T0295216

CONSTRUCTING QUANTUM MECHANICS

Constructing Quantum Mechanics Volume Two

The Arch: 1923–1927

Anthony Duncan and Michel Janssen

OXFORD
UNIVERSITY PRESS

Great Clarendon Street, Oxford, OX2 6DP,
United Kingdom

Oxford University Press is a department of the University of Oxford.
It furthers the University's objective of excellence in research, scholarship,
and education by publishing worldwide. Oxford is a registered trade mark of
Oxford University Press in the UK and in certain other countries

Published in the United States of America by Oxford University Press
198 Madison Avenue, New York, NY 10016, United States of America

British Library Cataloguing in Publication Data
Data available

Library of Congress Control Number: 2023930769

ISBN 978–0–19–888390–6

DOI: 10.1093/oso/9780198883906.001.0001

Printed and bound by
CPI Group (UK) Ltd, Croydon, CR0 4YY

Dedicated to the memory of Philip M. Stehle (1919–2013)
and Roger H. Stuewer (1934–2022)

Preface

This is the second of two volumes on the genesis of quantum mechanics in the first quarter of the twentieth century. For some general comments on our goals and approach in these volumes, see the preface of Volume 1. As we did for Parts I and II in Volume 1 (see Chapter 1), we begin Volume 2 with an overview, as non-technical as possible, of the contents of Parts III and IV (see Chapter 8).

As in Volume 1, we wrote many sections of Volume 2 with the idea that they be read in tandem with the original papers discussed in them. The papers most important for the development of both matrix and wave mechanics (covered in Part III) are available in English translations. The anthology edited by B. L. van der Waerden (1968) presents translations of the papers documenting the development of matrix mechanics (see Chapters 10–12). Schrödinger (1982) conveniently collects translations of the papers in which he introduced wave mechanics (see Chapter 14). Partial translations of De Broglie's (1924) dissertation and five key papers by Schrödinger can be found in an anthology on wave mechanics edited by Günther (or Gunter) Ludwig (1968). We refer to the papers included in van der Waerden (1968) and Schrödinger (1982) by using the page numbers of these volumes. Unfortunately, several of the papers in which the theory was developed further in 1926–1927 (covered in Part IV) have not yet been translated.

Our discussions of the original papers serve three purposes. First, they should enable a reader with a background in physics and mathematics comparable to that of an advanced physics undergraduate or beginning graduate student to follow the arguments and derivations in these papers.[1] Secondly, we explain how these papers are related to other papers upon which they build or to which they respond. Finally, we place these papers in the broader context of the debates over the developing quantum theory, in published papers, at public talks and conferences, and in private correspondence.

For this third task we relied heavily on editions of correspondence (as well as papers and manuscripts) of several prominent protagonists in our story. Especially important for Volume 2 were Niels Bohr's *Collected Works* (Bohr 1972–2008, especially Vols. 4 and 6), *The Collected Papers of Albert Einstein* (Einstein 1987–2021, Vols. 12–15), and the late Karl von Meyenn's edition of the correspondence of Wolfgang Pauli (in particular Vol. 1, Pauli 1979). The microfilms of the Archive for History of Quantum Physics (referred to as AHQP microfilm) are our source for unpublished correspondence. A finding aid prepared by Thomas S. Kuhn, John L. Heilbron, Paul Forman, and Lini Allen (1967) gives the reel and frame numbers for all letters included in this collection.

[1] To this end, we added several more web resources (cf. the preface of Volume 1): on Bohr's atomic theory of the early 1920s, on Thomas precession, on intensities of spectral lines in the old and the new quantum theory, on time-dependent perturbation theory in the Three-Man-Paper of Born, Heisenberg, and Jordan, on extremal principles, and on (Schrödinger's original handling of) the radial Schrödinger equation.

For the third task we also made heavy use of the massive six-volume history of quantum mechanics by Jagdish Mehra and Helmut Rechenberg (1982a, 1982b, 1982c, 1982d, 1987, 2000–2001). While Mehra and Rechenberg offer little help to those trying to understand the detailed arguments and derivations, their work remains invaluable in placing those arguments and derivations in a broader biographical and institutional context. Later reminiscences, although they need to be handled with care, were useful as well (see, e.g., Fierz and Weisskopf 1960; Rozental 1967; Heisenberg 1971). So were several biographies (both those written for a scholarly and those written for a general audience), such as David Cassidy (1991, 2009) on Werner Heisenberg, Max Dresden (1987) on Hans Kramers, Michael Eckert (2013) on Arnold Sommerfeld, Graham Farmelo (2009) and Helge Kragh (1990) on Paul Dirac, Nancy Greenspan (2005) on Max Born, Walter Moore (1989) on Erwin Schrödinger, Abraham Pais (1982a, 1991) on Einstein and Bohr, and, more recently, Ananyo Bhattacharya (2022) on John von Neumann. We were able to find on Wikipedia the dates and other biographical information for all but the most obscure characters.

We have also tried, as much as we could, to consider the sprawling secondary literature on the genesis of quantum mechanics. Especially important for Volume 2 were Guido Bacciagaluppi and Antony Valentini's (2009) edition of the proceedings of 1927 Solvay conference (with a detailed historical introduction) and, as for Volume 1, Olivier Darrigol's (1992) *From c-Numbers to q-Numbers*. Other important secondary sources were an edited volume (Badino and Navarro 2013), a pair of dissertations (Jordi Taltavull 2017; Jähnert 2016, since published as a book: Jähnert 2019) and a pair of papers (Joas and Lehner 2009; Blum et al. 2017) coming out of the quantum project in the context of which we started writing our book. For more information about this project, which ran from 2006 to 2013 and had its headquarters at the Max Planck Institute for History of Science in Berlin, see the preface to Volume 1 and the foreword by Alexander Blum, Christoph Lehner, and Jürgen Renn to Eckert (2020).

We also made use of our own earlier work. For dispersion theory and its role in the run-up to Heisenberg's (1925c) "reinterpretation" (*Umdeutung*) paper (see Chapter 10), we used Duncan and Janssen (2007). For Jordan's derivation of Einstein's fluctuation formula in the Three-Man-Paper (*Dreimännerarbeit*) of Born, Heisenberg, and Jordan (1926) (see Section 12.3.5), we used Duncan and Janssen (2008). For our discussion of "Kuhn losses" (see Section 15.2), we used Midwinter and Janssen (2013). For the Stark effect in wave mechanics (see Section 15.3.2), we used Duncan and Janssen (2014, 2015). For the statistical transformation theory of Jordan (1927b) and Dirac (1927a) and von Neumann's (1927a, 1927b) Hilbert space formalism (see Chapters 16 and 17), we used Duncan and Janssen (2013). Finally, for the arch-and-scaffold metaphor (see Chapter 18), we used Janssen (2019). For permission to reuse this material, we are grateful to Elsevier, publisher of *Studies in History and Philosophy of Modern Physics* (for Duncan and Janssen 2008, 2014); Springer, publisher of *Archive for History of Exact Sciences* (for Duncan and Janssen 2007); the Royal Danish Academy of Sciences and Letters (for Duncan and Janssen 2015); Edition Open Access (for Midwinter and Janssen 2013); and University of Minnesota Press (for Janssen 2019).

In addition to those friends and colleagues we already acknowledged in the preface of Volume 1, we thank Rudrajit (Rudi) Banerjee, Alex Blum, Victor Boantza, Cindy Cattell, Patrick Cooper, Paul Crowell, Mike Cuffaro, Rafael Fernandez, Michael Janas, Jim Kakalios, Alex Kamenev, Karl-Henning Rehren, and Kurt Schönhammer for helpful discussion. A significant part of this volume was written in Barcelona, where we enjoyed the company of and discussions with Enric Pérez. We thank Olival Freire Jr. for inviting us to contribute a brief synopsis of an important strand in our account of the genesis of quantum theory (i.e., the early history of quantization conditions) to *The Oxford Handbook of the History of Quantum Interpretations* (see Duncan and Janssen 2022, in Freire 2022). Olivier Darrigol once again deserves our special thanks for his detailed comments on drafts of several chapters of this volume. We are grateful to Laurent Taudin for drawing the "quantum cathedral" in Figure 18.1. One of us (MJ) thanks the Alexander von Humboldt Foundation for a Research Grant that provided generous financial support for work on this book.

We thank the Emilio Segrè Visual Archives of the American Institute of Physics (AIP), the Bibliothèque nationale de France, the Solvay Institutes, Brussels, the Pauli Archive at CERN, the Niels Bohr Archive in Copenhagen, the MIT Museum, the American Philosophical Society, and the Mathematisches Forschungsinstitut Oberwolfach for permission to use images in their collections for the plates in this volume. We thank Dominique Bogaerts, Urte Brauckmann, Joe DiLullo, Jennifer Hinneburg, Samantha Holland, Allison Rein, Rob Sunderland, Jens Vigen, and Ariel Weinberg for their help in obtaining these permissions.

It has once again been a pleasure to work with Sonke Adlung, our editor at Oxford University Press. We thank Michele Marietta for her careful copy-editing of the manuscript and Mark Ajin Millet for overseeing the production of this volume.

We dedicate this volume to the memory of Philip M. Stehle (1919–2013) and Roger H. Stuewer (1934–2022). Phil Stehle is responsible for one of us (AD) developing a deeper appreciation for and an interest in contributing to the history of physics. Phil Stehle is probably best known as one of the authors of a textbook on classical mechanics that is still in print (Corben and Stehle 1994, first published in 1950) but he also wrote an excellent textbook on quantum mechanics (Stehle 1966). He even wrote a book on the same topic as ours: *Order, Chaos, Order. The Transition from Classical to Quantum Physics* (Stehle 1994). Roger Stuewer, in addition to his many other contributions to the history of physics, created the position of a historian of physics embedded in the School of Physics and Astronomy at the University of Minnesota currently held by one of us. Roger Stuewer is probably best known for his authoritative publications on the history of nuclear physics, synthesized in his last monograph, *The Age of Innocence: Nuclear Physics between the First and Second World Wars* (Stuewer 2018). This is a topic beyond the scope of our book. But he also wrote the definitive account of the discovery of the Compton effect (Stuewer 1975), a classic in the history of physics on which our treatment of this topic is based (see Section 10.4). We were also able to consult many books in his impressive and extensive library, which he donated to the School of Physics and Astronomy at the University of Minnesota. Our book is written in the spirit and in fond memory of Phil Stehle and Roger Stuewer.

Contents

Part III Transition to the New Quantum Theory

Part IV The Formalism of Quantum Mechanics and Its Statistical Interpretation

List of Plates

8

Introduction to Volume 2

8.1 Overview

This is the second of two volumes on the genesis of quantum mechanics in the first quarter of the twentieth century. It covers the period 1923–1927 and tracks how what we now call the old quantum theory gave way to modern quantum mechanics. As in Volume 1 (see Chapter 1),[1] we begin with a detailed overview, as non-technical as possible, of the contents of Volume 2 (Chapter 8). Except for this introductory chapter and a short conclusion (Chapter 18), the chapters in this volume, like the ones in Volume 1, are divided into two parts, Part III and Part IV. Mirroring the appendix on the mathematics of the old quantum theory in Volume 1 (Appendix A), this volume has an appendix on the mathematics for the new quantum theory (Appendix C).

In Part III, consisting of seven chapters (Chapters 9–15), we cover the transition from the old to two distinct forms of the new quantum theory. In Chapter 9, we start with Bohr's (unsuccessful) attempts in the early 1920s to account for the periodicity of the table of elements, and the introduction of two ingredients, the first of which was indispensable for this task: Pauli's (1925b) exclusion principle and Uhlenbeck and Goudsmit's (1925) electron spin (after Pauli had talked Kronig out of publishing the idea). The next five chapters are devoted to the emergence of matrix mechanics (Chapters 10–12) and wave mechanics (Chapters 13–14). In Chapter 15, the final chapter of Part III, we revisit the successes and failures of the old quantum theory (see Chapters 6–7) and show how the new theory took care of the failures and put the successes on a more solid basis.

Matrix mechanics and wave mechanics resulted from different strands in the developments covered in Volume 1 (cf. Darrigol 2009). Matrix mechanics grew out of the atomic theory introduced by Bohr in his path-breaking 1913 trilogy (see Chapter 4) and elaborated by Sommerfeld and others, mostly working in the three main centers of the old quantum theory, Copenhagen, Munich, and Göttingen (see Chapters 5–7). After examining the roots of matrix mechanics in dispersion theory (Chapter 10), we

[1] In all further references to (sections, equations, and figures in) Parts I and II, Chapters 1–7, and Appendices A and B, we will no longer note explicitly that these are references to material in Volume 1 of our book.

Constructing Quantum Mechanics. Anthony Duncan and Michel Janssen, Oxford University Press.
 DOI: 10.1093/oso/9780198883906.003.0008

turn to the *Umdeutung* [reinterpretation] paper with which Heisenberg (1925c) laid the foundations of matrix mechanics (Chapter 11). We then examine its elaboration in the Two-Man-Paper of Born and Jordan (1925b) and the Three-Man-Paper [*Dreimänner-arbeit*] of Born, Heisenberg, and Jordan (1926), both coming out of Göttingen, as well as the paper in which Dirac (1925), working in Cambridge, introduced the closely related q-number theory (Chapter 12).

Wave mechanics, unlike matrix mechanics, grew out of work done by the pioneers of quantum theory *before* Bohr's 1913 trilogy (see Chapters 2 and 3). These pioneers—most importantly Planck, Einstein, and Ehrenfest—were not concerned with the structure of individual atoms, but rather with the statistics of large collections of simple systems modeled as harmonic oscillators. Instead of spectroscopy, they studied black-body radiation and the specific heat of solids at low temperatures—two areas where the equipartition theorem of the kinetic theory of gases of Maxwell and Boltzmann notoriously broke down. The year before the arrival of matrix mechanics, Indian physicist Satyendra Nath Bose (1924) proposed a new derivation of the Planck law for black-body radiation, using a new statistics for light quanta, which Einstein (1924, 1925a, 1925b) subsequently transferred to atoms in his quantum theory of the ideal gas. Einstein found that the density fluctuations in such a gas are given by a formula like the one he had found earlier for fluctuations in black-body radiation (1909a, 1909b), consisting of a wave and a particle term. This supported a suggestion made in the dissertation by the French physicist de Broglie (1924) to extend the wave-particle duality Einstein had proposed for radiation to matter. Drawing on this work by De Broglie, Bose, and Einstein (Chapter 13), and on the much older optical-mechanical analogy due to William Rowan Hamilton (1834, 1835) (Section 13.3), Schrödinger (1926c, 1926d, 1926f, 1926h), working in Zurich rather than in one of the centers of the old quantum theory, developed wave mechanics in a celebrated four-part article (Chapter 14).

In Part IV, consisting of just two chapters (Chapters 16–17), we cover the completion of the formalism of quantum mechanics and its probabilistic interpretation. By the middle of 1926, three authors, independently of one another, had already established that matrix and wave mechanics are intimately related (see Section 14.5). Schrödinger (1926e) and Eckart (1926) gave their "equivalence proofs" in published papers, and Pauli (1979, Doc. 131) gave his in private correspondence. In December 1926, Dirac (1927a) and Jordan (1927b) published, again independently of one another, what has come to be known as the Dirac–Jordan *statistical transformation theory* (Section 16.2). Their formalism unified all four versions of quantum theory that had meanwhile been proposed: matrix mechanics, wave mechanics, Dirac's q-number theory, and the operator calculus of Born and Wiener (1926, see Section 14.5). As the term "statistical" suggests, Dirac and Jordan generalized the probabilistic interpretation of Schrödinger's wave function proposed by Born (1926a, 1926b) in papers on collision processes, and by Pauli (1979, Doc. 143), true to form, in private correspondence (Section 16.1). As the term "transformation theory" suggests, canonical transformations familiar to Jordan and Dirac from the celestial mechanics used in the old quantum theory (see Appendix A) played a central role in the Dirac–Jordan formalism. These mathematical techniques, however, proved ill-suited to the task. This was clearly recognized by the young Hungarian mathematician Janos von Neumann, who came to work with Hilbert in Göttingen in 1927. Building on recent work by Hilbert and other mathematicians on what is now known as functional analysis (Appendix C), von Neumann (1927a) introduced the modern Hilbert space formalism in a paper presented to the Göttingen Academy in May 1927 and used it to formulate quantum mechanics in a way that steered clear of such mathematical monstrosities as the Dirac delta function, unavoidable in Dirac and Jordan's approach. Prompted in part by Heisenberg's (1927b) uncertainty paper (Section 16.3), von Neumann published two more papers later that year (von Neumann 1927b, 1927c). In the second installment of this 1927 trilogy, drawing on von Mises's ideas about probability (soon to be published in book form: von Mises 1928), von Neumann (1927b) reformulated the statistical interpretation of quantum mechanics in terms of his Hilbert space formalism. This paper only appeared after the Como and Solvay conferences of September and October 1927, which mark the beginning of the contentious and ongoing debate over the interpretation of quantum mechanics, a topic beyond the scope of our book. We therefore decided to cover Heisenberg's (1927b) uncertainty principle and Bohr's (1928b) complementarity principle (first presented at Como and Solvay) in Chapter 16, and to end with von Neumann's Hilbert-space formalism in Chapter 17.

In the short concluding Chapter 18, we elaborate and reflect on the arch-and-scaffold metaphor we used for the subtitles of Volumes 1 and 2 (Janssen 2019). As will become clear in the course of this volume, the "quantum revolution" of the mid-1920s, while a major turning point in the history of physics, does not follow the pattern of a "paradigm shift" familiar from Thomas S. Kuhn's (1962) classic *The Structure of Scientific Revolutions*, in which the old paradigm crumbles under the weight of an accumulation of anomalies and a new paradigm is erected on its ruins. The old quantum theory had run into serious difficulties by the early 1920s, but the way out of the resulting Kuhnian crisis was not, as Kuhn (1970, pp. 256–257) himself clearly recognized, to trash the old theory and start from scratch. A more apt image, our analysis suggests, is that the new quantum theory was built as an arch on the scaffold provided by the old quantum theory and, in turn, the classical mechanics and electrodynamics upon which it was built. Much of this scaffolding was discarded once the arch could support itself, making the discontinuity in its construction look, in hindsight, much more pronounced than it actually was.

8.2 Quantum theory in the early 1920s: deficiencies and discoveries (exclusion principle and spin)

By the middle of 1923, as we saw in Chapter 7, numerous serious deficiencies had been exposed in the formal framework of the old quantum theory, as developed in the aftermath of Sommerfeld's groundbreaking papers of 1916, in which the Bohr–Sommerfeld quantization rules were introduced and applied with remarkable success to several central problems of atomic physics. These deficiencies can broadly be divided into two categories. First, there were problems that seemed at least amenable to the accepted framework of the old quantum theory—formulation of a mechanical problem using the Hamilton–Jacobi technique, given that the problem allowed treatment in terms of a multiply periodic motion, followed by imposition of the quantization rules to the associated action variables, but in which the execution of the formalism simply led to an incorrect result—one in clear disagreement with the experimental data. The quintessential example of this type was the failure of the old theory to obtain the correct binding energy of the ground state of helium, despite the discovery of a perfectly "reasonable" multiply periodic coordinated motion for the two electrons consistent with classical mechanics.

The second set of failures—more numerous than the first—comprised those problems in atomic physics in which no single consistent treatment based on mechanical principles and the Bohr–Sommerfeld methodology could be found for the phenomena in question. The cluster of problems associated with the interpretation of the complex structure of optical spectral multiplets, and the explanation of the complicated rules that had arisen from the study of anomalous Zeeman multiplets belong to this second category. Other problems in this category crop up in the theory of optical dispersion and in the theory of the Stark effect. In all these cases, the root of the problem was the association of energy levels with definite classical orbits of charged (but non-magnetic) electrons in miniature solar systems.

The early 1920s saw an enormous increase in the interest in, and attention paid to, two more topics, which were eventually found to elude consistent treatment with the methods of the old quantum theory—the periodic nature of the chemical and optical properties of the elements, presumably to be explained by some shell-like arrangement of the electrons in larger atoms, and the appearance of unexplained "superfluous" lines in X-ray spectra, with spectral splittings with an inexplicable (on old quantum theory grounds) dependence on the atomic number Z. This cluster of problems would turn out to be deeply connected with the two final major discoveries of the quantum theory prior to the advent of modern quantum mechanics: the exclusion principle for atomic electron states proposed in December 1924 by Wolfgang Pauli, and the existence of an intrinsic electron spin angular momentum, first proposed by Ralph Kronig in January 1925, and independently rediscovered in November of the same year by George Uhlenbeck and Samuel Goudsmit. Ignorance of these two basic facts, quite apart from the (in hindsight) deep kinematical and dynamical defects of the old theory which would be cured by matrix and wave mechanics, explains a large part of the failure to arrive at a consistent treatment of complex spectra or the anomalous Zeeman effect within the framework of the old quantum theory. The discovery of the exclusion principle in late 1924, and shortly after, of electron spin, is our focus in Chapter 9. It is useful here, however, to begin our broad overview of the construction of modern quantum mechanics by describing the dramatic progress made in these two areas in the years leading up to the extraordinary developments of 1925.

8.3 Atomic structure *à la* Bohr, X-ray spectra, and the discovery of the exclusion principle

As we saw in Chapter 4, Bohr's earliest efforts in attempting to insert quantum principles into an understanding of atomic structure (post the Rutherford nuclear atom) were summarized in the memorandum of 1912 that he sent to Rutherford. Here, we can see him wrestling already with the problem of the configuration of electron orbits in an atom with more than one electron. These rudimentary considerations were much amplified and deepened in the second paper of his 1913 trilogy, entitled "Systems containing only a single nucleus" (but in general, more than one electron—see Section 4.5). The basic conclusions at this early stage of the old quantum theory were necessarily tentative, and indeed, only tenable given the rudimentary quantitative evidence available at the time concerning the detailed energetics of the quantized states of atoms. For example, even for the simplest multi-electron atom, helium, the total binding energy of the atom was only very roughly known. Bohr's earliest attempts in this direction, which we may term here "Bohr's first atomic theory", can be summarized in the following statements, as regards the configuration of electron orbits:

1. All the electrons circulate in a single plane in (roughly) concentric circular orbits. This is, of course, analogous to the solar system, in which the major planets orbit the sun in roughly (to within 6°) the same ("ecliptic") plane.
2. Every electron in the ground state of the atom has a quantized value of $h/2\pi$ for its orbital angular momentum.
3. For a given nuclear charge, there is a maximum number of electrons that can be accommodated in each of the concentric rings containing electrons. This number was determined by a mechanical stability condition with regard to small perturbations of the orbit perpendicular to the "ecliptic" plane of the atom.

Even in the absence of detailed quantitative information on electron binding energies, there was clear evidence from chemistry and optical spectroscopy of a "periodic" character in the properties of the elements, which had already been codified in Mendeleev's periodic table. Bohr's (1913b) stability arguments were an early, but (as Bohr himself recognized) far from successful, attempt to come to terms with this periodicity.

By the early 1920s, it was clear that none of these three principles could possibly be valid. For numerous reasons, including, but not limited to, the diamagnetism of the rare gases, the crystal structure of various elements, optical dispersion measurements, and (perhaps most critically) the extraordinary dual feature of the helium spectrum, which divided into two disconnected sets of lines separately connecting the terms of "orthohelium" and "parahelium", the picture of electrons circulating in coplanar orbits, as originally imagined by Bohr in 1913, had become completely untenable. The definitive abandonment of this picture in the case of helium came with the introduction of the non-coplanar Bohr–Kemble model of the ground state of helium (see Section 7.3), which was itself a natural descendant of the Sommerfeld–Landé model of helium, developed in the process of wrestling with the peculiarities of the helium spectrum.

Remarks at the end of a long paper entitled "On the series spectra of the elements" (Bohr 1920), submitted to *Zeitschrift für Physik* in the Fall of 1920, offer a glimpse of the methodology Bohr was to adopt in the years 1920–1923 in the study of atomic structure. Most of this paper is devoted to a description of the understanding which had been achieved of the remarkable efficacy of Rydberg-type formulas (see Appendix B) for the terms of the optical series of the elements (especially the alkali metals, such as sodium) using the methods of the old quantum theory (see Section 7.1), as well as a recapitulation of the applications of correspondence-principle ideas, for example, in the interpretation of spectra in the Stark effect (see Section 6.3). In the concluding section, Bohr apologizes for the lack of attention specifically to questions of atomic structure, and points forward to the need for a more sophisticated approach to the problem of the configuration of electronic orbits:

> In the foregoing I have intentionally not delved more closely into the question of the *structure of atoms and molecules*, although this subject is clearly intimately connected with the theory of the origin of spectra which we have considered ... On the occasion of the first treatment of this matter [Bohr's trilogy of 1913], the author also attempted to sketch the foundations of a theory for the structure of the atoms of the elements [Part II] and for the chemical combinations of molecules [Part III] ... I would like to take the opportunity here, however, to state that from the standpoint of more recent developments of the quantum theory ... many of the detailed special assumptions made in the [earlier] theory must be altered, on the basis of disagreement of the theory with experimental results which have been raised on numerous fronts. In particular it does not seem possible to justify the orienting assumption introduced at that time [1913], that in the ground state electrons move in geometrically particularly simple orbits, such as "electron rings." Considerations of the stability of atoms and molecules with regard to external influences, and on the possibility of construction of the atom by successive acquisition of single electrons, require us to insist, first, that the electron configurations to be considered are mechanically stable, and moreover possess a certain stability with regard to the requirements of the usual mechanics, and second, that the configurations to be used are so constituted that they may be arrived at by transitions from other stationary states of the atom. *These requirements are in general not fulfilled by such simple configurations as electron rings, and force us to look around for more complicated form of motion* (Bohr 1920, pp. 468–469, our emphasis).

In the three years following this work, Bohr delved more closely into the problem of atomic structure, specifically, in determining the mutual orientation of the orbits followed by electrons in multi-electronic atoms, and relating the proposed arrangement of atomic orbits to the observed chemical properties and optical spectra of the elements, with particular attention to the recurring periodicity of these properties with atomic number embodied in the periodic table. There would be two short *Nature* papers (Bohr 1921a, 1921c) as well as a series of longer papers, often elaborated versions of lectures Bohr presented at various venues—and often to audiences who were not atomic theorists, and could not therefore follow the more technical aspects of Bohr's program—such as Bohr (1921d), a greatly expanded version of a lecture in Copenhagen in October 1921, and Bohr (1922e), based on his legendary Wolfskehl lectures in Göttingen in June 1922, which became known as the "Bohr festival" (*Bohr Festspiele*) (Mehra and Rechenberg 1982a, sec. III.4, pp. 332–358).

The general nexus of ideas proposed in these papers has become known as "Bohr's second atomic theory" (see Kragh 1979, 2012, Ch. 7). In these works, Bohr attempted to develop and apply the ideas visible in embryo form in the quote given above. First, an electrically neutral multi-electronic atom with an atomic nucleus of charge $+Ze$ would be built up in stages by first putting a single electron into the lowest energy quantized orbit available for the bare nucleus, and then a second electron would be added in the lowest quantized energy state compatible with the execution of a multiply periodic motion[2] by both electrons. A third electron would then be added in the same way, allowed to settle down into an orbit which necessarily had to possess a high degree of "cooperation" with the motions of the preceding two electrons—and so on, until a full complement of Z electrons had been added to the atom, rendering it electrically neutral.

The multiply periodic character of the motion at each stage would allow for the application of the Bohr–Sommerfeld quantization principles, so that the orbit of each electron could be characterized by assigning the two quantum numbers familiar from the old theory: the principal quantum number n, the main determinant of the orbital size and energy, and the azimuthal (orbital angular momentum) quantum number k, so that one could speak of an electron in the state n_k (with $k = 1, 2, \ldots, n$).

Two guiding principles were supposed to be applied in the execution of this *Aufbau* ("build-up") process: (a) the quantum numbers of the previously added electrons would not be altered by the addition of subsequent electrons, although of course the detailed shape and size of the inner orbits would be presumed to change as new electrons were added, in conformance with the classical equations of motion; and (b) the final orbit reached by the latest electron added must correspond to an allowed transition as prescribed in the correspondence principle. Here, one should recall that the extended correspondence principle developed by Bohr and Kramers in the late teen years associated the probability of a quantum transition in which electromagnetic radiation was absorbed (or emitted) with the Fourier components of the motion of the absorbing (or emitting) electron. This association was quantitatively exact in the limit of large quantum numbers, but assumed to hold roughly for small quantum numbers. In particular, a transition disallowed for some symmetry reason (giving zero Fourier components for the corresponding emitted light quantum) was assumed to be disallowed for all (and not just large) quantum numbers, an extension that led to the remarkably successful application of the correspondence principle in deriving selection rules for optical transitions.

The guiding principles Bohr attempted to apply to his "second theory of atomic structure" seem fairly straightforward in principle, but turned out to be extremely slippery, and ultimately untenable, in practice, as we examine in detail in Chapter 9. His references to the correspondence principle as determining the choice of orbits for each newly added electron in the *Aufbau* procedure were unsupported by explicit calculations and for the most part expressed in vague and obscure language. A typical example is the third electron in lithium, which cannot (because of the exclusion principle, as we now know) occupy 1s ($n = 1, k = 1$) orbits like the first two, but which (according to Bohr) is prevented from joining the first two electrons in a 1s state because such a state cannot

[2] We remind the reader that a multiply periodic motion of N electrons is one in which the time dependence of each of the $3N$ Cartesian coordinates of the system of electrons can be written as a Fourier series combination involving at most $3N$ distinct frequencies.

be reached (by the third electron) from higher orbits by a transition compatible with correspondence principle ideas:

> Such a transition process, of which one must assume (as a more detailed examination of the possibilities of motion shows [this is never given]) that it would lead to an atomic state in which the third electron would appear as a equivalent participant in the collective motion of the three electrons, would in fact be of a quite different type as the transitions between stationary states connected with the emission spectrum of lithium, and in contrast to these would display no correspondence to the harmonic components of the motion of the atom (Bohr 1922b, p. 35).

Sentences of this sort abound in the sequence of articles and lectures mentioned above whenever the physico-chemical properties of an element required the addition of an electron in a new, less-bound orbit to the preceding ones. Appeals to the correspondence principle were therefore of critical importance in Bohr's attempts to understand the periodicity of these properties in terms of a shell structure (more precisely, a grouping into subclasses of similar orbits) of the electrons surrounding larger atoms. Inasmuch as Bohr's final (pre-exclusion principle) specification of electronic shell structures, in a paper co-authored with Dirk Coster (Bohr and Coster 1923, p. 344), still has an incorrect assignment of quantum numbers (as, for example, in the filled third shell for elements from copper onward, containing six each of s, p, and d electrons, instead of 2, 6, and 10, respectively), we know that these arguments must have been faulty, even if they existed in a more explicit mathematical form than is apparent from the published work. In Chapter 9 we discuss Bohr's arguments in more detail,[3] and demonstrate that atomic systems subject to the correspondence principle, but lacking spin and exclusion properties, simply do not behave as Bohr asserted.

8.3.1 Important clues from X-ray spectroscopy

In our discussion in Chapter 7 of the various attempts to deal with the appearance of "extra" lines—hence, extra quantum states—in the study of the complex optical spectra that occurred especially in larger atoms, we saw that the new levels were at first absorbed into the theory by the purely phenomenological ruse of introducing a new quantum number, Sommerfeld's "inner quantum number". A dynamical explanation for this new degree of freedom would eventually (in the few years remaining to the old quantum theory) come to be widely accepted: namely, that the splitting of Bohr–Sommerfeld levels associated with this new quantum number was due to the freedom to alter (discretely) the relative orientation of the orbital angular momentum of an outer valence electron (the quantum state of which was under consideration) to the orientation of the combined angular momentum of the atomic core (i.e., all the inner electrons). A further intriguing phenomenological observation would soon prove to play a critical

[3] See also the web resource, *Bohr's Second Theory and Atomic Aufbau before Spin and the Exclusion Principle.*

role—when a weak magnetic field was applied, a valence energy level would split into a number of close levels equal to twice the inner quantum number.

By 1920, the increasing accuracy and detail available from X-ray studies—especially, the studies of X-ray absorption edges by Gustav Hertz (the nephew of Heinrich Hertz) at the *Physikalisches Institut* of the University of Berlin—had shown quite clearly the existence of a similar set of "extra" levels in the atomic energy levels ("terms") associated with the innermost, and most tightly bound, electrons of larger atoms (all the way up to uranium, with $Z = 92$). The L levels, corresponding to principal quantum number 2, were found to be associated with a triplet of energy values, which were labeled by Sommerfeld as the L_1, L_2, and L_3 levels.

The increasing importance of X-ray studies, more specifically, of focusing attention on the innermost electrons, was precisely that such electrons should most closely approximate the behavior of single electrons in a hydrogenic atom (now of large nuclear charge) given the dominance of the nuclear electrostatic field, and the relative weakness of the influence of the other single electrons (each having a much smaller charge than the nucleus). Outer electrons were assumed to surround the inner ones (and nucleus) more or less symmetrically, and therefore, by a well-known theorem of classical electrostatics, exert no (or very little) net Coulomb force on the inner ones. An inner electron (say, an L electron with principal quantum number $N = 2$) would see almost the full effect of the atomic nucleus of charge $+Ze$, and the small screening effect of the other inner electrons could be taken into account quite accurately by replacing $Z \to Z_{\text{eff}} = Z - \sigma$, where the screening factor σ was typically of order unity (compared to Zs of interest in the X-ray studies between 50 and 90). The square root of the Bohr binding energy for electrons of a given principal quantum number, which was the quantity typically plotted as the vertical ordinate in X-ray term diagrams post-Moseley,[4] was therefore roughly proportional to Z for medium to large Z.

Sommerfeld's calculation of relativistic effects, giving rise to the famous fine structure formula (see Eqs. (6.51)–(6.53)), showed that the varying eccentricity of orbits of differing n (orbital angular momentum) for a fixed principal quantum number[5] N would lead to different relativistic corrections to the electron binding energy—corrections of order of magnitude $(v/c)^2$ relative to the leading energy term, where the typical velocity v of an electron circulating in a Bohr orbit around a nucleus of charge Ze satisfies $v/c \simeq Z\alpha$ (with $\alpha \simeq 1/137$ the Sommerfeld fine structure constant). The presence of such relativistic effects would therefore necessarily induce shifts in the inner electron term energies of order Z^4 (compared to the leading Z^2 Balmer term), or a divergence of order Z^3 between the roughly linear Moseley square-root-of-binding energy ordinates for orbits of differing n value (hence, orbital eccentricities).

[4] Moseley's X-ray data were typically displayed with the atomic number on the vertical axis (ordinate) and the square root of the term energy on the horizontal axis (abscissa). Cf. Figure 6.3.

[5] We apologize here for the confusing evolution of conflicting notations for electronic quantum numbers, with N, n (for principal and orbital angular quantum numbers) of the late teens giving way later to n, k (as in our discussion of Bohr's *Aufbau*), and subsequently n, l. We have tried to hold to the notation used by the authors being discussed as much as possible.

Let us first consider the simplest case, the "L" X-ray terms corresponding to inner electrons with principal quantum number $N = 2$, in atoms of large Z (> 50, say). Pairs of X-ray terms diverging for large Z from each other as the cube of the atomic number were indeed well established by the work of Gustav Hertz in early 1920, and were dubbed "regular doublets" by Sommerfeld. Then, there were the "extra" terms, which maintained a more or less constant (i.e., Z-independent) difference (rather than increasing as the cube of atomic number) in Moseley ordinate from one of Sommerfeld's "regular" terms! In fact, this situation is the one we should have anticipated from the fact that electron orbits of differing angular momentum would penetrate the innermost region of the atom (and be exposed to the full nuclear charge) to differing degrees, and hence, would have different screening constants. Two such terms would then have Moseley terms exhibiting a constant difference in the *leading* (i.e., Balmer) energy square root: $\sqrt{(Z-\sigma_1)^2} - \sqrt{(Z-\sigma_2)^2} \propto \sigma_1 - \sigma_2$ (hence, independent of Z). This would be much larger than, and should therefore always dominate, the Sommerfeld fine-structure shifts (with their extra factor of the square of α). The problem was that one found *both* the "regular" doublets, with a splitting that looked just like relativistic fine structure, and "irregular" doublets with a constant Moseley splitting—which was perfectly reasonable on the basis of the expected difference in nuclear charge screening for orbits of different shape (see Figure 9.2), but seemed to lack the expected relativistic corrections for orbits differing in angular momentum.

As mentioned earlier, an important reason for the increasing interest in X-ray studies at this point was the hope (already expressed by Sommerfeld in the first edition of *Atombau und Spektrallinien*; cf. Volume 1, p. 277) that the nuclear Coulomb attraction on an inner electron would so dominate the relatively small inter-electron repulsive interactions (which could be ignored for the outer electrons) that the motion of these electrons would be very close to that expected on the basis of the Bohr–Sommerfeld formalism. In particular, the peculiar complex spectra effects in optical spectra (i.e., in the energies of the outer electron orbits) presumed to arise from an interaction of the valence electron orbit with a magnetic field produced by the "core" (i.e., the nucleus plus inner electrons) should no longer be relevant, as the innermost electrons *were* the core, and the screening effects exerted on each other could apparently be accounted for very accurately simply by a small shift in the effective nuclear charge, $Z \to Z_{\text{eff}}$, as indicated above.

Nevertheless, the triplet of L ($N = 2$) terms, quintuplet of M ($N = 3$) terms, etc., clearly indicated that an additional degree of freedom was in play for these inner electrons as well, giving rise to "extra" stationary-state energy levels. And the Z-dependence of the extra splittings, as we have just seen, was inexplicable in terms of simple screening plus relativistic effects.

8.3.2 Electron arrangements and the emergence of the exclusion principle

These matters came to a head in an influential paper written in late 1922 by Bohr and Coster (1923) entitled "Röntgen [X-ray] spectra and the periodic system of the elements." Coster had worked on X-ray spectroscopy at Lund University with Manne

Siegbahn's group in the preceding two years before coming to visit the Niels Bohr Institute in Copenhagen in the summer of 1922. In fact, his dissertation, completed under the direction of Paul Ehrenfest at the University of Leyden earlier that year, had been precisely on the subject of "Röntgen spectra and the atomic theory of Bohr" (Coster 1922). Coster had at his fingertips all the latest results for X-ray term levels in a wide range of elements, which now represented a greatly expanded cornucopia of phenomenological information, compared to the relatively sparse results of Hertz just two years earlier. Bohr and Coster undertook an extensive review of the data and gave a detailed discussion of the various screening effects to be expected in the motion of an inner electron. This motion, while dominated by the very strong nuclear Coulomb field (we are thinking of atoms with a fairly large atomic number here), would also be affected occasionally by the penetration of an outer electron with a very elliptical orbit (i.e., low angular momentum) into the inner region, and more frequently by the presence of the electrostatic repulsion of the electrons of the same or smaller principal quantum number as the electron in question. Bohr and Coster asserted that these effects could be subsumed into screening shifts (such as the factor σ shifting the nuclear charge Z discussed above). Unfortunately, the appearance of extra "anomalous" (in the Bohr–Coster terminology) energy levels leading to the apparently incompatible simultaneous appearance of "screening" and "relativistic" doublets could not be reconciled with the fundamental tenet of the *Aufbau* program, which assumed that the addition to or subtraction from the atom of a single electron would not appreciably affect the orbits (or energies) of the remaining energies. Instead, as we discuss in more detail in Chapter 9, one was forced to assume that the removal of an inner electron resulted in an essential (*wesentliche*) change of the behavior of the other electrons in the same shell (or "group", in the Bohr–Coster language). In fact, the authors state

> Such a conception is necessitated by the theory if it is assumed that the completion of an electron group is essentially determined by the interaction of electrons within the various subgroups [of the shell] (Bohr and Coster 1923, p. 365).

The mysterious "interaction between subshells" referred to here can (with hindsight!) be regarded as a foreshadowing of the constraints that would, by the end of the next year, be recognized to arise from the exclusion principle, where the completion of the electron subshells within a shell (of given principal quantum number) is indeed determined by the previous occupancy of these subshells. But in the absence of any understanding of either the additional physical spin degree of freedom of the electrons or the completely non-classical constraints imposed on the electron states by exclusion, Bohr and Coster had to admit that "it is still quite unclear how a more detailed execution of this idea could provide an explanation for the separate appearance of screening- and relativity-doublets" (p. 365).

A few months after the publication of Bohr and Coster's paper, a major step toward the exclusion principle appeared in an article entitled "The distribution of electrons among atomic levels," submitted to the *Philosophical Magazine* by Edmund Stoner

(1924), a graduate student working with Rutherford and Fowler at the Cavendish Laboratory in Cambridge. In discussing Stoner's contribution, we should first note yet another shift in terminology in describing electronic configurations. Where Bohr and Coster speak of electron "groups" and "sub-groups", we find in Stoner the designation of electron "levels" (electrons sharing a given value of the principal quantum number n), which are subdivided into "sub-levels" characterized by a given value of the azimuthal (orbital angular momentum) quantum number k ($=1, 2, \ldots, n$).[6]

For Stoner the appearance of "extra" terms in X-ray spectra, which seemed very similar to the extra terms which had been known in optical "complex" spectra for decades, was more than a coincidence. The number of terms appearing in the X-ray data, just as in the optical case, suggested that each stationary state with an azimuthal number $k = 2, 3, \ldots$ (thus, the $p, d, \ldots$ states) was duplicated, with the two states distinguished by the numerological artifice of an extra "inner" quantum number (j for Stoner), which could take the two values $k - 1$ and k. The s states, with $k = 1$, on the other hand, appeared singly, with no extra "twin." Now in the optical case, numerous studies of the anomalous Zeeman effect had shown that in a weak magnetic field a stationary state with inner quantum number j splits into exactly $2j$ evenly spaced energy levels. Stoner suggested that this splitting revealed precisely the number of dynamically possible distinguishable electron states, not just for the outer light-emitting valence electrons relevant to the optical spectra, but also for the inner electrons occupying already filled levels (or "groups", or "shells", ...) in larger atoms.[7] Stoner refused to commit himself to any specific model-theoretic "explanation" for the existence of these states, except to accept the widely held proposition that the split energy levels revealed in the Zeeman effect corresponded to a space-quantized variation of the orientation of the orbits. He concludes the fifth chapter of his paper with the summary

> In brief, then, it is suggested that, corresponding roughly to the definite indication in the optical case, the number of possible states of the atom is equal to $2j$; so, for the X-ray sub-levels, $2j$ gives the number of possible orbits differing in orientation relative to the atom as a whole; *and that electrons can enter a sub-level until all the orbits are occupied* (Stoner 1924, p. 726, our emphasis).

Note that the term "sub-level" here indicates the specification of all three quantum numbers: principal (n), azimuthal (k), and inner (j). This remark is followed immediately in the next section of the paper by the assertion that all $2j$ electron states (for given n and k) have identical statistical weight, "for there is then one electron in each possible equally probable state" (p. 726). Although Stoner does not put it this way, it is very hard to see how all of this works if one does not also grant that more than one electron is in

[6] We are now adopting, in place of Sommerfeld's use of n for the azimuthal quantum number, the denotation k, which was increasingly used in the last few years of the old quantum theory. The modern notation n for the principal quantum number also becomes more usual at this time. The extra inner quantum number also exhibited a confusing abundance of different notations: for Bohr and Coster it was k_2 (with $k_1 = k$ the usual azimuthal quantum number), while Stoner, following Landé, uses j for the inner quantum number.

[7] It is important to realize that the relatively very small splittings induced by application of a weak magnetic field were not at this stage empirically accessible in the X-ray data.

fact *excluded* from occupying the same orbit, at least in the case of completed groups. However, it must be admitted that the statement of identical statistical weight for the various levels revealed after magnetic splitting cannot be taken to imply that each such level can be occupied at most by a single electron, only that the probability of occupancy on average is the same for the levels (in the weak field limit).

A simple example of how Stoner's configurations differ from Bohr and Coster's can be seen in the suggested distribution of electrons in the inert gas argon (atomic number $Z = 18$). In the Bohr scheme (using the n_k notation) there were two 1_1 ($1s$) states; four 2_1 ($2s$) and four 2_2 ($2p$) states; and four 3_1 ($3s$) and four 3_2 ($3p$) states. Stoner's assignments instead correspond exactly to the modern distribution of electrons: two $1s$, two $2s$, six $2p$ (divided into a sub-shell of two $j = 1$ and four $j = 2$ electrons), two $3s$, and six $3p$ electrons (again divided into two sub-shells of $j = 1, 2$). In the final part of his paper, Stoner gives some heuristic arguments to support his numerical assignments of electron orbits. The most convincing of these come from consideration of the intensity of X-ray lines arising from transitions of electrons out of different sub-levels, as in the comparison of $L_2 \to K$ and $L_3 \to K$ transitions, and from the relative absorption of X-rays by electrons in various sub-levels, as in the relative absorption of L_1, L_2, and L_3 electrons.

Stoner's paper was read with great enthusiasm by Wolfgang Pauli, who fully approved of the strategy of trying to combine information available from X-ray spectroscopy with the insights obtained from the (by now) multi-decade struggle with the mysteries of the anomalous Zeeman effect. The results of his study of the problem were published in a paper entitled "On the connection between the closing of electron groups in the atom with the complex structure of spectra," submitted on January 16, 1925 (Pauli 1925b).

After a brief review of the difficulties encountered by the old quantum theory in arriving at a satisfying—that is, dynamically consistent model-theoretic—account of the anomalous Zeeman effect for weak magnetic fields, Pauli turns to the most basic problem faced by the quantum theory in explaining the regularities of the periodic table: namely, the appearance of the period lengths 2, 8, 18, 32, ... (more concisely, the numbers $2n^2$ for $n = 1, 2, 3, 4, \ldots$), which suggest that the group of electrons with principal quantum number n is unable to accommodate the addition of any further electrons once there are, as in the case of the rare gases, already $2n^2$ electrons present. Pauli fully accepts Stoner's interpretation of this counting, reproducing both Bohr's and Stoner's proposed electron distributions for the rare gases from helium up to "radium emanation" (i.e., radon) (Pauli 1925b, Tables 1 and 2, respectively).

In common with Stoner, Pauli also proposes that critical insights into the filling of atomic electron groups could be obtained from the splittings observed in the optical Zeeman effect, but in Pauli's case, the important clues are to be found in the special case of large magnetic fields (Paschen–Back regime; cf. Section 7.2.3), where the structure of the splitting simplifies, as the interaction between the magnetic moment due to the orbital angular moment of a valence electron with that of the putative "atomic core" becomes negligible, and the magnetic energy shifts take on the form of a sum of two terms, one proportional to the component of the valence electron angular momentum in the direction (say, z) of the magnetic field, and the other to the z-component of the core magnetic momentum (enhanced by a factor of 2, reflecting the "double magnetism" of the core).

However, by this time in late 1924, Pauli had become convinced, on the basis of various calculations of the effects of relativity on the motion of inner "core" electrons, that the second term had to be associated with some completely non-classical and essentially mysterious "double-valuedness" (*Zweideutigkeit*) of single electrons in stationary states (Pauli 1925b, p. 768). The simplification of the dynamics of electron states in special circumstances was essential for both Stoner and Pauli: in the case of the former, the negligibility of inter-electron interactions for inner electrons dominated by the Coulomb field of a large nuclear charge; for the latter, as a result of the unimportance of the core–orbit (later spin–orbit) interaction once the external magnetic field became very large.

Pauli showed that the counting of Stoner—in particular, the division into sub-levels containing a number of electrons given by twice the inner quantum number—could be reproduced by introducing a set of four quantum numbers, which seemed relevant in describing the magnetically split levels of a single valence electron in the strong field (Paschen–Back) Zeeman effect. These were n, the principal quantum number, k_1, the azimuthal orbital angular momentum (thus $s, p, d, \ldots$ electrons for $k_1 = 1, 2, 3, \ldots$), k_2, the inner quantum number, which by now had been interpreted by Landé as equal to $j + 1/2$ (where j was the total angular momentum, valence plus core, of the atom), and finally, a magnetic quantum number $m_1 = -j, -j + 1, \ldots, j - 1, j$, which came into play once the levels were magnetically split, as it reflected the space-quantized component of the total angular momentum in the direction of the magnetic field.

To the classification of dynamically allowed electron states specified by the set (n, k_1, k_2, m_1), Pauli added an additional—and crucial—stricture not to be found, at least explicitly, in Stoner: the *prohibition* of double occupancy of any state specified by a unique value of these four quantum numbers. As we shall see, this prohibition, which would soon become known as the "Pauli exclusion principle" (*Paulisches Ausschliessungsprinzip*, or more simply, the *Pauli-Prinzip* or *Pauli-Verbot*), leads to exactly the same distribution of electron shells into sub-levels as Stoner's scheme. The eight electrons in the L shell of principal quantum number $n = 2$, for example, would be divided into three sub-levels containing two $2s$ electrons (with $k_1 = 1$, $k_2 = 1$, and $m_1 = \pm 1/2$), two $2p$ electrons (with $k_1 = 2$, $k_2 = 1$, and $m_1 = \pm 1/2$), and four $2p$ electrons (with $k_1 = 2$, $k_2 = 2$, and $m_1 = +3/2, +1/2, -1/2, -3/2$), respectively. This is clearly different from Bohr's scheme, which simply apportioned electrons in completed shells equally among the different k_1 values (thus, four each of $2s$ and $2p$ electrons).

Pauli's article concludes with the remark that, in contradistinction to Bohr's reliance on correspondence-principle arguments in arriving at proposed electronic arrangements in multi-electron atoms, arguments based on the "principle of invariance of statistical weights of quantum states" (p. 783) provide a much firmer foundation for the assignment of the quantum numbers to electrons in closed shells of atoms.[8] This principle is, in fact, a direct corollary of the second of the three "guiding principles" of the old quantum

[8] Pauli had made just this point in a much more explicit way in a letter to Bohr in early December 1924. See Section 9.1.

theory, Ehrenfest's adiabatic principle (cf. Section 5.2). For Pauli, Bohr's vague assertions that electrons could only enter certain correspondence-principle selected orbits in the "build-up" process of constructing an electrically neutral atom by adding electrons to the initially bare nucleus could easily lead to erroneous assignments. By contrast, the adiabatic stability of quantum states, whereby single stationary states revealed by the "dissection" of states induced by application of an initially weak magnetic field would transform continuously as the magnetic field was increased (without alteration of their discrete quantum number assignments) all the way to the strong field Paschen–Back end, led directly, with the addition of an exclusion postulate, given the *unique* identification of the electron state in terms of four quantum numbers, to the numerology observed in the periodic table.

Pauli's demotion of the correspondence principle in atom-building, to which he gave full credit as an "indispensable aid" (p. 770) in understanding the origin of selection rules for atomic transitions, but which he abandoned in his arguments for the exclusion principle, are connected with a deep allergy to classical-mechanical intuition, which he had developed by late 1924, primarily as a result of the enormous efforts expended on the Zeeman effect over many years by theorists, who were ultimately unable to arrive at a dynamically consistent (according to classical mechanics) description of the motion of an atomic electron in an applied magnetic field using the methods of the old quantum theory. He states quite explicitly that "it is not in fact a requirement of the correspondence principle to assign to each electron in a stationary state a uniquely determined orbit in the sense of the usual kinematics" (p. 770) Indeed, the word "orbit" (*Bahn*) appears only twice in the article, while "state" (*Zustand*) occurs twenty-six times. Pauli had clearly become convinced that any satisfactory new quantum theory would necessarily have to eschew the conventional visualizations of particle motion which had become ingrained over three centuries of dominance of Newtonian mechanics. We will see that, this attitude, while in many respects sanctioned by subsequent developments, would lead to an unfortunate rigidity—even hostility—in his initial reaction to the idea of an intrinsic electron spin.

8.3.3 The discovery of electron spin

Few important discoveries in the development of quantum theory can be dated with as much precision as that of electron spin, which we can confidently place a day or two following the arrival of the German physicist Ralph Kronig in Tübingen on January 7, 1925. Kronig at the time was a doctoral student at Columbia University in New York, and was in the middle of an extended tour of several important European loci of research in atomic theory, having already made stops in Cambridge and Leyden (where he worked with Samuel Goudsmit on the Zeeman effect). In Tübingen, Kronig would be hosted primarily by Alfred Landé, who had already made important contributions to the theory of the anomalous Zeeman effect and the helium spectrum (see Sections 7.3.2–7.3.3).

Shortly after his arrival in Tübingen, Kronig was handed a letter that Landé had just received from Pauli summarizing his recent progress in interpreting the closure of electron groups in atoms with the use of a fourth two-valued quantum number directly attributed to each valence electron *separately*, and not to a single putative atomic core. Kronig later claimed that "it occurred to me immediately that [the fourth quantum number] might be considered as an intrinsic angular momentum of the electron" (Kronig 1960, p. 20). Kronig did not leave the matter there, but immediately set out to reinterpret the puzzling relativistic doublets of X-ray spectroscopy model-theoretically with this new idea. We remind the reader that the existence of pairs of inner electron states which appeared to be split in energy by an amount given by the Sommerfeld fine-structure formula from 1916—in particular, with a splitting proportional to the fourth power of the effective nuclear charge Z_{eff}—which had initially seemed such a great success of the old quantum theory, had come to be seen as a serious difficulty once further energy levels were discovered with an energy splitting apparently purely due to screening (and not relativistic) effects. The critical problem was the absence of obvious screening effects in the relativistic doublets, which (Landé had argued) meant that these doublets corresponded to pairs of electron states with the same spatial orbits (thus, azimuthal quantum number $k = k_1$) and so the same screening, but different orientation of the electron orbital momentum to the core angular momentum. In other words, per Landé, these splittings would be analogous to the alkali metal doublet splittings which Heisenberg had calculated in the ancient days of late 1921 with his core model (cf. Section 7.1.2). The problem was that, in the core model, the doublet splittings, when computed with an effective nuclear charge Z_{eff} appropriate for an inner electron in a large atom, inevitably emerged with a dependence proportional to the cube, rather than fourth power, of Z_{eff}, which was the observed behavior.[9]

Kronig realized that with the transference of the fourth quantum number to an intrinsic spin of the electron (rather than the atomic core) engaged in the transition (optical or X-ray) of interest, the energy shifts due to the core–valence magnetic interaction would acquire the additional factor of effective nuclear charge, Z_{eff}, needed to convert the Heisenberg core-model Z_{eff}^3 dependence to the Z_{eff}^4 form observed in the relativistic doublets. The reason is quite simple. In the Heisenberg core model, doublet splittings arose because of the varying orientation of the magnetic moment associated with the atomic core with respect to the magnetic field generated by the circling charge $-e$ of the valence electron. In the new picture, which appears for the first time in Kronig's calculations of January 1925, the energy modifications to the Balmer–Bohr term levels arise because the magnetic moment due to the spin of the circulating electron assumes different orientations with respect to the magnetic field (in the frame of the moving electron) arising from the nuclear charge, which is $Z_{\text{eff}}e$: this magnetic field is therefore Z_{eff} larger than that felt by the core in the earlier model, where the magnetic field at the core was that generated by a single outer valence electron. The result obtained by Kronig was about a factor of 2 larger than the empirically observed splitting (on which more in Chapter 9), but the recovery of the correct Z-dependence was certainly a most encouraging development. Or so it would seem.

[9] Note that an energy correction of order Z^4 to the leading order Z^2 Bohr–Balmer energy will lead to a correction of order Z^3 in the Moseley ordinate, which reflects the square root of the total binding energy.

The explanation of the relativistic doublets in terms of the varying orientation of spin and a given orbital angular momenta ($k = 2$ for the L terms, for example) of the inner electrons in the X-ray spectra also cleared up the difficulty of the screening doublets, which simply corresponded to the energy shifts expected between states of the same principal quantum number but different orbital angular momentum (s versus p states in the L terms, for example).[10] As the screening effects appear in the leading Balmer–Bohr energy term, they dominate the energy difference and lead to a roughly constant offset in the Moseley ordinate, as we saw previously.

Kronig's idea seems to have met with a positive reception by his host Landé, who had struggled mightily with the problem of relativistic versus screening doublets in the previous year. Unfortunately, with Pauli's physical arrival in Tübingen, the whole development came to a screeching halt. According to Kronig's later recollections,[11] Pauli dismissed out of hand his attempts to associate his fourth quantum number with an explicit mechanical source—the angular momentum of an electron viewed as a ball of charge spinning on its own axis—as "a quite clever insight, but Nature is just not like that" ("*ein ganz witziges aperçu, aber so ist die Natur schon nicht*"). We admittedly do not know the specific criticisms Pauli leveled at the idea of an intrinsic electron spin angular momentum, but difficulties with reconciling a classically motivated picture of a ball of charge (with various assumed charge distributions) spinning fast enough to generate a spin $h/4\pi$ (or $\hbar/2$ in modern notation with $\hbar \equiv h/2\pi$) in a relativistically consistent way are not hard to find.[12] We shall see that obstacles of this kind emerged once again toward the end of the year when the idea of spin was independently discovered by Uhlenbeck and Goudsmit. In their case, the objections came from the master of classical electron theory, H. A. Lorentz.

Kronig took Pauli's criticism very seriously (as did Landé, who withdrew his previous support for Kronig's idea) and his calculation showing the agreement (apart from the annoying factor of 2) of the relativistic doublet splitting with a spin–orbit magnetic interaction of the electron was never written up and submitted for publication. However, Kronig does seem to have explained his calculation to Heisenberg in the course of his subsequent visit to the Bohr Institute in Copenhagen.[13] Heisenberg's awareness of the potential importance of the spin idea would play a significant role, as we shall see shortly, in the subsequent emergence, and final acceptance, of electron spin.

The definitive appearance of the idea of electron spin would come ten months after Kronig's ill-fated visit to Tübingen. In mid-October 1925, two graduate students of Paul Ehrenfest in Leyden, Samuel Goudsmit and George Uhlenbeck, submitted a short

[10] In modern notation, for the $N = 2$ L states, the relativistic doublets expose the energy difference of $2p_{3/2}$ and $2p_{1/2}$ states, which arises solely from the spin–orbit interaction (as they have the same orbital angular momentum, thus the same screening), while the pair $2p_{1/2}$, $2s_{1/2}$ correspond to different orbits, as they have different orbital angular momentum, and are therefore subject to the leading order screening effect.

[11] Interview with Kronig for the AHQP by John Heilbron in 1962.

[12] In his AHQP interview (see note 11), Kronig suggests that Pauli's objection might have concerned the excessively large electromagnetic mass which a spinning electron would necessarily develop if its intrinsic angular momentum was as large as $h/4\pi$. This is clearly a speculation on his part.

[13] In his AHQP interview (see note 11), Kronig recalled that this happened "on the ferry between Gedser and Warnemünde ... sometime in the Spring of 1925."

paper (less than two pages) to *Die Naturwissenschaften* in which they proposed a "different approach" to the problem of the fourth quantum number R previously used in the theory of the anomalous Zeeman effect to indicate the angular momentum of the atomic core, the magnetic interaction of which with both the valence electron and the external magnetic field was a crucial part of the phenomenology of the Zeeman effect, in particular, in the heuristic "derivation" of the formula for the Landé g-factor (cf. Sections 7.3.2–7.3.3). After describing the ascription by Pauli of a two-valued fourth quantum number to the valence electron itself (i.e., its removal from the atomic core), Goudsmit and Uhlenbeck suggest that, while the other three quantum numbers (principal, azimuthal, and inner) retained their previous interpretation, the quantum number R was to be ascribed directly "to the rotation of the electron itself" (Uhlenbeck and Goudsmit 1925, p. 954). There are no explicit calculations presented to support this hypothesis, but the authors do recognize in closing that the viability of this hypothesis rests crucially on its ability, on the basis of the "variable orientations of R to the orbital plane of the electron [hence, the direction of the orbital angular momentum vector K]," to yield a correct explanation of the relativity doublets.

Goudsmit and Uhlenbeck's paper had not had an entirely smooth trip from conception to publication, as Uhlenbeck had asked Lorentz to assess the dynamical consistency of an intrinsic electron spin, which, as the acknowledged master of classical electron theory, Lorentz was clearly in position to do. When Lorentz got back to Uhlenbeck, it was with the bad news that any interpretation of a spin angular momentum of an electron based on a model of a rotating charge distribution ran into conflicts with special relativity, as the surface of the electron sphere would be moving faster than the speed of light. Objections of this sort had also probably disturbed Pauli much earlier, when he dismissed Kronig's suggestion of a mechanical interpretation of his fourth quantum number. Fortunately for Goudsmit and Uhlenbeck, Ehrenfest, who was responsible for sending the paper in to the journal, was far more willing to allow his students to entertain "dangerous" ideas, and their short note, establishing for posterity their priority in the discovery of electron spin, proceeded to *Die Naturwissenschaften* unimpeded and was duly published on November 20.

The day after the paper appeared in print, Heisenberg wrote to Goudsmit enquiring about the factor of 2 disagreement in the spin–orbit coupling energy for inner electrons—the empirical splittings seemed systematically about twice as large as one would expect from a direct calculation of the coupling energy of a magnetic electron moving in the nuclear field. Everything else about this result (which, of course, goes back to Kronig) was encouraging, in particular the crucial correct Z^4 dependence that had been such a mystery in the atomic core models. The same issues were raised in a separate letter to Pauli sent the same day, in which Heisenberg asks for Pauli's opinion on the whole idea of a spinning electron. Heisenberg leaves no doubt that, for him, apart from the annoying extra factor of 2, the spin idea "eliminates all difficulties with a magic stroke" (Pauli 1979, Doc. 107). The stage was certainly set for a widespread acceptance of the idea of electron spin in the months to come. By the end of the year, Bohr, on the basis of conversations with Einstein explaining the relativistic origin of the spin–orbit coupling, had also accepted the idea of an intrinsic electron angular momentum. As we shall see in Chapter 9, the whole issue was settled fairly conclusively a few months later, as L. H. Thomas was able to show that in virtue of a rather subtle relativistic precession effect, the spin–orbit energy of a bound electron with an intrinsic magnetic moment in an external nuclear field was indeed reduced by exactly a factor of 2.

Pauli never publicly acknowledged the role he had played in robbing Kronig of the credit for one of the most spectacular discoveries in the history of quantum mechanics. To some degree, however, he did make amends in a letter of recommendation for Kronig, which had been solicited by Coster in an attempt to procure a position for Kronig in Groningen. This letter is quoted in a scientific biography of Kronig. Pauli wrote:

> In my opinion Mr. Kronig is in every respect, on the basis of his scientific achievements, ready for a professorship of theoretical physics ... All his work is characterized by great originality in the framing of the problem and a particular reliability and thoroughness in its execution. It is especially to be emphasized that, although there is no publication from his hand concerning this, at one point he came very close to the discovery of the spinning electron (Casimir 1996, p. 58).

Close, but no cigar, as the saying goes. Before proceeding to a survey of the remarkable evolution that quantum theory was already undergoing at the time of Uhlenbeck and Goudsmit's work, it is important to acknowledge that, although the introduction of the exclusion principle and electron spin were to prove indispensable in resolving the failures of the old theory (see Chapter 7), these ideas played practically no role in the initial radical transformations of the theory initiated by Heisenberg with his matrix mechanics, or a little later by Schrödinger in his wave mechanics. For Heisenberg, the indispensable workhorse on which to exercise the tenets of his "reinterpreted kinematics" was the one-dimensional anharmonic oscillator; for Schrödinger, it was the hydrogen atom, *sans* any fine structure or spin effects. In neither case did electron spin nor the Pauli exclusion principle play a role in the initial developments of the new formalisms. They were, however, crucial in the next stage, where the new methods were systematically applied to the resolution of the incorrect results and loose ends of the old quantum theory, most especially in the theory of the anomalous Zeeman effect and of the helium atom (see Chapter 15).

8.4 The dispersion of light: a gateway to a new mechanics

The study of visible light emitted from and absorbed by atoms was by far the most important source of empirical data constraining the evolution of atomic models following Bohr's introduction of the quantum into atomic theory in 1913. Apart from the study of emission and absorption of light, a third aspect of optical research that would turn out to have a profound impact on the development of modern quantum mechanics was *dispersion*—technically, the coherent scattering of light by the atoms (or molecules) comprising an extended homogeneous medium, typically a gas, liquid, or transparent solid.[14] Newton's experiments with prisms in the 1660s had established the essential facts underlying optical dispersion: a circular "pencil" of white light could be decomposed after passage through a prism into an elliptical spot displaying the colors of the spectrum; if a particular color (say, blue) was selected by passing the light through a small hole at the appropriate spot, the light remained blue in subsequent refractions through other prisms; and the light at the blue end of the spectrum was more bent (*more refrangible*, in Newton's terminology) than the light at the red end. In terms of the index of refraction n, which had already been introduced in quantitative accounts of refraction by Snell and Descartes, the existence of a variable refrangibility indicated that the index of refraction, which was a quantitative expression of the extent to which a light ray underwent bending in passage from air into a transparent medium, was a number greater than one that increased monotonically as one went from the red to the blue end of the spectrum in typical refractive media (such as prism glass).

With the advent of the wave theory 150 years later, and the ascription of color to the varying wavelengths λ (inversely related to frequency ν by $\lambda = c/\nu$, where c is the speed of light), the task of physical optics became clear. Any complete theory of the interaction of light with a material medium through which it passes must be capable of explaining quantitatively the functional dependence of the index of refraction on frequency, $n(\nu)$. By the close of the nineteenth century, Maxwell's electromagnetic theory of light was the dominant paradigm, and the interaction of light with matter was presumed to be determined by the effect of the oscillating electric field generated by a light wave as it passes over the constituent charged "ions" of opposite and, in bulk, cancelling signs in an electrically neutral medium. The point of view generally accepted for the underlying process can be seen in the following excerpt from the second edition of Drude's textbook on optics, in the first section of Chapter 5, "The Dispersion of Bodies:"

> One arrives at a theory which adequately represents the observed phenomena by introducing the assumption that the smallest parts ... of a body possess the possibility of characteristic vibrations. These [vibrations] are excited to a greater or lesser sense to the extent that their [natural] period lies nearer or further from that of the light vibrations impinging from outside. Such vibrations, stimulated by a light wave (i.e., an oscillating electrical force) are immediately understandable if we generalize the conception (made necessary by electrolysis) that each molecule of a body is composed out of positive or negative atoms, the so-called *ions*. These ions are not in general identical with those obtained in electrolysis and they are to be referred to by a different word, for example *electrons* (Drude 1906, pp. 362–363).

8.4.1 The Lorentz–Drude theory of dispersion

The idea that physical optics had become a study of the interaction of light with harmonically bound "mobile ions" within the atoms or molecules of a material substance had already been used by Lorentz a decade before the second edition of Drude's textbook

[14] See the dissertation of Marta Jordi Taltavull (2017) for a history of dispersion in the period of interest to us (1870–1925).

quoted here, to explain the normal Zeeman effect (cf. Section 7.2.1), or splitting of spectral lines when an external magnetic field was applied to the radiation emitted by excited atoms (e.g., in a heated or electrically excited sample). What concerns us here is a somewhat different instance of the interaction of light with matter: the absorption and coherent re-emission of light waves of a given frequency and phase by atoms in their normal (i.e., ground state). By the time of Drude, it was accepted that this process was fully described by the theory of Maxwellian light emission from bound charged particles (with some natural oscillatory frequency ν_0) undergoing forced simple harmonic motion as a consequence of the oscillating electric field (with frequency ν) they experience as a classical electromagnetic field passes over them. In elementary physics courses, one learns that the induced displacement of a simple harmonic oscillator of natural frequency ν_0 when subject to an external harmonic force of frequency ν is proportional to $1/(\nu^2 - \nu_0^2)$, if the harmonic oscillator is undamped. The fact that the oscillator responds with an amplitude that goes to infinity when the external frequency ν is tuned to the natural frequency ν_0 is an artifact of the neglect of damping: including damping introduces an additional term in the denominator that protects the induced displacement from actually becoming infinite. From the induced displacement of a charged "ion", one proceeds directly to the induced electric polarization and thence, by a well-known formula in classical electromagnetic theory, to a formula for the index of refraction.[15]

For now, we ignore the issue of damping, and refer to the effect of the term $1/(\nu^2 - \nu_0^2)$ as the Sellmeier *resonance pole* associated with the presence of oscillators of frequency ν_0, giving due homage to Wolfgang Sellmeier's work in 1872 in relating the index of refraction of light in media to forced simple harmonic motion of massive point particles. The obvious generalization, in which charged particles of various distinct natural oscillation frequencies ν_i (with $i = 0, 1, 2, \ldots$) are present, leads to a formula for the index of refraction as unity plus a sum of Sellmeier terms with resonance poles $C_i/(\nu^2 - \nu_i^2)$. Formulas of this type were indeed found to describe very accurately the measured index of refraction for many transparent media, including the case of "anomalous dispersion", which occurred when one moved from one side to the other of a resonance pole, thereby interrupting the normal continuous monotonic increase of the index of refraction with increasing frequency. Thus, in the closing years of classical physics, a perfectly reasonable "explanation" of the frequency dependence of the index of refraction $n(\nu)$, the core theoretical problem of optical dispersion, seemed at hand.

In one respect, however, the Lorentz–Drude derivation of the Sellmeier formula led to empirical difficulties. The location ν_i of the resonance poles did indeed agree with the observed location of discrete lines in the emission and absorption spectrum

[15] A full physical explanation of the dispersion phenomenon goes something like this: incident monochromatic light (of a fixed frequency ν) passing over a charged oscillator will induce a forced oscillation of the particle, with the same frequency and in a fixed phase relation to the incident wave. The oscillation of a charged particle necessarily results in emitted electromagnetic radiation of the same frequency, so the outgoing light from the particle is a superposition of the incident wave and the additional electromagnetic field produced by the particle itself. The coherent superposition of an incident wave with the emitted stimulated radiation of a uniform array of such charged oscillators results in a wave of the same frequency but, typically, smaller wavelength—that is, a wave of slower phase velocity c/n, with the index of refraction $n(\nu) > 1$.

of the dispersing medium, as one would expect if these frequencies corresponded to natural frequencies of oscillations of the mobile charges in the medium. On the other hand, the numerator factors C_i in the Sellmeier poles were, quite understandably, directly related to the number density $\mathfrak{N}_i$ of electrons capable of executing harmonic vibrations at the corresponding natural frequency ν_i, and could be immediately extracted from measurements by fitting the observed index of refraction to the Lorentz–Drude version of the Sellmeier formula. It turned out that the electron number densities extracted in this manner were usually orders of magnitude *smaller* than expected on the basis of the known densities of atoms in the samples under investigation. This problem remained a severe one through the years of the old quantum theory, and really was resolved properly only with the emergence of modern quantum theory.

8.4.2 Dispersion theory and the Bohr model

The advent of the Bohr model for atomic hydrogen in 1913, and its spectacular success in explaining the Balmer formula, meant at the same time a "Kuhn loss"[16] for the previous success of the classical electron theory approach to optical dispersion. Electrons were no longer thought to be bound by harmonic forces in atoms, at fixed equilibrium positions in the normal (i.e., ground) state of the atom, and then perturbed by the electromagnetic field of a passing light wave into sympathetic oscillations around these equilibrium points. Instead, the normal state of an atom consisted of a stationary nucleus surrounded by electrons executing a discrete set of selected quasi-Keplerian orbits. In fact, after Bohr's talk at the September 1913 meeting in Birmingham of the British Association for the Advancement of Science, where his new model for hydrogen received its first widespread exposure (see Volume 1, pp. 200–201), Lorentz had asked how Bohr's concepts of stationary states connected by quantum transitions could possibly be compatible with the classical dispersion theory (see Dresden 1987, p. 147). The Lorentz–Drude theory of dispersion had proven to be extremely successful at the empirical level, but clearly would have to be radically altered to take this new picture of atomic structure into account. In particular, the treatment of dispersion would have to be compatible with the new, utterly non-classical principle introduced by Bohr to explain the hydrogen spectrum: namely, that electrons in orbits corresponding to stationary states were subject to a *radiation veto*, preventing them from emitting Maxwellian electromagnetic radiation, despite the obvious presence of acceleration in these bounded orbits.

By early 1915, a new approach to dispersion had been introduced by Arnold Sommerfeld and Peter Debye (a former student of Sommerfeld). The rather ad-hoc procedure advocated by these authors was to insist on the application of classical electromagnetic radiation laws *for the small perturbations around Bohr orbits induced by an incident light*

[16] This is the phenomenon, discussed by Thomas S. Kuhn (1962, Ch. 9), that a new theory (or paradigm) cannot explain something that was explained by the theory (or paradigm) it replaced. See Section 15.2, and Midwinter and Janssen (2013).

wave of small amplitude passing over the atom, while the overall Bohr veto of transitions out of a stationary state orbit (on the time scale of the orbital period, or the period of the incident light) was maintained.

Dispersion measurements were available for hydrogen gas, which consists of hydrogen molecules, not separated atoms. Debye adopted Bohr's 1913 model of a hydrogen molecule, in which two electrons circulated around the line between two hydrogen nuclei, at opposite sides of a circular orbit, and subjected the electrons to the small electric field of an electromagnetic wave of frequency ν (Debye 1915).

It is hardly surprising that the resulting formula for the index of refraction obtained by Debye did not have the classic Sellmeier form, in two respects. In addition to resonant pole terms of Sellmeier form, $C_i/(\nu_i^2 - \nu^2)$, there were terms with resonant poles at complex frequencies, corresponding to unstable perturbations of the Bohr orbit. Moreover, the frequencies of the real Sellmeier poles ν_i were *completely unrelated* to the frequencies associated with lines in the spectrum of molecular hydrogen, which by the Bohr frequency condition were associated with the energy difference between Bohr stationary states (divided by Planck's constant). The Debye–Sommerfeld calculations instead related these frequencies to the Fourier components of the initial orbit being followed by the electron(s), and not to any other stationary state orbits which the electron(s) might occupy after a quantum jump. By 1922, Paul Epstein could assert confidently, after an extremely careful rederivation of the Sommerfeld–Debye results, using the full machinery of the canonical perturbation theory of multiply periodic systems, that, given the failure of the theory to yield locations of maximum dispersion and absorption as required by the "Kirchhoff principle"—namely, at the same frequencies corresponding to emission lines of the material—the "conclusion seems unavoidable" that the foundations of a hybrid dispersion theory of this sort were incorrect (Epstein 1922c; see Section 10.2.1).

Progress in reconciling the Bohr view of an atom in terms of electrons following a discrete set of classical paths, distinguished by appropriate quantization conditions, with the Lorentz–Drude–Sellmeier theory, which seemed at least phenomenologically to describe very well the passage of light through transparent media, would require the combination of two ingredients that became available by the end of the second decade of the twentieth century. The first ingredient was Bohr's correspondence principle, in the extended version covering both frequencies and intensities of transitions. We have already seen that the correspondence principle was (at best) a deceptive red herring in the search for the rationale underlying the periodic character of atomic structure as revealed in Mendeleev's arrangement of the elements. This periodicity in the end could only be explained once the Pauli exclusion principle had been asserted. However, the correspondence principle would turn out to be critically important in the developments leading up to the Kramers–Heisenberg theory of optical dispersion—one of the few quantitative predictions of the old quantum theory to survive completely intact the transition to modern quantum mechanics.

The second ingredient essential to progress in dispersion theory was Einstein's quantum radiation theory of 1916/1917 (see Section 3.6). Einstein's remarkable insight was to explain the emergence of the classical laws describing the emission and absorption of electromagnetic radiation from charged particles (at the macroscopic level) in terms of statistical phenomena (at the microscopic, atomic level).

For example, a bound electron executing an orbit would classically, because of the inevitable accelerations in its motion, *continuously* emit radiation, thereby losing energy and dropping to a lower state. Instead of this, Einstein proposed that electrons in an excited state k with energy E_k could make a quantum jump to a state i of lower energy E_i by emitting a light quantum with a frequency given by the Bohr frequency condition, i.e., $\nu_{ki} = (E_k - E_i)/h$. The probability per unit time of such an emission arose from two distinct sources: a *spontaneous emission probability* A^i_k, which was always present (even in the absence of external radiation), and a *stimulated emission probability* $B^i_k\rho$, proportional to the presence of electromagnetic radiation incident on the atom with energy density ρ, at the frequency of the light quanta emitted in the transition $k \to i$. Similarly, in the presence of electromagnetic radiation, the inverse process of absorption, in which an electron would use light energy absorbed from the environment to execute the reverse transition $i \to k$, would not occur continuously, as in classical theory, but randomly and stochastically, with a probability per unit time of an absorption event given by $B^k_i\rho$. The A and B coefficients appearing in Einstein's theory were not calculable with the tools at his disposal, but remarkably, using very general arguments of thermodynamic equilibrium, Einstein was able to recover both the Bohr frequency condition and the Planck black-body radiation law for the dependence of ρ on frequency on the basis of these very simple statistical assumptions.

In a paper entitled "The quantum-theoretic meaning of the number of dispersion electrons," Rudolf Ladenburg (1921), an experimental physicist in Breslau (now Wrocław), made a significant step forward in the interpretation of the Sellmeier formula by combining the Einstein radiation theory with an argument inspired by the correspondence principle equating two distinct approaches, one quantum and the other classical, to calculating the radiation emitted by a charged oscillator. By this point it was known that the numerator factors in the Sellmeier poles simply could not be associated directly with the "number of dispersion electrons" $\mathfrak{N}$, if this was to correspond to the number of valence electrons in the dispersing sample, which were presumably the electrons participating in the dispersion of incident light. Ladenburg showed that by equating the classical formula for the radiation rate of a charged Planckian oscillator immersed in ambient black-body radiation to the Einstein formula for spontaneous emission, one could show that the number of dispersion electrons $\mathfrak{N}$ was not equal to the actual number of electrons N, but rather given by N multiplied by a factor involving the Einstein A coefficient for the transition associated with the Sellmeier pole in question. This factor was in general a real number that was certainly not constrained to be equal to unity. However, while it could be extracted from measurements, this factor could not at this stage be predicted by theory, as it involved the Einstein A coefficient, the relation of which to the underlying quantum dynamics of atomic electrons was as yet unknown.

The year 1923 saw the tenth anniversary of Bohr's remarkable trilogy establishing the planetary Rutherford model, supplemented by quantum conditions on the electron orbits, as the basis for atomic theory. The journal *Die Naturwissenschaften* devoted an entire issue to articles commemorating Bohr's great contribution. Ladenburg, in collaboration with his Breslau colleague Fritz Reiche, a theoretical physicist, contributed an article on "Absorption, scattering, and dispersion in the Bohr atomic model" to this issue (Ladenburg and Reiche 1923). Bohr himself had long been puzzled by the infamous paradox whereby the poles in the well-established Sellmeier dispersion formula were located at the quantum transition frequencies, rather than the frequencies corresponding to actual Fourier components of the electron motion in a Bohr stationary state orbit. In a long paper entitled "On the application of the quantum theory to atomic structure," submitted in November 1922, Bohr had emphasized that "dispersion must be so conceived that the reaction of the atom on being subjected to radiation is closely connected with the unknown mechanism which is responsible for the emission of radiation on the transition between stationary states" (Bohr 1923d, p. 162). Although Bohr had no very definite ideas to suggest that would help to flesh out the quantitative nature of this connection, the problem of dispersion comes up on several occasions in the correspondence between Bohr and Ladenburg in the Spring of 1923, and may have been the impetus for Ladenburg's return to the subject, this time together with Reiche, in an attempt to deepen the connections he had already found in 1921 between the Einstein radiation theory and dispersion phenomena (Duncan and Janssen 2007, p. 585).

Ladenburg and Reiche (1923) fully adopted the guiding methodology of the sharpened correspondence principle, by which quantum transition frequencies and intensities) were asymptotically related to the Fourier frequencies and squares of Fourier amplitudes, respectively, of the classical motion of the radiating electron, in the event that the motion was multiply periodic (i.e., a superposition of a finite set of basic frequencies together with their harmonic overtones). This principle could be used to guess the analytic form of the Einstein spontaneous radiation A coefficient, which Ladenburg had shown two years earlier to be related to the numerator factor in the Sellmeier dispersion formula—and hence, to the number $\mathfrak{N}$ of dispersion electrons.

The extended/sharpened correspondence principle was most easily applied to the simplest dynamical system representing a bound charged particle, the Planckian simple harmonic oscillator of natural frequency ν, as in this case there were no harmonic overtones, just the fundamental frequency ν, which meant that only quantum transitions with unit change in the quantum number n of the stationary state were allowed, $\Delta n = \pm 1$. When the correspondence argument was applied to a system of N such oscillators (moving in an isotropic three-dimensional potential) in the quantum state n, the authors found that the number of dispersion particles needed to account for the numerator factor in the Sellmeier formula was $\mathfrak{N} = ((n+3)/3)N$. Evidently, this meant that the number of dispersion particles $\mathfrak{N}$ agreed with the *actual* number N of particles only when the system was in its ground state (all particles with $n = 0$). For a system with the oscillators all in an excited state $n > 0$, the "number of dispersion particles" was evidently a number (not necessarily an integer!) greater than the actual number of particles. This discrepancy would be resolved a year later with the derivation of a "new, improved" dispersion formula by Hans Kramers.

Ladenburg and Reiche, in their attempts to find a new language to describe the admittedly bizarre situation where the number of apparent "dispersion electrons" could not in general be brought into agreement with the actual number of radiating charged particles in the optical medium of interest, introduced a concept that would play a very important role in the 1924/1925 period, and which would lead to a final, and conceptually satisfactory, solution to the quantum dispersion problem for atomic models of the Bohr type. Namely, it was assumed that optical dispersion phenomena could be understood by treating the radiating charged particles in atoms as "substitute oscillators" (*Ersatzoszillatoren*). In other words, as far as phenomena of absorption, emission, or coherent scattering (dispersion) were concerned, atomic electrons circulating in multiperiodic, quasi-Keplerian stationary state orbits, for some reason, *acted as though* they were actually charged simple harmonic oscillators, with each Bohr transition $r \rightarrow s$ with emitted frequency ν_{rs} associated with a corresponding substitute oscillator with a natural frequency of oscillation equal to ν_{rs}. The peculiar values of the Sellmeier numerator factors (now connected to the Einstein spontaneous emission coefficient A_r^s) could now be adjusted on the assumption that the substitute oscillators were to be treated as having non-standard values for the electron charge e and/or mass m. The idea of such oscillators as the responsible entities for the emission of radiation from atoms was not in fact entirely new: in a response to a question from Langevin in the question session following his talk at the 1921 Solvay conference ("Notes on the theory of electrons"), H. A. Lorentz had speculated

> one could for example imagine that besides the atoms of Bohr a [radiating] gas contains true "vibrators" [oscillators] which could provisionally store the energy lost by the atom in passing from one stationary state to another; it would be necessary in this case for the quantity of [stored] energy be exactly the quantum corresponding to the proper frequency of the vibrator [i.e., $h\nu$ for an oscillator of natural frequency ν] ... Perhaps also the atom temporarily changes itself into a vibrator (Solvay 1923, p. 24).

8.4.3 Final steps to a correct quantum dispersion formula

The derivation of correct quantum theoretical expressions for the interaction of bound atomic electrons with the electromagnetic field would be accomplished, more or less independently, by two physicists working in different continents and in very different scientific environments. The first, Kramers, Bohr's trusted assistant in Copenhagen, we have already met on several occasions in connection with the Stark effect and the Bohr–Kemble model of helium (see Sections 6.3 and 7.4). The other physicist, the young American theorist John Van Vleck, we encountered in our account of the final (failing) attempts of the old quantum theory to deal with the binding energy of two electrons in the helium atom (see Section 7.4). Indeed, the extra technical equipment essential in their attack on the helium atom, a complete facility with canonical perturbation theory in classical mechanics, was the essential ingredient that allowed both Kramers and Van Vleck to go beyond the work of Ladenburg and Reiche, and arrive at formulas for emission, absorption, and (in the case of Kramers, and later, Heisenberg) both coherent (dispersion) and incoherent (Raman) scattering, formulas that survived completely intact the transition to modern quantum mechanics. Of course, the formulas in question contain elements (such as the Einstein coefficients) that could not be calculated from first principles until the full apparatus of the new theory was in place, but the form and structure of the results obtained by Van Vleck, Kramers, and Heisenberg are completely correct, and represent an inspired application of the correspondence-principle methodology, insofar as the authors managed to "guess" the correct results—remarkably, in the absence of a consistent underlying dynamical framework.

While Kramers was finishing his calculations on the binding energy of helium in the Bohr–Kemble model—he submitted his paper on the topic the last day of 1922 (Kramers 1923a)—he was also working on rotational band spectra in diatomic molecules, in collaboration with Pauli. They submitted their joint paper, "On the theory of band spectra," the first week of January 1923 (Kramers and Pauli 1923). Sometime in the next few months, probably by early May 1923 at the latest, when Reiche wrote to Kramers asking for help on the dispersion problem,[17] Kramers must have begun to think more intensely about the difficulties of reconciling the classical dispersion theory with Bohrian atomic models. One presumes after all that Kramers, working in Copenhagen at the Bohr Institute, and in almost daily contact with his mentor, would have shared Bohr's unease about the state of the quantum dispersion problem that Bohr had clearly described in the article on applications of quantum theory to atomic structure, which had been submitted in November of 1922. The conflict between central scientific contributions of Kramers' two greatest heroes, Lorentz and Bohr, would certainly have engaged him directly. Reiche's letter to Kramers on May 9, 1923, informing him of the upcoming publication of his article with Ladenburg in the Bohr commemorative issue of *Die Naturwissenschaften*, might well have been the additional impulse needed for Kramers to focus on the question of the Sellmeier formula—specifically, the problem of the "number of dispersion electrons" $\mathfrak{N}$ and the connection of $\mathfrak{N}$ to the Einstein spontaneous emission A coefficients.

What we do know with considerable certainty is that, by the time the young American physicist John Slater arrived in Copenhagen in late December of 1923, Kramers had derived, and could show to Slater, a generalized form of the Sellmeier–Lorentz–Drude dispersion formula, expressing the electric polarization of a material exposed to an incident monochromatic electromagnetic wave as a *difference of two sums*, the individual terms of the sums taking the classic Sellmeier form, with poles at frequencies corresponding to allowed quantum transitions of the material (as empirically established from the location of emission and absorption lines). For materials in which all the atoms were in their lowest possible energy state (which was the case in the vast majority of dispersion measurements), the sum of negative terms was absent, and Kramers' formula reduces to the usual Sellmeier form as a sum of poles with positive numerators.

How did Kramers arrive at this peculiar generalization of a formula that had survived 50 years of intense empirical testing? As we explain in greater detail in Chapter 10, the reconstruction of the exact chronology of Kramers' discovery of the correct quantum dispersion formula must remain to some degree speculative, as his earliest publication

[17] Reiche to Kramers, May 9, 1923 (Duncan and Janssen 2007, p. 588; see Section 10.3.3).

on the subject came several months after his discovery of the formula, and, somewhat misleadingly, was phrased in terms of a new theory by Bohr, Kramer, and Slater (1924a), of which it is really conceptually independent. Here we merely summarize what the best evidence shows, and follow the chronological sequence of events as best we can determine it.

It seems that by late 1923 Kramers had succeeded in deriving, using the technology of action/angle variables for multiperiodic classical systems, a formula for the coherent polarization of systems of mobile charged particles induced by an applied periodic sinusoidally varying electric field, such as that associated with an incident electromagnetic wave. As mentioned earlier, this polarization can be directly related to the (squared) index of refraction n^2 of the system. In the case of a system of charged particles subject to harmonic forces, and hence, executing harmonic oscillations around their equilibrium positions, the resulting formula for n^2 is then of Sellmeier form. Even for particles executing more general types of multiperiodic motion, however, Kramers found a formula giving n^2 as a sum of terms of Sellmeier type, with poles located at frequencies corresponding to the non-vanishing Fourier components of the classical motion. However, the formula had a critical feature of great importance in the correspondence-principle context where it would be applied: each Sellmeier pole appeared with a continuous derivative with respect to the action variables $\mathcal{J}_i$ of the system. Classically, of course, the action variables of a system were continuously variable quantities, so the derivative could be calculated according to the normal rules of calculus. In the old quantum theory, however, the allowed stationary states of the system were selected by insisting that each action variable was an integral multiple of Planck's constant, $\mathcal{J}_i = n_i h$. In fact the stationary states were identified by specifying the associated set of quantum numbers $\{n_i\}$.

In the correspondence limit of large quantum numbers $n_i \gg 1$, the action variables (such as the energy) changed relatively slowly in a quantum transition when the (large) n_i changed by numbers of order unity and one could asymptotically replace continuous derivatives with discrete ones. In the action/angle formalism, where a change of n_i by unity involved a change of the associated action variable $\mathcal{J}_i$ by h, one could therefore approximate the derivative $df(\mathcal{J}_i)/d\mathcal{J}_i$ of a function $f(\mathcal{J}_i)$ by the discrete difference operation $(f(\mathcal{J}_i=(n_i+1)h) - f(\mathcal{J}_i=n_i h))/h$. Kramers's derivation of a quantum dispersion rule involved the replacement of the action derivatives appearing in his classical formula with discrete differences, in which each Sellmeier pole involved a *difference* of two terms, in which the negative term was related to the positive term by a reduction of the quantum numbers by unity (or, in the event that there were Fourier components corresponding to higher harmonics, the corresponding small whole numbers). It is clear in retrospect that Kramers' work on dispersion in the Fall of 1923 must have involved the invocation of the following *Replacement Postulate*:

Postulate: *In classical formulas expressing the behavior of a multiperiodic system susceptible to description in terms of action/angle variables, each derivative of an action variable $\mathcal{J}_i$ is to be replaced by the corresponding discrete difference operation for the associated quantum number n_i, divided by Planck's constant.*

This apparently precise and concrete instruction represents in some sense the deepest "sharpening" that the correspondence principle would achieve prior to the arrival of modern quantum mechanics. It subsequently became known as the "Born replacement rule", as it appears explicitly in print for the first time in Born's paper entitled "On quantum mechanics" submitted in mid-June 1924 (Born 1924), but it was certainly invoked earlier independently—and used to great effect—by both Kramers and Van Vleck (Duncan and Janssen 2007, pp. 637–638).[18] Note, however, that the replacement rule, as natural as it seems in retrospect, required a certain amount of inspired guesswork, as the replacement of a continuous derivative by discrete differences can be accomplished in a number of distinct ways, which asymptotically coincide, but give quite different results when the proposed quantum formula is extended (as it must be, to be useful) from the regime of large quantum numbers down to that of small quantum numbers where the spectroscopic data available in the 1920s were focused.

The dispersion formula that Kramers found, after application of the replacement postulate, necessarily contained pairs of Sellmeier poles with positive and negative signs, as the translation of the action derivative in each of the terms in the classical formula into a discrete difference automatically generated such pairs. For example, if one considered dispersion of light through a material consisting (for simplicity) of charged Planckian resonators (oscillating harmonically), all in the $n = 1000$ quantum number state, the normal Ladenburg formula, which would contain a single Sellmeier pole with a numerator determined by the Einstein spontaneous emission coefficient for the transition from state 1001 to the given state $n = 1000$, would be supplemented by a similar term with a negative sign, and a numerator with the Einstein coefficient for the transition from $n = 1000$ to $n = 999$. Only in this way would the quantum formula merge into the classical, as the difference of two terms with action $\mathcal{J} = 1000h$ and $\mathcal{J} = 999h$ would approximate, as required by correspondence, a continuous derivative with respect to the action.

However, as Kramers realized, the extension of a formula of this type to the lowest quantum states resulted in an unpaired, purely positive, Sellmeier term for a system in which the resonators (or atoms) were all in their ground state $n = 0$, as there was no negative pole corresponding to the non-existing transition $n = 0$ to $n = -1$. A similar situation existed if the charged particles interacting with light were electrons in quantized Bohrian orbits. As essentially all quantitatively accurate measurements of optical dispersion (in gases, liquids, or transparent solids) were carried out with atoms in their ground state, the relevant dispersion formula—as had been verified empirically for decades, since the days of Sellmeier—would take the normal form, as rewritten by Ladenburg, with Sellmeier poles containing only positive numerator factors determined by the Einstein A coefficients for all possible transitions from higher energy states down to the ground (and therefore, lowest possible) state.

Kramers waited several months before publishing his new dispersion formula. We can only speculate as to the reason for this delay. It is certainly true that there was absolutely no empirical evidence for the need for negative terms in the dispersion formula. Such evidence came only toward the end of the 1920s from Ladenburg and other workers.

[18] In a letter to Born the following year (October 19, 1925), Van Vleck complained that he did not receive enough credit for his contributions in Born's publications (Duncan and Janssen 2007, pp. 569–571; cf. Chapter 10, note 22).

The only motivation for the new terms was the desire to exhibit a quantum result, which (at least in the regime of large quantum numbers) would merge with the classical formulas. This agreement was highly non-trivial, and was an impressive response to the long-standing paradox of the location of the poles of the Sellmeier formula at the quantum transition frequencies $\nu_{rs} = (E_r - E_s)/h$, rather than at the Fourier-component frequencies of the electron motion in a Bohr orbit. Of course, these frequencies did asymptotically coincide in the limit of large quantum numbers, so it seemed perfectly reasonable to start there and by a process of induction guess the form of an exact quantum formula that could be extended down to the interesting regime of small quantum numbers. However, the appearance of negative terms was extremely difficult to reconcile with classical intuition: what, for example, would a *negative* number of oscillators mean if one attempted to interpret the numerator factors literally in terms of a number of dispersion particles?

Kramers' first description of his new dispersion formula appears in a short paper entitled "The law of dispersion and Bohr's theory of spectra" (Kramers 1924a), submitted to *Nature* on March 25, 1924. The critical element of the calculation, as in the classical theory, was a determination of the polarization P induced in a bound charged electron by an incident electromagnetic wave of frequency ν. As motivation for the results to be stated, Kramers invokes the "paper by Bohr, Slater, and the writer" which had been submitted in January to *Philosophical Magazine*, entitled "The quantum theory of radiation" (Bohr, Kramers, and Slater 1924a) and which had by this time become the central object of attention for many atomic theorists attempting to resolve the numerous paradoxes of the old quantum theory as regards the description of the interaction of light with matter. The contents of this paper, which became known as the "BKS theory" for the initials of its authors, is described in more detail in Section 10.4. The theory represents an evolution of an idea of Slater, which he communicated to Kramers and Bohr on his arrival in Copenhagen in December 1923, whereby the absorption and emission of discrete point-like light quanta by bound atomic electrons were governed by a "virtual electromagnetic field" that was continually generated by "virtual oscillators", a new name for the "substitute oscillators", of Ladenburg and Reiche (1923), a set of charged particles moving with the *frequencies of allowed transitions*, and not the frequencies associated with the Bohr orbits.[19]

Bohr and Kramers took up Slater's idea with enthusiasm, but eliminated the Einsteinian point-like light quanta. Despite the evidence of the Compton effect, first reported in December 1922 and published a few months later (Compton 1923), Bohr still insisted on a completely continuous wave theory of light, which he argued (somewhat circularly) was required by the correspondence principle. The essential feature of the theory, which was never given a very precise quantitative form (the only equation in the BKS paper is the Bohr frequency condition $h\nu = E_1 - E_2$), was the loss of exact

[19] Contrary to what the terminology may suggest, the BKS theory did *not* introduce a distinction between "real" and "virtual" electromagnetic fields. It posited a new (and short-lived) conception of *any* electromagnetic field according to which the action of the field on charged particles is stochastic (see Section 10.4 for further discussion).

energy and momentum in individual atomic events. Instead, energy and momentum conservation was restored at the statistical level when one considered a large number of events—via the "communication" between atoms established by a virtual field.[20]

Kramers suggests in his first *Nature* note that the BKS methodology would allow a correspondence-principle approach to the phenomena of dispersion and (light) scattering. After stating the usual sum-over-poles Sellmeier expression for the induced polarization (with all positive terms), Kramers asserts

> The present state of the quantum theory does not allow a rigorous deduction of these laws. It is, however, possible to establish a very simple expression for *P*, which fulfills the condition, claimed by the correspondence principle, that, in the region where successive stationary states of an atom differ only comparatively little from each other, the interaction between the atom and the field of radiation tends to coincide with the interaction to be expected on the classical theory of electrons (Kramers 1924a).

After an explication of the notation to be used to describe the enumeration of all possible quantum transitions between the stationary state of the electron being illuminated and higher states (giving positive Sellmeier terms) and, if present, lower states (giving negative terms), the new dispersion formula is stated, without proof. In a second paper, submitted to *Nature* in July, Kramers (1924b) gives a very abbreviated outline of the derivation of his formula, starting with the classical formula giving the polarization in terms of derivatives with respect to action variables. Roughly contemporaneously, Born's paper "On quantum mechanics" appeared in *Zeitschrift für Physik*, giving a much more explicit derivation (using canonical perturbation theory) of the classical polarization formula, and its transformation (via the replacement rule) to the Kramers dispersion formula (Born 1924).

The virtual oscillators of the BKS theory provided a convenient terminology with which Kramers could justify the rather bizarre presence of negative terms in the dispersion formula. The purely formal character of virtual oscillators, the motion of which after all did not correspond to the actual stationary state orbits of the Bohr–Sommerfeld theory, could be used to explain the negative terms by assuming that the factor of squared oscillator charge divided by mass, e^2/m, which multiplies each term in the Sellmeier formula, while clearly positive for the normal "absorption" oscillators of the Ladenburg dispersion formula, could be negative for the "emission" oscillators, which became active if light was incident on bound electrons in an excited state, capable of dropping to a lower state. However, the historical evidence is incontestable that Kramers derived his generalized dispersion formula well—possibly months—before Slater arrived on the scene with his idea for a virtual field (generated by virtual oscillators) guiding light quanta.

[20] Despite its vagueness, the BKS theory fortunately satisfied one very important trait of any admissible scientific theory, refutability by empirical observation. By the Spring of 1925, the conservation of energy and momentum in individual Compton scattering events had been established by experiments by Bothe and Geiger, and by Compton and collaborators in Chicago, and the BKS theory was formally consigned to "as honourable a funeral as possible" by Bohr (see Section 10.4).

The negative terms in the Kramers dispersion formula turn out to resolve the disagreement noted by Ladenburg and Reiche between the actual number of charged oscillators N and the "number of dispersion electrons" $\mathfrak{N}$ as extracted from the numerator factor in the Sellmeier formula. We recall that these two numbers, which should logically be the same, actually differed by a factor of $(n+3)/3$ for light dispersed in a medium where all the (three-dimensional) oscillators were in the nth quantum state.[21] Thus, agreement obtained only for the situation in which all oscillators were in the ground state $n = 0$. The role of Kramers' "differential dispersion" formula in restoring sanity for the case where the oscillators were in an excited state $n > 0$ is most easily seen for the case of one-dimensional oscillators, where each quantum energy level is non-degenerate. Then, each positive Sellmeier term appears with a numerator proportional to $n + 1$ (from the Einstein spontaneous emission coefficient for the transition $n+1 \rightarrow n$), and is paired with a negative term with numerator proportional to n (for the transition $n \rightarrow n-1$). Meanwhile, the denominators of the two terms are the same, as the frequency for the transitions $n+1 \rightarrow n$ and $n \rightarrow n-1$ are equal, for a simple harmonic oscillator, to the natural oscillation frequency ν_0 of the oscillator. Thus, the discrepancy factor between $\mathfrak{N}$ (from the combined numerators) and N becomes $(n+1) - n = 1$: as expected, coincidence between classical and quantum radiation theories is achieved exactly at large quantum numbers and in virtue of the especially simple character of harmonic oscillators, remains exact even for small quantum numbers, all the way down to $n = 0$.

The missing ingredient (a pairing of positive and negative terms) needed to bring the Einstein quantum radiation theory into congruence with classical electromagnetic theory for large quantum numbers was also identified explicitly—and independently of Kramers—in Van Vleck's paper in the *Journal of the Optical Society of America* on the application of the correspondence principle to radiation theory (Van Vleck 1924a), submitted on April 7, 1924. The detailed calculations supporting the results for absorption announced in this paper were presented in October 1924 in *Physical Review* in an extended two-part paper (Van Vleck 1924b, 1924c). The first part was initially concerned purely with the reconciliation of the classical formulas for the rate of absorption by bound charged particles in multiperiodic orbits with the results expected on the basis of Einstein's quantum radiation theory, in the limit of large quantum numbers.

The classical Larmor formula for the rate of absorption of light energy for a bound charged particle immersed in an ambient electromagnetic field corresponding to blackbody radiation could be brought into agreement with the Einstein radiation theory by calculating the *differential absorption*—i.e., the difference between the directly absorbed radiation energy per unit time, *minus* the stimulated radiation emitted per unit time by the particle as a consequence of its immersion in the ambient field. The absorption and stimulated emission rates were both given by the Einstein B coefficients, which, in turn, are simply related to the A coefficients that determine the spontaneous emission rate of the charged particle (in the absence of an external field), the classical limit of which

[21] For a one-dimensional simple harmonic oscillator, this factor would be just $n + 1$.

was (of course) just the Larmor formula. Van Vleck was able to show the agreement of the quantum and classical radiation theories in the correspondence limit only if the stimulated emission energy loss predicted by Einstein's theory was correctly included.

At the end of the first part of his *Physical Review* paper,[22] Van Vleck mentions Kramers's notes in *Nature* containing the new dispersion formula, and a section (15) outlining the derivation of this result was added. Van Vleck's (1924b, 1924c) two-part paper is beautifully and clearly written, and provides a much more accessible introduction to the essence of the correspondence-principle approach to radiation than either of Kramers' two, very elliptical, notes in *Nature* or the slightly more detailed, but overly formal, paper by Born (1924), which also presents a derivation of the Kramers formula based on canonical perturbation theory and the replacement rule.[23]

8.4.4 A generalized dispersion formula for inelastic light scattering—the Kramers–Heisenberg paper

Following his receipt of the Nobel Prize in Physics in 1922, Bohr was able to secure expanded funding for his Copenhagen Institute for Theoretical Physics, not only from the Danish government, but also from the International Education Board (IEB) of the Rockefeller Foundation. Much of the funds obtained from the latter source were deployed as longer-term (up to one year) fellowships for visiting promising young scientists engaged in forefront research in atomic theory. Bohr had convinced Born in the summer of 1924 to allow him to "borrow" Heisenberg from his position as Born's assistant in Göttingen, bringing him to Copenhagen with the financial support provided by the IEB. By all accounts,[24] Heisenberg's first visit to the Bohr Institute had left him with an attitude of extreme hero-worship with respect to Bohr, combined with a decided sense of insecurity, if not inferiority, with respect to Kramers, Bohr's right-hand man at the Institute. Heisenberg returned to Copenhagen in mid-September to take up his IEB fellowship—having chosen, as letters to Bohr show, precisely a time at which Kramers would be abroad (he was attending the *Naturforscherversammlung*—the annual meeting of the Society of German Natural Scientists and Physicians—held in Innsbruck that year), leaving the path clear for a direct attack by Heisenberg on Bohr's attention—and affections.

In the event, Kramers returned to Copenhagen by the end of September, and Bohr's interaction with Heisenberg was gradually supplanted by Kramers, who (in any event)

[22] In a note added in print to the published version of his first short note on absorption in the *Journal of the Optical Society of America*, Van Vleck indicates that he had just seen Kramers' first *Nature* note on dispersion, published May 10. He writes: "By pairing together positive and negative terms in the Kramers formula, a differential dispersion may be defined resembling the differential absorption of the present article. It is found that this differential quantum theory dispersion approaches asymptotically the classical dispersion by the general multiply periodic orbit" (Van Vleck 1924a). Evidently Van Vleck had obtained, by the time his first note on absorption was published, the derivation of the Kramers formula, which he later presented in full detail in his *Physical Review* papers.

[23] See Duncan and Janssen (2007) for a detailed account of Van Vleck's involvement in this critical phase of the development of modern quantum mechanics.

[24] See, for example, Cassidy (1991, pp. 183–185) and Dresden (1987, pp. 264–266).

was far more involved than Bohr in the technical nitty-gritty of calculational atomic theory, of which Heisenberg was by now a master. After finishing a paper on applications of the correspondence principle to resonance fluorescence, Heisenberg became increasingly involved in conversations, and at some point detailed calculations, in dispersion theory. By this time, Kramers had finally decided to publish a detailed account of his correspondence-principle approach to the dispersion problem, leading to the dispersion formula, with the strange negative terms.

Of course, by this time, more or less detailed derivations of the Kramers dispersion formula by Van Vleck and Born were already in print. However, Kramers realized that the theory could be further generalized to deal with processes in which the optically active electrons were responsible for *incoherent* light scattering, as well as the purely coherent dispersive effects treated earlier. The possibility of processes in which an incoming light quantum could be absorbed by an atom and reemitted with a different energy/frequency, with the energy discrepancy compensated by a change in the energy level of an optically active electron, had recently been raised in a paper in *Die Naturwissenschaften* by Adolf Smekal (1923). At the classical level, the incident lightwave produced an induced dipole moment in each scattering electron with Fourier components at frequencies differing from that of the incident wave. The energy gained or lost by the electron would leave it in a different classical orbit. At the quantum level, for Smekal, who was a firm believer in the light-quantum hypothesis and in Einstein's quantum radiation theory, any such process in which energy and momentum of the initial atom plus incident light-quantum were equal to that of the final atom plus outgoing light-quantum was a possible observable phenomenon. These inelastic Smekal processes, which would be observed experimentally four years later by the Indian physicist C. V. Raman (1928), were a clear generalization of coherent dispersion, in which the atom was left unaltered and the scattered light was identical in frequency (and phase coherent with) the incident light. As Heisenberg began his discussions with Kramers, the latter was already thinking about possible generalizations of his dispersion theory. In line with the philosophy espoused by the BKS paper, which had not yet been refuted experimentally, Kramers substituted purely classical Maxwellian electromagnetic waves passing over quantized atoms for the unappetizing (to Bohr and Kramers) discrete light quanta of Einstein and Smekal.

Heisenberg's desire to work directly with Bohr, and his competitive urge to replace Kramers as Bohr's chief theoretical lieutenant, were not in the end required, as Bohr made it plain to his acolyte that the development of dispersion theory along the lines of Kramers's work was a matter of prime urgency—perhaps as this theory was now regarded as one of the main supporting pillars of the BKS scheme of virtual oscillators. Heisenberg thus ended up engaged in intense conversations with Kramers in the final months of 1924, culminating in the submission in early January 1925 of their seminal joint paper "On the dispersion of light by atoms" (Kramers and Heisenberg 1925). There is no question that Kramers was by far the most important contributor to this work. Kramers's biographer asserts that "he devised the methods and carried out most (but not all) of the calculations" (Dresden 1987, p. 273). Heisenberg's primary role was as a persistent and perceptive critic of the emerging results, subjecting them to all the conceptual tests of physical consistency he could devise.

The formula for inelastic light scattering which emerged from this work—now universally called the "Kramers–Heisenberg" formula—did display, in addition to the negative poles, which were by now accepted as essential consequences of quantum/classical correspondence, new "false resonances" at frequency locations, which seemed to have no interpretation in terms of quantum transitions of the optically active electrons. Heisenberg criticized the appearance of these resonances on physical grounds, while Kramers insisted on the necessity of their appearance as a consequence of the correspondence-principle replacement rule. Eventually, under the relentless pressure of Heisenberg's objections, Kramers cooked up an argument that suggested the false resonances would be cancelled by destructive interference from simultaneous spontaneous emission events. In fact, this argument is incorrect—the "false" resonances are indeed present, as indicated in the Kramers–Heisenberg formula—but their correct physical interpretation would have been premature at this stage, as it really depends on the understanding of rather subtle effects in quantum electrodynamics (see Section 10.5). Nonetheless, Heisenberg's participation in the discussions on inelastic light scattering was enough to cause Bohr to intercede directly with Kramers, and request the inclusion of Heisenberg's name on the paper, which was received by *Zeitschrift für Physik* on January 5, 1925.

The word *Streuung* in the title of the paper of Kramers and Heisenberg is translated as "dispersion" (of light) in the anthology by van der Waerden (1968) but is more accurately rendered as "scattering". Indeed, sections 3 and 4 of the paper are concerned explicitly with coherent and incoherent "scattered radiation" (*Streustrahlung*), respectively. The object of study here is the light radiated from bound atomic electrons as a result of the induced displacements of the electron position caused by an incident monochromatic electromagnetic wave. This displacement (via the corresponding induced dipole moment) results in radiated light which can be:

1. in the same direction, of the same frequency and in a fixed phase relation (coherent) to the incident light (the "secondary waves" of this type lead to a change of wavelength of the forward scattered light, hence, classical optical dispersion).
2. coherent light of the same frequency but a different direction to the incident wave (Rayleigh scattering, stronger for higher frequencies, hence, the blue color of the daytime sky).
3. light of a different frequency from the incident wave, with the electron changing its classical orbit or quantum stationary state (Raman–Smekal scattering). As we shall see shortly, the secondary waves produced by the scattering electrons are in random phase relation to the incoming wave from one atom to the next, so the net re-radiated light is the sum of destructively interfering contributions—the archetypal *incoherent* radiation—and consequently much weaker than the coherent, Rayleigh-scattering case.

Two years after modern quantum mechanics was already established and widely accepted, the extremely faint scattered radiation in this third category would first be detected by Raman (1928), winning him the 1930 Nobel prize for Physics.[25]

The main achievement of the Kramers–Heisenberg paper was to use the correspondence principle in the regime of large quantum numbers where classical formulas for light scattering were unambiguous to "obtain, quite naturally, formulae that contain only the frequencies and amplitudes which are characteristic for the transitions, while all those symbols which refer to the mathematical theory of periodic systems will have disappeared" (Kramers and Heisenberg 1925, p. 234). Specifically, the frequencies which appear as integer harmonics in the Fourier decomposition of the classical orbits (and which only in the limit of large quantum numbers correspond to the Bohr transition frequencies) "disappear" and, licensed by the correspondence principle, are simply replaced with the empirical quantum transition frequencies for all quantum numbers. Similarly, the Fourier coefficients of the classical orbits are replaced by complex quantities $\mathfrak{A}_q$ ("q" for "quantum"), which Kramers and Heisenberg call the "characteristic amplitudes for the transition under discussion." The absolute squares of these complex amplitudes are simply related (as Ladenburg had proposed four years earlier) to the Einstein spontaneous transition A coefficient for the corresponding transition. The phase of the amplitude is irrelevant (in both the classical and quantum formulas) in those cases in which an amplitude appears multiplied by its complex conjugate, which is the case both for dispersion (case 1) and elastic (Rayleigh, case 2) scattering.

But in general, and especially in the case of the inelastic component of the scattering (case 3), in which the final state of the electron differed from its initial state, the phase of the complex transition amplitude was critical. In the classical case, this phase simply corresponds to the phase difference between the incident wave and the motion of the particular electron being influenced by the wave. For coherent scattering, the scattered wave is given by the product of the amplitude for transition from the initial state to another "virtual" state, multiplied by the amplitude for the return of the electron from the virtual state to its initial state. These two amplitudes are complex conjugates of each other so the phase information disappears, and the induced scattered light is exactly in phase with the incident light.

But for inelastic (Smekal–Raman) scattering, the electrons execute a transition to a different state from their initial one (again, via an intermediate "virtual" state). Classically, electrons in different atoms can end up in their new orbits at completely uncorrelated phases, so the secondary waves induced by separate electrons are not in phase, and the resultant scattered light is incoherent and much weaker than in the coherent case (e.g., Rayleigh scattering). Similarly, in the quantum scattering formula, the product $\mathfrak{A}_{q'}\mathfrak{A}_q$ of characteristic amplitudes for a transition q from an initial state P to another state R, if followed by a quantum transition q' from the state R to a state Q (different from P) will contain an intrinsically random net phase, varying randomly from one scattering electron to another. The intrinsic, and necessary, complexity of the characteristic quantum amplitudes—inherited, via the correspondence principle,

[25] The Raman effect, as it has come to be known, was found essentially simultaneously and independently by two Russian physicists, Landsberg and Mandelstam (1928).

from their classical Fourier coefficient counterparts—would become one of the most essential features of the reinterpretation of these amplitudes as replacements for the classical electron coordinates in Heisenberg's revolutionary *Umdeutung* paper.

The final stage of dispersion theory precedent to Heisenberg's breakthrough of summer 1925 came in the form of a specialization of the Kramers formula for coherent light scattering, namely, its high frequency limit—as, for example, with hard X-rays. The examination of this asymptotic limit, in which the energy of the incoming light quanta far exceeded the binding energy of the atomic electrons (which could therefore be regarded as essentially free charges), would yield the final resolution of the puzzling disparities between the number of dispersion electrons extracted from the empirical data and the actual number of optically active electrons in the material under examination. The result obtained in this limit, found (independently it seems) by Werner Kuhn (1925, submitted in May) working in Copenhagen at the Bohr Institute and by Fritz Reiche and his doctoral student Willy Thomas in Breslau (Reiche and Thomas 1925, submitted in August), was a sum rule giving unity for the total "oscillator strength" (defined as a dimensionless number proportional to the corresponding Einstein A coefficient) for each optically active electron, provided transitions downward from the state of the electron (if it happens to be in an excited state) are counted negatively in the sum.[26] For a Planck resonator—a charged simple harmonic oscillator—in the quantum state n, the oscillator strength f_n^{n+1} for the absorption transition $n \to n+1$ was just $n+1$ (as Ladenburg and Reiche hypothesized), while the oscillator strength f_n^{n-1} for the emission transition $n \to n-1$ was n, so the Thomas–Kuhn or Thomas–Reiche–Kuhn sum rule (as it came to be known) corresponds to the trivial identity $n+1-n=1$.[27] The derivation provided by these three authors showed that a sum rule constrained the numerators of the dispersion formula even for the much more complicated multiply periodic systems for which the Kramers formula had been derived. This essentially put to bed the troublesome question of how to relate the numerator factors in dispersion formulas to the actual number of electrons involved in light scattering. Still, except for the extremely simple case of charges bound by harmonic potentials (which had been dealt with by Ladenburg and Reiche), there was no procedure for *calculating* individual characteristic quantum amplitudes from first principles for small quantum numbers—in particular, in the case of primary phenomenological interest, for atomic electrons bound by Coulombic electrostatic forces.

8.5 The genesis of matrix mechanics

8.5.1 Intensities, and another look at the hydrogen atom

In his contribution to a memorial volume for Pauli, published in 1960, Heisenberg offered, as its title announced, some "remembrances of the time of development of

[26] Van Vleck appears to have been the first to hit upon this sum rule but he failed to recognize its importance at the time (see Chapter 10, note 96).

[27] The reader will recall that the simple harmonic oscillator has the property that only transitions between adjacent quantum states are allowed.

quantum mechanics." After describing Pauli's discovery of the exclusion principle, as he became aware of it in December 1924, Heisenberg lists the main topics of discussions with Pauli in early 1925:

> Further important points of our discussions were at that point the Ornstein intensity rules for multiplets, as well as the Kramers dispersion theory. In both cases Pauli and I were in agreement that the transition from an only symbolically useful and thus only qualitatively correct model-based mechanics in the sense of the Bohr–Sommerfeld conditions to a real quantum mechanics was to be had by guesswork; that one had to mathematically sharpen the Bohr correspondence principle as far as necessary to arrive at the correct formulas. It was therefore perhaps possible someday, simply by clever guessing, to accomplish the transition to a complete mathematical framework of quantum mechanics (Fierz and Weisskopf 1960, p. 42).

Heisenberg's mention of the Ornstein rules connects to the other research project which absorbed his attention in the early part of his stay in Copenhagen in 1924/1925: the polarization of scattered light under conditions of resonant fluorescence, where the incident light is at just the right frequency to excite quantum transitions in the irradiated material (typically, a bulb containing sodium vapor), which is also immersed in a magnetic field. The re-radiated fluorescent light is then observed at various angles to determine its polarization, which is an indication of the magnetic quantum numbers of the quantum states to which the atoms in the vapor were excited by the incident light. Bohr was in the process of preparing a paper (Bohr 1924c, submitted in November to *Die Naturwissenschaften*) on this subject and must have had extended conversations with Heisenberg about it, as the latter also prepared and submitted a paper on the same subject at the end of November (Heisenberg 1925a).

At this time, in the Fall of 1924, the BKS theory was still the accepted wisdom at the Bohr Institute, and the discussion of light scattering was phrased (in the papers of both Bohr and Heisenberg) in terms of the virtual oscillators of this theory, which determined (in a probabilistic way) the absorption and emission of light by atomic electrons in Bohrian stationary states. A particular difficulty that occupied Bohr—and by osmosis, Heisenberg—was the subtle problem of the transition from a degenerate system (say, the atom in the absence of an external magnetic field, with stationary state electron orbits of equal energy with various orientations and shapes) to a non-degenerate system (say, the same atom with a weak applied magnetic field, where only a special selection of spatially quantized orbits were now allowed, with orientations discretely determined by the direction of the magnetic field).

This problem is connected at a deep level with the paradoxical situation of orbit ambiguity in the old quantum theory, where the application of the Bohr–Sommerfeld procedure led to orbits of completely different shapes (but the same energy) if the quantization rules were applied in different, but a priori equally acceptable, coordinate systems to a degenerate system. We encountered this problem in the context of the Stark effect (Section 6.3, p. 293). Lifting the degeneracy, by even a very weak magnetic field, would select a particular coordinate system, and the shape of the resulting orbits would depend very strongly on the direction of the field. This orbit instability seemed completely incompatible with the principle of "spectroscopic stability", which asserted that the spectral properties of optically active electrons—and therefore, presumably, their orbital behavior—varied smoothly and continuously when one gradually turned on a weak external field. The net result of Bohr's reasoning at this time was to demand a disconnection between the behavior of the virtual oscillators responsible for the interaction of the atom with radiation, and the underlying Bohrian orbits of the optically active electrons. Or, as Heisenberg would put it in his paper on fluorescence:

> the virtual oscillators are connected in only a very symbolic way with the motion of the electrons in stationary states (Heisenberg 1925a, p. 617).

The details of Heisenberg's correspondence-principle-inspired study of polarized fluorescent light need not concern us too much here, although the emphasis in this paper on a purely formal role for Bohrian orbits, and the abandonment of a detailed spatio-temporal picture of electrons orbiting nuclei in quasi-Keplerian paths, would be greatly amplified in the algebraically formulated quantum mechanics that Heisenberg would propose the following summer. Even though the BKS theory had been empirically refuted at that point—experiments by Bothe and Geiger (1924, 1925a, 1925b) and Compton and Simon (1925) showed that energy and momentum were conserved in individual scattering events and not just statistically as the BKS theory demanded—the concept of virtual oscillators, which had been imported from dispersion theory, continued to serve as a convenient placeholder to represent the state of quantum systems in lieu of orbits (Duncan and Janssen 2007, pp. 613–617).

An important ingredient of Heisenberg's discussion of resonant fluorescence, the use of the "Utrecht intensity sum rules" requires mention at this point, however, inasmuch as the consideration of spectral intensities would play an important role in the chronology of theoretical investigations that would lead Heisenberg to the introduction of matrix mechanics.[28] It had long been an ambition of atomic spectroscopists to bring the quantitative measurement of the intensities of spectral lines to a higher level of precision, which would perhaps give additional clues to atomic structure analogous to the spectacular progress (such as the Bohr model of hydrogenic atoms) that the much more precise measurement of spectral frequencies and wavelengths had achieved since the closing years of the nineteenth century. By the early 1920s, technological progress with measuring and calibrating precisely the blackening of photographic plates had allowed the Utrecht group led by Leonard Ornstein and Herman Burger (1924a, 1924b, 1924c) to obtain intensity measurements for resolvable spectral lines appearing in complex multiplets, and in some cases of the further split lines that appeared in the Zeeman effect when magnetic fields were applied, to a precision of a few percent—still far short, of course, of the precision to eight or nine decimal places in measurements of spectral wavelengths,

[28] This strand in Heisenberg's work in the months leading up to matrix mechanics is the focus of a paper by Blum et al. (2017) and the dissertation of one of its co-authors, Martin Jähnert (2016), which formed the basis for his book (Jähnert 2019).

but now accurate enough to start to formulate phenomenological sum rules that could rival the empirically successful "number mysticism" of Sommerfeld in dealing with the frequency splittings in the anomalous Zeeman effect, for example.

In fact, the problem of the relative intensity of magnetically split lines in resonance fluorescence is intimately related to the inelastic (Smekal) light scattering problem, which Heisenberg was discussing with Kramers at the same time in Copenhagen in the Fall of 1924. The emission of a light quantum with a frequency magnetically shifted from the incident light frequency could be regarded as a special case of Smekal (later, Raman–Smekal) light scattering, the quantitative treatment of which would culminate in the famous Kramers–Heisenberg formula. The intensities of spectral lines were, after all, directly connected to the Einstein spontaneous emission (A) coefficients, which appear in the numerators of the Sellmeier poles in the dispersion formula. What must have become clear once the Utrecht sum rules were applied successfully to explain the intensity ratios—as rational numbers, involving whole number ratios—for numerous spectral multiplets, both with and without magnetic splitting, was that the correspondence principle would require serious amplification (or "sharpening") to deal with the intensity problem. The reason is simply that in the large-quantum-number regime, the correspondence principle applied to the spectral intensity problem involved the classical Fourier coefficients of the electron orbits, which, in the simplest case, a single electron in a hydrogenic atom, had been found by Kramers in his doctoral thesis to be rather messy Bessel functions, the ratios of which were obviously not simple rational numbers. Thus, the correspondence principle route did not seem to be compatible for Bohr atoms in any simple way with the emerging numerology for spectral intensities, unlike the case for Planckian resonators (with simple harmonic motion), where the methods of Ladenburg and Reiche had succeeded, as discussed, in reconciling the numerators in the Kramers formula with sharpened correspondence-principle formulas for the Einstein coefficients.

After returning to Göttingen in April 1925, Heisenberg made a determined attack on the intensity-ratio problem for the simplest case relevant to observed atomic spectroscopic data: the hydrogen atom, where the explicit analytic formulas of Kramers could be examined in detail for large quantum numbers, and a simplified quantum result involving integer ratios obtained, that could then be extended, necessarily with inspired guesswork, to a formula that would hold all the way down to the low-quantum-number regime where actual spectroscopic data existed. In fact, using known asymptotic formulas for Bessel functions, Heisenberg was able to guess a sharpened version of the formula for the Einstein coefficient for allowed transitions in the hydrogen atom (i.e., those in which the electron transitioned from a state (n, k) to a lower one $(n - \tau, k - 1)$, where n and k are the principal and azimuthal quantum numbers, respectively, and $\tau = 1, 2, 3, \ldots$), which agreed asymptotically with Kramers' Bessel functions but satisfied the selection rules for atomic transitions, for all values of the quantum numbers. These formulas involved broken factorials (products of successive whole numbers, but not extending down to unity), so the appearance of whole numbers encouraged the hope that a derivation of Utrecht-style intensity relations was close.

What Heisenberg wished to do, but could not do at this stage, was to show that such a formula followed necessarily from a formalism in which classical derivatives were replaced by discrete differences, which would simultaneously result in the conversion of the unwanted Bessel functions to simpler algebraic functions of the quantum numbers. After all, this approach had led to the brilliant successes of the Kramers dispersion theory, the results of which were by this time broadly accepted in the theoretical community. In any case, Heisenberg's guess for hydrogen atom intensities was incorrect—although he did not know this at the time. The hydrogen-atom project was abandoned sometime in May, and Heisenberg returned (as we know from his correspondence with Kronig in early June) to the simpler problem of the one-dimensional anharmonic oscillator, with results that would alter the course of twentieth-century physics.

8.5.2 The *Umdeutung* paper

The frame of mind expressed in Heisenberg's correspondence in May and June 1925—primarily with Kronig, Kramers, and Pauli—concerning his struggles with the construction of a new quantum kinematics alternated between statements of optimism and admissions that the way ahead remained obscure and difficult. We cannnot construct the exact sequence of events with precision at this late stage: indeed, there are some inconsistencies in the later recollections of Heisenberg with the surviving letters he exchanged with his closest friends at the time (discussed in more detail in Chapter 11), which suggest that Heisenberg made somewhat less progress in his visit to Helgoland in June 1925 than later hagiographical accounts of this episode claim. But by the end of July, in the paper given to Born and submitted by him to *Zeitschrift für Physik*, the basic form of a new quantum-theoretic kinematics, which would eventually become the fully self-supporting arch of modern quantum mechanics, had emerged. The paper, entitled "Quantum-theoretical reinterpretation of kinematic and mechanical relations," is now universally referred to by historians of physics as the *Umdeutung*, or "reinterpretation" paper.[29] From here on, we use this appellation to refer to Heisenberg's extraordinary article as it points to the most critical feature of Heisenberg's achievement—the *preservation* of the formal structure of the equations of classical mechanics—while attaching a completely new *interpretation* to the kinematical symbols (coordinates, momenta, energy, etc.) appearing in these equations. The resulting dynamical formalism was applied (successfully) to the problem of the quantized energies of the one-dimensional anharmonic oscillator, and (not so successfully) to the problem of band spectra and Zeeman line intensities.

In the first section of his paper, Heisenberg introduces, via the correspondence principle, a new kinematical framework for describing mechanical systems in quantum theory. The coordinates of a particle, formerly represented each by real functions of time prescribing the path of the particle, were now replaced by assemblies (arrays) of complex numbers, which extend the notion of the Fourier components of the coordinate in the special case of periodic or multiply periodic systems. Let us consider for simplicity, as

[29] The term *Umdeutung* also appears in the title of a paper in which Sommerfeld (1922a) adapted Voigt's classical theory of the Zeeman effect to the old quantum theory (see Section 7.3.2, especially p. 344, note 41).

does Heisenberg, a one-dimensional system of a bound particle with a single coordinate $X(t)$, executing a periodic motion. The discrete stationary states of this particle are enumerated with the single integer n. The Fourier component $X_\alpha(n)$ (necessarily complex, as they are associated with the time dependence exp $(i\alpha\omega_n t)$, where $2\pi/\omega_n$ is the period of the motion in the nth state), which classically corresponds to the αth harmonic component of the motion of the particle in the nth quantum state, was associated, via the correspondence principle, with quantum transitions $n \to n + \alpha$. This association was now generalized to all states (not just large quantum number ones) by symmetrizing the relation between the initial and final state in a quantum transition, and replacing the Fourier components $X_\alpha(n)$ with the two-index arrays $X(n, m)$ (where here $m = n + \alpha$). The derivation of the Kramers dispersion relations had already shown that when squares of these coordinates occurred classically (as in the numerators of the Sellmeier poles), the resulting quantity involved a sum over "intermediate" states, so that the array element of X^2 corresponding to the transition $n \to m$ was a sum of the products of $X(n, p)$ (for the transition $n \to p$) times the amplitude $X(p, m)$ (for the transition $p \to m$), *summed over all states* p. This peculiar rule for multiplying assemblies of amplitudes corresponding to the coordinate of a particle's motion was in fact just matrix multiplication, but Heisenberg was, at the time of the *Umdeutung* paper, clearly innocent of the nomenclature and procedures of matrix algebra, although he clearly appreciated some of the unfamiliar properties of the multiplication of arrays by this rule—chief among them that the result of multiplication of two *different* arrays depended in general on the order of the arrays.

Heisenberg's strategy is summarized very directly at the beginning of section 2 of his paper: in "earlier theory" (i.e., what we now call "the old quantum theory") one proceeded in two steps. First, one solved the equation of motion, basically Newton's second law (acceleration = force divided by mass). Second, one imposed the Bohr–Sommerfeld quantization condition for the action variable $\mathcal{J}$ (= integral of momentum with respect to coordinate over a single cycle of the motion), setting it equal to a whole number times Planck's constant, $\mathcal{J} = nh$.

The new quantum mechanics would follow an analogous procedure, except that the coordinates and velocities appearing in the equation of motion would be replaced by two-index coordinate amplitude arrays labeled by a pair of quantum numbers. Squares, cubes, and so on of the coordinates or velocities appearing in the equation of motion would be interpreted using the rule for multiplying arrays suggested by former experience in the Kramers dispersion theory. The second element, the introduction of Planck's constant in the theory, was achieved by imposing a non-linear constraint on the coordinate arrays, which Heisenberg derived by a simple correspondence-principle argument (replacing action derivatives with discrete differences) applied to the Bohr–Sommerfeld quantization condition of the "earlier theory." But in essence, the two-step process was repeated as previously, but with a *reinterpretation*/*Umdeutung* of the symbols appearing in the equations. The application of this procedure to the Bohr–Sommerfeld quantization condition resulted in the Thomas(–Reiche)–Kuhn sum rule, found (by Thomas, prior to the *Umdeutung* paper) in dispersion theory, which inspired confidence that this was indeed the correct quantization condition in the new theory (Duncan and Janssen 2007, pp. 654 and 659).

Of course, Heisenberg had to check the validity of the results obtained by the new procedures, assuming that they could be applied unambiguously. As the systems that Heisenberg examined were the analytically more tractable one-dimensional oscillators (harmonic and anharmonic, with cubic and quartic anharmonic terms in the potential energy)—and not the empirically more interesting hydrogen (or larger) atom—he could only rely on the theoretical results obtained in the old theory. Quantized oscillations were indeed supposed to be responsible for the measurable vibrational energy levels of diatomic molecules, but the interpretation of the data was much more complicated than for the spectrum of hydrogenic atoms, for example. There were, however, elaborate formulas for the expansion in powers of the anharmonic coefficients of the quantized oscillator energy levels that Born had obtained in the old quantum theory using Hamilton–Jacobi methods, and could now be consulted in his "Lectures on the mechanics of the atom" (Born 1925, Sec. 42). Agreement at some level with these detailed results of the old theory would certainly be encouraging.

There was also the desire that the results of the new theory display internal consistency—in particular, the Bohr frequency condition relating the frequencies of light-quanta emitted in quantum transitions to the energy difference of the initial and final quantum states should be satisfied—and comply with time-honored mechanical theorems such as conservation of energy which, post-BKS, were once again presumed to hold exactly (and not merely statistically) at the microscopic level. The anharmonic oscillator models examined by Heisenberg in the *Umdeutung* paper could not be solved exactly, but he was able to verify conservation of energy and the Bohr frequency condition up to second order in the anharmonic coefficient, in a perturbative expansion in this parameter. Heisenberg does not in fact (notwithstanding the assertion made following Eq. (27) of *Umdeutung*) obtain a formula for the quantized energy of the quartic oscillator that exactly agreed with Born's result (although the two agree as regards the leading terms for large n), but his formula is absolutely correct in quantum mechanics, and can be regarded as a brilliant fulfillment of the "sharpening" strategy for the correspondence principle that had guided theoretical work in atomic physics for the previous five years.

In the final two pages of the *Umdeutung* paper, Heisenberg turns to the consideration of two important problems in spectroscopy intimately connected with the quantum mechanical treatment of rotational motion. The first was the rotationally quantized band spectra of molecules composed of two atoms, which arose from transitions between states with a different magnitude of angular momentum of the whole molecule, which could be modeled as a rigid body with a specified moment of inertia I, with rotational energy given, as in classical mechanics, by $\vec{L}^2/2I$, where $\vec{L}$ was the rotational angular momentum of the molecule about its center of gravity. Empirical observation had shown, as first noted by Adolf Kratzer (1923), that the quantized values of $\vec{L}^2$ seemed to be proportional to the squares of *half-integers* $(1/2, 3/2, 5/2, \ldots$ etc.). The second problem, also a prime example of the "numerological spectroscopy" pioneered by the Munich school, and now amplified by the intensity measurements and sum rules proposed by the Utrecht group, concerned the relative intensities of the split spectral lines appearing in the Zeeman effect. Quite explicit formulas had by now been proposed by various workers—Heisenberg refers to papers by Goudsmit and Kronig (1925) and

Hönl (1925)—for the relative intensities of these lines, all in agreement, and generally accepted as correct in the theoretical community.

Unfortunately, for rather technical reasons we cannot go into here,[30] Heisenberg's hitherto infallible intuition leads him astray in both of these problems. In the rotational molecular energy case, he makes the fundamental error, naturally inherited from the ways of thinking of the old quantum theory, that three-dimensional motion can be reduced to a two-dimensional problem, given that the diatomic molecule can classically be assumed to be rotating in a two-dimensional plane. This assumption is destined to fail, as "leakage" into a third dimension is inevitable in quantum mechanics. Moreover, Heisenberg also falls prey to the trap of "knowing the answer" one is trying to reproduce, and makes an incorrect argument, analogous to the "vanishing at the edges" one (see the web resource mentioned in note 30), which successfully yielded the (half-integral) zero point energy of the harmonic oscillator, in order to come up with the desired half-integral quanta. In the Zeeman intensities problem, he does indeed treat the problem correctly by formulating it directly in three spatial dimensions—the actual problem considered is that of a charged particle constrained to move on the surface of a sphere, where $x^2 + y^2 + z^2 = a^2$, in the presence of an external magnetic field—but incorrectly implements the "sharpened" formula for the quantization condition specifying the magnetic quantum number, which plays the role that the Thomas–Reiche–Kuhn sum rule had performed in the oscillator problem.[31]

Nonetheless, even at the remove of a century, Heisenberg's achievement in the *Umdeutung* paper is breathtaking in its intellectual audacity and depth of intuition. The algebraically clumsy formulation of his rethinking of quantum theory would soon be remedied, and the whole structure put on a much firmer conceptual and mathematical foundation, by the work of his Göttingen colleagues Max Born and Pascual Jordan, soon followed by his own participation in the famous *Dreimännerarbeit* (Three-Man-Paper) on matrix mechanics in late 1925.

8.5.3 The new mechanics receives an algebraic framing—Born and Jordan's Two-Man-Paper

In the absence of Heisenberg (away in England delivering a prearranged talk at Cambridge), Born and his former student Pascual Jordan wasted no time in taking the physical ideas of the *Umdeutung* paper, clumsily clothed in obscure algebraic manipulations, and reformulating them in a much cleaner and more powerful mathematical language (Born and Jordan 1925b). The increase in power and transparency the new mechanics gained by this transformation—which, it must be emphasized, did not alter its physical content in the slightest—is a classic example of the extraordinary efficacy of mathematics as the natural language for expressing the physical laws of nature. Born and

[30] See the web resource, *The Problem of Spectral Intensities in the Old and New Quantum Mechanics*, for a detailed commentary on Heisenberg's calculations and missteps in the final section of the *Umdeutung* paper.

[31] The constrained motion problem, if implemented correctly, would indeed have yielded the Goudsmit–Kronig–Hönl intensity rules, as we show explicitly in the web resource mentioned in note 30.

Jordan reiterated the basic principles enunciated by Heisenberg, but now in the language of matrices, as follows:

1. The coordinate and momentum of a particle (moving in one dimension), expressed as real functions of time in classical theory, are to be represented in quantum mechanics by matrices (square arrays of complex numbers, in general with an infinite number of rows and columns), with the elements of these matrices *complex* functions of time. Moreover—and this is the essence of Heisenberg's *Umdeutung*—the classical equation of motion for the particle, giving the rate of change of the momentum as a function of the coordinate (via the potential energy function), is presumed to remain formally valid, *by simply replacing the coordinate and momentum functions by their matrix versions.*
2. Planck's quantum of action is inserted in the theory by requiring the difference between the coordinate and momentum matrices multiplied in opposite order (the "commutator") to be $h/2\pi i$ times the identity matrix (which has the number 1 on the diagonal elements and zero elsewhere). This was Born and Jordan's formal matrix transcription of Heisenberg's use of the Thomas–Kuhn sum rule to introduce quantization into the theory.

The rephrasing of Heisenberg's ideas in the language of matrices proved extraordinarily felicitous, and not just from a purely conceptual point of view, but practically, as it enormously reduced the algebraic complexity of the calculations required to establish even the most basic results expected from the theory, for example, the proof that energy was conserved at a microscopic level. Even in the very simple systems studied by Heisenberg—one-dimensional anharmonic oscillators—he had only been able to establish this property by laborious calculations, and only to a low order in a perturbative expansion in the anharmonic coefficient. Using matrix technology, Born and Jordan, in the space of a few pages, were able to establish:

1. The quantization condition (giving the commutator of the position and momentum matrices as $h/2\pi i$ times the identity matrix), once enforced at time zero, is automatically preserved by the dynamics of the theory.
2. A new "sharpened" correspondence between classical and quantum theory (which would prove to be the final step in the multi-year process of elaboration of the correspondence principle) can be formulated simply in terms of a replacement of classical derivatives of a mechanical quantity (with respect to position or momentum) by commutators of that quantity (with the momentum or position matrix), times $2\pi i/h$.
3. The time rate of change of any mechanical quantity can be obtained by computing the commutator of the matrix of the quantity with the energy matrix, times $2\pi i/h$.

4. The energy matrix is automatically independent of time and diagonal. This was an exact result, extending Heisenberg's results for the lowest orders of perturbation theory.
5. The Bohr frequency condition follows immediately from considering the commutator of the energy and coordinate (or momentum) matrices. The interpretation of the diagonal elements of the energy matrix as the allowed stationary state energies is thereby assured.

In addition to these formal results, Born and Jordan checked Heisenberg's perturbative results for the stationary state energies of the quartic oscillator, and added a further calculation, evaluating the second-order energy for the cubic oscillator.

In the final section of their paper, Born and Jordan attempted to establish formally the connection between Heisenberg's coordinate matrix—or more specifically, the element of this matrix corresponding to the initial and final states of a quantum transition—and the Einstein A coefficient giving the spontaneous transition rate for that pair of states. Here, they fall victim, as Heisenberg had in his attack on rotational problems, to the trap of "knowing the answer," as the argument they give, essentially an attempt to reproduce a generalized version of the results of Ladenburg and Reiche, is fallacious, as we show in our discussion of the paper in Section 12.1. A fully consistent description of the quantum radiation process would have to wait for Dirac's work on quantum electrodynamics, which would appear two years later.

8.5.4 Dirac and the formal connection between classical and quantum mechanics

At about the same time that Born and Jordan were diligently transcribing Heisenberg's mysterious algebraic version of quantum mechanics into matrix language, a young graduate student of Ralph Fowler at Cambridge, began to study the manuscript of Heisenberg's *Umdeutung* paper, which had been mailed to him by Fowler with the handwritten request on the front page to evaluate and report on the paper ("What do you think of this? I would be glad to hear"). Paul Adrien Maurice Dirac had probably attended—although this appears to be in some doubt (Farmelo 2009, p. 82)—Heisenberg's talk on the empirical numerology of complex spectra and Zeeman effect splittings on July 28, to the Kapitza Club at Cambridge, but the new theory was only mentioned briefly at the end of the talk by Heisenberg, so the receipt of the manuscript from Fowler over a month later would have been Dirac's first real contact with the strange new way of thinking about quantum theory proposed by Heisenberg.

Dirac had studied engineering in Bristol, but his true love was mathematics, especially of the more abstract variety. He had therefore moved to Cambridge, beginning graduate studies in mathematical physics, of which the most active focus of activity at the time was clearly the problem of developing a consistent and quantitatively successful quantum theory of the atom. Dirac's background in mathematics included a thorough course in projective geometry, including the geometry of abstract spaces in which coordinates could be non-commuting numbers. He was also aware of Hamilton's introduction of quaternions, which served as a concrete example of a consistent number field, generalizing the complex numbers, but in which the multiplication of two elements was not in general commutative. On the other hand, it is not clear how familiar Dirac was at the time with the technology of matrix algebra, which Born and Jordan were so fruitfully exploiting in Göttingen. In fact, matrices are not mentioned in the paper "The fundamental equations of quantum mechanics," which Dirac would complete and submit (through Fowler) to the proceedings of the Royal Society early in November 1925 (Dirac 1925). The rule for composing two amplitude arrays which appears in the *Umdeutung* paper is referred to throughout Dirac's paper as "Heisenberg multiplication."

The formal property that was crucial to Dirac's remarkable achievement, and which he seems to have recognized as of central significance sometime after the beginning of the Fall term (October 1925), was the role of the commutator $AB - BA$ of two non-commuting quantities, such as the coordinate and velocity arrays of Heisenberg's paper. Although Heisenberg pointed quite explicitly to the fact that the result of multiplying the amplitude arrays $\mathfrak{A}$ and $\mathfrak{B}$ corresponding to two different dynamical quantities depended on the order, it never occurred to him to ponder the possible significance of the (in general, non-zero) difference $\mathfrak{A}\mathfrak{B} - \mathfrak{B}\mathfrak{A}$. By early October 1925, Dirac had succeeded, through calculations similar to those for incoherent scattering in Kramers and Heisenberg (1925), in deriving an expression for this "difference between the Heisenberg products," or, to use the later term, commutator.[32]

Dirac later recollected that, on a long Sunday walk in early October, he recognized in his expressions a strong resemblance to certain analytic expressions he had seen in the classic text of the time on analytical mechanics, E. T. Whittaker's *Treatise on the Analytical Dynamics of Particles and Rigid Bodies.* These combinations of derivatives of dynamical quantities, introduced more than a century earlier by Siméon Denis Poisson in a study of constants of the motion in problems of multi-body motion in classical mechanics, turned out to exactly match the expressions that Dirac had found for the elements of the commutator of any pair of dynamical quantities, for a system that could be formulated using the special set of coordinates and momenta called "angle" and "action" variables, as Dirac was able to check in Whittaker's book, once the library at Cambridge reopened on the Monday morning following Dirac's revelation. These "Poisson brackets" (for which Dirac uses exactly the commutator notation that would soon be adopted for Heisenberg's arrays, e.g., $[q_r, p_s]$) therefore corresponded precisely, modulo a factor of $ih/2\pi$, to the "difference of Heisenberg products."

A defining feature of Poisson brackets is that they preserve their form under canonical transformations, so that the particular angle and action variables employed in the treatment of a multiply periodic system can be replaced with *any* convenient set of coordinates q_r and momenta p_r. The corresponding quantum $qp - pq$ arrays would therefore once again reproduce the sacred $ih/2\pi$ factor. Dirac proposed that this correspondence

[32] The word commutator for this difference, which would soon become standard, does not appear in Dirac's early papers on quantum mechanics, nor, for that matter in Born and Jordan (1925a) or Born, Heisenberg, and Jordan (1926). We nevertheless employ it, anachronistically, as a convenient shorthand.

between classical Poisson brackets and quantum amplitude array commutators should be assumed to hold for a *general mechanical system*, and not just for the very special sub-class of multiply periodic systems for which he had managed to derive the connection between the classical Poisson brackets and the quantum commutator in the correspondence limit. Dirac's bold leap meant that the dynamics of a quantum system (which classically could be formulated directly in terms of Poisson brackets) could now be specified precisely for any mechanical system (including the cases of multi-electron atoms like helium, with which Dirac was much concerned at the time)—the shackles imposed by the restriction of quantum mechanics to *multiply periodic* systems treated by correspondence-principle arguments could finally and permanently be thrown off.

At a formal level, Dirac's first paper on quantum mechanics solves precisely the problem of specifying the dynamics of a multi-particle system of the sort one encounters in dealing with atoms and molecules. In it, the quantization conditions are correctly stated, and the problem of correct ordering of the variables in the Hamiltonian energy function (with which both Heisenberg and Dirac were explicitly concerned) is moot, as in atomic physics and quantum chemistry the basic Hamiltonian is a sum of kinetic and potential energy expressions where products of coordinate and momentum variables do not occur in the same term.

Technically, however, Dirac was certainly not in a position to solve the mathematical problem posed by his postulates, which would require the full machinery of Schrödinger's wave mechanics to become computationally manageable. The only system treated in Dirac's paper is the simple harmonic oscillator, so in this respect the paper is not an advance on the *Umdeutung* article, where at least the route to a systematic perturbative solution of more complicated systems had been indicated. But at a purely mathematical level, Dirac was the first to completely and correctly frame the principles underlying the quantum dynamics of atoms and molecules. The interaction of atomic systems with the electromagnetic field was not yet dealt with, of course—the correct way of incorporating electromagnetism into quantum theory would appear two years later, again in another spectacular theoretical *tour-de-force*, in Dirac's paper on quantum electrodynamics (Dirac 1927b).

8.5.5 The Three-Man-Paper [*Dreimännerarbeit*]—completion of the formalism of matrix mechanics

At roughly the same time that Dirac was perusing Heisenberg's paper and developing his own interpretation of the strange new ideas therein, Born and Jordan in Göttingen, now with the active collaboration at a distance of Heisenberg (who had returned from his travels but was stuck in Copenhagen finishing his fellowship duties at the Niels Bohr institute) were composing a long paper in which an explicit and comprehensive formulation of the principles of matrix mechanics would be presented, together with the application of these principles to the solution of many of the problems which had proved especially refractory in the old quantum theory. Heisenberg would be released (by request of Born) from his Copenhagen duties a little earlier than originally planned—he would return to Göttingen in late October, just in time to assist with the final preparation of the manuscript.

The Three-Man-Paper (*Dreimännerarbeit*) of Born, Heisenberg, and Jordan (1926), entitled "On quantum mechanics II," extends the formal description of the theory which had already appeared in the Two-Man-Paper of Born and Jordan (1925b), "On quantum mechanics," in several important ways. The treatment of one-dimensional systems (Chapter 1) is elaborated by providing (a) the correct connection between the matrix commutators of the quantum theory and the classical derivatives that appear in the equations of motion; (b) a precise correspondence between canonical transformations of the coordinate and momentum variables in the classical theory and a matrix similarity transformation in the quantum theory; (c) an application of the canonical matrix formalism to the solution of quantum energy levels order by order in their expansion in a perturbative parameter appearing in the energy matrix (for cases where the energy function is time-independent); and (d) an extension of the canonical matrix formalism to problems, such as light scattering, where the energy matrix contains explicit time dependence. In the last case, the authors were able to re-derive the Kramers–Heisenberg dispersion relations (both for elastic Rayleigh, and inelastic Raman, scattering), now on an entirely quantum-mechanical footing, and without recourse to the correspondence principle.

The second chapter of the paper proposes a generalization (first worked out by Heisenberg while still in Copenhagen) of the fundamental canonical commutation relations between position and coordinate variables to systems of several degrees of freedom. The argument is one of plausibility—with the "natural" proposed generalization one easily recovers the expected Hamiltonian equations of motion by exploiting the formal analogy between matrix commutators and classical derivatives (in this case, partial derivatives with respect to each coordinate or momentum variable). The beautiful Dirac derivation exhibiting the intimate relation between quantum commutators and classical Poisson brackets appears not to have occurred to the authors. Another important topic discussed in this section is the appearance of degeneracy in these systems—i.e., the existence of more than one quantum state with the same energy. This complicates the execution of perturbative calculations of the energy, but the authors are up to the job of designing the correct workaround, which turns out to be much the same "prediagonalization" of the energy matrix that one finds in modern texts.

Another topic explored in this second chapter of the paper, the intrinsic phase ambiguities present in the matrix representations of the kinematic variables, which seemed inescapable but unconnected to measurable physical quantities, appears at first sight (to the modern reader) to be a technical issue of little interest. Such ambiguities—which violated the underlying positivistic impulse that had guided Heisenberg in the *Umdeutung* paper, to restrict the theory to "relations between observable quantities"—were clearly disturbing to the authors of the Three-Man-Paper. In fact, such ambiguities would continue to be a confusing feature of the new quantum mechanics until two years later, when the concept of a quantum state represented by a ray vector in an abstract complex vector space emerged, and was put on a firm mathematical foundation, by John von Neumann (see Chapter 17, and Duncan and Janssen 2013). In this formulation, the multiplication of a state vector by a complex phase (i.e., a complex number of unit absolute value) was physically irrelevant. This freedom directly implies a corresponding fluidity in the values of the complex matrix elements of the kinematical variables, exactly the phase indeterminacy uncovered by Born and his colleagues in their examination of canonical transformations in the Three-Man-Paper.

The third chapter explores the algebraic mathematical substructure of matrix mechanics at a much more sophisticated level than that reached in either Heisenberg's *Umdeutung* paper, or in the formal reorganization achieved in the Born–Jordan follow-up article. It seems to have been written by Born, whose mathematical acumen and experience was clearly more advanced than that of either of his younger collaborators. As the first section of this chapter ("General method") indicates,

> behind the formalism of this perturbation theory [transposed from classical to matrix mechanics] there lurks a very simple, purely algebraic connection and it is well worth while to bring this into the limelight. Apart from the deeper insight into the mathematical structure of the theory, we thereby gain the advantage of being able to use the methods and results developed earlier in mathematics (Born, Heisenberg, and Jordan 1926, p. 348).

What "earlier mathematics" is Born referring to here? In fact, the mathematical structures coming to light in the new matrix-based quantum theory were as yet only incompletely accessible to the mathematical community. On several occasions in the history of theoretical physics, advances in mathematics have provided exactly the needed technical apparatus to express new physical concepts. Newton's development of calculus in order to facilitate his investigations of gravity and motion is a famous example. Another was the incorporation of the sophisticated methods of analytical dynamics (Hamilton–Jacobi theory, action/angle variables, etc.), previously of interest only to theoretical astronomers, into the quantum theory of the atom, initiated by Karl Schwarzschild in 1916 (see Section 5.1.4). In Einstein's development of general relativity, the differential calculus for Riemannian manifolds, as presented in the canonical text of Ricci and Levi-Civita (1901), played a central role (Reich 1994). And by the Fall of 1925, it had become clear to Born and his co-authors that the rigorous underpinnings of matrix mechanics would require exploitation of a branch of mathematical analysis under intense exploration at exactly the same time, and, by a remarkable coincidence, in the same place—Göttingen.

This field, combining elements of algebra and analysis, would eventually be dubbed "functional analysis."[33] The origins of functional analysis can be found in the work of Vito Volterra, Erik Ivar Fredholm, Carl Neumann, and others on linear integral equations, which stimulated David Hilbert, beginning in 1904 and continuing through 1910, to develop a detailed theory of these equations. Hilbert's formalism contained precisely the spectral theory of infinite dimensional matrices (in somewhat disguised form), which Born recognized as an essential tool for understanding how matrix mechanics

[33] The term "functional" to denote a linear operation on some well-defined set of functions goes back at least to Hadamard (1903). It would eventually encompass an enormous range of topics stemming from the core idea of a "function space," or set of suitably defined functions viewed as a topological space, the points of which are individual functions.

could serve as a *general* framework for quantum mechanical problems. In particular, the formalism was sufficiently general to include situations in which the allowed energies (eigenvalues[34] of the energy matrix) could assume discrete (as in the case of bound electrons in the hydrogen atom) as well as continuous (ionized electrons unbound from the nucleus but still subject to its Coulomb attraction) values.

Unlike the examples of Hamilton–Jacobi theory and differential geometry cited previously, which were available to physicists in a more or less finished state when they were recognized as useful tools for physics, the functional analysis needed for quantum mechanics was still under development in 1925. It would be another five years before the relevant elements (spectral theory of unbounded normal operators) would be provided by von Neumann, and thereafter incorporated explicitly into the foundations of quantum mechanics in his famous book (von Neumann 1932). In this respect, the mutual fertilization and synchronous co-development of physical ideas and underlying mathematical formalism that occurred in the late 1920s is more analogous to Newton's elaboration of Newtonian mechanics in close symbiosis with his discovery of the tools of differential calculus in the 1660s.

In 1925, the technology available to Born consisted mainly of Hilbert's results (1904–1910) for the spectra of bounded Hermitian forms/matrices (those for which there was a finite maximum absolute value for the eigenvalues), and the completion of this theory (explaining in particular how to deal with the non-normalizable eigenvectors of the matrix in the continuous spectrum case) by his student Ernst Hellinger (1910). Born simply had to assume—until von Neumann could fill the lacuna—that the basic results continued to hold for the obviously unbounded matrices needed to describe momentum, energy etc. in quantum mechanics. This is made explicit in a footnote in the Three-Man-Paper:

> Up till now, the theory of quadratic (or Hermitian) forms of infinitely many variables has been developed mainly for a special class ("bounded" forms) [references to Hilbert (1912) and Hellinger (1910)]. But here we are concerned just with non-bounded forms. We may nevertheless assume that in the main the rules run likewise (Born, Heisenberg, and Jordan 1926, p. 351, footnote 1).

The authors of the paper did not attempt to demonstrate the applicability of the Hilbert–Hellinger theory in the case that most obviously demanded it: the hydrogen atom, with its combination of a discrete and a continuous spectrum. The discrete spectrum can in fact be derived by a purely matrix-mechanical calculation, as Pauli (1926a) would demonstrate a few months later in a paper completed in mid-January 1926—a mathematical tour-de-force of a considerably higher order of complexity than any of the computations appearing in the Three-Man-Paper. Nevertheless, the basic point made in the third chapter of the paper remains correct (even if it is surprising to some modern physicists who sometimes assume that matrix mechanics is rigorously applicable only to systems like the harmonic oscillator with a purely discrete spectrum), namely, that the complete dynamics of *any mechanical Hamiltonian system* that can be formulated in

[34] See Section 2 in Appendix C for an introduction to the relevant linear algebra.

terms of a canonical set of position and momentum variables can be rephrased in the language of discrete (infinite) matrices, and that this language can, at least in principle, handle both bound (and energetically discrete) stationary states of a system and states where the system becomes unbound and the energy spectrum is continuous.

The fourth and final chapter of the paper, entitled "Physical applications of the theory," focuses almost entirely on the development of the algebraic properties of the matrices representing the components of the angular momentum vector in three spatial dimensions. After showing that, for a system defined by an energy function which is invariant (both kinetic and potential energy) under spatial rotations, this vector is conserved in time, the authors go on to derive, starting from the canonical commutation relations of the Cartesian coordinate and momentum matrices, the commutation relations for the three components $\mathbf{M}_x$, $\mathbf{M}_y$, and $\mathbf{M}_z$ of $\mathbf{M}$, the total orbital angular momentum of all the particles comprising the system. They also calculate the commutation relations of these matrices with the matrices $\mathbf{q}_x$, $\mathbf{q}_y$, and $\mathbf{q}_z$, which represent the coordinates of a light emitting electron (and therefore, modulo a factor of the electric charge, the electric dipole moment responsible for optical transitions). The algebraic properties of these commutators then imply very quickly the selection rules for radiation Bohr had intuited from correspondence-principle arguments in his seminal 1918 paper on line spectra (cf. Section 5.3). These properties are then applied to obtain at last the correct intensity ratios for the split lines in the Zeeman effect. These ratios had already been correctly guessed on the basis of correspondence-principle arguments by Goudsmit, Kronig, and Hönl, and Heisenberg had tried, but failed to reproduce them in the *Umdeutung* paper.[35] The second section, entitled "The Zeeman effect," consists of two paragraphs in which the authors basically admit that the whole concept of an atomic core (in which the supplementary angular momentum needed to account for the anomalous Zeeman effect was presumed to originate) was untenable; instead, the authors opine, "one might perhaps hope that the hypothesis of Uhlenbeck and Goudsmit might later provide a quantitative description of the above-mentioned phenomena." This would indeed come to pass, in a paper appearing a few months later, written by Heisenberg and Jordan (1926).

The most interesting part of this final chapter of the Three-Man-Paper paper is the third section, entitled "Coupled harmonic resonators—statistics of wave fields." This section was conceived and written entirely by Jordan, as testified by both his co-authors and emphasized frequently by Jordan in later years, who even came to regard it as "almost the most important contribution I ever made to quantum mechanics" (Jordan to van der Waerden, April 10, 1962, quoted in Duncan and Janssen 2008, p. 639). Moreover, it seems clear that he only persuaded Born and Heisenberg with some difficulty that his calculations were correct, and deserved inclusion in the paper, which the authors had self-consciously set about writing as the foundational text for a revolutionary reformulation of quantum theory, framed in as general terms as possible. However, the extension of the ideas of matrix mechanics to systems of infinitely many degrees of freedom, as any

[35] For a detailed explanation of Heisenberg's near-miss in his treatment of three-dimensional rotation, see the web resource, *The Problem of Spectral Intensities in the Old and New Quantum Theory*.

direct attempt to quantize waves in a box (such electromagnetic radiation in a cavity) would require, apparently gave Jordan's collaborators a distinct sense of mathematical vertigo.

Nonetheless, the results of Jordan's calculation of energy fluctuations in radiation, which modeled the electromagnetic field in a simplified way as standing waves confined to a finite interval on a one-dimensional line, were impressive. He managed to reproduce Einstein's famous formula from 1909 showing that the mean square energy fluctuations of the black-body radiation in a subvolume of a cavity is given by a simple additive combination of two terms that exactly agreed (in form) with the fluctuations one would expect if the field were a classical wave or if the field energy was instead distributed over discrete particles (Einstein 1909a, 1909b, see Section 3.4.3). The bizarre "particle-plus-wave" duality detected sixteen years earlier by Einstein could now be explained by Jordan as arising from a *single, unified dynamical framework*. One only had to subject the (infinitely many) Fourier components of the wave, now regarded as independent quantum mechanical "position" variables, to the same matrix reinterpretation that one gave to the coordinates of a Planckian resonator, for example, and the Einstein result emerged as if by magic.

The bold extension of a matrix mechanics originally conceived as an attempt to rationalize the behavior of a single light-emitting (or light scattering) atomic electron to a very different sort of system—a continuous field, mathematically a Hamiltonian system, but one with an infinite number of degrees of freedom—looks both backwards, to Einstein's remarkable energy fluctuation theorem, and forward, to the gradual development (by Jordan and Dirac, and then Heisenberg and Pauli, followed by many others) of modern relativistic quantum field theory. It does not play a big part in the further development of quantum mechanics in the period of interest to us, however (up to von Neumann's formalization of the theory in 1927), and so we leave our survey of the discovery of matrix mechanics here.[36]

8.6 The genesis of wave mechanics

8.6.1 The mechanical-optical route to quantum mechanics

The year before the developments of matrix mechanics in Göttingen, Einstein had read with great interest a short paper by the Indian physicist Satyendra Nath Bose (1924) that provided a new derivation of the Planck black-body radiation law, in which the statistical counting procedures introduced by Boltzmann in his seminal work on classical statistical mechanics were modified for light quanta. The distinct new method of counting introduced by Bose would soon be dubbed "Bose statistics", and later, after Einstein (1924, 1925a, 1925b) had suggested the application of the new counting also to a gas composed of non-interacting atoms, "Bose–Einstein" statistics. Remarkably, Einstein

[36] For detailed discussion of Jordan's derivation of Einstein's fluctuation formula, see Duncan and Janssen (2008) and Bacciagaluppi, Crull, and Maroney (2017).

found that the mean square density fluctuations in such a gas would take a form exactly analogous to his 1909 fluctuation formula for black-body radiation, which Jordan would derive from matrix mechanics in late 1925.

As in the case of black-body radiation, the fluctuation formula in Einstein's quantum theory for an ideal gas is a simple algebraic sum of a "particle" term and a "wave" term. The appearance of a wave term suggested to Einstein a connection with recent work of Paul Langevin's doctoral student in Paris, Louis de Broglie, in which particle motion was presumed to be associated with a concomitant wave phenomenon. The whole nexus of ideas appearing here in embryonic form would lead within a year to Schrödinger's discovery of wave mechanics. The origins of a wave interpretation of particle motion, however, lay almost a century back, with the mechanical-optical analogy developed by Hamilton in the 1830s, and in fact, even earlier, with the introduction of extremal principles, for ray optics by Fermat, and for mechanical systems, by Maupertuis (Joas and Lehner 2009). So our outline of this part of the story of quantum mechanics must include yet another, if brief, departure from strict chronological order.

In Volume 1 (see Appendix A), we noted the central role played in the development of the old quantum theory by the methods of classical analytical mechanics based on the reformulation of classical mechanics by William Rowan Hamilton in the early 1830s. The Hamiltonian approach to classical mechanics (further deepened by Jacobi, Staude, Stäckel, Delaunay, Poincaré, and others) would provide a well-defined scaffold capable of implementing the quantization postulates of the Bohr–Sommerfeld theory. But if we examine the roots of Hamilton's mechanics, which lay in his reformulation of classical *optics*, we find an alternative framework that would lead quite naturally to the work of de Broglie, and thence to the precise expression of quantum-mechanical behavior of material particles in Schrödinger's wave equation.

Hamilton's work on optics in the late 1820s, which precedes his reframing of classical mechanics now familiar to any physics undergraduate student, began from an investigation of the Fermat principle regulating the propagation of light on reflecting surfaces or through refracting transparent media. This "principle of least time" asserted that a light ray passing from point A to point B would take the route that would accomplish the transit in the least possible time. The principles of equal angles of incidence and reflection, as well as Snell's law for the bending of light passing through a plane separating two media of differing index of refraction, could easily be derived from this principle by geometrical reasoning. Given that the frequency of light remains constant as it propagates, Fermat's principle of least time can be equivalently expressed as the requirement that the number of wavelengths along the path from A to B be a minimum—or, more technically, that the integral of the inverse wavelength (which is proportional to the index of refraction) with respect to the path distance be minimal along the actual track of the ray.[37]

[37] For a detailed review of the history of extremal principles in optics and mechanics, see the web resource, *Extremal Laws in Classical Optics and Mechanics*.

More than eight decades after Fermat's enunciation of his minimum principle, the French mathematician Pierre Louis Maupertuis suggested that an analogous minimum principle—which he called "the principle of the least quantity of action"—might be invoked to understand the behavior of mechanical systems. The mysterious new "action" introduced by Maupertuis turned out to involve the product of mass, velocity, and distance traveled by the particle. Maupertuis was able to reinterpret the behavior of particles in a variety of elastic and inelastic collisions by interpreting in a rather flexible way the appropriate quantities to be employed in these algebraic components of the "action". The principle was put on a much more precise foundation by Euler and Lagrange—indeed, it served as the motivating force for the development by Lagrange of a general analytic formalism for deriving the dynamics of multiparticle systems, which to this day is taught to undergraduate students of classical mechanics. Hamilton's work on propagation of rays in the 1830s, and his adaptation of the optical formalism to the new "Hamiltonian" formulation of classical mechanics provided yet another example of the close formal analogies between optics and mechanics.

If we recognize the product of mass and velocity as momentum, the least action principle of Maupertuis and the Fermat extremal principle can be seen to be formally very similar indeed, the difference lying solely in the replacement of momentum in the mechanical case with inverse wavelength in the optical one. These two quantities (momentum and inverse wavelength) differ by a factor with the dimensions of momentum multiplied by distance, to which we not surprisingly now attach the name of "action". Of course, Maupertuis, Euler, Lagrange, and Hamilton did not have at their disposal a natural universal constant with dimensions of action that would allow a direct transcription between the optical and the mechanical minimal principles. This, post-Planck, de Broglie did possess.

De Broglie's (1924) doctoral thesis, which summarized and extended papers he had published previously in *Comptes Rendus*, was defended on November 25, 1924. It appeared in *Annales de Physique* in early 1925 (De Broglie 1925). The title of de Broglie's dissertation was rather vague, "Researches on the theory of quanta" [*Recherches sur la théorie de Quanta*], but the basic thrust of his ideas was made clear in the second sentence of the abstract of the thesis:

> Based on an understanding of the relationship between frequency and energy, we proceed in this work from the assumption of the existence of a certain periodic phenomenon of a yet to be determined character, which is to be attributed to each isolated energy parcel [*morceau isolé d'énergie*] and which depends on its proper [i.e., rest] mass by the equation of Planck–Einstein. In addition, relativity theory requires that uniform motion of a material particle be associated with propagation of a certain wave for which the phase velocity is greater than that of light (De Broglie 1925, p. 22).

The "equation of Planck–Einstein," as we soon discover, is a simple combination of two distinct expressions for energy: the Einsteinian expression for the relativistic energy of the particle at rest, $E = mc^2$, with m the rest mass of the particle; and the Planckian expression giving the quantum of energy for any quantized vibrational phenomenon in terms of its frequency ν, $E = h\nu$. De Broglie simply combines these to obtain the "Planck–Einstein" equation $h\nu = mc^2$, which, it hardly needs to be said, was never written down by either Planck or Einstein. The appeal to relativity amounts to the use of the Lorentz–Einstein transformation of the time variable in special relativity, which asserts that a simple harmonic universal vibration $\sin(2\pi\nu t)$ of frequency $\nu = mc^2/h$ associated with a particle at rest will appear to observers moving at speed v relative to the particle (which therefore appears to move in their frame, also with speed v) as a spatially variable *wave*, $\sin(2\pi\nu\gamma(t' - x'v/c^2))$, where $\gamma = 1/\sqrt{1 - v^2/c^2}$. Here, t' and x' refer to the time and spatial location at which the wave is examined, as measured by the observers with respect to which the particle is in motion with the subluminal speed $v < c$. Such a wave moves with the phase velocity (i.e., the speed of the individual peaks of the wave) c^2/v, which, as de Broglie points out, is faster than light. In the first chapter of his thesis, de Broglie shows that this (at first sight unacceptable) property is in fact harmless, as the transmission of physically measurable properties of a wave phenomenon were already known classically to be associated with the *group velocity*, rather than the phase velocity of the wave, that is, with the velocity with which the envelope of a localized wave bundle built from the superposition of waves of nearby frequencies moves. A short calculation shows that this group velocity[38] coincides exactly with the speed v (less than c!) of the material particle associated with the wave. The wave "to be attributed to any isolated parcel of energy" therefore plays the role of describing the transportation of the relativistic energy of a particle through space, if we associate the physical properties of the material particle with the envelope of the de Broglie wave, and not its individual peaks and dips.

In the second chapter of his thesis, de Broglie proposes a firm link between the optical and mechanical extremal principles described above. The connection is very simple: if one assigns to the "associated wave" a four component wave-vector k_μ, with the time component given by the frequency and the magnitude of the spatial part $|\vec{k}|$ by the inverse wavelength (times 2π), then, by the simple expedient of equating this wave vector to a four-component particle momentum $p_\mu = (h/2\pi)k_\mu$, Fermat's principle of least time becomes formally identical to the least action principle of Maupertuis. The connection between wave optics and mechanics—one that had been noticed 90 years earlier by Hamilton and exploited in his derivation of Hamiltonian dynamics—then becomes completely analogous to the connection between physical and geometrical optics. As de Broglie (1925, p. 22) puts it: "The rays of the wave are identical to the possible trajectories of the moving particle."

In the next chapter of his thesis, de Broglie shows that the Bohr quantization condition for angular momentum of an electron executing a circular orbit, and more generally, the Bohr–Sommerfeld quantization conditions for particles executing multiply periodic motions, can be interpreted as a "phase wave resonance" condition. The desired resonance occurs when the wave associated with the particle contains precisely a whole number of wavelengths when followed around a single cycle of any of the independent degrees of freedom of the multiply periodic motion. The phase

[38] We find the group velocity U by writing the wave in the form $\sin(\omega t - kx)$, and calculating the derivative $d\omega/dk$ (see Section 13.2 for details).

integrals that count this number turn out to be precisely the action phase integrals of the Bohr–Sommerfeld theory.

Of course, this observation, while it may give added support to the quantization conditions of the old quantum theory, does not get us any closer to a really new quantum dynamics based exclusively on the properties of de Broglie's guiding waves.[39] Although de Broglie attempts (in the second chapter) to discuss the problem of motion of a particle in a varying external potential, the analog of the behavior of a wave propagating through a medium of continuously varying index of refraction, he never succeeds in writing down an exact wave equation that would handle situations more general than those appropriate for the eikonal limit where ray optics merges with physical optics.[40] This step, in retrospect so obviously necessary, would be taken a year later by Schrödinger.

At about the same time as de Broglie was preparing the final version of his thesis, Bose's (1924) paper, which would play a key role in the emergence of wave mechanics, was approaching publication in *Zeitschrift für Physik*. Bose had sent his paper to Einstein, who personally translated it from English to German and forwarded it to the journal in July 1924, with the strong recommendation that it be published, as it represented "in my opinion, an important step forward."[41] Bose's paper, entitled "Planck's law and the light quantum hypothesis," contains the first self-contained derivation of the Planck black-body radiation law based entirely on the statistics of light quanta,[42] viewed as *indistinguishable* localized particles of energy $h\nu$ and momentum $h\nu/c$.[43] Nowhere is there any consideration of quantized Planckian resonators in thermal equilibrium with both an external heat bath and the radiation inside the cavity. Instead, the frequency distribution of the electromagnetic energy is derived on the basis of standard Boltzmannian ideas—the equilibrium distribution corresponds to that which yields the maximum number of microstates compatible with the prescribed total energy in the cavity. However, the derivation departs strikingly from Boltzmannian statistics in the counting procedure used for the microstates of the radiation field. According to Bose, the microstates of the radiation field are prescribed by specifying the occupation number of light quanta in each possible "cell", where a cell is defined by a region of volume h^3 in the combined spatial and momentum (hence, six-dimensional) phase space of the quanta, viewed as

[39] Such a theory would emerge a few years later in de Broglie's "pilot wave" theory, subsequently elaborated by David Bohm (see note 55).

[40] In the eikonal limit, the fractional variation of the index of refraction over a single wavelength is assumed to be very small. For small quantum numbers, the fractional variation of the Coulomb potential over a single de Broglie wavelength is not small.

[41] Einstein to Bose, July 2, 1924 (Einstein 1987–2021, Vol. 14, Doc. 279).

[42] We shall continue to use the terminology of "light quanta", instead of the obvious simpler choice "photon," which is now universally employed, as the latter term was introduced only somewhat later, in December 1926, by the physical chemist Gilbert Lewis (1926). Translations of de Broglie employing the term "photon", while not conceptually misleading as such, are therefore anachronistic.

[43] The earlier derivation of Planck's law by Debye (1910) (cf. Volume 1, p. 112, note 35) was based on the consideration of the statistics of the radiation field only, expressed in terms of (non-localized) standing wave modes *à la* Rayleigh and Jeans, where the critical counting formula for available microstates of the system (see Eq. (2.35)) referred to the integrally quantized energy assigned to each available standing wave mode, and not, as in the original Planckian derivation, to the energy assigned to material resonators in the cavity. Bose's derivation explicitly assumed that the electromagnetic radiation consisted of localized particles, to which coordinates and momenta could be assigned as in the classical treatment of ideal gases. This made its extension to a gas of massive atoms conceptually straightforward.

particles with a given position $\vec{r}$ and momentum $\vec{p}$ (with $|\vec{p}| = h\nu/c$). *It is not allowed, when the counting of microstates is carried out, to distinguish individual light quanta in a multiply occupied cell.* This merging of light quanta of the same momentum and energy is a distinctly non-classical phenomenon, but it is essential for Bose in his counting of the microstates compatible with a given total energy in the cavity, which then leads precisely to Planck's black-body formula.

The correct execution of Bose's new statistical method, which we must assume was not yet known to de Broglie when he defended his thesis, requires two important steps:

1. One must first identify the allowed modes of the radiation field, each of which can be separately excited with an integral number (zero or positive) of light quanta. The mode counting was originally performed by Rayleigh (then corrected by Jeans) simply by identifying the possible stationary waves of an electromagnetic field in a cubical volume (cf. Volume 1, p. 91). This counting is shown by Bose to be equivalent to the identification of phase space cells of volume h^3, once the Einsteinian attribution of momentum $h\nu/c = h/\lambda$ to each light quantum is accepted.
2. One then has to decide how to assign equal a-priori probability to distinguishable states of the radiation field. It is at this point that the strange loss of identity of individual light quanta occupying the same mode enters, in order to give the correct counting, and then, by Boltzmann's principle (relating entropy to the logarithm of the number of microstates), the correct expressions for entropy, energy distribution over frequency, etc.

In fact, in chapter 7 of his thesis, de Broglie succeeded perfectly in the execution of the first of these tasks—indeed, he went beyond Bose in obtaining the correct counting of modes both for light quanta and material point particles of non-zero mass, describing both situations in terms of "atoms" satisfying in all cases the correct relativistic relation between energy E and momentum p, that is, $E = \sqrt{p^2c^2 + m^2c^4}$, with momentum related to the wavelength of an associated wave via the relation $p = h/\lambda$. The second step, the identification and counting of distinct microstates, is certainly recognized by de Broglie as the essential element in obtaining Planck's black-body formula. To specify a definite method of identifying the allowed quantum states of a gas of such "atoms" (which embraces both the case of light quanta, with proper mass $m = 0$, as well as gases of massive atoms), de Broglie proposes the following hypothesis:

> *If two or more atoms have phase waves that exactly superpose* [i.e., coincide in phase] *so that it may be said that as a consequence they are transported by the same wave, their movements may no longer be considered entirely independent, and these atoms may no longer be treated as distinct entities in the calculation of probabilities.* The "undulatory" movement of these atoms therefore presents a type of coherence as a result of unspecifiable interactions, which are probably connected to a mechanism which renders unstable the movement of atoms where the phase wave is not stationary (De Broglie 1925, p. 116, emphasis in the original, our translation).

This hypothesis could (with some charity) be regarded as containing the essence of the Bose statistical method, provided one correctly understands the connection between the wave and particle aspects of the theory. Unfortunately, de Broglie's ideas are too vague and imprecise at this point, and in one respect, definitely incorrect—he is clearly under the impression that there is a *single* phase wave in physical space to which separate atoms contribute, much as separate charges all contribute to the Maxwellian electric and magnetic fields. Thus, the melding of several atoms contributing to the same standing wave as a single entity is attributed to the need for constructive interference of their respective contributions to the phase wave of the system. Nonetheless, de Broglie correctly identifies the number $n_\nu d\nu$ of "elementary stationary waves" in a given frequency interval $(\nu, \nu + d\nu)$ and then asserts that "each elementary wave can transport 0, 1, 2, ... atoms," with a relative probability for p atoms given by the canonical Boltzmann factor $\exp(-ph\nu/kT)$.

This sounds superficially like the Bose prescription, but in fact misses the correct statistical analysis given by Bose (and reiterated later by Einstein), as the Planck formula is reproduced by writing its high-frequency expansion (beginning with the Wien law term) as an infinite sum corresponding to "the mixture of an infinite number of gases each characterized by a whole number p such that the number of possible states of a quantity of gas [in an infinitesimal volume] with energy between $ph\nu$ and $ph(\nu + d\nu)$ is [inversely proportional to p]" (De Broglie 1925, p. 119). In other words, de Broglie simply "reverse engineers" the counting of states to reproduce the Planck black-body formula, without giving a correct statistical argument for identifying *à la* Boltzmann the number of *physically distinguishable* microstates compatible with the given macroscopic constraints.

We have yet another case here of the deceptive convenience of "knowing the desired answer". This becomes clear when de Broglie's vague reasoning fails him as soon as he attempts to obtain the result analogous to Planck's formula for a "gas in the ordinary sense of the word", namely, one in which the "atoms" have a finite rest mass m_0. Here, de Broglie's oversimplified application of the canonical ensemble leads him to neglect all states where an elementary stationary wave "transports" more than one atom, on the basis that the relative factor $\exp(-m_0c^2/kT)$ for each extra atom would be vanishingly small for any reasonable temperature. This factor is indeed tiny, but for the correct treatment of a Bose gas of massive atoms, the correct statistical treatment requires one to maximize the probability allowing for fluctuations of the number of atoms in the cavity by introducing a chemical potential, which then effectively removes the enormous relativistic rest energy from the exponential factor. By throwing away the multiple occupancy states, de Broglie effectively eliminates the terms that would give the characteristic interference terms in the behavior of a Bose gas that indicate the wave aspect of the quantum behavior of the atoms.

The correct extension of Bose statistics to the case of massive atoms was carried out by Einstein (1924, 1925a), clearly inspired by Bose's ideas, which Einstein had been championing. As already noted, Einstein (1925a), in his second paper on the quantum theory of monatomic ideal gases, showed that the mean square fluctuations of the number of such atoms in a subvolume was described by a formula of exactly the same algebraic form as the one he had obtained for black-body radiation in 1909, with the characteristic sum of purely wave-like and particle-like terms. Einstein then immediately references de Broglie's thesis: "E. [sic] De Broglie has shown in a very noteworthy paper how a material particle (resp. a system of material particles) can be associated with a (scalar) wave field" (Einstein 1925a, p. 9). In the footnote with the reference to this thesis, he adds: "In this dissertation one finds also a very striking geometrical interpretation of the Bohr–Sommerfeld quantum rule" (ibid.), an obvious reference to de Broglie's use of the standing wave condition to obtain the quantization conditions for closed orbits. Einstein then describes the de Broglie phase-wave ideas discussed previously, emphasizing de Broglie's crucial observation that the group velocity of these waves coincides with the actual (subluminal) velocity of the associated material particle. Einstein's enthusiasm for de Broglie's ideas would resonate with Erwin Schrödinger, who had precisely the technical and conceptual background needed to erect the stable arch of wave mechanics on the admittedly rickety scaffold provided by de Broglie's phase waves, with their motivating background of the Hamiltonian optical-mechanical analogy.

8.6.2 Schrödinger's wave mechanics

It is often said that truly ground-breaking work in theoretical physics is the province of the young—specifically, physicists in their mid-twenties, full of energy and enthusiasm for their subject, and unencumbered by the inherited prejudices of the previous generation. In the case of quantum theory, this apparent truism indeed holds for the majority of the iconoclasts in the development of the theory at critical junctures—for Einstein's introduction (1905) of light quanta at age 26, Bohr's first atomic theory (1913) at age 27, Pauli's work on the Zeeman effect and introduction of the exclusion principle (1924) at age 24, and perhaps most impressive of all, Heisenberg's *Umdeutung* paper (1925), at age 24.

Physicists with their twenties well behind them may nevertheless take comfort in some prominent counterexamples to the dictum just stated. Planck (born in 1858) was 42 when he published his groundbreaking work on black-body radiation in 1900. The central figure in the development of the old quantum theory, in the years from 1916 up to 1925, when the new mechanics appeared, was Arnold Sommerfeld, born in 1868. Finally, modern quantum mechanics would be unrecognizable without the concepts and techniques introduced in early 1926 by Erwin Schrödinger (born in 1887), at the ripe old age of 38. Schrödinger was an only child born into an upper-middle-class Viennese family, and showed his academic proficiency (always getting the top grades) already in the *Akademisches Gymnasium*, which he attended before enrolling in the Fall of 1906 at the University of Vienna, intending to study physics. Unfortunately, the University's preeminent star in theoretical physics, Ludwig Boltzmann, had resigned months earlier due to progressively worse mental illness and taken his own life in Italy in the summer preceding Schrödinger's matriculation.

Schrödinger's training in physics, received first through the lectures of Friedrich Hasenöhrl, appointed as Boltzmann's successor, and later, for his doctorate (on electrical conductivity), supervised by the experimentalist Franz Exner, was thoroughly classical in outlook, and in fact his published output prior to 1926—with the exception of two fine articles, one on penetrating electron orbits in 1921, and another in 1922 suggesting a phase quantization associated with these orbits (Schrödinger 1921, 1922)—gives scarcely a clue of the tremendous impact Schrödinger would shortly have on the construction of the modern framework of quantum mechanics.

However, there are at least two indications that make his later seminal work on wave mechanics appear as a natural extension of previous modes of thought. First, a very detailed examination of the work of Hamilton linking the equations governing ray optics and classical mechanics can be found in Schrödinger's notes from 1919. It is clear that Schrödinger was deeply impressed by the optical-mechanical analogies uncovered by Hamilton eighty years earlier. Secondly, the 1922 paper on phase quantization clearly anticipated, if only tentatively, de Broglie's interpretation of the Bohr–Sommerfeld quantization condition in terms of a periodicity property that was more characteristic of a wave phenomenon than of classical point particle motion.

The route leading to the remarkable series of papers in early 1926 that established the wave mechanical formulation of quantum theory can be traced with some confidence, and begins in 1925 with Schrödinger's investigations into the theory of quantum gases. Schrödinger was by this time teaching at the University of Zurich, following stints at Jena (as assistant to Wien), Stuttgart, and Breslau.

Following the appearance of Bose's 1924 paper and the papers of Einstein (1924, 1925a, 1925b) extending Bose's work to ideal monatomic quantum gases, the problem of the correct counting of quantum states in the derivation of the thermodynamics properties of such gases gained currency, with several theorists, in particular Planck and Schrödinger, entering the fray and proposing alternative (and incorrect) procedures for implementing Boltzmann's entropy formula. An exchange of letters between Schrödinger and Einstein culminated in the former accepting the results obtained by the latter. The particular element of this work of interest to us here was Einstein's emphatic promotion of de Broglie's work on a guiding wave regulating the quantum motion of particles, which Einstein had used as a useful conceptual prop in interpreting his rederivation of a "particle plus wave" formula for density fluctuations in the case of a quantum gas, precisely analogous to his famous formula from 1909 giving the energy fluctuations in black-body radiation, that is, in a gas of light quanta.

A second impetus to examine the work of de Broglie more carefully came from Peter Debye, at the nearby ETH (*Eidgenössische Technische Hochschule*) in Zurich, who invited Schrödinger to give a pedagogical account of de Broglie's ideas in the joint colloquium of the ETH and University of Zurich (Mehra and Rechenberg 1987, p. 420). Debye also seems to have suggested to Schrödinger around this time that it would be interesting to develop a precise wave equation that incorporated fully the particle–wave correspondences invoked by de Broglie. This appears to have happened in late November or early December of 1925.

In any event, by Christmas of 1925, Schrödinger had indeed succeeded in obtaining a fully relativistic wave equation for a charged particle moving in an external Coulomb field, such as that provided by the nucleus of a hydrogen atom. The calculations are preserved in three pages of notes entitled "H[ydrogen]-atom, characteristic vibrations."

They show that Schrödinger had obtained the correct equation describing the relativistic energy levels, *for a massive—but spinless—charged "electron"*. Schrödinger was able to show that the solutions of this equation, for a bound electron with total relativistic energy *less* than its rest energy mc^2, misbehave badly (i.e., diverge) when the distance r between the electron and the nucleus goes to zero or when it goes to infinity, *except* when the total relativistic energy E of the particle (related to the frequency of the associated wave by the Planck relation $E = h\nu$) took specific quantized values $E_{n,l}$, where n and l were whole non-negative integers, with n taking values $1, 2, 3, \ldots$ (the principal quantum number), and $l = 0, 1, \ldots, n-1$ identified with the angular momentum of the particle, which for any value of l could assume a maximum value of $lh/2\pi$ in any given direction.

Schrödinger obtained his relativistic wave equation for an electron in a Coulomb field by taking the usual wave equation, such as one finds for the components of the electric field of a free electromagnetic wave propagating at speed c, and replacing the speed of light c by the phase velocity V proposed by de Broglie for a massive charged particle immersed in an electrostatic potential. He expanded the energy levels $E_{n,l}$ associated with the well-behaved solution of this equation in inverse powers of the speed of light, with the encouraging result that the next-to-leading term (after the passive rest energy mc^2, which canceled in all energy differences) exactly coincides with the Balmer–Bohr energy levels. Unfortunately, the third-from-leading term, which in the old theory had given the fine structure associated with the Sommerfeld formula, gave a result different from Sommerfeld's. The discrepancy seemed to amount to simply a shift of 1/2 in the principal and angular momentum quantum numbers. In fact, the exact (unexpanded) energy formula that Schrödinger had obtained could be transformed into Sommerfeld's (unexpanded) fine structure formula by the same shift of 1/2. The failure of his new quantized energy formula to reproduce one of the greatest triumphs of the old quantum theory—which seemed at this point empirically unassailable—led Schrödinger to discard his relativistic wave equation, which, as a perfectly adequate description for charged *spinless* particles would be rediscovered a few months later and dubbed the *Klein–Gordon equation* (Mehra and Rechenberg 1987, p. 774).

Schrödinger made various attempts to modify his relativistic wave equation to bring the energy values of the quantized solutions into compliance with the Sommerfeld formula. By Christmas of 1925, however, he had (fortunately for his subsequent work) decided to ignore the failures of the theory at the post-Newtonian level (i.e., the relativistic corrections) and concentrate on developing the theory, by deepening the interpretation of the wave function, and applying the new wave mechanics to problems other than the archetypal hydrogen atom. This work commenced in full force over the Christmas/New Year vacation, which he took in the same Arosa mountain resort that, as a tuberculosis patient, he had already visited a few years earlier.

Three fairly complete notebooks have survived from this period in which we can trace Schrödinger's progress in parallel with the five foundational papers that appeared in the Winter and early Spring of 1926. There was a series of four papers, each with the title "Quantization as an eigenvalue problem" (Schrödinger 1926c, 1926d, 1926f, 1926h), interrupted (after the second paper) by a separate paper entitled "On the relation of the Heisenberg–Born–Jordan quantum mechanics to mine" (Schrödinger 1926e). In addition, one can follow fairly directly the progress Schrödinger made in Arosa in the critical two weeks between Christmas 1925 and the second week of January 1926, when he was back in Zurich, using both the surviving notebook and letters to Wien keeping him informed on his investigation of a new "vibration theory of the atom" (Schrödinger to Wien, January 8, 1926).[44]

Schrödinger's painstaking construction of a new wave-based quantum mechanics through the five articles cited above shows remarkable sure-footedness. There are a few notable stumbles along the way, to be sure, but taken together these papers contain a large part of the contents of any introductory text on non-relativistic quantum mechanics used today—except for matters concerning spin, and, of course, interpretational aspects of the wave function (such as the probability interpretation initiated by Born in mid-1926). Alas, the very first paper begins with one such misstep, as Schrödinger tries to recover the approximate version of his earlier fully relativistic wave equation, which gave the correct Bohr–Balmer terms but failed at the fine-structure level, in the framework of classical mechanics. He argues for this non-relativistic equation, not by reference to its origins in the fully relativistic one, but by analogy to the classical Hamilton–Jacobi equation. The action function S in the latter was reinterpreted as an exponential phase for a new wave function: the variation of a functional that would be exactly zero in the classical theory (via the Hamilton–Jacobi equation) could now be shown to give the desired wave equation—one for which the "proper values" (eigenvalues) of the energy, those for which the electron wave solutions behaved properly both near and far from the nucleus, agreed with the Bohr–Balmer values.

Unfortunately, not to put too fine a point on it, this argument is nonsense. Yet, it contained the germs of the much deeper, and fully correct, analysis of the Hamiltonian "optical-mechanical" connection, which Schrödinger would present in the second paper of the series. There he put his finger precisely on the way in which the "physical optics" represented by the wave equation made contact with the "geometrical optics" corresponding to well-defined rays following the classical mechanical paths determined by the Hamilton–Jacobi equation.

Returning to Schrödinger's first paper, it contains mainly a discussion of the analytic properties of the solutions of the non-relativistic wave equation (now universally called "Schrödinger's equation") for the hydrogen atom, for the cases where the total relativistic electron energy either exceeds mc^2—in which case the electron is unbound, and one finds oscillatory functions, damped by powers of the radial distance from the nucleus—or, more interestingly, the case where the energy is less than mc^2 by a binding energy, which must then assume definite quantized values for the associated solutions not to grow exponentially far from the nucleus. Although the techniques used by Schrödinger to establish the connection between quantization and asymptotic behavior of wave functions involved sophisticated applications of complex-function theory no longer to be found in standard textbooks, the results (reproduced by simpler power-expansion methods) are familiar to any modern student of quantum mechanics. The first paper also

[44] The sequence of events from December 1925 to the following Spring is laid out in detail in Mehra and Rechenberg (1987, ch. 3; this letter to Wien is quoted on p. 460).

discusses, in qualitative terms, the other great paradox of Bohr's 1913 breakthrough—the classically inexplicable appearance of line frequencies associated with *differences* of quantized energies, via the Bohr frequency condition. In a letter to Lorentz of June 6, 1926 (von Meyenn 2011, Doc. 76, p. 258), Schrödinger would call this picture of the excitation of light "*monstrous*" and "almost *inconceivable*" (see Volume 1, p. 16). Schrödinger speculates that the appearance of such frequencies can be attributed to "beats" between different temporal Fourier components of the wave function of the light-emitting electron.

In the three succeeding papers in the series, "Quantization as an eigenvalue problem," Schrödinger solves (correctly) a number of important problems that had preoccupied physicists in the preceding decade, during the sway of the old quantum theory (page references are to Schrödinger 1982):

1. As de Broglie had demonstrated for free particles, the superposition of waves close to the frequency $\nu = E/h$ associated with a particle of energy E, but now allowed to move through a potential energy field V produces a wave packet whose peak follows the expected classical path, provided the potential function V varies slowly over the extent of a single de Broglie wavelength of the particle. As Schrödinger emphasizes, this correspondence is exactly analogous to the reproduction of ray optics from physical optics (wave theory of light) in the eikonal approximation (index of refraction varying slowly over a wavelength). With this observation, the optical-mechanical analogy adumbrated almost a century earlier by Hamilton, and revived by de Broglie, receives a complete, and mathematically precise formulation (Paper II, Chs. 1 and 2).
2. The wave equation for the one-dimensional simple harmonic oscillator is solved, and the well-behaved solutions are shown to correspond exactly to the quantized eigenvalues (including the zero point energy) found by Heisenberg in his *Umdeutung* paper (Paper II, pp. 30–32).
3. The puzzling orbit ambiguities which plagued the old quantum theory are resolved. In our discussion of the Stark effect in Section 6.3, we pointed out that, for systems with degeneracy (distinct quantum states with the same energy), the Bohr–Sommerfeld procedure of identifying action variables by separating the Hamilton–Jacobi equation led to ambiguous sets of orbits (though not energies), since the Hamilton–Jacobi equation could be successfully separated in a variety of coordinate systems, giving quantized orbits which differed in shape and size depending on which coordinate system was used. It hardly seemed plausible that nature would display a preference for one particular coordinate system in such cases. Instead, as Schrödinger points out, while the proper functions obtained when his new wave equation was solved by separation of variables are indeed different for different choices of coordinate systems (like the orbits in the old theory), the allowed "possible state(s) of vibration" of the particle involved an "arbitrary ... linear aggregate" (i.e., linear combination, in modern terminology) of these proper solutions, and the set of such possible "aggregates" does not depend at all on which coordinate system is used to solve the Schrödinger equation (Paper II, pp. 33–34).

4. The problem of the quantized rotational energies of a rigid rotator is solved in both two and three spatial dimensions. Here, Schrödinger puts right an incorrect argument given by Heisenberg in his treatment of the rotator problem in the *Umdeutung* paper (see Section 8.5.2). Heisenberg had managed to obtain the desired "square of half-integers" for the rotational energies of a three-dimensional rotator (known empirically from molecular band spectra), *by using a two-dimensional rotator.* Schrödinger emphasizes that the behavior of quantum systems differs radically—just as that of waves—for different spatial dimensions. It is therefore critical that the problem under investigation be set up ab initio in the correct dimensions (Paper II, pp. 34–36).
5. The Schrödinger wave equation is extended from a single particle system to one containing multiple particles, by allowing the wave function to become a function of all the coordinate variables of the particles. Thus, the wave function, which had only depended on three coordinate variables x_1, y_1, z_1 for a single particle, becomes a function of the six coordinate variables $x_1, y_1, z_1, x_2, y_2, z_2$ for a system with two particles, and so on. This dependence of the wave function on more than three variables clearly meant that any attempt of a physical interpretation of the wave function as a conventional dynamical field, such as the electric, magnetic, or gravitational fields of classical theory, would have to be abandoned. With this extension, Schrödinger could reproduce the separation of center of mass motion and relative motion, and make a start on the empirically important problem of the simultaneous rotational and vibrational excitations of diatomic molecules. Of course, the generalization of the Schrödinger equation to more than one particle is a critical step, allowing the application of wave mechanics to atomic theory as well as quantum chemistry (Paper II, pp. 36–39).
6. Only a very small number of problems are susceptible to a complete analytic solution. In most cases, fortunately, the wave equation can be split into two parts, one "large but simple" part, which can be solved analytically, and the other "perturbation" part, "small but complicated," which can only be treated approximately. Technically, this is done by attaching a perturbation parameter (say, λ) to the perturbing piece, and then calculating the effect of the perturbation order by order in a formal expansion in the (presumably small) parameter λ. In his third paper on eigenvalue problems, Schrödinger shows how this formalism can be implemented in general, and then applies it to the problem of the energy shifts in the Stark effect, reproducing the (in hindsight) serendipitously correct results of Epstein (1916a) and Schwarzschild (1916) in the old theory. Additionally, however, Schrödinger was now in a position to calculate reliably the intensities and polarizations of the Stark lines, avoiding the arbitrariness of such calculations in the old theory on the basis of the correspondence principle (Paper III, chs. 1 and 2). Schrödinger's new predictions for the intensities would be confirmed within a few years by new and better experimental data (Foster and Chalk 1929).[45]

[45] For discussion, see Section 15.3.2 and Duncan and Janssen (2015, pp. 246–250).

7. In the final installment of his series on wave mechanics, Schrödinger turns to the critical issue of dynamics—the specification of the time development of a quantum system given some prescribed set of initial conditions. He had derived the original, time-independent version of the Schrödinger wave equation, valid only for conservative systems in which the classical energy function does not contain the time explicitly, by starting from the usual wave equation familiar from electromagnetic theory, in which both the space and time derivatives appear to second order. The time derivative term is then eliminated by assuming a strictly harmonic time dependence, with the frequency ν determined from the (conserved) energy by the Planck relation $\nu = E/h$. Once the system contains explicit time dependence—and in problems such as dispersion, in which the atom is immersed in an external oscillating electric field, this is clearly the case—Schrödinger quickly realized that this approach had to be modified, as the energy E of the system is no longer conserved.

   After trying various modifications, including a particularly baroque fourth-order equation, Schrödinger finally, and somewhat reluctantly, settled on his final proposal for the time evolution of the wave function in a non-conservative system. His reluctance arose from two aspects of the equation, which would have been extremely uncomfortable (or at the very least, unfamiliar) to a classically trained physicist. First, the wave equation was now only first order in time: thus, the specification of the wave function at an initial time was all that was necessary to determine the wave function at any future time, unlike the situation in classical mechanics where one needed to specify both the coordinates of a particle and the velocity components (i.e., the first time derivatives of the coordinates), given the second-order character of Newton's second law. The second uncomfortable feature of Schrödinger's time-dependent equation was that it implied that *the solutions for the wave function were unavoidably complex*. The dependence of the wave function on the full configuration space for a multiparticle system (rather than just the three dimensions of physical space) had already reduced expectations of a direct interpretation of the wave function as a physical field. The necessarily complex character of the wave function was just the final nail in the coffin for such an interpretation.

   Nevertheless, Schrödinger had no doubts that his first-order time-dependent equation had to be correct, as a straightforward perturbative calculation for the induced dipole moment of a bound electron subjected to an incident electromagnetic wave, starting from this equation, reproduced precisely the Kramers–Heisenberg dispersion formulas that had played such a critical role in the development of the matrix-mechanical formulation, and which were by now widely regarded as theoretically unassailable.

   In the last paragraphs of this fourth paper, Schrödinger makes an important advance toward a direct physical interpretation of the wave function. He shows that the absolute square ($\psi\psi^{\star}$) of the wave function, together with a vector field constructed from the wave function and its spatial gradient, satisfy a conservation equation exactly analogous to that satisfied by the electric charge density and electric current vector in classical electrodynamics. From this result, he infers that

"$\psi\psi^{\star}$ is a kind of *weight-function* in the system's configuration space ... the system exists, as it were, simultaneously in all the positions kinematically imaginable, but not 'equally strongly' in all." The explicit probabilistic interpretation of the (squared) wave function that would shortly be introduced by Born (1926a, 1926b) is not to be found here, but one might argue that Schrödinger came very close to it in spirit (Paper IV).

In March 1926, Schrödinger interrupted his sequence of papers on "Quantization as an eigenvalue problem" to publish an article that went a long way toward clarifying the relation between the wave mechanics he had developed, based entirely on the analytic properties of solutions of differential equations, and the Göttingen matrix mechanics, primarily algebraic in character. Remarkably, in all cases where the quantum system could be analytically solved in both approaches, the results (to this point, the eigenvalues of energy and angular momentum for these systems) were in complete agreement, even though, as Schrödinger points out, "the starting points, presentations, methods, and in fact the whole mathematical apparatus, seem fundamentally different" (Schrödinger 1926e, p. 45).

In this article, entitled "On the relation between the quantum mechanics of Heisenberg, Born and Jordan, and mine," Schrödinger provided an unambiguous transcription that allowed one to start from the full *discrete* set of energy eigenfunctions, found by solving the time-independent Schrödinger equation and applying the appropriate boundary conditions (typically, sane behavior of the solution at the origin and at infinity), and combine them pairwise in integrals with the coordinate variables to obtain infinite square arrays of numbers that, Schrödinger asserted, were just the matrix manifestations of the corresponding coordinate in the Göttingen approach. Likewise, the matrices associated with the momentum component of a particle could be found from the Schrödinger energy eigenfunctions by multiplying pairs of eigenfunctions, with one of the eigenfunctions differentiated in the direction of the desired momentum component (and multiplied by $h/2\pi i$). These matrices were then found to behave exactly as required in matrix mechanics: they satisfied the correct commutation relations with each other, and the energy matrix of the system, which was also found easily, by the same prescription, was diagonal, as required by the fundamental axioms of matrix mechanics.

Schrödinger's correspondences between the wave-mechanical solutions and the matrices of Heisenberg and his colleagues can be found in any modern introductory textbook on quantum mechanics. However, although correct as far as it goes, his discussion is a far cry from a rigorous proof of complete mathematical equivalence of the two theories.[46] Such a proof would be provided within a few years in the work of von Neumann (1927a, 1932), although the germs of a suitable theory can already be found in the Hilbert–Hellinger theory of infinite Hermitian forms and their connection with the solutions of Sturm–Liouville differential equations.

It so happened that Born had already deployed this machinery in his treatment of systems with both a discrete and continuous spectrum in the Three-Man-Paper (Born,

[46] For discussion, see Muller (1997–1999).

Heisenberg, and Jordan 1926, Ch. 3; see Sections 8.3.5 and 12.3.3). It would thus have been quite natural for Born to have arrived (in early 1926) at a much more complete analysis of the connection between matrix mechanics and an equivalent differential-theoretic formulation. Born, however, was occupied, at just this time, with giving lectures in the United States on the matrix-mechanical approach to quantum problems. The young American physicist Carl Eckart attended some of these lectures and, in early June, published a paper outlining many of the same arguments of Schrödinger's "equivalence" paper (Eckart 1926). In fact, even earlier than this, in a long letter to Jordan of April 12, 1926, Pauli (1979, Doc. 131) had also arrived at many of the same conclusions put in print by Schrödinger (1926e) and Eckart (1926).

The appearance of a new formulation of quantum mechanics based on the familiar analytic terrain of well-studied differential equations, rather than the unfamiliar algebraic one of non-commuting (and infinite-dimensional!) matrices, was greeted with a collective sigh of relief by the vast majority of physicists engaged in atomic theory. Schrödinger's approach was not only realized to be "equivalent" to Göttingen's matrix mechanics, but to be preferable to the latter for both conceptual and practical reasons. Conceptually, the wave theory provided an obvious candidate for the concept of "quantum state of a particle", which had been conspicuously missing from Heisenberg's theory, which was based on arrays corresponding to *pairs* of quantum states. In the new theory, one could simply identify the Schrödinger wave function of the particle at any given time with its quantum state. And from a practical point of view, where the matrix-mechanical formulation had given no general prescription for actually evaluating the matrix elements of the particle coordinates—long suspected, and, in the Three-Man-Paper, definitively shown to be the amplitudes in the Sellmeier numerators of dispersion theory, and the determinants of the intensities of spectral lines—the wave mechanics of Schrödinger immediately provided an unambiguous prescription for the calculation of these amplitudes, in terms of integrals of products of Schrödinger's wave functions. Indeed, in early May 1926, Schrödinger (1926f) was able to give explicit intensity formulas for the split lines of the Stark effect. These results were independently confirmed and extended by Pauli, who, we recall, had also independently (by early April) discovered the connection between Heisenberg's matrix amplitudes and overlap integrals of Schrödinger wave functions.

8.7 The new theory repairs and extends the old

Our journey through the old quantum theory in Volume 1, covering the progress made in applying the concepts of the Bohr atomic model of 1913 as extended by the guiding principles of Bohr–Sommerfeld quantization, the Ehrenfest adiabatic principle, and the correspondence principle, passed through a number of striking successes in accounting for atomic spectroscopy, but also ran into a number of embarrassing failures, which became increasingly problematic for the adherents of atomic planetary models by the early years of the 1920s. Even accomplishments that seemed at first unassailable—such as Sommerfeld's fine structure formula of 1916, and its astonishing success in dealing quantitatively with the doublets in X-ray spectra—would eventually encounter spectroscopic phenomena that simply could not be explained properly within the (relativistically improved) Bohr–Sommerfeld quantization framework. The problems of the anomalous Zeeman effect and the peculiarities of the helium spectrum stubbornly resisted attacks by the most sophisticated theorists, still wielding the conceptual weaponry of celestial mechanics, augmented by quantization of action variables.

The development of quantum mechanics—passing as it did from purely classical theory, through the early days of Planck–Einstein quantization of radiation and mechanical oscillators, then expanding in scope enormously with the Bohr model and its further elaboration, then undergoing a radical transformation with the appearance of matrix mechanics in Göttingen (1925) and wave mechanics (1926) in Zurich, followed two years later by the relativistic extension of the theory with the Dirac equation—provides an especially rich field for the appearance and study of "Kuhn losses" (Kuhn 1962; cf. Midwinter and Janssen 2013): phenomena which appeared to have been adequately described by an earlier theory turn out to resist explanation on the basis of a new, presumably "improved" theoretical framework. A classic example of such a loss is precisely the magical fine-structure formula of Sommerfeld mentioned earlier. Initially it seemed to do a marvelous job in explaining the fine structure of simple mono-electronic atoms (like hydrogen and ionized helium) and in bringing a huge range of X-ray phenomenology under control. However, with the arrival of matrix mechanics, the success of the Sommerfeld formula was at first a mystery, finally solved when it was understood that the formula was, in fact, formally identical to the one given by solving Dirac's relativistic wave equation, *once the quantum numbers appearing in the Sommerfeld formula were appropriately reinterpreted.*

Two other examples of Kuhn losses (which we treat in some detail in Section 15.2) involve dispersion theory and the theory of electric susceptibilities. The derivation of the classic Sellmeier dispersion formulas (which had been empirically successful from the late 1800s), which was unproblematic in classical electron theory (where the optically active electrons were assumed to be harmonically bound), became quite incomprehensible once Bohr atomic models became the foundation for atomic spectroscopy, a difficulty which was one of the primary motivations in the development of matrix mechanics, where the Kuhn loss was regained. Similarly, the analysis of the temperature dependence of electric susceptibilities, which had appeared satisfactory in classical theory, endured a Kuhn loss in the old quantum theory, which was only resolved once the techniques of the new quantum theories of 1925–1926 became available.

Returning to the successes and failures detailed in Volume 1, one clearly has two types of conceptual transformation to examine: (i) the apparent successes of the old theory (primarily in relativistic fine structure and the Stark effect) need to be understood in the light of their derivation from a manifestly inadequate—indeed, incorrect—theory; and (ii) the outright failures of the old theory (most prominently, the complex multiplet structure of electronic states in atoms, the anomalous Zeeman effect, and the structure of the helium spectrum) need to be remedied, as far as possible, by the purportedly superior new frameworks of matrix and wave mechanics.

In Chapter 15, we discuss in detail how the new theory, augmented by the critical element of intrinsic electron spin (absent from the old quantum theory), managed to arrive finally at a satisfactory understanding of the miraculous success of Sommerfeld's treatment of relativistically deformed Kepler orbits (of spinless electrons!) in obtaining the correct fine structure of the energy levels. Schrödinger's (1926f) and, independently, Epstein's (1926) solution of the Stark effect also showed that the success of the old theory was only temporary, valid to the first order of perturbation theory in the applied electric field, but failing at any further order. This was hardly surprising, as it was already known from the results obtained by Heisenberg in the *Umdeutung* paper that the quantized energy levels of an anharmonic oscillator obtained via Bohr–Sommerfeld quantization differed in detail from the matrix-mechanical ones in higher (second)-order perturbation theory, while agreeing at the lowest non-trivial order.

The analysis of fine structure in the new quantum mechanics (now including the Uhlenbeck–Goudsmit electron spin) also addressed the failure of the old theory to account for the complex structure of atomic line spectra (with new energy levels appearing as a consequence of possible variations in the relative orientation of the intrinsic spin and orbital angular momentum of the radiating electrons) and for the mysterious patterns of the anomalous Zeeman effect, which could now be perfectly explained by rather simple perturbative calculations, employing either matrix-mechanical or (more efficiently) wave-mechanical methods. The first paper to address, and resolve, both the issue of complex spectra and the mysteries of the anomalous Zeeman effect was submitted in the middle of March 1926 (Heisenberg and Jordan 1926). Using exclusively matrix-mechanical methods, the authors were able to reproduce almost completely the fine structure of hydrogenic atoms (induced by relativistic and spin–orbit effects), and the splittings (in the weak field limit) of the anomalous Zeeman effect. The only lacuna, in the case of the fine structure, was the failure to obtain the Darwin term (Darwin 1928), which appears only for *s* states, and is a result of intrinsically relativistic quantum effects, which would emerge naturally a while later in the Dirac equation.

With the further insights provided by Pauli in his introduction of the exclusion principle, the mysteries of the helium spectrum also began to dissolve, as Heisenberg (probably influenced by conversations with Fermi) finally deciphered the baffling segregation of the energy levels in the helium atom into the two non-communicating sectors of para- and ortho-helium. Heisenberg's work on helium in 1926 was soon followed by increasingly accurate perturbative calculations by Unsöld, and then by the variational ones (Kellner, Hylleraas) which are the forerunners of the Hartree–Fock (and related) schemes used to this day to accomplish astonishingly precise spectral calculations—now accurate, for the ground state energy of helium, with modern computational resources, to 40 significant digits! Quantitative studies of this sort, in which three-body systems such as helium could be treated with accuracy comparable to the spectroscopic precision available at the time, surely played a critical role in convincing more conservative physicists of the correctness of the strange new ideas introduced by Heisenberg, Schrödinger, and others. What was particularly impressive was that the new theory could control phenomena (once the appropriate mathematical techniques were brought to bear), such as the three-body problem, which were of nightmarish difficulty if treated purely classically.

8.8 Statistical aspects of the new quantum formalisms

The appearance of intrinsically stochastic processes in atomic physics, now regarded as perhaps the most characteristic feature of the new quantum-mechanical formalisms that emerged in the mid-1920s, actually precedes these developments by a full 25 years. To cite just a few examples, we have: the work of Rutherford and Soddy (1902) in which the phenomenon of radioactivity was attributed to absolutely random nuclear transformations, at a rate characteristic of each nucleus, giving rise to the famous exponential decay law; Planck's (1913, pp. 190–191) "second theory" of heat radiation, in which a Planckian oscillator (natural frequency ν) attempting to pass (upwards) through a quantized energy boundary $nh\nu$ would, with some fixed probability, either continue through the boundary (probability η) or lose all its energy (probability $1 - \eta$); Bohr's (1913a) introduction of quantum jumps of electrons in atomic stationary states in 1913; Einstein's (1916a, 1917a) work on the A and B coefficients that describe the emission and absorption of radiation quanta by an atomic system with discrete energy levels in a completely stochastic fashion (see Section 3.6); and finally, in the ill-fated theory of Bohr, Kramers, and Slater (1924a), where the virtual field acted in an explicitly probabilistic fashion to "guide" the transitions between stationary states of atomic electrons (see Section 10.4).

Still, it is somewhat surprising that the association of the square of the wave function with a probability density, usually to be found in the first few pages of any elementary text on quantum mechanics nowadays, is not to be found in any of the five papers of Schrödinger establishing the wave-mechanical formalism. In fact the word "probability" occurs only (twice) in the combination "transition probability", in connection with the long-sought relation between the strength (i.e., relative frequency) of the various possible quantum transitions of an atomic electron. There is an intriguing suggestion in Schrödinger's final paper that $\psi\bar{\psi}$[47] should be interpreted as "a kind of *weight-function* in the system's configuration space ... If we like paradoxes, we may say that the system exists, as it were, simultaneously in all the positions kinematically imaginable, but not 'equally strongly' in all" (Schrödinger 1926h, p. 120).

In hindsight, we can clearly see that a conceptually consistent incorporation of probabilistic/statistical ideas into quantum theory was accomplished almost exclusively by the Göttingen group responsible for the discovery and elaboration of matrix mechanics: Born, Heisenberg, Jordan, and, a little later, the mathematician John von Neumann, who came to Göttingen in 1926 at Hilbert's invitation, primarily to continue his work on the axiomatic foundations of mathematics in set theory. The other two main contributors, Pauli and Dirac, also belong, in this respect, to the matrix-mechanics camp.

The first tentative steps in the evolution of these ideas were taken by Born (1926a, 1926b), in two papers on the quantum mechanics of collision processes, submitted in

[47] In most of the early papers on quantum mechanics, bars ($\bar{\psi}$) are used to represent complex conjugates, which we would now write using asterisks, $\psi\psi^*$). We have largely adhered to the former notation throughout this book, to improve accessibility to the primary sources.

mid-1926. Born used the time-dependent wave-mechanical formulation of this problem, in which a particle is sent in along the x-axis toward a force field centered at the origin, and then is detected after scattering to emerge in some direction $\vec{k}'$, specified by the wave-vector of the outgoing particle, which far from the force center has a simple plane wave function $\varphi(\vec{k}') \exp(i\vec{k}' \cdot \vec{r})$. In the first of these two papers on collision processes, Born (1926a, p. 865) declared that the probability of observing the scattered particle in the direction of $\vec{k}'$ was "determined by the amplitude [in general, this is a complex number] $\varphi(\vec{k}')$." In a footnote added in proof, he amended this statement to identify the probability with the *absolute square* of the amplitude factor $\varphi(\vec{k}')$, a relation which has since been universally dubbed the "Born rule". The second paper considerably amplified the results of the first: in particular, a "conservation of probability" principle was confirmed, with the sum of probabilities for all possible final scattering states giving unity, as required for a consistent probability interpretation. In both papers, Born was quite clear that these probabilistic statements exhausted the predictive power of the theory: one could not make predictions for the outcome of a scattering on a case-by-case level, only statistically, as a statement about the fraction of scattered particles emerging in a given angular cone, for example.

One issue that became a real obstruction for the Göttingen theorists in the development of a correct statistical interpretation of the theory was connected with Bohr's original notion (going back to the 1913 trilogy) that atoms generally existed in one, and only one of the possible stationary states determined by the electronic dynamics, with occasional sudden (and effectively instantaneous) transitions between stationary states—the infamous "quantum jumps." By contrast, Schrödinger had quite happily written down expressions in which the wave function involved the superposition of energy eigenfunctions with different energies (expressions which he called in his first paper "a *pot-pourri* of its proper vibrations," Schrödinger 1926c, p. 11): indeed, such combinations were central to his derivation of the Kramers dispersion formula in wave mechanics. In an important paper on the reinterpretation of Ehrenfest's adiabatic principle (see Section 5.1) in wave mechanics, Born was at pains to asser that

> the statement that an atom can be at one and the same time in several stationary states not only runs counter to the whole Bohrian theory, but directly contradicts the natural, and so far never disputed, interpretation of the stationary states corresponding to points of the continuous spectrum . . . we shall therefore hold on to the Bohrian picture, that an atomic system is always only in *one* stationary state (Born 1926c, p. 170).

In his paper on the adiabatic theorem, Born considers the time development of an atomic system initially prepared in a definite stationary state (the nth, say) and then subjected to a temporary external influence (an external electric field, say, turned on at time $t = 0$ and off at time $t = T$). Application of Schrödinger's time-dependent equation then gave the wave function of the system after the removal of the external perturbation as a discrete Fourier sum of (in general) all stationary states, with the complex coefficients attached to each term (which, when squared and added, gave unity) interpreted by Born as determining the probability that the atom would be found to be in the corresponding stationary state after the perturbation. At this point, Born, in harmony with the "Bohrian picture," thought that the atom would at all times be in *some* stationary state and that the appearance of a superposition of different stationary states in the wave function simply allowed for a determination of the probability of the atom to be found in a particular one after it was "shaken" by the perturbation. This probability would be measured empirically by repeating the experiment a large number of times, starting from the same initial stationary state, applying the same temporary field, and taking note of the particular stationary state the atom ended up in at times $t > T$ in each case.

A few weeks after Born had completed and submitted his paper on the adiabatic theorem, Heisenberg (1926d) submitted a short note, entitled "Fluctuation phenomena and quantum mechanics," in which the Bohrian interpretation of the time evolution of atomic systems (via a sequence of quantum jumps at instantaneous temporal points between well-defined stationary states—a point of view which Heisenberg refers to as the "discontinuity theory") was subjected to scrutiny from the point of view of a matrix-mechanical examination of an idealized system of two atoms a and b, each capable of occupying one of two states, n and m, of energy E_n and E_m, respectively. One then imagines a weak coupling to be introduced allowing the atoms to exchange an energy quantum of magnitude $|E_n - E_m|$ (say, via a light quantum of appropriate frequency). At time zero the atoms are placed in different stationary states (say, atom a in n and atom b in m). A simple matrix mechanical calculation establishes that the energy matrix of the entire system (i.e., of atom a plus atom b plus the coupling energy term) has, in place of the two values E_n and E_m of the individual atoms, the diagonal values $(E_n+E_m)/2$. Quite reasonably (if one adopts the Bohr picture), this is interpreted as implying a dynamical process in which the two atoms randomly switch states in such a way that, over a sufficiently long period of time, each atom will be found to have spent 50% of its time in each of the two available stationary states (see Figure 16.1). Heisenberg then shows that *all* statistical properties of the energy of either atom (average value, mean square fluctuation, etc.) are in perfect harmony with the notion that the atoms are always in one or the other stationary state, spending on average half their time in each. Heisenberg concludes from this that "the fact of discontinuity [i.e., a discrete succession of quantum jumps between definite stationary states] is naturally incorporated in the system of quantum mechanics" (Heisenberg 1926d, p. 504). This conclusion, as we will see below, turned out to be wrong.

The clues to the emergence of a deeper understanding of the statistical content of quantum mechanics can be found in a short, but prescient, analysis in Born's (1926c) adiabatic-theorem paper. Let us suppose that an atomic system is already, at time zero, described by a wave function that is a linear superposition of several (perhaps infinitely many!) stationary states n, with the nth state associated with an amplitude coefficient c_n (a complex number the absolute square of which then corresponds to the probability that the atom is *actually* in state n). It could be, for instance, that the atom was previously in a known stationary state and then subjected to a perturbation. At time $t = 0$ a temporary external field is turned on, which is switched off permanently at time $t = T$. Born shows that at late times $t > T$ the atomic wave function is once again a linear superposition, but with each stationary state n associated with a different complex amplitude C_n, the square of which would again give the probability that the atom is in state n after time T. The new complex amplitudes are linearly related to the old ones by $C_n = \sum_m c_m\, b_{mn}$, where the matrix b_{mn} depends in detail, of course, on the particular perturbation applied between $t = 0$ and $t = T$. They can fortunately be squared and summed to give unity, preserving the interpretation of $|C_n|^2$ as the probability that the atom will be in a particular stationary state n after the perturbation is removed.

Now, it is apparent that the quantity $|b_{mn}|^2$ is just the probability that the perturbation induces a transition from state n to state m, if the atom is known to be in state n at time zero. As Born (1926c, p. 174) points out, if the quantum jumps undertaken by the atom were statistically independent, the joint probability of "being in some state m at $t = 0$" and "transitioning from m to n via the perturbation" would be $|c_m|^2\, |b_{mn}|^2$, and the total probability of being in any state m and arriving at state n after time T just the sum $\sum_m |c_m|^2\, |b_{mn}|^2$. This is *not* the same as the wave-mechanical result $|C_n|^2 = |\sum_m c_m\, b_{mn}|^2$ given earlier: the sum of absolute squares of complex numbers does not equal the absolute square of their sum—in general, these differ by cross-terms, the famous interference terms that would come to dominate the discussion of the statistical interpretation of quantum mechanics over the next year.

An important advance in the understanding of the connection between the kinematical and statistical aspects of the new theory was made by Pauli, in a long letter of October 19, 1926 to Heisenberg (Pauli 1979, Doc. 143). The considerations contained in this letter, which include prescient comments on the mutual uncertainty of position and coordinate values, were not published in their own right, but Pauli's letter did make the rounds at Bohr's institute in Copenhagen, where Heisenberg had taken up residence. In this letter, Pauli pointed out that a simple mathematical transformation could be used to derive the probability amplitude $\varphi(p)$ for a particle in momentum (p) space, starting from the wave function $\psi(q)$ in coordinate (q) space. Just as $|\psi(q)|^2 dq$ gave the probability for locating the particle between positions q and $q + dq$, $|\varphi(p)|^2 dp$ would give the probability that the particle would be found to have momentum between p and dp. The mathematical transformation required—a Fourier transformation—was completely independent of the particular dynamical environment of the particle (harmonic oscillator, electron in hydrogen atom, etc.): it was just a consequence of the kinematical relation between coordinate and momentum, in particular, of the fact that they are canonically conjugate quantities. A wide-reaching generalization of this observation of Pauli's, connecting (via a specified linear transformation, in this case, of integral form) the probability amplitudes written as functions of different kinematical variables, would be undertaken within a few months by Dirac (1927a) in Copenhagen and Jordan (1927b) in Göttingen, leading to essentially identical formalisms (in substance), now known collectively as the Dirac–Jordan statistical transformation theory. An obvious difference is in the notation: Jordan's is almost impossibly awkward, while Dirac's notation (with minor cosmetic alterations) would evolve into the almost universally employed bra-ket notation familiar to any modern physicist. It is probably for this reason that Jordan's paper, entitled "On a new foundation [*Neue Begründung*] of quantum mechanics,"[48] has been more or less forgotten by physicists, and largely ignored by historians of modern physics. Nevertheless, Jordan's

[48] We refer to this paper as *Neue Begründung* I. Its sequel, *Neue Begründung* II (Jordan 1927g), is discussed below.

paper pays far more attention to conceptual issues associated with the statistical content of the theory, so we emphasize it in our brief summary here of the Jordan–Dirac formalism, with occasional indications of the notational differences with Dirac's formulation.

In the opening pages of *Neue Begründung* I, Jordan gives Pauli full credit for stimulating his new formulation, which, "in close connection to Pauli's thoughts, attempts to base quantum mechanical laws on a few simple statistical assumptions" (1927b, pp. 810–811). These assumptions turn out not to be based on probabilities directly, but on *probability amplitudes*, which are complex numbers, the absolute square of which are to be interpreted as probabilities. The probabilities in question are conditional ones, and hence, refer to two separate kinematic quantities—arbitrary functions of position and momentum, $A(q,p)$ and $B(q,p)$, say—where the theory is supposed to assign a complex number $\varphi(a,b)$ whose *square* gives the conditional probability $\Pr(a|b)$ (in modern notation) that the kinematical quantity A will be found to take the value a if the system is already known to have the definite value b for the kinematical quantity B. Strictly speaking, for variables taking continuous values (such as position or momentum), we should really speak of a conditional probability density: $\Pr(a|b) = |\varphi(a,b)|^2 da$ gives the probability that, given the value b for B, the quantity A will be found to have a value between a and $a + da$.

Two very important examples, which recur throughout *Neue Begründung*, are (i) when $A = Q$ (the coordinate of a particle) and $B = H$ (the energy function), in which case the associated probability amplitude is just $\varphi(q,E) = \psi_E(q)$, the Schrödinger energy eigenfunction for a particle with energy E, and (ii) the amplitude relating canonically conjugate quantities (such as $A = Q$; $B = P$, the momentum), in which case the probability amplitude is just $\varphi(q,p) = \exp{(2\pi ipq/h)}$, the familiar plane-wave expression for a free particle wave function. We note here that all of these statements have precise correlates in Dirac's (1927a) paper. One simply has to replace $\varphi(a,b)$ with the expression (a/b), which would become $(a|b)$ in his famous textbook (Dirac 1930), and somewhat later the notation $\langle a|b\rangle$ found in modern texts (cf. Chapter 16, note 21).

The most important of Jordan's four postulates, which define and constrain probability amplitudes in quantum mechanics, is the third (Postulate C of section 2 of *Neue Begründung* I), in which Jordan (1927b, p. 813) states that "probabilities combine in an interfering fashion." Following the notations of the previous paragraph, this states that if we have *three* distinct kinematical variables A, B, and C, which can be measured to assume values a, b, and c, respectively, then the conditional *probability amplitude* $\varphi(a,c)$ for finding the value a for A given c for C, is a sum (or integral, for continuous quantities) of the amplitude products $\varphi(a,b)\,\varphi(b,c)$ over all possible values b of B. Only once these complex products are summed (over b) can they be squared to give the conditional probability $\Pr(a|c)$. The square of the sum (or integral) evidently produces cross-terms, which we recognize as appearing in exact analogy to the interference terms commented on by Born in his discussion of quantum transitions induced by transitory perturbations.

The genetic origins of the Jordan–Dirac formalisms in matrix mechanics are somewhat obscured by the fact that both authors emphasized quantities with completely continuous spectra. However, it should be realized that the Göttingen quantum theoreticians had by this time completely accustomed themselves to the idea that any function of two variables (in particular, the amplitudes $\varphi(a, b)$) could be thought of as two-indexed matrices, with either index being either discrete or continuous. In particular, Jordan's formalism takes over very literally the idea presented in the Three-Man-Paper of Born, Heisenberg, and Jordan (1926) that classical canonical transformations, from one set of suitable conjugate (q, p) pairs to another (q', p'), could be implemented in quantum mechanics by a similarity transformation generated by a transformation matrix S (e.g., $q' = SqS^{-1}, p' = SpS^{-1}$ in the Three-Man-Paper, where q, p, and S are discrete numerical matrices). In *Neue Begründung* I, these canonical transformation matrices take on a new form (they have continuous indices, and act as integral kernels, rather than discrete matrix multiplicands) and play a new role as the implementers of the transformations of probability amplitudes between different sets of kinematical variables.

Heisenberg, and later von Neumann, would object strenuously to Jordan's terminology "interference of probabilities" (and more generally, to any implication of a breakdown in classical probability theory induced by quantum mechanics) in his introduction of the third postulate C, with very good reason.[49] These objections were openly expressed in a letter from Heisenberg to Jordan dated March 7, 1927 (AHQP), after Heisenberg had (with much difficulty) finally mastered the technicalities of Jordan's transformation machinery. In this letter, Heisenberg indicates that he was working on a paper which would serve as "physical commentary" on the Jordan–Dirac theory, in which the formalism introduced could be used to give a precise mathematical constraint on the ability to simultaneously specify the p and q of a particle. He goes on to insist that the appearance of interference terms in the Jordan "interfering probabilities" rule (Postulate C of *Neue Begründung* I) "has nothing to do with the laws of probability," which continue to hold *without* "interference." Heisenberg repeats (with approval) in this letter Dirac's opinion: "all statistics is brought in only through our experiments." The fundamental dynamics of quantum theory, according to Dirac, is *not* intrinsically probabilistic (hence, indeterministic), but the introduction of probabilistic aspects in the specification and utilization of an experimental system will inevitably lead to predictions of a statistical nature:

> The notion of probabilities does not enter into the ultimate description of mechanical processes; only when one is given some information that involves a probability ... can one deduce results that involve probabilities (Dirac 1927a, p. 641).

Heisenberg finished the upcoming paper mentioned in his letter to Jordan, now universally known as "the uncertainty paper" (Heisenberg 1927b), in the first three weeks of March 1927, thereby avoiding the constant interruption and criticism by Bohr, who had departed for a month-long skiing holiday in Norway in mid-February. In Bohr's absence, there was frequent (and constructive) communication with Pauli in Hamburg.

[49] The full mathematical justification for Jordan's postulate C would depend on von Neumann's spectral theory for self-adjoint operators, hypothesized in the latter's "Mathematical foundation [*Mathematische Begründung*] of quantum mechanics" (von Neumann 1927a), which would appear in May 1927, and proved with full mathematical rigor two years later (von Neumann 1929). Modern students would simply refer to this postulate as the completeness relation for the eigenstates of a physical observable.

In the winter of 1926–1927, Bohr and Heisenberg had had intensive, practically daily, conversations on the "paradoxes of quantum theory." In Heisenberg's later recollections, the interpretation of the two-slit experiment and of the distinct paths generated by electrons passing through a cloud chamber (which seemed to contradict the spreading of the wave packet predicted by wave mechanics) were explored repeatedly in nightly discussions with Bohr in Heisenberg's room in the attic of the Bohr Institute.

These conversations frequently became quite heated, as Heisenberg continued to reject the use of the Schrödinger wave formulation, insisting on an analysis solely based on the clear mathematics (but obscure physical implications) of the new Jordan–Dirac theory, whereas Bohr was convinced that the injection of wave properties à la Schrödinger was an essential prerequisite to resolving the thorny epistemological issues raised by the new quantum mechanics. Fortunately for Heisenberg, Bohr's departure on holiday left him free to develop his ideas, which he explained in great detail to Pauli in a long letter sent on February 23, 1927 (Pauli 1979, Doc. 154). In this letter, Heisenberg summarizes (in itemized form) his new ideas on the "visualizable sense" (*anschaulichen Sinn*) of quantum mechanics. In point 1 of his itemized list, Heisenberg emphasizes the need for an operational definition, taking into account experimental constraints, in order to clearly define kinematic concepts such as position in quantum mechanics:

> What does one understand with the word [phrase]: "position of an electron?" This question is to be replaced in the well-known fashion by another: "how does one *determine* the position of an electron?" (Pauli 1979, Doc. 154, p. 376).

Heisenberg goes on to answer this question by using the example of a microscope "of sufficient resolution" used to "look at" the electron. In fact, this can (in principle) be done with arbitrarily high resolution by using light of sufficiently small wavelength—for example, gamma rays—but at the cost of disturbing the momentum to an increasingly large account as a result of the Compton scattering of the highly energetic gamma-ray quantum with the particle. In item 5, the famous momentum–position mutual uncertainty principle is stated precisely for the first time:

> If a quantity p is determinable with a precision which is characterized by a mean error p_1, then the canonically conjugate coordinate q is characterized by a mean error $q_1 \approx \frac{h}{2\pi p_1}$ (Doc. 154, p. 378).

The abstract for Heisenberg's (1927b) celebrated paper on the uncertainty principle clearly indicates Heisenberg's achievements in clarifying the physical, "visualizable" (*anschaulich*) content of the theory. The kinematical concepts of classical particle mechanics, position, speed, energy, and so forth, are to be given precise operational definitions, which imply that they can *individually* be determined with arbitrary precision on any given system by a suitable experimental setup. However, for canonically conjugate quantities (e.g., momentum-position, or energy-time), there is an inescapable limitation on the accuracy with which they can simultaneously be determined. Heisenberg announces that he will use the formalism of Jordan and Dirac (primarily, postulate C of *Neue Begründung* I) to provide a firm mathematical treatment of several thought experiments that show "how macroscopic processes can be understood on the basis of quantum mechanics."

The contents of this paper are analyzed in detail later in Section 16.3. Here, it suffices to say that, in his treatment of the position–momentum uncertainty relation, Heisenberg introduces the famous "gamma-ray-microscope" still used in many elementary texts on quantum mechanics to explain the mutual indeterminacy of p and q, whose uncertainties multiply to a quantity greater than or of the order of Planck's constant. In discussion of the energy–time uncertainty relation (stating a similar constraint on the uncertainties in energy and duration of a quantum process), the thorny issue of transitions between stationary states and quantum jumps is revisited. The paper also discusses several examples of experiments involving passage of atoms through a sequence of Stern–Gerlach magnets, another mainstay of many authors in introducing the fundamental superposition and interference features of quantum theory.[50]

In early June 1927, Jordan submitted *Neue Begründung* II (Jordan 1927g), a follow-up article to *Neue Begründung* I (Jordan 1927c), and introduced a number of cosmetic improvements (mostly notational) to the formalism of the first paper. Jordan also made an attempt to come to grips with the more general circumstance encountered in quantum mechanics, whereby physical observables could have a partly discrete spectrum (the energy of the hydrogen atom is an obvious example). The mathematical formalism needed for dealing properly with such situations was simply not yet available—or at least, unknown to physicists. Moreover, in both Dirac's and Jordan's formalism one encountered "improper functions", such as the famous "delta function" $\delta(a-b)$ (the probability amplitude $\varphi(a, b)$ for two continuous variables a and b in the special case that the associated kinematical variables were in fact the same (e.g., $A = B = Q$). This remarkable "function" had the property of vanishing for all non-zero values of its argument, but somehow integrating to unity from the single contribution at zero argument. The delta function, introduced formally and explicitly by Dirac (1927a), and its relatives (various derivatives, for example) was viewed with extreme suspicion by the mathematicians next door to the Göttingen physicists.[51]

There were several other difficulties with Jordan's reformulated formalism in *Neue Begründung* II. We only mention two here:

1. It is a central tenet of Jordan's discussion that a canonical transformation (of similarity type, e.g., $\alpha = TpT^{-1}$, $\beta = TqT^{-1}$) could be established between any canonical pair (p, q) and any other pair (α, β), in analogy with the situation in

[50] The first chapter of McIntyre's (2012) excellent introductory text is, in fact, titled "Stern–Gerlach experiments." Like several other authors using Stern–Gerlach magnets in this way (for a partial list, see Janas, Cuffaro, and Janssen 2022, p. 203, note 41), McIntyre credits the Feynman lectures for this approach (see Feynman, Leighton, and Sands 1964, Vol. 3, Ch. 5).

[51] A completely rigorous theory of generalized functions, or *distributions*, would be developed starting with the work of Sergei Sobolev in the late 1930s, and with much amplification, the work of Laurent Schwartz in the 1940s. Fortunately, the purely formal properties of the delta function used by physicists in the 1920s turned out to be perfectly reliable placeholders for von Neumann's rigorous spectral theory of 1929.

classical mechanics, where one could move freely, for example, between an initial set of Cartesian coordinates and momenta, and (for suitably multiply periodic systems) a more convenient set of conjugate action and angle variables $(\mathcal{J}, w)$. In doing this, Jordan assumed that one could go from variables that assumed purely continuous values, to others with partially discrete, partially continuous spectra. In fact, well-defined similarity transformations cannot alter the nature of the spectrum—the transformations envisaged by Jordan are in fact much more limited in scope than he supposes.

2. In his stubborn adhesion to canonical coordinate pairs, Jordan even ends up trying to interpret different Cartesian components of the spin observable (which do not commute) as somehow canonically conjugate to each other. Of course, the spin components (of which there are an odd number, namely 3!) cannot be neatly sorted into pairs of canonically conjugate kinematical quantities, unlike the ps and qs of classical point mechanics.[52]

Fortunately, only a few months after *Neue Begründung* II was published, the whole mathematical substructure of quantum mechanics, attached to a conceptually clear and mathematically precise probability interpretation, was put on a rigorous footing by John von Neumann in three remarkable papers. His Hilbert space formulation of the theory, together with his introduction of the use of density matrices to describe exactly the statistical content of the theory, remain to this day the indispensable basis for a clear understanding of quantum phenomena. The three installments of von Neumann's trilogy were presented (by Born) at two meetings of the Göttingen Academy of Sciences in 1927. The first, "Mathematical foundations [*Mathematische Begründung*] of quantum mechanics" (von Neumann 1927a), was presented on May 20; the remaining two, "Probability-theoretic construction [*Wahrscheinlichkeitstheoretischer Aufbau*] of quantum mechanics" (von Neumann 1927b) and "Thermodynamics of quantum-mechanical ensembles" (von Neumann 1927c), on November 11.

Between these two dates, two important international gatherings of quantum physicists occurred. The first was held in Como, Italy (September 11–20) to commemorate the centenary of the death of Alessandro Volta. The second, the fifth Solvay conference, was held in Brussels a little over a month later, October 24–29. Only von Neumann's first paper, on the mathematical structure of the theory, had been published before these conferences were held, and it did not excite much comment or interest among the most prominent quantum theorists in the Summer and Fall of 1927 (with the exception of Max Born, who mentioned von Neumann's work approvingly in discussion at both conferences). Instead, much of the discussion at these conferences focused on interpretational issues, which follow fairly directly from the conflict between the statistical/stochastic view (now supplemented by the epistemological implications of Heisenberg's uncertainty principle) advanced by the Göttingen camp, and the continuum wave theory, eschewing quantum jumps, whose principal proponents were

[52] For an extended analysis and criticism of Jordan's formalism, see our paper "(Never) mind your p's and q's: von Neumann versus Jordan on the foundations of quantum theory" (Duncan and Janssen 2013).

Schrödinger and de Broglie. Before turning to a description of von Neumann's great achievement in his Göttingen Academy trilogy, we briefly describe the interactions and conversations between the leaders of quantum theory in Como and Brussels.

8.9 The Como and Solvay conferences, 1927

Bohr, in contrast to the Göttingen theorists (especially Heisenberg), had enthusiastically welcomed Schrödinger's wave formalism as an important advance in the development of quantum mechanics—not just from a practical viewpoint (as an efficient tool for calculating transition amplitudes, for example), but for complementing the particle picture underlying matrix mechanics by a wave picture. Bohr in a sense placed himself in between the contesting camps of those theorists who had developed the matrix mechanics of Heisenberg into the much more general Jordan–Dirac statistical transformation theory, attached to an interpretational framework in which discontinuity (in particular, random quantum jumps) was essential, and the growing (and much more numerous) group of physicists who gladly accepted the Schrödinger formalism, with its familiar partial differential equations (reminiscent of Maxwellian electrodynamics) in which continuous evolution of the quantum state was emphasized. He argued vociferously with Schrödinger, on the occasion of the latter's visit to Copenhagen in October 1926 (famously sitting at Schrödinger's sickbed and bombarding him with objections), insisting on the need for a discontinuous aspect in wave mechanics; with equal intensity, he argued with Heisenberg in the Spring of 1927 on the necessity for the inclusion of wave aspects in the examination even of discontinuous quantum processes.

An important example of Bohr's insistence on the inclusion of both wave and particle features can be found in his criticism of Heisenberg's derivation (argument would be more accurate) of the momentum–position uncertainty relation, $\Delta p\ \Delta q \sim h$ (in modern notation) in his uncertainty paper. After reading Heisenberg's manuscript (which had already been submitted for publication), Bohr became increasingly uncomfortable with Heisenberg's analysis of the "gamma-ray microscope" thought experiment used to establish the uncertainty principle. Bohr pointed out that there were *two* critical inputs to the derivation of the mutual uncertainty relation: (i) the Compton-effect argument describing the transfer of momentum (and energy) from the gamma-ray quantum (photon) used to "observe" the electron and thereby determine its position; and (ii) the classical Abbe formula for the resolution of a microscope of a given aperture using light of a given wavelength. Bohr insisted that Heisenberg's analysis of the measurement processes used to operationally define kinematic concepts, such as position, momentum, energy etc., while in essence correct, did not sufficiently emphasize the dual character of elementary objects under study in atomic physics (primarily electrons, and light quanta/photons). He prevailed on Heisenberg to add a postscript to his paper (still in proof) admitting the shortcomings of his examination of the microscope, inasmuch as "the uncertainty in measurement is not based exclusively on the appearance of discontinuities, but rather ... on the requirement to treat properly the experiential requirements expressed both by the particle theory on the one hand, and the wave theory on the other" (Heisenberg 1927b, pp. 197–198).

Bohr had for some time planned to write an extensive paper treating the conceptual foundations of quantum mechanics, and an invitation to speak at an international conference in Como in mid-September, together with the intense arguments with Heisenberg on the proper treatment of measurements, added an element of urgency. He started in April 1927, continuing (with multiple revisions) until April 1928, when an article entitled "The quantum postulate and the recent development of atomic theory" was published simultaneously in German, in *Die Naturwissenschaften* (Bohr 1928a), and in English, in *Nature* (Bohr 1928b).[53] Bohr's lecture at Como was not reproduced exactly in later published proceedings (Bohr 1928c), but can be roughly reconstructed given extant manuscripts from the period July–September (including one written during the conference itself, which presumably most closely echoes Bohr's remarks). The word "complementarity" appears already in a manuscript of July 10, and increasingly becomes the underlying theme of Bohr's interpretational structure.

In a manuscript of September 13 sketching his lecture at Como (Bohr 1972–2008, volume 6, pp. 75–80 (transcription), pp. 81–88 (facsimile)), prepared when the conference was already underway (Mehra and Rechenberg 2000–2001, pp. 192–193), Bohr reproduces the Heisenberg uncertainty relations for momentum–position and energy–time by an ingenious use of wave theory, followed by the imposition of particle properties using the Planck (energy equals Planck's constant times frequency) and de Broglie (momentum equals Planck's constant times wave number) relations. From wave theory, one knows from the properties of Fourier integrals that a wave pulse limited to spatial extent Δq must be analyzable into a combination of monochromatic waves with a spread of wavenumber given by $1/\Delta q$. The de Broglie relation then immediately implies that the spread in momentum Δp must satisfy the Heisenberg relation $\Delta p \cdot \Delta q \sim h$. Similarly, the time extent of a wave pulse Δt is, by Fourier analysis, just the inverse of the spread in frequencies $\Delta \nu$ present in the pulse. Using the Planck relation $E = h\nu$ associating a wave property (frequency) to a particle one (energy), one thus finds the energy–time uncertainty relation $\Delta E \cdot \Delta t \sim h$.

Bohr also sketches a thought experiment, in which a beam of particles (either electrons or light quanta) passes through a small hole, and shows that the same uncertainty relations in properties of the emerging particles obtain, once we employ both the classical wave diffraction effects spreading the outgoing beam *and the Planck and de Broglie relations connecting wave to particle properties*. The need for dual argumentation of this kind, according to Bohr, reflects (quoting once again from his manuscript of September 13) "the complementary features of the apparent[ly] contrary claims of individuality [i.e., particle nature] and superposition [i.e., wave nature]" (Bohr 1972–2008, volume 6, p. 79). The concepts of a particle, as such, or a wave, as such, are just "abstractions according to quantum theory" (ibid.) Any analysis of actual interactions at the atomic level required the dual (or complementary) use of both wave and particle properties.

Bohr's talk at Como on September 16 did not, as Abraham Pais (1991, p. 315) put it, "bring down the house." The participants from the Göttingen–Copenhagen axis were

[53] The tortuous gestation of this famous paper of Bohr's is described in detail in Section 16.4.

in sympathy with Bohr's remarks, but the general sentiment was that Bohr was simply restating, in characteristically ponderous language, ideas that were current and already much discussed by theorists working at the cutting edge of quantum theory. People like Schrödinger and Einstein, who held quite different views on the necessity for a statistical interpretation of quantum measurement along the lines of Heisenberg and Bohr, were absent from the Como conference.

A month later, many of the participants at the Como conference would regather in Brussels for the fifth Solvay conference, devoted to "The quantum theory and the classical theories of radiation"—although the proceedings of the conference would later appear under the simplified title "Electrons and photons," using the term the American chemist Gilbert N. Lewis (1926) had recently introduced for the awkward "light-quantum" construction (cf. note 42). But now, the presence of de Broglie, Schrödinger, and Einstein would provide a significant counterweight to the emerging Bohrian dogma. The conference, which lasted from Monday October 24 to Saturday October 29, featured presentations by W. L. Bragg (Monday, on X-rays), A. H. Compton (Tuesday morning, on quantum aspects of electromagnetic radiation), de Broglie (Tuesday afternoon, on a deterministic pilot wave interpretation of the wave function), Born and Heisenberg (Wednesday morning, combined report, giving an overview of quantum mechanics), and Schrödinger (Wednesday afternoon, on wave mechanics). The general discussion on Thursday morning included extended remarks by Bohr, which seem to have largely echoed his lecture at Como.[54]

De Broglie's pilot-wave theory, which attempted to restore a deterministic underlying dynamics for particles "guided" by the spatio-temporal behavior of the Schrödinger wave function, was subjected to criticism in the discussion section (with some defense by Léon Brillouin), but left little impact on most in the audience, who clearly expected the "final theory" to emerge from some sort of amalgam of matrix and wave ideas.[55] The joint presentation of Born and Heisenberg presented a review of the basic ideas of matrix mechanics (as described in Chapter 13), supplemented by the statistical-transformation theory of Dirac and Jordan. The Schrödinger wave formalism was presented as a special case of the latter. Even at this point, the authors were still adhering firmly to the Bohrian precept: "a system can always be in only *one* quantum state" (Bacciagaluppi and Valentini 2009, p. 386). Such a state would be an eigenstate of the unperturbed Hamiltonian: the application of a perturbation then causes the system to transition randomly (the infamous "quantum jumps") between these eigenstates. The statistical content of the theory (and interpretation of the squared coefficients of the time-dependent wave function as transition probabilities) was then regarded as an inevitable consequence of the insistence that the system at any given time be in a definite stationary

[54] For a detailed account of the entire conference, see Bacciagaluppi and Valentini (2009). Here, we pass over the contributions of Bragg and Compton and concentrate on the lectures and discussion directly addressing quantum theory.

[55] The de Broglie pilot-wave theory would re-emerge two and a half decades later in Bohm's mechanics, which recruited a fairly large following, with adherents to be found to this day. See Bacciagaluppi and Valentini (2009, Pt. I, Ch. 2, pp. 27–79) for discussion of De Broglie's pilot-wave theory and Freire (2019) for a biography of Bohm.

state—albeit, one that can not be predicted in advance, except probabilistically. The connection of the Jordan–Dirac theory to the operational definitions of kinematic quantities, with the consequent emergence of the Heisenberg uncertainty relations, was described briefly.

Born and Heisenberg concluded their report with an account of various applications of the new quantum mechanics (cf. Chapter 15). Primarily, these concerned (i) the successful treatment, at long last, of the anomalous Zeeman effect, incorporating the Uhlenbeck–Goudsmit electron spin; and (ii) problems involving more than one particle, where the need for the supplementary constraints implied by Bose–Einstein (for photons) or Fermi–Dirac statistics (for electrons, and, probably, protons) were essential to understand atomic spectra for atoms with more than one electron. The first inklings of quantum field theory were mentioned at the close: Jordan's quantization of a field whose quanta satisfy Fermi–Dirac statistics, and Dirac's recent quantization of the electromagnetic field, allowing a first-principles derivation of the Einstein A and B radiation coefficients, as well as an improved dispersion theory in which damping terms appear in the Kramers–Heisenberg denominators.

Schrödinger's lecture basically reprised in highly abbreviated form the content of his four papers on wave mechanics. The time-dependent version of his (non-relativistic) wave equation was argued for on the basis of its success in obtaining (a) an understanding of the Bohr frequency condition; (b) the successful extraction of the dipole matrix elements that correctly predict the intensity and polarization of radiation emitted in atomic transitions; and (c) its ability to reproduce the Kramers–Heisenberg dispersion theory. However, Schrödinger was clearly ambivalent on the correct interpretation of the wave function itself. Born proposed that $\psi\psi^*$ be regarded as a probability density for finding an electron at a given position, while Schrödinger preferred to regard the square of the wave function (times the electron charge) as a physical charge density, the time variation of which can then be employed to calculate emitted dipole radiation. On the critical issue of discontinuity (quantum jumps), Schrödinger clearly took the position that isolated quantum systems should evolve (following his time-dependent equation) continuously and *causally*.

Schrödinger's lecture in Brussels was quite different in spirit from that of Born and Heisenberg, where one gets the impression that the theory is basically settled, with a few minor points of difficulty yet to be cleared up. In contrast, Schrödinger emphasized multiple points where his formalism encountered serious difficulties: for example, the interpretation of the wave function for N-particle systems (defined on a space of $3N$ variables, the joint coordinate space of all the particles, rather than ordinary three-dimensional space), the treatment of radiation reaction in atomic transitions, and the failure of the relativistic version of his equation (which he had originally found, and discarded, in December 1925) to reproduce the Sommerfeld fine-structure. These issues were brought up again in the discussion section—with inconclusive results, as the correct formalism for dealing with multiparticle quantum systems in terms of a quantum field defined in physical space-time had only recently (early October) been submitted for publication by Jordan and Klein (1927), and (with the exception of Pauli) was clearly not appreciated by the participants in Brussels. Jordan had not been invited. Dirac was present and his quantum electrodynamics, giving a consistent treatment of the radiation problem in terms of a fully quantized system of matter plus radiation, had already appeared in print six months before this Solvay conference (Dirac 1927b), but does not seem to have been fully digested by Schrödinger at this point, as he only responds in a rather vague fashion to Bohr's (correct) suggestion in the discussion that Dirac's approach would avoid certain inconsistencies in Schrödinger's treatment of radiation.

The general discussion on the fourth morning of the conference was dominated by extended remarks by Bohr. The discussion began with some introductory remarks by Lorentz, who expressed the desire that he would be "able to keep [my] deterministic faith for the fundamental phenomena ... must one elevate indeterminism to a principle?"—a touching *cri de coeur* from the last of the great classical physicists.[56] Bohr's remarks have been preserved in an extremely sketchy form in notes taken by O. W. Richardson, J.-É. Verschaffelt, and H. A. Kramers (Bacciagaluppi and Valentini 2009, Appendix, pp. 432–469). Bohr reprises the arguments employing Fourier wave analysis (described earlier), which quickly led to the uncertainty principles for momentum–position and energy–time once the de Broglie ($p = h/\lambda$) and Planck ($E = h\nu$) relations linking particle and wave properties were invoked. Both uncertainty relations were then illustrated by thought experiments: the gamma-ray microscope, now analysed with careful attention to both the Compton effect and the Abbe microscope resolution formula, was used to illustrate the momentum–position uncertainty formula. Heisenberg's Stern–Gerlach formulation of energy measurements was also described, this time with an emphasis on the loss of phase information once a measurement is used to isolate a definite stationary state. These remarks all prefigured discussions that would appear in the 1928 paper, which would represent Bohr's final summary of his thinking on quantum foundational issues in the period from 1926 to 1928 (Bohr 1928a, 1928b, 1928c).

The Solvay conference of 1927 is famous for the interactions between Bohr and Einstein, where apparent paradoxes induced by the application of quantum principles were identified by Einstein, and subsequently defanged by Bohr. These conversations (later recounted in some detail in Bohr 1949) occurred for the most part outside the formal framework of the conference meetings, but one interesting example survives in the discussion transcript from the morning of Thursday, October 27. Einstein points out that the interpretation of the de Broglie–Schrödinger wave function as a "complete theory of individual processes" makes it very hard to understand the process of detection of a single particle that passes through a small hole (of dimension much smaller than the de Broglie wavelength of the particle), and which then emerges as an almost spherical wave from the hole, thence passing on to a semi-spherical screen, where it is detected as a single localized spot. Einstein asserts that "the probability that *this* particle [up to the point of impact on the screen described by a spherical wave] is found at a given point, assumes an entirely peculiar mechanism of action at a distance, which prevents the wave continuously distributed in space from producing an action in *two* places on

[56] Lorentz died the following year. He had presided over the first five Solvay conferences. The proceedings of the fifth opened with a picture of him in Brussels in 1927 and a short obituary by Curie (Solvay 1928, pp. v–vi).

the screen" (Bacciagaluppi and Valentini 2009, p. 441). This argument, which clearly contains the embryo of the famous EPR thought experiment proposed eight years later (Einstein, Podolsky, and Rosen 1935), does not receive a very satisfactory response from Bohr.[57] Once again, the catch-all escape concept of complementarity is invoked: a causal space–time description "is based on observation without interference . . . if we speak of observations we play with a statistical problem . . . There are certain features complementary to the wave pictures (existence of individuals [i.e., particles])" (Bacciagaluppi and Valentini 2009, p. 442). This response does nothing to allay Einstein's concerns about the *non-local* aspects of the measurement process. The need for complementary descriptions of a quantum process at various times seems to be interpreted by Bohr as reflecting simply the inadequacy of classical concepts in describing quantum processes, conveniently allowing almost any peculiarity of the theory to be swept under the capacious rug of complementarity.

Bohr's complementarity ideas formed the nucleus of an interpretational structure that would be absorbed (consciously or unconsciously) by several subsequent generations of physicists—the "Copenhagen interpretation" of quantum mechanics.[58] The further development of these ideas is beyond the temporal scope of our story. It would necessarily lead into a discussion of the complex history of the quantum measurement problem, which remains a highly contentious subject, with many opposing camps and points of view. But it is certainly appropriate to point out that the basic antinomy at the heart of Bohr's attempts to construct a consistent quantum epistemology—the dual particle/wave behavior of elementary quantum systems—had in some sense already been resolved well before the time of the Solvay conference. In the final section of the Three-Man-Paper of Born, Heisenberg, and Jordan, written entirely by Jordan (but viewed at the time, and for a considerable period thereafter, with great suspicion by his coauthors), the apparently paradoxical simultaneously particle- and wave-like energy fluctuations (which Einstein had originally pointed out in 1909—cf. Section 3.4.3) were shown to emerge naturally and unambiguously in a toy model of the quantized electromagnetic field (Duncan and Janssen 2008). In other words, there was a unitary, consistent quantum-dynamical framework in which the physical properties of the system inescapably exhibited both particle and wave features. The further elaboration of quantum field theory—by Dirac, for the electromagnetic field, by Jordan, Klein, Wigner, and others, for matter fields (both bosonic and fermionic)—would also eliminate the disturbing need to introduce higher-dimensional coordinate spaces when dealing with the wave function of multiparticle systems.[59] Instead, with the introduction of "second quantization," one arrived at a situation, clearly much more adaptable to the viewpoint of special relativity, where

[57] Bohr's response to Einstein does not appear in the published proceedings in French, but are preserved, with some lacunae, in Verschaffelt's notes, a transcription of which can be found in Bacciagaluppi and Valentini (2009, pp. 483–497).

[58] For discussion of how "the Copenhagen interpretation" emerged, see, for example, Howard (2004, 2022) and Camilleri (2009a, 2009b, 2022).

[59] Exactly the advantage of quantum fields in this respect was emphasized by Pauli in the Thursday discussion at the Solvay conference (Bacciagaluppi and Valentini 2009, p. 443).

physical quantum fields were defined directly in the same four-dimensional space-time in which their classical precursors had operated.[60]

8.10 Von Neumann puts quantum mechanics in Hilbert space

The final steps in the construction of a self-supporting conceptual arch, which would allow physicists to discard the scaffold of inappropriate classical concepts still persisting to some degree as a hangover from the old quantum theory, was accomplished in the three papers by the young Hungarian mathematician John von Neumann[61] that we already mentioned at the end of Section 8.8 (von Neumann 1927a, 1927b, 1927c). The first paper contains a detailed exposition of the framework of functional analysis that von Neumann took over from the work of Hilbert and Hellinger, extending and adapting it for the needs of quantum theory. In particular, the states of a quantum system were now identified with vectors (more precisely, rays) in a separable[62] Hilbert space. Physical observables were associated with self-adjoint linear operators in this space, and measurements with projection operators in the space (more on this below). In the second paper, von Neumann presents a beautiful inductive construction of the basic principles of quantum mechanics, starting with a few extremely simple and natural postulates regarding the statistical features of quantum measurement, as they appear when one carries out a measurement (or sequence of measurements) on an ensemble of quantum systems. In the final paper, "Thermodynamics of quantum-mechanical ensembles," von Neumann went on to establish the foundations of modern quantum statistical mechanics.

Von Neumann's involvement with the development of quantum theory was to some extent fortuitous. He had come to Berlin in 1923 to study chemistry as a prelude to matriculating at the ETH in Zurich to pursue a degree in chemical engineering, but was simultaneously enrolled as a doctoral student in mathematics (his true love) in Budapest, with a thesis project involving the ambitious goal of providing an axiomatization of set theory free of Russellian paradoxes.

In Berlin he came into contact with Erhard Schmidt, a doctoral graduate (Göttingen 1905) of Hilbert. It was in fact Schmidt who had given the first example of what is now universally called a Hilbert space (Schmidt 1908). Here, the notion of a vector in "infinite-dimensional space" is introduced, as well as an inner product and metric (a notion of distance), with respect to which sequences of vectors which eventually become arbitrarily close ("Cauchy sequences") are guaranteed to converge to a vector in the space. In doing this, Schmidt had introduced for the first time an explicitly geometric

[60] For the early history of quantum field theory, see, for example, Darrigol (1986) or Duncan (2012, chapters 1–2).

[61] Von Neumann's given name was originally the Hungarian "Janos," but on moving to Germany he adopted the form "Johann," which then finally, on his emigration to the United States, became "John."

[62] That is, a space with an infinite but countable basis.

interpretation of the function spaces with which Hilbert had been dealing in his examination of the general theory of integral equations. Before long, von Neumann was in contact with Hilbert, discussing his ongoing work on set theory, which was of great interest to Hilbert in connection with his deep interest in the formal logical underpinnings of mathematics. In any event, by 1926 he was in Göttingen on a Rockefeller fellowship to study and work with Hilbert.

It is a historical coincidence of enormous consequence that, at just the time von Neumann arrived in Göttingen, Hilbert's interests had shifted to the unnervingly rapid developments occurring in the neighboring building, where the Institute for Theoretical Physics (inaugurated only four years earlier, with Born as its first director) housed the theorists directly responsible for the initiation and development of matrix mechanics. In the Fall of 1925 (after completing the Three-Man-Paper), Heisenberg was enlisted to give a colloquium to the mathematicians. Hilbert had already given a series of lectures on the old quantum theory in 1922–1923, with the assistance of Lothar Nordheim, a doctoral student of Born. Now he decided to update these lectures with the incorporation of the post-1925 developments in the winter semester of 1926–1927—once again with the assistance of Nordheim, who had been keeping track of all the developments in the field, up to and including Jordan's (1927b) *Neue Begründung* I (Sauer and Majer 2009).

The lecture notes for Hilbert's course were prepared by Nordheim, and became the basis for a paper submitted in April 1927, "On the foundations [*Grundlagen*] of quantum mechanics," in which von Neumann appears as a co-author (Hilbert, von Neumann, and Nordheim 1928, see Section 16.2.3). This paper has very little that goes beyond *Neue Begründung* I, but it helped improve the notation in *Neue Begründung* II (Jordan 1927g), where it is cited even though the paper only appeared in *Mathematische Annalen* early the following year. The inevitable appearance of "improper [*uneigentliche*] functions"—such as the Dirac delta "function" and its derivatives—in Jordan's theory (as well as in Dirac's) must have induced a certain degree of mathematical nausea in Hilbert and von Neumann. This is apparent from remarks of the latter in his next paper (von Neumann 1927a). Nevertheless, his collaboration with Hilbert and Nordheim on this paper allowed von Neumann to absorb the basic physical content of the new quantum mechanics, as it existed in early Spring of 1927, as well as to come up to speed on its most sophisticated mathematical expression to date.

Von Neumann's (1927a) masterful exposition of the mathematical foundations of quantum mechanics effectively completes the construction of quantum mechanics as a mathematically consistent physical theory. The qualifier "effectively" is added only to indicate that von Neumann had at this point surmised, but not yet completed, a rigorous proof of the fundamental spectral theorem on which the transformation theory of Dirac and Jordan relies: this lacuna would be filled two years later (von Neumann 1929). But the formal structure of quantum mechanics, as recognizable by any modern student who takes an advanced undergraduate or a core graduate course in the subject, is all there:[63]

[63] For a review of the essential mathematical technology needed to understand von Neumann's papers, see Appendix C, especially Section C.5.

1. (Chs. 1–4) The matrix-mechanical approach to quantum dynamics, in which the energy appears as a numerical matrix $H_{\mu\nu}$, with the allowed energies of the system w_μ corresponding to diagonal elements of the H-matrix (if it is already diagonal) or (if not diagonal) those elements after the matrix is reduced to diagonal form by a similarity transformation $H \to S^{-1}HS$, has a basic feature in common [*gemeinsame Grundzug*] with the wave-mechanical approach, in which the energy H is represented by a differential operator, and the allowed values of the energy w are those (real) numbers for which the Schrödinger equation $H\psi = w\psi$ has well-behaved solutions (i.e., everywhere finite, and bounded at large distance) for the functions ψ. This basic common feature is that physical observables are represented by linear operators acting on elements of a linear vector space. In the case of matrix mechanics, the linear operators (such as the one for energy H) are just matrices (in general infinite dimensional) acting on (infinite dimensional) vectors $(x_1, x_2, x_3, \ldots)$; in the case of wave mechanics, the linear operators are differential operators acting on functions satisfying certain regularity requirements, of exactly the kind that render them elements of a linear "function" space (infinite dimensional!) of the kind that Hilbert and collaborators had been studying for two decades.

   The space of infinite sequences of complex numbers $(x_1, x_2, x_3, \ldots)$ with finite (i.e., convergent) squared sum of the components, $\sum_n |x_n|^2 < \infty$, had already been studied in detail by Schmidt in 1908 (mentioned earlier). By now, von Neumann explains, the space of such sequences had been given the name "(complex) Hilbert space." This space, the simplest of all instantiations of a Hilbert space, is given the notation $\mathfrak{H}_0$ by von Neumann. For modern mathematicians, it goes by the name l^2. The space of (complex) functions on the real line $\psi(x)$ which are square-integrable (i.e., $\int |\psi(x)|^2 dx < \infty$) is also an infinite-dimensional linear vector space, which von Neumann here dubs $\mathfrak{H}$. For modern mathematicians, this space goes by the name L^2 (the "L" for Henri Lebesgue, who specified precisely the type of integration that must be employed to achieve a necessary completeness property of the function space). Famous theorems of Ernst Fischer (cf. Appendix C, Theorem C.5.7) and Frigyes Riesz (Theorem C.5.8) guarantee that the spaces $\mathfrak{H}_0$ (l^2) and $\mathfrak{H}$ (L^2) are structurally isomorphic. This isomorphism means that

> Even without the introduction of "continuous matrices" and "improper forms" [e.g., the delta function and its relatives] the different sequence spaces [i.e., l^2, and the machinery of matrix mechanics] and function spaces [i.e., L^2, and the machinery of wave mechanics] are in essence identical—even under the requirements of absolute mathematical rigor (von Neumann 1927a, p. 14)

In other words, *any problem that can be phrased in wave-mechanical language can be expressed, in a mathematically completely equivalent way, in matrix-mechanical language, and vice versa.*

2. (Ch. 5) The precise axioms defining a Hilbert space (fulfilled, of course by the concrete realizations of l^2 and L^2) are given in chapter 5 of the paper: it is a linear vector space, equipped with an inner product Q (for any two vectors f, g in the Hilbert space, $Q(f, g)$ is a complex number with the usual bilinearity properties), Cauchy sequences converge, and so on.[64]
3. (Chs. 6–8) A number of important lemmas (for example, the Parseval relation, Eq. (C.128)) that follow directly from the axioms are proven. Various essential components of operator calculus are now introduced, in the context of a general Hilbert space (and therefore applicable in either the matrix-mechanical or wave-mechanical framework). Most importantly, the concept of the adjoint of an operator is introduced, a sort of generalization of the process of complex conjugation for complex numbers. Only the *self-adjoint* operators are to be associated with physical observables (just as the result of a physical measurement must give a number equal to its complex conjugate, i.e., a real number). A special class of self-adjoint operators E (*Einzeloperator* or *E.Op.*) whose square is equal to themselves, $E^2 = E$, are introduced—the modern term is "projection operator"—and it is easily seen that the action of these operators on a vector is to project it onto a linear subspace of the Hilbert space. Once projected, a second projection leaves the vector unchanged—this follows from the idempotency property $E^2 = E$. These projection operators will turn out to be at the heart of von Neumann's theory of quantum measurement.
4. (Chapters 9–11) There is a detailed discussion of the *eigenvalue problem*, as formulated up to this point in quantum theory. The measured values of an observable, represented by a (self-adjoint) linear operator T, must fall in the spectrum of the operator, namely, the set of all (real) numbers l for which a state vector f in the Hilbert space exists satisfying $Tf = lf$: in this case f is called the eigenvector/eigenfunction of T with eigenvalue l. Von Neumann points out two immediate problems with this formulation: (i) The eigenfunctions are not uniquely determined by this definition: rather they are subject to a phase freedom $f \to e^{i\vartheta} f$ (or, in the case of degeneracy, where several fs have the same l, an arbitrary unitary mixing), which von Neumann finds unacceptable; and (ii) more seriously, for observables with a continuous spectrum, *there are no eigenvectors belonging to the Hilbert space at all*! The eigenfunctions of momentum, for example, are infinitely extended plane waves, which are not square-integrable functions. In this section of his paper, von Neumann gives an outline of the full spectral theory of self-adjoint operators, which he would complete in the next two years. Both problems (i) and (ii) are shown to be avoided by using projection operators, which are completely unambiguous and well defined in the Hilbert space of states. Projection operators (which project onto subspaces of the Hilbert space where the eigenvalues of the given operator lie in an arbitrary interval on the real line) are used to write integral expressions—the famous von Neumann "spectral resolution"—for the operators

[64] See Section C.5.4 for a detailed description of von Neumann's axiomatic prescription for a Hilbert space.

that are valid, whether for operators with a purely discrete or purely continuous spectrum, or for operators, such as the energy in the hydrogen atom, with both discrete and continuous eigenvalues.

5. (Chapters 12–15) The final sections of von Neumann's paper provide a comprehensive, and mathematically rigorous, transcription of the statistical content of the Jordan–Dirac transformation theory into the language of projection operators. The spectral theorem whereby an operator is written as an integral of projection operators over the spectrum of the operator is used throughout, and the conditional probabilities of the Jordan–Dirac theory appear as traces of products of appropriately chosen projection operators. This discussion prefigures the much more detailed and conceptually clean inductive construction of the theory which von Neumann would present in his next paper.

Von Neumann's "Probability-theoretic construction [*Wahrscheinlichkeitstheoretischer Aufbau*] of quantum mechanics" (von Neumann 1927b) is an intellectual achievement of the highest rank. It marks the closure of one chapter in the evolution of modern physics by providing a conceptually clear, and mathematically completely rigorous, formulation of the theory in which the formal structure and the statistical/phenomenological import of the theory are brought into complete harmony. It also marks the beginning of another, as yet unfinished, chapter, where the question of the completeness of the theory with regard to the description of the measurement process itself is brought into examination (the infamous "measurement problem," which has occupied so many physicists and philosophers in the century since the invention of the theory). In this book, we focus on the first of these chapters. We thus only follow the story of the construction of quantum mechanics up to the point where von Neumann embedded the theory firmly in Hilbert space, and established a precise connection between the objects in the theory and the statistical nature of the results obtained therefrom. His paper providing an "inductive derivation" of quantum mechanics, and of the "Born rule" central to its statistical interpretation, therefore serves as a natural endpoint for our story.

The formal ingredients of von Neumann's construction are threefold: (a) a specific physical system $\mathfrak{S}$, which is considered to be isolated from the external world, with the exception of occasional interventions from the outside designed to perform measurements; (b) ensembles consisting of a very large (usually infinite) number of exemplars of $\mathfrak{S}$, which can conveniently be labelled $\{\mathfrak{S}_1, \mathfrak{S}_2, \ldots\}$; and (c) a set of "physical quantities" $\mathfrak{a}$, $\mathfrak{b}$, etc., just the quantum versions of the kinematical variables of classical theory, which can be measured for each element of an ensemble. For concreteness, we may take for (a) the iconic hydrogen atom (at rest in its center of mass frame); for (b) a very large, perhaps infinite imaginary collection of such atoms, with the quantum state of each atom arbitrary; and for (c), the electron three-component coordinates, $(\mathfrak{q}_x, \mathfrak{q}_y, \mathfrak{q}_z)$, momenta $(\mathfrak{p}_x, \mathfrak{p}_y, \mathfrak{p}_z)$, and (discrete) spin component in the z-direction, $\mathfrak{s}_z$, as well as all possible functions of these. Of these three ingredients, the concept newly employed in an attempt to get to the heart of quantum theory is clearly that of an "ensemble"

[*Gesamtheit*], which had been recently emphasized by Richard von Mises,[65] director of the Institute for Applied Mathematics at the University of Berlin, as a basic tool in probability theory.

The basic ensemble that von Neumann considers, from which all others can be obtained by suitable "pruning," is the "elementary randomized ensemble" [*elementar ungeordnete Gesamtheit*] (p. 247),[66] infinite ensemble in which all possible states of the system occur with equal probability.[67] Any observable $\mathfrak{a}$ can be subjected to measurement on all members of a given ensemble, defining an average, or "expectation," value $\mathbf{E}(\mathfrak{a})$ of that observable in the given ensemble. Complete knowledge [*Kenntnis*] of an ensemble amounts to the association, *for each observable* $\mathfrak{a}$, of a real number giving the expectation value of the observable in that ensemble, that is,

$$\mathfrak{a} \longleftrightarrow \text{Expectation value of } \mathfrak{a} = \mathbf{E}(\mathfrak{a})$$

(p. 248). Von Neumann (pp. 249–250) now postulates (a) that the expectation value of a linear combination of observables is the corresponding linear combination of the expectation value of each observable (e.g., for our electron in a hydrogen atom, $\mathbf{E}(\text{total energy}) = \mathbf{E}(\text{kinetic energy}) + \mathbf{E}(\text{potential energy})$);[68] and (b) that the expectation value of an intrinsically positive observable (e.g., the kinetic energy) is always a positive number.

These very general considerations, of course, lead nowhere if we have no theoretical structure allowing the computation of expectation values in quantum mechanics:

> A theory is impossible as long as we are unable to find a formal equivalent, susceptible to calculation, for the physical observables $\mathfrak{a}, \mathfrak{b}, \mathfrak{c}, \ldots$ of the system $\mathfrak{S}$. But we know very well from quantum mechanics which mathematical structures are to be associated with physical quantities: they are the so-called linear symmetric [self-adjoint] functional operators, which are defined to act on complex-valued functions—the wave functions—in the state space of the system $\mathfrak{S}$, transforming them into other such functions (pp. 250–251).

The mathematical concepts and terminology of the *Mathematische-Begründung* paper are now invoked explicitly: for any physical system $\mathfrak{S}$ underlying an ensemble, there is

[65] Von Mises, in his papers on probability calculus in the late teens, as well as in his treatise *Probability, Statistics, and Truth* [*Wahrscheinlichkeit, Statistik und Wahrheit*] (von Mises 1928), more typically uses the term *Kollektiv* for an ensemble of similar systems, distinguished by individually measurable properties, with *Gesamtheit* reserved for the more general notion of a collection of mathematical values, coin tosses, or what have you.

[66] Page references here and below are to von Neumann (1927b).

[67] The notion of "equiprobable" becomes precise only once we have the Hilbert-space view (shortly to appear in von Neumann's paper) of the state of the system as a vector in Hilbert space—then, the elementary ensemble is simply an infinite enumerable set of unit vectors, randomly distributed on the unit sphere in Hilbert space.

[68] This linearity assumption becomes problematic when von Neumann's 1927 derivation of the Born rule is repurposed, as suggested by the book that grew out of his 1927 trilogy (von Neumann 1932), for an attempt to show that it is impossible to add hidden variables to the basic formalism of quantum mechanics. For discussion of this issue, which is beyond the scope of our book, see Bub (2010) and Dieks (2017).

an abstract Hilbert space $\mathfrak{H}$ in which each physical quantity acts as a "normal" operator,[69] defined as one for which the eigenvalue problem is solvable (at least in principle), allowing an expression of the operator as a sum over projections onto all possible eigenstates of the given quantity, multiplied by the corresponding eigenvalue—that is, the famous von Neumann "spectral resolution." Once one has a spectral resolution, it turns out to be trivial to define normal operators for more or less arbitrary functions of the basic physical quantities of the theory. The calculation of the variance (mean-square-deviation) of a quantity, for example, would require evaluating the expectation value of the quantity as well as of its square.

Von Neumann's step-by-step construction of the statistical content of quantum mechanics is explored in detail in Chapter 17. Here we merely summarize his main results. Every ensemble of quantum systems can be completely characterized by a complex matrix U (later to be called the "density matrix"), which is uniquely specified once a specific complete basis of states has been chosen for the underlying Hilbert space of the system. Once this matrix is known, the expectation value $\mathbf{E}(\mathfrak{a})$ of any physical quantity $\mathfrak{a}$ can be immediately calculated, if the matrix of the corresponding operator in the chosen basis is known.

The direct connection between the phenomenological/statistical formulation of the theory via ensembles and its underlying Hilbert space structure is finally made with the definition of a "pure [*rein* or *einheitlich*] ensemble," which von Neumann (pp. 255–256) defines as one that cannot be obtained by composing two other ensembles, unless they agree with each other (i.e., have identical density matrices, up to an overall irrelevant numerical factor). Von Neumann (p. 257) then proves that in this case, the density matrix is just the projection operator onto a unique vector in the Hilbert space: exactly the concept of a "pure state" with which modern quantum physicists are familiar.[70] In such an ensemble, the quantum state of every member $\mathfrak{S}_1, \mathfrak{S}_2, \ldots$ is identical, and can be identified with a unit vector φ in the Hilbert space. The expectation value of an arbitrary observable, realized as a normal operator S, is then given by the instantly familiar formula $\mathbf{E}(S) = Q(\varphi, S\varphi)$, or, in more modern notation (dropping the Q used by von Neumann to indicate an inner product), $\mathbf{E}(S) = (\varphi, S\varphi)$. This is, if one inserts the spectral resolution of S, easily seen to be the generalized Born rule of quantum mechanics.

Von Neumann had by this time read the Heisenberg uncertainty paper (it is cited on the second page of *Wahrscheinlichkeitstheoretischer Aufbau*), and was in complete agreement with Heisenberg that the ordinary probability calculus was perfectly applicable in quantum mechanics, as long as one was careful to specify the exact order and character of the measurements performed on a quantum system (more precisely, on an ensemble

[69] Two years later, von Neumann (1929) would identify the class of normal operators as those that commute with their adjoint, including of course all self-adjoint operators, as well as unitary operators, for which the adjoint coincides with the inverse.

[70] The more general ensembles considered by von Neumann, which can be considered as compositions of two statistically distinguishable sub-ensembles, correspond, in modern terminology, to "mixed states."

of such systems, the properties of which were determined by the experimental setup).[71] In particular, he asserted that, in general, "our knowledge of a system, the structure of the associated ensemble, is never described by specification of the state $[\varphi]$ ... ; rather, in general through the results of experiments carried out on the system" (p. 260). Such experiments typically are only capable of establishing that a given observable $\mathfrak{a}$ has a value in a prescribed interval I on the real line.

More generally, von Neumann shows that for a given physical system $\mathfrak{S}$ (say, the iconic hydrogen atom), and a maximal, or "complete," set of compatible (i.e., simultaneously measurable) observables $\mathfrak{a}_1, \mathfrak{a}_2, \ldots, \mathfrak{a}_m$, which correspond to normal operators $S_1, S_2, \ldots, S_m$, *which all commute with each other*, one can construct an ensemble $\mathfrak{S}'_1, \mathfrak{S}'_2, \ldots$ defined by the requirement that measurements of $S_1, S_2, \ldots$ can (a) be carried out simultaneously; and (b) in all cases lead to the value of $\mathfrak{a}_1$ being in an interval I_1, $\mathfrak{a}_2$ being in an interval I_2, etc.[72] In the case of a system with a completely discrete spectrum, the isolation of a single simultaneous eigenvector for all the observables in the commuting set, with suitable choice of the intervals $I_1, I_2, \ldots$, then leads directly to the identification of the $\mathfrak{S}'$ ensemble as a *pure* one, with an associated density matrix given by the projection operator onto a unique unit vector φ in the Hilbert space. Such a pure ensemble has the property that measurements of $S_1, S_2, \ldots, S_m$ give identical results on all members of the ensemble $\mathfrak{S}'$: the ensemble is "sharp" [*scharf*] with respect to all of the quantities in the maximally commuting set. A simple example of such a set, for ensembles of bound hydrogen atoms at rest, would be to define operators S_1, S_2, S_3, and S_4 whose eigenvalues correspond to the familiar quantum numbers n, l, m, s_z (principal, angular, magnetic, and spin quantum numbers of the electron, respectively[73]).

The restriction to pure ensembles in which all values of a maximal set of simultaneously measurable quantities are sharp is then relaxed to allow consideration of the more usual situation, in which a set of simultaneously measurable quantities $\mathfrak{a}_1, \mathfrak{a}_2, \ldots, \mathfrak{a}_m$ are known to take values (in the given ensemble) in specified intervals $I_1, I_2, \ldots, I_m$. The density matrix U for such an ensemble is shown by von Neumann to be simply a product of the projection operators $E = E_1(I_1)\, E_2(I_2) \ldots E_m(I_m)$ obtained from the spectral resolution of the normal operators associated with our set. The Jordan–Dirac transformation theory is beautifully encapsulated in the determination of the conditional probability of measuring *a different set of observables* (in general, *not* simultaneously measurable with the first!) $\mathfrak{b}_1, \mathfrak{b}_2, \ldots, \mathfrak{b}_n$ to have values in intervals $\mathcal{J}_1, \mathcal{J}_2, \ldots, \mathcal{J}_n$, with corresponding density matrix $F = F_1(\mathcal{J}_1)\, F_2(\mathcal{J}_2) \ldots F_n(\mathcal{J}_n)$. The answer is mathematically a simple trace operation: $\Pr(F|E) = \mathrm{Tr}(FE)$. All the (mathematically) correct consequences of the

[71] Earlier von Neumann (1927a, p. 46) and Hilbert, von Neumann, and Nordheim (1928, p. 5) had endorsed the idea that quantum mechanics changes the basic rules of probability, which was central to Jordan's (1927b) theory but was sharply criticized by Heisenberg (1927b, pp. 183–184, p. 196) in his uncertainty paper.

[72] The construction is very simple: one merely starts with the "elementary randomized ensemble" in which all states appear equiprobably, measures all the $\mathfrak{a}_i$ on each member of the ensemble, and discards all ensemble elements where the measured values do not lie in the desired intervals. The remaining systems $\mathfrak{S}'_1, \mathfrak{S}'_2, \ldots$ evidently satisfy the desired property.

[73] We are ignoring the degrees of freedom of the atomic nucleus, which for our purposes simply acts as an immovable Coulomb force center.

Jordan–Dirac statistical transformation theory follow essentially from this simple and transparent result.

With von Neumann's Hilbert-space framing of quantum mechanics—in particular, his identification of pure states with the results of simultaneous measurement of a maximal set of compatible (i.e., commuting) observables—the scaffold of the canonical formalism inherited from classical mechanics (via the old quantum theory), whereby the states of a system are naturally described in phase space, necessitating the choice of a full set of conjugate (coordinate, momentum) pairs, has finally been discarded: a self-supporting arch, mathematically solid, and with a clear connection to the statistical predictive content of the theory, has taken its place. We no longer need to find (and mind) "our *p*s and *q*s": rather, the new mechanics enjoins us to "find a maximal commuting set of observables". For *elementary particles*, quantum field theory demonstrates that this set is very simple indeed: one may take just the spatial momentum of the particle, and its spin (component in the z-direction, say; or, for massless particles, along the direction of motion).

In his final summation, von Neumann insists that his inductive derivation of the theory rests on only three basic postulates: (i) every measurement alters the measured object; (ii) if a measurement is done twice in rapid succession, the second measurement yields the same result as the first (thus, measurements must have the property of idempotency in the formalism); and (iii) physical quantities are represented in the theory by functional operators (i.e., linear operators in a Hilbert space), satisfying some simple formal requirements. From these principles "both quantum mechanics and its statistics follow inevitably" (p. 271).

His comments on the apparently acausal nature of quantum dynamics also display a refreshing clarity:

> The statistical, "acausal" nature of quantum mechanics arises from the (in principle!) inadequacy of measurement [here, Heisenberg's uncertainty paper is cited]; for a system left to itself (which is subjected to no measurements of any kind) will develop in time in a fully causal way: if one knows the state φ at time $t = t_0$, then one can calculate the state at any later time from the time-dependent Schrödinger equation [here, a footnote gives the explicit solution of said equation, $\varphi_t = \exp(2\pi i(t - t_0)H/h)\varphi_{t_0}$]. With regard to experiments however, the statistical character cannot be avoided: for every experiment there is an adapted state [or states], for which the result is unique (indeed, such a state is generated, if it did not previously exist, by the experiment); but for each state there are also "non-adapted" experiments, the performance of which destroys [*zertrümmert*] the state, replacing it by one adapted to the experiment (von Neumann 1927b, pp. 271–272).

This, as best as we can tell, is the first clear statement of the "collapse of the wave function" hypothesis familiar to every modern student of quantum theory.

Part III

Transition to the New Quantum Theory

9

The Exclusion Principle and Electron Spin

In Chapter 7, the final chapter of Volume 1, we discussed three major areas of atomic physics in which the old quantum theory encountered serious barriers to the incorporation and quantitative interpretation of empirical data using the conceptual machinery of the Bohr–Sommerfeld methodology: the complex structure of multiplets in atomic series spectra, the anomalous splitting patterns of series terms in the Zeeman effect, and finally, the failure of the theory to account for the spectrum of ortho- and para-helium, and ultimately, even the binding energy of two electrons in the ground state of helium, the simplest atom with more than one electron.

The failure of the old quantum theory to come to terms with these problems was of course partly due to the incorrect underlying kinematic framework of the theory, which would only be rectified by the development of the (equivalent) matrix and wave mechanical formalisms of Heisenberg and Schrödinger in late 1925 and early 1926. But the difficulties were also greatly magnified by the absence of two critical pieces of information that would have enormously simplified the job of the old quantum theory in producing a (at least qualitatively) satisfying account in all three cases, had the practitioners of the theory been aware of them. The first was the antisocial characteristic of electrons bound in atoms expressed by the Pauli exclusion principle, which eliminated a large number of a-priori possible multi-electron states, and was an absolute precondition to a real understanding of the spectra of multi-electron atoms, and in particular, the periodic aspects of these spectra as seen in the classification of the elements in the periodic table.[1] The second bit of missing information, closely connected to the first, was the existence of an intrinsic angular momentum, or spin, with a concomitant magnetic moment, attached to each electron, which is an absolutely critical ingredient in achieving a full understanding of either the complex multiplet structure or the anomalous Zeeman effect.

Both of these essential realizations, Pauli's formulation of the exclusion principle and the discovery of electron spin, occurred in late 1924 and early 1925, well before the

[1] For a historical analysis of the origins of the exclusion principle, see Heilbron (1983). See also van der Waerden (1960).

Constructing Quantum Mechanics. Anthony Duncan and Michel Janssen, Oxford University Press.
 DOI: 10.1093/oso/9780198883906.003.0009

new quantum mechanics of Heisenberg and Schrödinger emerged. They were therefore initially phrased in the language of the old quantum theory. But the acceptance and implementation of spin and the Pauli exclusion principle were essential ingredients in the successful resolution by the new theory of the "failures" of the old quantum theory discussed in Chapter 7, and serve as an appropriate starting point for our account in this volume of the emergence of modern quantum mechanics in 1925–1926.

9.1 The road to the exclusion principle

9.1.1 Bohr's second atomic theory

In the second part of his 1913 trilogy ("Systems containing only a single nucleus"), Bohr (1913b) addressed the problem of the structure of atoms containing more than one electron (see Section 4.5.2). The models considered in this article amounted to the simplest possible generalization of the circular orbits for hydrogen introduced in the first paper of the trilogy: concentric circular rings of equally spaced electrons. The *Aufbau* ("building up") procedure introduced here is one that Bohr would continue to employ for the next decade in his attempts to devise an explicit mechanical picture for all the atoms of the periodic table (see Figure 9.1):

> we shall assume that the cluster of electrons is formed by the successive binding by the nucleus of electrons initially at rest, energy at the same time being radiated away. This will go on until . . . the total negative charge on the bound electrons is numerically equal to the positive charge on the nucleus (Bohr 1913b, p. 476).

By imposing stability requirements for small displacements perpendicular to the ring (with stability with respect to displacements in the ring enforced by the requirement of fixed quantized values for the angular momentum of the electrons) Bohr could limit the number of electrons which could be stably "accommodated" in a single ring as a function of the nuclear charge. For example, neutral atoms with all electrons in a single ring could only exist for nuclear charge $Z \leq 7$ (i.e., up to nitrogen), atoms with $Z = 20$ could accommodate rings with up to ten electrons, $Z = 40$ up to thirteen electrons, and so on.

By 1920 Bohr had abandoned the notion of planar atoms with electrons circulating in concentric rings. At that point, with the experimental work of Franck and Knipping (1919), it had become clear that the simplest such model, with two electrons for helium, gave a value for the ionization potential of neutral helium that was about 20% too high. Moreover, such models offered no hope of explaining the striking numerical regularities of the periodic table, with recurring chemical properties after the suggestive periods of 2, 8, 18, etc., in atomic number. Furthermore, the three-dimensional structure of crystals seemed incompatible with such a simple geometry of coplanar electronic orbits (i.e., with the supposition of "flat atoms"). As we saw in Section 7.3, by late 1921,

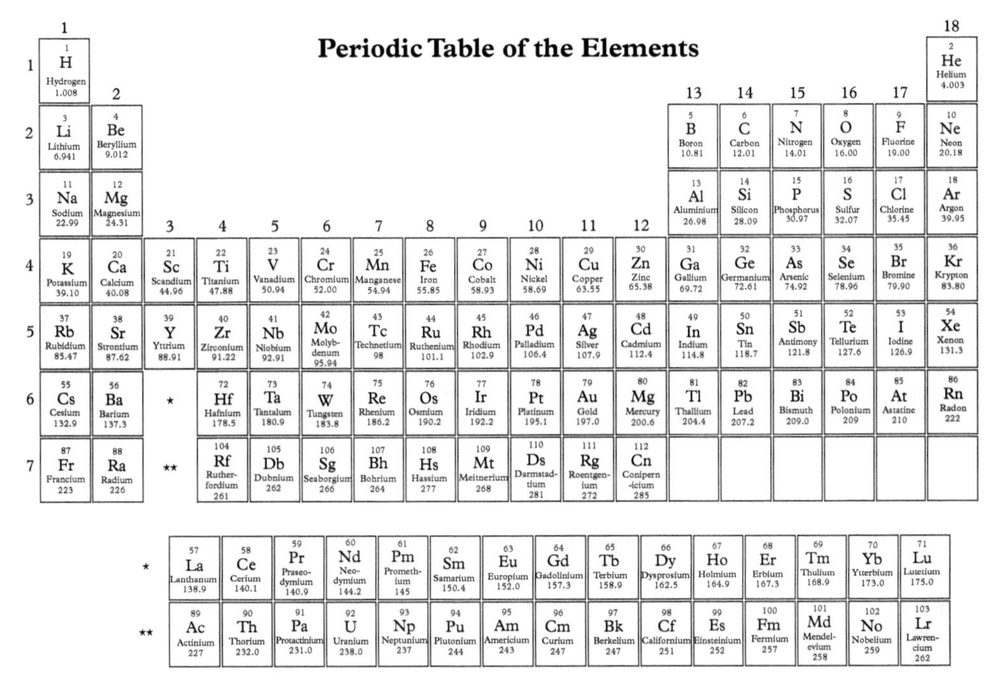

Figure 9.1 *The Periodic Table, in modern format.*

Bohr, influenced by the detailed calculations of Landé (which suggested that parahelium states involved electron orbits tilted with respect to one another) had adopted definitively the non-planar Kemble configuration of intersecting orbits for neutral helium, and Kramers was engaged in the detailed calculations of the ionization energy for this configuration. The era of fully three-dimensional atoms, with electrons orbiting the nucleus in a combination of (roughly) circular and elliptical, but non-coplanar, paths had arrived.

Over the next few years, Bohr devoted a great deal of effort to constructing a comprehensive theory of atomic structure for multi-electron atoms that would lead to properties consonant, at least qualitatively, with the chemical and physical periodicities manifest in the periodic table. This theory has come to be known as "Bohr's Second Atomic Theory" (see Kragh 1979, 2012, Ch. 7, pp. 271–312; see also Pais 1991, Ch. 10). The basic tool announced for this task was the *correspondence principle*, applied in the context of the *Aufbau* procedure, in which the neutral atom is built up in a stepwise fashion by successive addition of electrons to the bare nucleus. As mentioned in Section 8.3, Bohr gave lectures on this topic in various places and to various audiences. He worked out some of these in lengthy manuscripts. The most important of these are Bohr (1921d) (in Danish) based on a lecture in Copenhagen in October 1921, and Bohr (1922e), based on his Wolfskehl lectures in Göttingen of June 12–22, 1922—the "Bohr festival" (*Bohr*

Festspiele).[2] Bohr (1921d) was reprinted in book form (Bohr 1922a) and also published in German (Bohr 1922b, 1922c, Essay III), in English (Bohr 1922d, Essay III), as well as in French and Russian (Kragh 2012, p. 275). An English translation of the Wolfskehl lectures can be found, under the title "Seven lectures on the theory of atomic structure," in Bohr (1972–2008, Vol. 4, pp. 341–419).

In addition to Göttingen mathematicians (Courant, Hilbert, Runge), physicists (Born, Franck) and their students (Hund, Jordan, Rudolph Minkowski), many leading quantum physicists from other places attended Bohr's Wolfskehl lectures: Sommerfeld came from Munich, Pauli from Hamburg, Landé and Gerlach from Frankfurt, Ehrenfest from Leyden. Klein and Oseen traveled with Bohr from Copenhagen (Mehra and Rechenberg 1982a, pp. 344–345).[3] As Hund recalled in a volume in honor of Heisenberg: "Sommerfeld brought a blonde youngster [*einen blonden Jüngling*], whose interpretation of the anomalous Zeeman effect Bohr could not quite agree with. But he bravely [*tapfer*] defended himself in discussion and we looked at him with astonishment [*wir staunten ihn an*]" (Hund 1961, p. 2). After the third lecture, Bohr invited Heisenberg for a walk on Göttingen's Hainberg to further discuss a point the youngster had raised about Kramers and the quadratic Stark effect (Cassidy 2009, p. 100). They ended up walking and talking for about three hours. As Heisenberg wrote in his memoirs: "This walk was to have profound repercussions on my scientific career, or perhaps it is more correct to say that my real scientific career only began that afternoon" (Heisenberg 1971, p. 30; cf. Mehra and Rechenberg 1982a, pp. 357–358).

Heisenberg was not the only one on whom the event made an indelible impression. In his book on the history of quantum mechanics, Hund remembered that "Bohr spoke, as he often did, rather indistinctly [*undeutlich*], and he was barely audible in the back rows where the younger members were obliged to sit. This merely increased the excitement and interest" (Hund 1967, p. 106). In the reminiscences from which we quoted above, he elaborated:

> We had read a little in Sommerfeld's *Atombau und Spektrallinien* and in 1920 Debye had offered a course in quantum theory . . . but what Bohr presented sounded different, and we felt that it was something absolutely essential [*ganz wesentlich*]. The splendor of the event can no longer be conveyed today. For us it was as splendid as the Göttingen Händel *Festspiele* in those years (Hund 1961, p. 1; quoted in Mehra and Rechenberg 1982a, p. 345).

This comparison with the Händel *Festspiele* may have been the origin of the "designation [Bohr] 'Festspiele' surfacing shortly afterwards" (see p. 107 in the third edition of Hund 1967).

[2] Bohr also gave lectures in Cambridge and London in March 1922. A paper based on his Guthrie Lecture at the Physical Society of London, March 24, 1922, was published the following year (Bohr 1923b).

[3] See Volume 1, Plate 26, for a picture taken during the "Bohr Festival", showing Bohr with his hosts, Born and Franck, and his entourage, Oseen and Klein.

But it was not just the gullible youngsters in the back who were impressed; their elders in the front rows were too. Sommerfeld, who had been skeptical about the correspondence principle (which took center stage in these lectures), had already softened his stance on this principle after reading earlier papers on Bohr's "second atomic theory" (Mehra and Rechenberg 1982a, p. 359). In the preface to the third edition of *Atombau und Spektrallinien* (Sommerfeld 1922a), dated January 1922, he expressed confidence in Bohr's approach: "Bohr's recent far-reaching ideas will, indeed, add much that is new, but will not throw doubt on what now appears to be established" (Sommerfeld 1923, p. vii). Sommerfeld sent a copy of the new edition to Bohr. In a letter of April 30, 1922, thanking Sommerfeld for the book, Bohr wrote:

> I would like to express my gratitude for the friendly manner in which you have treated the work of my collaborators and myself. In the past years I have often felt very lonely in scientific matters, because my attempts to develop the principles of quantum theory systematically to the best of my ability have been received with very little sympathy ... My joy has therefore been so much greater, for I believe I noticed a change in the point of view contained in the latest edition of your book (Sommerfeld 2004, Doc. 55, quoted in Mehra and Rechenberg 1982a, p. 359).

All in all, Bohr was riding high in 1922. Sommerfeld's belated appreciation of the correspondence principle—which he would use in two papers with Heisenberg submitted in August (Sommerfeld and Heisenberg 1922a, 1922b)—and the success of the Göttingen *Festspiele* were followed in November by the news that he had been awarded the Nobel Prize in Physics for 1922. Finally, in his Nobel lecture on December 11, 1922, Bohr (1923d, p. 44) could announce that two of his collaborators, Coster and Hevesy, had just discovered hafnium, a new element predicted by his theory and named for the town where it was discovered (Mehra and Rechenberg 1982a, p. 370).

Bohr's winning streak, however, would not last. The fairytale brought to mind by the demise of his proud theory of the periodic table is "The emperor's new clothes" by his countryman Hans Christian Andersen. Bohr's colleagues had assumed that his correspondence-principle arguments were backed up by detailed calculations. They slowly came to the realization that they were not. We will see shortly that they *could not have been*, as a detailed quantum-mechanical treatment of a multi-electron atom containing distinguishable electrons (thus, not subject to the exclusion principle) reveals absolutely no evidence of the emerging shell structure, for which Bohr claimed to have found valid arguments based on the correspondence principle.

Although he continued to express faith in Bohr's program, Sommerfeld clearly understood, by the time he published the fourth edition of *Atombau und Spektrallinien*, that Bohr's correspondence-principle arguments were not supported by any calculations. In the third edition, he had assured his readers that considerations about the "natural system of elements," the topic of Ch. 2 of his book, "have received a much firmer foundation through the later views of Bohr [Sommerfeld refers to Bohr (1921a)], *which were accompanied by calculations*" (Sommerfeld 1923, p. 109, emphasis added; quoted and discussed in Seth 2010, p. 237; see also Kragh 1979, p. 161). In the preface of the fourth edition, dated October 1924, Sommerfeld cautioned that there would inevitably be exceptions to Bohr's atomic theory, adding: "We can expect such exceptions all the more, since *Bohr's system is not founded mathematically but conceived intuitively*" (Sommerfeld 1924, p. V, our emphasis).

A decade later, Kramers, Bohr's closest collaborator at the time, looked back on this period and wrote:

> It is interesting to recollect how many physicists abroad thought at the time of the appearance of Bohr's theory of the periodic system that it was extensively supported by unpublished calculations which dealt in detail with the structure of the individual atoms, whereas the truth was, in fact, that Bohr had created and elaborated with a *divine glance* a synthesis between results of a spectroscopic nature and of a chemical nature (Kramers 1935, p. 90; translation from Kragh 2012, p. 300, our emphasis).

A letter from Pauli written another two decades later shows just how badly Bohr's "second atomic theory" missed the mark. Pauli is responding to a letter from Léon Rosenfeld of May 24, 1954 (Pauli 1999, Doc. 1810), suggesting that Pauli submit some recollections about the connection between Bohr's work and his discovery of the exclusion principle for a volume in honor of Bohr's seventieth birthday (Pauli, Rosenfeld, and Weisskopf 1955). On May 31, 1954, Pauli replied:

> It is my suspicion that you ask me to write about Bohr's historical connexion with the exclusion principle because you do not know this connexion and you are curious to hear it. So I answer with giving you some personal informations about it together with the sincere assertion that I really wonder whether it is a nice thing to excavate a person's errors at his 70th birthday. But an historical article of me *would reveal it relentlessly*! (Pauli 1999, Doc. 1818).

Pauli then proceeds to list some serious problems in Bohr's theory, citing chapter and verse in the reprint of the German translation of Bohr (1921d) as Essay III in Bohr (1922c). He concludes by asking Rosenfeld some rhetorical questions: "But now: why tell all this old errors [sic] again on a 70th birthday? Why not let Bohr's paper of 1922 stay asleep? . . . What do you think now about the theme of my article? Do you still have such a curiosity about old times?" (Pauli 1999, Doc. 655). Whereas Pauli understandably wanted to draw the veil of charity over this whole production, we need to cover it in considerable detail if we want to understand the origins of Pauli's own exclusion principle.

Bohr's basic strategy in using the correspondence principle to explain the periodic table can be summarized as follows:

1. At any given stage the system (nucleus of charge Z with N electrons, $N \leq Z$) would execute a multiply periodic motion selected from among the vastly greater number of non-multiply-periodic (and as we now know, almost entirely "chaotic") motions that were classically available to a multi-electron atom. This required a high degree of "cooperation" in the motion of the separate electrons. The simplest example

is just the Kemble crossed-orbit configuration adopted for the ground state of helium, as discussed in Chapter 7. The multiply periodic motion would allow each electron to be separately subjected to Bohr–Sommerfeld quantization, and assigned identifying quantum numbers n and k (for the principal and azimuthal quantum numbers). An atomic configuration can then be specified by indicating with a superscript the number of electrons in each such stationary state n_k (e.g., $(1_1)^2(2_1)^4$ for carbon, $(1_1)^2(2_1)^4(2_2)^4$ for neon, all the way up to a putative noble gas at $Z = 118$, as given in Bohr (1922e, pp. 418–419).

2. The addition of the next electron to an atom with less than its full complement of electrons was assumed to leave the quantum numbers of the previously added electrons unchanged (the principle of "permanence of quantum numbers"). This can be thought of as a sort of extended adiabatic principle, valid if the absorption of the electron into the atomic ion could be accomplished by gradually switching on the Coulomb interaction of the new electron with the nucleus and previous electrons in such a way as to preserve the multiply periodic motion of the whole system at each stage.
3. The states accessible to the newly added electron would have to correspond to those reachable (from high quantum number states) by a succession of quantum jumps, with the emission of radiation at each jump constrained by the correspondence principle, and the electron finally settling into the state of this type with lowest energy. The symmetric ring states of 1913, in which electrons were symmetrically arranged in polyhedra on circular rings, were now excluded by correspondence principle reasoning, as incapable of absorbing or releasing dipole radiation (as the electric dipole moment of such an arrangement vanishes by symmetry). Of course, the application of the correspondence principle to states with low quantum numbers—the inevitable final stage of the atom building process in smaller atoms—is far from unambiguous, so much of what Bohr said in this regard amounted to pure (and, in fact, invalid) hand-waving, as we will soon see.

Bohr's use of the correspondence principle in restricting the ground-state configuration of atoms involves an important extension of the "deepening" of the principle to estimate spectral intensities, which Kramers had pioneered in his thesis (see Section 5.3). Not only was correspondence-principle reasoning to be applied for low quantum numbers, but also, it would be used to assert that the vanishing of a Fourier component in the *final stationary state* of an electron in a low orbit could be used to exclude the possibility of the electron in arriving at such a state (even if intermediate states existed leading to the final orbit with non-zero coefficients). For example, symmetrical arrangements of the electrons in the final state with vanishing dipole moment (albeit with non-zero quadrupole and higher moments) were excluded on the grounds that they could not have arisen by the final electron dropping into this state via dipole radiation (which was assumed to be the only possibility—yet another dubious aspect of the whole program). This excluded the 1913 model of helium with two electrons circling the nucleus at opposite points on a single circle[4] and was a primary motivation for Bohr's adoption of the non-planar Kemble model, in which the two electrons moved in a way which produced a non-zero oscillating dipole moment for the whole atom (see Figure 7.12).

At the early stages of the development of his "second theory", Bohr is clearly using the correspondence principle to propose specific orbital configurations for successively added electrons, but the application of the principle is always at a qualitative level (at least in his published works), with no detailed mathematical justification offered. A typical example can be found in his discussion of the addition of a third electron to a Li^+ ion to form neutral lithium in Bohr's unpublished notes for the 1921 Solvay conference, which in the end he was unable to attend (Bohr 1921b). The preceding two electrons were already supposed to occupy equivalent (i.e., of similar size and shape) $n = k = 1$ orbits, with the non-planar configuration of the Kemble model preferred over the original 1913 model of two electrons on opposite sides of the same circular orbit, even though, as we have seen, the ionization energy was incorrectly given (in opposite ways) in both cases (cf. Volume 1, pp. 189 and 373). The question then arises whether the third electron could also enter a $n = k = 1$ orbit equivalent to that of the first two electrons. In particular, Bohr makes the following argument, clearly inspired by the correspondence principle, to exclude a situation in which all three-electrons execute non-coplanar crossed, almost circular $n = k = 1$ orbits (i.e., a three-electron version of the Kemble model):

> we would in lithium . . . have to reject configurations of the latter type [i.e., non-coplanar, with three equivalent $n = k = 1$ electrons], from a consideration of formation of the atom by a spontaneous process of transition accompanied by radiation. In fact, no process—which bears any resemblance with a possible transition between two stationary states of a multiply periodic system—can be imagined, by which the configurations of the two first bound electrons would allow a third electron coming from outside to enter as an equivalent partner in the motion round the nucleus (Bohr 1921b, p. 143).

Vague arguments along these lines, never substantiated by clear and explicit calculations, are used by Bohr to claim the "exclusion of a state, in which this [third] electron moves in an orbit analogous with that of the electrons in the normal state of the helium atom" (Bohr 1921b, p. 145). Similarly, to prevent the addition of a ninth $n = 2$ electron in a ten-electron atom with nuclear charge +11, he argued that

> not only the inner group of orbits $n = 1$, but also the outer group of orbits $n = 2$, cannot take up further electrons by a process corresponding to the stages of binding of an eleventh electron . . . the inclusion of a further electron [in the outer group] would not show any resemblance with a process of transition between stationary states of a multiple periodic system (p. 153).

[4] It also excluded Sommerfeld's "ellipse club" (*Ellipsenverein*), in which electrons orbiting in ellipses were located at any given time symmetrically at the vertices of a regular polygon (Sommerfeld 1922b, pp. 612–613).

In a contemporary review of Bohr's correspondence-principle approach to atomic structure prepared for *Die Naturwissenschaften* by Kramers (1923b), one senses a certain degree of insecurity concerning the rigor of the Bohrian approach to electronic arrangements. The article is, in classic Bohr style, completely qualitative. There are only three equations: the Bohr frequency condition; the formula for the Rydberg constant; and the Balmer formula for hydrogen. The discussion of multi-electron atoms follows the strategy outlined above—electrons are added one by one to an initially bare nucleus, and the orbits accessible to the last added electron are restricted on the basis of vague arguments based on the supposed absence of the appropriate frequency components of this electron's motion, subsequent to falling into the lowest stationary state available to it. As Kramers admits, however, in arguing for the impossibility of adding a third electron to the previous two $n = 1$ electrons in order to arrive at neutral lithium:

> At present it has not been possible to seriously deepen this sort of argument concerning the closing of a [electron] group; the matter is mainly complicated by the failure of classical mechanical laws in the description of the motion (Kramers 1923b, p. 557).

One could certainly imagine, even in the conceptual framework envisaged by atomic theorists in 1921, in which electrons were distinguishable particles with no intrinsic spin angular momentum and bound by purely electrostatic attraction to a positively charged nucleus, that the constraints of mechanical stability and quantization of appropriate action variables of the motion might nevertheless lead to the emergence of a shell structure. In this structure, newly added electrons would prefer to occupy a stationary state given by a different (n, k) set of quantum numbers (corresponding to orbits further removed from the core of the atom) from the preceding electrons, in order to reduce the electrostatic repulsion energy, and allow the atom to assume a quantized state of minimum energy.

For such an imaginary world of spinless distinguishable electrons, however, wave-mechanical calculations show that the ground state of a "lithium atom" contains three electrons in exactly equivalent "orbits" (from the point of view of the old quantum theory). This is in a framework where the usual arguments suggesting classical radiative behavior for large quantum numbers—that is, a correspondence principle for radiation frequencies and intensities in the large quantum number limit—would still hold. This can be established by quantum-mechanical variational calculations for an assembly of charged, but, in keeping with the presuppositions of the old quantum theory, and in particular, with Bohr's second theory, distinguishable spinless electrons.[5] One finds, not surprisingly, that the ground state of atoms containing more than two electrons have the electrons occupying identical (1s) orbitals: there is *no evidence for the emergence of a*

[5] For a much more detailed account of Bohr's second atomic theory than we have space for here, as well as a detailed description of the variational calculations of atomic structure, and their Hartree self-consistent-field extensions, giving the results shown in Table 9.1, see the web resource, *Bohr's Second Theory and Atomic Aufbau before Spin and the Exclusion Principle.*

shell structure, on the basis of a dynamics determined solely, as in Bohr's second theory, by electrostatic interactions, and in the absence of a dynamically extrinsic exclusion principle.

Table 9.1 shows the binding energies of atoms built from distinguishable electrons employing a simple variational method, in which the energy is minimized, assuming that electrons occupy only *s* state ($l = 0$) orbitals, with zero angular momentum, which are the analogs in modern quantum mechanics of Bohr's $k = 1$ s-states. We use trial wavefunctions with independent screened nuclear charges z_1 and z_2 for the 1*s* and 2*s* states, which serve as the parameters to be varied in order to obtain the minimum (negative) energy E (or maximum binding energy $W = -E$). We have also checked these results by a complete Hartree consistent field calculation (see the web resource mentioned in note 5)—the more accurate binding energies obtained in this way are a few percent larger typically than the values given in the table, but the qualitative results are identical. The addition of more electrons in a 1*s* state to an atom is always energetically preferred to the addition of an electron in a 2*s* state. The correct configuration for a lithium atom in a world with such electrons would therefore be $(1s)^3$, as the $(1s)^2(2s)$ configuration is higher in energy by about 29 eV (see the third and fourth rows of Table 9.1), and therefore would correspond to an excited state of the atom.

Table 9.1 *Variational binding energy estimates for atoms* ($2 \leq Z \leq 6$) *built from* N_1 *1s,* N_2 *2s distinguishable (spinless) electrons.*

Z	N_1	N_2	$z_{1,\max}$	$z_{2,\max}$	$W_{\max}$(eV)
2	2	0	1.688	-	77.43
2	1	1	2.022	0.632	58.80
3	3	0	2.376	-	230.1
3	2	1	2.697	0.721	201.4
4	4	0	3.063	-	510.0
4	3	1	3.381	0.790	470.1
4	2	2	3.718	1.103	394.4
5	5	0	3.750	-	955.9
5	4	1	4.066	0.846	903.4
5	3	2	4.395	1.180	807.1
6	6	0	4.438	-	1606
6	5	1	4.753	0.893	1540
6	4	2	5.076	1.247	1422
6	3	3	5.415	1.567	1251

As we emphasized above, correspondence-principle arguments *for large quantum numbers* are just as applicable in this imaginary situation as in the real-world case of indistinguishable, spin-$\frac{1}{2}$ electrons, subject to the Fermi–Dirac statistics, which would emerge a few years later. Admittedly, in analyzing atomic *Aufbau* for these atoms (up to "carbon," with $Z = 6$), we are relying on methods not available to Bohr in 1921. The point is simply that it would be fruitless to try to uncover *valid* arguments, along the lines of the correspondence principle, based on explicit calculations, and leading to the shell structure of atoms and the regularities of the periodic table, as the actual quantum-mechanical calculations containing the correct physics for a world with electrons of this type, which are also obviously consistent with correspondence-limit constraints, show no signs of atomic shell structure.

As we saw earlier, leading theorists at the time appear to have been convinced that such calculations existed. As we now know, they did not—indeed, could not—exist, as the results of the actual quantum-mechanical calculations described above show. The shell structure of atoms in the real world simply cannot be properly accommodated in a theoretical framework lacking the additional constraints of the Pauli exclusion principle.

The slippery and unreliable nature of the arguments used by Bohr in the earlier stages of his second theory can be clearly seen in the evolution of his identification of principal quantum numbers for the successively added groups[6] of electrons in the *Aufbau* process. In a paper published in *Nature* in March 1921, Bohr (1921a) arrives at the following assignments for the first four rare gases (where here 2_1 means 2 electrons with $n = 1$, 8_2 eight electrons with $n = 2$, etc.):

$$\text{Helium } 2_1, \quad \text{Neon } 2_1\, 8_2, \quad \text{Argon } 2_1\, 8_2\, 8_2, \quad \text{Krypton } 2_1\, 8_2\, 18_3\, 8_2.$$

In a paper published in *Nature* in October 1921, however, Bohr (1921c) has changed the assignments to ensure that the principal quantum number increases uniformly by one unit in each successive row of the periodic table (which had not been the case in the earlier paper in *Nature*):

$$\text{Helium } 2_1, \quad \text{Neon } 2_1\, 8_2, \quad \text{Argon } 2_1\, 8_2\, 8_3, \quad \text{Krypton } 2_1\, 8_2\, 18_3\, 8_4.$$

In fact, in the evolution of Bohr's ideas concerning the orbital structures of multi-electron atoms, an increasingly central role was played by inductive arguments that relied on the available spectroscopic data (initially optical, then later, X-ray) to infer

[6] Bohr criticizes the use of the term "shell" as he points out that a group of electrons with the same principal quantum number n would be further divided into subgroups with different values of the azimuthal quantum number $k \leq n$, with the elliptical orbits with $k < n$ distributing the charge over a range of radii, in contrast to circular orbits of equal radius (and varying orientation) where the charge density would indeed lie on a thin shell. Accordingly, he speaks of "groups" of electrons of the same principal quantum number, which consist of subgroups of varying k values. Bohr (1921a) divides the electron groups democratically into equal numbers of electrons in each possible subgroup.

the appropriate quantum numbers to be assigned to electrons as they were sequentially added in the *Aufbau* process, rather than on "deductive" arguments motivated by correspondence-principle considerations.

Initially, Bohr relied heavily on the optical series, where the Rydberg spectral defects (cf. Section 7.1.1) could be related to the screening effects on the outer valence electron due to the inner (core) electrons.[7] The change in emphasis—away from vague correspondence-principle arguments, and toward inductive reasoning based on solid empirical evidence—can be seen clearly in a lengthy paper by Bohr (1921d) based on a lecture in October 1921 before a joint meeting of the Physical and Chemical Societies of Copenhagen. After a brief summary of the Kemble model for helium (and an advertisement of the Kramers calculations for this model that were underway but not yet finalized), Bohr turns to the problem of the third electron added to a bare nucleus of charge +3:

> We obtain direct information about the *binding of the third electron* from the spectrum of lithium. This spectrum shows the existence of a number of series of stationary states, where the firmness with which the last captured electron is bound is very nearly the same as in the stationary states of the hydrogen atom. These states correspond to orbits where k is greater than or equal to 2, and where the last captured electron moves entirely outside the region where the first two electrons move. But in addition this spectrum gives us information about a series of states corresponding to $k = 1$ in which the energy differs essentially [i.e., substantially] from the corresponding stationary states of the hydrogen atom (Bohr 1921d, pp. 89–90).

Bohr reasons that the strong screening effects seen for the lithium s ($k = 1$) states argue that these terms (corresponding to binding energies $1/1.6^2$, $1/2.6^2$, ... Rydbergs, rather than, as for hydrogen, $1/2^2$, $1/3^2$, ... Rydbergs) correspond to elliptical "dipping" (or "penetrating") orbits (for which necessarily $k < n$) that penetrate into the core region of the inner electrons and are therefore able to sample a less screened nuclear charge. In the October 1921 lecture Bohr assigns the lowest s-term of lithium, namely, the orbit occupied by the valence electron of the atom in its ground (normal) state to quantum numbers $n = 2$ and $k = 1$, for which he introduces the compact notation 2_1.

The assignment of principal quantum number $n = 2$ (with a spectral defect of 0.4) was further supported by detailed calculations of screening already published by Schrödinger in the September 1921 issue of *Zeitschrift für Physik* (Schrödinger 1921). Later in 1921, Bohr's more general screening arguments would lead him to alter Schrödinger's provisional assignment of the valence electron of sodium to principal quantum number $n = 3$ (Schrödinger had obtained $n = 2$), leading to the change in assignment of principal quantum numbers (from 2 to 3) in the third period of the periodic table evident in his second *Nature* paper.

[7] A detailed account of the analysis of dipping orbits and the effect of screening on spectral defects can be found in Kragh (1979) and Darrigol (1992, p. 153).

The assignment of orbits to the fourth, fifth, and six electrons added in the building up of beryllium, boron, and carbon suffered from the problem (Bohr 1921d, p. 91) that the spectra of the associated elements did not exhibit the simple series properties found in the alkali metals (Figure 9.1, column 1). Bohr once again had to rely on his "divine glance"—as Kramers put it—to infer that all of these electrons, like the single valence electron of lithium, should be assigned to $n = 2, k = 1$ elliptical orbits, as symmetrically disposed with respect to each other as was geometrically possible. This at least had the advantage of producing a tetrahedral arrangement of the major axes of these orbits in carbon, where the crystal structure of diamond was well known to exhibit precisely such a symmetry. The four electrons in the 2_1 orbits in carbon would execute a carefully timed ballet in which each electron was allowed to "visit" the inner nuclear region at separate (and equally spaced) times. The addition of the fifth electron in the $n = 2$ group (to form nitrogen) was assumed to require the assignment of the newcomer to a 2_2 circular orbit, as "the occurrence of five [2_1] orbits would so definitely destroy the symmetry in the interaction of these electrons that it is inconceivable that a process resulting in the accession of a fifth electron to this group would be in agreement with the correspondence principle" (Bohr 1921d, p. 94). And similarly for the sixth, seventh, and eighth electrons, all placed in carefully arranged circular 2_2 orbits to bring us to the exceptionally symmetric, and hence, stable, configuration of neon, with two $n = 1, k = 1$, four $n = 2, k = 1$, and four $n = 2, k = 2$ electrons.

In his Wolfskehl lectures of June 1922, Bohr covered similar terrain. In the sixth lecture (Bohr 1922e, pp. 397–406), he turned to the third and subsequent periods of the periodic table. The arguments made for the assignment of electrons to definite n_k orbits in the third shell for the elements from sodium ($Z = 11$) to argon ($Z = 18$) are entirely based on the empirical inputs provided by knowledge of the physical properties (such as atomic size, which could be estimated from gas theory), chemical behavior, and especially optical spectroscopy (arc spectrum for the neutral atom, spark spectra for the singly ionized one). The correspondence principle is *nowhere mentioned* in this lecture as giving the theoretical rationale for the choice of orbits. In tables included with these lectures, Bohr (1922e, pp. 418–419) gave explicit atomic structures for all the elements up to a putative noble gas at $Z = 118$, and in particular, for the as-yet undiscovered element 72—hafnium, first identified by Coster and Hevesy in Copenhagen in late 1922 in minerals containing zirconium. Bohr (1923d, p. 44) explicitly mentioned their work in his Nobel lecture on December 10, 1922 as a confirmation of his atomic structure procedure, which assigned to element 72 an outer shell structure $\ldots(5_1)^4(5_2)^4(5_3)^2(6_1)^2$, analogous to element 40, zirconium, with structure $\ldots(4_1)^4(4_2)^4(4_3)^2(5_1)^2$. Despite the plausibility of many of Bohr's assignments, in many cases made purely using the intuition associated with his "divine glance", the orbital assignments given in his tables are (beyond beryllium) all incorrect, inasmuch as they assign frequently 4, 6, or even 8 electrons to a single s-state orbit, violating the exclusion principle constraint introduced by Pauli just two years later.

9.1.2 Clues from X-ray spectra

We identified earlier Sommerfeld's derivation of the relativistic fine-structure formula for hydrogenic atoms (see Section 6.1) and the subsequent application of this formula in 1916 to the L-doublets in X-ray spectra (Section 6.2) as two of the most striking successes of the old quantum theory. As so often happened in the convoluted history of the development of modern quantum mechanics, the latter success was short lived. In fact, the inadequacies of Sommerfeld's relativistic interpretation of X-ray fine structure began to emerge once more accurate and detailed empirical data became available in 1920 and played an important role in the series of theoretical efforts, which were to lead to Pauli's introduction of the exclusion principle in late 1924.

At first sight it is perhaps unclear why the study of atomic X-ray spectra should provide useful clues for the understanding of the periodic physical and chemical properties of the elements manifested in the periodic table. Formulas of the type pioneered by Moseley (see, for example, Eq. (6.63)) show a very smooth (indeed linear) dependence of the square root of the frequency of X-ray spectral lines on the atomic number, in stark contrast to the periodic and recurrent characteristics of those chemical and physical properties dependent on the outermost electrons in the atom. But the evidence provided by X-ray investigations indicated certain very important features of multi-electron systems, which were to play a critical role along the way to the exclusion principle, which itself would be the key to understanding the structure of the periodic table.

Specifically, the interpretation of X-ray spectra using "K, L, M, ... " terms associated with electrons with principal quantum numbers 1, 2, 3, ... , taken together with the extremely slow and smooth dependence of the term values as the atomic number Z was increased implied that the addition of electrons at the periphery of the atom for high Z did not affect the pre-established arrangement of the inner electrons, which could be interpreted as occupying filled "rings", "levels", "shells" or "groups" (to illustrate the evolution of the language used by Sommerfeld and Bohr over the years from 1916 to 1923). This inner stability could be inferred, for example, from the stability of the screening number k (resp. l) for K (resp. L) electrons (cf. Section 6.2, p. 281), which remained essentially constant once Z was large—enough so that the changes to the atom as one increased the atomic number were due to additional electrons with orbits far enough out not to alter substantially the Coulomb forces felt by the inner electrons (apart, of course, from the steadily increasing electrostatic attraction of the nucleus as the nuclear charge was increased). The reasoning was already clearly expressed in the final paragraph of Sommerfeld's three-part paper of 1916:

> [from the independence of the screening numbers k, l from the atomic number Z] one may conclude that the inner construction of the atom for all Z above a certain limit is in essence the same (Sommerfeld 1916b, p. 166).

The term values assigned to electrons in the various groups corresponded directly to the binding energy of inner electrons, which were essentially unchanged by the addition of electrons much further out from the nucleus. This property was of course incorporated directly by Bohr in his assumption of the "permanence of quantum numbers" in the second atomic theory. Any quantitatively adequate mechanical model of a multi-electron atom should evidently address the task of (i) calculating these possible inner electron discrete energies on quantum principles; and (ii) explaining why the number of electrons $2N^2$ needed to fill a shell bore a simple relation to the principal quantum number (cf. Chapter 8, note 5) N for that shell, as was apparent from the periodic table.

We recall from the discussion of Sommerfeld's application of his relativistic fine-structure formula to X-ray spectra in Section 6.2 that a very convincing interpretation of the multiplets seen in the K and L series could be obtained by simply taking into account the dependence of an electron orbit energy on both the principal quantum number N and the azimuthal (angular momentum) quantum number n due to relativistic effects. The relativistic effects broke the degeneracy between orbits of the same principal quantum number N but different azimuthal quantum number n, with the smaller n (more eccentric) orbits lower in energy. This meant that the group of electrons with principal quantum number N should be divided into N sub-groups of distinct energy, corresponding to n values $1, 2, \ldots, N$. The K level ($N = 1; n = 1$) should be unique, there should be two L levels (with $N = 2; n = 1, 2$), three M levels (with $N = 3; n = 1, 2, 3$), and so on. In particular, the energy difference of the L-doublet arising from the transition from an electron in a given M-level to either of the two L-levels could be traced with impressive accuracy through an enormous range of atomic numbers using Sommerfeld's formula for the relativistic fine structure. The simplicity of this scheme would soon be destroyed by the discovery of additional lines in the X-ray spectrum that could only be explained by the positing of additional terms within each of the levels of a given principal quantum number. This development was just the analog, in the context of the high-frequency spectra of atoms, of the earlier discovery of complex structure in the lines of the optical spectrum. By the late teens (1916–1920) the direct identification of individual energy levels of inner electrons (rather than the differences corresponding to X-ray transition frequencies) by measurement of absorption limits had become a precise experimental tool. The idea was to directly identify the energy needed to liberate an inner electron from the atom (i.e., to promote it from a discrete energy level to a continuum, unbound state) by locating the frequency of *incident* X-rays at which their absorption by the element in question underwent a sudden increase.

By 1920, the study of these absorption edges for L electrons had progressed to the point where Gustav Hertz (1920) could announce definitively the existence of a third L term (or energy level for inner electrons with principal quantum number 2). The binding energies of the three levels (more precisely, the value of $\sqrt{\nu/R}$, where ν is the wavenumber $1/\lambda$ of the absorption edge, or $1/hc$ times the binding energy, and R is the Rydberg constant) as a function of atomic number can be seen clearly in Figure 9.2, taken from Hertz's paper. The new L_3 (A_3 in Hertz's notation) level corresponded to an electron state slightly more tightly bound than the previously known L_2 level (which, we recall, was interpreted by Sommerfeld as due to electrons in an elliptical orbit with $N = 2, n = 1$, with the L_1 level corresponding to the circular orbit $N = n = 2$). By the time the third edition of Sommerfeld's (1922b) "Bible" appeared in early 1922, inner electron terms had proliferated also to five (instead of three) for the M levels (principal quantum number $N = 3$), seven (instead of four) for the N levels (principal quantum number $N = 4$), and so on.

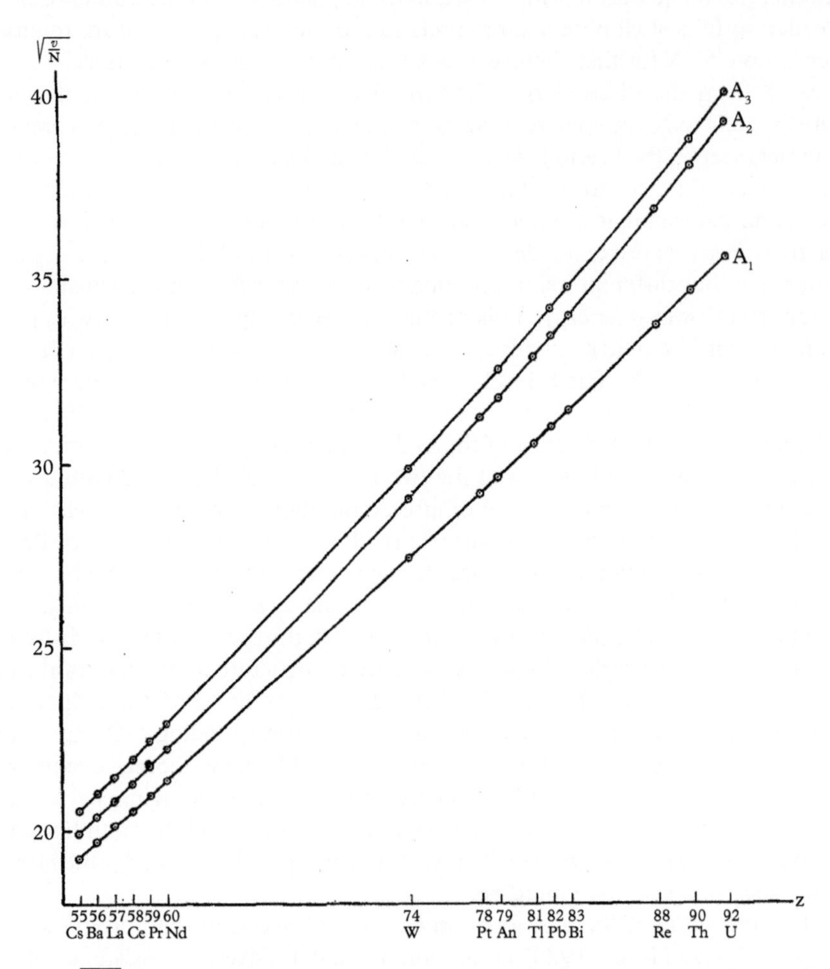

Figure 9.2 $\sqrt{\nu/R}$ *for the L triplet for (some) elements in the range* $55 \leq Z \leq 92$. *The notation* A_1, A_2, A_3 *corresponds to Sommerfeld's* L_1, L_2, L_3. *N is R, the Rydberg constant. Fig. 2 from Hertz (1920).*

The progressive divergence of the L_1 (A_1) and L_2 (A_2) terms seen in Figure 9.2 follows directly from the Sommerfeld fine structure analysis, which implies that the separation in $\sqrt{\nu/R}$ should grow roughly as Z^3. Introducing into the Sommerfeld formula (see Eq. (6.53)), with the notational changes $n \to N, l \to n$), independent screening constants σ

and s for the Balmer and fine-structure correction terms, one finds (with the Rydberg constant $R = e^4 m_0/2\hbar^2 hc$)

$$E = -Rhc\left\{\frac{(Z-\sigma)^2}{N^2} + \alpha^2\frac{(Z-s)^4}{N^4}\left(\frac{N}{n} - \frac{3}{4}\right) + O(\alpha^4)\right\} \tag{9.1}$$

for the energy E of an electron.[8] We may now calculate the term wavenumber $\nu/R = -E/(Rhc)$,

$$\frac{\nu}{R} = \frac{(Z-\sigma)^2}{N^2} + \alpha^2\frac{(Z-s)^4}{N^4}\left(\frac{N}{n} - \frac{3}{4}\right) + O(\alpha^4). \tag{9.2}$$

If we temporarily ignore the difference in screening constants σ and s (compared to the much larger Z) and take the square root, one easily sees that the splitting of $\sqrt{\nu/R}$ between the $n = 1$ (L_2) and $n = 2$ (L_1) terms should indeed grow roughly as the cube of the atomic number, as seen in the Hertz plot in Figure 9.2.[9] Pairs of levels related in the way implied by the fine-structure formula were termed by Sommerfeld "regular doublets", the archetypal example being the L_1, L_2 doublet. On the other hand, it is also apparent from a glance at Figure 9.2 that the splitting between the L_2 and L_3 levels is roughly *constant* over a wide range of atomic numbers. This splitting could therefore not have an origin in the relativistic fine structure, and such a term pair was dubbed by Sommerfeld an "irregular doublet" (Sommerfeld 1922b, p. 605).

The appearance of the "superfluous" (from the point of view of the Bohr–Sommerfeld picture of the atom) L_3 line in X-ray spectra was of course immediately realized to be analogous to the appearance of complex spectra (where doublets and triplets appeared in place of the expected singlet lines) in the optical regime Section 7.1). As noted in Section 7.1, the empirical data could be at least organized, if not theoretically understood, by the ruse of a new quantum number m, which Sommerfeld in the third edition of *Atombau und Spektrallinien* dubs the "ground quantum number" (Sommerfeld 1922b, p. 634) in a cautious attempt to distinguish it from the "inner quantum number" n_i introduced in the same work to make sense of the complex series line structure (Sommerfeld 1922b, p. 446). He did not of course know at this point that both "additional" quantum numbers had a common origin, in the as-yet undiscovered property of intrinsic electron spin. But the regularities observed by Hertz for the L-spectra, and soon extended to the M, N, and higher X-ray series, could be accounted for by assuming a doubling of all the electron states (at given principal quantum number) corresponding to non-circular orbits, and assigning two possible ground quantum number

[8] The need for separate screening constants in the Balmer and fine-structure terms was known from the empirical fits needed to produce both the individual $L_1(A_1)$ and $L_2(A_2)$ energy values, and their splitting as a function of Z in Figure 9.2.

[9] To see this, note that as α is small, we can, taking $s = \sigma$, and for not too large Z, expand

$$\sqrt{\frac{\nu}{R}} \simeq \frac{(Z-\sigma)}{N}\left(1 + \frac{\alpha^2}{2}\frac{(Z-\sigma)^2}{N^2}\left(\frac{N}{n} - \frac{3}{4}\right) + \ldots\right),$$

so that the difference of two $\sqrt{\nu/R}$ values with the same N but different n does indeed grow as $(Z-\sigma)^3 \simeq Z^3$.

Table 9.2 *Quantum number classification for X-ray terms (after Sommerfeld, Atombau, 3rd edn, p. 634). Symbols in parentheses indicate notation of Bohr and Coster (1923): "normal levels" of latter authors indicated in blue, "anomalous levels" in red. In Landé (1923b), the assignments are switched, with k_1 interpreted as the azimuthal, k_2 as an "inner" quantum number.*

Orbit	K	L_1	L_2	L_3	M_1	M_2	M_3	M_4	M_5
$N(n)$	1	2	2	2	3	3	3	3	3
$n(k_2)$	1	2	1	1	3	2	2	1	1
n_r	0	0	1	1	0	1	1	2	2
$m(k_1)$	1	2	2	1	3	3	2	2	1

values $m = n, n+1$ to an orbit with azimuthal quantum number n (for the non-circular cases with $n < N$). The possibilities emerging in this way for the first three X-ray series levels (K, L, and M) are shown in Table 9.2.

For the "regular" (later "relativistic") doublets, the beautiful agreement obtained between the term splittings and the $O(\alpha^2)$ fine structure term of Eq. (9.2) implied that the screening constant σ should be the same for the two levels in a regular doublet, so that the first (Balmer) term would cancel in the difference, leaving precisely the observed relativistic fine structure only (see note 9). By contrast, for the irregular doublets, the screening constant σ in the leading Balmer term in Eq. (9.2) would have to be different for the two states, so that the term splitting would arise from a change in the Balmer energy, and one would obtain the desired constant (i.e., Z-independent) displacement in the Hertz plot:

$$\Delta\left(\sqrt{\frac{\nu}{R}}\right) \approx \Delta\frac{Z-\sigma}{N} = -\frac{\Delta\sigma}{N}. \tag{9.3}$$

The observed behavior could therefore be reproduced on the assumption that the screening constants σ and s in Eq. (9.2) depend on the principal quantum number N and the new ground quantum number m, but not on the azimuthal quantum number n. The two states in a regular ("relativistic") doublet (e.g., (L_1, L_2), (M_1, M_2), and (M_3, M_4) in Table 9.2) would then have a splitting determined solely by relativistic effects, whereas the term splitting in irregular doublets (e.g., (L_2, L_3), (M_2, M_3), and (M_4, M_5) in Table 9.2) would be due to the varying screening effect in the leading Balmer term (due to the dependence of $\sigma(N, m)$ on m). For this reason, the irregular doublets would soon be redubbed "screening doublets".

The implications of the new levels appearing in X-ray spectra for mechanical models of multi-electron atoms in general, and specifically for Bohr's *Aufbau* tactic for interpreting the regularities in the periodic table, were the subject of study in an important paper by Bohr and Dirk Coster (1923), submitted to the *Zeitschrift für Physik* in November

1922. After reviewing the status of the latest available empirical data (on which Coster was a recognized expert), the authors turn their attention to the implications of the newly discovered X-ray terms for atomic structure (sec. 5, "General comparison of the classification of empirical X-ray levels with the theory of atomic structure"). They review the Sommerfeld classification of pairs of terms (doublets) into "relativistic" and "screening" doublets (Bohr and Coster used the same set of quantum numbers as Sommerfeld to identify terms, with the notational change $(N, n, m) \rightarrow (n, k_2, k_1)$, see Table 9.2) and directly address the difficulties of reconciling the data with (a) the screening effects to be expected on the basis of the usual electrostatic arguments; and (b) the selection rules for transitions based on the azimuthal quantum number ($\Delta n = \Delta k_2 = \pm 1$).

The basic problem was simply that the Bohr *Aufbau* depended on inserting electrons successively into groups of orbits characterized by the principal quantum number n, further subdivided into subgroups of orbits with the n possible values of the azimuthal (angular momentum) quantum number $k = n, n-1, \ldots, 2, 1$, corresponding to orbits of increasing eccentricity (and therefore of increasing penetration toward the nucleus, and decreasing screening of the nuclear charge by intervening electrons). If one simply ignored the terms with Sommerfeld's $m \neq n$ (Bohr and Coster's $k_1 \neq k_2$), leaving the levels labeled K, L_1, L_3, M_1, M_3, M_5 (and identified by blue symbols in Table 9.2), these remaining levels, now dubbed "normal" by Bohr and Coster, behaved exactly as one would expect on the basis of *Aufbau* principles. The energy differences of a $n = k = 2$ (circular) orbit and a $n = 2, k = 1$ (elliptical) orbit should after all contain both a screening effect due to the Balmer term in Eq. (9.1) *and* a relativistic splitting due to the second, fine structure term. Moreover, if only the normal levels are considered, the selection rules for transitions precisely correspond to the old Bohr–Sommerfeld requirement that $\Delta k (= k_1 - k_2) = \pm 1$.

Once the levels with $k_1 \neq k_2$ (now called "anomalous" levels, red symbols L_2, M_2, M_4 in Table 9.2) are included, however, a peculiar situation arose: the difference of energies of a normal level with the adjacent abnormal level just lower (in binding energy) seemed to come purely from the difference in screening constants in the leading Balmer term in Eq. (9.1), while the difference in energy of an anomalous level with the adjacent normal level just lower in binding energy was very well described by the fine-structure splitting arising from the second term in Eq. (9.1). The possibility of attributing the latter difference to two electron states with the same spatial behavior (orbit) and therefore the same screening constant σ, but differing in some intrinsic property of the electron, did not of course occur to the authors. After all, Sommerfeld's fine-structure theory of 1916 made absolutely no sense unless one considered the relativistic splitting to arise from a difference in energy of orbits of differing eccentricity—and therefore, necessarily differing screening behavior. The best Bohr and Coster could do in coming to terms with this paradoxical state of affairs was to abandon a basic tenet of the *Aufbau* procedure (on which Bohr's second atomic theory was based)—the stability of the remaining electrons in an inner group when one of their number is removed from the atom:

> This situation leads us to seek the origin of these anomalous levels in an intimate interaction of the electron motion between two neighboring subgroups of the excited atom,

> which effects a significant change of the harmonic interplay of the electron motion within *both* subgroups, if an electron is removed from *either* subgroup ... On the other hand it is still quite unclear how an explanation of the independent appearance of screening and relativity doublets can be given on the basis of a closer development of this observation (Bohr and Coster 1923, p. 365).

The quandary in which atomic theorists attempting to decipher X-ray spectra found themselves was best summarized by Landé (1923b), who asked a very simple question: what is the mechanical interpretation of the *two* quantum numbers k_1 and k_2, in relation to the *single* azimuthal quantum number k representing the orbital angular momentum of an inner electron ejected from an atom? Using the Bohr–Coster notation n, k_1, k_2 for the X-ray terms, and the notation n_k to identify an electron state with principal quantum number n and azimuthal quantum number k, Landé summarizes the central dilemma in the interpretation of screening versus relativistic doublets on the one hand, and normal versus anomalous term values on the other, as follows:

> If one wishes to go further in this question, one encounters immediately the choice: (a) either the "anomalous" $n(k_1, k_2)$ term arises from the removal of a n_{k_1} electron, or (b) from the removal of a n_{k_2} electron from the atom. The screening doublet splitting $n(k_1, k_2) - n(k_2, k_2)$ speaks in favor of (a) and against (b). The relativistic doublet splitting $n(k_1, k_2) - n(k_1, k_1)$ speaks in favor of (b) and against (a). But also for (a) and against (b) we have the selection rules $k_1 \rightarrow k_1 \pm 1$ and $k_2 \rightarrow k_2 \pm 1, k_2$, which suggest the interpretation of k_1 as an azimuthal, and k_2 as an "inner" quantum number (Landé 1923b, p. 186).

The additional weight of the selection rules—breaking the tie, as it were—settles the choice for Landé in favor of option (a). In the rest of the paper he develops an interpretation of the "inner quantum number" k_2 (equal to k_1 or $k_1 - 1$) in terms of two possible relative orientations of an unpaired (by the removal of the electron whose term value we are examining) inner electron (with angular momentum k_1) with respect to the innermost "helium core" of two $n = k = 1$ electrons (in crossed Kemble orbits, with unit net angular momentum). As in the case of the inner quantum number (Sommerfeld's n_i) for optical spectra, the quantum number k_2 is interpreted as the total resultant angular momentum of the helium core and the unpaired inner electron (with the fully aligned $k_1 + 1$ option excluded as in the optical case, leaving only the two possibilities $k_2 = k_1$ or $k_2 = k_1 - 1$).

The price paid for this view was severe, as the beautiful agreement of the relativistic doublet splitting with Sommerfeld's fine-structure formula now had to be regarded as essentially an accident. These splittings instead were assumed to be generated by magnetic effects, analogous to the explanation for alkali doublet splittings in Heisenberg's core (*Rumpf*) model (discussed in Section 7.1.2).

In two later papers, Landé (1924a, 1924b) explicitly attributed the splitting of Sommerfeld's "relativistic doublets" to the same mechanism responsible for the optical doublet structure in alkali metals: namely, the varying orientation of the orbit of the light-emitting valence electron with respect to the atomic core, with an energy difference generated by the effect of the inner magnetic field of the valence electron on the magnetic moment of the core (cf. the discussion of the Heisenberg core model in Section 7.1). Landé points out the remarkable formal similarity between the formulas for relativistic fine structure *à la* Sommerfeld and the magnetic-core/valence-orbit splitting assumed to be responsible for optical multiplets (Landé 1924a, p. 95; the formulas turn out to be identical for the case of hydrogen, $Z = 1$ (cf. Eqs. (6.53) and (7.28)). The rapid increase of the splitting of the relativistic X-ray doublets with atomic number (as compared to the splitting of the "screening" doublets), and the decrease (as the third inverse power) with principal quantum number was also reproduced by the core–orbit calculation.

The one fly in the ointment, Landé pointed out, was that the core–orbit calculation gave a splitting proportional to Z^3, which differed from the extremely well verified Z^4 dependence—or more accurately, when screening is taken into account, $(Z - s)^4$ dependence—of the relativistic fine structure formula.[10] Despite these difficulties, Landé insisted in the essential conclusion, which was to be of critical importance in the question of determining the number of electrons in complete shells following definite n_k orbits, namely, that the $L_1 - L_2$ energy difference was *not* due to electrons of different orbital angular momentum, occupying orbits of different eccentricity, but that both terms corresponded to electrons of the *same* angular momentum, that is, $n = k = 2$ (or, in modern notation, 2p) electrons.

9.1.3 The filling of electron shells and the emergence of the exclusion principle

Dramatic progress in unraveling the structure of the periodic table came in the summer of 1924 from a perhaps unlikely quarter and was crucially important for the discovery of the exclusion principle.[11] Edmund Stoner, an experimental research student of Rutherford's at Cambridge, had attended Bohr's lectures there in March of 1922 and had become enthralled with the problem of atomic structure (Cantor 1994, p. 281). Stoner's (1924) critical advance was to combine the insights that could be gleaned by study of X-ray spectra with those that had emerged from the (by now, almost three decades old) examination of the anomalous Zeeman effect in optical spectra. The classification of X-ray levels initiated by Sommerfeld, and further modified by Landé (Stoner references

[10] The discrepancy is due to the fact that in theories of the core–orbit variety, the optical splittings were due to the interaction of the magnetic moment of the core with a magnetic field of the electron at the core location, which varies inversely with the cube of the semi-major axis of the orbit ($= a_0/Z$ for a hydrogenic atom—see Eq. (7.26)), and hence, is of order Z^3 for an electron moving in an unshielded Bohr orbit. In the modern (post-electron spin) theory, the splittings are due to the interaction of the magnetic moment of the electron, due to its spin, with a magnetic field due to the electron motion relative to the nucleus of order Z^4 (see below, especially the derivation leading to (9.11)). The fact that the relativistic kinematic effects and the spin–orbit corrections—also ultimately relativistic in origin—have a very similar form, and combine to give precisely the Sommerfeld (1916) fine-structure formula (with the replacement $k \to j + \frac{1}{2}$, with j the total spin plus orbital angular momentum of the electron) is due to a remarkable numerical "accident" to which we return later. The extra symmetries present in the dynamics of a particle moving in a pure Coulomb potential—both classically and quantum-mechanically—are at the core of this unexpected success of the old theory.

[11] This part of the story is covered in Mehra and Rechenberg (1982a, pp. 665–684).

Table 9.3 *Stoner's classification for X-ray terms. The X-ray terms are given in the Bohr–Coster notation (Roman numerals, in reversed order from Sommerfeld, thus M_1 (Sommerfeld)→ M_V, etc.). Note the switch in assignment of azimuthal and inner quantum numbers from Sommerfeld's (Table 9.2), as proposed by Landé.*

Sub-level	K	L_I	L_{II}	L_{III}	M_I	M_{II}	M_{III}	M_{IV}	M_V
n	1	2	2	2	3	3	3	3	3
k	1	1	2	2	1	2	2	3	3
j	1	1	1	2	1	1	2	2	3

Landé 1923b, where the X-ray-optical doublet analogy is already discussed), clearly showed the presence of an additional quantum number, and therefore of a further subdivision of Bohr's electron n_k subgroups (characterized only by principal and azimuthal quantum numbers) in terms of the new inner quantum number j (following the notation of Bohr and Landé[12]) into sub-levels of Bohr's subgroups characterized by the triplet n, k, j (see Table 9.3).

The analogy emphasized by Landé between X-ray and optical doublets suggested to Stoner that the number of electrons that could be accommodated in a given sub-level was given by simply twice the inner quantum number j. This was, after all, just the number of distinct energy levels that emerged in optical spectra when a weak magnetic field was turned on. For example, the p_1 level in sodium (Figure 7.2) with inner quantum number $j = n_i = 2$, split into four distinct energy levels in a magnetic field (see Figure 7.5), while the p_2 level with $j = 1$, split into just two levels. The strong analogy between X-ray and optical doublet terms promoted by Landé suggested to Stoner that the energy levels of inner electrons would divide in just the same way under the influence of a weak magnetic field as those of an outer valence electron, with the multiplicity again given by twice the inner quantum number.

The result of this hypothesis, which Stoner went on to justify with X-ray emission and absorption data (as we will see) was an alteration in Bohr's proposed assignment of n and k quantum numbers already visible in the second and third periods of the periodic table. Bohr had relied on his "divine glance" (purportedly supported by the correspondence principle) to infer the atomic distribution $(1_1)^2(2_1)^4(2_2)^4$ for neon, with $Z = 10$ (where the notation $(n_k)^p$ implies the presence of p electrons in a n_k orbit). Heuristic symmetry considerations (numerical and geometrical) required for Bohr an equal distribution of electrons for given n among the different possible k values (see Kragh 1979, p. 159). Stoner's Table I, partially reproduced here as Table 9.3, shows

[12] The letter "j" for a quantum number associated with a total angular momentum (core plus valence electron) appears to have been first introduced by Bohr, in the fourth of his Wolfskehl lectures (Bohr 1922e, pp. 372–387). The use of $\mathcal{J}$ for total angular momentum appears first in Landé (1922), with the inner quantum number, still however defined to be integer in all cases, redubbed j in Landé (1923a).

that the second, or "L", shell should contain two electrons with $(n, k, j) = (2, 1, 1)$, two electrons with $(n, k, j) = (2, 2, 1)$, and four electrons with $(n, k, j) = (2, 2, 2)$: hence, only two electrons with azimuthal quantum number $k = 1$ (in modern notation, 2s) and *six* electrons altogether with $k = 2$ (2p electrons).

The empirical evidence adduced by Stoner in support of his electron distribution hypothesis involved (a) the intensities of X-ray emission lines; (b) the relative absorption of X-rays; (c) ionic paramagnetism of the elements in the third period of the periodic table; (d) chemical properties (valencies) and (e) optical spectra. Only in the first two cases was Stoner able to present convincing *quantitative* evidence for his proposed allotment of electrons to sub-levels, so we restrict our discussion to the X-ray data.

In the case of X-ray emission lines, Stoner points out that, for elements ranging from iron (Fe, $Z = 26$) to tungsten (W, $Z = 74$), the ratio of intensities of the K_{α_1} line, corresponding to the transition $L_{III}(= L_1) \to K$, to that of the K_{α_2} line, corresponding to $L_{II}(= L_2) \to K$, was practically constant and equal to 2 (to within a few percent). This strongly suggested that the number of electrons of the L_{III} type (with inner quantum number $j = 2$) in a closed L shell, available for transition to a vacated K level, was twice as many as those of the L_{II} type (with $j = 1$). In Bohr's scheme, where only four (not six) of the eight L electrons were assumed to have $n = k = 2$, a subdivision into two sub-levels with a relative occupancy of 2:1 was clearly not possible. In the case of X-ray absorption, a similar argument could be used, based on the relative strength of absorption from electrons in different sub-levels, which should be proportional to the number of electrons occupying each sub-level, and therefore available to absorb an incoming X-ray quantum (with a resulting transition to a higher or completely ionized state). After compensating for the effect of the different sub-level energies with a semi-empirical absorption formula (due to Louis de Broglie), Stoner once again found that the data supported a relative number of electrons in the $L_I : L_{II} : L_{III}$ sub-levels of 1:1:2, or an actual occupancy (given the fact that the total number of electrons in a closed L shell was 8) of 2, 2, and 4.

Stoner's hypothetical counting of sub-levels provides for the first time an explanation for the long-observed, and very suggestive, sequence of period numbers in the periodic table—2, 8, 18, 32, etc.—or in other words, the remarkably simple rule that the shell corresponding to principal quantum number n is filled once $2n^2$ electrons are present. For such a shell, the possible k values range from 1 to n, and for each value of k the allowed j values are k or $k - 1$, accommodating respectively $2k$ and $2k - 2$ electrons, in total $2(2k - 1)$. The total number of electrons in the shell characterized by principal quantum number n is therefore

$$\sum_{k=1}^{n} 2(2k - 1) = 2n^2. \tag{9.4}$$

The assumption implicit in Stoner's analysis, of course, is that the possibility of accommodating *more* than $2j$ electrons with inner quantum number j is somehow *excluded.* This suggests, though by no means implies, that one electron, and one electron only, could be present for each of the distinct energy levels manifested by the imposition of a weak magnetic field. It is, however, nowhere explicitly stated by Stoner that these electrons necessarily occupy separate levels, although there is hardly any other reasonable interpretation of his level filling hypothesis. Stoner's work was noticed immediately by Pauli (now back in Hamburg, having returned from his frustrating year at Bohr's Institute in Copenhagen, unsuccessfully wrestling with the anomalous Zeeman effect), who recognized the essential feature of the Stoner scheme, an exclusion principle, which emerges almost automatically once the proper identification of energy levels with quantum numbers can be made. By December 1924,[13] Pauli had prepared a paper (submitted to *Zeitschrift für Physik* in January 1925) entitled "On the connection of the closing of electron groups in the atom with the complex structure of spectra"—the famous "exclusion principle" paper (Pauli 1925b).

Actually, the statement of the exclusion principle here, while of enormous historical significance, is perhaps not the most conceptually significant feature of the paper. Rather, as we will see, the essential feature of Pauli's reasoning (indispensable for the "proper identification" alluded to above) is the definitive removal of the source of the complex multiplet structure of optical spectra from an assumed magnetic core, and its transferral to a "classically indescribable duality (*Zweideutigkeit*) of the quantum-theoretical properties of the valence electron" (Pauli 1925a, p. 765).

This change of viewpoint was certainly motivated by a calculation Pauli had done recently, and submitted to *Zeitschrift für Physik* in early December (Pauli 1925a), establishing the inescapable presence of relativistic corrections to the magnetic moment generated by the core electrons in high Z atoms due to the much larger velocities of the inner electrons, corrections which should undoubtedly affect the calculation of the Landé g-factor for such atoms. Data provided to Pauli by Landé showed absolutely no evidence of the expected Z-dependence of the g-factor in such cases. For Pauli, "the evident absence of such effects, as well as other arguments, argue against any essential participation of the inner, closed shells . . . on the emergence of the complex structure of optical spectra and their Zeeman effects" (Pauli 1925a, abstract). The only other place to look for the extra degree of freedom responsible for complex (i.e., multiplet) structure, and the further dissection of levels emerging in the anomalous Zeeman effect, was clearly in some intrinsic property of the valence (i.e., the outer, light-emitting) electron itself.

Pauli begins by reviewing the introduction of an inner quantum number in the classification of X-ray terms. He adopts the Landé notation (Landé 1923b): n for the principal quantum number, $k_1 = 1, 2, 3, \ldots, n$ for the azimuthal (orbital angular momentum) quantum number, and k_2 for the inner quantum number. From his previous analysis of the Zeeman effect (Pauli 1923, cf. Section 7.3.3), the magnetic quantum number m labeling the split levels of the Zeeman effect in an alkali atom is interpreted as giving the

[13] The considerations contained in Pauli's famous paper on the exclusion principle must have already matured at the latest by November 24, when, in a letter to Landé (Pauli 1979, Doc. 71), he described the attribution of a fourth quantum number to individual electrons, and stated the exclusion principle in more or less exactly the same terms as in the *Zeitschrift für Physik* paper. This letter would play an important role in the discovery of electron spin.

space-quantized allowed values[14] associated with a total angular momentum quantum number j, namely $m = -j, -j+1, \ldots, j-1, j$, or $2j+1$ possible values.[15] As emphasized by Stoner, the weak-field Zeeman multiplicity was given by just twice the inner quantum number, so one had $2j+1 = 2k_2$, or $j = k_2 - \frac{1}{2}$.

As we saw earlier (Section 7.2), Pauli had interpreted the levels emerging in the extreme strong field (Paschen–Back) limit as due to the combination of a normal Zeeman effect based on the component m of the total angular momentum along the field direction with an extra core contribution described by a new quantum number $\mu = \pm\frac{1}{2}$, giving a total magnetic energy (cf. Eq. (7.88))

$$V_{\text{mag,tot}} = h\Delta\nu_{\text{norm}}(m_1 \pm \tfrac{1}{2}) = h\Delta\nu_{\text{norm}}(m_1 + \mu) \equiv h\Delta\nu_{\text{norm}}\, m_2. \tag{9.5}$$

To accommodate the new quantum number, Pauli relabels, as indicated here, the old total magnetic quantum number m as m_1, and introduces an additional magnetic quantum number $m_2 = m_1 + \mu$.[16] The m_2 quantum number, as is apparent from Eq. (9.5), directly indicates the total magnetic moment of the atom (the component in the field direction, in units of the Bohr magneton) responsible for generating a magnetic energy when placed in a strong external magnetic field—strong enough to overwhelm all other magnetic effects due to the other components of the atom. In sum, a Zeeman level was now uniquely identified by *four* quantum numbers, either n, k_1, k_2, m_1, or equivalently by n, k_1, m_1, m_2.[17]

Pauli's assignment of quantum numbers is clearly motivated by experience with the behavior of the outermost electrons in an atom under magnetic influence, but he wastes no time in invoking Bohr's principle of "permanence of quantum numbers" to extend the classification to all the electrons in a multi-electron atom. After all, in Bohr's *Aufbau* procedure, what was once the outermost valence electron will subsequently become an inner one, as further electrons are added on the way to producing an electrically neutral atom, but one still characterized by the same quantum numbers it had prior to the addition of further electrons. In particular, although experimentally inaccessible (this would require a study of the Zeeman effect on X-ray spectra, not feasible at the time), the inner electrons are assumed to exhibit exactly the same multiplet and anomalous Zeeman properties as the valence electrons (Pauli 1925b, p. 767). The difficulties of a consistent dynamical interpretation of multiplet and anomalous Zeeman phenomena

[14] For simplicity, we omit throughout the qualifier "in units of $\hbar$", when describing angular momentum values in terms of quantum numbers.

[15] Of course, in his 1923 paper, this total angular momentum was still assumed to arise by combining the orbital angular momentum of the valence electron with the core angular momentum of $\frac{1}{2}$.

[16] For the modern reader, the identifications $m_1 = \mathcal{J}_z = L_z + S_z$, $\mu = S_z$ may be useful. The magnetic energy is proportional to $m_1 + \mu = L_z + 2S_z$, reflecting the "double magnetism" of the core/electron magnetic moment.

[17] Note that these two specifications of a state are not in one-to-one correspondence. In the first case, a n_{k_1} electron state is identified, in modern notation, by giving its $\mathcal{J}$ $(k_2 - \frac{1}{2})$ and $\mathcal{J}_z$ (m_1) quantum numbers, in the second, by specifying the $\mathcal{J}_z$ (m_1) and S_z $(m_2 - m_1)$ quantum numbers. In the modern theory, these alternative denotations represent different angular-momentum bases for describing electron states. The exclusion principle, of course, applies independently of the state basis used.

(cf. the discussion of the Landé–Pauli vector model in Section 7.2) are frankly admitted, but by now Pauli is quite determined *not* to be dragged into any specific commitments as to the actual geometry of electron orbits (either valence or core). Instead, he intends to focus specifically on the problem of the closure of electron groups:

> We will only be concerned with drawing conclusions about the number of possible stationary states of the atom in the presence of several equivalent electrons, and not with the orientation and arrangement of the levels (*Termwerte*) (Pauli 1925b, p. 771).

In connection with precisely the problem of electron shell closure, Pauli turns to the "essential progress" contained in the considerations advanced by Stoner. The agreement of Stoner's procedure with the observed periodic table shell closure numbers, as in Eq. (9.4), is viewed as an encouraging development. The Stoner level counting rules, motivated by the number of split levels appearing in the weak Zeeman effect for alkali atoms, is now reinterpreted by Pauli as following from the quantum number classification (n, k_1, k_2, m_1 or n, k_1, m_1, m_2) appropriate in the strong field limit. The fact that the same number of levels must appear in both cases is justified on "thermodynamic grounds" (in terms of the invariance of statistical weights under adiabatic transformations, i.e., from weak to strong fields).[18] But the new ingredient is not really the counting scheme, but the clear expression of the electron shell (and sub-shell) closure rules in terms of a *prohibition*:

> *There can never be two or more equivalent electrons in the atom, for which the value of all quantum numbers n, k_1, k_2, m_1 (or, equivalently n, k_1, m_1, m_2) in strong fields agree. If an electron is present in the atom with definite values of these quantum numbers (in an external field), this state is "occupied"* (Pauli 1925b, p. 776).

After stating this exclusion (of multiply occupied states) principle, Pauli examines some of the consequences, as a first check of consistency of the new rule with the empirical facts. The agreement with the famous sequence 2, 8, 18, 32, ... of the "natural system (of elements)" already achieved in Stoner's work is a given.

As a first check, for the alkali earths (Figure 9.1, column 2), the lowest triplet s term was known to correspond to a principal quantum number one larger than that corresponding to the outer shell in the normal (ground) state of the atom. For example, for beryllium (with ground state electronic configuration $(1s)^2(2s)^2$, where we now, finally, adopt the modern notation where $k_1 = 1, 2, 3, 4, \ldots$. levels are identified as $s, p, d, f, \ldots$ levels, preceded by the principal quantum number, replacing the Bohr n_k notation), the triplet s terms corresponded to situations in which the two electrons had to have different n values, as they necessarily had to have the same k_1 (=1), k_2 (also =1), and m_1 values, as the triplet contained values of the total m_1 of both electrons of $+1$ and -1, as well as zero, so the situations in which both electrons had $m_1 = +\frac{1}{2}$, or $m_1 = -\frac{1}{2}$ had to be allowed. The exclusion principle would then force the remaining quantum number—the

[18] In a footnote, Pauli insists on the validity of this invariance even independently of the validity of classical mechanics, a point of view to which Bohr had come by this point (Pérez and Pié Valls 2015).

principal—to be different for the two outer electrons. As the excited beryllium states giving rise to triplet *s* terms of the atom corresponded to one (L shell) electron remaining in the ground level with $n = 2$, the other, excited *s* electron would have to have $n \geq 3$, in agreement with observation.

A second example given by Pauli anticipates the analysis of atomic level quantum numbers that can be found in any modern text on atomic physics (see, for example, Bethe and Jackiw 1986, p. 119). The exclusion principle provides a "quick and dirty" glimpse of the set of possible total angular momentum $\mathcal{J}$ values available to the atom consistent with the prohibition of two electrons sharing the same quantum numbers. Pauli illustrates the procedure for the two 2*p* electrons in the incomplete subgroup of carbon: one constructs (in modern terminology) all possible "*m* sets", that is, possible values of the (m_1, m_2) assignments for each electron, avoiding duplication.[19] If one examines the possible *m* values for the two $n = 2, k_1 = 2$ (2*p*) electrons in the unfilled subshell of carbon, there are three possibilities to be considered:

1. Both electrons have total angular momentum $j = \frac{1}{2}$ $(k_2 = 1)$. By the exclusion principle, they must then have different m_1 values (i.e., one must have $+\frac{1}{2}$, the other $-\frac{1}{2}$), giving a total angular momentum $\mathcal{J}_z$ in the field direction (this is called $\overline{m}_1$ by Pauli) equal to zero. As no other values for $\mathcal{J}_z$ occur, Pauli concludes that this situation also corresponds to the quantum number $\mathcal{J} = 0$ for the net angular momentum of both electrons, and hence, of the atom (as the other electrons occupy symmetrically filled subgroups and their total angular momentum has already canceled to zero).
2. One electron has $j = \frac{1}{2}$, the other $j = \frac{3}{2}$ $(k_2 = 2)$. In this case there is no exclusion restriction as the two electrons are already distinguished by their k_2 value. As there are two possible m_1 values $(\pm\frac{1}{2})$ for the first electron, and four possible values $(\pm\frac{1}{2}, \pm\frac{3}{2})$ for the second, one obtains eight possible pair states, which can be sorted by their $\mathcal{J}_z = \overline{m}_1$ values into a $\mathcal{J} = 2$ series (with $\overline{m}_1 = +2, +1, 0, -1, -2$), and a $\mathcal{J} = 1$ series (with $\overline{m}_1 = +1, 0, -1$).
3. Both electrons have $j = \frac{3}{2}$. In this case, one has to exclude, from the sixteen a-priori possible pair combinations, the four cases where the electrons have equal m_1 values. The remaining twelve cases correspond to six possible values of $\mathcal{J}_z$ (as the interchange of quantum numbers of the two "equivalent" electrons is not regarded as producing a different state), namely $\pm 2, \pm 1, 0, 0$, which can be in turn divided into a $\mathcal{J} = 2$ series (with $\mathcal{J}_z$ values $+2, +1, 0, -1, -2$) and a $\mathcal{J} = 0$ singlet with $\mathcal{J}_z = 0$.

As Pauli (1925b, p. 781) immediately points out, "we have nothing to say about the grouping of these terms with regard to their size [energy] or interval relationships," in

[19] Taking a peek forward, we would now construct *m* sets by specifying (m_l, m_s) for each electron, where m_l is the z-component of the electron's orbital angular momentum $(0, \pm 1$ for *p* electrons), and $m_s = \pm\frac{1}{2}$ is the z-component of the spin angular momentum. Pauli's quantum numbers correspond to $m_1 = m_l + m_s$, $m_2 = m_l + 2m_s$.

contrast to their expected Zeeman splittings, for which a quantitative theory (at this stage still empirically based, of course) existed. In fact, we now know that the lowest five energy levels of the carbon atom have total angular momenta $\mathcal{J} = 0, 1, 2, 2$, and 0, respectively (in order of decreasing binding energy), exactly the set of possible values identified by Pauli.

We have previously commented on the misplaced enthusiasm with which Bohr attempted to squeeze the regularities of the periodic table, and in particular the rules for closure of electron shells (and sub-shells), out of correspondence-principle considerations. As shown in our discussion of Bohr's second atomic theory, this approach can certainly be recognized as incorrect. The inadequacy of correspondence arguments in determining atomic shell structure was already clear to Pauli in December 1924 as he was preparing his paper on the exclusion principle. Admittedly, the two fundamental guiding principles of the old theory—the adiabatic principle and the correspondence principle—had by this time parted company in the minds of most physicists, with the adiabatic principle losing ground as the evidence for a strictly multi-periodic atomic mechanics withered: with the failure of the Kemble model for helium, it would seem finally to have lost its relevance to the development of atomic theory. Indeed, rigorous proofs of the adiabatic principle on the basis of classical mechanics *à la* Burgers (1917) certainly depended on the multiple periodicity of the entire atomic system, which had more or less been abandoned by this point.

On the other hand, there was still no doubt that the correspondence principle, in the limited interpretation of providing a smooth transition from quantum theory to classical electromagnetic in the regime of large quantum numbers, could be relied on even for multi-electron atoms, especially as regards the derivation of selection rules for atomic transitions. But Bohr's attempts to use the correspondence principle as a means to enforce restrictions on individual electron states in low orbits (as opposed to transitions between states) was clearly recognized by Pauli as misguided. Instead, Pauli was quite sure that some extension of the adiabatic principle (which he terms a principle of "invariance of statistical weights") would preserve the counting of electron energy levels in the transition from the weak field region (relied on by Stoner) to the strong field Paschen–Back region, where his quantum number classification was valid, even though a *proof* of such a principle on a purely mechanical basis was clearly impossible. The evidence for Pauli's views on this point can be found both in the exclusion-principle paper itself, and in correspondence at the same time (early December 1924) with Bohr. At the very end of the exclusion-principle paper we find the following remarks:

> In general it should be remarked that the considerations given here which concern the transition from strong to weak or vanishing fields are based in principle on the invariance of statistical weights of the quantum states. However, there does not seem to be support for the point of view, on the basis of the results achieved here, as supposed by Bohr, that there exists a connection of the problem of closure of electron groups in the atom with the correspondence principle. The problem of a more precise justification of the general rule advanced here concerning the appearance of equivalent electrons in the atom will

> probably only be manageable after a further deepening of the fundamental principles of the quantum theory (Pauli 1925b, p. 783).

In a letter to Bohr of December 12, 1924, sent as Pauli was preparing his exclusion principle paper for submission (and in which he included a copy of the forthcoming paper), the same sentiments are expressed frankly and unapologetically. The ideas expressed are important as a summation of the general situation in the dying days of the old quantum theory, and deserve an extended excerpt. In describing the contents of the enclosed manuscript, Pauli begins by tossing a bone, as it were, to the other (than the correspondence principle) basic ingredient of Bohr's second atomic theory, the *Aufbau* principle, specifically, the stability of electron quantum numbers under successive addition of further electrons to a (not yet neutral) atom:

> I saw that I could rescue the *Aufbau* principle, and that with recourse to the work of E.C. Stoner (of which I became aware through the foreword to the new edition of the Sommerfeld book [the fourth edition of *Atombau und Spektrallinien*, Sommerfeld 1924, p. VI]) I could in an "unforced way" arrive at a general theory of the closing of electron groups in the atom.
>
> There is a question concerning which I would willingly hear your point of view. I have often said to you that in my opinion the correspondence principle has in truth absolutely nothing to do with the problem of closure of electron groups in the atom. You have always answered, that I was *too* critical about this. But now I believe that I am fairly sure of my view. The exclusion of certain stationary states (*not*: transitions), which is the matter of interest here, is rather to be compared with the exclusion of the states with $m = 0$ or $k = 0$ in the hydrogen atom than, for example, with the selection rule $\Delta k = \pm 1$. Do you still adhere to the supposed applicability of the correspondence principle in this case? I think, it is all much simpler in reality; one does not need at all to speak of a harmonic interplay.
>
> I am very enthusiastic about the work of Stoner. The more I read it, the more I like it. That was surely an eminently clever idea, to connect the number of electrons in closed subgroups with the number of Zeeman-terms of the alkali spectra!
>
> Now, one more important point. The formulation from which I proceed is quite certainly nonsense (*ein Unsinn*). For one can certainly not reconcile in this way the motion of a precessing central force orbit required by the correspondence principle with the relativistic doublet formula. This type of motion seems at the moment to be irreconcilable with my assumptions. But I believe that what I have done is *no greater* nonsense than the previous conception of the complex structure [i.e., series multiplet structure]. My nonsense is conjugate to the previous nonsense ... The physicist who succeeds in adding these two nonsenses will achieve the truth!
>
> ...
>
> The relativistic doublet formula seems to me to show without doubt that not only the concept of force, but also the kinematical concept of motion in the classical theory must necessarily undergo fundamental modifications. (For this reason I have stringently avoided the use of the term "orbit" in my work.) As this concept of motion lies at the heart of the correspondence principle, the clarification (of motion) deserves above everything else the efforts of theorists. I believe that the energy and (angular)-momentum values of the stationary states are something much more real than "orbits" (Pauli 1979, Doc. 74).

Pauli's aversion at this time to orbits, and more generally, to explicit mechanical models of any kind is remarkably prescient given the new theory that was to emerge within a year, in which precisely the *kinematic* underpinnings ("concept of motion") of quantum theory were to undergo a radical shift. Ironically, as we will see, this attitude also made him extremely resistant to the concept of electron spin, which he initially viewed as yet another attempt to interpret the anomalies of complex spectra and the Zeeman effect in terms of misguided analogies from classical mechanics.

9.2 The discovery of electron spin

In the early 1900s, Abraham, Lorentz, and others had modeled the electron as an extended spherical surface-charge distribution (Janssen and Mecklenburg 2007, cf. Volume 1, pp. 18–19). With the appearance and dramatic success of the Bohr model, however, the idea of attributing a spatial extension to electrons receded completely, as the successes of the old quantum theory were clearly based on the treatment of electrons as massive point-like negative charges. In 1920 Arthur Compton attempted, in a paper entitled "The magnetic electron" and published the following year (Compton 1921a), to resuscitate the notion of an extended electron, this time adding the feature of electron spin, giving rise to both an intrinsic angular momentum of an electron whose center of mass was at rest, as well as an intrinsic magnetic moment (due to the charge). The reintroduction of an extended structure for electrons was partly motivated by experiments on the scattering of hard X-rays and gamma rays, which indicated far less scattering than that expected on the basis of a point-like electron (Compton 1921b), and partly by theoretical arguments concerning the origin of permanent (as opposed to transient) diamagnetism. The mysterious features of complex spectra or the anomalous Zeeman effect, or of optical spectra more generally, are not mentioned in this connection. Compton's magnetic electron paper was in fact basically ignored, and certainly eclipsed by his much more influential discovery, on further pursuance of his experiments on X-ray scattering, of the Compton effect two years later.[20]

The discovery of electron spin in the modern sense, and in particular its connection to the complex structure of spectral multiplets, can be localized quite precisely in time, to a day or two after the arrival (on January 7, 1925) of Ralph Kronig in Tübingen. Kronig was born of an American father in Dresden in 1904, and educated until 1918 in Germany, later gaining a doctorate in experimental physics from Columbia University in New York; he was traversing Europe on a traveling fellowship, having previously visited Cambridge and Leyden. In Leyden, Kronig collaborated with Goudsmit, publishing a paper giving some semi-empirical rules for the intensities of lines in the Zeeman effect. Unlike Compton four years earlier therefore, Kronig was fully *au courant* with the

[20] For a detailed analysis of the experimental program that led Compton from a large-ring electron model to a large-sphere electron model to the discovery of the Compton electron, see Stuewer (1975).

intricacies of Zeeman spectroscopy, and it was consequently very natural that he should wish to visit Tübingen, where two masters of Zeeman spectroscopy, Paschen on the experimental side, and Landé on the theoretical, were resident at the time.

On his arrival in Tübingen, Kronig was informed that Pauli was expected on a visit two days later. More significantly, Landé showed Kronig a letter (Pauli 1979, Doc. 71) he had recently received from Pauli outlining the essential results of the exclusion principle paper (not yet in print). Pauli's assignment of four quantum numbers to each individual electron, and in particular, the reassignment of the fourth quantum number from the atomic core to a property of the electron itself, made an immediate impact on Kronig, as he recalled thirty five years later in his essay "The turning point," written for a memorial volume to Pauli (Fierz and Weisskopf 1960). As Kronig says, "it occurred to me immediately that it [i.e., the fourth quantum number] might be considered as an intrinsic angular momentum of the electron" (Kronig 1960, p. 20). He immediately set to work to rederive the doublet splitting formula for inner electrons associated with X-ray transitions on the basis of this new physical picture.

We saw above that Landé, in the Spring of 1924, argued forcefully for the interpretation of the relativistic X-ray doublets as analogous to optical doublets. In other words, the splitting of these level pairs, which spectroscopic evidence indicated had the same azimuthal quantum number (i.e., orbital angular momentum), was to be attributed to the same sort of interaction of the core magnetic moment with the inner magnetic field produced by the valence electron that had been used successfully in the Heisenberg core model to calculate the splittings of optical alkali doublets (cf. Section 7.1), rather than due, as originally proposed by Sommerfeld, to a relativistic fine-structure effect (which required the two levels to correspond to different angular momenta of the electron, with different orbital eccentricities). The problem was that this core–inner magnetic field picture, extended to inner electrons now experiencing a large shielded nuclear charge $Z_{\rm eff}$, led to a splitting which grew like $Z_{\rm eff}^3$, rather than the empirically verified $Z_{\rm eff}^4$, which, we recall (cf. Eq. (6.64)), emerges inescapably from the relativistic fine structure calculation, and was widely regarded as a great success of the old quantum theory. Kronig saw that the correct, fourth power, dependence on the (effective) nuclear charge emerged immediately in the new physical picture, where the level splittings were due to the interaction of the electron's intrinsic magnetic moment (due to its spin) with the magnetic field experienced by the electron due to its motion relative to the electrostatic field of the nucleus.

We can easily reproduce Kronig's result, as the calculations involved are fairly elementary. It was by now clear that the core magnetic moment, now transferred to the electron, was associated with an angular momentum of $\hbar/2$ but a magnetic moment of twice the classically expected value (Pauli's "double magnetism of the core"): thus, the electron should now be assigned a magnetic moment equal to a full Bohr magneton,

$$\mu_B = 2 \times \frac{e}{2mc} \times \frac{\hbar}{2} = \frac{e\hbar}{2mc}. \tag{9.6}$$

An electron moving with velocity $\vec{v}$ in a static electric field $\vec{E}$ experiences a magnetic field[21]

$$\vec{H} = \frac{1}{c}\vec{E} \times \vec{v}. \tag{9.7}$$

For an inner electron subject to an effective screened nuclear charge Z_{eff}, the electric field is that due to the nucleus, $\vec{E} = (eZ_{\text{eff}}/r^3)\,\vec{r}$, and this magnetic field becomes

$$\vec{H} = \frac{1}{c}\frac{eZ_{\text{eff}}}{r^3}\,\vec{r} \times \vec{v} = \frac{eZ_{\text{eff}}}{mcr^3}\,\vec{L}, \tag{9.8}$$

where $\vec{L} = m\vec{r} \times \vec{v}$ is the orbital angular momentum of the electron, of magnitude $k\hbar$ for an orbit of type n_k. The instantaneous "spin–orbit" energy splitting between the two states in which the magnetic moment of the electron is aligned or anti-aligned with this magnetic field is evidently

$$\Delta E_{\text{s.o.}} = 2\mu_B|\vec{H}| = \frac{e\hbar}{mc} \cdot \frac{eZ_{\text{eff}}}{mcr^3} \cdot k\hbar = \frac{ke^2\hbar^2 Z_{\text{eff}}}{m^2c^2}\frac{1}{r^3}. \tag{9.9}$$

The measured energy splitting should correspond to a time average of this quantity over an orbital cycle: for eccentric orbits with $k < n$, this requires a simple integral over the elliptical path of the electron, which is found to give

$$\langle\frac{1}{r^3}\rangle = \frac{1}{a^3(1-\epsilon^2)^{3/2}}, \tag{9.10}$$

where $a = (n^2/Z_{\text{eff}})\,a_0$ (with $a_0 = \hbar^2/me^2$ the Bohr radius for hydrogen) is the semi-major axis of the orbit, and ϵ the eccentricity, given by $\sqrt{1-\epsilon^2} = k/n$. Inserting Eq. (9.10) in Eq. (9.9), one finds

$$\Delta E_{\text{s.o.}} = Z_{\text{eff}}^4 \frac{me^8}{\hbar^4 c^2}\frac{1}{k^2n^3} = Z_{\text{eff}}^4 \frac{\alpha^4}{k^2n^3}mc^2. \tag{9.11}$$

The desired fourth power behavior for the relativistic X-ray doublets is clearly seen to emerge. The fact that the magnetic moment responsible for the splitting has been transferred from the core to the electron, which sees a magnetic field Z_{eff} times larger (due to motion relative to the nuclear electric field) than that seen by the atomic core in the

[21] This formula, completely familiar from modern-day relativistic treatments of electromagnetic theory, was not widely known at the time—for example, by Goudsmit and Uhlenbeck (Mehra and Rechenberg 1982a, p. 702) or even Bohr (Mehra and Rechenberg 2000–2001, p. 121). The same result is obtained by going to the rest frame of the electron, in which the circulating nuclear charge produces a magnetic field easily calculated by the Biot–Savart Law. In this approach, the analogy to Heisenberg's original derivation of doublet splitting in the core model becomes clear (see Section 7.1, Eqs. (7.21)–(7.27)).

previous model due to the circulation of an outer electron (of *unit* charge), supplies precisely the extra factor of nuclear charge needed to restore at least qualitative agreement with the data, with one annoying discrepancy.

Unfortunately, Kronig's formula (our Eq. (9.11)) runs into quantitative difficulties[22] by giving too large a result, by about a factor of 2.[23] As we show later, the correct quantum mechanical calculation of the spin–orbit splitting gives a very similar result to Eq. (9.11), with two differences: (a) the factor k^2 must be replaced by $k(k-1)$ (or $l(l+1)$, where $l = k-1$ is the orbital angular momentum quantum number in quantum mechanics); and (b) there is an overall missing factor of $\frac{1}{2}$, due to a subtle relativistic precession effect (the Thomas precession, discussed below). The second factor in particular was to prove a real obstacle to the full acceptance of the electron spin hypothesis until its origin was finally explained by L. H. Thomas, in a paper published in April 1926 in *Nature* (Thomas 1926). In any event, in Tübingen, Kronig's proposal met with a cold reception from Pauli on his arrival the next day in Tübingen, who complimented Kronig on his "cute idea" (*witziger Einfall*) but was adamant that it could not correspond to reality. Landé (who appears to have been initially enthusiastic) seems to have adopted Pauli's negative attitude in a later conversation alone with Kronig: the net result was that the discoverer of electron spin was discouraged from publishing anything on the subject that could later be regarded as establishing his "official priority" as the originator of one of the most critical ideas in the development of modern quantum theory.

There is a very interesting passage in a letter from Pauli to Kronig of October 9, 1925 (Pauli 1979, Doc. 100)—nine months after their meeting in Tübingen—in which the concept of electron spin seems to be present in all but name. The ideas presented in this letter are to some extent an amplification of the concept of "two-valued-ness" (*Zweideutigkeit*), which Pauli had introduced two years earlier in a letter to Landé of September 23, 1923 (Pauli 1979, Doc. 46). In this letter, Pauli had suggested that, in the treatment of the anomalous Zeeman effect, "every [angular] momentum is represented not by a single number, but by a pair of numbers: the momenta appear in a certain sense to be doubly valued" (Pauli 1979, Doc. 46, p. 122). In the letter to Kronig two years later, Pauli is describing a "substitute model" (*Ersatzmodell*) in terms of which a recent paper of Heisenberg's (1925b) on the anomalous Zeeman effect, in which Heisenberg had proposed various "schemata" (for interpreting the complex numerological mumbo-jumbo that had evolved to "explain" multiplets and Zeeman lines), could be reduced to a single unified framework. Pauli proposes the following:

> In general this point of view leads to the introduction of *a single* substitute model for an arbitrary atom, which contains all the Heisenberg schemata, and further amplifies these in regard to the question of connections between different systems of terms [i.e.,

[22] In his article for the Pauli memorial volume, Kronig (1960, p. 21) claims to have noticed the difficulty with getting too large a result immediately, on the day before Pauli's arrival.

[23] Of course this difficulty could be avoided by abandoning the factor of 2 assumed for the gyromagnetic ratio of the electron—a descendant of the double magnetism of the core— but only at the expense of destroying the understanding of the Zeeman phenomenology.

> energy levels]. This substitute model consists in associating to *every* electron *two* angular momentum vectors i and $\bar{k}$, of which i has the magnitude $\frac{1}{2}$, while $\bar{k}$ is valued as above [i.e., 0 for s, 1 for p, etc.—in other words precisely the modern angular momentum quantum number l] ... (for the Zeeman effect) one has to add to the interaction energies between the angular momenta of the electrons also magnetic energies proportional to the cosine of (the angle of) these vectors with the magnetic field direction. The magnetic "anomaly" is expressed in the fact that the vectors i of the electrons have a double magnetism, while the vectors k are magnetically normal. In very strong fields, when the electrons are independent of one another, every vector i or $\bar{k}$ has quantized components along the field direction, in particular
>
> $$m_i = +\tfrac{1}{2} \text{ or } -\tfrac{1}{2}; \quad m_{\bar{k}} = 0, \pm 1, \pm 2, \ldots, \pm \bar{k} \tag{9.12}$$
>
> The resulting angular momentum component parallel to the field is the sum over all electrons, the resulting magnetic component is $\sum(2m_i + m_{\bar{k}})$ (Pauli 1979, Doc. 100, p. 245).

With the attribution of the angular momentum i to an intrinsic electron spin (and it is very difficult to see what other physical quantity this could be referring to), as Kronig had suggested to Pauli just nine months earlier, we have precisely the treatment of the Paschen–Back effect in modern quantum mechanics. But there is not a single mention of Kronig's earlier suggestion either in this letter of Pauli's (to Kronig!) or in any other letters in Pauli (1979) for the period from January to November of 1925, when the paper of Uhlenbeck and Goudsmit appeared. There evidently remained, in the Fall of 1925, an extreme reluctance on Pauli's part to accept an explicit identification of the "vectors i of the electrons" (with their "double magnetism") with an intrinsic angular momentum of said electrons.

The formal public introduction of the electron spin hypothesis instead took place roughly 10 months after Kronig's unfortunate interaction with Pauli in Tübingen, this time in a short communication to *Die Naturwissenschaften* by two young Dutch physicists, Samuel Goudsmit and George Uhlenbeck. We saw earlier that Kronig, on his visit to Leyden, had in fact collaborated in late 1924 with Goudsmit on the intensities of Zeeman lines. One therefore is entitled to wonder whether Kronig may not have communicated his ideas on electron spin to Goudsmit at some point subsequent to the Tübingen visit. He certainly discussed the idea on his visit to Copenhagen (after visiting Göttingen and Berlin) with Kramers and Heisenberg. In particular, Heisenberg was certainly aware of Kronig's derivation of the doublet splitting formula, with its encouraging Z^4 dependence and equally discouraging extra factor of 2, and he immediately wrote to Goudsmit asking about the latter discrepancy the day after the *Naturwissenschaften* article appeared. However, there is no evidence that Kronig communicated the idea of electron spin directly to his erstwhile collaborator back in the Netherlands at any time between January and October of 1925.

In any event, to judge from the subsequent reminiscences of both men, the considerations that led Goudsmit and Uhlenbeck to entertain the notion of a spinning electron in late September of 1925 appear remarkably similar to those of Kronig in early January.

The sequence of events appears to have been something along the following lines.[24] In the summer of 1925, Goudsmit was dividing his time between Ehrenfest's group in Leyden and a stint as Zeeman's assistant in Amsterdam. Uhlenbeck, who had previously studied in Leyden but had interrupted his studies to go to Rome as a tutor for the children of the Dutch ambassador there, returned to the Netherlands and was assigned to Goudsmit by Ehrenfest for a crash course in atomic physics. Goudsmit had been immersed for years in the empirical intricacies of atomic spectroscopy, but his general knowledge of classical physics was quite sketchy. For Uhlenbeck, who was largely self taught, the problem was the opposite: he had a quite sophisticated grasp of general theoretical physics, but very limited acquaintance with the continuing flood of empirical knowledge being provided by the optical and X-ray spectroscopists.

In the course of his private lectures to Uhlenbeck, Goudsmit explained the new classification of electron states in terms of four quantum numbers, which had appeared in January in the (by now, widely read) paper of Pauli on the exclusion principle. For Uhlenbeck, four quantum numbers had to mean four *mechanical* degrees of freedom,[25] and, following Pauli, as all four degrees of freedom now had to be associated with the electron itself, the conclusion was inevitable that the fourth attribute (apart from the three spatial degrees of freedom) should be associated with an intrinsic rotation of the electron. In later reminiscences, both authors agreed that this step was taken by Uhlenbeck. But the realization that this simple assumption immediately clarified many previously mysterious features of complex spectra and the Zeeman effect was Goudsmit's contribution.

An essential feature of the electron spin hypothesis, unavoidable if the Paschen–Back phenomenology was to be understood, with its "double magnetism of the core", was that the spin of the electron (half integral to produce the half integral total angular momenta after combination with integral orbital angular momenta) had to be associated with twice the classically expected magnetic moment: in other words, the electron, with intrinsic angular momentum $\hbar/2$, had to nevertheless possess a magnetic moment of a full Bohr magneton (just as Kronig had assumed in his calculation of the spin–orbit splitting). When Uhlenbeck discussed this with Ehrenfest, he was referred to Abraham's 1902 paper in which a non-relativistic model of a rotating spherical electron with charge restricted to the surface in which the gyromagnetic ratio (magnetic moment/angular momentum) indeed turned out to have twice the classical value. The upshot of these discussions was that Ehrenfest encouraged the pair of young physicists to write a short note for submission to *Die Naturwissenschaften* summarizing their proposal. This was

[24] The highly abbreviated recounting of events presented here depends to some extent on the personal reminiscences of several of the participants, mainly Kronig (1960), Goudsmit (1971, 1976), and Uhlenbeck (1976). The article by van der Waerden (1960) in the Pauli memorial volume is also a useful resource, especially in its discussion of Pauli's resistance to the spin idea. Although there are slight variations in the detailed chronology of some of the events, the basic sequence of events is consistent in the several retellings of the discovery of spin. The discovery of spin is also covered in Mehra and Rechenberg (1982a, sec. VI.4, pp. 684–709).

[25] Uhlenbeck was a real aficionado of Boltzmann's statistical mechanics, so the concept of degrees of freedom, and the mechanical correlates thereof, was second nature to him. See his reminiscences in *Physics Today* (Uhlenbeck 1976).

duly done, and the short letter (one page in *Die Naturwissenschaften*) was handed to Ehrenfest for forwarding to the journal.

The paper of Uhlenbeck and Goudsmit (1925), which is commonly agreed to mark the permanent introduction of electron spin into quantum physics, almost underwent the fate of Kronig's ideas nine months earlier in Tübingen. In this case, the criticism came from an unexpected direction—the eminent theoretical physicist H. A. Lorentz, whom Uhlenbeck asked for assistance on the subject of electron models of the Abraham type. Lorentz, the founder of classical electron theory, had not surprisingly undertaken much more detailed calculations of the properties of such electron models, but had dealt with them fully relativistically, and the conclusion was inescapable that for a model of the Abraham type to yield the right angular momentum and gyromagnetic ratio, the surface velocity of the charge on the rotating electron would have to be an order of magnitude greater than the speed of light. Understandably, once Lorentz had shown Uhlenbeck these results, the latter immediately got cold feet and went to Ehrenfest, asking him to return the incriminating article. Fortunately for Uhlenbeck and Goudsmit, Ehrenfest replied that he had already sent it in for publication, adding the immortal comment "you are both young enough to allow yourself a stupidity!" (*Sie sind beide jung genug um sich eine Dummheit leisten zu können!*).

The paper of Uhlenbeck and Goudsmit, with the cumbersome title "Replacement of the hypothesis of unmechanical stress [*unmechanischen Zwang*] with a requirement regarding the inner behavior of each separate electron," was published in the November 20, 1925 issue of *Die Naturwissenschaften* (Uhlenbeck and Goudsmit 1925). After reviewing Pauli's objections to the standard core–vector model, and his arguments (in the exclusion principle paper) in support of the assignment of a fourth quantum number to each electron, Uhlenbeck and Goudsmit get to the point:

> It is now necessary to grant each electron, with its four quantum numbers, also four degrees of freedom. One can, for example, give the following interpretation to the quantum numbers:
>
> n and k remain, as previously, the principal- and azimuthal quantum numbers of each electron in its orbit.
>
> R [the Landé "Rumpf", or core, quantum number, equal to twice the spin] is to be associated with the rotation of the electron itself.
>
> The other quantum numbers maintain their old interpretation ... The electron must now display the still not understood property (of double magnetism), which Landé has ascribed to the atomic core (Uhlenbeck and Goudsmit 1925, p. 954).

On November 21, the day after the Uhlenbeck and Goudsmit paper appeared in *Die Naturwissenschaften*, Heisenberg dashed off two letters, to Goudsmit and Pauli, respectively, that raised the question of the unwanted factor of 2 in the spin–orbit coupling energy of an electron with a magnetic moment. Uhlenbeck and Goudsmit had not performed the calculation that Kronig had done immediately nine months earlier on the basis of a magnetic electron—in fact, Goudsmit later recalled that, even after the possibility of calculating this quantity was pointed out to them, they had no idea how to do the calculation. Thus, they were unaware that their new proposal had solved the problem of the Z^4 dependence of X-ray relativistic doublet splittings, but with a result roughly two times too big in magnitude. Heisenberg is clearly aware of Kronig's result (as indicated previously, he may in fact have been told directly by Kronig, who had spent several months in Copenhagen at the same time as Heisenberg earlier in the year), as he quotes the formula in his letter to Pauli. Without giving any formulas, Heisenberg asked Goudsmit if he (a) knew how to get rid of the factor of 2; and (b) whether he understood how to reconcile the apparent lack of relativistic effects (i.e., *à la* the Sommerfeld 1916 calculation) *in addition to* the spin–orbit splitting in the hydrogen spectrum.[26] In his letter of November 21 to Pauli, Heisenberg quotes the Kronig formula without derivation:

> I would like to hear what you, as a master of criticism, think about Goudsmit's note in *Naturwissenschaften.* In my opinion it is difficult for Goudsmit to get the relativistic formula; rather he would clearly get classically
>
> $$\Delta\nu\, h \left[= \frac{1}{c}\Delta E_{\text{s.o.}} \right] = \frac{RhZ^2}{n^3}\frac{\alpha^2 Z^2}{k^2} \cdot 2. \tag{9.13}$$
>
> [Here, R is the Rydberg constant, hence, $Rh = \frac{1}{2}\alpha^2 mc$, so this formula coincides with Eq. (9.11).] That k^2 appears in the denominator appears instead of $k(k-1)$ would not disturb anyone; but I just cannot get rid of the factor of 2 ... It is cute [*witzig* (cf. Pauli's reaction to Kronig's idea)] in Goudsmit's model that he can now get the Z^4 Law with magnetism, and for that reason I am not a-priori hostile to the G[oudsmit] theory. But I would like to hear your objections (Pauli 1979, Doc. 107).

Unfortunately, we do not have Pauli's response to this letter, in which Heisenberg essentially outlines the same story that Pauli had dismissed after Kronig proposed it in Landé's house in early January. But Pauli seems to have asked for a more explicit derivation of Eq. (9.13), for the requisite algebra (more or less along the lines given above for Eq. (9.11)) is provided three days later in another letter to Pauli (Pauli 1979, Doc. 108). One is inclined to conclude that Pauli had not even given poor Kronig the chance to explain his derivation of the Z^4 dependence in January, instead dismissing the electron–spin proposal out of hand.

The noisome factor of 2 was finally explained by L. H. Thomas, a Cambridge graduate visiting Copenhagen in the winter of 1925–1926. by the end of December, Bohr had become convinced of the truth of the Uhlenbeck–Goudsmit theory (after conversations in Leyden with Einstein who pointed out the inevitability on relativistic grounds of a spin–orbit interaction energy in this theory), so the problem of the extra factor of 2 was presumably an urgent matter of discussion at Bohr's institute. Thomas's (1926) paper, submitted to *Nature* in late February of 1926, explained the unwanted factor of 2 by pointing out that, due to a subtle relativistic effect arising from the acceleration of the

[26] As we will discuss in Chapter 15, with the full apparatus of quantum mechanics, and in particular with the correct identification of the orbital quantum number l, taking values $0, 1, 2, \ldots$, the intricate cooperation of kinematic relativistic and spin–orbit effects needed to yield the strange dichotomy of screening and relativistic doublet behavior in X-ray spectra was finally understood, at least for the states with non-zero l. The accidental degeneracy of states of the same total angular momentum j in hydrogenic atoms, and the gyromagnetic ratio of 2 for the electron, as well as the correct energy levels for the s ($l = 0$) states, was only fully understood with the relativistic Dirac treatment of the electron, in the famous Dirac equation.

electron, there was, in addition to the precession of the electron spin due to the magnetic field $\vec{H} = (1/c)\,\vec{E} \times \vec{v}$, a supplemental kinematical precession of one-half the size and opposite sign (now referred to as the "Thomas precession"), which effectively reduced the coupling energy by a factor of one half.[27]

As can be inferred from letters from Bohr of late 1925 and early 1926 (cf. Pais 1986, p. 279),[28] Pauli initially did not like Uhlenbeck and Goudsmit's version of the electron–spin idea any better than Kronig's: he referred to the idea as "a new Copenhagen heresy" (*eine neue Kopenhagener Irrlehre*).[29] He made his peace with the idea after Thomas had removed the discrepancy of a factor 2 but still felt that it was too closely attached to naive classical conceptions. He would always maintain—quite correctly—that electron spin is an intrinsically quantum-mechanical concept without a classical correlate, as opposed to orbital angular momentum for which a classical correspondence limit clearly exists, as one may obviously consider the case of large orbital angular momentum quantum numbers l (or k) $\gg 1$, whereas the spin angular momentum is, as it were, "stuck" at the intrinsic value $\frac{1}{2}(\hbar)$. In his Nobel lecture of December 13, 1946, Pauli endorsed a more controversial argument of Bohr's for the quantum nature of spin:

> Although at first I strongly doubted the correctness of this idea because of its classical mechanical character, I was finally converted to it by Thomas' calculation [Thomas 1926] on the magnitude of the doublet splitting. On the other hand, my earlier doubts as well as the cautious expression "classically non-describable two-valuedness" experienced a certain verification during later developments since Bohr was able to show on the basis of wave mechanics that the electron spin cannot be measured by classically describable experiments (as, for instance, deflection of molecular beams in external electromagnetic fields) and must therefore be considered as an essentially quantum mechanical property of the electron[30] (Pauli 1947, p. 1083; the same passage can also be found in Pauli 1946, pp. 1074–1075).

An amusing postscript to the tale of Kronig's unfortunate interaction with Pauli in early 1925 is provided by a letter of Thomas to Goudsmit of March 25, 1926, reproduced in part in Goudsmit (1971). Thomas writes:

> I think you and Uhlenbeck have been very lucky to get your spinning electron published and talked about before Pauli heard of it. It appears that more than a year ago Kronig believed in the spinning electron and worked out something; the first person he showed it to was Pauli. Pauli ridiculed the whole thing so much that the first person became also the last and no one else heard anything of it. Which all goes to show that the infallibility of the Deity does not extend to his self-styled vicar on earth (Goudsmit 1971).

[27] For a simple derivation of this factor, see the web resource, *Thomas Precession*.

[28] See Bohr to Ehrenfest, December 22, 1925 (Bohr 1972–2008, Vol. 5, p. 329) and Bohr to Pauli, February 20, 1926 (Pauli 1979, Doc. 121).

[29] The phrase comes from Bohr's letter to Pauli cited in the preceding note. The old Copenhagen heresy was presumably the BKS theory (Bohr, Kramers, and Slater 1924a, see Section 10.4).

[30] Pauli refers to a section in his contribution to the proceedings of the sixth Solvay conference held in Brussels in October 1930 (Pauli 1932, pp. 217–225). For (discussion of) related manuscripts on the "magnetic electron" by Bohr and Pauli, see Bohr (1972–2008, Vol. 6, pp. 305–349).

10

Dispersion Theory in the Old Quantum Theory

In parallel to the crucial developments outlined in Chapter 9, leading to the introduction of the exclusion principle and the discovery of electron spin, attempts were made to bring the phenomenologically successful but entirely classical treatment of optical dispersion going back to Lorentz and Drude under the aegis of the basic principles of the old quantum theory. Results of these efforts were to play a critical role in the reinterpretation of classical concepts employed by Heisenberg in his seminal work of 1925. In order to trace this development accurately, we need to review in some detail the change in understanding of the basic physics of dispersion of light in transparent media over the century leading up to the appearance of quantum mechanics in the 1920s.

10.1 Classical theories of dispersion

The spreading (or *dispersion*) of a ray of white light passed through a prism into a spectrum of visible rays of different colors was famously interpreted by Newton in 1666 as implying (a) the composite nature of white light, as a superposition of homogeneous components corresponding to each color in the spectrum; and (b) a varying *refrangibility* (degree of refraction, or bending) for the different colors of the spectrum, with the light rays at the red end of the spectrum the least, and those at the violet end the most, refrangible. In terms of the laws of refraction known from the work of Snell and Descartes, this amounts to a dependence of the index of refraction of the transparent material of the prism on the color of the light.

Until the emergence of the wave theory in the early nineteenth century, this dependence could not be expressed in a quantitatively precise way, although there were several attempts to do so on the basis of Newtonian corpuscular theories of light (see, e.g., Whittaker 1951, Vol. 1, p. 99). Once the idea that rays of light of a definite color (*homogeneous* rays) corresponded to waves of a definite wavelength λ and frequency ν (traveling with a speed $\lambda\nu$) took hold, one had the prerequisites for developing a precise physical theory of the phenomenon of optical dispersion. In particular, an adequate theory of optical dispersion would have to furnish a specific functional form for the dependence $n(\lambda)$

Constructing Quantum Mechanics. Anthony Duncan and Michel Janssen, Oxford University Press.
 DOI: 10.1093/oso/9780198883906.003.0010

(or $n(\nu)$) of the index of refraction on wavelength (or frequency) of the light for any given transparent medium, and a dynamical explanation for the variation in the propagation of light in that medium (as compared to *in vacuo*) implied by this functional form.

The history of attempts to develop a quantitative understanding of optical dispersion in the nineteenth century is somewhat complicated.[1] One may distinguish three main phases in the development of theories of dispersion in the period following the broad acceptance of the wave theory, as a result of the work of Thomas Young in the first decade of the nineteenth century. In the first stage, associated with the names of Fresnel, Cauchy, MacCullagh, and others, variations in the behavior of light traversing different transparent media were attributed to a variation in the character—either the density or the rigidity—of the aether (the universally premised carrier of the oscillatory motion constituting light) instigated by the presence of the material particles composing the transmitting medium. Considerations of this sort, augmented by explicit consideration of the coarse-grained character of the transmitting medium (see Whittaker 1951, Vol. 1, p. 123), even led to phenomenologically successful descriptions of dispersion. Cauchy, for instance (cf. Section 7.4, Eq. (7.111)), was able to derive a formula for the index of refraction as an expansion in the inverse squared wavelength,

$$n = 1 + A\left(1 + \frac{B}{\lambda^2} + O(1/\lambda^4)\right). \tag{10.1}$$

Such a formula accounted for the "normal" dispersion of many materials, in which the index of refraction decreased smoothly and monotonically with increasing wavelength.

In the second stage, associated with the names of Sellmeier, Kundt, Helmholtz, and others and lasting roughly from the late 1860s to 1890, the oscillations in the aether corresponding to light passing through a medium were dynamically coupled to the constituent matter particles of the medium, which were harmonically bound and subject to elastic restoring forces when displaced from their equilibrium position (in the absence of transient light). The effect of light passing through a transparent material could therefore be reduced to a study of the forced harmonic oscillations of massive point particles vibrating synchronously (hence, the German term, *Mitschwingungen*, or *co-vibrations*) with the aether wave traversing the material. An important and detailed theory of this type was developed by Sellmeier in 1872. Elementary physics tells us that the response of an undamped harmonic oscillator to a harmonic (sinusoidal) external force displays a resonance pole of the form $1/(\nu_0^2 - \nu^2)$ when the frequency ν of the external force (here, the light wave) equals the natural frequency ν_0 of oscillation of the matter particle (see Sections 10.1.1–10.1.2). Using energetic arguments, Sellmeier could relate the induced forced displacements of the material particles to the index of refraction, with the result that (with the assumption that the material contained a variety of matter particles

[1] For a more detailed account and references to the extensive primary and secondary literature, see Jordi Taltavull (2017). A useful contemporary account can be found in a review article on wave optics for the *British Association for the Advancement of Science* (Glazebrook 1886). Our brief account follows Duncan and Janssen (2007, sec. 3.1, pp. 573–576). For a concise history of dispersion theory from Descartes' rainbows and Newton's prisms to the end of the nineteenth century, see, for example, Darrigol (2012).

with different natural frequencies and densities) the squared index of refraction could be expressed as a sum of resonance poles

$$n(\nu)^2 = 1 + \sum_k \frac{C_k}{\nu_k^2 - \nu^2}. \tag{10.2}$$

For materials in which the resonance frequencies ν_k were all in the ultraviolet, and for light in the optical range with $\nu < \nu_k$, each resonance pole in the Sellmeier sum can be expanded

$$\frac{1}{\nu_k^2 - \nu^2} \simeq \frac{1}{\nu_k^2} + \frac{\nu^2}{\nu_k^4} + \ldots, \tag{10.3}$$

giving the expansion in powers of the squared inverse wavelength (as $\nu^2 = c^2/\lambda^2$, with c the speed of light) proposed by Cauchy, which was known to agree well with experiment.

The Sellmeier formula Eq. (10.2) was modified within a few years (in 1875) by Helmholtz to include a frictional damping term that has the salutary effect of introducing an additional (imaginary) term in the denominator, protecting the index of refraction from developing an unphysical infinity when the light frequency ν equals the natural frequency ν_i of one of the material oscillators. It was immediately recognized to provide a convincing explanation—and quantitative description—of the phenomenon of *anomalous dispersion*: the existence of materials in which the index of refraction was found to decrease with increasing frequency over some frequency interval (within the optical range). Leroux (1862) had already determined quantitative values for the index of refraction for iodine vapor and shown that red light was more, rather than less, refrangible than blue light. These results were confirmed and extended by the Danish physicist and Bohr's PhD advisor Christian Christiansen (1870, 1871) for fuchsine solutions (Whittaker 1951, Vol. 1, p. 261; for more detailed discussion, see Jordi Taltavull 2017, sec. 2.2.1). Sellmeier's formula provided an immediate explanation of the phenomenon, due to the change in sign of a resonance term as one goes from frequencies ν below to those above the resonance frequency ν_i of an oscillator.[2] In most solid transparent media the resonance frequencies are in the ultraviolet so one does not cross any resonance pole in traversing the entire range of the optical spectrum, giving the monotone behavior of the index of refraction characteristic of normal dispersion.

The third, and final, phase of classical dispersion theory arrived in the early 1890s. By this time, following Maxwell's great work unifying electric and magnetic phenomena, and his discovery that electromagnetic waves propagated at the speed of light, it was becoming clear that the influence of a medium on the transmission of light passing through it had to involve the excitation of charged particles ("mobile ions") in the medium by the oscillating electric field of the light wave.[3] The previous purely mechanically interpreted *Mitschwingungen* of Sellmeier were now understood in terms of the

[2] It was presumed that there would be numerous oscillators, capable of oscillations at a variety of natural frequencies, hence, the index i.

[3] For discussion, see, for example, Buchwald (1985, sec. 27.2) and Darrigol (2000, sec. 8.3).

forced oscillations of charged particles subjected to an oscillating electric field.[4] The oscillations of a charged particle would, according to Maxwellian theory, result in the emission of electromagnetic radiation coherent (i.e., of the same frequency and in a fixed phase relation) with the incident light. Theories of this type (differing in some details, but fundamentally in agreement on the basic physical mechanism) were developed in the 1890s by Helmholtz (1892, 1893) and Lorentz (1892), based on an earlier electromagnetic theory of dispersion by Lorentz (1878, see Jordi Taltavull 2017, sec. 2.3.3.1). This approach to the treatment of optical dispersion was further developed and popularized by Drude (1900) in *Lehrbuch der Optik*, his influential textbook on optics.[5] The classical Lorentz–Helmholtz–Drude approach to dispersion therefore begins with an examination of a charged harmonically bound particle influenced by an external oscillating electric field.

Before turning to a detailed explanation of the treatment of dispersion in terms of charged harmonic oscillators, it may be worthwhile to explain to the modern reader, inevitably imbued with the "planetary" picture of atoms as miniature solar systems with charged electrons circulating around a positive nucleus, why the idea of harmonically bound charged particles in atoms or molecules seemed (in the 1890s) not only plausible, but indeed inescapable. The evidence of spectroscopy was clear: excited atoms or molecules emitted light with a discrete spectrum consisting of a superposition of homogeneous components, that is, light of definite wavelength and frequency. The only mechanical system known to exhibit vibrations of a definite frequency, independent of the amplitude of the oscillation, was the simple harmonic oscillator. Any other classically bound system would necessarily emit a continuous spectrum, as the frequency of the oscillation would vary with its amplitude. It was therefore very natural for people to adopt a model in which the charged "mobile ions" (by the mid-1890s, "electrons") responsible for dispersion should be able to vibrate harmonically around equilibrium positions in the atoms/molecules containing them. A quote from a short article by Arthur Schuster in *Nature* in 1895 makes this point of view clear:

> If a molecule is capable of sending out a homogeneous vibration, it means that there must be a definite position of equilibrium of the "electron". If there are several such positions, the vibrations may take place in [i.e., with] several periods ... The probability of an electron oscillating about one of its positions of equilibrium need not be the same in all cases. Hence a [spectral] line may be weak not because the vibration has a smaller amplitude, but because fewer molecules give rise to it (Schuster 1985).[6]

[4] The Lorentz force induced by the magnetic field of the electromagnetic wave on the moving charged particles could be neglected in a first approximation, provided the velocity of the particles was much smaller than the speed of light.

[5] The writing and reception of Drude's book are analyzed in Jordi Taltavull (2013).

[6] The term "electron" (initially "electrine") had been introduced by George Johnstone Stoney as a universal basic quantity of electricity in 1874. The need for such an "atom of electricity" was further championed by Helmholtz as a natural explanation of the phenomena of electrolysis. Talk about electrons, as charged particles which were constituents of atoms, clearly preceded by two decades the discoveries of J. J. Thomson in the late 1890s (see Keller 1983, p. 34).

The term "dispersion electrons", which becomes ubiquitous in the discussion of optical dispersion in the following two decades, can be taken to refer to exactly the situation suggested here—electrons bound within atoms that (under presumably small displacements from equilibrium) execute harmonic vibrations of well-defined frequency when disturbed.

Let us now return to an outline of the classical dispersion theory, as it was developed up to 1913, when the introduction of the Bohr atomic model made quantal considerations indispensable in describing the interaction of light and matter.

10.1.1 Damped oscillations of a charged particle

First, we consider the simplest case possible—a particle of mass m moving in a harmonic potential in one dimension, subject to Newton's Law:

$$m\ddot{x} = -kx = -m\omega_0^2 x. \tag{10.4}$$

The spring constant k of the harmonic potential has been re-expressed in terms of $\omega_0 = \sqrt{k/m}$. As the general solution of this differential equation,

$$x_0(t) = \mathrm{Re}(A\,e^{i\omega_0 t}) = C\cos(\omega_0 t + \varphi) \quad (\text{with } A = Ce^{i\varphi}), \tag{10.5}$$

shows, ω_0 is the natural oscillatory (angular) frequency of the particle. We have introduced a subscript 0 to indicate that we are treating the case of an isolated oscillator, with no external (e.g., electric) forces applied. At this point, the amplitude C and phase φ are arbitrary real constants, determined by the initial conditions of the problem (e.g., x and $\dot{x}$ at time $t = 0$).

In fact, if we now assume our particle also possesses electric charge $-e$ (we anticipate the use of this formalism for negatively charged electrons by displaying the sign explicitly), then the acceleration $\ddot{x}$ of the particle results in the loss of energy by emitted electromagnetic radiation, with the instantaneous radiated power given by the Larmor formula,[7] which in Gaussian units takes the form

$$-\frac{dE}{dt} = \frac{2}{3}\frac{e^2\ddot{x}^2}{c^3}. \tag{10.6}$$

One can model the radiative energy loss phenomenologically by including a retarding frictional force $F_{\text{fric}} = -R\dot{x}$ on the right-hand side of Eq. (10.4), with the friction

[7] See, for example, Griffiths (1999, Eq. (11.61)).

constant R adjusted to lead to the average power loss per period implied by the Larmor formula. Thus, we now solve, instead of Eq. (10.4), the *damped harmonic equation*

$$m\ddot{x} = -m\omega_0^2 x - R\dot{x}. \tag{10.7}$$

Inserting here the usual complex Ansatz $x_0(t) = A\, e^{i\omega_1 t}$, one finds the quadratic equation,

$$m\omega_1^2 - iR\omega_1 - m\omega_0^2 = 0, \tag{10.8}$$

for ω_1, with the solution[8]

$$\omega_1 = \sqrt{\omega_0^2 - \frac{R^2}{4m^2}} + \frac{iR}{2m}. \tag{10.9}$$

It turns out (see Eq. (10.18)) that for problems of optical dispersion, with ω_0 corresponding to optical frequencies, the second term in the square root is *much* smaller than the first (by some sixteen orders of magnitude!), and we can therefore write more simply

$$\omega_1 = \omega_0 + \frac{iR}{2m}, \tag{10.10}$$

giving the general solution of Eq. (10.7) as

$$x_0(t) = \mathrm{Re}\left(Ae^{-Rt/2m\, +\, i\omega_0 t}\right) = Ce^{-Rt/2m} \cos\left(\omega_0 t + \varphi\right). \tag{10.11}$$

Unsurprisingly, the effect of the retarding frictional force is simply to introduce an exponential damping of the harmonic oscillations. The fact that, in cases of interest, $R/2m \ll \omega_0$ simply indicates that the fractional loss of amplitude due to the damping is far less than unity over a single period of the oscillation. The average energy $\overline{U}$ of the oscillator[9] over a cycle involves the square of the position $x_0(t)$ and velocity $\dot{x}_0(t)$ and therefore decays exponentially as

$$\overline{U}(t) = \overline{U}(0)e^{-Rt/m} = \overline{U}(0)e^{-t/\tau}, \tag{10.12}$$

where we introduced the decay time $\tau \equiv m/R$. The rate of loss of energy (i.e., energy radiated) by an oscillator at any given time is therefore

$$-\frac{d\overline{U}}{dt} = \frac{\overline{U}}{\tau}. \tag{10.13}$$

[8] We pick the sign in the quadratic formula to ensure $\omega_1 \to \omega_0$ as $R \to 0$.

[9] To maintain consistency with the notation frequently used at the time, we use U rather than E to denote oscillator energy.

The work done by the frictional force over a single period $T = 1/\nu = 2\pi/\omega$ is

$$\begin{aligned}\int_0^{2\pi/\omega_0} R\dot{x}_0^2\,dt &\approx RC^2 \int_0^{2\pi/\omega_0} \omega_0^2 \sin^2(\omega_0 t + \varphi)\,dt \\ &= \frac{1}{2} R\omega_0^2 C^2 \frac{2\pi}{\omega_0} + O(R^2),\end{aligned} \tag{10.14}$$

where we have neglected the small quantity R in the damping exponential in Eq. (10.11) in calculating the integral over a single period. The radiated energy emitted over one period is given the Larmor formula (Eq. (10.6)), which, in the same approximation, works out to

$$\frac{2}{3}\frac{e^2}{c^3} \int_0^{2\pi/\omega_0} C^2 \omega_0^4 \cos^2(\omega_0 t + \varphi)\,dt = \frac{1}{3}\frac{e^2\omega_0^4}{c^3} C^2 \frac{2\pi}{\omega_0}. \tag{10.15}$$

Equating the right-hand sides of Eqs. (10.14) and (10.15), we find

$$R = \frac{2}{3}\frac{e^2}{c^3}\omega_0^2. \tag{10.16}$$

This gives us an explicit expression for the decay time $\tau = m/R$ in Eq. (10.12):

$$\tau = \frac{3mc^3}{2e^2\omega_0^2}. \tag{10.17}$$

For typical optical transitions, $\hbar\omega_0$ is a few eV, $mc^2 \approx 500000$ eV (for an electron), and $e^2/\hbar c$ (Sommerfeld's fine-structure constant) $= 1/137$. Hence, the ratio[10]

$$\frac{R}{2m\omega_0} = \frac{1}{3}\frac{e^2}{mc^3}\omega_0 = \frac{1}{3}\frac{e^2}{\hbar c}\frac{1}{mc^2}\hbar\omega_0 \tag{10.18}$$

is typically of order 10^{-8} and completely negligible.

The phenomenological friction constant R is usually replaced by the related quantity

$$\Gamma \equiv R/m \tag{10.19}$$

with the dimension of frequency, which (for reasons that will shortly become clear) is called the "full width at half-maximum."

The fact that our damped oscillator involves an accelerated charged particle means that, according to classical electromagnetic theory, it will emit electromagnetic radiation

[10] It may appear somewhat peculiar to use Planck's constant in the calculation of the purely classical quantity τ but this allows us to use units (eV) familiar to the modern reader and to avoid having to insert explicit values for e, m, c, and ω_0.

(indeed, just this radiation is responsible for the damping factor introduced phenomenologically via the Larmor formula). We obtain the emission spectrum of the oscillator—the relative intensity of light of different frequencies ω—by performing a Fourier analysis of the coordinate of the particle $x_0(t)$ given in Eq. (10.11). According to classical theory, the intensity of emitted radiation at frequency ω is proportional to the square of the corresponding Fourier component. We imagine putting our oscillator in motion at time $t = 0$ and examine the radiation emitted subsequently. In other words, we write[11]

$$x_0(t) = Ae^{-\frac{1}{2}\Gamma t + i\omega_0 t}\,\theta(t) = \int f(\omega)\, e^{i\omega t}\, \frac{d\omega}{2\pi}, \tag{10.20}$$

so that

$$\begin{aligned} f(\omega) &= \int_0^{+\infty} e^{-i\omega t} A e^{-\frac{1}{2}\Gamma t + i\omega_0 t}\, dt \\ &= \int_0^{+\infty} A e^{-\left(\frac{1}{2}\Gamma + i(\omega - \omega_0)\right) t}\, dt \\ &= \frac{A}{\frac{1}{2}\Gamma + i(\omega - \omega_0)}. \end{aligned} \tag{10.21}$$

The emitted radiation will have an intensity profile proportional to the functional form

$$|f(\omega)|^2 = f(\omega) f^*(\omega) = \frac{|A|^2}{(\omega_0 - \omega)^2 + \frac{1}{4}\Gamma^2}, \tag{10.22}$$

which is called a "Lorentzian". If $\Gamma \ll \omega_0$, the function displays a sharp peak at frequencies close to the natural frequency of oscillation of our oscillator. The emitted light will therefore consist of a sharp (though not infinitely so) bright line centered on the natural oscillator frequency. The width of the line, or (more precisely) the *full width at half-maximum*, is easily seen to be just the quantity Γ associated with the radiation damping.[12]

What complicated the correct interpretation of empirical measurements of optical dispersion (and in particular of the shape of the spectral lines observed both in absorption and emission) in the first three decades following the introduction of the electromagnetic picture outlined above was that, in many cases, the natural damping due to loss of energy by radiation was swamped by an unrelated source of damping—the effect of random collisions between the molecule (or atom) containing the radiating electron and other molecules (or atoms). Such collisions would interrupt the smooth damped oscillatory decay indicated in Eq. (10.20) and inject a random phase into the subsequent oscillations. The net effect of frequent collisions of this kind were shown by Lorentz

[11] The theta function switches on the oscillations at time zero.

[12] One finds $|f(\omega_\pm)|^2 = |f(\omega_0)|^2/2$ for $\omega_\pm = \omega_0 \pm \Gamma/2$.

(1909, sec. 120) to be phenomenologically equivalent to an additional source of damping leading to an increase in the value of R (or Γ). The effect on the line shape of this supplementary source of damping is called "collisional broadening", and its presence (the magnitude of which varied with the temperature and density of the medium being studied) frequently made it difficult, in the early days of spectroscopy, to separate the universal component of the line width due purely to radiative damping, which was independent of the atomic environment.[13]

10.1.2 Forced oscillations of a charged particle

Electrons in realistic materials are, of course, able to move in all three spatial dimensions, so the Lorentz–Drude theory necessarily involved electrons bound harmonically in three dimensions to a force center, the positions of equilibrium. We make the simplifying assumption that the binding is isotropic (equal spring constant in all three dimensions). We now consider the response of such a harmonically bound electron to a monochromatic electromagnetic wave polarized in the x-direction, that is, one in which the electric field is in the x-direction, taking the value

$$E_x(t) = \mathcal{E}e^{i\omega t} \tag{10.23}$$

at the position of the electron. The angular frequency ω of the applied electromagnetic wave is freely tunable, and distinct from the natural frequency of oscillation ω_0 of the oscillator. As usual, we use a complex formalism here to simplify the solution of the forced harmonic equation that this field induces. We obtain the physical fields and particle displacements in the end by taking the real part (which must also be a solution as the equations of motion are linear and only involve real coefficients).

The presence of the electric field results in an additional force $-eE_x(t)$ in the x-direction acting on our negatively charged electron. Eq. (10.7), the x-component of equations of motion of the electron, thus picks up an extra term and becomes

$$m\ddot{x} + m\Gamma\dot{x} + m\omega_0^2 x = -e\mathcal{E}e^{i\omega t}, \tag{10.24}$$

where we used Eq. (10.19) to replace R with $m\Gamma$. As the equations for the y and z coordinates are unchanged, the additional displacement induced by the presence of the electric field vanishes for these components. For the x coordinate, the general solution of the inhomogeneous equation Eq. (10.24) takes the form

$$x(t) = x_0(t) + \Delta x_{\text{coh}}(t), \tag{10.25}$$

where $x_0(t)$ is just the general solution given in Eq. (10.11) of the homogeneous equation, Eq. (10.7), and the displacement $\Delta x_{\text{coh}}(t)$ is the shift in coordinate due

[13] The two sources of damping described here—radiative and collisional—also correspond to the two possible destinations of the light energy absorbed by a charged oscillator: either as reradiated electromagnetic energy, or into random thermal motion (i.e., heat) of the molecules with which it collides. Discussions of absorption spectra thus necessarily involve both modes.

(and directly proportional) to the applied electric field. This displacement must satisfy the full inhomogeneous equation

$$m\frac{d^2\Delta x_{\rm coh}}{dt^2} + m\Gamma\frac{d\Delta x_{\rm coh}}{dt} + m\omega_0^2\Delta x_{\rm coh} = -e\mathcal{E}e^{i\omega t}. \tag{10.26}$$

As we demonstrate shortly, this part of the solution is undamped and oscillates with exactly the applied frequency ω of the applied electric field. It is *coherent* with the applied field. Hence, the subscript "coh". It also corresponds to the "steady state" part of the solution, as the homogeneous solution $x_0(t)$ dies away exponentially in time.

To solve Eq. (10.26), we make the obvious Ansatz

$$\Delta x_{\rm coh}(t) = Be^{i\omega t}. \tag{10.27}$$

Inserting this in Eq. (10.26), one finds (canceling an overall factor of $e^{i\omega t}$)

$$B\left(-m\omega^2 + im\Gamma\omega + m\omega_0^2\right) = -e\mathcal{E}, \tag{10.28}$$

from which it follows that

$$B = \frac{e\mathcal{E}}{m}\frac{1}{\omega^2 - \omega_0^2 - i\Gamma\omega}. \tag{10.29}$$

With the determination of B, we have an explicit expression for the coherent contribution to the particle's displacement. An important aspect of the classical theory needs to be emphasized at this point: *while the homogeneous (and exponentially damped) part of the solution $x_0(t)$ depends on initial conditions* (through the two-parameter complex constant A in Eq. (10.11)), *the steady state coherent part $\Delta x_{\rm coh}(t)$ induced by the applied field is completely fixed by the properties of the oscillator and the applied field.* This means that the classical theory can be used to provide a complete treatment of the dispersion as well as the absorption spectrum of a medium consisting of Lorentz–Drude oscillators that depends only on the coherent response to an applied field. On the other hand, the classical theory is unable to deal with the *emission* (i.e., bright-line) spectrum of a set of radiating oscillators in the same unambiguous way, as this clearly requires a detailed understanding of the continuously variable way in which the oscillators were excited in the first place.[14]

[14] Remarkably, in quantum mechanics (post-1925) the restriction of allowed excited states of electrons to a discrete enumerable set (as opposed to the continuous infinity possible classically), together with a precise understanding of the initial and final states in an atomic transition, means that we are once again in a position to make specific quantitative predictions concerning not just the frequency, but also the intensity of bright emission lines from distinct excited states. See also the discussion of the restriction to Zeeman absorption lines in the classical Voigt theory, Section 7.2.2, p. 328.

The induced dipole moment vector $\vec{d}$ due to the applied electric field $\vec{E}$ is therefore in the same direction as the field (here, the x-direction) with magnitude

$$-e\Delta x_{\text{coh}} = -\frac{e^2}{m}\frac{1}{\omega^2 - \omega_0^2 - i\Gamma\omega}\mathcal{E}e^{i\omega t}. \tag{10.30}$$

The induced polarization $P(t)$ is defined as the induced dipole moment per unit volume. If there are $\mathfrak{N}_0$ oscillators per unit volume with natural oscillation frequency ω_0, the induced polarization is thus given by[15]

$$P(t) = \frac{e^2}{m}\frac{\mathfrak{N}_0}{\omega_0^2 - \omega^2 + i\Gamma\omega}\mathcal{E}e^{i\omega t} \equiv \chi_e E_x(t), \tag{10.31}$$

where χ_e is the "electric susceptibility," which by definition is the proportionality constant between the induced polarization and the applied field. Hence, in this case

$$\chi_e = \frac{e^2}{m}\frac{\mathfrak{N}_0}{\omega_0^2 - \omega^2 + i\Gamma\omega}. \tag{10.32}$$

More generally, if the material is supposed to consist of a variety of oscillators with natural frequencies ω_k, damping constants Γ_k, and number densities $\mathfrak{N}_k$, one has the obvious extension of Eq. (10.32):

$$\chi_e = \frac{e^2}{m}\sum_k \frac{\mathfrak{N}_k}{\omega_k^2 - \omega^2 + i\Gamma_k\omega}. \tag{10.33}$$

The dielectric constant ε (or, to be more precise, the "relative electric permittivity", $\varepsilon/\varepsilon_0$ in SI units, where ε_0 is the electric permittivity of the vacuum) of an (imaginary) material composed of charged oscillators of this type is directly related to the susceptibility via:

$$\varepsilon = 1 + \chi_e. \tag{10.34}$$

The index of refraction is given by $n = \sqrt{\varepsilon\mu}$, where μ is the relative magnetic permeability, which we henceforth set to unity, as most of the materials of interest in questions of

[15] The induced polarization in a medium where the polarizable molecules/atoms are sufficiently dense so that an individual electron sees both the external applied field and the electric field due to all nearby (induced) dipoles satisfies a more complicated relation, known as the Clausius–Mossotti relation. We assume that our assembly of oscillators is of sufficiently low density, as in a dilute gas, for example, that the field seen by each oscillator is just that due to the applied external field. We also assume that the susceptibility (which in Gaussian units is dimensionless) is small compared to unity ($\chi_e \ll 1$).

optical dispersion are non-magnetic. Thus, the refractive index is directly related to the susceptibility derived above via

$$n = \sqrt{1 + \chi_e} \approx 1 + \tfrac{1}{2}\chi_e. \tag{10.35}$$

Since the imaginary part of χ_e is always negative,[16] we define

$$n = n_r - i n_i, \tag{10.36}$$

with

$$n_r = 1 + \tfrac{1}{2}\text{Re}(\chi_e), \quad n_i = -\tfrac{1}{2}\text{Im}(\chi_e). \tag{10.37}$$

If we neglect the damping factors Γ_k in Eq. (10.33) (along with χ_e^2), we see that the Lorentz–Drude theory in fact reproduces the phenomenologically successful Sellmeier formula Eq. (10.2):

$$n^2 = 1 + \chi_e \approx 1 + \sum_k \frac{C_k}{\nu_k^2 - \nu^2}, \tag{10.38}$$

where we have gone over from angular to cyclic frequencies ($\omega = 2\pi\nu$, etc.), and the coefficients C_k in the Sellmeier formula are now given by

$$C_k = \frac{e^2}{4\pi^2 m}\mathfrak{N}_k. \tag{10.39}$$

10.1.3 The transmission of light: dispersion and absorption

Consider the passage of a plane electromagnetic wave of angular frequency ω, with its electric field polarized in the x-direction, propagating in the z-direction through a

[16] Consider Eq. (10.32) (the generalization to Eq. (10.33) is straightforward). Multiplying both numerator and denominator by the complex conjugate of the denominator, we find

$$\begin{aligned}\chi_e &= \frac{\mathfrak{N}_0 e^2}{m} \frac{\omega_0^2 - \omega^2 - i\Gamma\omega}{(\omega_0^2 - \omega^2 + i\Gamma\omega)(\omega_0^2 - \omega^2 - i\Gamma\omega)} \\ &= \frac{\mathfrak{N}_0 e^2}{m} \frac{\omega_0^2 - \omega^2 - i\Gamma\omega}{(\omega_0^2 - \omega^2)^2 + \Gamma^2\omega^2}.\end{aligned}$$

From this last expression, we read off that

$$\text{Re}(\chi_e) = \frac{\mathfrak{N}_0 e^2}{m} \frac{\omega_0^2 - \omega^2}{(\omega_0^2 - \omega^2)^2 + \Gamma^2\omega^2}, \quad \text{Im}(\chi_e) = -\frac{\mathfrak{N}_0 e^2}{m} \frac{\Gamma\omega}{(\omega_0^2 - \omega^2)^2 + \Gamma^2\omega^2}.$$

medium consisting of a dilute assembly of charged oscillators as described in Section 10.1.2. The electric field $E_x(z,t)$ of such a wave satisfies the wave equation

$$\frac{\partial^2 E_x(z,t)}{\partial z^2} = \frac{\varepsilon\mu}{c^2}\frac{\partial^2 E_x(z,t)}{\partial t^2}. \tag{10.40}$$

We see that the effect of the factors ε and μ, the electric permittivity and the magnetic permeability, is to replace the speed of light c by an altered phase velocity $v_p = c/n$ with the index of refraction n given by $\sqrt{\varepsilon\mu}$, or more simply, for our non-magnetic medium with $\mu = 1$,

$$n = \sqrt{\varepsilon}. \tag{10.41}$$

As n is complex, it is not immediately obvious that c/n is velocity but this is clarified once we introduce a complex Ansatz for our plane wave:

$$E_x(z,t) = A\,e^{i\omega t - ikz}. \tag{10.42}$$

Inserting this into Eq. (10.40), one finds $k^2 = \varepsilon\omega^2/c^2$ or

$$k = n\frac{\omega}{c} = (n_r - in_i)\frac{\omega}{c}, \tag{10.43}$$

where in the last step we used Eq. (10.36). We can thus rewrite Eq. (10.42) as

$$E_x(z,t) = A\,e^{-n_i\omega z/c}\,e^{i\omega(t - n_r z/c)}, \tag{10.44}$$

where, as usual, we obtain the physical electric field by taking the real part (converting the complex exponential to a cosine). This corresponds to an oscillating wave with phase velocity c/n_r undergoing exponential damping as it progresses in the positive z-direction.

The damping is due to *absorption* of light energy by the oscillators, which must extract energy from the wave to maintain their oscillations in the presence of frictional terms. The damping is regulated by the imaginary part n_i of the refractive index (sometimes written as κ).[17] For this reason, it is called the "absorption coefficient", and is also known as the "extinction coefficient". The real part n_r regulates the wave's phase velocity (and through the relation $v_g^{-1} = \partial k/\partial\omega$, its group velocity) and is responsible for the usual phenomena of optical dispersion, the frequency dependence of the index of refraction.

[17] As we showed in note 16, n_i is proportional to the friction constant Γ.

Inserting Eq. (10.33) for χ_e into Eq. (10.37) for n_r and n_i, we obtain explicit expressions for the frequency dependence of the real and imaginary parts of the refractive index in the Lorentz–Drude theory:

$$n_r = 1 + \frac{e^2}{2m}\sum_k \frac{\mathfrak{N}_k(\omega_k^2 - \omega^2)}{(\omega_k^2 - \omega^2)^2 + \Gamma_k^2\omega^2}, \tag{10.45}$$

$$n_i\ (= \kappa) = \frac{e^2}{2m}\sum_k \frac{\mathfrak{N}_k\Gamma_k\omega}{(\omega_k^2 - \omega^2)^2 + \Gamma_k^2\omega^2}. \tag{10.46}$$

For the low damping situation envisioned here, with $\Gamma_k \ll \omega_k$, the absorption coefficient n_i exhibits strong narrow peaks at the natural oscillator frequencies ω_k. For $|\omega - \omega_k| \gg \Gamma_k$, absorption is negligible. For frequencies $\omega \approx \omega_k$, n_i is proportional to

$$\begin{aligned}\frac{1}{(\omega_k^2 - \omega^2)^2 + \Gamma_k^2\omega^2} &\approx \frac{1}{(\omega_k + \omega)^2(\omega_k - \omega)^2 + \Gamma_k^2\omega_k^2} \\ &\approx \frac{1}{4\omega_k^2(\omega_k - \omega)^2 + \Gamma_k^2\omega_k^2} \\ &\propto \frac{1}{(\omega_k - \omega)^2 + \frac{1}{4}\Gamma_k^2}.\end{aligned} \tag{10.47}$$

This last expression is just the Lorentzian that we encountered earlier in the emission spectrum of damped oscillators (see Eq. (10.22) with $\omega_0 = \omega_k$ and $|A|^2 = 1$). It has a maximum at ω_k and a frequency width at half-maximum of Γ_k, which means that, for lower damping, the resonance peaks in the absorption coefficient are sharper. Peaks in the absorption coefficient will result in the appearance of dark lines in white light passed through a medium consisting of an assembly of Lorentz–Drude oscillators. Our analysis shows that the *frequency profile* of these dark lines is the same (in reverse) as the Lorentzian line shape of the bright emission lines emitted by these oscillators, once excited and allowed to radiate their energy away freely.

An example of the frequency dependence of the real and imaginary parts of the refractive index is given in Fig. 10.1, where we consider a toy example where there are just two resonance poles, with resonance frequencies (in some units) of $\omega_1 = 50$ and $\omega_2 = 100$. Notice that the real part n_r, giving the "normal" refractive index, generally (i.e., in between successive resonance poles) increases with frequency (or decreases with increasing wavelength) of the light: this is *normal dispersion* (red light bends less than blue light in a prism). In particular, for most solid transparent dispersive media we commonly encounter, the resonance frequencies are in the ultraviolet, so we are to the

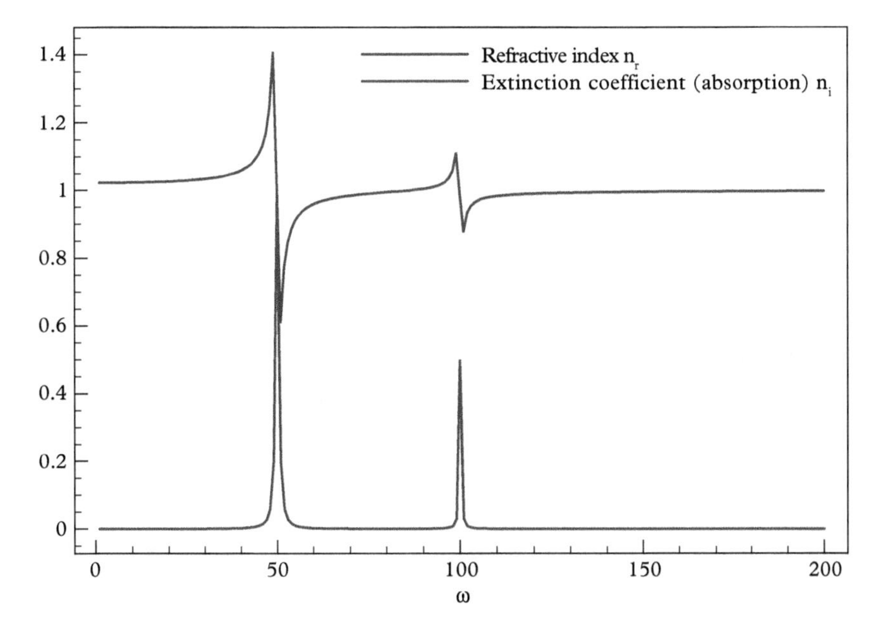

Figure 10.1 *Example of dispersion and absorption curves for system with two resonance poles ($\omega_1 = 50, \omega_2 = 100$).*

left of both resonance poles in the figure, in a region of normal dispersion. However, in the frequency intervals in the immediate neighborhood of the resonance poles, the index of refraction *decreases* with increasing frequency. This is called *anomalous* dispersion.[18] This phenomenon provided an important impetus for the development of a new generation of dispersion theories in the 1860s in terms of harmonically bound particles emitting and absorbing light (see Section 10.1).

10.1.4 The Faraday effect

In Section 10.1.2, we analyzed the interaction of *linearly* polarized light with an assembly of charged oscillators. In this section, we adapt this analysis to the case of *circularly* polarized light. This is important for understanding the rotation of the plane of polarization in a magnetic field discovered by Faraday in 1845. As was realized by the late 1890s, this Faraday effect is closely connected to the dispersion theory of Lorentz and Drude. In fact, an important method for determining the parameters $\mathfrak{N}_k$ in the dispersion formula (see Eq. (10.45)) was through measurements of the Faraday effect (see Section 10.1.5).

[18] The term was originally used for the phenomenon discovered by Leroux (1862) that a prism filled with iodine vapor refracted red light more than blue (see Section 10.1).

Suppose a light wave is passing through the medium in the z-direction in the presence of a magnetic field $\vec{H}$ in that same direction. At the origin, where we imagine our charged oscillator to be located, the electric field of this light wave is given by

$$\vec{E} = \mathcal{E}\big(\cos(\omega t), \pm\sin(\omega t), 0\big). \tag{10.48}$$

Here and below the upper and lower signs are for right and left circularly polarized light, respectively. The Lorentz force $-(e/c)\,\vec{v}\times\vec{H}$ exerted by this magnetic field (where $\vec{v}$ is the velocity of the oscillator) gives rise to additional terms in the equations of motion for the oscillator (cf. Eq. (10.24)):

$$m\ddot{x} + m\Gamma\dot{x} - \frac{eH}{c}\dot{y} + m\omega_0^2 x = -e\mathcal{E}\cos(\omega t), \tag{10.49}$$

$$m\ddot{y} + m\Gamma\dot{y} + \frac{eH}{c}\dot{x} + m\omega_0^2 y = \mp e\mathcal{E}\sin(\omega t). \tag{10.50}$$

There is no effect of either the electric or the magnetic field on the motion of the electron in the z-direction, so we omit the z-component of the equations of motion. Introducing the complex coordinate $\zeta_\pm(t)$, defined as

$$\zeta_\pm(t) = x(t) + iy(t), \tag{10.51}$$

and adding Eq. (10.49) to i times Eq. (10.50), we can combine the two equations for the real coordinates x and y into one equation for the complex coordinate ζ:

$$m\ddot{\zeta}_\pm + m\Gamma\dot{\zeta}_\pm + i\frac{eH}{c}\dot{\zeta}_\pm + m\omega_0^2\zeta_\pm = -e\mathcal{E}e^{\pm i\omega t}. \tag{10.52}$$

Inserting the Ansatz $\zeta_\pm = B_\pm e^{\pm i\omega t}$ (cf. Eqs. (10.27)–(10.28)), we find

$$B_\pm\left(m\big(\omega_0^2 - \omega^2 + i\Gamma\omega\big) \mp \frac{eH}{c}\omega\right) = -e\mathcal{E}. \tag{10.53}$$

Comparison with Eq. (10.31) shows that the presence of the magnetic field results in an additional (real) term

$$\mp\frac{eH\omega}{mc} \tag{10.54}$$

in the denominator of the dispersion formula.

The generalization to the case with several oscillators is obvious. Equation (10.33) and Eqs. (10.38)–(10.39) for the susceptibility χ_e and the index of refraction n, respectively, should be replaced by

$$\chi_{e,\pm} = n_{\pm}^2 - 1 = \frac{e^2}{m} \sum_k \frac{\mathfrak{N}_k}{\omega_k^2 - \omega^2 \mp \dfrac{eH\omega}{mc} + i\Gamma_k \omega}. \tag{10.55}$$

Setting the damping terms Γ_k to zero, we see that Eq. (10.55) has resonance poles whenever

$$\omega_k^2 = \omega^2 \left(1 \pm \frac{eH}{\omega mc}\right). \tag{10.56}$$

In the vicinity of these poles (i.e., when $\omega \approx \omega_k$), we thus have

$$\omega = \omega_k \mp \frac{eH}{2mc}, \tag{10.57}$$

and we recover the Zeeman shifts for circularly polarized light traveling in the direction of the magnetic field originally found by Lorentz (see Eqs. (7.39)–(7.40)).

A simple analysis[19] shows that the rotation angle ϑ of the linear polarization after passing through a length l in the medium is given by

$$\vartheta = \frac{\omega l}{2c}(n_- - n_+). \tag{10.58}$$

For small magnetic fields H, $n_- - n_+$ is proportional to H. Faraday had found and Émile Verdet had confirmed in the 1850s that this angle is proportional to H and l. The proportionality "constant" $\mathcal{V}$—in scare quotes because it actually varies, unsurprisingly, with frequency—is known as the "Verdet constant". Combining the equation $\vartheta = \mathcal{V}lH$ with Eq. (10.58), we obtain a dispersion relation for $\mathcal{V}$. The effect is most pronounced in the immediate neighborhood of the resonance poles, $\omega \approx \omega_k$, where the shift in Eq. (10.54) induces the largest change in the index of refraction. Many of the most accurate determinations of the number of dispersion electrons $\mathfrak{N}_k$ in the first two decades of the twentieth century involved just such measurements, as the rotation of the polarization plane could easily be measured with Nicol prisms.

10.1.5 The empirical situation up to ca. 1920

Increasingly accurate measurements of dispersion in the neighborhood of spectral lines were carried out in the first two decades of the twentieth century, allowing one to extract the constant C_k in Eq. (10.38) associated with a given spectral line, and therefore with

[19] See the web resource, *Classical Dispersion Theory.*

a particular type of bound electron, capable of elastic oscillation at frequency ν_k. In the work of Drude, it was supposed that such electrons should be identified with the more loosely bound electrons presumed to be involved in chemical activity of the given atom/molecule,[20] that is, with the valence electrons. Thus, in the case of hydrogen or the alkali metals, there should be one such electron per radiating atom. Measurements by J. J. Hallo (1902) in Amsterdam of the Faraday effect, which is governed, as we saw in Section 10.1.4, by closely related formulas of Sellmeier type, indicated that the number density $\mathfrak{N}_k$ of electrons associated with the sodium D line was only 1/200 of the expected number of radiating atoms. This puzzling situation was summarized by Voigt in his treatise *Magneto- und Elektro-Optik*:

> Through somewhat uncertain considerations one has acquired a notion of the number of sodium molecules [i.e., atom] which are radiating in a colored flame [i.e., a flame into which sodium has been introduced]. Comparison [with the number $\mathfrak{N}_k$] leads to the surprising result that only a small fraction of these molecules participate in emission or absorption. It thus appears that the mere presence of sodium in the flame is insufficient to ensure radiation, but rather that the sodium molecule must first find itself in a particular condition, unknown to us, which only a small fraction of molecules actually assumes (Voigt 1908, p. 144).

An even larger discrepancy was found by Rudolf Ladenburg and Stanislaw Loria in Breslau in experiments on dispersion of the H_α Balmer line of hydrogen. For the number of electrons per cc, they found $\mathfrak{N} = 4 \cdot 10^{12}$,

> while the number of molecules of hydrogen, which are available in the given unit of volume under the relevant experimental conditions and temperature can be calculated by kinetic theory to be $2 \cdot 10^{17}$, i.e., one finds a single "dispersion electron" for every 50,000 molecules (Ladenburg and Loria 1908, p. 865).

Ironically, these results would later turn out to be highly flawed, primarily due to the crudeness of the experimental measurements. It was extremely difficult to determine the number of atoms actually participating in the observed phenomena. In the case of the hydrogen H_α line, one needed to know precisely what fraction of the hydrogen atoms were excited to the second Bohr orbit, which serves as the lower state for this line. For the sodium vapor measurements, one needed an accurate determination of the density (or pressure) of the sodium atoms in the sample under study, from which the number of atoms (in this case, in the ground state) could be established.

At this point, the terminology "dispersion electron" was introduced by Ladenburg, bowing to the by now inescapable implication that, under the tenets of classical Lorentz–Drude dispersion theory, only a privileged class of electrons, in general considerably less than the actual number of valence electrons present in the radiating atoms, were

[20] In the literature of the time, the term "molecule" is frequently used to refer to both atoms and combinations of atoms.

apparently capable of participating in the phenomena associated with dispersion (and magneto-rotation of light).[21]

10.2 Optical dispersion and the Bohr atom

With the appearance, and, within a few years, fairly widespread acceptance, of the Bohr model in 1913, it became obvious that the whole Lorentz–Drude approach to dispersion had to be critically reassessed. It was by now abundantly clear that if atoms consisted of electrically bound planetary systems, with the valence electrons corresponding to the outermost and most loosely bound "planets", the derivation of the (admittedly empirically successful) Sellmeier-type formulas for the refractive index in terms of elastically bound charges no longer held water. The first attempts to incorporate the insights of Bohr into dispersion theory were made in late 1914 and 1915, first by Peter Debye, a former student of Sommerfeld, and then by Sommerfeld himself. The "Sommerfeld–Debye" approach to dispersion that emerged from these efforts was based on a peculiar hybrid type of theory, to which we now turn.

10.2.1 The Sommerfeld–Debye theory

The physical basis for dispersion in the classical theory can be summarized as follows. The passage of an electromagnetic field over a bound charged particle will induce (via the electric field of the wave) a supplementary acceleration of the particle of the same frequency and locked in a definite phase relation to the incident wave. By classical electromagnetic theory, such accelerations must result in the emission of a secondary electromagnetic wave, coherent to the incident one. The superposition of the incident and secondary wave corresponds to the observed refracted wave (of the same frequency but typically shorter wavelength as a wave passed from vacuum into a transparent medium). The problem in the Bohr model is that bound electrons in a stationary state (and in most cases one was considering the dispersion of light in materials in which the atoms were unexcited, with all the electrons in their "normal," i.e., lowest, stationary states) were specifically prohibited from emitting radiation, even though they manifestly were undergoing acceleration as they circled the atomic nucleus.

The admittedly ad-hoc escape from this dilemma adopted by Debye and Sommerfeld was to assert that the much smaller supplementary accelerations induced by a light wave passing over an atom (in comparison to the large accelerations of the electron in its Bohr orbit) should be treated according to classical electromagnetic theory: they would be assumed to generate a secondary wave that would act to modify the incoming wave exactly as in the classical Lorentz–Drude theory. In practice, this involved calculating

[21] An alternative escape route was to assume that all valence electrons participated in dispersion but that, for some reason, they had a different charge-to-mass ratio e/m than that measured in cathode ray experiments. Yet another proposal was that they would act in dispersive phenomena as though they had charge xe and mass xm, which would preserve the charge-to-mass ratio, but compensate for the disparity in the expected value of $\mathfrak{N}$ via the factor x, as the dispersion involves the ratio of charge *squared* to mass (see Section 10.2.2).

the deviations to the orbit of an electron in a Bohr stationary state induced by a small periodic electric field (and calculated to first order in the magnitude of the field) and evaluating the polarization per unit volume induced by these deviations, from which the index of refraction follows (see Eqs. (10.34) and (10.35)).

The strategy outlined here was applied first by Debye (1915) to optical dispersion in a gas of molecular hydrogen. The model of the hydrogen molecule adopted by Debye was that originally proposed by Bohr in his Rutherford memorandum of 1912 (see Figure 4.2b in Volume 1). The two electrons circulated in the same sense (and opposite to each other) on a circle, with the line between the nuclei piercing the center of the circle and perpendicular to the plane of the electron orbits. The effect of a small periodic electric field applied to the electrons resulted in displacements of their orbits and a corresponding induced dipole moment. There was no inclusion of damping effects—either of radiative or collisional origin. Not surprisingly, given the Coulomb binding of the electrons, already moving in a circular orbit in the absence of the field, the formula obtained by Debye for the index of refraction did not have the classic Sellmeier form. Instead, one found a complicated function of the frequency ω (of the light) which could be separated into three terms that had the Sellmeier resonance pole form $1/(\omega_k^2 - \omega^2)$, where the resonance frequencies ω_k correspond to eigenfrequencies of the perturbed motion associated with stable perturbations of the orbit, plus two more terms where the resonance poles were at complex values, corresponding to unstable perturbations (as $e^{\pm i\omega_k t}$ will contain exponentially increasing terms if ω_k is complex):

$$n^2 - 1 \propto \frac{C_1}{\nu_1^2 - \nu^2} + \frac{C_2}{\nu_2^2 - \nu^2} + \frac{C_3}{\nu_3^2 - \nu^2} + \left(\frac{C_4 + D_4\nu}{(\nu_4 - \nu)(\nu_4^\star - \nu)} + [\nu \to -\nu] \right). \tag{10.59}$$

The resonance frequencies ν_1, ν_2, ν_3, and $\mathrm{Re}(\nu_4)$ were in the ultraviolet so one could expand in powers of the light frequency ν, keeping only the first two terms, which could be compared with empirical data in the visible optical range. Rather impressively, Debye found quite good agreement with experiment, which clearly lent some credence to this admittedly artificial procedure. In a long paper written in 1917, Sommerfeld (1918) extended Debye's calculations to measurements of magnetic rotation (the Faraday effect) and to other diatomic atoms, specifically oxygen and nitrogen. The results of both Sommerfeld and Debye were confirmed (given the strategy of the approach) in a much streamlined treatment of the problem (using canonical perturbation theory) by Epstein in the final installment of a trilogy entitled "Perturbation theory in the service of quantum theory" (Epstein 1922a, 1922b, 1922c). The subtitle of this third installment was "Critical comments on dispersion theory."

The central problem with the tactic of Debye and Sommerfeld—quite apart from its conceptual artificiality—was apparent from the beginning. The resonance poles in the Sellmeier formula were known to be coincident with the frequencies of the sharp spectral lines emitted by radiating atoms. In Bohr's theory, however, these frequencies were given by the Bohr frequency condition, and had nothing to do with the frequencies to be found in a Fourier resolution of the mechanical orbit of the radiating electron. As Bohr explained to the Swedish physicist Carl Wilhelm Oseen (who also felt a fundamental incompatibility between the Lorentz–Drude and Bohr theories) in a letter of December 20, 1915, if the characteristic frequencies involved in dispersion

> are determined by the laws for quantum emission, the dispersion cannot, whatever its explanation, be calculated from the motion of the electrons and the usual electrodynamics, which does not have the slightest connection with the frequencies considered (Bohr 1972–2008, Vol. 2, p. 337).

Bohr (1972–2008, Vol. 2, pp. 433–461) repeated this criticism at some length in a paper intended for publication in 1916 but withdrawn after the appearance of Sommerfeld's pathbreaking work introducing phase integral quantization, emphasizing that any full understanding of dispersion would have to involve the same mechanism as the discontinuous quantum jumps between stationary states giving rise to sharp spectral lines. Much the same point was made six years later by Epstein:

> [T]he positions of maximal dispersion and absorption [in the formula Epstein derived] do not lie at the position of the emission lines of hydrogen but at the position of the mechanical frequencies of the model ... *the conclusion seems unavoidable to us that the foundations of the Debye–Davysson* [sic][22] *theory are incorrect* (Epstein 1922c, pp. 107–108).

10.2.2 Dispersion theory in Breslau: Ladenburg and Reiche

The outlines of a new approach to reconciling the classical treatment of optical dispersion in terms of atoms continuously absorbing and re-emitting electromagnetic radiation and the quantum view in which atoms jump between discrete stationary states (losing or gaining energy by emitting or absorbing radiation in accordance with the Bohr frequency condition) emerged at the end of the second decade of the twentieth century. This approach depended critically on elements of Einstein's new quantum theory of radiation, such as the A and B coefficients, which one could use without buying into light quanta (Einstein 1916a, 1916b, 1917a; see also Section 3.6). It also involved the correspondence principle, initially in a form related to Bohr's ideas (only in spirit and not in detail) whereby quantities appearing in the (phenomenologically highly successful) classical Lorentz–Drude theory were shown to correspond to—in fact, be quantitatively equivalent to—analogous quantities in Einstein's theory. A qualitative outline of this strategy appeared in a paper on fluorescence by Otto Stern and Max Volmer in 1919:

> In connection with the absorption of light, according to the classical theory it proceeds by an equal absorption of light by all the molecules, whereas in the quantum theory

[22] Clinton Davisson (1916) had proposed a theory of dispersion similar to that of Debye (1915) and Sommerfeld (1918). Epstein had moved to Pasadena in 1921 and, like Van Vleck (see Chapter 8, note 18), may have felt that European physicists did not properly acknowledge the contributions of those working in the United States (cf. Duncan and Janssen 2007, p. 580).

> (according to the formulation of Einstein and Bohr) the majority of the molecules do not absorb and remain unchanged (in state *a*), and only a few molecules receive energy, but then only in absolutely equal quanta of amount $h\nu$, whereby they transition to a state *b*. The *decay process* proceeds according to the classical theory by gradual radiation of energy by all the molecules, but in the quantum theory occurs when from time to time a molecule from the state *b* springs back to the state *a* emitting light, in the same way and following the same statistical law by which the atoms of radioactive materials decay. The decay time in the classical theory is the time in which the [previously] absorbed energy has radiated away to the $1/e$ level, in the quantum theory on the other hand it is the average lifetime of an atom in the [excited] state *b* [at this point, there is a reference to Einstein (1917a)] (Stern and Volmer 1919, p. 186).

A year later, this approach was applied with quantitative precision to an analysis of absorption at spectral lines from the point of view of quantum theory by Christian Füchtbauer (1920), working in Friedrich Paschen's group in Tübingen. Füchtbauer was concerned with the absorption of incident light by an atom, resulting in the excitation of the atom from its ground state to an excited state higher in energy by $h\nu$, where the incident light was of uniform intensity i over a range of frequencies around ν encompassing the entire width of the spectral absorption line corresponding to the transition in question. He showed that one could relate the total light energy A absorbed per unit time by a given number of the atoms to an integral over the extinction coefficient κ of the absorption line, which, in the classical theory, is given by an inverted Lorentzian (see Section 10.1.3). The point of the calculation was to extract the proportionality constant P between A and the light intensity i. This probability was then associated to the probability per unit time that an atom exposed to the light beam would make a quantum jump (*Quantensprung*) per unit intensity of the light. Füchtbauer, in effect, tried to measure Einstein's (1917a) B coefficient, which he cited at this point (Füchtbauer 1920, p. 322).

In a seminal paper, following a strategy similar to that of Stern, Volmer, and Füchtbauer, the Breslau experimentalist Rudolf Ladenburg (1921) addressed the thorny issue of, as it says in the title of the paper, "The quantum-theoretical interpretation of the number of dispersion electrons," $\mathfrak{N}$ appearing in the numerator of the Sellmeier formula (see Eqs. (10.38)–(10.39)). Again, the basic idea was to accept the structural form of the classical Sellmeier formula as given, with the resonance poles located at frequencies corresponding to discrete quantum transitions in the Bohr model (and given by Bohr's frequency condition). By connecting the notion of classically radiating dispersion electrons associated with each of the resonance frequencies underlying this formula to the demands of Einstein's radiation theory, one arrives at an interpretation of the numerator factor in the Sellmeier formula containing the troublesome $\mathfrak{N}$, in terms of Einstein's A coefficients for spontaneous emission. As Ladenburg wrote on the introduction of his paper:

> According to classical electron theory, the absorption of isolated spectral lines is characterized above all by the number $\mathfrak{N}$ of dispersion electrons per unit volume, the "dispersion

> constant", apart from the frequency ν and the damping coefficient ... In quantum theory, on the other hand, the absorption is produced by the transition of the molecules from a state i to a state k, and the strength of the absorption is determined by the probability of such transitions $i \to k$. This result follows from Einstein's well-known considerations and has recently been used by Füchtbauer in calculations connected with his absorption measurements on alkali vapours. Einstein's above-mentioned theory ... now leads to an important relation between this probability factor and the probability of the spontaneous (reverse) transition from state k to state i. It will be shown that it is this latter probability ... which takes the place of the dispersion constant $\mathfrak{N}$, so that it is directly obtainable from absorption measurements, as well as from emission measurements and those of anomalous dispersion and magnetic rotation at and near the spectral lines (Ladenburg 1921, p. 139).

As we will see below, Ladenburg's results turned out to be a critical ingredient in the ultimate resolution of the problem of optical dispersion within the framework of the old quantum theory.

Ladenburg's argument starts on the classical side, with the observation that the rate of radiative energy loss for $\mathfrak{N}$ "oscillatory electrons" (i.e., dispersion electrons) of natural frequency $\nu_0 = \omega_0/2\pi$ is given by (see Eq. (10.13))

$$\mathcal{J}_{\mathrm{El}} = \frac{\overline{U}\mathfrak{N}}{\tau} = \overline{U}\mathfrak{N}\left(\frac{8\pi^2 e^2 \nu_0^2}{3mc^3}\right), \tag{10.60}$$

where in the last step we used Eq. (10.17) for the decay time τ. $\mathcal{J}_{\mathrm{El}}$ is Ladenburg's somewhat peculiar notation for the energy radiation rate from the dispersion electrons of natural frequency ν_0 according to *classical electron theory*. This quantity will soon be compared with the corresponding result obtained on the basis of Einstein's quantum radiation theory.

Ladenburg now focuses on the system considered by Einstein in his quantum radiation theory papers, an assembly of quantized atoms/molecules immersed in a container filled with black-body radiation in equilibrium with them at some temperature T. In the course of his investigations in the late 1890s into irreversible radiation processes, Planck (1897–1899, Pt. V, p. 575) had derived a simple relation between the average energy $\overline{U}$ of a charged three-dimensional oscillator and the ambient black-body energy density $\rho_0 = \rho(\nu_0)$:

$$\overline{U} = \frac{3c^3}{8\pi\nu_0^2}\rho_0. \tag{10.61}$$

This expression differs by a factor 3 from Eq. (2.7) in Volume 1 for a one-dimensional oscillator: for an oscillator in three dimensions, the energy can be written as a sum of three identical one-dimensional oscillators. Inserting Eq. (10.61) for $\overline{U}$ into Eq. (10.60) for $\mathcal{J}_{\rm El}$, one finds

$$\mathcal{J}_{\rm El} = \left(\frac{3c^3}{8\pi\nu_0^2}\rho_0\right)\mathfrak{N}\left(\frac{8\pi^2e^2\nu_0^2}{3mc^3}\right) = \frac{\pi e^2}{m}\mathfrak{N}\rho_0. \tag{10.62}$$

Ladenburg's next step is to calculate the absorbed energy per unit time A_Q at frequency ν_0, but now calculated according to Einstein's quantum theory of radiation (hence, the subscript Q).[23] Consider two quantum states i and k of our atoms/molecules with an energy gap of

$$E_k - E_i = h\nu_{ik} = h\nu_0, \tag{10.63}$$

so that transitions between these two states involve light quanta of the resonance frequency of the dispersion electrons under consideration. The probability per unit time that an atom immersed in black-body radiation of energy density $\rho(\nu)$ absorbs a single quantum of frequency ν_0 and undergoes an upward transition $i \to k$ is given by

$$B_i^k\rho(\nu_{ik}) = B_i^k\rho_0 \tag{10.64}$$

(cf. Eq. (3.87)). Hence, the rate of energy absorption A_Q of N_i such atoms is

$$A_Q = h\nu_0 N_i B_i^k \rho_0. \tag{10.65}$$

Einstein also considered the probability (again per unit time per atom) of spontaneous emission, leading to a transition $k \to i$, given by the coefficients A_k^i, and the probability of emission stimulated by the ambient black-body radiation, given by $B_k^i\rho(\nu_{ik})$. He also took into account the "statistical weight" p_i of the state i (Eq. (3.84)), which we can think of in terms of degeneracy as the number of equiprobable states with the same energy E_i. Thermal equilibrium (keeping the number of atoms N_i in any given state i constant) requires that

$$p_i B_i^k = p_k B_k^i; \tag{10.66}$$

while the Wien displacement law requires that

$$A_k^i = \alpha\nu_{ik}^3 B_k^i \tag{10.67}$$

[23] We use the notation for the Einstein coefficients that we use in our discussion of Einstein's quantum radiation theory in Section 3.6, which differs slightly from Ladenburg's.

(Eqs. (3.94) and (3.98), respectively). Agreement with the Wien law for high frequencies, finally, requires that the proportionality constant α is equal to $8\pi h/c^3$. Combining these results, we arrive at

$$B_i^k = \frac{p_k}{p_i}\frac{c^3}{8\pi h\nu_0^3}A_k^i. \tag{10.68}$$

Inserting this expression for B_i^k into Eq. (10.65), we find

$$A_Q = N_i\frac{p_k}{p_i}A_k^i\frac{c^3}{8\pi\nu_0^2}\rho_0. \tag{10.69}$$

We once again remind the reader that the spectral frequency ν_0 associated with the dispersion electrons under consideration corresponds to an actual quantum transition $i \to k$ of our N_i atoms.

In thermal equilibrium, the quantum rate of absorption of energy A_Q at frequency ν_0 must be balanced exactly by the emission at this frequency, for which we now take the *classical* result in Eq. (10.62). Setting $\mathcal{J}_{\rm El}$ and A_Q equal to each other,

$$\frac{\pi e^2}{m}\mathfrak{N}\rho_0 = N_i\frac{p_k}{p_i}A_k^i\frac{c^3}{8\pi\nu_0^2}\rho_0, \tag{10.70}$$

we obtain an explicit expression for the number of dispersion electrons in terms of Einstein's spontaneous emission coefficient for the transition of interest:

$$\mathfrak{N} = N_i\frac{p_k}{p_i}A_k^i\frac{mc^3}{8\pi^2e^2\nu_0^2}. \tag{10.71}$$

Ladenburg's result Eq. (10.71) only related the number of dispersion electrons associated with a given transition to the spontaneous decay coefficient for the upper state of the transition, but gave no insight into how to calculate independently the latter. Two years later, he would at least partially eliminate this shortcoming in work with his Breslau colleague, the theoretical physicist Fritz Reiche, discussed later. For now, Ladenburg contented himself with using his result to draw some rough quantitative conclusions concerning the relative values of the decay coefficient for hydrogen (he found, for instance, that the probability for transitions in H_α was about four times that for H_β) and for alkali metals, where it was already known that the transition probability, and hence, the number of associated dispersion electrons, was much greater for the first line of the principal series than for the succeeding ones.

In the final section of his paper, Ladenburg addresses the perennial issue of the absolute number of dispersion electrons, and its relation to the number of valence electrons (one per atom for both hydrogen and alkalis). The problem here was what it had been from the early measurements of Hallo (1902) and Wood (1904): it was very difficult to pin down precisely the number of atoms in the appropriate quantum state in the given experimental sample. For the sodium D lines the total number of dispersion electrons $\mathfrak{N}_{D_1+D_2}$ (from both lines of the doublet) was variously found to be $0.41N$ (Gouy 1920), $1.16N$ (Senftleben 1915) or $0.21N$ (Fuchtbauer and Hofmann 1914), where N was the number of sodium atoms in the sample. The only general area of agreement was in the relative number of dispersion electrons associated with the D_2 versus D_1 lines (about a factor of 2, $\mathfrak{N}_{D_2} \approx 2\mathfrak{N}_{D_1}$).

The experimental situation would be substantially improved in a matter of months. Ladenburg's doctoral student in Breslau, Rudolph Minkowski, a nephew of Hermann Minkowski, carried out a careful investigation of magnetic rotation (the Faraday effect) in sodium vapor, in the immediate neighborhood of the principal series D lines, but now (finally, in contrast to all earlier measurements) under conditions of uniform density and temperature, which allowed the actual number N of dispersing sodium atoms to be calculated (using also measurements performed by Haber and Zisch (1922) in Berlin) and compared with the number of D_1 and D_2 dispersion electrons (Ladenburg and Minkowski 1921). The result was in one respect gratifying: the total number of dispersion electrons from both D lines, $\mathfrak{N}_{D_1+D_2}$, now agreed with the number of dispersing sodium atoms N, to within a few percent. Of course, as indicated previously, this agreement only held for the first member of the alkali principal series (D lines): the higher members showed rapidly decreasing dispersion and magnetic rotation.

Ladenburg's (1921) paper met with a rather muted response from the community of physicists intensely involved in developing the implications of the Bohr atomic model (Duncan and Janssen 2007, pp. 584–586). The forcible, and admittedly ad hoc, wedding of the classical formula Eq. (10.62) with its quantum counterpart Eq. (10.69) must have seemed unconvincing at the time, especially since the Bohr atomic theory provided absolutely no understanding of the resonance pole structure, with poles located at spectral line frequencies and not at mechanical frequencies of the orbiting electrons. The Ladenburg approach required one to accept the denominator structure of the Sellmeier equation as given (with no classical explanation given the mechanics of the Bohr atom), but apply classical radiation theory to the numerator factor containing the mysterious number $\mathfrak{N}$. Moreover, without an independent way to calculate, estimate, or directly measure the decay coefficients A^i_k, there was really no way to check the validity of Ladenburg's Eq. (10.71). Two years later, Ladenburg, in collaboration with Reiche, his friend and colleague in Breslau, published an important amplification of the ideas advanced in his 1921 paper. The paper (entitled "Absorption, scattering and dispersion in the Bohr atomic model") was prepared as a contribution to a special issue of *Die Naturwissenchaften* to commemorate the tenth anniversary of Bohr's 1913 trilogy. Not surprisingly, the theoretical tenets of Bohr's approach (not referred to in any detail in the 1921 paper) here move to center stage—in particular, the consequences of Bohr's correspondence principle, which the authors characterize as follows:

> We say, with *Bohr*, that the radiated frequency ν in the transition $n' \to n''$ "corresponds" to the harmonic component $(n' - n'')\nu_n$ [where ν_n, with $n \approx n', n''$, is the fundamental frequency in the mechanical motion of the orbit, with $n' - n'' \ll n', n''$] (Ladenburg and Reiche 1923, p. 586).

This is followed by a statement of Bohr's extension of this basic idea to the calculation of transition *intensities* in the limit of large quantum numbers, namely, the asymptotic agreement of "the probability factor A^i_k of the relevant quantum transition and the squared amplitude of the 'corresponding' component (in the initial, final or intermediate orbits)" (ibid.).

This correspondence had been applied with some success to the Stark effect by Kramers (1919) in his dissertation (see Section 6.3). It is particularly straightforward to implement in the case of a quantized harmonic oscillator. This is the only system in which the frequency of the motion $\omega = 2\pi\nu$ is independent of amplitude, so the only transitions allowed correspond to $n' - n'' = \pm 1$, where the energy of state n is $U_n = nh\nu$ (we ignore the zero-point energy), and the intensity of the spontaneous decay $n+1 \to n$, from which we may determine the coefficient A^n_{n+1}, is determined by the coefficient C in Eq. (10.5). Of course, the same ambiguity that plagued Kramers' work remains: should we use the amplitude C associated with the initial state $n + 1$, the final state n, or some average of the two? If we assume that one should take the amplitude of the initial state, then C is determined by equating

$$\begin{aligned} U_{n+1} &= \frac{1}{2}m\dot{x}(t)^2 + \frac{1}{2}m\omega^2 x(t)^2 \\ &= \frac{1}{2}mC^2(2\pi\nu)^2\Big(\sin^2(\omega t) + \cos^2(\omega t)\Big) \\ &= 2m\pi^2\nu^2 C^2, \end{aligned} \tag{10.72}$$

with $(n + 1)h\nu$. The relevant squared amplitude can thus be set equal to

$$C^2 = \frac{(n+1)h}{2m\pi^2\nu}. \tag{10.73}$$

On the classical side, such an oscillator radiates energy at a rate given by the Larmor formula, Eq. (10.6). Averaged over a single cycle, this gives an amount of energy

$$\frac{2e^2}{3c^3}C^2(2\pi\nu)^4 \cdot \frac{1}{2} = \frac{16\pi^4 e^2\nu^4}{3c^3}C^2, \tag{10.74}$$

which should be equated on the quantum side to $A^n_{n+1}h\nu$. We thus have:

$$A^n_{n+1}h\nu = \frac{16\pi^4 e^2\nu^4}{3c^3}C^2. \tag{10.75}$$

Using Eq. (10.73) for C^2, we find the following expression for this Einstein spontaneous emission coefficient:

$$A^n_{n+1} = \frac{1}{h\nu}\frac{16\pi^4 e^2\nu^4}{3c^3}\frac{(n+1)h}{2m\pi^2\nu} = \frac{8\pi^2 e^2\nu^2}{3mc^3}(n+1). \tag{10.76}$$

With an independent evaluation of the emission coefficient in hand, we are now in a position to apply Ladenburg's (1921) result Eq. (10.71) to obtain the ratio of the number of "dispersion oscillators (electrons)" to actual oscillators (which in this case means the number of oscillators in the lower quantum state $i = n$ in Eq. (10.71), with the higher quantum state $k = n + 1$):

$$\frac{\mathfrak{N}}{N} = \frac{p_{n+1}}{p_n} A^n_{n+1} \frac{mc^3}{8\pi^2 e^2 \nu^2} = \frac{p_{n+1}}{p_n} \frac{n+1}{3}. \tag{10.77}$$

To complete the calculation, we need to know the degeneracy factor p_n, that is, the number of quantum states of energy $nh\nu$ (where $n = 0, 1, 2, \ldots$). For the quantized version of the isotropic three-dimensional oscillators considered in classical dispersion theory, the quantum degeneracy factor is[24]

$$p_n = \frac{1}{2}(n+1)(n+2). \tag{10.78}$$

Applying this result in Eq. (10.77) we find

$$\frac{\mathfrak{N}}{N} = \frac{(n+2)(n+3)}{(n+1)(n+2)} \frac{(n+1)}{3} = \frac{n+3}{3}. \tag{10.79}$$

Of course, the usual objection to correspondence arguments of this kind holds: in principle, we should only expect them to be valid asymptotically, for large quantum numbers. Nevertheless, in the spirit of Bohr, one crosses one's fingers and extends the result to low quantum numbers, which perhaps is somewhat more plausible in this case, given the extreme simplicity of the harmonic oscillator (where the transition frequencies are the same for low or high quantum numbers).

At first sight, the result Eq. (10.79) is encouraging: for an assembly of quantum "dispersion" oscillators in the ground state $n = 0$, the number of "dispersion" and actual oscillators in fact agree, so the embarrassing issue of a fractional number of oscillators is avoided. Of course, this agreement depends on us having chosen the initial state as determining the amplitude C to use in Eq. (10.74), rather than the final state or some average of the two.

On closer examination, however, the result is decidedly puzzling. One would expect the agreement between quantum and classical theory to be best for large quantum numbers: after all, in classical theory the dispersion formulas in Eqs. (10.31)–(10.33) make no reference to the amplitude of the classical oscillations, which for large electric fields would be correspondingly large, and certainly would not correspond to oscillators in the quantum ground state. Eq. (10.79), however, clearly implies that the number of "dispersion" oscillators $\mathfrak{N}$ would be larger than the number N of actual oscillators once

[24] Labeling the quantum state of a three-dimensional oscillator by (n_x, n_y, n_z), one needs to count the number of states with $n_x + n_y + n_z = n$. There is a single ground state, (0,0,0), three first excited states with $n = 1$, (1,0,0), (0,1,0), and (0,0,1), six states with $n = 2$, etc.

the oscillators are excited into higher levels with $n > 0$. This puzzle would be resolved within a year by Kramers, who introduced a dispersion formula with an additional negative term for excited states, which precisely restores the agreement between the number of dispersion and actual oscillators in the case of an assembly of quantized isotropic three-dimensional harmonic oscillators, whatever their state of excitation.

Ladenburg and Reiche's paper introduces two concepts that would play a central role in the development of the quantum theory of radiation in the coming years: first, the notion of a dimensionless *oscillator strength* (used to the present day to quantify transition rates) associated with each pair of quantum states of an atomic system; and, second, the concept of a "substitute oscillator" (*Ersatzoszillator*) as an interpretation of the non-equality of N and $\mathfrak{N}$, which by now appeared to be the case in essentially all atomic transitions.

Note that the quantity $mc^3/8\pi^2e^2\nu_0^2$ appearing in Eq. (10.71) is just one-third the classical decay time τ for a three-dimensional oscillator of charge e, mass m vibrating at frequency ν_0 (see Eq. (10.17) for τ). As the spontaneous emission coefficient A_k^i has dimensions of inverse time, the quantity f_k^i (later called the "oscillator strength") defined as

$$f_k^i \equiv \tau A_k^i \tag{10.80}$$

is dimensionless, and determines (together with the degeneracy ratio of the transition) the fractional relation between the number of dispersion electrons $\mathfrak{N}$ associated with the transition $i \to k$ and the number of atoms N_i in the lower state i (Ladenburg and Reiche 1923, p. 588):

$$x \equiv \frac{\mathfrak{N}}{N_i} = \frac{1}{3}\frac{p_k}{p_i}f_k^i. \tag{10.81}$$

This changes the puzzling fact in classical dispersion theory that the number of dispersion electrons extracted from fits of the Sellmeier formula seemed in many cases to be only a small fraction of the available supply of optically active electrons from a bug to a feature, allowing us to relate the mismatch to the presence of an intrinsically quantum-mechanical factor, the dimensionless oscillator strength, determined by Einstein's spontaneous emission coefficient, which was under no *a priori* constraint to yield an integer value for $\mathfrak{N}$.[25]

The second notion introduced by Ladenburg and Reiche, the idea of "substitute oscillators", would carry through into, and play a central role in, the BKS radiation theory of Bohr, Kramers, and Slater (1924a, see Section 10.4), universally referred to, after the initials of the authors, as the BKS theory. The point is a simple one. A glance at

[25] An important sum rule involving these oscillator strengths would be derived two years later by Kuhn (1925) and by Reiche and Thomas (1925). This sum rule would play a central role in Heisenberg's (1925c) *Umdeutung* paper (see Section 10.5).

the basic dispersion formula Eq. (10.38) shows immediately that the discrepancy, if any, between the number of actual electrons and dispersion electrons (expressed by the ratio x introduced in Eq. (10.81)) can be eliminated by assuming that the dispersion electrons associated with any given term in the Sellmeier formula *behave as though* they are classical oscillators having a charge xe and a mass xm (thereby maintaining the measured charge-to-mass ratio of free electrons). The appearance of the factor e^2/m in the Sellmeier formula would then produce the needed factor x to account for the discrepancy between $\mathfrak{N}$ and N_i. Of course, electrons responsible for different resonance poles (i.e., different quantum transitions) in the Sellmeier formula would have different non-standard charges and masses. Note that another possible route for explaining the discrepancy that only electrons present in some randomly selected fraction of atoms participated in dispersion, but with canonical mass and charge, was excluded by the observation that "the $\mathfrak{N}$ oscillators must be uniformly distributed among the N spatial points [location of the atoms], as otherwise the outgoing secondary waves would lead to false interferences" (Ladenburg and Reiche 1923, pp. 589–590). The introduction of substitute oscillators was clearly a purely formal maneuver.[26] It nevertheless strongly influenced the BKS theory.

10.3 The correspondence principle in radiation and dispersion theory: Van Vleck and Kramers

The confusing status of the phenomenologically successful Sellmeier–Lorentz–Drude formula—capable of describing dispersion quantitatively in an enormous variety of materials—in the old quantum theory underwent a dramatic transformation in the year following the appearance of the paper of Ladenburg and Reiche (1923). These authors had, it turns out, correctly identified the two critical physical ingredients in formulating a dispersion theory for Bohr atoms that would give formulas for the refractive index structurally identical to the Sellmeier form while incorporating central features of the quantum theory. These ingredients were: first, the Einstein stochastic quantum theory of radiation (with the A and B coefficients playing an essential role); and second, Bohr's correspondence principle. They would lead to a quantum dispersion formula (now called the Kramers–Heisenberg formula, but derived first by Kramers alone) that survived intact the transition from the old quantum theory to modern quantum mechanics.

But for this correct quantum dispersion theory to emerge, a third ingredient was necessary—a sophisticated technical grasp of the methods of canonical perturbation theory in classical mechanics. Familiarity with these techniques among quantum theorists had grown rapidly following the introduction of the Bohr–Sommerfeld–Schwarzschild approach to quantization in 1916, with its reliance on the use of action/angle variables in multiply-periodic systems (see Section 1.3.2, Section 5.1.4, and Appendix A). Not

[26] It is reminiscent of the notion of Fresnel drag, introduced in the nineteenth century, which had ether dragged along in transparent media with a velocity dependent on the index of refraction, which, precisely because of dispersion, would vary with the frequency of the light (Duncan and Janssen 2007, p. 574).

surprisingly, the two physicists who would contribute first, and most decisively, to the development of an adequate quantum dispersion theory, Hans Kramers in Copenhagen and John Van Vleck in Minneapolis, were both adept practitioners of the black arts of Hamiltonian perturbation theory applied to multiply periodic dynamical systems. In fact, in the previous two years, both theorists had applied such techniques to the extremely complex task of working out the three-body dynamics of neutral helium, leading (independently) to the same binding energy results that represented one of the most glaring failures of the old quantum theory (see Section 7.4). Initially the emphasis of the two was somewhat different: Van Vleck was concerned with using the correspondence principle to connect Einstein's results for emission and absorption in quantized atomic systems with the corresponding classical formulas for charged particles engaged in multiply periodic motion, while Kramers was engaged in using the correspondence principle to understand, and then generalize, Ladenburg's hypothesis for the origin of the numerator factors in the Sellmeier dispersion formula.

The chronology of the development of dispersion theory in the year from the Fall of 1923 to late 1924 is somewhat complicated by the fact that both Kramers and Van Vleck initially published papers in which results were announced without any detailed derivations. Kramers' (1924a) first note on dispersion was submitted to *Nature* in late March 1924, and published in May; Van Vleck's (1924a) note on absorption was submitted to the *Journal of the Optical Society of America* in early April, and published in July. A second note by Kramers (1924b) (in response to criticism of Gregory Breit, Van Vleck's colleague in Minneapolis at the time) submitted in late July and published in *Nature* at the end of August contains only the barest outline of the derivation of his new dispersion relation, which he had stated without proof in his first communication in May. Van Vleck became aware of Kramers' *Nature* notes sometime between his own first note on absorption in April,[27] equally lacking in specifics, and the appearance in print of a much longer and more detailed pair of papers in October (Van Vleck 1924b, 1924c). Very rapidly, he was able to use exactly the same type of correspondence arguments he had previously applied to Einstein's radiation theory in the context of absorption to derive the new dispersion formula stated in Kramers' first note. Also entering the fray in this period was Max Born in Göttingen, who published yet another correspondence-principle-inspired derivation of the Kramers dispersion formula, using a slightly different version of canonical perturbation theory than Van Vleck, in late August (Born 1924). Van Vleck would thus have been unaware of Born's paper.

A further complication in clearly delineating the conceptual evolution of dispersion theory—which would eventually culminate, via the exhaustive analysis of the new quantum dispersion theory by Kramers and Heisenberg in late 1924, in Heisenberg's reinterpretation (*Umdeutung*) of classical dynamical variables in terms of complex arrays—was the emergence in the Spring of 1924 of a new theory of the interaction of radiation and matter—the notorious BKS theory (Bohr, Kramers, and Slater 1924a). Its germination was occasioned by a set of ideas introduced by the American physicist

[27] This was presumably before June 19, the date of submission of the *Physical Review* two-part paper.

John Slater, who arrived at the Bohr institute in Copenhagen around Christmas 1923 on a fellowship. Slater's idea was that the substitute oscillators of Ladenburg and Reiche, now redubbed "virtual oscillators", should be regarded as generating a "virtual electromagnetic field" even when the system was in a stationary state, and moreover, that this field contained all frequencies corresponding to possible transitions of the atom from the given stationary state to other ones. Initially, Slater viewed this field as determining the transition probabilities for the stochastic photon transitions of Einstein's theory, but Bohr and Kramers, both violently opposed to the notion of light quanta (cf. note 72), while enthusiastically adopting much of Slater's conception, insisted on stripping the theory of all mention of light quanta. The result was a theory in which energy and momentum could only be conserved statistically, but not in individual atomic processes.[28]

The trouble the BKS theory poses for attempts to decipher the evolution of a new quantum dispersion theory in 1923–1924 is that Kramers' new dispersion theory was presented under the general umbrella of the BKS theory. We return to this issue later but it should be emphasized right away that the correspondence-principle derivation of the quantum dispersion relation *preceded and was independent of* the development of the BKS theory (Dresden 1987, pp. 144–146, pp. 220–222; see also Darrigol 1992, p. 225).[29] In particular, as Slater later recalled, Kramers already had the dispersion relation in hand when Slater arrived in Copenhagen in December 1923.[30] In Van Vleck's case, it is perfectly clear that his derivation of a correspondence principle for absorption (the results of which were clearly stated in the April note to the *Journal of the Optical Society of America*, and which proceeds along completely similar lines to his later derivation of the quantum dispersion formula) must have preceded any knowledge of the BKS theory. We therefore present the correspondence-principle arguments of Van Vleck and Kramers first *with no reference to the BKS theory*, relegating discussion of the rise and fall of the latter to Section 10.4.

[28] The BKS theory did at least have the virtue of being explicitly disprovable: it survived for rather less than a year, with experiments of Bothe and Geiger (1924, 1925a, 1925b) in Berlin, and Compton and Simon (1925) in Chicago dealing the death blow by April 1925. Their experiments showed that energy and momentum are conserved not just statistically in Compton scattering, but also event by event (see Section 10.4).

[29] Dresden and Darrigol thus corrected Jammer (1966), who wrote that BKS "was the point of departure of Kramers's detailed theory of dispersion" (p. 184). Dresden (1987, p. 221) provides a helpful chronology of events in 1923–1925 related to BKS and dispersion theory.

[30] This can be inferred from a letter that Slater wrote on his way back from Copenhagen to his former fellow graduate student Van Vleck, in which he gives a rather unflattering picture of Bohr's institute: "Don't remember just how much I told you about my stay in Copenhagen. The paper with Bohr and Kramers [proposing the BKS theory] was got out of the way the first six weeks or so—written entirely by Bohr and Kramers. That was very nearly the only paper that came from the institute at all the time I was there; there seemed to be very little doing. Bohr does very little and is chronically overworked by it ... Bohr had to go on several vacations in the spring, and came back worse from each one ... Kramers hasn't got much done, either. You perhaps noticed his letter to Nature on dispersion; *the formulas & that he had before I came, although he didn't see the exact application*; and except for that he hasn't done anything, so far as I know. They seem to have too much administrative work to do. Even at that, I don't see what they do all the time. Bohr hasn't been teaching at all, Kramers has been giving one or two courses" (quoted in Duncan and Janssen 2007, pp. 564 and 588, emphasis added).

10.3.1 Van Vleck and the correspondence principle for emission and absorption of light

Van Vleck's work was carried out in the context of a charged particle executing general multiply periodic motion, where the dynamics was assumed separable, and amenable to the introduction of a complete set of action variables

$$\mathcal{J}_1, \mathcal{J}_2, \ldots, \mathcal{J}_u$$

with their conjugate angle variables

$$w_1 = \nu_1 t, \; w_2 = \nu_2 t, \; \ldots, \; w_u = \nu_u t,$$

with u the number of degrees of freedom of the system (so $u = 3$ for a single charged particle moving in three dimensions). The classical formulas for emission and absorption of light of a charged particle, specifically, a bound electron in some Bohr orbit, could be calculated to the lowest non-trivial order in the electric field of the emitted or absorbed light using canonical perturbation theory. In the case of absorption especially, this required a tour-de-force application of these perturbation techniques. We describe Van Vleck's results here for the much simpler case of a one-dimensional simple harmonic oscillator, as the critical new ingredient, the formalization and application of correspondence-principle ideas, can be quickly appreciated in this simpler system.

The correspondence-principle result for emission which Van Vleck (1924b, Eqs. (8) and (9)) derives is already contained, in the special case of a harmonic oscillator, in the work of Ladenburg and Reiche (1923; see also Eq. (10.74)). For a one-dimensional simple harmonic oscillator the Fourier sum representing the general multiply periodic motion,

$$x(t) = \sum_{\tau=1}^{\infty} X_\tau \cos(\tau\omega t), \tag{10.82}$$

degenerates to the single term $\tau = 1$ corresponding to oscillation at the fundamental frequency. The classical radiation emission rate is then given by Eq. (10.74) (with a slight change of notation, that is, X_1 instead of C):

$$\frac{\Delta E_{\text{emiss}}}{\Delta t} = \frac{16\pi^4 e^2 \nu^4}{3c^3}(X_1^r)^2. \tag{10.83}$$

The superscript r refers to a classical oscillator with an amplitude corresponding to an oscillator energy $rh\nu$ (i.e., in the rth excited quantum state). In the case considered by

Ladenburg and Reiche, the emitting state has $r = n + 1$ and Eq. (10.73) for $C = X_1^r$ gives

$$(X_1^r)^2 = \frac{(n+1)h}{2m\pi^2\nu}. \tag{10.84}$$

Like Ladenburg and Reiche, Van Vleck now imposes the correspondence-principle requirement that, *for high quantum numbers*, the classical spontaneous emission rate given by Eq. (10.83) should agree with the quantum transition rate to the lower state s (here the immediately lower state $s = n$), which (according to Einstein's radiation theory) is

$$A_r^s h\nu_{rs} = A_{n+1}^n h\nu. \tag{10.85}$$

Setting this equal to the right-hand side of Eq. (10.83), using Eq. (10.84) for X_1^r, and solving for A_{n+1}^n, we recover Eq. (10.74):

$$A_{n+1}^n = \frac{1}{h\nu}\frac{16\pi^4 e^2\nu^4}{3c^3}\frac{(n+1)h}{2m\pi^2\nu} = \frac{8\pi^2 e^2\nu^2}{3mc^3}(n+1). \tag{10.86}$$

For the more general case of multiply periodic motion in three dimensions, Van Vleck writes the obvious generalization relating the classical Fourier amplitudes for the three Cartesian coordinates

$$X_{\vec{\tau}}^r,\ Y_{\vec{\tau}}^r,\ Z_{\vec{\tau}}^r$$

for motion in the state r, now specified by three non-zero integers $\vec{\tau} = (\tau_1, \tau_2, \tau_3)$ (not necessarily equal to unity), to the Einstein emission coefficient A_r^s from the higher state r to a lower state s. The frequency associated with a given choice of τ_1, τ_2, τ_3 is the sum $\tau_1\nu_1+\tau_2\nu_2+\tau_3\nu_3$. The emission coefficient for two states r and s with an energy difference

$$E_r - E_s = h\nu_{rs} = h(\tau_1\nu_1 + \tau_2\nu_2 + \tau_3\nu_3)$$

is given by (van Vleck 1924b, Eq. (9)):

$$A_r^s = \frac{16\pi^4 e^2\nu_{rs}^3}{3hc^3}(D_{\vec{\tau}}^r)^2, \tag{10.87}$$

with $(D_{\vec{\tau}}^r)^2 \equiv (X_{\vec{\tau}}^r)^2 + (Y_{\vec{\tau}}^r)^2 + (Z_{\vec{\tau}}^r)^2$.

So far, Van Vleck seems to have simply produced a straightforward extension of the results of Ladenburg and Reiche for a harmonic oscillator to a general multiply periodic system. The next result, however, relating the classical and quantum expressions for the rate of light energy absorbed by a quantum system (in the correspondence limit) is highly non-trivial and requires complete facility with the intricacies of canonical perturbation theory. The calculation for absorption is considerably more involved than the analogous one for the polarization of the system (leading to the dispersion formula), as the perturbation theory for the action variables must be carried out to second order in the electric field of the light wave. The result, however, is of great interest, as it introduces a new feature, which would be critical in the dispersion case as well: the necessity for including stimulated emission (or "negative absorption" as it was called at the time) in order to attain agreement at high quantum numbers between the classical formula and the quantum expression given by Einstein's theory.

Section 10.3.2 illustrates the application of general canonical methods with a full derivation of the Kramers dispersion formula, which also contains "negative absorption" terms (see Sections 10.3.2–10.3.3). In rest of this section we illustrate Van Vleck's result for absorption only for the simple case of a one-dimensional harmonic oscillator.

We already derived the classical formula for the rate of emission of light energy of $\mathfrak{N}$ charged three-dimensional oscillators of natural frequency ν_0 immersed in a radiation field characterized by an energy density $\rho(\nu)$ (see Eq. (10.62)). In equilibrium, this must be equal to the net rate of absorption of energy by the oscillating particle. In one dimension (which means that the expression in Eq. (10.62) has to be divided by 3) and for a single oscillator ($\mathfrak{N}$=1), we thus have

$$\frac{\Delta E_{\text{abs,cl}}}{\Delta t} = \frac{\pi e^2}{3m}\rho(\nu_0). \tag{10.88}$$

In the corresponding quantum radiation description *à la* Einstein, the net light energy absorbed (per unit time) by a quantum oscillator in a given quantum state s is the difference between the light energy absorbed, accompanied by a transition to a higher quantum state r, *minus* the light energy lost via stimulated emission (or "negative absorption"), accompanied by a transition to a lower quantum state t. In the Einstein theory, these rates are determined by the B coefficients. In the case of a one-dimensional harmonic oscillator, the only higher state r possible is the one just above s, while the lower state t must be the one just below s.[31] For the net, or *differential*, rate of absorption we have (van Vleck 1924b, Eq. (15)):

$$\frac{\Delta E_{\text{abs,qu}}}{\Delta t} = h\nu_{rs}\,\rho(\nu_{rs})B_s^r - h\nu_{st}\,\rho(\nu_{st})B_s^t. \tag{10.89}$$

The relation between the B coefficients and the A coefficient is given by Eq. (10.68). For an oscillator in a state s with energy $nh\nu_0$, the state t has energy $(n-1)h\nu_0$, the state r energy $(n+1)h\nu_0$. The transition frequencies are $\nu_{rs} = \nu_{st} = \nu_0$. Using Eq. (10.76) for the relevant B coefficients, we find

$$\begin{aligned} B_s^r = B_n^{n+1} = B_{n+1}^n &= \frac{\pi e^2}{3hm\nu_0}(n+1), \\ B_s^t = B_n^{n-1} = B_{n-1}^n &= \frac{\pi e^2}{3hm\nu_0}n. \end{aligned} \tag{10.90}$$

[31] This is implied by the correspondence principle as the only Fourier component of the motion corresponds to the principal harmonic.

Inserting these values in Eq. (10.89) one finds

$$\frac{\Delta E_{\text{abs,qu}}}{\Delta t} = h\nu_0 \rho(\nu_0) \frac{\pi e^2}{3hm\nu_0}(n+1-n) = \frac{\pi e^2}{3m}\rho(\nu_0), \tag{10.91}$$

which agrees exactly with the classical result in Eq. (10.88).

The correspondence between classical and quantum behavior in this case is so straightforward that one is inclined to ascribe it to the extraordinarily simple character of a one-dimensional harmonic oscillator, where only the simplest transitions (to neighboring states) are allowed, all with the same transition frequency, the natural frequency of the oscillator. Moreover, the correspondence obtains not just at high quantum numbers—the agreement between the classical and quantum formulas persists down to the lowest quantum states as well. What Van Vleck was able to show was that a careful calculation of the generalization of Eq. (10.88) to a charged particle undergoing arbitrary three-dimensional multiply periodic motion (amenable to analysis in action/angle variables) could be brought into congruence with Eq. (10.89), the quantum formula for the differential rate of absorption, *in the limit of high quantum numbers*. This is just the sort of generalization needed in order to deal with electron motion in the case of the Bohr atomic model, which involved motion in a Coulomb (or Coulomb-like) potential, rather than in the simpler harmonic potential.

As we mentioned earlier, the classical perturbation theory required in the general multiply periodic case is quite involved, as the perturbation calculation must be carried to second order. The arguments used to establish the agreement, however, are identical in spirit to those employed by Kramers in the derivation of his dispersion formula (Duncan and Janssen 2007, p. 560, note 17; pp. 638–639). As the classical perturbation calculation needed for dispersion is only a first-order one, we describe the formalization of the correspondence principle introduced by Kramers and Van Vleck in the general multiply periodic case in the simpler case of dispersion. In preparation for this, Section 10.3.2 covers the perturbation theory needed to calculate the displacement of a particle in periodic or multiply periodic motion due to a periodic external applied force. Unfortunately, this involves—at intermediate stages—a number of complicated formulas, although the final result is refreshingly simple. Given the importance of the dispersion formula for subsequent developments, however, we decided to present its derivation in detail, treating the one-dimensional case before tackling the multi-dimensional one and offering a generous helping of intermediate steps. Hopefully, this will make it easier for the reader to absorb the derivation without having to take out pen and paper.

10.3.2 Dispersion in a classical general multiply periodic system

We first carry out the derivation of the dispersion formula for the special case of a periodic (but not necessarily harmonic—i.e., pure sinusoidal) one-dimensional system, as the formula for the general multiply periodic case is identical in form, with the only difference being the obvious introduction of appropriate summations over the multiple canonical degrees of freedom in the latter case. Our dispersion electron is therefore a particle of mass m, charge $-e$, subject to a symmetric potential $V(x) = V(-x)$ allowing bounded periodic motion with a period T and fundamental frequency $\nu_0 = 1/T$. We consider potentials more general than the simple-harmonic (quadratic in x) one, which means that we consider cases in which period and frequency depend on the amplitude of the oscillations of the particle.[32] In general, any periodic function can be written as a Fourier series, so in particular for the coordinate of the particle we must have

$$x_0(t) = \sum_{\tau} A_\tau e^{2i\pi\tau w}, \tag{10.92}$$

where $w = \nu_0 t$ and A_τ are complex coefficients, with the summation index τ running over all positive and negative integers. We may assume that $A_0 = 0$ by placing the center of the force and the time-average of the periodic motion at the origin. We impose the constraint $A_{-\tau} = A_\tau^\star$ to guarantee that the coordinate $x_0(t)$ is a real number.[33] Alternatively, one can employ the purely real form

$$x_0(t) = \sum_{\tau>0} X_\tau \cos(2\pi\tau\nu_0 t + \gamma_\tau), \tag{10.93}$$

where $X_\tau^2 = 4\,|A_\tau|^2$ and γ_τ is a phase angle for each harmonic (the phase of the complex number A_τ).

The Hamiltonian of our particle in canonical coordinates x and $p = m\dot{x}$ is given by

$$H_0 = \frac{p^2}{2m} + V(x). \tag{10.94}$$

By a suitable time-independent canonical transformation, one can treat the system instead in terms of a new momentum $\mathcal{J}$ and coordinate w, the so-called action and angle variables.[34] In terms of the new (canonical) variables, the Hamiltonian is independent of the angle variable w, so Hamilton's equations take the form

$$\frac{\partial H_0}{\partial w} = -\dot{\mathcal{J}} = 0, \tag{10.95}$$

$$\frac{\partial H_0}{\partial \mathcal{J}} = \dot{w} = \nu_0(\mathcal{J}). \tag{10.96}$$

[32] A concrete example would be the anharmonic oscillator with $V(x) = ax^2 + bx^4$. See Section A.2.4 for the more elementary example of a simple harmonic oscillator. See also Duncan and Janssen (2007, secs. 5.1 and 6.2).

[33] Using that $A_0 = 0$ and $A_{-\tau} = A_\tau^\star$, we can write Eq. (10.92) as:

$$x_0(t) = \sum_{\tau=-\infty}^{\infty} A_\tau e^{2i\pi\tau w} = \sum_{\tau=1}^{\infty} \left(A_\tau e^{2i\pi\tau w} + A_\tau^\star e^{-2i\pi\tau w}\right) = 2\,\mathrm{Re}\left(\sum_{\tau=1}^{\infty} A_\tau e^{2i\pi\tau w}\right).$$

[34] See Section A.2.3 for an introduction to the method of action/angle variables.

We now submit our oscillating particle (of charge $-e$) to the influence of an external electric field in the x-direction of fixed frequency ν described by $E_x(t) = \mathcal{E}\cos(2\pi\nu t)$. The frequency ν is unconnected to the natural frequency ν_0 of the particle. The field is switched on at time $t = 0$, so the Hamiltonian now becomes

$$H = H_0 + e\mathcal{E}x(t)\cos(2\pi\nu t)\,\theta(t), \tag{10.97}$$

where $\theta(t)$ is the theta function (defined as $\theta(t) = 0$ for $t < 0$ and $\theta(t) = 1$ for $t > 0$). Since we are only doing the calculation here to first order in $\mathcal{E}$, we can use the Fourier expansion of the unperturbed coordinate $x_0(t)$ in Eq. (10.92) for $x(t)$ in Eq. (10.97). For $t > 0$, we will get extra terms in Hamilton's equations compared to Eqs. (10.95)–(10.96). Eq. (10.95) gets replaced by

$$\begin{aligned}\dot{\mathcal{J}} &= -\frac{\partial H}{\partial w}\\ &= -e\mathcal{E}\frac{\partial}{\partial w}\left(\sum_\tau A_\tau e^{2i\pi\tau w}\right)\cos(2\pi\nu t)\\ &= -2i\pi e\mathcal{E}\sum_\tau \tau A_\tau e^{2i\pi\tau w}\cos(2\pi\nu t).\end{aligned} \tag{10.98}$$

Eq. (10.96) gets replaced by

$$\begin{aligned}\dot{w} &= \frac{\partial H}{\partial \mathcal{J}}\\ &= \frac{\partial H_0}{\partial \mathcal{J}} + e\mathcal{E}\frac{\partial}{\partial \mathcal{J}}\left(\sum_\tau A_\tau e^{2i\pi\tau w}\right)\cos(2\pi\nu t)\\ &= \nu_0(\mathcal{J}) + e\mathcal{E}\sum_\tau \frac{\partial A_\tau}{\partial \mathcal{J}} e^{2i\pi\tau w}\cos(2\pi\nu t).\end{aligned} \tag{10.99}$$

We now calculate the shifts $\Delta\mathcal{J}(t)$ and $\Delta w(t)$ induced by the electric field. Since there are no shifts before $t = 0$, we obtain the accumulated shifts at any later time by integrating the shifts in $\dot{\mathcal{J}}$ and $\dot{w}$ from 0 to t. For $\Delta\mathcal{J}(t)$, we note that $\dot{\mathcal{J}}$ vanishes as long as the electric field is switched off, so Eq. (10.98) gives the full shift $\Delta\dot{\mathcal{J}}$ induced by the field. With $\Delta w(t)$ we need to be more careful. In addition to the second term in the expression for $\dot{w}$ in Eq. (10.99), we need to extract the shift to first order in $\mathcal{E}$ in the term $\nu_0(\mathcal{J})$ coming from the first-order shift $\Delta\mathcal{J}$ in $\mathcal{J}$. The first-order shift in $\dot{w}$ is thus given by

$$\Delta\dot{w} = \frac{\partial \nu_0}{\partial \mathcal{J}}\Delta\mathcal{J} + e\mathcal{E}\sum_\tau \frac{\partial A_\tau}{\partial \mathcal{J}} e^{2i\pi\tau w}\cos(2\pi\nu t). \tag{10.100}$$

We are interested in the displacement $\Delta x_{\mathrm{coh}}(t)$ induced by and coherent with (i.e., of the same frequency as) the applied field (cf. Eq. (10.25)), to first order in the (by assumption small) electric field $\mathcal{E}$. It is critical at this point to realize[35] that we may proceed to obtain a general formula for this coherent response of our system to a periodic applied force *without specifying the specific form of the initial Hamiltonian* H_0 (or, for that matter, the explicit form of the canonical transformation yielding the action/angle variables $\mathcal{J}$ and w for this Hamiltonian). The success of the correspondence-principle approach to quantization employed by Kramers and Van Vleck hinges on obtaining the most universal characterization possible of the displacement of a multiply periodic system induced by a periodic external force. The first-order (in $\mathcal{E}$) calculation of the particle displacement can be carried out in many different ways. One can, for example, try to find action/angle variables for the new Hamiltonian Eq. (10.97), in terms of which the coordinate shift can be calculated. This is the approach taken by Born (1924) and, a few months later, by Kramers and Heisenberg (1925). In our view, the method used by Van Vleck (1924b, 1924c) is much simpler, inasmuch as it does not require intimate familiarity with the technology of canonical perturbation theory (Duncan and Janssen 2007, pp. 634–635).

In Van Vleck's version of the calculation, one continues to use the same time-independent canonical transformation leading to $\mathcal{J}$ and w above (and in terms of which the new Hamiltonian is equal to the old one, re-expressed in terms of the new variables), and instead calculates the perturbations $\Delta\mathcal{J}(t)$ and $\Delta w(t)$ in the action/angle variables $\mathcal{J}$ and w induced by the applied field by performing a time integral of the rates of change $\Delta\dot{\mathcal{J}}$ and $\Delta\dot{w}$ induced by the applied electric field of the incident light wave. One then uses the results to find the corresponding shift in the original physically relevant Cartesian coordinate $x(t) = x_0(t) + \Delta x(t)$. To first order, $\Delta x(t)$ is given by

$$\begin{aligned}\Delta x &= \frac{\partial x}{\partial \mathcal{J}}\Delta\mathcal{J} + \frac{\partial x}{\partial w}\Delta w \\ &= \sum_{\tau'}\left(\frac{\partial A_{\tau'}}{\partial \mathcal{J}}\Delta\mathcal{J} + 2\pi i\tau' A_{\tau'}\Delta w\right) e^{2\pi i\tau' w}. \qquad (10.101)\end{aligned}$$

Once the full shift Δx has been found, its coherent part can easily be extracted.

To find the first-order shifts $\Delta\mathcal{J}(t)$ and $\Delta w(t)$ at any time $t > 0$, we integrate their time derivatives from 0 to t. It will be convenient to convert the cosines to complex exponentials prior to the integration. Thus, for the shift in $\mathcal{J}$, we first rewrite Eq. (10.98) for $\dot{\mathcal{J}} = \Delta\dot{\mathcal{J}}$ as

$$\begin{aligned}\Delta\dot{\mathcal{J}} &= -i\pi e\mathcal{E}\sum_{\tau}\tau A_\tau e^{2i\pi\tau w}\left(e^{2i\pi\nu t} + [\nu \to -\nu]\right) \\ &= -i\pi e\mathcal{E}\sum_{\tau}\tau A_\tau\left(e^{2i\pi(\tau\nu_0+\nu)t} + [\nu \to -\nu]\right). \qquad (10.102)\end{aligned}$$

[35] As Van Vleck (1924c, p. 350) explicitly noted in a passage quoted toward the end of this section.

Here and below "$[\nu \to -\nu]$" is shorthand for "the same term with ν replaced by $-\nu$." Integrating this equation from 0 to t, we find:

$$\Delta\mathcal{J}(t) = \frac{e\mathcal{E}}{2}\sum_{\tau}\tau A_{\tau}\Big(\frac{1-e^{2i\pi(\tau\nu_0+\nu)t}}{\tau\nu_0+\nu} + [\nu \to -\nu]\Big). \tag{10.103}$$

Inserting this result into Eq. (10.100) for $\Delta\dot{w}$, we find

$$\begin{aligned}\Delta\dot{w} &= \frac{e\mathcal{E}}{2}\sum_{\tau}\Big\{\frac{\partial A_{\tau}}{\partial\mathcal{J}}e^{2i\pi(\tau\nu_0+\nu)t} + \tau\frac{\partial\nu_0}{\partial\mathcal{J}}A_{\tau}\frac{1-e^{2i\pi(\tau\nu_0+\nu)t}}{\tau\nu_0+\nu}\Big\}\\ &\quad + [\nu \to -\nu]).\end{aligned} \tag{10.104}$$

Integrating this equation from 0 to t, we find:

$$\begin{aligned}\Delta w(t) &= \frac{e\mathcal{E}}{4\pi}\sum_{\tau}\Big\{i\frac{\partial A_{\tau}}{\partial\mathcal{J}}\frac{1-e^{2i\pi(\tau\nu_0+\nu)t}}{\tau\nu_0+\nu}\\ &\quad + \tau\frac{\partial\nu_0}{\partial\mathcal{J}}A_{\tau}\frac{2\pi(\tau\nu_0+\nu)t - i(1-e^{2i\pi(\tau\nu_0+\nu)t})}{(\tau\nu_0+\nu)^2}\Big\}\\ &\quad + [\nu \to -\nu].\end{aligned} \tag{10.105}$$

The action and angle shifts $\Delta\mathcal{J}$ and Δw given by Eqs. (10.103) and (10.105) can now be inserted into Eq. (10.101) for Δx to obtain the following rather formidable expression for the perturbation in the oscillator coordinate induced by the electric field (to first order in the strength $\mathcal{E}$ of the field):

$$\begin{aligned}\Delta x(t) &= \frac{e\mathcal{E}}{2}\sum_{\tau,\tau'}\Big\{\Big(\tau\frac{\partial A_{\tau'}}{\partial\mathcal{J}}A_{\tau} - \tau'\frac{\partial A_{\tau}}{\partial\mathcal{J}}A_{\tau'}\Big)\frac{1-e^{2i\pi(\tau\nu_0+\nu)t}}{\tau\nu_0+\nu}\\ &\quad + \tau\tau' A_{\tau}A_{\tau'}\frac{\partial\nu_0}{\partial\mathcal{J}}\frac{2\pi i(\tau\nu_0+\nu)t + 1 - e^{2i\pi(\tau\nu_0+\nu)t}}{(\tau\nu_0+\nu)^2}\Big\}e^{2\pi i\tau'\nu_0 t}\\ &\quad + [\nu \to -\nu].\end{aligned} \tag{10.106}$$

There are two basic types of exponential time dependence appearing in this expression. There are terms that only involve the natural frequency harmonics $\tau'\nu_0$ of the oscillator itself (multiplied by a linear function of time $A+Bt$), and terms containing the frequency ν of the external electric field (i.e., of the light wave being dispersed), which all appear in the form of exponential $e^{2\pi i(\nu+(\tau+\tau')\nu_0)t}$. The former terms are of no interest for light scattering or dispersion, as they contain no reference to the frequency of the incoming light. The latter terms again fall into two categories. If $\tau + \tau' \neq 0$, they correspond to oscillations at a frequency $\nu + (\tau + \tau')\nu_0$ in which the original light frequency has been shifted by some harmonic of the charged particle motion.[36] However, for the subset of terms with $\tau + \tau' = 0$ (i.e., $\tau' = -\tau$), the displacement of our charged particle (and hence, the induced dipole moment, and polarization of an assembly of particles) is coherent, and in phase[37] with the incoming wave. This last situation is the one relevant for dispersion.

We may therefore obtain the desired coherent contribution to the particle displacement by selecting just the subset of terms with $\tau' = -\tau$, whereupon the double sum in Eq. (10.106) reduces to a single sum:

$$\Delta x_{\text{coh}} = -\frac{e\mathcal{E}}{2}\sum_{\tau}\left\{\tau\frac{\partial(A_\tau A_{-\tau})}{\partial\mathcal{J}}\frac{e^{2\pi i\nu t}}{\tau\nu_0+\nu} - A_\tau A_{-\tau}\tau\frac{\partial(\tau\nu_0)}{\partial\mathcal{J}}\frac{e^{2\pi i\nu t}}{(\tau\nu_0+\nu)^2}\right\} + [\nu \to -\nu]. \tag{10.107}$$

This expression contains, as expected classically, poles when the light frequency ν matches one of the mechanical frequencies $\tau\nu_0$ (for $\tau < 0$ in the displayed term, $\tau > 0$ in the term indicated by $[\nu \to -\nu]$) present in the Fourier decomposition of the particle's motion. This is just as expected from our experience with the Sellmeier formula and its descendants. Note that for the very special case of a harmonic oscillator, where the natural frequency is independent of the amplitude (and hence, the action variable $\mathcal{J}$), the second term containing *double* poles (which are *not* present in the Sellmeier formula) drops out, as $\partial\nu_0/\partial\mathcal{J} = 0$.

The generalization of Eq. (10.107) to a multi-periodic conditionally periodic system—amenable to analysis in action/angle variables—with u degrees of freedom is straightforward.[38] Any given coordinate of the system $x(t)$ now has the multiply periodic Fourier expansion (cf. Eq. (10.92))[39]

$$x(t) = \sum_{\vec{\tau}} A_{\vec{\tau}} e^{2\pi i\vec{\tau}\cdot\vec{w}}, \tag{10.108}$$

with the corresponding real (cosine) expansion

$$x(t) = \sum_{\vec{\tau},\vec{\tau}\cdot\vec{\nu}>0} X_{\vec{\tau}}\cos(2\pi\vec{\tau}\cdot\vec{\nu}t + \gamma_{\vec{\tau}}). \tag{10.109}$$

[36] We show later that these classical terms have a quantum counterpart, proposed theoretically by Smekal (1923) as a generalization of light absorption and emission in Einstein's radiation theory, and experimentally confirmed five years later by Raman (1928). The Smekal–Raman effect corresponds to the inelastic scattering of light: the emitted light differs in frequency from the absorbed light. Dispersion of light corresponds to elastic light scattering, in the special case where the emitted "secondary" light is of the same frequency and collinear with the incident light (forward scattering).

[37] For $\tau + \tau' = 0$, the complex phases of A_τ and $A_{\tau'}$ cancel. For inelastic scattering, where $\tau + \tau' \neq 0$, this does not happen: the resulting radiation is *incoherent*, with phases changing randomly from one scattering electron to the next.

[38] See Duncan and Janssen (2007, pp. 648–651) for a detailed derivation.

[39] The notation is greatly simplified by combining the assembly of u coordinates into single u-dimensional vectors: thus, $\vec{w} = (w_1, w_2, ..., w_u)$, and similarly for action coordinates $\vec{\mathcal{J}} = (\mathcal{J}_1, \mathcal{J}_2, ..., \mathcal{J}_u)$, and mode harmonic indices $\vec{\tau} = (\tau_1, \tau_2, ..., \tau_u)$.

(cf. Eqs. (10.92)–(10.93)). As before, the complex amplitudes in Eq. (10.108) must satisfy

$$A_{-\vec{\tau}} = A^{\star}_{\vec{\tau}} \tag{10.110}$$

to ensure that the sum is a real number and they are related to the real amplitudes $X_{\vec{\tau}}$ via

$$X^2_{\vec{\tau}} = 4A_{\vec{\tau}}A_{-\vec{\tau}}. \tag{10.111}$$

The u independent angle variables have been abbreviated into a single vector,

$$\vec{w} \equiv \vec{\nu}t \equiv (\nu_1, \nu_2, \ldots, \nu_u)\, t, \tag{10.112}$$

as have the integer coefficients specifying a particular harmonic of the system,

$$\vec{\tau} \equiv (\tau_1, \tau_2, \ldots \tau_u). \tag{10.113}$$

We assume that the Hamilton–Jacobi equation for the unperturbed Hamiltonian is separable[40] with action variables

$$\vec{\mathcal{J}} \equiv (\mathcal{J}_1, \mathcal{J}_2, \ldots \mathcal{J}_u) \tag{10.114}$$

conjugate to the angle variables $\vec{w}$. The single term $\tau\nu_0$ appearing in Eq. (10.107) is replaced by $\vec{\tau} \cdot \vec{\nu}$ and the action derivative $\tau(\partial/\partial\mathcal{J})$ by the corresponding multi-dimensional generalization

$$\sum_i \tau_i \frac{\partial}{\partial \mathcal{J}_i} \equiv \vec{\tau} \cdot \vec{\nabla}_{\mathcal{J}}. \tag{10.115}$$

Using the notation introduced in Eqs. (10.108)–(10.115), we can easily generalize Eq. (10.107) from one to many degrees of freedom:

$$\begin{aligned} \Delta x_{\text{coh}} &= -\frac{e\mathcal{E}}{2} \sum_{\tau} \left\{ \vec{\tau} \cdot \vec{\nabla}_{\mathcal{J}} \left(A_{\vec{\tau}} A_{-\vec{\tau}} \right) \frac{e^{2\pi i \nu t}}{\vec{\tau} \cdot \vec{\nu} + \nu} \right. \\ &\qquad \left. - A_{\vec{\tau}} A_{-\vec{\tau}}\, \vec{\tau} \cdot \vec{\nabla}_{\mathcal{J}} \left(\vec{\tau} \cdot \vec{\nu} \right) \frac{e^{2\pi i \nu t}}{(\vec{\tau} \cdot \vec{\nu} + \nu)^2} \right\} \\ &\quad + [\nu \to -\nu]. \end{aligned} \tag{10.116}$$

The resonance poles once again occur at values $\vec{\tau} \cdot \vec{\nu}$ corresponding to the mechanical frequencies of the problem, those present in the Fourier decomposition of the particle

[40] For discussion of these notions, see Sections A.2.1 and A.2.3.

motion, and *not* at the frequencies corresponding to the quantum transitions of the Bohr model. Moreover, in general, except for purely harmonic problems, there are the peculiar non-Sellmeier double poles in the second term in curly brackets, that is, divergences of the form $1/(\vec{\tau}\cdot\vec{\nu}+\nu)^2$.

We may return to a purely real notation by further simplifying this result to explicitly eliminate imaginary terms (Δx_{coh} is, of course, real). The terms in Eq. (10.116) involving $\sin(2\pi\nu t)$ vanish, as can be seen with the help of the identities

$$\sum_{\vec{\tau}} \tau_j \left(\frac{1}{\vec{\tau}\cdot\vec{\nu}+\nu} - \frac{1}{\vec{\tau}\cdot\vec{\nu}-\nu}\right)\cdot(\text{even function of } \vec{\tau}) = 0$$

$$\sum_{\vec{\tau}} \tau_j\tau_k \left(\frac{1}{(\vec{\tau}\cdot\vec{\nu}+\nu)^2} - \frac{1}{(\vec{\tau}\cdot\vec{\nu}-\nu)^2}\right)\cdot(\text{even function of } \vec{\tau}) = 0.$$

Thus Eq. (10.116) simplifies to

$$\begin{aligned}\Delta x_{\text{coh}} &= -\frac{e\mathcal{E}}{2}\cos(2\pi\nu t)\sum_{\vec{\tau}}\left\{\vec{\tau}\cdot\vec{\nabla}_J\left(\frac{A_{\vec{\tau}}A_{-\vec{\tau}}}{\vec{\tau}\cdot\vec{\nu}+\nu}\right)+[\nu\to-\nu]\right\}\\ &= e\mathcal{E}\cos(2\pi\nu t)\sum_{\vec{\tau}}\vec{\tau}\cdot\vec{\nabla}_J\left(\frac{\vec{\tau}\cdot\vec{\nu}A_{\vec{\tau}}A_{-\vec{\tau}}}{\nu^2-\vec{\tau}\cdot\vec{\nu}^2}\right).\end{aligned} \tag{10.117}$$

Using Eq. (10.111), we may go over to the cosine form of the expansion (cf. Eq. (10.109)), summing over only positive values of $\vec{\tau}\cdot\vec{\nu}$ (with a factor of 2):

$$\Delta x_{\text{coh}} = \frac{e\mathcal{E}}{2}\cos(2\pi\nu t)\sum_{\vec{\tau},\,\vec{\tau}\cdot\vec{\nu}>0}\vec{\tau}\cdot\vec{\nabla}_J\left(\frac{\vec{\tau}\cdot\vec{\nu}X_{\vec{\tau}}^2}{\nu^2-\vec{\tau}\cdot\vec{\nu}^2}\right). \tag{10.118}$$

The electric dipole moment induced by the electric field is obtained by multiplying this result by the electric charge, and the polarization by a further factor of the number density of charged oscillators.

Kramers must already have arrived at a formula equivalent to the one Eq. (10.118) by the Fall of 1923. It is a prerequisite to the correspondence-principle argument leading to his quantum dispersion formula, which, according to Slater (see note 30), he already had in hand in early December 1923. This classical formula is first explicitly stated in the literature in Kramers' (1924b) second note to *Nature* (dated July 22, 1924), albeit in highly abbreviated form. Explicit derivations by Born (1924) and, independently, Van Vleck (1924c), appeared in print in the Summer and Fall of 1924.

In the simplest case, a one-dimensional oscillator executing simple harmonic motion with frequency ν_0, the formula for Δx_{coh} simplifies as follows. The sum over $\vec{\tau}$ degenerates to a single term $\tau = 1$ (as there are no overtones in the Fourier expansion), the dot product $\vec{\tau} \cdot \vec{\nu}$ turns into ν_0 (which is independent of the single action variable $\mathcal{J}$), and $\vec{\tau} \cdot \vec{\nabla}_{\mathcal{J}}$ turns into $\partial/\partial\mathcal{J}$. For the amplitude $X_{\tau=1}$ we can use Eq. (A.191):[41]

$$X_{\tau=1} = \frac{1}{\pi}\sqrt{\frac{\mathcal{J}}{2m\nu_0}}. \tag{10.119}$$

With these simplifications, Eq. (10.118) turns into

$$\begin{aligned} \Delta x_{\text{coh}} &= \frac{e\mathcal{E}}{2}\cos\left(2\pi\nu t\right)\frac{\nu_0}{\nu^2-\nu_0^2}\frac{\partial}{\partial\mathcal{J}}\left(\frac{\mathcal{J}}{2m\pi^2\nu_0}\right) \\ &= \frac{e\mathcal{E}}{m}\frac{1}{\omega^2-\omega_0^2}\cos\left(\omega t\right), \end{aligned} \tag{10.120}$$

in agreement with Eqs. (10.27)–(10.29).[42] Of course, as we noted in Appendix A (p. 424), the use of sophisticated methods of canonical perturbation theory to obtain this simple result is akin to killing a fly with a sledgehammer. The response of a harmonic oscillator to a periodic applied force is much more readily obtained simply by solving Newton's equations, as we did in Section 10.1. However, although such methods are physically transparent, they depend on an explicit treatment of the equations of motion of a specific and completely specified Hamiltonian. The advantage of the canonical approach employing action/angle variables is that it allows us to obtain general formulas for the perturbation in the coordinate(s) of the system *completely independently of the specific nature of the dynamics.* Even when action/angle methods are employed, as in Epstein's treatment of dispersion discussed briefly in Section 10.2.1 in connection with the Sommerfeld–Debye theory, one may specialize too soon to special cases and miss the powerful and general result Eq. (10.118), which is in an ideal form to make the transition to quantum theory under the guidance of the correspondence principle.

As Van Vleck put it:

> If we were to study the perturbations in the motion produced by the incident wave purely with the aid of [Newton's second law] it would be impossible to make further progress without specializing the form of the potential function ... However, it is quite a different story when we seek to compute the perturbations ... in the "angle variables" w_1, w_2, w_3 and their conjugate momenta $\mathcal{J}_1, \mathcal{J}_2, \mathcal{J}_3$ ... In fact by using them rather than x, y, z, which is the essential feature of the present calculation, the periodic properties of the system come to light even without knowing the form of [the potential] (Van Vleck 1924c, p. 350).

[41] Equation (A.191) says that $A_{\tau=1} = \frac{1}{2}\sqrt{\frac{\mathcal{J}}{\pi m\omega_0}}$. Using Eq. (10.111), which in this case reduces to $X_{\tau=1}^2 = 4A_{\tau=1}^2$ or $X_{\tau=1} = 2A_{\tau=1}$, and $\omega_0 = 2\pi\nu_0$, we arrive at Eq. (10.119). For a simple derivation of Eq. (10.119), see Duncan and Janssen (2007, pp. 628–630, Eqs. (20)–(32)).

[42] We have ignored damping in our treatment of the dispersion problem in this section, which means that we set $\Gamma = 0$ in Eq. (10.29).

In particular, the feature of independence of specific dynamics extends to the associated quantum dispersion formula and explains why dispersion in much more complicated systems such as multi-electron atoms with electrons bound with Coulomb forces is described by Sellmeier formulas of the same structure as those for the dynamically much simpler three-dimensional isotropic harmonic oscillators at the heart of the classical Lorentz–Drude theory. The deep reason for this remarkable formal coincidence would only become apparent once the discovery and application of matrix mechanics revealed the underlying dynamical structure of quantum theory.

10.3.3 The Kramers dispersion formula

In the Fall of 1923, Kramers turned to the study of dispersion theory in the context of the old quantum theory.[43] In a letter of May 9, 1923, Reiche had already contacted Kramers with a request for assistance in deciphering some of the puzzling features he had uncovered with Ladenburg in their attempt to reconcile Bohr's correspondence principle with the properties of dispersion. At some point between this request and the end of the year, Kramers derived a new dispersion formula that had the standard Sellmeier form (with Ladenburg's 1921 reinterpretation of the numerator in terms of Einstein A coefficients) for systems in the normal (ground) state, but contained new *negative* terms for dispersing systems that were in excited quantum states.

As mentioned in note 30, we know from a letter Slater wrote to Van Vleck that Kramers had found this new formula by the time Slater arrived in Copenhagen in December 1923. Kramers, however, only published his formula in a note submitted to *Nature* in March and published in July 1924 without giving the reasoning that had led him to it (Kramers 1924a). In a second note to *Nature*, submitted in July and published in August 1924 (Kramers 1924b), he supplied a qualitative outline of his reasoning but it would not be till late 1924 that he started working on a paper (with Heisenberg) containing all the details of the argument and extending the formula to inelastic processes, that is, the Raman effect (Kramers and Heisenberg 1925). By that time, explicit derivations by Born (1924) and Van Vleck (1924b, 1924c) of the new dispersion formula (for the elastic, coherent dispersion case) based on the correspondence principle had already appeared in print.[44]

Our account of Kramers' discovery of the new formula must therefore be somewhat speculative, as there are no surviving contemporary written indications of his first attack on the problem in the Fall of 1923. Nevertheless, the indications given in the second

[43] For another account of Kramers' work on dispersion theory and references to both primary and secondary sources on the topic, see Konno (1993).

[44] In a letter to Kramers of April 9, 1924, Reiche wrote that he had been able to reconstruct the derivation of the new dispersion formula with the help of papers by Epstein (1922c) and Born and Pauli (1922). With Kramers' blessing, Reiche originally made this derivation part of a paper with Ladenburg in *Die Naturwissenschaften* (Ladenburg and Reiche 1924) but took it out when the editors asked the authors to shorten their paper. On June 5, 1924, Kramers had written to Reiche that "it will probably be a while before I have time to write [a detailed article on his dispersion formula]; because of lack of time [cf. note 30] I have not thought through many details and I consequently would not mind it at all if your note [with a derivation of the formula] appears first" (Duncan and Janssen 2007, p. 589–590).

Nature note are fairly clear, and the independent derivations given by Born and Van Vleck are so similar to each other and to these indications by Kramers, that we can reconstruct his reasoning with a considerable degree of confidence.

We shall, with Kramers (in his notes in *Nature*) make two simplifying assumptions at the outset. The charged oscillators under consideration, although moving in three dimensions, will be assumed to be restricted to motion in the direction of polarization (say, x) of the light impinging on them. They can thus be considered as one-dimensional oscillators, which results in an additional factor of 3 in Eq. (10.71). In addition, the statistical weights p_i (energy degeneracy) are assumed to be equal to 1 for all the states (as would be the case for one-dimensional oscillators). Thus, the effective number density of dispersion electrons $\mathfrak{N}$ in a state s is related to the actual number density of oscillators N_s and the Einstein spontaneous emission coefficient A_r^s for a transition from a higher state r to a lower state s by a formula of the kind found by Ladenburg (1921, cf. Eq. (10.71)):[45]

$$\mathfrak{N}_r = N_s A_r^s \frac{3mc^3}{8\pi^2 e^2 \nu_{rs}^2}, \tag{10.121}$$

where ν_{rs} is the quantum transition frequency given by the Bohr frequency condition, that is, $h\nu_{rs} = E_r - E_s$.[46]

We now consider light of angular frequency $\omega = 2\pi\nu$ impinging on an assembly of oscillators all in the quantum state s. Let $\omega_r = 2\pi\nu_{rs}$ be the angular frequency associated with the transition frequency ν_{rs}.[47] Including a sum over all possible higher states r, we find the following formula (in real form) for the resulting polarization (cf. Eq. (10.31))[48]

$$\begin{aligned} P_s(t) &= \frac{e^2}{m}\sum_r \frac{\mathfrak{N}_r}{\omega_r^2 - \omega^2}\,\mathcal{E}\cos(\omega t) \\ &= \frac{3c^3}{8\pi^2} N_s \sum_r \frac{A_r^s}{\nu_{rs}^2(\omega_r^2 - \omega^2)}\,\mathcal{E}\cos(\omega t) \\ &= \frac{3c^3}{32\pi^4} N_s \sum_r \frac{A_r^s}{\nu_{rs}^2(\nu_{rs}^2 - \nu^2)}\,\mathcal{E}\cos(\omega t). \end{aligned} \tag{10.122}$$

This formula, of classic Sellmeier form, contains only simple poles (divergences of the form $1/(\nu - \nu_{rs})$) when the frequency of the incident light approaches the resonance

[45] To facilitate the comparison with the notation used by Van Vleck in his treatment of absorption, which we followed in Section 10.3.2, and his derivation of the polarization formula, we relabeled the indices i and k used by Ladenburg as s and r, respectively.

[46] Note that the subscript r in $\mathfrak{N}_r$ indicates that, for a given state s, which the dispersing electrons subject to the incident light are occupying, $\mathfrak{N}_r$ "virtual" oscillators must be associated with each of the higher states r for which a quantum transition $r \to s$ is allowed.

[47] In the following, the state s is held fixed as we consider all possible transitions to other states r—accordingly, we suppress the indication of s in the ω subscript.

[48] Recall (see note 42) that, throughout our discussion of the Kramers dispersion theory, we ignore damping effects and thus set $\Gamma = 0$.

frequencies, associated with the absorption spectral lines of the system in quantum theory and with the frequencies present in the Fourier decomposition of the mechanical motion in classical theory. By contrast, Eq. (10.116), the classical formula for the coherent displacement Δx_{coh}, also contains *double* poles (divergences of the form $1/(\nu-\nu_{rs})^2$). These double poles are also implicit in Eq. (10.118), the simpler final formula, where they arise once the action derivatives are applied to the mechanical frequencies $\vec{\tau}\cdot\vec{\nu}$ in the (simple-pole) denominator.

At some point in 1923, Kramers must have realized—as Van Vleck did in his study of absorption (see Section 10.3.1)—that the derivatives with respect to continuous classical action variables $\mathcal{J}_i$ implied discrete differences once quantization had been imposed via $\mathcal{J}_i = n_i h$. Unlike the case with a continuous derivative, the application of a discrete difference operation to a term containing a single Sellmeier resonance pole would not result in a double pole, but rather in the algebraic difference between two simple pole terms. The non-Sellmeier double poles of the classical theory would thus be replaced by an equally non-canonical feature: simple poles of Sellmeier form, but with a *negative* coefficient. Such terms, of course, could not be given a classical interpretation, unless one wished to introduce particles of negative mass or imaginary charge (which would change the sign of e^2/m in the first line of Eq. (10.122)).

In the regime of very high quantum numbers the difference between taking derivatives with respect to an action variable and applying the discrete difference operation described above would be small. The vast majority of empirical studies of dispersion, however, involved atomic systems in the ground state. Even in the few exceptional cases where excited states were studied (as in the Balmer lines of hydrogen), the quantum numbers of the accessible states were quite low. To arrive at a quantum formula that would merge satisfactorily with the classical result Eq. (10.118), Kramers therefore had to leave the empirically accessible regime completely and embark on a purely conceptual thought experiment.

Imagine the dispersion of light by a set of oscillators in a very highly excited state s, with an infinite number of possible higher states r and a very large (though finite) number of lower states t. Of course, in the situation usually under empirical study, where s refers to the ground state of the system, the set of lower states would be empty. A simple generalization of Eq. (10.122), based on the work of Ladenburg (1921), containing negative simple-pole contributions is

$$P_s(t) = \frac{3c^3}{32\pi^4} N_s \left(\sum_{r>s} \frac{A_r^s}{\nu_{rs}^2(\nu_{rs}^2 - \nu^2)} - \sum_{t<s} \frac{A_s^t}{\nu_{st}^2(\nu_{st}^2 - \nu^2)} \right) \mathcal{E} \cos(2\pi\nu t). \tag{10.123}$$

In the correspondence limit where the state s is specified by very large quantum numbers n_1, n_2, and n_3 (for a particle in three dimensions), we can associate the states r and t in the sums in Eq. (10.123) symmetrically and pairwise with the central state s:

$$\begin{aligned} r &\to (n_1 + \tau_1, n_2 + \tau_2, n_3 + \tau_3), \\ t &\to (n_1 - \tau_1, n_2 - \tau_2, n_3 - \tau_3), \end{aligned} \tag{10.124}$$

with the integers τ_1, τ_2, and τ_3 chosen so that $\vec{\tau} \cdot \vec{\nu} > 0$ (where ν_1, ν_2, and ν_3 are the fundamental frequencies of the assumed multiply periodic motion), as in Eq. (10.118). We now assume that $\vec{\tau} \cdot \vec{\nu} \ll \vec{n} \cdot \vec{\nu}$, so transitions $r \to s \to t$ correspond to *very slight relative changes* in the classical orbits. In this limit we may expect that difference quotients in the quantum system will approximate well to derivatives in the classical theory. This leads to a new formal interpretation of Bohr's correspondence principle. Given that quantization is accomplished in the old quantum theory by the simple expedient of setting the action variables $\mathcal{J}_i = n_i h$, we may associate discrete differences in the quantum case with action derivatives in the classical theory as follows:[49]

$$\delta_{\vec{\tau}} F(\vec{n}) \equiv F(\vec{n}) - F(\vec{n} - \vec{\tau}) \quad \longrightarrow \quad h\vec{\tau} \cdot \vec{\nabla}_{\mathcal{J}} F(\vec{\mathcal{J}}). \tag{10.125}$$

As the transition $s \to t$ in the negative terms in Eq. (10.123) correspond to the positive ones involving the transition $r \to s$ with the replacement $\vec{n} \to \vec{n} - \vec{\tau}$, the Kramers polarization formula may be written, at least for the high terms for which the correspondence limit applies, as

$$P_s(t) = \frac{3c^3}{32\pi^4} N_s \mathcal{E} \cos(2\pi\nu t) \sum_{\vec{\tau}, \vec{\tau}\cdot\vec{\nu}>0} \delta_{\vec{\tau}} \left(\frac{A_r^s}{\nu_{rs}^2(\nu_{rs}^2 - \nu^2)} \right). \tag{10.126}$$

Using Eq. (10.87) to replace A_r^s by $(D_{\vec{\tau}}^r)^2$, with $D_{\vec{\tau}}^r = X_{\vec{\tau}}^r$ and $Y_{\vec{\tau}}^r = Z_{\vec{\tau}}^r = 0$, we can rewrite this as

$$P_s(t) = \frac{1}{2} e^2 N_s \mathcal{E} \cos(2\pi\nu t) \sum_{\vec{\tau}, \vec{\tau}\cdot\vec{\nu}>0} \frac{1}{h} \delta_{\vec{\tau}} \left(\frac{\nu_{rs}(X_{\vec{\tau}}^r)^2}{\nu_{rs}^2 - \nu^2} \right). \tag{10.127}$$

In this equation, we must take $\nu_{rs} = \vec{\tau} \cdot \vec{\nu}$, as the difference in energy $h\nu_{rs}$ between the state r with quantum numbers $(n_1 + \tau_1, n_2 + \tau_2, n_3 + \tau_3)$ and the state s with quantum numbers (n_1, n_2, n_3), which is just $h\vec{\tau} \cdot \vec{\nu}$. Employing the correspondence prescription Eq. (10.125) we obtain

$$P_s(t) = \frac{1}{2} e^2 N_s \mathcal{E} \cos(2\pi\nu t) \sum_{\vec{\tau}, \vec{\tau}\cdot\vec{\nu}>0} \vec{\tau} \cdot \vec{\nabla}_{\mathcal{J}} \left(\frac{\vec{\tau} \cdot \vec{\nu} (X_{\vec{\tau}}^r)^2}{(\vec{\tau} \cdot \vec{\nu})^2 - \nu^2} \right), \tag{10.128}$$

which agrees with the classical equation Eq. (10.118) for the displacement Δx_{coh}, once we have multiplied by a further factor of electric charge $-e$ and number density N_s to obtain the polarization.

[49] This was already known in the case when F is the Hamiltonian and was, in fact, a key ingredient of Bohr's correspondence principle (Darrigol 1992, p. 227, Eqs. (210) and (211)). Kramers, his biographer relates (Dresden 1987, p. 158), explained the formal association of action derivatives with discrete quantum differences to Slater at some time on or around January 17, 1924. He had presumably needed this already in the Fall of 1923 in his derivation of the new dispersion formula. The formal rule associating action derivatives to discrete difference operations is also discussed by Born (1924, sec. 3) and by Van Vleck (1924b), who found this rule independently of Kramers and of each other (Duncan and Janssen 2007, pp. 638–639).

Kramers' new dispersion formula, Eq. (10.123), differed from Ladenburg's by the additional negative terms, which are present only when the atoms in the dispersing medium are in an excited state. This meant that it was extremely difficult to establish experimentally the presence of the additional terms. The excited atomic states were typically produced by electrical discharge, where it was not possible to determine accurately the population of the different quantum levels. Heroic efforts of Ladenburg and collaborators established by the late 1920s the existence of a negative contribution attributable to the negative dispersion terms, but a precise quantitative comparison with the corresponding terms in the Kramers formula was not really possible given the technology available at the time (see Kopfermann and Ladenburg 1928). Nevertheless, the cogency of the correspondence-principle argument, which gave a precise agreement with the highly non-trivial classical formula, together with the already widespread acceptance of the existence of "negative absorption" (stimulated emission) processes in Einstein's (1916a, 1917a) quantum theory of radiation meant that this purely theoretical improvement gained widespread, if not universal, acceptance from the start.[50]

The need to calculate a "differential polarization", including negative terms associated with transitions to lower states for electrons already in excited states, would of course have been immediately apparent to Van Vleck, given his experience with exactly the same issue in his correspondence-principle approach to absorption, once he saw the new dispersion formula in Kramers' first note to *Nature.* Ladenburg and Reiche had become aware of this point somewhat earlier, as a letter of Reiche to Kramers of December 28, 1923, makes clear. Kramers had apparently criticized results like Eq. (10.79) in Ladenburg and Reiche (1923), presumably because of the strange implication that a quantum oscillator in an excited state would correspond to more than one dispersion oscillator. As Ladenburg and Reiche noted, the number of dispersion and actual oscillators would only agree for systems in the lowest quantum state, $n = 0$. But the two concepts should agree in the classical limit of large quantum numbers n, exactly where Eq. (10.79) gave $\mathfrak{N} \gg N$. In fact, it is precisely the inclusion of the negative terms in the Kramers formula that solves this problem.[51]

Let us see how this works in the case of dispersion from N_s one-dimensional harmonic oscillators of natural frequency ν_0, in state s (where s is now an integer). The sums in Eq. (10.123) reduce to a single positive and negative term

$$P_s(t) = \frac{3c^3}{32\pi^4} N_s \left(\frac{A_{s+1}^{s}}{\nu_0^2(\nu_0^2 - \nu^2)} - \frac{A_s^{s-1}}{\nu_0^2(\nu_0^2 - \nu^2)} \right) \mathcal{E} \cos(2\pi\nu t). \tag{10.129}$$

[50] Gregory Breit advanced an alternative formula for dispersion, criticizing Kramers' version: the second *Nature* note of Kramers is in fact a response to Breit's (1924) criticism. In his NRC bulletin on the old quantum theory two years later, Van Vleck (1926a, sec. 49, pp. 156–159) briefly discussed three alternative theories of dispersion (all using light quanta) by Smekal, Charles Galton Darwin, and Karl F. Herzfeld, but concluded that Kramers' theory "furnishes by far the most satisfactory theory of dispersion." That this theory was not universally accepted even at that late date is illustrated by a review of Van Vleck's bulletin. Commenting Van Vleck's opinion, Arthur E. Ruark (1926) wrote: "Many readers will not agree with the author's conclusion . . . the reviewer believes that a final solution cannot be achieved until we have a much more thorough knowledge of the dispersion curves of monatomic gases and vapors" (quoted in Midwinter and Janssen 2013, pp. 155–156).

[51] This point is appreciated, and clearly exhibited in an explicit calculation for the harmonic oscillator, in a letter from Reiche to Kramers of December 28, 1923.

From Eq. (10.76), we have

$$A^s_{s+1} = \frac{8\pi^2 e^2 \nu_0^2}{3mc^3}(s+1), \; A^{s-1}_s = \frac{8\pi^2 e^2 \nu_0^2}{3mc^3}s. \tag{10.130}$$

Inserting these values for the spontaneous emission coefficients, Eq. (10.129) becomes

$$P_s(t) = N_s \frac{e^2}{4\pi^2 m} \frac{1}{\nu_0^2 - \nu^2} \mathcal{E} \cos(2\pi\nu t) = N_s \frac{e^2}{m} \frac{1}{\omega_0^2 - \omega^2} \mathcal{E} \cos(\omega t). \tag{10.131}$$

The dependence on s has completely disappeared, and we see that there is a complete coincidence with the classical polarization result for all states s, and in particular for arbitrarily large quantum numbers (see Eq. (10.31), with $N_s = \mathfrak{N}_0, \Gamma = 0$, and taking the real part to get the physical polarization). The coincidence is of course crucially dependent on the new negative term(s) in the Kramers formula. Admittedly, the persistence in this case for low quantum numbers in the agreement of classical and quantum results, arrived at by application of the correspondence principle in the regime of high quantum numbers, is an artifact of the extremely simple dynamics of a harmonic oscillator.

The coincidence of the number of dispersion electrons associated with a given frequency of light and the actual number of optically active electrons just derived is in fact a *unique* property of the simple harmonic oscillator, and fails to hold in any other situation—for example, once anharmonicity is present. Certainly it does not hold for the mechanical motions appropriate for Bohrian atoms. The essential feature is that, for simple harmonic oscillators, transitions only occur from a given quantum state n to the immediate neighboring state $n \pm 1$, and all with the same frequency ν_0, the natural frequency of the oscillator. Once transitions are possible to more than one state from the given dispersing state (the state s above), the identity between the dispersion numbers $\mathfrak{N}$ appearing in the numerators of the Sellmeier poles and the actual number of dispersing electrons N_s is destroyed. Instead, a more general sum rule, discovered independently by Willy Thomas (1925) in Breslau and Werner Kuhn (1925) in Copenhagen, comes into play, in which all available transitions (up and down from the given state, with minus signs for the "negative dispersing" terms) must be taken into account (see Section 10.5). For the special case of the one-dimensional harmonic oscillator, this sum degenerates to just the two terms in Eq. (10.129), with the simple result that the number of dispersion and actual oscillators agree.

As discussed in the introduction to Section 10.3, both in the preliminary notes to *Nature* of March and July 1924, in which Kramers (1924a, 1924b) first published his dispersion formula, and in the detailed exposition and extension of his ideas in a paper co-authored with Heisenberg over the Christmas break of 1924 and published the following March (Kramers and Heisenberg 1925), the Kramers dispersion theory is presented in the context of the BKS theory from early 1924. Its rise and (rapid) fall is the topic of the next section. Here we remind the reader that we have Slater's testimony (see note 30) that Kramers had already found his dispersion formula *before* Slater arrived in Copenhagen in December 1923. Since the BKS theory was only assembled *after* Slater's arrival, Kramers' work on dispersion covered in this section clearly predates it. As our analysis has shown and as our discussion in the next section confirms, the correspondence-principle arguments that led Kramers to his quantum dispersion theory are logically independent of the BKS theory. The same is true for the further elaboration of Kramers' dispersion theory in his paper with Heisenberg, which we examine in detail in Section 10.5. Within a little over a year, the BKS theory was decisively refuted in experiments in X-ray scattering experiments of the kind that had led Compton to the discovery in late 1922 of the effect named after him (Bothe and Geiger 1925b; Compton and Simon 1925). By April 1925 it was widely accepted that the BKS theory had become untenable. Despite its association with the BKS theory, however, Kramers' dispersion theory survived this refutation unscathed.

10.4 Intermezzo: the BKS theory and the Compton effect

This section takes a closer look at the BKS theory, the controversial quantum theory of radiation of Bohr, Kramers, and Slater. The paper in which they introduced this theory is signed "Institute for Theoretical Physics, Copenhagen, January 1924" but only appeared in the *Philosophical Magazine* in April 1924 (Bohr, Kramers, and Slater 1924a, p. 162).[52] However, the main thrust of the BKS theory was known to the community of physicists working in this area well before the paper was published. Both the paper and the theory are best described as programmatic. The only equation appearing in the paper is the Bohr frequency condition, $h\nu = E_1 - E_2$ (ibid., p. 162, eq. (1)). Many aspects of the theory remained to be worked out but never were during its short lifespan.

We start our discussion of the BKS theory by examining some of the considerations that went into its formulation. This takes us back to the early 1920s. In late 1922, the discovery of the Compton effect provided strong evidence for Einstein's light quanta and made it particularly challenging to formulate a quantum theory of radiation in which radiation is still conceived as consisting of classical electromagnetic waves rather than particles. Bohr, Kramers, and Slater tried to pull this off by turning these waves into what they called "virtual radiation" emitted by "virtual oscillators" (the substitute oscillators familiar from dispersion theory) determining probabilities (according to some unspecified rule) of electrons jumping from one orbit to another. This mechanism forced them to give up strict energy conservation, which turned out to be the theory's undoing.[53]

[52] The paper would also be published in German translation (Bohr, Kramers, and Slater 1924b). There is an extensive historical literature on the BKS theory. See, for example, Klein (1970, pp. 23–39), Stuewer (1975, pp. 291–305; 2014, pp. 161–165), Hendry (1981), the dissertation of Neil Wasserman (1981), Pais (1982a, Ch. 22, pp. 416–422; see also Pais 1991, sec. 11(d), pp. 232–239), Mehra and Rechenberg (1982a, secs. V.2 and V.5), Dresden (1987, pp. 159–215), the editorial note by Klaus Stolzenburg (1984) in Bohr's collected papers, Darrigol (1992, pp. 214–224), and Kragh (2012, pp. 325–337). Our discussion follows Duncan and Janssen (2007, sec. 4, pp. 597–617).

[53] This stochastic element did earn the BKS theory a place in (the discussion about) Paul Forman's (1971) classic paper, "Weimar culture, causality, and quantum theory, 1918–1927: Adaptation by German physicists and mathematicians to a hostile intellectual environment." Embracing acausality, Forman argued, was part of such adaptation. Even though none of the three authors of the BKS paper were German, their theory did

Using the Compton effect, which had already spelled trouble for the wave theory, Bothe and Geiger (1925b) and Compton and Simon (1925) showed that energy is strictly conserved in the relevant processes, thereby conclusively disproving the one definite prediction of the BKS theory. Given that the theory only lasted a little over a year, its impact on the development of quantum theory was limited and transient. This makes it easy to forget both how prominent it was during its short lifetime and how radically Bohr was willing to modify the earlier theory less than two years before it gave way to modern quantum mechanics. We will see how these modifications actually brought the theory in closer alignment with the correspondence principle.

While embraced enthusiastically by some, the BKS theory was met with profound skepticism by others. Among the supporters were Heisenberg, Born, Ladenburg, Reiche, and Schrödinger.[54] The most prominent skeptics were Einstein and Pauli. In a well-known passage from a letter to Born and his wife, Einstein wrote:[55]

> The idea that an electron ejected by a light ray can choose *of its own free will* the moment and direction in which it will fly off, is intolerable to me. When it comes to that, I would rather be a shoemaker or even an employee in a gambling casino than a physicist. My attempts to give quanta a form one can grasp have failed again and again, it is true, but I am far from giving up hope.[56]

Einstein's remarks point to two of three key features of the BKS theory: the rejection of light quanta and the abandonment of strict energy conservation. The third is the introduction of "virtual radiation", which is responsible for the violation of energy conservation.

Adopting a statistical law of energy conservation was the price Bohr and company were willing to pay to avoid light quanta and preserve the wave theory of light. Like most of his contemporaries (see Chapter 3), Bohr had been a staunch opponent of Einstein's light-quantum hypothesis from the beginning. In the first installment of his 1913 trilogy, for instance, he wrote that he wanted to show that his theory

> may afford a simple basis of representing a number of experimental facts which cannot be explained by help of the ordinary electrodynamics, and that *the assumptions used do*

have a considerable following in Germany for reasons that may have included cultural factors. We focus on factors internal to the physics of the day. There is an extensive literature on the Forman thesis. See, for example, Carson, Kojevnikov, and Trischler (2011), which grew out of a 2007 conference devoted to the Forman thesis, reviews of that volume by Camilleri (2012) and Seth (2013), and the contributions by Kojevnikov (2022) and Forman (2022) himself to the *Oxford Handbook on the History of Interpretations of Quantum Mechanics* (Freire 2022).

[54] Schrödinger (1924c) published a favorable review of the theory in *Die Naturwissenschaften* in September 1914 (Stuewer 1975, pp. 297–299).

[55] For further discussion of Einstein's reaction to the BKS theory (and the Compton effect), see Einstein (1987–2021, Vol. 14, Introduction, sec. II, pp. xl–xlv).

[56] Einstein to Max and Hedi Born, April 29, 1924 (Vol. 14, Doc. 240), quoted and discussed, for example, in Klein (1970, p. 32).

> *not seem to be inconsistent with* experiments on phenomena for which a satisfactory explanation has been given by the classical dynamics and *the wave theory of light* (Bohr 1913a, p. 19, our emphasis).

The photoelectric effect is one of the examples he gave: "Obviously, we get . . . the same expression for the kinetic energy of an electron ejected from an atom by photo-electric effect as that deduced by Einstein" (Bohr 1913a, p. 17). Over the next few years, Bohr's resistance to light quanta only hardened, even as he helped himself to elements of Einstein's (1916a, 1917a) quantum theory of radiation, notably the A and B coefficients. Bohr felt that light quanta could not be reconciled with the correspondence principle, with which he hoped "to trace the analogy between the quantum theory and the ordinary theory of radiation as closely as possible" (Bohr 1918, p. 4, quoted and discussed in Section 1.3.4).

In the draft of a letter composed in the summer of 1919 but never actually sent, Bohr already contemplated giving up strict energy conservation to avoid light quanta.[57] Another hint that he was considering such a move can be found in comments he prepared for the third Solvay conference held in Brussels in April 1921. These comments were presented by Ehrenfest as illness prevented Bohr from attending himself. Light quanta, Bohr conceded, seem to be the only way to account for the photoelectric effect but then he added the tantalizing clause, "if we stick to the unrestricted applicability of the ideas of energy and momentum conservation" (Bohr 1923a, pp. 241–242). A letter written early the following year by Bohr's Leyden colleague and Solvay spokesman confirmed the impression given by this conditional statement. On January 17, 1922, Ehrenfest wrote to Einstein that Bohr was "much more willing to give up the energy and momentum theorems (in their classical form) for elementary atomic processes, and to maintain them only statistically, than to 'lay the blame on the ether' " (Einstein 1987–2021, Vol. 13, Doc. 24). Bohr finally made a more definite statement himself in paper completed in November 1922 and published in the spring of 1923, the first (and, as it turned out, only) part of a series, "On the application of the quantum theory to atomic structure" (Bohr 1923c). After reviewing the difficulties facing the light-quantum hypothesis, he concluded that a "general description of the phenomena, in which the laws of the conservation of energy and momentum retain in detail their validity in their classical formulation, cannot be carried through" (Bohr 1923a, p. 40).[58]

While Bohr looked for ways to avoid light quanta, Einstein looked for ways to confirm them. The letter from Ehrenfest to Einstein was written in connection with a proposal by Einstein in late 1921 of what he thought was a crucial experiment, involving the emission of light from canal rays, to decide between the light-quantum hypothesis and the wave theory of light (Einstein 1922a). Hans Geiger and Walther Bothe immediately set out to perform this experiment at the *Physikalische Reichsanstalt* in Berlin and found what Einstein expected on the basis of the light-quantum hypothesis. With this result, Einstein told Hedwig and Max Born in a letter of December 30, 1921, "it is proven

[57] Bohr to Charles G. Darwin, July 1919, reproduced in full by Stolzenburg (1984, p. 15).
[58] The quotations in this paragraph can all be found in Klein (1970, pp. 19–22).

beyond doubt that the undulatory field has no real existence and that Bohr emission is an instantaneous process in the true sense of the word. This is my strongest scientific experience in years" (Einstein 1987–2021, Vol. 12, Doc. 345). He informed Ehrenfest of these developments in a letter of January 11, 1922 (Einstein 1987–2021, Vol. 13, Doc. 13). In his response of January 17, 1922, the incredulous recipient told his friend: "If your light experiment really turns out anti-classically—I mean, after both theoretical and experimental criticism—well, you know, then you give me the spooks" (Einstein 1987–2021, Vol. 13, Doc. 24).[59] In a follow-up letter two days later, however, the same day that Einstein gave a lecture about the experiment at the Prussian Academy (Einstein 1987–2021, Vol. 13, Doc. 29), Ehrenfest showed that Einstein had used a phase velocity where he should have used a group velocity in his calculation for the wave theory. If this error was corrected, Ehrenfest showed, the wave theory predicted the same result as the light–quantum hypothesis (Einstein 1987–2021, Vol. 13, Doc. 30). After some further back-and-forth, Einstein conceded the point and withdrew his claim in another lecture at the Prussian Academy two weeks later, on February 2, 1922 (Einstein 1922b). Looking back on this episode a few months later, in May 1922, Einstein wrote to Born that he "had committed a colossal blunder" [*einen monumentalen Bock geschossen*], but consoled himself with the thought that "the only thing that helps against blundering [*Böcke-Schiessen*] is death" (Einstein 1987–2021, Vol. 13, Doc. 190).[60]

By the end of 1922, despite this embarrassing episode at the beginning of the year, the tide was finally turning in favor of Einstein's light-quantum hypothesis. In December 1922, Arthur Holly Compton, an American experimentalist at Washington University in St. Louis, reported the effect (soon to be named after him) in a paper he submitted to *Physical Review*, where it appeared in March of the following year (Compton 1923).[61] It won him the 1927 Nobel Prize. Compton measured the decrease in frequency of X-rays scattered by graphite as a function of the scattering angle. He found a formula that could accurately account for this effect by treating the interaction of the X-rays with their target as collisions between light quanta (with energy $h\nu$ and momentum $h\nu/c$) and electrons in the graphite and applying the relativistic conservation laws of energy and momentum. Examining the data on X-ray scattering published by Compton (1922) a few months earlier, Debye (1923) independently arrived at the same interpretation in terms of collisions between light quanta and elections (Stuewer 1975, pp. 234–237).[62]

[59] Klein (1970, p. 10) translates this last clause, *dann bist du mir unheimlich geworden*, as "then you will have become really *uncanny* to me." It seems odd to apply the term "uncanny" to a person but the same can be said about the term "unheimlich" in German.

[60] "Einen Bock schiessen" (literally: shooting a ram) is standard German idiom for committing a blunder. For discussion of this episode, see Klein (1970, sec. 3, pp. 8–13). A few years later, Einstein (1926a) proposed another experiment to decide between a wave and a particle theory of light. Emil Rupp claimed to have confirmed the prediction of the particle theory but his results later turned out to have been fraudulent (van Dongen 2007a).

[61] For the authoritative account of the research program, begun in 1916, that eventually led Compton to his discovery, see Stuewer (1975). Unlike Millikan (see Section 3.3.3), Compton did not embark on this research program motivated to prove or disprove Einstein's light quantum hypothesis. For a concise discussion of the role of the Compton effect both in getting Einstein's quantum theory of radiation accepted and in bringing down the BKS theory, see Stuewer (2014, pp. 157–165).

[62] See also Wheaton (1983, pp. 285–286).

It so happened that Sommerfeld, unaware of his former student Debye's interest in these X-ray scattering experiments, was on a lecture tour in the United States when Compton announced his results (Eckert 2013, p. 270). Sommerfeld spent Christmas 1922 with another former student, Jacob Kunz, a professor at the University of Illinois in Urbana. As he wrote to his wife, Sommerfeld considered making the seven-hour journey to St. Louis to visit Compton but could save himself the trouble "since the man I wanted to speak to came to Urbana for ten hours."[63] On January 21, 1923, Sommerfeld reported to Bohr from California: "The most interesting thing I have experienced scientifically in America . . . is a work of Arthur Compton in St. Louis. After it the wave theory of Röntgen-rays will become invalid" (Sommerfeld 2004, Doc. 65, quoted in Stuewer 1975, p. 241). He told Bohr that it was not entirely clear yet whether the result would hold up but wanted to let him know that "eventually we may expect a completely fundamental and new lesson" (ibid.).

Sommerfeld championed Compton's work wherever he went. Perley Ason Ross, an experimental physicist at Stanford, learned about Compton's results from Sommerfeld's lectures at Berkeley and set out to do his own experiments to see whether he could confirm them (Stuewer 1975, pp. 241, 245). He found conflicting results at first but by early June his data agreed with Compton's findings (Ross 1923a). In September, he could report additional measurements supporting Compton's (Ross 1923b). Bothe (1923, 1924) in Berlin and C. T. R. Wilson (1923a, 1923b) in Cambridge also confirmed Compton's results (Stuewer 1975, pp. 242–243).

The experiments by Ross in particular seem to have dispelled the doubts Sommerfeld expressed in his letter to Bohr. After his return to Munich, on October 9, 1923, Sommerfeld wrote to Compton that

> [y]our discovery of the change in wavelength of Röntgen rays keeps the scientific world in Germany extremely busy . . . [I]n my book [*Atombau und Spektrallinien*], the 4th edition of which I am now preparing, I have inserted a section in the first chapter on the quantum structure of light; I there discuss the "Doppler effect and the Compton effect" [Sommerfeld 1924, pp. 52–56] . . . After the beautiful experiments of Ross, there can be no doubt that your observation and theory are completely accurate . . . I was asked to discuss your results in the *Physikalische Zeitschrift*. I have not yet had time to do it and was very surprised to see Debye's note on it [Debye 1923], which in essence agrees with your theory . . . I wrote to Debye that you naturally have the priority not only in the experiments, but also in the theory (Sommerfeld 2004, Doc. 68).[64]

Sommerfeld was probably the first to use the term "Compton effect" in print (Stuewer 1975, p. 249). Whether or not he coined the term, his prominent discussion of the "Compton effect" in the fourth edition of the "Bible" of atomic theory is probably

[63] Arnold to Johanna Sommerfeld, December 28–29, 1922, quoted by Eckert (2013, p. 270).
[64] A full translation of this letter is given in Stuewer (1975, p. 247).

what ensured that the name stuck and that the effect did not become known as the Compton–Debye effect.[65]

In the first edition of *Atombau und Spektrallinien*, Sommerfeld (1919, p. 382, note) had explicitly rejected Einstein's (1916a, 1917a) quantum theory of radiation (see Section 3.6.2). In the preface to the fourth edition, he did an about-face:

> Whereas I formerly sought to uphold the wave theory for the pure propagation processes as long as possible, the Compton Effect forces me more and more to accept the extreme theory of light quanta (Sommerfeld 1924, p. VII, quoted in Eckert 2013, p. 285)

That Sommerfeld gave Compton's work his papal blessing was probably at least partially responsible for European physicists not paying much attention to a controversy in which Compton would soon find himself embroiled in the United States. In October 1923, right around the time of the letter from Sommerfeld to Compton, William Duane, professor of bio-physics at Harvard, and George L. Clark, a National Research Council Fellow working in Duane's laboratory, announced that they had been unable to reproduce Compton's results (Clark and Duane 1923a, 1923b). This touched off what Compton, decades later, described as "the most lively scientific controversy that I have ever known" (Compton 1961, p. 818).[66] It was not until late 1924 that Duane finally backed down. As noted by Stuewer (1975, p. 273), the end of the Duane–Compton controversy meant that Compton's experimental findings were now accepted everywhere. However, the "conceptual turmoil" (ibid.) over how to make sense of them was just beginning.

The most important alternative to the account of the Compton effect based on Einstein's light-quantum hypothesis was the account based on the BKS theory,[67] in which radiation consists of waves rather than particles, and the Compton effect comes out as a Doppler shift. Compton himself explored this option. The relativistic conservation laws of energy and momentum gave him the recoil velocity of the electron after its interaction with the X-ray radiation. He then calculated the velocity the electron should have if the difference in frequency between absorbed and reemitted radiation were simply a Doppler shift. These two velocities are not the same. They are not in the same direction and they have different magnitudes, given as fractions β and β' of the velocity of light, respectively (Compton 1923, p. 487, Eqs. (6) and (7)). Compton, however, did not conclude that the Compton effect therefore cannot be a Doppler effect. Instead, noting that there is a simple relation between β and β', he wrote: "so far as the effect on the wave-length is concerned, we may replace the recoiling electron by a scattering electron moving in the direction of the incident beam" with a velocity of just the right size to make the observed frequency shift come out as a Doppler shift (Compton 1923,

[65] Debye did not complain and appears to have made his peace with this turn of events because, as he said when interviewed for the AHQP project in the 1960s, he had only produced the theory, whereas Compton had both worked out the theory and carried out the experiments (Stuewer 1975, p. 237).

[66] Quoted by Stuewer (1975, p. 249), who discusses the ensuing controversy in detail (ibid., pp. 249–273).

[67] For discussion of other attempts to account for the Compton effect, see Stuewer (1975, pp. 288–291).

p. 487).[68] The authors of the BKS paper seized upon this option of using two different velocities to account for the Compton effect.

In the BKS theory atoms are represented by two sets of quantities. In addition to the orbits of the original Bohr–Sommerfeld theory there are virtual oscillators for every pair of orbits with characteristic frequencies determined by the energy difference between them.[69] As the authors explicitly acknowledge, these virtual oscillators are just the substitute oscillators (*Ersatzoszillatoren*) of dispersion theory under a new name. They cite both Ladenburg (1921) and Ladenburg and Reiche (1923) at this point (Bohr, Kramers, and Slater 1924a, pp. 163–164). Treating the electrons in Compton's experiment in similar terms, the BKS theory explained the Compton effect as a Doppler shift by assigning the recoil velocity to the electron and the velocity required to get the right Doppler shift to the virtual oscillator associated with the electron. The authors of the BKS paper recognized that this was an unusual move (to put it mildly) but were undaunted:

> That in this case the virtual oscillator moves with a velocity different from that of the illuminated electrons themselves is certainly a feature strikingly unfamiliar to the classical conceptions. In view of the fundamental departures from the classical space-time description, involved in the very idea of virtual oscillators, it seems at the present state of science hardly justifiable to reject a formal interpretation as that under consideration as inadequate (Bohr, Kramers, and Slater 1924a, p. 173).

We already noted that the BKS theory is incompatible with strict energy conservation. We now see that it is also incompatible with "the classical space-time description." This raises the question of how the authors arrived at this peculiar theory and what they (and more than a few of their readers) found so appealing about it.

The BKS paper was triggered by an idea of its third and most junior author. John C. Slater, a graduate student at Harvard in the same cohort as Van Vleck,[70] earned

[68] For analysis of Compton (1923), see Stuewer (1975, pp. 223–232).

[69] The phrase "virtual orchestra" for the set of all virtual oscillators associated with an atom comes from a later paper by Landé (1926, p. 456), who used it to describe matrix mechanics rather than the BKS theory. The author of a book on the new physics aimed at a general audience still used this imagery to describe matrix mechanics in the 1930s:

> The state of an atom should no longer be described by the unobservable position and momentum of its electrons, but by the measurable frequencies and intensities of its spectral lines ... Regardless of the nature of the real musicians who play the optical music of the atoms for us, Heisenberg imagines assistant or auxiliary musicians [*Hilfsmusiker*]: every one plays just one note at a certain volume. Every one of these musicians is represented by a mathematical expression, q_{mn}, which contains the volume and the frequency of the spectral line ... These auxiliary musicians are lined up in an orchestra [*Kapelle*] according to the initial and final states n and m of the transition under consideration. The mathematician calls such an arrangement a "matrix" (Zimmer 1934, pp. 161–162; quoted in Duncan and Janssen 2007, p. 616).

[70] On the role that young American theorists like Van Vleck and Slater played in the transition from the old quantum theory to matrix and wave mechanics, see, for example, Coben (1971), Schweber (1986),

his PhD in 1923 and was awarded a Sheldon Fellowship to visit some of the leading centers of physics in Europe. He spent the fall of 1923 at the Cavendish Laboratory in Cambridge. Around Christmas, he traveled on to Copenhagen, where he would spend the first half of 1924 in Bohr's institute. Shortly after his arrival, Slater suggested to Bohr and Kramers that one way to reconcile the wave and particle properties of light might be to introduce a virtual electromagnetic field equivalent to that generated by an assembly of classical charged oscillators with frequencies equal to all possible atomic transitions, which would guide—probabilistically—the emission and absorption of light quanta.[71] Bohr and Kramers liked the idea of a virtual field of this kind, but stripped it of all reference to light quanta.[72] Against his better judgment, as he later insisted, Slater went along with this.[73] As he told his family[74] and reiterated in a letter to Van Vleck on his way back to the United States (see note 30), Slater had no hand in the actual writing of the BKS paper. In a short letter sent to *Nature* a week after the paper had been submitted, Slater, talked out of its original version by Bohr and Kramers, couched his idea in proper BKS terms:

> Any atom may, in fact, be supposed to communicate with other atoms all the time it is in a stationary state, by means of a virtual field of radiation originating from oscillators having the frequencies of possible quantum transitions and the function of which is to provide for the statistical conservation of energy and momentum by determining the probabilities for quantum transitions (Slater 1924, p. 307).

In Bohr's (1913a) original theory and its further developments by Sommerfeld and others, atoms only emit and absorb radiation, real and in the form of classical waves, when they jump from one stationary state to another. In the new theory, radiation, still in the form of classical waves but now somehow virtual rather than real, is produced while atoms stay in their stationary states. The BKS paper fails to make it clear in exactly

Sopka (1988), and Duncan and Janssen (2007, section 2, pp. 560–571). On Slater's education and early work, see Wasserman (1981, Ch. 5).

[71] Slater probably did not realize that Einstein and De Broglie had proposed similar ideas before. For discussion, see Hendry (1981, p. 199) and Darrigol (1992, p. 218).

[72] In his biography of Kramers, Dresden (1987, pp. 170, p. 290) relates the story of how, in 1921, Bohr had likewise talked Kramers out of the idea of light quanta. Kramers, it seems, had actually come close to anticipating the Compton effect. However, as Dresden describes the situation, Bohr subjected Kramers "to an unrelenting series of interminable and acrimonious discussions in which the tenuous ... character of the photon notion was examined in excruciating detail ... After or during these discussions, which left Kramers exhausted, depressed, and let down, Kramers got sick and spent some time in the hospital ... Kramers did not only acquiesce to Bohr's views; he made Bohr's arguments his own ... Kramers' complete reversal is a perfect example of the fanatical adherence recent converts often exhibit to their new-found faith" (Dresden 1987, p. 290).

[73] As documented by Dresden (1987, pp. 168–171) and Schweber (1990, pp. 350–356), this led to a lingering resentment on Slater's part against Bohr and his Copenhagen institute.

[74] Pais (1991, p. 235) quotes from six letters from Slater to his family of January 1924, which allow us to follow how Slater came around to Bohr and Kramers's point of view. On January 6, he grudgingly wrote that he was willing to let Bohr and Kramers "have their way," as he had concluded that "the part they believe is the only part that leads to any results anyway." On January 18, he reported that he had "finally become convinced that the way they want things, without the little lump carried along on the waves ... is better." On January 22, he wrote that the "paper is just about done. Prof. Bohr has done all the writing, but it suits me just fine."

what sense the radiation field through which atoms are supposed to "communicate" with each other is "virtual". As the German translation of the BKS paper was being prepared, Bohr wrote to Pauli, anxious to get his endorsement of "the words 'communicate' and 'virtual', for after lengthy consideration, we have agreed here on these basic pillars of the exposition."[75] Pauli, who had not yet seen the paper, replied: "On the basis of my knowledge of these two words (which I definitely promise you not to undermine), I have tried to guess what your paper may deal with. But I have not succeeded."[76]

So what does it mean for radiation to be "virtual"?[77] In Slater's original conception, the radiation might be called virtual because light quanta are primary and the radiation is there only to guide them. In the BKS theory, however, there are no light quanta, there is only radiation. In follow-up papers by both Kramers and Slater, it is made clear that the "virtual" radiation through which atoms "communicate" with each other is on the same footing as external radiation.[78] It is hard to see how this could be otherwise. After all, it should be possible to have interference between these two types of radiation. As it says in the BKS paper, "we shall assume that [illuminated atoms] will act as secondary sources of virtual wave radiation which interferes with the incident radiation" (1924a, p. 167) A few pages later, the authors refer to external radiation as "incident virtual radiation" (Bohr, Kramers, and Slater 1924a, p. 173). In the final paragraph they talk about the "(virtual) radiation field" produced by ordinary antennas (Bohr, Kramers, and Slater 1924a, p. 175). The concluding sentence (which has Bohr written all over it) shows how the authors struggled with their own terminology:

> It will in this connexion be observed that the emphasizing of the "virtual" character of the radiation field, which at the present state of science seems so essential for an adequate description of atomic phenomena, automatically loses its importance in a limiting case like that just considered [i.e., a classical antenna], where the field, as regards its observable interaction with matter, is endowed with all the attributes of an electromagnetic field in classical electrodynamics (Bohr, Kramers, and Slater 1924a, p. 175).

Subsequent expositions of BKS by Slater and Kramers removed much of the tentativeness of this passage.

In a lengthy paper written in late 1924 and published in April 1925, Slater described more clearly than in the BKS paper how to picture the interaction between matter and radiation in the BKS theory. The one new element of the BKS theory, he wrote, is

> that the wavelets sent out by an atom in connection with a given transition were sent out, not as a consequence of the occurrence of the transition, but as a consequence of the existence of the atom in the stationary state from which it could make that transition.

[75] Bohr to Pauli, February 16, 1924 (Bohr 1972–2008, Vol. 5, p. 409).
[76] Pauli to Bohr, February 21, 1924 (Vol. 5, p. 412).
[77] The discussion below closely follows Duncan and Janssen (2007, pp. 604–607)
[78] The BKS paper suggests that there are two types of radiation, leading some commentators astray. Dresden (1987, p. 179), for instance, refers to the "somewhat vague, tenuous relation between the virtual field and the real electromagnetic field." However, we are by no means the first to set the record straight: see, for instance, Darrigol's (1992, pp. 219–221) discussion of "virtuality" in the BKS theory.

> On this assumption, the stationary state is the time during which the atom is radiating or absorbing; the transition from one state to another is not accompanied by radiation, but so far as the field is concerned, merely marks the end of the radiation or absorption characteristic of one state, and the beginning of that characteristic of another. The radiation emitted or absorbed during the stationary state is further not merely of the particular frequency connected with the transition which the atom is going to make; it includes all the frequencies connected with all the transitions which the atom could make (Slater 1925a, p. 398).

This picture, Slater pointed out, is incompatible with strict energy conservation:

> Although the atom is radiating or absorbing during the stationary states, its own energy does not vary, but changes only discontinuously at transitions ... It is quite obvious that the mechanism becomes possible only by discarding conservation (Slater 1925a, p. 399).

Slater's portrayal of the BKS theory agrees with the exposition given by Kramers and Helge Holst in the German edition of a popular book on Bohr's atomic theory originally published in Danish (Kramers and Holst 1922). In the section, "Bohr's new conception of the fundamental postulates," added to the German edition, the authors explained that the BKS theory breaks with one of the basic tenets of Bohr's original theory, namely, that atoms only emit light when one of its electrons makes a transition from, to use his example, the second to the first stationary state. "According to the new conception, radiation with frequency ν_{21} is still tied to the possibility of a transition to the first state, but it is assumed that the emission takes place during the entire time the atom is in the second state" (Kramers and Holst 1925, p. 135). Another difference is that "if the atom is in the third state, it will simultaneously emit the frequencies ν_{32} and ν_{31} until it either jumps to the second or to the first state" (ibid.). The authors emphasize the advantage of the new conception in regard to the correspondence principle, which had appealed instantly to Bohr and Kramers when Slater explained his ideas to them in December 1923:

> This situation shows that the new conception is closer to the classical electron theory than the old one; the simultaneous emission of two frequencies mentioned above has its counterpart in that an electron moving on an ellipse emits both its fundamental tone and its first overtone ... while earlier one had to assume that these two frequencies were produced by different transitions in different atoms. It is a welcome consequence, especially from the point of view of the correspondence principle, that the radiation emitted by a single atom contains all the frequencies that correspond to possible transitions; for in the border region of large quantum numbers the radiation demanded by the quantum theory will now merge very smoothly with the radiation demanded by the classical theory (Kramers and Holst 1925, pp. 135–136).

The term "virtual radiation" no longer occurs in these expositions of the theory by Slater and Kramers. Their presentations make it clear that the BKS theory does not introduce two different kinds of radiation, real and virtual, but a new picture of the interaction between radiation and matter, which is different both from the classical picture and from

Einstein's light-quantum picture. In the BKS picture, the strength of the electromagnetic radiation of some frequency ν gives the probability (according to some unspecified rule) that an electron jumps between two orbits with an energy difference $h\nu$.[79]

This stochastic element—transferred from atoms to electrons in X-ray scattering experiments, as in the BKS explanation of the Compton effect—proved to be the Achilles heel of the BKS theory. Experiments by Bothe and Geiger (1925b) in Berlin and by Compton and Alfred W. Simon (1925) in Chicago showed that energy-momentum is strictly conserved in Compton scattering (i.e., event by event) and not just statistically.[80] The detection of a scattered electron almost always coincided with the detection of a light quantum, which went against the BKS picture that light is emitted and absorbed continuously, whereas the electron changes its energy and momentum only at discrete intervals. Of course, radiation is detected via its effect on electrons in some detector. In the BKS picture, radiation only determines the probability of an electron absorbing energy. The crucial difference between the BKS prediction and the light-quantum prediction is that according to the latter there is a perfect correlation between detection of a scattered electron and detection of a light quantum, whereas the former predicts no such correlation.

The experiments that eventually disproved the theory began shortly after the BKS paper was published (see Bothe and Geiger 1924), but the final verdict did not come in until the following year. Bothe and Geiger (1925a, 1925b) published their results in April 1925. The paper by Compton and Simon (1925) is signed June 23, 1925, and appeared in September 1925.[81] On April 17, 1925, Geiger sent Bohr a letter giving him a heads-up. Bohr's reaction has been preserved in a letter he wrote to Ralph Fowler four days later: "there is nothing else to do than to give our revolutionary efforts as honourable a funeral as possible" (quoted, e.g., in Stuewer 1975, p. 301).[82] His co-authors took the fall of BKS harder. So did other supporters of the theory, such as Ladenburg, Reiche, and Born. By contrast, Einstein and Pauli, the theory's most vocal critics, rejoiced. In a letter to Ehrenfest, Einstein dryly noted: "We both had no doubt about it."[83]

When Slater found out about the experimental refutation of the BKS theory, he reverted to his original position as he made clear in another letter to *Nature* (dated July 25, 1925): "The simplest solution to the radiation problem then seems to be to return to the view of a virtual field to guide corpuscular quanta" (Slater 1925). Kramers and

[79] As Heisenberg, who was very much enamored of the BKS theory, put it in one of his AHQP interviews, radiation is a "half reality" in this new picture in that it only determines the probabilities of quantum transitions in matter (see Duncan and Janssen 2007, pp. 603–604, 607).

[80] See Stuewer (1975, pp. 299–302) and Stolzenburg (1984, pp. 75–80).

[81] See, for example, Stuewer (1975, pp. 299–302) for discussion of these experiments,

[82] Darrigol (1992, pp. 249–252) has argued that, prior to receiving the definitive notice of the Bothe–Geiger results contradicting the BKS theory (in the third week of April 1925), Bohr had already begun to doubt the main tenets of BKS, given the problems encountered in reconciling the theory with the Ramsauer effect, in which a beam of electrons (with energy near 1 eV) was seen to pass apparently without any interaction through a noble gas. Bohr certainly accepted the results of Bothe and Geiger immediately: on April 21, in a letter to Geiger, he stated: "I was completely prepared that our proposed point of view ... should turn out to be incorrect."

[83] Einstein to Ehrenfest, August 18, 1925 (Einstein 1987–2021, Vol. 15, Doc. 49; quoted, e.g., in Klein 1970b, p. 35, and Stuewer 1975, p. 303).

Bohr agreed: "we think that Slater's original hypothesis contains a good deal of truth."[84] The following year, Bohr mentioned in passing in a letter to Slater that he had "a bad conscience in persuading you to our view." Slater told him not to worry about it,[85] but he remained resentful (see note 73).

Pauli was delighted to see the BKS theory collapse. In a letter to Kramers, he called it "a magnificent stroke of luck" and berated his colleague for "the reactionary Copenhagen Putsch, propagandized by you to fanatical excess!"[86] Pauli was at pains to emphasize that the Kramers dispersion theory is independent of the BKS theory, even though Kramers (1924a, 1924b) only published his dispersion formula *after* the BKS theory and even though Kramers and Heisenberg explicitly say in the abstract of their joint paper on the Kramers dispersion theory that it is based "on an extension of the point of view, recently put forward in a paper by Bohr, Kramers and Slater" (Kramers and Heisenberg 1925, p. 223). In the next and final line of this abstract, they write: "the conclusions, should they be confirmed, would form an interesting support for this [i.e., the BKS] interpretation" (ibid.). Pauli took strong exception to these statements. In a paper that would not be published until later in 1925, he added a footnote saying "that the formulae of [Kramers and Heisenberg 1925] used here are independent of the special theoretical interpretation concerning the detailed description of the radiation phenomena in the quantum theory taken as a basis by them [i.e., BKS]" (Pauli 1925c, p. 5). As he explained in the letter to Kramers, "if I had not added the footnote in question, it would also have been true that the conclusions of my paper, should they be confirmed, 'would form an interesting support for this interpretation.' This impression I had, of course, to counteract!"

Pauli's clear separation of the Kramers dispersion theory and the BKS theory[87] serves as a reminder that the latter only played a limited role in the developments that led to matrix mechanics.[88] As long as we think of the Kramers dispersion theory as part and parcel of the BKS theory, it may look as if matrix mechanics replaced a decisively refuted theory. Once we recognize that the Kramers dispersion theory was developed before and independently of the BKS theory, we see that matrix mechanics grew naturally out of an eminently successful earlier theory. The BKS theory and its experimental refutation then become a sideshow distracting from the main plot line, which runs directly from dispersion theory to matrix mechanics. A corollary to this last observation is that the acceptance of the light-quantum hypothesis was irrelevant to the development of matrix mechanics. Compton scattering provided convincing evidence for the light-quantum hypothesis and against the BKS theory, but it had no bearing on dispersion theory. The

[84] Kramers to Harold Urey, July 16, 1925, quoted in Stolzenburg (1984, p. 86).

[85] Bohr to Slater, January 28, 1926; Slater to Bohr, May 27, 1926 (Bohr 1972–2008, Vol. 5, pp. 68–69).

[86] Pauli to Kramers, July 27, 1925 (Pauli 1979, Doc. 97, pp. 232–234; Bohr 1972–2008, Vol. 5, p. 87).

[87] Like the BKS theory, however, the Kramers dispersion theory implied that the atom and its virtual oscillators do not move together. As Kramers and Heisenberg note in passing in their elaboration of the Kramers dispersion theory, "[w]e shall not discuss in any detail the curious fact that the centre of these spherical waves moves relative to the excited atom" (Kramers and Heisenberg 1925, p. 229).

[88] This and the following statements are the conclusions we reached in our examination of the BKS theory in Duncan and Janssen (2007, p. 613).

Kramers dispersion theory crucially depended on Einstein's A and B coefficients but not on the theory of light quanta in which these coefficients were introduced. In the paper by Kramers and Heisenberg (1925), to which we turn next, dispersion—coherent light scattering—is still being discussed in terms of waves rather than light quanta.

10.5 The Kramers–Heisenberg paper and the Thomas–Reiche–Kuhn sum rule: on the verge of *Umdeutung*

In September 1924, Werner Heisenberg arrived in Copenhagen at the Bohr Institute to take up a six-month-long leave of absence from his position as *Privatdozent* and assistant to Max Born in Göttingen. His stay in Copenhagen, which had been arranged a few months earlier in the Spring of 1924 on a brief visit to Copenhagen, was funded by the Rockefeller Foundation. Heisenberg had been extremely impressed, indeed captivated, by Bohr's approach to physics (and more general issues of culture[89]) in his earlier visit, and hoped on his return to learn much more from, and presumably to collaborate with, the great man. As it was, Heisenberg ended up spending most of his time in the Fall of 1924 in discussions with Kramers, Bohr's 'deputy' at the Institute. This collaboration, which must have begun soon after Kramers' return to Copenhagen in late September, would lead to a joint paper on quantum dispersion, which was completed (with the actual writing done almost entirely by Kramers) by December 1924 (Kramers and Heisenberg 1925).

The object of the Kramers–Heisenberg paper was twofold. First, Kramers wanted to give a detailed account of the correspondence principle ideas that had led him to the new dispersion formula. He had so far only published his dispersion formula in two short notes to *Nature* (Kramers 1924a, 1924b), providing no derivation in the first and only the barest of outlines of one in the second. Meanwhile, Born (1924) and Van Vleck (1924b, 1924c) had published detailed derivations of the Kramers dispersion formula (with minor differences of strategy) so the derivation in the first part of Kramers and Heisenberg's paper served mainly to establish notation and strategy for the second, totally new, part of the paper. Here, the authors generalized the approach that had led to a quantum dispersion formula, where the relevant process was *elastic (and coherent) scattering* (with the incident and re-radiated light from a quantized atomic system having the same frequency and a fixed phase relation), to deal with *inelastic (and incoherent) scattering*, where the perturbing influence of light of frequency ν on bound electrons would result in the radiation of light of a different frequency ν', bearing no fixed phase relation to the incident light.

The possibility of such inelastic processes had already been raised a year earlier by Adolf Smekal (1923) in Vienna. Smekal, who (unlike Bohr and Kramers) fully accepted

[89] See Cassidy (2009, Ch. 10) and Dresden (1987, pp. 261–263) for accounts of this period in Heisenberg's life.

Einstein's light-quantum hypothesis, proposed a generalization of the probabilistic processes of emission and absorption considered in Einstein's quantum theory of radiation, in which a light quantum is either absorbed or emitted by the atom or molecule, with a concomitant jump between quantum states governed by the Bohr frequency condition. Smekal proposed the existence of processes in which a light quantum of frequency ν could be absorbed by an atom or molecule in a quantum state m with energy E_m accompanied by the emission of a light quantum of a *different* frequency ν', leaving the atom in a final quantum state n with energy E_n. Since the change in kinetic energy of the atom/molecule due to recoil effects is usually completely negligible, energy conservation simply requires that

$$E_m + h\nu = E_n + h\nu'. \tag{10.132}$$

In the event that $E_n > E_m$ (the only possibility if the initial state m is the ground state), the emitted light must be of lower frequency than the incident light, that is, $\nu' < \nu$. The frequency difference is of the same kind as in fluorescence (where the absorption and emission events are separated by a very long time on atomic scales), which had been studied in the nineteenth century by George Stokes and was therefore dubbed a "Stokes shift" (with the displaced line of the emitted radiation a "Stokes line").[90] In the more unusual case where the atom/molecule is initially in an excited state, one could also have $E_n < E_m$ and $\nu' > \nu$, with the resulting displacement to the blue called an "anti-Stokes shift".

The main new contribution of the Kramers–Heisenberg paper was to provide a detailed correspondence-principle derivation of the scattered light intensity in the inelastic case, for general initial states (either ground or excited) of the bound electrons responsible for the scattering. Unlike Smekal, however, and in accordance with Bohr's views (see Section 10.4), they treated radiation as consisting of electromagnetic waves rather than light quanta. It is probable that the underlying calculations had already been carried out by Kramers before Heisenberg's arrival in Copenhagen and that much of Heisenberg's input concerned the physical interpretation of the results. In particular, some of the terms contained in the new generalized formula seemed quite inexplicable on physical grounds (Kramers would label these terms "false resonances") and led to heated arguments between Heisenberg and Kramers. In fact, the formulas obtained by Kramers and Heisenberg were correct and survived the transition to modern quantum mechanics fully intact. But the proper interpretation of the "false resonances" turned out to be impossible without a full acceptance of light quanta and the general framework provided by the new quantum mechanics.

[90] In the paper in which he first introduced his light-quantum hypothesis, Einstein (1905, sec. 7) had listed the "Stokes rule of fluorescence" (the emitted light is always of lower frequency than the incident light) as one of three phenomena supporting the hypothesis (see Section 3.3.3).

We mentioned, in the course of our derivation of the classical polarization formula Eq. (10.118), the existence of terms in the induced displacement $\Delta x(t)$ of a frequency differing from that of the incident light. Such terms were visible in the expression Eq. (10.106) and involve the exponential behavior

$$e^{2\pi i((\tau+\tau')\nu_0+\nu)t} \tag{10.133}$$

in the one-dimensional case, with $\tau + \tau' \neq 0$. These terms dropped out when we restricted ourselves to terms with $\tau' = -\tau$. Now these terms, corresponding to frequencies $\nu' = \nu + (\tau + \tau')\nu_0 \neq \nu$, and thereby exhibiting a Stokes shift, are precisely the ones of interest.

To understand Kramers and Heisenberg's results, we return to three dimensions, where the Fourier components are labeled by an integer vector $\vec{\tau}$ and the interesting terms have time dependence

$$e^{(2\pi i(\vec{\tau}+\vec{\tau}')\cdot\vec{\nu}+\nu)t}, \tag{10.134}$$

with $\vec{\tau} \neq -\vec{\tau}'$. It will also be convenient to consider the case in which the electric field $\vec{\mathcal{E}}$ of the incident light of frequency ν is in an arbitrary direction. The Hamiltonian in Eq. (10.97) thus changes slightly to

$$H = H_0 + e\vec{\mathcal{E}}\cdot\vec{x}(t)\cos(2\pi i\nu t)\theta(t), \tag{10.135}$$

with

$$\vec{x}(t) = \sum_{\vec{\tau}} \vec{A}_{\vec{\tau}}\, e^{2\pi i\vec{\tau}\cdot\vec{w}}. \tag{10.136}$$

It is easy to see that the three-dimensional version of Eq. (10.106), with action derivatives replaced by difference terms, will involve terms of the form[91]

$$\sum_{\vec{\tau},\vec{\tau}'} \left(\frac{\vec{\mathcal{E}}\cdot\vec{A}_{\vec{\tau}}\,\vec{A}_{\vec{\tau}'}\, e^{2\pi i((\vec{\tau}+\vec{\tau}')\cdot\vec{\nu}+\nu)t}}{\nu + \vec{\tau}\cdot\vec{\nu}} + [\nu \to -\nu] \right). \tag{10.137}$$

Such terms give rise to outgoing radiation with a frequency $\nu + (\vec{\tau}+\vec{\tau}')\cdot\vec{\nu}$, that is, to a Stokes shift of $(\vec{\tau}+\vec{\tau}')\cdot\vec{\nu}$. Terms of the form $\vec{\tau}\cdot\vec{\nu}$ or $\vec{\tau}'\cdot\vec{\nu}$ correspond to the component frequencies of the classical motion, and therefore to Bohr quantum-transition frequencies via the correspondence limit. The Stokes shift must therefore be regarded as due to

[91] The steps leading up to Eq. (10.106) make it clear that the electric field $\vec{\mathcal{E}}$ appears in $\Delta\mathcal{J}$ and Δw, and therefore in a dot product with $\vec{A}_{\vec{\tau}}$.

a double transition of the dispersing electron from an initial state P to a final state Q via some intermediate state R, with $\vec{\tau} \cdot \nu$ or $\vec{\tau}' \cdot \vec{\nu}$ associated with the transition $P \rightarrow R$ or $R \rightarrow Q$.

We focus on a special case that caused considerable discomfort for Kramers and Heisenberg (1925, pp. 245–249). First, if the energy E_P of the initial state P is less than the energy E_Q of the final state Q, where

$$E_Q - E_P \equiv h\nu^\star, \tag{10.138}$$

we have a Stokes shift to the red, with the scattered light appearing at a frequency $\nu - \nu^\star$. This occurs in the correspondence limit when $(\vec{\tau} + \vec{\tau}') \cdot \vec{\nu} = -\nu^\star$. There are three possibilities for the energy E_R of the intermediate state R,

$$E_R > E_Q, \quad E_Q > E_R > E_P, \quad E_R < E_P, \tag{10.139}$$

which Kramers and Heisenberg labeled R_a, R_b, and R_c, respectively (see Figure 10.2a). They closely examined the third possibility, in which there is a decrease in energy going from the (excited) initial state P to the intermediate state R_c (transition 5 with $h\nu_5 = E_P - E_{R_c}$) followed by a (larger) increase in energy going from R_c to the final state Q (transition 6 with $h\nu_6 = E_Q - E_{R_c}$).

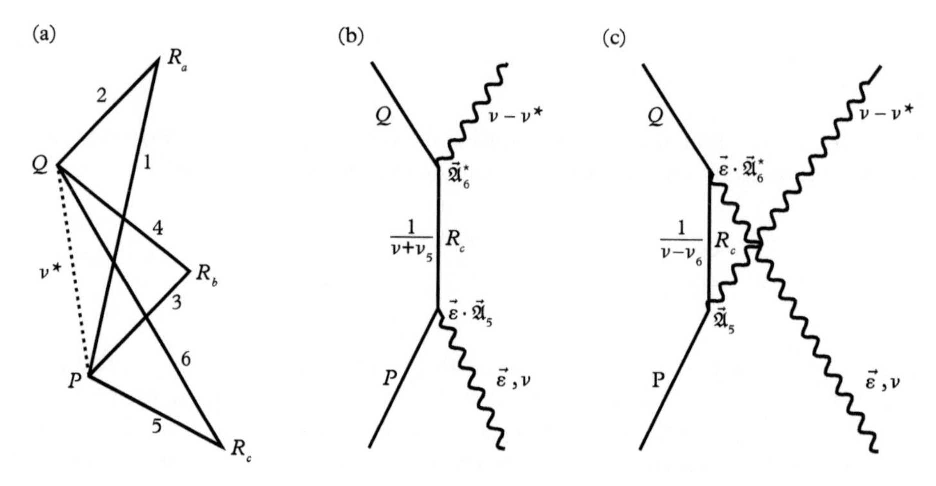

Figure 10.2 *(a) Stokes line transitions, (b) direct inelastic scattering graph, (c) crossed inelastic scattering graph. Figure (a) is (our rendition of) Fig. 4 in Kramers and Heisenberg (1925, p. 243). Figures (b) and (c) are our own diagrams, illustrating the situation from a modern point of view.*

We can identify two terms of the general form of Eq. (10.137) which contribute to transitions to the final state via such an intermediate state R_c. If we choose

$$\begin{aligned} \vec{\tau} \cdot \vec{\nu} &= \frac{E_P - E_{R_c}}{h} = \nu_5, \\ \vec{\tau}' \cdot \vec{\nu} &= \frac{E_{R_c} - E_Q}{h} = -\nu_6, \\ (\vec{\tau} + \vec{\tau}') \cdot \vec{\nu} &= \frac{E_P - E_Q}{h} \equiv -\nu^\star, \end{aligned} \tag{10.140}$$

and

$$\vec{A}_{\vec{\tau}} = \vec{\mathfrak{A}}_5, \quad \vec{A}_{\vec{\tau}'} = \vec{\mathfrak{A}}_6^\star \tag{10.141}$$

(recall the conjugation property in Eq. (10.110)), we get the term[92]

$$\frac{\vec{\mathcal{E}} \cdot \vec{\mathfrak{A}}_5 \, \vec{\mathfrak{A}}_6^\star}{\nu + \nu_5}. \tag{10.142}$$

Similarly, if we choose

$$\begin{aligned} \vec{\tau} \cdot \vec{\nu} &= \frac{E_{R_c} - E_Q}{h} = -\nu_6, \\ \vec{\tau}' \cdot \vec{\nu} &= \frac{E_P - E_{R_c}}{h} = \nu_5, \\ (\vec{\tau} + \vec{\tau}') \cdot \vec{\nu} &= \frac{E_P - E_Q}{h} = -\nu^\star, \end{aligned} \tag{10.143}$$

and

$$\vec{A}_{\vec{\tau}} = \vec{\mathfrak{A}}_6^\star, \quad \vec{A}_{\vec{\tau}'} = \vec{\mathfrak{A}}_5 \tag{10.144}$$

we get the term

$$\frac{\vec{\mathcal{E}} \cdot \vec{\mathfrak{A}}_6^\star \, \vec{\mathfrak{A}}_5}{\nu - \nu_6}. \tag{10.145}$$

These are the final two terms in Kramers and Heisenberg (1925, p. 242, Eq. (40)). The term in Eq. (10.142) displays no resonance pole (both ν and ν_5 are positive) and corresponds to the absorption of energy from the incident wave, leading to a state even

[92] If we wish to select for outgoing light polarized in a direction $\vec{\mathcal{E}}'$, Eqs. (10.142) and (10.145) will contain dot products $\vec{\mathcal{E}}' \cdot \vec{\mathfrak{A}}_6^\star$ and $\vec{\mathcal{E}}' \cdot \vec{\mathfrak{A}}_5$, respectively.

further removed in energy from the intermediate state R_c than the initial state. It is paired (via the replacement $\nu \to -\nu$ in Eq. (10.137)), with a term displaying a resonance pole $1/(\nu_5 - \nu)$ as ν approaches ν_5, which is physically understandable as incident radiation of frequency ν_5 can induce stimulated emission, with a transition of atomic state from P to R_c.

In the quasi-classical view of dispersion that still held sway at this point, resonances occurred when the incoming light directly made an electron jump to a higher quantum state (via absorption) or to a lower quantum state (via stimulated emission or "negative absorption"). In either case the resonance frequency corresponded to the Bohr frequency associated with the corresponding transition. The problem noted by Kramers and Heisenberg (with considerable dismay) was that the pole at $\nu \to \nu_6$ in Eq. (10.145) does not correspond to any transition available to the initial state P! Instead, it corresponds to the transition between the intermediate state R_c and final state Q. There seemed to be no physical explanation for a resonance occurring at frequency $\nu = \nu_6$ when light of frequency ν irradiates an atom in state P. The appearance of such terms in the total amplitude for Stokes scattering led Kramers to introduce the term "false resonances".

The presence of this peculiar term led to heated arguments between Heisenberg and Kramers (Dresden 1987, pp. 241–242). Heisenberg insisted on their indispensability, given their origin in the same type of correspondence-principle arguments that had led to the Kramers dispersion formula. Kramers maintained that the amplification at $\nu \approx \nu_6$ should be canceled by some other effect. Kramers succeeded in cooking up a complicated argument, which made it into the published paper (Kramers and Heisenberg 1925, pp. 247–249), suggesting that there would be destructive interference between the inelastic scattered wave arising from the term in Eq. (10.145) in the limit $\nu \to \nu_6$ and the always present spontaneous radiation of frequency ν_5, which coincides with $\nu - \nu^\star$ when ν approaches ν_6 (see Eq. (10.143)). In fact, this argument is specious, as there is no reason for the amplitudes of the two contributions to be exactly equal and with opposite phase. Heisenberg was right—the "false resonances" were there—but Kramers was also right in that the conventional intuition about resonances could not account for their presence.

The proper interpretation of these "false resonances" only came with the new quantum mechanics and with the full integration of the photon concept in Dirac's (1927b) seminal paper on quantum electrodynamics. The basic idea can be explained simply if, instead of the diagrams used by Kramers and Heisenberg, we help ourselves to a graphical representation of light scattering akin to Feynman diagrams (see Figures 10.2b and 10.2c). These diagrams, unlike those of Kramers and Heisenberg, show the temporal order (with time increasing in the upward vertical direction) of the transitions $P \to R_c$ and $R_c \to Q$ and the radiation absorbed and emitted in the process. The lower and upper squiggly lines represent the incoming and outgoing radiation, respectively. Kramers and Heisenberg still thought of both as continuous waves, albeit carrying definite amounts of energy $h\nu$ and $h(\nu - \nu^\star)$, respectively. Within a year, with the definitive experimental refutation of the BKS theory, it would be recognized that these processes actually involve a single incoming and a single outgoing photon. Regardless of how one conceives it, the incoming light interacts with the electron in state P, inducing a transition to a different state Q via the intermediate state R_c, accompanied by the emission of light with amplitudes $\vec{\mathfrak{A}}_5$ or $\vec{\mathfrak{A}}_6^\star$ and the absorption of light polarized in the direction of $\vec{\mathcal{E}}$ with amplitudes $\vec{\mathcal{E}}{\cdot}\vec{\mathfrak{A}}_5$ or $\vec{\mathcal{E}}{\cdot}\vec{\mathfrak{A}}_6^\star$. Products of these amplitudes show up in the numerators of the terms in the dispersion formula given in Eqs. (10.142) and (10.145). Eq. (10.142) has $\vec{\mathcal{E}}\cdot\vec{\mathfrak{A}}_5\vec{\mathfrak{A}}_6^\star$ and corresponds to the process depicted in Figure 10.2b. Equation (10.145) has $\vec{\mathcal{E}}{\cdot}\vec{\mathfrak{A}}_6\vec{\mathfrak{A}}_5^\star$ and corresponds to the process depicted in Figure 10.2c. One finds the numerators in these two terms by multiplying the amplitudes assigned to the two vertices in these figures. The denominators in these terms are assigned to the line connecting the vertices. One finds those by taking the initial energy of the system and subtracting the energy of the whole system (the electron and, in Figure 10.2c, the two photons) at a time when the system is in the intermediate state R_c. For what can be called the "direct scattering", depicted in Figure 10.2b, we thus find:

$$E_P + h\nu - E_{R_c} = h(\nu + \nu_5), \tag{10.146}$$

where we used Eq. (10.140). Apart from a factor h, this is just the denominator of the term in Eq. (10.142). For what we can be called the "crossed scattering" depicted in Figure 10.2c, we find

$$E_P + h\nu - (E_{R_c} + h\nu + h(\nu - \nu^\star)) = h(\nu_6 - \nu), \tag{10.147}$$

where we used Eq. (10.143). This is just (h times) the denominator of the term in Eq. (10.145) with the peculiar resonance pole at $\nu = \nu_6$.

It should be emphasized that exactly the same pair of direct and crossed processes occurs in the simplest case of coherent scattering (dispersion), with the electron initially in the ground state (so that no negative dispersion terms are present). In this case the denominator for Figure 10.2b (with $P = Q$ and R an excited state with $E_R - E_P = h\nu_0 > 0$) is $1/(\nu_0 - \nu)$, while the crossed graph in Figure 10.2c gives $1/(\nu_0 + \nu)$. Combining the two gives the factor $\nu_0/(\nu_0^2 - \nu^2)$, clearly visible in Eq. (10.118) (with $\vec{\tau} \cdot \vec{\nu} = \nu_0$) or, for the one-dimensional oscillator, in Eq. (10.120). So there is no question that the crossed process must be included to recover even the most basic Sellmeier form of the dispersion formula.

The trouble comes when we think about the physical interpretation of a process like that depicted in Figure 10.2c, especially for the dispersing electron being in the ground state P. Even before the incoming photon has arrived, bringing the energy needed to restore the system to the ground state, the outgoing photon has been emitted! This seems to require a remarkable prescience on the part of the emitting electron. What if there had been no incoming photon? The explanation lies in one of the peculiarities of the quantum electrodynamics introduced by Dirac. The emission of "virtual photons" is a constant fact of life for an electron, even one completely isolated from all other matter or light. The emission almost always is followed, after a very short time, by reabsorption of the emitted virtual photon, and this ongoing process actually modifies (renormalizes) the mass of the electron from the value appearing in the fundamental Hamiltonian of the theory to the actual physically measured mass. On rare occasions, the reabsorption is interrupted by the arrival of a photon from the external world, and if the arriving photon has just the right energy to restore energy conservation for the whole system, the originally emitted photon is allowed to escape. Of course, this process can also occur inelastically, with the incoming photon having just the right energy $h\nu'$ to allow the originally emitted photon to escape, as long as the electron makes an appropriate energy-conserving quantum jump to a new state Q, satisfying the Smekal condition $E_P + h\nu = E_Q + h\nu'$ in Eq. (10.132). If in addition the incoming photon is at the ("false") resonance frequency for an absorption line of the intermediate state R, there will be a large amplification (limited of course by damping, not considered here) of the amplitude characteristic of resonant behavior. This is exactly what happens in Figure 10.2c, where if ν is close to ν_6 the absorption of the late arriving incoming photon (promoting the lower state R to the final state Q) is resonant. This very peculiar, on the surface acausal, type of resonant interaction was clearly not within the conceptual range of Kramers and Heisenberg in late 1924, and it is not difficult to see that it must have led to considerable perplexity—and, apparently, to some acrimonious discussions.[93]

Just as had been the case for the negative dispersion terms in the Kramers formula Eq. (10.123), the generalized formulas of Kramers and Heisenberg for inelastic light scattering arose in a purely theoretical context. They gave a quantitative and precise description of the phenomena hypothesized a year earlier by Smekal (1923). Actual observation of inelastically scattered light came three years later, in February 1928 with the discovery by C. V. Raman of a shift in frequency of light passed through a liquid when a filter is used to block out the much more intense elastically (Rayleigh) scattered light.[94]

That this "Raman–Smekal" effect is much weaker than the elastically scattered Rayleigh component (in which the radiating electrons return to their original state, with frequency shift $\nu^\star = 0$) is primarily because it is intrinsically incoherent. At the classical level, this incoherence goes back to the complex amplitudes $A_{\vec{\tau}}$ in the numerator terms encoding the relative phase information of the mechanical motion of the electron with the incoming electric field of the incident light. These phases vary randomly from one electron to another, which leads to destructive interference except in the elastic case, such as for dispersion, which singles out the elastically scattered light in the forward direction, that is, the direction of the incident light. In that case, the complex number $A_{\vec{\tau}}$ is always accompanied by its complex conjugate $A^\star_{\vec{\tau}} = A_{-\vec{\tau}}$ (see Eq. (10.117)), so that the phase information drops out. For the incoherent (Raman–Smekal) terms examined by Kramers and Heisenberg, by contrast, the amplitudes multiplied in the numerator

[93] The presence of false resonances is no longer considered a troublesome issue a few years after this, for example, in Breit's review article on quantum dispersion theory (Breit 1932), fully based on Dirac's quantum electrodynamics (Dirac 1927b). Breit refers to the false resonances as "resonance of . . . the scattered frequency with the atom" (p. 572). By this time the quantum electrodynamic formalism of Dirac, which leads inescapably to the Kramers–Heisenberg formula, was completely accepted, even if a clear physical intuition of the meaning of the various terms along the lines described here would only come later.

[94] The same effect was found essentially simultaneously and independently by two Russian physicists (Landsberg and Mandelstam 1928).

terms (such as $\vec{\mathfrak{A}}_6^\star$ and $\vec{\mathfrak{A}}_5$ in Eqs. (10.142) and (10.145)) are different. The phases of the multiplied amplitudes do not cancel, but rather play a real physical role in determining the net intensity of the process.

The physical importance of the complex phases in the amplitudes appearing in the generalized Kramers–Heisenberg expression did not go unheeded by Heisenberg, who later referred to them as an important stimulus for his reinterpretation (*Umdeutung*) of the amplitudes $\vec{\mathfrak{A}}$.[95] Classically, these amplitudes were the Fourier components of the mechanical motion, while in the quantum dispersion theory they were associated with pairs of quantum states, corresponding to the Einstein coefficient for the particular transition under consideration. As we discuss in Chapter 11, the conceptual shift adopted by Heisenberg amounted to accepting the assembly of such amplitudes, for all possible transitions, as a full replacement in quantum theory of the classical concept of position $\vec{X}$. Moreover, the summations over intermediate states of products of complex amplitudes appearing in the Kramers–Heisenberg formula already suggest the new type of multiplication of amplitude arrays, which would be central to the new quantum kinematics that Heisenberg (1925c) would propose in his pivotal *Umdeutung* paper—even the commutator structure is present in embryo (in the high frequency limit).

The transition to the new quantum mechanics that emerged in late 1925 was preceded by one final important development in the area of dispersion theory: the sum rules found by Werner Kuhn (1925), working in Copenhagen, and independently, by Willy Thomas (1925) in Breslau. The latter subsequently expanded his paper in a joint work with his doctoral supervisor (Reiche and Thomas 1925), whence the modern appellation "Thomas–Reiche–Kuhn sum rule."[96] These papers were all prepared in the late spring and early summer of 1925—in the case of the papers of Kuhn and Thomas, early enough for Heisenberg (1925c, p. 268) to refer to them in his *Umdeutung* paper, which was submitted to *Zeitschrift für Physik* at the end of July. We describe the simpler approach used by Kuhn, which connects the large-frequency limit of the dispersion formula to the classical result of Thomson for the rate of energy loss due to elastic Thomson (free particle) scattering.

Our first step is to observe that in the limit where the frequency $\omega = 2\pi\nu$ of incident light far exceeds the natural frequency ω_0 of a classical oscillator, the latter can be neglected, and the particle acts as though the elastic force is absent, that is, as a free particle. In the quantum case, the appropriate limit is one in which the incident light frequency ν is much greater than the Bohr transition frequencies ν_{rs} and ν_{st} in Eq. (10.123), the Kramers formula for the induced dipole moment for N_s in state s. We may therefore set both $\nu_{rs}^2 - \nu^2$ and $\nu_{st}^2 - \nu^2$ equal to $-\nu^2$. For one oscillator in state s, Eq. (10.123) then reduces to

$$P_s(t) = -\frac{3c^3}{32\pi^4\nu^2}\left(\sum_r \frac{A_r^s}{\nu_{rs}^2} - \sum_t \frac{A_s^t}{\nu_{st}^2}\right) \mathcal{E}\cos(2\pi\nu t). \tag{10.148}$$

[95] Cf. Duncan and Janssen (2007, p. 555, notes 4 and 5, and pp. 614–615).

[96] Van Vleck seems to have been the first to have found this sum rule as he mentioned in his NRC bulletin on the old quantum theory (van Vleck 1926a, p. 152; see Duncan and Janssen 2007, pp. 595–596). He did not emphasize the result at the time because he thought it was problematic but later took pride in having been the first to hit upon it (Roger Stuewer, private communication, see Duncan and Janssen 2007, p. 668).

Using Eq. (10.80), we replace the Einstein coefficients A_r^s and A_s^t for spontaneous emission by the Ladenburg and Reiche (1923) "oscillator strengths" f_r^s and f_s^t divided by the classical decay time $\tau = (3mc^3/8\pi^2 e^2 \nu_0^2)$ (see Eq. (10.17)), replacing the characteristic frequency ν_0 by ν_{rs} and ν_{st}, respectively. Inserting

$$A_r^s = \left(\frac{8\pi^2 e^2}{3mc^3}\right) \nu_{rs}^2 f_r^s, \quad A_s^t = \left(\frac{8\pi^2 e^2}{3mc^3}\right) \nu_{st}^2 f_r^s \tag{10.149}$$

into Eq. (10.148), we see that ν_{rs} and ν_{st} drop out and we are left with

$$P_s(t) = -\frac{e^2}{4\pi^2 \nu^2 m}\left(\sum_r f_r^s - \sum_t f_s^t\right) \mathcal{E} \cos(2\pi\nu t). \tag{10.150}$$

The Larmor formula (see Eq. (10.6)) gives us the average rate of energy lost by an oscillating dipole moment $P_s(t)$:

$$-\frac{dE}{dt} = \frac{2}{3c^3}\overline{\ddot{P}_s(t)^2}, \tag{10.151}$$

where the bar denotes a time average. Differentiating Eq. (10.150) twice, we find that

$$\ddot{P}_s(t) = \frac{e^2}{m}\left(\sum_r f_r^s - \sum_t f_s^t\right) \mathcal{E} \cos(2\pi\nu t). \tag{10.152}$$

As $\overline{\cos^2(2\pi\nu t)} = 1/2$, Eq. (10.151) thus gives

$$-\frac{dE}{dt} = \frac{1}{3}\frac{e^4}{m^2 c^3}\mathcal{E}^2\left(\sum_r f_r^s - \sum_t f_s^t\right)^2. \tag{10.153}$$

The classical formula for the energy loss rate due to Thomson scattering off a free particle of charge e and mass m is

$$-\left.\frac{dE}{dt}\right|_{\text{Thom}} = \frac{e^4 \mathcal{E}^2}{3m^2 c^3}. \tag{10.154}$$

Comparing Eqs. (10.153) and (10.154), we conclude that

$$\sum_r f_r^s - \sum_t f_s^t = 1. \tag{10.155}$$

The Kuhn sum rule, with its further elaborations by Reiche and Thomas, represents the final resolution of the puzzling disparity between the actual number of optically active electrons and the number of "dispersion electrons" phenomenologically extracted from the numerators in the Sellmeier formula. The sum rule makes clear that each such electron, in a given quantum state s, distributes its dispersive effect fractionally. Each possible transition upwards (to a state r in the first sum) is associated with a positive term, the "oscillator strength" for that transition, while downward transitions to a state t, if possible, appear in the second sum, with a negative sign. Yet, the sum over all possible transitions yields exactly unity. This sum rule would play a critical role in Heisenberg's *Umdeutung* paper.

With a view to simplifying our discussion of Heisenberg's breakthrough in Chapter 11, it will be convenient to rewrite the sum rule Eq. (10.155) in a different form. We consider (with Heisenberg) the special case of one-dimensional *anharmonic* oscillators. These are just the systems for which we developed canonical perturbation theory in Section 10.3.2. For such an oscillator in its nth quantum state, the "orbit" according to the old quantum theory is described by a periodic Fourier sum just as in classical mechanics (cf. Eq. (10.92))[97]

$$x^{(n)}(t) = \sum_{\alpha} A_\alpha(n)\, e^{i\alpha\omega_n t}, \tag{10.156}$$

or, in real rather than complex form,

$$x^{(n)}(t) = \sum_{\alpha} X_\alpha(n) \cos(\alpha\omega_n t + \gamma_n). \tag{10.157}$$

The overtone integer α is summed over both positive and negative values in the complex form Eq. (10.156), but only over positive values in Eq. (10.157). The amplitudes in these two equations are related via

$$A_\alpha(n) = \frac{1}{2} X_\alpha(n) e^{i\gamma_n}. \tag{10.158}$$

Note that the quantized orbit being considered (the nth one) is explicitly indicated in the amplitudes $A_\alpha(n)$ and $X_\alpha(n)$, the phase factors γ_n, and the frequency ω_n of the periodic motion, which, for anharmonic oscillators, depends on the amplitude of the motion. The presence of overtones with arbitrary integer values of α implies, according to the correspondence principle, transitions from the initial state n to all higher $n+\alpha$ and lower $n-\alpha$ states. We may therefore rewrite Eq. (10.155) as

$$1 = \sum_{\alpha>0} \left(f^{n}_{n+\alpha} - f^{n-\alpha}_{n}\right). \tag{10.159}$$

[97] We use the index α, rather than τ, as in Section 3.2, to simplify comparison with Heisenberg's work, as well as to avoid confusion with the decay time τ involved in the definition of the oscillator strength.

We now switch back from oscillator strengths to Einstein A coefficients for spontaneous emission (cf. Eqs. (10.148 and (10.150)). Using Eq. (10.80), we obtain (cf. Eq. (10.149))

$$f^n_{n+\alpha} = \left(\frac{3mc^3}{2e^2}\right)\frac{A^n_{n+\alpha}}{\omega^2_{(n+\alpha,n)}}, \quad f^{n-\alpha}_n = \left(\frac{3mc^3}{2e^2}\right)\frac{A^{n-\alpha}_n}{\omega^2_{(n,n-\alpha)}}, \tag{10.160}$$

where $\omega_{(n+\alpha,n)}$ and $\omega_{(n,n-\alpha)}$ are the frequencies associated with the transitions $n+\alpha \to n$ and $n \to n-\alpha$, respectively. In the correspondence limit these are both equal to $\alpha\,\omega_n$. In this limit, as elucidated by Ladenburg and Reiche (1923), the Einstein coefficient $A^n_{n+\alpha}$ is connected to the Fourier amplitude $X_\alpha(n)$ via

$$h\alpha\frac{\omega_n}{2\pi}A^n_{n+\alpha} = \frac{e^2\alpha^4\omega_n^4}{3c^3}|X_\alpha(n)|^2 = \frac{4e^2\alpha^4\omega_n^4}{3c^3}|A_\alpha(n)|^2, \tag{10.161}$$

where in the last step we used Eq. (10.158) (cf. Eq. (10.75) for A^n_{n+1} with $X_\alpha(n)$ for C and $\alpha\,\nu_n = \alpha\,\omega_n/2\pi$ for ν). We thus find that

$$A^n_{n+\alpha} = \frac{1}{h}\frac{8\pi e^2\alpha^3\omega_n^3}{3c^3}|A_\alpha(n)|^2, \tag{10.162}$$

and therefore, from the expression for $f^n_{n+\alpha}$ in Eq. (10.160) with $\omega_{(n+\alpha,n)} = \alpha\,\omega_n$, that

$$f^n_{n+\alpha} = \frac{3mc^3}{2e^2}\frac{1}{h}\frac{8\pi e^2\alpha\omega_n}{3c^3}|A_\alpha(n)|^2 = \frac{1}{h}4\pi m|A_\alpha(n)|^2\alpha\omega_n. \tag{10.163}$$

Inserting Eq. (10.163) into the sum rule Eq. (10.159), we obtain the Thomas–Reiche–Kuhn sum rule in a form appropriate for the one dimensional systems under consideration:

$$h = 4\pi m\sum_\alpha\Big(|A_\alpha(n)|^2\alpha\,\omega_n - |A_\alpha(n-\alpha)|^2\alpha\,\omega_n\Big). \tag{10.164}$$

Of course, the Bohr transition frequency $\omega(n+\alpha,n)$ from state $n+\alpha$ to state n is only equal to $\alpha\,\omega_n$ in the correspondence limit of very large n, while the sum in Eq. (10.165) extends over all α, and therefore inevitably to low quantum numbers. The obvious extrapolation—or, in the language of the time, "sharpening"—of the correspondence principle would lead us to replace Eq. (10.164) by

$$h = 4\pi m\sum_{\alpha>0}\Big(|A_\alpha(n)|^2\,\omega(n+\alpha,n) - |A_\alpha(n-\alpha)|^2\,\omega(n,n-\alpha)\Big). \tag{10.165}$$

As we see in Chapter 11, this constraint is the one adopted by Heisenberg as the critical "point of the lever" to introduce Planck's constant into the reinterpretation of mechanics he introduced in the summer of 1925.

11

Heisenberg's *Umdeutung* Paper

11.1 Heisenberg in Copenhagen

When Heisenberg arrived in Copenhagen in September 1924 (on leave from Göttingen, courtesy of a Rockefeller Foundation fellowship arranged by Bohr), his primary intention was to collaborate with Bohr.[1] We saw in Section 10.5 that he ended up spending much of his time working with Kramers on the paper extending the dispersion relation for elastic light scattering to inelastic (Smekal–Raman) processes. However, conversations with Bohr presumably led him to work also on another topic of great current interest: the relative intensities of spectral lines in resonant fluorescent scattering. This work would lead to a paper submitted to *Zeitschrift für Physik* at the end of November (Heisenberg 1925a), a month before the Kramers–Heisenberg (1925) paper on dispersion was completed.

While a student in Munich, Heisenberg had already worked on the problem of intensities with his doctoral advisor, Arnold Sommerfeld.[2] In a paper submitted in August 1922, Sommerfeld and Heisenberg (1922b) had approached the problem from the point of view of the correspondence principle, obtaining the relative intensities of spectral lines from the squared Fourier components of the orbit of the radiating electron corresponding to the given transition. The geometry of the electron orbit was specified in the core model in which both the angular momentum of the core and of the radiating electron precessed around the conserved total angular momentum (for the geometry of the atomic model in question, see Figure 7.7).

By the spring of 1924, improvements in the experimental approach to intensity measurements of spectral lines in the Utrecht group led by Leonard Ornstein had provided a wealth of new experimental data on line intensities, which (at admittedly much lower precision) supplemented the information provided by precisely measured line wavelengths. Although the determination of the absolute intensities was both experimentally

[1] For Heisenberg's own reminiscences of this period, see, for example, his autobiography (Heisenberg 1971, pp. 59–62), his contribution to Rozental's (1967) volume on Bohr (Heisenberg 1967, pp. 95–100), and the interviews with him for the AHQP project.

[2] For detailed discussion, see the web resource, *The Problem of Spectral Intensities in the Old and New Quantum Theory*. The importance of the problem of intensities has been emphasized by MacKinnon (1977, 1982), Darrigol (1992), and, more recently, Jähnert (2016, 2019) and Blum *et al.* (2017).

Constructing Quantum Mechanics. Anthony Duncan and Michel Janssen, Oxford University Press.
 DOI: 10.1093/oso/9780198883906.003.0011

and theoretically challenging,[3] the measured relative intensities appeared to follow some rather simple arithmetical regularities (or *Gesetzmässigkeiten*).

An example of the type of rules extracted from the data is the relation between the intensities of the two components of a sharp series doublet in the alkali spectra, as stated by Burger and Dorgelo (1924) in a paper submitted in early March 1924. Sharp series alkali lines arise from a transition from a higher singlet s state to a lower doublet p state. The p states consist of two quantum levels labeled by Sommerfeld's inner quantum number $n_i = 1, 2$, where the state with $n_i = 1$ is split into two components by a magnetic field, the state with $n_i = 2$ into four components. The Sommerfeld inner quantum number was numerically identical to Landé's $\mathcal{J}$ quantum number, interpreted as the total angular momentum of the atom (vector sum of the angular momentum of the core and of the series electron), and equal to half the multiplicity of the state (its "statistical weight") once it had been split by a magnetic field in the anomalous Zeeman effect.[4] The first rule stated by Burger and Dorgelo asserts that the total intensity of lines corresponding to transitions from a (unique) higher singlet state to the two lower p states in a doublet is in the ratio of the inner quantum numbers of the latter (i.e., 2:1)—that is, in the ratio of the statistical weights of the final state. The same 2:1 ratio holds for the relative intensities of the inverse p-s transitions in the lines of the principal series of the alkalis.

Moreover, in cases (as in the diffuse series, with d-p transitions) where the upper state was not a singlet but a very closely spaced multiplet that was not spectroscopically resolved, the same intensity ratios between final states differing by inner quantum number were found. Thus, the rule could be generalized to the sum of intensities for spectral lines arising from a multiplet of unresolved initial states going to a set of (separated) final states distinguished by inner quantum number. Rules of this type were extended in a number of ways by the Utrecht group and shown in some cases to lead to an essentially unique determination of the relative intensities for some complex multiplets (e.g., for the three allowed transitions in certain doublet to doublet multiplets).

In some cases (e.g., transitions between triple p and d states) the Utrecht sum rules did not uniquely determine the individual intensities, but did when supplemented by an "integer rule" [*Ganzzahligkeits-Regel*] (Sommerfeld 1924, p. 654), whereby integer values were associated to the intensity of each spectral line, with the sum of all intensities equal to the (integer) number of possible transitions between distinct initial and final states (as revealed by applying a magnetic field, for example). This sort of numerology was of course very familiar to atomic spectroscopists studying the Zeeman effect, long accustomed to Runge denominators and frequency shifts given by rational fractions of the classical Lorentz shift. For Sommerfeld, it was an indication that real

[3] In the former case it required a determination of the absolute number of atoms in the initial quantum state of the given line transition, in the latter a first-principles calculation of the associated Einstein spontaneous emission A coefficient, which as yet was only accessible in the correspondence limit, and for atoms for which a believable multiply periodic mechanical model was available.

[4] These quantum numbers are related to the total angular momentum quantum number j in modern quantum mechanics by $\mathcal{J} = n_i = j + \frac{1}{2}$. Thus, the denominator of the Landé g-factor, $\mathcal{J}^2 - \frac{1}{4} = j(j+1)$. The statistical weight—the number of possible magnetic quantum numbers m, $-j \leq m \leq +j$—is $2j + 1 = 2\mathcal{J}$.

progress toward a quantitative theoretical description of line intensities in atomic spectra would have to go beyond the continuous analytic formulas for Fourier components of classical orbits obtained from the correspondence principle: a truly quantum-theoretic description based *ab initio* on integer quantum numbers was needed instead.

As pointed out a few weeks later by Ornstein and Burger (1924a), the fact that the relative intensities were in proportion to the multiplicity of states (of different magnetic quantum number) revealed by imposing a magnetic field suggested that the Einstein emission A coefficient from each (non-degenerate) initial state to each of the (non-degenerate) final states was independent of the inner quantum number of either, depending only on the azimuthal (orbital) quantum number k of the states involved ($k = 1$ for s, $k = 2$ for p, etc.). Ornstein and Burger interpreted this rule along Bohrian lines:

> According to the correspondence principle, this means that the Fourier coefficients of the initial orbit, which has the same form and therefore the same Fourier coefficients for the various initial levels [of a complex multiplet of given inner quantum number], determines the total decay probability (Ornstein and Burger 1924a, p. 45).

According to the current conception of atomic structure at the time, this meant that the orientation of the core angular momentum with respect to the orbital angular momentum of the radiating series electron was irrelevant in determining the quantum transition rate of the latter. All that mattered was the orbit of the radiating electron, determined by the principal quantum number n and azimuthal quantum number k.

In August 1924, Ornstein and Burger (1924b) extended their treatment of intensity relations in complex multiplets to the case of magnetically split lines, in the regime of weak magnetic field (where the magnetic splitting is very small compared to the splittings within a complex multiplet, due to varying inner quantum number). Here, the sum rules based on statistical weight were supplemented by a further condition originating from the requirement that the total light emitted by each complex multiplet level in the limit of zero magnetic field was unpolarized, and should remain so for very weak magnetic fields (up to terms of first order in the field strength). The dissection of each multiplet level into a set of closely spaced levels by the magnetic field (distinguished by the magnetic quantum number m) also produces differing polarization in the light observed transverse to the magnetic field, with transitions having $\Delta m = \pm 1$ transversely polarized (to the field direction), those with $\Delta m = 0$ parallel polarized. The unpolarized character of the light when the magnetic field is switched off implies that the total intensity of transverse and parallel transitions must be equal, giving another constraint. Arguments of this sort allowed Ornstein and Burger to determine the relative intensities of the four transverse and two parallel polarized lines of the magnetically split D_2 sodium line, and of the two transverse and two parallel lines of the split D_1 line (cf. Figure 7.5).

By the time Heisenberg got to Copenhagen in September, the intensity problem had taken a new turn. Recent experiments of Wood and Ellett (1923) had shown a remarkable sensitivity of the polarization of resonant fluorescent light arising from a linearly polarized incident beam to the direction of observation of the scattered light, and to the presence of a weak external magnetic field. For example, if the light from a sodium arc lamp (essentially all the yellow D_1 and D_2 lines) is polarized linearly (say, in the z direction) and then enters (say in the x direction) a bulb containing sodium vapor immersed in a magnetic field aligned with the direction (z) of the electric field of the incident light, resonant light scattering occurs as the spectral profile of the incoming light overlaps the (now magnetically split) frequencies needed to induce transitions from sodium atoms in the ground state to the first excited p_1 and p_2 states (associated with the D_1 and D_2 lines), and the D_2 line observed in the perpendicularly (y) scattered light is found to be strongly polarized, while the D_1 line is unpolarized. A number of authors (Breit, Gaviola and Pringsheim, and Joos, among others) had already addressed this problem theoretically, with differing methods and results.

Bohr was apparently preoccupied with the problem of polarization in resonant fluorescence at the time of Heisenberg's arrival in Copenhagen, as he submitted a paper to *Die Naturwissenschaften* on the subject at the beginning of November (Bohr 1924c). This paper is completely qualitative in character. No calculations or quantitative results are presented: rather, Bohr uses the opportunity afforded by the observed effects to examine carefully the interpretation of the virtual oscillators of the BKS theory (see Section 10.4) in connection with the classical motions presupposed by the old quantum theory.

The essential point is one we encountered earlier in our discussion of the Stark effect in Section 6.3: for degenerate systems (e.g., the hydrogen atom in the absence of external fields), the discrete assembly of quantized orbits selected by the Bohr–Sommerfeld procedure depends on the coordinate system used to separate the problem. This orbit ambiguity is removed once the degeneracy is lifted by the imposition of an external field, at which point an essentially unique coordinatization, compatible with separation of the problem in action/angle variables, comes into play, and the orbits are uniquely specified (up to, of course, the freedom to rotate them continuously in space). The virtual oscillators of BKS are presumably associated with a transition from one fully specified orbit to another, so in the case of degenerate systems, as Bohr says

> we must be prepared for the situation that, the behavior of a degenerate atom, as far as radiation is concerned, is not only determined by its motion in the corresponding stationary state, but requires a further specification of the virtual oscillators (Bohr 1924c, p. 1115).

In the case of resonance fluorescence, this further specification is (at least in part) determined by the properties of the incident light, in particular, by its linear polarization in the experiments of Wood and Ellett. Polarized incident light at the frequency corresponding to an upward transition from the ground state would therefore lead to the preferential activation of a subset of excited states. Once an external magnetic field is imposed (in the same direction as the polarization of the incident light), the degeneracy of the excited states is lifted (by the Zeeman effect), and the selection of the virtual oscillators with vibrations appropriate to the direction of polarization of the incident light will lead to selective population of excited states, and therefore to a particular polarization of the light emitted when those states later decay.

The basic lesson of Bohr's note, which was clearly absorbed by Heisenberg in conversations with Bohr after his arrival in Copenhagen, and further developed in his paper on fluorescent polarization, entitled "On an application of the correspondence principle to the question of polarization of fluorescent light" and submitted a few weeks later (Heisenberg 1925a),[5] is that (except in the limit of very large quantum numbers where unvarnished correspondence-principle ideas hold) the virtual oscillators responsible for the interaction of atoms with radiation cannot be connected directly with spatially defined motions of the radiating electron. As Heisenberg puts it:

> The correspondence principle in general, apart from the limit of high quantum numbers, allows only approximate conclusions concerning the intensity and state of vibration of the virtual oscillators associated with the possible transition processes. The reason for this is that the virtual oscillators are only connected in a very symbolic way with the motion of electrons in stationary states. This circumstance is particularly prominent when we are concerned with a degenerate problem, as then, as Bohr [1924c] has more closely examined, the virtual oscillators possess a higher degree of freedom than the motion in the stationary states affords (Heisenberg 1925a, p. 617).

Clearly, the increasingly precise spectral information, now available for intensities as well as frequencies of spectral lines, demanded some extension of the correspondence principle, which was (a) quantitatively precise, and (b) applicable to the low quantum numbers relevant to the quantum transitions actually being studied. At this time (late 1924), such extensions were fairly widely referred to as requiring a "sharpening" [*Verschärfung*] of the correspondence principle (although present usage would suggest "broadening" would be more appropriate). The sharpening referred to by Heisenberg requires that the virtual oscillators behave in some respects in very close analogy to classical systems, while at the same time the linkage of each virtual oscillator to the quantized Bohr orbits of the radiating electron is greatly loosened. As Heisenberg puts it, "the virtual oscillators determine the radiation of the atom, in a certain sense independently of the motion of the electrons in the corresponding stationary state" (Heisenberg 1925a, p. 618).

The required sharpening of the correspondence principle starts from the observation that the frequency and intensity of radiated electromagnetic energy from a bound system of accelerated charges, in classical theory, depends continuously on the parameters in the energy function (Hamiltonian) of the system. The parameters considered here may either be associated with internal effects in the atom, for example, the coupling between the magnetic moment of the core and the magnetic field generated by the valence electron (believed in the core model to be responsible for the splitting of complex multiplets and the appearance of the inner quantum number), or to externally applied fields, such

[5] MacKinnon (1977) emphasizes the importance of this paper for the developments leading up to Heisenberg's (1925c) *Umdeutung* paper. Heisenberg agreed. Commenting on a draft of MacKinnon's article, he wrote to the author in July 1974: "I was especially glad to see that you noticed how important the paper on the polarization of fluorescent light has been for my further work on quantum mechanics. Actually, in Copenhagen I felt that this paper contained the first step in which I could go beyond the views of Bohr and Kramers" (quoted in MacKinnon 1977, p. 149, note 29). See also MacKinnon (1982, pp. 192–200).

as electric fields in the Stark effect, or magnetic fields in the Zeeman effect. In all of these cases, the term in question is (i) quantitatively a small perturbation, and (ii) lifts (at least partially) a degeneracy that holds in the absence of core–valence coupling or external fields. In the case of externally imposed fields, this continuity requirement had become known as Bohr's "principle of spectroscopic stability". Clearly, the ability to obtain consistent and precise results for the line spectra of atoms generated in a wide variety of experimental arrangements (where small uncontrolled external electric and magnetic fields were almost always present) suggested this sort of insensitivity of spectra to small perturbations. The example originally used by Bohr to illustrate this principle turns out to be relevant in the present case (of resonant fluorescence) as well:

> In fact, from a consideration of the necessary "stability" of spectral phenomena, it follows that the total radiation of the components, in which a spectral line, which originally is unpolarised, is split up in the presence of a small external field, cannot show characteristic polarisation with respect to any direction (Bohr 1918, p. 85).

While spectroscopic stability is eminently plausible in the classical context, where continuously variable orbits change only slightly under the imposition of a small external field, it is not at all obvious in the quantum case, as the appearance of a small external field can alter discontinuously the set of Bohr–Sommerfeld quantized electron paths once the problem is separated in a coordinate system adapted to the direction of the new field, however weak. Thus, in the transition from a degenerate problem (zero external field) where a multiplicity of coordinate systems are available in which the problem separates into action/angle variables (with differing sets of quantized orbits in each case) to a non-degenerate one where a particular coordinate system is required to effect the separation, and the quantization procedure becomes essentially unique, one should expect in general a discontinuous change in the physical properties of the quantized atom, even for very small imposed fields. This clearly runs counter to all experience. One is led to conclude that the virtual oscillators responsible for radiation, in the transition from degenerate to non-degenerate situations, must somehow act in a coordinated fashion which, in Heisenberg's words, "allows the analogy of the virtual oscillators with the classical radiation quantities to be carried through more sharply than one would expect from a consideration of the stationary states alone" (Heisenberg 1925a, p. 618).

The first application of the "sharpened" correspondence principle in Heisenberg's paper concerns the Ornstein–Burger rule stated above for complex multiplets (in the absence of a magnetic field): the intensity for a transition from a given (non-degenerate) state in a complex multiplet to any allowed final state—in other words, the total radiation rate for the radiating electron in such an initial state—is independent of the inner quantum number of the initial state, that is, of the total atomic angular momentum, which in turn depends on the relative orientation of the core and valence orbital angular momentum (we are still in the era of unknown electron spin!). First, let us consider the case in which the interaction of the valence electron with the core (via the magnetic field generated by the electron at the core) is neglected. Classically the total radiation rate of the valence electron can then only depend on its orbital characteristics—in this case, completely determined by principal quantum number n and azimuthal quantum number k, but not on the angle between the core and valence angular momenta (which determines the total angular momentum $\mathcal{J}$ by vector addition). Thus, the radiation rate of an alkali principal series electron in any of the four p_1 (with $\mathcal{J} = 2$) or of the two p_2 states (with $\mathcal{J} = 1$) must be equal, in this approximation (in which also the p_1 and p_2 states are degenerate). This classical feature is then transferred by "sharpened correspondence" to the quantized system, leading to the conclusion that the total intensity for radiation from all p-states of a given inner quantum number (or $\mathcal{J}$ value) must be proportional to the statistical weight $2\mathcal{J}$ only, neglecting corrections of first order in the neglected core–valence coupling.[6]

In Heisenberg's second application of sharpened correspondence, to the polarization of resonant fluorescent light, an explicit reference is made to the principle of spectroscopic stability, inasmuch as one is now dealing with a case where an external (magnetic) field is responsible for lifting the degeneracy of the system. In this case, spectroscopic stability was invoked to assert that the net polarization of the light emitted by a degenerate set of excited states is only altered slightly (i.e., to first order in the field) when a weak external field is applied. This behavior is clearly expected in the classical case, and the correspondence principle then implies that the virtual oscillators selected for transitions in the presence of the field must be just such as to preserve the claimed insensitivity.

The advantage of using resonant scattering, with the incident light tuned to the frequencies corresponding to the Bohr transitions responsible for producing the excited states of interest (e.g., in the Wood–Ellett experiments cited above, yellow sodium D light, capable of exciting sodium atoms in the ground state to the $3p_1, 3p_2$ excited multiplets), is that one can selectively populate[7] a subset of excited states of known magnetic quantum number by appropriate choice of the polarization of the incident light. In the experimental configuration described above, the magnetic field in the z-direction results in space quantization of the sodium atoms being irradiated, which therefore have magnetic quantum numbers $m = -\frac{3}{2}, -\frac{1}{2}, +\frac{1}{2}, +\frac{3}{2}$ for the p_1 state, and $m = -\frac{1}{2}, +\frac{1}{2}$ for the p_2 state (see Figure 11.1: cf. Figure 7.3(b)). On the other hand, the incident light, linearly polarized in the z-direction, can only effect transitions obeying the selection rule $\Delta m = 0$. As the ground state has a valence electron in a s state (with $m = \pm\frac{1}{2}$), only $m = \pm\frac{1}{2}$ levels of the excited p_1 state are populated, and only those states give rise to the outgoing scattered light on de-excitation. The net degree of polarization (parallel–transverse/parallel+transverse intensity) can then be found using the relative intensities found by Ornstein and Burger, given as integers next to the transition arrows in Figure 11.1. Heisenberg concludes that the degree of polarization of the D_2 line should be $(4 + 4 - 1 - 1)/(4 + 4 + 1 + 1) = 0.6$, while for the D_1 line one gets complete depolarization, agreeing (at least qualitatively) with the experimental results.

[6] The energy splitting between the p_1 and p_2 lines due to the "core–valence coupling" is about a thousand times smaller than the transition energy for the D_1 or D_2 lines, so this a good approximation.

[7] In the absence of external fields, the original situation envisaged by the Utrecht sum rules, one is considering multiplet line intensities of light generated in arc discharges, under the assumption of Maxwell–Boltzmann equilibrium, with $\exp \Delta E/kT \approx 1$, so all multiplet levels were equally populated to a good approximation, and the observed line intensities could be directly associated with the Einstein spontaneous emission (A) coefficient.

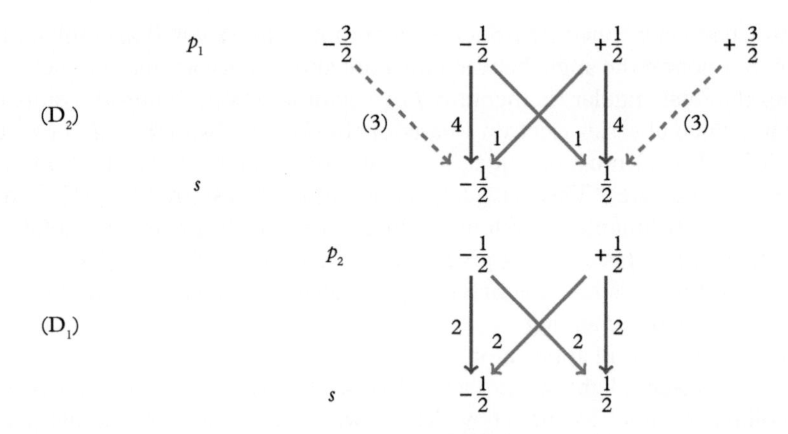

Figure 11.1 *Sodium D line transitions (blue lines: parallel polarization; red lines: transverse polarization). Relative intensities are indicated by integers adjacent to transition arrows.*

11.2 A return to the hydrogen atom

Heisenberg's stay in Copenhagen allowed him time and opportunity to wrestle with problems that were at the leading edge of atomic theory. In his work with Kramers, and also in his attack on the polarization problem in resonant scattering, the nature of the peculiar virtual oscillators invoked in the BKS theory and the complicated issue of their relation to the picture of electrons following Bohr–Sommerfeld orbits was clearly of prime importance. These problems were connected at the hip, as it were, by the need to find a reliable and quantitatively accurate extension—"sharpening"—of the correspondence principle (thus, valid for small as well as large quantum numbers) to the calculation of the Einstein spontaneous emission A coefficients, which determined the strength of atomic transitions, and also appeared directly, as Ladenburg had shown, in the numerators of the Sellmeier terms in the Kramers–Heisenberg formula.

The appearance of rational fractions (ratios of whole numbers) in intensity relations, as pioneered by the Utrecht group, was particularly puzzling. Ladenburg and Reiche had shown that the appearance of integer intensity ratios was not surprising in the case of simple harmonic oscillators, where the squared Fourier components of the motion—that is, the Einstein A coefficient—for a transition from state n to state $n-1$ was proportional to n (for large n) by correspondence, and could simply be assumed to remain so for all values of n. The appearance of such rational fractions for electrons in Bohrian orbits was far less obvious, where, even for the simplest atom, hydrogen, the Fourier components (as demonstrated by Kramers (1919) in his thesis) were messy Bessel functions with arguments involving the integer quantum numbers. Multi-electron atoms had the additional complication of the interaction of the valence orbits with the atomic core (and each other), making the situation even worse.[8]

[8] For a detailed account of the problem of intensities of atomic spectral lines, from the Bohr–Sommerfeld theory up to Heisenberg's *Umdeutung*, see the web resource, *The Problem of Spectral Intensities in the Old and New Quantum Theory.*

On his return to Göttingen in early April 1925, Heisenberg decided to tackle the simplest possible dynamical problem involving electrons in a realistic atomic environment, the hydrogen atom, in which the complications of an atomic core, complex spectra, inner quantum numbers, etc., were not relevant. Nevertheless, the system did not have the deceptive simplicities (evenly spaced energy levels, only unit changes of quantum number, the same Bohr frequency for all transitions, etc.) of the simple harmonic oscillator, so it might show the way to a more generally applicable procedure for extending correspondence-principle arguments. Here, the hydrogen atom was the obvious system of choice, as Kramers (1919) had already done the hard work of finding exact analytic expressions for the Fourier components of the motion, which would directly give the Einstein coefficients in the large quantum number limit.[9]

Already while in Copenhagen Heisenberg had had discussions with Pauli (presumably during the latter's visit in the latter half of March 1925) and Kronig, where the possibility of using a one-dimensional anharmonic oscillator as a convenient toy model—more complicated than the simple harmonic case, but not of course directly relevant to atomic models—was examined, with a view to obtaining intensity formulas for allowed quantum transitions via use of the "sharpened" correspondence principle. This system offered an ideal "warm-up" for the real goal of understanding hydrogen atom intensities. Although the classical motion of such an oscillator, obeying an equation of motion

$$\ddot{x} + \omega_0^2 x + \lambda x^2 = 0, \tag{11.1}$$

was not analytically solvable (unlike the case for the hydrogen atom), the model could be solved in a perturbative way by expanding in the anharmonic coupling λ. And physically the model was much closer to the situation in actual atomic spectra than the simple harmonic case, in that an infinite number of transitions from any given state were allowed—in fact, transitions between any pair of stationary states.[10] Classically, this corresponded to the fact that the motion was bounded and periodic (as long as the amplitude was not too large), with $x(t) = x(t + T)$, implying a Fourier expansion of the particle coordinate containing the fundamental frequency $\omega = 2\pi/T$ as well as *all* its overtones:

$$x(t) = \sum_{\tau=0}^{\infty} X_\tau \cos(\tau\omega t). \tag{11.2}$$

If one substitutes this Fourier series into the equation of motion Eq. (11.1), and sets the Fourier coefficient of each harmonic term (i.e., the constant term, coefficient of $\cos(\omega t)$, $\cos(2\omega t)$, etc.) on the left to zero, one finds a series of non-linear equations

[9] For a detailed introduction to Kramers's beautiful analysis of the Fourier properties of the Kepler problem, see the web resource, *Quantized Orbits in the Old Quantum Theory and Wave Mechanics.*

[10] The transitions in the anharmonic oscillator correspond to changes in the *principal* quantum number in the hydrogen atom case, for which there is no selection rule.

relating the coefficients X_τ. To facilitate a perturbative expansion in powers of the small coupling parameter λ, it is convenient to extract the leading power of λ from each coefficient, so with malice aforethought, we set

$$\begin{aligned} X_0 &= \lambda a_0, \\ X_1 &= a_1, \\ X_2 &= \lambda a_2, \\ X_3 &= \lambda^2 a_3, \ldots \text{etc.} \end{aligned} \tag{11.3}$$

The classical amplitude a_1 (which would be the only non-vanishing coefficient in the simple harmonic limit with $\lambda = 0$) should be regarded as set *ab initio*: it determines the overall amplitude of the motion, and therefore takes on a discrete set of values once quantization is imposed. With these definitions one finds the following relations among the other $a_i, i \neq 1$ coefficients:

$$\omega_0^2 a_0 = -\frac{1}{2} a_1^2 + \ldots \tag{11.4}$$

$$(4\omega^2 - \omega_0^2) a_2 = \frac{1}{2} a_1^2 + \ldots \tag{11.5}$$

$$(9\omega^2 - \omega_0^2) a_3 = -a_1 a_2 + \ldots, \text{etc.}, \tag{11.6}$$

where the ellipses indicate terms of higher order in λ, or involving higher order coefficients. These relations allow one to solve recursively for the coefficients $a_0, a_2, a_3, \ldots$ (hence, $X_0, X_2, X_3, \ldots$) once a_1 is specified. The notation here is deficient in one important respect: the Fourier coefficients a_τ refer to the Fourier decomposition of the classical orbit associated with a specific quantized stationary state—specifically, the state n. A more explicit notation, writing $a_\tau^{(n)}$, for example, would make this clear. For notational simplicity, we have not done this, but remind the reader at this point that a specific state, characterized by a definite integer n, is implicitly under consideration in what follows.

By the usual correspondence reasoning (cf. Section 5.3) the presence of the τ-th overtone would correspond (for large quantum numbers) to transitions that changed the quantum number n associated with the discrete energy levels by an amount $\pm\tau$. As all the X_τ in Eq. (11.2) are in general non-zero, all transitions are therefore allowed, as in the case of the principal quantum number for the hydrogen atom. Also, the transition from state n to $n - \tau$ (say) would be associated with a Bohr transition frequency $\omega(n, n-\tau) = (E_n - E_{n-\tau})/\hbar$, which here is not a simple multiple of the fundamental frequency ω (unless, of course $\lambda = 0$).[11] So, if the question of appropriate sharpening

[11] The simplifying notation $\hbar \equiv h/2\pi$ does not appear in any of the papers of Heisenberg, Born, and Jordan in 1925, but we use it frequently to simplify the algebra.

of the correspondence-principle results for intensities and frequencies of quantum transitions could be settled unambiguously in this model—and with it, the still-mysterious paradox of discrepant Fourier and Bohr transition frequencies—there was a strong hope that an analogous procedure would work for the hydrogen atom, and ultimately for more complex atoms as well.

The amplitude of the anharmonic oscillations, and therefore the conserved energy of the motion, can be fixed by choosing a value for X_1—we saw that the equation of motion then determines (in an expansion in λ) the other coefficients X_0, X_2, X_3, etc. Now let us imagine that $X_1(= a_1)$ has been chosen to select the state with principal quantum number n (where it is convenient to assign $n = 0$ to the ground state)—by imposing a Bohr–Sommerfeld condition, for example. The squared coefficient X_1^2 then determines (as in the simple harmonic case) the Einstein spontaneous emission coefficient for transitions $n \rightarrow n-1$ (with frequency $\omega(n, n-1)$). For $\lambda = 0$, the harmonic oscillator case, this quantity was found by Ladenburg and Reiche (cf. Eq. (10.63)) to be proportional to n, vanishing, as it must, for transitions from the ground state $n = 0$ to the non-existent lower $n = -1$ state, so that we have $X_1 \propto \sqrt{n}$. The recursive argument above shows that the term X_τ can be written as a combination of products $X_m X_n$ with $m+n = \tau$ (see, for example, Eqs. (11.5)–(11.6)), so eventually one has (asymptotically, for large quantum numbers) $X_\tau \propto X_1^\tau \propto \sqrt{n^\tau}$. This formula was "sharpened" into a quantum-theoretically acceptable one by setting

$$X_\tau \propto \sqrt{n(n-1)\cdots(n-\tau+1)}, \tag{11.7}$$

which agrees with the classical result $\sqrt{n^\tau}$ in the asymptotic correspondence regime where $n \gg \tau \gg 1$, but vanishes if $\tau > n$ (leading to a non-existent final state with negative quantum number).

The "vanishing at the edge" of quantum transition amplitudes, necessitated by the limited range of various quantum numbers in atomic systems (and thus, the non-existence of transitions into or out of impermissible states) was by the Spring of 1925 a well-known springboard to the derivation of intensity relations in the Zeeman effect, particularly in the work of Hönl (1925) and Hönl and Sommerfeld (1925).[12] This Ansatz is rendered at least somewhat plausible by the fact that the factor $n(n-1)\cdots(n-\tau+1)$ also appears in the joint probability for the successive allowed transitions $n \rightarrow n-1 \rightarrow n-2 \rightarrow \cdots \rightarrow n-\tau+1 \rightarrow n-\tau$ in the simple harmonic case. The arguments presented here were probably developed already in Copenhagen in the Spring of 1925 by Heisenberg, in conversations with Pauli and Kronig, and are outlined in some detail in a letter of Heisenberg to Kronig dated June 5, 1925 (Fierz and Weisskopf 1960, pp. 23–25).[13]

[12] See the web resource, *The Problem of Spectral Intensities in the Old and New Quantum Theory*, Section 8. See also Jähnert (2019, Ch. 4).

[13] In the cited letter to Kronig, Heisenberg says: "This formula [Eq. (11.7)] is indeed the one previously discussed (discussions with Pauli), and I could imagine that it provides a general law for the calculation of intensities." The reference here is presumably to discussions of Heisenberg with Pauli and Kronig in Copenhagen in early 1925.

Once back in Göttingen in April 1925, Heisenberg tried to derive an analogous "sharpened" formula for intensities in the hydrogen atom spectrum, valid for small quantum numbers, once again beginning with the Fourier components of the classical motion. Analytical expressions for these components had been derived by Kramers in his 1919 doctoral dissertation. As Kramers had pointed out in his discussion of the intensities of lines in the Stark effect, the correspondence principle in its unvarnished form did not specify uniquely for low-lying states whether the relevant Fourier coefficient determining the intensity of a given transition was associated with the Fourier component of the motion in the initial orbit, the final orbit, or some average of the two (or of all orbits in between). In fact, Kramers was unable to find an averaging procedure that would satisfy exactly the requirement of spectroscopic stability, whereby the sum of intensities of lines split in the Stark effect would go over continuously to the intensity of the single line in the absence of electric field. Heisenberg rather ambitiously set out to find (really, to guess) an analytic expression for intensities that would agree asymptotically with the classically calculated components at large quantum number, while respecting spectroscopic stability.

Heisenberg's starting point was the formula for the Fourier expansion of $x + iy$ of the electron in the hydrogen atom once the problem is separated in polar coordinates (Kramers 1919, p. 299). Kramers had found, with some minor changes of notation,

$$x + iy = \text{const} + \sum_{\tau \neq 0} C_\tau e^{i\tau\omega t}, \tag{11.8}$$

$$C_\tau = \frac{a}{2\tau}\left(\left(1 + \frac{k}{n}\right)\mathcal{J}_{\tau-1}(\tau\varepsilon) - \left(1 - \frac{k}{n}\right)\mathcal{J}_{\tau+1}(\tau\varepsilon)\right), \tag{11.9}$$

where a is the semi-major axis of the orbit and $\varepsilon = \sqrt{1 - k^2/n^2}$ the eccentricity. Here, as usual at the time, n represents the principal, k the azimuthal (orbital angular momentum) quantum number, with $1 \leq k \leq n$.

In the correspondence limit where one assumes $n, k \gg 1$, $n - k \ll n$, and $\varepsilon \ll 1$ (i.e., almost circular large orbits), the Bessel functions in Eq. (11.9) can be expanded and the leading term becomes

$$C_\tau \simeq \frac{a}{\tau!}\left(\frac{\tau}{\sqrt{2n}}\right)^{\tau-1}(n-k)^{(\tau-1)/2} \quad \Rightarrow \quad C_\tau^2 \propto (n-k)^{\tau-1}. \tag{11.10}$$

This result is obviously analogous to the behavior $C_\tau^2 \propto n^\tau$ (for large n) found in the anharmonic oscillator case.

The asymptotic form can be modified (or better, "sharpened") in a fashion analogous to that used in the anharmonic oscillator to impose the lower bound of the spectrum. Here, the corresponding quantum-theoretical constraints are that $k \leq n$ for any stationary state with principal quantum number n and azimuthal quantum number k. The Fourier component C_τ in Eq. (11.9) is associated by the correspondence principle with the emission of circularly polarized light (in the z direction), which induces a drop in the principal quantum number n by τ, and by the usual selection rule, in the azimuthal quantum number by one: in other words, the transition $(n,k) \rightarrow (n-\tau, k-1)$. This component therefore must vanish unless $k-1 \leq n-\tau$, that is, unless $n-k \geq \tau-1$. For $n-k = 0, 1, \ldots, \tau-2$, we expect the component to be associated with forbidden transitions and to be zero. This constraint is incorporated by Heisenberg with the introduction of broken factorials as in the anharmonic oscillator case:

$$(n-k)^{\tau-1} \rightarrow (n-k)(n-k-1)(n-k-2)\cdots(n-k-\tau+2). \tag{11.11}$$

This modification is innocuous in the correspondence limit where $n-k \gg \tau, n \gg 1$ (it agrees asymptotically with Eq. (11.10)), but is presumed to hold even for low quantum numbers. It ensures that the amplitude "vanishes at the edge" once $n-k$ becomes equal to, or less than, $\tau-2$. Unfortunately, Heisenberg's guess is incorrect—the actual quantum transition intensities for the hydrogen atom are considerably more complicated expressions, as one can see by consulting a classic text on atomic spectra published ten years later (Condon and Shortley 1963, p. 133).

Heisenberg was in any case unable to make further progress along these lines. Apparently, he did not regard guesswork of the sort responsible for Eq. (11.11) as convincing if it was not supported by a precise implementation of the correspondence principle in which classical derivatives were replaced by discrete quantum differences—just the sort of procedure that had led to precise and reliable dispersion formulas of the Kramers type. He seems to have spent several weeks in hopes of finding such discrete difference equations for Fourier components which quantum-theoretically would involve rational functions of the quantum numbers but asymptotically would agree with the Bessel functions found by Kramers. This project was abandoned early in May, at which point Heisenberg returned to the much simpler problem of the anharmonic oscillator—the subject of his conversations in Copenhagen with Pauli and Kronig—with remarkable and, in the words of Steven Weinberg (1992, pp. 67–68), "magical" results.

11.3 From Fourier components to transition amplitudes

In Göttingen, having failed to establish an unambiguous and theoretically convincing "sharpening" of the correspondence-principle results for intensities of lines in the hydrogen atom, Heisenberg returned to the simple case of a one-dimensional anharmonic oscillator, which he had already discussed with Pauli and Kronig in Copenhagen. In a letter to Kronig dated June 5, 1925 (reprinted in Fierz and Weisskopf 1960, pp. 23–25), Heisenberg returns to the problem of intensities in this model, introducing a "quantum-theoretic transformation" [*quantentheoretische Verwandlung*], which would soon evolve into the central formal element of Heisenberg's matrix mechanics.

From Eqs. (11.2)–(11.3) it is clear that the time dependence $\cos(\omega t)$ associated with the coefficient a_1 (or X_1) can be broken into complex terms $e^{+i\omega t}$, associated with an upward transition $n-1 \rightarrow n$, and $e^{-i\omega t}$, associated with a downward transition

$n \to n-1$. In the quantum theory these transitions are associated with amplitudes that Heisenberg now writes as $a(n-1, n)$ and $a(n, n-1)$, with the associated frequencies $\omega(n, n-1) = (E_n - E_{n-1})/\hbar = -\omega(n-1, n)$. More generally, the quantum transition $n \to m$ is associated with a frequency given by the Bohr frequency condition $\omega(n, m) = (E_n - E_m)/\hbar$, which, in the general anharmonic case, need not be proportional to the integer difference $n - m$ one finds in the simple harmonic case.

The problem now arises: how should we write a "quantum-theoretic transformation" for classical non-linear relations such as Eq. (11.6), obtained by substituting the Fourier expansion of the coordinate $x(t)$ into the classical equation of motion? The essential objective, of course, is to obtain relations that hold all the way down to low quantum numbers, in the "sharpened" sense of the correspondence principle. In the cited letter to Kronig, Heisenberg suggests writing Eq. (11.5) quantum-theoretically as

$$3\omega_0^2 a_2(n, n-2) = \frac{1}{2} a_1(n, n-1) a_1(n-1, n-2) + \ldots \tag{11.12}$$

where the lowest order approximation $\omega \approx \omega_0$ has been used to replace $4\omega^2 - \omega_0^2$ by $3\omega_0^2$. Similarly, Eq. (11.6) is rewritten as

$$\begin{aligned} 8\omega_0^2 a_3(n, n-3) &= -\frac{1}{2}\Big(a_1(n, n-1) a_2(n-1, n-3) \\ &\quad + a_2(n, n-2) a_1(n-2, n-3)\Big) + \ldots \end{aligned} \tag{11.13}$$

Where do these mysterious—but suggestive!—equations in Heisenberg's letter to Kronig come from? Our direct documented evidence of Heisenberg's train of thought from this period (late-May/early-June 1925) is scanty: in fact, we have to rely almost entirely on his correspondence with Kronig (letters of May 8 and June 5). Nevertheless, there are clear motivations for these maneuvers in Heisenberg's work over the period from his collaboration with Kramers on dispersion theory in the late Fall of 1924, up to the point of his re-engagement with the anharmonic oscillator in May 1925.

The first, and most critical point, is that the right-hand sides of Eqs. (11.12)–(11.13) must satisfy the correspondence principle. Indeed, the work of Ladenburg and Reiche allows us to associate a_1 (recall that this is the Fourier coefficient in the nth state) with the transition amplitude $a(n, n-1)$, which in the limit of large quantum numbers is approximately equal to $a(n-1, n-2)$, so that the right-hand side of Eq. (11.12) is equivalent to the classical $a_1^2/2$ of Eq. (11.5) in the correspondence limit. A similar argument (with $a_2(n-1, n-3) \approx a_2(n, n-2)$ and $a_1(n-2, n-3) \approx a_1(n, n-1)$ shows that the right-hand side of Eq. (11.13) agrees with that of Eq. (11.6) in the correspondence limit.

This brings us to the specific choice of, and sum over, states in the quantum versions of these formulas. The terms non-linear in the a_τ coefficients arise classically from the

Fourier expansion of the $x(t)^2$ term in the equation of motion Eq. (11.1). If we use the complex version of the Fourier expansion Eq. (11.2),

$$x(t) = \sum_{\tau=-\infty}^{\infty} A_\tau e^{i\tau\omega t}, \tag{11.14}$$

we see that the squared coordinate possesses a Fourier expansion of the form

$$x(t)^2 = \sum_{\tau=-\infty}^{\infty} B_\tau e^{i\tau\omega t}, \tag{11.15}$$

where the new coefficients B_τ are given as a sum of quadratic terms in the A_τ,

$$B_\tau = \sum_{\tau_1} A_{\tau_1} A_{\tau-\tau_1}. \tag{11.16}$$

The coefficient B_τ, if it appears in a classical expression as a contribution to an A_τ coefficient, must be associated with a transition $n \to n-\tau$, and should therefore be equal, once quantized, to a sum over the product of transition amplitudes for $n \to n-\tau_1$ and $n-\tau_1 \to n-\tau_1-(\tau-\tau_1) = n-\tau$. The appearance of a sequence of states linked by single transitions into and out of an intermediate state would have been familiar to Heisenberg from his work on the inelastic Raman dispersion formula with Kramers.[14] Furthermore, this "sharpened" replacement of amplitudes gives a time dependence consistent with the Ritz combination principle for frequencies (valid for all transitions, of course, and not just in the correspondence limit), as $\omega(n, n-\tau) = \omega(n, n-\tau_1) + \omega(n-\tau_1, n-\tau)$, whence the time dependent factor $\exp(i\omega(n, n-\tau)t)$ associated with the transition $n \to n-\tau$ is indeed the product $\exp(i\omega(n, n-\tau_1)t) \cdot \exp(i\omega(n-\tau_1, n-\tau)t)$ associated with the sequence of transitions $n \to n-\tau_1$ followed by $n-\tau_1 \to n-\tau$. It is crucial for the success of Heisenberg's program that this consistency holds (via the Ritz combination principle) for arbitrary energy quantized systems—not just for the one-dimensional anharmonic oscillator, but for Hamiltonians in higher dimensions with potentials leading to discrete bound states.

Now, the coefficient a_1 can only characterize transitions where the state quantum number changes by one, so the contribution to a_2 (after quantum transcription $a_2(n, n-2)$) from the x^2 term must be a sum that degenerates to a single term, with $\tau_1 = 1$,

$$\begin{aligned} B_2 &= \sum_{\tau_1} A_{\tau_1} A_{2-\tau_1} \\ &\to \sum_{\tau_1} a_1(n, n-\tau_1) a_1(n-\tau_1, n-2) \\ &= a_1(n, n-1) a_1(n-1, n-2), \end{aligned} \tag{11.17}$$

[14] See Eq. (10.137) and following discussion, and Figure 10.2, showing the $P \to R$, $R \to Q$ sequential transitions, and the total scattering amplitude given by a sum over the intermediate state R.

exactly the combination appearing in Eq. (11.12). The right arrow here (and in Eq. (11.18)) expresses the "quantum-theoretic transformation" that is the essential crux of Heisenberg's new sharpening of the correspondence principle.

Similarly, in Eq. (11.13), the transition $n \to n-3$ induced by the transition amplitude $a_3(n, n-3)$ is obtained by the transformation

$$\begin{aligned} B_3 &= \sum_{\tau_1} A_{\tau_1} A_{3-\tau_1} \\ &\to \sum_{\tau_1} a_{\tau_1}(n, n-\tau_1) a_{3-\tau_1}(n-\tau_1, n-3) \\ &= a_1(n, n-1) a_2(n-1, n-3) + a_2(n, n-2) a_1(n-2, n-3), \end{aligned} \tag{11.18}$$

which explains the right-hand side of Eq. (11.13). The coefficient of $1/2$ is required so that the correspondence limit yields the classical result Eq. (11.6). It is apparent that with this Ansatz for the quantum-theoretical amplitudes, the transition amplitude $a_\tau(n, n-\tau)$ that changes the quantum number from n to $n-\tau$ can be reduced recursively using the non-linear equations arising from the equation of motion to a product of a_1 factors, specifically

$$\begin{aligned} a_\tau(n, n-\tau) &\propto a_1(n, n-1) a_1(n-1, n-2) \cdots a_1(n-\tau+1, n-\tau) \\ &\propto \sqrt{n(n-1)(n-2) \cdots (n-\tau+1)}, \end{aligned} \tag{11.19}$$

a result Heisenberg communicated to Kronig in the June 5 letter, reminding him that it had already been discussed with Pauli (and Kronig, presumably when all three were in Copenhagen in March 1925). As we saw above, it was this partial ("broken") factorial expression that had inspired Heisenberg to search for general intensity formulas for hydrogen after returning to Göttingen.

11.4 A new quantization condition

At around the time his letter to Kronig of June 5, 1925, Heisenberg was suffering from a severe bout of hay fever, and decided to seek relief in the largely vegetation-free island of Helgoland, in the North Sea.[15] He left for Helgoland on June 6, arriving by ferry at the

[15] As Heisenberg's biographer notes, "[w]hat exactly happened on that barren, grassless island during the next ten days has been the subject of much speculation and no little romanticism" (Cassidy 2009, p. 142). Heisenberg's own account in his autobiography (Heisenberg 1971, pp. 60–61) has undoubtedly contributed to this mystique. After calculating through the night, Heisenberg wrote: "I was far too excited to sleep, and so, as a new day dawned, I made for the southern tip of the island, where I had been longing to climb a rock jutting out into the sea. I now did so without too much trouble and waited for the sun to rise" (Heisenberg 1971, p. 61). He went on to emphasize, however, that "what I saw that night in Heligoland was admittedly not very much more than [another] sunlit rock edge I had glimpsed in the autumn of 1924" (Heisenberg 1971, pp. 61–62, cf. note 23).

island the following day. In the next ten days, Heisenberg worked on extending the ideas in his letter to Kronig. In particular, it was obviously necessary to introduce Planck's constant by some form of quantization condition, as the application of the equation of motion to the transition amplitudes, as described to Kronig, gave relations between the amplitudes but was insufficient to pin down their absolute magnitudes, which clearly depended on Planck's constant.

After spending about ten days on Helgoland, Heisenberg returned to Göttingen via Hamburg, where he stopped off to discuss his ideas with Pauli. On June 21, 1925, back in Göttingen, he wrote a short letter to Pauli, in which he thanked him for his hospitality and indicated that "my attempts to fabricate a quantum mechanics proceed only slowly, but I am not inclined to worry about how far I have distanced myself from the theory of conditionally periodic systems [i.e., the notion of discretely selected classical orbits]" (Pauli 1979, Doc. 91).

A few days later, on June 24, 1925, Heisenberg sent Pauli a much more detailed letter (Pauli 1979, Doc. 93), which contains the quantization condition that would assume a central role in the *Umdeutung* paper of July, as well as a derivation of the quantized energies of the simple harmonic oscillator, including the zero-point energy, namely $E_n = (n + \frac{1}{2})h\nu$. There are however some confusing errors in the formulas Heisenberg gives, which makes it tricky to assess exactly how much Heisenberg had actually accomplished in his trip to Helgoland.

The part of Heisenberg's letter of June 24 concerned with his new approach to quantization begins with the following comments:

> I have almost no desire to write about my own work, because everything is still unclear, and I only vaguely grasp how it will turn out; but perhaps the underlying thoughts are nonetheless correct. The basic assumption is: in the calculation of any quantity, such as energy, frequency, etc., only relations between quantities which are in principle controllable should appear. (In this respect the Bohr theory seems to me much more formal than the Kramers dispersion theory.) (Pauli 1979, Doc. 93).

These remarks, which clearly echo statements along much the same lines in Kramers and Heisenberg (1925) (cf. Section 10.5), are followed by the statement of the equation of motion for a simple harmonic oscillator[16]

$$\ddot{q} + \omega_0^2 q = 0. \tag{11.20}$$

A solution to this is then written, entirely in terms of an amplitude and frequency that depends on two indices (labeling an initial and final state):

$$q = a(n, n-1)\, e^{i\omega(n,n-1)t}. \tag{11.21}$$

Inserting Eq. (11.21) into Eq. (11.20), one concludes that in this case $\omega(n, n-1) = \omega_0$.

[16] The frequency in this equation should be ω_0, rather than ω as Heisenberg writes, as is clear from what follows.

Of course, for an anharmonic oscillator, with equation of motion Eq. (11.1), the periodic classical motion would be associated with an angular frequency ω, which differs from the frequency ω_0 appearing in the equation of motion, and the motion would involve all harmonics in general, and not just the fundamental frequency, so that the classical particle coordinate and momentum have a Fourier expansion

$$q(t) = \sum_{\tau} a_\tau e^{i\tau\omega t}, \quad p(t) = m\dot{q} = im \sum_{\tau} a_\tau \tau\omega e^{i\tau\omega t}. \tag{11.22}$$

The classical action variable $\mathcal{J}$, given by the Sommerfeld phase integral $\oint p\dot{q}dt$, can now be expressed in terms of the classical components a_τ, as follows:[17]

$$\begin{aligned}
\mathcal{J} &= \int_0^{2\pi/\omega} im \sum_{\tau} a_\tau \tau\omega e^{i\tau\omega t} \cdot i \sum_{\tau'} a_{\tau'} \tau' \omega e^{i\tau'\omega t} dt \\
&= -m \sum_{\tau\tau'} \delta_{\tau,-\tau'} a_\tau \tau\omega a_{\tau'} \tau' \omega \cdot \frac{2\pi}{\omega} \\
&= 2\pi m \sum_{\tau} \tau a_\tau^2 \tau\omega.
\end{aligned} \tag{11.23}$$

Differentiating this equation with respect to $\mathcal{J}$, one finds

$$1 = 2\pi m \sum_{\tau} \tau \frac{\partial}{\partial \mathcal{J}} \left(a_\tau^2 \tau\omega \right). \tag{11.24}$$

Note that in the general anharmonic case the frequencies will depend on the amplitude of the motion (controlled by $\mathcal{J}$) and therefore must be subjected to differentiation by $\mathcal{J}$.

The usual rule from dispersion theory for converting classical to quantum equations, namely

$$\tau \frac{\partial}{\partial \mathcal{J}} \rightarrow \frac{1}{h}\Delta_\tau, \tag{11.25}$$

$$\tau\omega \rightarrow \omega(n+\tau, n), \tag{11.26}$$

$$a_\tau^2 \rightarrow a(n+\tau, n)^2, \tag{11.27}$$

is now applied to Eq. (11.24), to obtain

$$h = 2\pi m \sum_{\tau} \left(a(n+\tau, n)^2 \omega(n+\tau, n) - a(n, n-\tau)^2 \omega(n, n-\tau) \right). \tag{11.28}$$

[17] Note that $a_\tau = a^*_{-\tau}$ is required to ensure that $q(t)$ and $p(t)$ are real. This means that Eq. (11.23) should really contain $a_\tau a_{-\tau} = |a_\tau|^2$. The absolute magnitude in the square is carelessly dropped by Heisenberg, but must be regarded as implied in the subsequent equations.

The sum over τ here runs over all positive and negative integer values. Using the symmetry relations $\omega(n+\tau,n) = -\omega(n,n+\tau), a(n,m) = a(m,n)$,[18] we can eliminate the negative terms by doubling the prefactor:

$$h = 4\pi m \sum_{\tau>0} \Big(a(n+\tau,n)^2\omega(n+\tau,n) - a(n,n-\tau)^2\omega(n,n-\tau)\Big). \qquad (11.29)$$

Referring to Eq. (10.165), we see that this is just the Thomas–Reiche–Kuhn sum rule in disguise (with the relabeling $\tau \to \alpha$, and with the obvious association $a(n+\tau,n) \to A_\alpha(n)$).

Heisenberg would not have seen, at the time of his letter to Pauli in late June, the published versions of either the Thomas or Kuhn paper on the sum rule, but he had overlapped with Kuhn throughout his stay in Copenhagen, during which time there were (according to Mehra and Rechenberg 1982b, pp. 245–247) discussions on dispersion theory leading to sum rules of this type. In any event, the non-linear constraint Eq. (11.29) would prove to be the critical ingredient introducing Planck's quantum into the new theory that was taking shape.

Returning to the June 24 letter to Pauli, we see that Heisenberg specializes his argument by considering the case of the simple harmonic oscillator, in which only the fundamental frequency ω_0 appears, so τ is restricted to the single value 1 in Eq. (11.29). Rather annoyingly, Heisenberg changes the normalization of his amplitudes by a factor of 2 at this point, writing (in parentheses) instead of Eq. (11.22),[19]

$$q = \frac{1}{2}\left[a(n,n-1)\, e^{i\omega_0 t} + a(n,n+1)\, e^{-i\omega_0 t}\right]. \qquad (11.30)$$

The factor of $1/2$ changes the sum rule (now with only a single term) to

$$h = \pi m\left[a(n,n+1)^2 - a(n,n-1)^2\right]\omega_0. \qquad (11.31)$$

This equation can be rearranged to reveal a recursion relation for the transition amplitude $a(n,n+1) = a(n+1,n)$:

$$a(n+1,n)^2 = a(n,n-1)^2 + \frac{h}{m\pi\omega_0}. \qquad (11.32)$$

Iterating this relation, one finds

$$a(n+1,n)^2 = a(0,-1)^2 + \frac{(n+1)h}{m\pi\omega_0}. \qquad (11.33)$$

[18] In general, one has $a(n,m) = a(m,n)^*$, analogous to $A_\tau = A^*_{-\tau}$ classically. Here, Heisenberg assumes zero complex phase, and writes ordinary squares instead of absolute magnitudes squared, which does not affect the final result.

[19] As pointed out by van der Waerden, there is an obvious error in this formula in the letter, where Heisenberg writes $a(n,n-1)$ instead of $a(n,n+1)$ in the second term. We have also replaced ω by the ω_0 appropriate in this case.

If we label the ground state as the state with quantum number $n = 0$, as is conventional, then $a(0, -1)$ must equal zero, as there is no transition to a state $n = -1$. Thus, one finds (replacing $n \to n - 1$)

$$a(n, n-1)^2 = \frac{nh}{m\pi\omega_0}, \tag{11.34}$$

equivalent to the result obtained previously by Ladenburg and Reiche (cf. Eq. (10.73)), which the reader will recall involves a specific, but at the time arbitrary, choice of emission amplitude (namely, that of the initial state, rather than the final, or some average of initial and final state) in applying the correspondence principle to the harmonic oscillator.

Heisenberg next turns to the evaluation of the quantized energy levels of the harmonic oscillator. By this time, the collapse of the BKS theory was complete, and the necessity for enforcing conservation principles at the microscopic level, and in particular, that of energy conservation, was generally accepted. It was certainly not obvious that inserting the expression for the coordinate q and momentum $m\dot{q}$—now in terms of two-index transition amplitudes rather than Fourier components, and with the new multiplication rule—into the formula for the energy would lead to a quantity independent of time. Moreover, the formula Heisenberg gives for the energy in the letter to Pauli is highly problematic:[20]

> The energy is (the squares of $\dot{q}$ and q are again symbolically meant):
>
> $$W = \frac{m}{2}(\dot{q}^2 + \omega_0^2 q^2) \tag{11.35}$$
>
> $$= \frac{m}{2}\left(-\omega_0^2\,\frac{(a(n,n-1)\,e^{i\omega_0 t} - a(n,n+1)\,e^{-i\omega_0 t})^2}{4} + \omega_0^2\,\frac{(a(n,n-1)\,e^{i\omega_0 t} + a(n,n+1)\,e^{-i\omega_0 t})^2}{4}\right) \tag{11.36}$$
>
> $$= \frac{m}{2}\omega_0^2\,\frac{a(n,n+1)^2 + a(n,n-1)^2}{2} \tag{11.37}$$
>
> $$= \frac{(n+\frac{1}{2})\omega_0 h}{2\pi}. \tag{11.38}$$
>
> (Pauli 1979, Doc. 93).

The classical expression for the oscillator energy W, Eq. (11.35), and the final result for its quantized values, Eq. (11.38), are obviously correct, but it is (at first sight) difficult to make sense of Eq. (11.36), which, taken literally, does not lead to the (correct) result Eq. (11.37).

[20] A number of obvious copying errors in the handwritten letter have been corrected, as discussed by van der Waerden (1968, pp. 25–27). The equation numbers are ours.

The problem goes back to Eq. (11.30), which shows that Heisenberg's notation is still enmeshed to some degree in classical thinking, as he writes the particle coordinate $q(t)$ as an algebraic sum of different Fourier components, corresponding to distinct quantum transitions, rather than as a two-indexed array of distinct transitions, with a special multiplication rule that must be employed to obtain the product of two such arrays, in particular, the squared coordinate q^2 and velocity $\dot{q}^2$ in Eq. (11.35). This is presumably what Heisenberg means when he alerts Pauli to the fact that "the squares [in Eq. (11.35)] are symbolically meant." Heisenberg must already, in his brief visit to Hamburg a few days earlier, have explained to Pauli how the result Eq. (11.37) can actually be obtained from Eq. (11.35), along the lines he had already described to Kronig in the letter of June 4.

The quantity called W in Eq. (11.35), and evaluated in Eq. (11.38), should really be regarded as an element $W(n,n)$ of an array of energies, calculated by the array multiplication rule visible, for example, in Eq. (11.17), except that here, in the calculation of the (n,n) element of the squares q^2 and $\dot{q}^2$, array elements are chosen so that the leftmost index is n, the rightmost is n, and the intermediate states coincide. As the only non-zero transition amplitudes link states differing by one in quantum number, only four elements of $q(t)$ are relevant, namely

$$\begin{aligned} q(t) \quad \rightarrow \quad & \frac{1}{2}\Big\{a(n,n-1)\,e^{i\omega_0 t},\ a(n,n+1)\,e^{-i\omega_0 t}, \\ & a(n+1,n)\,e^{i\omega_0 t},\ a(n-1,n)\,e^{-i\omega_0 t}\Big\}, \end{aligned} \tag{11.39}$$

and similarly, four elements of the velocity array

$$\begin{aligned} \dot{q}(t) \quad \rightarrow \quad & \frac{i\omega_0}{2}\Big\{a(n,n-1)\,e^{i\omega_0 t},\ -a(n,n+1)\,e^{-i\omega_0 t}, \\ & a(n+1,n)\,e^{i\omega_0 t},\ -a(n-1,n)\,e^{-i\omega_0 t}\Big\}. \end{aligned} \tag{11.40}$$

The (n,n) element of the squared coordinate array q^2 is then found to arise from only two possible terms (cf. Eq. (11.18)):

$$\begin{aligned} q^2(n,n) &= \frac{1}{2}a(n,n-1)\,e^{i\omega_0 t}\cdot\frac{1}{2}a(n-1,n)\,e^{-i\omega_0 t} \\ &\quad + \frac{1}{2}a(n,n+1)\,e^{-i\omega_0 t}\cdot\frac{1}{2}a(n+1,n)\,e^{i\omega_0 t} \\ &= \frac{1}{4}\big(a(n,n-1)^2 + a(n,n+1)^2\big), \end{aligned} \tag{11.41}$$

where we have used the symmetry property $a(m,n) = a(n,m)$ (see note 18). Similarly, for the squared velocity,

$$\dot{q}^2(n,n) = \frac{1}{4}\omega_0^2\big(a(n,n-1)^2 + a(n,n+1)^2\big). \tag{11.42}$$

Combining Eq. (11.41) and Eq. (11.42), we arrive at Eq. (11.37). The quantized energies for the simple harmonic oscillator given in Eq. (11.38) then follow using Eq. (11.34).

Note that the multiplication of terms according to the new multiplication rule for the coordinate and velocity arrays results in the cancellation of the exponentials carrying the time dependence: the resulting expression for the energy array element $W(n,n)$ is time independent, consistent with the post-BKS requirement of microscopic conservation of energy.[21] It was certainly not obvious at this point that this would continue to be the case for more complicated models (e.g., anharmonic oscillators, to all orders in the anharmonic couplings). However, Heisenberg did inform Pauli in this letter of the generalization of Eq. (11.38) to the quartic anharmonic oscillator (with equation of motion $\ddot{q}+\omega_0^2 q+\lambda q^3=0$, corresponding to an energy function $\frac{m}{2}(\dot{q}^2+\omega_0^2 q^2+\lambda q^4/2)$), up to terms of first order in the anharmonic coupling:

> For the anharmonic oscillator one obtains in essence your form "B":[22]
>
> $$E=\left(n+\tfrac{1}{2}\right)h\nu+\beta\left(n^2+n+\tfrac{1}{2}\right),\ \left[\beta=\frac{3\lambda h^2}{32m\omega_0^2\pi^2}\right] \tag{11.43}$$
>
> I would be very grateful if you could inform me in writing of any arguments which favor these formulas. Apart from the formulation of the quantum condition I am not really satisfied with the whole scheme. The strongest objection seems to me that the energy, written as a function of q and $\dot{q}$ does not in general need to be a constant, even when the equations of motion are fulfilled; this is finally due to the fact that the product of two Fourier series is not uniquely defined—but I don't wish to bore you any more with such matters (Pauli 1979, Doc. 93).

Heisenberg is pointing here to an inconvenient feature of his new multiplication rule for the amplitude arrays associated with kinematical quantities (like q or $\dot{q}$): the result depends in general on the order of multiplication,

$$\sum_{n'} a(n,n')\,b(n',n'')\neq\sum_{n'} b(n,n')\,a(n',n''). \tag{11.44}$$

Thus, an unambiguous quantum-theoretical transcription of classical expressions could not be provided unless a definite (physically motivated) prescription for the ordering of the factors was given.

[21] One can also verify, though Heisenberg does not comment on this in the Pauli letter, that all the off-diagonal elements of the energy ($W(m,n)$, with $m\neq n$) are zero.

[22] The reference to Pauli's "form B" is unclear: perhaps Heisenberg is referring to a private conversation with Pauli, in his recent visit in Hamburg, or to correspondence from Pauli, no longer extant.

Why was this relevant to the issue of energy conservation? For the anharmonic oscillator discussed in the letter to Pauli, with classical equation of motion $\ddot{q}+\omega_0^2 q+\lambda q^3 = 0$, the conservation of energy follows directly from the equation of motion, as

$$\frac{d}{dt}\left[\frac{m}{2}\left(\dot{q}^2+\omega_0^2 q^2+\lambda q^4/2\right)\right] = m\dot{q}\left(\ddot{q}+\omega_0^2 q+\lambda q^3\right) = 0. \tag{11.45}$$

This conclusion—the equivalence of energy conservation and the equation of motion—no longer follows automatically if the quantities $q(t)$, $\dot{q}(t)$, and $\ddot{q}(t)$ are not simply classical functions of time, but rather arrays subject to a non-commutative multiplication rule. The question then arises how to justify the rearrangement of different non-commuting arrays appearing in a product, such as $\dot{q}q + q\dot{q}$ in $d(q^2)/dt$, in order to reproduce the classical demonstration of energy conservation.

It turns out that in the class of models (one-dimensional anharmonic oscillators) selected by Heisenberg as test cases for his new theory the quantum-theoretical equations of motion are fortunately still compatible with a conserved (quantum-theoretical) energy function. Of course, in the calculation of the energy itself the ambiguity is absent, as only squares $(q^2, \dot{q}^2)$ appear, and the product rule is clearly unambiguous if one is multiplying an array with itself. Heisenberg seems to have realized this shortly after his June 24 letter to Pauli.

In a postcard sent June 29, 1925, Heisenberg assures Pauli of his confidence in the energy calculation for the simple harmonic oscillator (ignoring the issue of compatibility of the equations of motion with energy conservation in more general cases—*audentis fortuna iuvat*):

> As far as the energy calculation in my quantum-mechanical formalism is concerned, it is as unambiguous as the multiplication of Fourier series. In other words *if* one believes in the quantum-theoretical transformation of this multiplication, and *further* in the form of the energy $p^2/2m + \omega^2 q^2$, then one must also grant $W = (n + \frac{1}{2})h\nu$. In the meantime I have made some, but not much, progress, and am once again heartily convinced that this quantum mechanics is really right, for which Kramers accuses me of optimism (Pauli 1979, Doc. 94).

Later recollections of Heisenberg[23] suggest that, during his stay in Helgoland, he had already managed, by employing his new methodology of multiplying amplitude arrays, to calculate the quantized energy of the simple harmonic oscillator, and likewise (to some order of the perturbative expansion in the anharmonic parameter) the quantized energies in anharmonic cases in which a q^3 or q^4 term was added to the harmonic energy function. Moreover, these later recollections suggest that he had already realized that the temporal constancy of the surviving (i.e., non-zero) energy array elements was an important check on the validity of his procedure, had checked this constancy for the harmonic case as well as, up to some order in the anharmonic terms, for the anharmonic

[23] See, for example, Heisenberg (1971, p. 61; cf. note 15) and an AIP oral history interview of February 22, 1963 with Thomas S. Kuhn.

cases. However, his letters to Pauli—*subsequent* to Helgoland—betray an uneasiness and tentativeness about the new approach that are hard to reconcile with the confidence one would expect if all the results presented in the *Umdeutung* paper to come were already in hand. It seems more likely that many of the calculations presented in detail in the paper he handed to Born in mid-July were only completed after his return from Helgoland, in late June and early July 1925. This paper was completed in the first week of July, and, after being sent (July 9) to Pauli for comments, was entrusted to Born for examination, and (if favorably assessed) transmission to *Zeitschrift für Physik* for publication.

11.5 Heisenberg's *Umdeutung* paper: a new kinematics

On July 29, 1925, a full month after Heisenberg had sent the postcard to Pauli expressing confidence in his new approach, Born submitted Heisenberg's paper, entitled "On a quantum-theoretic reinterpretation [*Umdeutung*] of kinematic and mechanical relations," to *Zeitschrift für Physik*. The abstract of this paper was short and succinct: "In this work we attempt to obtain foundations for a quantum-theoretical mechanics, based exclusively on relations between quantities which are observable in principle" (p. 261).[24] This paper undoubtedly deserves a place among the handful of contributions to quantum theory that fundamentally altered the terms of discourse used by physicists to address the problems of atomic physics.

The introduction consists of a single paragraph in which Heisenberg first outlines the sources of failure of the old quantum theory, before indicating in general terms the direction in which the new ideas, capable of surmounting the inadequacies of the old theory, could be found. It represents the first systematic exposition of Heisenberg's motivations for his new approach to which we have access, and is therefore worth quoting in some detail:

> It is well known that the formal rules [i.e., Bohr–Sommerfeld quantization rules], which are generally used in the quantum theory for the calculation of observable quantities (such as the energy in the hydrogen atom), are subject to the serious objection that they contain, as an essential component, relations between quantities which are in principle unobservable (for example, the position, or the period of the orbit of an electron), and that consequently these rules lack any intuitive physical underpinning, unless one relies on the hope that quantities which as of now are unobservable might later become accessible to experimental determination. This hope might be justified if the rules in question were internally self-consistent and applicable to a precisely delineated range of quantum-theoretical problems. Experience shows however that only the hydrogen atom and the Stark effect of this atom are amenable to treatment with these formal rules of quantum theory, and that already fundamental difficulties arise with the problem of "crossed

[24] Unless noted otherwise, all references in this section and the next are to the translation of Heisenberg's (1925c) paper in van der Waerden (1968) even though we shall sometimes deviate slightly from the translations in this anthology.

> fields" (the hydrogen atom in electric and magnetic fields of different direction), that further the reaction of atoms to periodically changing fields can certainly not be described by the aforesaid rules, and that finally an extension of the quantum rules to the treatment of atoms with several electrons has proven to be impossible (p. 261).

Here, Heisenberg points to three areas in which the old quantum theory had encountered serious obstructions—a far from exhaustive list (for example, the anomalous Zeeman effect is not mentioned here). The problem of crossed fields had become a salient one following the discussion by Bohr (1918, pp. 93–94), and had been reexamined by a number of theorists (Epstein, Klein, and Lenz, among others) in 1923 and 1924. The most critical difficulty here was that, in the presence of crossed fields, an adiabatic process could be devised in which permissible Bohr–Sommerfeld quantized electron orbits could be continuously deformed into nuclear-crossing ones (with zero orbital angular momentum),[25] which were explicitly forbidden in the old quantum theory (for example, in the treatment of the Stark effect (cf. Section 6.3 and Bohr 1918, pp. 27, 75).

The second failure pointed to here was the peculiar inability of the Bohr–Sommerfeld quantization procedure to handle cases of an imposed time varying external electric field—even a very slowly varying one—as in cases of dispersion of light, when these methods had proved so successful in the limit of zero frequency (i.e., static fields, the Stark effect). Instead, a wholly different approach had been required—the Kramers dispersion theory (see Chapter 10)—to handle such problems.

Finally, there was the obvious failure of the old theory to provide any quantitatively adequate treatment of the energy levels of even helium, the simplest atom with more than one electron. Heisenberg continues:

> It has become customary to characterize the failure of the quantum-theoretical rules, which in essence involved an application of classical mechanics, as a deviation from classical mechanics. Such a characterization can hardly be regarded as sensible however if one considers that the (already quite generally valid) Einstein–Bohr frequency condition represents such a complete rejection of classical mechanics, or rather (from the standpoint of wave theory) of the kinematics underlying classical mechanics, that classical mechanics cannot be considered at all valid in even the simplest quantum-theoretical problem. In this circumstance it would seem more advisable to completely abandon any hope for an observation of previously unobservable quantities (such as position or orbital period of an electron), admitting at the same time that the partial agreement [e.g., for the hydrogen atom and the Stark effect] of the aforesaid quantum rules with experience is more or less accidental, and to try to develop a quantum-theoretical mechanics, analogous to classical mechanics, in which only relations between observable quantities appear (pp. 261–262).

It is of course only possible to understand precisely what Heisenberg is getting at here after seeing how these rather vague conceptions are implemented, and used to arrive at concrete results, in the rest of the *Umdeutung* paper.

[25] The crossed field problem is discussed in detail by Pauli (1926a, p. 163) in his *Handbuch* article.

But intimations of the new approach were already clearly apparent in the letters to Kronig and Pauli described previously. On the one hand, the idea of quantized orbits (taking seriously a classical function of time $q(t)$ or $x(t)$ as providing a meaningful position for an electron at any given time) had morphed into the use of arrays of (in general complex) numbers $a(n, m)$ directly associated with the numerators of observable Sellmeier terms in dispersion theory, and only connected asymptotically via the correspondence principle with their classical Fourier component origins.

On the other hand, the new theory was to maintain a *precise analogy* with classical mechanics insofar as one insisted that the new array quantities would formally obey exactly the same equations of motion as their classical counterparts. Heisenberg concludes his introduction by pointing to the recent work on dispersion theory, which he regards as the direct antecendent to his new approach:

> Apart from the frequency condition, one can regard the most important first steps to such a quantum-theoretical mechanics as provided by the Kramers dispersion theory [Kramers (1924a, 1924b)] and the works extending this theory [Born (1924), Kramers and Heisenberg (1925), Born and Jordan (1925a)]. In the following we wish to establish some new quantum-mechanical relations and use them for the complete treatment of some special problems. We will restrict ourselves in this to problems of a single degree of freedom (p. 262).[26]

Heisenberg begins the argument for reinterpreting classical expressions by pointing to the fact that, in classical electromagnetic theory, the calculation of arbitrary higher order (beyond dipole) moments of the radiation emitted by an electron involve increasingly complicated products of the coordinate and velocity vectors of the electron. By contrast, in optical dispersion, or dipole radiation, only the electron coordinate enters, and is completely specified by its Fourier components, which are given a quantum-theoretical transcription in terms of the associated transition amplitude array. For the calculation of electric quadrupole, magnetic dipole, and further higher multipole components of the radiation, one had to develop an unambiguous prescription for associating a quantum-theoretical quantity (presumably a two-indexed array) with an arbitrary product of coordinate and velocity terms. The development of such a prescription would, as we shall see, cause considerable discomfort in the early months of matrix mechanics.

In line with the positivistic approach indicated in the introduction,[27] Heisenberg proposes to begin by identifying the "observable quantities" that are to play an exclusive role in the new quantum mechanics. They are just the quantities that can be directly extracted from the formulas encountered in the Kramers dispersion theory, and its

[26] In fact, in the discussion of Zeeman intensities that concludes the paper, a system with two degrees of freedom is studied.

[27] For a detailed discussion of Heisenberg's philosophical attitudes, and the extent to which the procedures introduced in the *Umdeutung* paper reflect a serious commitment to Machian ideas, see Mehra and Rechenberg (1982b, sec. V.2). For another detailed analysis of the calculations in the paper, see Aitchison, McManus, and Snyder (2004).

extensions in the Kramers–Heisenberg paper: namely, the quantum transition frequencies and transition amplitudes appearing in dispersion formulas (e.g., ν_5, ν_6 and $\vec{\mathfrak{A}}_5$, $\vec{\mathfrak{A}}_6$ in Eqs. (10.142) and (10.145)), and therefore extractable in principle from experimental measurements. Both the classical and quantum versions of these quantities are then presented side by side, to emphasize the formal analogy between classical theory and its new quantum-theoretical *Umdeutung*. For the frequencies (Heisenberg uses at various points both angular ω and cyclic ν frequencies, somewhat confusingly) one has

$$\text{Classical}: \ \nu(n,\alpha) = \alpha\nu(n) = \alpha\frac{1}{h}\frac{dW}{dn}, \tag{11.46}$$

$$\text{Quantum}: \ \nu(n,n-\alpha) = \frac{1}{h}(W(n) - W(n-\alpha)). \tag{11.47}$$

Note that in Eq. (11.46) we are explicitly working in the regime of large quantum numbers where the energy levels are to a good approximation evenly spaced.[28] So the qualification "classical" here refers to the correspondence limit treatment of a periodic system, with classical motions selected and labeled via the integer n according to the usual quantization rules. This even spacing of the energy levels only holds down to small quantum numbers for the simple harmonic oscillator: once anharmonicity is present, the frequency of the (still periodic) motion depends on the amplitude, and must be reflected by an explicit dependence on the quantum number, as in $\nu(n)$.

By contrast, the Bohr frequency condition expressed in Eq. (11.47) is exact for all quantum numbers. The frequencies in Eqs. (11.46)–(11.47) obviously satisfy classical and quantum versions of the Ritz combination principle:

Classical:

$$\nu(n,\alpha) + \nu(n,\beta) = \nu(n,\alpha+\beta), \tag{11.48}$$

Quantum:

$$\nu(n,n-\alpha) + \nu(n-\alpha,n-\alpha-\beta) = \nu(n,n-\alpha-\beta). \tag{11.49}$$

For the amplitudes, the correspondence between classical Fourier components and quantum transition amplitudes is expressed (using the now rather antiquated Fraktur fonts common at the time)

$$\text{Classical: } \mathrm{Re}(\mathfrak{A}_\alpha(n)\, e^{i\omega(n)\alpha t}), \tag{11.50}$$

[28] Thus, the classical equation giving the frequency as the action derivative of the energy, $\nu = dW/d\mathcal{J}$ (cf. Eq. (A.163)), corresponds to Eq. (11.46) once the action is rescaled via $\mathcal{J} = hn$ in terms of a (still continuous) state label n. The point of this way of writing the classical equation is to emphasize the formal similarity to the quantum version, which follows immediately in Eq. (11.47), which is the Bohr frequency condition. However, the appearance of the integer offset α in the classical equation also presupposes that one is thinking about the quantized system in the correspondence limit, where energy levels are equally spaced (for $\alpha \ll n$).

$$\text{Quantum: } \mathrm{Re}(\mathfrak{A}(n, n-\alpha)\, e^{i\omega(n,n-\alpha)t}). \tag{11.51}$$

The next question—for Heisenberg, a purely kinematic one, although it would soon have profound dynamical consequences—is to settle on a composition rule allowing one to construct quantum-theoretic quantities (for example, the energy, or the more complicated expressions involving products of coordinates and velocities appearing in the calculation of the higher radiation multipoles) which involve products of the basic two-index amplitudes appearing in Eq. (11.51). At this point, the time dependence indicated in this equation takes on a critical role, once we insist that it hold for an arbitrary kinematical quantity, expressed as an assembly of two-index amplitudes. For example, if the coordinate $x(t)$ is associated quantum-theoretically with $\mathfrak{A}(n, n-\alpha)\, e^{i\omega(n,n-\alpha)t}$, then $x(t)^2$ should be associated with a two-index array given by

$$\begin{aligned}
&\textstyle\sum_\alpha \mathfrak{A}(n, n-\alpha)\, e^{i\omega(n,n-\alpha)t} \cdot \mathfrak{A}(n-\alpha, n-\beta)\, e^{i\omega(n-\alpha,n-\beta)t} \\
&\quad = \textstyle\sum_\alpha \mathfrak{A}(n, n-\alpha)\, \mathfrak{A}(n-\alpha, n-\beta)\, e^{i\omega(n,n-\beta)t} \\
&\quad \equiv \mathfrak{B}(n, n-\beta)\, e^{i\omega(n,n-\beta)t},
\end{aligned} \tag{11.52}$$

where the cyclic frequencies used earlier are replaced by angular ones. The time dependence in the first and third lines matches precisely because of the Ritz frequency combination condition Eq. (11.49) (with some relabeling of indices). As Heisenberg states: "in fact, this type of combination [i.e., multiplication] is an almost necessary consequence of the frequency combination rules" (p. 265).

In an attempt to take into account systems where the spectrum has a continuous portion (for example, the unbound states in the hydrogen atom), Heisenberg writes versions of his composition rules where the discrete sums are replaced by continuous integrals. However, the examples given in the paper—two versions of the one-dimensional anharmonic oscillator, and the rigid rotator—have purely discrete spectra, and it is certainly apparent throughout that there is no obvious generalization of Heisenberg's methodology to this more general situation.[29]

In the case just considered, where the square of a quantity (in this case $x(t)^2$) is being considered, the order of multiplication is not an issue, and the result obtained is unambiguous. However, as Heisenberg had indicated at the outset of his exposition of his new kinematics, in general one also needs a quantum-theoretical version of quantities like $x(t)\,\dot{x}(t)$ and $v(t)\,\dot{v}(t)$ (where $v(t) = \dot{x}(t)$), for example, in calculating the time derivative

[29] Within a few months, Pauli (1926a) would employ matrix-mechanical techniques to derive the discrete quantized energies of the bound electron in the hydrogen atom. The calculation is a *tour de force* of algebraic sleight of hand, which perhaps only Pauli could have accomplished at this time. Pauli's techniques are also restricted to the discrete bound spectrum. The case of continuous spectra was first addressed, in a general way, in the Three-Man-Paper [*Dreimännerarbeit*] of Born, Heisenberg, and Jordan (1926), where the theory of bounded infinite Hermitian forms (developed by Hilbert and collaborators) was invoked—see Section 12.3.3.

of an energy function involving x^2 or v^2, or in evaluating higher multipoles in radiation theory. In this case, the composition rule is clearly non-commutative, as in general

$$\begin{aligned}\mathfrak{C}(n, n-\beta) &= \sum_{\alpha} \mathfrak{A}(n, n-\alpha)\, \mathfrak{B}(n-\alpha, n-\beta) \\ &\neq \sum_{\alpha} \mathfrak{B}(n, n-\alpha)\, \mathfrak{A}(n-\alpha, n-\beta).\end{aligned} \tag{11.53}$$

In cases like this, Heisenberg proposes the use of a symmetric product rule, where $v(t)\,\dot{v}(t)$ is replaced by $(v\dot{v}+\dot{v}v)/2$, thereby preserving the Leibniz rule for differentiation of a product.

If one relabels the indices in Eq. (11.53) in an obvious way, setting $p = n - \alpha$ and $m = n-\beta$, one finds that the composition rule of arrays $\mathfrak{A}$ and $\mathfrak{B}$ (in that order) becomes

$$\mathfrak{C}(n, m) = \sum_{p} \mathfrak{A}(n, p)\mathfrak{B}(p, m), \tag{11.54}$$

which any undergraduate student in the physical sciences nowadays will immediately recognize as the rule for matrix multiplication. The word "matrix" does not appear anywhere in the *Umdeutung* paper: indeed Heisenberg seems blissfully unaware of even the simplest aspects of what we now term "linear algebra". It is apparent that this particular mathematical toolbox was simply not a common ingredient in the training most theoretical physicists received in the mid-1920s. The situation is reminiscent of the one we encountered in Volume 1 with respect to the action/angle formalism and Hamilton–Jacobi theory, where even a physicist such as Sommerfeld with a magnificent preparation in the classical mathematical physics of the late-nineteenth and early-twentieth century had to lean on the knowledge of an astronomically trained Schwarzschild before coming to the realization that his quantized phase integrals were just the action variables of the celestial mechanicians.

In fact, some physicists—in particular, Heisenberg's colleagues in Göttingen, Max Born, and especially Pascual Jordan[30]—were already familiar with the essentials of matrix algebra, and were soon to reformulate Heisenberg's ideas in a much more transparent form using this technology. Remarkably, Pauli, despite being well-versed in tensor calculus,[31] failed to recognize that Heisenberg's non-commutative multiplication rule was just the rule for matrix multiplication. He never once mentioned matrices in his extant correspondence in the period from June to September, despite many letters

[30] As Born recalled in an obituary for the mathematician Otto Toeplitz, who had been a fellow student at Breslau: "[Toeplitz] insisted on my learning matrix calculus properly and occasionally refreshed my knowledge when we were together again as young teachers in Göttingen. This turned out to be an advantage to me, first in developing Minkowski's form of relativity, then for the study of vibrations in crystals, which are determined, indeed, by a quadratic form of a (practically) infinite number of variables, but decidedly at the birth of quantum mechanics" (Born 1940, quoted in Mehra and Rechenberg 1982c, p. 38). Jordan had assisted Courant in writing Courant and Hilbert (1924).

[31] At just 21 years of age, Pauli had, after all, written the article on relativity for the *Encyclopedia of the Mathematical Sciences* (Pauli 1921a).

from Heisenberg, especially in June and July. And Heisenberg takes pains, in a letter of September 18, 1925 (Pauli 1979, Doc. 98), to remind Pauli of the definition of matrix multiplication, and basic properties thereof (with a reference to Courant–Hilbert). Had Pauli known about matrices, one presumes he would have told Heisenberg by the end of June that his peculiar rules for multiplying amplitude arrays simply correspond to matrix multiplication. Armed with this information, it seems plausible that Heisenberg would have been able to anticipate many of the results of the Born–Jordan paper which would appear later in the Fall of 1925.

In his discussion of the new rules for constructing the kinematic quantities appropriate for a fully quantum-theoretical description of atomic phenomena, Heisenberg puts particular emphasis in the *Umdeutung* paper (as he would also frequently do in later recollections of this period) on the physical significance of the relative phases of the complex array elements appearing in these quantities. This is perhaps the first point in which the critical role played by complex quantities in quantum mechanics becomes apparent, in contradistinction to classical mechanics, where the use of complex expressions is a convenience rather than a necessity. We should first recall the physical significance of the complex phase in classical dispersion theory. In the Fourier expansion for the coordinate (one dimensional for simplicity—cf. Eq. (10.92))

$$x_0(t) = \sum_{\tau} A_\tau e^{2i\pi\tau w}, \quad w = \nu_0 t, \tag{11.55}$$

each complex Fourier component A_τ contains a phase γ_τ (i.e., $A_\tau = |A_\tau|\, e^{i\gamma_\tau}$), which encodes the point in the periodic trajectory that this component of the motion reaches at time $t = 0$. The reality of the coordinate $x(t)$ requires the symmetry property

$$A_{-\tau} = A_\tau^\star \tag{11.56}$$

in the Fourier sum, but in general the coefficients A_τ must be allowed to be complex numbers. Their phase can be expected to vary randomly from a radiating (or dispersing) electron in one atom to another such electron, in the same or another atom. One must take account here of the fact that an incoming light wave can interact simultaneously with electrons in many atoms. If the effect on the light being studied is dependent on these phases, the overall influence of the matter on the electromagnetic radiation will involve an incoherent sum with typically destructive cancellation between the scattered light from different electrons. For the coherent part of the polarization however, as in Eq. (10.116), only the product $A_\tau A_{-\tau} = A_\tau A_\tau^\star$ appears, in which the dependence on the random phases γ_τ disappears: in other words, optical dispersion is a *coherent* process in which the scattered light from electrons in different atoms combines coherently to produce the dispersed wave, which maintains a fixed phase relation with the incoming light.

As was emphasized by Kramers and Heisenberg (1925) in their work on incoherent (Smekal–Raman) scattering, the lack of dependence on the complex γ_τ phases in the coherent (elastic scattering) case is decidedly not the case once the scattering is inelastic, as the electron is left in a final state different from the initial one, and the relative phase of the initial and final electron motion can vary randomly from one atom to the next. For example, in Eq. (10.142), the amplitudes $\mathfrak{A}_5$ and $\mathfrak{A}_6$ have a random relative phase, which does not cancel in the product $\mathfrak{A}_5\mathfrak{A}_6^\star$. Moreover, it was clear from the analysis of the incoherent Smekal–Raman terms that the final scattering amplitude involved a sum over intermediate states with unavoidable interference effects due to the varying phase of the component terms in the sum—just as in the sum over intermediate states in Heisenberg's definition of array multiplication in Eq. (11.53) The importance of phase considerations had also been brought home to Kramers and Heisenberg in their (mistaken) attempts to elude the "false resonances", as discussed in Section 10.5.

The qualification "relative phase" in the preceding (purely classical) discussion is important. There is clearly a physically irrelevant global shift in the phases of the Fourier coefficients corresponding to the freedom to shift the origin of the time coordinate. But it was clear to Heisenberg that the physically relevant phase dependence of the classical coefficients should extend to the complex transition amplitudes in the quantum case. What was, of course, not clear at this point was the proper physical interpretation of these phases, given that the concept of multiply periodic orbital motion, with a clear geometrical meaning to the phase of each harmonic component of the motion at any given time, had been abandoned. As Heisenberg says:

> A geometrical interpretation of such quantum-theoretical phase relations in analogy with those of classical theory seems at present scarcely possible (p. 265).

The issue of phases of the complex amplitudes would be addressed in greater detail in the Two-Man paper of Born and Jordan (1925b), discussed in Chapter 12. However, a full understanding of the phase issue would have to wait for the Hilbert space formulation of quantum theory developed by von Neumann two years after the *Umdeutung* paper (see Chapter 17). Also, and perhaps more surprisingly (given its later importance), the quantum constraint (later called "hermiticity") corresponding to the reality condition Eq. (11.56) for the classical amplitudes,

$$A(n, n-\tau) = A(n-\tau, n)^\star, \tag{11.57}$$

was not emphasized at all by Heisenberg either in the weeks preceding the *Umdeutung* paper, or in the paper itself.

11.6 Heisenberg's *Umdeutung* paper: a new mechanics

In the next section of his paper, Heisenberg addresses the construction of a new, fully quantum-theoretical, dynamical theory, which he proposes to realize in two steps (pp. 266–267):

1. Integration of the equation of motion

$$\ddot{x} + f(x) = 0. \tag{11.58}$$

2. Determination of the constants for periodic motion through a Bohr–Sommerfeld quantization condition

$$\oint p dq = \oint m\dot{x} dx = \mathcal{J}(= nh). \tag{11.59}$$

The underlying guiding philosophy here is the desire to obtain a quantum formalism "corresponding as closely as possible to that of classical mechanics" (hence, the preservation of the classical equations at least in form) but where, in keeping with the positivistic impulse to deal only with observable quantities, one must replace "the quantities $\ddot{x}$ and $f(x)$ by their quantum-theoretical representatives" (p. 267) as described in the preceding kinematic section of the paper.

Heisenberg's second step indicates the route that he is going to use to introduce Planck's constant, and hence, explicit quantization, into his theory. The argument here reproduces exactly the result Heisenberg outlined to Pauli in his letter of June 24, 1925, (see Section 11.4) leading to the sum rule Eq. (11.29), with a relabeled index of summation α (instead of τ):[32]

$$h = 4\pi m \sum_{\alpha>0} \Big(a(n+\alpha, n)^2 \omega(n+\alpha, n) - a(n, n-\alpha)^2 \omega(n, n-\alpha) \Big). \tag{11.60}$$

At this point Heisenberg explains the connection of this formula to the high-frequency limit of the Kramers dispersion formula, namely the Thomas–Reiche–Kuhn sum rule. So, in a sense, the constraint Eq. (11.60) is hardly an original contribution of Heisenberg. However, his use of it in combination with a radically new interpretation of the quantities appearing in the classical equations of motion is, of course, the key to the remarkable conceptual breakthrough contained in the *Umdeutung paper*.

The objectives Heisenberg hopes to achieve employing his two steps are clearly stated:

> one could regard equations [11.58] and [11.60] as a satisfactory solution, at least in principle, of the dynamical problem if it were possible to show that this solution agrees with (or at any rate does not contradict) the quantum-mechanical relationships which we know at present. It should, for instance, be established that the introduction of a small perturbation into a dynamical problem leads to additional terms in the energy, or frequency, of the type found by Kramers and Born—but not of the type given by classical theory. Furthermore, one should investigate whether equation [11.58] in the present quantum-theoretical form would in general give rise to an energy integral $\frac{1}{2}m\dot{x}^2 + U(x)$=const., and whether the energy so derived satisfies the condition $\Delta W = h\nu$ (p. 269).

[32] There is a slight error in the formula as written in the paper, as Heisenberg interchanges the order of the arguments of the frequency arrays, writing $\omega(n, n+\alpha)$ instead of $\omega(n+\alpha, n)$, etc., which introduces an incorrect sign.

In short, the theory should provide a conserved energy—in the new kinematics, as an assembly of two-indexed quantities—with the diagonal elements $W(n,n) \equiv E_n$ satisfying the Bohr frequency condition $E_n - E_m = h\nu(n,m)$.

To examine these issues, Heisenberg turns to the application of the new theory to two mechanical models which, although one-dimensional and structurally simple, are analytically non-trivial even in the classical context. First, he examined the cubic anharmonic oscillator, defined by equation of motion

$$\ddot{x} + \omega_0^2 x + \lambda x^2 = 0, \tag{11.61}$$

and with energy

$$W = \frac{1}{2}m\dot{x}^2 + \frac{1}{2}m\omega_0^2 x^2 + \lambda\frac{m}{3}x^3. \tag{11.62}$$

This model is physically not very sensible, as the potential energy is unbounded below, and the motion is periodic and bounded only if the amplitude is less than a critical value. The asymmetry of the potential also means that the classical time average position of the oscillating particle is not zero. Nevertheless, one can work in the regime of periodic motions and examine the motion (both classically, as we did previously (cf. Eqs. (11.1)–(11.3)), and, as we shall see, quantum mechanically) via a perturbative expansion in the small parameter λ.

The second model, the quartic anharmonic oscillator, is physically more sensible, as the energy is bounded below (indeed, it is strictly positive). The equation of motion is

$$\ddot{x} + \omega_0^2 x + \lambda x^3 = 0, \tag{11.63}$$

with energy

$$W = \frac{1}{2}m\dot{x}^2 + \frac{1}{2}m\omega_0^2 x^2 + \lambda\frac{m}{4}x^4. \tag{11.64}$$

If we remove the anharmonicity by setting $\lambda = 0$, in either model, we return of course to the simple harmonic oscillator, which Heisenberg had already completely solved with his new methods, presumably during or even before his visit to Helgoland. We have already described the solution as he outlined it to Pauli in the June 24 letter. The transition amplitudes, frequencies, and quantized energy levels in this case were found to be (cf. Eqs. (11.34)–(11.38)):

$$a(n,n-1) = a(n-1,n) = \sqrt{\frac{nh}{m\pi\omega_0}}, \tag{11.65}$$

$$a(n,m) = 0, \quad n \neq m \pm 1, \tag{11.66}$$

$$\omega(n,n-\alpha) = \alpha\,\omega_0, \tag{11.67}$$

$$W(n,n) = (n+\frac{1}{2})\hbar\omega_0, \quad W(n,m) = 0 \text{ for } n \neq m. \tag{11.68}$$

As the (n, m) array element of any kinematical quantity constructed from $x(t)$ and $\dot{x}(t)$ according to Heisenberg's array multiplication rule was forced to have the time dependence $e^{i\omega(n,m)t}$, for a system with a non-degenerate spectrum (like the one-dimensional oscillator systems examined in the *Umdeutung* paper) where $\omega(n, m) \neq 0$ if $n \neq m$, the principle of microscopic energy conservation (newly reinforced after the collapse of the BKS theory) required that the energy array $W(n, m)$ be *diagonal* once the arrays corresponding to the coordinate $x(t)$ and velocity $\dot{x}(t)$ had been calculated (by first calculating the transition amplitudes $a(n, m)$) and inserted into the expression for the total energy. The diagonal elements $W(n, n)$ obviously corresponded to the quantized energy in the nth stationary state. In general, Heisenberg was aware that a full demonstration of the conservation of energy principle in his new mechanics required a proof that all off-diagonal elements of the energy array vanish. As we shall see, he was only able to provide limited evidence in support of this requirement.

In his examination of the anharmonic models Heisenberg resorts to a recursive procedure to obtain the formal expansion of the amplitudes $a(n, n-\alpha)$ and frequencies $\omega(n, n-\alpha)$ in increasing powers of the anharmonic coupling constant λ. We begin with the cubic anharmonic oscillator. Referring to Eq. (11.3) for the classical model, we see that, if we assume (by correspondence arguments) that the same power dependence applies to the classical Fourier component as to the associated quantum amplitude, the leading power of λ appearing in the transition amplitudes $x(n, m)$ corresponding to the particle coordinate is as follows[33]

$$\begin{aligned} x(n,n) &\simeq X_0 \simeq O(\lambda), \\ x(n,n-1) &\simeq X_1 \simeq O(1), \\ x(n,n-2) &\simeq X_2 \simeq O(\lambda), \\ x(n,n-3) &\simeq X_3 \simeq O(\lambda^2), \text{ etc.} \end{aligned} \tag{11.69}$$

To zeroth order in λ (the simple harmonic limit), the only non-vanishing elements are, from Eq. (11.65),

$$x(n,n-1) = x(n-1,n) = \frac{1}{2}\sqrt{\frac{nh}{m\pi\omega_0}}. \tag{11.70}$$

If we examine the $(n, n-\alpha)$ element of the arrays appearing in the cubic oscillator equation of motion Eq. (11.61), we find, recalling the time-dependence $e^{i\omega(n,m)t}$ associated with $x(n, m)$,

$$(-\omega(n,n-\alpha)^2 + \omega_0^2)\, x(n,n-\alpha) + \lambda \sum_{\beta} x(n,n-\beta)\, x(n-\beta,n-\alpha) = 0, \tag{11.71}$$

[33] We shall work with the amplitudes $x(n, m)$ directly associated with the particle coordinate, rather than with Heisenberg's $a(n, m)$ as the connection to the equation of motion is more direct. The comparison with the equations of the *Umdeutung* paper simply involves inserting appropriate factors of 2 and λ, as $x(n, m) = \lambda^p a(n, m)\, e^{i\omega(n,m)t}/2$, where $p = 1$ for $m = n$, $p = 0$ for $m = n-1$, etc.

where the time dependence $e^{i\omega(n,n-\alpha)t}$ appearing in every term (in the last via the Ritz frequency combination condition $\omega(n,n-\beta)+\omega(n-\beta,n-\alpha)=\omega(n,n-\alpha)$) has been divided out.

We now examine the consequences of Eq. (11.71) by successively putting $\alpha=0,1,2,\ldots$, in each case keeping only terms with the lowest power of the coupling λ. For example, for $\alpha=0$, we obtain (as $\omega(n,n)=0$, and using the symmetry property $x(n,n-\beta)=x(n-\beta,n)$)

$$\begin{aligned}\omega_0^2 x(n,n) &= -\lambda\sum_\beta x(n,n-\beta)^2 \\ &= -\lambda(x(n,n-1)^2+x(n,n+1)^2)+O(\lambda^2).\end{aligned} \tag{11.72}$$

This equation, reinserting appropriate factors of 2 and λ, appears as the first of Heisenberg's Eq. (19):

$$\omega_0^2 a_0(n)+\frac{1}{4}\left[a^2(n+1,n)+a^2(n,n-1)\right]=0. \tag{11.73}$$

Inserting Eq. (11.70) in Eq. (11.72) and solving for $x(n,n)$, one finds

$$x(n,n)=-\lambda\frac{\hbar}{2m\omega_0^3}(2n+1), \tag{11.74}$$

confirming that $x(n,n)$ is indeed of order λ (plus higher orders, of course) as claimed in Eq. (11.69). The associated classical Fourier coefficient X_0 has the physical interpretation of being the time average position of the oscillator (see Eq. (11.2)). In modern quantum mechanics, it is simply the expectation value of the particle position in the nth stationary state. It is non-zero in this case simply because the potential energy function is asymmetric around the origin.

Next, taking $\alpha=1$ in Eq. (11.71), we find

$$\begin{aligned}\left(-\omega(n,n-1)^2+\omega_0^2\right)x(n,n-1) &= -\lambda\{x(n,n)x(n,n-1) \\ &\quad + x(n,n-1)x(n-1,n-1) \\ &\quad + x(n,n+1)x(n+1,n-1) \\ &\quad + x(n,n-2)x(n-2,n-1) \\ &\quad + \ldots\}.\end{aligned} \tag{11.75}$$

Referring to Eq. (11.69), we see that the contents of the curly braces are at least of first order in λ, whence we conclude

$$-\omega(n,n-1)^2+\omega_0^2=O(\lambda^2), \tag{11.76}$$

corresponding to the second of Heisenberg's Eq. (19) on p. 270 (which has zero on the right-hand side, discarding terms of higher than first order in λ).

Finally, setting $\alpha = 2$, we find

$$\begin{aligned}(-\omega(n,n-2)^2+\omega_0^2)\,x(n,n-2) &= -\lambda\{x(n,n)x(n,n-2)\\ &\quad + x(n,n-1)x(n-1,n-2)\\ &\quad + x(n,n+1)x(n+1,n-2)\\ &\quad + x(n,n-2)x(n-2,n-2)\\ &\quad + \ldots\}. \end{aligned} \tag{11.77}$$

To leading order, only the second term in the curly braces (of zeroth order in λ) survives, so to the desired order this reduces to

$$(-\omega(n,n-2)^2+\omega_0^2)x(n,n-2) = -\lambda x(n,n-1)x(n-1,n-2), \tag{11.78}$$

corresponding to the third of Heisenberg's Eq. (19):

$$(-\omega(n,n-2)^2+\omega_0^2)a(n,n-2)+\frac{1}{2}a(n,n-1)a(n-1,n-2) = 0. \tag{11.79}$$

From Eq. (11.76) we know that $\omega(n,n-2) = 2\omega_0 + O(\lambda^2)$, so Eq. (11.78) allows us to determine the $x(n,n-2)$ amplitudes to lowest order:

$$\begin{aligned}-3\omega_0^2 x(n,n-2) &= -\lambda x(n,n-1)x(n-1,n-2)\\ &= -\lambda\frac{h}{4m\pi\omega_0}\sqrt{n(n-1)},\end{aligned} \tag{11.80}$$

whence

$$x(n,n-2) = \lambda\frac{\hbar}{6m\omega_0^3}\sqrt{n(n-1)} + O(\lambda^2), \tag{11.81}$$

agreeing with the power counting indicated in Eq. (11.69). A simple recursive argument shows that the leading contribution (in powers of λ) to $x(n,n-\tau)$, for $\tau > 0$, displays the "vanishing at the edges" feature (cf. Eq. (11.7)), which Heisenberg had employed in conversations with Pauli in Copenhagen:

$$\begin{aligned}x(n,n-\tau) &\propto \lambda^{\tau-1}\sqrt{n(n-1)(n-2)\cdots(n-\tau+1)}\\ &= \lambda^{\tau-1}\sqrt{\frac{n!}{(n-\tau)!}}\end{aligned} \tag{11.82}$$

(cf. p. 271, Eq. (21)).

The calculation of the quantized energies in the cubic anharmonic oscillator amounts to the evaluation of the energy, written by Heisenberg simply as W, but more explicitly the diagonal element of the energy array

$$W(n,n) = \frac{1}{2}m\dot{x}^2(n,n) + \frac{1}{2}m\omega_0^2 x^2(n,n) + \frac{m\lambda}{3}x^3(n,n), \tag{11.83}$$

where the squares and cubes are calculated by the array multiplication rule. With the power counting rules of Eq. (11.69), the reader may easily verify that there are no terms of order λ in these diagonal elements. Consequently, the quantized energies, neglecting terms of order λ^2, reduce to just those of the simple harmonic case, the evaluation of which we discussed in connection with the June 24 letter to Pauli:

$$W(= W(n,n)) = (n + \frac{1}{2})h\omega_0/2\pi \text{ (terms of order } \lambda^2 \text{ have been excluded)}, \tag{11.84}$$

which is Eq. (23) in the *Umdeutung* paper. Although the results obtained by Heisenberg for the cubic model were encouraging, the only new result was the establishment of the necessity for the zero-point energy, following from a "vanishing at the edge" argument.[34] Otherwise, they merely reproduce, to the given order, the all too-familiar properties of the quantized simple harmonic oscillator.

Much more extensive results are presented for the second model examined by Heisenberg: the quartic oscillator with total energy function Eq. (11.64). The cubic term present in the equation of motion (arising from the quartic x^4 term in the potential energy),

$$\ddot{x} + \omega_0^2 x + \lambda x^3 = 0, \tag{11.85}$$

results in only odd harmonics appearing in the Fourier expansion of the classical motion:

$$x(t) = X_1 \cos(\omega t) + X_3 \cos(3\omega t) + X_5 \cos(5\omega t) + \cdots \tag{11.86}$$

By the usual correspondence-principle arguments, this suggests that the only non-vanishing coordinate array amplitudes are $x(n,n-1)$, $x(n,n-3)$, etc. (and of course the associated conjugate quantities $x(n-1,n) = x(n,n-1)^*$, etc.). The quantum-theoretical equation of motion, analogous to Eq. (11.71) for the cubic oscillator, is

$$\begin{aligned}&(-\omega(n,n-\alpha)^2 + \omega_0^2)\, x(n,n-\alpha)\\ &= -\lambda \sum_{\beta,\gamma} x(n,n-\beta)\, x(n-\beta,n-\gamma)\, x(n-\gamma,n-\alpha) = 0.\end{aligned} \tag{11.87}$$

The evaluation of the diagonal elements of the energy array $W(n,n)$ up to and including terms of order λ^2 requires calculation of the frequencies $\omega(n,n-1)$ to the same order, while the off-diagonal matrix elements $x(n,n-1)$ and $x(n,n-3)$ are only needed up to

[34] The non-existence of a state with $n = -1$, implying $x(n,-1) = 0$, leads to the vanishing of $x(n,n-\tau)$ for $\tau = n+1$, which is apparent in the broken factorial of Eq. (11.82).

order λ. We do not reproduce all the algebra here, but give a simple example to illustrate the type of calculations needed, which are tedious, but ultimately perfectly straightforward. Taking $\alpha = 1$ in Eq. (11.87), and using Eq. (11.70), we find, neglecting terms of order λ^2,

$$
\begin{aligned}
\left(-\omega(n,n-1)^2+\omega_0^2\right)x(n,n-1) &= \frac{1}{2}\sqrt{\frac{nh}{m\pi\omega_0}}\left(\omega_0^2-\omega(n,n-1)^2\right)\\
&= -\lambda\Big\{x(n,n+1)x(n+1,n)x(n,n-1)\\
&\qquad + x(n,n-1)x(n-1,n-2)x(n-2,n-1)\\
&\qquad + x(n,n-1)x(n-1,n)x(n,n-1)\Big\} + O(\lambda^2)\\
&= -\frac{\lambda}{8}\left(\frac{h}{m\pi\omega_0}\right)^{3/2}\Big\{(n+1)\sqrt{n}+(n-1)\sqrt{n}+n\sqrt{n}\Big\} + O(\lambda^2)\\
&= -\frac{3\lambda}{8}\left(\frac{h}{m\pi\omega_0}\right)^{3/2} n\sqrt{n} + O(\lambda^2),
\end{aligned}
\tag{11.88}
$$

from which we conclude

$$
\omega(n,n-1)^2 = \omega_0^2 + \frac{3\lambda h}{4m\pi\omega_0}\,n + O(\lambda^2),
\tag{11.89}
$$

and, taking the square root,

$$
\omega(n,n-1) = \omega_0 + \frac{3\lambda h}{8m\pi\omega_0^2}\,n + O(\lambda^2).
\tag{11.90}
$$

The term of order λ in this equation is the second term on the right-hand side of Eq. (24) on p. 272 of Heisenberg's paper.

Heisenberg's result for the quantized energies of the quartic anharmonic oscillator, through terms of order λ^2 is impressively elaborate. After giving the formulas for the frequencies $\omega(n,n-1)$ and amplitudes $a(n,n-1), a(n,n-3)$, he states (pp. 272–273)

> The energy, defined as the constant [i.e., time-independent] term in the expression
>
> $$\frac{1}{2}m\dot{x}^2 + \frac{1}{2}m\omega_0^2x^2 + \frac{1}{4}m\lambda x^4,$$
>
> (I could not prove in general that all periodic terms actually vanish, but this was the case for all terms evaluated) turns out to be
>
> $$
\begin{aligned}
W[= W(n,n)] \; &= \frac{(n+\frac{1}{2})h\omega_0}{2\pi} + \lambda\frac{3(n^2+n+\frac{1}{2})h^2}{8\cdot 4\pi^2\omega_0^2 m}\\
&\quad - \lambda^2\frac{h^3}{512\pi^3\omega_0^5 m^2}\left(17n^3+\tfrac{51}{2}n^2+\tfrac{59}{2}n+\tfrac{21}{2}\right)
\end{aligned}
$$
>
> [this is Heisenberg's Eq. (27)].

A critical test (and valuable algebraic check!) of the whole procedure is that it reproduces the Bohr frequency condition, in some respects the most characteristic (and puzzling) feature of quantum theory since the Bohr atom of 1913,

$$W(n,n) - W(n-1,n-1) = \frac{h}{2\pi}\omega(n,n-1), \tag{11.91}$$

(this is a slightly rearranged version of the unnumbered equation following Heisenberg's Eq. (27) on p. 273). One easily verifies, substituting Heisenberg's Eq. (27) in Eq. (11.91), and keeping terms up to first order in λ, our earlier result Eq. (11.90) for the frequency $\omega(n,n-1)$. In fact, Heisenberg's Eq. (27) turns out to be precisely the correct result, obtained of course much more efficiently using the technical arsenal of modern quantum theory (specifically, Rayleigh–Schrödinger perturbation theory).[35]

Referring to his Eq. (27), Heisenberg remarks

> This energy can also be calculated using the *Kramers–Born* procedure, by taking the term $(m\lambda/4)\,x^4$ as a perturbation to the harmonic oscillator. One then comes once again exactly to the result Eq. (27), which appears to me to give remarkable support to the underlying quantum-mechanical equations (p. 273).

Heisenberg here somewhat misstates the actual situation: his new results do agree in their leading terms for large quantum numbers with those obtained by Hamilton–Jacobi perturbation theory in the old quantum theory, but there are discrepancies for finite values of n. The agreement is improved by including the zero-point energy $\frac{1}{2}\hbar\omega_0$ in the leading term, which is an ad hoc amendment of the original rules of the old theory. We will illustrate this point in the order λ term in Heisenberg's Eq. (27), using the classical perturbation theory result from Born's 1924 *Atommechanik*. There we find, to first order in the perturbation bx^4, an energy given in terms of powers of the action variable $\mathcal{J}$ (Born 1925, p. 294),

$$W = \nu_0\mathcal{J} + \frac{3}{2}b\frac{\mathcal{J}^2}{(2\pi)^4\nu_0^2m^2}. \tag{11.92}$$

Substituting $\nu_0 = \omega_0/2\pi$, $\mathcal{J} = (n+\frac{1}{2})h$, $b = m\lambda/4$, we find to first order in λ,

$$W = \frac{(n+\frac{1}{2})h\omega_0}{2\pi} + \lambda\frac{3(n^2+n+\frac{1}{4})h^2}{8\cdot 4\pi^2\omega_0^2 m}, \tag{11.93}$$

which agrees with the first two terms in Heisenberg's eq. (27), apart from the constant term $\frac{1}{4}$ replacing $\frac{1}{2}$ in the parenthesis of the first order term. Note that this level of agreement requires the inclusion of the zero-point energy in the zeroth order term.

Unfortunately, Heisenberg's parenthetic remark that he "could not prove in general that all periodic terms actually vanish" but that "this was the case for all terms evaluated,"

[35] See, for example, Powell and Crasemann (1962, p. 388).

does not tell us what exactly he checked in this context. Two letters to Pauli in late June (after the Helgoland trip) make it clear that the issue of energy conservation had been of great concern to Heisenberg (see Section 11.4). As he had told Pauli in his letter of June 24: "I am not really satisfied with the whole scheme . . . The strongest objection seems to be that the energy . . . does not in general need to be a constant" (Pauli 1979, Doc. 93).

If Heisenberg had, while in Helgoland, really verified explicitly the (highly nontrivial!) vanishing of the off-diagonal (and therefore explicitly time-dependent, via periodic factors $e^{i\omega(n,m)t}$) elements of the energy array to the same order of perturbation theory with which he treated the diagonal elements (i.e., up to order λ for the cubic, order λ^2 for the quartic oscillator), we would have expected a much more enthusiastic tone of voice in his letter of June 24 to Pauli. In later reminiscences, Heisenberg leaves no doubt that he had convinced himself while in Helgoland of the validity of energy conservation for some nontrivial cases.[36] Speaking of his work in Helgoland, in his 1971 memoir, he says

> I noticed that there was no guarantee that the new mathematical scheme could be put into operation without contradictions. In particular, it was completely uncertain whether the principle of conservation of energy would still apply, and I knew only too well that my scheme stood or fell by that principle. Other than that, however, several calculations showed that the scheme seemed quite self-consistent. Hence I concentrated on demonstrating that the conservation law held, and one evening I reached the point where I was ready to determine the individual terms in the energy table, or, as we put it today, in the energy matrix, by what would now be considered an extremely clumsy series of calculations. When the first terms seemed to accord with the energy principle [i.e., conservation], I became rather excited, and I began to make countless arithmetical errors. As a result, it was almost three o'clock in the morning before the final result of my computations lay before me. The energy principle had held for all the terms, and I could no longer doubt the mathematical consistency and coherence of the kind of quantum mechanics to which my calculations pointed (Heisenberg 1971, p. 61, right before the passage quoted in note 15).

On this basis, Mehra and Rechenberg (1982b, pp. 258–259) conclude that Heisenberg did in fact compute in Helgoland, through order λ^2, the $W(n, n-2)$ and $W(n, n-4)$ off-diagonal elements of the energy array, verifying that the numerous terms that contribute to these elements cancel exactly to that order.[37] The vanishing of farther off-diagonal terms ($W(n, n-6)$, etc.) would require working to higher than second order, which Heisenberg clearly did not do. Heisenberg may, in fact, have misremembered exactly how much he accomplished in Helgoland four decades earlier. This would also help explain the lukewarm and tentative tone of his letters to Pauli *after* his stay in Helgoland regarding the new formulation. In any event, he presumably had performed further

[36] A complete demonstration for all off-diagonal matrix elements would have required, using Heisenberg's methods, a calculation to all orders of perturbation theory, which was clearly an intractable proposition, given the elementary algebraic tools employed by Heisenberg in June 1925.

[37] For the explicit algebra verifying the vanishing of the energy elements $W(n, n-1)$, $W(n, n-2)$, $W(n, n-3)$ to order λ in the cubic oscillator, see Aitchison et al. (2004, p. 1378).

checks of energy conservation by the time of handing his paper to Born, with the request to submit it for publication if Born approved of the contents, before leaving Göttingen on July 9 for visits to Leyden and Cambridge.

In the final pages of the *Umdeutung* paper, Heisenberg turns to two topics that were of great current interest in spectroscopy: the infrared band spectra of diatomic molecules, and the Zeeman intensity multiplet rules developed recently (see Goudsmit and Kronig 1925; Hönl 1925; Sommerfeld and Hönl 1925). Here, unfortunately, Heisenberg fell victim to the trap that has ensnared many a scientist: the peril of knowing in advance the answer one wishes to get. We shall see before too long that the necessary quantum-mechanical technology—the correct treatment of rotation and angular momentum in three dimensions—was simply not available to Heisenberg at this point,[38] and his attempts to squeeze out the desired formulas for three-dimensional rotational systems by working in analogy to the one-dimensional oscillator problem led him astray—although in the case of the Zeeman intensities, he does come tantalizingly close to getting it right.

Let us briefly survey the experimental situation on band spectra, as described in *Atombau und Spektrallinien* (Sommerfeld 1924, p. 706). Here, we are dealing with quantum transitions of a diatomic molecule (such as HCl) which can simultaneously change the vibrational quantum number n (corresponding to oscillations along the radial vector connecting the two nuclei) and/or the rotational quantum number m (corresponding to rotations of the molecule around its center of mass) by a single unit, with an emitted light quantum taking away the net change in energy (rotational plus vibrational). The change in vibrational energy is just $h(n' - n)\nu_0$ where ν_0 is the natural frequency of oscillation if the two nuclei are separated radially by a small displacement from the equilibrium configuration. The rotational kinetic energy is calculated by taking the two atom molecule to be a rigid rotator of moment of inertia I around the center of mass. Classically, the rotational kinetic energy would therefore be

$$W_{\mathrm{rot}} = \frac{1}{2I}\vec{L}^2, \tag{11.94}$$

where $\vec{L}$ is the rotational angular momentum of the whole molecule. In the old quantum theory, one assumed that this angular momentum (assumed without loss of generality to be in the z-direction) was quantized by the usual quantum conditions

$$L_z = m\hbar, \quad \vec{L}^2 = m^2\hbar^2, \quad m = 0, \pm 1, \pm 2, \ldots \tag{11.95}$$

Thus, in a transition in which the vibrational quantum number dropped by a single unit and simultaneously there was a change by a single unit (up or down) of the rotational

[38] The correct treatment of angular momentum in quantum mechanics would be laid out in full in Born, Heisenberg, and Jordan (1926), discussed in Chapter 12.

quantum number, the emitted (or, in the case of the more normally measured absorption spectrum, the absorbed) light frequency would be

$$\nu = \nu_0 + B(m'^2 - m^2), \;\; m' = m \pm 1, \;\; B = \frac{h}{8\pi^2 I}. \tag{11.96}$$

Note that the first (vibrational) term is typically much larger than the second (rotational), so that we get a positive frequency ν with either sign in the second term.

Unfortunately, the measured infrared absorption spectra of diatomic gases showed clearly that Eq. (11.96) did not accord with the empirical situation. Instead, the data suggested that the rotational energy shift (the B term in the equation) corresponded to a difference of squares of *half integer* quantum numbers $m^* = \frac{1}{2}, \frac{3}{2}, \frac{5}{2}, \ldots$ etc., rather than a difference of squares of integer angular momentum quantum numbers $m = 0, 1, 2, \ldots$ This phenomenological rule was first proposed by Kratzer in 1922 and published the following year (Kratzer 1923). Sommerfeld (1924) emphasized Kratzer's rule in his review of the situation: it was yet another case of half-integral quanta appearing in the old quantum theory, with no clear physical explanation, and was accepted by Sommerfeld as just another case where empirical numerology took precedence over real physical understanding.[39]

Taking a peek ahead to the understanding of angular momentum, which would emerge from matrix mechanics before the end of 1925, we now realize that the problem arose from the fact that the squared orbital angular momentum in three dimensions is quantized as

$$\vec{L}^2 = l(l+1)\hbar^2 = \left(\left(l + \tfrac{1}{2}\right)^2 - \tfrac{1}{4}\right)\hbar^2, \quad l = 0, 1, 2, \ldots. \tag{11.97}$$

The quantum mechanical treatment of motion differs essentially from that in classical mechanics, or for that matter, in the old quantum theory, in that classical motion in a two-dimensional plane can be calculated ignoring completely the presence of an orthogonal dimension, whereas in modern quantum mechanics the presence of a third dimension essentially alters the dynamics (e.g., by the unavoidable "spreading" of the wave function into the third dimension). As the factor of $-\frac{1}{4}$ cancels when energy differences are taken, Eq. (11.97) explains Kratzer's phenomenological half-integer rule.

In the old quantum theory, it was perfectly reasonable to imagine the motion of a rigid rotator as contained entirely in a two-dimensional plane, with angular momentum given

[39] Kratzer had published an "explanation" of the half integral values by assuming that the net angular momentum of the electrons in the diatomic molecule contributed $\hbar/2$ to the orbital angular momentum, apart from the angular momentum associated with the rotation of the molecule as a whole. As we do here, Sommerfeld (1924, p. 713) relegated this rationalization to a footnote.

by an integer multiple of $\hbar$, as in Eq. (11.95). With this intuition, Heisenberg assumed that the rotational structure of band spectra can be examined by modeling a rigid rotator as a particle of mass m attached to a rigid massless rod of length a, and executing rotations around the origin with angular momentum ω, whence the quantization condition

$$nh = 2\pi m a^2 \omega. \tag{11.98}$$

Imitating the route in Eqs. (11.24)–(11.27) used to reach the quantization condition Eq. (11.28), one goes from

$$h = \frac{d}{dn}\left(2\pi m a^2 \omega\right) \tag{11.99}$$

to[40]

$$h = 2\pi m a^2 \left(\omega(n+1, n) - \omega(n, n-1)\right). \tag{11.100}$$

As for the harmonic oscillator, this provides a recursion relation for the frequency array $\omega(n, n-1)$, which can be solved up to an additive constant

$$\omega(n, n-1) = \frac{h(n + \text{const.})}{2\pi m a^2}. \tag{11.101}$$

Heisenberg's motivation in this calculation is to recover a half-integral quantization rule (in analogy to the result he had obtained for the energy levels of the simple harmonic oscillator) where the undetermined constant in Eq. (11.101) is fixed (at zero) by observing that the transition amplitude from the ground state $n = 0$ to the non-existent state $n = -1$ should vanish. Heisenberg repeats exactly the same argument here, assuming that the array element $\omega(n, n-1)$ corresponds also in this case to a transition amplitude, which should vanish at $n = 0$ as a consequence of the non-existence of a state with quantum number -1.[41] This gives

$$\omega(n, n-1) = \frac{hn}{2\pi m a^2}. \tag{11.102}$$

[40] For a more detailed account of Heisenberg's treatment of two- and three-dimensional rotators in his *Umdeutung* paper, see the web resource, *The Problem of Spectral Intensities in the Old and New Quantum Mechanics.*

[41] As mentioned, the "vanishing at the edge" argument used here by Heisenberg was frequently employed in the study of multiplet intensity rules (see also Jähnert 2016, 2019), and would return in the correct analysis of angular momentum in three dimensions in Born, Heisenberg, and Jordan (1926), but at the "outer edge": namely, by insisting that the magnitude $|m|\hbar$ of the z-component of angular momentum $L_z = m\hbar$ not exceed the maximum value prescribed by the angular momentum quantum number l, namely, $l\hbar$.

The next step is to construct the diagonal element of the energy array in the usual way:

$$\begin{aligned} W(n,n) &= \frac{ma^2}{2}\omega^2(n,n) \\ &= \frac{ma^2}{2}\Big(\omega(n,n-1)\omega(n-1,n)+\omega(n,n+1)\omega(n+1,n)\Big) \\ &= \frac{h^2}{8\pi^2 ma^2}\left(n^2+(n+1)^2\right) \\ &= \frac{h^2}{4\pi^2 ma^2}\left(n^2+n+\tfrac{1}{2}\right) \\ &= \frac{h^2}{4\pi^2 ma^2}\left(\left(n+\tfrac{1}{2}\right)^2+\tfrac{1}{4}\right). \end{aligned} \tag{11.103}$$

The desired answer has been achieved: the rotational energy involves a square of half-integers rather than integers, as empirically required in molecular band spectra.

Unfortunately, the whole argument makes no sense. In fact, there is no "ground state" in this problem: states of negative n simply correspond to angular momentum in the negative z-direction, if the particle is constrained to rotate in the x-y plane. A transition in Eq. (11.96) with initial state $m = 0$ and final state $m' = -1$ is perfectly possible energetically, as the gain in rotational energy is more than compensated by the vibrational energy quantum $h\nu_0$. A correct treatment of the two-dimensional rotator problem in matrix mechanics would recognize both the angular momentum L_z and energy $W = L_z^2/(2ma^2)$ as diagonal arrays, with $L_z(n,n) = n\hbar$ and $W(n,n) = n^2\hbar^2/(2ma^2)$ (with n integer!). The desired half-integral numbers appear, once we treat the angular momentum $\vec{L}$ properly in three dimensions, as a set of three non-commuting arrays, as we shall see in Chapter 12 in our discussion of the Three-Man-Paper of Born, Heisenberg, and Jordan (1926).

We saw at the beginning of this chapter that the subject of relative intensities of spectral lines had become the focus of much theoretical work in the year leading up to Heisenberg's *Umdeutung* paper, partly as a response to the much-improved empirical data provided by the Utrecht group led by Ornstein. By the Summer of 1925, the earlier sum rules of Ornstein, Burger, and Dorgelo had been "sharpened" to the point that explicit formulas were proposed for the relative intensities of the individual lines in complex multiplets, in zero magnetic field (Sommerfeld and Honl 1925), as well as for the separated lines of the anomalous Zeeman effect in the presence of a magnetic field (Goudsmit and Kronig 1925; Hönl 1925).[42]

In the final two pages of his paper, Heisenberg attempts to fashion amplitude equations, which possess as solutions the expressions found by the latter authors for the Zeeman line intensities. The model used is a massive particle constrained to move on the surface of a sphere of radius a in three dimensions,[43] with the total angular momentum

[42] For further details, see the web resource, *The Problem of Spectral Intensities in the Old and New Quantum Theory*.

[43] The idea of an electron orbiting on a sphere may seem hopelessly unrealistic in a calculation of Zeeman intensities, but in fact the relative intensity formulas depend only on the angular component of the motion of

vector $\vec{L}$ (corresponding to quantum number l) precessing slowly around the z axis, with the component along the z axis characterized by magnetic quantum number m. The attempt once again displays the perils of "knowing the answer". The usual *Umdeutung* strategy is employed: one introduces doubly indexed arrays x, y, z, which generalize the Fourier components of the Cartesian coordinates of a massive charged particle constrained to move on the surface of a sphere of radius a. The arrays for x, y, and z must therefore satisfy $x^2 + y^2 + z^2 = a^2$. The doubly periodic motion of such a particle once a weak magnetic field H is imposed, inducing a precession with angular frequency ω_H of the orbital angular momentum vector $\vec{L}$ around the z-axis (direction of $\vec{H}$), on top of the orbital motion of frequency ω, corresponding to the orbital angular momentum l, means that the indices are specified with the quantum number pair (l, m).[44]

Aided by the intensity formulas of Goudsmit, Kronig, and Hönl, Heisenberg attempts to guess the coordinate array elements by imposing three constraints:

1. The diagonal array elements of $x^2 + y^2 + z^2$ must equal a^2.
2. The off-diagonal array elements of $x^2 + y^2 + z^2$ must be zero.
3. Quantization is introduced via the requirement that the z-component of the orbital angular momentum of the particle be $mh/2\pi$, by reinterpretation (*Umdeutung*) of the classical expression for this requirement in terms of Fourier amplitudes. This condition plays the role of the Kuhn sum rule constraint in the previous oscillator calculations.

The first two constraints are correct, but Heisenberg makes an incorrect guess for the third—his quantization condition (analogous to the Kuhn sum rule in the oscillator case) is correct in the correspondence limit, but not for small quantum numbers—a salutary reminder of the treacherous ground tread by practitioners of the "sharpened correspondence principle." Moreover, his guessed solution amplitudes only satisfy approximately the first constraint—again, the diagonal elements of $x^2 + y^2 + z^2$ turn out to be $(2l + 3)/(2l + 2)$ times a^2, which is correct in the correspondence limit, but not for finite l. His (incorrect) amplitudes satisfy exactly his second and (incorrect) third condition, arising from the fixing of L_z. The "vanishing at the edges" condition

the radiating electron—the details of the radial motion (later, radial wave function) are irrelevant, as are the specific choices made for the binding potential: Coulomb, screened Coulomb, etc. We are therefore at liberty to pick a spherically symmetric binding potential, which localizes the electron to move on an arbitrarily thin shell of radius $r = a$, say, by making the potential very large and negative for $r \approx a$. The model chosen by Heisenberg is therefore a perfectly fine substitute for the actual atomic system. His incorrect results are not the fault of the model chosen, but rather of a flawed execution of the correspondence arguments.

[44] Heisenberg uses the notation n for l, common in the old quantum theory. Space quantization of the precessional motion around the direction of the magnetic field is imposed via the magnetic quantum number m, giving the component of the angular momentum along the z-axis. Of course, despite Heisenberg's assertion at the end of the introduction to his paper that only problems involving a single degree of freedom would be treated, he is here addressing a doubly periodic system with two independent degrees of freedom. For complete details of Heisenberg's calculations, identifying precisely where he goes off the tracks in his intensity calculations, see the web resource mentioned in note 42.

emphasized by Sommerfeld and Hönl is not fully satisfied either by Heisenberg's formulas, despite his assertions to the contrary.[45] A correct matrix mechanical treatment of the Zeeman amplitudes, precisely agreeing with the Goudsmit–Kronig–Hönl formulas, would be given in the Three-Man-Paper a few months later, which, tellingly, references these authors, but not Heisenberg's *Umdeutung* paper calculations.

Heisenberg's paper ends with the following remark:

> Whether a method to determine quantum-theoretical data using relations between observable quantities, such as that proposed here, can be regarded as satisfactory in principle, or whether this method in the end represents far too rough an approach to the problem of constructing a theoretical quantum mechanics, an obviously very involved problem at the present, can be decided only by a more intensive mathematical investigation of the method which has been very superficially employed here (p. 276).

The desired mathematical investigation would follow within the next two years, first in the matrix reformulation of Heisenberg's ideas by Born and Jordan, then in the connection to the Hellinger–Hilbert theory of infinite quadratic forms, and finally in the formulation of the theory in infinite dimensional Hilbert space by John von Neumann. A complete and consistent physical interpretation of the new conceptual structure for which Heisenberg laid the foundation would take a lot longer—and, perhaps, is still ahead of us.

[45] Again, the reader is referred to the web resource mentioned in note 42 for a detailed treatment of the algebra required here.

12

The Consolidation of Matrix Mechanics: Born–Jordan, Dirac and the Three-Man-Paper

12.1 The "Two-Man-Paper" of Born and Jordan

On July 9, 1925, Heisenberg wrote to Pauli from Göttingen, promising to send him the manuscript of the *Umdeutung* paper, and requesting Pauli to return the same within a few days (we are in the days well before the xerox machine or pdf files):

> It is really my conviction that an interpretation of the Rydberg formula along the lines of circular or elliptical orbits in *classical* geometry does not have the slightest physical sense, and all of my miserable efforts are directed towards the complete removal, and suitable replacement of the concept of orbits, which one cannot in any case observe. Consequently I am taking the liberty of sending you shortly the manuscript of my work, because I believe that it contains, at least in its critical, i.e., negative, part real physics. I have indeed a very bad conscience doing this, as I have to ask you to return it in 2–3 days, so that I can either complete it or burn it in the last days of my stay here (Pauli 1979, Doc. 96).

Heisenberg had previously been invited to visit Leyden (by Ehrenfest) and Cambridge (by Fowler, and later by Kapitza), as in both places interest in spectroscopy, and in particular the mysteries of the anomalous Zeeman effect, was high, and Heisenberg's work in this area was known and respected. As subsequent events show, Heisenberg did not burn his paper on recovering it from Pauli, but finished it in a few days and handed it to Born sometime in the third week of July, before leaving for Leyden, with the request to read it and forward it for publication if he (Born) found it of sufficient merit. Heisenberg would be absent from Göttingen until the middle of October (having taken a vacation in Bavaria and returned to Copenhagen to finish the required time of his fellowship, after his visits to Leyden and Cambridge). The next steps in the extension and formalization of Heisenberg's ideas were taken by Max Born and his assistant Pascual Jordan, in Göttingen, leading to the Two-Man-paper, "On quantum mechanics"

Constructing Quantum Mechanics. Anthony Duncan and Michel Janssen, Oxford University Press.
 DOI: 10.1093/oso/9780198883906.003.0012

(*Zur Quantenmechanik*), which was submitted to *Zeitschrift für Physik* on September 27, 1925 (Born and Jordan 1925b).

Max Born, who re-enters our story here on the verge of his most significant contributions to theoretical physics, was born to an upper middle-class (on his mother's side, quite wealthy) family in Breslau, East Prussia (now Wrocław, Poland) in 1882, three years after Einstein and three years before Bohr. His primary school and Gymnasium years were spent in Breslau, and he attended the University of Breslau (obtaining a degree in mathematics in 1904), before going to Göttingen, where he initially hoped to carry out doctoral studies in pure mathematics. In his first class, he came to David Hilbert's attention, who assigned him the task of compiling the official class notes for Hilbert's course (on mechanics). Hilbert's assigned thesis topic in pure mathematics (the transcendentality of the roots of Bessel functions) proved too difficult for Born, who ended up working on a topic in applied mathematics—the theory of elastic media—which was a favorite of Felix Klein, the second of the three mandarins of mathematics at Göttingen who would have a great influence (in Klein's case, not altogether positive[1]) on Born, the third being Hermann Minkowski, the originator of the geometrical four-dimensional space–time picture for special relativity. After holding positions at the University of Berlin (1915–1919) and the University of Frankfurt (1919–1921), Born returned to the scene of his doctoral work in Göttingen as a full (*Ordentlicher*) professor, becoming a year later the first director of the new Institute for Theoretical Physics—basically, the renamed "mathematical physics" section of the Physics faculty of the University of Göttingen. Under Born's direction, this new institute emerged as a third important center in the development of quantum theory alongside the Sommerfeld school in Munich and Bohr's institute in Copenhagen.[2]

At this point, Pascual Jordan,[3] another key protagonist in our account, makes his first appearance. Jordan, as we will see, played a leading role in the development of quantum theory in the two remaining years leading up to its formal culmination in John von Neumann's work. Jordan was born in Hanover, Germany on October 18, 1902. Although less than a year younger than Heisenberg, and only two and a half years younger than Wolfgang Pauli, his academic position and visibility in mid-1925 was far below these two "young Turks" who had already, with the introduction of matrix mechanics and the exclusion principle, made profound contributions to quantum theory. Jordan entered the University of Göttingen in early summer of 1922, after an unsatisfying first year at the Technical University in Hanover. Although physics was already his primary interest, he seems to have been far more diligent in his attendance of Richard Courant's mathematics lectures than in the required (but early morning) physics lectures of Robert Pohl (Mehra and Rechenberg 1982c, p. 47). Jordan acquired

[1] For further details on Born's complicated relations with his Göttingen mentors, see Ch. 2 (especially p. 31) in Greenspan's (2005) biography on Born, from which we are drawing here.

[2] See *Establishing Quantum Physics in Göttingen* (Schirrmacher 2019) and *Establishing Quantum Physics in Munich* (Eckert 2020). For Göttingen, see also Mehra and Rechenberg (1982a, secs. III.1 and III.2, pp. 262–313). For Copenhagen, see Robertson (1979).

[3] For biographical information on Jordan, see, for example, Mehra and Rechenberg (1982c, sec. I.4, pp. 44–57).

a very thorough grounding in the mathematical foundations of theoretical physics, in part by the simple expedient of assisting Courant in the revision and correction of the seminal text of Courant and Hilbert (*Methods of Mathematical Physics*), the first edition of which appeared in 1924 (with an acknowledgement in the foreword to Jordan, among other prominent mathematicians, as a "true helper . . . in the bitter work of corrections"). Jordan's physics interests were stimulated at the very beginning of his Göttingen years by attending the legendary Bohr *Festspiele* (Bohr's Wolfskehl lectures) in June 1922 (see Section 9.1.1), where he also met Pauli and Heisenberg for the first time. Bohr's lectures on his second atomic theory were the stimulus for an intensive study of the sophisticated classical analytical dynamics which was still the required technical machinery for quantum theory in the early 1920s.

Jordan soon got to know Born, whom he helped complete a long-delayed encyclopedia article on crystal dynamics. By mid-1924 Jordan had taken up and completed—under Born's supervision—a paper proposing an alternative (to Einstein's) quantum theory of radiation, in which the "extreme corpuscular" point of view of Einstein's theory was somewhat softened by avoiding the assignment of a definite momentum $h\nu/c$ to light quanta of frequency ν. This work served also as his doctoral research topic (Jordan 1924). After receiving his doctorate, Jordan stayed on in Göttingen as Born's assistant, collaborating with him on an extension of dispersion theory to the scattering off atoms of incoming non-periodic light pulses (Born and Jordan 1925a), submitted to the *Zeitschrift für Physik* in the second week of June 1925, just as Heisenberg was taking his famous Helgoland excursion.

The precise origins of the highly fruitful collaboration of Jordan, then only 22 years old, and Born (who was 20 years Jordan's elder) on matrix mechanics are a bit uncertain, as there are various versions of the story in the secondary literature. According to van der Waerden (1968, pp. 36–37), Born had read Heisenberg's new paper by July 15, and had immediately been impressed and fascinated by Heisenberg's bold proposal.[4] It is not completely clear whether Born recognized immediately the underlying mathematical structure of matrix algebra in Heisenberg's manipulations, but it is clear that he right

[4] In his AHQP interview in 1960 with Ewald, Born stated that "Heisenberg brought me this paper and I read it through and was fascinated and began to brood over it day and night." In his later commentary on a letter he wrote to Einstein on July 15, 1925, Born asserts that Heisenberg gave him the *Umdeutung* paper on July 11 or 12, and that he had read it, and was convinced of its essential correctness, by the time of his letter to Einstein a few days later. In his letter to Einstein, Born indeed refers to a paper of Heisenberg as "rather mystifying but . . . certainly true and profound" (Born 1971, p. 84). Unfortunately, from the context of this remark in the letter (and in particular, to its use by Hund (1925) in discussing complex multiplets in a paper sent to *Zeitschrift der Physik* on August 20), it appears that the paper referred to was actually an earlier paper on multiplet structure and the Zeeman effect (Heisenberg 1925b) and not the *Umdeutung* paper (Heisenberg 1925c), which would surely have been discussed in more explicit terms in his letter to Einstein, if Born had actually managed to digest its contents by July 15. Born also maintains in his annotation of this letter that he had probably established the equivalence of Heisenberg's quantization condition to the equality of the diagonal matrix elements of the commutator of the momentum and position arrays (to $h/2\pi i$) in these first few days, and that his conviction of the validity of Heisenberg's new scheme derived from its simple relation to matrix calculus (Born 1971, p. 87). The confusing discrepancies here simply serve to emphasize the caution required in taking the recollections of the founders of quantum mechanics at face value more than four decades after the pivotal points in its development.

away felt the need for a collaborator with powerful mathematical talents. Apparently, Born met Pauli on the train from Göttingen to Hanover on July 19 (where there was a meeting of the German Physical Society) and tried to interest him in the project, but was rebuffed by Pauli, who accused Born of wanting to spoil Heisenberg's physical reasoning with "tedious and complicated formalism" (Born 1978, p. 218).[5] Again, following van der Waerden's account, Born asked Jordan on the next day to join him in clarifying the structure of Heisenberg's new theory.[6]

Whatever the precise sequence of events leading to his collaboration with Born, it is fairly clear that, by the time Jordan offered his assistance to Born, the latter had already succeeded in translating Heisenberg's expressions into matrix notation, and had even recognized that Heisenberg's quantization condition (equivalent, as we saw, to the Thomas–Reiche–Kuhn sum rule) could be re-expressed as the statement that the diagonal elements of the commutator[7] of the arrays representing momentum p ($= m\dot{q}$) and position q were all equal to $\hbar/i$. Many of the remaining results in their paper (including the demonstration that all the off-diagonal elements of the commutator of p and q vanish) are due to Jordan. In the following, we describe the main results of Born and Jordan's Two-Man-Paper (1925b) without attempting to ascribe each one to a particular author.

Born and Jordan begin with an introduction laying out the objectives of their study:

> Heisenberg has expressed the physical concepts which have guided him in such a clear way, that any remarks aiming to complete them would be superfluous. But from a formal, mathematical perspective his considerations, as he himself emphasizes, are only at an initial stage . . . Benefiting from the circumstance that we have been able to familiarize ourselves with his considerations already in a nascent form, we have tried after the completion of his investigations to clarify the formal mathematical content of his assumptions, and present some of our results here (Born and Jordan 1925b, pp. 277–278).

Born and Jordan, in the paper's first chapter (entitled "Matrix calculation") present a short introduction to the basics of matrix algebra, and introduce a definition for the partial derivative of a function, given as a sum of products of matrices, with respect to a particular matrix appearing in the products.

In the second chapter (entitled "Dynamics"), they lay out a quantum-theoretical dynamics based on matrix representations[8] of the coordinate $\mathbf{q}$ and momentum $\mathbf{p}$ of

[5] This same anecdote is related, for example, in Greenspan's (2005, pp. 125–126) biography of Born.

[6] According to Jammer (1966, p. 209), Jordan happened to be in the same compartment as Born on the train to Hanover, and overheard him discussing his struggles with Heisenberg's work with an (unnamed) colleague, and it was after the train had arrived in Hanover that Jordan "introduced himself" to Born and offered his help. As Born and Jordan were already very well acquainted, and had collaborated on a paper by this point, this version does not make a lot of sense. On the other hand, Jammer's version, placing Jordan in the train compartment with Born and a third colleague (whose name is missing due to a gap in the tape) agrees with that of Born in his 1960 AHQP interview with Ewald. However, Born's recollections 35 years after the event are faulty in several respects (cf. note 4): for example, he places Heisenberg in Denmark and Finland in the summer of 1925!

[7] The commutator $[A, B]$ of two matrices A, B is defined as $AB - BA$.

[8] Bold letters are used throughout the Born–Jordan paper to denote matrix quantities. For a brief history and introduction to matrix algebra, see Appendix C.

a particle moving in one dimension, and show that the classical Hamiltonian equations of motion carry over unchanged in form to quantum mechanics once the kinematical quantities they contain are regarded as matrices. It is also shown that the (matrix) Hamiltonian equations can be derived from a variational principle based on a matrix version of the Lagrangian. The Heisenberg quantization condition is shown to be equivalent to the statement that the diagonal elements of the matrix $\mathbf{pq} - \mathbf{qp}$ (the commutator of $\mathbf{p}$ and $\mathbf{q}$) are all equal to $\hbar/i$. It is then shown that the commutator is time independent, whence it is purely diagonal, and equal to $\hbar/i$ times the identity matrix. The role of the energy matrix in giving the time rate of change of arbitrary functions of the phase coordinates $\mathbf{g}(\mathbf{p}, \mathbf{q})$ is explained, from which the time independence of the energy matrix itself (the energy law," or *Energiesatz*) follows immediately, as an exact result, and without the need for tedious perturbative calculations along the lines of the *Umdeutung* paper.

Chapter 3 of their paper presents a detailed examination of the simple harmonic and cubic anharmonic oscillators employing the new matrix language, although the technology of the perturbative calculations in the latter case is basically equivalent to that of Heisenberg. The calculation for the energy of the cubic anharmonic oscillator is carried to the second order in the coupling λ, one stage further than Heisenberg. In their treatment of the anharmonic oscillator Born and Jordan pay explicit attention to the complex phase freedom in the transition amplitudes.[9]

In chapter 4 (entitled "Remarks on electrodynamics") there is a first attempt to treat the electromagnetic field, both on its own and in interaction with a charged system. The aim is to put on a firm basis Heisenberg's tacit assumption that the transition probabilities in atomic processes are determined by (i.e., proportional to) the absolute squares of the coordinate matrix elements $\mathbf{q}(nm)$. Although some of the insights in this last section are correct, the arguments given are (as in the final section of Heisenberg's *Umdeutung* paper) the least-satisfying part of this otherwise extraordinary paper. A first stab at a correct quantum-mechanical treatment of the electromagnetic field (in a one-dimensional toy model) would be given by Jordan in the final section of the Three-Man-Paper of Born, Heisenberg, and Jordan, to which we turn shortly.

Born in later recollections (AHQP interview with Ewald 1960) claimed that he only recognized after a few days of "brooding" over Heisenberg's paper that the composition rule for arrays was just matrix multiplication, and that he was then able to rewrite Heisenberg's expressions in matrix form. He was unable however to establish conservation of the energy from the matrix equations of motion. It was this difficulty (he recalls) that he was overheard discussing by Jordan in the train to Hanover. While Born's knowledge of matrix technology was rudimentary at this point (it took him several days to even recognize them in Heisenberg's work), Jordan had assisted in the preparation and correction of Courant and Hilbert's (1924) book *Methods of Mathematical Physics*, in which matrices already appear in the first chapter. Jordan had also consulted the German

[9] The origin and extent of the phase freedom of quantum amplitudes would only become completely clear once the Hilbert space formulation of quantum mechanics had been introduced by von Neumann two years later.

translation of Bôcher's (1907) *Introduction to Higher Algebra.*[10] It should be noted that both of these texts only discussed matrices whose elements were real numbers, whereas complex matrices appear from the outset in matrix mechanics. Presumably, Jordan had encountered the necessary generalizations in the mathematical literature (in particular, he was aware of the definition of a *Hermitian* matrix), while assisting in the preparation of the Courant–Hilbert text.

For the most part, we follow the notation used by Born and Jordan, with the exception of the commutator and trace, where we use modern notation. For example, the elements of a matrix $\mathbf{a}$ would now be written using subscripts, with $\mathbf{a}_{nm}$ representing the element in the nth row and m'th column of the matrix $\mathbf{a}$, whereas the same element is written by Born and Jordan using parentheses, as $a(nm)$ (the matrix elements themselves are just numbers, and are not rendered in bold type). Matrix addition and multiplication are defined in the usual way:

$$\mathbf{a} = \mathbf{b} + \mathbf{c} \quad \text{means} \quad a(nm) = b(nm) + c(nm), \tag{12.1}$$

$$\mathbf{a} = \mathbf{bc} \quad \text{means} \quad a(nm) = \sum_{k=0}^{\infty} b(nk)c(km). \tag{12.2}$$

Note that we are here dealing with infinite square arrays, with the rows and columns indexed by $0, 1, 2, \ldots$. As the mere act of multiplying two infinite matrices involves in general the calculation of an infinite sum (e.g., in the anharmonic oscillator case), convergence questions naturally arise, and Born and Jordan studiously avoid them. One proceeds as though all the properties present for finite matrices remain valid in the infinite dimensional case—in particular, associativity, $\mathbf{a}(\mathbf{bc}) = (\mathbf{ab})\mathbf{c}$, and distributivity, $\mathbf{a}(\mathbf{b} + \mathbf{c}) = \mathbf{ab} + \mathbf{ac}$. In general, of course, as Heisenberg had already realized, matrix multiplication is not commutative:

$$\mathbf{ab} \neq \mathbf{ba} \quad \text{(in general)}. \tag{12.3}$$

The identity matrix 1 is simply the diagonal matrix with all diagonal elements equal to 1, or, using the Kronecker symbol,

$$1(nm) = \delta_{nm}. \tag{12.4}$$

A matrix $\mathbf{a}$ is said to have an inverse matrix $\mathbf{a}^{-1}$ if

$$\mathbf{a}^{-1}\mathbf{a} = \mathbf{a}\mathbf{a}^{-1} = 1. \tag{12.5}$$

We will usually be dealing with matrices that involve kinematical variables like position or momentum that are functions of time. If the array elements $b(nk), c(km)$ in Eq. (12.2)

[10] See Jammer (1966, p. 207) and Mehra and Rechenberg (1982c, p. 59).

are (implicitly) time dependent, one easily establishes by taking the time derivative on the right

$$\frac{d}{dt}(\mathbf{bc}) = \dot{\mathbf{b}}\mathbf{c} + \mathbf{b}\dot{\mathbf{c}}, \tag{12.6}$$

with an obvious generalization to products of more than two matrices.

Two further notational issues should be addressed here. For the trace of a matrix (the sum of its diagonal elements) Born and Jordan write $D(\mathbf{a}) = \sum_k a(kk)$. We instead employ the modern notation $\mathrm{Tr}(\mathbf{a})$. Similarly, for the commutator of two matrices, Born and Jordan write

$$\mathbf{ab} - \mathbf{ba} = \begin{vmatrix} \mathbf{a} \\ \mathbf{b} \end{vmatrix}$$

whereas we use the more convenient modern notation $\mathbf{ab} - \mathbf{ba} \equiv [\mathbf{a}, \mathbf{b}]$.

The classical Hamilton equations of motion involve derivatives of the energy function (Hamiltonian) $H(p, q)$ with respect to the kinematical variables p and q. As these are now all matrices, one clearly needs in the quantum case a precise definition for (partial) differentiation of a function of matrices with respect to a particular matrix appearing therein. Born and Jordan introduce a symbolic matrix derivative by using the trace of the derived function, as follows

$$\frac{\partial F(\mathbf{x}, \mathbf{y}, \ldots)}{\partial \mathbf{x}}(mn) \equiv \frac{\partial}{\partial x(nm)} \mathrm{Tr}(F(\mathbf{x}, \mathbf{y}, \ldots)). \tag{12.7}$$

With this definition, we have the natural result

$$\frac{\partial \mathbf{x}^2}{\partial \mathbf{x}}(mn) = \frac{\partial}{\partial x(nm)} \sum_{pq} x(pq)x(qp) = x(mn) + x(mn) = 2x(mn), \tag{12.8}$$

and by similar manipulations,

$$\frac{\partial \mathbf{x}^N}{\partial \mathbf{x}}(mn) = N\mathbf{x}^{N-1}(mn). \tag{12.9}$$

The Born–Jordan rule for matrix differentiation leads to a peculiar sum of cyclic products in the case where the function being differentiated contains products of non-commuting matrices (e.g., terms like $\mathbf{p}^2\mathbf{q}^2$ involving both momentum and position). The only physical systems examined in the Born–Jordan paper involve Hamiltonian functions of the form $H(\mathbf{p}, \mathbf{q}) = \mathbf{p}^2/2m + V(\mathbf{q})$, with V a sum of powers of $\mathbf{q}$, so Eq. (12.9) suffices for our purposes.

The second chapter of the Born–Jordan paper is entitled "Dynamics." It erects the basic framework for the treatment of dynamics—in particular, the deterministic time evolution—of quantum systems. The commutation relation of position and momentum in terms of Planck's constant appears here for the first time as the point at which the quantum of action is inserted into the theory. The basic postulates are built up systematically, initially adhering very closely to the arguments in Heisenberg's *Umdeutung* paper. One begins by specifying the time development of the matrices associated with position and momentum:

$$\mathbf{q} = (q(nm)\, e^{2\pi i\nu(nm)t}), \quad \mathbf{p} = (p(nm)\, e^{2\pi i\nu(nm)t}). \tag{12.10}$$

The frequencies $\nu(nm)$ are presumed to be the Bohr condition frequencies for light emitted in a transition between quantum states n and m. For the first time in the development of matrix mechanics, a requirement that would later acquire central importance in the theory is explicitly stated. The matrices in Eq. (12.10) are required to be *Hermitian*: matrix elements related by interchange of row and column index must be complex conjugates,

$$q(nm) = q(mn)^{\star}. \tag{12.11}$$

This condition needs only be imposed at a single time, say $t = 0$, as the anti-symmetry property of the frequency arrays,

$$\nu(nm) = -\nu(mn), \tag{12.12}$$

then ensures its validity for all subsequent time:

$$q(nm)\, e^{2\pi i\nu(nm)t} = (q(mn)\, e^{2\pi i\nu(mn)t})^{\star} \tag{12.13}$$

Apart from pointing out that the Hermiticity requirement ensures that the product $q(nm)q(mn)$, which determines the transition probability between the states n and m, is a positive real number, the authors make no further reference to this critical property in the rest of the paper. In fact, it is central to the new theory, in particular because: (i) it ensures the reality of the diagonal elements of any Hermitian matrix W (as $W(nn) = W(nn)^{\star}$, taking $m = n$ in Eq. (12.11)); and (ii) it is easily seen that a power of a Hermitian matrix is Hermitian, so the Hermiticity of the kinetic energy $\mathbf{p}^2/2m$ and potential energy $V(\mathbf{q})$, given as a sum of powers of $\mathbf{q}$, is guaranteed, whence also the reality of the diagonal elements of the energy matrix identified by Heisenberg with the energies of the stationary states of the system. We see later, in our discussion of Born, Heisenberg and Jordan (1926), that the Hermiticity requirement would allow the theory to be connected with a highly developed area of mathematics, which would be crucial in the rigorous mathematical formulation of modern quantum mechanics.

The Ritz combination principle appears in the form

$$\nu(jk) = \nu(jl) + \nu(lk) \Rightarrow \nu(jk) + \nu(kl) + \nu(lj) = 0. \tag{12.14}$$

We saw in our discussion of the *Umdeutung* paper that this condition ensures that any kinematic observable built out of products of the coordinate and position arrays in

Eq. (12.10) will have exactly the same time dependence of its matrix elements as the basic kinematic variables themselves. For example,

$$(\mathbf{qp})(jk) = \sum_l q(jl)\, e^{2\pi i\nu(jl)t} \cdot p(lk)\, e^{2\pi i\nu(lk)t} = (\sum_l q(jl)p(lk))\, e^{2\pi i\nu(jk)t}. \qquad (12.15)$$

The property Eq. (12.14) implies that there exist real numbers W_n, with dimension energy, such that

$$h\nu(nm) = W_n - W_m. \qquad (12.16)$$

Recalling that the overall zero of energy can be chosen freely, as only differences of energy are physically meaningful, we may choose for example,

$$W_n \equiv h\nu(n0), \qquad (12.17)$$

thereby setting the zero of energy to coincide with the energy of the state labeled by index 0 (as $\nu(nn) = 0$). The Bohr frequency condition Eq. (12.16) then follows immediately with the help of Eqs. (12.11) and (12.14). Of course, so far we have not shown that the quantities W_n have anything to do with the allowed energies of the system—they are simply numbers extracted from the assumed frequencies of oscillation of the matrix elements of the kinematic quantities.

There is one other property of the transition frequencies $\nu(nm)$ that is implicitly assumed in this section of the paper,[11] but critical for further developments, namely,

$$n \neq m \Rightarrow \nu(nm) \neq 0. \qquad (12.18)$$

This then implies that two distinct stationary states must have different energies: the system is non-degenerate. A very important consequence of this is that any conserved quantity of the theory $\mathbf{Q}$, $\dot{\mathbf{Q}} = 0$ must have a diagonal matrix, as any non-zero off-diagonal element $Q(nm), n \neq m$ of the matrix $\mathbf{Q}$ will have a non-trivial periodic time dependence proportional to $e^{2\pi i\nu(nm)t}$, with $\nu(nm) \neq 0$ when $n \neq m$. The frequencies $\nu(nn)$ are of course automatically zero from Eq. (12.11), so diagonal elements of any kinematical quantity are guaranteed to be constant in time. In matrix mechanics, the conservation (time independence) of a kinematical quantity is inextricably linked to the diagonality of the associated matrix. An associated property frequently exploited in the early years of matrix mechanics concerns the time-average of (the matrix of) any kinematical quantity, which can be obtained by simply dropping the off-diagonal matrix elements, which oscillate periodically, and keeping only the diagonal ones, which are automatically time independent.

[11] Born and Jordan (1925b, p. 877) explicitly assert the non-degeneracy postulate in their discussion of the simple harmonic oscillator in sec. 5. It can be shown that the discrete spectrum of any one dimensional particle moving in a potential $V(x)$ admitting bound states (so $V(x) \to +\infty, x \to \pm\infty$) has a non-degenerate spectrum: every allowed energy corresponds to one, and only one, quantum state.

The crucial connection of the W_n to the quantized energies of the system is established in a series of steps. First, one defines a diagonal matrix **W** with the quantities W_n along the diagonal:

$$W(nm) = W_n \delta_{nm}. \tag{12.19}$$

For any kinematical quantity **g** with the time-dependence Eq. (12.10) the time-derivative $\dot{\mathbf{g}}$ has matrix elements

$$\dot{g}(nm) = 2\pi i \nu(nm) g(nm) = \frac{2\pi i}{h}(W_n - W_m) g(nm). \tag{12.20}$$

On the other hand, the commutator $[\mathbf{W}, \mathbf{g}] = \mathbf{Wg} - \mathbf{gW}$ has matrix elements

$$(Wg - gW)(nm) = \sum_k (W_n \delta_{nk} g(km) - g(nk) W_k \delta_{km}) = (W_n - W_m) g(nm). \tag{12.21}$$

Comparing the right-hand sides of Eqs. (12.20) and (12.21), we find

$$\dot{\mathbf{g}} = \frac{2\pi i}{h}(\mathbf{Wg} - \mathbf{gW}) = \frac{2\pi i}{h}[\mathbf{W}, \mathbf{g}]. \tag{12.22}$$

This equation represents the fundamental dynamical law of the new quantum mechanics, playing a role in the new theory corresponding to Hamilton's equations in classical mechanics. The quantum correlates of the classical Hamilton equations of motion are assumed to be formally identical to their classical counterparts,[12] so one writes (Born and Jordan 1925b, Eq. (35)):

$$\dot{\mathbf{q}} = \frac{\partial \mathbf{H}}{\partial \mathbf{p}}, \quad \dot{\mathbf{p}} = -\frac{\partial \mathbf{H}}{\partial \mathbf{q}}. \tag{12.23}$$

Here, the Hamiltonian matrix **H** is differentiated with respect to the momentum and coordinate matrices **p**, **q** upon which it depends using the symbolic matrix differentiation rule Eq. (12.7). This definition can be applied to Hamiltonians containing arbitrary combinations of **p** and **q**, in an arbitrary (but prescribed) order, but we restrict our

[12] Born and Jordan also show that the matrix Hamitonian equations can be derived from a matrix version of the classical Hamilton action principle, where one obtains the classical equations of motion by extremizing $\int (p\dot{q} - H(q, p))dt$ with respect to $p(t), q(t)$. This derivation plays no further role in the development of the new quantum procedures, so we have not included it here.

discussion to Hamiltonian energy functions of a one-dimensional particle of mass m moving in a potential $V(q)$, so

$$\mathbf{H} = \frac{1}{2m}\mathbf{p}^2 + V(\mathbf{q}), \tag{12.24}$$

for which the equations of motion take the form

$$\begin{aligned}\dot{\mathbf{q}} &= \mathbf{p}/m,\\ \dot{\mathbf{p}} &= -\frac{\partial V(\mathbf{q})}{\partial \mathbf{q}}.\end{aligned} \tag{12.25}$$

The only systems considered in detail in the paper are harmonic and anharmonic oscillators, for which $V(\mathbf{q})$ is a (finite) sum of powers of $\mathbf{q}$.

So far, Planck's constant has entered the theory in an artificial way, allowing us to define quantities W_n with the dimension of energy from the (directly observable) transition frequencies $\nu(nm)$. If we can establish the connection of these quantities to the energy matrix, we will have derived the Bohr frequency condition from the new theory. The physical introduction of Planck's constant into the theory involves an argument similar to that of Heisenberg (cf. Eqs. (11.22)–(11.24)), but more general as the authors do not assume $p = m\dot{q}$ (i.e., a kinetic energy of the form $p^2/2m$). One again begins with the classical phase integral

$$\mathcal{J} = \oint p\,dq = \int_0^{1/\nu} p\dot{q}\,dt. \tag{12.26}$$

Introducing the Fourier expansions for a periodic system

$$p = \sum_{\tau=-\infty}^{\infty} p_\tau e^{2\pi i\nu\tau t}, \quad q = \sum_{\tau=-\infty}^{\infty} q_\tau e^{2\pi i\nu\tau t}, \tag{12.27}$$

and performing the phase integral,

$$\begin{aligned}\mathcal{J} &= \int_0^{1/\nu} \sum_{\tau,\tau'} p_{\tau'} e^{2\pi i\nu\tau' t} \cdot 2\pi i\nu\tau q_\tau e^{2\pi i\nu\tau t}\,dt\\ &= \sum_{\tau,\tau'} 2\pi i\nu\tau p_{\tau'} q_\tau \int_0^{1/\nu} e^{2\pi i\nu(\tau+\tau')t}\,dt.\\ &= \sum_{\tau,\tau'} 2\pi i\nu\tau p_{\tau'} q_\tau \frac{1}{\nu}\delta_{\tau',-\tau}\\ &= 2\pi i \sum_{\tau} \tau p_{-\tau} q_\tau.\end{aligned} \tag{12.28}$$

Differentiating both sides with respect to the action $\mathcal{J}$, one finds the classical result (Born and Jordan 1925b, Eq. (36)):

$$1 = 2\pi i \sum_{\tau=-\infty}^{\infty} \tau \frac{\partial}{\partial \mathcal{J}}(q_\tau p_{-\tau}). \tag{12.29}$$

One now applies the usual correspondence argument linking continuous derivatives with respect to the action to discrete differences:

$$\sum_{\tau=-\infty}^{\infty} \tau \frac{\partial}{\partial \mathcal{J}}(q_\tau p_{-\tau}) \to \frac{1}{h} \sum_{\tau=-\infty}^{\infty} (q(n+\tau, n)p(n, n+\tau) - q(n, n-\tau)p(n-\tau, n)). \tag{12.30}$$

Relabeling the index $n+\tau$ as k in the first term and $n-\tau$ as k in the second term, and using Eq. (12.29), the expression on the right becomes

$$\frac{1}{h}\sum_k (p(n,k)q(k,n) - q(n,k)p(k,n)) = \frac{1}{h}(pq - qp)(n,n) = \frac{1}{2\pi i}, \tag{12.31}$$

or[13]

$$(pq - qp)(n,n) = \frac{h}{2\pi i} = \frac{\hbar}{i}. \tag{12.32}$$

In other words, the *diagonal* elements of the commutator of the momentum and position matrices $[\mathbf{p}, \mathbf{q}]$ are all equal to the constant number $\hbar/i$. At this point, Born and Jordan point out that this simple result actually implies that one must be dealing with an *infinite* number of stationary states. Indeed, if the state index n had only a finite range (say, from 1 to N), with the matrices $\mathbf{q}$ and $\mathbf{p}$ finite-dimensional $N \times N$ matrices, we would find a contradiction by taking the trace

$$\mathrm{Tr}([\mathbf{p}, \mathbf{q}]) = \mathrm{Tr}(\mathbf{pq} - \mathbf{qp}) = 0 = N \cdot \frac{\hbar}{i} \Rightarrow N = 0. \tag{12.33}$$

This contradiction can only be avoided if the matrices under consideration are infinite dimensional.

Our discussion so far takes us up to the critical section 4 of the paper, entitled "Consequences. energy- and frequency laws" (Born and Jordan 1925b, pp. 870–875). The laws referred to here are: first, the law of conservation of energy, which we recall served as a critical check for Heisenberg on the validity of his quantum transcription of the classical equations of motion; and second, the Bohr frequency condition linking the transition frequencies to energy differences of stationary states. These important results

[13] To avoid a plague of πs we henceforth frequently, and anachronistically, use Dirac's version of Planck's constant, $\hbar = h/2\pi$.

rely on a "sharpened quantum condition" specifying completely the matrix commutator $\mathbf{d} \equiv \mathbf{pq} - \mathbf{qp}$, and not just its diagonal elements, which we obtained above.

The required sharpening depends on the validity of the quantum Hamilton equations of motion Eq. (12.23). Using these equations, we find

$$\begin{aligned}\dot{\mathbf{d}} &= \dot{\mathbf{p}}\mathbf{q} + \mathbf{p}\dot{\mathbf{q}} - \dot{\mathbf{q}}\mathbf{p} - \mathbf{q}\dot{\mathbf{p}} \\ &= -\frac{\partial \mathbf{H}}{\partial \mathbf{q}}\mathbf{q} + \mathbf{p}\frac{\partial \mathbf{H}}{\partial \mathbf{p}} - \frac{\partial \mathbf{H}}{\partial \mathbf{p}}\mathbf{p} + \mathbf{q}\frac{\partial \mathbf{H}}{\partial \mathbf{q}} \\ &= \mathbf{q}\frac{\partial \mathbf{H}}{\partial \mathbf{q}} - \frac{\partial \mathbf{H}}{\partial \mathbf{q}}\mathbf{q} + \mathbf{p}\frac{\partial \mathbf{H}}{\partial \mathbf{p}} - \frac{\partial \mathbf{H}}{\partial \mathbf{p}}\mathbf{p}.\end{aligned} \tag{12.34}$$

It turns out that with the definition of symbolic matrix differentiation given by Born and Jordan (cf. Eq. (12.7)), the quantity appearing on the final right-hand side of Eq. (12.34) vanishes identically, as long as the Hamiltonian can be expressed as a sum of products (in any order) of $\mathbf{p}$ and $\mathbf{q}$. In the special case of interest here, with $\mathbf{H} = \mathbf{p}^2/2m + V(\mathbf{q})$, one obtains

$$\dot{\mathbf{d}} = \mathbf{q}V'(\mathbf{q}) - V'(\mathbf{q})\mathbf{q} + \mathbf{p}\frac{\mathbf{p}}{m} - \frac{\mathbf{p}}{m}\mathbf{p} = 0, \tag{12.35}$$

where we have used the fact that $V'(\mathbf{q})$ is a sum of powers of $\mathbf{q}$ to conclude that the first and second terms cancel (as do the third and fourth).

We saw earlier that any time-independent matrix constructed from the fundamental kinematical variables of the theory $\mathbf{q}$, $\mathbf{p}$ must be a diagonal matrix. We may therefore write, using Eq. (12.32),

$$\mathbf{d} = [\mathbf{p}, \mathbf{q}] = \frac{h}{2\pi i}\mathbb{1}, \tag{12.36}$$

which we will regard as the fundamental quantization condition of the new theory: in fact, it is the *only* place where Planck's constant needs to be inserted to obtain the characteristic features of quantum theory.[14] According to Born and Jordan (1925b, p. 292), "all further results depend" on Eq. (12.36).

The very simple structure of the fundamental commutation relation Eq. (12.36), giving the commutator of momentum and position matrices as a pure number (multiplying the *identity* matrix), is the key to the role it plays in the tight (one might say, *sharpened*) correspondence between the derivatives in phase space appearing in classical mechanics and the commutators appearing ubiquitously in the new quantum mechanics. The

[14] It is widely accepted that this result, in particular the conservation of $[\mathbf{p}, \mathbf{q}]$ and consequent vanishing of the off-diagonal elements, is due to Jordan.

connection is easily seen if we recall the algebraic identity for the commutator of a given matrix with a product of matrices:

$$[\mathbf{a}, \mathbf{bcd}\cdots] = [\mathbf{a}, \mathbf{b}]\mathbf{cd}\cdots + \mathbf{b}[\mathbf{a}, \mathbf{c}]\mathbf{d}\cdots + \mathbf{bc}[\mathbf{a}, \mathbf{d}]\cdots + \ldots \tag{12.37}$$

This identity thus implies, for $\mathbf{a} = \mathbf{p}, \mathbf{b} = \mathbf{c} = \mathbf{d} = \ldots \mathbf{q}$,

$$[\mathbf{p}, \mathbf{q}^n] = \frac{\hbar}{i} n\mathbf{q}^{n-1}. \tag{12.38}$$

For a classical function of position expressible as a power series, $V(q) = \sum_n a_n q^n$, the associated quantum matrix thus satisfies

$$[\mathbf{p}, V(\mathbf{q})] = \sum_n a_n[\mathbf{p}, \mathbf{q}^n] = \frac{\hbar}{i}\sum_n a_n n\mathbf{q}^{n-1} = \frac{\hbar}{i}V'(\mathbf{q}). \tag{12.39}$$

Similarly, for a classical function of momentum expressible as a power series, $K(p) = \sum_n a_n p^n$, the quantum version satisfies (using $[\mathbf{q}, \mathbf{p}] = -\hbar/i\mathbb{1}$)

$$[\mathbf{q}, K(\mathbf{p})] = \sum_n a_n[\mathbf{q}, \mathbf{p}^n] = -\frac{\hbar}{i}\sum_n a_n n\mathbf{p}^{n-1} = -\frac{\hbar}{i}K'(\mathbf{p}). \tag{12.40}$$

The action of commutators in the quantum theory can therefore be associated with classical phase space derivatives by the following correspondence:

$$[\mathbf{p}, \cdots] \rightarrow \frac{\hbar}{i}\frac{\partial}{\partial \mathbf{q}}\cdots \tag{12.41}$$

$$[\mathbf{q}, \cdots] \rightarrow -\frac{\hbar}{i}\frac{\partial}{\partial \mathbf{p}}\cdots. \tag{12.42}$$

For a classical Hamiltonian given as a sum of a function $K(p)$ of p and a function $V(q)$ of q—in particular for our one-dimensional particle subject to a potential, and with Hamiltonian $H = p^2/2m + V(q)$—the quantum Hamiltonian matrix will have, using the equations of motion Eq. (12.23),

$$[\mathbf{H}(\mathbf{p}, \mathbf{q}), \mathbf{q}] = \frac{\hbar}{i}\frac{\partial \mathbf{H}}{\partial \mathbf{p}} = \frac{\hbar}{i}\dot{\mathbf{q}}, \tag{12.43}$$

$$[\mathbf{H}(\mathbf{p}, \mathbf{q}), \mathbf{p}] = -\frac{\hbar}{i}\frac{\partial \mathbf{H}}{\partial \mathbf{q}} = \frac{\hbar}{i}\dot{\mathbf{p}}. \tag{12.44}$$

If we instead commute the Hamiltonian with an arbitrary function $g(\mathbf{p}, \mathbf{q})$, written as a sum of products of the $\mathbf{p}, \mathbf{q}$ matrices, we can see with a glance at Eq. (12.37) that the result is a sum of terms where the Hamiltonian effects a time derivative as it passes through each appearance of a position **q** or momentum **p** matrix. The resulting sum of terms is simply, apart from the factor $\hbar/i$ appearing in each term, the total time derivative of the given function:

$$[\mathbf{H}(\mathbf{p},\mathbf{q}),\mathbf{g}(\mathbf{p},\mathbf{q})]=\frac{\hbar}{i}\frac{d}{dt}\mathbf{g}(\mathbf{p},\mathbf{q}). \qquad (12.45)$$

If for the general function **g** we insert the Hamiltonian itself, we find

$$[\mathbf{H}(\mathbf{p},\mathbf{q}),\mathbf{H}(\mathbf{p},\mathbf{q})]=\frac{\hbar}{i}\frac{d}{dt}\mathbf{H}(\mathbf{p},\mathbf{q})=0\Rightarrow\dot{\mathbf{H}}=0. \qquad (12.46)$$

This result asserts the exact conservation of the energy matrix (Born and Jordan's *Energiesatz*), a very powerful result when we remember the tedious algebraic manipulations undertaken by Heisenberg to check this by explicit calculation just to low orders of perturbation theory. The conservation of the energy matrix also implies, as we have seen, that it must be a diagonal matrix. The diagonal elements $H(nn)$ must clearly then give the quantized energies of the stationary states of the system.

If we examine the nm matrix element of Eqs. (12.43) and (12.44), and recall that the right-hand sides must have time-dependence $e^{2\pi i\nu(nm)t}$, we find, using Eq. (12.45),

$$H(nn)q(nm)-q(nm)H(mm) = \frac{\hbar}{i}2\pi i\nu(nm)q(nm), \qquad (12.47)$$

$$H(nn)p(nm)-p(nm)H(mm) = \frac{\hbar}{i}2\pi i\nu(nm)p(nm). \qquad (12.48)$$

For allowed transitions, with $q(nm), p(nm)\neq 0$, we can divide by the position and momentum matrix elements, in both cases obtaining the Bohr frequency condition (Born and Jordan's *Frequenzsatz*)

$$h\nu(nm)=H(nn)-H(mm). \qquad (12.49)$$

We now know that the energies W_n previously defined purely operationally in terms of the observable transition frequencies (cf. Eq. (12.17)) are simply the quantized energies of the system $H(nn)$ up to an overall constant shift. The matrix of W appearing in the dynamical equation Eq. (12.22) is just the Hamiltonian matrix, shifted by a constant (multiple of the identity matrix), so Eqs. (12.22) and (12.45) are equivalent.

The final pages of section 4 are devoted to examining the form of the quantum equations of motion Eqs. (12.43) and (12.44) if the Hamiltonian contains terms with products of the (non-commuting) **p** and **q** matrices, such as $\mathbf{p}^s\mathbf{q}^r$. It is shown that a particular reordering of these matrices is necessary if the Hamiltonian is to obey these equations of motion (with the derivatives with respect to **p** and **q** interpreted as in Eq. (12.7)). It turns out that all of this would play hardly any role in the further development of quantum mechanics: the physical systems of importance in atomic physics only rarely have Hamiltonians in which these ordering problems pose any difficulty. We therefore pass over these remaining considerations in section 4 of the paper and move on to chapter 3, entitled "Examination of the anharmonic oscillator."

Born and Jordan begin their examination of the anharmonic oscillator (here, restricted to the cubic anharmonic case) by referring to Heisenberg's (1925c) treatment of this system in the *Umdeutung* paper (the equation numbers are ours):

> The anharmonic oscillator with
>
> $$\mathbf{H} = \frac{1}{2}\mathbf{p}^2 + \frac{\omega_0^2}{2}\mathbf{q}^2 + \frac{1}{3}\lambda\mathbf{q}^3 \tag{12.50}$$
>
> has already been carefully examined by Heisenberg. Nevertheless this system will here be subjected to a further investigation, in particular with the aim of establishing the *most general* solution of the fundamental equations in this case. If the fundamental equations of the theory [i.e., the quantum equations of motion and the quantization condition—now implemented with the $[\mathbf{p}, \mathbf{q}]$ commutator] are really complete and not in need of amplification, the absolute values $|q(nm)|$, $|p(nm)|$ of the elements of $\mathbf{q}$ and $\mathbf{p}$ will be *uniquely* determined by them, and it will be important to establish this in the case of the oscillator Eq. (12.50). In this connection one expects that with regard to the phases φ_{nm}, ψ_{nm} in
>
> $$q(nm) = |q(nm)|\, e^{i\varphi_{nm}}, \quad p(nm) = |p(nm)|\, e^{i\psi(nm)} \tag{12.51}$$
>
> there is still a remnant indeterminacy. For the statistics of the interaction, for example, of quantized atoms with external radiation fields it will be of fundamental significance to determine exactly the degree of this indeterminacy (Born and Jordan 1925b, pp. 296–297).

The complex phases in question had typically been ignored by Heisenberg (see note 9), and products like $q(n, n-1)q(n-1, n) = q(n, n-1)q(n, n-1)^\star = |q(n, n-1)|^2$ written as ordinary squares $q(n, n-1)^2$ of a real number. Born and Jordan (as was Heisenberg) were aware of the fact that in incoherent scattering processes (e.g., Raman–Smekal transitions) where the atom is left in a final state different from the initial state, the relative phase of the initial and final electron motion is classically physically significant, which suggests that the quantum description corresponding to the classical one should exhibit a similar sensitivity to phases that are not determinable from the fundamental equations, but simply accidents of the experimental situation in which light scattering is observed from many atoms with uncorrelated classical motions (or stationary states, in the quantum case). The question here is whether the phase freedom available in the quantum case exactly mimics that found in the classical case.[15]

Born and Jordan begin their study of one-dimensional oscillators in section 5 by setting $\lambda = 0$ and returning to the simplest case, a simple harmonic oscillator, which Heisenberg had of course already solved exactly. They make heavy work in the next

[15] See also the discussion following Eq. (11.55)).

few pages of showing that the quantum equations of motion and quantization condition together uniquely determine the diagonal elements of the energy matrix W_n, up to an obvious permutation symmetry corresponding to a reordering of the indices which label the stationary states. Once the ordering of states is chosen so that the energy elements W_n increase monotonically with the index, all ambiguity is resolved and one has $W_n = (n + 1/2)h\nu_0$ (which the authors call the "normal form" of the general solution). However, the quantum equations and quantization condition only determine the *absolute magnitude* of the coordinate matrix elements, of which the only non-vanishing ones take the form

$$|q(n, n+1)|^2 = |q(n+1, n)|^2 = \frac{h}{8\pi^2\nu_0}(n+1), \tag{12.52}$$

which clearly leaves the phases φ_{nm} in Eq. (12.51) undetermined.[16] In section 6, Born and Jordan turn to the examination of the cubic anharmonic oscillator Eq. (12.50), which Heisenberg had introduced as the simplest example of his new dynamical scheme. They rederive, in a somewhat altered notation, the recursion relations allowing the stepwise determination of the energy and coordinate matrix elements $q(nm)$ in a perturbative expansion in λ. The only new results (going beyond Heisenberg's results in the *Umdeutung* paper) concern their discussion of the phase freedom of the off-diagonal elements $q(nm), n \neq m$ (the diagonal elements are all real by the Hermitian property Eq. (12.11)), and their calculation of the quantized energies W_n up to terms of order λ^2 (Heisenberg had only calculated the first order contribution, which vanishes for the cubic oscillator). With regard to the phases, the conclusion is quite simple:

> Their phases [of the higher order contributions to $q(nm)$ elements)] are determined by those of the harmonic oscillator [i.e., the lowest order terms] (Born and Jordan 1925b, p. 881).

In particular,

$$\begin{aligned} q(n+1, n) &= |q(n+1, n)|\, e^{i\varphi_n}, \\ q(n+2, n) &= |q(n+2, n)|\, e^{i(\varphi_n + \varphi_{n+1})}, \\ q(n+3, n) &= |q(n+3, n)|\, e^{i(\varphi_n + \varphi_{n+1} + \varphi_{n+2})}, \quad \text{etc.} \end{aligned} \tag{12.53}$$

The phases φ_n appearing here are just the ones assumed at the lowest order (for the simple harmonic oscillator with $\lambda = 0$). Switching on the cubic term changes the magnitudes of the (real) quantized energies and the (in general complex) coordinate matrix elements, but leaves the phases unchanged.[17] The proper interpretation of this phase

[16] The relation $\mathbf{p} = m\dot{\mathbf{q}}$ actually requires equality of the phases ψ_{nm} and φ_{nm} in Eq. (12.51).

[17] Born and Jordan are not able to verify this phase stability in the case of the second-order contribution to $q(n+1, n)$, but it is easy to check using the formulas of modern (Rayleigh–Schrödinger) perturbation theory (Born and Jordan 1925b, p. 881). The modern reader will recognize the phase φ_n as the arbitrary relative phase between the quantum states $|n\rangle$ and $|n+1\rangle$, in Dirac notation, at some initial time.

freedom really requires an understanding of the quantum state as a ray in a complex Hilbert space, which would emerge two years later in von Neumann's work. However, Born and Jordan's treatment of complex phases, and the adoption of the fundamental quantization condition Eq. (12.36), with the prominent presence of i on the right side, testify to the growing awareness that in quantum theory, in contrast to classical mechanics, complex numbers play an unavoidable and central role.

The second new result obtained by Born and Jordan is the calculation of the quantized energies to second order in the cubic anharmonic parameter λ (Heisenberg had already performed the equivalent calculation for the quartic oscillator in his *Umdeutung* paper, perhaps the most impressive explicit calculational result of that paper). They find

$$W_n = h\nu_0(n+\frac{1}{2}) - \lambda^2\frac{5}{3}\frac{C^2}{\omega_0^2}(n(n+1)+\frac{17}{30}) + \cdots, \quad C = \frac{h}{8\pi^2\nu_0}. \tag{12.54}$$

This result is then compared with that obtained from classical Hamilton–Jacobi perturbation theory, which gives the energy for the cubic oscillator (with $m = 1$), up to terms of second order (Born 1925, p. 294):

$$W(\mathcal{J}) = \mathcal{J}\nu_0 - \frac{5}{12}\frac{\lambda^2}{(2\pi)^6\nu_0^4}\mathcal{J}^2 + \cdots \tag{12.55}$$

If we quantize the classical action variable $\mathcal{J} = (n+1/2)h$ by including the zero-point term now known to be present in the leading term, this becomes

$$\begin{aligned} W(\mathcal{J}) &= h\nu_0(n+\frac{1}{2}) - \lambda^2\frac{5}{3}\frac{C^2}{\omega_0^2}(n+\frac{1}{2})^2 + \cdots \\ &= h\nu_0(n+\frac{1}{2}) - \lambda^2\frac{5}{3}\frac{C^2}{\omega_0^2}(n(n+1)+\frac{1}{4}) + \cdots . \end{aligned} \tag{12.56}$$

Just as was the case for the quartic oscillator, the old quantum theory result, *provided we include the zero-point energy in the quantization of the action variable* $\mathcal{J}$, almost exactly agrees with the exact quantum result. The discrepancy (the factor 17/30 appearing in the exact quantum energy, instead of 1/4) is an additive constant which is subdominant in the correspondence limit of large quantum numbers n. As the discrepant constant is independent of the quantum number, it disappears in differences of energies, so that the quantum Bohr frequency $\nu_{qu}(n, n-1) = (W_n - W_{n-1})/h$ is almost identical to the classical frequency ν_{cl}, obtained by differentiating Eq. (12.55) with respect to the action variable $\mathcal{J}$,

$$\nu_{cl} = \frac{\partial W}{\partial \mathcal{J}} = \nu_0 - \frac{5}{6}\frac{\lambda^2}{(2\pi)^6\nu_0^4}\mathcal{J} + \cdots = \nu_0 - \frac{5}{6}\frac{\lambda^2}{(2\pi)^6\nu_0^4}nh + \cdots = \nu_{qu}, \tag{12.57}$$

but now provided we *omit* the zero-point energy and write $\mathcal{J} = nh$. In any case, all we can really expect in these comparisons is that the exact quantum result and the result obtained by classical perturbation theory subjected to Bohr–Sommerfeld quantization (with or without a zero-point term) should agree asymptotically for large quantum numbers. This means that we can only expect agreement for the terms with the highest powers of the large quantum number n.

In the final chapter of this ground-breaking paper, entitled "Remarks on electrodynamics,"[18] Born and Jordan make a first attempt to adapt the fundamental equations of Maxwellian electrodynamics to the new conceptual framework, applied so far only to the dynamics of a massive charged particle moving in one dimension. The basic objective here is to justify Heisenberg's identification of the absolute square of the coordinate matrix elements $|q(nm)|^2$ as the critical quantity determining the transition probabilities for the quantum transition (they actually use the word "jump" (*Sprung*)) between quantum states n and m. They apologize at the outset for the purely provisional character of their considerations, and promise to examine the relation of the new theory to the theory of light quanta in a future publication.

The discussion begins in a promising way. The authors observe that cavity (black-body) radiation can be regarded as a system of infinitely many uncoupled simple harmonic oscillators. The simple harmonic oscillator has an exceedingly simple structure. In particular, the energy conservation principle follows directly from the (linear) equations of motion, without the need for specifying the commutators of position and momentum via the quantization condition. From

$$\mathbf{H} = \frac{1}{2}(\mathbf{p}^2 + \omega_0^2\mathbf{q}^2) \tag{12.58}$$

follows immediately

$$\begin{aligned}\dot{\mathbf{H}} &= \frac{1}{2}(\dot{\mathbf{p}}\mathbf{p} + \mathbf{p}\dot{\mathbf{p}} + \omega_0^2\dot{\mathbf{q}}\mathbf{q} + \omega_0^2\mathbf{q}\dot{\mathbf{q}} \\ &= \frac{1}{2}\omega_0^2(-\mathbf{q}\mathbf{p} - \mathbf{p}\mathbf{q} + \mathbf{p}\mathbf{q} + \mathbf{q}\mathbf{p}) = 0.\end{aligned} \tag{12.59}$$

Note that energy conservation can be established here without the need to commute momenta and coordinate matrices, thereby requiring knowledge of the fundamental commutator Eq. (12.36). Accordingly, it is quite plausible that, once the electric and magnetic field vectors $\vec{\mathfrak{E}}, \vec{\mathfrak{H}}$ are reinterpreted as (triplets of) matrices, the local conservation principle of the electromagnetic energy density $\mathfrak{W}$ embodied in the Poynting theorem (with $\vec{\mathfrak{S}}$ the Poynting vector giving the local energy flux)

$$\dot{\mathfrak{W}} = \vec{\nabla} \cdot \vec{\mathfrak{S}}, \quad \vec{\mathfrak{S}} \equiv \frac{c}{8\pi}(\vec{\mathfrak{H}} \times \vec{\mathfrak{E}} - \vec{\mathfrak{E}} \times \vec{\mathfrak{H}}), \tag{12.60}$$

[18] This chapter is omitted in the translation included in the anthology by van der Waerden (1968) to which we have been referring.

follows from Maxwell's equations, without the need to commute electric and magnetic field components past each other. Total energy conservation follows of course by integrating the preceding equation over all space and using Gauss's theorem, assuming that the flux vanishes at infinity. Born and Jordan also verify the analogous result for local conservation of the angular momentum flux.

So far, so good. Alas, in section 8, in an attempt to connect the probability of a radiative transition to the square of coordinate matrix elements on the basis of the new theory, Born and Jordan fall victim to the same peril that had led Heisenberg to an erroneous calculation in his treatment of the rotator: knowing the desired answer in advance. Let us recall some of our earlier results concerning this connection, which goes back to the work of Ladenburg and Reiche (1923). For a simple harmonic oscillator, these authors identified, for an oscillator in the $n+1$ state, the coordinate amplitude C (in the cosine Fourier expansion, corresponding to the coordinate matrix element $q(n+1,n) = C/2$ in the complex exponential expansion (cf. Eqs. (10.92) and (10.93)), by Eq. (10.73):

$$C^2 = \frac{(n+1)h}{2m\pi^2\nu} \Rightarrow |q(n+1,n)|^2 = \frac{(n+1)h}{8m\pi^2\nu}. \tag{12.61}$$

The rate of energy radiation in the transition $n+1 \to n$ is given in Eq. (10.74),

$$A^n_{n+1}h\nu = \frac{16\pi^4 e^2\nu^4}{3c^3}C^2 = \frac{64\pi^4 e^2\nu^4}{3c^3}|q(n+1,n)|^2, \tag{12.62}$$

or, if we wish to consider the transition $n \to n-1$, and use the new matrix notation for the frequency $\nu(n,n-1)$ associated with this transition, the spontaneous energy radiation rate takes the form

$$A^{n-1}_n h\nu = \frac{64\pi^4 e^2\nu(n,n-1)^4}{3c^3}|q(n,n-1)|^2. \tag{12.63}$$

Presumably, this formula gives the template for the radiative energy rate in more general situations, for example, anharmonic oscillators, where transitions take place for $n \to k$, for any $k < n$, and

$$A^k_n h\nu = \frac{64\pi^4 e^2\nu(n,k)^4}{3c^3}|q(nk)|^2. \tag{12.64}$$

Born and Jordan attempt to derive Eq. (12.64) by reinterpreting the Larmor formula Eq. (10.6) for the rate of radiation of energy by a classical charged particle:

$$P(t) = \frac{2e^2}{3c^3}(\ddot{\vec{q}})^2. \tag{12.65}$$

Assuming the particle to be executing oscillations in a fixed direction, the classical coordinate vector $\vec{q}$ can be replaced by a single coordinate matrix $\mathbf{q}$. Furthermore, as we are interested in the time-averaged rate of radiation (which we henceforth denote with an overbar), the off-diagonal matrix elements of the squared acceleration matrix that are proportional to complex periodic exponentials average to zero, leaving only the time-independent diagonal elements:

$$\begin{aligned}\overline{P}(t)(n,n) &= \frac{2e^2}{3c^3}\ddot{\mathrm{q}}^2(n,n)\\ &= \frac{2e^2}{3c^3}\sum_k \ddot{q}(nk)\ddot{q}(kn)\\ &= \frac{2e^2}{3c^3}\sum_k (2\pi i\nu(nk))^2 e^{2\pi i\nu(nk)t}q(nk)\cdot(2\pi i\nu(kn))^2 e^{2\pi i\nu(kn)t}q(kn)\\ &= \frac{32\pi^4 e^2}{3c^3}\sum_k \nu(nk)^4|q(nk)|^2, \qquad (12.66)\end{aligned}$$

where the time dependence disappears as expected, as $\nu(nk) = -\nu(kn)$. This looks tantalizingly close to the desired expression Eq. (12.64), but there is a serious problem, as Born and Jordan recognize:

> Here we wish only to check, whether the radiation is really determined by the quantities $|q(nk)|^2$; the expression Eq. (12.66) shows that that is the case, but at the same time we see that the quantity written here is not the total outgoing spontaneous radiation from a stationary state. For spontaneous transitions always take place to states of lower energy, i.e., to states of lower quantum number for suitably labeled states (Born and Jordan 1925b, pp. 887–888).

Born and Jordan are simply pointing out that the sum over k in Eq. (12.66), for a fixed value of the initial state n, contains values of the final state quantum number k which are both larger as well as smaller than n, whereas the total radiation rate from state n should only contain contributions from final states k with $k < n$. Moreover, the coefficient appearing in Eq. (12.66) is apparently off by a factor of two from the desired result in Eq. (12.64). Born and Jordan suggest a clever fix to both of these problems, admitting quite openly that their solution is of a "completely formal nature". One forms, instead of the time-averaged radiation rate, the *trace* (i.e., sum of diagonal matrix elements) of the matrix reinterpretation of Eq. (12.65):

$$\mathrm{Tr}\left(\frac{2e^2}{3c^3}\ddot{\mathrm{q}}^2\right) = \frac{32\pi^4 e^2}{3c^3}\sum_{nk}\nu(nk)^4|q(nk)|^2 \qquad (12.67)$$

$$= \sum_n \frac{64\pi^4 e^2}{3c^3}\sum_{k<n}\nu(nk)^4|q(nk)|^2. \qquad (12.68)$$

In arriving at the second line here, we have relied on the fact that the double sum in the first line only contains off-diagonal terms (as $\nu(nk) = 0$ if $n = k$), and that each term with

$k > n$ is paired with an equal term with $k < n$ (as $\nu(kn) = -\nu(nk), |q(kn)| = |q(nk)|$). The correct prefactor now appears in each term of the sum over n. These terms are thus identified with the total spontaneous radiation rate from the state n to all lower states k, agreeing with the known result Eq. (12.64).

Of course, there is no clear physical argument guiding the choice of the trace operation here, other than the desire to arrive at the attractive goal of Eq. (12.64). The correct treatment of spontaneous radiation in the new quantum mechanics would have to wait another two years for Dirac's (1927b) quantum electrodynamics paper, in which the electromagnetic field as well as the kinematical variables associated with the radiating particle is subjected to quantization, and the spontaneous radiation rate calculated in terms of a process in which both the quantum state of the particle and that of the electromagnetic field change simultaneously.

For modern physicists, the paper of Born and Jordan is vastly more readable than the seminal work from which it springs, Heisenberg's *Umdeutung* paper. This is partly because the intimate knowledge of dispersion theory of which theorists in Copenhagen and Göttingen could boast in the period in question has evaporated from the standard training one receives nowadays in physics, even at the graduate level. In their extension of Heisenberg's perturbative oscillator calculations, the methods and notations are hardly an improvement over those in the *Umdeutung* paper. But their rephrasing of the general principles of the theory in terms of matrices, and the introduction of the quantum commutation relation of position and momentum, instantly reveals a clear formal structure that all modern students of quantum mechanics can easily recognize. The achievement was summarized succintly (if somewhat snidely) by a remark made by Alfred Landé during a lunch break in his AHQP interview with Thomas Kuhn and John Heilbron in 1962. Landé told his interviewers: "Heisenberg stammered something. Born made sense of it" (quoted in Duncan and Janssen 2007, p. 557).

12.2 The new theory derived differently: Dirac's formulation of quantum mechanics

At the same time that Born and Jordan were reformulating and extending the formal framework of Heisenberg's ideas in Göttingen, 23-year-old Cambridge graduate student Paul Adrien Maurice Dirac was reprocessing Heisenberg's ideas in a different, and conceptually considerably deeper, direction. Dirac arrived at a formulation of quantum mechanics that not only emphasized the formal connection to classical theory just as powerfully as Heisenberg had (in insisting in the validity of the classical equations of motion, once the constituent kinematic quantities had been suitably reinterpreted), but also showed how to remove the annoying reliance on a multiply periodic classical formulation, instead producing a quantum theory defined by a formal analogy to classical Hamiltonian dynamics valid for arbitrary classical systems, multiply periodic or not.[19]

[19] For analysis of the development of Dirac's version of quantum mechanics, see Mehra and Rechenberg (1982d), Kragh (1990), and Darrigol (1992).

Dirac matriculated (at the age of 16) in the engineering department of the University of Bristol, expecting to pursue a career as an engineer. It seems clear that his original interests, which definitely lay more in the direction of pure mathematics rather than in a "practical profession" such as engineering, had been suppressed, probably because the only perceivable outcome of a degree in mathematics lay in teaching, and Dirac was completely lacking in the sort of social skills that would make persistent contact with the uninitiated bearable. As it turned out, by the time Dirac graduated with his engineering degree in 1921, the British economy was in depression and there were in any case no jobs available for a newly minted engineer. Dirac took the entrance examination to Cambridge in June of 1921 and passed easily, but was not awarded enough financial aid to make attendance possible. Instead, he returned to Bristol with the intention of taking an undergraduate degree for the second time, but this time in mathematics. He managed this easily in two years, and in 1923 won two scholarships, which together were just about able to support his attendance at Cambridge for graduate studies.[20]

When he arrived at Cambridge, Dirac had already studied a number of topics that would turn out to be of critical importance in his work on quantum mechanics two years later. In his "applied mathematics" courses in Bristol, he had studied classical mechanics, in particular in its Hamiltonian formulation, and had also taken a course in the rudiments of atomic theory—that is, the old quantum theory. Relativity (special) became a particular focus of interest for Dirac at this time: later, his first response to a new physical theory would frequently be an attempt to "relativize" the theory, if it did not already take the requirements of special relativity into account. He had become intrigued by Hamilton's *quaternions*, an extension of the complex number field to include quantities a, b, which failed to satisfy the commutation relation $ab = ba$. In "pure mathematics" he had studied (in the excellent course taught by Peter Fraser) projective geometry, a formulation of geometry that also admitted non-commuting quantities in some of its more recherché extensions. This acquaintance with objects with unusual algebraic qualities would stand Dirac in good stead when he encountered Heisenberg's new kinematics in September 1925. Like Heisenberg (but unlike Jordan), however, Dirac does not seem to have been aware of the technology of matrix algebra: in any case, matrices are not mentioned[21] in his first two papers on quantum theory.

As discussed, Heisenberg had been invited by Ralph Fowler to lecture in Cambridge at the end of July. His talk (entitled "Term-zoology and Zeeman-botany") concerned the panoply of empirically motivated rules that had evolved to deal with complex multiplets and their Zeeman splittings—apparently, his own recent work contained in the *Umdeutung* paper was not mentioned, except perhaps in passing.[22] Heisenberg's talk was

[20] For a detailed account of the life of this "strangest man", see the biography of the same name by Farmelo (2009). See also Kragh (1990).

[21] In his paper on the hydrogen atom, Dirac (1926a) cites the Born–Jordan paper in which matrices are ubiquitous, so he is clearly aware of the matrix representation of the theory by this time. Nonetheless, he scrupulously avoids any reference to matrices.

[22] There is some uncertainty on this point, as Heisenberg later claimed to have discussed his *Umdeutung* work in Cambridge (see, e.g., Mehra and Rechenberg 1982b, p. 321, especially footnote 334).

presented at the Kapitza club, a regular informal seminar organized by Peter Kapitza. Dirac was presumably present, though neither later recalled having been introduced to the other. Instead, Dirac found out about Heisenberg's breakthrough a month later, early in September, when Fowler forwarded the proofs of Heisenberg's paper to him, with a scribbled request: "What do you think of this. I shall be glad to hear" (Farmelo 2009, p. 83). By early November, Dirac was able to hand his first paper on quantum mechanics, entitled "The fundamental equations of quantum mechanics," to Fowler, who, as a member of the Royal Society, was able to secure rapid acceptance and publication in the proceedings of the Royal Society (Dirac 1925).

Dirac's paper begins with a short introduction, which reviews in the first paragraph the basic tenets of the old quantum theory: the validity of classical mechanics for the stationary states, selected by specific quantum conditions from the manifold of all possible classical motions, the Bohr frequency condition for the radiation emitted in transitions between such states, and the Correspondence Principle (so capitalized by Dirac), which requires classical electromagnetic theory to agree asymptotically (for large quantum numbers)—as Dirac (1925, p. 307) puts it, "give the right results"—with motions selected by the quantum conditions of the old quantum theory.

The second paragraph of Dirac's introduction deserves to be quoted in its entirety:

> In a recent paper Heisenberg puts forward a new theory which suggests that it is not the equations of classical mechanics that are in any way at fault, but that the mathematical operations by which physical results are deduced from them require modification. *All* the information supplied by the classical theory can thus be made use of in the new theory (Dirac 1925, p. 307).

Here, Dirac is making explicit an assumption about the new theory which is certainly implicit in Heisenberg's work, but never addressed as clearly as Dirac has here: namely, that the dynamical equations of quantum mechanics—or the energy function (Hamiltonian) defining the dynamics—do not contain additional terms that vanish as Planck's constant goes to zero, and therefore would be empirically invisible at the classical level, but would have to be taken into account to determine correctly the behavior of systems at the atomic level. Such terms were certainly *a priori* possible, and the assertion of their absence, so that the classical equations were *all* the information needed to understand atomic systems, once the symbols appearing in them were appropriately interpreted, is a very strong statement indeed. In fact, the essence of this remarkable paper is Dirac's insistence in preserving *as much as possible* of the formal structure of classical mechanics in the new theory, in particular, the specification of the time evolution of classical systems in (p, q) phase space, as given in the Hamiltonian formulation, by first-order equations of motions expressed in terms of Poisson brackets.[23]

The second section of Dirac's paper, entitled "Quantum algebra," begins with the following expression for the multiple (complex) Fourier expansion for a given coordinate

[23] For a review of the essential properties of Poisson brackets needed here, see Section A.2.2.

in a classical system with u degrees of freedom undergoing multiply periodic motion:

$$\begin{aligned} x &= \sum_{\alpha_1\alpha_2\ldots\alpha_u} x(\alpha_1\alpha_2\ldots\alpha_u)\exp i(\alpha_1\omega_1+\alpha_2\omega_2+\ldots+\alpha_u\omega_u)t \\ &\equiv \sum_{\alpha} x_\alpha \exp i(\alpha\omega)t. \end{aligned} \tag{12.69}$$

This is equivalent to Eq. (10.108) in our previous discussion of classical perturbation theory of multiply periodic motion:

$$x(t) = \sum_{\vec{\tau}} X_{\vec{\tau}}(\vec{\mathcal{J}})\, e^{2\pi i\vec{\tau}\cdot\vec{w}}, \quad \vec{\tau} = (\tau_1, \tau_2, \ldots, \tau_u), \tag{12.70}$$

where the angle variables w_i are related to the cyclic frequencies ν_i associated with each degree of freedom by $w_i = \nu_i t$, $i = 1, 2, \ldots, u$, and we have now explicitly indicated the dependence of the amplitude $X_{\vec{\tau}}(\vec{\mathcal{J}})$ on the u action variables $\vec{\mathcal{J}} = (\mathcal{J}_1, \mathcal{J}_2, \ldots, \mathcal{J}_u)$ conjugate to the angle variables $\vec{w} = (w_1, w_2, \ldots, w_u)$. We have used the capitalized letter X to represent the Fourier amplitudes of $x(t)$ as we shall shortly be using analogous labeling for another kinematical coordinate $y(t)$ (with Fourier amplitudes $Y_{\vec{\tau}}$). It should be emphasized that these coordinates ($x(t), y(t)$, etc.) can represent either position or momentum variables: in a multiply periodic system, all the phase space coordinates have a multiple Fourier expansion, as do arbitrary functions of them.

In Dirac's paper, the integer indices identifying Fourier components (or, via the correspondence principle, quantum transitions from a given initial state) are denoted (following Heisenberg's *Umdeutung* paper) by early letters of the Greek alphabet ($\alpha, \beta, \ldots$ etc.), which correspond to the notation using τ, τ' that appears in our previous analysis of the work on dispersion theory of Kramers, Born, and Heisenberg. We use this latter notation (although it differs from that in Dirac's published paper) because, apart from the fact that it is familiar from earlier discussions in this book, it is actually the notation Dirac uses in the rough calculations that led him directly from formulas appearing in the Kramers–Heisenberg paper to his famous equivalence between commutators and Poisson brackets.

In the remainder of this section Dirac reviews ("following Heisenberg," p. 309) the kinematic correspondence formulas introduced by Heisenberg (1925c) in section 1 of the *Umdeutung* paper (cf. Eqs. (11.46–11.51) and Eqs. (11.52) and (11.53)). The reasoning (though not the notation) is more or less exactly that used by Heisenberg, although the formula for multiplication of quantum amplitudes is given in a notation instantly recognizable (to modern eyes) as matrix multiplication:

$$xy(nm) = \sum_k x(nk)y(km). \tag{12.71}$$

Nonetheless, Dirac never mentions the word "matrix" in this paper—the aspects of matrix algebra relevant to his arguments are derived directly when necessary with explicit index summations. Although we can be certain that Dirac was familiar with some rather sophisticated areas of contemporary pure mathematics (most famously, projective geometry), he does not seem to have known, or seen the need to use, the basic features of matrix algebra exploited, for example, in the Born–Jordan paper. Throughout this paper, Dirac simply refers to matrix multiplication as either "quantum multiplication", or as the "Heisenberg product". The fact that such a product is not in general commutative,

$$xy(nm) \neq yx(nm), \tag{12.72}$$

is pointed out, together with the reassuring statement that the "Heisenberg product" is associative and distributive. Having stated these properties, Dirac then says

> We are now able to take over each of the equations of motion of the system into the quantum theory provided we can decide the correct order of the quantities in each of the products. Any equation deducible from the equations by algebraic processes not involving the interchange of the factors of a product, and by differentiation and integration with respect to t, may also be taken over into the quantum theory. In particular, the energy equation may be thus taken over (Dirac 1925, p. 310).

The reference to the "energy equation" here suggests that Dirac has in mind energy functions (Hamiltonians) of the restricted form $H(q,p) = K(p)+V(q)$, where the kinetic energy part $K(p)$ only depends on momentum variable(s), while the potential energy $V(q)$ only depends on coordinate(s). As we saw earlier, Heisenberg, (and especially) Born, and Jordan were much more concerned in trying to develop a formalism general enough to encompass situations in which coordinates and momenta appear multiplied together, in which the ordering problem cannot be avoided.

Like Born and Jordan, Dirac introduces a definition of "quantum differentiation" in the third section of his paper. He wishes to find the "most general quantum operation d/dv" satisfying linearity and the Leibniz rule:[24]

$$\frac{d}{dv}(x+y) = \frac{d}{dv}x + \frac{d}{dv}y, \tag{12.73}$$

$$\frac{d}{dv}(xy) = \frac{d}{dv}x \cdot y + x \cdot \frac{d}{dv}y. \tag{12.74}$$

Note that the variable v here can be either a scalar variable (such as the time t), or a "quantum variable" (i.e., a matrix). Dirac then assumes that the linearity condition

[24] Dirac's definition of quantum differentiation differs from the Born–Jordan rule Eq. (12.7), which does not obey the Leibniz rule in general. For example, $\partial/\partial p\,(pqpq) = 2qpq$ according to the Born–Jordan rule, while splitting $pqpq = (p)(qpq)$ and applying the Leibniz rule would give $\partial/\partial p\,(p)(qpq) = qpq + pq^2$, which is not the same if p and q are non-commuting matrices. For functions $K(p)$ consisting only of powers of p, or functions $V(q)$ only involving powers of q, Born–Jordan differentiation gives the natural result, namely, $K'(p)$ or $V'(q)$, so the Leibniz rule is obviously satisfied.

Eq. (12.73) is satisfied by writing the derivative as a linear combination of the matrix elements of the differentiated quantity

$$\frac{dx}{dv}(nm) = \sum_{n'm'} a(nm; n'm')x(n'm'). \tag{12.75}$$

The imposition of the Leibniz rule Eq. (12.74) then restricts the coefficients $a(nm; n'm')$ in such a way that the linear combination exhibited in Eq. (12.75) reduces to taking the commutator of the differentiated matrix x with another matrix A,[25] so that

$$\frac{dx}{dv}(nm) = \sum_{n'm'} a(nm; n'm')x(n'm') = (xA - Ax)(nm) = [x, A](nm) \tag{12.76}$$

a result that can easily be seen to satisfy the linearity requirement, as well as the Leibniz rule, by the commutator identity

$$[A, xy] = Axy - xyA = Axy - xAy + xAy - xyA = [A, x]y + x[A, y]. \tag{12.77}$$

In fact, the form Eq. (12.75) (leading to Eq. (12.76)) is not "required" by Eqs. (12.73) and (12.74), as Dirac claims. If we take v to be a scalar parameter on which the matrix elements of x depend, it is certainly not the case that the result of the differentiation can always be expressed as a commutator, as that would require (assuming finite matrices for simplicity), taking the trace,

$$\mathrm{Tr}\left(\frac{dx}{dv}\right) = \mathrm{Tr}(xA - Ax) = 0, \tag{12.78}$$

which can hardly be true in general, if the dependence on v of the $x(nm)$ matrix elements does not take a very particular form.

Of course, Dirac already knows that for the time derivative of the kinematic variables in Heisenberg's new theory (cf. Eqs. (12.19) and (12.20)),

$$\frac{d}{dt}x(nm) = i\omega(nm)x(nm), \quad \omega(nm) = \Omega(n) - \Omega(m), \tag{12.79}$$

which implies, defining the matrix A as a diagonal matrix with $A_{nn} = -i\Omega(n)$,

$$\frac{d}{dt}x(nm) = -A(nn)x(nm) + x(nm)A(mm) = [x, A], \tag{12.80}$$

so the derivative with respect to time does take the particular form assumed by Dirac. Indeed, we already saw that, given the fundamental commutator Eq. (12.36) of position

[25] Our notation differs from Dirac in that he always writes quantum commutators explicitly as "the difference of Heisenberg products," for example, $xa - ax$. The usual square bracket notation [,], which we have been using, is reserved by Dirac for the classical Poisson bracket. We instead use curly brackets, { , } for Poisson brackets, to distinguish them from their quantum counterparts.

and momentum, derivatives with respect to p and q are in fact realized via commutators, as we see in Eqs. (12.41) and (12.42). But this basic commutator has not yet been derived by Dirac—it will emerge from his Poisson bracket formulation in section 4 of his paper—so his reasoning here seems a bit out of place. Nevertheless, the fundamental importance of non-commutativity was clear to Dirac from the outset, and in particular, the relevance of the quantum commutator $xy-yx$ of two "quantum variables/quantities," which quantifies the extent of the non-commutativity.

In section 4 of his paper, entitled "The quantum conditions," Dirac applies correspondence-principle methods to derive a relation between the classical concept of Poisson brackets[26] and the commutator of quantum variables. This relation allows Planck's constant to be introduced into the theory in a way that mimics the formal structure of classical mechanics (in its Hamiltonian phase space formulation) in an extremely close way. Dirac's approach, in contrast to the path followed by Heisenberg (and extended by Born and Jordan), has the merit that the extension to systems with more than one degree of freedom is immediate and unproblematic.

The route taken by Dirac in arriving in his quantum reinterpretation of Poisson brackets can fortunately be reconstructed in detail, as Dirac's hand-written notes leading to the result have survived.[27] The starting point is a graphical representation of quantum transitions found in Kramers and Heisenberg's (1925) paper, in their discussion of the dispersion theory formula for inelastic (Raman–Smekal) light scattering (see Figure 12.1).[28] In the Kramers-Heisenberg paper, this diagram expresses the need to calculate the amplitude for transitions between states n and $n-\tau-\tau'$ as a sum of the amplitude for transitions via two different intermediate states, namely as a sum of the amplitude for $n \to n-\tau' \to n-\tau-\tau'$ and for $n \to n-\tau \to n-\tau-\tau'$, that is, by transition 2 followed by 1, or by transition 3 followed by 4. Let us consider any two functions of the phase space coordinates (i.e., functions of the position and/or momentum coordinates, which for suitable multiply periodic systems might just be the action and angle variables of the system) $x(t)$ and $y(t)$ in a multiply periodic system, with $x(t)$ given by Fourier

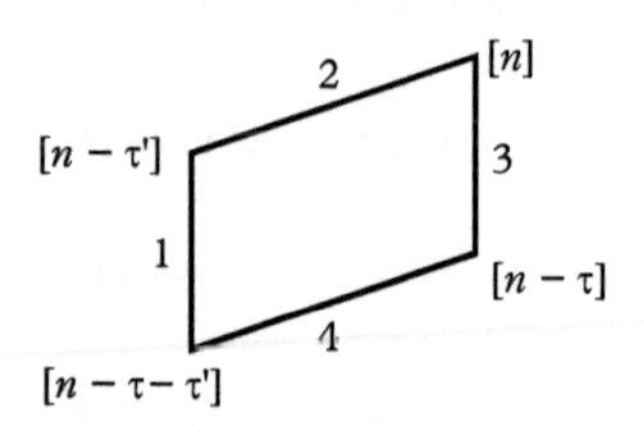

Figure 12.1 *Quantum transition diagram in Dirac's notes, taken from Fig. 2 in Kramers and Heisenberg (1925, p. 238).*

[26] For a review of Poisson brackets, see Section A.2.2.

[27] For a detailed discussion, see Darrigol (1992, p. 312).

[28] The labeling of the states in the Kramers–Heisenberg diagram is slightly different from that used by Dirac: one obtains the states as labeled by Dirac by replacing $n \to n-\tau-\tau'$ in Kramers and Heisenberg (1925).

expansion Eq. (12.70), and $y(t)$ similarly given (with Fourier amplitudes $Y_{\vec{\tau}}$). In the correspondence limit, the stationary states are specified by integral quantization of the action variables, so we can identify states $\vec{n}$ and $\vec{m}$ by setting $\mathcal{J}_r = n_r h, r = 1, 2, \ldots, u$ in state n and $\mathcal{J}_r = m_r h$ in state m. Transitions between these two states can occur either by

$$n_r \rightarrow n_r - \tau_r \rightarrow n_r - \tau_r - \tau_r' \equiv m_r, \tag{12.81}$$

or

$$n_r \rightarrow n_r - \tau_r' \rightarrow n_r - \tau_r - \tau_r' \equiv m_r, \tag{12.82}$$

corresponding to routes 3,4 or 2,1 in Figure 12.1. We use vector notation to indicate the multi-dimensionality of the system: thus, the state defined by quantum numbers $n_1, n_2, \ldots, n_u$ can be denoted simply by $\vec{n}$. The commutator of the quantum variables x and y are given by Eq. (12.71):

$$(xy - yx)(\vec{n}\vec{m}) = \sum_{\vec{k}} (x(\vec{n}\vec{k})y(\vec{k}\vec{m}) - y(\vec{n}\vec{k})x(\vec{k}\vec{m})). \tag{12.83}$$

In the first term on the right-hand side we relabel the vector summation index $\vec{k} = \vec{n} - \vec{\tau}$ and define $\vec{\tau}' = \vec{n} - \vec{m} - \vec{\tau}$, while in the second term on the right the summation variable is relabeled $\vec{k} = \vec{n} - \vec{\tau}'$ with $\vec{\tau} = \vec{n} - \vec{m} - \vec{\tau}'$. With these substitutions, Eq. (12.83) becomes

$$\begin{aligned}(xy - yx)(\vec{n}\vec{m}) &= \sum_{\vec{\tau}+\vec{\tau}'=\vec{n}-\vec{m}} \Big(x(\vec{n}, \vec{n} - \vec{\tau})\, y(\vec{n} - \vec{\tau}, \vec{n} - \vec{\tau} - \vec{\tau}') \\ &\qquad - y(\vec{n}, \vec{n} - \vec{\tau}')\, x(\vec{n} - \vec{\tau}', \vec{n} - \vec{\tau} - \vec{\tau}') \Big)\end{aligned} \tag{12.84}$$

Adding and subtracting the term

$$\sum_{\vec{\tau}+\vec{\tau}'=\vec{n}-\vec{m}} x(\vec{n}, \vec{n} - \vec{\tau})\, y(\vec{n}, \vec{n} - \vec{\tau}') \tag{12.85}$$

and rearranging terms, we can write this as:

$$\begin{aligned}(xy - yx)(\vec{n}\vec{m}) &= \sum_{\vec{\tau}+\vec{\tau}'=\vec{n}-\vec{m}} \Big(\Big\{ x(\vec{n}, \vec{n} - \vec{\tau}) - x(\vec{n} - \vec{\tau}', \vec{n} - \vec{\tau} - \vec{\tau}') \Big\} y(\vec{n}, \vec{n} - \vec{\tau}') \\ &\qquad - x(\vec{n}, \vec{n} - \vec{\tau}) \Big\{ y(\vec{n}, \vec{n} - \vec{\tau}') - y(\vec{n} - \vec{\tau}, \vec{n} - \vec{\tau} - \vec{\tau}') \Big\} \Big).\end{aligned} \tag{12.86}$$

In the correspondence limit, where $|\vec{\tau}|, |\vec{\tau}'| \ll |\vec{n}|$ and $|\vec{n}| \gg 1$, we can replace the individual terms occurring in the sum on the right-hand side of this equation by Fourier components (see Eq. (12.70)),

$$x(\vec{n}, \vec{n} - \vec{\tau}) \to X_{\vec{\tau}}(\vec{\mathcal{J}})\, e^{2\pi i \vec{\tau}\cdot\vec{w}}; \tag{12.87}$$

and differences between them by changes in action variables (see (10.125)),

$$x(\vec{n}, \vec{n} - \vec{\tau}) - x(\vec{n} - \vec{\tau}', \vec{n} - \vec{\tau} - \vec{\tau}') \quad \to \quad h\vec{\tau}' \cdot \vec{\nabla}_{\mathcal{J}} X_{\vec{\tau}}(\vec{\mathcal{J}})\, e^{2\pi i \vec{\tau}\cdot\vec{w}}. \tag{12.88}$$

This last substitution is just the classic maneuver used in the construction of the Kramers dispersion formula guided by the correspondence principle (see Chapter 10). Using these two substitutions, we can, in the correspondence limit, replace the expression on the first line of Eq. (12.86) by:

$$h\vec{\tau}' \cdot \left(\vec{\nabla}_{\mathcal{J}} X_{\vec{\tau}}(\vec{\mathcal{J}})\, e^{2\pi i \vec{\tau}\cdot\vec{w}}\right) \, Y_{\vec{\tau}'}(\vec{\mathcal{J}})\, e^{2\pi i \vec{\tau}'\cdot\vec{w}}. \tag{12.89}$$

Using that $\vec{\tau}' \cdot \vec{\nabla}_{\mathcal{J}} = \sum_r \tau_r' \, \partial/\partial \mathcal{J}_r$ and that

$$\frac{\partial}{\partial w_r} e^{2\pi i \vec{\tau}\cdot\vec{w}} = 2\pi i \tau_r \, e^{2\pi i \vec{\tau}\cdot\vec{w}}, \tag{12.90}$$

we can rewrite this as

$$\begin{aligned} & h \sum_r \frac{\partial}{\partial \mathcal{J}_r} \left(X_{\vec{\tau}}(\vec{\mathcal{J}})\, e^{2\pi i \vec{\tau}\cdot\vec{w}}\right) Y_{\vec{\tau}'}(\vec{\mathcal{J}})\, \tau_r' \, e^{2\pi i \vec{\tau}'\cdot\vec{w}} \\ & \quad = \frac{h}{2\pi i} \sum_r \frac{\partial}{\partial \mathcal{J}_r} \left(X_{\vec{\tau}}(\vec{\mathcal{J}})\, e^{2\pi i \vec{\tau}\cdot\vec{w}}\right) \frac{\partial}{\partial w_r} \left(Y_{\vec{\tau}'}(\vec{\mathcal{J}})\, e^{2\pi i \vec{\tau}'\cdot\vec{w}}\right). \end{aligned} \tag{12.91}$$

We can similarly replace the expression on the second line of Eq. (12.86) by:

$$\frac{h}{2\pi i} \sum_r \frac{\partial}{\partial \mathcal{J}_r} \left(Y_{\vec{\tau}'}(\vec{\mathcal{J}})\, e^{2\pi i \vec{\tau}'\cdot\vec{w}}\right) \frac{\partial}{\partial w_r} \left(X_{\vec{\tau}}(\vec{\mathcal{J}})\, e^{2\pi i \vec{\tau}\cdot\vec{w}}\right). \tag{12.92}$$

The correspondence between commutator array elements and Fourier components is now found by inserting the expressions in Eqs. (12.91) and (12.92) for the right-hand side of Eq. (12.86). In the correspondence limit, the commutator $(xy-yx)(\vec{n}\vec{m})$ can thus be written as:

$$\frac{h}{2\pi i}\sum_{\vec{\tau}+\vec{\tau}'=\vec{n}-\vec{m}}\sum_{r}\left(\frac{\partial}{\partial \mathcal{J}_r}\left(X_{\vec{\tau}}(\vec{\mathcal{J}})\,e^{2\pi i\vec{\tau}\cdot\vec{w}}\right)\frac{\partial}{\partial w_r}\left(Y_{\vec{\tau}'}(\vec{\mathcal{J}})\,e^{2\pi i\vec{\tau}'\cdot\vec{w}}\right)\right.$$
$$\left.-\frac{\partial}{\partial \mathcal{J}_r}\left(Y_{\vec{\tau}}(\vec{\mathcal{J}})\,e^{2\pi i\vec{\tau}\cdot\vec{w}}\right)\frac{\partial}{\partial w_r}\left(X_{\vec{\tau}'}(\vec{\mathcal{J}})\,e^{2\pi i\vec{\tau}'\cdot\vec{w}}\right)\right), \tag{12.93}$$

where we have relabeled $\tau \leftrightarrow \tau'$ in the second line. The rule Eq. (11.16) for the classical Fourier components of a product of two (here *multiply* periodic) quantities in our case takes the form

$$(xy)_{\vec{n}-\vec{m}} = \sum_{\vec{\tau}} X_{\vec{\tau}} Y_{\vec{n}-\vec{m}-\vec{\tau}} = \sum_{\vec{\tau}+\vec{\tau}'=\vec{n}-\vec{m}} X_{\vec{\tau}} Y_{\vec{\tau}'}. \tag{12.94}$$

Comparing Eqs. (12.93) and (12.94), we see that, in Dirac's words (italicized in the original), *the difference between the Heisenberg products of two quantum quantities is equal to $ih/2\pi$ times their Poisson bracket expression.* In other words,[29]

$$(xy-yx)(\vec{n}\vec{m}) \rightarrow \frac{ih}{2\pi}\{x(t),y(t)\}_{\vec{n}-\vec{m}}, \tag{12.95}$$

where we recall the definition of the Poisson bracket of any two kinematical quantities in terms of action/angle variables $(\mathcal{J}_r, w_r)$ in a multiply periodic system with u degrees of freedom:

$$\{x(t),y(t)\} \equiv \sum_{r=1}^{u}\left(\frac{\partial x}{\partial w_r}\frac{\partial y}{\partial \mathcal{J}_r} - \frac{\partial y}{\partial w_r}\frac{\partial x}{\partial \mathcal{J}_r}\right). \tag{12.96}$$

Classical Poisson brackets have the fundamental property of invariance under canonical transformations,[30] so we also have, for *any set of conjugate canonical variables* p_r, q_r,

$$\{x(t),y(t)\} \equiv \sum_{r=1}^{u}\left(\frac{\partial x}{\partial q_r}\frac{\partial y}{\partial p_r} - \frac{\partial y}{\partial q_r}\frac{\partial x}{\partial p_r}\right). \tag{12.97}$$

As x and y in Eq. (12.97) can be any functions on the classical phase space (p_r, q_r), they can be chosen to be just the canonical position and momentum variables themselves.

[29] We remind the reader that Dirac uses square brackets to indicate the Poisson bracket, and writes out quantum commutators explicitly, as $xy - yx$, etc. We have retained the more usual notation, using square brackets for commutators, curly brackets for Poisson brackets.

[30] See Section A.2.2.

We then obtain from Eq. (12.97) the simple results

$$\{q_r, q_s\} = \{p_r, p_s\} = 0, \tag{12.98}$$

$$\{q_r, p_s\} = \delta_{rs}, \tag{12.99}$$

implying for the corresponding quantum variables, via Eq. (12.95), the commutators

$$[q_r, q_s] = [p_r, p_s] = 0, \tag{12.100}$$

and, more interestingly,

$$[q_r, p_s] = \frac{ih}{2\pi}\delta_{rs} \cdot 1. \tag{12.101}$$

Here, 1 represents the unit matrix, with matrix elements $1(\vec{n}\vec{m}) = \delta_{\vec{n}\vec{m}}$ (corresponding to the fact that the Fourier expansion of a constant quantity contains only the zeroth mode $\vec{\tau} = 0$, implying $\vec{m} - \vec{n} = 0$ for the associated transitions). It is at this point that we can, with Dirac, adopt the very natural hypothesis that the commutation relations Eqs. (12.100) and (12.101) hold for *any mechanical system with Hamiltonian dynamics*—those in which canonical coordinates (q_r, p_r) and a Hamiltonian function $H(q_r, p_r)$ exist satisfying Hamilton's equations of motion—and not just for the very particular subset of Hamiltonian systems, which undergo multiply periodic motion amenable to separation in action/angle variables, although of course the transition from the old quantum theory to the new quantum mechanics via the correspondence principle required that we start with exactly such multiply periodic systems.

The commutation relations Eq. (12.101), of course, include the single nontrivial relation Eq. (12.36) obtained by Born and Jordan. But it provides the complete set of quantization conditions needed for the analysis of a system of arbitrarily many degrees of freedom. And such systems, by assumption, are not required to be multiply periodic at the classical level, as the successful application of the new quantum mechanics to multi-electron atomic systems in the next few years would confirm (and as Dirac surmises in the final paragraph of his paper).

For a system of one degree of freedom, Dirac rederives Heisenberg's quantum condition (Eq. (16) of the *Umdeutung* paper; cf. Eq. (11.60)) by looking at the constant (i.e., diagonal) element of the q, p commutator:

$$\begin{aligned}\frac{ih}{2\pi} &= (qp - pq)(nn)\\ &= (qm\dot{q} - m\dot{q}q)(nn)\\ &= m\sum_k (q(nk)\dot{q}(kn) - \dot{q}(nk)q(kn)).\end{aligned} \tag{12.102}$$

Using that $\dot{q}(mn) = i\omega(mn)\, q(mn)$ and that $\omega(nm) = -\omega(mn)$ (see Eqs. (12.20) and (12.12)), we can rewrite this as:

$$\begin{aligned}\frac{ih}{2\pi} &= im\sum_k (q(nk)\omega(kn)q(kn) - \omega(nk)q(nk)q(kn)) \\ &= 2im\sum_k q(nk)q(kn)\omega(kn). \qquad (12.103)\end{aligned}$$

It follows that

$$4\pi m\sum_k q(nk)q(kn)\omega(kn) = h. \qquad (12.104)$$

In this formula the two terms in Heisenberg's version of the quantization condition, Eq. (11.60), correspond to $k = n + \alpha$ and $k = n - \alpha$ (with $\alpha > 0$) in Eq. (12.104).

The dispersion-theoretic origins of both Heisenberg's and Dirac's derivation of a quantum condition should by now be abundantly clear: the superiority of the Dirac approach to Heisenberg's lies in his use of the correspondence-principle arguments associated with the transitions displayed in Figure 12.2 for the *inelastic* Raman-type processes (in which $\vec{m} \neq \vec{n}$, i.e., $\vec{\tau} + \vec{\tau}' \neq 0$) that allowed Dirac to obtain *all* (and not just the diagonal) matrix elements of the commutator, whereas Heisenberg employed the Thomas–Reiche–Kuhn sum rule, which is already a special case of the elastic scattering ($\vec{m} = \vec{n}$) dispersion relation.[31]

In section 5 of his paper, entitled "Properties of the quantum Poisson bracket expressions," Dirac explores the consequences for the quantum dynamics (time dependence) of a system specified classically by a Hamiltonian function $H(p_r, q_r)$. Classically, the Hamiltonian equations of motion can be expressed simply in terms of the Poisson brackets

$$\dot{p}_r = \{p_r, H\}, \quad \dot{q}_r = \{q_r, H\}, \qquad (12.105)$$

which the reader can easily verify by referring to the definition of the Poisson bracket, Eq. (12.96). Of course, on translating these equations to quantum theory via the correspondence Eq. (12.95), the order of non-commuting coordinate and momentum operators (if they appear multiplied in a single term) in H must be specified to obtain an unambiguous result. In Dirac's words:

> These equations [i.e., Eq. (12.105)] will be true on the quantum theory for systems for which the orders of the factors of products occurring in the equations of motion are unimportant. They may be taken to be true for systems for which the orders are important if one can decide upon the orders of the factors in H (Dirac 1925, p. 317).

[31] Dirac did not of course know at this point that Born and Jordan had also succeeded in extending this result to the vanishing of the off-diagonal elements of the commutator, in their case, by appealing to the time independence of the fundamental commutator.

The quantum transcription of the equations of motion are, using Eq. (12.95), and using $\hbar = h/2\pi$ for convenience,[32]

$$\dot{p}_r = -\frac{i}{\hbar}[p_r, H]; \quad \dot{q}_r = -\frac{i}{\hbar}[q_r, H]. \tag{12.106}$$

These equations are not explicitly written by Dirac, who curiously continues to employ Poisson brackets and their properties, specifically the bilinearity and Leibniz properties, and the cyclic Jacobi identity:

$$\{x+y, z\} = \{x, z\} + \{y, z\}, \tag{12.107}$$

$$\{xy, z\} = \{x, z\}y + x\{y, z\}, \tag{12.108}$$

$$\{x, y, z\} \equiv \{\{x, y\}, z\} + \{\{y, z\}, x\} + \{\{z, x\}, y\} = 0. \tag{12.109}$$

These equations can be verified starting from the definition of Poisson brackets, although Eq. (12.109) requires some effort. In contrast, as Dirac points out, the Jacobi identity is much more easily checked if the Poisson brackets "are replaced by the differences of Heisenberg products $(xy - yx)$" (Dirac 1925, p. 317). These algebraic properties, valid both for Poisson brackets and for commutators, are sufficient to extend the equations of motion (either the classical Eq. (12.105) or the quantum Eq. (12.106)) to any function $x(t)$ of the phase space variables (p_r, q_r). Thus, the classical formula $\dot{x} = \{x, H\}$ takes the quantum form (not written by Dirac)

$$\dot{x} = -\frac{i}{\hbar}[x, H] = \frac{2\pi i}{h}[H, x], \tag{12.110}$$

which we recognize as Eq. (12.22) of Born and Jordan, with the quantity x replaced by $\mathbf{g}$ and H by $\mathbf{W}$ in their notation. Classical conservation principles carry over naturally to the quantum version: any quantity A with vanishing time derivative, $\dot{A} = 0$, clearly satisfies, from the preceding equation,

$$[A, H] = 0. \tag{12.111}$$

Of course, H commutes with itself, and is conserved. Moreover, if we have two conserved quantities A_1 and A_2, satisfying $[A_1, H] = [A_2, H] = 0$, then it is a simple consequence of

[32] The absorption of the ubiquitous factor of 2π into Planck's constant became standard after Dirac's follow-up paper, in which he continues to write h, defined as "a real universal constant, equal to $(2\pi)^{-1}$ times the usual Planck's constant" (Dirac 1926a, p. 418). At some later point, a bar was introduced to distinguish the two, $\hbar = h/2\pi$. As noted, we will help ourselves to either version, with the aim of simplifying formulas where possible.

the Jacobi identity (Eq. (12.109), with Poisson brackets replaced by commutators) that their commutator is also conserved,

$$[[A_1, A_2], H] = 0 \Rightarrow \frac{d}{dt}[A_1, A_2] = 0. \qquad (12.112)$$

Indeed, this was Poisson's original motivation for introducing the brackets carrying his name (see Section A.2.2).

For any classical multi-periodic system expressible in action/angle variables (or, in the language of Dirac's time, "uniformizing" variables) $\mathcal{J}_r, w_r, r = 1, 2, \ldots, u$, one can define a new set of canonical conjugate variables,[33]

$$\xi_r = \sqrt{\frac{\mathcal{J}_r}{2\pi}}\, e^{2\pi i w_r}, \quad \eta_r = -i\sqrt{\frac{\mathcal{J}_r}{2\pi}}\, e^{-2\pi i w_r}. \qquad (12.113)$$

These complex canonical variables are closely related to real canonical variables (involving sines and cosines) $\xi^0 = -\frac{1}{\sqrt{2}}(\eta + i\xi)$ and $\eta^0 = \frac{1}{\sqrt{2}}(\xi + i\eta)$, originally introduced by Poincaré, and used by Born in his discussion of the canonical perturbation of a degenerate system (1925, p. 317).

Classically, multiplying ξ_r into the Fourier expansion of a multiply periodic quantity, such as Eq. (12.70),

$$x(t)\left(= \sum_{\vec{\tau}} X_{\vec{\tau}}(\vec{\mathcal{J}})e^{2\pi i \vec{\tau}\cdot\vec{w}}\right) \rightarrow \xi_r x(t), \qquad (12.114)$$

evidently corresponds to a shift upward by unity of the τ_r index in the complex exponential. The quantum correlate of this amounts to saying that the associated coordinate array element $\xi_r(\vec{n}\vec{m})$ is zero unless $n_r = m_r + 1$, with no alteration in the other components $s \neq r, n_s = m_s$. Similarly, $\eta_r(\vec{n}\vec{m})$ is zero unless $n_r = m_r - 1$, with $n_s = m_s$ for $s \neq r$.

Henceforth, to simplify the notation, we drop the subscripts r, s, etc., and focus our attention on a single separated canonical pair, e.g., the rth one, with r fixed, as the canonical transformations we use only involve a single canonical pair $\mathcal{J}_r, w_r$, and the associated quantum transitions only affect a single quantum number (among the u numbers $n_1, n_2, \ldots, n_u$ which define a stationary state).

In the final section of his paper, entitled "The stationary states," Dirac turns to the question of the quantized energies of a classically multiply periodic system. He derives the Bohr frequency condition exactly along the lines described previously for the Born–Jordan paper, by taking the (nn) element of $Hq - qH = i\hbar\dot{q}$. He then turns to the question of the quantized stationary state energies (and action variable values) for particular

[33] The reader may easily verify that ξ_r, η_r are canonical, either by evaluating their Poisson brackets with $w_r, \mathcal{J}_r$, or more directly, evaluating the Jacobian $\partial(\xi_r, \eta_r)/\partial(w_r, \mathcal{J}_r) = 1$.

dynamical systems, for example, the simple harmonic oscillator. The calculations are done using the ξ, η variables defined above.

Since the only non-zero elements correspond to $n = m + 1$ (resp. $n = m - 1$) for $\xi(nm)$ (resp. $\eta(nm)$), we have

$$\begin{aligned}(\xi\eta)(nn) &= \sum_k \xi(nk)\eta(kn) = \xi(n, n-1)\eta(n-1, n) \\ &= \eta(n-1, n)\xi(n, n-1) = \sum_k \eta(n-1, k)\xi(k, n-1) \\ &= (\eta\xi)(n-1, n-1). \end{aligned} \tag{12.115}$$

On the other hand, the (nn) diagonal element of the commutator of ξ and η is just $ih/2\pi$ times their Poisson bracket, which is 1, so

$$(\xi\eta)(nn) = (\eta\xi)(nn) + i\frac{h}{2\pi}, \tag{12.116}$$

or, using Eq. (12.115), with $n \to n+1$,

$$(\xi\eta)(nn) = \xi\eta(n+1, n+1) + i\frac{h}{2\pi}, \tag{12.117}$$

yielding the recursion relation

$$(\xi\eta)(nn) = (\xi\eta)(n-1, n-1) - i\frac{h}{2\pi}, \tag{12.118}$$

which is solved by setting

$$(\xi\eta)(nn) = -i\frac{h}{2\pi}(n + \text{const.}). \tag{12.119}$$

If there is a lowest state, conventionally assigned quantum number 0, such that $\xi(n.n-1)$ vanishes when $n = 0$, then by Eq. (12.115) we must take the constant here to be zero, whence

$$(\xi\eta)(nn) = -i\frac{h}{2\pi}n. \tag{12.120}$$

This then gives, using Eq. (12.116),

$$(\eta\xi)(nn) = -i\frac{h}{2\pi}(n+1). \tag{12.121}$$

It may already have occurred to the reader that the definitions Eq. (12.113) are immediately problematic once we go over to the quantum theory, as they involve products of non-commuting quantities (namely, the action $\mathcal{J}_r$ and angle w_r), in which case ordering ambiguities necessarily arise. For example, if we define the quantum η variable with the complex exponential on the left (with $\xi = \sqrt{\frac{\mathcal{J}}{2\pi}}e^{2\pi i w}$ as before),

$$\eta = -ie^{-2\pi i w}\sqrt{\frac{\mathcal{J}}{2\pi}}, \tag{12.122}$$

then we immediately find $\xi\eta = -i\mathcal{J}/2\pi$, so, using Eq. (12.120), the action variable is quantized as $\mathcal{J} = nh$, the familiar rule from the old quantum theory. Of course, $\xi\eta$ is classically indistinguishable from $\eta\xi$ or $(\xi\eta + \eta\xi)/2$. As Dirac points out, which of these versions to use requires "a detailed investigation of any particular dynamical system" (1925, p. 318). In fact, for the classical simple harmonic oscillator, where the coordinate and momentum operators are related to the action $\mathcal{J}$ and angle w variables by

$$q = \sqrt{\frac{\mathcal{J}}{m\pi\omega}}\sin(2\pi w), \quad p = \sqrt{\frac{m\mathcal{J}\omega}{\pi}}\cos(2\pi w), \tag{12.123}$$

one finds that the classical quantity a, and its complex conjugate $a^\star$, defined by[34]

$$a \equiv \sqrt{\frac{m\omega\pi}{h}}q + i\sqrt{\frac{\pi}{m\omega h}}p, \quad a^\star = \sqrt{\frac{m\omega\pi}{h}}q - i\sqrt{\frac{\pi}{m\omega h}}p, \tag{12.124}$$

manifestly have no ordering problems when interpreted quantum mechanically, but are simply related to Dirac's ξ, η:

$$a = -\sqrt{\frac{2\pi}{h}}\eta, \quad a^\star = -i\sqrt{\frac{2\pi}{h}}\xi, \tag{12.125}$$

a fact that is not explicit in Dirac's paper.

With this realization, one easily finds that, to obtain the Hamiltonian function $H = p^2/2m + m\omega^2 q^2/2$ in the quantum theory (with p, q in Eq. (12.124) non-commuting), one must write

$$H = \frac{1}{2}(aa^\star + a^\star a)\hbar\omega = \frac{i\omega}{2}(\xi\eta + \eta\xi), \tag{12.126}$$

[34] The reader schooled in modern quantum mechanics may recognize here the appearance of the famous raising and lowering operators, which implement transitions up or down of the quantum number by unity.

which settles the issue of how to define the ordering for the case of a simple harmonic oscillator. The stationary state energies are then given as the (nn) elements of H, namely, using Eqs. (12.120) and (12.121),

$$H(nn) = \hbar\omega\left(n + \frac{1}{2}\right), \tag{12.127}$$

corresponding to half-integral quantization of the action variable, $\mathcal{J} = (n+1/2)h$. Dirac appears not to realize the relation of his ξ, η variables to the underlying q, p, so he simply assumes the ordering $((\xi\eta + \eta\xi)/2)$ to reproduce Heisenberg's result, which contains the zero-point energy.

Dirac's remarkable paper concludes with a paragraph that points to the further development of the theory. We reproduce it here in its entirety, with some comments:

> Up to the present we have considered only multiply periodic systems. There does not seem to be any reason, however, why the fundamental equations (11) [our Eq. (12.95)] and (12) [our Eqs. (12.100) and (12.101)] should not apply as well to non-periodic systems, of which none of the constituent particles go off to infinity, such as a general atom.[35] One would not expect the stationary states of such a system to classify,[36] except perhaps where there are pronounced periodic motions,[37] and so one would have to assign a single number n to each stationary state according to an arbitrary plan.[38] Our quantum variables [Born and Jordan's matrices] would still have harmonic components, [the (nm) element of any kinematical quantity has the time dependence $\exp i\omega(nm)t$] each related to two ns, and Heisenberg multiplication [matrix multiplication] could be carried out exactly as before. There would thus be no ambiguity in the interpretation of equations

[35] This extrapolation is a rather bold one, given that the whole derivation leading to the Poisson–commutator connection is so firmly based on the correspondence-limit analysis of multiply periodic systems, as we pointed out previously. Of course, Dirac's equations (11) and (12) no longer carry any direct reference to an underlying multiply periodic dynamics—they can be applied to any system possessing a Hamiltonian dynamics specified in terms of conjugate coordinates and momenta. They can be regarded as the self-supporting *arch* from which the previous supporting structure, or *scaffold*, of multiply periodic dynamics, has been discarded. Dirac restricts himself to bound systems ("of which none of the constituent particles go off to infinity") so that the stationary states form a discrete, enumerable set, which can be labeled by integer quantum number(s).

[36] Dirac presumably means here that the stationary states would not be specified by u quantum numbers $n_1, n_2, \ldots, n_u$ for a system of u degrees of freedom, as in the multiply periodic case. In fact, the configurations that serve as the starting point of approximative schemes used in modern atomic theory, such as Hartree–Fock, do begin with the assigning of a set of four quantum numbers (principal quantum number, orbital angular momentum, magnetic orbital, and spin quantum numbers) for each electron, subject, of course, to exclusion requirements, which is actually rather similar to the situation in multiply periodic systems, which real atoms most decidedly are not.

[37] Dirac here comes perilously close to relapsing into the mode of thinking of the old quantum theory, in which electrons execute visualizable orbital "motions".

[38] The bound stationary states of any atom certainly form an enumerable set, so one *could* label them with a single discrete index n, although this is not necessary.

(12) [our Eqs. (12.100) and (12.101)] or of the equations of motion.[39] (Dirac 1925, p. 320).

Dirac's paper ends with an acknowledgement of gratitude to R. H. Fowler "for many valuable suggestions in the writing of this paper" (p. 320). Fowler, who at this point had added to his previous interest in stellar astrophysics and statistical mechanics an active participation in theoretical atomic spectroscopy,[40] can be regarded as playing a roughly analogous role with Dirac as Pauli had played with Heisenberg. The analogy is *very* rough: Fowler's knowledge of the deeper issues in quantum theory simply cannot be compared with Pauli's. However, as a Fellow of the Royal Society, he could ensure that Dirac's paper was promptly accepted and published at the beginning of December 1925, roughly simultaneously with the paper of Born and Jordan.

12.3 The "Three-Man-Paper" of Born, Heisenberg, and Jordan

Immediately after handing the final version of his *Umdeutung* paper to Born in Göttingen in mid-July 1925, with instructions to forward the article for publication only if Born found it satisfactory, Heisenberg departed for visits to Leyden and Cambridge. By the beginning of August Heisenberg had returned to his home town of Munich, which would serve as his base for various hiking excursions in the Alps. He had by this time become aware of the rapid development of his theory Born and Jordan were undertaking in Göttingen. He could not return immediately to Göttingen after his summer vacation as he still had to spend time in Copenhagen to fulfill the terms of his Rockefeller fellowship, which required his presence at the Niels Bohr Institute. Around September 11, Heisenberg was back in Copenhagen, and fully engaged in working on the new theory, and with the new methods of matrix algebra that Born and Jordan had introduced (details of which he had already asked Jordan to send him in late August).

Once in Copenhagen, Heisenberg had no intention of letting the grass grow under his feet. Within a few days, in a flurry of letters to Jordan (September 13, 16, and 21, 1925; AHQP microfilm) and Pauli (September 18, 1925; Pauli 1979, Doc. 98), he announced important new results. He informed Jordan that he had been able to reproduce the dispersion theory of Kramers, and its extension to inelastic scattering as worked out in the Kramers–Heisenberg paper, from the new formalism, and in the letter of September 21, the generalization of the canonical quantum commutator relation ($[p, q] = -i\hbar$) to

[39] Ambiguity in the ordering of ps and qs within a single term in the quantum Hamiltonian is the only potential source of difficulty, and Dirac is fortunate here that such ambiguities do not appear in the physically important terms entering atomic Hamiltonians such as those for the electron–nuclear Coulomb potential, inter-electron electrostatic repulsion, spin–orbit interaction, electric (Stark) or magnetic (Zeeman) perturbations, etc.

[40] Fowler (1925a, 1925b) published two papers in this area, exploring line intensities of band spectra and summation rules for intensities (cf. Jähnert 2019, p. 273).

multi-dimensional systems. In the much longer letter to Pauli of September 18, Heisenberg spells out in considerable detail the basic postulates of the theory (now referred to ubiquitously as *Quantenmechanik*—quantum mechanics) as formulated in the Born–Jordan paper, and showed how energy conservation, the Bohr frequency condition, and the equations of motion (formally identical to the classical ones) arise naturally from the quantum condition

$$\mathfrak{p}\mathfrak{q} - \mathfrak{q}\mathfrak{p} = \frac{h}{2\pi i}\mathfrak{e}, \tag{12.128}$$

where the unit matrix $\mathfrak{e}$ is rather confusingly referred to as the "unit determinant."[41]

Heisenberg's letter to Pauli concludes with a section on perturbation theory, in which he lays out a matrix analog of the classical perturbation theory (for multiply periodic systems separable in action/angle variables) which had been developed three years earlier by Born and Pauli (1922). Here, the objective was to calculate the shifts in the quantized energies of a system defined by a Hamiltonian expandable in a small parameter λ:

$$\mathfrak{H} = \mathfrak{H}_0 + \lambda\mathfrak{H}_1 + \lambda^2\mathfrak{H}_2 + \ldots \tag{12.129}$$

In the classical case, the corresponding perturbative expansion of classical orbit energies of multiply periodic systems had been done by expanding a generating function of type $F_2(q, P) = F_2(w, \mathcal{J})$ (see Section A.2.1), which effects a transformation from an unperturbed set of action/angle variables $w_k^0, \mathcal{J}_k^0$ to new variables $w_k, \mathcal{J}_k$ in terms of which (order by order in the expansion in λ) the Hamiltonian would be a function only of the new action variables $\mathcal{J}_k$. Heisenberg, in his letter to Pauli, asserts that the quantum analog of this procedure amounts to subjecting the fundamental coordinate and momentum quantum variables to a *canonical transformation*, which in the quantum case amounts to a matrix similarity transformation:

$$\mathfrak{q}_0 \quad \rightarrow \quad \mathfrak{q} = e^{\mathfrak{S}}\mathfrak{q}_0 e^{-\mathfrak{S}} \tag{12.130}$$

$$\mathfrak{p}_0 \quad \rightarrow \quad \mathfrak{p} = e^{\mathfrak{S}}\mathfrak{p}_0 e^{-\mathfrak{S}} \tag{12.131}$$

where the matrix quantity $\mathfrak{S}$ has an expansion in powers of λ

$$\mathfrak{S} = \mathfrak{S}_0 + \lambda\mathfrak{S}_1 + \lambda^2\mathfrak{S}_2 + \ldots \tag{12.132}$$

with the $\mathfrak{S}_i$ chosen order by order so that the transformed Hamiltonian is diagonal in each order of λ. The diagonal matrix elements then give the shifts of the quantized

[41] It appears that it was only after his return to Göttingen in October that Heisenberg was able to delve into the technology (and correct terminology) of matrix algebra, by studying Bôcher's (1911) *Introduction to Higher Algebra*, and Ch. 1 of Courant and Hilbert's (1924) text on mathematical methods of physics (Mehra and Rechenberg 1982c, pp. 93–94). Heisenberg uses both bold type and Fraktur notation to indicate matrix quantities in this letter, which requires an impressive level of calligraphy.

energies to each order of the perturbation theory. Moreover, Heisenberg informed Pauli (without proof) that the change of the elements of the coordinate matrix $\mathbf{q}$ to first order in λ due to Eq. (12.130) corresponded exactly to the zero frequency limit of the Kramers–Heisenberg dispersion formula, while the general dispersion formula (for time varying perturbing field) could also be obtained by a simple modification of the procedure to handle time-dependent perturbations. The perturbative expansion used in the Three-Man-Paper (1926) is slightly different from that given by Heisenberg in his letter to Pauli, which caused some disagreement across the Copenhagen/Göttingen axis. In fact both are valid, and the choice comes down to a matter of convenience.

Over the course of the next month (mid-September to mid-October) the collaboration was carried out by an exchange of letters between Heisenberg in Copenhagen and Born and Jordan in Göttingen, who were still partially preoccupied with writing up the final version of their Two-Man-Paper. It appears that Heisenberg was the first to propose the correct generalization of the quantum condition Eq. (12.128) for systems with several degrees of freedom, which we saw was independently obtained by Dirac by his correspondence principle "Poisson brackets to commutators" argument (see Section 12.2). In any case, a justification for the commutators in Eqs. (12.100)–(12.101), independent of correspondence considerations, can be found by considering the two competing definitions of matrix differentiation given in the Three-Man-Paper (which became a second source of disagreement between the authors before its final preparation), which was probably the basis for Heisenberg's proposal of the extended commutation rules for multidimensional systems.

We recall that Born and Jordan had given a rather peculiar definition, Eq. (12.7), of the derivative of a matrix function of matrix arguments, in terms of the trace of the function being differentiated. The result was that a strange cyclic reordering of the matrices appeared in the result. For example, according to Born and Jordan, if we consider a function

$$F(\mathbf{p},\mathbf{q}) = \mathbf{p}^2\mathbf{q}^2\mathbf{p}\mathbf{q}^3\mathbf{p}^3\mathbf{q}, \tag{12.133}$$

then the derivative with respect to the coordinate matrix $\mathbf{q}$ is

$$\frac{\partial F(\mathbf{p},\mathbf{q})}{\partial \mathbf{q}} \equiv 2\mathbf{q}\mathbf{p}\mathbf{q}^3\mathbf{p}^3\mathbf{q}\mathbf{p}^2 + 3\mathbf{q}^2\mathbf{p}^3\mathbf{q}\mathbf{p}^2\mathbf{q}^2\mathbf{p} + \mathbf{p}^2\mathbf{q}^2\mathbf{p}\mathbf{q}^3\mathbf{p}^3. \tag{12.134}$$

This definition turned out to be useful in formulating a matrix version of the extremal action principle of classical mechanics allowing one to derive the Hamiltonian equations from the extremization of (a matrix version of) $\int(p\dot{q} - H(p,q))dt$. However, as Heisenberg pointed out to Born in a letter in early October, this definition is not compatible with the need to maintain the equations of motion when canonical transformations are made from one set of conjugate variables, $(\mathbf{p},\mathbf{q})$, to another, $(\mathbf{P},\mathbf{Q})$. In order for this to work, one needs to maintain at all times the correspondences Eqs. (12.41)–(12.42) between matrix derivatives and commutators, thereby guaranteeing the validity of

Eqs. (12.43)–(12.44). The basic commutation rule $[\mathbf{p}, \mathbf{q}] = -i\hbar\mathbf{1}$ means that, returning to our example above,

$$\frac{i}{\hbar}[\mathbf{p}, \mathbf{p}^2\mathbf{q}^2\mathbf{p}\mathbf{q}^3\mathbf{p}^3\mathbf{q}] = 2\mathbf{p}^2\mathbf{q}\mathbf{p}\mathbf{q}^3\mathbf{p}^3\mathbf{q} + 3\mathbf{p}^2\mathbf{q}^2\mathbf{p}\mathbf{q}^2\mathbf{p}^3\mathbf{q} + \mathbf{p}^2\mathbf{q}^2\mathbf{p}\mathbf{q}^3\mathbf{p}^3, \tag{12.135}$$

exactly in accordance with the *natural* definition

$$\frac{\partial F(\mathbf{p}, \mathbf{q})}{\partial \mathbf{q}} \equiv \lim_{\varepsilon \to 0} \frac{F(\mathbf{p}, \mathbf{q} + \varepsilon\mathbf{1}) - F(\mathbf{p}, \mathbf{q})}{\varepsilon}, \tag{12.136}$$

but in disagreement with Eq. (12.134). The point is simply that the commutation of the $\mathbf{p}$ matrix through a product of $\mathbf{p}$s and $\mathbf{q}$s results in a sum of terms where each $\mathbf{q}$ matrix is removed in turn, leaving the other matrices in the product *in situ*, which is exactly what Eq. (12.136) accomplishes. Note that we have introduced a bold division bar in the left-hand side (as in the Three-Man-Paper) to distinguish between the original Born–Jordan and the "natural" definition Eq. (12.136). In effect, the earlier definition of Born and Jordan (which agreed with the "natural" one for simple Hamiltonians of the form $F(p) + G(q)$) was superseded for all practical purposes by the one championed by Heisenberg, which was exclusively employed throughout the Three-Man-Paper.

Given the various disagreements, both stylistic and substantive, between the authors, it is not surprising that Born wrote to Bohr asking him to allow Heisenberg to return to Göttingen a little earlier than prescribed by his fellowship requirements, especially as Born had previously committed to visit the Massachusetts Institute of Technology in the US for the winter semester 1925–1926, and was planning to leave at the beginning of November. Heisenberg returned to Göttingen in the third week of October, which allowed for just two weeks of intense consultation with his two coauthors in preparing a final draft of the paper. The essential content of the paper was agreed upon (starting from a draft produced by Jordan) by the time of Born's departure on November 2, but there remained a certain amount of rewriting, with Heisenberg doing the major part of the proof-reading (Mehra and Rechenberg 1982c, p. 101). The completed paper, entitled "On quantum mechanics. II," was received by *Zeitschrift für Physik* on November 16, 1925. As the Roman numeral in the title indicates, this Three-Man-Paper was conceived as the sequel to Born and Jordan's (1925b) Two-Man-Paper.

The Three-Man-Paper consists of an introduction and four chapters, of which the first three amplify and deepen the formal framework constructed by Born and Jordan to implement the physical ideas of Heisenberg's (1925c) *Umdeutung* paper, while chapter 4 is devoted to physical applications of the theory. The major new conceptual contributions presented in the first three chapters were: (a) the extension of the formalism to systems of arbitrarily many degrees of freedom; (b) the introduction of a quantum analog of the classical notion of canonical transformations; (c) the application of quantum canonical transformations to the development of a quantum-theoretical perturbation theory (including the treatment of degenerate systems); and, finally, (d) the connection of the formalism with the theory (due primarily to Hilbert and Hellinger) of quadratic forms of infinitely many variables, which allowed a unified treatment of systems with discrete and continuous spectra.

The physical applications of chapter 4 were: (a) the development of a full matrix-mechanical theory of angular momentum in three dimensions; (b) a preliminary application of the new angular momentum technology to the Zeeman effect; and (c) a derivation in a simple one dimensional model (a quantized string) of Einstein's (1909a, 1909b) formula for the mean square energy fluctuation in black-body radiation (cf. Section 3.4.3). This last result was due entirely to Jordan, and corresponds to the first application of the new methods to a system with infinitely many degrees of freedom. It is, in fact, the first construction of what we now call a "quantum field", albeit an extremely simple one.

The motivations adduced in the introduction for the new theory echo motifs expressed previously in the *Umdeutung* and Born–Jordan paper (of which the present paper is presented as the concluding second part). The Machian/positivistic reliance on "observable quantities" returns as the motivation for adopting an abstract schema at the admitted sacrifice of intuitive transparency:

> The point of departure for the desired theory was the conviction that it would not be possible to gain control over the difficulties which have met us at every point in recent years before we had at our disposal a mathematical system of relations between (in principle) observable quantities which was of similar simplicity and unity to the system provided by classical mechanics. Such a system of quantum-theoretical relationships between observable quantities will certainly display the defect, in comparison with the quantum theory used up to now [the old quantum theory], that it cannot be directly interpreted in a geometrically intuitive way, as the motions of electrons cannot be described in terms of the current concepts of space and time (Born, Heisenberg, and Jordan 1926, p. 322).[42]

Despite the lack of intuitive appeal, the authors suggest that the new theory displays a feature which at first sight seems contradictory to its admitted reliance on a non-classical description of motion. Namely, the basic guiding principle which underlies both the preceding two papers (the *Umdeutung* and Born–Jordan papers) is identified as an attempt to represent a quantum mechanics, which

> should be as similar as possible to classical mechanics, as one could possibly hope. In this context we merely recall the validity of the conservation principles of energy and angular momentum, and of the form of the equations of motion (Ch. 1, sec. 2). This similarity of the new theory with classical theory also precludes any question of a separate correspondence principle alongside the new theory: rather, the latter can itself be regarded as an exact formulation of Bohr's correspondence considerations. In the further development of the theory, an important task will lie in the closer investigation of the nature of this correspondence and in the description of the manner in which symbolic quantum geometry goes over into visualizable classical geometry (p. 322).

[42] Unless noted otherwise, all page references in the remainder of this chapter are to the English translation of the Three-Man-Paper in van der Waerden (1968).

Evidently, the procedure previously employed in the old quantum theory, where classical theory (in the correspondence regime of large quantum numbers) was used as a starting point for inducing features of the as yet unknown underlying quantum theory, was now to be reversed: the new theory, thought of as a *precise* statement of quantum principles, would be the starting point, and one would try to see how a classical world could emerge deductively from this underlying mechanics.

The optimism of the first two paragraphs of the introduction gives way in the third and fourth to the admission that the new approach has not yet been able to resolve various central problems of atomic physics. Singled out for attention are the proper description of collision processes and the complex of questions surrounding the anomalous Zeeman effect. The recent proposal of Uhlenbeck and Goudsmit (1925) that an intrinsic spin angular momentum be attributed to each electron is mentioned as providing at least a qualitative solution to the problems arising both in the treatment of complex multiplets and in the Zeeman splittings, although it is admitted that it was not yet clear that an adequate quantitative treatment of these phenomena would be possible on the basis of the new theory. The treatment of angular momentum in chapter 4 of the paper is indicated here as strengthening the "hope of finding a quantitative interpretation at some later date" (p. 324). The introduction to the Three-Man-Paper ends with a description of the treatment of a "well-known statistical problem" (ibid.) with the new methods: the quantization of the standing wave modes of cavity radiation (in an enclosure surrounded by reflecting walls). The new treatment of this problem is claimed to obtain the correct value of the mean square energy fluctuation, which Einstein (1909a, 1909b) had shown in 1909 to consist of a sum of two terms, understandable as due to a particle or wave interpretation of the radiation, respectively. Here, the claim is made that both terms (which together are necessary to assure compatibility with the Planck black-body law) arise naturally once the calculation is performed according to the rules of the new matrix mechanics (Duncan and Janssen 2008).

12.3.1 First chapter: systems of a single degree of freedom

The presentation of the new theory begins in chapter 1, section 1, entitled "Foundational principles" (*Grundprinzipien*). There is no beating around the bush this time: the essentials of the new theory are given succinctly and directly in four postulates (see points I through IV on pp. 325–327):

1. Any quantum-theoretical quantity **a** (be it a coordinate, a momentum, or an arbitrary function of both) is represented by the collection of quantities

$$a(nm)\, e^{2\pi i\nu(nm)t}, \tag{12.137}$$

where $a(nm)$ is a complex number specified for all pairs of stationary states n, m, and $\nu(nm)$ the frequency associated with a radiative quantum transition from state n to state m. The complex exponential time dependence of each array amplitude is identical for all quantities; the assembly of complex numbers indicated in Eq. (12.137) can be regarded as a matrix.

2. Functions of matrix quantities generated algebraically from individual quantum-theoretical quantities are formed by the application of the usual rules of matrix addition and multiplication.
3. Two possible definitions of (partial) differentiation of a matrix function with respect to any one of its (matrix) arguments are given: they correspond to the definitions Eq. (12.136) and Eq. (12.7). In fact, only the first of these (distinguished by the boldfont) will be employed in the rest of the Three-Man-Paper, for the reasons discussed previously.
4. The interpretation of the product of non-commuting quantities is to be rendered unambiguous by the introduction of the fundamental quantum relation

$$\mathbf{pq} - \mathbf{qp} = \frac{h}{2\pi i}\mathbf{1}. \tag{12.138}$$

This equation is in fact the only point in the enumeration of the fundamental principles of the theory at which Planck's constant enters. As the authors comment, the usual commutative property of kinematic variables in classical mechanics is seen to return if we set $h = 0$ in Eq. (12.138).

The first section of the paper concludes with the observation that differentiation (of the first kind) with respect to coordinate (resp. momentum) quantities is equivalent to commutation with momentum (resp. coordinate) quantities (cf. Eqs. (12.41)–(12.42)):

$$\mathbf{fq} - \mathbf{qf} = \frac{h}{2\pi i}\frac{\partial \mathbf{f}}{\partial \mathbf{p}} \tag{12.139}$$

$$\mathbf{pf} - \mathbf{fp} = \frac{h}{2\pi i}\frac{\partial \mathbf{f}}{\partial \mathbf{q}} \tag{12.140}$$

and section 2 of the paper ("The canonical equations, energy principle and frequency condition") reprises the results obtained previously by Born and Jordan for the quantum matrix transcription of the classical Hamilton formalism. The matrix form of Hamilton's equations, Eqs. (12.23), together with the time dependence generated by the diagonal matrix **W** associated with the quantum transition frequencies (cf. Eqs. (12.17)–(12.22)), are shown to lead to the energy (conservation) principle $\dot{\mathbf{H}} = 0$, and the conclusion that the Hamiltonian matrix must be diagonal, with diagonal elements equal to those of the **W** matrix, up to a shift, $H_n = W_n +$ const. The canonical equations are then seen to follow directly from Eqs. (12.41)–(12.42), which in turn derive from the fundamental commutator relation Eq. (12.138).

Section 3, entitled "Canonical transformations," breaks new ground, as it introduces concepts which would prove indispensable for the development of a practical perturbation theory, vastly more efficient than Heisenberg's awkward recursive computations in the *Umdeutung* paper. The introduction of a quantum version of the canonical transformations of classical theory is also in harmony with the authors' avowed desire, expressed in the introduction, to produce a quantum theory "as similar as possible to classical mechanics."

Canonical transformations can be defined in various ways in classical Hamiltonian theory. Starting with a classical system (of one degree of freedom) defined in terms of a momentum/coordinate pair (p, q), one can define a canonical transformation as a change to a new pair of phase space variables $(P(p, q), Q(p, q))$ which leaves the form of the equations of motion invariant. Thus if we re-express the Hamiltonian function $H(p, q)$ in terms of the new variables, so that $\tilde{H}(P, Q) = H(p, q)$, then the Hamiltonian equations should take the same form as the original ones:

$$\frac{\partial H}{\partial p} = \dot{q}, \ \frac{\partial H}{\partial q} = -\dot{p} \Rightarrow \frac{\partial \tilde{H}}{\partial P} = \dot{Q}, \ \frac{\partial \tilde{H}}{\partial Q} = -\dot{P} \tag{12.141}$$

By this point it should be apparent that this will only work in the quantum context if the matrices associated with the new canonical variables satisfy the fundamental commutation relation Eq. (12.138),[43]

$$\mathbf{PQ} - \mathbf{QP} = \frac{h}{2\pi i}\mathbf{1}. \tag{12.142}$$

At the quantum level, an obvious way to preserve the form of the $(\mathbf{p}, \mathbf{q})$ commutation relation is to construct the new canonical quantities as matrix similarity transforms of the old ones:

$$\mathbf{P} = \mathbf{SpS}^{-1}, \ \mathbf{Q} = \mathbf{SqS}^{-1}, \tag{12.143}$$

which then implies

$$\begin{aligned} [\mathbf{P}, \mathbf{Q}] &= \mathbf{SpS}^{-1} \cdot \mathbf{SqS}^{-1} - \mathbf{SqS}^{-1}\mathbf{SpS}^{-1} \\ &= \mathbf{S}[\mathbf{p}, \mathbf{q}]\mathbf{S}^{-1} \\ &= \frac{h}{2\pi i}\mathbf{S1S}^{-1} = \frac{h}{2\pi i}\mathbf{1}. \end{aligned} \tag{12.144}$$

The matrix $\mathbf{S}$, the only constraint on which (for the time being) is that it be invertible, would then represent the quantum analog of the generating function of the canonical transformation in classical theory. Any function $\mathbf{f}$ of the original canonical coordinates

[43] This is also clear from the alternative classical definition of canonical transformations, as changes of variable leaving the Poisson brackets invariant. With Dirac's transcription of Poisson brackets to commutators, the requirement stated here also follows directly. However, the notion of Poisson brackets, and their correspondence to commutators is nowhere explicitly stated in the Three-Man-Paper, although Heisenberg mentioned the similarity of the generalized commutation relations for several degrees of freedom to the "bracket symbols . . . of classical mechanics," in a letter to Jordan dated September 29, 1925 (AHQP, microfilm).

$(\mathbf{p},\mathbf{q})$ expandable as a (possibly infinite) sum of products of powers of $(\mathbf{p},\mathbf{q})$ would then be similarity transformed by the same matrix to the same function of the new coordinates,

$$\mathbf{Sf}(\mathbf{p},\mathbf{q})\mathbf{S}^{-1} = \mathbf{f}(\mathbf{SpS}^{-1},\mathbf{SqS}^{-1}) = \mathbf{f}(\mathbf{P},\mathbf{Q}) \tag{12.145}$$

The authors *presume*, but admittedly do not prove, that the transformations Eq. (12.143) are in fact the most general canonical transformations allowed in the quantum theory.

The practical significance of the matrix similarity definition of a canonical transformation given here for actual problems in quantum theory arises from the fact, which by now must have been fully appreciated by all three authors of the Three-Man-Paper, that such similarity transformations are exactly what is needed to effect the conversion of a large class of matrices, initially not of diagonal form, to purely diagonal matrices. For the energy matrix in particular, the ability to find canonical coordinates in terms of which the energy matrix was diagonal was the prerequisite to identifying the stationary state energies of the system, which are just the diagonal elements of the transformed matrix. In other words, if the Hamiltonian matrix $\mathbf{H}$ of a system is initially expressed in terms of variables $(\mathbf{p}_0,\mathbf{q}_0)$, such that $\mathbf{H}(\mathbf{p}_0,\mathbf{q}_0)$ is not diagonal, the solution of the system at the quantum level amounts to finding a transformation matrix $\mathbf{S}$ such that

$$\mathbf{p} = \mathbf{Sp}_0\mathbf{S}^{-1}, \quad \mathbf{q} = \mathbf{Sq}_0\mathbf{S}^{-1}, \tag{12.146}$$

and

$$\mathbf{H}(\mathbf{p},\mathbf{q}) = \mathbf{SH}(\mathbf{p}_0,\mathbf{q}_0)\mathbf{S}^{-1} = \mathbf{W} \tag{12.147}$$

is a diagonal matrix. The (real) numbers W_n on the diagonal of the transformed matrix are then the desired stationary state energies of the system.

In almost all cases the canonical transformation effecting the diagonalization of the Hamiltonian matrix is not analytically obtainable. Instead, one has recourse to approximations, usually involving the formal expansion of $\mathbf{S}$, and of the desired stationary state energies W_n, in powers of a small coupling constant λ. In the *Umdeutung* paper the expansion of the energies in powers of the anharmonic coupling was done by first obtaining the coupling expansion of the coordinate matrix $\mathbf{q}$, which satisfies the equation of motion arising from the full Hamiltonian $\mathbf{H}$ by an awkward recursive procedure that implemented the equations of motion at each order. Section 4 presents a new procedure for obtaining the matrix elements of $(\mathbf{q},\mathbf{p})$ and the diagonal matrix elements W_n, which is extremely efficient and greatly reduces the effort needed to obtain the desired perturbative expansions. Indeed, at first order, the results for the anharmonic oscillator, and even the Kramers–Heisenberg dispersion relations, become a matter of a few lines of algebra.

One begins, as usual, with a splitting of the full energy (matrix) function of the theory $\mathbf{H}$ into (a) a part $\mathbf{H}_0$ which is, one hopes, "large," that is, quantitatively the most significant, but analytically solvable, and (b) a term, or series of terms $\lambda\mathbf{H}_1+\lambda^2\mathbf{H}_2+\ldots$, which are quantitatively smaller, i.e., contain powers of a small coupling λ, but are analytically not solvable (i.e., reducible to diagonal form by an explicit canonical transformation). A similar expansion is made for the desired canonical transformation **S** in Eq. (12.147). Thus, one has

$$\mathbf{H} = \mathbf{H}_0 + \lambda\mathbf{H}_1 + \lambda^2\mathbf{H}_2 + \ldots \tag{12.148}$$

$$\mathbf{S} = 1 + \lambda\mathbf{S}_1 + \lambda^2\mathbf{S}_2 + \ldots \tag{12.149}$$

$$\mathbf{W} = \mathbf{W}_0 + \lambda\mathbf{W}_1 + \lambda^2\mathbf{W}_2 + \ldots \tag{12.150}$$

Of course, the assumption is that at zeroth order in λ, we already possess explicit matrices $(\mathbf{p}_0, \mathbf{q}_0)$ such that $\mathbf{H}_0(\mathbf{p}_0, \mathbf{q}_0) = \mathbf{W}_0$ is diagonal with known diagonal entries. Thus, at lowest order we do not need to perform any canonical transformation, and **S** is simply the unit matrix. The matrices $\mathbf{H}_1, \mathbf{H}_2, \ldots$ are written as functions of the known $(\mathbf{p}_0, \mathbf{q}_0)$, as these matrices satisfy the fundamental commutation relations Eq. (12.138). Also, for the moment we are assuming that the only time dependence in the perturbing matrices $\mathbf{H}_1(\mathbf{p}_0, \mathbf{q}_0), \mathbf{H}_2(\mathbf{p}_0, \mathbf{q}_0), \ldots$ arises through the known time-dependence implicit in $(\mathbf{p}_0, \mathbf{q}_0)$ via Eq. (12.137). In other words, there is no *explicit* time-dependence term in the Hamiltonian function, arising from, say, an externally imposed oscillating field. This case, obviously essential for dispersion theory, is discussed in due course. In fact, as the total energy is conserved, once we have it in diagonal form the time dependence disappears completely, and must do so for each order in λ separately. Of course, in the individual terms contributing to the energy, time-dependent oscillatory factors (involving the *unperturbed* frequencies $\nu_0(mn)$) will appear.[44]

From Eq. (12.149) it follows that

$$\mathbf{S}^{-1} = 1 - \lambda\mathbf{S}_1 + \lambda^2(\mathbf{S}_1^2 - \mathbf{S}_2) + \ldots. \tag{12.151}$$

Inserting Eqs. (12.148)–(12.151) in Eq. (12.147), and equating the coefficients of each power of λ on the left and right side, one finds a sequence of equations

$$\begin{gathered} \mathbf{H}_0 = \mathbf{W}_0 \\ \mathbf{S}_1\mathbf{H}_0 - \mathbf{H}_0\mathbf{S}_1 + \mathbf{H}_1 = \mathbf{W}_1 \\ \mathbf{S}_2\mathbf{H}_0 - \mathbf{H}_0\mathbf{S}_2 + \mathbf{H}_0\mathbf{S}_1^2 - \mathbf{S}_1\mathbf{H}_0\mathbf{S}_1 + \mathbf{S}_1\mathbf{H}_1 - \mathbf{H}_1\mathbf{S}_1 + \mathbf{H}_2 = \mathbf{W}_2 \end{gathered} \tag{12.152}$$

[44] It may appear counterintuitive that the calculation of the energy shifts due to $\mathbf{H}_1, \mathbf{H}_2$ etc., involves terms that contain only the time dependence of the unperturbed system. Readers familiar with modern quantum mechanics will recognize the use here of what is now called the "interaction picture" (originally due to Dirac), in which the observables (or, in this context, matrices) of the theory are given the time dependence associated with the unperturbed Hamiltonian, with the state vectors carrying the time dependence arising from the perturbing terms. Of course, as the concept of a state vector is still two years in the future, the usage of the interaction picture here is hardly explicit. One can avoid any confusion by simply setting the time t to zero everywhere, eliminating the exponential oscillation factors: as the total energy is conserved, we can calculate it at any time, and once the matrix of **H** is diagonal, to the desired order in λ, we can be sure that the matrix elements give the correct energy at any other time t. The perturbation procedure carried out in this way now goes under the rubric of "time-independent perturbation theory", for obvious reasons.

and, in general, for the coefficient of λ^r one finds

$$\mathbf{S}_r\mathbf{H}_0 - \mathbf{H}_0\mathbf{S}_r + \mathbf{F}_r(\mathbf{H}_0, \ldots, \mathbf{H}_r, \mathbf{S}_1, \ldots, \mathbf{S}_{r-1}) = \mathbf{W}_r. \tag{12.153}$$

Note that all the matrices $\mathbf{H}_1, \mathbf{H}_2, \ldots$ appearing here are specified as functions of, and built out of, the zeroth order canonical quantities $(\mathbf{p}_0, \mathbf{q}_0)$. Hence, they are explicitly known. If Eq. (12.153) allows the determination of $\mathbf{S}_r$ for a given value of r, then at each stage the $\mathbf{F}_r$ is given in terms of already determined matrices, so is known.[45]

To solve for the energy $\mathbf{W}_r$ in the rth order in Eq. (12.153), we first note that the matrix $\mathbf{S}_r\mathbf{H}_0 - \mathbf{H}_0\mathbf{S}_r$ has no diagonal matrix elements:

$$\begin{aligned}(\mathbf{S}_r\mathbf{H}_0 - \mathbf{H}_0\mathbf{S}_r)(nm) &= \sum_p \left(S_r(np)\, W_0(p)\, \delta_{pm} - W_0(n)\, \delta_{np}\, S_r(pm)\right) \\ &= \big(W_0(m) - W_0(n)\big)\, S_r(nm) \\ &= h\nu_0(mn)\, S_r(nm),\end{aligned} \tag{12.154}$$

which vanishes for $m = n$. Thus, the diagonal matrix elements of $\mathbf{W}_r$ on the right-hand side of Eq. (12.153) derive entirely from the matrix $\mathbf{F}_r$ on the left. The effect of time averaging any matrix quantity in matrix mechanics is simply to eliminate the off-diagonal matrix elements (assuming, as here, that $\nu_0(nm) \neq 0$ if $n \neq m$, i.e., that the zeroth order system is non-degenerate), as such matrix elements always contain a non-trivial

[45] The perturbative procedure being followed here has a close analogy in classical perturbation theory, where one seeks a canonical transformation from action/angle variables $(\mathcal{J}_0, w_0)$ (indices suppressed) of the unperturbed Hamiltonian H_0 to action/angle variables $(\mathcal{J}, w)$ of the full Hamiltonian H. The expansion of the latter in $(\mathcal{J}_0, w_0)$ is

$$H = H_0(\mathcal{J}_0) + \lambda H_1(\mathcal{J}_0, w_0) + \lambda^2 H_2(\mathcal{J}_0, w_0) + \ldots$$

It is only when we use $(\mathcal{J}, w)$, the "uniformizing" variables for the full Hamiltonian H, that the action variables are constants, the angle variables are linear in time, and the terms in the expansion of H only depend on the action variable:

$$W(\mathcal{J}) = W_0(\mathcal{J}) + \lambda W_1(\mathcal{J}) + \lambda^2 W_2(\mathcal{J}) + \ldots$$

In the quantum theory, expressing the full Hamiltonian in terms of $\mathcal{J}$ corresponds to diagonalization—knowledge of $\mathcal{J}$, for a non-degenerate system, amounts to knowledge of H_0, and writing all the higher terms as functions of H_0 in matrix terms means putting them in diagonal form. The generating function can be expanded $S = S_0 + \lambda S_1 + \lambda^2 S_2 + \cdots$, with $W_1 = \{H_0, S_1\} + H_1$ (Born and Pauli 1922, p. 140; Born 1925, p. 287). The correspondence with the second equation in Eq. (12.152) is apparent from the Poisson bracket correspondence with commutators (modulo a factor of $\hbar/i$), and of S_1 with the order λ part of the F_2 classical generating function (modulo the same factor—see the web resource, *Time-Dependent Perturbation Theory in the Three-Man-Paper*, for a detailed account of the connection between classical generating functions and their quantum counterparts).

exponentially oscillating term $e^{2\pi i\nu_0(nm)t}$, which time averages to zero. Thus Born *et al.* simply write

$$\mathbf{W}_r = \overline{\mathbf{F}}_r. \tag{12.155}$$

where the bar over **F** indicates a time average. Consequently, the rth order contribution to the nth quantized energy is simply given by the diagonal matrix element $F_r(nn)$. Of course, the task remains of ensuring that the canonical transformation has done its job, and that the left-hand side of Eq. (12.153) is diagonal. To accomplish this, using Eq. (12.154), we merely have to choose

$$S_r(mn) = \frac{F_r(mn)}{h\nu_0(mn)}, \quad m \neq n, \quad S_r(mm) = 0 \tag{12.156}$$

The result for the first-order perturbation of the energy of any stationary state is particularly simple, as we see from the second line of Eq. (12.152) that $\mathbf{F}_1 = \mathbf{H}_1$, so

$$\mathbf{W}_1 = \overline{\mathbf{H}}_1 \Rightarrow W_1(n) = H_1(nn), \tag{12.157}$$

and the first order energy shift due to any perturbation is simply the corresponding diagonal element of the perturbation matrix. Also, from Eq. (12.156), we have

$$S_1(mn) = \frac{H_1(mn)}{W_0(m) - W_0(n)} = \frac{H_1(mn)}{h\nu_0(mn)}, \quad m \neq n, \quad S_1(mm) = 0. \tag{12.158}$$

All of the calculations to this point have only depended on the assumed invertibility of the canonical transformation matrix **S**. The fundamental property of *Hermiticity* of the kinematic variables (cf. Eq. (12.11)) is now invoked to restrict the matrix **S** to satisfy the orthogonality requirement[46]

$$\mathbf{S}\,\mathbf{S}^{\mathrm{T}\star} = 1, \tag{12.159}$$

where the T superscript indicates the transpose operation, $S^{\mathrm{T}}(mn) = S(nm)$, and the $\star$ superscript complex conjugates all matrix elements. In modern language, this condition defines the *unitary* property of **S**: in the terminology used by Born *et al.* (introduced in chapter 3), **S** is *orthogonal* (a term used in the modern era for the special case where the

[46] The reader may find it useful to consult Appendix C, which reviews the relevant linear algebra for finite-dimensional vector spaces, and the essentials of the functional analysis needed in the infinite dimensional case.

matrix is purely real). In any case, the orthogonal/unitary property for the full matrix **S** means, to order λ, that

$$(1+\lambda\mathbf{S}_1+\ldots)(1+\lambda\mathbf{S}_1+\ldots)^{\mathrm{T}\star}=1+\lambda(\mathbf{S}_1+\mathbf{S}_1)^{\mathrm{T}\star}+O(\lambda^2)=1, \tag{12.160}$$

or

$$\mathbf{S}_1{}^{\mathrm{T}\star}=-\mathbf{S}_1. \tag{12.161}$$

But this is indeed satisfied by our first order result in Eq. (12.158):

$$S_1^{\mathrm{T}\star}(mn)=S_1^\star(nm)=\frac{H_1^\star(nm)}{W_0(n)-W_0(m)}=-\frac{H_1(mn)}{W_0(m)-W_0(n)}=-S_1(mn) \tag{12.162}$$

Why is this property of the canonical transformation matrix important? For the first time in the Three-Man-Paper, the "Hermitian character" of the kinematical variables of the theory, for example, for **q**,

$$\mathbf{q}^{\mathrm{T}\star}=\mathbf{q}\iff\mathbf{q}^\star=\mathbf{q}^{\mathrm{T}}, \tag{12.163}$$

is mentioned as an important property, which, in addition to the fundamental commutation relations, should be preserved under canonical transformations. The central importance of the Hermiticity property is further clarified in the mathematical developments that appear in chapter 3. For the time being, we note that the orthogonality (i.e., unitarity) property of **S** is sufficient to ensure this, as, if we begin with Hermitian **q** in Eq. (12.143), then the transformed coordinate matrix satisfies

$$\mathbf{Q}^\star=\mathbf{S}^\star\mathbf{q}^\star\mathbf{S}^{\star-1}=(\mathbf{S}^{\mathrm{T}})^{-1}\mathbf{q}^{\mathrm{T}}\mathbf{S}^{\mathrm{T}}=\mathbf{Q}^{\mathrm{T}}, \tag{12.164}$$

with a similar result for any other initially Hermitian kinematic variable subjected to the canonical transformation **S**.

The second order result for the perturbed energy is a little trickier, as we need, from the third line of Eq. (12.152), the diagonal element of $\mathbf{F}_2$, which is the matrix

$$\mathbf{F}_2=\mathbf{H}_0\mathbf{S}_1^2-\mathbf{S}_1\mathbf{H}_0\mathbf{S}_1+\mathbf{S}_1\mathbf{H}_1-\mathbf{H}_1\mathbf{S}_1+\mathbf{H}_2. \tag{12.165}$$

We therefore find

$$\begin{aligned}F_2(nn) &= \sum_l\Big(W_0(n)\,S_1(nl)\,S_1(ln)-S_1(nl)\,W_0(l)\,S_1(ln)\\ &\qquad +S_1(nl)\,H_1(ln)-H_1(nl)\,S_1(ln)\Big)+H_2(nn)\\ &= \sum_l\Big(\big(W_0(n)-W_0(l)\big)S_1(nl)\,S_1(ln)+S_1(nl)\,H_1(ln)\\ &\qquad -H_1(nl)\,S_1(ln)\Big)+H_2(nn).\end{aligned} \tag{12.166}$$

Using Eq. (12.158), we can rewrite this as:

$$\begin{aligned} F_2(nn) &= \sum_l \Big(H_1(nl)S_1(ln) + S_1(nl)H_1(ln) \\ &\qquad - H_1(nl)S_1(ln)\Big) + H_2(nn) \\ &= \sum_l \Big(S_1(nl)H_1(ln)\Big) + H_2(nn) \\ &= \sum_{l\neq n} \left(\frac{H_1(nl)H_1(ln)}{h\nu_0(nl)}\right) + H_2(nn). \end{aligned} \tag{12.167}$$

As the $\mathbf{S}_1$ matrix is purely off-diagonal, the only terms in the sum over l are for $l \neq n$ (so the vanishing denominator $\nu_0(nn)$ is avoided). The formulas in Eq. (12.157) and Eq. (12.167) vastly simplify the computation of results such as eq. (27) for the quartic oscillator in the *Umdeutung* paper (Heisenberg 1925c, p. 273, quoted following Eq. (11.90)).

The change in the coordinate matrix due to the matrix canonical transformation Eq. (12.146), with $\mathbf{S} = 1 + \lambda\mathbf{S}_1 + \ldots$, is

$$\mathbf{q}_0 \to \mathbf{q} = \mathbf{q}_0 + \lambda(\mathbf{S}_1\mathbf{q}_0 - \mathbf{q}_0\mathbf{S}_1) + \ldots \tag{12.168}$$

Next, the first order coordinate shift due to an electric field in a one-dimensional system (a charged anharmonic oscillator, say), with corresponding shift in the Hamiltonian $\mathbf{H}_1 = e\mathcal{E}\mathbf{q}_0$, and with the matrix of $\mathbf{S}_1$ given as in Eq. (12.158), is calculated and shown to agree with the zero frequency limit of the Kramers–Heisenberg dispersion formula, where the perturbing Hamiltonian is now an explicit function of time:

$$\mathbf{H}_1 = e\mathcal{E}\mathbf{q}_0 \cos(2\pi\nu_0 t). \tag{12.169}$$

As Born *et al.* proceed in the following section to discuss the problem of time-dependent perturbations in general, thereby obtaining the Kramers–Heisenberg formula for general frequency ν_0, we defer this result for a short while, as it will follow immediately from the more general result.

In chapter 1, section 5 of the Three-Man-Paper, entitled "Systems, in which the time appears explicitly in the energy function," Born *et al.* attempt to generalize the perturbation theory of time-independent Hamiltonians to time-dependent Hamiltonians of the form

$$\mathbf{H} = \mathbf{H}_0(\mathbf{q}_0, \mathbf{p}_0) + \lambda\mathbf{H}_1(\mathbf{q}_0, \mathbf{p}_0, t) + \lambda^2\mathbf{H}_2(\mathbf{q}_0, \mathbf{p}_0, t) + \ldots, \tag{12.170}$$

where the perturbing terms $\mathbf{H}_1, \mathbf{H}_2$, etc., depend on time not only through their dependence on the coordinate and momentum variables $(\mathbf{q}_0, \mathbf{p}_0)$, but also explicitly, as for example in Eq. (12.169). The objective in this part of the paper is simply stated: one wishes to obtain a generalization of the perturbation expansion Eq. (12.152) in the case where the Hamiltonian contains explicit time dependence as in Eq. (12.170). However, in this case, the target of interest is the shift in the coordinate matrices of an electron induced by the external time-dependent perturbation—specifically, the oscillating electric field of an impinging electromagnetic wave.

The prescription proposed by Born, Heisenberg, and Jordan for dealing with a time-dependent perturbation is inspired by the corresponding treatment of a time-dependent perturbation in classical Hamilton–Jacobi theory. They do not give a detailed argument for this prescription, and indeed the derivation of their formulas is rather subtle, so we shall refer the reader to our proposed reconstruction of their argument.[47] The end result is easily stated. The shift in the electron coordinate matrix induced by the time-dependent harmonic perturbation matrix $\mathbf{H}_1(t)$ is generated by a time-dependent canonical transformation matrix $\mathbf{S}_1(t)$, via the equation (analogous to Eq. (12.168))

$$\Delta\mathbf{q} = \mathbf{q}_1 = \lambda(\mathbf{S}_1\mathbf{q}_0 - \mathbf{q}_0\mathbf{S}_1), \tag{12.171}$$

where $\mathbf{S}_1$ must satisfy

$$\mathbf{S}_1\mathbf{H}_0 - \mathbf{H}_0\mathbf{S}_1 - \frac{\hbar}{i}\frac{\partial\mathbf{S}_1}{\partial t} + \mathbf{H}_1 = 0. \tag{12.172}$$

This is rather similar to the corresponding equation for the time-independent problem, as seen in the middle equation of Eqs. (12.152), with the replacement $\mathbf{W}_1 \to (\hbar/i)\,\partial\mathbf{S}_1/\partial t$ (see note 47).

The analogy with classical perturbation theory, where the coordinate for a particle moving in a multiply periodic system is perturbed by a small external periodic force (such as the sinusoidal electric field of an electromagnetic wave, with frequency ν_0), led Born *et al.* to suggest that the elements of the coordinate matrices in matrix mechanics, for example, which initially have the time dependence $q(mn)\exp(2\pi i\nu(mn)t)$, should undergo a shift which contains additionally the Fourier harmonics of frequency ν_0 inherited from the external perturbation. The arrays of quantum mechanics—whether coordinate, momentum, or energy—would thus acquire an additional index, as in

$$X(mn)\exp(2\pi i\nu(mn)t) \to \sum_{\tau} X(mn,\tau)e^{2\pi i[\nu(mn)+\tau\nu_0]}. \tag{12.173}$$

[47] See the web resource, *Time-Dependent Perturbation Theory in the Three-Man-Paper.*

The multiplication of two such variables would then involve a hybrid[48] of the quantum "Heisenberg multiplication" and the classical formula for combining Fourier series (cf. Eq. (11.16)):

$$XY(mn,\tau) = \sum_{k,\tau'} X(mk,\tau-\tau')Y(kn,\tau'). \tag{12.174}$$

Let us now see how this works, in the case of optical dispersion, the problem that served as a primary stimulus in the development of matrix mechanics. We start with an unperturbed (one-dimensional) system for which the Hamiltonian $\mathbf{H}_0$ is time independent. This need not be a simple harmonic oscillator: indeed this is not assumed in the derivation of the Kramers dispersion relations. There we needed only a periodic motion, so we are at liberty to take $\mathbf{H}_0$ to be a charged cubic or quartic oscillator (or combination of both), and $q_0(mn)$ may be a complicated matrix (for example, it may have no non-zero matrix elements), itself to be determined by an approximative perturbative procedure. The perturbing Hamiltonian in this case is just the interaction energy of an electric field $\mathcal{E}\cos(2\pi\nu_0 t)$ impinging on a particle of charge e,

$$\mathbf{H}_1(t) = e\mathcal{E}\mathbf{q}_0\cos(2\pi\nu_0 t) = \frac{e\mathcal{E}}{2}\mathbf{q}_0(e^{2\pi i\nu_0 t} + e^{-2\pi i\nu_0 t}), \tag{12.175}$$

or, in the notation of Eq. (12.173), for the matrix elements

$$H_1(mn,+1) = H_1(mn,-1) = \frac{e\mathcal{E}}{2}q_0(mn). \tag{12.176}$$

As indicated in Eq. (12.172), the calculation of the coordinate shift induced by H_1 requires first that we find $\mathbf{S}_1$ such that, for all pairs of initial and final states m, n,

$$(\mathbf{S}_1\mathbf{H}_0 - \mathbf{H}_0\mathbf{S}_1 - \frac{\hbar}{i}\frac{\partial \mathbf{S}_1}{\partial t} + \mathbf{H}_1)(mn) = 0. \tag{12.177}$$

The solution given by Born *et al.* is

$$\begin{aligned} S_1(mn) &= \frac{e\mathcal{E}}{2}q_0(mn)\left(\frac{e^{2\pi i\nu_0 t}}{E_m^{(0)} - E_n^{(0)} + h\nu_0} + \frac{e^{-2\pi i\nu_0 t}}{E_m^{(0)} - E_n^{(0)} - h\nu_0}\right) \\ &= \frac{e\mathcal{E}}{2h}q_0(mn)\left(\frac{e^{2\pi i\nu_0 t}}{\nu_0(mn) + \nu_0} + (\nu_0 \to -\nu_0)\right), \end{aligned} \tag{12.178}$$

[48] This hybrid multiplication is due to the fact that the light-scattering is not being treated fully quantum-mechanically: the electron(s) are indeed quantum-mechanical, but the electromagnetic field is still a fully classical object. This is "semi-classical" quantum electrodynamics. In a fully quantum-mechanical treatment, the electromagnetic field is quantized, and the state of the full system—electron plus field—is indicated by specifying both the discrete electron state, and the Fock state of the electromagnetic field, that is, how many photons of each frequency and momentum are present. The multiplication formula for observables then becomes fully "Heisenberg" in form—that is, matrix multiplication.

where $E_m^{(0)}$ denotes the stationary state energies of $\mathbf{H}_0$, and $E_m^{(0)} - E_n^{(0)} = h\nu_0(mn)$.[49] Let us check that this indeed works.

From Eq. (12.178) we find first

$$
\begin{aligned}
(\mathbf{S}_1\mathbf{H}_0 - \mathbf{H}_0\mathbf{S}_1)(mn) &= (E_n^{(0)} - E_m^{(0)})S_1(mn) \\
&= -\frac{e\mathcal{E}}{2}\nu_0(mn)q_0(mn)\left(\frac{e^{2\pi i\nu_0 t}}{\nu_0(mn)+\nu_0} + (\nu_0 \to -\nu_0)\right).
\end{aligned}
\tag{12.179}
$$

Similarly,

$$
\frac{\hbar}{i}\frac{\partial S_1}{\partial t}(mn) = \frac{e\mathcal{E}}{2}\nu_0 q_0(mn)\left(\frac{e^{2\pi i\nu_0 t}}{\nu_0(mn)+\nu_0} - (\nu_0 \to -\nu_0)\right).
\tag{12.180}
$$

Combining these results,

$$
\begin{aligned}
(\mathbf{S}_1\mathbf{H}_0 - \mathbf{H}_0\mathbf{S}_1 - \frac{\hbar}{i}\frac{\partial \mathbf{S}_1}{\partial t})(mn) &= \frac{e\mathcal{E}}{2}q_0(mn)\Big(\frac{-(\nu_0(nm)+\nu_0)\,e^{2\pi i\nu_0 t}}{\nu_0(mn)+\nu_0} \\
&\qquad + (\nu_0 \to -\nu_0)\Big) \\
&= -\frac{e\mathcal{E}}{2}q_0(mn)\left(e^{2\pi i\nu_0 t} + e^{-2\pi i\nu_0 t}\right) \\
&= -H_1(mn),
\end{aligned}
\tag{12.181}
$$

where the last line follows by taking the (mn) matrix element of $\mathbf{H}_1$ in Eq. (12.175). This establishes the desired result Eq. (12.177).

Having determined S_1, we can proceed to the determination of the coordinate shift

$$
\mathbf{q}_1 \equiv \lambda(\mathbf{S}_1\mathbf{q}_0 - \mathbf{q}_0\mathbf{S}_1)
\tag{12.182}
$$

induced by the applied field, to first order, as given in Eq. (12.171). First, by matrix multiplication we find

$$
(S_1 q_0)(mn) = \frac{e\mathcal{E}}{2h}\sum_k\left(\frac{q_0(mk)q_0(kn)}{\nu_0(mk)+\nu_0}e^{2\pi i\nu_0 t} + (\nu_0 \to -\nu_0)\right),
\tag{12.183}
$$

$$
(q_0 S_1)(mn) = \frac{e\mathcal{E}}{2h}\sum_k\left(\frac{q_0(mk)q_0(kn)}{\nu_0(kn)+\nu_0}e^{2\pi i\nu_0 t} + (\nu_0 \to -\nu_0)\right).
\tag{12.184}
$$

[49] Rather confusingly, the Three-Man-Paper notation, which we are following, uses ν_0 for the frequency of the applied electric field and $\nu_0(mn)$ for the Bohr transition frequencies of the H_0 system. These two frequencies, of course, have nothing to do with each other.

Hence, writing

$$q_1(mn;t) = q_1(mn,+1)\, e^{2\pi i\nu_0 t} + q_1(mn,-1)\, e^{-2\pi i\nu_0 t}, \tag{12.185}$$

we find

$$q_1(mn,+1) = \frac{e\mathcal{E}}{2h}\sum_k \left(\frac{q_0(mk)q_0(kn)}{\nu_0(mk)+\nu_0} - \frac{q_0(mk)q_0(kn)}{\nu_0(kn)+\nu_0}\right). \tag{12.186}$$

We obtain $q_1(mn,-1)$ from $q_1(mn,+1)$ by changing the sign of ν_0.

For the case relevant to dispersion (elastic light scattering) the initial and final states are the same, $m = n = s$ (we have relabeled the state to facilitate comparison with earlier results), and we obtain,

$$\begin{aligned} q_1(ss,+1) &= \frac{e\mathcal{E}}{2h}\sum_k |q_0(sk)|^2 \left(\frac{1}{\nu_0(sk)+\nu_0} + \frac{1}{\nu_0(sk)-\nu_0}\right) \\ &= \frac{e\mathcal{E}}{h}\sum_k \frac{\nu_0(sk)|q_0(sk)|^2}{\nu_0(sk)^2-\nu_0^2}. \end{aligned} \tag{12.187}$$

The polarization $P_s(t)$ induced by the applied field is the induced dipole moment per unit volume, so we must multiply this result by the electron charge $-e$ and number of dispersing electrons in state s per unit volume N_s and obtain (putting back the exponential time dependence factor, and absorbing the negative sign in the electron charge in $\nu_0(ks) = -\nu_0(sk)$)

$$\begin{aligned} P_s(t) &= \frac{e^2\mathcal{E}}{h} N_s \sum_k \frac{\nu_0(ks)|q_0(ks)|^2}{\nu_0(ks)^2-\nu_0^2} e^{2\pi i\nu_0 t} + (\nu_0 \to -\nu_0) \\ &= \frac{2e^2}{h} N_s \left(\sum_{r>s} \frac{\nu_0(rs)|q_0(rs)|^2}{\nu_0(rs)^2-\nu_0^2} - \sum_{t<s} \frac{\nu_0(st)|q_0(st)|^2}{\nu_0(st)^2-\nu_0^2}\right) \mathcal{E}\cos(2\pi\nu_0 t). \end{aligned} \tag{12.188}$$

where we have separated the sum over states k into states above, $k = r > s$, and below, $k = t < s$, the dispersing state s.

This result can be compared with our previous statement of the Kramers dispersion formula, Eq. (10.123), which we repeat here verbatim:

$$P_s(t) = \frac{3c^3}{32\pi^4} N_s \left(\sum_{r>s} \frac{A_r^s}{\nu_{rs}^2(\nu_{rs}^2-\nu^2)} - \sum_{t<s} \frac{A_s^t}{\nu_{st}^2(\nu_{st}^2-\nu^2)}\right) \mathcal{E}\cos(2\pi\nu t). \tag{12.189}$$

The Einstein coefficients A_r^s appearing here are given in the correspondence limit by Eq. (10.162):

$$A_{n+\alpha}^n = \frac{1}{h}\frac{8\pi e^2(\alpha\omega_n)^3}{3c^3}|A_\alpha(n)|^2. \tag{12.190}$$

Here, the transition $r = n + \alpha \to s = n$ corresponds in the correspondence limit to a transition (angular) frequency $\alpha\omega_n = 2\pi\alpha\nu_n$ which in turn becomes $2\pi\nu_0(rs)$ in the notation of the Three-Man-Paper. The Fourier amplitudes $A_\alpha(n)$ are, by *Umdeutung*, transition matrix elements $q_0(n+\alpha, n) = q_0(rs)$. Thus, rewriting the previous equation in the notation of the Three-Man-Paper,

$$A_r^s = \frac{64\pi^4 e^2 \nu_0(rs)^3}{3c^3 h}|q_0(rs)|^2. \tag{12.191}$$

Inserting this result in Eq. (12.189), and with the further obvious changes of notation $\nu_{rs} \to \nu_0(rs), \nu \to \nu_0$, we see that the new result Eq. (12.188) is indeed identical to the dispersion relation proposed and (by correspondence methods) derived by Kramers. The generalization of the Kramers (elastic) dispersion formulas to the inelastic (Raman–Smekal) case as carried out by Kramers and Heisenberg just corresponds to taking the initial and final states in Eq. (12.186) to be different, $m \neq n$.

The final application of the time-dependent perturbation theory developed in section 5 of the Three-Man-Paper brings us full circle, as the high-frequency limit of the dispersion formula (in this case, in its general form Eq. (12.186), for arbitrary initial and final states m, n) is shown to be intimately connected with classical mechanics via the fundamental commutation relation of the coordinate and momentum matrices. The momentum matrix $p_0(mn; t)$ is simply related to the coordinate matrix if the latter refers to a Cartesian coordinate, for which we have the simple relation $p = \mu\dot{q}$,[50] in which case

$$q_0(mn; t) \propto e^{2\pi i\nu_0(mn)t} \Rightarrow p_0(mn; t) = 2\pi i\mu\nu_0(mn)q_0(mn; t), \tag{12.192}$$

so, stripping off the time-dependence factors, we may replace $\nu_0(mn)q_0(mn)$ by $p_0(mn)/(2\pi i\mu)$. Returning to Eq. (12.186), we can rewrite the first order coordinate

[50] We employ the symbol μ here for the particle mass, to avoid confusion with the use of m to identify a stationary state of the particle.

shift as follows

$$
\begin{aligned}
q_1(mn,+1) &= \frac{e\mathcal{E}}{2h}\sum_k \left(\frac{q_0(mk)q_0(kn)}{\nu_0(mk)+\nu_0} - \frac{q_0(mk)q_0(kn)}{\nu_0(kn)+\nu_0} \right) \\
&= \frac{e\mathcal{E}}{2h}\sum_k \frac{(\nu_0(kn)+\nu_0-(\nu_0(mk)+\nu_0))q_0(mk)q_0(kn)}{(\nu_0(mk)+\nu_0)(\nu_0(kn)+\nu_0)} \\
&= \frac{e\mathcal{E}}{2h}\sum_k \frac{q_0(mk)\nu_0(kn)q_0(kn)-\nu_0(mk)q_0(mk)q_0(kn)}{(\nu_0(mk)+\nu_0)(\nu_0(kn)+\nu_0)} \\
&= \frac{e\mathcal{E}}{4\pi i\mu h}\sum_k \frac{q_0(mk)p_0(kn)-p_0(mk)q_0(kn)}{(\nu_0(mk)+\nu_0)(\nu_0(kn)+\nu_0)}.
\end{aligned}
\tag{12.193}
$$

We already can recognize in the numerator of the last expression the germ of the matrix multiplications leading to the commutator of q_0 and p_0. The only fly in the ointment is the dependence on the summation index k through the transition frequencies in the denominator. We now suppress this dependence by considering the limit where the frequency ν_0 of the applied electric field is much greater than any of the transition frequencies: $\nu_0 \gg \nu_0(mk), \nu_0(kn)$. Physically, this is the limit in which the motion of the electron is overwhelmingly determined by the applied field, and it "forgets" that it is bound in an atom, or with elastic forces. Accordingly, we may neglect $\nu_0(mk)$ and $\nu_0(kn)$ in the denominator factors of Eq. (12.193), and find that the induced coordinate shift (now putting back the time dependence due to the applied field) becomes

$$
\begin{aligned}
q_1(mn;t) &= q_1(mn,+1)\,e^{2\pi i\nu_0 t} + q_1(mn,-1)\,e^{-2\pi i\nu_0 t} \\
&= -\frac{e\mathcal{E}}{2\pi i\mu h\nu_0^2}(\mathbf{p}_0\mathbf{q}_0 - \mathbf{q}_0\mathbf{p}_0)(mn)\cos(2\pi\nu_0 t).
\end{aligned}
\tag{12.194}
$$

The fundamental quantum relation $\mathbf{p}_0\mathbf{q}_0 - \mathbf{q}_0\mathbf{p}_0 = (h/2\pi i)\,1$, for which the authors employ for the first time the now-common terminology "commutation relation" (*Vertauschungsrelation*), now reduces this result to a solution of Newton's second law for an unbound particle of mass μ subject only to the electric force $-e\mathcal{E}\cos(2\pi\nu_0 t)$,

$$
q_1(t) = \frac{e\mathcal{E}}{4\pi^2\mu\nu_0^2}\cos(2\pi\nu_0 t) \Rightarrow \mu\ddot{q}_1 = -e\mathcal{E}\cos(2\pi\nu_0 t), \tag{12.195}
$$

a most satisfying result. As the authors state:

> This finding indicates that in fact the quantum-mechanical commutation relation . . . ultimately entails the fact that for sufficiently high frequencies the electron behaves on scattering like a free electron (p. 338).

12.3.2 Second chapter: foundations of the theory of systems of arbitrarily many degrees of freedom

The second chapter, dealing with the treatment of systems of more than one degree of freedom (i.e., having more than one conjugate pair of coordinate and momentum), is divided into two sections, and each addresses questions that naturally arise in problems with a multi-dimensional coordinate space. The first is the need to generalize the fundamental commutation relation for a single degree of freedom, Eq. (12.138) (which may be supplemented with the trivial relations $[\mathbf{q}, \mathbf{q}] = [\mathbf{p}, \mathbf{p}] = 0$), to systems with several coordinates and conjugate momenta. The second is the presence of degeneracy in the discrete stationary states of the system.[51]

Before introducing the expanded set of quantum conditions/commutation relations needed in the quantum treatment of a system classically described by a $2f$ dimensional phase space $(q_1, \ldots q_f, p_1, \ldots p_f)$, the authors address a noisome notational question that immediately arises: how to deal with transitions between states specified by a set of integer quantum numbers $n_1, \ldots n_f$ without having to resort to an awkward tensorial notation, for example, $q(n'_1 \ldots n'_f n_1 \ldots n_f)$, instead of the simple matrix notation (with the associated, and by now well-understood, matrix algebra) $q(mn)$. The point made earlier in Born and Jordan's Two-Man-Paper is here reiterated: the ordering of the discrete states in the quantum transcription of a classical mechanical system is a purely conventional matter. None of the essential physics depends on how we choose to order the discrete states, although there are obvious issues of convenience in selecting, for example, an ordering that corresponds to monotonically increasing energies (so the first state is the ground state of the system). In the multi-dimensional case, one has simply to imagine ordering the states in an arbitrary, but well-defined way, so that each quantum (stationary) state is associated with a unique integer N (the capitalization here is our notation; Born *et al.* continue to use lower case letter indices). One could, for example, order the states by enumerating first the finite set of states for which $|n_1 + \ldots + n_f|$ assumes its minimum value (say, all $n_i = 0$ labeled as state $N = 0$), then all the states with $|n_1 + \ldots + n_f| = 1$, and so on. The quantum matrix variables of the theory would then take the usual form, depending on just two indices, for example, $q(MN)$.

The proposed generalization of the fundamental commutation relation, which appears as equation (3) of chapter 2, section 1, prescribes the algebra for a set of $2f$ coordinate and momentum matrices $\mathbf{q}_1 \ldots \mathbf{q}_f, \mathbf{p}_1, \ldots \mathbf{p}_f$

$$\mathbf{p}_k\mathbf{q}_l - \mathbf{q}_l\mathbf{p}_k = \frac{h}{2\pi i}\delta_{kl}\,1, \tag{12.196}$$

$$\mathbf{p}_k\mathbf{p}_l - \mathbf{p}_l\mathbf{p}_k = 0, \tag{12.197}$$

[51] The bound states of a one-dimensional potential can be proven to be non-degenerate. In higher dimensions, the appearance of degeneracy—multiple quantum states with the same energy—is almost ubiquitous.

$$\mathbf{q}_k\mathbf{q}_l - \mathbf{q}_l\mathbf{q}_k = 0. \qquad (12.198)$$

Heisenberg had already proposed these relations shortly after his return to Copenhagen. In a postcard to Jordan of September 21, 1925 (Mehra and Rechenberg 1982c, p. 97), he pointed out that the relations followed trivially if the variables were completely separated in the dynamics (e.g., if the Hamiltonian was a sum of Hamiltonians depending on each conjugate pair separately $H = H_1(p_1, q_1) + H_2(p_2, q_2) + \ldots$), but that otherwise he did not understand how to obtain these very natural conditions from classical theory. The beautiful correspondence argument of Dirac, which unambiguously produces Eqs. (12.196)–(12.198) by transcription of classical Poisson brackets, simply did not occur to Heisenberg, or either of his co-authors, at this time, or indeed later in the final preparation of their paper, which evidently occurred before they had a chance to see Dirac's paper.[52]

There is much beating around the bush in chapter 2, section 1 of the Three-Man-Paper, as the authors confessedly lack a convincing "derivation" of the multi-dimensional commutation relations, natural though they appeared. Even in the absence of the Dirac argument, this seems a bit unnecessary: it was absolutely essential to preserve the validity of the Hamiltonian equations of motion, as transcribed to the quantum theory, now for a system with several conjugate pairs,

$$\dot{\mathbf{q}}_k = \frac{\partial \mathbf{H}}{\partial \mathbf{p}_k}, \quad \dot{\mathbf{p}}_k = -\frac{\partial \mathbf{H}}{\partial \mathbf{q}_k}. \qquad (12.199)$$

where we recall that the bold font in the derivative indicates the new "natural" definition of partial matrix differentiation adopted in the Three-Man-Paper. This definition simply produces the derivative of a term involving a product of matrix p_is and q_is, with respect to a particular (say) q_k, as a sum of terms in which that q_k is successively removed at each position at which it occurs without altering the order of the other matrices in the product. It is perfectly clear that the proposed commutation rules Eqs. (12.196)–(12.198) are *sufficient* to ensure this desired result: for example, on commutation with $\mathbf{p}_k$, each $\mathbf{q}_k$ that is encountered is removed and replaced by $h/2\pi i$, thus

$$\mathbf{p}_k\mathbf{f}(\mathbf{q}_1\ldots\mathbf{q}_f, \mathbf{p}_1\ldots\mathbf{p}_f) - \mathbf{f}(\mathbf{q}_1\ldots\mathbf{q}_f, \mathbf{p}_1\ldots\mathbf{p}_f)\mathbf{p}_k = \frac{h}{2\pi i}\frac{\partial \mathbf{f}}{\partial \mathbf{q}_k}, \qquad (12.200)$$

$$\mathbf{q}_k\mathbf{f}(\mathbf{q}_1\ldots\mathbf{q}_f, \mathbf{p}_1\ldots\mathbf{p}_f) - \mathbf{f}(\mathbf{q}_1\ldots\mathbf{q}_f, \mathbf{p}_1\ldots\mathbf{p}_f)\mathbf{q}_k = -\frac{h}{2\pi i}\frac{\partial \mathbf{f}}{\partial \mathbf{p}_k}. \qquad (12.201)$$

Indeed, it is very hard to see how else one could achieve the equivalence of commutation with each of the canonical variables with the corresponding partial derivative operation.

The authors describe a number of additional results that can be regarded as providing "circumstantial evidence" for the utility (if not the validity) of the assumed multi-dimensional commutation relations, to wit:

[52] Dirac's paper is referenced on the first page of the published Three-Man-Paper, as a note added in proof.

1. The equations of motion imply the (necessary) vanishing of the time derivative of $\mathbf{p}_k\mathbf{q}_k - \mathbf{q}_k\mathbf{p}_k(=h/2\pi i)$, as follows,[53]

$$\begin{aligned}
\frac{d}{dt}(\mathbf{p}_k\mathbf{q}_k - \mathbf{q}_k\mathbf{p}_k) &= \sum_k(\dot{\mathbf{p}}_k\mathbf{q}_k + \mathbf{p}_k\dot{\mathbf{q}}_k - \dot{\mathbf{q}}_k\mathbf{p}_k - \mathbf{q}_k\dot{\mathbf{p}}_k) \\
&= -\frac{\partial \mathbf{H}}{\partial \mathbf{q}_k}\mathbf{q}_k + \mathbf{p}_k\frac{\partial \mathbf{H}}{\partial \mathbf{p}_k} - \frac{\partial \mathbf{H}}{\partial \mathbf{p}_k}\mathbf{p}_k + \mathbf{q}_k\frac{\partial \mathbf{H}}{\partial \mathbf{q}_k} \\
&= \frac{2\pi i}{h}(-\mathbf{p}_k\mathbf{H}\mathbf{q}_k + \mathbf{H}\mathbf{p}_k\mathbf{q}_k + \mathbf{p}_k\mathbf{H}\mathbf{q}_k - \mathbf{p}_k\mathbf{q}_k\mathbf{H} \\
&\quad -\mathbf{H}\mathbf{q}_k\mathbf{p}_k + \mathbf{q}_k\mathbf{H}\mathbf{p}_k + \mathbf{q}_k\mathbf{p}_k\mathbf{H} - \mathbf{q}_k\mathbf{H}\mathbf{p}_k) \\
&= \frac{2\pi i}{h}(\mathbf{H}(\mathbf{p}_k\mathbf{q}_k - \mathbf{q}_k\mathbf{p}_k) + (\mathbf{q}_k\mathbf{p}_k - \mathbf{p}_k\mathbf{q}_k)\mathbf{H}) \\
&= \frac{2\pi i}{h}(\mathbf{H}\frac{h}{2\pi i} - \frac{h}{2\pi i}\mathbf{H}) = 0.
\end{aligned} \quad (12.202)$$

2. The fundamental commutation relations are preserved under a large class of matrix canonical transformations of the type used in the preceding section to implement both time-independent and time-dependent versions of perturbation theory, that is, if $\mathbf{p}_k, \mathbf{q}_k$ satisfy the fundamental multi-dimensional commutation relations specified in Eqs. (12.196–12.198), then so do

$$\mathbf{q}'_k \equiv \mathbf{S}\mathbf{q}_k\mathbf{S}^{-1}, \quad \mathbf{p}'_k \equiv \mathbf{S}\mathbf{p}_k\mathbf{S}^{-1}, \quad (12.203)$$

   where $\mathbf{S}$ is any invertible matrix.

3. One would certainly expect (and require) the commutation relations to be valid for any set of Cartesian coordinate/momentum pairs obtained from another such set by an orthogonal rotation, described by a matrix a_{kl}, satisfying $\sum_h a_{kh}a_{lh} = \delta_{kl}$, and indeed, setting

$$\mathbf{q}'_k = \sum_l a_{kl}\mathbf{q}_l, \quad \mathbf{p}'_k = \sum_l a_{kl}\mathbf{p}_l, \quad (12.204)$$

[53] For some reason Born *et al.* include a sum over k in the statement of this result, which is unnecessary: the identity holds for each value of k separately. The argument here really establishes that, if the commutator is a pure number (commuting with the Hamiltonian matrix), then it must be a constant.

one immediately verifies, using Eq. (12.196),

$$\begin{aligned} \mathbf{p}'_k\mathbf{q}'_l - \mathbf{q}'_l\mathbf{p}'_k &= \sum_{hj} a_{kh}a_{lj}(\mathbf{p}_h\mathbf{q}_j - \mathbf{q}_j\mathbf{p}_h) \\ &= \frac{h}{2\pi i}\sum_h a_{kh}a_{lh} = \frac{h}{2\pi i}\delta_{kl}. \end{aligned} \tag{12.205}$$

In chapter 2, section 2 of the paper the authors examine some subtleties of degenerate systems, which become commonplace once one begins to examine systems with more than a single degree of freedom. This problem was identified very early in the development of the old quantum theory. As we saw in Sections 1.3.2 and 5.1.3, Sommerfeld (1915a) realized that the Balmer energies in hydrogen actually corresponded to a degenerate set of stationary states, with n distinct orbits of varying eccentricity at each value of the principal quantum number n having the same energy. Degeneracy was also the rule rather than the exception for atoms with a non-zero total angular momentum, in the absence of external fields, as rotational invariance meant the physically distinct states, corresponding to varying orientation of the total angular momentum vector, would have the same energy.

As had been the case in the treatment of degenerate systems in classical canonical perturbation theory, there were several problems, both fundamental and practical, to be faced in the application of quantum methods to such systems. Once it was possible for $\nu(nm)$ to be zero for distinct states $n \neq m$, the link between time independence of a kinematical quantity A and its diagonality is lost, as we clearly have, for the corresponding off-diagonal matrix element that

$$\frac{d}{dt}\left(A(nm)\, e^{2\pi i\nu(nm)t}\right) = \frac{d}{dt}A(nm) = 0 \tag{12.206}$$

for $n \neq m$. In particular, it is no longer a priori the case that the energy matrix is diagonal, which then makes the derivation of the frequency condition impossible. Another problem, of a more practical nature, is that the time-independent perturbation theory developed in the preceding chapter becomes singular. In particular, the equation giving the first order contribution to the canonical transformation effecting the diagonalization of the Hamiltonian (cf. Eq. (12.158)),

$$S_1(mn) = \frac{H_1(mn)}{W_0(m) - W_0(n)} = \frac{H_1(mn)}{h\nu_0(mn)}, \quad m \neq n \tag{12.207}$$

develops an infinity for pairs of states (n, m) such that $H_1(mn) \neq 0$ with $W_0(m) = W_0(n)$.

To restore the validity of the frequency condition, the authors adopt a supplementary condition, in addition to the fundamental commutation relations, the requirement that the Hamiltonian of the system always be in (or be brought to) diagonal form:

$$\mathbf{H} = \mathbf{W} = \text{Diagonal Matrix.} \tag{12.208}$$

Even if this has been achieved, however, there remains a disturbing ambiguity in the underlying kinematical variables $(\mathbf{p}, \mathbf{q})$ due to the existence of non-trivial canonical transformations $\mathbf{S}$ that leave the Hamiltonian matrix intact (and diagonal) but produce non-trivial changes in these variables. In fact, a certain freedom of choice in the $\mathbf{S}$ exists even for totally non-degenerate systems. For the set of quantum numbers n_i for which $W_{n_i} \neq W_{n_j}$, $n_i \neq n_j$, the requirement that the Hamiltonian matrix remains stable under a canonical transformation $\mathbf{S}$ implies that, for $i \neq j$,

$$\begin{aligned} \mathbf{SWS}^{-1} = \mathbf{W} \quad &\Rightarrow \quad (SW)(n_i n_j) = (WS)(n_i n_j) \\ &\Rightarrow \quad (W_{n_i} - W_{n_j}) S(n_i n_j) = 0 \\ &\Rightarrow \quad S(n_i n_j) = 0. \end{aligned} \tag{12.209}$$

This means that the $\mathbf{S}$ matrix is diagonal in the sector of non-degenerate states. We saw previously that the canonical transformation matrices are in fact orthogonal (in modern terminology, unitary), satisfying (cf. Eq. (12.159)) $\mathbf{S}\,\mathbf{S}^{\mathrm{T}\star} = 1$, which for a diagonal matrix means

$$S(n_i n_i) S(n_i n_i)^\star = 1 \quad \Rightarrow \quad S(n_i n_i) = e^{i\varphi_{n_i}}. \tag{12.210}$$

So if we transform the kinematical variables $(\mathbf{p}, \mathbf{q})$ to $(\mathbf{p}', \mathbf{q}')$ with $\mathbf{S}$,

$$\mathbf{p}' = \mathbf{SpS}^{-1}, \quad \mathbf{q}' = \mathbf{SqS}^{-1}, \tag{12.211}$$

we have, in the non-degenerate sector,

$$p'(n_i n_j) = e^{i(\varphi_{n_i} - \varphi_{n_j})} p(n_i n_j) \tag{12.212}$$

$$q'(n_i n_j) = e^{i(\varphi_{n_i} - \varphi_{n_j})} q(n_i n_j). \tag{12.213}$$

This phase freedom is the latest manifestation of the situation first diagnosed in Born and Jordan's Two-Man-Paper (1925b; cf. note 17):

> The indeterminacy appearing here means therefore an arbitrariness of *phase constants*; and in fact we have obtained here a general proof of the presumption previously raised in I [Born and Jordan 1925a], that in every problem, for every state n, one phase φ_n remains undetermined (p. 345).

That this indeterminacy in fact exhausts the arbitrariness in the quantum solution of a totally non-degenerate problem—one in which no two distinct quantum states have the same energy—is argued on the basis of the perturbation theory of chapter 1, section 4: once the phase freedom is admitted at the zeroth order of the perturbation theory (i.e., in the $\mathbf{p}_0, \mathbf{q}_0$), no additional phase ambiguities occur as the perturbation expansion is performed to higher order.

So far, we have been discussing the ambiguities that arise even for non-degenerate systems. If a degeneracy occurs, where, for example, a finite number N of states, labeled by quantum numbers $n_1, \ldots n_N$, have exactly the same energy E (not shared by any other states), then it is easy to see that we may choose $S(n_i n_j)$, with $1 \leq i, j \leq N$, to constitute an arbitrary $N \times N$ unitary matrix, which will clearly execute a non-trivial change in the kinematical variables $(\mathbf{p}', \mathbf{q}')$ in Eq. (12.211). Such a transformation matrix will leave the Hamiltonian matrix unchanged in the N-dimensional degeneracy sector, as the latter is simply, by assumption, a multiple of the identity matrix there, which commutes with an arbitrary unitary matrix. However, such a unitary matrix will *not* commute with the kinematical variables $(\mathbf{p}, \mathbf{q})$, leading to new solutions $(\mathbf{p}', \mathbf{q}')$, which differ substantially from the initial ones, despite the fact that they correspond to exactly the same values for stationary state energies. Born *et al.* comment on this feature explicitly:

> It seems that this lack of uniqueness lies in the nature of the matter. Degenerate systems clearly possess a *lability* (*Labilität*), on account of which finite alterations of the coordinates can be produced by arbitrarily small perturbations, and which finds its mathematical expression in the fact that in the complete absence of perturbations the solution of the dynamical equations remains partially undetermined. Naturally, for each actual single atom the coordinates determining the physical properties of the system, in particular the transition probabilities, are always uniquely fixed either by external perturbations or by the prehistory of the system (p. 345).

The point being made here is that an initial degeneracy of states due to some symmetry of the system (e.g., the equality of energies of states of different magnetic quantum number due to the rotational invariance of an atomic system in the absence of external fields) can typically be lifted by the application of a very small external field. At this point the new fundamental requirement expressed in Eq. (12.208), that the system be always represented in terms of states in which the Hamiltonian matrix is diagonal, will require that we immediately apply a canonical transformation—by no means one close to the identity operation, in general—which diagonalizes the Hamiltonian, revealing the presence of slightly different energies in the previously degenerate sector. For example, the $\mathbf{S}$ matrix needed to diagonalize the energy matrix when a very small magnetic field in the x-direction is imposed on a previously field-free atom differs by a finite amount from that needed if a very small field is imposed in the z-direction. Mathematically, the previous freedom to pick an arbitrary $N \times N$ unitary matrix to transform the variables has been reduced to the much smaller phase freedom of N complex phases, one for each state. This reduction occurs instantaneously and discontinuously. It is certainly disconcerting, and no less so for the authors' attempt to defuse the situation by appealing to unspecified "external perturbations" or the "prehistory" of the system.

The dynamical instability, or "lability", noted by Born *et al.* has a precise correlate in a corresponding difficulty of the old quantum theory. The quantization of a degenerate multiply periodic system in terms of action variables suffers from an essential ambiguity of the spatial orbits determined by the quantization.[54] For example, a one-electron atom, where the electron experiences a pure Coulomb potential due to the nucleus, is a degenerate system, with the associated property that the Hamilton–Jacobi problem can be separated in a number of different coordinate systems (polar, parabolic, elliptical, etc.). The spatial orbits obtained by quantization of the action variables in these different coordinate systems are essentially different from one another (e.g., in the size of their semi-major axes, eccentricities, etc., hence, are not geometrically congruent) although the associated energies are the same, independent of coordinate choice.

Once an external field is imposed (e.g., a small electric field in the Stark effect) the choice of coordinate systems available for separation of the problem is greatly reduced (e.g., in the Stark effect, one is restricted to the parabolic coordinate system, up to trivial redefinitions). Of course, in the very first stages of the old quantum theory, the spectroscopically available energies (or term values) were of primary interest as one did not have—indeed, as the new theory would make clear, *could* not have—empirical access to the spatial structure of the orbits. A little later, the orbital structure would also (indirectly) become empirically relevant, via the correspondence principle, as spectral line intensities could be related to Fourier components of the orbital motion. The orbit ambiguity problem was noted already in 1916 by Epstein in his paper on the Stark effect (see Section 6.3). After determining the quantized energies of an electron subject to a Coulomb potential and small electric field (to first order in the latter), Epstein comments:

> As we have already mentioned, the orbits selected by such quantum conditions do not agree with those which appear for vanishing [electric field] *E*. Even though this does not entail, as seen above, any shift of the series lines, the idea that the shape and orientation of the stationary orbits should be changed in such a drastic fashion by the preferred direction given by an ever so small external field seems to us inadmissible. The elucidation of this apparent paradox is to be expected from a theory in which relativity and an external field are simultaneously taken into account (Epstein 1916a, p. 507).

Unfortunately, the solution proposed in the last sentence, wherein the problem of a particle bound by a Coulomb potential is solved including both effects of relativity and of an external field, would not be realized, for the simple reason that the Hamiltonian containing both effects does not allow for separation of the associated Hamilton–Jacobi equation *in any coordinate system*. Both Sommerfeld (1919, pp. 502–503), who inclined to Epstein's view that the inclusion of further corrections would resolve the ambiguity, and Bohr (1918, p. 23), who simply asserted that the appearance of small external fields would lead to spontaneous quantum jumps as the system readjusted itself to the instantaneously appropriate set of selected orbits, were well aware of this problem, but did not

[54] For a detailed discussion of the orbit ambiguity problem in the old quantum theory—and its ultimate resolution in wave mechanics (Schrödinger 1926d, pp. 33–34, quoted in Section 14.4.2)—see Duncan and Janssen (2015, sec. 2.3).

seem to regard it as a fundamental flaw in the conceptual framework of the old quantum theory.

In fact, despite having dispensed with the untenable notion of well-defined orbital trajectories for particles in the stationary states of atoms, the new quantum mechanics, at least in the version being developed in the early autumn of 1925, had inherited the same disturbing sensitivity to small changes in external conditions that had afflicted the old theory. The source of the problem lies in the insistence that the states of the atomic system always correspond to a diagonal Hamiltonian matrix, as in Eq. (12.208).[55] Later, we will see later that this error (perhaps forgivable given the absence at this point of a suitable mathematical conception of the state in terms of a vector space admitting linear combinations of states) would be addressed directly, and repaired, in Schrödinger's new wave formulation of quantum mechanics.

The discussion of degenerate systems in chapter 2, section 2, concludes with a discussion of time-independent perturbation theory in the presence of degeneracy, intended to forestall the appearance of singular terms. The technique used, where at each level of the perturbation theory a pre-diagonalization of the energy matrix in any remaining degenerate sectors is performed, should be familiar to students of modern quantum mechanics, and is not of great importance for the rest of our story, so we pass on quickly to the central section of the paper, in which for the first time we see the appearance of algebraic techniques that would soon form the mathematical core of the new theory.

12.3.3 Third chapter: connection with the theory of eigenvalues of Hermitian forms

The third chapter of the Three-Man-Paper contains the germs of the mathematical framework of the new theory eventually brought to formal completion by von Neumann (1927a; see Chapter 17). The precursors of the transformation theory of Jordan (1927b, 1927g) and Dirac (1927a) can also be detected just beneath the surface here (see Chapter 16). The level of mathematical expertise required here is substantially higher than that deployed heretofore in the development of matrix mechanics—in fact, results are needed from the cutting edge of mathematical research in the area of (what would come to be called) functional analysis in the first decade of the twentieth century. Characteristically, this section of the paper was due almost entirely to Born, who was fascinated with the idea that a deep reliance on invariance principles under linear transformations underlay the most fundamental laws of physics (Mehra and Rechenberg 1982c, sec. III.4).[56] The material in this third chapter of the paper was elaborated and

[55] This requirement is closely connected to another tenet of the "Bohrian picture"—later dubbed the "discontinuity theory"—which asserted that electrons were always in stationary states of the atom, except for the discrete moments in time when they randomly underwent a quantum jump from one stationary state to another. As we will see in Chapter 16, this (erroneous) picture persisted a full two years after the discovery of matrix mechanics and played an important role in the initial struggles to develop a consistent probabilisitic interpretation of the theory.

[56] This casts doubt on the following anecdote related by Condon (who was visiting Göttingen in the summer of 1926) in later reminiscences: "Hilbert was having a great laugh on Born and Heisenberg and the Göttingen

written up by Born in the final weeks before his departure from Göttingen (en route to the United States) at the end of October 1925.

The first section of chapter 3, entitled "General methods," introduces the basic motivation as follows:

> In the preceding the solution of the fundamental equations of quantum theory were developed in the closest possible correspondence to the classical theory. Behind this formalism of perturbation theory however a very simple, and purely algebraic, connection lies concealed, and it is very profitable to bring this into the light. For apart from the deeper insight into the mathematical structure of the theory one also reaps thereby the advantage of being able to exploit previously developed methods and results of mathematics. In this way we shall reach a new definition of the energy constants ("terms") which remains valid in the case of aperiodic motions, hence in the case of continuously variable indices (p. 348).

The "previously developed methods" referred to primarily point to the development of the theory of bilinear forms in infinitely many variables over the preceding two decades, primarily by David Hilbert, and with important contributions by Ernst Hellinger.[57] The fundamental problem posed in the new theory had already been addressed in chapter 2: in general, the energy matrix $H(q_0, p_0)$ as expressed in terms of coordinate and momentum matrices q_0, p_0 satisfying the fundamental commutation relations (say, those of a simpler system that could be completely solved) was not a diagonal matrix, and one was then faced with the task of finding a similarity transformation to new coordinate and momentum matrices q, p such that the Hamiltonian $H(q, p) = SH(q_0, p_0)S^{-1}$ is a diagonal matrix. Moreover, the matrices in question were typically infinite-dimensional ones. Hilbert and Hellinger's work was in fact precisely concerned with finding such transformations, although expressed in the equivalent language of bilinear forms, rather than matrix transformations.[58]

theoretical physicists because when they first discovered matrix mechanics they were having, of course, the same kind of trouble that everybody else had in trying to solve problems and to manipulate and to really do things with matrices. So they had gone to Hilbert for help and Hilbert said the only times he had ever had anything to do with matrices was when they came up as a sort of by-product of the eigenvalues of the boundary-value problem of a differential equation. So if you look for the differential equation which has these matrices you can probably do more with that. They had thought it was a goofy idea and that Hilbert didn't know what he was talking about. So he was having a lot of fun pointing out to them that they could have discovered Schrödinger's wave mechanics six months earlier if they had paid a little more attention to him" (Condon 1962, p. 276; quoted without a source in Reid 1986, p. 182; see also Condon 1954, p. 202). According to Born (1940), he had been familiar with matrices since his student days in Breslau and had encountered the connection between matrices and quadratic forms in his work on vibrations in crystals before the advent of matrix mechanics (see Chapter 11, note 30).

[57] It should be noted that Born was personally acquainted with, and a friend of, both Otto Toeplitz and Ernst Hellinger, with whom he overlapped both as students in Breslau and later in Göttingen. Toeplitz had already encouraged Born in Breslau to study the theory of algebraic equations (as well as matrix calculus; see note 56), and became (with Hellinger) an expert in the theory of infinite quadratic forms, which Born then encountered in his work with von Kármán on crystal lattices in 1912–1913. Thus, of all the physicists engaged in the development of quantum theory in 1925, Born was probably the best equipped to see the emerging relations with previous mathematical work (see Mehra and Rechenberg 1982c, Ch. 1).

[58] See Appendix C for an introduction to the linear algebra and elementary functional analysis important for understanding this section of the Three-Man-Paper.

The connection between the two problems is in fact very easy to see. For every matrix **a**—and for the time being, we shall assume these to be finite-dimensional, avoiding convergence issues in the associated index sums—defined by an array of matrix elements $a(nm)$ one can define an associated bilinear form $A(xy)$ of variables $x_1, x_2, \ldots$ and $y_1, y_2, \ldots$ by setting

$$A(xy) = \sum_{nm} a(nm) x_n y_m. \tag{12.214}$$

The matrices representing kinematical variables in matrix mechanics are all *Hermitian*, which means that the transpose matrix is equal to the complex conjugated one, or equivalently, that the process of taking the transpose and complex conjugation brings us back to the original matrix:

$$\mathbf{a}^T = \mathbf{a}^\star, \; a(mn) = a(nm)^\star \iff \mathbf{a}^{T\star} = \mathbf{a}. \tag{12.215}$$

For purely real matrices, the Hermitian requirement reduces to the symmetry property $a(mn) = a(nm)$. For Hermitian matrices, the special case of quadratic forms where we set $y_n = x_n^\star$ is particularly important,

$$A(xx^\star) = \sum_{nm} a(nm) x_n x_m^\star, \tag{12.216}$$

as for arbitrary complex x_n, the form defined in Eq. (12.216) is real:

$$A(xx^\star)^\star = \sum_{nm} a(nm)^\star x_n^\star x_m = \sum_{nm} a(mn) x_m x_n^\star = A(xx^\star). \tag{12.217}$$

Linear changes of variable, defined by a transformation matrix $v(ln)$, such as

$$x_n = \sum_l v(ln) y_l, \tag{12.218}$$

result in the appearance of a new quadratic form

$$\begin{aligned} B(yy^\star) \equiv A(xx^\star) &= \sum_{nm} a(nm) \sum_l v(ln) y_l \sum_k v(km)^\star y_k^\star \\ &= \sum_{kl} b(lk) y_l y_k^\star, \quad b(lk) = \sum_{nm} v(ln) a(nm) v(km)^\star. \end{aligned} \tag{12.219}$$

The change from x to y variables therefore corresponds to the matrix transformation

$$\mathbf{b} = \mathbf{v}\mathbf{a}\mathbf{v}^{T\star}. \tag{12.220}$$

The sort of transformation described here has the fortunate property of leaving the Hermitian property (essential for the kinematical quantities of matrix mechanics) of the form invariant:

$$\mathbf{b}^T = \mathbf{v}^\star \mathbf{a}^T \mathbf{v}^T = \mathbf{v}^\star \mathbf{a}^\star \mathbf{v}^T = \mathbf{b}^\star. \tag{12.221}$$

The transformations $\mathbf{v}$ with $\mathbf{v}^{T\star} = \mathbf{v}^{-1}$ are called *orthogonal* (in modern terminology, *unitary*, with "orthogonal" reserved for purely real matrices and forms), and clearly correspond, as we see from Eq. (12.220), to a subclass of the similarity type canonical transformations $\mathbf{a} \to \mathbf{SaS}^{-1}$ discussed earlier in the paper. Such orthogonal transformations are just those which leave the *identity form* $E(xx^\star) = \sum_n x_n x_n^\star$ invariant. The unit matrix 1 is given by matrix elements $1(nm) = \delta_{nm}$, so

$$\begin{aligned} E(xx^\star) &= \sum_n x_n x_n^\star = \sum_{nm} x_n 1(nm) x_m^\star = E(yy^\star) \\ &\iff \mathbf{v}\mathbf{v}^{T\star} = 1 \iff \mathbf{v}^{T\star} = \mathbf{v}^{-1}. \end{aligned} \tag{12.222}$$

A fundamental theorem of linear algebra states that, in the finite-dimensional case, with variables $x_1, x_2, \ldots x_N$ and $\mathbf{a}$ a square $N \times N$ Hermitian matrix, there always exists a (complex) orthogonal matrix $\mathbf{v}$ such that[59]

$$\mathbf{vav}^{T\star} = \mathbf{W}, \quad W(nm) = W_n \delta_{nm}, \quad W_n = W_n^\star, \tag{12.223}$$

or equivalently, in terms of forms,

$$A(xx^\star) = \sum_n W_n y_n y_n^\star. \tag{12.224}$$

Such a transformation of a general quadratic form to a sum of squares was usually referred to at the time as a "principal axis transformation".

Obviously, the generalization of this result to the case of infinite dimensional matrices (or forms which are quadratic functions of infinitely many variables) is of critical importance to the mathematical viability of the new theory. Indeed, we saw previously that the fundamental quantization conditions of the new theory, which specify the commutation relations of the coordinate and momentum variables in matrix form, cannot be realized with finite matrices. Also, in the cases of greatest physical interest (oscillator problems, the Coulomb problem, with or without external fields) there were inevitably an infinite number of distinct stationary states.

Fortunately for the authors of the Three-Man-Paper, precisely this sort of generalization had been the concern of Hilbert and Hellinger in the fairly recent past (1904–1910). In a groundbreaking series of six papers ("Foundations of a general theory of linear integral equations, I–VI") between 1904 and 1910, Hilbert had developed a general

[59] See Section C.3 for a review of the relevant linear algebra.

theory of linear integral equations by exploiting the close relationship of such equations to the behavior of bilinear and quadratic forms of an infinite number of variables. In the first paper, Hilbert (1904) introduced the terms "eigenvalues" (*Eigenwerte*) and "eigenfunctions" (*Eigenfunktionen*) in connection with the theory of linear integral equations. In Hilbert (1906a), the terms "point spectrum" (p. 169) and "continuous spectrum" (p. 172) appear. The use of terminology associated with optics in the context of function theory seems to go back to a paper of Wilhelm Wirtinger (1897), which compares the bunching of characteristic frequencies of a vibrating string into "band-spectra", with the gaps between adjacent bands disappearing in the limit of infinite string length (giving an uninterrupted continuous spectrum). Hilbert may well have read this paper, though he does not credit Wirtinger with the introduction of a terminology that must now seem extraordinarily prescient, as the intimate connection of infinite dimensional forms/matrices to atomic spectra was of course completely unknown in 1906 (not to speak of 1897). Hilbert's results were further extended by Hellinger (1910), who in particular delineated the rigorous treatment of the eigenfunctions in the case of bounded forms with both a discrete and continuous spectrum.

Hilbert and Hellinger's results were of great interest to Born *et al.* as they had been able to show rigorously that, for a particular class of infinite-quadratic forms, the transformation of Eq. (12.223) for finite matrices could be generalized to the infinite-dimensional case, with one new wrinkle: the appearance in some cases of a *continuous spectrum*, where the eigenvalues W_n, obviously discrete in the finite dimensional case, now formed a continuous set of values. In general, both types of spectrum, point and continuous, would appear in the same form. This would correspond physically to the appearance (for example, in atoms) of both bound states with discrete energies, and unbounded states of continuously variable energy in which one or more electrons were detached from the atom (eventually escaping to infinity) but still subject to the Coulomb attraction of the remnant charged ion. In relation to the classic result Eq. (12.223), the authors state

> For infinite matrices an analogous statement holds in all cases studied heretofore; however it can occur that the index n on the right hand side, in addition to a sequence of discrete values, also runs through a continuous range of values, corresponding to an integral component [in Eqs. (12.224) and (12.218)].
>
> The quantities W_n are called "eigenvalues", they constitute in their entirety the "mathematical" spectrum of the form, consisting of a "point-" and "continuous-" spectrum. This is, as we shall see, identical with the "term-spectrum" of physics, while the "frequency spectrum" arises by taking differences.
>
> This principal-axis-transformation now gives us directly the solution of our dynamical problem: a transformation $(p_0 q_0) \rightarrow (pq)$ should be found which leaves [the fundamental commutation relations Eqs. (12.196)–(12.198)] invariant and simultaneously brings the energy into a diagonal matrix (pp. 583–584).

In the second section of chapter 3, the authors re-examine time-independent perturbation theory, recovering in a more streamlined way the results of chapter 2. The principal axis transformation for an energy matrix **H** is developed as follows.[60] One attempts to find solutions for the linear equations

$$\sum_l H(kl)x_l = Wx_k. \tag{12.225}$$

Note that an arbitrary finite proportionality factor can be used to multiply all the x_k without altering the validity of this equation.

In modern language we would say that the application of the matrix **H** to the column vector of (complex) coefficients x_k produces the same vector, scaled by a real number W. W is then said to be the eigenvalue corresponding to the eigenvector (x_k). In particular, if W should correspond to the nth discrete stationary state energy W_n, we may specify this in the eigenvector notation by adding n as a column specifier:

$$\sum_l H(kl)\, x_{ln} = W_n\, x_{kn}. \tag{12.226}$$

Notice that we are now using (following the Three-Man-Paper) a peculiar hybrid notation, with some indices (e.g., in $H(kl)$) indicated within parentheses, and others (e.g., in x_{kn}) as subscripts, in line with modern practice. Changing n to m and taking the complex conjugate, we obtain

$$\sum_l H^\star(kl)x^\star_{lm} = \sum_l H(lk)x^\star_{lm} = W_m x^\star_{km}. \tag{12.227}$$

If we multiply Eq. (12.226) by $x^\star_{km}$, Eq. (12.227) by x_{kn}, subtract the two and sum over k, we find

$$(W_n - W_m)\sum_k x^\star_{km}x_{kn} = 0. \tag{12.228}$$

This equation states that either $W_n = W_m$, or $\sum_k x^\star_{km}x_{kn} = 0$ (or both). The latter quantity is just the complex inner product of the eigenvectors associated to W_n and W_m, and we may say that *eigenvectors corresponding to different eigenvalues are orthogonal.* Through a suitable choice of the proportionality factor mentioned above one can further arrange that, for $m = n$,

$$\sum_k x_{kn}x^\star_{kn} = 1. \tag{12.229}$$

As the authors point out, this (together with $\sum_k x^\star_{km}x_{kn} = 0$ for $m \neq n$) means that "the x_{kn} form an orthogonal matrix, $\mathbf{S} = (x_{kn})$" (p. 354). In modern terminology,

[60] We have slightly rearranged the equations to make them more recognizable for the modern reader.

Eq. (12.229) says that the (squared) norm of the column vector of complex coefficients x_{kn}, with $k = 1, 2, 3, \ldots$ and for fixed n, is unity.

There are two problems with these statements, the resolution of which is postponed to the third section of chapter 3. First, it may well be the case that there are solutions of Eq. (12.225) for which the norm $\sum_k x_{kn}x^{\star}_{kn}$ of the corresponding eigenvector is infinite so that the vector cannot be normalized to unity by applying a scalar factor. This typically occurs when the eigenvalue W can take on a continuous range of values. Secondly, one is by no means guaranteed that the number of eigenvectors, labeled by index m, n, etc., is "as large" as the infinite number of rows and columns (labeled above by k, l) of the energy matrix. For example, one might be dealing with a system with only a finite number of bound states, so that the array x_{kn} has an infinite number of rows but a finite number of columns.

In the second section of chapter 3, however ("Application to perturbation theory"), Born *et al.* are mainly concerned with executing the principal axis transformation *staying within the sector of discrete energy eigenvalues, corresponding to normalizable eigenvectors.* The essential property to be relied on, which follows from the discussion above, is that, for all m and n corresponding to discrete eigenvalues

$$\sum_k x_{kn}x^{\star}_{km} = \delta_{mn}. \tag{12.230}$$

Thus, if one takes the variables x_k and y_k, related to one another via

$$x_k = \sum_m x^{\star}_{km}y_m, \tag{12.231}$$

and forms the usual quadratic form

$$\begin{aligned}\sum_{kl} H(kl)\, x_k x^{\star}_l &= \sum_{kl}\sum_{mn} H(kl)\, x^{\star}_{km}x_{ln}y_m y^{\star}_n \\ &= \sum_{mn}\sum_{k} W_n\, x^{\star}_{km}x_{kn}y_m y^{\star}_n \\ &= \sum_m W_m\, y_m y^{\star}_m,\end{aligned} \tag{12.232}$$

where in the last step we used Eq. (12.230), we see that the quadratic form in the highly restricted set of variables x_k in Eq. (12.231) is indeed reduced to diagonal form by the transformation obtained by stacking the discrete eigenvectors of the matrix **H** into a (not necessarily square) array. Of course, in cases like the one-dimensional oscillators that were the bread and butter of early matrix mechanics, there are only discrete energy stationary states (as the potentials rise to infinity, disallowing unbounded motion), so in this case the discussion needs no amplification.

The new algebraic approach is applied to the development of perturbation theory for the eigenvalues and eigenvectors of an energy matrix of the form

$$H(kl) = \delta_{kl} W_l^0 + \lambda H_1(kl) + \lambda^2 H_2(kl) + \dots. \tag{12.233}$$

We focus our attention on the perturbative expansion of the energy of the nth state, which at zeroth order is just W_n^0. The components of the corresponding eigenvector are x_{kn}. As the energy is assumed diagonal at the lowest order, the corresponding eigenvector is simply a column vector with an arbitrary number (we shall constrain this number below by imposing normalization conditions) in the nth slot and zero elsewhere:

$$x_{kn}^0 = y_n^0 \delta_{kn}. \tag{12.234}$$

If we impose the normalization condition Eq. (12.229), then to lowest order in λ, we must take $|y_n^0|^2 = 1$, or $y_n^0 = e^{i\varphi_n^0}$, with φ_n^0 an undetermined phase constant.

We follow the calculations through the first order only. Inserting Eq. (12.233) and

$$W_n = W_n^0 + \lambda W_n^1 + \dots, \quad x_{kn} = x_{kn}^0 + \lambda x_{kn}^1 + \dots \tag{12.235}$$

into Eq. (12.226) and collecting terms of order λ, we find

$$\sum_l \left(\delta_{kl} W_l^0 x_{ln}^1 + H_1(kl) x_{ln}^0 \right) = W_n^1 x_{kn}^0 + W_n^0 x_{kn}^1. \tag{12.236}$$

Noting that the first sum on the left-hand side gives $W_k^0 x_{kn}^1$ and rearranging terms, we can rewrite this as

$$\begin{aligned} (W_n^0 - W_k^0) x_{kn}^1 &= -W_n^1 x_{kn}^0 + \sum_l H_1(kl) x_{ln}^0 \\ &= -W_n^1 \delta_{kn} y_n^0 + H_1(kn)\, y_n^0, \end{aligned} \tag{12.237}$$

where in the last step we used Eq. (12.234). For $k = n$, this implies

$$W^1 = H_1(nn), \tag{12.238}$$

while for $k \neq n$, we have

$$x_{kn}^1 = -\frac{H_1(kn)\, y_n^0}{W_k^0 - W_n^0}, \quad k \neq n. \tag{12.239}$$

The component x_{nn}^1 is undetermined at this order—but once again, the imposition of the normalization condition Eq. (12.229) will constrain it.

We begin to see the results of Eqs. (12.157) and (12.158) emerging directly from the new algebraic interpretation. The previous calculations had implicitly made the phase choice that the diagonal elements of the canonical transformation matrix **S** were real, and equal to unity to lowest order in λ (as one wrote $\mathbf{S} = 1 + \lambda \mathbf{S}_1 + \ldots$), so in particular, $y_n^0 = 1$. In fact, the general phase freedom available in the diagonalization problem is easily understood by looking at the orthogonalization condition Eq. (12.230), which remains satisfied if we replace x_{kn} by $e^{i\varphi_n} x_{kn}$. In particular, the φ_n can obviously be chosen to remove the complex phase in each diagonal element x_{nn}. As we remarked earlier, this phase freedom would later take on a deeper meaning in terms of the ultimate framing of the quantum states of the theory as rays in a Hilbert space.

For non-degenerate systems, the determination of the energies through first-order terms requires no further diagonalization than that already performed to render the zeroth order energy diagonal: as shown, the first order shift in each non-degenerate level W_n^0 is just $H_1(nn)$. This is no longer the case once degeneracy is present in the lowest-order energy. In the final part of chapter 3, section 2, the authors discuss degenerate perturbation theory from the new algebraic angle.

We saw above that in the non-degenerate case, the freedom in the allowed orthogonal transformations at the lowest order level (i.e., the transformations leaving the diagonal energy matrix $\delta_{kl} W_l^0$ invariant) amounted to a complex phase factor for each state. This freedom expands considerably once degeneracy is present. To make the analysis as simple as possible, we assume that there is a single set of r degenerate levels (in the Three-Man-Paper, these are assumed to be the states labeled $n, n+1, \ldots, n+r-1$) with energy W_n^0, with no other degeneracies present in the system. Let us use m and n to indicate the r degenerate levels and k and l to refer to all the other states, differing in energy from W_n^0 (and from each other). The orthogonal matrix corresponding to assembling the eigenvectors of $\mathbf{H}_0$ in columns, as described above, will reduce, as previously, to complex phase factors for the non-degenerate levels

$$x_{kl}^0 = e^{i\varphi_l^0} \delta_{kl}, \tag{12.240}$$

while within the degenerate sector, at lowest order (i.e., as far as H_0 is concerned) we are free to choose an arbitrary $r \times r$ unitary matrix, as the energy matrix H_0 is a multiple of the identity (by the single energy W_n^0), so any set of orthogonal normalized vectors can be chosen as the energy eigenvectors in this sector:

$$\sum_{n'} x_{n'n}^0 x_{n'm}^{0\star} = \delta_{nm}. \tag{12.241}$$

Also, the zeroth-order transformation matrix $\mathbf{x}^0$ has no matrix elements connecting the degenerate with the non-degenerate levels:

$$x_{km}^0 = x_{mk}^0 = 0. \tag{12.242}$$

The non-trivial unitary matrix x_{nm}^0 can now be chosen to diagonalize the energy matrix through first order by simply picking the transformation that diagonalizes the first-order energy (non-diagonal) matrix $H_1(nm)$, as in the degenerate sector the zeroth-order energy is a multiple of the identity matrix, $H_0(nm) = W_n^0\delta_{nm}$, and is left unchanged by an arbitrary unitary rotation.

The specific matrix x^0_{nm} to be used will depend on the nature of the first-order perturbing energy H_1 (which may be due to a weak magnetic field in the z-direction, or a weak electric field in the x-direction, etc.). However, the sum of squares of coordinate matrix elements connecting a specific non-degenerate state k to all degenerate levels m is independent of the perturbation H_1. If we perform the canonical transformation induced by the x^0 unitary rotation on the coordinate matrix element,

$$q(mk) = \sum_{nl} x^0_{mn}\, q_0(nl)\, x^{0\star}_{kl} = e^{-i\varphi^0_k} \sum_{n} x^0_{mn}\, q_0(nk), \tag{12.243}$$

one easily verifies with the help of Eq. (12.241) (and with some index relabeling) that

$$\begin{aligned}\sum_m q(mk)\, q^\star(mk) &= \sum_{mnn'} x^0_{mn} x^{0\star}_{mn'}\, q_0(nk)\, q^\star_0(n'k) \\ &= \sum_{nn'} \delta_{nn'}\, q_0(nk)\, q^\star_0(n'k) \\ &= \sum_n q_0(nk)\, q^\star_0(nk).\end{aligned} \tag{12.244}$$

This exactly reproduces the intuition that had led Bohr to propose the idea of spectroscopic stability: the imposition of a very tiny external field, of whatever kind (i.e., whatever the Hermitian H_1), should lead to a total intensity of spectral transitions (proportional in dipole radiation to the square of the associated coordinate matrix element) from a given state k to a set m of now (very slightly) split levels, which is essentially the same as the previous total intensity from the given state to any of the spectroscopically indistinguishable (since degenerate) n states. It is apparent from the preceding discussion that these observations apply to weak field situations, as they were obtained within the context of first order perturbation theory. As the authors point out, Eq. (12.244) "provides a mathematical representation of the so-called spectroscopic stability, which has played an important part in the more recent theories of fine-structure intensities" (p. 358; cf. Section 11.1) They promise to return to these intensities in chapter 4 of their paper.

12.3.4 Third chapter (cont'd): continuous spectra

In the final section of chapter 3 of their paper, Born *et al.* attempt to show that their discrete version of quantum mechanics, based on infinite-dimensional Hermitian matrices, is sufficiently general to encompass the common situation in Nature, where the same system displays both discretely quantized quantum states, as well as states of continuous energy. This is of course the case for the hydrogen atom, where the states in which the electron is bound to the nucleus (classically described by elliptical or circular orbits) have discretely quantized energy levels (with the total energy negative), but the states with positive total energy, in which the electron eventually recedes to infinite distance from the nucleus (classically described by hyperbolic orbits), have arbitrary, continuously variable total (positive) energy. The results described in this section are fully correct and rigorous, and anticipate many aspects of the transformation theory of Jordan (1927b, 1927g) and Dirac (1927a), which would emerge in the next two years, and even the final completion of the formalism by von Neumann (1927a) in terms of self-adjoint operators in Hilbert space (see Chapters 16 and 17). What obscures this connection is that the theoretical framework used—the Hilbert–Schmidt–Hellinger theory of bilinear forms—is no longer well known (not even to mathematicians, at least in its original form) and that assertions that would be obvious (even trivial) to modern physicists when phrased in the now-ubiquitous Dirac notation look strange and unfamiliar in the notation of the time.

The possibility of a continuous spectrum (i.e., set of eigenvalues) arising from an infinite matrix with a denumerably infinite set of rows and columns may seem strange, but in fact, just this circumstance had already been encountered in the mathematical literature in the two decades preceding the Three-Man-Paper, and in fact, the seminal work in this nascent field of functional analysis had been done precisely in Göttingen, so Born and his co-authors would have been able to consult the world experts in the field by walking a short distance. The results that were directly of interest for the new matrix mechanics were the theorems of Hellinger on the spectral theory of bounded Hermitian forms. We provide a quick exposition of Hellinger's results in Appendix C, taking advantage of the spectral theory (in Dirac notation) that can be found in any modern textbook on quantum mechanics. To maintain chronological fidelity in our story, however, we only illustrate the main points of the Hellinger theory, sticking as much as possible to the original notation, using a simple, analytically solvable example that makes the emergence of a continuous spectrum from a discrete matrix completely transparent.

Hellinger's spectral theory was developed for the case of *bounded* (equivalently, *continuous*; see Appendix C) Hermitian forms,

$$H(xx^{\star}) = \sum_{kl} H(kl)\, x_k x_l^{\star}, \tag{12.245}$$

which take real values bounded by a finite positive number B for all unit vectors:

$$H(xx^{\star}) \leq B, \qquad \sum_{k=1}^{\infty} |x_k|^2 = 1. \tag{12.246}$$

The existence of such a bound is related to the boundedness of the spectrum of the matrix with components $H(kl)$—simply put, there is a maximum absolute value for the eigenvalues. For many of the kinematic quantities of interest in quantum mechanics (momentum, energy, position, etc.) the spectrum is unbounded, as we must physically allow for arbitrarily large energies, momenta, etc. Fortunately for the authors of the Three-Man-Paper, the essential results of Hellinger's theory for bounded forms carry through to the type of matrices needed in quantum mechanics, as von Neumann (1929) would show with his spectral theorem for densely defined self-adjoint operators (bounded or unbounded) in Hilbert space. In fact, the theory of Hermitian forms had been most fully developed by Hilbert for an even more special subclass of bounded forms—the *completely continuous* (or *compact*) bilinear forms. These are based on infinite-dimensional matrices that are in some sense the most precise analogs to the finite Hermitian matrices, which automatically satisfy Eq. (12.223). The infinite-dimensional, completely continuous forms can be shown to possess a purely point spectrum, so that a linear transformation of variables can always be found which reduces the form to an infinite diagonal (sum of pure squares) expression (cf. Eq. (12.232)):[61]

$$\sum_{kl} H(kl) x_k x_l^\star = \sum_n W_n y_n y_n^\star. \tag{12.247}$$

Hellinger's theory asserts that for the more general class of bounded forms, Eq. (12.247) must be generalized to include the possibility of a continuous spectrum, with the discrete sum on the right-hand side supplemented by a continuous integral:

$$\sum_{kl} H(kl) x_k x_l^\star = \sum_n W_n y_n y_n^\star + \int W(\varphi)\, y(\varphi)\, y^\star(\varphi)\, d\varphi. \tag{12.248}$$

The appearance of a continuous integral component on the right-hand side is associated with the existence of eigenvector solutions $x_k(W)$ of Eq. (12.225)

$$\sum_l H(kl) x_l(W) = W x_k(W), \tag{12.249}$$

where the eigenvalue W can assume a continuous range of values. The arguments leading to Eq. (12.228) can be repeated verbatim in this case, whence,

$$(W' - W'') \sum_k x_k(W')\, x_k^\star(W'') = 0. \tag{12.250}$$

Unlike the case for the discrete point spectrum, this equation implies an extremely singular dependence of the orthogonality product $\sum_k x_k(W')\, x_k^\star(W'')$ on the continuously variable eigenvalues W' and W''. As Born *et al.* put it, "this function . . . becomes wildly irregular, if it exists at all" (p. 360). After all, it must vanish identically for $W' \neq W''$, but for $W' = W''$ one typically obtains ∞! The "eigenvectors" for points in the continuous spectrum are not normalizable in the usual way (cf. Eq. (12.229)): the sum of squares of the components $x_k(W)$ typically diverges and we cannot therefore make this sum equal to unity by scaling it with a finite normalization factor.

[61] Completely continuous forms/operators do not play a significant role in the early stages of quantum mechanics, and therefore do not concern us in this volume: however, they would become significant tools in the resolution of some important difficulties in multi-particle scattering theory in the 1960s.

As we illustrate with a simple example of Hellinger's theory, the continuous functions $y(\varphi)$ appearing in Eq. (12.248), which turn out to be linear combinations of the x_k variables on the left-hand side, are determined in terms of the continuous eigenfunctions $x_k(W)$ as follows. Assuming the continuous spectrum to begin at the value W_0, one defines a positive function $\varphi(W)$:

$$\varphi(W) \equiv \sum_k \left| \int_{W_0}^{W} x_k(W')\, dW' \right|^2 . \tag{12.251}$$

It can be shown that the function $\varphi(W)$, which is manifestly positive, is also a monotone increasing function of W. The functions $y(\varphi)$ appearing in the spectral representation Eq. (12.248) are then linear combinations of the original x_k variables, as follows,

$$y(\varphi)d\varphi = \sum_k x_k(W)dW\, x_k. \tag{12.252}$$

We illustrate these relations and the essential features of the Hellinger theory with a simple example, using only methods and notation available at the time of the Three-Man-Paper (see Appendix C for a modern treatment in Dirac notation). The complex Hermitian aspects of the theory are irrelevant for our purposes, so we start with a real bilinear form $\sum_{mn} H(mn)\, x_m x_n$, based on a real symmetric matrix $H(mn) = H(nm)$, given by

$$\mathbf{H} = \begin{pmatrix} -2 & 1 & 0 & 0 & 0 & 0 & \cdots \\ 1 & -2 & 0 & 0 & 0 & 0 & \cdots \\ 0 & 0 & 2 & -1 & 0 & 0 & \cdots \\ 0 & 0 & -1 & 2 & -1 & 0 & \cdots \\ 0 & 0 & 0 & -1 & 2 & -1 & \cdots \\ 0 & 0 & 0 & 0 & -1 & 2 & \cdots \\ \vdots & \vdots & \vdots & \vdots & \vdots & \vdots & \ddots \end{pmatrix}, \tag{12.253}$$

corresponding to the bilinear form

$$H(xx) = -2(x_{-1}^2 + x_{-2}^2 - x_{-1}x_{-2}) + 2\sum_{k=1}^{\infty} x_k(x_k - x_{k+1}), \tag{12.254}$$

where for reasons that will shortly become apparent, we have labeled the initial variables with an index set $-2, -1, 1, 2, 3, \ldots$ (rather than, more conventionally, $1, 2, 3, \ldots$). We will show that the spectrum of $\mathbf{H}$ consists of a point spectrum containing the discrete eigenvalues $-3, -1$, and a continuous spectrum consisting of the entire open interval (0,4).

A glance at the matrix $\mathbf{H}$ makes it clear that the two-dimensional sector corresponding to the index values $-2, -1$ does not communicate with the rows and columns labeled by positive index values $1, 2, 3, \ldots$ The 2×2 real symmetric matrix at the top left is trivial to diagonalize: the eigenvalues are -3, corresponding to the unit eigenvector $\frac{1}{\sqrt{2}}(1, -1, 0, 0, \ldots)$ and -1, corresponding to the unit eigenvector $\frac{1}{\sqrt{2}}(1, 1, 0, 0, \ldots)$. In the language of bilinear forms, the introduction of new variables y_1 and y_2, defined as

$$y_1 \equiv \frac{1}{\sqrt{2}}(x_{-2} + x_{-1}), \quad y_2 \equiv \frac{1}{\sqrt{2}}(x_{-2} - x_{-1}), \tag{12.255}$$

reduces the first terms in Eq. (12.254) to quadratic form:

$$-2(x_{-1}^2 + x_{-2}^2 - x_{-1}x_{-2}) = -3y_1^2 - y_2^2 \equiv W_1 y_1^2 + W_2 y_2^2. \tag{12.256}$$

The remaining part of $\mathbf{H}$ (corresponding to rows and columns indexed $1, 2, 3, \ldots$, and the infinite sum contribution to $H(xx)$ in Eq. (12.254)) can be treated completely separately and, as we will now see, possesses a completely continuous spectrum, corresponding to the real open interval (0,4).[62] From the trigonometric identity

$$2\sin(k\vartheta) - \sin((k-1)\vartheta) - \sin((k+1)\vartheta) = 2(1 - \cos(\vartheta))\sin(k\vartheta) \tag{12.257}$$

we see that the matrix $\mathbf{H}$ has, for every value $0 < \vartheta < \pi$, an eigenvector with eigenvalue

$$W(\vartheta) = 2(1 - \cos(\vartheta)) \tag{12.258}$$

and components

$$x_k(W) = C(\vartheta)\sin(k\vartheta), \quad k = 1, 2, 3, \ldots \tag{12.259}$$

where $C(\vartheta)$ is an arbitrary normalization factor (which can be chosen independently for each distinct eigenvector labeled by ϑ, hence, can be an arbitrary function of ϑ, non-zero on the interval $0 < \vartheta < \pi$). The reader can readily verify this by writing out the action of part of the matrix in Eq. (12.253) on these eigenvectors:

$$\begin{pmatrix} 2 & \cdots & 0 & 0 & 0 & 0 & 0 & \cdots \\ \vdots & \ddots & \vdots & \vdots & \vdots & \vdots & \vdots & \cdots \\ 0 & \cdots & -1 & 2 & -1 & 0 & 0 & \cdots \\ 0 & \cdots & 0 & -1 & 2 & -1 & 0 & \cdots \\ 0 & \cdots & 0 & 0 & -1 & 2 & -1 & \cdots \\ \vdots & \vdots & \vdots & \vdots & \vdots & \vdots & \vdots & \ddots \end{pmatrix} \begin{pmatrix} \sin\vartheta \\ \vdots \\ \sin((k-1)\vartheta) \\ \sin(k\vartheta) \\ \sin((k+1)\vartheta) \\ \vdots \end{pmatrix} = W \begin{pmatrix} \sin\vartheta \\ \vdots \\ \sin((k-1)\vartheta) \\ \sin(k\vartheta) \\ \sin((k+1)\vartheta) \\ \vdots \end{pmatrix}.$$

[62] We have chosen this particular infinite-dimensional matrix in part because it appears naturally in the treatment of one-dimensional crystal dynamics already treated by Born and von Kármán (1912), where the calculation of the frequencies of vibration involves the diagonalization of just this matrix. Born and von Kármán (1912, p. 298) explicitly cite Toeplitz's theory of infinite dimensional systems of this type, a special case of the more general Hilbert–Hellinger theory.

The kth row of this matrix equation reproduces Eq. (12.257).

As expected, the eigenvector is not square-normalizable. For any ϑ between 0 and π,

$$\sum_{k=1}^{\infty} \sin^2(k\vartheta) = \infty. \tag{12.260}$$

The continuous spectrum is simply the range of values assumed by $W(\vartheta)$ in Eq. (12.258), namely, the open interval $(0, 4)$. The angle parameter ϑ is related to the "energy" eigenvalue W by

$$\vartheta(W) = \arccos(1 - W/2). \tag{12.261}$$

The eigenvalues of $\mathbf{H}$ are bounded in magnitude by 4, so we are dealing with a bounded Hermitian matrix/form, to which the Hellinger theory applies.

To confirm the Hellinger spectral theorem Eq. (12.248) for this example, we shall need a result from the classical theory of Fourier series. The set of eigenfunctions $\sin(k\vartheta)$, with $k = 1, 2, \ldots$, form a complete set of orthogonal functions on the interval $0 < \vartheta < \pi$. In particular, completeness amounts to the statement that, for any continuous functions $f(\vartheta)$ and $g(\vartheta)$:[63]

$$\sum_{k=1}^{\infty} \int_0^{\pi} f(\vartheta) \sin(k\vartheta) d\vartheta \int_0^{\pi} g(\vartheta') \sin(k\vartheta') d\vartheta' = \frac{\pi}{2} \int_0^{\pi} f(\vartheta) g(\vartheta) d\vartheta. \tag{12.262}$$

The first step in verifying the Hellinger theory is the calculation of the Hellinger "basis function" $\varphi(W)$ defined in Eq. (12.251). Changing variables in the integrals from W to ϑ (see Eqs. (12.258) and (12.261))

$$\begin{aligned} \varphi(W) &= \sum_k \int_0^{\vartheta(W)} C(\vartheta) \sin(k\vartheta) 2 \sin(\vartheta) d\vartheta \int_0^{\vartheta(W)} C(\vartheta') \sin(k\vartheta') 2 \sin(\vartheta') d\vartheta' \\ &= \frac{\pi}{2} \int_0^{\vartheta(W)} C(\vartheta)^2 4 \sin^2(\vartheta) d\vartheta. \end{aligned} \tag{12.263}$$

This implies $d\varphi = C(\vartheta)^2\, 2\pi \sin^2(\vartheta) d\vartheta$ or

$$\frac{d\varphi}{d\vartheta} = 2\pi C(\vartheta)^2 \sin^2(\vartheta). \tag{12.264}$$

By definition (where we used Eq. (12.258) to set $dW = 2\sin(\vartheta) d\vartheta$),

$$y(\varphi) d\varphi = \sum_k x_k(W) dW x_k = C(\vartheta) \sum_k \sin(k\vartheta) \cdot 2 \sin(\vartheta) d\vartheta\, x_k, \tag{12.265}$$

[63] In modern notation, this integral property (which is how completeness is stated in Hilbert–Hellinger theory) is more compactly expressed using the Dirac delta function, $\sum_{k=1}^{\infty} \sin(k\vartheta) \sin(k\vartheta') = (\pi/2)\, \delta(\vartheta - \vartheta')$.

or, using Eq. (12.264),

$$y(\varphi) = \frac{1}{\pi C(\vartheta)\sin(\vartheta)}\sum_k x_k \sin(k\vartheta). \tag{12.266}$$

The continuous part of the Hellinger spectral representation Eq. (12.248) is therefore

$$\begin{aligned}
&\int W(\varphi)y(\varphi)^2 d\varphi \\
&= \int \frac{2(1-\cos(\vartheta))}{\pi^2 C(\vartheta)^2 \sin^2(\vartheta)} \sum_k x_k \sin(k\vartheta) \sum_l x_l \sin(l\vartheta)\, 2\pi C(\vartheta)^2 \sin^2(\vartheta) d\vartheta \\
&= \sum_{kl} x_k x_l \frac{2}{\pi}\int_0^\pi 2(1-\cos(\vartheta))\sin(k\vartheta)\sin(l\vartheta)d\vartheta.
\end{aligned} \tag{12.267}$$

Note that the dependence on the arbitrarily chosen normalization factor $C(\vartheta)$ has disappeared at this point. Using the orthogonality property,

$$\frac{2}{\pi}\int_0^\pi \sin(k\vartheta)\sin(l\vartheta)d\vartheta = \delta_{kl}, \tag{12.268}$$

and the identity Eq. (12.257), we find

$$\frac{2}{\pi}\int_0^\pi 2(1-\cos(\vartheta))\sin(k\vartheta)\sin(l\vartheta)d\vartheta = 2\delta_{kl} - \delta_{k-1,l} - \delta_{k+1,l}. \tag{12.269}$$

This is recognizable as the portion of the $H(kl)$ matrix corresponding to index values $k, l = 1, 2, 3, \ldots$, which we saw earlier gave a completely continuous spectrum. Inserting this result in Eq. (12.267), one recovers the (continuous spectrum part of) the original quadratic form Eq. (12.254):

$$\sum_{kl} x_k x_l (2\delta_{kl} - \delta_{k-1,l} - \delta_{k+1,l}) = 2\sum_k x_k(x_k - x_{k+1}). \tag{12.270}$$

The Hellinger spectral theory implies (at least for bounded Hermitian matrices) a transformation from an initially fully discrete description of the (non-diagonal) energy matrix to a new, partially discrete, partially continuous description with rows and columns labeled by the stationary states of the theory. The transformation is effected by a quasi-orthogonal "matrix" with both discrete and continuous column indices:

$$\mathbf{S} = (x_{kn}, x_k(W)dW). \tag{12.271}$$

The orthogonality properties of this generalized type of transformation are discussed in more detail in Appendix C. Here, we just outline the application made of these ideas by

Born *et al.* in the Three-Man-Paper. As usual, we begin with an initial description of the theory in terms of the (fully discrete) coordinate and momentum matrices $q_0(kl), p_0(kl)$ associated with a fully solvable initial energy matrix $H_0(kl)$, where H_0 has a fully discrete spectrum (it could be, for example, an oscillator Hamiltonian). The actual system under consideration, with energy matrix $H = H_0 + \lambda H_1$, say,[64] is in general not fully solvable except in a perturbative expansion in λ. Nevertheless, the calculation of, for example, dipole transition intensities in the full theory will require transforming the discrete $q_0(kl)$ matrix to the coordinate matrix appropriate for the new Hamiltonian H, which may have a partially continuous spectrum. Thus, the indices for the new coordinate matrix $\mathbf{q} = \mathbf{S}\mathbf{q}_0\mathbf{S}^{-1}$, will be partially discrete and partially continuous, and the matrix elements fall into four categories, namely,

$$q(mn),\ q(m, W')dW',\ q(W, n)dW,\ q(W, W')dWdW', \tag{12.272}$$

which the authors describe as corresponding to four different types of transitions between stationary states in the Bohr theory:

> There are four types of "transitions" which to some extent furnish a simple analog to the "transitions" postulated hitherto in the theory of the hydrogen atom, viz. (1) from ellipse to ellipse; (2) from ellipse to hyperbola; (3) from hyperbola to ellipse; (4) from hyperbola to hyperbola (p. 363).

In modern terms, these quantities are just the dipole matrix elements needed for the calculation of transition probabilities (1) between two bound electron states; (2) for ionization of a bound electron to an unbound one; (3) for capture of an unbound electron to a bound state; and (4) for radiative transitions between unbound electron states induced by a Coulomb potential of a nearby hydrogen nucleus (for example). As Born *et al.* explicitly acknowledge, the application of the Hellinger spectral theory is not really justified in matrix mechanics, given the fact that the important kinematic variables (position, momentum, angular momentum, energy, etc.) are represented by unbounded Hermitian matrices—indeed, the corresponding classical quantities can assume arbitrarily large values—while the Hellinger theory specifically assumes bounded forms. This shortcoming of the available mathematical technology would be definitively resolved within a few years with von Neumann's proof of the spectral theorem for unbounded self-adjoint operators in Hilbert space, so that by 1930, the necessary functional analysis forming the rigorous mathematical basis for quantum mechanics (in either its matrix or wave-mechanical formulation) was in place, as presented in von Neumann's (1932) classic text. However, the form of the spectral representation Eq. (12.248), and the essential

[64] Note that by appropriate choice of H_1 we can convert an arbitrary initial Hamiltonian H_0 to the desired final one H. Of course, the rapidity of convergence of perturbation approximations will be strongly affected by the choice of H_0. For example, the energy matrix for the hydrogen atom can be written in terms of the coordinate and momentum matrices appropriate for a three-dimensional isotropic harmonic oscillator, with a completely discrete spectrum: the resulting discrete matrix (complicated and non-diagonal!) will have both the usual discrete Bohr point spectrum of bound states and a continuous spectrum corresponding to the unbound states.

content of the Hellinger theory, remains valid even in the unbounded case, so the discussion of continuous spectra in chapter 3, section 3 of the Three-Man-Paper is in fact perfectly correct (as far as it goes), and contains the germs of the much more transparent and physically intuitive transformation theory, which would be developed within two years by Jordan and Dirac (see Chapter 16).

12.3.5 Fourth chapter: physical applications of the theory

In the first section of chapter 4 of the Three-Man-Paper, entitled "Laws of conservation of momentum and angular momentum; intensity formulae and selection rules," a correct treatment of angular momentum in three-dimensional systems in quantum theory finally makes an appearance, fully twelve years after Bohr (and Nicholson) first injected the quantization of angular momentum into atomic theory in his treatment of the hydrogen atom spectrum (see Chapter 4). The contents of this section can be attributed to joint work of Jordan and Heisenberg (based on later recollections of Jordan; see Mehra and Rechenberg 1982c, p. 157). The critical new ingredient making this analysis possible was the introduction of the multi-dimensional commutation relations for position and momentum, Eqs. (12.196)–(12.198). The angular momentum vector $\vec{M}$ for a particle with momentum $\vec{p} = (p_x, p_y, p_z)$ and coordinate $\vec{q} = (q_x, q_y, q_z)$ is given by the cross-product $\vec{p} \times \vec{q}$ (note that the sign is opposite to the modern definition), or, for a set of particles labeled by the index k, the total angular momentum is given by three matrices,

$$\mathbf{M}_x = \sum_k (\mathbf{p}_{ky}\mathbf{q}_{kz} - \mathbf{q}_{ky}\mathbf{p}_{kz}), \tag{12.273}$$

$$\mathbf{M}_y = \sum_k (\mathbf{p}_{kz}\mathbf{q}_{kx} - \mathbf{q}_{kz}\mathbf{p}_{kx}), \tag{12.274}$$

$$\mathbf{M}_z = \sum_k (\mathbf{p}_{kx}\mathbf{q}_{ky} - \mathbf{q}_{kx}\mathbf{p}_{ky}). \tag{12.275}$$

Note that, although products of coordinate and momentum matrices occur in each term, the order is immaterial, as the two matrices always involve different Cartesian components of the corresponding vector, which therefore commute. Also, of course, the matrices for different particles automatically commute. A short calculation, based on Eq. (12.196), gives the commutator of two different components of the angular momentum, for example,

$$\mathbf{M}_x\mathbf{M}_y - \mathbf{M}_y\mathbf{M}_x = \frac{h}{2\pi i}\mathbf{M}_z, \tag{12.276}$$

with the two other independent commutators obtained by cyclic permutation of the Cartesian coordinates ($x \to y, y \to z, z \to x$). Note the change in sign of the factor on the right-hand side, which in modern quantum mechanics is $ih/2\pi = -h/2\pi i$, which goes back to the peculiar sign choice adopted in the Three-Man-Paper for the definition of angular momentum (which uses a "left-hand rule", rather than the usual "right-hand rule" in modern texts).

The first physical application of the matrix-mechanical treatment of angular momentum in the Three-Man-Paper concerns the derivation of angular momentum selection rules that had previously been arrived at by a combination of empirical guesswork and correspondence-principle arguments. Let $\mathbf{q}_{lz}$ be the matrix representing the z-component of the coordinate of the lth electron (for example, the light-emitting valence electron in optical spectra), and $\mathbf{M}_z$ the total angular momentum of all the atomic electrons, as indicated by the sum in Eq. (12.275). Then

$$\mathbf{q}_{lz}\mathbf{M}_z - \mathbf{M}_z\mathbf{q}_{lz} = \mathbf{q}_{lz}(\mathbf{p}_{lx}\mathbf{q}_{ly} - \mathbf{q}_{lx}\mathbf{p}_{ly}) - (\mathbf{p}_{lx}\mathbf{q}_{ly} - \mathbf{q}_{lx}\mathbf{p}_{ly})\mathbf{q}_{lz} = 0, \quad (12.277)$$

as $\mathbf{q}_{lz}$ commutes with $\mathbf{p}_{lx}$ and $\mathbf{p}_{ly}$. If we assume that the dynamics is such that $\mathbf{M}_z$ is conserved (i.e., no net torque in the z-direction on the electrons) and that the system is non-degenerate, then $\mathbf{M}_z$ becomes a diagonal matrix (with the diagonal eigenvalues M_{zn}) and the (nm) matrix element of Eq. (12.277) becomes

$$q_{lz}(nm)(M_{zn} - M_{zm}) = 0. \quad (12.278)$$

Recall that at this point the emission of light by atomic electrons is still considered a semi-classical process, in which the dipole moment $e\vec{q}_l$ of the lth electron emits classical Maxwellian radiation eliciting a transition between states m and n as though the associated virtual oscillator were a charged particle with dipole moment $eq_l(nm)$. If in this transition, $M_{zn} \neq M_{zm}$, then Eq. (12.278) implies $q_{lz}(nm) = 0$, so the emitted spherical radiation must be polarized as though it originates in the oscillation of a charged particle in the x–y plane.

The transverse (x, y) components of the dipole moment can be similarly examined, through the calculation of

$$\begin{aligned}\mathbf{q}_{lx}\mathbf{M}_z - \mathbf{M}_z\mathbf{q}_{lx} &= \mathbf{q}_{lx}(\mathbf{p}_{lx}\mathbf{q}_{ly} - \mathbf{q}_{lx}\mathbf{p}_{ly}) - (\mathbf{p}_{lx}\mathbf{q}_{ly} - \mathbf{q}_{lx}\mathbf{p}_{ly})\mathbf{q}_{lx} \\ &= (\mathbf{q}_{lx}\mathbf{p}_{lx} - \mathbf{p}_{lx}\mathbf{q}_{lx})\mathbf{q}_{ly} \\ &= i\frac{h}{2\pi}\mathbf{q}_{ly}, \end{aligned} \quad (12.279)$$

and similarly,

$$\mathbf{q}_{ly}\mathbf{M}_z - \mathbf{M}_z\mathbf{q}_{ly} = -i\frac{h}{2\pi}\mathbf{q}_{lx}. \quad (12.280)$$

Examining the (nm) matrix elements of these equations, one finds

$$q_{lx}(nm)(M_{zm} - M_{zn}) = i\hbar q_{ly}(nm), \tag{12.281}$$
$$q_{ly}(nm)(M_{zm} - M_{zn}) = -i\hbar q_{lx}(nm). \tag{12.282}$$

Thus, if the quantum number associated with M_z does not change in a transition, the associated radiation must have $q_{lx}(nm) = q_{ly}(nm) = 0$, that is, the emitted light is linearly polarized in the z-direction. By inserting Eq. (12.282) in Eq. (12.281), and using $q_{l\perp}$ to indicate either x or y component of the coordinate matrix, one finds

$$\{(M_{zm} - M_{zn})^2 - \hbar^2\}\, q_{l\perp}(nm) = 0. \tag{12.283}$$

The appearance of z-polarized light (requiring $q_{lz}(nm) \neq 0$) is therefore associated (via Eq. (12.278)) with the selection rule $\Delta M_z = 0$, while circularly polarized light (requiring $q_{l\perp}(nm) \neq 0$) is associated, via Eq. (12.283), with the selection rule $\Delta M_z = \pm\hbar$. Together, the selection rule $\Delta M_z = 0, \pm\hbar$ reproduces the Bohr prescription originally justified on the basis of correspondence-principle arguments (cf. Section 5.3).

In the following pages of chapter 4, section 1, Born *et. al.* develop many of the familiar formulas of angular momentum algebra, which can be found in any undergraduate text on quantum mechanics. As the arguments used are basically the same as those found in modern texts,[65] we briefly summarize the results:

1. The (total) angular momentum components commute with the square (e.g., $\mathbf{q}_l^2 = \mathbf{q}_{lx}^2 + \mathbf{q}_{ly}^2 + \mathbf{q}_{lz}^2$) of three-vector kinematical quantities:

$$\mathbf{q}_l^2\mathbf{M}_z - \mathbf{M}_z\mathbf{q}_l^2 = \mathbf{p}_l^2\mathbf{M}_z - \mathbf{M}_z\mathbf{p}_l^2 = 0, \tag{12.284}$$

   and, in particular,

$$\mathbf{M}^2\mathbf{M}_z - \mathbf{M}_z\mathbf{M}^2 = 0. \tag{12.285}$$

2. The "raising" (resp. "lowering") matrices $\mathbf{M}_x + i\mathbf{M}_y$ (resp. $\mathbf{M}_x - i\mathbf{M}_y$) have vanishing matrix elements except for transitions in which the eigenvalue of $\mathbf{M}_z$ changes by $+1$ (resp. -1).
3. The possible eigenvalues of $\mathbf{M}^2$ take the form $j(j+1)\hbar^2$, where $j = m_{\max}$ ($m_{\max}$ the largest eigenvalue of $\mathbf{M}_z$) must be integer or half-integer to ensure symmetry of the spectrum of $\mathbf{M}_z$ around zero.

The final pages of chapter 4, section 1 of the Three-Man-Paper provide an awkward and messy derivation of the selection rule for the magnitude j of total angular

[65] See, for example, (Griffiths 2005, sec. 4.3).

momentum (where $\mathbf{M}^2 = j(j+1)\hbar^2$).[66] The outcome of the calculations are that the coordinate matrices for any individual electron $\mathbf{q}_{lx}, \mathbf{q}_{ly}, \mathbf{q}_{lz}$ only connect stationary states subject to the selection rule $\Delta j = 0, \pm 1$. The ratio of the corresponding matrix elements for varying values of the magnetic quantum number m are then shown to agree with the Goudsmit–Kronig–Hönl intensity rules previously derived from correspondence-principle arguments. As we pointed out in Chapter 11, Heisenberg's (1925c) "derivation"[67] in his *Umdeutung* paper of these intensity formulas for split Zeeman lines contained errors, which can be traced back to the incorrect execution of correspondence-principle ideas in the final pages of his paper. The new, correct, formulas indeed "agree with the intensity formulae derived from correspondence considerations" (p. 374), but the reference given at this point is to the work of Goudsmit and Kronig (1925), and Hönl (1925)—Heisenberg's misfires in *Umdeutung* are passed over in charitable silence.

The very brief (two paragraphs) section 2, entitled simply "The Zeeman effect," is clearly an attempt to acknowledge the importance of the Uhlenbeck–Goudsmit electron spin hypothesis, of which the authors had become aware during the preparation of the Three-Man-Paper, but the implications of which they had not yet had a chance to explore. The first paragraph discusses one last hope to account for the mysteries of the anomalous Zeeman effect (specifically, the apparent failure of the Larmor precession theorem,[68] which leads inevitably to the normal Zeeman effect) by invoking effects quadratic in the strength of the magnetic field, which are explicitly neglected in the derivation of the Larmor theorem. Such effects might be more important than expected if there were close connections imposed by quantum theory between lower states and very high bound (or even unbound) orbits. But in the second paragraph, the authors admit that this proposal is hardly likely to produce a satisfactory explanation for the rich phenomenology of the anomalous Zeeman effect. Instead, they return to the Uhlenbeck–Goudsmit–proposal, which they had previously mentioned in the introduction to the paper:

> For the nuclear atom it does not seem to be entirely excluded that the intimate connection between innermost and outermost orbits leads to findings which differ somewhat from the normal Zeeman effect. However, we must emphasize that a whole set of weighty reasons speak against the possibility of explaining the anomalous Zeeman effects on this basis. Rather, one might perhaps hope that the hypothesis of Uhlenbeck and Goudsmit might later provide a quantitative description of the above-mentioned phenomena (p. 375).

[66] The selection rule for j is an immediate consequence of the Wigner–Eckart theorem, which would emerge within a few years.

[67] In fact, as we pointed out in Chapter 11, Heisenberg derived formulas of similar appearance, but not identical to the correct Goudsmit–Kronig–Hönl ones.

[68] The Larmor theorem referred to here asserts that the entire effect of an imposed magnetic field on a system of charged particles, of identical charge/mass ratio, *to first order in the magnetic field*, amounts to an additional uniform precession of the system around the axis of the field.

This emblematic failure of the old quantum theory, it would thus seem, remained yet to be solved with the new methods—but not for long. By the middle of March 1926, Heisenberg and Jordan (1926) submitted a paper to *Zeitschrift für Physik* in which the essential features of the anomalous Zeeman effect (for alkali doublet spectra) were shown to follow simply and directly from the new theory (see Section 15.3.1).

The final section (chapter 4, section 3) of this remarkable paper was written entirely by Pascual Jordan, and plays an unusual Janus-like role in the history of quantum theory.[69] On the one hand, it avowedly looks backward, to the second fluctuation theorem of Einstein (1909a; 1909b; cf. Section 3.4.3), that exhibited the extraordinary dual presence of particle and wave terms in the formula for energy fluctuations in black-body radiation. Jordan's derivation of this dual behavior on the basis of a single, unified dynamical framework—that provided by the new matrix mechanics—served as an additional point of support for the admittedly radical new approach to mechanics, especially impressive as it was quite unrelated to the dispersion-theoretic investigations that had provided the primary initial motivation for quantum mechanics. On the other hand, the technique used by Jordan here also looks forward, inasmuch as it is based on the quantization of a one-dimensional string, a system of infinitely many degrees of freedom, using matrix-mechanical methods, and as such represents the first example of a quantum field theory, albeit a very simple one—a one-dimensional free field.

The title of the section, "Coupled harmonic resonators. Statistics of wave fields," gives a clue as to the methodology and objectives of the calculation. Jordan begins by reminding the reader of a well-known property of classical systems such as crystal lattices, or, of more immediate relevance, the "vibrations of an elastic body idealized to a continuum or finally, of an electromagnetic cavity" (p. 375). Such systems possess a Hamiltonian energy function that can be transformed by appropriate choice of coordinates into a system of *uncoupled* harmonic oscillators. Such oscillators are, of course, the simplest system amenable to analytic treatment by the new methods discussed in the first three chapters of the Three-Man-Paper. After discussing previous attempts to understand radiative energy fluctuations, in which the quantal nature of the electromagnetic field is not imposed *ab initio*, Jordan states

> in each such treatment of cavity radiation by quantum theory hitherto, one encountered the fundamental difficulty that although it led to Planck's law of radiation, it did not yield the correct mean square deviation of energy in an element of volume. One thus finds that a consistent treatment of the natural vibrations of a mechanical system or an electromagnetic cavity in accordance with past theory leads to most serious contradictions. This caused us to hope that the modified kinematics which forms an inherent feature of the theory proposed here would yield the correct value for the interference fluctuations (pp. 376–377).

In way of preparation for our account of this last remarkable, and still under-appreciated, portion of the Three-Man-Paper, we briefly review the original results of

[69] For discussion of this paper, see Duncan and Janssen (2008) and Bacciagaluppi, Crull, and Maroney (2017).

Einstein, which motivated Jordan's investigation. The energy fluctuations present in a fixed subvolume V of a closed container within which electromagnetic radiation (say, of frequency ν) is present can be calculated first under the assumption that the radiation consists of a large number of randomly moving localized "light quanta" each of energy $h\nu$, where on average $\overline{n}$ quanta are present in the subvolume, so that the mean energy in the subvolume is $\overline{E}_V = \overline{n}h\nu$. In this case the root-mean-square (RMS) fluctuation in the number of quanta in the subvolume is $\sqrt{\overline{n}}$, and the consequent mean square energy fluctuation is evidently

$$\overline{\Delta E^2} = (\sqrt{\overline{n}}h\nu)^2 = h\nu\overline{E}_V. \tag{12.286}$$

This relation between the average mean energy in a region and the mean deviation from this average is characteristic of energy that is partitioned into discrete localized units, as would be the case for particles of a gas, for example.

On the other hand, in a purely classical wave model of the electromagnetic radiation in a container, the energy in a subvolume fluctuates from time to time as a result of varying constructive and destructive interference effects, and the typical mean square energy fluctuation is proportional to the average energy *squared*, $\overline{\Delta E^2} \propto \overline{E}^2$. This is perfectly reasonable as the result of a random interference between two waves of a typical amplitude A gives a resultant with an amplitude also of order A. In fact, all that is needed for this result to hold is that the fields comprising the wave be a random superposition of monochromatic components (of given wavelength and direction), with totally uncorrelated phase differences between different components.

It is important to realize that both of these classical fluctuation results—for the mean square fluctuation in the mean number of particles (say, of a gas) in a subvolume, or of the mean square energy of the electromagnetic field in a subvolume of a cavity—depend only on a very general condition of disordered motion in both cases. In particular, it is not necessary, either in the particle or the wave case, that the system be in thermal equilibrium with a heat bath at some temperature T. For a system of point particles moving independently and bouncing elastically off the walls of their container, it is only necessary that the initial locations and direction of velocities be random for the mean number fluctuation in a subvolume to equal the square root of the average occupancy of the subvolume.[70] For example, the square root property applies even if all the particles have the same speed (but are randomly oriented in direction and initial position)—very far indeed from a thermal (Maxwell–Boltzmann) distribution characteristic of a system at a well-defined temperature. Similarly, for the energy of waves, (Lorentz 1916) established that the mean square energy was proportional to the square of the average energy (again, in a subvolume) simply by assuming an adequate incoherence of the Fourier components, with random phase differences between the different modes of the wave

[70] This is most easily established with particles moving in a finite one-dimensional segment; the generalization to several dimensions follows immediately if the particles are non-interacting as each Cartesian coordinate behaves independently.

field. We will see that an analog of this more general property of random, but not necessarily thermal, systems reappears in the quantum case, in Jordan's calculations of the energy fluctuations of a quantized string, leading to the puzzling (at first sight) absence of any reference to a temperature T in his results.

The second part of the title of this section, "Statistics of wave fields," points to the really novel contribution of this part of the Three-Man-Paper, as it marks the beginning of two extraordinarily important fields of modern physics: quantum statistical mechanics, which lies at the heart of condensed matter physics; and quantum field theory, which is the structural framework for our most fundamental theories of the interaction of matter at the smallest scales. First, in order to execute a thoroughly quantum-mechanical treatment of a thermal system such as black-body radiation, Jordan had to address the question of the status of the new statistics, which had only a year earlier been introduced by Bose to derive the Planck formula, and later extended to monatomic gases by Einstein, *when the new kinematics for describing quantum systems was fully taken into account.*

Jordan's approach to reconciling matrix mechanics with Bose statistics[71] amounted simply to taking the re-expression of electromagnetic energy as a sum of decoupled harmonic oscillators—basically, the standing wave eigenmodes[72] of the electromagnetic field confined to a cavity—seriously from a matrix-mechanical point of view, where the canonical momentum and coordinate matrix from each independent eigenmode of frequency ν satisfied the fundamental commutation relation, while the matrices for independent eigenmodes commuted with each other, just like the matrices representing independent coordinate/momentum variables in any other system with multiple degrees of freedom. A state of the electromagnetic field was then completely, and uniquely, specified by giving the energy quantum number for the quantum harmonic oscillator associated with each mode. Thus, a state in which there are n_1 light quanta of frequency ν_1, n_2 light quanta of frequency ν_2, etc., should be denoted simply by giving the list $n_1, n_2, \ldots$, and counted *once* in any statistical computation. The n_1 quanta (as we now say, photons) are in no way separable into distinct entities, but counted simply as a unit, with unit statistical weight accorded to the state once the occupation numbers, as it were, of each eigenmode are specified. The state of the whole field would then have a total energy $h(n_1\nu_1 + n_2\nu_2 + \ldots)$, *provided the zero-point energy $h\nu_i/2$ for each oscillator* is dropped.[73] Once the possible eigenmodes are identified, the enumeration of the possible quantum states becomes a trivial task.

We have already seen in Chapter 3 (cf. Eqs. (3.9)–(3.13)) how these modes are identified and counted in a three-dimensional cubical cavity of sides L and volume $V = L^3$. The upshot is that one finds $z_\nu^{3D} V\Delta\nu$ modes in the frequency interval $(\nu, \nu + \Delta\nu)$, where

[71] See Section 13.5 for a full discussion of the Bose and Einstein papers.

[72] These eigen-modes are also sometimes referred to as "eigenvibrations" or "cells" by Jordan.

[73] Jordan proposes that this part of the energy be called the "thermal energy", which is misleading in that the description being advanced has nothing whatsoever in principle to do with thermal equilibrium. Some justification for this terminology may be nevertheless be given by noting that Jordan's thermal energy, missing the (infinite) zero-point contribution, is the part subject to change when heat energy is added or subtracted from the electromagnetic energy. The reappearance of an infinite zero-point energy is unpleasantly reminiscent of, but quite different from, the ultraviolet catastrophe of classical radiation theory.

$z_\nu^{3D} = 8\pi\nu^2/c^3$. A similar calculation carried out for standing waves confined to a one-dimensional line interval of length L gives $z_\nu^{1D} L\Delta\nu$ modes in this interval, with $z_\nu^{1D} = 2/c$. In either case, the application of Bose statistics to the situation where the electromagnetic field is in thermal equilibrium with the external environment at temperature T (e.g., the material walls of the cavity) gives an average thermal energy of the Planck form

$$\overline{E} = \frac{z_\nu h\nu}{e^{h\nu/kT} - 1} \cdot V, \tag{12.287}$$

with the volume V to be interpreted as the length L in the case of a wave confined to a one-dimensional line interval $0 < x < L$.

On well-known Boltzmannian principles (cf. Volume 1, pp. 121–123), the mean-square fluctuation in the energy in a subvolume V of thermal radiation is simply related to the temperature derivative of this average energy, provided, as Jordan says, "there is communication between a volume V and a very large volume such that waves having frequencies which lie within a small range ν to $\nu + d\nu$ can pass unhindered from one to the other" (p. 379). Indeed, Einstein had found in 1909 that the mean square energy fluctuation for black-body radiation in the frequency range $(\nu, \nu + \Delta\nu)$, in a subvolume V, using thermodynamic formulae of unimpeachable validity, was (cf. Eq. (3.77)), if one used the Planck distribution formula,

$$\overline{\Delta E^2} = \frac{\overline{E}^2}{z_\nu V} + h\nu\overline{E}, \tag{12.288}$$

where $\overline{E}$ is the average energy in the given range in the subvolume. Evidently, this result is a linear combination of terms that are individually of exactly the type expected for either waves or particles, additively superposed. It is exactly this extraordinary—even paradoxical—result that provides the impetus for Jordan's seminal idea in this last section of the Three-Man-Paper. What was needed was a direct and unambiguous connection between the requirements of the new quantum kinematics and this peculiar hybrid result for energy fluctuations, typical of the schizophrenic form—both particles, and waves—that quantum physics had increasingly taken since Einstein's 1905 photoelectric paper. Jordan's achievement in the closing pages of the Three-Man-Paper was to show that Einstein's fluctuation formula emerges from a straightforward application of Heisenberg's *Umdeutung* procedure to classical waves, and that the two terms in the formula, at first sight quite irreconcilable with each other, arise from a unified dynamical framework, once the new quantum kinematics is imposed on the classical system. To simplify the calculations, without losing the features essential for displaying the desired particle–wave duality, Jordan simply imagines a classical wave realized on a one-dimensional string of length L held stationary at both ends. The general displacement of the string can then be written as a function $u(x,t)$ constrained by $u(0,t) = u(L,t) = 0$.[74] Such a function can be written as a Fourier series,

$$u(x,t) = \sum_{j=1}^{\infty} q_j(t)\sin(k_j x), \quad k_j = \frac{j\pi}{L}. \tag{12.289}$$

The energy of the wave is the sum of kinetic and potential energy terms that can be chosen (by appropriate choice of the mass per unit length and tension in the string) to take the form[75]

$$H = \frac{1}{2}\int_0^L \left((\frac{\partial u}{\partial t})^2 + (\frac{\partial u}{\partial x})^2\right) dx = \sum_j (\frac{p_j^2}{2m} + \frac{1}{2}m\omega_j^2 q_j^2), \tag{12.290}$$

with $m = L/2$ and $p_j = m\dot{q}_j$. We see that, as claimed earlier, the energy function for the classical wave can be re-expressed as a sum of independent simple harmonic oscillators, labelled by the index j, all of "mass" $L/2$, and with the jth oscillator having a natural frequency of $\omega_j = k_j = j\pi/L$.[76] Jordan proposes, in accordance with the "new kinematics", that the momentum and coordinate variables p_j, q_j for each classical standing wave mode be promoted to matrices $\mathbf{p}_j, \mathbf{q}_j$ in the usual way, conforming to the canonical commutation relation

$$\mathbf{p}_j\mathbf{q}_j - \mathbf{q}_j\mathbf{p}_j = \frac{h}{2\pi i}\cdot 1. \tag{12.291}$$

The Hamiltonian H of the string, as given in Eq. (12.290), lacks any explicit time dependence, and the system as a whole therefore satisfies conservation of energy. Accordingly, the matrix version $\mathbf{H}$ of H must be a diagonal matrix, where the rows and columns are identified by specifying the set of quantum numbers $(n_1, n_2, \ldots)$, which determines the energy contribution arising from each of the independent oscillators,[77] with the diagonal matrix element in the state "n", which Jordan uses as a shorthand for the full set $(n_1, n_2, \ldots)$, given by $h(n_1\omega_1 + n_2\omega_2 + \ldots)/2\pi$. These diagonal matrix elements are time independent, and the matrix as a whole has no non-zero off-diagonal matrix elements. The question of energy fluctuations for the string as a whole is therefore moot: as energy is conserved, there are none.

If one considers, however, as Einstein had done in his study of energy fluctuations in black-body radiation in 1909, the energy in a subvolume of the cavity, one is led to

[74] We have altered in some minor ways the notation to make the results more readable for the modern reader.

[75] The summation on the right-hand side is easily obtained by inserting the Fourier expansion Eq. (12.289), and using the orthogonality relations $\int_0^L \sin(k_j x)\sin(k_l x)dx = m\delta_{jl}$, where $m = L/2$.

[76] The wave equation $u'' = \ddot{u}$ corresponds to a choice $c = 1$ for the "speed of light".

[77] Note that for states of the string with finite energy, at most a finite number of these n_j can be non-zero, and the stationary states of the string can be ordered lexicographically in any convenient way. The number n can then be considered as a single ordinal integer n encoding the set of all states $(n_1, n_2, \ldots)$, ordered in any suitable way (cf. Section 12.3.2).

investigate, in the case of the string, the energy contained in a proper subsegment of the string, say, the segment $0 < x < a$ with $a < L$, given classically by

$$H_a = \frac{1}{2}\int_0^a \left((\frac{\partial u}{\partial t})^2 + (\frac{\partial u}{\partial x})^2 \right) dx, \qquad (12.292)$$

and it is then easy to see that the resulting matrix version $\mathbf{H}_a$ contains a non-trivial time dependence. The algebra is somewhat tedious in this case, and we refer the reader to Duncan and Janssen (2008), where Jordan's calculations are dissected and explained in detail. The important point is that it is now possible to compute a non-trivial mean square fluctuation in the energy E_a of this portion of the string.

Jordan interprets the "mean" in "mean energy" or "mean square energy" to refer to: (a) the time average of the respective quantity (in this case, the energy in the subinterval $0 < x < a$), which in matrix mechanics amounts to eliminating the off-diagonal elements of the corresponding matrix, and considering only the diagonal matrix elements; and (b) examining the resulting diagonal matrix element for a stationary state n of the whole string Hamiltonian specified by prescribing the mode excitation numbers $n_1, n_2, \ldots$ for all the allowed eigenvibrations of the string. Since the calculations are somewhat lengthy and are given in detail in Duncan and Janssen (2008), we simply state Jordan's results here.[78] For the mean energy of the subinterval, one finds the unsurprising result

$$\overline{E} \equiv H_a(n,n) = \frac{a}{L}H(n,n). \qquad (12.293)$$

For the desired mean-square-fluctuation (i.e., the diagonal (n,n) matrix element of the square of the energy fluctuation matrix $\mathbf{\Delta} \equiv \mathbf{H}_a - \overline{E}$) Jordan obtains (using the canonical commutation relations of the $\mathbf{p}_j$, $\mathbf{q}_j$, as indicated),

$$\overline{\Delta^2} \;\; (= \mathbf{\Delta}^2(n,n)) = h\nu\overline{E} + \frac{\overline{E}^2}{z_\nu V}, \quad \text{(Three–Man–Paper, equation(55))}, \qquad (12.294)$$

where $z_\nu = z_\nu^{1D} = 2/c = 2$ (as the speed c of the wave on the string was chosen to be unity), and the "volume" V is just the length a of the subinterval being considered. This is the one-dimensional version of Einstein's fluctuation theorem, demonstrating that the squared energy fluctuations are simply an additive result of contributions of pure particle and pure wave form, respectively. The appearance of the quantum particle term (with the characteristic presence of h) is a direct consequence of the non-commutativity of the coordinate and momentum variables evident in Eq. (12.291).

This (at first sight) extremely gratifying result becomes, after a little thought, rather surprising, as nowhere in the calculation does the concept of a thermalized system, characterized by a definite temperature T, appear. Einstein's fluctuation theorem explicitly

[78] For a more easily digestible version of Jordan's calculation, streamlined by the use of the formalism of raising and lowering operators, see Duncan (2012, pp. 21–26).

relies on thermal arguments (cf. Section 3.4.3), which relate the mean square energy fluctuation to a temperature derivative of a thermal mean energy (Eq. (3.71)). On the other hand, as we saw previously, the classical results for the root mean square (RMS) fluctuation of particle number in a subregion, or of classical wave energy in a subvolume of a cavity, only depend on a certain degree of randomness in their respective systems: random distribution and direction of velocity of the particles ("molecular chaos") in the former case, decoherence (phase randomness) of the different Fourier components of the waves in the latter. The acquisition of a thermal distribution of the modes by equilibration with a heat bath at a definite temperature is *consistent* with such randomness, but is not *necessary* to obtain the "particle" and "wave" terms of Einstein's formula, in a classical context. The surprising aspect of Jordan's result is that it refers not to a thermal ensemble, but rather to the mean energy, and mean squared energy of a subvolume of the wave system, *in a definite stationary state n of the string*, specified by fixing the quantized occupation numbers $n_1, n_2, \ldots$ of all the modes of vibration (which are just the energy excitation quantum numbers of each mode—in today's terminology, the "photon numbers" for each allowed frequency). In the language of modern quantum physics, Jordan has calculated the quantum energy dispersion of the subvolume energy in a pure state of the string. Where is the desired "randomness" that would seem to be necessary to yield a fluctuation formula containing the dual particle/wave structures of the Einstein formula?

A complete understanding of the success of Jordan's calculation in reproducing a fluctuation formula of manifestly thermal roots in a clearly non-thermal situation requires a brief anachronistic excursion in our story, employing results and concepts unavailable to Jordan in 1925. The essential ideas were introduced two years later in Dirac's (1927b) seminal paper on quantum electrodynamics. There, operators N_r, $r = 1, 2, \ldots$ are introduced that represent the number of excitations in each eigenmode of the field (whose eigenvalues are thus just Jordan's $n_1, n_2, \ldots$, etc,. for the one-dimensional wave). One can also introduce operators θ_r, which represent the *phase* of the associated (rth) Fourier component. It turns out that the number and phase operators satisfy canonical commutation relations completely analogous to those of coordinate and momentum for particles. The quantum state of an electromagnetic field, in which the excitation values N_r of each Fourier mode is exactly specified, will necessarily have a completely unspecified (infinite dispersion) in the associated phase θ_r, exactly analogous to the situation exploited by Lorentz classically to derive the mean square energy fluctuation formula, $\overline{\Delta E^2} = \overline{E}^2/(z_\nu V)$, for classical waves.[79]

The pure state used by Jordan to calculate a mean square energy dispersion is therefore *maximally phase dispersed*, just the situation needed in the classical context to obtain the typical fluctuation formulas in either the particle or the wave case. It is not surprising therefore that Jordan was able to recover a formula with apparently thermal origins without any reference to thermal equilibration, which is sufficient, but *not necessary*, for

[79] We are here appealing to the association between non-commutativity of observables and their mutual uncertainty that would emerge in Heisenberg's famous uncertainty principle paper (Heisenberg 1927b), and be further deepened in Jordan's axiomatization of quantum principles (Jordan 1927b).

the derivation of the mean square fluctuation formula.[80] If Jordan was surprised that he had managed to reproduce Einstein's avowedly thermal result without once mentioning a temperature T (or heat bath), he gives no indication of that in the concluding pages of the Three-Man-Paper.

Actually, Jordan would later point to his quantization of a wave field with some pride as an important contribution to the development of quantum theory—in fact, in a letter to van der Waerden in 1961, he insisted that "what the Three-Man-Paper says about energy fluctuations in a field of quantized waves is, in my opinion, almost the most important contribution I ever made to quantum mechanics" (quoted, e.g., in Duncan and Janssen 2008, p. 639). It is also clear that his co-authors on the Three-Man-Paper did not share his judgment on this matter, and seem to have regarded the calculations presented in the final section with considerable suspicion. Heisenberg was particularly troubled by the extension of matrix-mechanical principles to a system with an infinite number of degrees of freedom, where the total zero-point energy would be infinite—an infinity which, as we saw, Jordan simply subtracted out and ignored throughout his calculation. In fact, the whole calculation can be restricted to only the (finite number of) modes in a narrow frequency interval, so the worries about the total zero-point energy are really misplaced. Although there is no direct evidence on this front, one imagines that Born and Heisenberg may also have been puzzled—as the discussion in the preceding paragraphs describes—by the appearance of Einstein's *thermal* fluctuation formula from a calculation where there is no whiff of heat baths, statistical ensembles *à la* Boltzmann, Planck, Bose, or Einstein, but simply the calculation of the energy dispersion in a fixed stationary state of the string. For more details of the uneasy acceptance of the Jordan calculation by his closest collaborators—all the way up to an article in 1939 by Born and Klaus Fuchs criticizing Jordan's fluctuation result, and later retracted entirely—we refer the reader to the discussion in Duncan and Janssen (2008, pp. 640–642).

[80] Exactly the same situation holds in the case of Planck's famous formula Eq. (2.7) for the average energy of a harmonic resonator in equilibrium with an ambient black-body field: the formula can be derived just employing phase decoherence of the Fourier modes of the field, with no reference to thermal equilibrium. See the web resource, *Classical Dispersion Theory.*

13

De Broglie's Matter Waves and Einstein's Quantum Theory of the Ideal Gas

13.1 De Broglie and the introduction of wave–particle duality

In 1923 Louis de Broglie, a doctoral student at the University of Paris, introduced a concrete realization of the optical–mechanical analogies circulating at least from the time of Maupertuis (and made very explicit in the work of Hamilton), by directly associating a wave phenomenon with the motion of material particles. De Broglie's theory was outlined in some short communications to *Comptes Rendus* in 1923, and laid out in much greater detail in his doctoral thesis, which he defended on November 25, 1924. In the introduction to his thesis de Broglie summarizes his motivation for this remarkable move:

> Guided by the idea of a general relation between the concepts of frequency and energy, we propose in the present work the existence of a periodic phenomenon, of a nature still to be determined, which should be associated with every isolated parcel of energy, and which depends on its mass according to the equation of Planck and Einstein [$h\nu = E = mc^2$]. The theory of relativity then leads us to associate to the uniform motion of every material point [particle] the propagation of a certain wave, of which the phase is propagated in space more rapidly than the speed of light (Ch. 1).
>
> In order to generalize this result to the case of non-uniform motion, one is led to propose a proportionality between the momentum world vector of a material particle and a vector characteristic of the propagation of the wave, of which the time component is the frequency. The application of Fermat's principle to the wave then becomes identical to the principle of least-action applied to the particle. The rays of the wave are identical to the possible trajectories of the particle (Ch. 2) (De Broglie 1924, p. 1).[1]

[1] Here and in the following, the page numbers refer to the English translation of De Broglie's thesis by Kracklauer but unless noted otherwise the translations are our own. The thesis was published in *Annales de Physique* (De Broglie 1925). An abridged translation can be found in Part Two of Ludwig (1968, pp. 73–93), with a historical introduction in Part One.

Constructing Quantum Mechanics. Anthony Duncan and Michel Janssen, Oxford University Press.
 DOI: 10.1093/oso/9780198883906.003.0013

13.2 Wave interpretation of a particle in uniform motion

In chapter 2 of his thesis, de Broglie introduces the basic framework for his theory, in the simplest possible context—that of a freely moving particle, which is to be associated with a monochromatic wave, of definite frequency and wavelength. The starting point is the "Planck–Einstein" equation,

$$h\nu_0 = m_0 c^2. \tag{13.1}$$

This relation represents, as de Broglie puts it, "the basis of our system, valid, as are all hypotheses, to the extent that the consequences deduced from them are found to hold" (p. 8).

In Eq. (13.1), a particle of rest mass m_0, in its rest frame, and hence, with relativistic energy m_0c^2, is to be associated with a periodic phenomenon of frequency ν_0, with the Planck relation "energy = Planck's constant times frequency" determining the frequency in terms of the rest energy. In this way each material particle can be regarded as a little "clock", and as every beginning student of relativity learns, such a clock when placed in motion with speed v (say, along the x direction) is seen by a stationary observer to be slowed down by the factor $\sqrt{1-\beta^2}$, where $\beta \equiv v/c$. The phase accumulated by the moving particle when it moves from the origin to coordinate $x = \beta ct$ (t as measured in the rest frame) is therefore

$$\varphi_1 = 2\pi\nu_0\sqrt{1-\beta^2}\frac{x}{\beta c}. \tag{13.2}$$

The question now arises as to the specific form (frequency and wavelength) of the wave motion to be associated with such a particle. De Broglie determines this by appealing to a "theorem of phase harmony" (p. 8): one examines the phase of a wave, which would be a spatially independent uniform periodic phenomenon in the rest frame, but would appear to have a spatially and temporally varying character in a frame moving relative to this with speed v, by the usual Lorentz transformation for time:

$$\varphi_2 = 2\pi\nu_0 t_0 = 2\pi\nu_0\frac{(t-\beta x/c)}{\sqrt{1-\beta^2}}. \tag{13.3}$$

The phase of this wave at time $t = x/\beta c$ (when the particle "clock" has arrived at x) is then found to agree with that registered by the "slowed-down particle clock" of Eq. (13.2),

$$\varphi_2 = 2\pi\frac{\nu_0\left(\dfrac{1}{\beta c}-\dfrac{\beta}{c}\right)x}{\sqrt{1-\beta^2}} = 2\pi\nu_0\sqrt{1-\beta^2}\frac{x}{\beta c} = \varphi_1. \tag{13.4}$$

As was the case for the fundamental relation Eq. (13.1), this rather obscure mode of reasoning is to be considered acceptable only to the extent that it leads to desired consequences. At first sight, this hardly seems to be the case, as a wave with the phase behavior given by Eq. (13.3) would seem to move with a superluminal phase velocity $c/\beta = c^2/v > c$. In fact, as de Broglie reminds us, it was well known from classical theory that the physical effects of wave phenomena depend not on the *phase velocity* of a monochromatic wave, but on the *group velocity* of the wave-packets obtained from superposition of such waves. Let us briefly review how these two quite different conceptions of the velocity of a wave are related.

The monochromatic wave described by the phase function appearing in Eq. (13.3) corresponds to some (co)sinusoidally varying quantity, which for algebraic convenience we can express as the real part of a complex exponential:

$$\cos(2\pi\nu_0\gamma t - 2\pi\nu_0\gamma\beta x/c) = \mathrm{Re}\, e^{-i(2\pi\nu_0\gamma t - 2\pi\nu_0\gamma\beta x/c)} = \mathrm{Re}\, e^{ikx - i\omega t}, \tag{13.5}$$

where $\gamma \equiv 1/\sqrt{1-\beta^2}$, and the wavenumber k, wavelength λ, and angular frequency ω are defined as

$$k = 2\pi\nu_0\gamma\beta/c \equiv 2\pi/\lambda, \tag{13.6}$$

$$\omega = 2\pi\nu_0\gamma \equiv 2\pi\nu. \tag{13.7}$$

Note that the kinematical information for the associated material (and massive, cf. Eq. (13.1)) particle is inserted through its velocity v, contained in $\beta = v/c$ (and thence, in γ). As usual, the phase velocity of such a wave—in this one-dimensional case, the speed with which a crest of the wave moves—is just $\lambda\nu = \omega/k = c/\beta$, which is greater than the speed of light c, as material particles must move with a speed $v < c$, that is, with $\beta < 1$. Note that the frequency ω is determined as a non-linear function of the wavenumber k: writing $\omega_0 \equiv 2\pi\nu_0$, we find from Eqs. (13.6)–(13.7),[2]

$$\omega(k) = \sqrt{\omega_0^2 + c^2k^2}. \tag{13.8}$$

Now let us suppose that waves of different wavenumber are superposed, with an amplitude $A(k)$ associated with the component of wavenumber k, and with $A(k)$ a smooth function strongly peaked around the value $k = \bar{k}$. One could, for example, take the Gaussian function

$$A(k) = A\exp(-(k-\bar{k})^2/2\Delta^2). \tag{13.9}$$

[2] Substituting ω_0 for $2\pi\nu_0$ in Eqs. (13.6) and (13.7) and squaring both equations, we find that $c^2k^2 = \omega_0^2\gamma^2\beta^2$ and $\omega^2 = \omega_0^2\gamma^2$. Subtracting the first from the second equation, we arrive at

$$\omega^2 - c^2k^2 = \omega_0^2\gamma^2(1-\beta^2) = \omega_0^2,$$

from which Eq. (13.8) immediately follows.

Defining $\omega(\bar{k}) = \bar{\omega}$, we can approximate the superposed wave by a Taylor expansion of the exponent around $\bar{k}$, as the amplitude prefactor $A(k)$ is by assumption strongly peaked around $k = \bar{k}$:

$$\begin{aligned}\int A(k)e^{i(kx-\omega(k)t)}\,dk &\approx e^{i(\bar{k}x-\bar{\omega}t)}\int A(k)e^{i\left((k-\bar{k})x-\frac{d\omega}{dk}\big|_{k=\bar{k}}(k-\bar{k})t\right)}\,dk \\ &= e^{i(\bar{k}x-\bar{\omega}t)}F_{\text{env}}\left(x-\frac{d\omega}{dk}\bigg|_{k=\bar{k}}t\right).\end{aligned} \tag{13.10}$$

Here, F_{env} is just the Fourier transform of $A(k)$. The monochromatic wave $e^{i(\bar{k}x-\bar{\omega}t)}$ (at a wavenumber and frequency determined by the peak at $k = \bar{k}$ of the amplitude function $A(k)$) is therefore modulated by an envelope $F_{\text{env}}(x-v_g t)$ moving with a "group velocity" v_g, given by

$$v_g = \frac{d\omega}{dk}\bigg|_{k=\bar{k}} = \frac{c^2\bar{k}}{\omega(\bar{k})}, \tag{13.11}$$

where in the last step we used Eq. (13.8). For example, if $A(k)$ is the Gaussian function cited above, the envelope function (its Fourier transform) would also be a Gaussian, namely, $F_{\text{env}} \propto \exp\left(-\Delta^2(x - v_g t)^2/2\right)$, with its peak at $x = v_g t$ moving with the group speed v_g.

Now substituting Eqs. (13.6)–(13.7) (where the wavenumber and frequency now refer to the values relevant for the peak wavenumber $\bar{k}$) in Eq. (13.11), we find

$$v_g = \frac{c^2 k}{\omega} = \frac{c\omega_0\gamma\beta}{\omega_0\gamma} = c\beta = v. \tag{13.12}$$

This is just the speed (now *less* than the speed of light) to be associated with the material particle corresponding to the de Broglie wave, or, in de Broglie's words, "*the group velocity of phase waves equals the velocity of its associated body*" (De Broglie 1924, p. 11). This is an eminently reasonable result, as it was already known classically that the transfer of energy (or, indeed, any physically measurable quantity) occurs with group velocity of the wave. The concentration of phase waves in the small region where the envelope is sizable can be associated with what we would now call a "wave packet" (De Broglie called it an "energy packet").

Using $\hbar\omega_o = h\nu_0 = m_0c^2$, the fundamental hypothesis for a particle at rest (see Eq. (13.1)), in combination with Eq. (13.8) (again, dropping the overbars to simplify the notation), we find that $\hbar$ times ω, the angular frequency associated with the moving particle, is given by

$$\hbar\omega(k) = \sqrt{\hbar^2\omega_0^2 + c^2\hbar^2k^2} = \sqrt{m_0^2c^4 + c^2\hbar^2k^2}, \tag{13.13}$$

which corresponds exactly to the relativistic energy equation,

$$E(p) = \sqrt{m_0^2 c^4 + c^2 p^2}, \tag{13.14}$$

if we identify $\hbar k$ with the momentum p of the particle (of rest mass m_0). Here, the equivalence of group velocity to particle velocity occurs via the simple relativistic identity $dE(p)/dp = pc^2/E = v$. This identification between spatial momentum and wavenumber is actually only introduced explicitly by de Broglie in the second chapter of his thesis, entitled "The principles of Maupertuis and Fermat." To understand what de Broglie is doing here, a digression into classical mechanics and optics is useful.

13.3 Classical extremal principles in optics and mechanics

The differential analytic approach to mechanical systems that was initiated by Newton's *Principia* and later extensively developed by Lagrange, Hamilton, Jacobi, and others lies at the foundation of the classical mechanics that supports the development of the old quantum theory, as we saw in Volume 1.[3]

However, an alternative, more global—and distinctly teleological—point of view concerning the motions of bodies goes much further back in time.[4] Its antecedents can be traced back to Aristotle and Hero of Alexandria. The notion that mechanical and optical phenomena were guided by an extremal principle reemerged in the seventeenth and eighteenth centuries, in the work of Fermat and Maupertuis. The underlying idea, the necessity to accomplish the motion as efficiently as possible, on the basis of some criterion suggested by the phenomena but often given a post hoc metaphysical or theological justification, was essentially the same as that proposed by Aristotle or Hero.

In the 1660s Fermat proposed that optical phenomena (both of reflection and refraction of light) could be subsumed under a single extremal principle, namely, that the route taken by a light ray from point A to point B is one that requires the least time. In particular, if the time taken by the light in the actual route is compared with that for paths that deviate slightly from it, the change in time is of second order in the magnitude of the small deviation.[5] In a medium of refractive index n, the speed of light (more exactly, the phase velocity) is given locally by c/n, so the time increment dt for the light

[3] See Appendix A for a review of the essentials of this approach.

[4] See the web resource, *Extremal Laws in Classical Optics and Mechanics*, for more detailed discussion.

[5] This principle of least time is closely related to Huygens's principle, proposed shortly after Fermat's work, in which the propagation of light is explained in terms of the superposition of spherical waves originating at any given time from the crest (wave front) of the propagating wave. In fact, Huygens was able to show the equivalence of his wave theory to Fermat's principle, with the light rays determined by the normals to the wave front at each point. The hegemony of the particle interpretation of light advanced by Newton more or less contemporaneously had not yet been established.

to advance a distance ds is $n\,ds/c$ and Fermat's principle can be expressed in the language of variational calculus simply as

$$\delta \int_A^B \frac{n}{c}\, ds = 0 \quad \text{or} \quad \delta \int_A^B n\, ds = 0, \tag{13.15}$$

where the speed of light in vacuo c is a fixed constant which can be multiplied out in the second version. The law of equality of angles of incidence and reflection, and Snell's law relating the angle of incidence to the angle of refraction, can be established quickly with this principle.[6]

About eighty years after Fermat had proposed his least time principle for light, Maupertuis proposed (in a speech to the *Académie des Sciences in Paris* in 1744) a "principle of the least quantity of action," where the action was something of the form $m \cdot v \cdot s$, with m the mass, v the speed, and s the distance covered by the moving body. Maupertuis applied this principle with inconsistent interpretations of the speed and distance components of this product, yet in each case—elastic and inelastic collisions of two particles, and the refraction of light particles (we are in the era of Newtonian corpuscular theory)—he obtained the correct results[7] by suitable massaging of the algebra. Maupertuis's principle was put on a precise, and as we now know, technically correct footing almost immediately (i.e., later in 1744) by Euler, who showed that the correct dynamics of a point particle moving in two dimensions could be obtained by demanding that the integral of the mass times velocity over the path followed by the particle be a minimum, which implies the stationarity condition

$$\delta \int mv\, ds = 0, \tag{13.16}$$

where s is the length covered along the chosen path. The variation in Eq. (13.16) has to be interpreted as comparing paths that maintain conservation of energy, with the particle subjected to forces given by the gradient of a potential function of position $V(\vec{r})$. Thus, one picks a definite total energy $E = \frac{1}{2}mv^2 + V$, and determines v at any point along any of the paths considered in the variation by setting $mv = \sqrt{2m(E-V)}$ at any point in the path integral. Euler surmised, and Lagrange was able to prove (in 1760), that the principle extended to a system of several particles (possibly moving in three dimensions). In modern notation, this would mean[8]

$$\delta S_0 \equiv \delta \int_{\vec{q}_i}^{\vec{q}_f} \sum_n p_n \dot{q}_n dt = \delta \int_{\vec{q}_i}^{\vec{q}_f} \sum_n p_n dq_n = 0. \tag{13.17}$$

[6] See the web resource, *Extremal Laws in Classical Optics and Mechanics*.

[7] In the optical case, Snell's law was obtained, but as the action was proportional to the speed, rather than the time or inverse speed, the incorrect inference was drawn that the ratio of refractive indices was proportional to the ratio of speeds, so that light particles *speeded up* on entering a denser optical medium.

[8] We denote the action S_0 here with a subscript zero to distinguish it from the action quantity introduced by Hamilton, which we shortly discuss.

Here, q_n and p_n are just the generalized coordinates and associated canonical momenta of the set of particles considered, and $\vec{q}_i$ and $\vec{q}_f$ are shorthand for the assembly of values of the q_n at the initial and final states of the particles. In those cases where the dynamics is represented by motion in a time-independent conservative potential $V(q_n)$, and with a kinetic energy quadratic in the first-order time derivatives of the q_n, the Lagrangian takes the form

$$\mathcal{L} = T - V = \frac{1}{2}\sum_{mn} \dot{q}_m K_{mn} \dot{q}_n - V(q_n), \tag{13.18}$$

where the "mass matrix" K_{mn} may itself depend on the coordinates q_n.[9] The canonical momenta are given by

$$p_m = \frac{\partial \mathcal{L}}{\partial \dot{q}_m} = \sum_n K_{mn} \dot{q}_n. \tag{13.19}$$

Thus, we find that the Maupertuis–Euler–Lagrange action can be written simply as the time integral of (twice) the kinetic energy:

$$\begin{aligned} S_0 &= \int_{\vec{q}_i}^{\vec{q}_f} \sum_n p_n \dot{q}_n dt \\ &= \int_{\vec{q}_i}^{\vec{q}_f} \sum_{mn} \dot{q}_m K_{mn} \dot{q}_n dt = \int_{\vec{q}_i}^{\vec{q}_f} 2T\, dt. \end{aligned} \tag{13.20}$$

In the late 1820s and early 1830s, the Irish mathematician Hamilton carried out a series of investigations in geometrical optics, culminating in the "Third supplement" (Hamilton 1837, originally presented in 1832) to his article "Theory of systems of rays" (Hamilton 1828).[10] In this supplement, Hamilton laid out a comprehensive mathematical theory of the geometrical structure of light rays obeying Fermat's principle in situations that involve reflection from (stationary) mirrors and/or motion through a medium of varying refractive index $n(\vec{r})$. The seminal character of this work can hardly be overemphasized, as it led not only to Hamilton's remarkable reformulation of classical mechanics, which would prove central to the development of the old quantum theory (as described in Volume 1), but also to an optical–mechanical analogy that would later provide one of the inspirations for Schrödinger's discovery of wave mechanics.

In his work on optics, Hamilton found that Fermat's integral Eq. (13.15) could be used to define a "characteristic function" V, which determines the direction of rays originating from any point $\vec{r}_A$ (denoted simply A in Eq. (13.15)) to any other point $\vec{r}_B$ (the final point in the Fermat path integral). The characteristic function was simply

[9] See the web resource, *Extremal Laws in Classical Optics and Mechanics.*
[10] See Hankins (1980, Ch. 4) for discussion of this paper and its supplements.

the Fermat path integral itself, but now integrated *along the actual path* $\vec{r}(s)$ for which the integral is stationary:

$$V(\vec{r}_B, \vec{r}_A) = \int_A^B n(\vec{r}(s))\, ds. \tag{13.21}$$

Let us suppose that this function is explicitly known. We can use it to determine the direction, as a unit vector $\hat{r}_B$, of the light ray passing through any point $\vec{r}_B$ which originated at some fixed source at $\vec{r}_A$ by simply taking the gradient of V:

$$\hat{r}_B = \frac{1}{n(\vec{r}_B)} \frac{\partial V(\vec{r}_B, \vec{r}_A)}{\partial \vec{r}_B}. \tag{13.22}$$

Alternatively, we can say that the rays originating from any $\vec{r}_A$ are normal at any point to the surface of constant V (with respect to varying the coordinate $\vec{r}_B$). As V measures the optical path length (or time of transit of the light, dividing by the speed of light c *in vacuo*), the existence of such a V is easily seen to be equivalent to Fermat's principle. From the point of view of the wave theory, and Huygens's principle, the surfaces of constant V represent the wave fronts for waves originating at the point source $\vec{r}_A$. Hamilton called these surfaces of constant V (or, of equal optical path length, or transit time from the point of origin) "surfaces of constant action".

Over the next couple of years, Hamilton presented the reformulation of mechanics for which he is most famous, published in two parts—"On a general method in dynamics" (Hamilton 1834) and "Second essay on a general method in dynamics" (Hamilton 1835). In the first essay, by direct imitation of the methods he had used in the preceding years to reformulate geometric optics, he introduced a mechanical analog of his optical characteristic function in Eq. (13.21). To simplify the notation and the algebra, we restrict our discussion to the dynamics of a single particle of mass m moving in three dimensions under the influence of a time-independent potential function $V(\vec{r})$ (not to be confused with the optical characteristic function above). The total energy is then given by $H = T + V$.[11] Maupertuis's principle asserts that the action (cf. Eq. (13.17))

$$S_0 = \int_{\vec{r}_i}^{\vec{r}_f} \vec{p} \cdot d\vec{r} \tag{13.23}$$

is stationary at the correct physical path taking the particle from initial location $\vec{r}_i$ to final location $\vec{r}_f$ with total energy E, *when the variation is performed over all energy-conserving paths of the same energy E connecting $\vec{r}_i$ to $\vec{r}_f$.* Along such paths the magnitude

[11] We do not adhere to Hamilton's notation for mechanical quantities, which differs almost completely from modern usage, and which would thoroughly confuse the present-day reader.

of the momentum is just $|\vec{p}| = \sqrt{2m(E - V)}$ so that the path integral becomes (writing $|d\vec{r}| = ds$, and using the fact that $d\vec{r}$ is parallel to $\vec{p}$)

$$S_0 = \int_{\vec{r}_i}^{\vec{r}_f} \sqrt{2m(E - V(\vec{r}))}\, ds. \tag{13.24}$$

For every prescribed value of $(\vec{r}_i, \vec{r}_f)$ and E, the evaluation of this integral at its stationary point defines a function of these variables, which we also denote by S_0:

$$S_0(\vec{r}_i, \vec{r}_f, E) = \int_{\vec{r}_i}^{\vec{r}_f} \sqrt{2m(E - V(\vec{r}))}\, ds \Bigg|_{\text{stat}}. \tag{13.25}$$

Hamilton considered this function to be the exact analog of his characteristic function Eq. (13.21) for optics, which was also calculated by a stationarity principle applied to integrals along spatial paths. Note that the result of the variational procedure is to determine a spatial path $\vec{r}(s)$ connecting the start and end positions (e.g., segments of Kepler ellipses in the central force problem). The time dependence can be recovered by calculating the speed $v(s) = \sqrt{2(E - V)/m}$ at each point on the path and integrating $t = \int ds/v(s)$. A closer examination of the variational calculation[12] shows that if the initial and final positions, as well as the energy (held fixed along each path considered) are allowed to vary, the variation of S_0 in Eq. (13.23) gives[13]

$$\delta S_0 = \int_{\vec{r}_i}^{\vec{r}_f} \delta(\vec{p} \cdot d\vec{r}) = \vec{p}_f \cdot \delta\vec{r}_f - \vec{p}_i \cdot \delta\vec{r}_i + t\,\delta E. \tag{13.26}$$

Of course, the Maupertuis–Euler–Lagrange principle of least (or rather, stationary) action $\delta S_0 = 0$ follows from this as a special case once the initial and final points, as well as the energy of the particle, are not allowed to vary.

Hamilton was able to derive from the action principle a new extremal law, by considering the variation of the time integral of the Lagrangian $\mathcal{L} = T - V = 2T - H$:

$$\delta S \equiv \delta \int_{t_i}^{t_f} \mathcal{L}\, dt = \delta \int_{t_i}^{t_f} 2T\, dt - \delta \int_{t_i}^{t_f} H\, dt. \tag{13.27}$$

The first variation on the right-hand side is just the variation of the Maupertuis integral found in Eq. (13.26) (cf. Eq. (13.20)), while the variation of the time integral of the energy function H arises from two sources: (a) the variation of the chosen energy (conserved along each path) $H = E$; and (b) the variation in the temporal endpoints t_i and t_f

[12] See the web resource, *Extremal Laws in Classical Optics and Mechanics*.
[13] Ibid.

of the time integral that arise as we are considering only energy-conserving paths. The net result is[14]

$$\delta S = \vec{p}_f \cdot \delta\vec{r}_f - \vec{p}_i \cdot \delta\vec{r}_i - E\,\delta t, \tag{13.28}$$

where $t = t_f - t_i$ is the elapsed time along the path. In fact, the time integral of the Lagrangian defines a new function $S(\vec{r}_i, \vec{r}_f, t)$, now called Hamilton's principal function, which is just the Legendre transform of S_0 with respect to the energy variable (as $\partial S_0/\partial E = t$ and $\partial S/\partial t = -E$):

$$S(\vec{r}_i, \vec{r}_f, t) = \int \mathcal{L}\, dt = S_0(\vec{r}_i, \vec{r}_f, E) - Et. \tag{13.29}$$

Hamilton then showed that the integral $\int \mathcal{L}\, dt$, when varied with respect to arbitrary smooth paths $\vec{r}(t)$, fixed at the spatial *and temporal* endpoints $(\vec{r}_i, t_i)$ and $(\vec{r}_f, t_f)$, but not necessarily energy conserving, was stationary precisely for paths satisfying the Lagrange equations of motion,

$$\frac{d}{dt}\frac{\partial \mathcal{L}}{\partial \dot{\vec{r}}} = \frac{\partial \mathcal{L}}{\partial \vec{r}}, \tag{13.30}$$

which for our case where $\mathcal{L} = \frac{1}{2}m\dot{\vec{r}}^2 - V(\vec{r})$ just amounts to Newton's second law, $m\ddot{\vec{r}} = -\partial V/\partial \vec{r} = \vec{F}$. The variational principle

$$\delta \int \mathcal{L}(\vec{r}, \dot{\vec{r}})\, dt = 0 \tag{13.31}$$

was dubbed "Hamilton's principle" in the nineteenth century, but, as it appears much more frequently in modern discussions of extremal principles in mechanics than the earlier Maupertuis law, it is usually referred to as *the* least action principle, and the term "action" has also been appropriated by Hamilton's principal function S, the integral of the Lagrangian. Nevertheless, both versions of the extremal condition appear prominently in de Broglie's arguments, and will also appear in Schrödinger's construction of wave mechanics—hence, the rather lengthy digression here.

There is a further aspect of the action formulation of mechanics, of which we shall later have need: its intimate connection with the Hamilton–Jacobi formulation of mechanics, which indeed originally sprang from Hamilton's study of least action principles. By fixing the origin of the particle's classical motion at some arbitrary point of space and time (say, $\vec{r}_i = 0$ and $t_i = 0$), we may regard the action function S in Eq. (13.29)

[14] Ibid.

as a function of the path endpoint and time, $\vec{r} \equiv \vec{r}_f$ and $t \equiv t_f$, where now Eq. (13.28) implies

$$\frac{\partial S(\vec{r}, t)}{\partial \vec{r}} = \vec{p}, \quad \frac{\partial S(\vec{r}, t)}{\partial t} = -E. \tag{13.32}$$

Here, the momentum $\vec{p}$ is the actual momentum at the point in the correct physical path starting at the origin with conserved energy E. Thus, for a single particle of mass m moving in the conservative potential $V(\vec{r})$ (so $H = \vec{p}^{\,2}/2m + V(\vec{r}) = E$), we have

$$\begin{aligned} -\frac{\partial S(\vec{r}, t)}{\partial t} = E &= \frac{1}{2m}\left|\frac{\partial S(\vec{r}, t)}{\partial \vec{r}}\right|^2 + V(\vec{r}) \\ &= \frac{1}{2m}\left|\frac{\partial S_0(\vec{r})}{\partial \vec{r}}\right|^2 + V(\vec{r}), \end{aligned} \tag{13.33}$$

where in the last step we have defined a spatial function $S_0(\vec{r})$ (cf. Eqs. (13.25–13.26)) in analogy to the procedure used to define $S(\vec{r}, t)$ above, namely, by setting $\vec{r}_i = 0$ and $\vec{r} = \vec{r}_f$, and suppressing the dependence on the conserved energy E. Eq. (13.33) is just the Hamilton–Jacobi equation (in both time-dependent and time-independent forms), which we introduced in Volume 1 by starting from the generating function for canonical transformations of a certain type.[15] This interpretation of the action integral (at its extremal and therefore physical value) plays a central role in Schrödinger's development of wave mechanics.

The extension of the variational principles just discussed to the case of a particle undergoing relativistic motion is not difficult. The relativistic momentum of a particle of rest mass m_0 moving with velocity $\dot{\vec{r}} = \vec{v}$ is

$$\vec{p} = \frac{m_0 \vec{v}}{\sqrt{1 - v^2/c^2}} = m_0 \gamma \vec{v}, \tag{13.34}$$

which can be obtained as the usual derivative $\partial \mathcal{L}/\partial \dot{\vec{r}}$ by choosing the Lagrangian

$$\mathcal{L} = -m_0 c^2 \sqrt{1 - v^2/c^2}. \tag{13.35}$$

[15] See Sections A.1.3 and A.2.1. The history of the Hamilton-Jacobi method is plagued by a proliferation of different (though often confusingly similar) notations used for the same quantity. In Hamilton's (1834) famous essay, "On a general method in dynamics," the abbreviated action S_0 is called the "characteristic function" and denoted V (with $-U$ used for the potential energy), while the quantity normally called the action nowadays, S, is called by Hamilton the "principal function", and denoted also by the letter S, with the definition $S = V - tH$, with H the energy function. In modern treatments of classical mechanics, as in Goldstein, Poole, and Safko (2002), the abbreviated action is denoted W. For Schrödinger, the notations for W and S are interchanged, so $S = W(q, t)$, $S_0 = S(q)$. We stick to S and S_0 as unambiguous representations of the action in the time-dependent (Hamilton) and time-independent (Maupertuis) version of the extremal action principle, respectively.

The least action principle in Hamilton's formulation (cf. Eq. (13.31)) then takes the form

$$\begin{aligned} 0 &= \delta \int \mathcal{L}\, dt \\ &= \delta \int -m_0 c^2 \sqrt{1 - \frac{1}{c^2}\dot{\vec{r}}^2}\, dt \\ &= \delta \int -m_0 c^2 \sqrt{dt^2 - \frac{1}{c^2} d\vec{r}^2} \\ &= \delta \int -m_0 c \sqrt{(cdt)^2 - d\vec{r}^2}. \end{aligned} \tag{13.36}$$

The quantity appearing inside the final square root is just the (squared) invariant space–time interval ds^2 along the infinitesimal segment $dx^\mu = (cdt, dx, dy, dz)$ (with $\mu = 0, 1, 2, 3$) of the world-line of the particle (or, alternatively, the elapsed proper time as measured in the particle's rest frame), so the Hamiltonian action principle takes the simple form

$$0 = \delta \int -m_0 c\, ds, \tag{13.37}$$

which asserts that the particle executes a trajectory which is a geodesic (shortest proper time path) in space–time. This is evidently a straight line for a free particle in flat Minkowski space, but the principle remains true in general relativity for test particles, which follow the appropriate curved paths in the presence of a gravitational field.

The addition of external electrostatic and magnetic forces can be accomplished as follows. Let our particle now possess a charge e and be subject to a four-vector electromagnetic potential $A_\mu = (\varphi, A_i)$ (with $i = 1, 2, 3$). Here, φ and A_i represent the electrostatic potential and the spatial components of the vector potential, respectively. We need to extend the Lagrangian so that the equations of motion reflect the presence of the Lorentz force

$$\vec{F}_{\text{Lor}} = \frac{d\vec{p}}{dt} = e\vec{E} + \frac{e}{c}\vec{v} \times \vec{B}, \quad \vec{E} = -\vec{\nabla}\varphi, \quad \vec{B} = \vec{\nabla} \times \vec{A}. \tag{13.38}$$

We can accomplish this by extending the free Lagrangian in Eq. (13.35) to the expression

$$\mathcal{L} = -\left(m_0 c^2 + e u^\mu A_\mu\right) \sqrt{1 - v^2/c^2}, \tag{13.39}$$

where $u^\mu \equiv dx^\mu/ds$ is the four-velocity, related to the four-momentum by $p^\mu = m_0 c u^\mu$. Note that $u^\mu u_\mu = 1$, and $u_\mu dx^\mu = ds$. The action integral $\int \mathcal{L}\, dt$ now becomes the covariant expression

$$S = \int \mathcal{L}\, dt = \int -\left(m_0 c u_\mu + \frac{e}{c} A_\mu\right) dx^\mu \equiv -\int \mathbf{p}_\mu dx^\mu. \tag{13.40}$$

This expresses the standard prescription for including electromagnetic effects by replacing the kinematical momentum $p_\mu = m_0 c u_\mu$ by the canonical momentum

$$\mathbf{p}_\mu \equiv p_\mu + \frac{e}{c} A_\mu. \tag{13.41}$$

Hamilton's principle then asserts that the physical trajectory satisfies the stationarity condition

$$\delta S = \delta \int \mathbf{p}_\mu dx^\mu = 0. \tag{13.42}$$

Here, the variation is carried out over all space–time trajectories with fixed initial and final space–time coordinates x_i^μ and x_f^μ, respectively.

In particular, if we specialize to the situation where only electrostatic forces are present by setting $\vec{A} = 0$, the Lagrangian in Eq. (13.39) becomes

$$\begin{aligned} \mathcal{L} &= -\left(m_0 c^2 + e u^0 A_0\right)\sqrt{1 - v^2/c^2} \\ &= -m_0 c^2 \sqrt{1 - v^2/c^2} - e\varphi(\vec{r}), \end{aligned} \tag{13.43}$$

where we used that $u^0 = \gamma = 1/\sqrt{1 - v^2/c^2}$. The Lagrangian equation of motion now becomes

$$\begin{aligned} \frac{d}{dt}\frac{\partial \mathcal{L}}{\partial \vec{v}} &= \frac{d}{dt}\left(\frac{m_0 \vec{v}}{\sqrt{1 - v^2/c^2}}\right) \\ &= \frac{\partial \mathcal{L}}{\partial \vec{r}} \\ &= -e\vec{\nabla}\varphi(\vec{r}) = e\vec{E}, \end{aligned} \tag{13.44}$$

which is just the relativistic version of Newton's second law for a charged particle in the presence of an electric field $\vec{E}$ derived from an electrostatic potential $\varphi(\vec{r})$. The Hamiltonian in this case is just the total relativistic energy:[16]

$$\begin{aligned} H &= \dot{\vec{r}} \cdot \vec{p} - \mathcal{L} \\ &= \vec{v} \cdot \frac{m_0 \vec{v}}{\sqrt{1 - v^2/c^2}} + m_0 c^2 \sqrt{1 - v^2/c^2} + e\varphi \\ &= \frac{m_0 c^2}{\sqrt{1 - v^2/c^2}} + e\varphi. \end{aligned} \tag{13.45}$$

The least action principle in its original (Maupertuis–Euler–Lagrange) form is obtained by varying the restricted action S_0. In the absence of an electromagnetic field the principle is expressed by

$$\delta S_0 = \delta \int \vec{p} \cdot \dot{\vec{r}} dt = \delta \int m_0 c u_i dx^i, \tag{13.46}$$

where the variation of this purely spatial line integral must be carried out over trajectories that conserve energy, as discussed above in the non-relativistic case. In the presence of an electromagnetic field, Eq. (13.46) needs to be replaced by:

$$\delta S_0 = \delta \int \mathbf{p}_i dx^i = 0. \tag{13.47}$$

13.4 De Broglie's mechanics of waves

In the second chapter of his thesis, de Broglie sets out to generalize his phase wave hypothesis for free particles to the case where the particle is subject to a field of force. In particular, he restricts his attention to the effect of electromagnetic fields on the motion of a charged particle. The motivation and strategy are expressed very clearly in the introduction to this chapter:

> The phase wave which accompanies the movement of a particle ... possesses properties which depend on the nature of the particle, given that its frequency, for example, is determined by the total energy. It then seems natural to suppose that if a field of force acts on the movement of a particle, it will also act on the propagation of its phase wave. Guided by the idea of a fundamental identity of the principle of least action with that of Fermat, I have been led from the outset of my studies to suppose that, for a given value of the total energy and consequently of the frequency of its phase wave, the dynamically possible trajectories of the one [the particle] coincide with the possible rays of the other [the phase wave] (De Broglie 1924, p. 15).

[16] The first two terms on the right side in the second line of Eq. (13.45) can be combined as $(m_0 v^2 + m_0 c^2(1 - v^2/c^2))/\sqrt{1 - v^2/c^2} = m_0 c^2/\sqrt{1 - v^2/c^2}$.

In the second section of the chapter, de Broglie reviews the two extremal principles of mechanics: the Maupertuis purely spatial principle of least action, and the spatio-temporal version expressed in Hamilton's principle. The treatment is restricted to the non-relativistic case, and for Lagrangians that take the form $T - V$ (kinetic energy minus potential energy). The required generalization of these two principles to the case of a charged particle moving relativistically in an electromagnetic field is presented in section 2.3 of the thesis, more or less along the lines we have described in Section 13.3.[17]

In the fourth section of the chapter, entitled "Propagation of waves; principle of Fermat," the aim is to "study the propagation of a sinusoidal [i.e., periodic] phenomenon by a method parallel to that of the last two sections" (p. 21). Fermat's principle is presented first in its Hamiltonian (spatio-temporal) form, as the extremal condition for a phase function $\varphi(\vec{r}, t)$:

$$\delta \int_P^Q d\varphi = 0, \tag{13.48}$$

where P and Q are two points in *space–time*, and, in line with the intuitions expressed in Huygens's principle, if a ray associated with the wave phenomenon passes from P to Q, one requires that the accumulated phase change along any infinitesimally displaced neighboring path can differ from that of the actual ray only by quantities of second order in the displacement, as otherwise the superposition of waves implied by Huygens's principle would lead to destructive interference. For electromagnetic waves traveling in vacuo, the phase of the wave must be a Lorentz scalar, as the physical space–time locations where maxima and minima occur, for example, are independent of the frame of observation. In this case the phase change induced by a displacement dx^μ takes the form

$$d\varphi = k_\mu dx^\mu \tag{13.49}$$

($2\pi O_i dx^i$ in de Broglie's notation), where k_μ are the components of a four-vector called the "world wave vector" by de Broglie (1924, p. 21), with $k_0 = 2\pi\nu$ and $|\vec{k}| = 2\pi/\lambda$.

In general in optics we consider rays of light passing through refractive media, so in particular the spatial components $k_i(\vec{r})$ (with $i = 1, 2, 3$) will acquire a spatial dependence in the inertial frame in which the refracting medium is at rest, as the local phase velocity will depend on the local value of the refractive index $n(\vec{r})$: specifically $V(\vec{r}) = c/n(\vec{r}), |\vec{k}| = 2\pi\nu/V(\vec{r})$ in that frame. The frequency ν, by contrast, remains

[17] There are some minor differences of notation: for example, de Broglie denotes the four-momentum of a particle by $\mathcal{J}_i$ with $i = 1, 2, 3, 4$, where we have used p_μ with $\mu = 0, 1, 2, 3$. For waves, the four-dimensional wave vector is $2\pi O_i$ for de Broglie, whereas we use k_μ.

constant (in the given frame) throughout the transit of the light.[18] This means that in the optical case the spatio-temporal (i.e., Hamiltonian) extremal principle given in Eq. (13.48) is trivially equivalent to the purely spatial Fermat version. As the temporal part of the inner product in Eq. (13.49) integrates just to $k_0(t_Q - t_P)$, and $k_0 = 2\pi\nu$ is constant throughout, and by assumption the points P and Q are related by the transit of a light ray, the extremal property applies simultaneously to the temporal part of the inner product and to the spatial part, as in Eq. (13.15). Hence, the least time principle

$$0 = \delta \int_P^Q 2\pi\nu dt = \delta\Big(2\pi\nu(t_Q - t_P)\Big) \tag{13.50}$$

is equivalent to the spatial version of Fermat's principle

$$0 = \delta \int_A^B \vec{k} \cdot d\vec{x} = \delta \int_A^B \frac{2\pi\nu}{c} n(\vec{r})\, ds, \tag{13.51}$$

where A and B are the spatial points associated with space–time events P and Q. For optics, the passage between the Maupertuis and Hamilton versions of the least action principle is much more direct and obvious than for mechanics.[19]

The explicit connection between the properties of a material point particle (in particular the four-momentum p_μ) and of the phase wave associated with it (the "world wave vector" k_μ, with dimensions of inverse distance) is provided by de Broglie in section 2.5 of his dissertation ("Extension of the quantum relation"). A proportionality between these two four-vectors will require a scaling constant with the dimension of momentum times distance (i.e., action). In fact, the fundamental relation $h\nu = m_0c^2$ (cf. Eq. (13.1)) on which de Broglie bases his entire theory already gives the relation between the temporal component of these vectors for a free particle at rest, namely $k_0 = (2\pi/h)p_0$.[20] This relation is extended in the obvious way to the spatial components of the momentum and wave vector:

$$k_i = \frac{2\pi}{h} p_i, \quad i = 1, 2, 3. \tag{13.52}$$

[18] We recall from our discussion of dispersion that refractive phenomena correspond to elastic scattering of light, in which the frequency, or alternatively, the energy of the light quanta, is unchanged.

[19] In fact, for de Broglie's space–time formulation of Fermat to make sense, the phase difference φ accumulated between P and Q must by definition be zero, as the associated ray must be regarded as "riding along with the wave", so that the statement $\delta \int d\varphi = 0$ becomes a triviality. The separate statements that the temporal and spatial parts of the integrated inner product of Eq. (13.49) be extremal, are then equivalent to each other and to Fermat's principle, which determines the propagation of the wave fronts in the eikonal approximation to physical optics. See the web resource, *Extremal Laws in Classical Optics and Mechanics*, for further details.

[20] We have $k_0 = 2\pi\nu$, $E = m_0c^2 = p_0$. In de Broglie's notation, the relation is $O_4 = (1/h)\,\mathcal{J}_4$. De Broglie's wave vector O differs from the modern definition by a factor of 2π. See Eq. (13.49).

With this identification, the principle of Fermat and the least action principle of Maupertuis–Euler–Lagrange become not merely analogous, but *identical*:

$$\delta \int_A^B k_i \, dx^i = 0 \quad \Leftrightarrow \quad \delta \int_A^B p_i \, dx^i = 0. \tag{13.53}$$

De Broglie puts it as follows:

> The principle of Fermat applied to the phase wave is identical to the principle of Maupertuis applied to the particle in motion; the dynamically possible trajectories of the particle are identical to the possible rays of the wave (De Broglie 1924, p. 23).

Some special cases of particular interest are discussed in section 2.6, "Particular cases: discussions," for example, a free particle of energy E, a charged particle (also with total energy E) moving in an electrostatic potential φ, and then, most generally, a charged particle in the presence of both electric and magnetic fields (described by the electrostatic and vector potentials φ and $\vec{A}$).[21] In this case, the four-momentum must be modified in the standard way (cf. Eq. (13.40)):

$$p_\mu \to \mathbf{p}_\mu \equiv p_\mu + \frac{e}{c} A_\mu. \tag{13.54}$$

For the free particle ($A_\mu = 0$), the quantum condition for the time component of the phase gives

$$E = h\nu = m_0 c^2 / \sqrt{1 - v^2/c^2}, \tag{13.55}$$

while the phase velocity V (the product of the frequency ν and wavelength λ, where $|\vec{k}| = 2\pi/\lambda$) is related to the spatial momentum by

$$V = \nu\lambda = \nu \cdot 2\pi/|\vec{k}| = \nu h/p = E/p = \frac{c^2}{v}, \tag{13.56}$$

where we have used Eq. (13.52) to relate $|\vec{k}|$ to $|\vec{p}\,| = m_0 v/\sqrt{1 - v^2/c^2}$. As we saw in Section 12.1.1, this gives a phase velocity greater than the speed of light, as the material particle has velocity $v < c$. However, the transmission of physical quantities (such as energy) occurs with the group velocity of the envelope of the wave-packets formed from superposition of phase waves. This (subluminal!) velocity turns out to be exactly coincident with that of the material particle associated with the wave (cf. Eq. (13.12)).

[21] Once again, we have modified de Broglie's notation in accord with modern usage: de Broglie uses W for the total conserved energy, ψ for the electrostatic potential, and $a_{x,y,z}$ for the components of the vector potential.

The situation becomes considerably more complicated once the particle is subject to an electrostatic potential. Now Eqs. (13.55)–(13.56) are replaced by

$$E = h\nu = \frac{m_0 c^2}{\sqrt{1 - v^2/c^2}} + e\varphi(\vec{r}), \tag{13.57}$$

$$V = \nu\lambda = \frac{h\nu}{p} = \frac{c^2}{v}\frac{E}{E - e\varphi(\vec{r})}. \tag{13.58}$$

Given the energy E, one determines v at every spatial point from Eq. (13.57) (as in the Maupertuis principle). This value of v is then inserted in Eq. (13.58) to obtain the varying phase velocity V as a function of spatial position $\vec{r}$. However, there is no attempt to provide a precise wave equation to determine both amplitude and phase of the wave associated with the particle. In particular, there is no prescription given for determining the direction of the normal to the wave front at each point, as Eqs. (13.57)–(13.58) only involve the magnitude of the momentum, not its direction. The situation is even worse once a vector potential (giving a magnetic field) is present, as the phase velocity V depends on the angle between the vector potential $\vec{A}$ and the normal to the wave front at each spatial point, analogously to the situation in the propagation of light in anisotropic dispersive media.

In our discussion of Schrödinger's work in Chapter 14, we show that it is precisely at this point that de Broglie missed the opportunity to develop a wave equation capable of handling the dynamics of a charged particle subject to electrostatic forces (such as an electron in a hydrogen atom), by uniquely determining both the amplitude and phase of the periodic phenomenon to be associated with the particle. The road leading back from classical Maxwellian electromagnetic theory to ray optics begins with the wave equation for the electric field components, passing via an eikonal approximation (the assumption that the properties of the transmitting medium change very little fractionally over a single wavelength of the light) to an equation for the phase that implies both Huygens's principle and the Fermat principle of least time (strictly speaking, optical length). De Broglie, by contrast, seems much more concerned, in his discussions of the implications of a varying phase velocity, with the difficulties of treating non-uniform motion in special relativity: nowhere in the thesis is there the suggestion that what is needed here is an analog of physical optics, based on a precise wave equation, underlying the "ray optics" of particle motion in de Broglie's theory.

Given the approach taken by de Broglie, it is not surprising that, with one important exception, he was unable to use his new ideas to illuminate any of the burning questions of atomic physics that had dominated theoretical physics in the years since the Bohr model. In chapter 3 of his thesis ("Quantum stability conditions for trajectories"), de Broglie reviews the Bohr–Sommerfeld quantization conditions, alluding to Einstein's

(1917b) reformulation of these conditions in terms of a single integral (invariant under coordinate transformations) over a closed orbit of the particle

$$\oint \vec{p} \cdot d\vec{q} = nh. \tag{13.59}$$

We do not discuss here Einstein's (1917b) formulation of the Bohr–Sommerfeld quantization conditions in terms of conservative[22] vector fields $\vec{p}(\vec{r})$. Suffice it to say that, after pointing out that the orbit integral Eq. (13.59) is identical to the quantity to be extremized in Maupertuis's principle, de Broglie shows that his phase wave interpretation of particle motion also allows an explanation, in terms of a resonance condition, of the allowed quantized values nh appearing on the right-hand side of the action integral. If we assume that for a closed orbit the associated phase wave must experience a whole number of wavelengths on returning to the starting point—as de Broglie (1924, p. 28) puts it, "the length of the channel must be resonant with the wave"—then the loop integral of $\nu/V = 1/\lambda$ must have a (non-zero) integer value:

$$\oint \frac{\nu}{V} dl = n. \tag{13.60}$$

Since $\nu/V = p/h$ (see Eq. (13.58)), this means that

$$\oint \vec{p} \cdot d\vec{r} = \oint p\, dl = nh. \tag{13.61}$$

For the case of circular orbits, this of course reduces to the original Bohr quantization condition $2\pi Rp = 2\pi L = nh$ or $L = n\hbar$ for the orbital angular momentum of an electron in the nth circular orbit of the hydrogen atom.

In chapter 4 of his thesis de Broglie extends his stability criterion to the case of two-body motion, where an electron and a nucleus both execute circular motion around the common center of mass. The resonance condition is then shown to apply to both phase waves—that of the nucleus as well as that of the electron.

Succeeding chapters of the thesis discuss a number of topics with very little connection to the phase wave hypothesis; chapter 5 explores the idea of treating light quanta in a manner similar to material particles by introducing a very small mass for the photon, and chapter 6 offers discussion and critique of various theories of X- and γ-ray diffusion.

Finally, in chapter 7, "Quantum statistical mechanics," de Broglie returns to the consideration of phase waves, now in the context of the statistical-mechanical treatment of atoms in a gas. In section 7.2 (after reviewing some basic elements of classical statistical thermodynamics in section 7.1), de Broglie begins his discussion of a "new conception of gas equilibrium" as follows:

[22] Derivable from a potential function—$p_i = \partial S/\partial q_i$, where S is just the Hamilton–Jacobi function—hence, invariant under continuous deformation of the path. See also Volume 1, Section 5.1.5.

> If moving atoms of a gas are accompanied by waves, the container [of the gas] must then contain a pattern of standing waves. We are naturally drawn to consider how within the notions of black body radiation developed by M. Jeans [i.e., Jeans's work on what is now known as the Rayleigh–Jeans law (see, e.g., Volume 1, pp. 89–92 and pp. 112–114)], these phase waves forming a standing pattern (that is, with respect to a container) as the only stable situation, can be incorporated into the study of thermodynamic equilibrium. This is somehow an analog to a Bohr atom, for which stable trajectories are defined by stability conditions [such as the requirement that the circular orbit of an electron contain a whole number of wavelengths of the associated phase wave] such that unstable waves would be regarded as unphysical (De Broglie 1924, p. 61).

The idea that the phase wave of free particle(s) contained in a box must be subject to boundary conditions is familiar in our time to any beginning student of quantum mechanics. The standing wave condition for a particle moving in one dimension and confined to the interval $0 < x < L$ would imply waves of the form $\sin(n\pi x/L)$, vanishing at $x = 0$ and $x = L$. The value for the wavenumber $k = 2\pi/\lambda$ of allowed phase wave modes is thus restricted to integral multiples of π/L. The number Δn of such modes between $k = n\pi/L$ and $k + \Delta k = (n + \Delta n)\,\pi/L$ is thus given by $\Delta n = (L/\pi)\,\Delta k$. Inserting the de Broglie quantum condition $\lambda = h/p = h/m_0 v$ into $k = 2\pi/\lambda$, we see that $k = 2\pi m_0 v/h$.[23] The number of modes in a velocity interval Δv is thus given by

$$\Delta n = \frac{L}{\pi}\Delta k = \frac{2m_0 L}{h}\Delta v. \tag{13.62}$$

In fact, de Broglie's formula differs from Eq. (13.62) by a factor of 2, as he imposes periodic boundary conditions ($\psi(x) = \psi(x+L)$), rather than the correct Dirichlet boundary conditions ($\psi(x) = 0$ for $x = 0, L$). In generalizing this formula to three dimensions, however, de Broglie arrives (via a calculation not given in the thesis) at the correct counting of modes. Labeling the standing wave modes by a wavenumber vector $\vec{k} = (\pi/L)\,\vec{n}$, with $\vec{n} \equiv (n_x, n_y, n_z)$, and spatial momentum given by the de Broglie quantum condition $\vec{p} = \hbar\vec{k}$, one finds, returning to relativistic mechanics,[24]

$$p = |\vec{p}\,| = \frac{\hbar\pi}{L}|\vec{n}| = \frac{h}{2L}n. \tag{13.63}$$

Using

$$h\nu = \frac{m_0 c^2}{\sqrt{1 - v^2/c^2}} \quad \text{and} \quad p = \frac{m_0 v}{\sqrt{1 - v^2/c^2}}, \tag{13.64}$$

[23] As de Broglie frequently returns to the relativistic formulas, he is careful to always indicate the rest mass m_0 with a subscript. Here we are dealing (for the moment) with the non-relativistic case.

[24] The reader is cautioned to distinguish carefully between the unfortunately too similar symbols used for the frequency ν of the phase wave and the velocity v of the associated particle.

we arrive at

$$p = \frac{hv}{c^2}\nu. \tag{13.65}$$

We now want to find the relation between Δp and $\Delta\nu$. We do this by combining two relations, one between $\Delta\nu$ and Δv and one between Δp and Δv. The former is given by

$$h\Delta\nu = \frac{d}{dv}\left(\frac{m_0c^2}{\sqrt{1-v^2/c^2}}\right)\Delta v = \frac{m_0 v}{(1-v^2/c^2)^{3/2}}\,\Delta v; \tag{13.66}$$

the latter by

$$\begin{aligned}\Delta p &= \frac{d}{dv}\left(\frac{m_0 v}{\sqrt{1-v^2/c^2}}\right)\Delta v\\ &= \left(\frac{m_0}{\sqrt{1-v^2/c^2}} + \frac{m_0v^2/c^2}{(1-v^2/c^2)^{3/2}}\right)\Delta v\\ &= \frac{m_0}{(1-v^2/c^2)^{3/2}}\,\Delta v.\end{aligned} \tag{13.67}$$

Combining these two relations, we find that $h\Delta\nu = v\Delta p$ or

$$\Delta p = \frac{h}{v}\,\Delta\nu. \tag{13.68}$$

This yields a general expression for the number of modes $n(\nu)\Delta\nu$ in the frequency interval $(\nu, \nu+\Delta\nu)$:

$$n(\nu)\Delta\nu = \frac{1}{8}\cdot 4\pi n^2\Delta n = \frac{4\pi L^3}{h^3}p^2\Delta p, \tag{13.69}$$

where in the second step we used Eq. (13.63) to set $n^2\Delta n = (8L^3/h^3)\,p^2\Delta p$. Using $p = vE/c^2 = vh\nu/c^2$, and $\Delta p = \Delta E/v = h\Delta\nu/v$, we can rewrite this as

$$n(\nu)\Delta\nu = \frac{4\pi L^3}{h^3}\cdot\frac{v^2h^2\nu^2}{c^4}\cdot\frac{h}{v}\Delta\nu = 4\pi L^3\frac{v}{c^4}\nu^2\Delta\nu. \tag{13.70}$$

Using the expressions earlier obtained for phase velocity $V = c^2/v$ and group velocity (which agrees with the velocity of the particle) $U = v$, de Broglie writes this as[25]

$$n(\nu)\Delta\nu = \gamma\; 4\pi L^3\frac{1}{UV^2}\,\nu^2\Delta\nu. \tag{13.71}$$

[25] In fact, the expression (7.2.5) in De Broglie (1924, p. 62) is for the number per unit spatial volume, so the box volume L^3 is absent.

Here, γ is a polarization factor, equal to 1 for longitudinal waves with a single degree of freedom, 2 for light with two independent transverse polarizations for any given wavevector. For the massless case of a gas consisting of light quanta, $U = V = c$, and we recover the familiar Rayleigh–Jeans mode factor $8\pi\nu^2/c^3$ for the number density, which gives the Rayleigh–Jeans law when multiplied by the classical equipartition energy kT. Similarly, by taking the non-relativistic limit, the mode number density is seen to involve the factor (using $h\nu \approx m_0c^2 + p^2/2m_0 + \cdots, h\Delta\nu = p\Delta p/m_0$)

$$4\pi L^3 \frac{v}{c^4}\nu^2\Delta\nu \approx \frac{4\pi}{h^3}L^3p^2dp \simeq \int \frac{dxdydzdp_xdp_ydp_z}{h^3}, \tag{13.72}$$

reproducing the old Planck association of quantum states with cells of volume h^3 in phase space.

Simply attaching the classical Maxwell–Boltzmann factor

$$e^{-E/kT} = e^{-h\nu/kT} \tag{13.73}$$

and the energy per mode $h\nu$ to the mode number expression Eq. (13.71) leads to Wien's Law, as de Broglie points out, and not the desired Planck expression. Evidently, simply counting modes of standing waves in the standard way as indicated by Rayleigh and Jeans, and attaching a quantized energy $h\nu$ to each mode in the quantum case (or the equipartition factor kT in the classical case) could not lead to the correct result. Instead, as de Broglie says, "To get Planck's Law *a new hypothesis is needed* ... this hypothesis can be formulated as follows:"

> *If two or more light quanta have phase waves that exactly coincide, then since they are carried by the same wave their motion cannot be considered independent, and these light quanta must be treated as identical when calculating probabilities.* Motion of these light quanta "as a wave" exhibits a sort of coherence of inexplicable origin, but probably is such that out-of-phase motion is rendered unstable (De Broglie 1924, p. 64; emphasis in the original).

This pregnant phrase suggests that de Broglie might well have succeeded in developing the correct method of counting microstates for gases obeying Bose–Einstein statistics a year or more before the idea was published by Bose and then extended by Einstein. Of course, the notion that the waves for separate atoms should be directly superimposed and therefore be subject to constructive or destructive interference is incorrect—but the basic idea of treating multiply occupied modes as a single "complexion" (in Boltzmann's language), where the atoms are essentially indistinguishable and therefore cannot be labeled, is the essential ingredient in the new statistics developed by Bose and Einstein. In the case of light quanta, de Broglie is able to reproduce the Planck formula by applying the canonical distribution formula for each mode separately, so the relative probability of having p light quanta in a particular mode of frequency ν is $e^{-ph\nu/kT}$, which results in the mean occupation number for that mode of

$$\frac{\sum_{p=1}^{\infty} pe^{-ph\nu/kT}}{\sum_{p=0}^{\infty} e^{-ph\nu/kT}} = \frac{1}{e^{h\nu/kT} - 1}, \tag{13.74}$$

which yields, when multiplied by the density of modes given in Eq. (13.71) (with of course, $U = V = c$), Planck's distribution formula for black-body radiation.

The indistinguishability of individual light quanta is built into this calculation, as is the fact that the total number of light quanta is allowed to vary freely (as the distribution of light quanta in a given mode is considered independently of all the other modes). De Broglie's attempt to extend this result to massive particles (still treated by full relativistic kinematics) fails for the simple reason that he does not introduce a Lagrange multiplier to control the total number of atoms N (i.e., a chemical potential μ), which in a realistic gas is a fixed number, unlike the case for light quanta (which have effectively chemical potential zero). The analog of Eq. (13.74) for massive atoms then reduces to the first term $p = 1$ only, as terms with more than one atom in a given mode are suppressed by a huge factor $e^{-m_0c^2/kT}$ due to the rest energy.[26] As in the case of the wave equation, de Broglie missed here another opportunity to anticipate an important development in quantum theory, in this case, the dramatic progress in the theory of quantum gases, which would be made in the following year by Bose and Einstein.

13.5 The new statistics of Bose and Einstein and Einstein's quantum theory of the ideal gas

A subject of intense discussion around this time was the troublesome conflict between two apparently secure tenets of classical thermodynamics, namely, the requirement that the entropy be an extensive quantity,[27] and the requirement of Nernst's theorem that the entropy of systems in internal equilibrium vanish at absolute zero. This conflict seemed to follow inexorably if Boltzmann statistics were employed. The conflict with the extensivity requirement, known as the *Gibbs paradox*, could be avoided if the usual Boltzmann factor for the number of micro-configurations of a system of N identical particles compatible with a given state of specified extensive macroscopic properties were divided by $N!$, indicating that configurations obtained by simply exchanging two of the identical particles were not be considered as different for the purposes of counting states.

[26] In the correct treatment of a Bose–Einstein gas, which would emerge in the following year, the Planck factor $1/(e^{h\nu/kT} - 1)$ is replaced by $1/(e^{(E-\alpha)/kT} - 1)$, where in the non-relativistic limit the energy becomes $E \approx m_0c^2 + p^2/2m_0$, and the Lagrange multiplier α removes the inert rest energy by simply defining $\alpha = m_0c^2 + \mu$, where μ is the usual chemical potential used to adjust the number density of atoms in the gas. See the excellent article by Darrigol (1991) for an in-depth examination of the development of quantum statistical mechanics in early quantum theory.

[27] This property asserts that if the extensive properties of a system, such as volume, internal energy, and number of particles, are scaled by a factor λ, without changing the intensive parameters (pressure, temperature, chemical potential, etc.), the entropy also scales by the same factor λ.

Unfortunately, this then resulted in a violation of Nernst's theorem (Nernst 1906; cf. Section 3.5).

There were numerous attempts to navigate a safe course between the Scylla of the Gibbs paradox and the Charybdis of Nernst's theorem violation in the period leading up to 1925, with Planck, Einstein, and Schrödinger weighing in on the subject on several occasions. The issue of counting of states was clearly intimately bound up with the matter of identifying the allowed *distinct* quantized energy states of the particles of a gas (with any associated degeneracy, as we have seen, identified with the a priori probability of the given energy state).[28] Schrödinger addressed both of these issues in 1925, the first in a paper entitled "Remarks on the statistical entropy function of an ideal gas" (Schrödinger 1925),[29] the second in a paper on "The energy steps of the ideal monatomic gas model" (Schrödinger 1926a). Both were attempts to come to terms with recent work by Planck (1925a) on quantized gases, as well as the recent work of Einstein (1924, 1925a, 1925b) extending a paper by the Indian physicist Satyendra Nath Bose (1924), which gave a new derivation of the Planck blackbody radiation law, to the case of monatomic ideal gases. Prior to the work of Bose and Einstein, these papers all gave different—and incorrect—results for the thermodynamic functions and the form of the ideal (quantum) gas law, and even after the appearance of Einstein's work, it took quite some time for Schrödinger to fully absorb the lessons of the new quantum statistics proposed by Bose and Einstein. By the time he had finally done so, in a paper entitled "On the Einstein gas theory" (Schrödinger 1926b), he had also become acutely sensitized to the work of de Broglie and the need for an undulatory mechanism to fully understand the properties of gases in the new statistics.

The essential conceptual content of the new gas statistics, and its resolution of the Nernst/entropy-additivity conflict, was beautifully explained by Einstein in the second of his two seminal papers on monatomic ideal gases (Einstein 1925a).[30] Einstein begins the statistical analysis of such a gas by first introducing a very fine discretization of the energy for each atom of the gas, with $0 = E_0 < E_1 < \ldots < E_\nu < E_{\nu+1} < \ldots$, where $\Delta E_\nu = E_{\nu+1} - E_\nu \ll E_\nu$ is a very small increment, but still large enough to contain a very large number z_ν of quantized energy levels of the atom. The number of such states was evaluated (as per Bose) by dividing the volume of phase space occupied by the energy shell $E_\nu < E < E_{\nu+1}$ by h^3. Using the step function $\theta(x)$ (with $\theta(x) = 0$ for $x < 0$ and $\theta(x) = 1$ for $x \geq 0$), we can write z_ν as

$$z_\nu = \frac{1}{h^3} \int \theta(E - E_\nu)\, \theta(E_{\nu+1} - E)\, d^3x\, d^3p. \tag{13.75}$$

[28] The intricate history of quantum statistics in the first two and a half decades of the twentieth century has been lucidly described in a pair of articles by Darrigol (1988, 1991).

[29] This paper was presented as the same session of the Prussian Academy in July 1925 as a paper on the same topic by Planck (1925b).

[30] At this point the role of the Pauli exclusion principle for electrons, and more generally, the need for (as it later was called) Fermi–Dirac statistics in systems composed of identical particles with half-integral spin, was not understood—a quantum theory of gases specifically implied a treatment of an assembly of identical, non-interacting monatomic *bosons*, such as photons, in the massless case, or rare gases of even atomic weight, such as helium 4, in the massive case.

where the product of the two θ-functions equals 1 for values of E between E_ν and $E_{\nu+1}$ and vanishes for all other values. Using that $\int d^3x = V$ and that the energy shell consists of concentric spheres of area $4\pi p^2$, we find that

$$\begin{aligned} z_\nu &= \frac{4\pi V}{h^3} \int \theta(E - E_\nu)\, \theta(E_{\nu+1} - E)\, p^2\, dp \\ &= \frac{4\pi V}{h^3} \int_{E_\nu}^{E_{\nu+1}} \frac{p(E)^2}{E'(p)}\, dE \\ &\approx \frac{4\pi V}{h^3} \frac{p(E_\nu)^2}{E'(p(E_\nu))}\, \Delta E_\nu. \end{aligned} \tag{13.76}$$

Here, $E'(p) = dE(p)/dp$ depends on the kinematics employed for the atoms. Most generally, for atoms of rest mass m_0,

$$E(p) = \sqrt{m_0^2 c^4 + p^2 c^2}. \tag{13.77}$$

Hence, $E'(p) = c^2 p / E(p)$. Using this result and noting that $E(p(E_\nu)) = E_\nu$, we find that

$$\frac{p(E_\nu)^2}{E'(p(E_\nu))} = \frac{E_\nu p(E_\nu)}{c^2} = \frac{1}{c^3} E_\nu \sqrt{E_\nu^2 - m_0^2 c^4}, \tag{13.78}$$

where we used Eq. (13.77) to write

$$p(E_\nu) = \frac{1}{c}\sqrt{E_\nu^2 - m_0^2 c^4}. \tag{13.79}$$

Substituting Eq. (13.78) into Eq. (13.76), we arrive at

$$z_\nu = \frac{4\pi V}{h^3 c^3} E_\nu \sqrt{E_\nu^2 - m_0^2 c^4}\, \Delta E_\nu. \tag{13.80}$$

This expression is equivalent to the one obtained by de Broglie by counting standing wave modes of the phase waves associated with gas atoms.[31]

Eq. (13.80) contains both the Rayleigh–Jeans result for massless photons (absent a factor of 2 for polarization) and Einstein's result for non-relativistic massive atoms. We recover the former if we set $m_0 = 0$ and $E_\nu = h\nu$ (now with ν interpreted as the frequency) in Eq. (13.80):

$$z_\nu = \frac{4\pi V}{h^3 c^3} h\nu \sqrt{(h\nu)^2}\, h\,\Delta\nu = \frac{4\pi V}{c^3} \nu^2 \Delta\nu \tag{13.81}$$

[31] See Eq. (13.70), with $E_\nu = h\nu$, $V = L^3$, $v/c^4 = p/(c^2 E_\nu)$, and $pc = \sqrt{E_\nu^2 - m_0^2 c^4}$.

(cf. Section 3.2, Eq. (3.13)). Note that Planck's constant drops out of this equation. For the case of non-relativistic massive atoms, we first use Eq. (13.79) to rewrite Eq. (13.80) with $E_\nu = m_0c^2$ as

$$z_\nu = \frac{4\pi V}{h^3}\, m_0\, p(E_\nu)\, \Delta E_\nu. \tag{13.82}$$

The energy E_ν in this equation is the non-relativistic kinetic energy of atoms with mass m_0, so we can substitute $\sqrt{2m_0E}$ for $p(E_\nu)$. We thus arrive at:

$$z_\nu = 2\pi\frac{V}{h^3}(2m_0)^{3/2}E_\nu^{1/2}\Delta E_\nu. \tag{13.83}$$

Except for the subscript ν, which Einstein omits, this is Eq. (2a) in Einstein (1925a, p. 5).

The macroscopic volume V can always be chosen sufficiently large so that the number of available modes z_ν is a number very much larger than one, even for a very small energy increment ΔE_ν. In fact, we need to assume it so large that the occupation number of the available modes in the νth slice, n_ν, is also much greater than 1, allowing us to use Stirling's approximation $n_\nu! \approx n_\nu^{n_\nu}$ throughout (and likewise for z_ν).

Einstein distinguishes two ways of determining the number W of microscopic states to be associated with a given macroscopic state of a gas of N identical atoms, where the total energy is constrained to be E. We assume that the possible states of each atom are labeled by the integer index ν as described, with z_ν states available in the νth energy slice. Einstein considers the evaluation of W using (a) the new statistics of Bose and (b) classical Boltzmann statistics. We invert the order of presentation, considering the classical case first.

First, we emphasize a point that, in retrospect, seems obvious, but is not sufficiently underscored in many discussions of quantum statistics in the old quantum theory—indeed, it may have been imperfectly understood by the practitioners of the theory up to the early 1920s. If we assign well-defined paths (or orbits) to particles, as is certainly the case in the Bohr model and its post-1913 extensions, the particles become *ipso facto* uniquely distinguishable from each other. We need only examine the system at some taking note of the spatial location (say $t=0$), taking note of the spatial location $\vec{r}_i^{(0)}$ of the i particle at that time. As each particle can in principle be followed along its subsequent path, these vectors serve as unambiguous labels for the particles at all subsequent times, save for a set of situations of relative measure zero where two particles are at exactly the same spatial point initially. If we then ask how many microscopic configurations of N such distinguishable particles correspond to having n_ν of the (in our case) atoms with energy in the νth energy slice, where evidently $\sum_\nu n_\nu = N$, we can compute the desired number by first noting that N objects can be divided into groups containing $n_1, n_2, \ldots$ objects each in $N!/(n_1!n_2!\ldots.) = N!/\prod_\nu n_\nu!$ ways. Once this partition has been effected, we must still decide which of the available z_ν each of the distinguishable n_ν atoms should be placed in the νth energy slice. This can be done in $z_\nu^{n_\nu}$ ways for the νth slice. Combining these factors we find the classical number of "complexions" (in Boltzmann's terminology common at the time) of the whole gas to be

$$W_{\text{Boltz}} = \frac{N!}{\prod_\nu n_\nu!} \prod_\nu z_\nu^{n_\nu}. \tag{13.84}$$

Note that the requirement that the entropy (or the logarithm of W) vanish at absolute zero requires that $W = 1$ in this limit: there is a single lowest energy quantum state, which we can label $\nu = 0$, with $z_0 = 1$ and $n_0 = N, n_\nu = 0$ for $\nu \neq 0$, that is, all atoms occupy this single lowest energy quantum state, which then gives $W = N!/n_0! = 1$. This result requires the presence of the $N!$ factor in the numerator in Eq. (13.84).

The new counting procedure introduced by Bose (for a gas of light quanta) assumes that the constituent particles of the gas are indistinguishable, in some fundamental, and classically incomprehensible, way, so that there is no way of attaching labels—as we did using the world lines of the atoms—to each atom, in the case of a material gas composed of a single type of atom, as we pick a particular energy level for that atom. Consequently, a state of the gas is specified simply by stating the total number of atoms n_ν occupying each of the z_ν energy levels in each of the energy slices. The combinatorics of this case is easily accomplished using the "dots and bars" trick of Ehrenfest described in Section 2.5 in connection with Planck's derivation of his black-body law. One finds that the number of ways of assigning occupation numbers summing to n_ν to each of the z_ν states in the νth slice is $(n_\nu + z_\nu - 1)!/(n_\nu!(z_\nu - 1)!)$.[32] The total number of complexions available to the gas that we find using this Bose statistics is thus

$$W_{\text{Bose}} = \prod_\nu \frac{(n_\nu + z_\nu - 1)!}{n_\nu!(z_\nu - 1)!}. \tag{13.85}$$

This quite different formula will, unsurprisingly, lead to quite different thermodynamic consequences for the equilibrium behavior of the gas. Note that this formula, like the Boltzmannian Eq. (13.84), satisfies $W_{\text{Bose}} = 1$ at absolute zero (and therefore zero entropy) if only the lowest quantum state (of degeneracy $z_0 = 1$) is occupied ($n_0 = N, n_\nu = 0$ for $\nu \neq 0$). Thermal equilibrium of the gas implies the maximization of the entropy, given by the famous relation due to Boltzmann and carved into his tombstone (see Volume 1, Plate 6) but first written in the now-familiar form by Planck (see Section 2.5):

$$S = k \ln W. \tag{13.86}$$

[32] We note here that no mystery whatsoever attaches to the use of this formula in the context in which Planck applied it: there, one was counting the number of ways of assigning n_ν energy quanta to a set of z_ν identical harmonic oscillators. It clearly makes no sense to try to distinguish one component of the energy of an oscillator from another.

In the classical case this gives, using throughout Stirling's approximation $\ln n! \approx n \ln n$,

$$S_{\rm Boltz} = k\Big(N \ln N + \sum_{\nu} (n_\nu \ln z_\nu - n_\nu \ln n_\nu)\Big). \tag{13.87}$$

This expression must now be maximized subject to the constraints $\sum_\nu n_\nu = N$ and $\sum_\nu n_\nu E_\nu = E$. Introducing Lagrange multipliers α for the total number of atoms and β for the energy, we must maximize (with respect to the now freely variable n_ν—the inert piece $kN \ln N$ may be ignored here) the expression

$$\sum_{\nu} (n_\nu \ln z_\nu - n_\nu \ln n_\nu - \alpha\, n_\nu - \beta\, n_\nu E_\nu). \tag{13.88}$$

Differentiating Eq. (13.88) with respect to n_ν, one finds the maximum condition (using an overbar to indicate the distribution $\overline{n}_\nu$ for which the entropy assumes its maximum value) for each ν,

$$\ln z_\nu - \ln \overline{n}_\nu - 1 - \alpha - \beta E_\nu = 0. \tag{13.89}$$

This can be rewritten as

$$\ln \frac{\overline{n}_\nu}{z_\nu} = -\alpha - 1 - \beta E_\nu. \tag{13.90}$$

It follows that

$$\overline{n}_\nu = C z_\nu e^{-\beta E_\nu}, \quad \text{with } C \equiv e^{-\alpha - 1}. \tag{13.91}$$

This is just the Maxwell–Boltzmann distribution, with the Lagrange multiplier β determined by the temperature, $\beta = 1/kT$, and α chosen (through the constant C) to ensure that the total number of atoms is adjusted to the desired quantity, $\sum_\nu \overline{n}_\nu = N$.

With $\overline{n}_\nu$ substituted for n_ν, Eq. (13.87) gives the entropy of the system in thermal equilibrium at temperature T:

$$S^{\rm equil}_{\rm Boltz} = k\Big(N \ln N - \sum_{\nu} \overline{n}_\nu \ln \frac{\overline{n}_\nu}{z_\nu}\Big). \tag{13.92}$$

Using Eq. (13.90) in the sum on the right-hand side, we arrive at

$$\begin{aligned} S^{\rm equil}_{\rm Boltz} &= k\Big(N \ln N + \sum_{\nu} \overline{n}_\nu (\alpha + 1 + \beta E_\nu)\Big) \\ &= k\Big(N \ln N + (\alpha + 1)N + \beta E\Big). \end{aligned} \tag{13.93}$$

This expression clearly displays the dilemma faced by the classical formalism in implementing both entropy additivity and the Nernst theorem, which requires the presence of the $N!$ term in W, and therefore the $N\ln N$ term in Eq. (13.93). On the other hand, the extensivity property of the entropy is ruined by precisely this term, due to the logarithm.[33] Removing this term, say, by dividing by the factor $N!$ in Eq. (13.84), would reduce the entropy at absolute zero to the nonsensical negative value $-kN\ln N$.

The Boltzmann expression for the number of microstates, $W = e^{S/k}$, allows us to find the mean square fluctuation $(\Delta n_\nu)^2$ around the maximizing value $\overline{n}_\nu$ for each ν. It can be done in the same way that Einstein (1909a, 1909b) derived his famous formula of the mean square fluctuation of energy and momentum in black-body radiation (see Section 3.4.3). First, we expand the entropy in Eq. (13.87) in a Taylor series around the maximum (where the first derivative is zero):

$$\begin{aligned} S_{\text{Boltz}}(n_\nu) &= S_{\text{Boltz}}(\overline{n}_\nu) + \frac{1}{2}\left(n_\nu - \overline{n}_\nu\right)^2 \left.\frac{\partial^2 S_{\text{Boltz}}}{\partial n_\nu^2}\right|_{n_\nu = \overline{n}_\nu} + \dots \\ &= S_{\text{Boltz}}(\overline{n}_\nu) - \frac{k}{2}\frac{\left(n_\nu - \overline{n}_\nu\right)^2}{\overline{n}_\nu} + \dots . \end{aligned} \tag{13.94}$$

We use this result to calculate the probability $W(n_\nu)dn_\nu = e^{\Delta S(n_\nu)/k_B}dn_\nu$ of a fluctuation $\Delta n_\nu \equiv n_\nu - \overline{n}_\nu$ (cf. (Eq. 3.64)). Using Eq. (13.94) for $\Delta S(n_\nu) \equiv S_{\text{Boltz}}(\overline{n}_\nu) - S_{\text{Boltz}}(n_\nu)$, we can write $W(n_\nu)$ as

$$W(n_\nu) \propto e^{\Delta S(n_\nu)/k} = \exp\left(-\frac{1}{2}\frac{(n_\nu - \overline{n}_\nu)^2}{\overline{n}_\nu}\right). \tag{13.95}$$

A simple Gaussian integral (cf. Eq. (3.70)) then shows that the mean square fluctuation in the number of occupied modes in the ν slice is

$$(\Delta n_\nu)^2 = \overline{n}_\nu, \tag{13.96}$$

which agrees with the usual $\sqrt{n}$ fluctuation for the root mean square fluctuation, which we are accustomed to when dealing with the behavior of classical particles being scattered randomly into and out of the given energy slice. In terms of the total energy in this energy slice, $\mathcal{E}_\nu \equiv n_\nu E_\nu$, the fluctuations take the form

$$(\Delta\mathcal{E}_\nu)^2 = E_\nu \overline{\mathcal{E}}_\nu. \tag{13.97}$$

In the case of a photon gas, with mode energy $E_\nu = h\nu$, this is just the energy fluctuation formula displaying particle-like behavior that we find for black-body radiation in the Wien limit (see Eq. (3.75)).

We turn now to the second case examined by Einstein, namely, the entropy calculation for the case of a gas of (once again massive) atoms subject to Bose statistics,

[33] The coefficients $\alpha+1$ and β in second and third term in Eq. (13.93) are intensive variables, depending on the number density and temperature of the gas, respectively, so that these terms behave as extensive quantities, due to their linear dependence on N and E.

where $S_{\text{Bose}} = k \ln W_{\text{Bose}}$, with W_{Bose} given in Eq. (13.85). Once again, using Stirling's approximation, we have

$$S_{\text{Bose}} = k \sum_{\nu} \big((n_\nu + z_\nu) \ln (n_\nu + z_\nu) - z_\nu \ln z_\nu - n_\nu \ln n_\nu \big). \tag{13.98}$$

As previously in the case of Boltzmann statistics, this quantity must be maximized subject to the constraints $\sum_\nu n_\nu = N$ and $\sum_\nu n_\nu E_\nu = E$. In conformance with modern notation, we redefine the Lagrange multiplier attached to the total number of atoms as $\alpha \equiv -\beta\mu$, with $\beta = 1/kT$ as before, while μ, now called the "chemical potential" must be adjusted to tune the total number of particles to the desired amount. In analogy with Eqs. (13.88)–(13.89), we need to differentiate the expression

$$\sum_{\nu} \big((n_\nu + z_\nu) \ln (n_\nu + z_\nu) - z_\nu \ln z_\nu - n_\nu \ln n_\nu + \beta\mu n_\nu - \beta n_\nu E_\nu \big) \tag{13.99}$$

with respect to n_ν and set the result to 0, to obtain the extremal (hence, equilibrium) value at $n_\nu = \overline{n}_\nu$:

$$\ln (\overline{n}_\nu + z_\nu) + 1 - \ln \overline{n}_\nu - 1 + \beta\mu - \beta E_\nu = 0. \tag{13.100}$$

It follows that

$$\ln \Big(1 + \frac{z_\nu}{\overline{n}_\nu}\Big) = \beta(E_\nu - \mu), \tag{13.101}$$

in which we readily recognize the now famous *Bose–Einstein distribution*:

$$\overline{n}_\nu = \frac{z_\nu}{e^{\beta(E_\nu - \mu)} - 1}. \tag{13.102}$$

We now calculate the equilibrium value of the entropy. To this end, we first rewrite Eq. (13.98) as

$$\begin{aligned} S_{\text{Bose}} &= k \sum_{\nu} \Big((n_\nu + z_\nu) \ln n_\nu \Big(1 + \frac{z_\nu}{n_\nu}\Big) - z_\nu \ln z_\nu - n_\nu \ln n_\nu \Big) \\ &= k \sum_{\nu} \Big((n_\nu + z_\nu) \ln \Big(1 + \frac{z_\nu}{n_\nu}\Big) + z_\nu \ln \frac{n_\nu}{z_\nu} \Big). \end{aligned} \tag{13.103}$$

Substituting $\overline{n}_\nu$ for n_ν in this expression and using Eqs. (13.101) and (13.102), we find:

$$\begin{aligned} S_{\text{Bose}}^{\text{equil}} &= k\sum_\nu \Big((\overline{n}_\nu + z_\nu)\beta(E_\nu - \mu) - z_\nu \ln\left(e^{\beta(E_\nu-\mu)} - 1\right)\Big) \\ &= k\Big(\beta E - N\beta\mu + \sum_\nu z_\nu \Big(\ln e^{\beta(E_\nu-\mu)} - \ln\left(e^{\beta(E_\nu-\mu)} - 1\right)\Big)\Big) \\ &= k\Big(\beta(E-\mu N) - \sum_\nu z_\nu \ln\left(1 - e^{-\beta(E_\nu-\mu)}\right)\Big). \end{aligned} \tag{13.104}$$

This expression displays the desired extensivity property of the entropy, which depends linearly on the extensive variables of the system, the total energy E, the total number of atoms N, and the volume V (through z_ν, as can be seen in Eq. (13.76)). Apart from this, the entropy depends only on intensive variables, the temperature (through β), and the chemical potential μ. We already saw that Nernst's theorem is also guaranteed by Eq. (13.85), so, as Einstein emphasizes, the classically mysterious Bose counting procedure naturally incorporates all the desired thermodynamic behavior without contradiction. As Einstein says,

> For these reasons I believe that the calculational method [given by Bose's statistical Ansatz] must be considered preferable, even if there are no a priori reasons for the adoption of this method with respect to others (Einstein 1925a, p. 7).

The mean square fluctuation in the number, or energy, of atoms in a given energy slice can be calculated as in the Boltzmann case. The second derivative of the entropy function in Eq. (13.98) now gives (instead of $-k/\overline{n}_\nu$, in the Boltzmann case)

$$\left.\frac{\partial^2 S_{\text{Bose}}}{\partial n_\nu^2}\right|_{n=\overline{n}} = k\left(\frac{1}{\overline{n}_\nu + z_\nu} - \frac{1}{\overline{n}_\nu}\right) = -k\frac{z_\nu}{\overline{n}_\nu(\overline{n}_\nu + z_\nu)}. \tag{13.105}$$

The corresponding Gaussian integral then gives

$$(\Delta n_\nu)^2 = \overline{n}_\nu + \frac{\overline{n}_\nu^2}{z_\nu}, \tag{13.106}$$

or, equivalently,

$$\frac{(\Delta n_\nu)^2}{\overline{n}_\nu^2} = \frac{1}{\overline{n}_\nu} + \frac{1}{z_\nu}. \tag{13.107}$$

The mean square fluctuation of the energy $\mathcal{E}_\nu = n_\nu E_\nu$ of the atoms occupying the ν slice is then given by

$$(\Delta\mathcal{E}_\nu)^2 = E_\nu\,\overline{\mathcal{E}}_\nu + \frac{\overline{\mathcal{E}}_\nu^2}{z_\nu}. \tag{13.108}$$

This expression displays the remarkable sum of linear and quadratic dependence on the average energy $\overline{\mathcal{E}}_\nu = \overline{n}_\nu E_\nu$, which Einstein had already found in the case of black-body radiation (see Eq. (3.79)) and which led him to predict that "the next phase of the development of theoretical physics will bring us a theory of light that can be interpreted as a kind of fusion of the wave and emission [i.e., particle] theories" (Einstein 1909b, pp. 482–483).[34]

Einstein follows his derivation of the fluctuation formula for a Bose gas with a description of de Broglie's phase wave hypothesis for massive particles. The motivation is clear: de Broglie's work shows how "a wave field can be assigned to a system of material particles" (Einstein 1925a, p. 9) and the appearance of terms corresponding to interference phenomena in the density fluctuation formulas for a gas of massive atoms would certainly support a hypothesis of this kind. In a footnote, Einstein mentions de Broglie's derivation of the Bohr–Sommerfeld quantization rules on the basis of a "very interesting [*sehr bemerkenswerte*] geometrical interpretation" (ibid.) His enthusiasm for de Broglie's ideas is clear: "I will examine this interpretation more closely, as I believe that we are dealing here with more than a mere analogy" (ibid.).

The argument presented is identical to the procedure de Broglie used to arrive at Eq. (13.4): if a particle at rest is associated with a spatially independent "everywhere synchronous oscillation" (in Einstein's words), with frequency given by $h\nu_0 = m_0c^2$, then the same oscillation viewed in an inertial frame in which the particle is moving with speed v in the x direction corresponds to a "wavelike process" of the form[35]

$$u(x,t) = A\sin\left(2\pi\nu_0\frac{t - vx/c^2}{\sqrt{1 - v^2/c^2}}\right). \tag{13.109}$$

Replacing ν_0 by m_0c^2/h in this expression, we get

$$u(x,t) = A\sin\frac{2\pi}{h}(Et - px) = A\sin(\omega t - kx), \tag{13.110}$$

where $E = m_0c^2/\sqrt{1 - v^2/c^2}$ (resp. $p = m_0v/\sqrt{1 - v^2/c^2}$) are the usual relativistic energy (resp. momentum) expressions, related to the wave quantities ω (resp. k) by

$$E = \hbar\omega,\ p = \hbar k,\ k = \frac{2\pi}{\lambda},\ \omega = 2\pi\nu,\ \nu = \frac{\nu_0}{\sqrt{1 - v^2/c^2}}. \tag{13.111}$$

[34] The situation discussed by Einstein in 1909 concerned the fluctuations in the energy in a frequency range of thermal black-body radiation in a subvolume of a larger container, but the formula obtained is characteristic of a much more general set of fluctuation results, which in general only depend on a sufficient level of phase randomness or decorrelation between the particle states. Here, it holds for the fluctuations in the number of atoms in a narrow energy range—but in the whole volume of the system—given that the atomic system as a whole is in thermal equilibrium at a definite temperature.

[35] Einstein writes only the sine function that follows, without attaching an amplitude factor A, or assigning a specific notation for the wave field. Equation (13.109) follows directly from the Lorentz transformation equation for time.

It is clear that the wave characteristics of frequency and wavenumber are alternatively equally well expressed in terms of particle properties energy and momentum. Moreover, as de Broglie had discovered, the group velocity of the associated waves (i.e., the speed with which signals, or wave-packets, propagate) is given by

$$v_g = \frac{\partial \omega}{\partial k} = \frac{\partial E}{\partial p} = \frac{\partial}{\partial p}\sqrt{p^2c^2 + m_0^2c^4} = \frac{pc^2}{E} = v \tag{13.112}$$

and agrees precisely with the speed with which the associated particle is seen to be moving (i.e., the relative speed of the two frames in which the particle is at rest or in motion with speed v). This speed, of course, is completely different from the phase velocity $V \equiv \omega/k = E/p = c^2/v > c$, which cannot represent the transmission of physical effects.

After reviewing de Broglie's phase wave/particle connection, Einstein connects this approach to the new gas statistics giving the fluctuation formula Eq. (13.106) with the following remark:

> One now sees that a scalar wave-field can be assigned to such a gas, and I have convinced myself by calculation that the mean square fluctuation of this wave-field is $1/z_\nu$ [for the part of the field] corresponding to the energy interval ΔE examined above (Einstein 1925a, p. 10).

Here, Einstein is referring to the *classical* component of the fluctuation formula, namely, the final term in the second formula in Eq. (13.106). It is not clear at this point exactly to which calculation Einstein is referring. In principle, one can directly calculate the classical fluctuations due to the interference of superposed standing-wave modes of a general wave field of the form of Eq. (13.109).[36] Alternatively, one can use the fact that the scalar wave field here can be decomposed into independent harmonic oscillator normal modes, just as in the electromagnetic case (in the Rayleigh–Jeans calculation), the only difference being an altered dispersion relation (dependence of frequency on wavenumber) corresponding to massive particles, which simply changes the mode counting number z_ν to the form displayed in Eq. (13.80) (which reduces to the Rayleigh–Jeans form setting $m_0 = 0$). Classical equipartition then gives

$$\overline{\mathcal{E}}_\nu = z_\nu kT, \tag{13.113}$$

[36] In Duncan and Janssen (2008), we reconstructed the calculation for the massless case in the one-dimensional model used by Jordan in the final section of the *Dreimännerarbeit* (Born, Heisenberg, and Jordan 1926). See Section 12.3.5.

and applying the well-known formula for the mean square energy fluctuation, which Einstein himself had derived in two different ways before (see Eqs. (3.16)–(3.20) and (3.61)–(3.71)):

$$\overline{(\Delta \mathcal{E}_\nu)^2} = kT^2 \frac{d\overline{\mathcal{E}_\nu}}{dT} = (kT)^2 z_\nu = \frac{\overline{\mathcal{E}_\nu}^2}{z_\nu}. \tag{13.114}$$

One is inclined to think, however, that this latter argument, simple as it is, would have been mentioned explicitly by Einstein. In any case, within a year of Einstein's second paper on gas theory, as noted in our discussion of the final section of the Three-Man-Paper, Jordan would fill the remaining gap and produce a derivation of Eq. (13.108) (in the massless case, and for a one-dimensional system), including both particle and wave terms, based on a fully consistent quantum-mechanical formalism (cf. note 36).

At the conclusion of the section on fluctuations in his second paper on Bose gases, after his review of the de Broglie theory, Einstein (1925a, p. 10) comments on a paradox, related to the Gibbs paradox and known as the *mixing paradox*, which he had already discussed briefly at the conclusion of his first paper. At that point he had conceded he had been unable to resolve it (Einstein 1924, p. 267). There was no doubt that thermodynamics requires the additivity of entropy in the case in which two gases each composed of a distinct type of atom are mixed in a single container. Classically, this results from the fact that the microstates available to the first gas (numbering, say, W_1 for a given macrostate) can be distinguished from those available to the second (numbering W_2), so that the entropy of the mixed system, $S = k \ln (W_1 W_2) = k \ln W_1 + k \ln W_2$, is the sum of the separate gas entropies.

The question arises how to understand this result once the states of the gas are interpreted in terms of modes of a single wave field. For Einstein, as for de Broglie, as we saw earlier (cf. quote following Eq. (13.73)), this wave field is conceived as a physical disturbance analogous to the electromagnetic field, which (in some sense) guides the motion of the particles of the gas. Einstein suggests in the second paper that the characteristic interference effects visible in the fluctuation result Eq. (13.108) are noticeable only if the interfering modes are close in frequency and phase velocity. Two distinguishable atoms (or molecules) will necessarily have different masses, resulting in enormously different frequencies (as the rest frequency, for example, is $\nu = m_0 c^2/h$). Such components of the wave field are therefore to be treated separately—the language used by Einstein (1924, p. 267) is that each type of molecule would have its own particular "cells", and can therefore be treated statistically as if it were the only type present.

None of this makes much sense: it is difficult to see how to separate in a precise way different sets of Fourier components (however close or far separated in frequency) of a single wave to be treated distinctly in the statistical analysis of states of the system. Instead, the waves needed for a proper quantum-mechanical description of a multi-particle system propagate, as we see, in a higher-dimensional ($3N$-dimensional for N particles) configuration space, and the imposition of independent statistics for different types of particles is achieved via symmetry requirements imposed on this higher-dimensional wave function. There does indeed exist in Einstein's (and de Broglie's) sense a single universal wave function describing a multiparticle system, but its interpretation and properties are quite different from the classically inspired waves—basically, scalar fields taking real values at each point in physical three-dimensional space—these physicists were imagining in late 1924 and early 1925.[37]

[37] We want to emphasize that the wave field/function referred to here, naturally generalizing de Broglie's phase wave for a single particle, is *not* a relativistic quantum field of the type that would emerge in the late 1920s and 1930s—we are referring here to the non-relativistic multiparticle Schrödinger wave function $\psi(\vec{r}_1, \vec{r}_2, \ldots, \vec{r}_N, t)$ of an assembly of N particles. This function in general takes complex values and is defined in the higher dimensional configuration space which specifies the simultaneous location of all the particles, rather than in ordinary three-dimensional space.

14

Schrödinger and Wave Mechanics

14.1 Schrödinger: early work in quantum theory

Erwin Schrödinger was born on August 12, 1887 into an upper middle class family in Vienna, Austria.[1] Only two years younger than Niels Bohr, his seminal contributions to the development of modern quantum mechanics would arrive almost thirteen years after Bohr conceived of his model of the hydrogen atom, initiating the explosive growth of the old quantum theory. His academic background was as promising as could be expected in Vienna in the closing years of the Austro–Hungarian empire—attendance at the Akademisches Gymnasium (the most secular of Vienna's secondary schools aimed at students aspiring to University degrees) until the age of eighteen, always at the top of his class, and with a clear facility for mathematics and physics, followed by matriculation at the University of Vienna in the Fall of 1906, where he just missed the lectures of Boltzmann, who had committed suicide a few months earlier.

At the University, Schrödinger followed the course of lectures in theoretical physics offered by Friedrich Hasenöhrl, Boltzmann's successor, but also engaged in experimental work, receiving his doctorate (Dr. Phil.) in May 1910 for a thesis summarizing his experimental work on electrical conductivity (which was supervised by the pre-eminent experimentalist in Vienna at the time, Franz Exner). Official appointment to a university position required the submission of a *Habilitationschrift*, a post-doctoral thesis which would establish the author's facility for independently conceived and executed research. In Schrödinger's case, this entailed a turn (in 1912) to purely theoretical research, interestingly, on a topic that had concerned Bohr just a few years earlier in his doctoral work in Copenhagen: the diamagnetism of metals.

Despite having an excellent background in classical theoretical physics, Schrödinger's early work betrays the somewhat provincial character of Viennese physics in the first two decades of the twentieth century. The most senior and accomplished physicists at the University of Vienna concentrated on research topics at the peripheryof the new

[1] For more details on Schrödinger's early years, and the cultural milieu in which he grew up, see Moore (1989) and Mehra and Rechenberg (1987). For another account of the evolution of Schrödinger's thinking, see Wessels (1979).

Constructing Quantum Mechanics. Anthony Duncan and Michel Janssen, Oxford University Press.
 DOI: 10.1093/oso/9780198883906.003.0014

physics ushered in by the development of quantum theory and special relativity. Much of Schrödinger's early theoretical work is completely classical in content, encompassing subjects such as the melting point of crystals, the emergence of wave phenomena from assemblies of coupled oscillators, or even capillary pressure in bubbles. His first paper specifically deploying the quantum theory was a two-part review (Schrödinger 1917) on specific heats, in which the (by this time well-known) Einstein–Debye theory was described, and an attempt made to summarize the state of the art for specific heats of gases, which had been a sore point in classical statistical mechanics since the days of Maxwell. Schrödinger's treatment of the latter was confused and incorrect (hardly surprising since the correct treatment of the specific heat of even as simple a gas as molecular hydrogen necessitated an understanding of wave-function exchange symmetry which was only possible after the new quantum mechanics was developed in 1925/26 (Gearhart 2010)).

In the Fall of 1920 Schrödinger assumed a position as full professor in Stuttgart. The first edition of Sommerfeld's (1919) "Bible" of atomic physics, *Atombau und Spektrallinien*, had appeared the previous year, and provided a mother lode of information on the most pressing current problems in the application of quantum theory to atomic spectroscopy, and which was especially useful for physicists who were not based at Copenhagen and Munich, the two main centers of research in quantum theory. Schrödinger studied *Atombau* assiduously and by the following January had submitted a paper to *Zeitschrift für Physik* addressing the thorny issue of penetrating orbits in the old quantum theory (Schrödinger 1921). By this time, the great successes of the old quantum theory (for example, the relativistic fine structure and the Stark effect) had all been achieved. Indeed, they had followed in swift succession the introduction of the orbit quantization rules by Sommerfeld and others in 1915–1916. By 1920, the old quantum theory was facing several problems, which stubbornly resisted resolution.

The problem treated by Schrödinger was somewhat peripheral, and far from an original subject of enquiry in the sense that it hearkened back to Sommerfeld's (1916c) earlier work on screening effects and spectral defects—the effect of the screening of the nuclear charge by inner electron shells on the quantized energy of an outer "valence" electron (see Section 7.1.1). The consequent shift of the term energy from its hydrogenic Bohr–Balmer value (the spectral defect) was particularly prominent for the alkali *s*-states, where it typically amounted to as much as a half unit in the effective principal quantum number. The problem was that the *s* orbits (with azimuthal quantum number 1, and varying non-zero radial quantum numbers) were all quite eccentric, with the perihelion distance to the nucleus so small that they would necessarily penetrate the full shell of electrons corresponding to the next lower principal quantum number, thereby experiencing a much increased effective nuclear charge, and a severe distortion of the outer (Keplerian) elliptical orbit. Schrödinger was able to arrive at an approximate solution by treating the inner shell as a uniform spherical distribution, in which case the valence electron orbit switched between two Keplerian ellipses (of different semi-major axis and eccentricity) as it passed in and out of the inner shell. Quantization of this more complicated orbit then led to a quantitative value for the spectral defect of the *s* state of lithium (0.26), unfortunately not very close to the observed value (0.59).

Schrödinger's next significant work on the quantum theory contains some intriguing—and prescient—speculations on an alternative interpretation of the quantization rules of the old quantum theory. It anticipates in some respects de Broglie's treatment of phase waves in the presence of an electromagnetic potential, and was, by an amusing coincidence, written in Arosa, the same Swiss resort (specializing in treatment of pulmonary conditions—in Schrödinger's case, probably a mild case of tubercolosis) where Schrödinger would, a little over three years later, produce his seminal paper on quantization from the wave equation that now bears his name.

The 1922 Arosa paper, "On a remarkable property of quantized orbits of a single electron" (Schrödinger 1922), has its origins however in a set of ideas far removed from quantum theory, namely, Hermann Weyl's (1918) attempt to construct a variant of general relativity that incorporates the physics of electromagnetism in the metrical structure of space–time, thereby producing a unified geometrical picture of both gravitation and electromagnetism.[2] Weyl transferred the role of the metric tensor $g_{\mu\nu}$ in general relativity[3] in prescribing the magnitude of vectors undergoing parallel transport to the electromagnetic potential A_μ, while leaving directional information encoded in the metric tensor, which effectively becomes a tensor density. In particular, the length l of a space–time interval under parallel transport is "re-calibrated" (in Weyl's language) under an infinitesimal displacement by $dl = -lA_\mu dx^\mu/\gamma$ (where γ is a *real* constant with dimensions of action), or if the interval is transported over a finite path C from point A to point B, a finite rescaling

$$l \to l\exp\left(-\frac{1}{\gamma}\int_C A_\mu dx^\mu\right). \tag{14.1}$$

As in the case of the usual parallel transport in affine Riemannian geometry, the resulting rescaling is path-dependent, with the variation in rescaling factors directly dependent on the non-vanishing of the "curvature" tensor $F_{\mu\nu} = \partial_\mu A_\nu - \partial_\nu A_\mu$.

Weyl's unified geometrical theory of gravitation and electromagnetism differs essentially from Einstein's general relativity (in particular, it leads to unacceptable field equations of fourth order; see Bergmann 1942, p. 253), and is now of purely historical interest. Schrödinger's investigation of the Weyl scale factor is of interest to us because it contains a clear premonition of the emergence of periodic *complex* phase factors, dependent on the electromagnetic potential A_μ, and exactly of the form suggested two years later by de Broglie. Much later, these phase factors induced by an electromagnetic potential would become famous in the Aharanov–Bohm effect. In his paper, Schrödinger examines a number of special cases (Kepler orbit in Coulomb potential, then the weak field Zeeman and Stark effects, and finally, relativistic effects absent external fields) and shows that the exponent in the Weyl scale factor Eq. (14.1) in all cases assumes the form

[2] Einstein (1918) appended a short note to Weyl's paper criticizing this theory, about which the two men corresponded extensively in 1918 (Einstein 1987–2021, Vol. 8B). For further discussion of Weyl's theory, see, for example, Weyl (1922, secs. 35–36, pp. 282–312) and Bergmann (1942, Ch. XVI).

[3] We use a notation for space–time indices and for the electromagnetic four-potential more familiar to modern readers: for example, both Weyl and Schrödinger write φ_i for A_μ.

nh/γ, $n = 1, 2, 3, \ldots$ when the integral is extended over a period (appropriately chosen) of the motion.

Let us illustrate this for the first two examples given by Schrödinger. Consider an electron executing a quantized Kepler orbit in the static Coulomb potential of a hydrogen atom, so that the electromagnetic four-potential has only a time component $\varphi(\vec{r}) = e/|\vec{r}|$ (where the positive quantity e is the magnitude of the electron charge). The presence of the electrostatic potential then induces a potential energy function $V = -e\varphi$ for the negatively charged electron. The virial theorem for this potential relates the average kinetic T and potential V energies, or, in terms of an integral over a single orbital period τ,

$$\tau\overline{T} = \int_0^\tau T(t)dt = -\frac{1}{2}\int_0^\tau V(t)dt = -\frac{1}{2}\tau\overline{V}. \tag{14.2}$$

The quantization theorem of the old quantum theory in its most elementary form derives from Ehrenfest's adiabatic theorem, which asserts that in any system obtainable by adiabatic transformation of a harmonic oscillator, the kinetic energy integrated over a period must be an integer multiple of $h/2$, $\tau\overline{T} = nh/2$ (cf. Eq. (5.59)). Accordingly,[4]

$$\int_0^\tau e\varphi dt = 2\int_0^\tau T dt = nh. \tag{14.3}$$

In other words, viewed as an integral along the world-line of the orbiting electron, the line integral $\int eA_\mu dx^\mu$, where $A_0 = \varphi/c$ ($A_i = 0, i = 1, 2, 3$) and $x^0 = ct$, taken for a single orbital period, is quantized in integer multiples of Planck's constant.[5] In the Weyl theory, by Eq. (14.1), this would entail a recalibration of the scale factor determining distances by $e^{-nh/\gamma}$ after a single orbital period of an electron, a fundamentally absurd proposition unless γ is chosen suitably large.

As Schrödinger openly confesses, there is a more immediate conceptual problem with this result, however: it clearly assumes a particular choice of the zero level of the electrostatic potential, namely, that it vanish at infinity. Normally, such a choice is purely a matter of convention, but here, as the result depends crucially on use of the virial theorem, which in turn assumes that the potential energy be a pure power of r, the addition of a floating constant to $\varphi(r)$ is precluded.[6]

[4] In the Bohr model, the average kinetic energy for principal quantum number n is $me^4/(2\hbar^2n^2)$, with orbital period $\tau = 2\pi n^3\hbar^3/(me^4)$, giving $\tau\overline{T} = nh/2$.

[5] The presence of this electric Aharanov–Bohm phase factor, and its role in inducing measurable interference effects, has been directly verified experimentally (van Oudenaarden et al. 1998). In the case considered here, there is no magnetic field, so the spatial components $A_i, i = 1, 2, 3$ of the electromagnetic potential are zero.

[6] More generally, the components of the electromagnetic potential (the electrostatic potential φ and the vector potential $\vec{A}$) are all gauge-dependent quantities, as the replacement $\varphi \to \varphi - \partial\Lambda(\vec{r}, t)/\partial t$, $\vec{A} \to \vec{A} + \nabla\Lambda(\vec{r}, t)$ should leave the physics of the problem unchanged, where Λ is an arbitrary (differentiable) function of space and time. Such a transformation will in general ruin the proposed quantization of the line integral of the potential. The issue of gauge dependence of the result is not raised by Schrödinger—in the wave mechanics that would emerge later, the effect of a gauge transformation is known to result in an unphysical, hence unobservable, phase shift of the wave function (Baym 1969).

Undeterred by this, Schrödinger presses ahead, next considering the Zeeman effect, where he finds that for a constant magnetic field in the z-direction, $\vec{H} = H\hat{z}$, and an electron in an orbit with z-component of angular momentum $L_z = m\hbar$, the spatial line integral of the vector potential over the world-line executed by the particle in a complete Larmor precession period $\tau_L = 2\pi/\omega_L$ ($\omega_L = eH/2mc$) is quantized by

$$\int \vec{A} \cdot d\vec{r} = mh. \tag{14.4}$$

After considering a number of other examples of quantization results of this kind, Schrödinger concludes his paper by suggesting that "it is difficult to believe that these results are just an accidental mathematical consequence of the quantum conditions, without any deeper physical meaning." On the other hand, he admits that "it is more than questionable that the electron really carries around a "line element" [subject to rescaling by the potentially enormous factor $e^{nh/\gamma}$] during its motion."

As for the mysterious constant γ, with dimensions of action, Schrödinger remarks that there are only two obvious choices. One could choose $\gamma \approx e^2/c$ (whence $h/\gamma \approx 2\pi/\alpha, \alpha = 1/137$), giving an exponential rescaling of the order of e^{1000} per orbital period in the Weyl theory, clearly physically nonsensical.

Alternatively, one could choose γ to be the *purely imaginary* number $\hbar/i$, in which case the Weyl scale factor Eq. (14.1) becomes a complex phase, which in virtue of the quantization condition reproduces unity, $e^{nh/\gamma} = e^{2\pi i n} = 1$ extended over an appropriate path. Of course, this complex interpretation is no longer compatible with a metrical interpretation in terms of distances that are necessarily real. On the other hand, such a phase factor is perfectly reasonable once one adopts a mechanics based on the description of particle motion in terms of periodic waves. This possibility almost certainly was not at the front of Schrödinger's mind at this point in 1922, but on encountering de Broglie's work three years later, the appearance of quantization rules in wave-like complex exponential factors would certainly have primed him for acceptance of a wave approach, especially given the fact that he had already in 1919 studied (and preserved in detailed notes) the Hamiltonian analogy between optics and mechanics, discussed in Section 13.3.

14.2 Schrödinger and gas theory

Schrödinger had maintained a lively interest in gas theory from the time of his two-part review on specific heats (Schrödinger 1917), and this continued in his early Zurich years with papers on gas degeneracy and mean free paths and on the specific heat of hydrogen (Schrödinger 1924a, 1924b).

Schrödinger's interest in the treatment of gases in quantum theory continued in his papers on the statistics of entropy and the energy states of a monatomic ideal gas (Schrödinger 1925, 1926a), led him to study with great interest Einstein's papers on the quantum statistical mechanics of ideal gases subject to Bose statistics. He had presumably met and perhaps discussed with Einstein the issue of gas statistics at the Innsbruck *Naturforscherversammlung* in September 1924, around the time of publication of Einstein's (1924) first paper on gas theory. Schrödinger had little time in the Fall of 1924 for research, given his teaching duties in Zurich (Mehra and Rechenberg 1987, p. 385). A letter to Einstein of February 5, 1925 (before seeing Einstein's second gas theory paper) indicates that he simply did not believe the results of Einstein's application of Bose statistics to a monatomic gas (Einstein 1987–2021, Vol. 14, Doc. 433). In particular, Schrödinger claimed that the Bose distribution Eq. (13.102), with the atomic occupation number $n_\nu \propto 1/(e^{\beta(E_\nu-\mu)}-1)$, disagreed with the Maxwell–Boltzmann distribution $e^{-\beta E_\nu}$ that Einstein had used to such great effect in his radiation papers on the A and B coefficients in 1917. This complaint is unjustified—Einstein's arguments in 1917 should be imagined to take place in a regime (certainly the one relevant to gas discharge tubes whence the spectroscopic information of the time was gleaned) analogous to the Wien limit of black-body radiation, in which $e^{\beta(E_\nu-\mu)} \gg 1$ so that the *atomic* occupation numbers took on precisely the Maxwell–Boltzmann form $n_\nu \propto e^{-\beta E_\nu}$ used by Einstein in his discussion of emission and absorption of radiation.[7] In fact, Schrödinger suggested that the deviations from Maxwell–Boltzmann behavior were the result of an incorrect counting of states in the situation where there were "multiply occupied cells".[8] In his response to Schrödinger on February 28, 1925 (Einstein 1987–2021, Vol. 14, Doc. 446), Einstein emphasized the correctness of his arguments, and pointed Schrödinger to two further papers on the Bose gas statistics problem, one already published, the other forthcoming (Einstein 1925a, 1925b).

It is clear, in fact, that the approach of Bose, and its utilization in developing a quantum theory of ideal gases by Einstein, did not receive immediate approval and acceptance by the most prominent physicists working on this problem. Throughout 1925, Planck and Schrödinger continued to work on gas theory, developing distinct (and incorrect) approaches to both the identification of the quantum states of a gas (which was treated as a single quantized dynamical system, with the canonical phase space sliced into cells in varying ways) and the counting procedure used to compute an entropy.[9]

By the end of October 1925, Schrödinger seems to have finally digested the importance—and correctness—of Einstein's gas theory, and also, of the importance of de Broglie's work, which had been highlighted so prominently in Einstein's second paper. Another stimulus for Schrödinger's study of de Broglie appears to have been Peter Debye, then at the ETH, which shared a physics colloquium with the University of Zurich, where Schrödinger was. Also a student at the ETH at the time, Felix Bloch

[7] In contrast, in Einstein's A and B coefficients papers [cf. Vol. 1, p. 135] papers, the possibility of multiple occupancy of *light quanta modes* is most assuredly *not* discounted—as the success of his "miraculous" derivation of the Planck radiation law would indicate.

[8] Again, Einstein's counting procedure, as presented in Section 13.5, is perfectly correct: in the A and B coefficients papers, the gas atoms (even if bosons) are overwhelmingly to be found in singly occupied atomic states.

[9] The back and forth on these issues between Planck, Einstein, and Schrödinger can be followed in more detail in Mehra and Rechenberg (1987, Ch. 3).

recalled years later that Debye coaxed Schrödinger into giving a colloquium explaining ideas in de Broglie's thesis, which had by this point appeared in print in the *Annales de Physique*.[10]

A letter to Einstein of early November 1925 makes it clear that Schrödinger had read de Broglie's thesis (either in its original form or in the version published in *Annales de Physique*) and absorbed the lessons of Einstein's discussion thereof in his second gas paper:

> A few days ago I read with the greatest interest the ingenious thesis of Louis de Broglie, which I finally got hold of; with it section 8 of your second paper on degeneracy has also become clear to me for the first time. The de Broglie interpretation of the quantum rules seems to be related to my note [Schrödinger (1922)] where a remarkable property of the Weyl "gauge factor" along every quasi-period is shown ... Naturally de Broglie's consideration in the framework of his comprehensive theory is altogether of far greater value than my single statement, which I did not know what to make of at first (Schrödinger to Einstein, November 3, 1925, Einstein 1987–2021, Vol. 15, Doc. 103).[11]

Schrödinger's conversion to a wave theory of gas molecules is explicitly stated in his paper "On Einstein's gas theory" (Schrödinger 1926b), which he submitted to *Physikalische Zeitschrift* in mid-December, and which he presumably was working on at the time of or shortly after the 3 November letter to Einstein. In the introduction to this paper, Schrödinger points out that the essential point of difference in the new statistics employed by Bose and Einstein from that of Boltzmann can be summarized as simply the application to gas molecules of the same statistics as in the treatment of "light-atoms" (i.e., light quanta, or photons) in Planck's radiation law. The latter statistics can be seen as a "natural" one by the simple expedient of applying the statistical arguments to the "aether-resonators" (the discretely enumerated standing wave modes of Rayleigh and Jeans), rather than to their associated particles, the light quanta/atoms.[12] The treatment of the five energy elements contributing to a mode excited to the fifth energy level as completely indistinguishable from one another does far less violence to a classically oriented intuition than the assertion that five light quanta (especially if thought of as localized entities) are completely indistinguishable entities. As Schrödinger put it himself (with some interpretive parenthetical remarks added):

> The transition from the natural to the Bose statistics can always be replaced by exchanging the roles of "manifold of energy states" [specification of numbers of particles in definite energy states] with "manifold of bearers of those states" [specification of degree of excitation of quantized standing wave modes associated with a given particle energy state]. One must therefore simply form a picture of a gas in accordance with the picture of cavity radiation which does not yet correspond to the extreme light quantum conception [of localized particles]; then the natural statistics—the convenient Planck method of state summation—will lead to the Einstein gas theory. This amounts to nothing else than

[10] See, for example, Moore (1989, p. 192) and Mehra and Rechenberg (1987, p. 420).
[11] Quoted in Moore (1989, p. 192) and Mehra and Rechenberg (1987, p. 415).
[12] This approach had already been adopted by Debye (1910), in his derivation of the Planck Law.

> taking seriously the de Broglie–Einstein wave theory of moving particles, according to which these [particles] are nothing more than a sort of "foam-crest" of a wave radiation underlying the universe (Schrödinger 1926b, p. 95).

Schrödinger's paper on Bose–Einstein ideal gases does not really contain anything new—the results he obtains are either identical to or special cases of results already obtained by de Broglie and Einstein. The methodology used to evaluate the constrained state sums (in modern language, the partition sum) is adapted from Darwin and Fowler (1922), who used a complex integral formalism to extract from an unconstrained sum over gas configurations only the terms where the total number of atoms occupying the distinct quantized modes is fixed at some macroscopically specified number N. The approximate saddle point evaluation of this integral avoids in an elegant way the explicit invocation of Stirling's approximation, but is of course valid under the same circumstances which allow use of the latter.

In the third section of the paper ("Determination of the frequency spectrum") the density of modes factors are derived by counting standing wave modes, and Schrödinger arrives at formulas equivalent to de Broglie's Eq. (13.71) and Einstein's Eq. (13.80). In section 4, expressions for the average occupation number of a single "cell' (unique quantum state, rather than all states in an energy interval, as in Einstein), and the mean square fluctuations of this number, are obtained from the partition function obtained from the Darwin–Fowler method and shown to agree with Einstein's formulas (by setting $z_\nu = 1$).

Section 5 ("On the possibility of representing molecules or light quanta through interference of plane waves") contains some intriguing anticipations of the quantum mechanics of wave-packets (which Schrödinger calls "signals") which every first-year student of quantum mechanics must learn. Schrödinger points out that such wave pulses, localized in all three spatial dimensions, can indeed be constructed in de Broglie's theory, by superposing waves of differing frequencies and differing directions of the wave vector.[13] Schrödinger is perfectly aware that such a signal/pulse/wave packet does not maintain its shape—rather, precisely because of the presence of various frequencies and wavelengths in the signal, it must spread and become increasingly less localized in the course of time, a phenomenon perfectly understood already at the classical level for optical signals. Such a property would seem to be unacceptable for a material particle such as an atom or molecule. The section ends with the hope that a quantum-mechanical escape route from this "classical wave law" can somehow be found to elude the undesirable spreading. As we now know, the spreading of a particle's wave packet is an ineluctable consequence of the dynamical laws of quantum theory, based on linear superposition of states—the algebraic structure of linear wave superposition is exactly the same in classical and quantum theory.

[13] This is just the three-dimensional generalization of the one-dimensional wave packet constructed in Eq. (13.10).

14.3 The first (relativistic) wave equation

There is a story, told by Bloch in later reminiscences (cf. note 10), that Debye had suggested to Schrödinger, on the occasion of his colloquium lecture on the de Broglie theory (probably toward the end of November 1925), that the appropriate way to study the physics of wave phenomena was in terms of a precisely formulated wave equation. The results of Schrödinger's first attempts to implement such a program have been preserved in a three-page set of notes headed "H-atom. Characteristic vibrations" (*H-Atom. Eigenschwingungen*—see Figure 14.1).[14] We describe here the contents of this memorandum, with minor changes of notation to accommodate the notations used previously, and for convenience of a modern reader.

In three dimensions, a plane wave with wavelength λ and frequency ν (and the wave vector defining the normal to the wave fronts given by $\vec{k}$ with $|\vec{k}| = 2\pi/\lambda$) can be described by a sinusoidal function analogous to the one-dimensional one written down by Einstein (cf. Eq. (13.110)),

$$\psi(\vec{r}, t) = A \sin\left(\vec{k} \cdot \vec{r} - 2\pi\nu t\right), \tag{14.5}$$

where we have now adopted Schrödinger's famous usage of the Greek letter ψ (see Figure 14.1), to denote the wave function (A is for the moment an unimportant constant giving the amplitude of the wave). The phase velocity of such a wave is just the product of the wavelength and the frequency, $V = \lambda\nu$. The wave function thus satisfies the equation

$$\Delta\psi = -|\vec{k}|^2\psi = -\frac{4\pi^2}{\lambda^2}\psi = -\frac{4\pi^2\nu^2}{V^2}\psi \quad \left(= \frac{1}{V^2}\frac{\partial^2\psi}{\partial t^2}\right), \tag{14.6}$$

where, as usual, Δ is the Laplacian defined as $\partial^2/\partial x^2 + \partial^2/\partial y^2 + \partial^2/\partial z^2$. The final equality gives the wave equation in the form perhaps most familiar to the reader (e.g., with $V = c$ for electromagnetic waves). Schrödinger assumes throughout an Ansatz in which the wave function is a product of a function of space and a periodic function of time (with frequency ν), so that the right-hand side can be replaced by the fourth term in the preceding equation, and the wave function is now considered only a function of space, $\psi(\vec{r})$. This amounts simply to considering "standing waves" in which the space and time behavior of the wave are separated and multiplied, as opposed to traveling waves as in Eq. (14.5).[15]

For a free particle, the associated wave has phase velocity given, as we saw earlier, by c^2/v. But for a de Broglie wave associated not to a free particle, but to one moving under

[14] These notes are also discussed by Mehra and Rechenberg (1987, pp. 423–433), who refer to it as the "Memorandum."

[15] Schrödinger uses u, rather than V, for phase velocity. We also employ $\hbar = h/2\pi$, rather than Planck's original constant h, throughout to reduce the appearance of annoying factors involving π. We continue to write pedantically the rest mass as m_0—Schrödinger uses m_e.

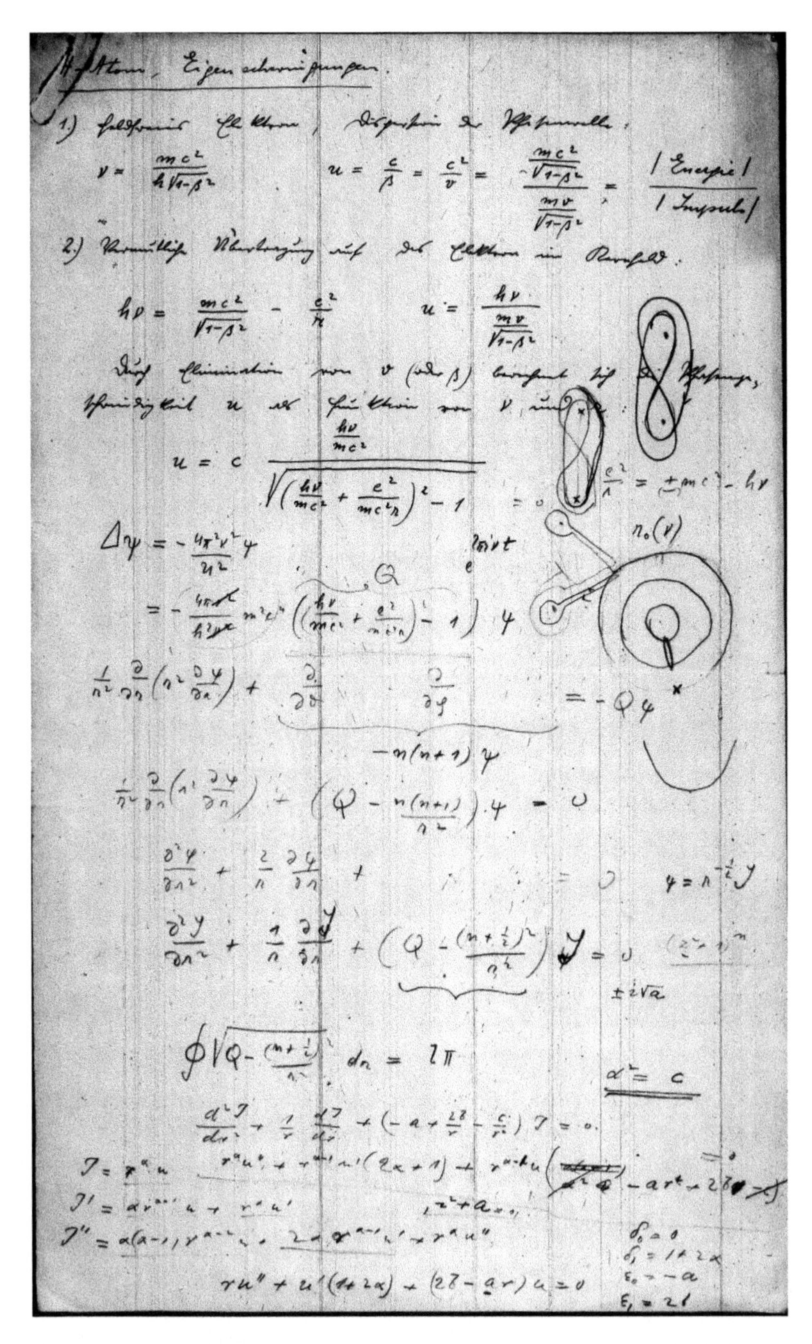

Figure 14.1 *First page of Schrödinger's three-page memorandum (December 1925), "H-atom. Characteristic vibrations" (H-Atom. Eigenschwingungen).*

the influence of an electrostatic potential energy $e\varphi(\vec{r}) = -e^2/r$ (electron in a hydrogen atom), we have instead, as de Broglie found in his thesis (cf. Eqs. (13.57)–(13.58)),

$$E = h\nu = \frac{m_0 c^2}{\sqrt{1 - v^2/c^2}} - \frac{e^2}{r}, \tag{14.7}$$

$$V = \nu\lambda = \frac{c^2}{v}\frac{h\nu}{h\nu + e^2/r}. \tag{14.8}$$

Solving Eq. (14.7) for the prefactor c^2/v,[16]

$$\frac{c^2}{v} = c\frac{h\nu + e^2/r}{\sqrt{\left(h\nu + e^2/r\right)^2 - m_0^2 c^4}}, \tag{14.9}$$

and substituting this into Eq. (14.8), we obtain for the phase velocity

$$V = c\frac{\dfrac{h\nu}{m_0 c^2}}{\sqrt{\left(\dfrac{h\nu}{m_0 c^2} + \dfrac{e^2}{m_0 c^2 r}\right)^2 - 1}}. \tag{14.10}$$

This rather unpromising expression is then inserted into the wave equation Eq. (14.6) to obtain

$$\Delta\psi = -\frac{4\pi^2\nu^2}{V^2}\psi = -Q(r)\psi, \tag{14.11}$$

[16] Rearranging Eq. (14.7),

$$\frac{m_0 c^2}{\sqrt{1 - v^2/c^2}} = h\nu + \frac{e^2}{r},$$

squaring and taking the inverse of both sides, we find

$$\frac{1 - v^2/c^2}{m_0^2 c^4} = \frac{1}{\left(h\nu + e^2/r\right)^2}.$$

It follows that

$$\frac{v^2}{c^2} = 1 - \frac{m_0^2 c^4}{\left(h\nu + e^2/r\right)^2} = \frac{\left(h\nu + e^2/r\right)^2 - m_0^2 c^4}{\left(h\nu + e^2/r\right)^2}.$$

Taking the inverse and the square root on both sides, we arrive at the expression for c/v used in Eq. (14.9).

with

$$Q(r) \equiv \frac{m_0^2 c^2}{\hbar^2} \left(\left(\frac{h\nu}{m_0 c^2} + \frac{e^2}{m_0 c^2 r} \right)^2 - 1 \right). \tag{14.12}$$

Note that, despite its rather complicated appearance, $Q(r)$ is simply a quadratic function of $1/r$.[17]

Since we are dealing with a problem with rotational symmetry—Q only depends on r and not on the polar angles ϑ and φ—we write the Laplacian operator Δ in spherical coordinates,

$$\Delta\psi = \frac{1}{r^2}\frac{\partial}{\partial r}\left(r^2 \frac{\partial \psi}{\partial r}\right) + \frac{1}{r^2 \sin\vartheta}\frac{\partial}{\partial \vartheta}\left(\sin\vartheta \frac{\partial \psi}{\partial \vartheta}\right) + \frac{1}{r^2 \sin^2\vartheta}\frac{\partial^2 \psi}{\partial \varphi^2}, \tag{14.13}$$

and separate the radial dependence of the wave function from its angular dependence by writing

$$\psi(r, \vartheta, \varphi) = R(r)\, Y(\vartheta, \varphi). \tag{14.14}$$

With the help of Eqs. (14.13) and (14.14), we can rewrite (r^2/ψ) times $\Delta\psi + Q(r)\psi$, the expression set equal to 0 in Eq. (14.11), as

$$\frac{r^2}{\psi}\Big(\Delta\psi + Q(r)\psi\Big) = \frac{1}{R(r)}\frac{d}{dr}\left(r^2 \frac{dR(r)}{dr}\right) + r^2\, Q(r)$$

$$+ \frac{1}{Y(\vartheta, \varphi)}\left(\frac{1}{\sin\vartheta}\frac{\partial}{\partial \vartheta}\left(\sin\vartheta \frac{\partial Y(\vartheta, \varphi)}{\partial \vartheta}\right) + \frac{1}{\sin^2\vartheta}\frac{\partial^2 Y(\vartheta, \varphi)}{\partial \varphi^2}\right). \tag{14.15}$$

The two terms on the right-hand side on the first line depend only on r while the two terms on the second line depend only on ϑ and φ. Since the sum of all four must be zero, the sums of the first two and the sum of the second two separately must have equal and opposite values $\pm\lambda$, independent of r, ϑ, and φ:

$$\frac{1}{R(r)}\frac{d}{dr}\left(r^2 \frac{dR(r)}{dr}\right) + r^2\, Q(r) = -\lambda, \tag{14.16}$$

$$\frac{1}{Y(\vartheta, \varphi)}\left(\frac{1}{\sin\vartheta}\frac{\partial}{\partial \vartheta}\left(\sin\vartheta \frac{\partial Y(\vartheta, \varphi)}{\partial \vartheta}\right) + \frac{1}{\sin^2\vartheta}\frac{\partial^2 Y(\vartheta, \varphi)}{\partial \varphi^2}\right) = \lambda. \tag{14.17}$$

[17] If we reinsert the time dependence, as a monochromatic $e^{-2\pi i \nu t}$ factor, the wave function $\psi(\vec{r}, t)$ is seen to satisfy the equation $((i\hbar\partial_t + e^2/r)^2 - \hbar^2 c^2 \vec{\nabla}^2 + m_0^2 c^4)\psi = 0$. This would become known as the Klein–Gordon equation (with the Coulomb potential switched off), though Schrödinger clearly deserves part of the credit for its discovery.

Eq. (14.17), the angular equation for $Y(\vartheta, \varphi)$, is known to have solutions which are (a) single-valued, and (b) non-singular for all values of ϑ (i.e., $0 \leq \vartheta \leq \pi$) and φ (i.e., $0 \leq \varphi < 2\pi$), only if

$$\lambda = -l(l+1), \tag{14.18}$$

where l is a non-negative integer, $l = 0, 1, 2, \ldots$ Moreover, for a given value of l there are $2l + 1$ distinct solutions labeled by an index running between $-l$ and l in steps of unity,

$$m = -l, -l+1, \ldots -1, 0, 1, \ldots, l-1, l. \tag{14.19}$$

The resulting solutions are the famous spherical harmonics, $Y_{lm}(\vartheta, \varphi)$, well known to physicists of Schrödinger's and subsequent generations:

$$Y_l^m(\vartheta, \varphi) = C_{lm} P_l^m(\vartheta) e^{im\varphi}, \quad -l \leq m \leq +l. \tag{14.20}$$

Here, $P_l^m(\vartheta)$ is the associated Legendre function, which turns out to be a polynomial in powers of $\sin\vartheta$ and $\cos\vartheta$. Note that the single-valuedness criterion imposes the constraint that m be an integer, so that $e^{im(\varphi+2\pi)} = e^{im\varphi}$, while the criterion that l be integer arises from the demand that $P_l^m(\vartheta)$ not exhibit singularities at $\vartheta = 0$ and $\vartheta = \pi$, that is, at the purely arbitrarily chosen "north" and "south" poles of the spherical coordinate system.

The angular dependence of the possible well-behaved solutions having been determined, we turn to Eq. (14.16), the radial equation for $R(r)$. Inserting Eq. (14.18) for λ, we can write this linear second-order *ordinary* differential equation as

$$\frac{1}{R(r)}\left(2r\frac{dR(r)}{dr} + r^2\frac{d^2R(r)}{dr^2}\right) + r^2\, Q(r) = l(l+1). \tag{14.21}$$

Multiplying by R/r^2 and rearranging terms, we arrive at:

$$\frac{d^2R(r)}{dr^2} + \frac{2}{r}\frac{dR(r)}{dr} + \left(Q(r) - \frac{l(l+1)}{r^2}\right) R(r) = 0. \tag{14.22}$$

Given Eq. (14.12) for $Q(r)$, this equation takes the form

$$\frac{d^2R(r)}{dr^2} + \frac{2}{r}\frac{dR(r)}{dr} - \left(A - \frac{2B}{r} + \frac{C}{r^2}\right) R(r) = 0, \tag{14.23}$$

where

$$A = \frac{1}{\hbar^2 c^2}\left(m_0^2 c^4 - (h\nu)^2\right), \tag{14.24}$$

$$B = \frac{e^2}{\hbar^2 c^2}(h\nu), \tag{14.25}$$

$$C = l(l+1) - \frac{e^4}{\hbar^2 c^2}. \tag{14.26}$$

The differential equation Eq. (14.23) has a regular singular point at $r = 0$ (i.e., the coefficient function of dR/dr has at most a single pole there and the coefficient of R at most a double pole). The singular behavior (if any) at $r = 0$ can therefore be extracted, leaving a residual function given by a power series convergent for all r (as the only other singularity of the equation is an irregular one at infinity). In his three pages of notes dealing with the relativistic equation, Schrödinger carries out this procedure familiar to modern students of quantum mechanics. Examining the differential equation for $R(r)$ near $r = 0$, we only need to retain the terms with the most singular dependence on r at $r = 0$. Introducing the function $R_0(r)$ to represent $R(r)$ near $r = 0$, we can thus write Eq. (14.23) near $r = 0$ as

$$\frac{d^2 R_0(r)}{dr^2} + \frac{2}{r}\frac{dR_0(r)}{dr} - \frac{C}{r^2}R_0(r) = 0, \tag{14.27}$$

which has the solution $R_0(r) = r^p$, with $p = -1/2 \pm \sqrt{C + 1/4}$.[18] We must take the positive sign in the square root as otherwise the radial function diverges as r goes to zero.

We now write $R(r)$ as

$$R(r) = r^p U(r), \tag{14.28}$$

where $U(r)$ should now be an entire function of r, as the singularity at $r = 0$ has been removed and the only remaining singularity is at infinity.

[18] Inserting $R_0(r) = r^p$ into Eq. (14.27), we find

$$p(p-1)r^{p-2} + \frac{2}{r}p\,r^{p-1} - \frac{C}{r^2}r^p = 0,$$

which means that p has to satisfy the quadratic equation $p^2 + p - C = 0$. The two roots of this equation are

$$p_{1,2} = -\frac{1}{2}\left(1 \pm \sqrt{1+4C}\right).$$

Substituting Eq. (14.28) into Eq. (14.23), we find[19]

$$r\frac{d^2U(r)}{dr^2} + (2p+2)\frac{dU(r)}{dr} + (2B - Ar)U(r) = 0. \tag{14.29}$$

Introducing

$$\alpha \equiv A, \quad \beta \equiv 2B, \quad \gamma \equiv 2p+2 = 1 + \sqrt{1+4C}, \tag{14.30}$$

we can rewrite this equation as

$$0 = r\frac{d^2U(r)}{dr^2} + \gamma\frac{dU(r)}{dr} + (\beta - \alpha\, r)U(r). \tag{14.31}$$

This is a special case of Laplace's differential equation, which is characterized by having linear functions of the independent variable r in the coefficients of all three terms of the differential equation. A general integral representation of the solutions of this equation was given by Laplace himself, in terms of what is now called a Laplace transform:[20]

$$U(r) = \int_L dz\, e^{zr} w(z). \tag{14.32}$$

For a function $U(r)$ satisfying a second-order differential equation of Laplace type, the Laplace transform "weight" function $w(z)$ takes a particularly simple form: with an

[19] Inserting

$$R = r^p U, \quad \frac{dR}{dr} = r^p\frac{dU}{dr} + p\,r^{p-1}U, \quad \frac{d^2R}{dr^2} = r^p\frac{d^2U}{dr^2} + 2p\,r^{p-1}\frac{dU}{dr} + (p-1)p\,r^{p-2}U$$

into Eq. (14.23), we find:

$$\begin{aligned} r^p\frac{d^2U}{dr^2} &+ 2p\,r^{p-1}\frac{dU}{dr} + (p-1)p\,r^{p-2}U \\ &+ \frac{2}{r}\left(r^p\frac{dU}{dr} + p\,r^{p-1}U\right) - \left(A - \frac{2B}{r} + \frac{C}{r^2}\right)r^pU = 0. \end{aligned}$$

Dividing by r^{p-1} and regrouping terms, we can rewrite this as

$$r\frac{d^2U}{dr^2} + (2p+2)\frac{dU}{dr} + \left(\frac{(p-1)p}{r} + \frac{2p}{r} - Ar + 2B - \frac{C}{r}\right)U.$$

The terms with $1/r$ in the expression multiplying U cancel since $(p-1)p + 2p - C = p^2 + p - C = 0$ (see note 18).

[20] The reader whose facility in complex analysis may be a bit rusty may safely skip the next few pages, and go directly to Schrödinger's results Eqs. (14.49) and (14.50), which express the conditions for the radial function $U(r)$ to be well behaved when $r \to 0$ and $r \to \infty$.

appropriate choice of the contour of integration L in the complex plane, it is the product of two pure powers centered at two singular points,

$$w(z) = (z - c_1)^{\alpha_1 - 1} (z - c_2)^{\alpha_2 - 1}. \tag{14.33}$$

In general, the real numbers α_1 and α_2 are not integers, and the weight function is an analytic function with two branch point singularities at the (typically real) points $z = c_1$ and $z = c_2$. In this particular case, it turns out that

$$c_1 = -c_2 = \sqrt{\alpha}, \tag{14.34}$$

while for the powers at the branch singularities one has

$$\alpha_1 = \frac{1}{2}\left(\gamma + \frac{\beta}{\sqrt{\alpha}}\right), \quad \alpha_2 = \frac{1}{2}\left(\gamma - \frac{\beta}{\sqrt{\alpha}}\right). \tag{14.35}$$

Solving for γ and β, we find that

$$\gamma = \alpha_1 + \alpha_2, \qquad \beta = \sqrt{\alpha}\,(\alpha_1 - \alpha_2). \tag{14.36}$$

If we insert the integral representation Eq. (14.32) into the differential equation Eq. (14.31), using Eqs. (14.33)–(14.34), we find:

$$\begin{aligned} 0 &= \int_L dz \left(rz^2 + \gamma z + \beta - \alpha r\right) e^{zr} w(z) \\ &= \int_L dz \left(rz^2 + \gamma z + \beta - \alpha r\right) e^{zr} \left(z - \sqrt{\alpha}\right)^{\alpha_1 - 1} \left(z + \sqrt{\alpha}\right)^{\alpha_2 - 1}. \end{aligned} \tag{14.37}$$

Making the substitutions

$$\left(rz^2 - \alpha r\right) e^{zr} = \left(z^2 - \alpha\right) \frac{de^{zr}}{dz} = \left(z - \sqrt{\alpha}\right)\left(z + \sqrt{\alpha}\right) \frac{de^{zr}}{dz} \tag{14.38}$$

and (cf. Eq. (14.36))

$$\begin{aligned} \gamma z + \beta &= (\alpha_1 + \alpha_2) z + (\alpha_1 - \alpha_2)\sqrt{\alpha} \\ &= \alpha_1\left(z + \sqrt{\alpha}\right) + \alpha_2\left(z - \sqrt{\alpha}\right), \end{aligned} \tag{14.39}$$

and reordering the various factors in the integrand, we can rewrite Eq. (14.37) as:

$$\begin{aligned}
0 &= \int_L dz\,(z-\sqrt{\alpha})^{\alpha_1-1}(z+\sqrt{\alpha})^{\alpha_2-1} \\
&\quad \times \left((z-\sqrt{\alpha})(z+\sqrt{\alpha})\frac{d}{dz} + \alpha_1(z+\sqrt{\alpha}) + \alpha_2(z-\sqrt{\alpha})\right) e^{zr} \\
&= \int_L dz \left((z-\sqrt{\alpha})^{\alpha_1}(z+\sqrt{\alpha})^{\alpha_2}\frac{de^{zr}}{dz} \right. \\
&\quad \left. + (z+\sqrt{\alpha})^{\alpha_2}\frac{d}{dz}(z-\sqrt{\alpha})^{\alpha_1}e^{zr} + (z-\sqrt{\alpha})^{\alpha_1}\frac{d}{dz}(z+\sqrt{\alpha})^{\alpha_2}e^{zr} \right) \\
&= \int_L \frac{d\hat{w}}{dz}dz, \qquad (14.40)
\end{aligned}$$

where

$$\hat{w}(z) \equiv e^{zr}(z-\sqrt{\alpha})^{\alpha_1}(z+\sqrt{\alpha})^{\alpha_2}. \qquad (14.41)$$

The contour integral of the derivative of the function $\hat{w}$ along some as-yet undetermined path L is automatically zero if the path L is either: (a) a closed loop on the Riemann surface of the branched function $\hat{w}$ (on which the function is everywhere single-valued); or (b) connects two points (on the Riemann surface) where $\hat{w}$ has identical values, for example, zero. In practice, the general theory of linear second-order differential equations tells us that there are at most two linearly independent solutions of Eq. (14.31), so once we find two paths that give functions that are not simply multiples of each other, our job is complete.

Schrödinger was completely familiar with the procedures described here, as he had at his disposal, in the critical months of December 1925 and January 1926, the standard text on ordinary differential equations at the time, Ludwig Schlesinger's *Introduction to the Theory of Differential Equations with One Independent Variable*, in which the Laplace equation was treated in detail along the lines just described (Schlesinger 1900, sec. 49).[21]

He was also completely familiar with Sommerfeld's treatment—in chapter 6 of *Atombau und Spektrallinien* (Sommerfeld 1924) for instance—of the relativistic fine structure of the hydrogen atom, which constituted one of the most spectacular successes (both theoretical and experimental) of the old quantum theory (see Section 6.1).

The Sommerfeld result clearly had to be reproduced by any new theory of the hydrogen atom. It is clear that Schrödinger was paying close attention to the Sommerfeld

[21] The methods used to study the Laplace equation in Schlesinger's text, making full use of complex function theory, in particular a rather sophisticated analysis of complex integral contours on a multi-sheeted Riemann surface, are not at all familiar to modern students, and cannot be found in the texts used even for graduate level quantum theory, which almost exclusively employ the less elegant, but technically straightforward, power series expansion method. For this reason we have relegated most of the technical discussion of Schrödinger's early work on the radial equation to a web resource, *Solving the Radial Schrödinger Equation*, where the interested reader can find Schrödinger's arguments dissected in gory detail.

calculation while developing his relativistic wave equation, because the notation he uses for the coefficients A, B, and C appearing in Eq. (14.23) closely parallels Sommerfeld's notation for the terms appearing in the radial momentum p_r,[22] the phase integral of which, once subjected to Bohr–Sommerfeld quantization, leads directly to the relativistic fine structure:

$$\oint p_r\, dr = n'h \text{ with } p_r = \sqrt{A_{\text{Som}} + 2\frac{B_{\text{Som}}}{r} + \frac{C_{\text{Som}}}{r^2}}. \tag{14.42}$$

In fact, the radial phase integral is written explicitly on the first page of Schrödinger's three-page note on the relativistic problem.[23] In Eq. (14.42), n' is Sommerfeld's notation for the radial quantum number, $n' = 0, 1, 2, \ldots$ and we have added the subscript "Som" (for "Sommerfeld") to the coefficients simply to distinguish them from Schrödinger's. The coefficient A and B are the same, apart from an overall factor of $\hbar^2$ and a sign factor introduced by Schrödinger to make all the coefficients positive in the case of bound states (with energy $h\nu < m_0c^2$):

$$A_{\text{Som}} = -\hbar^2 A, \quad B_{\text{Som}} = \hbar^2 B. \tag{14.43}$$

Sommerfeld's coefficient C_{Som} is given by

$$C_{\text{Som}} = \frac{e^4}{c^2} - \hbar^2 n^2, \tag{14.44}$$

where n is the azimuthal quantum number (angular momentum) in Sommerfeld's notation. The corresponding coefficient used by Schrödinger is[24]

$$C = -\frac{1}{\hbar^2}\left(\frac{e^4}{c^2} - \hbar^2 l(l+1)\right). \tag{14.45}$$

The radial integral quantization in Eq. (14.42), leading to Sommerfeld's famous fine structure formula, gave, on performing the integral,

$$\oint p_r\, dr = 2\pi\left(\frac{B_{\text{Som}}}{\sqrt{-A_{\text{Som}}}} - \sqrt{-C_{\text{Som}}}\right) \approx \left(\frac{B}{\sqrt{A}} - \sqrt{C}\right) h, \tag{14.46}$$

[22] This notation goes back to Sommerfeld's treatment of screening effects in alkali atoms in 1916 (cf. Eqs. (7.4)–(7.7)), where the radial phase integral is seen to contain the factor $B/\sqrt{A} - \sqrt{C}$. In a letter to Sommerfeld of January 29, 1926, the day Schrödinger's (1926c) first paper on wave mechanics was received by *Annalen der Physik*, its author told Sommerfeld that the "characteristic $-B/\sqrt{A} + \sqrt{C}$ suddenly shone out from the exponents α_1 and α_2 like a Holy Grail" (Sommerfeld 2004, Doc. 92, p. 238, or von Meyenn 2011, p. 172; quoted by Mehra and Rechenberg 1987, p. 462).

[23] See the equation in Figure 14.1 toward the bottom of the page involving a loop integral $\oint \sqrt{Q - \ldots}\, dr$. The 2π on the right is missing a factor of the radial quantum number n'.

[24] In fact, Schrödinger uses n for l in his notes, so the factor $n(n+1)$ appears in the radial equation. We have switched to the modern notation to avoid confusion with the principal quantum number, for which Schrödinger uses l—just the reverse of the assignments common now!

where the final quasi-equality sign $\approx$ signals that Sommerfeld's n^2 (the square of the azimuthal quantum number, i.e., the quantum number for orbital angular momentum) has been replaced by Schrödinger's $l(l+1)$. Schrödinger was initially excited to find that a very similar quantity was appearing as the power α_2 of $z + \sqrt{\alpha}$ in Eq. (14.41). Using the definitions of α, β, and γ in Eq. (14.30), we have

$$\begin{aligned}\alpha_2 &= \frac{1}{2}\left(\gamma - \frac{\beta}{\sqrt{\alpha}}\right)\\ &= \frac{1}{2}\left(1 + \sqrt{1+4C} - 2\frac{B}{\sqrt{A}}\right).\end{aligned} \tag{14.47}$$

In fact, Schrödinger was able to show[25] that the radial function $U(r)$ was well behaved for large r (as well as when $r \to 0$, as we have already separated off any singular behavior there) only if the parameter α_2 was a negative integer or zero (quantization!), and the path L chosen to encircle only this singularity, which was now a pole $(1/(z+\sqrt{\alpha})^{|\alpha_2-1|})$, rather than a branch point in $w(z)$ (cf. Eq. (14.33)). Setting $\alpha_2 = -n'$ with $n' = 0, 1, 2, \ldots$, we thus rewrite Eq. (14.47) as

$$\frac{1}{2} + \sqrt{C + \frac{1}{4}} - \frac{B}{\sqrt{A}} = -n'. \tag{14.48}$$

Using the definitions of A, B, and C in Eqs. (14.24)–(14.26) and solving for $E = h\nu$, we find[26]

$$E = h\nu = \frac{m_0 c^2}{\sqrt{1 + \alpha^2/\mathfrak{N}^2}}, \tag{14.49}$$

[25] The presence of the factor e^{zr} in Eq. (14.32) means that the integration path L has to remain in the left half plane $\mathrm{Re}(z) < 0$ to avoid the appearance of terms that would blow up exponentially as $r \to +\infty$, which appear in e^{zr} as soon as z acquires a positive real part. The need for the singularity on the left side at $z = -\sqrt{\alpha}$ to be a pure pole, rather than a branch point, arises from the need to ensure regularity of $U(r)$ as $r \to 0$. For complete details on Schrödinger's analysis of the radial equation using the Laplace–Schlesinger method, see the web resource, *Solving the Radial Schrödinger Equation*.

[26] We can rewrite

$$-n' = \tfrac{1}{2} + \sqrt{C + \tfrac{1}{4}} - \frac{B}{\sqrt{A}} = \tfrac{1}{2} + \sqrt{l(l+1) - \alpha^2 + \tfrac{1}{4}} - \frac{\alpha E}{\sqrt{m_0^2 c^4 - E^2}},$$

(with $\alpha = e^2/\hbar c$ the fine structure constant) as

$$\frac{\alpha E}{\sqrt{m_0^2 c^4 - E^2}} = n' + \tfrac{1}{2} + \sqrt{\left(l + \tfrac{1}{2}\right)^2 - \alpha^2}.$$

The right-hand side is just the quantity $\mathfrak{N}$ introduced in Eq. (14.50). Squaring both sides and rearranging terms we arrive at $\alpha^2 E^2 = \mathfrak{N}^2\left(m_0^2 c^4 - E^2\right)$, from which Eq. (14.49) immediately follows.

where $\alpha = e^2/\hbar c$ is the fine structure constant and

$$\mathfrak{N} \equiv n' + \tfrac{1}{2} + \sqrt{\left(l+\tfrac{1}{2}\right)^2 - \alpha^2}. \tag{14.50}$$

As we will see below, $\mathfrak{N}$ reduces to the usual principal quantum number N in the non-relativistic limit. Unfortunately for Schrödinger, this formula, though very similar in appearance, differs from Sommerfeld's result for the relativistic energy, which has the same form as Eq. (14.49), but with $\mathfrak{N}$ replaced by

$$\mathfrak{N}_{\text{Som}} = n' + \sqrt{n^2 - \alpha^2}, \tag{14.51}$$

where $n = 1, 2, 3, \ldots$ is the angular momentum quantum number of the old quantum theory.

Schrödinger's ambitious attempt to reproduce the empirically well-established formula of Sommerfeld—the basis not only for the understanding of the fine structure of hydrogen and ionized helium, but also for his beautiful organization of the L terms in X-ray spectra—had obviously failed. As a comparison between Eqs. (14.50) and (14.51) shows, Schrödinger found *half-integers* where integers were required.

The leading fine structure can be found by expanding Eqs. (14.49) and (14.50) in powers of the fine structure constant α. In this way we find the relation between $\mathfrak{N}$ and the principal quantum number N already alluded to

$$\begin{aligned} \mathfrak{N} &= n' + \tfrac{1}{2} + \left(l+\tfrac{1}{2}\right)\left(1 - \frac{\alpha^2}{2\left(l+\frac{1}{2}\right)^2} + \ldots\right) \\ &= n' + l + 1 - \frac{\alpha^2}{2(l+\frac{1}{2})} + \ldots \\ &= N - \frac{\alpha^2}{2(l+\frac{1}{2})} + \ldots . \end{aligned} \tag{14.52}$$

Using this expression in the expansion of Eq. (14.49) in powers of α, we find

$$\begin{aligned} E &= m_0c^2\left(1 - \frac{\alpha^2}{2\mathfrak{N}^2} + \frac{3}{8}\frac{\alpha^4}{\mathfrak{N}^4} + \ldots\right) \\ &= m_0c^2 - \frac{\alpha^2 m_0c^2}{2N^2} + \frac{\alpha^4 m_0 c^2}{2N^3}\left(\frac{3}{4N} - \frac{1}{l+\frac{1}{2}}\right) + \ldots . \end{aligned} \tag{14.53}$$

This result is, in fact, perfectly correct for charged *spinless* particles (Baym 1969), but the absence of electron spin, which enters the energies precisely at the first non-leading relativistic correction, inevitably doomed Schrödinger's efforts to reproduce the Sommerfeld fine structure formula.[27]

[27] The Uhlenbeck–Goudsmit proposal of electron spin (see Section 9.2) had been mentioned, but not applied, in the Three-Man-Paper of Born, Heisenberg, and Jordan, submitted in late November 1925. There

It is easy to understand why Schrödinger would not publish an analysis of his relativistic wave equation yielding energies in disagreement with the Sommerfeld formula. This would inevitably have looked like a step backward from the old quantum theory, especially since it was based on the still rather vague conceptual notions introduced by de Broglie. Presumably, however, Schrödinger realized at once that the energies he had obtained at least correctly reproduced the Bohr–Balmer energy levels in the non-relativistic limit. Expanding Eq. (14.50) for $\mathfrak{N}$ in inverse powers of the speed of light—or, equivalently, in powers of $\alpha = e^2/\hbar c$—we find

$$\mathfrak{N} = n' + \tfrac{1}{2} + (l + \tfrac{1}{2})\sqrt{1 - \frac{\alpha^2}{(l+\frac{1}{2})^2}} = N + O(\alpha^2), \tag{14.54}$$

where $N = n' + l + 1$ is just the Bohr principal quantum number.[28] Inserting $\alpha = e^2/\hbar c$ in Eq. (14.53), we find

$$E = h\nu = m_0c^2 - \frac{m_0 e^4}{2\hbar^2 N^2} + O(\alpha^4). \tag{14.55}$$

Once the rest energy of the electron is removed, the remaining energy is just the negative quantity associated with the term energies of the Bohr–Balmer formula. Schrödinger communicated this result in a letter to Wien of December 27, 1925 (von Meyenn 2011, pp. 162–165),[29] including the rest energy of the electron in the "term frequency" ν_n to be associated with the nth level:

$$\nu_n = \frac{m_0c^2}{h} - \frac{R}{n^2}, \tag{14.56}$$

where $R = 2\pi^2 m_0 e^4/h^3$ is the Rydberg constant. By this time Schrödinger had clearly realized that his theory based on "vibrating system ... which has as its eigenfrequencies the term frequencies of the hydrogen atom" (Schrödinger to Wien, December 27, 1925) was only going to work at the non-relativistic level. Over the next few weeks, he introduced that approximation at the outset and attempted to complete the mathematical analysis of the resulting wave equation (which was simpler than, but in many ways

is no mention of this development in Schrödinger's (1926c) January paper in *Annalen*. Of course, it would still not have been clear to Schrödinger that this new ingredient was relevant, as Sommerfeld had apparently managed to get the right result (as we now know, serendipitously) without any electron spin! Unfortunately for Schrödinger, the fortuitous cancellations of errors that allowed Sommerfeld to obtain the correct result in the old theory no longer worked in the new one, which contained automatically the half integer Maslov index, the absence of which had previously been cancelled by the lack of electron spin in the old theory. The relativistic fine structure in the final term of Eq. (14.53) was rederived—with the exception of the s states, which were only correctly described once the Darwin term is included (automatically present in the Dirac equation)—by Heisenberg and Jordan (1926) using matrix-mechanical methods in ignorance of Schrödinger's earlier attack, in December 1925, on the relativistic problem (see Section 13.1). For a comparison with Sommerfeld's original fine structure formula from 1916, see Eq. (6.53).

[28] This corresponds to the old quantum theory relation $N = n' + n$, where n is the (previous notation for) azimuthal quantum number, related to the new angular quantum number l by $n = l + 1$.

[29] Quoted and discussed in Mehra and Rechenberg (1987, pp. 460–461).

formally similar to, the relativistic one). In particular, he needed to compute a full set of physically and mathematically acceptable solutions to the radial equation for the non-relativistic problem, at least for bound electrons (i.e., the case where $h\nu - m_0c^2 < 0$).

14.4 Four papers on non-relativistic wave mechanics

The exact chronology of Schrödinger's progress on the development of wave mechanics in December 1925 and January 1926 has been the subject of considerable controversy,[30] but there is general acceptance that the three-page note containing the quantization result for the relativistic wave equation (now known as the Klein–Gordon equation—see Eqs. (14.11) and (14.12)—and note 17) was completed by mid-December. Schrödinger made some fruitless efforts to modify the relativistic equation to bring it into compliance with the "Holy Grail" (cf. note 22) of the Sommerfeld formula, but within a few weeks, and certainly before Christmas, when he left for a two-week vacation in the Arosa resort where he had spent several months in 1922, he had made a virtue of necessity, and accepted the need to first explore the new formalism in a non-relativistic framework, before attempting a frontal attack on the relativistic problem.

The progress of his work in late December and early January can be followed in some detail, as three quite detailed notebooks have survived, in which we can scrutinize how Schrödinger developed much of the contents of the famous series of four papers on wave mechanics, "Quantization as an eigenvalue problem, Parts I–IV" (Schrödinger 1926c, 1926d, 1926f, 1926h), as well as his paper, "On the relation of the Heisenberg–Born–Jordan quantum mechanics to mine" (Schrödinger 1926e).[31] The first notebook in particular (entitled "Eigenvalue problem of the atom") contains, in its first 20 pages, material that would be incorporated directly into the first of these papers, submitted to *Annalen der Physik* in the last week of January, in addition to preliminary attempts to deal with dispersion theory and the Zeeman effect using the new theory.

This notebook begins with a section headed "Without relativity (first approximation)." Terms that are suppressed by an inverse power of the speed of light are dropped, beyond the first correction to the rest energy (which is retained although it plays no physical role as it is constant throughout and simply supplies an enormous energy offset), which represents the usual non-relativistic kinetic energy. As the rest energy is kept, the frequency ν of the atomic "vibrations" is primarily determined by the rest mass, via

[30] See Mehra and Rechenberg (1987, Ch. 3) for a detailed account.

[31] We use the English translations of these papers in Schrödinger (1982) as our basis for quotations from them. We refer to the four "communications" (*Mitteilungen*) of "Quantization as an eigenvalue problem," either as parts or installments of one paper or as four papers in a series of papers. With the exception of the third communication, (abridged) translations of these papers (including the one comparing matrix and wave mechanics) can also be found in Part Two of Ludwig (1968), with a historical introduction in Part One.

$h\nu = mc^2 + \ldots$, whereas the normal physics of the hydrogen atom is to be found in the relatively much smaller energetic terms in $h\nu - mc^2$.[32]

The relativistic equations Eqs. (14.7)–(14.11) are simplified since we can assume that both $h\nu - mc^2$ and e^2/r are much smaller than mc^2, whence the function $Q(r)$ introduced in Eqs. (14.11)–(14.12) now becomes

$$\begin{aligned} Q(r) &= \frac{m^2c^2}{\hbar^2}\left(\left(\frac{h\nu}{mc^2} + \frac{e^2}{mc^2 r}\right)^2 - 1\right) \\ &= \frac{m^2c^2}{\hbar^2}\left(\left(1 + \frac{h\nu - mc^2 + e^2/r}{mc^2}\right)^2 - 1\right) \\ &\approx \frac{2m}{\hbar^2}\left(h\nu - mc^2 + \frac{e^2}{r}\right). \end{aligned} \tag{14.57}$$

The wave equation now becomes

$$\Delta\psi = -Q(r)\psi = -\frac{2m}{\hbar^2}\left(h\nu - mc^2 + \frac{e^2}{r}\right)\psi. \tag{14.58}$$

The radial equation Eq. (14.23) is exactly as before, but with the parameters A, B, and C (cf. Eqs. (14.24)–(14.26)) now having the values[33]

$$A = \frac{2m}{\hbar^2}(mc^2 - h\nu), \quad B = \frac{me^2}{\hbar^2}, \quad C = l(l+1). \tag{14.59}$$

Writing $h\nu - mc^2 = E$ in Eq. (14.58), one recognizes the Schrödinger equation (for a hydrogen atom) in a form familiar to all first-year students of quantum mechanics.

14.4.1 Quantization as an eigenvalue problem. Part I

The first part of Schrödinger's four-part paper, "Quantization as an eigenvalue problem," was received by *Annalen der Physik* on January 27, 1926. The derivation of Eq. (14.58) in this first installment bears no evidence of its genetic origins in the phase waves of de Broglie. Instead, Schrödinger (1926c, p. 1) begins with the classical Hamilton–Jacobi equation[34] for the Coulomb problem of a single electron:

$$\frac{1}{2m}|\vec{\nabla}S|^2 + V(\vec{r}) = \frac{1}{2m}|\vec{\nabla}S|^2 - \frac{e^2}{r} = E. \tag{14.60}$$

The classical Hamilton–Jacobi function $S(\vec{r})$ has dimensions of action. A new function $\psi(\vec{r})$ can therefore be introduced as an exponential of the Hamilton–Jacobi function S,

[32] In keeping with our turn to a non-relativistic formalism, we henceforth simplify the notation by denoting the rest mass of the electron, with Schrödinger, simply as m, dropping the zero.

[33] We arrive at this identification of A, B, and C if we write $Q(r) - l(l+1)/r^2$ as $A - 2B/r + C/r^2$ as we did in going from Eq. (14.22) to Eq. (14.23) but now use Eq. (14.57) rather than Eq. (14.11) for $Q(r)$.

[34] For an introduction to the Hamilton–Jacobi theory, see Section A.2.1.

once the latter is divided by a quantity K with dimensions of action, or equivalently, we can set:

$$S(\vec{r}) = K \ln \psi(\vec{r}). \tag{14.61}$$

which then implies the equation for ψ

$$\frac{K^2}{2m}|\vec{\nabla}\psi|^2 - \frac{e^2}{r}\psi^2 - E\psi^2 = 0. \tag{14.62}$$

So far everything is still perfectly classical—we simply have changed the dependent variable in the Hamilton–Jacobi equation. Schrödinger now makes the remarkable assertion that the transition from classical to quantum mechanics amounts to the requirement that the quantity on the left of Eq. (14.62) not be exactly zero, but rather an extremum when integrated over all space and subjected to small variations of ψ. In other words, Eq. (14.62) is replaced by the variational principle

$$\delta \int \left(\frac{K^2}{2m}|\vec{\nabla}\psi|^2 - \frac{e^2}{r}\psi^2 - E\psi^2 \right) d^3r = 0. \tag{14.63}$$

The variational derivative of this expression yields just the Schrödinger equation in Eq. (14.58) (with $h\nu - mc^2 = E$), once we identify the constant K with Planck's constant (divided by 2π), $\hbar$.

As this argument, though it gets the right answer, misses the essential point of the optical–mechanical analogy, which Schrödinger would soon analyze correctly in his second paper on wave mechanics, we do not discuss it further here, other than to point out that at various points Schrödinger appeals to the need for the surface term (arising from integration by parts) to vanish at infinity to justify the boundary conditions imposed on $\psi(\vec{r})$ as $r \to \infty$. The logic here is not very convincing either as one could simply assert that $\delta\psi(\vec{r})$ vanishes sufficiently rapidly at infinity (to suppress the behavior of ψ) to eliminate any troublesome boundary terms. All of this really just amounts at this stage to an ex post facto justification for an equation that Schrödinger already knows gives the right results (i.e., the Balmer formula), at the non-relativistic level.[35] However, Schrödinger's familiarity (apparent already in his notebooks from the immediate post-war years) with

[35] With the formalism of modern quantum mechanics, the eigenvalue equation expressed by the Schrödinger equation can also be seen to be equivalent to the requirement that the expectation value of the energy is locally extremal with respect to small variations of the state vector when the latter is an eigenvector of the energy. The variational postulate (eq. (3) in Schrödinger 1926c) proposed by Schrödinger is exactly equivalent to this local extremization of the expectation value of the energy subject to the constraint that the wave function ψ be normalized (with the energy eigenvalue E the associated Legendre constraint parameter). In a note added in proof, Schrödinger essentially writes down (what we now call) the expectation value of the Hamiltonian operator, and correctly states that the wave equation amounts to the extremal condition applied to this integral quantity, subject to the constraint that $\int \psi^2 d^3r = 1$. The square-integrability of ψ is indeed the correct condition for a physical wave function, although the physical justification in terms of a probability interpretation and the full mathematical underpinnings of this innocent sounding requirement in terms of Hilbert space were yet to come.

the Hamiltonian analogy between optics and mechanics would shortly lead him to a much deeper understanding of the connection—via the Hamilton–Jacobi function and associated variational principles—between the new wave mechanics and its "geometric-optical" limit as classical mechanics. This connection is explored in detail in the second part of Schrödinger's four-part paper on wave mechanics, discussed in Section 14.5.

For now, however, we return to the first part (Schrödinger 1926c). The mathematical analysis of the non-relativistic wave equation Eq. (14.58) is by now essentially complete.[36] As before, the angular dependence of $\psi(r,\vartheta,\varphi) = R(r)\,Y_l^m(\vartheta,\varphi)$ is given by the spherical harmonics, where $l = 0, 1, 2, \ldots$ and the φ dependence is given by the single-valued function $e^{im\varphi}$, with m taking on the $2l+1$ integer values $-l, \ldots, 0, \ldots, l$. *The appearance of quantized integer values for the quantum numbers l and m here is a consequence of the demand that the angular part of the wave function be non-singular and single-valued on the sphere.*

The radial part $R(r)$ of the wave function is written as the product of r^l and a function $U(r)$, which is known from the singularity structure of the differential equation to be an entire function of r (cf. Eq. (14.28)). The associated Taylor series at $r = 0$ has infinite radius of convergence:

$$R(r) = r^l U(r), \quad U(r) = \sum_p a_p r^p \quad (< \infty, \ \forall r). \tag{14.64}$$

The other possible behavior of $U(r)$ as $r \to 0$, namely, the singular behavior $U(r) \simeq r^{-l-1}$, is excluded throughout as unphysical. Schrödinger also excludes any divergent behavior (in particular, exponential growth of $U(r)$ as $r \to +\infty$). *These requirements will lead to the quantization of the remaining, as yet unspecified, quantum number, the principal quantum number n.*

Consider the integral Laplace representation given in Eqs. (14.32)–(14.33).[37] Using Eq. (14.34) for c_1 and c_2, Eq. (14.35) for α_1 and α_2, and Eq. (14.30) for the parameters α, β, and γ in terms of A, B, and C (now given by Eq. (14.59)), we find

$$c_1 = -c_2 = \sqrt{\alpha} = \sqrt{A} = \sqrt{\frac{-2mE}{\hbar^2}} \tag{14.65}$$

[36] As previously, we occasionally depart slightly from Schrödinger's notation to make life easier for the modern reader. Despite having used ψ for the wave function in his notebook, Schrödinger for some reason switches to χ in the paper. We continue to use ψ. For the power prefactor r^ρ in Eq. (14.28), Schrödinger uses r^α. The Planck constant, divided by 2π, is written K throughout. And most disconcertingly, the letters used for azimuthal and principal quantum numbers (n and l, respectively) are just the reverse of modern notation, to which we adhere.

[37] Once again, we advise the reader that Schrödinger's analysis of the radial equation, which we are describing scrupulously in the interests of historical accuracy, involves nontrivial applications of complex function theory. The reader wishing to avoid an intricate detour into the complex plane is encouraged to proceed directly to the itemized results, beginning with Eq. (14.72).

(where once again we used that $E = h\nu - mc^2$) and[38]

$$\alpha_1 = \frac{me^2}{\hbar\sqrt{-2mE}} + l + 1, \qquad \alpha_2 = -\frac{me^2}{\hbar\sqrt{-2mE}} + l + 1. \tag{14.66}$$

Schrödinger establishes several crucial properties of the radial solutions to the new non-relativistic wave equation. The discussion is based on analysis of the two basic solutions of Eq. (14.29) (divided by r), which now takes the form

$$\begin{aligned} 0 &= \frac{d^2U}{dr^2} + \frac{2(l+1)}{r}\frac{dU}{dr} + \left(\frac{2B}{r} - A\right)U \\ &= \frac{d^2U}{dr^2} + \frac{2(l+1)}{r}\frac{dU}{dr} + \frac{2m}{\hbar^2}\left(E + \frac{e^2}{r}\right)U. \end{aligned} \tag{14.67}$$

The technique used by Schrödinger to analyze the solutions of this equation comes straight from chapters 49–52 Schlesinger's (1900, pp. 192–210) book on ordinary differential equations, which he consulted throughout the period of gestation of the first installment of "Quantization as an eigenvalue problem."

As explained in Section 14.3, solutions to equations of the Laplace form could be written as Laplace transforms of the analytic function (cf. Eq. (14.41))

$$\hat{w}(z) = e^{zr}(z - c_1)^{\alpha_1 - 1}(z - c_2)^{\alpha_2 - 1}, \tag{14.68}$$

where for general (non-integer) values of the exponents $\alpha_1 - 1$ and $\alpha_2 - 1$, the function is single valued on a Riemann surface with two branch cuts, which we can take as extending from $-\infty$ to $c_2 < 0$ along the negative real axis, and from $c_1 > 0$ to $+\infty$ along the positive real axis (see Figure 14.2). Of particular importance to us is the behavior of $U(r)$ at large positive r.[39] The integration contour L in the Laplace representation must be chosen so that the function $\hat{w}(z)$ is single valued along L and returns to the same value at the end of the contour as it had at the beginning (cf. Eq. (14.40)). In practice, this requires choosing L to be either a closed loop *on the Riemann surface*, or to begin and

[38] Eq. (14.59) tells us that $A = -2mE/\hbar^2$, $B = me^2/\hbar^2$, and $C = l(l+1)$. Eq. (14.30) (with p replaced by l) tells us that $\alpha = A$, $\beta = 2B$, and $\gamma = 1 + \sqrt{1+4C} = 2l + 2$. This last equality follows from $1 + 4C = 4\left(\frac{1}{4} + l^2 + l\right) = 4\left(l + \frac{1}{2}\right)^2$. Substituting these relations into Eq. (14.35) for α_1 and α_2, we arrive at Eq. (14.66):

$$\alpha_1 = \frac{1}{2}\left(\gamma + \frac{\beta}{\sqrt{\alpha}}\right) = \frac{1}{2}\left(2l + 2 + \frac{2B}{\sqrt{A}}\right) = l + 1 + \frac{me^2}{\hbar^2\sqrt{-2mE/\hbar^2}},$$

$$\alpha_2 = \frac{1}{2}\left(\gamma - \frac{\beta}{\sqrt{\alpha}}\right) = \frac{1}{2}\left(2l + 2 - \frac{2B}{\sqrt{A}}\right) = l + 1 - \frac{me^2}{\hbar^2\sqrt{-2mE/\hbar^2}}.$$

[39] As r is the radial coordinate, the only physically relevant values for this variable are positive real ones.

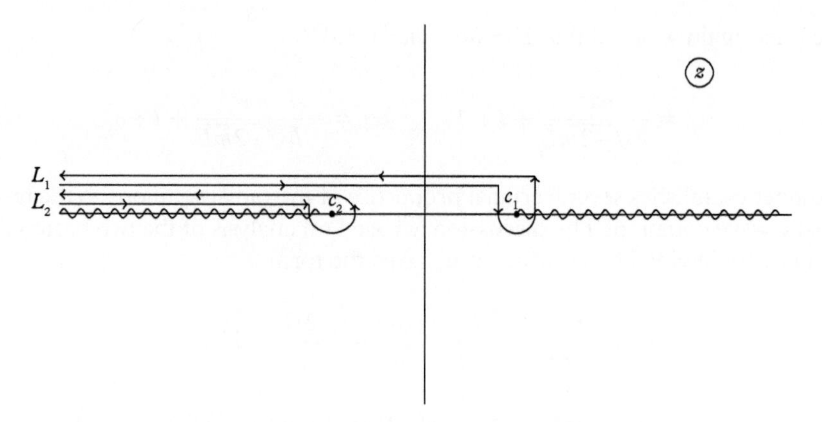

Figure 14.2 *Integration contours for Laplace solution of the Schrödinger radial equation.*

end at infinity, in a direction such that the exponential factor e^{zr} vanishes, so that one is integrating the derivative of a function that vanishes at both the start and the end of the path. For $r > 0$, this means that L should come in from $-\infty$ and return to $-\infty$. If L does not encircle a singularity, the line integral can be contracted to a null path by Cauchy's theorem and would simply give zero. Thus, as one finds (p. 198), there are two obvious linearly independent solutions, which can be written down immediately, given that we are interested in $U(r)$ for positive real r,

$$U_1(r) = \int_{L_1} e^{zr} \left(z - \frac{\sqrt{-2mE}}{\hbar} \right)^{\alpha_1 - 1} \left(z + \frac{\sqrt{-2mE}}{\hbar} \right)^{\alpha_2 - 1} dz, \tag{14.69}$$

and

$$U_2(r) = \int_{L_2} e^{zr} \left(z - \frac{\sqrt{-2mE}}{\hbar} \right)^{\alpha_1 - 1} \left(z + \frac{\sqrt{-2mE}}{\hbar} \right)^{\alpha_2 - 1} dz, \tag{14.70}$$

where the contours L_1 (resp. L_2) encircle once the branch point at $z = c_1 = \sqrt{-2mE/\hbar^2}$ (resp. at $z = c_2 = -\sqrt{-2mE/\hbar^2}$), as shown in Figure 14.2, before returning to $-\infty$ along the negative real axis (above the left-hand cut), where the integral converges exponentially. Any other solution of Eq. (14.67) must be a linear combination of these two.

Schlesinger (1900) showed that the leading asymptotic behavior of these functions for large positive r is

$$U_i(r) \simeq A_i e^{c_i r} r^{-\alpha_i} (1 + O(1/r)), \quad i = 1, 2, \tag{14.71}$$

where the coefficients A_i are given explicitly in terms of c_i and α_i. These coefficients, however, are of no further interest to us. The exponential behavior $e^{c_i r}$ is intuitively clear, as the contours L_1 and L_2 can be pulled leftward (while still encircling their respective branch points) so that the integration variable z reaches its maximum real part right at the branch point, where $z = c_i$. For large r the integral is dominated by the value of the integrand at the right-most point along the contour, that is, precisely at the branch points. In particular, if α_1 is non-integral, then the integral defining U_1 is trapped encircling the branch point at c_1 (the contour begins and ends on different sheets of the Riemann surface), and the corresponding growing exponential behavior of $e^{c_1 r}$ for $c_1 > 0$ as $r \to \infty$ would lead to a wave amplitude growing indefinitely as one goes away from the nucleus, which is clearly a physically unacceptable situation.

With this technology at his disposal, Schrödinger establishes the following:[40]

1. For positive energy $E > 0$, corresponding to unbound electrons (and therefore hyperbolic paths in the old quantum theory), the exponents α_1 and α_2 in Eqs. (14.69)–(14.71) are pure imaginary—in this case, the branch points in Figure 14.2 should actually be located on the imaginary axis. Defining $E \equiv (\hbar^2/2m)k^2$, with k real positive, we may set $c_1 = +ik$ and $c_2 = -ik$ (cf. Eq. (14.65)), so the asymptotic behavior of $U_i(r)$ in Eq. (14.71) becomes[41]

$$U_i(r) \simeq A_i r^{-(l+1)} e^{\pm i(kr - \ln(r)/ka_0)}, \quad a_0 \equiv \frac{\hbar^2}{me^2}. \tag{14.72}$$

   Returning to the original radial wave function $R(r) = r^l U(r)$, we see that for unbound electrons the asymptotic behavior assumes the characteristic complex oscillatory form for traveling (spherical) waves, multiplied by $1/r$, and this behavior holds *for all values of the positive energy E*, so that there is no quantization condition in this case. These solutions therefore correspond to the *continuous spectrum* (ionized states) of the hydrogen atom.

2. For the bound state situation, with negative energy, $E < 0$, Schrödinger first proves that if the (now real) quantity $me^2/\hbar\sqrt{-2mE}$ appearing in the exponents α_i in Eq. (14.66) is non-integer, then the solution regular at $r = 0$ (which we have assumed $U(r)$ to be) must be a linear combination of both U_1 and U_2, with non-vanishing coefficients.[42] As we saw previously, the behavior of U_1 at large r is

[40] Again, we have taken the liberty to rephrase Schrödinger's notation in a somewhat more compact and easily recognizable form for the modern reader, occasionally stating the results in more explicit form than he does. For a complete derivation of all the relevant conclusions of the analysis, see the web resource, *Solving the Radial Schrödinger Equation.*

[41] Using that $E \equiv (\hbar^2/2m)k^2$ and $a_0 \equiv \hbar^2/me^2$, we can rewrite $me^2/\hbar\sqrt{-2mE}$, the first term in the expressions for α_i in Eq. (14.66), as $\pm i/a_0 k$. We thus have $\alpha_i = l + 1 \pm i/a_0 k$. The factor $r^{-\alpha_i}$ in Eq. (14.71) for $U_i(r)$ can then be written as

$$r^{-\alpha_i} = r^{-(l+1\pm i/a_0 k)} = r^{-(l+1)} e^{\mp i \ln(r)/a_0 k}.$$

Using this expression along with $e^{c_i r} = e^{\pm ikr}$, we arrive at Eq. (14.72).

[42] See the web resource, *Solving the Schrödinger Radial Equation*, for details of the proof.

unacceptable, so we conclude that *the only acceptable solutions for negative energy must have* $\alpha_2 = -n'$, $n' = 0, 1, 2, \ldots$ (cf. Eqs. (14.48) and (14.66)), leading to

$$\frac{me^2}{\hbar\sqrt{-2mE}} = n, \quad n = 1, 2, 3, \ldots, \tag{14.73}$$

with $n = n' + l + 1$, which of course is precisely the Bohr–Balmer formula

$$E = -\frac{me^4}{2\hbar^2 n^2}. \tag{14.74}$$

With this quantization condition in place, we now have for the exponents $\alpha_1 - 1$ and $\alpha_2 - 1$ appearing in the function $w(z)$ in Eq. (14.33) in the integral representation in Eq. (14.32):

$$\alpha_1 - 1 = n + l, \qquad \alpha_2 - 1 = -n + l. \tag{14.75}$$

3. Finally, it remains to be seen what constraints exist, if any, between the values assumed by the principal quantum number n and the azimuthal (angular) quantum number l. There are two possibilities: $l \geq n$ or $l < n$. We will show that only the latter possibility is allowed. Suppose that $l \geq n$. In that case, Eq. (14.75) tells us that the exponent $\alpha_1 - 1$ is a positive integer while $\alpha_2 - 1$ is either zero or a positive integer. Thus, the function $w(z) = e^{zr}(z - c_1)^{\alpha_1 - 1}(z - c_2)^{\alpha_2 - 1}$ in the integral representation Eq. (14.32) loses its branch points and becomes an everywhere analytic (entire) function. The contour L must be chosen so that the integral in Eq. (14.40) vanishes. This can be done simply by choosing an open contour beginning and ending at zeroes of the function inside the derivative. The only such zeroes are (i) negative real infinity (for positive real r), (ii) at $z = c_1$, (iii) at $z = c_2$. Two linearly independent solutions can be obtained by picking for the contour L

   (a) the straight line from $-\infty$ to $z = c_2 < 0$, in which case the solution vanishes exponentially for large r but is singular at $r = 0$, or
   (b) the straight line from $c = c_2 < 0$ to $z = c_1 > 0$, in which case the solution is regular at $r = 0$ but blows up exponentially for $r \to \infty$.

Physically acceptable solutions therefore cannot be found in this case.[43]

If $n > l$, the exponent $\alpha_1 - 1$ is a positive integer while $\alpha_2 - 1$ is a negative integer. Hence, the Laplace transform function $w(z)$ is entire except for a pole at $z = c_2$ on the left real axis. In this case we can use the L_2 contour indicated in Figure 14.2: the associated function U_2 we already know to be well behaved (falling exponentially) as $r \to \infty$, so the only question that remains is regularity as $r \to 0$. There is no branch point in this case and the contour can be deformed into

[43] Again, the detailed arguments supporting these conclusions can be found in the web resource, *Solving the Schrödinger Radial Equation.*

a circle enclosing the pole of $e^{zr}(z-c_1)^{n+l}/(z-c_2)^{n-l}$ at $z=c_2=-\sqrt{-2mE}/\hbar$. The Cauchy residue theorem then gives

$$U_2(r) \propto \frac{d^{n-l-1}}{dz^{n-l-1}}(e^{zr}(z-c_1)^{n+l})\Big|_{z=c_2} \tag{14.76}$$

$$= e^{-\sqrt{-2mE}r/\hbar}\sum_{k=0}^{l-n-1} a_k\left(\frac{\sqrt{-2mE}}{\hbar}r\right)^k, \tag{14.77}$$

that is, a polynomial of order $l-n-1$ in r multiplying the decreasing exponential $e^{-\sqrt{-2mE}r/\hbar}$. This solution clearly satisfies both regularity at zero and damped behavior at infinity. Reinserting the extra factor of r^l (cf. Eq. (14.64)), we see that the radial part $R(r)$ of the wave function ψ has the desired behavior $R(r) \propto r^l$ as r goes to zero.

This completes the explicit solution of the wave equation for bound electrons, in terms of products of spherical harmonics and radial functions that are products of exponentials and polynomials, and which, as Schrödinger would later—consulting the book by Courant and Hilbert (1924)—discover, were associated with Laguerre. The number of linearly independent solutions for a given principal quantum number n (and therefore, of the same energy at the leading non-relativistic approximation, and in the absence of external fields) is, given the constraint $0 \leq l \leq n-1$, and the presence of $2l+1$ different m values in the spherical harmonic for a given l,

$$\sum_{l=0}^{n-1}(2l+1) = n^2. \tag{14.78}$$

As Schrödinger put it:

> *The discovered solution has exactly* $[2l+1]$ *arbitrary constants for any permissible* $[(l, n)]$ *combination; and therefore for a prescribed value of* $[n]$ *has* $[n^2]$ *arbitrary constants* (Schrödinger 1926c, p. 7; emphasis in the original).[44]

Schrödinger surmises at this point that the found set of solutions is complete, that is, "that no proper [eigen] value has escaped us," but admits that his statement is based on "experience of similar cases," rather than mathematical proof (ibid.)

The second section of the paper (p. 8) describes some qualitative properties of the solutions. The azimuthal quantum number in the old theory (k in the fourth edition of Sommerfeld's (1924) *Atombau und Spektrallinien*) is analogous to $l+1$, and the splitting of a single n level into $2l+1$ sub-levels is associated with solutions with varying numbers of node lines (where the wave function vanishes) on the sphere. The radial wave function

[44] Recall that Schrödinger's use of n and l is just the opposite of the modern convention we are following.

can be shown (this is stated but not proven) to have $n-l-1$ positive roots, corresponding to "node-spheres", that is, spherical surfaces on which the wave function vanishes identically. Finally, Schrödinger asserts that the value of the elliptical semi-major axis in the old theory for principal quantum number n is of the order of magnitude of the radial distances for which the radial function is sizeable.[45]

In the third and final section of the paper, Schrödinger (1926c, pp. 9–11) discusses some of the implications of a quantum mechanics based on wave dynamics. He emphasizes that the characteristic feature of the quantum theory—what he calls the "whole-numbers requirement" (*Ganzzahligkeitsforderung*)—has its origin in natural mathematical requirements (single-valuedness and finiteness) imposed on a spatial function. Further mathematical development of the theory is not discussed at this time, and Schrödinger admits that his hesitancy to do so stems from uncertainty that the physical results to be obtained would be both empirically verified, and, crucially, go beyond those already obtained in the old theory. In particular, he confesses that the relativistic treatment of the Kepler problem leads to the appearance of half-integer radial and azimuthal quantum numbers. As we saw earlier, these are not present in the (empirically successful) result obtained by Sommerfeld, and lead to an incorrect fine structure for the hydrogen atom.

Schrödinger's only reference to the origins of his theory in the work of de Broglie, via the Einstein gas theory, appears at this point. He comments on the appearance of whole numbers via the requirement of an integral number of de Broglie waves along each "period" of an electron's path (i.e., closed orbit), but asserts that he differs from de Broglie in considering standing ("stationary proper vibrations") rather than traveling waves. In fact, we saw earlier that, in his treatment of gas theory, de Broglie had also considered standing waves. The claim that he had "lately shown that the Einstein gas theory can be based on the consideration of such stationary proper vibrations, to which the dispersion law of de Broglie's phase waves has been applied" (Schrödinger 1926c, p. 9), is also a bit misleading, as Schrödinger's (1926b) paper on Bose–Einstein gases contains a different mathematical treatment of the problem (using the Darwin–Fowler method) but no results that are not already to be found in Einstein's work (see Section 13.5). Here, Schrödinger is perhaps guilty of a certain amount of "nostrification" (a term common in mathematical circles in Göttingen at the time for appropriating, if not downright plagiarizing, someone else's results).

The most interesting part of this final section concerns Schrödinger's attempt to come to terms with one of the most mysterious aspects of the old theory: the appearance of monochromatic emitted light of a frequency unrelated to any mechanical orbital frequencies in a quantum transition, as expressed in the Bohr frequency condition. Thus, when an electron makes a quantum "jump" from a state with quantized energy E_1 to

[45] This is most easily seen for the state with maximum angular momentum, $l = n - 1 \approx n$, for large n, corresponding in the old theory to a circular orbit. The radial function $r^n e^{-r/na_0}$ is a highly peaked function of r with a maximum at $r_{\max} = n^2 a_0$, exactly the radius of the n Bohr circular orbit.

one with a (lower) quantized energy E_2, light is emitted with a frequency ν_{12} given by the Bohr frequency condition

$$h\nu_{12} = E_1 - E_2. \tag{14.79}$$

In Schrödinger's (and de Broglie's) theory, the individual quantized energies E_1 and E_2 are themselves associated with "term frequencies" $E_1 = h\nu_1$ and $E_2 = h\nu_2$, so the Bohr frequency condition has a distinct similarity to the classical phenomenon of beats, whereby two periodic phenomena when suitably combined can lead to the appearance of a third, vibrating with the difference frequency. The non-linear dependence of the relativistic energy (involving a square root) on the energy measured in atomic phenomena might seem at first sight to invalidate this beat frequency interpretation, but as Schrödinger points out, in the non-relativistic regime relevant for atomic phenomena, the expansion of the relativistic energy gives a very large rest mass term, which is constant and universal (and therefore eliminated from difference frequencies). For a free electron with momentum p, for example, the relativistic energy is

$$\sqrt{m^2c^4 + p^2c^2} = mc^2 + E + O(E^2/mc^2), \tag{14.80}$$

where $E = p^2/2m$ in the second term is the usual non-relativistic kinetic energy. The third term, quadratic in energy, is much smaller (typically by five orders of magnitude) than the second, and we therefore find that if the appropriate vibration frequency of the electron is proportional to the full relativistic energy (as assumed by Schrödinger all along), then the beat differences of two such electron vibrations are to a very good approximation given by just the differences of the non-relativistic energy arising from the second term (modified by a potential energy term, if present), making the form of the Bohr frequency condition understandable. Of course, the precise mechanism by which the beats arise could not have been anticipated by Schrödinger at this point—it would remain for Dirac (1927b) to construct the necessary formalism in his seminal paper on quantum electrodynamics in the following year. But the following quote contains more than a whiff of the final resolution:

> It is hardly necessary to emphasize how much more congenial it would be to imagine that at a quantum transition the energy changes over from one form of vibration to another, than to think of a jumping electron. The changing of the vibration form can take place continuously in space and time, and it can readily last as long as the emission process lasts empirically (experiments on canal rays by W. Wien[46]); nevertheless, if during this transition the atom is placed for a comparatively short time in an electric field which alters the proper frequencies, then the beat frequencies are immediately changed as well, and this just for as long as the field is present (Schrödinger 1926c, pp. 10–11).

There is nothing in this statement that a modern quantum mechanician could reasonably object to, provided we interpret "vibration form" as the Schrödinger wave function (in

[46] See Wien (1919, 1921, 1924; cf. Mehra and Rechenberg 1987, p. 500)

either the Schrödinger or interaction/Dirac picture of time development) of the light emitting electron.

The paper ends with the comment that in the case of degeneracy (distinct states with the same quantized energy—the rule rather than the exception for atomic systems) an electron wave of a single frequency will typically contain a combination of all the proper (i.e., eigen-) functions corresponding to the corresponding energy:

> Under all circumstances, I believe, the proper functions, which belong to the same frequency, are in general all simultaneously stimulated. Multipleness of the proper values corresponds, namely, in the language of the previous theory to *degeneration* (Schrödinger 1926c, p. 11).

The intuition expressed here, would, of course, soon be explicitly incorporated into the theory in the guise of the fundamental precept of *linear superposition of quantum states.* A more explicit recognition of this most characteristic defining feature of quantum theory (as compared to classical theory) would come in the final section of Schrödinger's next paper, which as we shall soon see, also resolved the troubling orbit ambiguity problem of the old quantum theory (cf. Volume 1, p. 293).

14.4.2 Quantization as an eigenvalue problem. Part II

The second installment of Schrödinger's four-part paper on wave mechanics (Schrödinger 1926d), received by *Annalen der Physik* on February 23, 1926, a mere four weeks after the first, marks a considerable advance over the first, both conceptually and in the new applications and methods it introduces (such as, for example, perturbation theory). In particular, the connection of the new theory to Hamilton's development of mechanics in analogy to geometrical optics is explored in detail, allowing Schrödinger to replace the admittedly ad hoc motivation for the wave equation as a variational principle related (in an obscure way) to the Hamilton–Jacobi equation, with a much more clearly motivated connection along exactly the lines proposed by Hamilton almost a century earlier.[47]

The first section of the paper (entitled "The Hamiltonian analogy between mechanics and optics") is centered on a discussion of the Hamilton–Jacobi equation for a conservative mechanical system, defined by generalized coordinates q_k and momenta p_k, a kinetic energy $T(q_k, p_k)$, and potential energy function $V(q_k)$. In its time-dependent form, this equation takes the form[48]

$$\frac{\partial S}{\partial t} + T\left(q_k, \frac{\partial S}{\partial q_k}\right) + V(q_k) = 0. \tag{14.81}$$

[47] For discussion of Schrödinger's use of the optical–mechanical analogy and a review of the historical literature on this topic, see Joas and Lehner (2009). A detailed account of Hamilton's life and work can be found in Hankins (1980).

[48] We remind the reader that Schrödinger uses the notation W for S, and S for S_0. Schrödinger's second paper makes extensive use of the Hamilton–Jacobi procedure in classical mechanics—we suggest that the reader consult Appendix A, especially Sections A.1.3 and A.2.1, for a quick review of this material.

A simple special case of this equation—for a particle of mass m moving in three dimensions in a potential—was already discussed earlier (cf. Eq. (13.33)). For a conservative system, the time-dependence in the equation can be separated by setting

$$S(q_k, t) = -Et + S_0(q_k), \tag{14.82}$$

whence

$$T\left(q_k, \frac{\partial S_0}{\partial q_k}\right) = E - V(q_k). \tag{14.83}$$

Schrödinger (again, following Hamilton) associates solutions to the Hamilton–Jacobi equation for a general conservative system of this type with wave fronts defined as surfaces of constant $S(q_k, t)$, moving in a Riemannian metric space defined by a path distance element determined by the kinetic energy part of the Lagrangian (cf. Eq. (13.18))

$$ds^2 = 2T(q_k, \dot{q}_k)dt^2 = \sum_{k,l=1}^{3} K_{kl}(q_k)dq_k dq_l. \tag{14.84}$$

Evidently, $ds^2 = 2(E - V(q_k))dt^2$, and the "distance" so defined is not necessarily related directly to the position of the particle(s) comprising the system. However, we restrict our discussion to a single particle moving in three dimensions, as this is the only case to which most of the subsequent discussion can really be applied. In this case, the kinetic energy $T = mv^2/2$, so $ds^2 = 2Tdt^2 = mv^2dt^2 = mdl^2$, where dl is the displacement of the particle in time dt. Thus, ds represents (apart from a square root of the mass factor) the actual physical distance corresponding to the given change in coordinates, at least in this simple case.

As we saw earlier, the Hamiltonian analogy between optics and mechanics is based on the recognition that, as a consequence of Fermat's principle of least time—which determines the paths traced by rays of light subject to either reflection or (variable) refraction (or both)—a set of surfaces can be constructed at which the rays of light originating at a single point arrive simultaneously, and which are pierced normally by these rays. These surfaces are analogous to equipotentials in electrostatics, that is, they are surfaces for which a function $S(\vec{r})$ is constant,[49] and for which the normal directions to the surface at any point $\vec{r}$ are just the directions of the gradient $\vec{\nabla}S(\vec{r})$. By construction, this function must be proportional to the elapsed time for the ray to arrive from the point of origin to $\vec{r}$, or equivalently, to the integral of the refractive index μ (previously, in Chapter 13, n, cf. Eq. (13.15)) along the path from the point of origin to $\vec{r}$ (as the time required for

[49] Hamilton uses V for this function. We reserve V for the potential energy function.

light to advance a distance ds is $\mu ds/c$). One may therefore simply set $|\vec{\nabla}S(\vec{r})| = \mu(\vec{r})$, or equivalently,

$$|\vec{\nabla}S(\vec{r})|^2 = \mu(\vec{r})^2. \tag{14.85}$$

This is formally identical to the Hamilton–Jacobi equation for $S_0(\vec{r})$ in mechanics (cf. Eq. (13.33)), with $2m(E - V(\vec{r}))$ playing the role of the squared refractive index $\mu(\vec{r})^2$.

In Schrödinger's (1926d, section 1) revival of the Hamiltonian analogy, the mechanical function $S(\vec{r}, t)$ is reinterpreted, up to a proportionality constant, as the phase of an associated wave function, with the surfaces of constant S identified with the wave fronts. For a free particle, this is no more or no less than de Broglie's phase wave, but with the explicit association of the phase with the Hamilton–Jacobi function, Schrödinger is in possession of a crucial additional element of generality: the ability to handle the dynamics of particles subject to potentials.

The optical–mechanical analogy is further deepened by showing that the Fermat principle for the rays associated to the wave fronts constructed from S is *formally* identical to the Maupertuis least action principle for particle dynamics. In this, as we saw in Chapter 13, Schrödinger had been anticipated by de Broglie. To do this, one first calculates the phase velocity associated with wave fronts defined by $S(\vec{r}, t) = S_0(\vec{r}) - Et = \text{const}$. If we follow a given wave front along (in the normal direction) for a time dt, it will cover a distance (in the sense of Eq. (14.84)) of ds given by

$$Edt = |\vec{\nabla}S_0|ds, \tag{14.86}$$

from which it follows that

$$ds = \frac{Edt}{\sqrt{2(E - V)}}, \tag{14.87}$$

where we used that $|\vec{\nabla}S_0| = \mu(\vec{r}) = \sqrt{2m(E - V)}$ with $m = 1$. This corresponds to a phase velocity

$$u = \frac{ds}{dt} = \frac{E}{\sqrt{2(E - V)}}. \tag{14.88}$$

Schrödinger (1926d, p. 17, Eq. (7)) now writes:

$$\begin{aligned} 0 = \delta \int_{P_1}^{P_2} \frac{ds}{u} &= \delta \int_{P_1}^{P_2} \frac{ds\sqrt{2(E - V)}}{E} \\ \text{“}=\text{”}\ \delta \int_{t_1}^{t_2} \frac{2T}{E} dt &= \frac{1}{E}\,\delta \int_{t_1}^{t_2} 2Tdt. \end{aligned} \tag{14.89}$$

The first two equalities simply express the Fermat principle for the rays constructed from wave fronts moving with the phase velocity given in Eq. (14.88). However, in the third equality, which we put in quotation marks, ds has been reinterpreted as the distance covered by a mechanical point particle[50] with kinetic energy $\frac{1}{2}(ds/dt)^2 = E - V$ in time dt, and the variational symbol δ in the last two terms must be carried out with the energy held fixed (which means that the initial and final times t_1 and t_2 must necessarily be allowed to vary), so that the final identity reflects the Maupertuis principle. It is apparent that some legerdemain is required to go from the left-most identity (which states $\delta \int dt = 0$) to the right-most ($\delta \int 2T dt = 0$). The discrepancy really comes down, as Schrödinger (1926d, pp. 17–20, Eqs. (9), (11), and (13)) points out, to the difference between the phase velocity u of the wave used in the Fermat principle, proportional to the inverse square root of the kinetic energy, and the group velocity $v = ds/dt = \sqrt{2T}$ with which the "image point of the mechanical system" (Schrödinger 1926d, p. 17) associated with the wave moves, which involves the square root of the kinetic energy.

We return to this issue of "images" in our discussion of the next section of Schrödinger's (1926d) paper, as it is central to his conception of the new mechanics. The first section concludes with the explicit declaration of his motivation in introducing a wave theory:

> *we know today, in fact, that our classical mechanics fails for very small dimensions of the path and for very great curvatures.* Perhaps this failure is in strict analogy with the failure of geometrical optics, i.e., "the optics of infinitely small wavelengths", that becomes evident as soon as the obstacles or apertures are no longer great compared with the real, finite, wavelength . . . Then it becomes a question of searching[51] for an undulatory mechanics, and the most obvious way is the working out of the Hamiltonian analogy on the lines of undulatory [i.e., physical] optics (Schrödinger 1926d, p. 18).

The next section of Schrödinger's (1926d, section 2) paper, entitled "'Geometrical' and 'undulatory' mechanics," lies at the conceptual core of Schrödinger's attempt to induce the correct form of the undulatory (wave) theory ("physical optics"), which would correspond to the classical mechanics of point particles (viewed as the corresponding ray or "geometrical optics"), as expressed in the language of Hamilton–Jacobi theory. The fundamental intuition is taken over from the first section: the dimensionless phase of the wave process must be proportional to the Hamilton action function $S(q_k, t)$, with the proportionality constant having dimensions of inverse action. The obvious choice—one which, for obvious reasons, was not available to Hamilton—giving a universal transcription from classical to quantum phenomena, is to use the inverse of Planck's constant $\hbar$ as this constant.

The use of $\hbar = h/2\pi$ rather than simply h, or some other multiple of h, clearly is justified ex post facto by the procedure leading to the correct form of the wave equation already found in the first of Schrödinger's (1926c) four-part paper on wave mechanics.

[50] The argument is carried out for a particle of unit mass, hence, the dimensional discrepancies in the equations.

[51] At this point a footnote is inserted with a reference to Einstein's (1925a) second paper on the quantum theory of the ideal gas.

The radial solutions Schrödinger found there for the bound states of the hydrogen atom were all real, and he continues to write periodic functions in their real sinusoidal (rather than complex exponential) form:

$$\sin\left(\frac{2\pi S(q_k,t)}{h} + \text{const.}\right) = \sin\left(-\frac{2\pi Et}{h} + \frac{2\pi S_0(q_k)}{h} + \text{const.}\right). \qquad (14.90)$$

The fundamental "Einstein" equation with which de Broglie began his considerations now follows, as the frequency of the wave can be read off directly,

$$\nu = \frac{E}{h}, \qquad (14.91)$$

while the wavelength is given locally, in a region of space where the potential is varying very little over distances of the order of the wavelength, by the (local) phase velocity $u = E/\sqrt{2m(E-V)}$ (restoring the particle mass m treated as unity previously; cf. Eq. (14.88)) divided by the frequency,

$$\lambda = \frac{u}{\nu} = \frac{E}{\sqrt{2m(E-V)}} \cdot \frac{h}{E} = \frac{h}{\sqrt{2m(E-V)}}. \qquad (14.92)$$

Note that whereas the frequency can only be determined by making an absolute choice for the energy E (i.e., by eliminating the additive ambiguity typically present in the total energy of a system), the wavelength involves $E - V$, the kinetic energy, which is unambiguous.[52] For a single non-relativistic particle of mass m, $\sqrt{2m(E-V)} = p = mv$, so Eq. (14.92) is simply the de Broglie relation between wavelength and momentum. Moreover, the ratio of the wavelength to the radius of a Bohr orbit, for example, is $h/(mva)$ (with $v \simeq \alpha c$), so Schrödinger is able to point out that this ratio is of the order of unity (it is actually 2π in the ground state of the Bohr model), which confirms that the breakdown of the "geometrical optics" picture must occur precisely at the typical atomic dimensions encountered with bound electrons in atoms.

As emphasized previously by de Broglie—and re-emphasized by Einstein (1925a) in his second paper on the quantum theory of the ideal gas—the connection of the wave behavior to particle motion is found by identifying the velocity of the latter with the group velocity v_g, the velocity of the peak of the envelope of the wave packet obtained by superimposing monochromatic waves of frequency close to the mechanical particle

[52] De Broglie—and Schrödinger in his first attack on the hydrogen atom problem—had insisted on eliminating the ambiguity by taking E to be the total relativistic energy of the particle, including the rest energy.

energy divided by h. This group velocity is given by $\partial\omega/\partial k$ (with $\omega = 2\pi\nu, k = 2\pi/\lambda$, cf. Eq. (13.11)). Using the de Broglie relation $1/\lambda = p/h = \sqrt{2m(E-V)}/h$, one finds:

$$\begin{aligned} v_g &= \frac{\partial\omega}{\partial k} \\ &= \frac{\partial\nu}{\partial\lambda^{-1}} \\ &= \frac{dE}{d\sqrt{2m(E-V)}} \\ &= \left(\frac{d\sqrt{2m(E-V)}}{dE}\right)^{-1} \\ &= \left(\frac{1}{2}\sqrt{\frac{2m}{E-V}}\right)^{-1} = \sqrt{\frac{mv^2}{m}} = v, \end{aligned} \tag{14.93}$$

where in the final line we used that $E - V = T = mv^2/2$.

Schrödinger comments on this result with a direct reference to the work of de Broglie:

> the velocity of the system point is that of a *group of waves*, included within a small range of frequencies (signal-velocity [i.e., group velocity]). We find here again a theorem for the "phase waves" of the electron, which M. de Broglie had derived, with essential reference to the relativity theory, in those fine researches,[53] to which I owe the inspiration for this work. We see that the theorem in question is of wide generality, and does not arise solely from relativity theory, but is valid for every conservative system of ordinary mechanics (Schrödinger 1926d, p. 20).

The continued validity of the de Broglie connection between group and particle velocity in the non-relativistic domain is hardly surprising, but given Schrödinger's careful avoidance of the touchy issue of the validity of the theory in the relativistic domain, occasioned by his failure to obtain the correct fine structure from a fully relativistic equation, it is also not surprising that he takes the opportunity to emphasize that his arguments are justified in the non-relativistic context, which after all is exactly the regime in which the familiar Hamilton–Jacobi formalism is to be brought to bear.

The central part of section 2 of this paper consists of an extended examination of the connection between the Hamilton–Jacobi equation and the behavior of wave-packets linked with particle motion via the interpretation of the Hamilton–Jacobi function $S(q_k, t)$ (or $W(q_k, t)$ in Schrödinger's notation), divided by Planck's constant $\hbar$, as the phase of the wave. The argument is carried out for a general conservative system with coordinates q_k, with $k = 1, 2, \ldots, n$. The germ of the reasoning is however already to be found in our earlier discussion of the group velocity of a one-dimensional wave-packet

[53] At this point a footnote is inserted with a reference to the published version of de Broglie's (1925) thesis.

(cf. Eq. (13.10)). We therefore present Schrödinger's argument in the context of a one-dimensional system (a free particle of mass m), which is completely explicit and in which the basic idea is not lost in an algebraic thicket. Although Schrödinger does not make this connection, it will be apparent that the argument is really at heart yet another application of Bohr's extraordinarily fruitful correspondence principle.

In general, for a system with n degrees of freedom, the Hamilton–Jacobi equation (cf. Eq. (13.33))

$$T\Big(\frac{\partial S}{\partial q_k}, q_k\Big) + V(q_k) = -\frac{\partial S}{\partial t} \tag{14.94}$$

is a first-order (but non-linear) partial differential equation with n integration constants, which we may denote $\alpha_1, \alpha_2, \ldots, \alpha_n$.[54] Thus, a general solution should strictly be written $S(q_k, t; \alpha_k)$, and a *specific classical motion* is selected by picking definite values for the α_k (which can also be regarded as the new conserved momenta after executing the canonical transformation generated by S—see below). For a conservative system, as we saw previously, the time-dependence is trivial, as $S(q_k, t) = -Et + S_0(q_k)$. It is conventional to choose the first integration constant to be just the conserved energy: $\alpha_1 = E$. For a one-dimensional free particle, with $q_1 = x$, $T = p^2/2m$, and $V = 0$, this is the only integration constant (other than the physically irrelevant additive one), and we have,

$$\frac{1}{2m}\left(\frac{\partial S}{\partial x}\right)^2 = E = \alpha_1, \tag{14.95}$$

from which it follows that $S_0 = \sqrt{2m\alpha_1}\, x$. The Hamilton–Jacobi function is thus given by

$$S(x, t; \alpha_1) = S_0(x) - Et = \sqrt{2m\alpha_1} x - \alpha_1 t = px - \frac{p^2}{2m} t, \tag{14.96}$$

where we have rewritten the dependence on $\alpha_1 = E$ as a dependence on the (also conserved) momentum p of the particle.

Previous studies of Debye (1909), Laue (1914), and Sommerfeld and Runge (1911), cited here by Schrödinger (1926d, p. 18, p. 21), had established in the optical case the proper treatment of light "signals", in which the energy of an electromagnetic pulse was confined to a small spatial region. This is achieved by constructing a superposition of monochromatic plane waves $e^{i(\vec{k}\cdot\vec{r}-\omega t)}$ (with $\omega = c|\vec{k}|$) for varying directions and magnitudes of the wave-vector $\vec{k}$ (the latter implying varying frequency of the wave). We have already seen how to do this in the one-dimensional case (see Eq. (13.10)).

In the new wave mechanics, Schrödinger is concerned with finding the appropriate transcription in wave terms of the classical situation in which a particle located at a

[54] The equation involves $n+1$ independent variables, $q_1, \ldots, q_n$ and t, but the overall additive freedom in the solution, $S \to S+$ const., is physically insignificant, reducing the number of relevant integration constants to n. See Appendix A for further details on Hamilton–Jacobi theory.

certain point P at a given initial time passes through a point Q at a later time. In the new wave mechanics, one should demand that (at least in some limit where classical notions are approximately valid) the center of the associated wave signal (or wave packet, in the more common modern usage), if located at P at the initial time, will indeed be located at the classically determined (via the Hamilton–Jacobi equations) location Q at the later time.

With the new insight that, for the waves associated with particle motion, the phase of the monochromatic components of the wave should be identified with $S(q_k, t)/\hbar$, we may modify our previous one-dimensional construction in Eq. (13.10) with a superposition chosen to be located at $x = x_0$ at time $t = 0$:

$$\int A(p) e^{\frac{i}{\hbar}\left(p\,(x-x_0) - (p^2/2m)\,t\right)}\,dp. \tag{14.97}$$

Here, $A(p)$ is a smooth function peaked at some central momentum p_0. Provided the exponential factor oscillates many times in a momentum range over which the amplitude factor $A(p)$ varies sensibly, the stationary phase approximation can be applied, which states that the integral is dominated by the region of stationary phase, that is, the region where the derivative of the exponent with respect to the integration variable vanishes when the amplitude factor reaches its maximum value at p_0—in this case, when $x - x_0 - (p_0/m)\,t = 0$. As Schrödinger puts it:

> care is taken that all members of the infinitely small n-dimensional continuum of sets of waves meet together at time t in the point P in exactly agreeing phase (Schrödinger 1926d, p. 22).

Thus, the wave packet is peaked at $x = x_0 + (p_0/m)\,t$, namely, at x_0 at $t = 0$ and moving with velocity p_0/m so that the peak arrives at the desired classical value at time t. This condition (the origins of which in Hamilton–Jacobi theory are discussed more completely later) can alternatively be phrased as the requirement that the first derivative of the Hamilton–Jacobi function S (given by Eq. (14.96)) with respect to the integration constant(s) (in this case there is just one, α_1), *evaluated at the central value corresponding to the classical particle whose motion we wish to describe wave-theoretically*, should be independent of time,

$$\left(\frac{\partial S}{\partial \alpha_1}\right)_0 \equiv \frac{\partial S}{\partial \alpha_1}\left(x = x_0, t = 0\right)\Big|_{\alpha_1 = p_0^2/2m} = \frac{m x_0}{p_0}$$

$$= \frac{\partial S}{\partial \alpha_1} = \frac{\partial S}{\partial \alpha_1}\left(x, t\right)\Big|_{\alpha_1 = p_0^2/2m} = \frac{m}{p_0}\left(x - \frac{p_0}{m}t\right), \tag{14.98}$$

which is of course equivalent to the equation of motion $x = x_0 + (p_0/m)\,t$. Eq. (14.98) simply expresses the enforcement of the stationary phase condition at all times, in order to determine the point at which the monochromatic components of the packet coincide in phase and therefore add constructively to produce the peak amplitude of the pulse. The generalization of this requirement to an n-dimensional system is expressed by Schrödinger as follows:[55]

> The point Q [our point x] is defined as a function of time by the n equations
>
> $$\frac{\partial W}{\partial \alpha_1} = \left(\frac{\partial W}{\partial \alpha_1}\right)_0, \ldots, \frac{\partial W}{\partial \alpha_n} = \left(\frac{\partial W}{\partial \alpha_n}\right)_0 \tag{14.99}$$
>
> continues to be a point of agreeing phase for the whole aggregate of wave sets (Schrödinger 1926d, p. 24).[56]

The connection to the Hamilton–Jacobi procedure—or more accurately, to a particular version of the general canonical approach now given this name—is then indicated as follows:

> Since the system of equations (14.99) agrees with the known second system of equations of Jacobi, we have thus shown: *The point of phase agreement for certain infinitesimal manifolds of wave systems, containing n parameters, moves according to the same laws as the image point of the mechanical system* (Schrödinger 1926d, p. 24; emphasis in the original).

As the modern reader is unlikely to be familiar with the content and usage of "Jacobi's second system of equations", a brief digression may be in order here.[57] As discussed in Appendix A (pp. 393–402), the action function (in our notation $S(q_k, t; \alpha_k)$ in the one-dimensional example $S(x, t; \alpha_1)$) also plays the role of a generating function of a canonical transformation from a set of old coordinates (q_k, p_k) to new ones (Q_k, P_k). There are four versions of such a function depending on which of the initial and final canonical coordinates are used to specify the function. In our case we are dealing with a $F_2(q_k, P_k)$-type transformation, corresponding to the time-independent part $S_0(q_k; \alpha_k)$ of the action. Here, the α_k are the integration constants of the (time-independent) Hamilton–Jacobi equation satisfied by S_0, which will be identified with the new (conserved) momenta P_k. There is also the option to include time-dependence in the

[55] Schrödinger's approach here should be recognizable to modern physicists familiar with the Feynman–Dirac formulation of quantum mechanics in terms of sums of the phase factor $e^{iS/\hbar}$ over all the paths leading from the chosen initial point P to a given arrival point Q at some later time. In the correspondence limit in which $\hbar$ is very small compared to the value of the action S, this "sum over histories" (or "path integral") can also be evaluated by stationary phase methods, and turns out to be dominated by paths close to the ones for which the exponent is extremal. These are exactly the classical paths determined by Hamilton's principle, which are the ones given by the Hamilton–Jacobi equations Eq. (14.99). The Schrödinger and Feynman–Dirac formulations are, however, complementary to each other in one sense: in the former, classical behavior is recovered from a wave system by examining the constructive/destructive interference effects giving rise to wave "pulses" (or packets), viewed as focal "image points" of the wave phenomenon, whereas in the latter, the quantum behavior is regarded as arising from a superposition of complex amplitudes, each of which corresponds to a specific classical path of a point particle.

[56] Our Eq. (14.99) is Eq. (17) in Schrödinger's paper. We remind the reader that Schrödinger's W is our S.

[57] Readers immune to the charms of Hamilton–Jacobi theory may safely jump at this point to the paragraph following Eq. (14.106), without missing any essential component of our discussion.

canonical transformation, as is clearly the case if we work with $S(q_k, t; \alpha_k) = F_2(q_k, P_k, t)$. The new and old coordinates and Hamiltonians are related by (cf. Eqs. (A.82)–(A.83), and following paragraph):

$$p_k = \frac{\partial F_2(q_k, P_k, t)}{\partial q_k}, \tag{14.100}$$

$$Q_k = \frac{\partial F_2(q_k, P_k, t)}{\partial P_k}, \tag{14.101}$$

$$\tilde{H} = H + \frac{\partial F_2}{\partial t}. \tag{14.102}$$

In the present case, the time-dependent action function

$$F_2(q_k, P_k, t) = S(q_k, t; \alpha_k = P_k) = S_0(q_k; \alpha_k) - Et, \tag{14.103}$$

so the new Hamiltonian is $\tilde{H} = H - E$, which *vanishes identically* as we are dealing with an energy conserving system. Accordingly, the new Hamiltonian equations of motion become completely trivial,

$$\frac{\partial \tilde{H}}{\partial P_k} = \dot{Q}_k = 0, \tag{14.104}$$

$$\frac{\partial \tilde{H}}{\partial Q_k} = -\dot{P}_k = 0. \tag{14.105}$$

In other words, not only the new momenta (the P_k, identified with the automatically constant integration constants α_k) are temporally constant, but even the new coordinates Q_k, which are, using the action notation,

$$Q_k = \frac{\partial F_2(q_k, P_k, t)}{\partial P_k} = \frac{\partial S(q_k, t; \alpha_k)}{\partial \alpha_k} = \text{const.} \tag{14.106}$$

We have already seen above, using the free particle example as an illustration, that the equations of motion can indeed be rephrased as constancy of the derivatives $\partial S / \partial \alpha_k$ (cf. Eqs. (14.98) and (14.99)). This ends our brief detour into Hamilton–Jacobi theory.

As Schrödinger emphasizes, the constancy of the phase derivatives (interpreting S now as the phase of the associated wave) is just the required condition for locating the point of maximum amplitude of the wave-packet at any given time, which occurs because the superposition of monochromatic waves with nearby α values occurs with stationary phase at just such "image points", where the waves interfere constructively, rather than destructively. The whole situation is summed up as follows:

> What I now categorically conjecture is the following: The true mechanical process is realised or represented in a fitting way by the *wave processes* in q-space, and not by the motion of *image points* in this space. The study of the motion of image points, which is

> the object of classical mechanics, is only an approximate treatment, and has, as such, just as much justification as geometrical or "ray" optics has, compared with the true optical process. A macroscopic mechanical process will be portrayed as a wave signal of the kind described above, which can approximately enough be regarded as confined to a point, compared with the geometrical structure of the path. We have seen that the same laws of motion hold exactly for such a signal or group of waves as are advanced by classical mechanics for the motion of the image point. This manner of treatment, however, loses all meaning where the structure of the path is no longer very large compared with the wavelength or indeed is comparable with it. Then we *must* treat the matter strictly on the wave theory, i.e., we must proceed from the *wave equation* and not from the fundamental equations of mechanics, in order to form a picture of the manifold of the possible processes. These latter [classical] equations are just as useless for the elucidation of the micro-structure of mechanical processes as geometrical optics is for explaining the *phenomena of diffraction* (Schrödinger 1926d, p. 25).

The use of the term "image point" here for the location of the particle—but now corresponding to the center, or maximum-amplitude point of the wave-packet—is a direct inheritance from the Debye–Sommerfeld analysis of light pulses in the optical arena. The analogy with optics also brings out the relation of the new quantum wave formalism to classical mechanics: it is exactly analogous to the relation of physical optics, based on the proper treatment of electromagnetic waves using Maxwellian electrodynamics, to geometrical (or ray) optics, which is valid only in circumstances where "path dimensions, and especially the radius of curvature of the path, are very great compared with the wavelength" (Schrödinger 1926d, p. 20). The latter condition is clearly not in effect in the problems of atomic dynamics, where the Bohr radius is *inevitably* comparable to the de Broglie wavelength, as atoms in their "normal" (i.e. ground state) have their electrons in states of low quantum numbers, precisely the case where the spatial extension of the wave function is of the same order as the de Broglie wavelength.

The optical analogy—which Schrödinger (1926d, p. 27) calls "*very* strict" (*sehr eng*)—with the clear physical distinction between ray and physical (wave) optics, serves as a very useful expedient for Schrödinger in another respect, namely, in providing a clear explanation for the failures of the orbit picture of atomic dynamics in the old theory. From the description of wave packets in terms of combinations of monochromatic waves, it is clear that the complete phase agreement that obtains at the center of the wave-packet only diminishes continuously and very gradually away from the center, at least in circumstances where the width of the wave-packet encompasses many wavelengths of the underlying phase wave. Accordingly,

> the "system path" in the sense of classical mechanics, i.e., the point of exact phase agreement, will completely lose its prerogative, because there exists a whole continuum of points before, behind, and near the particular point, in which there is almost as complete phase agreement, and which describe totally different "paths" ... In *this* sense do I interpret the "phase waves" which, according to de Broglie, accompany the path of the electron; in the sense, therefore, that no special meaning is to be attached to the electronic path itself (at any rate, in the interior of the atom), and still less to the position of the electron on its path (Schrödinger 1926d, p. 26).

The following sentences, in which Schrödinger summarizes his dismissal of the old quantum theory in three basic precepts (which remain to this day perfectly correct, if in some cases vague, assertions in the context of modern quantum theory), with a nod in the third case to the alternative approach of Heisenberg, Born, Jordan, and Dirac, his first acknowledgement in print of the methodology of matrix mechanics:

> And in this sense I explain the conviction, increasingly evident today, *firstly*, that real meaning has to be denied to the *phase* of electronic motions in the atom; *secondly*, that we can never assert that the electron at a definite instant is to be found on *any definite one* of the quantum paths, specialised by the quantum conditions; and *thirdly*, that the true laws of quantum mechanics do not consist of definite rules for the *single path*, but that in these laws the elements of the whole manifold of paths of a system are bound together by equations, so that apparently a certain reciprocal action exists between the different paths (p. 26).[58]

The rather vague notion of a "certain reciprocal action" (where "action" here refers just to a coupling, not to the action in the technical Hamiltonian sense) between different paths is Schrödinger's oblique way of referring to the primacy of matrix elements, each involving two distinct stationary states, in Heisenberg's reinterpretation of the classical dynamical equations of motion. A few pages later, Schrödinger refers again, this time more explicitly, to the work of the Göttingen group:

> In its *tendency*, Heisenberg's attempt stands very near the present one, as we have already mentioned. In its method, it is so totally different that I have not yet succeeded in finding the connecting link ... The strength of Heisenberg's programme lies in the fact that it promises to give the *line intensities*, a question that we have not approached as yet. The strength of the present attempt—if I may be permitted to pronounce thereon—lies in the guiding, physical point of view, which creates a bridge between the macroscopic and microscopic mechanical processes, and which makes intelligible the outwardly different modes of treatment which they demand (Schrödinger 1926d, p. 30).

In the next three weeks these vague references would be replaced by a detailed (though not complete) understanding of the mathematical connections between the matrix and wave formulations of quantum mechanics.

Going back a few pages, we find the explicit form of the non-relativistic wave equation written down as the culmination of Schrödinger's reflections on the optical analogy described above. He arrives at it simply by inserting the formula Eq. (14.88) for the phase velocity of a de Broglie wave into the standard wave equation (p. 27)

$$\Delta\psi - \frac{1}{u^2}\ddot{\psi} = \Delta\psi + \frac{4\pi^2\nu^2}{u^2}\psi = 0, \tag{14.107}$$

[58] At this point, Schrödinger inserts references to Heisenberg (1925c), Born and Jordan (1925b), Born, Heisenberg, and Jordan (1926), Dirac (1925), and Bohr (1926).

where a separable time-dependence of the wave function $\psi(\vec{r},t) = e^{2\pi i\nu t}\psi(\vec{r})$ has been assumed.[59]

This Ansatz can hardly be unique, as it is (in the present paper) the result of correspondence-principle reasoning, which only requires that in the limit where the action of the mechanical motion far exceeds Planck's constant ($S \gg h$ in the phase factor), the ray optics limit of the theory should reproduce the classical paths arising from (say) Hamilton–Jacobi theory. In particular, in this limit slowly varying differences in the real amplitude prefactor of the wave function $\psi(\vec{r}) = A(\vec{r})\sin(S_0(\vec{r})/\hbar)$ are not distinguishable. In fact, Schrödinger's hypothesis (of the identity of the phase of the wave process with the classical Hamilton action function) clearly fails to specify unambiguously the concomitant amplitude information necessary to obtain a unique wave function. There must therefore be a multitude of wave equations giving the same geometrical limit.

In fact, in the optical case, it had already been understood for thirty years, since the eikonal theory of Bruns (1895), that the correspondence of the wave equation to the ray picture obtains in the approximation that the amplitude factor changes by a very small factor over a wavelength of the phase wave. Thus, if the relative change in $A(\vec{r})$ in a path element $d\vec{s}$ is much smaller than the corresponding change of phase

$$\frac{1}{A}\vec{\nabla}A(\vec{r})\cdot d\vec{s} \ll \vec{\nabla}(S_0/\hbar)\cdot d\vec{s}, \tag{14.108}$$

then we can neglect the spatial dependence of $A(\vec{r})$ in calculating the Laplacian in the wave equation, which then becomes

$$\begin{aligned}\vec{\nabla}\cdot\vec{\nabla}(Ae^{iS_0/\hbar}) &\approx A(\vec{r})\vec{\nabla}\cdot(\frac{i}{\hbar}\vec{\nabla}S_0e^{iS_0/\hbar}) \\ &= \frac{i}{\hbar}A(\vec{\nabla}^2S_0)e^{iS_0/\hbar} - \frac{1}{\hbar^2}A|\vec{\nabla}S_0|^2e^{iS_0/\hbar}.\end{aligned} \tag{14.109}$$

In the situation envisaged here, S_0 is very large in units of $\hbar$, so we may neglect the term $\vec{\nabla}^2S_0/\hbar$ in comparison to the term $|\vec{\nabla}S_0|^2/\hbar^2$, which is quadratic in $S_0/\hbar$. Thus, we find (cf. Eq. (13.33))

$$\begin{aligned}\vec{\nabla}\cdot\vec{\nabla}(Ae^{iS_0/\hbar}) &\approx -\frac{1}{\hbar^2}A\,|\vec{\nabla}S_0|^2\,e^{iS_0/\hbar} \\ &= -\frac{2m}{\hbar^2}(E-V(\vec{r}))\,Ae^{iS_0/\hbar}.\end{aligned} \tag{14.110}$$

[59] It is remarkable, given our present understanding of the inextricable way in which complex structure is woven into the formalism of quantum theory, that, with the exception of the asymptotic formula (17) in paper I (from Schlesinger), this is the first time the imaginary quantity i has appeared explicitly in either of Schrödinger's first two papers. Schrödinger's preoccupation with the radial wave functions of the bound electron in hydrogen in the first paper accounts for the absence of i, as these can all be chosen real with no loss of generality. The spherical harmonics, typically written in complex form (for the φ dependence), do not appear explictly in either paper. Elsewhere in the paper, the time dependence of the wave function is written in purely real form (see, e.g., Schrödinger 1926d, p. 19, Eq. (10), $\sin(-(2\pi Et/h) + (2\pi S(q_k)/h) + \text{const.}))$.

The right-hand side of this equation coincides with the term in Eq. (14.107) from the time derivative $-(4\pi^2\nu^2/u^2)\,\psi$ (with $\psi = Ae^{iS_0/\hbar}$) if we identify $E = h\nu$ and the phase velocity $u = E/\sqrt{2m(E-V)}$, exactly as in Eq. (14.88) (reinserting the mass m). This simple deductive argument, leading from the wave theory back to ray optics, must surely have been known to Schrödinger in early 1926. As it is, he simply proposes, in the interests of "striving for simplicity" (Schrödinger 1926d, p. 27), that we accept as valid, *without approximation*, the wave equation obtained by substituting the local phase velocity obtained previously by consideration of the moving wave fronts of the Hamilton function in Eq. (14.107), thereby obtaining

$$\Delta\psi + \frac{8\pi^2 m}{h^2}(E-V)\,\psi = \Delta\psi + \frac{2m}{\hbar^2}(E-V)\,\psi = 0, \qquad (14.111)$$

which is the non-relativistic time-independent Schrödinger equation familiar to every beginning student of quantum mechanics. There is no surprise that we have ended up here, of course: Schrödinger had already shown that exactly this equation is up to the job of reproducing the Bohr–Balmer energies, once appropriate regularity conditions are imposed. There remains the critical issue of identifying in the new formalism the *specific characteristic feature of the quantum theory—the appearance of discrete quantities where classical theory would normally demand continuous ones.* Of course, the quantum theory is already present in some sense in Eq. (14.111), inasmuch as Planck's constant is present. But the analog of the insistence of explicit quantum conditions in the old theory to select dynamically allowed motions is far from obvious in a wave formalism based on continuous, even differentiable functions. The solution here is just the one proposed in the previous paper:

> [I]t turns out that equation (18) [our Eqs. (14.107) and (14.111)] *carries within itself the quantum conditions.* It distinguishes in certain cases, and indeed in those where experience demands it, *of itself,* certain frequencies or energy levels as those which alone are possible for stationary processes, without any further assumption, other than the almost obvious demand that, as a physical quantity, the function ψ must be single-valued, finite and continuous throughout configuration space (Schrödinger 1926d, p. 28).

The "almost obvious demand" stated here by Schrödinger would receive within two years an essential sharpening and conceptual clarification in von Neumann's formulation of quantum theory employing the Hilbert spaces of function theory. But the basic idea is certainly correct, and it allows Schrödinger to point to the essential distinction between his new theory and the one it was in the process of supplanting:

> The definition of the quantum levels *no longer takes place* in two separated stages: (1) Definition of all paths dynamically possible. (2) *Discarding* of the greater part of those solutions and the selection of a few by special postulations; on the contrary, the quantum levels are *at once* defined as the *proper values* of equation (18) [our Eqs. (14.107) and (14.111)], which *carries in itself its natural boundary conditions* (Schrödinger 1926d, pp. 28–29).

In the final section of the paper, Schrödinger (1926d, sec. 3, pp. 30–40) turns to a number of explicit examples where his wave equation can either be solved exactly or in a systematic approximative scheme. First comes the simple harmonic oscillator, which proved such a fertile testing ground for Heisenberg (1925c) in his *Umdeutung* paper. The Schrödinger wave equation for a one-dimensional harmonic oscillator, with potential energy $V(q) = \frac{1}{2}m\omega^2 q^2 = 2\pi^2 m\nu^2 q^2$, after conversion to dimensionless variables $x = 2\pi\sqrt{m\nu/h}\, q$ (here q is the actual physical Cartesian coordinate of the oscillator), is easily seen to coincide with the Hermite equation

$$\frac{d^2\psi}{dx^2} + \left(\frac{2E}{h\nu} - x^2\right)\psi = 0. \tag{14.112}$$

By now, Schrödinger has the first edition of Courant and Hilbert's (1924) *Methods of Mathematical Physics* close at hand,[60] where the "proper" (now, "eigen") solutions are discussed in detail. They are the orthogonal Hermite functions

$$\psi_n(x) = e^{-x^2/2}H_n(x),\ n = 0, 1, 2, \ldots, \tag{14.113}$$

where the $H_n(x)$ are n-order polynomials, and the n function corresponds to the proper (eigen-)value $E_n = (n + \frac{1}{2})h\nu$, including the half-integral zero-point energy which had been found previously by Heisenberg. As Schrödinger (1926d) points out: "It is remarkable that our quantum levels are *exactly* those of Heisenberg's theory" (p. 31). Once again, as in the case of the hydrogen atom treated in Part. I of his paper on quantization as an eigenvalue problem (Schrödinger 1926c), the selection of a discrete, quantized set of possible values for the energy is brought back to the boundary conditions imposed on the wave function. For positive values of the energy E, the phase velocity $u = E/\sqrt{2(E - V)}$ is real as long as we are in the coordinate region where $E > V(x)$, which is bounded by the classical turning points of the motion. But outside of this region, u evidently becomes imaginary, which translates into real exponential behavior (for general E, *both* rising and falling), rather than the oscillatory behavior that holds between the turning points. Only a very careful choice of E—specifically, the selection of one of the values E_n—will prevent the "function represented by such an equation [from] growing beyond all bounds" (Schrödinger 1926d, p. 33).

This observation is followed by four paragraphs that may easily be passed over without realizing their enormous importance in resolving one of the most disturbing features of the old quantum theory. In our discussion of the Stark effect in Section, 6.3 (p. 293), we pointed out that for degenerate systems (such as the hydrogen atom or the isotropic three-dimensional harmonic oscillator), where the resolution of the problem led

[60] In his notebook entitled "Eigenvalue problem of the atom," Courant–Hilbert is explicitly referred to on p. 45, for the solution (on p. 261), of the equation $y'' + (1 - x^2)y + \lambda y = 0$, where the requirement that the solutions not grow exponentially at infinity is satisfied for $\lambda = 0, 2, 4, 6, \ldots$. As $2E/h\nu = \lambda + 1$, we find immediately the desired eigenvalues. The cumbersome contour-integral Laplace method, previously used by Schrödinger, following Schlesinger (1900), for the hydrogen atom's radial wave functions (and applicable also in this case) is thereby avoided.

to identity of (or a rational relation between) the individual frequencies of the separated coordinates, the selection of the particle orbits was ambiguous: the problem can be separated in a variety of distinct coordinate systems (e.g., polar or parabolic), and while the quantized energies obtained are the same irrespective of the coordinate system used, the orbital paths followed by the particle are geometrically different (in mathematical terms: not congruent). For the Coulomb problem, for example, one obtains Kepler ellipses of different eccentricities if one uses parabolic coordinates rather than polar coordinates. There can hardly be a deep preference shown by Nature for one rather than another coordinate system!

If the system is not degenerate, we are forced to a particular choice of coordinates in which the problem separates, and the ambiguity does not arise. Let us consider, for example, a three-dimensional oscillator with three distinct frequencies, so

$$V(q_1, q_2, q_3) = 2\pi^2 m\left(\nu_1^2 q_1^2 + \nu_2^2 q_2^2 + \nu_3^2 q_3^2\right), \qquad (14.114)$$

where $(q_1, q_2, q_3) = (x, y, z)$. In the new wave theory, provided ν_1, ν_2, and ν_3 are all distinct and incommensurable,

> nothing is changed [in comparison to the treatment of a one-dimensional system]. ψ is taken as the *product* of functions [thus, $\psi(q_1, q_2, q_3) = \psi_1(q_1)\,\psi_2(q_2)\,\psi_3(q_3)$], each of a single coordinate, and the problem splits up into just as many separate problems of the type treated above as there are coordinates present ... No proper value (for the whole system) [i.e., value of energy $E = E_1 + E_2 + E_3$] is multiple, if we presume that there is no rational relation between the ν-values [i.e., if we assume that there are no integers n_1, n_2, and n_3 for which $n_1\nu_1 + n_2\nu_2 + n_3\nu_3 = 0$] (Schrödinger 1926d, p. 33).

If the system is degenerate, however, as for the isotropic oscillator with $\nu_1 = \nu_2 = \nu_3 = \nu$, one may again write the solution for $\psi(q_1, q_2, q_3)$ as a product of one-dimensional solutions, but *not in a unique fashion.* For example, the coordinates (q_1, q_2, q_3) may be identified with an arbitrary set of orthogonal Cartesian (x, y, z) coordinates. Classically the orbits obtained in the old theory by performing such a rotation and then imposing integer quantization of the motion in each of the rotated directions clearly will result in a completely different set of allowed orbits. Or one might choose to employ polar, rather than Cartesian, coordinates, which also separate the problem in the case of an isotropic oscillator. Just as in the old quantum theory, in the new wave mechanics the quantized proper values for the energies obtained are the same irrespective of the coordinate system used (e.g., $E = (n_1 + n_2 + n_3 + \frac{3}{2})h\nu$ for any Cartesian system), while the associated wave functions clearly differ. But there is one critical difference between the old, classically oriented, treatment and the new theory:

> Only in one point is there a not unwelcome formal difference. If we applied the Sommerfeld–Epstein quantum conditions *without* regard to a possible degeneracy then we always got the same energy levels, but reached different conclusions as to the paths permitted, according to the choice of coordinates. Now that is *not* the case here. Indeed we come to a completely different system of proper [eigen] functions, if we, for

> example, treat the vibration [i.e., wave] problem corresponding to unperturbed Kepler motion [Coulomb potential] in *parabolic* coordinates instead of the polars used in Part I [Schrödinger (1926c)]. However, it is not just the *single proper vibration* [i.e., associated wave function $\psi_{n_1,n_2,n_3}(q_1, q_2, q_3, t)$, including its time-dependence] that furnishes a *possible state of vibration*, but an arbitrary, finite or infinite, *linear aggregate* [i.e., arbitrary linear combination] of such vibrations. And as such the proper functions found in any second way may always be represented; namely, they may be represented as linear aggregates of the proper functions found in an arbitrary way, provided the latter form a *complete* system (Schrödinger 1926d, pp. 33–34).

In other words, one may solve for the wave function of the electron in a hydrogen atom (unperturbed by an electric or magnetic field), with a particular quantized Bohr–Balmer energy, in either polar or parabolic coordinates. Although the individual eigenfunctions obtained depend on the coordinate system chosen, the eigenfunctions of one system can always be written as linear combinations of the eigenfunctions of the other (of the same energy).[61] This extremely important passage not only shows how the new theory circumvents the troubling ambiguity of paths (orbits) in the old theory, but it introduces for the first time, in a completely explicit way, a central new conceptual component of modern quantum mechanics, not to be found in the old quantum theory preceding it, namely, the notion of *linear superposition of quantum states*.

Schrödinger concludes his discussion of the harmonic oscillator in section 3.1 by addressing the question of "how the energy is really distributed among the proper vibrations" (Schrödinger 1926d, p. 34). He proclaims himself agnostic as to whether the system (between quantum transitions, clearly) must have a definite energy value, or whether it is possible for the system to exist in a superposition of different energy states. In other words, does the linear superposition property that must necessarily hold for states degenerate in energy (in order for the physics in different coordinate systems to be equivalent) extend to states of different energy? This is a question he admits will "have to be faced [at] some time" (p. 34). In fact, a few months later, Schrödinger (1926g) would submit a short paper to *Die Naturwissenschaften*, entitled "The continous transition from micro- to macro-mechanics" (published early July 1926), in which a specific superposition of harmonic oscillator stationary state proper(eigen-)functions was constructed and shown to correspond to a "wave group" (wave packet, in modern language) of breadth much smaller than the amplitude A of the motion, centered at $x = A\cos(2\pi\nu t)$, which bobs back and forth precisely as classical simple harmonic motion would require, while maintaining its integrity and form.[62] In other words, the recovery of particle motion in the classical sense from the new wave theory precisely requires the employment of superpositions of wave solutions of Schrödinger's equation of different frequency and,

[61] How this works explicitly can be found in Duncan and Janssen (2014, sec. 5, pp. 76–77) on the Stark effect in the old and the new quantum theory.

[62] The states constructed by Schrödinger (1926g) in his *Naturwissenschaften* article are the direct ancestors of the *coherent states* of modern quantum optics.

therefore, energy. In particular, such states cannot be stationary states of definite energy at the quantum level.[63]

The final two sections of this paper (secs. 3.2 and 3.3) address applications of the new theory to rigid rotator problems in two and three dimensions. In the two-dimensional case, one considers a rigid body of moment of inertia A constrained to rotate around the z-axis. The single generalized coordinate relevant for the specification of the state of the system is the azimuthal angle φ measuring rotations around the z-axis, so we are looking for a wave function $\psi(\varphi)$. If we consider as the simplest model of a rotator a particle of mass m constrained to rotate in the xy plane (at $z = 0$) at a distance a from the origin, the three-dimensional Laplacian Δ of Eq. (14.111) simplifies to

$$\frac{1}{a^2}\frac{d^2\psi(\varphi)}{d\varphi^2} + \frac{8\pi^2 mE}{h^2}\psi(\varphi) = 0, \tag{14.115}$$

or, in terms of the moment of inertia $A = ma^2$ (I in modern notation),

$$\frac{1}{A}\frac{d^2\psi(\varphi)}{d\varphi^2} + \frac{8\pi^2 E}{h^2}\psi(\varphi) = 0. \tag{14.116}$$

This harmonic equation has simple real (co)sinusoidal solutions

$$\psi = \sin\left(\sqrt{8\pi^2 EA/h^2}\,\varphi\right) \quad \text{or} \quad \psi = \cos\left(\sqrt{8\pi^2 EA/h^2}\,\varphi\right). \tag{14.117}$$

The single-valuedness requirement for the wave function that $\psi(\varphi) = \psi(\varphi + 2\pi)$ constrains the allowed energies to the quantized values

$$E_n = \frac{n^2 h^2}{8\pi^2 A}, \quad \text{with } n = 0, 1, 2, \ldots. \tag{14.118}$$

It is curious that Schrödinger still scrupulously avoids the complex exponential form (e.g., $e^{2\pi in\varphi}$) for the wave solutions, using real sines and cosines instead. There seems little doubt that he still considers the wave function as representing a field with a direct physical interpretation, which should therefore take real values. At this point, given the freedom to construct real solutions by taking linear combinations of complex ones (the basic wave equation is of course *linear*, a fact of critical importace both practically and conceptually), there is no loss of generality.

The result Eq. (14.118) coincides with the result of the old quantum theory, going back to Ehrenfest's (1913) paper on diatomic molecules (see Section 5.2.1). It disagrees with the rotational energies found in band spectra, which require half-integer values for the quantum number n. As we saw in Section 11.6, Heisenberg (1925c) attempted to

[63] Schrödinger's hesitancy on this matter in his second paper is a little puzzling, as with a little thought, it is apparent that, *for a free particle*, the construction of wave-packets already implies the superposition of states of different momentum, hence, energy. Apparently, Schrödinger initially thought that the same degree of superposition might not be allowed for bound as for free particles.

repair the problem in the last section of the *Umdeutung* paper, but with an incorrect argument.[64] The critical point leading to Heisenberg's error is noted by Schrödinger: whereas it is perfectly correct in classical physics to impose constraints on the motion of a particle (such as requiring the particle motion to be confined to a two-dimensional plane) by reducing ab initio the number of degrees of freedom in the equations of motion, this is impermissible in the new wave theory, as the "wave process fills the *whole* of the phase space ... it is well known that even the *number* of dimensions in which a wave process takes place is very significant" (Schrödinger 1926d, p. 35). The quantized rotational energies of a rotator in physical three-dimensional space will be quite different than those of a rotator in a physical two-dimensional plane.

The three-dimensional rigid rotator is the subject of section 3.3 of this paper (Schrödinger 1926d, pp. 35–36). Now the full angular dependence of the three-dimensional Laplacian must be taken into account, and the wave function becomes a function of the two angles appearing in polar coordinates, $\psi(\vartheta, \varphi)$. The wave equation is

$$\frac{1}{\sin\vartheta}\frac{\partial}{\partial\vartheta}\left(\sin\vartheta\frac{\partial\psi}{\partial\vartheta}\right) + \frac{1}{\sin^2\vartheta}\frac{\partial^2\psi}{\partial\varphi^2} + \frac{8\pi^2 AE}{h^2}\psi = 0. \tag{14.119}$$

The regularity and single-valuedness of the solutions to this equation, just as in the treatment of the hydrogen atom in Part I (Schrödinger 1926c), leads to the requirement that the energy take the quantized value[65]

$$E_l = \frac{h^2}{8\pi^2 A}l(l+1), \quad \text{with } l = 0, 1, 2, \ldots \tag{14.120}$$

The rotational energy is given by $\vec{L}^2/2A$, where $\vec{L}$ is the angular momentum vector of the rotator (now three dimensional), so this result reproduces the quantization rule obtained in the Three-Man-Paper (Born, Heisenberg, and Jordan 1926), namely, $\vec{L}^2 = l(l+1)\hbar^2$. As $l(l+1)$ can be written $(l+\frac{1}{2})^2-\frac{1}{4}$, one now has the mysterious "square of half-integers" rule that must be employed when computing band spectra frequencies (given, by the Bohr frequency condition, as differences of the quantized energies, in which the $-1/4$ factor cancels). Section 12.3.5 described how the authors of the Three-Man-Paper provided a correct quantum-mechanical treatment of angular momentum in three dimensions but they did not make the connection to half integral quantization in band spectra.

The final section of the paper (Schrödinger 1926d, sec. 3.4, pp. 36–40), entitled "Non-rigid rotator (diatomic molecule)," examines the wave equation for a molecular system composed of two atoms, of mass m_1 and m_2, respectively, treated as point particles subject to a harmonic potential energy function $V = 2\pi^2\nu^2(r - r_0)^2$, where r is the distance between the atoms and r_0 the equilibrium separation. The molecule therefore displays both rotational and vibrational excitations, and not surprisingly, the solution

[64] See Section 11.6 and, for further details, the web resource, *The Problem of Spectral Intensities in the Old and New Quantum Theory*.

[65] Schrödinger continues to use *n* for the azimuthal quantum number, a hangover from the old quantum theory—we have here switched to the modern notation for the angular momentum quantum number *l*.

of the wave equation reveals a spectrum (in the approximation of small vibrations $r - r_0 \ll r_0$) of the form

$$E_{l,n} = E_0 + B\frac{h^2 l(l+1)}{8\pi^2 A} + C\left(n + \tfrac{1}{2}\right)h\nu, \tag{14.121}$$

that is, a superposition of rotational and harmonic oscillator quantized energies. Here, A is the moment of inertia of the molecule, as before, and B and C are constants depending on l that express the centrifugal distortion of the radial behavior and hence, of the vibrational and rotational energies.[66] Semi-empirical formulas of this type had been employed previously in the old theory in the analysis of band spectra (see, e.g., Sommerfeld 1924). The two new theoretical elements introduced in this section by Schrödinger are again ideas that would play enormously important roles in the further development of quantum mechanics:

1. The extension of the Schrödinger equation from a single particle to a multiparticle system, which proceeds in complete analogy to the extension of the Hamilton–Jacobi equation for a system of three (say, Cartesian) coordinates (x_1, y_1, z_1), to one in which the system is specified by $3n$ such variables, for example, for the diatomic molecule with $n = 2$, $(x_1, y_1, z_1, x_2, y_2, z_2)$. The wave function for such a system (like the Hamilton–Jacobi function in the classical formalism) must now be a function of all six coordinates, $\psi(x_1, y_1, z_1, x_2, y_2, z_2)$. The Schrödinger wave equation becomes

$$\frac{1}{m_1}\left(\frac{\partial^2\psi}{\partial x_1^2} + \frac{\partial^2\psi}{\partial y_1^2} + \frac{\partial^2\psi}{\partial z_1^2}\right) + \frac{1}{m_2}\left(\frac{\partial^2\psi}{\partial x_2^2} + \frac{\partial^2\psi}{\partial y_2^2} + \frac{\partial^2\psi}{\partial z_2^2}\right) + \frac{8\pi^2}{h^2}(E - V)\psi = 0. \tag{14.122}$$

Schrödinger shows how, in analogy to the completely similar procedure used in classical mechanics, the center of mass and relative motion can be separated by changing to coordinates

$$\xi = \frac{m_1 x_1 + m_2 x_2}{m_1 + m_2}, \quad \eta = \frac{m_1 y_1 + m_2 y_2}{m_1 + m_2}, \quad \zeta = \ldots \tag{14.123}$$

and

$$x = x_1 - x_2, \quad y = y_1 - y_2, \quad z = z_1 - z_2, \tag{14.124}$$

[66] As above, we have used l for the rotational quantum number and n for the vibrational one. Rather confusingly for modern readers, Schrödinger does exactly the reverse! Schrödinger's calculations for the non-rigid rotator employ a harmonic approximation for the vibrational modes of the diatomic molecule that would later be improved by Erwin Fues in Zurich, working with the assistance of Schrödinger, in the earliest paper using wave mechanics not authored by Schrödinger (Fues 1926a). The problem of the terms and line intensities of band spectra arising from nonrigid diatomic molecules had been attacked somewhat earlier using matrix mechanical methods by Lucy Mensing (1926).

with

$$\psi = f(x, y, z)\, g(\xi, \eta, \zeta). \tag{14.125}$$

The equation for the center of mass wave function g is a free one (no potential term) and represents the motion of a free particle of mass $M = m_1 + m_2$, with unquantized oscillatory solutions in an unbounded space, or quantized modes as in the Einstein–Bose gas theory if the molecules are confined in a box. The equation for the relative wave function f retains the potential energy function giving rise to quantized vibrational and rotational levels.

The full equation Eq. (14.122) is the archetype for all future applications of the Schrödinger equation to multiparticle systems in atomic physics (multi-electron atoms) and quantum chemistry. At this point, a potential difficulty arises for a direct interpretation of the wave function as a physical field (analogous, say, to electric and magnetic fields), which we have argued earlier seems to be the point of view adopted at this stage by Schrödinger. Such a field should be formulated directly in the three-dimensional space of the physical world (with, of course, an additional dependence on the time variable if we are interested in the dynamical evolution of the system). Instead, the wave function for two particles depends on *six* variables—the wave function is staged on the abstract *configuration space* of the system, which for n particles is $3n$-dimensional. This is the first sign that the proper physical interpretation of the wave function will not be a trivial matter. Schrödinger refers to this in a single, somewhat elliptical, sentence:

> The direct interpretation of this wave function of *six* variables in *three*-dimensional space meets, at least initially, with difficulties of an abstract nature (Schrödinger 1926d, p. 39).

2. The introduction of a systematic perturbation scheme for approximating systems in which the wave equation cannot be solved analytically. Here, the actual scheme, which would be developed fully in Schrödinger's (1926f) analysis of the Stark effect in the third installment of his four-part paper, is merely pointed to here, as a necessary tool in improving the calculation of band spectra:

> If, however, we are going to push the calculation as far as is necessary for the fineness of band structure, then we must make use of the theory of the *perturbation of proper* [eigen] *values and functions*, that is, of the alteration experienced by a definite proper [eigen] value and the appertaining proper[eigen] functions of a differential equation, when there is added to the coefficient of the unknown function in the equation a small "disturbing term". This "perturbation theory" is the complete counterpart of that of classical mechanics, except that it is simpler because in undulatory mechanics we are always in the domain of *linear* relations. As a first approximation we have the statement that the perturbation of the proper value is equal to the perturbing term averaged "over the undisturbed motion" (Schrödinger 1926d, pp. 39–40).

Schrödinger closes the paper by remarking that he has already applied this technique to the problem of the Stark effect (small applied constant electric field) and that his results agree to the first order with those obtained by Epstein (1916a) in the old quantum theory (see Section 6.3).

14.4.3 Quantization as an eigenvalue problem. Part III

Schrödinger interrupted his series on quantization viewed as an eigenvalue problem with a paper in which he uncovered the intimate connection between the continuous analytic formalism he had developed and the more abstract algebraic methods of Heisenberg et al. (Schrödinger 1926e). We discuss this paper in Section 14.5, but since it does not have a very direct impact on the two remaining papers dealing purely with wave mechanics, we continue here with some brief comments on the third communication of the series "Quantization as an eigenvalue problem" (Schrödinger 1926f). In this paper, received by *Annalen der Physik* on May 10, 1926, the formalism of time-independent perturbation theory, for both the non-degenerate and degenerate cases, is fully developed, and then applied to the derivation of the first-order energy shifts in the Stark effect. We return to the specific application to the Stark effect in Chapter 15, as one example of how one of the great successes of the old quantum theory would be reinforced, modified, and reinterpreted in the new theory. The other obvious application of perturbation theory in atomic physics—and one in which, as we saw in Volume 1, the old theory had notoriously failed—was in the calculation of the splittings in atomic levels by an external static magnetic field. With regard to this problem, as well as to the incorrect solution given by his relativistic version of the wave equation, Schrödinger refers briefly to the recent proposal of Goudsmit and Uhlenbeck in the introduction to this third installment:

> The application (not yet completed) to the *Zeeman effect* will naturally be of much greater interest [than the analysis of the Stark effect] ... It was already mentioned in Part I [Schrödinger (1926c)] that the relativistic hydrogen atom may indeed be treated without further discussion, but that it leads to "half-integral" azimuthal quanta, and thus contradicts experience. Therefore, "something must still be missing". Since then I have learnt *what* is lacking from the most important publications of G. E. Uhlenbeck and S. Goudsmit, and then from oral and written communications from Paris (P. Langevin) and Copenhagen (W. Pauli), viz., in the language of the theory of electronic orbits, the *angular momentum* of the electron round its axis, which gives it a *magnetic moment* ... In the present paper, however, the taking over of the idea is not yet attempted (Schrödinger 1926f, pp. 63–64).

The resolution of the Zeeman effect (as well as the complex multiplet structure, and the relativistic fine structure problem) in the new quantum mechanics are discussed in Chapter 15. Schrödinger did not play a central role in these developments. The "perturbation method", as defined by Schrödinger in this paper, allows for the approximate solution of boundary value problems that are "sufficiently closely related to a directly soluble problem" (p. 62). In particular, one imagines that the directly soluble problem amounts to determining the eigenvalues E_k and eigenfunctions $u_k(x)$ of a self-adjoint operator L of Sturm–Liouville type:[67] type,

$$L u_k(x) + E_k \, \rho(x) \, u_k(x) = 0, \tag{14.126}$$

where $\rho(x)$ is a real, positive-valued measure function and the action of L on a function is defined as:

$$L y(x) \equiv \frac{d}{dx}\left(p(x)\,\frac{dy}{dx}\right) - q(x)\,y(x). \tag{14.127}$$

with $p(x)$ and $q(x)$ otherwise unspecified (real) coefficient functions of the differential equation.[68] One supposes that the problem in question, if it involves a multi-dimensional configuration space (e.g., the three dimensions of the hydrogen atom problem, possibly perturbed by an external electric field), has already been dissected by separation in an appropriate coordinate system (e.g., spherical coordinates, or parabolic coordinates if an electric field is present), so that one is dealing with a collection of ordinary differential equations of Sturm–Liouville type. The reader may easily verify, using integration by parts, the self-adjointness property of L

$$\int f(x)\, L g(x)\, dx = \int g(x)\, L f(x)\, dx, \tag{14.128}$$

which holds *provided the boundary terms, arising in the integration by parts whereby the derivatives in L are transferred from $g(x)$ to $f(x)$, can be neglected.* This is precisely the point where the appropriate "vanishing at the edge" or "vanishing at infinity" conditions need to be imposed, which we have seen previously are responsible for the emergence of quantization in the eigenvalue spectrum of the equation.

The self-adjointness of L is the essential property ensuring the orthogonality of the eigenfunctions (here assumed to form a discrete set $u_k(x)$, with eigenvalues E_k), by a standard argument. From

$$L u_k(x) + \rho(x)\, E_k\, u_k(x) = 0, \tag{14.129}$$

it follows that

$$\int u_l(x)\, L u_k(x)\, dx = -E_k \int \rho(x)\, u_k(x)\, u_l(x)\, dx. \tag{14.130}$$

[67] For a quick review of Sturm–Liouville theory, see Dennery and Krzywicki (1967, Ch. 4, sec. 12).

[68] For example, in the case of the simple harmonic oscillator, with an appropriately rescaled dimensionless variable, $p(x) = \rho(x) = e^{-x^2}$, $q(x) = 0$, and the functions $u_k(x) = H_k(x)$ that satisfy the differential equation $L u_k(x) + \lambda_k \, \rho(x)\, u_k(x) = 0$ are the orthogonal Hermite polynomials, relative to the measure $\rho(x)$, with eigenvalues $\lambda_k = 2k = 0, 2, 4, \ldots$. The actual wave functions $\psi_k(x)$ that satisfy the differential equation $\psi_k'' + (1 + \lambda_k - x^2)\psi_k = 0$ (cf. Eq. (14.112)) are $\psi_k = e^{-(x^2/2)}\, H_k(x)$.

From

$$L\,u_l(x) + \rho(x)\,E_l\,u_l(x) = 0, \tag{14.131}$$

it likewise follows that

$$\int u_k(x)\,L\,u_l(x)\,dx = -E_l \int \rho(x)\,u_l(x)\,u_k(x)\,dx. \tag{14.132}$$

Subtracting Eq. (14.130) from Eq. (14.132) and using that L is self-adjoint (see Eq. (14.128)), one finds

$$(E_k - E_l) \int \rho(x)\,u_k(x)\,u_l(x)\,dx = 0. \tag{14.133}$$

If the spectrum in non-degenerate (i.e. $k \neq l$ implies $E_k - E_l \neq 0$),

$$\int \rho(x)\,u_k(x)\,u_l(x)\,dx = 0, \quad k \neq l, \tag{14.134}$$

that is, the eigenfunctions $u_k(x)$ form an orthogonal set with respect to the measure $\rho(x)$. Multiplying them by appropriate constants, we can ensure that they are normalized (i.e., for all k, $\int \rho(x)\,u_k^2(x)\,dx = 1$). They then form an orthonormal set with respect to the measure $\rho(x)$:

$$\int \rho(x)\,u_k(x)\,u_l(x)\,dx = \delta_{kl}. \tag{14.135}$$

Now we suppose that a small perturbation $-\lambda\,r(x)$ is added to the operator L. The dimensionless parameter λ is a useful tool for keeping track of the order of approximation, as we imagine formally expanding both the eigenvalues and eigenfunctions of the perturbed differential equation in powers of λ. In this paper, Schrödinger (1926f) only considered perturbations to first order in the small quantity λ:[69]

$$\tilde{E}_k = E_k + \lambda\,\epsilon_k, \tag{14.136}$$

$$\tilde{u}_k(x) = u_k(x) + \lambda\,v_k(x). \tag{14.137}$$

[69] Schrödinger uses the notation E_k^* and u_k^* for the perturbed eigenvalues and eigenfunctions, respectively. As the equations and the wave functions are real throughout—Schrödinger has at this point yet to come to terms with the intrinsic complexity of the theory (still viewing the wave function as underlying a physical field which, like the fields of electromagnetism, should ultimately be a real quantity)—there is no possibility of confusion for him with the operation of complex conjugation. We prefer to reserve the asterisk for complex conjugation, and instead employ a tilde to indicate the effect of a small perturbation.

Inserting $L - \lambda r$ for L, $\tilde{u}_k(x)$ for $u_k(x)$, and $\tilde{E}_k$ for E_k in the basic eigenvalue equation Eq. (14.126) and suppressing the x-dependence of u_k, v_k, ρ, and r for a moment, we have

$$(L - \lambda r)\left(u_k + \lambda v_k\right) + \left(E_k + \lambda \epsilon_k\right) \rho \left(u_k + \lambda v_k\right) = 0. \tag{14.138}$$

Using that u_k is a solution of Eq. (14.126) and collecting terms of order λ (and restoring the x-dependence of u_k, v_k, ρ, and r), we arrive at

$$L v_k(x) + \rho(x)\, E_k\, v_k(x) = \Big(r(x) - \epsilon_k\, \rho(x)\Big)\, u_k(x), \tag{14.139}$$

which is an inhomogeneous differential equation for the as-yet unknown perturbation $v_k(x)$ of the k eigenfunction $u_k(x)$ (which is presumed known explicitly by solution of the unperturbed problem). Multiplying both sides by $u_k(x)$ and integrating, one has

$$\int u_k(x)\Big(L v_k(x) + \rho(x)\, E_k\, v_k(x)\Big)\, dx = \int \Big(r(x) - \epsilon_k\, \rho(x)\Big)\, u_k^2(x)\, dx. \tag{14.140}$$

Using that L is self-adjoint (see Eq. (14.128)), we can rewrite the left-hand side as:

$$\int v_k(x)\Big(L u_k(x) + \rho(x)\, E_k\, u_k(x)\Big)\, dx, \tag{14.141}$$

which vanishes since $u_k(x)$ is a solution of Eq. (14.126) for the eigenvalue E_k. Eq. (14.140) thus reduces to

$$0 = \int r(x)\, u_k^2(x)\, dx - \epsilon_k \int \rho(x)\, u_k^2(x)\, dx. \tag{14.142}$$

Since the u_k form an orthonormal set (see Eq. (14.135)), the integral in the second term equals 1 and we arrive at

$$\epsilon_k = \int r(x)\, u_k^2(x)\, dx. \tag{14.143}$$

This then gives us the desired result for the first-order perturbation in the eigenvalue E_k.

As Schrödinger puts it,

> This simple formula expresses the perturbation of the eigenvalue (of first order) in terms of the perturbing function $r(x)$ and the unperturbed eigenfunction $u_k(x)$. If we consider that the eigenvalue of our problem signifies mechanical energy or is analogous to it, and that the eigenfunction u_k is comparable to "motion with energy E_k", then we see in (7″) [our Eq. (14.143)] the complete parallel to the well-known theorem in the perturbation theory of classical mechanics, viz. the perturbation of the energy, to a first approximation, is equal to the perturbing function, averaged over the unperturbed motion (Schrödinger 1926f, pp. 66–67).

Given that the connection between wave and matrix mechanics had already been established by Schrödinger in the equivalence paper discussed in Section 14.5—submitted for publication almost two months earlier—it should have been clear to Schrödinger that this result coincides exactly with the matrix mechanical result $W_1(n) = H_1(nn)$ (see Eq. (12.157)) in Born, Heisenberg, and Jordan 1926). He does not however comment on this correspondence (and does not cite the Three-Man-Paper) in this third paper in the series on quantization as an eigenvalue problem. The second section of Schrödinger's (1926f) paper, entitled "Several independent variables (partial differential equation)," examines the treatment of eigenvalue problems where the operator L acts in a multi-dimensional configuration space. Such problems frequently lead to the appearance of *degeneracy*, that is, the existence of several linearly independent eigenfunctions associated with a single eigenvalue E_k. Schrödinger discusses in detail the subtleties which arise in implementing perturbation theory when degeneracy is present. Again, the results are equivalent (given the explicit connections uncovered in the meantime to the matrix mechanical method) to those obtained in the Three-Man-Paper for degenerate systems. The applications made of this formalism to (sec. 3) the energy shifts in the Stark effect, and (sec. 4) to the calculation of intensities and polarizations of the spectral lines split in the Stark effect, are discussed in Chapter 15 (cf. Section 15.3.2), as part of our review of the reinterpretation and resolution of the high points (both successes and failures) of the old quantum theory in the new (and still at this point dualistic) formalism.

14.4.4 Quantization as an eigenvalue problem. Part IV

The fourth, and final, installment of "Quantization as an eigenvalue problem" (Schrödinger 1926h), received by *Annalen der Physik* on June 23, 1926, is of great conceptual and historical importance. It introduces the critical element of time dependence that must be included for any mechanical theory that purports to contain an adequate dynamical component. The added formalism is essential in using the wave-mechanical approach to address the problem of optical dispersion, thereby closing the circle with the physics that had led to the matrix mechanics, which Schrödinger by now knew (see Section 14.5) to be intimately connected to his wave theory. There are also interesting discussions of the relativistic wave equation for a single electron including an electromagnetic potential (but not spin), and a final section with intriguing speculations concerning the physical interpretation of the wave function—or, as Schrödinger (1926h, p. 120) calls it, "the field scalar". In a letter to Wien of June 18, 1926 (von Meyenn 2011, pp. 274–275),[70] shortly before submitting this paper, Schrödinger indicated that his primary motivation was in extending the theory to deal with the coupling of charged systems (including those with more than one electron) to the electromagnetic field. The fact that the electromagnetic field in question (for example, in problems of optical dispersion) would typically be time dependent meant that the theory had to now assume a fully

[70] Quoted and discussed by Mehra and Rechenberg (1987, p. 620).

dynamical form, and in particular be able to deal with non-conservative systems where the Hamiltonian was not a constant of the motion. As he noted

> [W]e have always postulated up to now that the potential energy V is a pure function of the co-ordinates and does *not* depend explicitly on the time. There arises, however, an urgent need for the extension of the theory to *non-conservative* systems, because it is only in that way that we can study the behavior of a system under the influence of prescribed external forces, e.g., a light wave, or a strange atom flying past (Schrödinger 1926h, p. 103).

Schrödinger had in fact started already in December of 1925 from a wave function of both space and time. This is the wave function in Eq. (14.5), which satisfies a wave equation of standard form $\Delta\psi = (1/V^2)\partial^2\psi/\partial t^2$ (cf. Eq. (14.6)). In such a wave equation the time dependence is always expressed through a second derivative with respect to time. For a conservative system where the total energy E is fixed, Schrödinger writes the time dependence as[71]

$$\psi \simeq \text{real part of } \left(e^{\pm 2\pi i Et/h}\right), \tag{14.144}$$

so that the wave function ψ necessarily satisfies

$$\frac{\partial^2\psi}{\partial t^2} = -\frac{4\pi^2 E^2}{h^2}\psi. \tag{14.145}$$

In order to arrive at a more general equation—the "real wave equation," as Schrödinger (1926h, p. 103) puts it—for the wave function $\psi(\vec{r}, t)$, capable of dealing with situations where the energy is not conserved, we must clearly eliminate the energy E from Eq. (14.145). With the help of the time-independent (non-relativistic) equation used in the preceding papers (cf. Eq. (14.111) with the mass m stubbornly set to unity throughout this fourth and final installment), which can be written as

$$\left(\nabla^2 - \frac{8\pi^2}{h^2}V\right)\psi = -\frac{8\pi^2}{h^2}E\psi, \tag{14.146}$$

the energy factor in Eq. (14.145) can be eliminated in favor of a fourth-order differential operator,

$$\left(\nabla^2 - \frac{8\pi^2}{h^2}V\right)^2\psi = \frac{64\pi^4 E^2}{h^4}\psi = -\frac{16\pi^2}{h^2}\frac{\partial^2\psi}{\partial t^2}. \tag{14.147}$$

The question arises whether this equation, derived in the context of conservative systems, can be assumed valid for non-conservative systems as well (for example, if the

[71] The insistence on taking the real part of the complex exponential indicates Schrödinger's continuing interpretation of the wave function as a physical, ultimately observable, field. However, he will shortly be forced to the acceptance of an intrinsically complex wave function. Moreover, his discussion of the physical interpretation of the wave function in the final section suggests a significant evolution in his thought.

potential energy V depends on time as well as space). Schrödinger apparently had tried to apply this fourth-order equation directly to such systems (probably the optical dispersion problem treated in the next section), and had encountered problems "which seem to arise from the term in $\partial V/\partial t$" (Schrödinger 1926h, p. 104). In fact, a much simpler equation, second order in space derivatives but only first order in time, can be obtained if the energy is eliminated via

$$\frac{\partial \psi}{\partial t} = \pm \frac{2\pi i}{h} E\psi, \tag{14.148}$$

which follows from Eq. (14.144). Substituting the expression given by Eq. (14.148) for $E\psi$ into Eq. (14.146), we arrive at

$$\nabla^2\psi - \frac{8\pi^2}{h^2} V\psi = -\frac{8\pi^2}{h^2}\left(\pm \frac{h}{2\pi i}\frac{\partial \psi}{\partial t}\right) = \pm \frac{4\pi i}{h}\frac{\partial \psi}{\partial t}, \tag{14.149}$$

or, rearranging terms,

$$\nabla^2\psi - \frac{8\pi^2}{h^2} V\psi \mp \frac{4\pi i}{h}\frac{\partial \psi}{\partial t} = 0. \tag{14.150}$$

The presence of an explicit factor of i here means that the solutions are intrinsically complex. Hence, one cannot simply take the real (or imaginary) part of a solution of this equation and obtain a new solution. Instead, the two equations presented here connect the complex solution ψ obtained with one choice of sign with the complex conjugate solution $\psi^\star$ ($\overline{\psi}$ in Schrödinger's notation) satisfied by the equation with the other sign. The modern convention is to take the wave function ψ to satisfy Eq. (14.150) with the positive sign before the imaginary term. Schrödinger had at this point found that the perturbation theory developed from this equation reproduced perfectly the dispersion theory results that were by now widely accepted. However, one can easily see that this equation differs from the fourth-order equation (by terms involving $\partial V/\partial t$), so that in the non-conservative case the two are decidedly not equivalent. For the rest of this paper (and indeed, from this point on generally), Schrödinger exclusively employs the much simpler Eq. (14.150) as the general expression for the time development of a quantum system in wave mechanics. The sign choice he adopts (negative sign before the imaginary term)—which we follow here, to facilitate comparison with the original paper—means however that his wave function ψ and conjugate wave function $\psi^\star$ are interchanged with respect to modern usage.[72]

In section 2, Schrödinger (1926h, pp. 104–110) shows how perturbations that explicitly contain the time may be dealt with, using his new dynamical equation Eq. (14.150). This will allow him to make contact with the Kramers dispersion theory—the source

[72] Thus, for example, the time dependence of the wave function involves factors $e^{+2\pi i E_k t/h}$ (cf. Schrödinger 1926h, p. 106, Eq. (10)).

from which the matrix mechanics of his competitors in Göttingen had emerged. We simplify the algebra slightly (e.g., by using $\hbar = h/2\pi$ where convenient, and by representing the Hamiltonian differential operator symbolically where possible).

The calculation is performed rather generally, as the coordinate x for Schrödinger can represent the full configuration space of a multi-electron system (i.e., $x = (\vec{r}_1, \vec{r}_2, \ldots, \vec{r}_n)$), but we will imagine that monochromatic light of frequency ν is incident on a single optically active electron of charge e. The unperturbed system is subject to a Hamiltonian

$$H_0 = -\frac{\hbar^2}{2m}\nabla^2 + V_0(\vec{r}), \tag{14.151}$$

and the full Hamiltonian is

$$H = H_0 + A(\vec{r})\cos(2\pi\nu t), \tag{14.152}$$

with

$$A(\vec{r}) = -e\Big(\vec{\mathcal{E}} \cdot \vec{r}\Big) = -e\left(\mathcal{E}\vec{\epsilon}_i \cdot \vec{r}\right), \tag{14.153}$$

where $\vec{\mathcal{E}} = \mathcal{E}\vec{\epsilon}_i$ is the electric field amplitude and $\vec{\epsilon}_i$ is a unit polarization vector for the incident plane polarized light wave. It is presumed that the unperturbed problem has been completely solved, giving a complete set of eigenvalues E_k and orthonormal eigenfunctions $u_k(\vec{r})$ (cf. Eq. (14.135)):

$$H_0\, u_k(\vec{r}) = E_k\, u_k(\vec{r}), \quad \text{with} \int \rho(\vec{r})\, u_k(\vec{r})\, u_l(\vec{r})\, d\vec{r} = \delta_{kl}. \tag{14.154}$$

The dynamical wave equation Eq. (14.150) can be written succinctly as

$$H(t)\psi = -i\hbar\frac{\partial\psi}{\partial t}, \tag{14.155}$$

which, for our problem, becomes

$$\Big(H_0 + A(\vec{r})\cos(2\pi\nu t)\Big)\psi(\vec{r}, t) = -i\hbar\frac{\partial\psi}{\partial t}. \tag{14.156}$$

We assume that the incident light impinges on an atomic electron in a definite state u_k (for normal dispersion, as opposed to Raman dispersion, this would be the ground state), and we write the full solution,

$$\psi(\vec{r}, t) = u_k(\vec{r})e^{iE_k t/\hbar} + w(\vec{r}, t), \tag{14.157}$$

as a sum of the unperturbed solution plus a perturbation w, which we calculate to first order in the applied field strength $\mathcal{E}$. Inserting Eq. (14.157) in Eq. (14.156), neglecting terms of second order (i.e., products of w and A), and suppressing the dependence of u_k and w on $\vec{r}$, we find

$$H_0\, u_k\, e^{iE_k t/\hbar} + H_0 w + A\cos\left(2\pi\nu t\right) u_k\, e^{iE_k t/\hbar} \tag{14.158}$$

for the left-hand side and

$$-\,i\hbar\frac{\partial}{\partial t}\left(u_k\, e^{iE_k t/\hbar} + w\right) = E_k\, u_k\, e^{iE_k t/\hbar} - i\hbar\dot{w} \tag{14.159}$$

for the right-hand side. Since u_k is an eigenfunction of H_0 with eigenvalue E_k, the first term in Eq. (14.158) cancels against the first term on the right-hand side of Eq. (14.159). Regrouping the remaining terms and using Euler's formula for $\cos\left(2\pi\nu t\right)$, we find

$$H_0 w + i\hbar\dot{w} = -\frac{1}{2}A\, u_k\left(e^{2\pi i(E_k+h\nu)t/h} + e^{2\pi i(E_k-h\nu)t/h}\right). \tag{14.160}$$

This inhomogeneous equation has solutions for $w(\vec{r}, t)$, which are the sum of a homogenous solution of the form $w = C\, u_k\, e^{iE_k t/\hbar}$, which can be absorbed into the unperturbed component in Eq. (14.157), and the part relevant to our enquiry, namely the linear response to the incident light, and coherent with it, which can clearly be written as the sum of two parts, corresponding to the two exponential time behaviors in Eq. (14.160):

$$w_{\text{coh}} = w_+(\vec{r})\, e^{2\pi i(E_k+h\nu)t/h} + w_-(\vec{r})\, e^{2\pi i(E_k-h\nu)t/h}. \tag{14.161}$$

Inserting this Ansatz into Eq. (14.160), one finds

$$\left(H_0 - (E_k \pm h\nu)\right) w_\pm(\vec{r}) = -\frac{1}{2}A(\vec{r})\, u_k(\vec{r}). \tag{14.162}$$

Both $w_\pm$ and $A\, u_k$ may be expanded in a complete set of eigenfunctions (the possible presence of a non-trivial measure factor ρ is suppressed here and in the following for simplicity). We write

$$w_\pm(\vec{r}) = \sum_n c_n^\pm\, u_n(\vec{r}), \quad A(\vec{r})\, u_k(\vec{r}) = \sum_n a'_{kn}\, u_n(\vec{r}), \tag{14.163}$$

with

$$a'_{kn} = \int A(\vec{r})\, u_k(\vec{r})\, u_n(\vec{r})\, d\vec{r} = -e\,\mathcal{E}\, a_{kn}, \tag{14.164}$$

where we used that $A(\vec{r}) = -e\,\mathcal{E}\left(\vec{\epsilon}_i \cdot \vec{r}\right)$ (see Eq. (14.153)) and introduced

$$a_{kn} \equiv \int u_k(\vec{r})\, u_n(\vec{r})\left(\vec{\epsilon}_i \cdot \vec{r}\right) d\vec{r}. \tag{14.165}$$

Inserting the expansions in Eq. (14.163) in Eq. (14.162) and using that u_n is an eigenfunction of H_0 with eigenvalue E_n, we find

$$\sum_n \Big(E_n - (E_k \pm h\nu)\Big)\, c_n^{\pm}\, u_n(\vec{r}) = -\frac{1}{2}\sum_n a'_{kn}\, u_n(\vec{r}). \tag{14.166}$$

Comparing coefficients of u_n on the left- and right-hand sides, we conclude that

$$c_n^{\pm} = \frac{1}{2}\frac{a'_{kn}}{E_k - E_n \pm h\nu}. \tag{14.167}$$

We now have the desired solution for the wave function including the terms induced (to first order) by the incident light wave (cf. Schrödinger 1926h, p. 107, Eq. (16)),

$$\psi(\vec{r}, t) = u_k(\vec{r})e^{iE_k t/\hbar} + \frac{1}{2}\sum_n a'_{kn}u_n(\vec{r})(\ldots), \tag{14.168}$$

where the expression between parentheses is given by

$$(\ldots) = \frac{e^{i(E_k+h\nu)t/\hbar}}{E_k - E_n + h\nu} + \frac{e^{i(E_k-h\nu)t/\hbar}}{E_k - E_n - h\nu}. \tag{14.169}$$

Schrödinger still regards the solution Eq. (14.168), which is clearly a complex quantity, as an intermediate quantity:

> The real or the imaginary part of (16) [our Eqs. (14.168)–(14.169)] can be considered as the *real* solution. In the following, however, we will operate with the complex solution itself (Schrödinger 1926h, p. 108).

The reason for this concession to the intrinsic complexity of the wave function—inevitable once the fourth-order real equation Eq. (14.147) had been factored into two complex equations, which were now to be regarded as separately determining the time evolution—becomes clear when we consider the following paragraph, where an (at first sight to modern eyes) innocent assumption, essential for further progress in the dispersion problem, is made:

> To see the significance that our result [i.e., Eqs. (14.168)–(14.169)] has in the theory of dispersion, we must examine the radiation arising from the simultaneous existence of the excited forced vibrations [the terms in Eq. (14.169)] and the free vibration already present [first term on the right-hand side of Eq. (14.168)]. For this purpose, we form, following the method we [at this point, a footnote is inserted, on which more below] have always adopted up to now [*bisher*]—a criticism follows in section 7 [again, more on this criticism below]—the product of the wave function (16) [our Eqs. (14.168)–(14.169)] and its conjugate, i.e., the norm [actually, norm squared] of the complex wave function ψ (Schrödinger 1926h, p. 108).

Schrödinger now interprets the quantity $\psi\psi^\star$ (once multiplied by the electric charge e), which we are here taking for a system of a single electron, and which is therefore a function simply of space and time coordinates $\vec{r}$ and t, as the *electrical charge density*. Using Eq. (14.168), we can schematically write $\psi\psi^\star$ as

$$\Big(u_k e^{iE_k t/\hbar} + \tfrac{1}{2}\textstyle\sum_n a'_{kn} u_n (\ldots)\Big)\Big(u_k e^{-iE_k t/\hbar} + \tfrac{1}{2}\textstyle\sum_n a'_{kn} u_n (\ldots)^\star\Big), \tag{14.170}$$

where $(\ldots)$ is given by Eq. (14.169). We can thus write

$$\psi\psi^\star = u_k(\vec{r})^2 + \delta(\psi\psi^\star), \tag{14.171}$$

with $\delta(\psi\psi^\star)$ the shift due to the incident light wave. Neglecting terms of second order in the perturbing electric field (or, equivalently, in the matrix elements a'_{kn}), we see from Eq. (14.170) that $\delta(\psi\psi^\star)$ is the sum of two terms that are each other's complex conjugate, which means that $\delta(\psi\psi^\star)$ is twice the real part of one of these. Schematically:

$$\delta(\psi\psi^\star) = \mathrm{Re}\Big(u_k e^{iE_k t/\hbar} \sum_n a'_{kn} u_n (\ldots)^\star\Big). \tag{14.172}$$

Inserting Eq. (14.169) for $(\ldots)$, we find that

$$\begin{aligned}\delta(\psi\psi^\star) &= \mathrm{Re}\left(u_k \sum_n a'_{kn}\, u_n \left(\frac{e^{-2\pi i\nu t}}{E_k - E_n + h\nu} + \frac{e^{2\pi i\nu t}}{E_k - E_n - h\nu}\right)\right)\\ &= \sum_n u_k\, a'_{kn}\, u_n \cos(2\pi\nu t) \frac{2(E_k - E_n)}{(E_k - E_n)^2 - h^2\nu^2}.\end{aligned} \tag{14.173}$$

Inserting this expression for $\delta(\psi\psi^\star)$ into Eq. (14.171), we arrive at the equation given by Schrödinger (1926h, p. 108, Eq. (17)):

$$\psi\psi^\star = u_k(\vec{r})^2 + 2\cos(2\pi\nu t) \sum_n \frac{(E_k - E_n)\, a'_{kn}\, u_k(\vec{r})\, u_n(\vec{r})}{(E_k - E_n)^2 - h^2\nu^2}. \tag{14.174}$$

Schrödinger's interpretation of $\psi\psi^\star$, indispensable in establishing contact with Kramers' seminal formula, clearly indicates the first point at which he grudgingly accepts the need for complex ingredients in his theory: the electrodynamically required (and directly physical) concept of charge density, shortly to be converted into a time-dependent dipole moment, is seen to depend on the sum of the squares of the real and imaginary part of a complex wave function determined by an ineluctably complex wave equation, Eq. (14.156).

The assertion that this Ansatz for the charge density is the one he has used "up to now" (*bisher*) is a bit disingenuous. In the second paper, incorrectly, as Schrödinger now admits, the real part of the quantity $\psi\, \partial\psi^\star/\partial t$ had been proposed as the charge density

(Schrödinger 1926d, p. 60), while in the third paper, the intensity of the lines split in the Stark effect is discussed entirely in terms of real energy eigenfunctions of the Hamiltonian for the hydrogen atom.[73] At the conclusion of this fourth paper, Schrödinger would again complain about the need for the use of complex functions:

> Meanwhile, there is no doubt a certain awkwardness in the use of a *complex* wave function. If it were unavoidable *in principle*, and not merely a facilitation of the calculation, this would mean that there are in principle *two* wave functions, which must be used *together* in order to obtain information on the state of the system. This somewhat disagreeable inference allows [i.e., is compatible with], I believe, the much more agreeable interpretation that the state of the system is given by a real function and its time derivative (Schrödinger 1926h, p. 123).

It is clear that Schrödinger is still clinging to the hope that, as in classical mechanics and classical electrodynamics, the dynamical evolution of systems in the quantum theory will be determined by real equations of second order in the time variable, the initial conditions for which will require the specification of a real function and its time derivative at the starting time. He admits, however, that his attempt to provide such a dynamics, as embodied in the fourth-order equation Eq. (14.147), is unsatisfactory, as he had been unable to find an appropriate generalization for the non-conservative case (i.e. for a Hamiltonian with explicit time dependence). The first-order character of quantum dynamics, requiring only the specification of the state (and not also a time derivative) at the start to determine the future evolution of the system, and the inevitable presence of complex structure, would soon be accepted broadly by the community of quantum theorists.

Returning now to the problem of dispersion, we construct, following Schrödinger, the dipole moment due to the perturbation $A(\vec{r})\cos(2\pi\nu t)$ in Eq. (14.152). We may compute the component of this induced dipole moment in any direction $\vec{\epsilon}_f$ by multiplying the electron charge density perturbation $-e\,\delta(\psi\psi^\star)$ by $\vec{\epsilon}_f\cdot\vec{r}$ and integrating over space. In fact, for the purposes of comparison with the Kramers dispersion formula, we are interested in the polarization along the direction of the incident (plane-polarized) electric light vector, so we may take $\vec{\epsilon}_f = \vec{\epsilon}_i$. One thus obtains[74]

$$\begin{aligned}\int -e\,\delta(\psi\psi^\star)\,\vec{\epsilon}_i\cdot\vec{r}\,d\vec{r} &= -2e\cos(2\pi\nu t)\int\sum_n \frac{(E_k-E_n)\,a'_{kn}\,u_k\,u_n}{(E_k-E_n)^2-h^2\nu^2}\,\vec{\epsilon}_i\cdot\vec{r}\,d\vec{r}\\ &= 2e^2\mathcal{E}\cos(2\pi\nu t)\sum_n\frac{(E_k-E_n)(a_{kn})^2}{(E_k-E_n)^2-h^2\nu^2},\end{aligned}\qquad(14.175)$$

[73] For example, states of well-defined magnetic quantum number m, which would unavoidably have complex wave functions proportional to $e^{im\varphi}$, are avoided: instead of the usual spherical harmonics, Schrödinger uses only wave functions with real azimuthal angular dependence $\cos(m\varphi)$ or $\sin(m\varphi)$ (cf. Schrödinger 1926f, p. 70).

[74] Schrödinger considers a somewhat more general case, with $\vec{\epsilon} = \hat{z}\vec{\epsilon}_f = \hat{y}$, which would be necessary if one were interested in elastic light scattering, for example.

where in the second step we used that $a'_{kn} = -e\mathcal{E}a_{kn}$ and $\int u_k u_n (\vec{\epsilon}_i \cdot \vec{r})\, d\vec{r} = a_{kn}$ (see Eqs. (14.164)–(14.165)).

The corresponding result in matrix mechanics can be obtained from Eq. (12.187), which leads directly to the Kramers dispersion formula. In the notation of the Three-Man-Paper of Born, Heisenberg, and Jordan (1926), our matrix element a_{kn} would be written $q_0(kn)$, while the differences of the lowest-order energy eigenvalues are re-expressed in terms of frequencies, viz. $E_k - E_n = h\nu_0(kn)$. The first-order shift in the dipole moment of state k is

$$e\,q_1(kk;t) = e\,q_1(kk,+1)e^{2\pi i\nu t} + e\,q_1(kk,-1)e^{-2\pi i\nu t}. \tag{14.176}$$

With some minor changes of notation,[75] we find, using Eq. (12.187), that the matrix-mechanics result may be rewritten as

$$e\,q_1(kk;t) = 2\frac{e^2\mathcal{E}}{h}\cos(2\pi\nu t)\sum_n \frac{\nu_0(kn)|q_0(kn)|^2}{\nu_0(kn)^2 - \nu^2}. \tag{14.177}$$

Setting $a_{kn} = q_0(kn)$, we see that this result agrees with Eq. (14.175). The identification of a_{kn} with $q_0(kn)$ is exactly the mapping between wave and matrix mechanics, which Schrödinger (1926e) had already established in the paper discussed in Section 14.5 (and which had already appeared in *Annalen* early May)—with one telling difference. The complex character of the matrix elements recognized very early by the Göttingen group—an inheritance from the examination of inelastic light scattering in the Kramers–Heisenberg paper, discussed in detail in Chapter 10—which manifests itself in the appearance of the absolute square $|q_0(kn)|^2$, is nowhere to be seen in the Schrödinger formula, which is written entirely in terms of real matrix elements obtained by integrations involving only real-energy eigenfunctions u_k. It is evident that at this point Schrödinger still sees no need for the appearance of complex wave functions in any *essential* way.

In sections 3–6 of this final paper in the series on quantization as an eigenvalue problem, Schrödinger (1926h, pp. 110–120) deals with a number of ancillary issues raised by the possibility of treating time dependence in his new theory. These are: (a) the extension of dispersion theory for normal (ground state) atoms to the inelastic case (i.e., Raman scattering, treated in Kramers and Heisenberg 1925; see also Section 10.5) (b) degenerate systems, where multiple solutions can be found with the same eigenvalue; (c) the treatment of systems with a continuous, as well as discrete, spectrum; (d) (sec. 4) resonant light scattering; (e) (sec. 5) non-periodic time-dependent perturbations; and (f) inclusion of a electromagnetic four-potential, in a relativistically invariant fashion (thereby recovering the time-dependent version of the relativistic wave equation Eq. (14.11) which he had previously analysed, but now with a magnetic vector potential added). The treatment of these matters is mostly qualitative, with the exception of the

[75] Our ν is ν_0 in the Three-Man-Paper and the summation index k in the Three-Man-Paper is changed to n in Schrödinger's result.

generalization of the dispersion formula to include a continuous spectrum (via an integral) in the sum over states in Eq. (14.168), and the relativistic wave equation of section 6 (now called the *Klein–Gordon* equation).

The discussion of Raman scattering lays great emphasis on the simultaneous presence of multiple energy eigenstates in the temporal development of the wave function, already apparent in the first-order solution Eqs. (14.168)–(14.169).[76] Considerations based on a specific time development of an atomic system are, he claims, simply impossible in the matrix mechanical formalism of Heisenberg *et al.*:

> As far as I see, the above-mentioned dispersion theory of Heisenberg, Born and Jordan does *not* allow of such reflections as we have just made, in spite of its great formal similarity to the present one. For it only considers *one* way in which the atom reacts to incident radiation. It conceives of the atom as a timeless entity, and up till now is not able to express in its language the undoubted fact that the atom can be in *different* states at different times, and thus, as has been proved, reacts in different ways to incident radiation (Schrödinger 1926h, pp. 111–112).

This paragraph is followed by a reference (in a footnote) to the concluding passage of a paper by Heisenberg in *Mathematische Annalen*, submitted in December 1925, in which he summarized, for more mathematically inclined readers, the basic tenets of matrix mechanics.

> A considerable difficulty [of matrix mechanics] consists of the fact that time plays a different role, and is formally handled differently, from the spatial coordinates. The formal character of the time coordinate in the mathematical structure of the theory is particularly clearly a consequence of the fact that the temporal development of an event has up to now no definite sense in the theory, and that the concept of earlier or later can hardly be defined exactly (Heisenberg 1926a, p. 705).

Although Schrödinger here leaves the incorrect impression that the Göttingen approach was incapable of handling correctly the calculation of inelastic light scattering—the dispersion-theoretic treatment of which can already be found in the paper by Kramers and Heisenberg (1925)—he has certainly put his finger here on a defect of the matrix-mechanical approach, at least as it existed in mid-1926. The modern student of quantum mechanics will recognize in the phrase “atom as a timeless entity” an important feature of the representation of quantum dynamics in what is now called “Heisenberg representation”, in which all the time dependence is transferred to the operators, with the state vector fixed for all time. However, the absence of the very concept of a state (later, a vector in Hilbert space) in the initial formulations of matrix mechanics clearly makes an unambiguous description of the dynamics of more complicated processes very difficult, as Heisenberg himself confesses, in the passage quoted above.

[76] Schrödinger’s discussion necessarily comes in advance of Dirac’s (1927b) correct treatment of the interaction of photons and atoms that would appear in the following year. In particular, some of the statements he makes, asserting the necessity for the simultaneous presence of the initial and final states in Raman scattering, are incorrect.

In the final section of this fourth installment, with the title "On the physical significance of the field scalar," Schrödinger (1926h, sec. 7, pp. 120–123) struggles with the conceptual content of the new theory, in particular, the physical interpretation of the wave function. The very term used for this quantity—"field scalar"—indicates the extent to which Schrödinger is still thinking in analogy to classical field theory—specifically, Maxwellian electrodynamics. The *direct* analogy with the physical fields of electrodynamics has, admittedly, been shaken by two disturbing features of wave mechanics:

1. The wave function, in harmony with its genetic descent from the Hamilton–Jacobi function, is not in general, like normal field quantities, a function on the physical space–time in which actual processes evolve. Instead, if there are N particles, the wave function becomes a function of $3N$ spatial and one time variable, for example, $\psi(\vec{r}_1, \vec{r}_2, \ldots, \vec{r}_N, t)$. If the first particle has charge e_1, the second e_2, and so on, the interpretation previously advanced (in the discussion of dispersion in section 2) had assigned the quantity

$$\rho_1(\vec{r}) = e_1 \int \psi\psi^\star(\vec{r}, \vec{r}_2, \ldots, \vec{r}_N, , t)\, d\vec{r}_2 \ldots d\vec{r}_N \tag{14.178}$$

   to the charge density due to the first particle, with similar expressions—integrating the absolute square of the wave function $\psi\psi^\star$ over the position of all the other particles, and multiplying by the charge of the particle in question—for the charge density due to particles 2, 3, etc. This rather complicated prescription meant that Schrödinger was forced to adopt a considerably more abstract interpretation of the wave function than he had initially hoped.
2. The first-order (in time derivatives) character of the time evolution equation, essential for the derivation of charge conservation and for the derivation of the Kramers dispersion formula, in agreement with the results of matrix mechanics, comes with a high price: the wave function is now inescapably complex, with the real and imaginary parts intertwined in an intimate fashion. This clearly increases the difficulty of making a direct identification of the wave function with a physically accessible field.

The new conception required by these peculiar (to classical thinking) features is clearly described in the second paragraph of this concluding section:

> $\psi\psi^\star$ is a kind of *weight-function* in the system's configuration space [i.e., the space $\vec{r}_1, \vec{r}_2, \ldots, \vec{r}_N$]. The *wave-mechanical* configuration of the system is a *superposition* of many, strictly speaking of *all*, point-mechanical configurations kinematically possible. Thus each point-mechanical configuration contributes to the true wave-mechanical configuration with a certain *weight*, which is given precisely by $\psi\psi^\star$. If we like paradoxes, we may say that the system exists, as it were, simultaneously in all the positions kinematically imaginable, but not "equally strongly" in all (Schrödinger 1926h, p. 120).

If we were to replace the word "weight" with "probability" we would have exactly the interpretation shortly to be advanced by Born (1926a, 1926b; see Section 16.1), and which has survived to the present time. However, Schrödinger strongly resisted the attachment of a stochastic component to the framework of quantum mechanics, so the exact meaning of the "weight" prescribed by the wave function should probably be restricted at this point to the specific practical application that Schrödinger makes of it—as a route to the calculation of the "very real electrodynamically effective fluctuations of the electric space-density" (Schrödinger 1926h, p. 120). This interpretation was just the one needed for the results he had obtained in dispersion theory.

The interpretation of $\psi\psi^\star$ (times the electric charge) as a charge density is directly connected, via charge conservation, to the invariance in time of the total weight of all the "kinematically imaginable" configurations available to each particle in the system. We present Schrödinger's results for the simplified case where there is only a single particle present, described in Cartesian coordinates by a wave function $\psi(\vec{r}, t)$, satisfying the wave equation given in Eq. (14.150)

$$\frac{\partial\psi(\vec{r}, t)}{\partial t} = \frac{h}{4\pi i m}\left(\nabla^2 - \frac{8\pi^2 m}{h^2}V\right)\psi(\vec{r}, t) \tag{14.179}$$

(where we restored the factor m for the particle mass, previously set to unity), and its complex conjugate, $\psi^\star(\vec{r}, t)$, similarly satisfying

$$\frac{\partial\psi^\star(\vec{r}, t)}{\partial t} = -\frac{h}{4\pi i m}\left(\nabla^2 - \frac{8\pi^2 m}{h^2}V\right)\psi^\star(\vec{r}, t). \tag{14.180}$$

Multiplying the first of these two equations by $\psi^\star$, the second by ψ, and adding the two, we obtain

$$\frac{\partial}{\partial t}(\psi\psi^\star) = \frac{h}{4\pi i m}\left(\psi^\star\nabla^2\psi - \psi\nabla^2\psi^\star\right) = -\vec{\nabla}\cdot\vec{\mathcal{J}}(\vec{r}, t). \tag{14.181}$$

This is the famous *continuity equation*,[77] which guarantees that we can regard $\psi\psi^\star$ as a density of some "conserved stuff" that flows with a vector current density $\vec{\mathcal{J}}$, defined as

$$\vec{\mathcal{J}} \equiv \frac{ih}{4\pi m}(\psi^\star\vec{\nabla}\psi - \psi\vec{\nabla}\psi^\star). \tag{14.182}$$

In particular, the total amount of the "stuff' represented by $\psi\psi^\star$ is conserved in time, as

$$\frac{d}{dt}\int \psi\psi^\star d\vec{r} = -\int \vec{\nabla}\cdot\vec{\mathcal{J}}(\vec{r}, t)d\vec{r} = 0. \tag{14.183}$$

The final equality is guaranteed if the current vector vanishes sufficiently fast at infinity, as it is the integral of a gradient over all space. The conservation of the normalization

[77] See, for example, Griffiths (1999, Eq. (5.29)).

integral $\int \psi\psi^{\star} d\vec{r}$ in time thus ensures that the wave function, once normalized to unity (say at time zero), preserves its normalization thereafter, as a consequence of the time-dependent Schrödinger equation, and, equally importantly, *the boundary conditions of appropriate vanishing of the wave function at infinity*. Precisely the latter condition, as we saw already in our discussion of the first installment of this four-part paper, was Schrödinger's asserted requirement for the emergence of quantized eigenvalues in the new theory. After a derivation of the continuity equation in the more generalized context of a system of N particles, and allowing for the use of generalized coordinates (rather than the Cartesian ones employed above), Schrödinger closes this final paper in the series with some remarks on the undesirable appearance of complex quantities, which we have already commented on (cf. the discussion following Eq. (14.174)).

14.5 The "equivalence" paper

The series of four papers, "Quantisation as an eigenvalue problem," which Schrödinger produced over the course of the first half of 1926, was interrupted in mid-March, after the first two, by a separate paper, received by *Annalen der Physik* on March 18, entitled "On the relation between the quantum mechanics of Heisenberg, Born and Jordan, and mine" (SchrÖdinger 1926e). In a footnote at the beginning of the paper, Schrödinger, citing De Broglie (1925) and Einstein (1925a, p. 9 ff.) explained that:

> My theory was inspired by L. de Broglie ... and by brief yet infinitely far-seeing remarks by A. Einstein ... I did not at all suspect any relation to Heisenberg's theory at the beginning. I naturally knew about his theory, but was discouraged [*abgeschreckt*] if not repelled [*abgestoßen*] by what appeared to me as very difficult methods of transcendental algebra, and by the want of visualizability [*Anschaulichkeit*] (Schrödinger 1926e, p. 46, quoted, e.g., in Jammer 1966, p. 272).

In the meantime, however, it had become clear to him that in all the problems that had proved amenable to solution by both the matrix-mechanical approach of Heisenberg *et al.* and his own wave theory, the physical results obtained were identical.[78]

In particular, both theories reproduced naturally, indeed, inescapably, the peculiar half-integral quanta that had seemed so mysterious in the old quantum theory—the zero-point energy of the harmonic oscillator, and the half-integral rotational quanta in band spectra arising in three dimensions from the $l(l+1) = (l+1/2)^2 - 1/4$ squared angular momentum eigenvalue.

This agreement in results was in itself a mystery, for as Schrödinger points out, in the two theories,

> starting points, presentations, methods, and in fact the whole mathematical apparatus, seem fundamentally different. In Heisenberg's work the classical continuous variables are

[78] By mid-January, Pauli (1926a), in one of his most virtuosic calculations, had succeeded in deriving the spectrum of the hydrogen atom using matrix-mechanics methods.

replaced by systems of discrete numerical quantities (matrices) ... The authors themselves describe the theory as a "true theory of a discontinuum." On the other hand, wave mechanics shows just the reverse tendency: it is a step from classical point-mechanics towards a *continuum theory* (Schrödinger 1926e, p. 45).

As late as February 22, 1926, Schrödinger wrote to Wien that he could not find the connection between matrix and wave mechanics:

With privy councillor [*Geheimrat*] Sommerfeld I am convinced that there must be a close inner connection. It must lie pretty deep though, because Weyl, who has made a very thorough study of Heisenberg's theory and has developed it further himself and to whom I gave my first manuscript to read [i.e., Schrödinger 1926c], cannot find it. At which point I gave up looking for it myself (von Meyenn 2011, p. 186).

Shortly thereafter, Schrödinger must have hit upon the connection he was looking for. In the paper of March 18, he promises, a "very intimate *inner connection*" between the two theories existed: one could even, "from the formal mathematical standpoint ... speak of the *identity* of the two theories" (Schrödinger 1926e, p. 46).[79] The connection is to be established in two steps:

1. A mapping is established that assigns to each kinematic quantity (function of the coordinates and momenta) a numerical matrix satisfying the formal properties posited in matrix mechanics such as the commutation relations. This requires the choice of a complete orthogonal system of functions having as their domain the $3N$-dimensional configuration space (i.e., the spatial q_k coordinates for a system of N particles). Such a choice is of course not unique—the matrices associated with each kinematic quantity are therefore also not uniquely determined at this stage. This takes up sections 2 and 3 of the paper.
2. If a particular complete orthogonal set of functions is chosen—namely, the eigenfunctions arising from the solution of the Schrödinger wave equation—then the matrices corresponding to position and momentum, together with the matrix (suitably symmetrized) constructed from them representing the Hamiltonian (total energy) of the system, satisfy the algebraic matrix equations that embody the classical Hamiltonian equations of motion, again in agreement with the formulation of the theory in the papers of Born, Heisenberg, and Jordan. In this special case, the matrix representing the Hamiltonian function is diagonal. This is done in section 4 of the paper.

The modern student of quantum mechanics encounters the ideas Schrödinger develops in this paper early on in his or her introductory course on the subject. The canonical coordinates of a system, (q_l, p_l) (coordinates and momenta), are replaced by linear

[79] For a critical evaluation of Schrödinger's claim to have demonstrated the mathematical equivalence of wave and matrix mechanics, see Muller (1997–1999).

operators, $(\mathbf{q}_l, \mathbf{p}_l)$, acting on functions $\psi(q_1, .., q_{3N}) = \psi(q_k)$ in the $3N$-dimensional configuration space:

$$\mathbf{q}_l \, \psi(q_k) \equiv q_l \, \psi(q_k), \tag{14.184}$$

$$\mathbf{p}_l \, \psi(q_k) \equiv K \frac{\partial}{\partial q_l} \psi(q_k), \tag{14.185}$$

where K is a constant with dimensions of action, to ensure dimensional consistency of the second equation. Kinematic functions $F(q_k, p_k)$ must be written as sums of unambiguously ordered products of functions of the coordinates and momenta, given that the order in which a given coordinate, q_l, and the derivative with respect to that coordinate, $\partial/\partial q_l$, occurs affects the result. Of course, in many cases, where the function of interest is a Hamiltonian of the form $H = T(p_k) + V(q_k)$, there are no ordering ambiguities and the corresponding linear operator can be written down immediately.

The association of a matrix with each such linear differential operator requires the introduction of a complete orthogonal system of functions,[80] a concept treated in detail in the first edition of Courant and Hilbert's (1924) *Methods of Mathematical Physics*, which in early 1926 had replaced Schlesinger's (1900) text on differential equations as Schrödinger's *vade mecum* for mathematical questions. The theory of series expansions of arbitrary functions (Courant and Hilbert 1924, Ch. 2) in an orthogonal set of functions took on its most powerful and general form for the solutions of ordinary differential equations of Sturm–Liouville type (Courant and Hilbert 1924, Ch. 5, sec. 9, pp. 238–239), for which an infinite set of normalized orthogonal solutions $u_n(x)$ exist that satisfy (cf. Eq. (14.134)):

$$\int \rho(x) \, u_m(x) \, u_n(x) \, dx = \delta_{nm}, \tag{14.186}$$

where $\rho(x)$ (the "density function") is a positive real function arising from the specific form of the coefficients in the underlying differential equation.[81] Using the single variable x to represent the assembly of coordinate variables q_k, we now associate a matrix $\mathbf{F}_{kl}$ with the kinematic function $F(q_k, p_k)$ by defining[82]

$$\mathbf{F}_{kl} = \int \rho(x) \, u_k(x) \, \big(\mathbf{F} \, u_l(x)\big) \, dx. \tag{14.187}$$

It is obvious from this definition that the matrix corresponding to the sum of two kinematical quantities F and G is the sum of the corresponding matrices. That the mapping

[80] See Section C.5.

[81] For example, for the Hermite equation discussed previously in connection with the harmonic oscillator, $u_n(x) = H_n(x)$ and the density function is $\rho(x) = e^{-x^2}$ (cf. Eq. (14.113)).

[82] We have departed slightly from Schrödinger's notation to emphasize the connection with the matrix-mechanical formalism discussed in Chapter 12. In particular, we use bold type to indicate operators/matrices.

also preserves multiplication is less obvious—indeed, this requires a special additional property of the orthogonal set $u_m(x)$, namely, *completeness* (Courant and Hilbert 1924, p. 36). This basically assures that every wave function (satisfying appropriate boundary conditions) can be expanded in a series $\sum_n c_n u_n(x)$. In other words, there are no "missing" orthogonal functions.

In the differential operator **F** in the integral Eq. (14.187), thought of as acting to the right on the function $u_l(x)$, all occurrences of the momentum p_m are replaced by the differential operator $K\partial/\partial q_m$ (acting to the right).[83] Schrödinger is careful to emphasize the need for maintaining the order with which coordinates and momenta are arranged in the algebraic expressions for a given kinematical quantity—here, the trail had already been blazed by Born and Jordan, and much of what Schrödinger says in this connection can be traced back directly to the paper by Born and Jordan (1925b) six months earlier.

In order to establish contact with the matrix-mechanics program of Born, Heisenberg, and Jordan (1926), the special kinematical quantity $p_l q_l - q_l p_l$ must be associated with the matrix $(\hbar/i)\,1$, where 1 is the unit matrix. Evidently, acting on any function $u_k(x)$ (recall, x is shorthand for $q_1, q_2, \ldots$), the operator corresponding to $p_l q_l - q_l p_l$ must produce

$$(\mathbf{p}_l \mathbf{q}_l - \mathbf{q}_l \mathbf{p}_l)\, u_k(x) = K\left(\frac{\partial}{\partial q_l}(q_l\, u_k(x)) - q_l \frac{\partial u_k}{\partial q_l}\right) = K u_k(x), \tag{14.188}$$

whence

$$\int \rho(x)\, u_i(x)\, (\mathbf{p}_l \mathbf{q}_l - \mathbf{q}_l \mathbf{p}_l)\, u_k(x) = K \int \rho(x)\, u_i(x)\, u_k(x)\, dx, \tag{14.189}$$

which, on account of the orthonormality relation in Eq. (14.186), is equal to $K\delta_{ik}$. Accordingly, agreement with the fundamental canonical commutation relation of Heisenberg *et al.* is assured by setting $K = \hbar/i$. Just as in the paper by Born and Jordan (1925b), the momentum–position commutation relation implies, with a suitable definition of partial differentiation,[84] that commutation of a kinematic quantity F with $\mathbf{p}_l$ is equivalent to differentiation with respect to q_l (up to a factor of $K = \hbar/i$):

$$\mathbf{p}_l F - F \mathbf{p}_l = K \frac{\partial F}{\partial q_l} = \frac{\hbar}{i} \frac{\partial F}{\partial q_l}. \tag{14.190}$$

Similarly, commutation of the operator $\mathbf{q}_l$ (i.e., multiplication by q_l) with F is equivalent to $-K(\partial F/\partial p_l)$ (cf. Eqs. (12.41) and (12.42)).

[83] Schrödinger writes $\mathbf{F}u_l(x)$ as $[\mathbf{F}, u_l(x)]$. Here, the square brackets *could* be interpreted as a commutator, as in modern usage, but it is clear from the subsequent development that Schrödinger merely means to use this to indicate action of an operator **F** on a function. Where commutators occur, such as in Eq. (14), they are written out explicitly, for example, $p_l F - F p_l$.

[84] Namely, when partially differentiating an expression with respect to q_l, one examines the expression for every occurrence of q_l, which is then removed, leaving the order of the remaining quantities in the expression intact.

In section 4 of his "equivalence" paper, Schrödinger (1926e, pp. 52–57) establishes the connection between the dynamical content of the Heisenberg matrix theory—that is, the matrix equations that are formally identical to Hamilton's equations of motion and express time evolution in the theory—to a particular specialization of his wave-mechanical theory, specifically, the choice of a particular set of orthogonal wave functions to be used in the calculation of matrix elements. First, after some remarks on the need to choose a definite symmetrized ordering of p's and q's in the expression for the Hamiltonian $H(q_k, p_k)$ (given that these will now be represented either by non-commuting matrices in matrix mechanics, or non-commuting differential operators in wave mechanics), Schrödinger reminds the reader of the dynamical postulate first introduced by Born and Jordan (1925b; cf. Eq. (12.23)):

> Then the authors postulate that the *matrices* $(q_l)_{ik}$, $(p_l)_{ik}$ shall satisfy an infinite system of equations, as "equations of motion" ...
>
> $$\left.\begin{aligned} \left(\frac{dq_l}{dt}\right)^{ik} &= \left(\frac{\partial H}{\partial p_l}\right)^{ik} \\ \left(\frac{dp_l}{dt}\right)^{ik} &= \left(-\frac{\partial H}{\partial q_l}\right)^{ik} \end{aligned}\right\} \qquad \begin{aligned} l &= 1, 2, \ldots, n \\ i, k &= 1, 2, 3, \ldots \end{aligned}$$
>
> (Schrödinger 1926e, p. 53, Eq. (18)).

Here, the inductive approach used by Born and Jordan is recalled, in that the time derivative symbol d/dt appearing on the left here is to be interpreted formally, *after specification of an infinite series of numbers* ν_1, ν_2, ν_3, ν_4, ... (of course, with dimensions of inverse time, i.e., frequency) as multiplication of the $(ik)^{\text{th}}$ matrix element by $2\pi i(\nu_i - \nu_k)$. In other words, the equations of motion above really mean (cf. Eq. (14.190)):[85]

$$(\nu_i - \nu_k)\,(\mathbf{q}_l)^{ik} = \frac{1}{h}\left(\mathbf{H}\,\mathbf{q}_l - \mathbf{q}_l\,\mathbf{H}\right)^{ik}, \tag{14.191}$$

$$(\nu_i - \nu_k)\,(\mathbf{p}_l)^{ik} = \frac{1}{h}\left(\mathbf{H}\,\mathbf{p}_l - \mathbf{p}_l\,\mathbf{H}\right)^{ik}. \tag{14.192}$$

The question now arises: how can we ensure that the matrices as defined previously by Schrödinger—by computing matrix elements using elements of an orthogonal set of functions, as in Eq. (14.187)—satisfy these equations? There are two important issues to be settled here:

1. First, the explicit form of the matrices appearing in these equations of motion clearly depends on the choice of the set of orthogonal functions used to compute the individual matrix elements. Schrödinger asserts that agreement with the

[85] Again, we depart here from Schrödinger's notation by using bold type for matrices, as in Born and Jordan (1925b). The indices *ik* are erroneously omitted on the right-hand side of the following equation (Schrödinger 1926e, p. 53, Eq. (18′)).

program of Born, Heisenberg, and Jordan (1926) is obtained by choosing the orthogonal functions $u_i(x)$ to be eigenfunctions of the linear differential operator corresponding to the energy function H of the system

$$-H\psi(x)+E\psi(x)=0 \tag{14.193}$$

(the functions $u_i(x)$ are solutions of this equation for eigenvalues E_i). The problem must be formulated in such a way—by appropriate choice of a density function–that the operator becomes self-adjoint. Here, the mathematical background provided by Courant and Hilbert (1924, see, e.g., p. 239) is essential.

2. This choice is intimately connected with the second issue, which was also of great concern to Born, Heisenberg, and Jordan. In fact, the transition from classical kinematical variables to the corresponding (operatorial) quantum ones is necessarily ambiguous, as the classical variables q_k and p_k can be permuted at will without altering the classical quantity they compose, but a similar permutation at the quantum level alters the differential operator obtained, if non-commuting coordinate and momentum variables are interchanged. Not only must one choose the Hamiltonian differential operator so that the differential equation becomes a self-adjoint one, with an orthogonal set of eigenfunctions, but some argument must be advanced for the choice of a specific ordering of the algebraic components of H so that the operator so obtained is *unique*. Schrödinger asserts that when this is done, the operator to be chosen in Eqs. (14.191) and (14.192) is exactly the wave operator of his original equation, Eq. (14.111), that is, $H=-(h^2/8\pi^2m)\,\Delta+V$.

The first matter is easily settled. We assume a set of orthogonal eigenfunctions of H (as differential operator) have been obtained, and normalized to unity (thus, satisfying Eq. (14.186)). The matrix of the energy function $\mathbf{H}$ with respect to this orthonormal set is

$$\mathbf{H}^{kl} = \int \rho(x)\,u_k(x)\,H\,u_l(x)\,dx \tag{14.194}$$

$$= E_l\int \rho(x)\,u_k(x)\,u_l(x)\,dx = E_l\,\delta_{kl}, \tag{14.195}$$

clearly in diagonal form. The same calculation originally given by Born and Jordan (1925b) can thus be repeated here

$$(\mathbf{H}\,\mathbf{q}_l)^{ik} = \sum_m \mathbf{H}^{im}(\mathbf{q}_l)^{mk} = E_i\,(\mathbf{q}_l)^{ik}, \tag{14.196}$$

$$(\mathbf{q}_l\mathbf{H})^{ik} = \sum_m (\mathbf{q}_l)^{im}\mathbf{H}^{mk} = E_k\,(\mathbf{q}_l)^{ik}, \tag{14.197}$$

which is easily seen to agree with Eqs. (14.191) and (14.192), with the identification $E_i=h\nu_i$. This argument does not strictly require the discrete set of eigenfunctions $u_i(x)$

to be *complete*. There may be a continuous spectrum as well (in most problems of atomic physics, there is). But the argument correctly determines the time dependence of the subsector of discrete matrix elements of the coordinate and momentum operators. We return later to the issue of completeness, and its impact on "equivalence proofs" of matrix and wave mechanics.

The second matter—the correct ordering of non-commuting factors, and the associated question of "symmetrization" of the energy function H—raises considerably more subtle issues. We saw in Chapter 12 that this problem caused considerable difficulties also for the Göttingen matrix theoreticians. In fact, Schrödinger considered (correctly) that this problem had not been completely settled by Born, Heisenberg, and Jordan. Although an unambiguous definition of differentiation of a matrix function had been proposed that ensured consistency of the classical and quantum versions of the Hamilton equations of motion, there still was no clear prescription for deciding on a given ordering of operators in the quantum theory, starting with the classical expression for H.

Schrödinger gave a simple one-dimensional example of the ordering problem (ignoring dimensionful factors of mass and frequency), the harmonic oscillator. The two expressions for H, he observed,

$$H_a = \frac{1}{2}(p^2 + q^2), \quad H_b = \frac{1}{2}\left(\frac{1}{f(q)}\, pf(q)p + q^2\right) \tag{14.198}$$

are classically identical (as $f(q)$ and p commute when regarded as simple functions), but quantum-mechanically distinct.[86] Now in either case, the transcription from classical to quantum theory should be unambiguous: one simply replaces p by $(\hbar/i)\partial/\partial q$, wherever it occurs. But which one gives the correct quantum treatment, as the classical form is ambiguous?

The discussion given here by Schrödinger (1926e, pp. 55–56), where the special case of classical Hamiltonians of the form $T + V$ (kinetic plus potential, with the potential term independent of momenta) is treated, shows that the quantum Hamiltonian *can* be specified uniquely, and in such a way that the resulting differential operator is self-adjoint, thereby allowing the theory access to the well-developed theorems on the existence (and, with appropriate boundary conditions, completeness) of an orthogonal set of eigenfunctions of the same, which of course lies at the heart of Schrödinger's proposed correspondence between the wave and matrix theories.

The specification given by Schrödinger also has the advantage that (a) invariance of the theory under arbitrary point transformations (change of coordinate variables q_i to $q'_m(q_i)$ with $i, m = 1, 2, \ldots, n$) is guaranteed, and (b) the energy operator agrees with the original Schrödinger wave equation Eq. (14.111) in Cartesian coordinates. Cartesian coordinates play a special role in the discussion precisely because the expression for the energy operator in these coordinates does not present any ordering ambiguities: there are no terms in which coordinates and their derivatives appear in a single term together.

[86] They can easily be seen to differ by terms proportional to h, using the canonical commutation relations.

The theory also automatically generates the correct density function $\rho(x)$ with respect to which the coordinate integrals must be taken.

The set of dynamical problems examined by Schrödinger (1926e, pp. 55–57) in section 4 of his "equivalence" paper are defined in a phase space (q_k, p^k) with $k = 1, 2, \ldots, N$ by a classical Hamiltonian of the form[87]

$$H = T(q, p) + V(q) = \frac{1}{2}\sum_{kl} p^k g_{kl}(q)\, p^l + V(q). \tag{14.199}$$

Here, q without a subscript refer to the whole set $(q_1, q_2, \ldots, q_N)$ of coordinate variables defining the "configuration space" (thus, $dq = dq_1 dq_2 \ldots dq_N$). Similarly, p without a subscript refers to the whole set $(p_1, p_2, \ldots, p_N)$ of momentum variables. This form encompasses a very wide range of problems in non-relativistic quantum mechanics—essentially all problems in atomic physics with Coulomb (or other) potentials, as well as applied electric fields, as in the Stark effect (but without spin and magnetic fields). Also, no assumption is made as to the coordinate system employed for the calculations (e.g., spherical coordinates for the hydrogen atom, or parabolic coordinates for the Stark effect).

The wave equation determining the eigenfunctions of energy for such a system is then posited to be the extremal condition for the quadratic functional

$$\mathcal{J}_1 \equiv \int \left\{ \hbar^2 T\left(q, \frac{\partial\psi}{\partial q}\right) + \psi^2 V(q) \right\} \Delta^{-1/2} dq, \tag{14.200}$$

subject to the condition

$$\mathcal{J}_2 \equiv \int \psi^2 \Delta^{-1/2} dq = 1. \tag{14.201}$$

Here, $\Delta = \det(g)$ is the determinant of the quadratic form $g_{kl}(q)$ appearing in Eq. (14.199). The partial differential equation resulting from the extremal condition

$$\delta(\mathcal{J}_1 - E\mathcal{J}_2) = 0, \tag{14.202}$$

(where E is the Lagrange multiplier for the condition Eq. (14.201)) is then the wave equation for the system, wherein all ordering issues of the derivatives are already implicitly settled. Note that Schrödinger is still working in the context of purely real wave functions and operators: there is still no sign of the conjugate wave function $\psi^\star$ that would signal the transition to a complex space of functions.

First, let us check that the problem so formulated is indeed invariant with respect to point transformations, allowing us to reformulate the wave equation in whatever coordinate system is most convenient for the problem at hand, with no change in the physical

[87] The associated Lagrangian is $\mathcal{L} = \frac{1}{2}\dot{q}_k g^{kl}(q)\dot{q}_l - V(q)$, where $g^{kl}(q)$ is the inverse matrix of the symmetric matrix $g_{kl}(q)$. The canonical momenta are $p^k = \partial\mathcal{L}/\partial\dot{q}_k$, and $\dot{q}_k = g_{kl}p^l$.

content. We define the transformation for the quadratic form $g_{kl}(q)$ as in standard differential geometry: under a change of coordinates to $q'_m(q)$, with $m = 1, 2, \ldots, N$,

$$g'_{mn}(q') = \sum_{k,l=1}^{N} \frac{\partial q'_m}{\partial q_k} g_{kl}(q) \frac{\partial q'_n}{\partial q_l}, \tag{14.203}$$

with

$$\det(g') = \left| \frac{\partial q'}{\partial q} \right|^2 \det(g), \tag{14.204}$$

where $|\partial q'/\partial q|$ is the Jacobian of the transformation from the old to the new variables. The point transformation invariance of $\mathcal{J}_1$ is now easily checked. First, with a view to the kinetic energy term $T(q, \partial\psi/\partial q)$ in $\mathcal{J}_1$, we transform the expression (cf. Eq. (14.199))

$$\int \sum_{k,l=1}^{N} \left(\frac{\partial \psi(q)}{\partial q_k} g_{kl}(q) \frac{\partial \psi(q)}{\partial q_l} \right) \frac{1}{\sqrt{\det(g)}} \, dq \tag{14.205}$$

from q to q':

$$\int \sum_{k,l=1}^{N} \sum_{m,n=1}^{N} \left(\frac{\partial \psi(q')}{\partial q'_m} \frac{\partial q'_m}{\partial q_k} g_{kl}(q) \frac{\partial q'_n}{\partial q_l} \frac{\partial \psi(q')}{\partial q'_n} \right) \frac{1}{\sqrt{\det(g)}} \left| \frac{\partial q}{\partial q'} \right| dq'. \tag{14.206}$$

Given Eqs. (14.203) and (14.204), this reduces to

$$\int \sum_{m,n=1}^{N} \left(\frac{\partial \psi(q')}{\partial q'_m} g'_{mn}(q') \frac{\partial \psi(q')}{\partial q'_n} \right) \frac{1}{\sqrt{\det(g')}} \, dq', \tag{14.207}$$

which shows that the expression in Eq. (14.205) retains its form under transformation to new coordinates. To establish the invariance of the potential energy term in $\mathcal{J}_1$ and of the normalization condition $\mathcal{J}_2$ is straightforward and is left to the reader.

The extremal condition in Eq. (14.202) can be written as

$$\delta \int \left\{ \frac{1}{2} \hbar^2 \sum_{k,l=1}^{N} \left(\frac{\partial \psi}{\partial q_k} g_{kl} \frac{\partial \psi}{\partial q_l} \right) + \Big(V(q) - E \Big) \psi^2 \right\} \Delta^{-1/2} \, dq = 0. \tag{14.208}$$

Upon partial integration, this turns into[88]

$$\int \left\{ -\hbar^2 \Delta^{1/2} \sum_{k,l=1}^{N} \frac{\partial}{\partial q_k} \left(\Delta^{-1/2} g_{kl} \frac{\partial \psi}{\partial q_l} \right) + 2\Big(V(q) - E\Big)\psi \right\} \delta\psi \, \Delta^{-1/2} \, dq = 0. \quad (14.209)$$

As the variation must vanish for an arbitrary $\delta\psi(q)$, we conclude that

$$H\psi - E\psi = 0, \quad (14.210)$$

where the operator H is defined as

$$H = -\frac{\hbar^2}{2} \Delta^{1/2} \sum_{k,l=1}^{N} \frac{\partial}{\partial q_k} \left(\Delta^{-1/2} g_{kl}(q) \frac{\partial}{\partial q_l} \right) + V(q) \quad (14.211)$$

(cf. Schrödinger 1926e, p. 56, Eq. (31), with $g_{kl} = \delta_{kl}$). Of course, if we are dealing with a single particle of mass m in Cartesian coordinates $(q_1, q_2, q_3) = (x, y, z)$, we have $g_{kl} = (1/m)\delta_{kl}$ and Eq. (14.210) reduces to the original Schrödinger equation Eq. (14.111) (in this case the density factor Δ is constant and drops out of Eq. (14.211)). But it is apparent that the energy operator H, as given in Eq. (14.211), is uniquely defined inasmuch as the ordering of derivative factors with respect to coordinate functions is completely specified. The invariance under point transformations assures us that this is the correct form of the wave equation in an arbitrary coordinate system obtainable from Eq. (14.111) in Cartesian coordinates by a change of coordinate variables.[89]

[88] Subtracting the integral in Eq. (14.208) from

$$\int \left\{ \frac{1}{2}\hbar^2 \sum_{k,l=1}^{N} \left(\frac{\partial(\psi + \delta\psi)}{\partial q_k} g_{kl} \frac{\partial(\psi + \delta\psi)}{\partial q_l} \right) + \Big(V(q) - E\Big)(\psi + \delta\psi)^2 \right\} \Delta^{-1/2} \, dq$$

and neglecting all term with $\delta\psi^2$, we arrive at

$$\int \left\{ \frac{1}{2}\hbar^2 \sum_{k,l=1}^{N} \left(\frac{\partial \delta\psi}{\partial q_k} g_{kl} \frac{\partial \psi}{\partial q_l} + \frac{\partial \psi}{\partial q_k} g_{kl} \frac{\partial \delta\psi}{\partial q_l} \right) + 2\Big(V(q) - E\Big)\psi \, \delta\psi \right\} \Delta^{-1/2} \, dq.$$

Since $g_{lk} = g_{kl}$, we can replace the two sums in this expression by taking one of them twice. Integration by parts then gives Eq. (14.209). We are assuming here that the boundary terms arising in this integration by parts vanish. In the case of bound atomic systems, this is ensured by vanishing of ψ as the coordinates go to spatial infinity.

[89] As an explicit example, consider the point transformation taking us from Cartesian coordinates $q_k = (x, y, z)$ of a single particle in three dimensions, to spherical coordinates $q'_m = (r, \vartheta, \varphi)$. The initial quadratic form/metric appearing in the kinetic energy, setting the mass $m = 1$ for simplicity, is $g_{kl} = \delta_{kl}$, or $g = \text{diag}(1, 1, 1)$. Performing the transformation Eq. (14.203), we find $g' = \text{diag}(1, 1/r^2, 1/(r^2 \sin^2\vartheta))$. The Jacobian is $|\partial q'/\partial q| = \Delta^{1/2} = \sqrt{\det(g')} = 1/r^2 \sin\vartheta$. The reader may now verify that the differential operator appearing in the first term of Eq. (14.209) exactly reproduces the Laplacian in spherical coordinates, with the proper ordering of derivatives and coordinate factors. This procedure is, to the authors' knowledge, by far the least-painful method for obtaining the otherwise mystifying spherical-coordinate Laplacian.

Just as importantly, the differential operator so obtained can be shown to be *self-adjoint* with respect to the density function $\rho = \Delta^{-1/2}$ appearing in the coordinate integrals. This property says that, for arbitrary functions $\chi(q)$ and $\psi(q)$ (not necessarily solutions of the Schrödinger equation, but satisfying suitable boundary conditions allowing the neglect of boundary terms in integration by parts), the differential operator H appearing in the Schrödinger equation $H\psi = E\psi$ should satisfy

$$\int \chi(q)\,[H\psi(q)]\,\rho(q)\,dq = \int \psi(q)\,[H\chi(q)]\,\rho(q)\,dq, \tag{14.212}$$

where the square brackets indicate that the derivatives in H act only on ψ and χ, not on the density function ρ, which should be regarded as built into the integration measure. This self-adjointness property trivially obtains for the potential energy term in H in Eq. (14.211). We now check it for the kinetic energy term. Replacing H in Eq. (14.212) by just this kinetic part, we arrive at

$$-\frac{\hbar^2}{2}\int \chi(q)\,\Delta^{1/2}\sum_{k,l=1}^{N}\frac{\partial}{\partial q_k}\left(\Delta^{-1/2}g_{kl}(q)\frac{\partial\psi(q)}{\partial q_l}\right)\Delta^{-1/2}dq. \tag{14.213}$$

Integrating by parts twice, we can rewrite this first as

$$\frac{\hbar^2}{2}\int \sum_{k,l=1}^{N}\left(\frac{\partial\chi(q)}{\partial q_k}\Delta^{-1/2}g_{kl}(q)\frac{\partial\psi(q)}{\partial q_l}\right)dq, \tag{14.214}$$

and then as

$$-\frac{\hbar^2}{2}\int \sum_{k,l=1}^{N}\psi(q)\frac{\partial}{\partial q_l}\left(\Delta^{-1/2}g_{kl}(q)\frac{\partial\chi(q)}{\partial q_k}\right)dq. \tag{14.215}$$

Regrouping terms and inserting factors $\Delta^{1/2}$ and $\Delta^{-1/2}$, we see that this is just Eq. (14.213) with ψ and χ interchanged, thus proving the self-adjointness property for the kinetic energy term:

$$-\frac{\hbar^2}{2}\int \psi(q)\,\Delta^{1/2}\sum_{k,l=1}^{N}\frac{\partial}{\partial q_l}\left(\Delta^{-1/2}g_{kl}(q)\frac{\partial\chi(q)}{\partial q_k}\right)\Delta^{-1/2}dq. \tag{14.216}$$

Note that in these integrations by parts, we must assume that the wave functions $\psi(q)$ and $\chi(q)$ satisfy appropriate boundary conditions (e.g., that they vanish at spatial infinity if the dynamics takes place in an unbounded region), so that the boundary terms can be neglected. *It can hardly be emphasized enough how critical the self-adjointness property—which in turn depends on the imposition of appropriate boundary conditions—is to the program of implementing discrete quantum phenomena in a continuous differential framework!* The whole apparatus of complete orthogonal systems of functions elaborated in Courant and Hilbert (1924) depends on the self-adjointness property of the wave operators whose eigenvalues represent physical quantities like the energy. The emergence of quantization is directly connected to the imposition of appropriate regularity and asymptotic conditions on the solutions of the wave equation, as Schrödinger (1926c) had already emphasized in his first paper on quantization as an eigenvalue problem. In fact, *if we ignore the behavior of the solutions at infinity*, solutions exist for *arbitrary* energy E (positive or negative) for either the Hermite equation, which we encountered in the case of the one-dimensional oscillator (cf. Eq. (14.112)), or the radial Laguerre equation, which we encountered in the case of the hydrogen atom (cf. Eq. (14.67)). Only by requiring the solutions to remain finite at spatial infinity do we thin out the solution space to only those functions corresponding to the quantized energy eigenvalues (in the case of the bound-state spectrum) required in the quantum theory. It is hardly an overstatement to assert that in the wave theory, *quantization equals boundary conditions* (which in turn assures self-adjointness).

Section 12.3.3 showed that the *Hermitian* property (corresponding to self-adjointness for differential operators[90]) of infinite quadratic forms played an important role in Born's chapter 3 of the Three-Man-Paper (Born, Heisenberg, and Jordan 1926), where the Hilbert–Hellinger theory of bounded quadratic forms was brought to bear to explain the appearance of both discrete and continuous spectra in the matrices representing physical quantities in matrix mechanics. This theory (not covered in any detail in Courant and Hilbert 1924) does not seem to have been at Schrödinger's disposal at this point (although the Three-Man-Paper was).[91] The mapping Eq. (14.187) from wave functions to discrete matrices preserves the matrix algebra for multiplication of two kinematic quantities only if the discrete orthogonal function set $u_m(x)$ in Eq. (14.186) are complete, as discussed previously. In the case of the hydrogen atom, or indeed for any atomic system, the discrete energy eigenfunctions corresponding to bound stationary states obviously are not complete, as they cannot take into account (indeed, are orthogonal to) all the unbound wave functions corresponding to positive energy eigenvalues $E > 0$. As Schrödinger admits,

> We conclude with a general observation on the whole formal apparatus of secs. 2, 3, and 4. The basic orthogonal system was regarded as an absolutely *discrete* system of functions. Now in the most important applications this is *not* the case. Not only in the hydrogen atom but also in heavier atoms the wave equation (31) [see our Eqs. (14.210) and (14.211)] must possess a continuous eigenvalue spectrum as well as a line [discrete] spectrum. The former manifests itself, for example, in the continuous *optical* spectra which adjoin the limit of the series. It appeared better, provisionally, not to burden the formulae and the line of thought with this generalization, though it is indeed indispensable. The chief aim of this paper is to work out, in the clearest manner possible, the formal

[90] See Appendix C for an introduction to the necessary functional analysis.

[91] Born, Heisenberg, and Jordan (1926) is referred to in a footnote in the introduction of Schrödinger (1926e, p. 45).

> connection between the two theories, and this is certainly not changed, in any essential point, by the appearance of a continuous spectrum (Schrödinger 1926e, p. 57).

These remarks throw an interesting light on the complementary areas of mathematical expertise exhibited by the developers of matrix and wave mechanics, respectively. While the Göttingen group (and Born in particular) were fully informed on the treatment of infinite matrices with *both* discrete and continuous spectra—a theory that had developed naturally from the analysis of linear integral equations and culminated in the work of Hilbert and Hellinger in the first decade of the twentieth century—the important conceptual ingredient of a linear vector space as the theater of operations of the matrix quantities was missing in the Göttingen formalism. In Schrödinger's theory, the linear space, now identified as a space of functions that were either eigenfunctions of differential operators or linear combinations thereof, plays a central role, insofar as the central physical object is a wave function determining the dynamics of the particle (or particles) comprising the system. However, Schrödinger is clearly unaware at this point of the proper manner of handling completeness issues in systems with both a discrete and continuous spectrum. Both theories are limited also at this point by their emphasis on the energy operator: for the matrix mechanicians, all kinematic quantities are referred to a basis in which the energy matrix is diagonal, while for Schrödinger the only differential operator subjected to detailed examination is the one appearing in the wave equation (i.e., the Hamiltonian differential operator).

In the final section of his "equivalence" paper, Schrödinger (1926e, sec. 5) attempts, not quite successfully, to delineate the exact nature of the connection between the two theories (matrix mechanics and wave mechanics). and finally, to propose an argument for the emergence of the Bohr frequency condition from the time-dependence of the electric moment, as calculated from the time dependent spatial charge density of an electron. Both of these attempts turn out to be flawed, for reasons we now explain.

First, Schrödinger discusses the implications of an exact mathematical equivalence of matrix and wave mechanics. The introduction to this argument shows Schrödinger's acute awareness of the epistemological principles at the center of the discourse we now associate with the "Vienna Circle" (*Wiener Kreis*):[92]

> Today there are not a few physicists who, like Kirchhoff and Mach, regard the task of physical theory as being merely a mathematical description (*as economical as possible*) of the empirical connections between observable quantities, i.e., a description which reproduces the connection, as far as possible, without the intervention of unobservable elements. On this view, mathematical equivalence has almost the same meaning as physical equivalence. In the present case there might perhaps appear to be a certain superiority in the matrix representation because, through its stifling of intuition, it does not tempt us to form space-time pictures of atomic processes, which must perhaps remain uncontrollable (Schrödinger 1926e, p. 58).

[92] See Sigmund (2017) for an engaging discussion of the Vienna Circle and references to the extensive literature on the topic.

This philosophical aside is followed by the assertion that his new theory is fully equivalent to the one of Heisenberg *et al.*. Not only can the matrices of matrix mechanics be computed once the eigenfunctions of the wave theory have been determined (via Eq. (14.187), but also knowledge of the matrices allows reconstruction of the wave functions. The argument given by Schrödinger for this, based on moment theorems that determine a function under certain conditions from knowledge of all its moments (integrals with arbitrary powers of the independent variable), is incomplete and faulty, as is discussed in detail in Muller (1997–1999) (see note 79). Nonetheless, Schrödinger was in essence correct: all the essential elements are already present in the Hilbert–Hellinger theory to establish the desired reconstruction of wave functions from matrix representations, even before this theory was extended by von Neumann to cover the physically essential case of unbounded self-adjoint operators.[93]

In this final section, Schrödinger also addresses the perennial question of the correct interpretation of the Bohr frequency condition. A few months later, Schrödinger was to write to Lorentz, describing the appearance of light emission frequencies associated with differences of energies (rather than the individual energies of electrons in stationary states) as something "*monstrous* [*Ungeheuerliches*] ... and really almost *inconceivable* [*undenkbar*]."[94] In this paper, he proposes (incorrectly) that the charge density at location x and time t associated with a wave function $\psi(x,t)$ be given as the real part of $\psi\,\partial\psi^*/\partial t$ (Schrödinger 1926e, p. 60, Eq. (36)). The desired difference frequencies then appear in the Fourier components of the charge density, if ψ is written as a linear combination of all the stationary states available for transitions—a hypothesis very close in spirit to the "virtual oscillators" of the discredited BKS theory (Bohr, Kramers, and Slater 1924a; see Section 10.4). The correct treatment of the charge density and electric moment of a bound electron (see Section 12.2.3) would appear three months later in Schrödinger (1926h), his fourth paper on wave mechanics, in which he would develop time-dependent perturbation theory and correctly derive the essential results of dispersion theory, thereby bringing his theory into direct contact with the crucible of the matrix-mechanics approach.

One is rightfully impressed by the synthesis achieved by Schrödinger (1926e) in his "equivalence" paper in exposing the underlying structural relationships between two theories, which at first sight seem to have almost no points of conceptual contact, but nevertheless, after non-trivial calculations, lead to consonant results in a variety of important problems of atomic dynamics. In fact, more or less the same results were independently obtained in the spring of 1926 by at least two other physicists, Wolfgang Pauli (at the time, visiting the Bohr Institute in Copenhagen), and Carl Eckart, who was then a National Research Fellow at the California Institute of Technology. Indeed, once the surface anatomy of the matrix-mechanical and wave-mechanical formulations had

[93] See Appendix C for a more thorough discussion of the equivalence question.

[94] Schrödinger to Lorentz, June 6, 1926 (Lorentz 2008, Doc. 413, p. 615; English translation in Klein 1967, p. 61). We already quoted this passage in Volume 1 (p. 16), where we juxtaposed it with a similar comment by Planck (1920, p. 19) in his Nobel lecture.

been properly understood by theorists with an adequate mathematical background, the connecting underlying tissue could easily be exposed.

Pauli's discussion of the matrix–wave connection can be found in a lengthy letter to Jordan of April 12, 1926 (Pauli 1979, Doc. 131). It appears that Pauli at this time had carefully studied Schrödinger's (1926c) first paper on wave mechanics (having been alerted by Sommerfeld to its existence in early February). By the time he arrived in Copenhagen for the Easter holidays, he could report to Bohr that Schrödinger's scheme led to results agreeing with those obtained by matrix methods (Mehra and Rechenberg 1987, p. 655). The procedure followed in Pauli's letter to Jordan is close in spirit to Schrödinger's: one starts with the wave-mechanical theory and moves from there toward the matrix theory by defining matrix elements in terms of wave functions and showing that these matrices satisfy the posited properties in the matrix theory. The derivation of the relativistic version of the wave equation, and then its non-relativistic approximation, is briefly sketched. At this point Pauli had probably not had access to Schrödinger's (1926d) second paper on wave mechanics. Nonetheless, with his customary perspicacity, he comments

> The difference between the old quantum theory of periodic systems and Schrodinger's quantum mechanics based on Ansatz (5) [$\Delta\psi + (2m/K^2)(E - V)\psi$, with $K = \hbar$[95] and some minor changes of notation] is, from the point of view of the de Broglie radiation, the same as that between geometrical optics and wave optics. For small wavelengths of the de Broglie radiation one can in fact make the usual substitution
>
> $$\psi = e^{iS/K}. \tag{14.217}$$
>
> If S/K is large then one obtains, following Debye [1909], from (5) the Hamilton–Jacobi differential equation for S [cf. our Eqs. (14.108 and 14.110)] (Pauli 1979, Doc. 131, pp. 317–318).

We have here, in miniature, the basic idea of the optical–mechanical connection discussed at length in Schrödinger's (1926d) second paper on wave mechanics. Pauli next introduces a complete set of orthogonal and normalized eigenfunction solutions of the wave equation (now taken to be one-dimensional for simplicity, i.e., $d^2\psi/dx^2 + (2m/K^2)(E - V)\psi = 0$), with $\psi_1, \psi_2, \psi_3, \ldots$ corresponding to the discrete set of eigenvalues $E_1, E_2, E_3, \ldots$, respectively:

$$\int_{-\infty}^{+\infty} \psi_n \psi_m \, dx = \delta_{nm}. \tag{14.218}$$

The aura of Courant and Hilbert (1924) hangs over this entire procedure, although this by-now standard text is not mentioned explicitly.

[95] The standard usage $\hbar = h/2\pi$ has not yet taken hold. Also, we use V here for the potential energy function, rather than Pauli's E_{pot}.

Pauli proceeds to introduce the matrices for position and momentum. Noting that "an arbitrary function of x can be expanded in the ψ_n" (Pauli 1979, p. 318), he considers such an expansion of $x\,\psi_n(x)$ to introduce the position matrices:

$$x\,\psi_n(x) = \sum_m x_{nm}\,\psi_m(x) \quad \text{with } x_{nm} = \int_{-\infty}^{+\infty} x\,\psi_n\,\psi_m\,dx. \tag{14.219}$$

He introduces the momentum matrices via:

$$iK\frac{\partial\psi_n}{\partial x} = \sum_m (p_x)_{nm}\psi_m(x), \tag{14.220}$$

with

$$(p_x)_{nm} = iK\int_{-\infty}^{+\infty} \frac{\partial\psi_n}{\partial x}\psi_m dx. \tag{14.221}$$

Pauli then asserts (leaving the proof, which, as in Schrödinger's case, requires the use of the completeness property, to Jordan) that the resulting matrices

> satisfy the equations of the Göttingen mechanics, namely,
>
> $$\mathbf{p}_x x - x\mathbf{p}_x = -iK, \quad \frac{1}{2m}\mathbf{p}_x^2 - \mathbf{V}(x) = \mathbf{E}\ \text{(Diagonal matrix)}.$$
>
> (Pauli 1979, p. 318).

The association of matrices to functions (albeit only functions $F(x)$ of the coordinate x) is then asserted to follow the same rule as for the coordinate itself in Eq. (14.219), with $F(x)$ replacing x in the integral.

Pauli informs Jordan that, in addition to these results, he has also calculated, using Schrödinger's method: (a) the harmonic oscillator; (b) the rigid rotator; (c) the Zeeman intensity rules of Hönl (1925) and Kronig (1925a, 1925b) (involving overlap integrals of spherical harmonics); (d) the translation into wave-mechanical language of the perturbation theory based on principal axis transformations, as in Born, Heisenberg, and Jordan (1926); and (e) the transition probabilities (squares of dipole matrix elements) in hydrogen. As Schrödinger remarked in amazement in a letter to Sommerfeld on April 28, 1926, Pauli had "discovered everything again . . . in a tenth of the time which I needed for it!" (Sommerfeld 2004, Doc. 97, p. 251).[96] As with Schrödinger, however, the connections found by Pauli between the two new frameworks for quantum theory fall far short of a demonstration of complete *equivalence*. In particular, the proper treatment of quantities admitting both a discrete and continuous spectrum was absent, as both Pauli,

[96] Quoted in Mehra and Rechenberg (1987, p. 655), who give the date of the letter as April 18. Von Meyen (2011, p. 219) follows the editors of Sommerfeld (2004).

in his letter to Jordan, and Schrödinger, in his "equivalence" paper (Schrödinger 1926e, p. 57), admit.[97]

The paper by Eckart (1926),[98] submitted to *Physical Review* in early June, reflects knowledge of the work of the Göttingen physicists (up to Born, Heisenberg, and Jordan 1926), of Dirac's (1925, 1926a) first two papers, as well as of a paper (which we shall briefly discuss below) by Born and Norbert Wiener (1926) of the integral equation version of quantum mechanics proposed by Cornel Lanczos (1926), and of Schrödinger's (1926c) first paper on wave mechanics. Schrödinger's (1926d) second paper had also been published at this point but Eckart does not explicitly refer to it. Schrödinger's (1926e) "equivalence" paper only reached Eckart after his own paper had been submitted, and is acknowledged in a note added in proof as "containing all the essential results" of his own paper.

Eckart's paper is genetically related most directly to the work which Born did at the Massachusetts Institute of Technology (MIT) in collaboration with Norbert Wiener in November and December of 1926.[99] Born, in fulfillment of a long-delayed visit to the Physics Department of MIT, arrived in Cambridge in early November, and, beginning in mid-November, gave a series of 20 lectures on atomic dynamics culminating in the latest work on matrix mechanics, including the latest breakthrough—Pauli's (1926a) *tour de force* calculation of the hydrogen atom bound state spectrum using matrix mechanics. These lectures would appear shortly in both English and German (Born 1926d, 1926e) and served to bring a number of young American physicists quickly up to date in the recent developments on quantum theory in Göttingen.

While giving these lectures, Born entered into a collaboration with Wiener, a young (32 year-old) mathematician at MIT whose work had ranged from mathematical logic to Brownian motion and boundary value problems. Born's primary motivation in this work was to extend the reach of the matrix-mechanical scheme to aperiodic problems, specifically, those in which the spectrum was continuous so that a representation of the motion in terms of discrete matrices was apparently not appropriate. Indeed, the simplest possible problem in particle dynamics, the motion of a free particle, falls into this category. The historical interest of the Born–Wiener paper lies in the emphasis placed on the formal character of non-commuting linear differential operators—but, unlike the case of Schrödinger (1926e) in his "equivalence" paper, the operators appearing here are derivatives with respect to time, not spatial coordinates, with the operator $D = d/dt$ associated (up to a factor) with the (diagonal) energy matrix $(2\pi i/h)\, W$. The fundamental time dependence of the coordinate matrix elements

$$q(mn)\, e^{2\pi i\, \nu(nm)\, t}, \quad \text{with } \nu(nm) = \frac{W_n - W_m}{h}, \tag{14.222}$$

[97] We saw earlier, in our discussion of the Three-Man-Paper, that the Göttingen group already had at their disposal a rigorous treatment of the general case, including continuous spectra, in matrix mechanics. The only missing ingredient there was the generalization to unbounded operators, provided a few years later by von Neumann. See Appendix C, where we show that this generalization, though physically important, is to some degree superfluous, from a purely mathematical standpoint.

[98] See Jammer (1966, pp. 275–276) for a brief discussion of Eckart's approach.

[99] For a more elaborate discussion of this work by Born and Wiener, see Gimeno, Xipell, and Baig (2021).

meant that the energy matrix could be replaced by the differential operator $(h/2\pi i)\, D$. Indeed, applied to any matrix function of time, the operator $Dq(t) - q(t)D$ is clearly the same as $\dot{q}(t)$, which in turn is just $(2\pi i/h)\,(Wq-qW)$ (cf. Eq. (12.22)). The operator calculus based on the time-derivative operator D acting in concert with both periodic and more general matrix functions of time could just about be tortured into an extremely circuitous solution of the eigenvalue problems for a simple harmonic oscillator (giving the usual discrete spectrum as found by Heisenberg) and for a free particle (with a continuous spectrum), but unfortunately Born and Wiener completely missed the point that a far more useful representation of the abstract matrix algebra of the Göttingen theory would be obtained by replacing the kinematical quantities in that theory by linear differential operators based on the spatial coordinates of the system, as done by Schrödinger (and independently, by Pauli and Eckart).

This point was not however missed by Eckart (1926), who of course had the advantage of having studied both the matrix and wave incarnations of quantum mechanics, as well as having attended lectures which Born gave at Caltech following his visit to MIT. The starting point for Eckart were the canonical commutation relations,

$$P_iQ_j - Q_jP_i = \frac{h}{2\pi i}\delta_{ij}, \quad Q_iQ_j - Q_jQ_i = P_iP_j - P_jP_i = 0, \tag{14.223}$$

with $i, j = 1, 2, \ldots, n$. The realization of these algebraic constraints in terms of differential operators is recognizable to any modern-day student of elementary quantum mechanics: acting on functions $\psi(q_1, q_2, \ldots, q_n)$ (cf. Eqs. (14.184)–(14.185)),

$$P_j\psi \equiv \frac{h}{2\pi i}\frac{\partial}{\partial q_j}, \quad Q_j\psi \equiv q_j\psi. \tag{14.224}$$

Here, the critical shift has been made from operators acting on functions of time only, as in Born and Wiener (1926), to operators acting on functions of the spatial coordinates (and potentially, for the time-dependent problem, of the time also). In any case, it is immediately clear that Schrödinger's wave equation amounts to just the eigenvalue equation for the energy, with the momenta replaced by spatial derivatives in the Hamiltonian energy function, which Eckart (1926, p. 720, Eq. (17)) writes in the form

$$\frac{1}{\psi}\, H\Big(\frac{h}{2\pi i}\frac{\partial}{\partial q_j}, q_j\Big)\, \psi = W. \tag{14.225}$$

Eckart then shows, by reasoning that parallels exactly that used by Pauli in his letter to Jordan, that the coordinate operators Q are associated with matrices, once a complete set of energy eigenfunctions are introduced as an expansion basis for an arbitrary function ψ. If $\psi_n(q)$ (with $n = 1, 2, \ldots$) is such a basis, then

$$Q\,\psi_n(q) = q\,\psi_n(q) = \sum_k Q(nk)\,\psi_k, \tag{14.226}$$

with powers of Q corresponding to the appropriate powers of the matrix $Q(nk)$. Eckart (1926, pp. 721–722) similarly extracts the matrix $P(nk)$ of the momentum operator P by examining the expansion of $P\psi_n = (h/2\pi i)\,\psi_n$ in the basis of energy eigenfunctions. Eckart's paper ends with a solution of the wave equation for the one-dimensional simple harmonic oscillator, recovering the discrete eigenvalue sequence $W = (n+1/2)\,(h\omega/2\pi)$, provided the solutions are suitably behaved (1926, p. 724).

14.6 Reception of wave mechanics

In contrast to the admiring, but frequently baffled, reception the previous Fall of the matrix theory of Heisenberg and his Göttingen collaborators, the new approach developed by Schrödinger was soon welcomed enthusiastically, especially by senior physicists of the generation before Schrödinger. The formal basis of the theory, in boundary value problems of partial differential equations of a structure very familiar to a classically trained physicist, turned out to be far more agreeable to the vast majority of atomic theorists in the 1920s than the strange algebraic gymnastics of the matrix approach. Once Schrödinger's papers began to appear in print, Planck and Einstein in Berlin were among the first to greet the new approach with admiration—one might even say relief—at having an alternative to the Göttingen formalism. The Göttingen theorists (especially Heisenberg) received the new approach with far less enthusiasm, perhaps not surprisingly, as it clearly constituted a serious competitor to their hard-won formalism.[100]

The first prominent physicists to become aware of the new theory were, not surprisingly, Wien and Sommerfeld in Munich. Wien had been in fairly regular contact with Schrödinger since asking him in May 1925 to contribute to an article on color theory in an upcoming volume on experimental physics (Mehra and Rechenberg 1987, p. 447–450). Schrödinger also submitted several articles in 1925 to *Annalen der Physik*, of which Wien was the editor, and returned to this journal for the entire series of articles developing wave mechanics and showing its relation to matrix mechanics. In submitting these articles, Schrödinger asked Wien to make sure that Sommerfeld would look them over before publication (Mehra and Rechenberg 1987, p. 534). So Wien and Sommerfeld were probably the first to see Schrödinger's (1926c) first paper on wave mechanics in late January 1925, well before it became available in published form.[101]

In a letter to Schrödinger on February 6, 1926 (von Meyenn 2011, p. 177), Wien acknowledged receipt of this paper and congratulated him on developing a theory in which—in contrast to the matrix theory, the fundamental quantities of which only concerned pairs of states—the temporal change in state of a quantum system could be addressed. Wien's reservations about the theory concerned the treatment of light absorption in the theory, the derivation in the theory of the blackbody formula, and in

[100] For other discussions of the early reception of wave mechanics, see, for example, Mehra and Rechenberg (1987, pp. 534–538, 617–636), and Moore (1989, pp. 209–210, 220–229).

[101] In letters of December 27, 1925 and January 8, 1926, Schrödinger had already informed Wien of his progress toward a "vibration theory" of atoms. For quotations of passages from these letters, see Mehra and Rechenberg (1987, p. 460 and p. 465, respectively).

particular, the rather ad hoc way in which Planck's constant was introduced (Mehra and Rechenberg 1987, pp. 534–535).

Sommerfeld similarly, in a letter of February 3, 1926, written in response to one from Schrödinger of January 29, 1926,[102] and to the manuscript of Schrödinger's (1926c) first paper on "Quantization as an eigenvalue problem," had some vigorous complaints, especially concerning the hand-waving arguments for the basis of the Bohr frequency condition in the paper, as arising from "beats" in the wave processes of the new theory. But Sommerfeld's primary complaint was the standard positivistic Machian criticism—whereas Heisenberg had been careful only to include those elements in the theory that had a direct connection to observable quantities, Schrödinger's theory contained a suspicious amount of "large, unobservable ballast". It was certainly unclear at this point (and would remain so throughout Schrödinger's series of five substantial papers on wave mechanics produced in the first half of 1926) what the physical significance of the very detailed and specific temporal (e.g. very large oscillation frequencies, if one starts from $E = h\nu$, with E including the rest energy m_0c^2 of the particle) and spatial properties (spherical harmonics, nodes, etc.) of the wave function could be. Schrödinger's response to Sommerfeld, in a letter of February 20, 1926 (Sommerfeld 2004, Doc. 96; von Meyenn 2011, pp. 178–183), emphasized that the path to his wave equation was as predetermined for a wave theory (given the starting point of the Planck–Einstein equation $\nu = E/h$) as the Keplerian orbits of electrons in an atomic theory based on particle mechanics.

The response from the senior echelon of theorists in Berlin—Planck and Einstein—was more complimentary and enthusiastic from the beginning, that is, from the period starting in March when the first paper appeared in print in *Annalen der Physik*. Planck's comments, in a postcard dated April 2 and a letter of May 24, 1926 (Mehra and Rechenberg 1987, pp. 624–627), were not very specific but indicate that he was taking the trouble to study the new theory carefully.[103] He seems to have been most impressed by the significant role of the action function in the new wave theory, which is hardly surprising, given his early emphasis on the primacy of the quantization of action rather than energy in quantum theory. He also showed Schrödinger's papers to Einstein, who, despite listing some reservations, wrote in the margin of a letter to Schrödinger of April 16, 1926: "the idea of your article shows real genius" (Einstein 1987–2021, Vol. 15, Doc. 256). Einstein had at first misinterpreted or misremembered the wave equation being used by Schrödinger and complained about its failure to separate properly for two uncoupled systems. This confusion was soon cleared up and ten days later Einstein wrote (in the translation of Klein (1967, p. 24)) to Schrödinger that

[102] Sommerfeld (2004, Docs. 92 and 93), discussed in Mehra and Rechenberg (1987, pp. 537–539). These two letters can also be found in von Meyenn (2011, pp. 170–175)

[103] Schrödinger's correspondence with Planck, Einstein, and Lorentz, mostly from 1926, can be found in Przibram (1963), which was translated into English by Klein (1967). As Mehra and Rechenberg's (1987, p. 624) note, this volume unfortunately does not include Schrödinger's correspondence with Wien and Sommerfeld, nor, for that matter, his correspondence with Pauli. Schrödinger's correspondence of this period can be found in von Meyenn (2011) as well as in editions of the correspondence of Pauli (1979), Einstein (1987–2021, Vol. 15), Sommerfeld (2004), and Lorentz (2008).

> I am convinced that you have made a decisive advance with your formulation of the quantum condition, just as I am equally convinced that the Heisenberg–Born route is off the track (Einstein to Schrödinger, April 26, 1926, Einstein 1987–2021, Vol. 15, Doc. 267; von Meyenn 2011, p. 217).

The most detailed and penetrating early response to Schrödinger's new theory clearly came from Pauli, who, by early April, had independently discovered the connection between wave and matrix mechanics, which Schrödinger (1926e) would present in detail in his "equivalence" paper. In early February, Pauli was already aware (via Sommerfeld) of Schrödinger's (1926c) first paper on wave mechanics. In a letter to Sommerfeld of February 9, 1926, he refers to this work as "very interesting ... perhaps really not so crazy" (Pauli 1979, Doc. 120; Sommerfeld 2004, Doc. 95). He may even have seen Schrödinger's (1926c) second paper (published April 6) in Copenhagen before the letter of April 12 to Jordan discussed earlier, in which he lays out the mapping between wave functions and matrices, and shows (for fully discrete systems) that the matrices so defined obey the "equations of the Göttingen mechanics" (Pauli 1979, Doc. 131). This letter shows clearly that Pauli was remarkably free from ideological adherence to any predetermined approach to quantum theory. He was quite willing to abandon the dogma of the matrix approach (of which his calculation of the hydrogen spectrum using matrix techniques had shown him to be an absolute master) if a potentially equivalent, and calculationally more transparent and efficient, schema were to present itself. Pauli's (1930) famously caustic review of Born and Jordan's (1930) book *Elementare Quantenmechanik*, based exclusively on matrix mechanics, reflects his long-standing aversion to overly formal approaches to physical problems.[104] In any event, Pauli was perhaps the most prominent member of the younger generation of physicists to appreciate from the very outset the significance of wave mechanics.

Another important response to the wave theory came from Holland in late May, in the form of a detailed letter from H. A. Lorentz, the grand old master of classical physics who had received proofs of the first two installments of "Quantization as an eigenvalue problem" and the "equivalence" paper (Schrödinger 1926c, 1926d, 1926e) from the author, and had obviously taken the time to read all three articles carefully.[105] In an insightful letter, Lorentz quickly gets to the point, clearly identifying some of the salient features (both pro and con) of the two competing approaches to quantum physics represented by wave and matrix mechanics:

> If I had to choose now between your wave mechanics and the matrix mechanics, I would give the preference to the former, because of its greater intuitive clarity, so long as one has to deal with the three coordinates x, y, z. If, however, there are more degrees of freedom, then I cannot interpret the waves and vibrations physically, and I must therefore decide in favor of matrix mechanics. But your way of thinking has the advantage for this case too that it brings us closer to the real solution of the equations; the eigenvalue problem is the same in principle for a higher dimensional q-space as it is for a three-dimensional

104 For further discussion of Born and Jordan's book and Pauli's review, see Section 18.1.

105 For discussion of Lorentz's response to wave and matrix mechanics, see Kox (2013, sec. 8, pp. 166–168).

> space (Lorentz to Schrödinger, May 27, 1926, Lorentz 2008, Doc. 412, pp. 603–604; see also von Meyenn 2011, pp. 238–246).

The affinity in attitude between Lorentz and Schrödinger on the appropriate physical interpretation of the wave function is apparent in this quote: the originator of wave mechanics had also been troubled by the fact that the wave function $\psi(x_1, y_1, z_1, x_2, y_2, z_2, t)$ for, say, two particles could no longer be treated as a physically meaningful scalar field in real space(–time), as it was defined on the higher-dimensional configuration space specifying the location of both particles. In matrix mechanics, on the other hand, the spatial description adhered more closely to the basic structure of classical mechanics, inasmuch as a separate set of three matrices $\mathbf{x}_i, \mathbf{y}_i, \mathbf{z}_i, i = 1, 2, ..., N$ were introduced to correspond to the spatial coordinates in physical 3-space of each of the N particles comprising the system, maintaining thereby a clear structural similarity to classical mechanics.

There was, however, as Lorentz also saw immediately, a high price to be paid for the matrix representation of classical phase space in the Göttingen approach:

> There is another point in addition where your methods seem to me to be superior. Experiment acquaints us with situations in which an atom persists in one of its stationary states for a certain time, and we often have to deal with quite definite transitions from one such state to another. Therefore we need to be able to represent these stationary states, every individual one of them, and to investigate them theoretically. Now a matrix is the summary of all possible transitions and it cannot at all be analyzed into pieces. In your theory, on the other hand, each of the states corresponding to the various eigenvalues E plays its own role (Lorentz to Schrödinger, May 27, 1926; Lorentz 2008, Doc. 412, p. 604).

The critical new feature introduced in the wave-mechanical formulation—the concept of an *individual quantum state*—is clearly exposed here by Lorentz, even though the full implementation of this idea in the Hilbert state space of von Neumann lay more than a year in the future. In the remainder of his letter, Lorentz gives an itemized list with seven entries providing comments/criticisms on the contents of the three papers. Here, we focus on two of the difficulties identified by Lorentz in his remarks, namely:

1. The problem of the spreading of wave packets (items 3 and 4).
2. The origins of the Bohr frequency condition ($h\nu_{fi} = E_i - E_f$), as a consequence of either "beats" or "combination tones" (items 5, 6, and 7).

With regard to the first of these, Lorentz points out that the dispersive character of the Schrödinger wave equation—the fact that waves of different wavelengths move with different phase velocities—will inevitably lead to the spreading of wave-packets, even ones of initially atomic size, to arbitrarily large dimensions. The wave packet representing a single free electron (cf. Eq. (13.10)) is a superposition of components of well-defined frequency and wavelength that increasingly will lose phase agreement with each other over the course of time. One can show that, after a sufficient time, the growth in the width of the associated wave-packet is asymptotically linear in the time. In fact, the wave-packet of an electron liberated from an atom will spread to macroscopic dimensions in the course of milliseconds.[106] In his response to Lorentz, in a letter of June 6, 1926 (Lorentz 2008, Doc. 413; von Meyenn 2011, pp. 252–261), and in his *Naturwissenschaften* article published the following month (Schrödinger 1926g), Schrödinger pointed to the very agreeable property of Gaussian packets in the harmonic oscillator problem that they retain their initial width indefinitely, with the center of the packet oscillating, as one would expect according to the classical equation of motion. As for the behavior of *free* electrons, which clearly suffered from the spreading disease, he suggested, that perhaps

> "free" electrons do not permanently keep their identities at all in the usual sense? That speaking of individual electrons in a bundle of cathode rays perhaps means only that the bundle has a certain "granular" structure, in just the same way that many phenomena have made this plausible for a bundle of light rays, where in *both* cases neither a pure wave description nor a pure particle description exactly reaches the truth (Schrödinger to Lorentz, June 6, 1926; Lorentz 2008, Doc. 413, p. 614).

To Lorentz's comment that the concept of a wave-packet would necessarily lose its applicability for a bound electron in a hydrogen atom (for which the wave length is of the order of the size of the orbit), Schrödinger cautioned that a direct identification of particular stationary orbits with the eigenvibrations (solutions of definite energy) of his wave theory was illegitimate:

> One should *not* set the *individual* proper oscillations of the wave theory in parallel with the *individual* stationary orbits of the Bohr theory. For if one does that the transition from micromechanics to macromechanics by means of the correspondence principle is absolutely impossible. One can see how for large quantum number the individual Bohr orbits are built up by a superposition of very many proper oscillations which are relatively closely adjacent to one another (Schrödinger to Lorentz, June 6, 1926; Lorentz 2008, Doc. 413, p. 614).

It is indeed possible to concoct localized wave-packets that follow the Keplerian Bohr orbits by superposing wave solutions over a large range of principal n and angular quantum numbers l and m, with the average value of these quantum numbers much larger than unity (the original correspondence limit). Unlike the case of the Gaussian wave-packets in the simple harmonic oscillator, however, these wave packets inevitably spread out in the course of time to occupy a full three-dimensional region of the order of the atomic size. So Schrödinger's response does not really address Lorentz's qualms, specifically, that the "unavoidable blurring of a wave packet does not seem to me to be very

[106] The conceptual machinery (collapse of the wave function, decoherence due to interactions with the macroscopic environment, etc.) required to assimilate this fact still lay in the future. The question of wave-packet spreading resumed its relevance in the very practical setting of the detection of neutrinos from the supernova SN1987A. The arrival of these neutrinos in a short burst, despite spreading of their wave packets *in the neutrino rest frames* to dimensions of the order of 10^{18} meters, is perfectly understandable once relativistic frame effects are taken into account (Tzara 1988).

suitable for representing things to which we want to ascribe a rather permanent individual existence" (Lorentz to Schrödinger, May 27, 1926; Lorentz 2008, Doc. 412, p. 606).

The second main issue raised by Lorentz was the origin of the Bohr frequency condition in the new wave theory. In his "equivalence" paper, Schrödinger (1926e) proposed interpreting the real part of $e\psi\,\partial\psi^{\star}/\partial t$ as a charge density. As noted, Schrödinger (1926h) made the correct identification of the charge density with $e\psi\,\psi^{\star}$ in his fourth paper on wave mechanics and used that as the basis for a perturbative derivation of the Kramers dispersion formula. If the wave function $\psi(x,t)$ was composed of a superposition of energy eigenfunctions, with the appropriate complex exponential time dependent factor for each energy,

$$\psi(x,t) = \sum_k c_k u_k(x) e^{2\pi i E_k t/h}, \tag{14.227}$$

then a charge density constructed from the product of ψ and the time derivative of $\psi^{\star}$ would necessarily be a superposition of terms with time dependence $e^{2\pi i(E_k - E_m)t/h}$, that is, with exactly the monochromatic oscillatory behavior expected for a light wave/quantum emitted in the transition between the kth and m state. Lorentz (2008, Doc. 412, p. 608) called this time dependence a *combination tone*: it arises from the non-linear dependence of the charge density on the wave function, which contains the individual stationary state frequencies superposed *linearly*. Such a purely linear superposition also produces *beats*. For instance, in the example given by Lorentz, if we take real parts to get cosines rather than exponentials, we have

$$\begin{aligned} & a_1 \cos\left(2\pi\nu_1 t + b_1\right) + a_2 \cos\left(2\pi\nu_2 t + b_2\right) \\ &= (a_1 + a_2)\cos\left(2\pi\frac{\nu_1+\nu_2}{2}t + \frac{b_1+b_2}{2}\right)\cos\left(2\pi\frac{\nu_1-\nu_2}{2}t + \frac{b_1-b_2}{2}\right) \\ &\quad + (a_2 - a_1)\sin\left(2\pi\frac{\nu_1+\nu_2}{2}t + \frac{b_1+b_2}{2}\right)\sin\left(2\pi\frac{\nu_1-\nu_2}{2}t + \frac{b_1-b_2}{2}\right). \end{aligned}$$

Once the wave function contains at least two distinct components with frequencies ν_1 and ν_2, the combination can be re-expressed as a very fast[107] (co-)sinusoidal oscillation with frequency $(\nu_1+\nu_2)/2$, but modulated by the much slower difference frequency $(\nu_1-\nu_2)/2$. This difference frequency is still only one half of the Bohr transition frequency (assuming we set $E_1 = h\nu_1$ and $E_2 = h\nu_2$ as usual), so a further non-linear process is required in any case to demodulate this signal and produce the desired monochromatic oscillation of the emitted light with frequency $\nu_1 - \nu_2$. Schrödinger's response to Lorentz on the whole matter of beats versus combination tones throws an interesting light on the desperation which the Bohr frequency condition still evoked in 1926 (cf. note 94):

[107] It is clear that both Lorentz and Schrödinger imagine the basic frequencies as containing a part m_0c^2/h due to the unavoidably present, and very large, rest energy m_0c^2.

> You discuss the question of the explanation of radiation by means of beats or by means of difference [combination] tones in a very penetrating way that is also instructive for me. I must frankly admit that up to now I have not made enough of a conceptual distinction between these two things. I was so extremely happy to have arrived at a picture in which at least *something* really takes place with that frequency which we observe in the emitted light that, with the rushing breath of a hunted fugitive, I fell upon this something in the form in which it *immediately* offered itself, namely as the amplitudes periodically rising and falling with the beat frequency ... The frequency discrepancy in the Bohr model, by contrast, seems to me (and has indeed seemed to me since 1914) to be something so *monstrous*, that I should like to characterize the excitation of light in this way as really almost *inconceivable*. Between the *alternatives* of beats or difference tones, however, I obviously declare myself for the latter (Schrödinger to Lorentz, June 6, 1926; Lorentz 2008, Doc. 413, p. 615).

At the time of this letter Schrödinger was completing the fourth and final communication of the series "Quantization as an eigenvalue problem" (Schrödinger 1926h), so at least as far as the Kramers dispersion theory, Schrödinger already had (or would shortly have) at his disposal the time-dependent perturbation theory that would yield a formula for light scattering in which resonance poles appear in the correct places (cf. Eq. (14.175)), that is, at the frequencies corresponding to differences of quantized energies, and not, as would be natural in the Bohr model, at the frequencies of the orbital motions. The calculation in this fourth communication already contains the non-linearity (through the occurrence of the square of the wave function in the calculation of the dipole moment) necessary for the appearance of combination tones. A full understanding of the emission of light quanta, with frequencies given by the Bohr condition, in spontaneous quantum jumps between stationary states of atoms, would have to wait another nine months, when the whole matter was definitively settled in Dirac's (1927b) seminal paper on quantum electrodynamics.

Among the physicists who not only studied Schrödinger's work, but also actively tried to extend it to new problems—in the case of Pauli even anticipating some of Schrödinger's results—were Schrödinger's assistant in Zurich, Erwin Fues (1926a, 1926b), who concentrated on the application of the theory to diatomic molecules and their band spectra, Ivar Waller (1926), in Göteborg and Copenhagen, who worked on the Stark effect and the singly-ionized hydrogen molecule, and Paul Epstein (1926), at Caltech, who re-analysed the Stark effect in the new wave mechanics (see Section 15.3.2). A particularly active strand of research—here the main actors were Pauli, Gregor Wentzel, Fues, and Schrödinger himself—concerned the computation of matrix elements as overlap integrals of Schrödinger wave functions, which was universally agreed to be the royal road to the computation of spectral intensities. But here we leave our account of the genesis of wave mechanics: the story of the multitude of successful applications of Schrödinger's theory, and its further interaction, and eventual fusion with the matrix mechanics of Heisenberg *et al.*, properly belongs to our review of the resolution of the archetypal failures and successes of the old quantum theory in the new quantum mechanics, which follows in Chapter 15. In Chapter 16, we return to the debate about the interpretation of Schrödinger's wave function.

15

Successes and Failures of the Old Quantum Theory Revisited

Part Two of Volume 1 described the evolution of the quantum theory in the twelve years between the Bohr model and the appearance of matrix mechanics as guided by three important principles: the action quantization rules, now ascribed to Bohr and Sommerfeld (but independently discovered and elaborated by a number of other physicists), Ehrenfest's adiabatic principle, and Bohr's correspondence principle. The first and third of these in particular should perhaps more properly be called *methodologies*, as their application in many cases was more of an art than a science. In any case, these ideas were initially applied with great success by Sommerfeld and coworkers, leading within a few months in 1916 to the Sommerfeld formula for the relativistic fine structure of hydrogen and ionized helium, the extension of this formula to the understanding of X-ray spectra, and a successful derivation of the energy splittings in the Stark effect (by Schwarzschild and Epstein).

In other areas, however, these principles proved wholly inadequate, leading to the failure of the old quantum theory to account for several empirical features of atomic spectra, sometimes even at a qualitative level. In particular, the theory gave no explanation for the appearance of complex multiplet structure in non-hydrogenic atoms (atoms with more than one electron), where energy levels that should be singlets according to the theory were observed to be narrow doublets or triplets. The further splitting of these levels by an applied external magnetic field (the anomalous Zeeman effect) led to a rich phenomenology of varying patterns, which also proved impregnable to the theoretical assaults of the old quantum theory. Finally, the energy levels of the helium atom—both the appearance of a duplicate set of levels (parahelium and orthohelium), and the binding energy of the ground state itself—proved to be impossible to explain on the basis of the Bohr theory, even with the amplifications afforded by the three guiding principles of the old quantum theory.

Even in the areas where the old theory seemed to provide successful descriptions of the spectroscopic phenomena however, cracks soon began to appear in the edifice built by Sommerfeld's Munich school on the foundations of the 1913 Bohr model. The simple structure of the X-ray spectra that had seemed to fall perfectly in line with the fine structure formula of 1916 was found within a few years to reveal additional complex structure,

Constructing Quantum Mechanics. Anthony Duncan and Michel Janssen, Oxford University Press.
 DOI: 10.1093/oso/9780198883906.003.0015

just as in the case of the optical spectra. The anomalous Zeeman patterns stubbornly resisted explanation, despite increasingly elaborate contortions of the theory to account for the data. And the problem of the helium spectrum ultimately proved to be the most resistant of all to explanation on the basis of discretely selected electronic orbits following classically computable paths. About the only element of the theory that seemed to survive the barrage of increasingly accurate and comprehensive spectroscopic information was Sommerfeld's relativistic fine structure formula—and that only for single-electron atoms, of which only ionized helium could provide accurate empirical support.

The transition from classical physics to the old quantum theory, or from the old quantum theory to the new one which emerged from the complementary approaches of Heisenberg and Schrödinger, entailed a number of examples of what have come to be known as "Kuhn losses"—situations in which phenomena apparently correctly (or at least, adequately) treated in the old theory now proved problematic, or incorrectly described, by the new one, if only temporarily. The resolution of the role and interpretation of Sommerfeld's fine structure formula provides an example of just such a loss. Two other examples are the Kuhn losses associated with dispersion theory (in moving from classical to the old quantum theory), and the behavior of the temperature dependent contribution to electric susceptibilities (given essentially correctly in classical theory, but suffering a Kuhn loss in the old quantum theory, before a correct description emerged in the new quantum mechanics).

In this chapter we examine the extent to which the initial successes of the old quantum theory had to be reinterpreted in the context of the new theories initiated by Heisenberg and Schrödinger. This discussion inevitably connects to an explanation of how the most dramatic failures of the old theory (complex multiplet structure, in both optical and X-ray spectra, the anomalous Zeeman effect, and the helium atom) were first resolved in the post-1925 epoch. Thus, the chapter is divided into sections on fine structure, an intermezzo on Kuhn losses, external field problems (the Zeeman and Stark effects), and the helium atom.

15.1 Fine structure 1925–1927

Section 9.2 described Uhlenbeck and Goudsmit's (1925) proposal to attach a fourth *physical* degree of freedom to the electron, corresponding to the fourth quantum number that Pauli attributed to the electron in his introduction of the exclusion principle. This degree of freedom would be interpreted as a rotation of the electron on its own axis—in other words, the presence of non-zero angular momentum, even in the rest frame of the electron. The authors indicate (at the end of their extremely concise note) that the model had to satisfy two requirements: (a) the double magnetism previously attributed to the core must now be applied to the electron spin angular momentum itself, yielding a proportionality factor between magnetic moment and spin twice the normal (for orbital angular momentum) value; and (b) the explanation of the "relativity doublet" paradox of X-ray spectra (cf. Section 9.1) should now be sought in the varying orientation of the spin angular momentum with respect to the orbital angular momentum. In other words, the postulate of electron spin should explain the splittings—previously attributed to relativistic effects arising from different orbital shapes (known to give the famous Z^4 dependence of the Sommerfeld fine structure formula)—instead in terms of a spin–orbit coupling energy.

A quick calculation of this energy using Bohr–Sommerfeld methods quickly shows the important difference between the old core model of Heisenberg (cf. Section 7.1.2) and the new picture where magnetic effects were attributed directly to the electron. In the core model the splitting energy

$$\Delta E = 2\mu_R H_i = \frac{e\hbar}{2mc} H_i \tag{15.1}$$

is due to a magnetic atomic core aligning or anti-aligning with an internal magnetic field H_i due to an outer valence electron circulating around the core (which is placed at a focus of the corresponding Bohr–Sommerfeld ellipse, cf. Eqs. (7.25)–(7.26)),

$$H_i = \frac{2\pi I}{ac(1-\epsilon^2)}, \tag{15.2}$$

where I is the effective current in the loop formed by the electron's orbit, a the semi-major axis, and ϵ (with $\sqrt{1-\epsilon^2} = k/n$) the eccentricity (with k the orbital angular momentum in units of $\hbar$, n the principal quantum number). The current is e/T, where T is the orbit period, and for an atom with a core of effective charge $Z^\star$, $a = n^2 a_0/Z^\star$, where n is the principal quantum number and a_0 is the Bohr radius for hydrogen, and (Kepler's third law) $T = 2\pi m a^2\sqrt{1-\epsilon^2}/k\hbar$. When these results are inserted, a little algebra yields (cf. Eq. (7.27)):

$$\Delta E = \frac{1}{2} mc^2 \alpha^4 \frac{Z^{\star 3}}{n^3 k^2}. \tag{15.3}$$

On the one hand, this displays the troublesome third-power dependence on the effective nuclear charge that Landé had puzzled over, in conflict with the well-established fourth-power dependence seen in X-ray spectra. On the other hand, the formula displays an unmistakable similarity to the splitting (now between orbits of different angular momentum, hence, eccentricity, and arising from relativistic effects!) arising from the fine structure term in the Sommerfeld formula (cf. Eqs. (6.53) and (7.28)),

$$\Delta E = \frac{1}{2} mc^2 \alpha^4 Z^{\star 4} \left(\frac{1}{n^3(k-1)} - \frac{1}{n^3 k} \right) = \frac{1}{2} mc^2 \alpha^4 Z^{\star 4} \frac{1}{n^3 k(k-1)}. \tag{15.4}$$

We have already seen how this result is changed if the magnetic moment is transferred from the core to the electron, in our description of Kronig's introduction of electron spin in early 1925 (cf. Section 9.2). We re-derive this result in a slightly different way here, which isolates more clearly the origin of the shift in dependence on the effective charge of the core. In the new picture, the magnetic moment of the core is transferred to the electron, with a factor of 2 reflecting the double magnetism (which was by now unavoidable) while the magnetic field H_c at the location of the electron can be viewed as due to the motion of the electron through the electrostatic field of the core, thought of as executing an orbit around the electron (in the latter's rest frame), and therefore producing a magnetic field

$$H_c = \frac{2\pi I}{ac(1-\epsilon^2)}, \quad I = \frac{Z^\star e}{T}, \tag{15.5}$$

with a, ϵ, I, and T as before (cf. Eqs. (7.25) and (7.26)). Inserting these, one finds, instead of Eq. (15.3), a spin–orbit interaction splitting

$$\Delta E = 2 \times \frac{1}{2} mc^2 \alpha^4 \frac{Z^{\star 4}}{n^3 k^2}. \tag{15.6}$$

This now agrees with the Sommerfeld splitting Eq. (15.4) (with the correct fourth-power dependence on $Z^\star$), with the exception of an extra factor of 2, and the (less important qualitatively) change $k^2 \to k(k-1)$.

Section 9.2 described how Heisenberg (who was well aware of Landé's struggles in 1924 to reconcile the Z^3 behavior of the core model with the Z^4 law of X-ray doublets), on seeing Uhlenbeck and Goudsmit's paper in the November 20 issue of *Naturwissenschaften,* immediately wrote to Pauli pointing out the (very desirable) feature of the spin hypothesis in yielding the correct Z^4 behavior (Pauli 1979, Doc. 107). However, there was the (less-desirable) fact that the spin–orbit coupling energy came out too big by a factor of 2. Moreover, in the case of optical spectra, if the explanation for the fine structure was now to be found in a spin–orbit coupling, what role was there for Sommerfeld's relativity corrections? The splittings observed in hydrogen and ionized helium continued to agree exactly with the Sommerfeld relativistic fine structure after all, and there was the extremely puzzling similarity of the algebraic form of the splitting obtained *either* by calculating relativistic corrections to orbit energies *or* by considering the interaction of the magnetic moment of a spinning electron with the magnetic field seen by the electron due to its orbital motion through the electrostatic field of the nucleus (or core). These two physical mechanisms seemed to have nothing to do with each other, while the actual empirical fine structure required one or the other, but not both. Despite the many confusing aspects of the magnetic electron proposal, Heisenberg indicated to Pauli that he wanted "to put this model once through the matrix-mill and see whether the *g*-formulae, intensities, etc. come out correctly" (Pauli 1979, Doc. 107).

Pauli was no more sympathetic to the idea of electron spin when queried by Heisenberg in late November 1925 than he had been when presented with essentially the same idea by Kronig in January of that year. He was extremely unwilling to appeal to simple visualizable mechanical models, such as the spinning of an electron on its own axis, to explain the quintessentially quantum mechanical "double-valuedness" that lay at the heart of the problems of complex spectra and the anomalous Zeeman effect (see Section 9.2 and Pais 1986, p. 279). Bohr, however, became an early convert, after meeting

Goudsmit and Uhlenbeck in Leyden in early December (at the jubilee in honor of the 50th anniversary of Lorentz's doctorate), and a conversation there with Einstein emphasizing the natural emergence of a spin–orbit coupling energy using relativistic reasoning seems to have completed his conversion (Mehra and Rechenberg 1982c, p. 202). On his way back to Copenhagen, Bohr subsequently visited Göttingen and Berlin, touching base with Heisenberg and Pauli, respectively. Heisenberg, in concert with Jordan, had already set about applying matrix-mechanical techniques to the evaluation of both relativistic and spin–orbit effects, and to a re-evaluation of the anomalous Zeeman effect on the basis of the new magnetic electron hypothesis. In Berlin Pauli seems to have at least conceded to Bohr that it was worth examining the implications of electron spin when treated in a matrix-mechanical scheme.

Over the next few months, first Pauli and Heisenberg, and then Jordan in close collaboration with Heisenberg, struggled with the problem of fine structure and the Zeeman effect for hydrogenic atoms in the matrix-mechanical framework. The main technical difficulty was in evaluating the diagonal matrix elements of the matrices corresponding to inverse powers (specifically $1/r^2, 1/r^3$) of the radial coordinate r for the electron.[1] These matrix elements were needed in the calculation of the first-order perturbative shifts induced by relativistic corrections, and, if the Goudsmit–Uhlenbeck hypothesis were accepted, in the quantitative evaluation of the perturbative energy shift due to the spin–orbit interaction. By comparison, the evaluation of the magnetic perturbation energy (due to an applied magnetic field) turned out to be fairly trivial in the new theory. The results of these endeavors were revealed in an article submitted the middle of March 1926 and published in early June (Heisenberg and Jordan 1926).

It is convenient to collect at the outset the various relevant terms in the electronic energy analyzed by Heisenberg and Jordan in their article. One imagines an electron subject to a Coulomb potential $-Ze/r$, in the presence of an external magnetic field $\vec{\mathfrak{H}}$, with an electron to which an orbital angular momentum[2] $\vec{\mathfrak{l}}$ and, in accordance with Uhlenbeck–Goudsmit, a spin-angular momentum $\vec{\mathfrak{s}}$ is ascribed, and with the first post-Newtonian relativistic corrections to the kinetic energy included. The Hamiltonian of the system thus takes the form

$$H = H_0 + H_1 + H_2 + H_3 = \frac{\vec{p}^2}{2m} - \frac{Ze^2}{r} + H_1 + H_2 + H_3. \qquad (15.7)$$

Here, H_0 is the unperturbed Hamiltonian, and H_1, H_2, and H_3 represent the external magnetic field perturbation, the spin–orbit energy, and the leading relativistic corrections, respectively.

[1] At the time of these investigations, Heisenberg and his collaborators were, at least initially, unaware of the new formalism introduced by Schrödinger, in which the desired matrix elements would be given by straightforward integrals of products of wave functions of fairly simple analytic form. The contortions needed to obtain these matrix elements in the absence of this knowledge were, as we will soon see, extremely painful.

[2] $\vec{k}$ in the notation of Heisenberg and Jordan (1926)—we employ the more readily recognizable modern notation here.

Taking these in order, we find that the interaction energy of the electron with the external magnetic field is given by (cf. Eq. (7.98))

$$H_1 = \frac{e}{2mc}\vec{\mathfrak{H}} \cdot (\vec{\mathfrak{l}} + 2\vec{\mathfrak{s}}). \tag{15.8}$$

The doubled gyromagnetic ratio needed for the spin contribution to the magnetic moment is present in the second term, the Uhlenbeck–Goudsmit replacement for the former "double magnetism of the core".

Next, there is the spin–orbit interaction energy. The motion of the electron through an external electrostatic field $\vec{E} = Ze\vec{r}/r^3$ (Coulomb field of the positively charged nucleus) means that in its rest frame the electron experiences a magnetic field $\vec{\mathfrak{H}}_i = (1/c)\vec{E} \times \vec{v}$ where $\vec{v}$ is the velocity of the electron.[3] The magnetic energy $-\vec{\mu}_s \cdot \vec{\mathfrak{H}}_i$ associated with the interaction of the (doubled) spin magnetic moment $\vec{\mu}_s = -(e/mc)\,\vec{\mathfrak{s}}$ (minus sign from the electron charge) with this magnetic field is thus given by

$$\frac{e}{mc}\vec{\mathfrak{s}} \cdot \left(\frac{Ze}{cr^3}\vec{r} \times \vec{v}\right) = \frac{Ze^2}{mc^2r^3}\vec{\mathfrak{s}} \cdot (\vec{r} \times \vec{v}) = \frac{Ze^2}{m^2c^2r^3}\,\vec{\mathfrak{s}} \cdot \vec{\mathfrak{l}}, \tag{15.9}$$

where we have introduced the orbital angular momentum $\vec{\mathfrak{l}} = \vec{r} \times \vec{p} = m\vec{r} \times \vec{v}$. By the time Heisenberg and Jordan submitted their paper, it had become apparent that this result needs an extra factor of one half, due to a subtle relativistic effect, which was discovered by L. H. Thomas (1926).[4] The absence of this factor had plagued discussion of spin–orbit (or core–orbit) interactions for quite some time, at one point evoking from Pauli (letter to Heisenberg, 31 January 1926) the cry, evoking Shakespeare's Richard III, "A half, a half, a kingdom for the factor $\frac{1}{2}$!" Inserting the long-desired factor, one has for the spin–orbit contribution to the Hamiltonian:

$$H_2 = \frac{Ze^2}{2m^2c^2r^3}\,\vec{\mathfrak{s}} \cdot \vec{\mathfrak{l}}. \tag{15.10}$$

[3] This is a classic relativistic effect, and is probably what Einstein pointed out to Bohr at the Lorentz jubilee conference in December 1925, when "he explained that this [spin–orbit] coupling was an immediate consequence of the theory of relativity," as Bohr related to Kronig in a letter of 26 March, 1926. For the relativistic transformation of the electric field, see Griffiths (1999), Eq. (12.109)—in Gaussian units the magnetic field $\vec{B}$ becomes $\vec{\mathfrak{H}}/c$. We use the Fraktur font here to avoid confusion between the magnetic field $\mathfrak{H}$ and the Hamiltonian H.

[4] For an elementary derivation, see the web resource, *Thomas Precession*. According to Mehra and Rechenberg (1982c, p. 273), the Thomas correction factor of $\frac{1}{2}$ was actually inserted into the calculation of the H_2 term by Jordan in late February, after Heisenberg's departure on vacation. Pauli's initially vociferous objections to the Thomas factor (pp. 270–272), finally evaporated in early March, in a letter to Bohr (Pauli 1979, Doc. 127). Shortly thereafter, the Heisenberg–Jordan paper was submitted (by Jordan), and was received by *Zeitschrift für Physik* on March 16.

Finally, there is the relativistic correction term. Expanding the relativistic expression for kinetic energy, we find

$$\begin{aligned}\sqrt{m^2c^4+p^2c^2}-mc^2 &= mc^2\sqrt{1+\frac{p^2}{m^2c^2}}-mc^2\\ &= mc^2\left(1+\frac{p^2}{2m^2c^2}-\frac{1}{8}\frac{p^4}{m^4c^4}+\ldots\right)-mc^2\\ &= \frac{p^2}{2m}-\frac{1}{8}\frac{p^4}{m^3c^2}+\ldots\\ &= \frac{p^2}{2m}-\frac{1}{2mc^2}\left(\frac{p^2}{2m}\right)^2+\ldots. \end{aligned} \tag{15.11}$$

The final term in the last line here is the desired leading relativistic perturbation, while the first term is of course already included in the unperturbed Hamiltonian H_0. It is to be evaluated, as per the usual prescription for first order perturbation theory, with respect to the unperturbed theory. Since we can write

$$\frac{p^2}{2m}=H_0+\frac{Ze^2}{r}, \tag{15.12}$$

we can replace H_0 in this term by the unperturbed (Balmer) energy E_0 (W_0 in the Heisenberg–Jordan paper) of the electron state of which we are computing the energy shift, and write

$$H_3=-\frac{1}{2mc^2}\left(E_0^2+\frac{2Ze^2E_0}{r}+\frac{Z^2e^4}{r^2}\right). \tag{15.13}$$

By the end of December 1925, Heisenberg and Jordan had established that the perturbations H_1 and H_2 (with the as yet unknown diagonal matrix element of the $1/r^3$ matrix, needed for the evaluation of the spin–orbit energy, left embedded in an undetermined parameter λ, which also conveniently contained the still mysterious missing factor of $\frac{1}{2}$) gave rise to exactly the algebraic structure of the phenomenological Voigt–Sommerfeld theory for the anomalous Zeeman effect for hydrogenic and alkali atoms, including the characteristic square roots describing the transition to the Paschen–Back effect (see Sections 7.2–7.3). But a quantitative comparison of the theory with experiment, as well as a calculation of the zero-field doublet structure, necessitated evaluation of the diagonal elements of the matrices for the inverse powers of the radial coordinate $1/r$, $1/r^2$, and $1/r^3$ appearing in H_2 and H_3.

Reversing the order of the discussion of Heisenberg and Jordan (1926), we first examine their calculation of the energy for zero external magnetic field, setting $H_1 = 0$, returning in the next section to the Zeeman effect arising from this term. This represents the first efforts made in the new quantum mechanics to come to terms with the existence of complex structure in spectra of multi-electron atoms—a phenomenon we identified in Volume 1 as the first of three great failures of the old quantum theory. The practitioners of matrix mechanics addressed this problem first in the calculation of the fine structure for hydrogenic atoms (a single electron, with the nuclear charge $+Ze$, Z integer), or for the valence electron levels in alkali atoms, assuming that replacing Z by some effective (non-integral) nuclear charge $Z^\star$ is appropriate in the latter case.

It is apparent that the calculation of the energy shift due to the spin–orbit (H_2) and relativistic (H_3) terms requires the diagonal matrix elements (in modern language, the expectation values) of $1/r, 1/r^2$, and $1/r^3$ in the Balmer levels with respect to which the unperturbed Hamiltonian is a diagonal matrix (with eigenvalues $E_0 = -Z^2me^4/(2\hbar^2n^2)$). The evaluation of these quantities involves a somewhat complicated story stretching from Heisenberg's first encounter with the Goudsmit–Uhlenbeck proposal in late November 1925 to March of 1926, when the resulting article was completed and submitted (by Jordan).

In the evaluation of the first order perturbation shifts we must begin by reminding the reader that in the matrix mechanics of any periodic system the time average of a dynamical quantity eliminates off-diagonal matrix elements, provided the system is non-degenerate (cf. Eq. (12.154)). In fact, the time-average of quantities, written with an overbar (e.g. $\overline{1/r}, \overline{1/r^2}$, etc.) are used throughout the Heisenberg–Jordan paper to represent what we would nowadays call the expectation value of that quantity in the unperturbed state. The state in question here is a stationary state of the unperturbed hydrogenic atom described by Hamiltonian H_0, with a definite value of the principal quantum number n, and angular quantum number $l, 0 \leq l < n$. To evaluate the diagonal matrix element of $1/r$ in any stationary state corresponding to quantum number n, Heisenberg and Jordan employ the virial theorem for the Coulomb potential, which states (with $E_{\rm kin} = \vec{p}^{\,2}/2m, E_{\rm pot} = -Ze^2/r$),

$$\overline{E_{\rm kin}} = -\frac{1}{2}\overline{E_{\rm pot}} \Rightarrow E_0 = \overline{E_{\rm kin}} + \overline{E_{\rm pot}} = \frac{1}{2}\overline{E_{\rm pot}} = -\frac{1}{2}Ze^2\overline{\frac{1}{r}}, \tag{15.14}$$

where the overbars indicate time averages. For an atom in a state characterized by principal quantum number n, we have $E_0 = -Z^2me^4/(2\hbar^2n^2)$, so dividing the second equation by $Ze^2/2$, we find

$$\overline{\frac{1}{r}} = \frac{Zme^2}{\hbar^2n^2} = \frac{Z}{n^2a_0}, \tag{15.15}$$

where $a_0 = \hbar^2/(me^2)$ is the Bohr radius for the hydrogen atom. The virial theorem had always worked perfectly in the old quantum theory, and its continued validity in the new quantum mechanics was accepted tacitly and without further comment by Heisenberg and Jordan.

The evaluation of the diagonal elements/time averages of the remaining inverse radial powers $1/r^2, 1/r^3$ appearing in Eqs. (15.10)–(15.13) turned out to be much more problematic, and was the subject of intense effort by Pauli and Heisenberg for the two months leading up to the end of January 1926.[5] In a letter to Heisenberg of January 31, 1926, Pauli indicated that the sum of the spin–orbit and relativistic shifts reproduced the Sommerfeld formula, apart from a missing factor of one half in the spin–orbit term:

> There is a very peculiar circumstance here [footnote: You will naturally regard the following as a quite stupid remark. I agree with you fully in that. However I don't consider it completely excluded that somebody can make a clever remark out of it!]. If one calculates with one half of the magnetic energy E_M [i.e., the spin–orbit part of the Hamiltonian] (what that means, I do not know), then the sum $E_R + \frac{1}{2}E_M$ [thus, $\frac{1}{2}E_M$ is H_2 above, E_R the relativistic energy correction H_3 in the given stationary state] has exactly the properties that one needs ... the values of the terms with the same j fall *exactly* together, and the difference of $E_R + \frac{1}{2}E_M$ for terms with adjacent j corresponds exactly to the Sommerfeld formula. Accordingly: "*A half, a half, a kingdom for the factor* $\frac{1}{2}$"! (Pauli 1979, pp. 283–284).

What this letter makes clear is that, although not yet aware of the origin of the Thomas factor of $\frac{1}{2}$, Pauli and Heisenberg were by now in the possession of the correct expressions for the expectation values of $1/r$, $1/r^2$, and $1/r^3$, which would be needed for Pauli to draw the conclusions stated here. Of course, we now know that the ideal machinery (see note 13) for calculating these expectation values—Schrödinger's wave mechanics—was being developed simultaneously in Zurich as Pauli and Heisenberg struggled with the problem (in Hamburg and Göttingen, respectively, but with frequent visits and exchange of letters).

We begin with the evaluation of the inverse radial square $\overline{1/r^2}$, where the time average is taken with respect to the unperturbed system (electron in a Coulomb field). If we consider the classical motion of an electron around a nucleus of charge $+Ze$ in two dimensions, with the plane described in (r, φ) polar coordinates, the orbital angular momentum p_φ, which is a constant of the motion, is given by

$$p_\varphi = mr^2\dot{\varphi}. \tag{15.16}$$

This can be inverted to give an equation for $1/r^2$:

$$\frac{1}{r^2} = \frac{m}{p_\varphi}\dot{\varphi}. \tag{15.17}$$

Here, we imagine that the state in question (for which we wish to evaluate the expectation value of $1/r^2$) has been assigned a definite value for the azimuthal quantum number associated with p_φ, which can therefore be regarded as a fixed scalar number. The matrix

[5] The algebraic gymnastics needed to obtain the required matrix elements by purely matrix-mechanical reasoning, which occupy the next few pages, are unimportant for later developments. The essential results are Eq. (15.40) (for the diagonal matrix elements of $1/r^2$) and Eq. (15.44) (for $1/r^3$), which, together with Eq. (15.15) for $1/r$, determine the needed diagonal matrix elements of H_2 and H_3. The reader may resume the main theme of our story in the paragraph following Eq. (15.44).

of $1/r^2$ is therefore proportional to the matrix of the angular velocity $\dot{\varphi}$. It is important to recall that in matrix mechanics, only periodic time dependence (of the form $e^{2\pi i(E_m-E_n)t/h}$) is allowed in the matrix elements. The angle φ itself classically has a periodic piece superimposed on a linear term: $\varphi(t) = \omega t +$ periodic terms. For example, if the motion is elliptical, the angular velocity goes up near the perihelion and down at aphelion, averaging to $2\pi/T$, where $T = 2\pi/\omega$ is the period of the orbit:

$$\overline{\dot{\varphi}} = \frac{1}{T}\int_0^T \dot{\varphi}(t)dt = \frac{2\pi}{T}. \tag{15.18}$$

In the time derivative, the linear dependence on time disappears and the periodic terms remain, so $\dot{\varphi}$ has a perfectly well-defined matrix with only periodic time dependence.[6] Evidently, the desired time-averaged value of $1/r^2$ (automatically a diagonal matrix), can be obtained once the corresponding average of $\dot{\varphi}$ is known:

$$\overline{\frac{1}{r^2}} = \frac{m}{p_\varphi}\overline{\dot{\varphi}}. \tag{15.19}$$

In the end, the desired $\overline{\dot{\varphi}}$ was obtained using an ingenious mapping between the two- and three(spatial)-dimensional versions of the quantized Coulomb problem.

Staying with the Coulomb problem in two dimensions for the time being, we recall that the classical Hamiltonian

$$H^{(2D)} = \frac{1}{2m}\left(p_r^2 + \frac{p_\varphi^2}{r^2}\right) - \frac{Ze^2}{r} \tag{15.20}$$

can be regarded as a one-dimensional periodic radial system once some fixed value for the angular momentum p_φ is chosen. A canonical transformation from (r, p_r) to angle/action variables $(w, \mathcal{J})$,[7] can now be found such that

$$H^{(2D)} = -\frac{2\pi^2 me^4 Z^2}{(\mathcal{J} + 2\pi p_\varphi)^2}. \tag{15.21}$$

Here, $\mathcal{J} = \oint p_r dr$, the radial action variable, is a constant of the motion, and the Hamiltonian equation of motion implies

$$\dot{w} = \frac{\partial H^{(2D)}}{\partial \mathcal{J}} = \frac{4\pi^2 me^4 Z^2}{(\mathcal{J} + 2\pi p_\varphi)^3} \equiv \nu = \text{const.} \tag{15.22}$$

The angle variable $w(t)$ is therefore a linear function of time, and by the general theory of action/angle variables, it increments by exactly unity over a single orbital period

[6] In fact, as is apparent from Eq. (15.17), the matrix of $\dot{\varphi}$ is proportional to the product of the (diagonal) matrix of p_φ and the (not diagonal) matrix of $1/r^2$. All of the off-diagonal matrix elements of this matrix contain the usual periodic time-dependence factors that are posited in matrix mechanics.

[7] For an introduction to the use of action/angle variables, see Section A.2.

(the range over which the loop integral in $\mathcal{J} = \oint p_r dr$ is taken). Thus, ν is the cyclic frequency of the system. The angle variable $w(t)$ (associated with the radial coordinate $r(t)$) should be carefully distinguished from the original *physical* angle variable $\varphi(t)$—the latter is in general not constant in time, for example in elliptical orbits where the angular velocity increases toward perihelion and decreases toward aphelion. The angle variable $w(t)$ can be regarded as a "smoothed out" version of $r(t)$, increasing *uniformly* (whence the original alternative term for action/angle coordinates, as "uniformizing variables") in time, so as to increment by unity after each orbital period. On average, of course, $\varphi(t)$ increases by 2π per orbital period, or by a factor of 2π times the increase for $w(t)$, so

$$\overline{\dot{\varphi}} = 2\pi\dot{w} = 2\pi\nu. \tag{15.23}$$

In the old quantum theory, the desired average of the angular velocity would be obtained directly from Eq. (15.22) by the Bohr–Sommerfeld quantization prescriptions

$$\mathcal{J} = \oint p_r dr = n_r h, \quad 2\pi p_\varphi = 2\pi \cdot k\frac{h}{2\pi} = kh, \tag{15.24}$$

where n_r and k are the radial and azimuthal quantum numbers, respectively. Inserting these values in Eqs. (15.22)–(15.23), and defining the principal quantum number as $n = n_r + k$, one finds

$$\overline{\dot{\varphi}} = 2\pi\frac{4\pi^2 me^4 Z^2}{n^3 h^3} = \frac{8\pi^3 me^4 Z^2}{n^3 h^3}. \tag{15.25}$$

Given $\overline{\dot{\varphi}}$, $\overline{1/r^2}$ would follow immediately from Eq. (15.19):

$$\overline{1/r^2} = \frac{8\pi^3 m^2 e^4 Z^2}{n^3 h^3 p_\varphi}. \tag{15.26}$$

The question in December 1925 was whether these results continued to be valid in the new matrix mechanics. It was a problem with which Heisenberg and Pauli would wrestle for the next two months.

There were two essential sticking points in the derivation and use of Eqs. (15.25)–(15.26) in the context of the new matrix version of quantum theory which had to be confronted by Pauli and Heisenberg. First, it was essential to translate the action/angle formalism applied in the foregoing argument to the new matrix formalism; and, secondly, one had to properly interpret the quantum number associated with the angular momentum p_φ in the *three-dimensional* quantized hydrogen(ic) atom problem, the quantized spectrum of which clearly differed completely from the two-dimensional case in the new quantum mechanics, even though the treatment of the two cases gave identical results in the old quantum theory (where the plane of the orbit could always be described with x, y or r, φ coordinates, and motion in the third direction ignored without affecting the quantization procedure in any essential way).

The first difficulty has already been alluded to above: in matrix mechanics, the representation of a kinematic variable by a discrete matrix presupposes that the associated variable has a periodic dependence on time. On the other hand, for aperiodic motion, such as that of a free or unbound particle, or even for angle variables in periodic motion, the kinematic variables have a component that is monotone increasing. In the case of the position coordinate of a free particle or the angle variables of a multi-periodic system, the dependence on time is linear. The operator mechanics of Born and Wiener (1926, see Section 14.5) had been introduced to take care of just such cases: for example, by introducing linear differential (and integral) operators in the time, such as $D \equiv d/dt$, one could posit a canonical commutation relation between the (clearly aperiodic) time variable t and the energy operator H, now written as $H = (h/2\pi i)D$: thus,

$$Ht - tH = \frac{h}{2\pi i}, \tag{15.27}$$

as a consequence of $Dt - tD = 1$.

The linearly time-dependent angle variables and time-independent action variables could similarly be treated *à la* Born and Wiener by introducing operator versions satisfying

$$\mathcal{J}w - w\mathcal{J} = \frac{h}{2\pi i}. \tag{15.28}$$

The Hamiltonian equations associated with this new canonical pair would then imply, given that the energy is just a function of the action variable, $H = H(\mathcal{J})$,

$$\dot{w} = \frac{2\pi i}{h}[H(\mathcal{J}), w] = \omega(\mathcal{J}) = \text{const.} \tag{15.29}$$

Heisenberg also showed that the exponentiated version of these angle variables, $e^{i\tau w}$, $\tau = 0, \pm 1, \pm 2, \ldots$, had a time dependence corresponding to a Bohr transition between states with action $\mathcal{J}_n = n(h/2\pi)$ and $\mathcal{J}_{n-\tau} = (n - \tau)(h/2\pi)$ (for a system of a single degree of freedom), given by (see Pauli 1979, Doc. 117):

$$e^{2\pi i(H(\mathcal{J}_n) - H(\mathcal{J}_{n-\tau}))t/h}. \tag{15.30}$$

By the end of January, Heisenberg and Pauli had convinced themselves that, at least for a system of a single degree of freedom (such as the radial motion in the Coulomb problem), the action/angle formalism that had been applied classically in the old quantum theory could be brought over virtually intact to the new matrix/Born–Wiener operator formalism, thereby justifying the steps leading to Eq. (15.26).[8]

[8] In early March, Pauli read a paper by Dirac (1926a) in which the same issues were explored: the translation into matrix-mechanical terms of action/angle formalism. Dirac avoids the difficulty of dealing with aperiodic angular variables by studying the commutation relations of the action variables $\mathcal{J}_r$ with well-defined periodic

The second difficulty faced by Heisenberg and Pauli, which we alluded to previously, concerned the interpretation of the angular momentum quantum number associated with p_φ in Eq. (15.26). The angular momentum p_φ was that associated with the electron in the two-dimensional version of the Coulomb problem—that is, it corresponded to the total angular momentum in the physically relevant three-dimensional problem. What Heisenberg and Pauli realized was that the Coulomb problems in two and three dimensions could be written in a form that gave Hamiltonians of identical form, which could be mapped onto each other simply by picking an appropriate value for p_φ. To see how this works, let us begin with the two-dimensional problem, in Cartesian coordinates (x, y), where the (lowest-order) Hamiltonian is clearly

$$H_0^{(2D)} = \frac{1}{2m}\left(p_x^2 + p_y^2\right) - \frac{Ze^2}{r}, \quad r \equiv \sqrt{x^2 + y^2}. \tag{15.31}$$

Here p_x, p_y, x, y are matrices satisfying the canonical commutation relations

$$\begin{aligned} [p_x, x] &= [p_y, y] = \frac{h}{2\pi i}, \\ [p_x, y] &= [p_y, x] = [p_x, p_y] = [x, y] = 0. \end{aligned} \tag{15.32}$$

If one defines new matrices

$$p_r = \frac{1}{2}\left(\vec{p} \cdot \frac{\vec{r}}{r} + \frac{\vec{r}}{r} \cdot \vec{p}\right), \tag{15.33}$$

$$p_\varphi = x p_y - y p_x, \tag{15.34}$$

one finds that they are canonically conjugate to the polar coordinate variables $r = \sqrt{x^2 + y^2}$ and $\varphi = \arctan\,(y/x)$:

$$\begin{aligned} [p_r, r] &= [p_\varphi, \varphi] = \frac{h}{2\pi i}, \\ [p_r, \varphi] &= [p_\varphi, r] = [p_r, p_\varphi] = [r, \varphi] = 0. \end{aligned} \tag{15.35}$$

quantities like $e^{i\tau_r w_r}$, which have a perfectly fine matrix representation, without any need for introducing Born–Wiener type differential operators. Pauli shortly after expressed his disappointment to Kramers that he had wasted a lot of time on this only to be beaten to the punch by Dirac (see Mehra and Rechenberg 1982c, pp. 252–253). Dirac's paper also had a treatment of the hydrogen-atom Hamiltonian, in its two-dimensional version Eq. (15.36), with however a complicated and not quite complete derivation of the Balmer–Bohr eigenvalues. We recall that this had already been accomplished by Pauli in late 1925 (Pauli 1926a).

Using these new kinematic variables, our original Hamiltonian matrix Eq. (15.31) takes the form[9]

$$H_0^{(2D)} = \frac{1}{2m}\left(p_r^2 + \frac{1}{r^2}\left(p_\varphi^2 - \frac{h^2}{16\pi^2}\right)\right) - \frac{Ze^2}{r} \tag{15.36}$$

A similar argument shows that the three-dimensional Coulomb Hamiltonian,

$$H_0^{(3D)} = \frac{1}{2m}\left(p_x^2 + p_y^2 + p_z^2\right) - \frac{Ze^2}{r}, \quad r \equiv \sqrt{x^2 + y^2 + z^2}, \tag{15.37}$$

with momentum and coordinate matrices satisfying the obvious extension of Eqs. (15.32), can be re-expressed in terms of a radial momentum matrix p_r (with $[p_r, r] = \frac{h}{2\pi i}$) and the squared orbital angular momentum matrix $\vec{l}^2 = L_x^2 + L_y^2 + L_z^2$ (where $[L_x, L_y] = i\frac{h}{2\pi}L_z$ etc.), with the properties derived in the Three-Man-Paper:[10]

$$H_0^{(3D)} = \frac{1}{2m}\left(p_r^2 + \frac{\vec{l}^2}{r^2}\right) - \frac{Ze^2}{r}. \tag{15.38}$$

[9] Although this can be demonstrated by careful matrix algebra, paying attention to the commutation relations in Eqs. (15.32) and (15.35), as in Dirac (1926a), the modern reader will probably find it much easier to obtain Eq. (15.36) by using the differential representations

$$p_x = \frac{\hbar}{i}\frac{\partial}{\partial x}, p_y = \frac{\hbar}{i}\frac{\partial}{\partial y} \quad \text{and} \quad p_r = \frac{\hbar}{i}r^{-1/2}\frac{\partial}{\partial r}r^{+1/2}, p_\varphi = \frac{\hbar}{i}\frac{\partial}{\partial \varphi},$$

which clearly satisfy Eqs. (15.32) and (15.35). The differential identity

$$\left(r^{-1/2}\frac{\partial}{\partial r}r^{+1/2}\right)^2 = \left(\frac{\partial}{\partial r} + \frac{1}{2r}\right)^2 = \frac{\partial^2}{\partial r^2} + \frac{1}{r}\frac{\partial}{\partial r} - \frac{1}{4r^2},$$

together with the well known form for the two-dimensional Laplacian in polar coordinates,

$$\Delta = \frac{\partial^2}{\partial x^2} + \frac{\partial^2}{\partial y^2} = \frac{\partial^2}{\partial r^2} + \frac{1}{r}\frac{\partial}{\partial r} + \frac{1}{r^2}\frac{\partial^2}{\partial \varphi^2},$$

implies that the kinetic energy part of the Hamiltonian becomes

$$-\frac{\hbar^2}{2m}\Delta = \frac{1}{2m}(p_r^2 + (p_\varphi^2 - h^2/16\pi^2)/r^2),$$

whence Eq. (15.36).

[10] Again, the matrix manipulations required are somewhat painful: it is easiest to use the differential representations

$$p_r = \frac{\hbar}{i}r^{-1}\frac{\partial}{\partial r}r^{+1},\ p_r^2 = -\hbar^2\left(\frac{\partial^2}{\partial r^2} + \frac{2}{r}\frac{\partial}{\partial r}\right) \quad \text{and} \quad \vec{l}^2 = -\hbar^2\left(\frac{1}{\sin\vartheta}\frac{\partial}{\partial\vartheta}\left(\sin\vartheta\frac{\partial}{\partial\vartheta}\right) + \frac{1}{\sin^2\vartheta}\frac{\partial^2}{\partial\varphi^2}\right).$$

Eq. (15.38) then follows from the usual expression for the three-dimensional Laplacian (cf. Section 14.5, note 89) in spherical coordinates. For the angular momentum properties, see Section 12.3.5 (Born, Heisenberg, and Jordan (1926) use the notation M_x, M_y, M_z for the components of a general angular momentum vector).

As shown by Born, Heisenberg, and Jordan (1926), the allowed quantized value for $\vec{l}^2$ are $l(l+1)\,(h/2\pi)^2$ with $l = 0, 1, 2, \ldots$, so once we restrict our attention to states of a definite magnitude of orbital angular momentum, this can be written as[11]

$$H_0^{(3D)} = \frac{1}{2m}\left(p_r^2 + \frac{h^2}{4\pi^2}\frac{l(l+1)}{r^2}\right) - \frac{Ze^2}{r}. \tag{15.39}$$

Once p_φ in Eq. (15.36) and l in Eq. (15.39) are assigned numerical values, the only matrices contained by the two Hamiltonians are p_r and r, with identical commutation relations $[p_r, r] = -ih/2\pi$, so the problem of finding the eigenvalues is formally the same. In fact, if we set $p_\varphi = mh/2\pi$ and $m = l + \frac{1}{2}$, the two Hamiltonians become identical.[12] Of course, a physical (integer) value for m in the two-dimensional problem corresponds to an unphysical (half-integer) value for l in the three-dimensional problem, and vice versa, so the physical spectra of the two theories do not coincide, as we would expect (in contradistinction to the situation in the old quantum theory).

To summarize, from the extant correspondence between Heisenberg and Pauli, we know that by the end of January 1926 both physicists had convinced themselves that the time average/diagonal matrix element of $1/r^2$ in a one-dimensional Hamiltonian of the form Eq. (15.36) describing only radial motion, with p_φ at this point a purely numerical parameter affecting the net effective potential energy, would be given by the formula of the form of Eq. (15.26). Of course, for the problem of an electron actually restricted to two-dimensional motion in a Coulomb potential, the parameter p_φ corresponded physically to the orbital angular momentum and would have to take the allowed physical values of $mh/2\pi = m\hbar$, with m integer. However, there is no reason to suppose that if the parameter p_φ is continuously varied to non-integer values that the value of $\overline{1/r^2}$ in the stationary states of the one-dimensional radial potential $(p_\varphi^2 - \hbar^2/4)/2mR^2 - Ze^2/r$ would not continue to be given by Eq. (15.26). In particular, if p_φ is continuously varied and allowed to assume half-integral values, $p_\varphi = (l + \frac{1}{2})\hbar$, the Hamiltonian Eq. (15.36) becomes formally identical to Eq. (15.39), which describes the physically interesting situation of an electron in a three-dimensional Coulomb potential with azimuthal angular momentum l. Following Heisenberg and Jordan (1926, p. 276), we may therefore conclude that

$$\overline{1/r^2} = \frac{8\pi^3 m^2 e^4 Z^2}{n^3 h^3 p_\varphi} = \frac{16\pi^4 m^2 e^4 Z^2}{n^3 h^4 (l + 1/2)}. \tag{15.40}$$

Fortunately for Heisenberg and Jordan, this formula would be completely confirmed by the much easier computation of moments of radial powers using Schrödinger's wave

[11] Heisenberg and Jordan (1926) still use the notation k from the old quantum theory for the azimuthal quantum number. To ease the path for the modern reader, we use l instead.

[12] Inserting these values for p_φ and m into the expression $p_\varphi^2 - h^2/16\pi^2$ in Eq. (15.36), we find

$$m^2\frac{h^2}{4\pi^2} - \frac{h^2}{16\pi^2} = \frac{h^2}{4\pi^2}\left(m^2 - \frac{1}{4}\right) = \frac{h^2}{4\pi^2}\left(\left(l + \frac{1}{2}\right)^2 - \frac{1}{4}\right) = \frac{h^2}{4\pi^2}\,l(l+1),$$

which is just the corresponding expression in Eq. (15.39).

mechanics.[13] The comparative ease of the wave-mechanical calculations, involving simple radial integrals, needed to confirm the results of the preceding pages was surely more than enough to convince most theoretical physicists of the practical, if not the conceptual, superiority of the Schrödinger methodology.

The final quantity that remains to be determined in evaluating the fine structure was the time average of the stationary state value of $1/r^3$, which appears in the spin–orbit Hamiltonian H_2 (see Eq. (15.10)). First, we observe that the time average/diagonal matrix element of the commutator of any dynamical variable with our one-dimensional (and non-degenerate) Hamiltonian vanishes (cf. e.g. Eq. (12.154)). In particular, we have

$$\overline{[p_r, H^{(3D)}]} = 0 = \overline{\left[p_r, \frac{1}{2m}\left(p_r^2 + \frac{h^2}{4\pi^2}\frac{l(l+1)}{r^2}\right) - \frac{Ze^2}{r}\right]}, \tag{15.41}$$

from which it follows that

$$0 = -\frac{h^2 l(l+1)}{4m\pi^2}\,\overline{1/r^3} + Ze^2\,\overline{1/r^2}. \tag{15.42}$$

Rearranging this, we find

$$\overline{1/r^3} = \frac{4m\pi^2 Ze^2}{h^2 l(l+1)}\,\overline{1/r^2}, \tag{15.43}$$

and, inserting Eq. (15.40) for $\overline{1/r^2}$:

$$\overline{1/r^3} = \frac{64m^3\pi^6 e^6 Z^3}{h^6 n^3 l(l+\frac{1}{2})(l+1)}. \tag{15.44}$$

As can be inferred from his letter to Pauli of January 7, 1926 (Pauli 1979, Doc. 115), this result initially caused Heisenberg considerable consternation. He realized that for *s*-states, with $l = 0$, the average of $1/r^3$ would diverge, due to the presence of the factor of l in the denominator. By the time of the Heisenberg–Jordan paper of March, this concern had evaporated, presumably as he had realized that the evaluation of $1/r^3$ is only needed for the spin–orbit term, which is proportional to $\vec{\mathfrak{l}}\cdot\vec{\mathfrak{s}}$, and is therefore presumably absent anyway for states with zero orbital angular momentum. The authors did not know at this point that there is another intrinsically relativistic contribution *only for s-states* important for the full determination of the fine structure (at the same order as the other terms being considered here)—the "Darwin term" that emerged from C. G. Darwin's analysis of the Dirac equation two years later.

We have assembled the quantities needed to follow the derivation of the fine structure for hydrogenic (single electron) atoms given in Heisenberg and Jordan (1926). First,

[13] In other words, by evaluation of the appropriate radial integrals $\overline{1/r^n} = \int_0^\infty R_{nl}^2 r^{-n} r^2 dr$, with R_{nl} appropriately normalized Schrödinger radial wave functions for the Coulomb problem.

we define the total angular momentum vector $\vec{j} = \vec{l} + \vec{s}$ as the sum of orbital and spin components of the electron angular momentum. The dot product $\vec{s} \cdot \vec{l}$ can be seen to take definite quantized values once the total angular momentum is assigned a definite value, $\vec{j}^2 = j(j+1)h^2/(4\pi^2)$. Since

$$\vec{j}^2 = \vec{l}^2 + \vec{s}^2 + 2\vec{s} \cdot \vec{l} = \left(l(l+1) + \tfrac{1}{2}\left(\tfrac{1}{2}+1\right)\right)\frac{h^2}{4\pi^2} + 2\vec{s} \cdot \vec{l}, \tag{15.45}$$

it follows that

$$\vec{s} \cdot \vec{l} = \frac{\left(j(j+1) - l(l+1) - \frac{3}{4}\right)h^2}{8\pi^2}. \tag{15.46}$$

Heisenberg and Jordan had already realized by the end of 1925 that this formula contained the key to the Landé γ-factor,[14] defined as

$$\gamma = \frac{\mathcal{J}^2 + \frac{1}{4} - R^2 - K^2}{2KR}, \tag{15.47}$$

which gave the relative size of the splittings within the multiplets of complex spectra, according to the core model of the old quantum theory (cf. Volume 1, p. 355). The numerator factor here can be seen to be identical with the factor in parenthesis in Eq. (15.46) with the identification $\mathcal{J} = j + \frac{1}{2}$, $K = l + \frac{1}{2}$, and $R = s + \frac{1}{2}$ (with $s = \frac{1}{2}$). The correct explanation of the weak-field Zeeman effect, and the associated Landé g-factor, as well as the special case of the doublet Zeeman effect for all field values (Voigt–Sommerfeld theory), had also been worked out by Christmas 1925. As Heisenberg wrote to Pauli on December 24:

> Bohr's optimism with regards to the Goudsmit theory has influenced me so much that I would gladly believe in a magnetic electron. As agreed, we [presumably, Heisenberg and Jordan] have applied ourself to the quantum-mechanical calculations of this theory, but for "lack of time and spirit" have not gotten past rather trivial results. The formulas for g and γ are quite trivial, the sum rules almost so, it comes out as expected: for the doublets one gets the Voigt formula (Pauli 1979, Doc. 112).

Section 15.2 addresses the resolution of the anomalous Zeeman effect (doublet splitting theory, and more generally, the Landé g-factor). But first, we return to the problem of fine structure in the absence of external fields.

Ignoring for the time being the external magnetic field term H_1 in the Hamiltonian in Eq. (15.7), we first calculate the perturbative energy shift $\Delta E_2(n, l, j)$ to first order due to the spin–orbit Hamiltonian H_2 in a stationary state of the unperturbed Hamiltonian. Here $n = 1, 2, 3, \ldots$ is the principal quantum number; $l \neq 0$ is the angular momentum quantum number (for $l = 0$ the value of $\vec{s} \cdot \vec{l}$ must be zero—see Eq. (15.46) with $l = 0$

[14] Now universally called the Landé g-factor.

and $j = \frac{1}{2}$); and $j = \frac{1}{2}, \frac{3}{2}, \frac{5}{2}, \ldots$ Starting with Eq. (15.10) for H_2 and using Eq. (15.44) for $\overline{1/r^3}$ and Eq. (15.46) for $\vec{\mathfrak{s}} \cdot \vec{\mathfrak{l}}$ (with $s = \frac{1}{2}$), we arrive at

$$\begin{aligned}
\Delta E_2(n, l, j) &= \frac{Ze^2}{2m^2c^2}\overline{1/r^3}\,(\vec{\mathfrak{s}} \cdot \vec{\mathfrak{l}}) \\
&= \frac{Ze^2}{2m^2c^2} \cdot \frac{64m^3\pi^6e^6Z^3}{h^6n^3l(l+\frac{1}{2})(l+1)} \cdot \frac{\left(j(j+1) - l(l+1) - \frac{3}{4}\right)h^2}{8\pi^2} \\
&= \frac{8Z^4\pi^4me^8}{h^4c^2} \cdot \frac{1}{n^3} \frac{j(j+1) - l(l+1) - s(s+1)}{l(l+1)(2l+1)}.
\end{aligned} \tag{15.48}$$

Note that this term is present only for non-zero angular momentum, $l = 1, 2, 3, \ldots$ (*p*, *d*,... states), as it originates in a spin–orbit term proportional to $\vec{\mathfrak{l}} \cdot \vec{\mathfrak{s}}$. This avoids the singularity in the final term for $l = 0$ (originating in the time average of $1/r^3$) that had bothered Heisenberg.

We now turn to the first order energy shift $\Delta E_3(n, l)$ induced by the first post-Newtonian relativistic correction Hamiltonian H_3 in Eq. (15.13). Inserting the zeroth order Bohr–Balmer energy (see, e.g., Eq. (15.51)) into this expression along with the stationary state averages for $1/r$ in Eq. (15.15) and $1/r^2$ in Eq. (15.40), we find, after a bit of algebra,

$$\begin{aligned}
\Delta E_3(n, l) &= -\frac{1}{2mc^2}\left(E_0^2 + \frac{2Ze^2E_0}{r} + \frac{Z^2e^4}{r^2}\right), \\
&= \frac{8\pi^4Z^4me^8}{h^4c^2} \cdot \frac{1}{n^3}\left(\frac{3}{4n} - \frac{1}{l+\frac{1}{2}}\right).
\end{aligned} \tag{15.49}$$

This term only depends on the quantum numbers n and l: it is unconcerned with spin effects.[15]

The rather cumbersome numerical pre-factor appearing in these formulas becomes more compact, and more understandable, if it is rewritten in terms of the fine-structure constant $\alpha = e^2/\hbar c = 2\pi e^2/hc$,

$$\frac{8\pi^4Z^4me^8}{h^4c^2} = \frac{1}{2}\alpha^4Z^4mc^2. \tag{15.50}$$

The Balmer–Bohr energy,

$$E_0 = -\frac{2\pi^2Z^2me^4}{h^2n^2} = -\frac{1}{2}\alpha^2Z^2mc^2n^2, \tag{15.51}$$

[15] The relativistic correction here is identical with that obtained earlier by Schrödinger in his first relativistic treatment of the hydrogen atom in December 1925—see Eq. (14.53).

can therefore be regarded as an α^2 correction to the relativistic rest energy mc^2, with the fine structure appearing at $O(\alpha^4)$, with subsequent relativistic corrections arising from expansion of the Sommerfeld formula at orders α^6, α^8, etc.

Combining the relativistic and spin–orbit corrections, we find

$$\Delta(E_2+E_3)(n,l,j) = \frac{\alpha^4 Z^4 mc^2}{2n^3}\left(\frac{3}{4n} - \frac{1}{l+\frac{1}{2}} + \frac{j(j+1)-l(l+1)-\frac{3}{4}}{l(l+1)(2l+1)}\right). \tag{15.52}$$

It must be admitted that this looks nothing at all like the very compact, and empirically successful, Sommerfeld fine-structure formula (cf. Eq. (6.53)). But, as Heisenberg and Jordan point out, we must evaluate the contents of the parenthesis for the doublets arising from the combining of electron spin angular momentum $\frac{1}{2}$ with orbital angular momentum l to yield the two possible values for the total angular momentum quantum number $j = l + \frac{1}{2}$ or $j = l - \frac{1}{2}$, for which a short calculation shows that $j(j+1) - l(l+1) - \frac{3}{4} = l$ or $-(l+1)$, respectively. For $j = l + \frac{1}{2}$, one finds

$$\begin{aligned}\Delta(E_2+E_3)(n,l,j=l+\tfrac{1}{2}) &= \frac{\alpha^4 Z^4 mc^2}{2n^3}\left(\frac{3}{4n} - \frac{1}{j} + \frac{1}{2j(j+\frac{1}{2})}\right)\\ &= \frac{\alpha^4 Z^4 mc^2}{2n^3}\left(\frac{3}{4n} - \frac{1}{j+\frac{1}{2}}\right).\end{aligned} \tag{15.53}$$

Similarly, for $j = l - \frac{1}{2}$, one finds

$$\begin{aligned}\Delta(E_2+E_3)(n,l,j=l-\tfrac{1}{2}) &= \frac{\alpha^4 Z^4 mc^2}{2n^3}\left(\frac{3}{4n} - \frac{1}{j+1} - \frac{1}{(j+\frac{1}{2})2(j+1)}\right)\\ &= \frac{\alpha^4 Z^4 mc^2}{2n^3}\left(\frac{3}{4n} - \frac{1}{j+\frac{1}{2}}\right).\end{aligned} \tag{15.54}$$

When expressed in terms of the j quantum number, one sees that a single formula gives the correct energy as a function only of j, not of l. States with different $l \neq 0$ values, but the same total angular momentum j, are therefore degenerate, exactly as pointed out by Pauli in his letter to Heisenberg at the end of January. Since $j + \frac{1}{2}$ can only take integer values $1, 2, 3, \ldots$ (in other words, exactly the values assigned to the azimuthal quantum number k in the old theory), we see that the new theory has reproduced, at least for states with l not equal to zero, the leading fine structure shifts derived in 1916 by Sommerfeld on the basis of a purely relativistic treatment (*absent* any electron spin effects).[16] One has

[16] The identity of Eqs. (15.53) and (15.54) with the second (fine structure) term in the Sommerfeld result Eq. (6.53) is seen once we replace l in the latter formula (=k, in the old quantum theory, taking values 1, 2, 3, etc.) with $j + \frac{1}{2}$, and $e^4/\hbar^2$ in the prefactor with $\alpha^2 c^2$.

of course to *reinterpret* the meaning of the quantum numbers appearing in the formula in accordance with the new physical content, but from a quantitative point of view, the energy shifts predicted at order α^4, with the exception of the *s*-states for which $l = 0$, are just those that by now had been empirically confirmed:

> The formula [i.e., Eq. (15.54)] reproduces fully the empirical facts. In particular, the absence of k [the orbital angular momentum quantum number l] in [Eq. (15.54)] implies that the "screening doublets" are explained by the Uhlenbeck–Goudsmit theory. Moreover, the splittings of the magnetic doublet agree with those obtained from the Sommerfeld fine structure formula (Heisenberg and Jordan 1926, p. 277).

We shall see shortly that Heisenberg and Jordan overstate their achievement here, as the result obtained for the *s*-states ($l = 0$) *does not* agree with the Sommerfeld fine structure formula, and would produce, if their calculation is carried to its logical conclusion, levels at an incorrect location. However, taking their statement at face value, it is clear how the new quantum theory, amplified by the Uhlenbeck–Goudsmit spin hypothesis, has resolved the baffling problems of "screening" versus "relativistic" doublets in X-ray spectra (cf. Section 9.1).

Recall that the evidence from the L triplet in X-ray spectra (caused by the fine structure of the $n = 2$ states in transitions from principal quantum number 3 to 2) showed that, instead of a single doublet with a splitting due to the differing relativistic corrections to the energy of the circular (p-state, $k = 2$) and elliptical (s-state, $k = 1$) states at $n = 2$, as well as (for the X-ray case, in a multi-electron atom) a splitting due to screening effects for the orbits of different shape, there was a third state (A_2 in Figure 9.2, L_2 in Table 9.2) with a splitting from the *s*-state (A_3 in Figure 9.2), which was independent of Z, indicating a purely screening effect (from the Balmer term in the energy), while the splitting from the p-state (A_1 in Figure 9.2) behaved exactly the same way as the fine structure term (of order α^4) in the Sommerfeld formula, with a cubic dependence on Z (corresponding to the expected Z^4 dependence of the relativistic fine structure).

In the new theory, the explanation of these three levels was obvious. At principal quantum number $n = 2$, we can have orbital motion corresponding to either an s ($l = 0$) state or a p state ($l = 1$). But the intrinsic spin of the electron can also combine with the orbital angular momentum $l = 1$ in the p state to give total angular momentum of either $j = \frac{3}{2}$ or $j = \frac{1}{2}$. For the s state, only the value $j = \frac{1}{2}$ is possible. *Except for s states* Eq. (15.54) implies that states of differing j but identical l (namely, the A_1/L_1 and A_2/L_2 states) have identical screening properties (having the same "orbits" or, in Schrödinger's language, the same spatial wave functions) but differing fine structure, thus giving rise to the splittings expected from the Sommerfeld formula. However, states of differing l but the same j (the A_2/L_2 and A_3/L_3 states) will be split already at the Balmer level by screening effects. The appearance of the new physical phenomenon of electron spin as suggested by Uhlenbeck and Goudsmit was therefore the key to understanding an empirical situation in which additional states would appear, allowing screening effects and relativistic effects to manifest themselves separately, with characteristically different Z dependence.

Unfortunately, the calculation of Heisenberg and Jordan missed a crucial ingredient, essential to obtaining the correct energy shift for just the s states with $l = 0$, for which a spin–orbit term is absent (as $\vec{l} \cdot \vec{s}$ must take the value zero for any state with $l = 0$). For such states H_2 is zero, and the energy shift is given by the relativistic effects only:

$$\begin{aligned} \Delta E_3(n, l = 0) &= \frac{1}{2}\alpha^4 Z^4 mc^2 \frac{1}{n^3}\left(\frac{3}{4n} - \frac{1}{l + \frac{1}{2}}\right) \\ &= \frac{1}{2}\alpha^4 Z^4 mc^2 \frac{1}{n^3}\left(\frac{3}{4n} - 2\right), \end{aligned} \tag{15.55}$$

which differs from Eq. (15.54) (and from the Sommerfeld formula, which it reproduces) in having a -2 in the final parenthesis in Eq. (15.55), instead of a -1 (from $-1/(j + \frac{1}{2})$ with $j = \frac{1}{2}$).

Not a whisper of this discrepancy is to be found in the Heisenberg–Jordan paper, which ends with the assertion "one can regard the results of our calculations as important supports for the Compton–Unhlenbeck–Goudsmit[17] hypothesis on the one hand, and for quantum mechanics [i.e., matrix mechanics] on the other." What is even more strange, as critical and probing a theorist as Pauli does not seem to have regarded the calculations of the Heisenberg–Jordan paper as anything less than a full confirmation of the Sommerfeld fine structure formula—with, of course, the new interpretation of the quantum numbers appearing therein demanded by the new quantum mechanics and the spin hypothesis (transferring the magnetic properties from the atomic core to the valence electron(s)). It didn't hurt, of course, that Pauli had been the first to point out to Heisenberg (in the letter of January 31, 1926) the "dumb observation" that the spin–orbit energies (multiplied by the mysterious factor of $\frac{1}{2}$) plus relativistic corrections gave the doublet splittings found in the Sommerfeld formula; or, that the anomalous Zeeman effect for alkali atoms worked out perfectly (see Section 15.2), for *arbitrary field strengths*, once the zero field splittings were inserted (whether or not one could correctly derive these from the theory). On May 8, we find in a letter of Pauli to Wentzel:

> With the hypothesis of the magnetic electron and with quantum mechanics the doublet spectra (including the hydrogen spectrum) together with the fine structure and anomalous Zeeman effect are really put in order (*kommen . . . wirklich in Ordnung*) (Pauli 1979, Doc. 133).

The missing ingredient was first correctly identified two years later in Darwin's explicit solution of the Dirac equation for an electron in a Coulomb potential:

> The extra terms . . . rectify one of the earlier defects, for with my equations the s-levels of hydrogen fell in the wrong place though all others were correct . . . Since $f(0)$ [the radial wave function at the origin] vanishes unless $k = 0$ [$l = 0$], all levels other than s-levels are unaffected, and a more detailed calculation shows that the s-levels now fall in the right place (Darwin 1928, p. 663).

[17] Note that while Compton (1921a) is given credit for the spin hypothesis, Kronig is not (cf. Chapter 9).

Discussion of the new term in the Hamiltonian of the electron takes us somewhat beyond our announced terminus of 1927 for the development of the non-relativistic quantum theory, but is necessary to explain the point at which the empirical success of the Sommerfeld fine structure formula from 1916 could truly be confirmed, from the new point of view. The "Darwin term" only affects the *s*-states, so its absence did not affect the agreement of the Heisenberg–Jordan derivation of doublet splittings involving the $l \neq 0$ $(p, d, f, \ldots)$ states, which thus agreed with the 1916 formula. Its origin, however, is both quantum-mechanical and intrinsically relativistic, so it could only really be understood once a start had been made on the construction of a relativistic version of quantum mechanics.

Once both relativity and quantum mechanics are present, an additional phenomenon presents itself, which subtly alters the energy of a bound electron at the $O(\alpha^4)$ level relevant for the fine structure effects being considered here. In *relativistic* quantum mechanics the position of an electron cannot be precisely specified (even at the cost of the inevitably increasing momentum uncertainty) beyond the level of about the Compton wavelength $\hbar/mc$ of the electron. Thus, an electron at radial distance r "sees" not precisely the Coulomb potential $V(r) = -Ze^2/r$, but this quantity smeared out over a ball of radius of the order of $r_c \equiv \hbar/mc$, within which the electrostatic potential can be Taylor expanded $V(r) \to V(r) + Cr_c^2 \Delta V(r) + \ldots$, where Δ is the three-dimensional Laplacian.[18] As the Laplacian applied to the Coulomb potential gives the charge density of the nucleus, regarded as a point, one finds that the Hamiltonian of the electron acquires an additional "Darwin" term, which amounts effectively to a repulsive point-like potential energy spike at the nuclear position (taking the value 1/8 for C, following from the Dirac equation):

$$H_{\text{Darwin}} = \frac{1}{8}\frac{\hbar^2}{m^2c^2}\Delta\left(\frac{-Ze^2}{r}\right) = \frac{1}{8}\frac{\hbar^2}{m^2c^2} \cdot 4\pi Ze^2\delta^3(\vec{r}). \tag{15.56}$$

The perturbative energy shift caused by this term, to first order, is best determined using the techniques of wave mechanics (which had been almost universally adopted by 1928 by atomic theorists). One has only to evaluate the expectation value of H_{Darwin} in a normalized hydrogenic wave function $\psi_{nlm}(\vec{r})$:

$$\begin{aligned}\Delta E_{\text{Darwin}} &= \frac{1}{8}\frac{4\pi Z\hbar^2e^2}{m^2c^2}\int |\psi_{nlm}(\vec{r})|^2\delta^3(\vec{r})d^3r \\ &= \frac{\pi Z\hbar^2e^2}{2m^2c^2}|\psi_{nlm}(0)|^2\end{aligned}$$

[18] The linear terms in the Taylor expansion vanish when averaged over all angles around the central point. The constant C is a dimensionless number of order unity, the exact value of which, 1/8, is given by the Dirac theory. The relativistic blurring of the electron's position is sometimes pictorially attributed to an unavoidable "*Zitterbewegung*" ("shiver motion") of the electron.

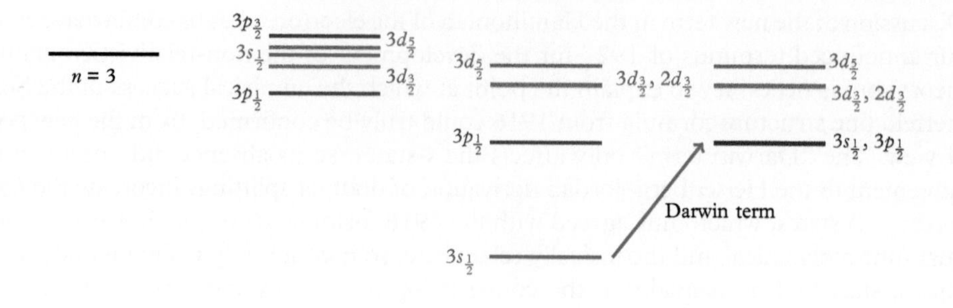

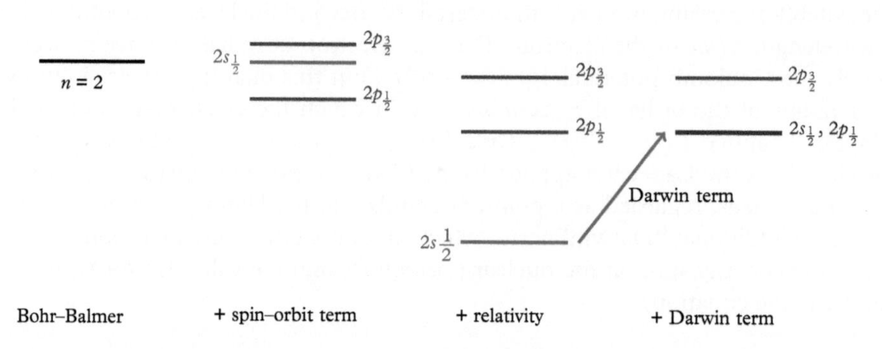

Figure 15.1 *Fine structure of the $n = 2$ and $n = 3$ levels in hydrogen. The separation of $n = 2, 3$ levels is not to scale (it should be roughly $1/\alpha^2 \approx 20000$ greater). Relative fine structure splittings are to scale for each principal quantum number: however, the scale of the $n = 3$ splittings as shown should be reduced by the factor 8/27 relative to the $n = 2$ splittings to reflect the overall $1/n^3$ falloff of the fine structure (cf. Eq. (15.54)). Non-degenerate s, p, d levels are shown in red, blue, and green, respectively, with degenerate levels indicated by thicker lines. The arrows show the role of the Darwin term in restoring degeneracy of s and p levels. The levels are denoted using the spectroscopic notation that became current in the late 1920s: $n(l)_j$, where (l) is s for $l = 0$, p for $l = 1$, etc.*

$$
\begin{aligned}
&= \frac{Z^4 m e^8}{2\hbar^4 c^2} \frac{1}{n^3} \delta_{l0} \delta_{m0} \\
&= \frac{1}{2} \alpha^4 Z^4 m c^2 \cdot \frac{1}{n^3} \delta_{l0} \delta_{m0}. \qquad (15.57)
\end{aligned}
$$

In the last line we used the expression for the hydrogenic wave function at the origin, $\psi_{nlm}(0) = (1/\sqrt{\pi})\,(Zme^2/n\hbar^2)^{3/2}\delta_{l0}\delta_{m0}$.[19]

[19] The relevant formulas can be found in Messiah (1966, App. B).

The addition of this term converts the troublesome -2 in Eq. (15.55) to the correct $-1 = 1/(j + \frac{1}{2})$, where for s-states with $l = 0$, we must have $j = \frac{1}{2}$. The degeneracy required by the Sommerfeld formula between states of equal j but differing l values—in this case, between the s and p states with $j = \frac{1}{2}$—is thereby restored. Figure 15.1 illustrates the critical role of the Darwin term in restoring the correct spectral degeneracies in the spectrum of hydrogen and singly ionized helium, where the appearance of the complex doublet spectra is indicated by progressively including the spin–orbit, relativity, and Darwin corrections. It is obvious from the third column of the figure that, without the Darwin term, false extra lines would be predicted at completely incorrect positions, for example in the p to s principal series transitions.

Dirac's (1928) four-component relativistic wave equation for the electron in early 1928 was solved independently for the case of a Coulomb potential by Gordon (1928) and Darwin (1928). The result given by Gordon (for hydrogen, $Z = 1$) for the total relativistic energy E (including the rest energy) is

$$\frac{E}{mc^2} = \left(1 + \frac{\alpha^2}{(n' + \sqrt{j'^2 - \alpha^2})^2}\right)^{-1/2}, \tag{15.58}$$

where $j' = j + \frac{1}{2} = 1, 2, 3, \ldots$ and $n' = n - j'$ (n the principal quantum number). This result is formally identical with Sommerfeld's 1916 result (cf. Eqs. (6.31) and (6.48)), where the azimuthal quantum number $k = 1, 2, 3, \ldots$ of the old theory has now been replaced by the total quantum number j augmented by one half (and therefore assuming the same set of possible non-zero integer values).

That this rather elaborate mathematical expression has retained exactly its form, if not the interpretation of the symbols it contains, has always seemed somewhat miraculous. The contentious (and in our view, somewhat semantic) disputes over the extent to which this coincidence is "accidental" (Yourgrau and Mandelstam 1979, pp. 113–115) or indicative of profound correspondences in the physics of the problem as treated in the old and new quantum theories (Biedenharn 1983) will not concern us here—the interested reader can follow the arguments on either side in the cited articles. Note, however, that the expansion of Eq. (15.58) in powers of the fine structure constant only gives the correct fine structure up to order α^4 (i.e., the first post-Newtonian term)—beyond this point, truly quantum-electrodynamic (i.e., quantum field theoretic) effects appear that are not contained in the Dirac equation. The first such effect is the famous Lamb shift, which appears at order α^5, a term clearly not contained in the Dirac–Gordon–Darwin expression Eq. (15.58), which only contains even powers of the fine structure constant.

The discussion of fine structure so far has been confined to hydrogenic (i.e., single-electron) atoms, of which the only experimentally available exemplars (in the 1920s) were hydrogen itself and singly ionized helium, He^+. The fine structure of the spectra of alkali atoms, with a single valence electron, could be roughly understood as it amounted qualitatively to a screened version of the hydrogenic case. The screening of the nuclear charge by the inner electrons could be taken into account by replacing the integer nuclear charge Z by an effective value Z_{eff} (or Z^*). The effective charge is a function of the angular momentum l, with the valence electron seeing a smaller effective charge for higher values of l, as it spends more time further away from the nucleus. The resulting energy shifts (as they directly affect the Balmer term in the energy) are typically much larger than the fine structure discussed above. In particular, the effect of the Darwin term in shifting the s-states is completely drowned in the much larger shift of s states away from $p, d, \ldots$ states induced by nuclear charge screening for atoms with more than one electron. Systematic attacks on the problem of evaluating spectral terms in multi-electron atoms including the effects of screening began with Hartree's wave-mechanical studies in late 1927 of electronic energies in non-Coulombic central fields (Hartree 1927a, 1927b, 1928).

15.2 Intermezzo: Kuhn losses suffered and recovered

The story of the fine structure of the hydrogen spectrum, from Sommerfeld (1916a) to Heisenberg and Jordan (1926) to Darwin (1928), makes for an instructive example of what is known in history and philosophy of science circles as a "Kuhn loss". The concept was introduced in *The Structure of Scientific Revolutions* (Kuhn 1962, Ch. 9; it was obviously only named after Kuhn later). Kuhn pointed out that it is not uncommon in instances of theory (or paradigm) change that there are phenomena that could be explained (to the satisfaction of the relevant community) in the old discarded theory but not in the newly adopted one. In the excitement over the new opportunities afforded by the new theory, such losses tend to get swept under the rug. The fine structure formula for hydrogen is an example of a Kuhn loss sustained and then recovered. Sommerfeld (1916a) had found the correct formula in the old quantum theory on the basis of an argument that looked convincing to his contemporaries. The formula Heisenberg and Jordan (1926) found in the new quantum theory agreed with Sommerfeld's in *most but not all* cases—it gives the wrong energies for the s-states of hydrogen (see Figure 15.1). Yet, in the next two years, nobody—not even Pauli—seems to have objected to Heisenberg and Jordan's claim, quoted in Section 13.1, that their formula "reproduces *fully* the empirical facts" (1926, p. 277, our emphasis). We have been unable to find a single mention in publications or extant letters of 1926–1927 of the problem with hydrogen's s-states. Given how many other problems were solved by the analysis of Heisenberg and Jordan it is perhaps not too surprising that this one remaining problem appears to have been ignored until Gordon (1928) and Darwin (1928), with the crucial input of the Dirac equation (1928), actually solved it.

Before turning to additional examples of problems tackled successfully by Heisenberg and Jordan (1926)—including the anomalous Zeeman effect over which the old quantum theory had come to grief—we briefly discuss two other Kuhn losses, both suffered in the transition from classical physics to the old quantum theory and recovered in the subsequent transition to the new quantum mechanics. The first of these occurred in dispersion theory (see Chapters 10 and 11, and Jordi Taltavull 2017); the second in the theory of electric susceptibility, an area that had not attracted much attention during the days of the old quantum theory but would help launch work in solid state physics based on the new quantum mechanics in the 1930s (see Midwinter and Janssen 2013).

In the classical dispersion theories of Sellmeier, Helmholtz, Lorentz, and Drude, the dispersion formula has poles at the resonance frequencies of small oscillators assumed to be present in large quantities in optical media. These resonance frequencies were chosen to coincide with the absorption frequencies of the optical medium. With this choice, the classical theory could account for the empirical fact that, in narrow intervals around those absorption frequencies, dispersion is anomalous (i.e., the index of refraction decreases rather than increases with frequency in those intervals).

In the old quantum theory, however, the dispersion formula had poles at the *orbital* frequencies of the electrons taking over the role of the oscillators in the emission and absorption of light. In what was widely seen as the most radical departure from classical theory, Bohr (1913a) had severed the relation between orbital and radiation frequencies in his quantum model of the atom (see, e.g., Volume 1, pp. 15–16 and p. 178). Except in the limit of high quantum numbers, these two frequencies no longer agreed. A direct consequence of this was that the poles in the dispersion formulas proposed by Sommerfeld (1915b), Debye (1915), and Davisson (1916) on the basis of the old quantum theory were at the wrong frequencies. Given its successes in other areas, this problem for the old quantum theory was largely ignored. As we saw at the end of Section 10.2.1, while Bohr already drew attention to the problem in a letter of December 1915 and an unpublished paper of 1916, it was not until 1922 that the problem was prominently mentioned in print (Epstein 1922c, pp. 107–108; see Section 10.2.1). It was solved two years later by Kramers (1924a, 1924b). Like the Sellmeier-type formula of classical dispersion theory, the Kramers dispersion formula has poles at the absorption frequencies of the optical medium (see Section 10.3.3), while fully incorporating the correspondence-principle requirement of merging with classical results in the limit of large quantum numbers. The Kramers dispersion formula was quickly incorporated in the short-lived BKS theory (Bohr, Kramers, and Slater 1924a), and then, more permanently, in the new matrix mechanics of Heisenberg, Born, and Jordan as well as in Schrödinger's wave mechanics (Heisenberg 1925c; Born, Heisenberg, and Jordan 1926; Schrödinger 1926h). And thus this Kuhn loss, suffered by the old quantum theory, was recovered in the new quantum theory.

Our third and final example of a Kuhn loss involves a formula due to Langevin (1905) and Debye (1912) for the electric susceptibility of diatomic gases such as hydrogen chloride,

$$\chi = N\left(\alpha + \frac{\mu^2}{3kT}\right), \tag{15.59}$$

where N is the number of molecules, α is a constant, μ is the permanent electric moment of a molecule of the gas, k is Boltzmann's constant, and T is the temperature. The first term in this Langevin–Debye susceptibility formula comes from the induced moment of the molecule, resulting from the deformation of the molecule by the external electric field. The second term comes from the alignment of the permanent moment of the molecule with the field. Thermal motion will frustrate this alignment, which is reflected in the inverse proportionality to the temperature T.

The Kuhn loss occurred in the second term of this susceptibility formula. If this term is written as $C(\mu^2/kT)$, with C a dimensionless parameter, classical theory gives the value $C = 1/3$, in agreement with the empirical data. Quantum mechanics gives that same value, except at very low temperatures (van Vleck 1932, p. 185, p. 197). Other than that, the factor $1/3$ is a remarkably robust prediction of both theories. It is true for different models of diatomic gas molecules and many other systems. In the old quantum theory, however, the value of C differed from $1/3$ and depended sensitively on which system one considered. In the case of diatomic molecules modeled as dumbbells, Pauli (1921b) and Pauling (1926) (submitted in February 1926 but still based on the old quantum theory) found values much greater than $1/3$, that is, 1.54 and 4.57, respectively.

What lay behind these discrepancies[20] is that Pauli and Pauling followed the old quantum theory's standard prescription to quantize the angular momentum of their dumbbells,

$$L = l\hbar, \quad L_z = m\hbar, \tag{15.60}$$

with l taking on either integer values $1, 2, 3, \ldots$ (Pauli) or half-integer values $\frac{1}{2}, \frac{3}{2}, \frac{5}{2} \ldots$ (Pauling) and m running from $-l$ to l in both cases. As usual in the old quantum theory, the state $l = m = 0$ is forbidden. The problem was quickly resolved once it became clear that this prescription should be replaced by:

$$L^2 = l(l+1)\hbar^2, \quad L_z = m\hbar, \tag{15.61}$$

with l taking on integer values *including* $l = 0$ and m running from $-l$ to l as before. As discussed in Section 12.3.5, these relations were first derived in the Three-Man-Paper (Born, Heisenberg, and Jordan 1926, Ch. 4). They were applied to the case of diatomic molecules by Lucy Mensing (1926), who had obtained her PhD in Hamburg working with Pauli and Wilhelm Lenz,[21] and by David Dennison (1926), who had joined the faculty at the University of Michigan in 1922 after six years at Bohr's institute in Copenhagen.

Inspired by Dennison's paper, which he read in manuscript as assistant editor of *Physical Review* (edited by John Tate, Van Vleck's senior colleague at the University of Minnesota), Van Vleck (1926b) appears to have been the first to use matrix mechanics to derive a formula for the temperature-dependent term in the electric susceptibility of diatomic gases and show that the classical value $C = 1/3$ is restored in the new theory. By the time his paper appeared in *Nature*, however, after a delay of several months (Fellows 1985, p. 109), the same result had already been obtained and published by others (Kronig 1926; Mensing and Pauli 1926). That the solution to this problem was

[20] For the detailed calculations, see Midwinter and Janssen (2013, sec. 7.5.4, pp. 178–189)

[21] Mensing gave up her career in physics when she married a fellow physicist in 1930 (Müenster 2020).

found so quickly and by different authors working independently of one another just goes to show how natural this solution was once the new rules for the quantization of angular momentum were available.[22]

The derivation of the temperature-dependent term in the susceptibility formula proceeds in two steps. First, the time average of the component of the permanent moment of a single molecule in the direction of the external field needs to be computed. This quantity then needs to be averaged over a thermal ensemble of many such molecules. The problem responsible for the deviations from $C = 1/3$ in the old quantum theory occurred in the first step.

As usual, we take the external field (of strength $\mathcal{E}$) to be in the z-direction. To find the time average (indicated by an overbar) of the z-component of the electric moment μ of a molecule we then have to average over the cosine of the angle between μ and the z-axis:

$$\overline{\mu_z} = \mu \overline{\cos \vartheta}. \tag{15.62}$$

Pauli (1921b, p. 324) and Pauling (1926, p. 570) did a perturbative calculation to find a formula for $\overline{\cos \vartheta}$ in classical theory in an energy regime where the rotational energy $E = L^2/2I$ (with L the magnitude of the angular momentum and I the moment of inertia of the dumbbell used to model diatomic atoms) is much greater than its electrical energy $\mu\mathcal{E}$ (note that, in quantum theory, E only depends on l and not on m). We focus on Pauli's paper. His result, in our notation, is:

$$\overline{\cos \vartheta} = \frac{\mu\mathcal{E}I}{2\overline{L^2}}\left(3\overline{L_z^2/L^2} - 1\right), \tag{15.63}$$

where L_z is the z-component of the dumbbell's angular momentum. Pauli obtained the corresponding equation in the old quantum theory by setting $\overline{L^2} = l^2\hbar^2$ and $\overline{L_z^2} = m^2\hbar^2$ (cf. Eq. (15.60)):

$$\overline{\cos \vartheta} = \frac{\mu\mathcal{E}I}{2\hbar^2 l^2}\left(\frac{3m^2}{l^2} - 1\right). \tag{15.64}$$

Classically, the ratio (m^2/l^2) corresponds to the time average of $(L_z/L)^2$, which is $\overline{\cos^2 \vartheta}$ for the *unperturbed* system. This average is equal to $1/3$, which means that for the *full perturbed* system (i.e., including the terms with the electric field in the Hamiltonian) $\overline{\cos \vartheta}$ vanishes in the classical theory. According to classical theory, in other words, there is no

[22] Van Vleck (1971, p. 7) later called it a "quadruple tie" adding a paper by Charles Manneback (1926), who, however, actually cited Mensing and Pauli (1926) and claimed priority only for having derived the result in wave rather than matrix mechanics (Manneback 1926, p. 564). These 1926 papers all dealt with the special case in which gas molecules are modeled as rotating dumbbells. The result was derived in full generality the following years in a series of papers by Van Vleck (1927a, 1927b, 1928) that formed the basis for his book on susceptibilities (Van Vleck 1932). Debye (1928) discussed these problems in the old quantum theory and their resolution in the new theory in his contribution to the Como conference of September 1927 (cf. Section 16.4).

contribution to the susceptibility *at all* from molecules in the energy regime for which Eq. (15.64) was derived.

As Pauli (1921b, p. 324) noted, this fits with the conclusion drawn earlier by Alexandrow (1921) that it is only the molecules in the lowest energy states that contribute to the susceptibility. As Pauli also noted, however, the lowest energy state in the old quantum theory ($l = m = 0$) is forbidden. In the old quantum theory, we thus have the strange situation that there are "*only such orbits present that according to the classical theory do not give a sizable contribution to the electrical polarization*" (Pauli 1921b, p. 325; emphasis in the original). Pauli went on to show that, contrary to the situation in the classical theory, the thermal ensemble average of $\overline{\cos\vartheta}$ as given in Eq. (15.64) does *not* vanish in the old quantum theory. Hence, he concluded, in the old quantum theory the temperature-dependent term in the susceptibility formula comes not from molecules in the low-energy states, but rather from those in the high $E \gg \mu\mathcal{E}$ energy regime in which Eq. (15.64) holds. It therefore should not surprise us, Pauli argued, that the old quantum theory does not reproduce the factor $1/3$ of the Langevin–Debye formula. This then is a Kuhn loss suffered in the transition from the classical theory to the old quantum theory.

We should note, however, that the situation is not as clear-cut as in the case of the Kuhn loss suffered in dispersion theory. Whereas there was abundant empirical evidence to suggest that the poles of the dispersion formula had to be at the absorption frequencies of the optical media, strong empirical evidence that C must be equal to $1/3$ in the susceptibility formula only emerged around the time that this Kuhn loss was recovered. The Kuhn loss in the case of susceptibilities is a loss of a theoretical feature that in hindsight proved to be empirically correct. The most persuasive argument against the results of Pauli and Pauling, accordingly, was that they deviated from the classical result even at high temperatures, where the old quantum theory was expected to agree with the classical theory (Midwinter and Janssen 2013, pp. 141–142). Van Vleck already emphasized this in his 1926 article in *Nature*. He reiterated the point in his book six years later: "the correspondence principle led us to expect usually an asymptotic connection of the classical and quantum results at high temperatures" (van Vleck 1932, p. 107).

In 1926, Pauli teamed up with his former student Mensing to derive the susceptibility formula in matrix mechanics. They wrote Eq. (15.62) in the form

$$\overline{\mu_z} = \alpha(l, m)\,\mathcal{E}. \tag{15.65}$$

Comparing this equation to Eq. (15.64), we see that in the old quantum theory, $\alpha(l, m)$ is given by $(\mu/\mathcal{E})$ times the expression for $\overline{\cos\vartheta}$. In the new quantum theory, Mensing and Pauli (1926, p. 512) found that it is given by

$$\frac{2\mu^2 I}{3\hbar^2}, \qquad \frac{2\mu^2 I}{\hbar^2}\frac{1}{(2l-1)(2l+3)}\left\{\frac{3m^2}{l(l+1)} - 1\right\}, \tag{15.66}$$

for $l = 0$ and $l \neq 0$, respectively.[23] With these results, the Kuhn loss is recovered. The only contribution to the temperature-dependent term in the Langevin–Debye formula in

[23] Note that, for $l \gg 1$, the second expression in Eq. (15.66) reduces to (μ/E) times the one on the right-hand side of Eq. (15.64), the corresponding expression in the old quantum theory.

Eq. (15.2) comes from the first term in Eq. (15.66), the one for $l = 0$, which was forbidden in the old quantum theory. Except at very low temperatures, this contribution is the same as the temperature-dependent term in the classical formula (Midwinter and Janssen 2013, p. 185). The thermal ensemble average over the second term in Eq. (15.66), for $l \neq 0$, simply vanishes. The easiest way to see this is that averaging $m^2/(l(l+1))$ gives $\overline{L_z^2/L^2}$, which in the new quantum mechanics, as in classical theory, is simply 1/3.[24] As Mensing and Pauli (1926, p. 512) noted with obvious relief: "*Only the molecules in the lowest state will therefore give a contribution to the temperature-dependent part of the dielectric constant*" (emphasis in the original). Van Vleck (1926b, p. 227) similarly pointed out that the "remarkable result is obtained that only molecules in the state of lowest rotational energy make a contribution to the polarisation. This corresponds very beautifully to the fact that in the classical theory only molecules with energies less than $[\mu\mathcal{E}]$ contribute to the polarisation." After giving the result in the new theory, he commented: "This is a much more satisfactory result than in the older version of the quantum theory, in which both the calculations of Pauli with whole quanta ... and of Pauling with half quanta yielded results diverging from the classical Langevin theory even at high temperatures" (p. 227).

15.3 External field problems 1925–1927

In this section we revisit two problems that played a critical role in the development of quantum theory. Both concern the changes in atomic spectra induced by the application of external fields to the atom: the Zeeman effect, observed first by Pieter Zeeman in 1896 in the broadening of sodium lines in the presence of an applied magnetic field; and the Stark effect, first detected in 1913 independently by Johannes Stark and Antonino Lo Surdo, in the Balmer lines of hydrogen when an external electric field is applied. The Stark effect succumbed relatively early, in 1916 (at least from the point of view of the observed splittings of the energy levels), to the ministrations of the old quantum

[24] We can also verify directly that the ensemble average of the second expression in Eq. (15.66) vanishes. This ensemble average will be proportional to

$$\sum_{m=-l}^{l} \left\{ \frac{3m^2}{l(l+1)} - 1 \right\}.$$

We can readily evaluate this sum using the well-known sum-of-squares formula $\sum_{i=1}^{n} i^2 = (1/6)n(n+1)(2n+1)$ and noting that $\sum_{i=-n}^{n} i^2 = 2\sum_{i=1}^{n} i^2$. Using that

$$3\sum_{m=-l}^{l} m^2 = l(l+1)(2l+1),$$

(which confirms that $\overline{L_z^2/L^2} = 1/3$), we arrive at:

$$\sum_{m=-l}^{l} \left\{ \frac{3m^2}{l(l+1)} - 1 \right\} = \frac{l(l+1)(2l+1)}{l(l+1)} - (2l+1) = 0.$$

theory, applied by Schwarzschild and Epstein, and was considered a great success of this theory (cf. Section 6.3). The Zeeman effect, in its non-Lorentzian "anomalous" manifestations (which, despite the anomalous appellation, actually were much more frequently observed than the "normal" Lorentz triplet splitting), ultimately turned out to resist all attempts at a consistent explanation in the terms of the old quantum theory (cf. Sections 7.2–7.3). We turn first to the anomalous Zeeman effect, the first successful treatment of which, using the combined machinery of the new quantum mechanics and the Uhlenbeck–Goudsmit spin hypothesis, was given by Heisenberg and Jordan in the same paper discussed above in connection with the doublet fine structure of hydrogen.

15.3.1 The anomalous Zeeman effect: matrix-mechanical treatment

In section 3 of their paper, Heisenberg and Jordan (1926) set out to resolve, using the new matrix mechanics augmented with the physical hypothesis of electron spin, the thorniest problem of the old quantum theory, the anomalous Zeeman effect. One needs here to calculate the additional splittings induced in the complex spectral multiplets produced by relativistic and spin–orbit effects once a constant and homogeneous magnetic field $\vec{\mathfrak{H}}$ is applied to the atomic system. A typical example are the sodium D lines, which arise from principal series transitions of the valence electron from $3p$ to $3s$ states (i.e., from $n = 3, l = 1$ to $n = 3, l = 0$).[25] As we are concerned only with the splittings for a multiplet of states with a given value of quantum numbers n, l that are initially degenerate with respect to the parts H_0 and H_3 (the relativistic correction term) of the full Hamiltonian Eq. (15.7), the splittings of interest arise from the terms H_1 and H_2, so we consider only the "Zeeman Hamiltonian" given by the spin-dependent parts of the full Hamiltonian,

$$H_Z = H_1 + H_2 = \frac{e}{2mc}\vec{\mathfrak{H}} \cdot (\vec{\mathfrak{l}} + 2\vec{\mathfrak{s}}) + \frac{Ze^2}{2m^2c^2r^3}\vec{\mathfrak{s}} \cdot \vec{\mathfrak{l}}. \tag{15.67}$$

To avoid unnecessarily complicated formulas, we define some simple parameters that encapsulate the relative strengths of the external magnetic field and the spin–orbit splittings discussed in the previous section. We also introduce dimensionless versions $\vec{s}, \vec{l}$ of the angular momentum operators $\vec{\mathfrak{s}}, \vec{\mathfrak{l}}$, by extracting a factor of $h/2\pi$:

$$\vec{\mathfrak{s}} = \frac{h}{2\pi}\vec{s}, \quad \vec{\mathfrak{l}} = \frac{h}{2\pi}\vec{l}. \tag{15.68}$$

The magnetic field will as usual be chosen to lie in the z-direction, with $\vec{\mathfrak{H}} = \mathfrak{H}\hat{z}$. Following the notation of Heisenberg and Jordan, define

[25] Of course, the treatment given by Heisenberg and Jordan is explicitly valid only for single-electron atoms, so there is no screening, and the lines of interest involve transitions between states of differing principal quantum number.

$$\mu \equiv \frac{e}{2mc}\mathfrak{H} \cdot (\frac{h}{2\pi}). \tag{15.69}$$

For the spin–orbit terms, we define, for a given value of n and l, a parameter λ (implicitly dependent on n, l, cf. Eq. (15.44)) by

$$\lambda \equiv \frac{Ze^2}{2m^2c^2}(\frac{h}{2\pi})^2 \overline{\frac{1}{r^3}}. \tag{15.70}$$

Both μ and λ have dimensions of energy. The "Zeeman Hamiltonian" Eq. (15.67) now takes the much simpler form

$$\begin{aligned} H_Z &= \mu(l_z + 2s_z) + \lambda\vec{s}\cdot\vec{l} \\ &= \mu(l_z + 2s_z) + \lambda(s_z l_z + \tfrac{1}{2}s_- l_+ + \tfrac{1}{2}s_+ l_-). \end{aligned} \tag{15.71}$$

Here, the famous "raising" (s_+, l_+) and "lowering" (s_-, l_-) matrices have been introduced:

$$s_\pm \equiv s_x \pm i s_y, \quad l_\pm \equiv l_x \pm i l_y. \tag{15.72}$$

The non-vanishing matrix elements of these quantities had already been determined, using the commutation relations of the angular momentum operators, in Born, Heisenberg, and Jordan (1926, Ch. 4, Eq. (25)):[26]

$$s_\pm(m_s', m_s) = \sqrt{s(s+1) - m_s m_s'}\; \delta_{m_s', m_s \pm 1}, \tag{15.73}$$

$$s_z(m_s', m_s) = m_s \delta_{m_s, m_s'}, \tag{15.74}$$

with $s = \frac{1}{2}$, $m_s = \pm\frac{1}{2}$, and, similarly,

$$l_\pm(m_l', m_l) = \sqrt{l(l+1) - m_l m_l'}\; \delta_{m_l', m_l \pm 1}, \tag{15.75}$$

$$l_z(m_l', m_l) = m_l \delta_{m_l, m_l'}, \tag{15.76}$$

where $l = 0, 1, 2, \ldots$ and for given l, m_l takes one of the $2l+1$ values $-l, -l+1, \ldots, l-1, l$.

The calculation of the energy shifts induced by H_Z is a problem in degenerate perturbation theory, insofar as there exist pairs of states that have the same energy with respect to the unperturbed Hamiltonian (which in this case is H_0, or, including relativistic effects $H_0 + H_3$), but which are connected by non-vanishing matrix elements

[26] Note that the modern convention for labeling row and column indices is reversed in the Three-Man-Paper. To avoid some awkwardness of notation, we follow Heisenberg and Jordan here in displaying row and column numbers in parentheses, rather than in subscripts. Simple derivations of these matrix elements can be found in any modern text on quantum mechanics: see, for example, McIntyre (2012, p. 360).

of H_Z. The prescription explained in the Three-Man-Paper for handling such situations must therefore be applied: the first order perturbative energy shifts are obtained by diagonalizing the matrices of H_Z in each degenerate set of unperturbed states. The latter may be characterized by definite values of $m_s(=\pm\frac{1}{2})$, the eigenvalue of s_z, and of $m_l(=-l,-l+1,\ldots,l-1,+l)$, the eigenvalue of l_z. Each such state has of course a definite value $m=m_s+m_l$ for the z component of the total angular momentum $j_z=s_z+l_z$. Note that H_Z in Eq. (15.71) does not connect states with different values of m—the term with the raising matrix s_+ for m_s has the lowering matrix l_- for m_l, and vice versa. So we can divide up the $2(2l+1)$ possible unperturbed states into (i) two isolated (i.e., unconnected with other states via H_Z) states with $m_l=+l, m_s=+\frac{1}{2}, m=l+\frac{1}{2}$ and $m_l=-l, m_s=-\frac{1}{2}, m=-l-\frac{1}{2}$, respectively, and (ii) $2l$ pairs of states with quantum numbers $m_l, m_s=-\frac{1}{2}$ and $m_l-1, m_s=+\frac{1}{2}$ (both with $m=m_l-\frac{1}{2}$).

For the two isolated states (case (i)), we obtain directly the energy shift as the diagonal matrix element of H_Z,

$$\Delta E_Z(m=l+\tfrac{1}{2}) = \mu(l+1)+\lambda l/2, \tag{15.77}$$

$$\Delta E_Z(m=-l-\tfrac{1}{2}) = -\mu(l+1)+\lambda l/2, \tag{15.78}$$

where in the λ term, only the $l_z s_z$ part of the spin–orbit matrix $\vec{l}\cdot\vec{s}$ contributes to the desired diagonal matrix element (the other terms containing raising and lowering matrices with vanishing matrix elements at the extremal "edge" values of m_l, m_s).

On the other hand, for case (ii), each $(m_l,-\frac{1}{2}),(m_l-1,+\frac{1}{2})$ degenerate pair, corresponding to a given m value (i.e., eigenvalue of $j_z=l_z+s_z$) of $m_l-\frac{1}{2}$, gives rise to a 2×2 matrix, $H_Z(m)$, the elements of which are easily seen to be (with the help of Eqs. (15.71) and (15.73)–(15.76)):

$$\begin{pmatrix} \mu(m-\frac{1}{2})-\frac{\lambda}{2}(m+\frac{1}{2}) & \frac{\lambda}{2}\sqrt{l(l+1)-(m+\frac{1}{2})(m-\frac{1}{2})} \\ \frac{\lambda}{2}\sqrt{l(l+1)-(m+\frac{1}{2})(m-\frac{1}{2})} & \mu(m+\frac{1}{2})+\frac{\lambda}{2}(m-\frac{1}{2}) \end{pmatrix}. \tag{15.79}$$

The desired energy shifts are just the eigenvalues of this matrix, obtained by solving the secular equation (here quadratic) $\det(E-H_Z(m))=0$. After a little algebra, the secular equation reduces to a quadratic equation,

$$E^2-(2\mu m-\frac{\lambda}{2})E+\mu^2(m^2-\tfrac{1}{4})-\lambda\mu m-\frac{\lambda^2}{4}l(l+1)=0, \tag{15.80}$$

with solutions

$$E=\mu m-\frac{\lambda}{4}\pm\frac{\mu}{2}\sqrt{1+\frac{2\lambda m}{\mu}+\frac{\lambda^2}{\mu^2}(l+\tfrac{1}{2})^2}. \tag{15.81}$$

In the old quantum theory, the phenomenology of Zeeman splittings in the doublets of alkali spectra, which had been perfected by Voigt and Sommerfeld (cf. Section 7.3), was expressed using a dimensionless parameter v that indicated the relative magnitudes of the zero field doublet splitting (written $h\Delta\nu_0$) and the normal Lorentz–Zeeman splitting $h\Delta\nu_{\text{norm}}$ (cf. Eq. (7.69)). The doublet splitting between a $j = l + 1/2$ and a $j = l - 1/2$ state due to the spin–orbit Hamiltonian

$$\lambda\vec{s}\cdot\vec{l} = \frac{\lambda}{2}\left(\vec{j}^2 - \vec{l}^2 - \vec{s}^2\right) = \frac{\lambda}{2}\Big(j(j+1) - l(l+1) - s(s+1)\Big) \tag{15.82}$$

is evidently

$$\frac{\lambda}{2}\Big((l+\tfrac{1}{2})(l+\tfrac{3}{2}) - (l-\tfrac{1}{2})(l+\tfrac{1}{2})\Big) = \lambda\,(l+\tfrac{1}{2}), \tag{15.83}$$

while we have here written $h\Delta\nu_{\text{norm}}$ simply as μ (Eq. (15.69)). Thus, reviving the Voigt–Sommerfeld parameter, we define

$$v \equiv \frac{\Delta\nu_0}{\Delta\nu_{\text{norm}}} = \frac{\lambda}{\mu}(l+\tfrac{1}{2}). \tag{15.84}$$

Eliminating λ in favor of v in Eq. (15.81), one finds[27]

$$E = \mu\left(m - \frac{v}{4(l+\frac{1}{2})} \pm \frac{1}{2}\sqrt{1 + \frac{2mv}{l+\frac{1}{2}} + v^2}\right). \tag{15.85}$$

This expression precisely reproduces the empirically confirmed Zeeman energy shifts of the phenomenological Voigt–Sommerfeld theory (cf. Volume 1, Table 7.2).[28]

The last section of chapter 3 of the Heisenberg–Jordan paper deals with the relative intensities of the Zeeman lines for the doublet system, predictions for which had already been made in the phenomenological Voigt–Sommerfeld theory. A preliminary examination of these intensity relations had already been made in the Three-Man-Paper, before the importance of the Uhlenbeck–Goudsmit spin hypothesis had been appreciated, and incorporated into the matrix-mechanical framework. Considerations of space (not to say, patience of the reader) proscribe a proper treatment of this topic here, so we simply refer the interested reader to an extensive web resource dealing in detail with the whole question of spectral line intensities.[29]

[27] Eqs. (15.80) and (15.81) correspond precisely to Eqs. (19) and (20) in Heisenberg and Jordan (1926). This paper contains typographical errors: missing factors of $\frac{1}{4}$ and $\frac{1}{2}$ in the second and third terms on the right-hand side, respectively, of the first line of Eq. (23).

[28] In comparing Eq. (15.85) with the second and third lines of Table 7.2, corresponding to $m = +\frac{1}{2}$ and $m = -\frac{1}{2}$, respectively, one must remember that the shifts in this table are measured relative to the midpoint of the zero field doublet, corresponding to $\frac{1}{2}(\lambda/2 + (-\lambda)) = -\lambda/4 = -\mu v/6$, and that the values in the table are in units of $\Delta\nu_{\text{norm}} = \mu$. Dividing the energy in Eq. (15.85) by μ (after including $-\mu v/6$), one obtains the expressions in Table 7.2.

[29] See the web resource, *The Intensity Problem in the Old and New Quantum Theory*.

The Voigt–Sommerfeld theory had successfully treated (albeit in a semi-empirical fashion) the entire progression of the Zeeman splittings of alkali doublet (and singlet) lines from weak magnetic fields ($\mu \ll \lambda$) through the intermediate field region (where non-linear dependence on the fields appeared due to the square root expressions), to the Paschen–Back regime of strong magnetic fields ($\mu \gg \lambda$). It was also important for the new theory to reproduce the variety of Zeeman patterns found for weak magnetic field, and not just for the alkali doublets, but for a whole range of elements in other columns of the periodic table, for example, the alkaline earths in the second column (see Figure 9.1), where triplets appeared instead of doublets. It was apparent for the alkaline earths that the Zeeman splittings involved Landé g-factors where the "S" quantum number, instead of the value $s = 1/2$ appropriate for the magnitude of spin- angular momentum in the alkali case, should be 0 or 1, as though the energy levels were associated with states where the two valence electrons were coupled in total spin angular momentum to either zero or one (in units of $\hbar$).

The calculation of energy levels for weak magnetic fields amounted to a first-order perturbation-theory calculation in which the diagonal matrix elements of H_1 (the "small perturbation" induced by a weak external magnetic field) were needed with respect to stationary states chosen to make H_2 (the spin-orbit term) a diagonal matrix. We already saw in our treatment of the zero-field problem that H_2 (as well as H_0 and the relativistic correction term H_3) is diagonal with respect to stationary states characterized by the quantum numbers n, j, and l. The diagonal matrix elements of $H_2 = \lambda \vec{s} \cdot \vec{l}$ in terms of these quantum numbers are obtained from Eq. (15.46) as[30]

$$H_2(nlj, nlj) = \frac{1}{2}\lambda \left(j(j+1) - l(l+1) - s(s+1)\right). \tag{15.86}$$

The problem now was to evaluate the diagonal matrix element (or, according to the dynamical postulates of matrix mechanics, the time average) of the external magnetic field term

$$H_1 = \mu(l_z + 2s_z) = \mu(j_z + s_z) \tag{15.87}$$

in a stationary state characterized by n, l, j, and, of course, also the magnetic quantum number m corresponding to the z-component of the total angular momentum, j_z, as the degeneracy of the $2j + 1$ states with total magnetic quantum number $j_z = m = -j, -j+1, \ldots, j-1, j$ is lifted by the external field term. The first term on the right in the preceding equation is already a diagonal matrix, as

$$j_z(nljm', nljm) = m\delta_{m'm}. \tag{15.88}$$

The question was how to calculate the diagonal matrix element of the second term, μs_z, involving the non-conserved spin angular momentum (as $[H_2, s_z] \neq 0$).

[30] We remind the reader that latin lower case letters, such as j, l, s, refer to the dimensionless angular momenta, differing from the physical quantities, written in Fraktur font, by a factor of $\hbar = h/2\pi$: $\vec{\mathfrak{j}} = \hbar\vec{j}$, etc.

At this stage of affairs, Heisenberg and Jordan were still feeling their way forward through a set of problems for which the appropriate technical apparatus in the new theory often had not yet been developed. They frequently resorted to intuition that had evolved over the decade of dominance of the old quantum theory. We saw an example of this above in Heisenberg's resorting to the action/angle formalism in the calculation of the time averages of the inverse radial powers needed for the evaluation of the fine structure. Similarly, to evaluate the needed matrix elements for the weak field Zeeman effect, Heisenberg and Jordan appeal to the methods of the Pauli–Landé vector model (see Volume 1, pp. 358–361). Classically, the spin–orbit coupling $\vec{s}\cdot\vec{l}$ will induce a rapid precession of both the spin and orbital momentum vectors around the *conserved* total angular momentum vector $\vec{j}$ (which remains so, of course, even if the central potential terms in H_0 and H_3 are included as well). On the other hand, the component $(1/\vec{j}^2)\,(\vec{s}\cdot\vec{j})\,\vec{j}$ of the spin angular momentum $\vec{s}$ along the direction of $\vec{j}$ is constant, and should therefore be a diagonal matrix. Indeed, an easy calculation gives

$$\begin{aligned} \vec{s}\cdot\vec{j} &= \frac{1}{2}\left(\vec{j}^2+\vec{s}^2-(\vec{j}-\vec{s})^2\right) \\ &= \frac{1}{2}\left(j(j+1)+s(s+1)-l(l+1)\right), \end{aligned} \tag{15.89}$$

which shows that $\vec{s}\cdot\vec{j}$ is diagonal once s, j, l are specified.

The picture of the vector model also suggests that the time average (and hence, the diagonal matrix elements) of the component of $\vec{s}$ transverse to the fixed $\vec{j}$ is zero, so that the time average of $\vec{s}$ reduces just to the aforesaid component. In other words,

$$\bar{\vec{s}} = \frac{1}{2j(j+1)}\left(j(j+1)+s(s+1)-l(l+1)\right)\vec{j} \tag{15.90}$$

reduces to

$$\overline{s_z} = \frac{1}{2j(j+1)}\left(j(j+1)+s(s+1)-l(l+1)\right) j_z, \tag{15.91}$$

from which we get the desired (weak-field) Zeeman shift as the diagonal matrix element

$$\begin{aligned} \Delta E_Z &= H_1(jlm, jlm) \\ &= \mu m\left(1+\frac{j(j+1)+s(s+1)-l(l+1))}{2j(j+1)}\right) \\ &= g\mu m, \end{aligned} \tag{15.92}$$

with $-j \leq m \leq +j$ and

$$g \equiv 1+\frac{j(j+1)+s(s+1)-l(l+1))}{2j(j+1)}. \tag{15.93}$$

The correct form for the Landé g-factor (cf. Eq. (7.104)) has at last been obtained on the basis of a dynamically consistent physical framework. A vast amount of the previously baffling anomalous Zeeman phenomenology, with the proliferation of strange splitting patterns dependent on the column of the periodic table where the particular element sat (through the number of optically active valence electrons), could now be understood on the basis of a formula with a basis in physical reasoning. The essential ingredient needed for the generalization of Eq. (15.93) from the mono-electronic hydrogenic (H and He^+), or quasi-hydrogenic alkali metals, to elements in columns 2,3, etc. of the periodic table had already been provided in 1924 by Russell and Saunders (1925), with the proposal that in situations where several valence electrons were present, optical transitions involved quantum jumps of several valence electrons at once. For the case of the alkaline earths in column 2, as the authors explained in the abstract of their paper:

> It follows that *both valence electrons may jump at once* from outer to inner orbits, while the net energy lost is radiated as a *single quantum*—the atom acting as a whole (Russell and Saunders 1925, p. 38).

This idea ("Russell–Saunders coupling") would turn out to be the key to understanding the lighter atoms in the periodic table: the Zeeman weak field splittings are obtained by first ordering the energy levels according to the values of S, the spin quantum number associated with the vector sum of all the spins of the valence electrons, and L, the total orbital angular momentum quantum number of all the valence electrons (small letters continued to be used for the single valence electron case, i.e., hydrogenic or alkali metal atoms). These quantities would all be conserved with respect to a zeroth order Hamiltonian H_0 containing the nuclear Coulomb potential and the electron-electron electrostatic repulsion terms. The levels would then be split into terms with different values of the total angular momentum $\mathcal{J}$ (with possible values $\mathcal{J} = |L - S|, \ldots, L + S$) by the spin–orbit terms for the individual valence electrons, and the Zeeman splittings of each of these given by a Landé factor

$$g(SL\mathcal{J}) = 1 + \frac{\mathcal{J}(\mathcal{J}+1) - L(L+1) + S(S+1)}{2\mathcal{J}(\mathcal{J}+1)}. \tag{15.94}$$

The "correct" derivation of the weak field Zeeman effect in the new quantum mechanics, as it is found in today's texts, depends on a technical tool, the Wigner–Eckart theorem, which emerged a few years later in the work of Eugene Wigner (1927) and Carl Eckart (1930).[31] We briefly outline how this works, as it is at the core of the success of the vector model still found in modern texts in "derivations" of the Landé g-factor.[32] The Wigner–Eckart theorem states that the triplet of matrices representing the Cartesian components of any kinematic spatial vector quantity $\vec{v}$ (special cases of which are

[31] For an elementary derivation, see, for example, McIntyre (2012, p. 399).
[32] See, for example, Eisberg and Resnick (1985, sec. 10-6).

the coordinate $\vec{q}$, momentum $\vec{p}$, orbital angular momentum $\vec{l}$, spin angular momentum $\vec{s}$, or the total angular momentum $\vec{j}$), satisfying (as shown in the Three-Man-Paper) the commutation relations with the total angular momentum matrices j_x, j_y, j_z

$$[j_x, v_y] = i\frac{h}{2\pi}v_z, \quad [j_y, v_z] = i\frac{h}{2\pi}v_x, \quad [j_z, v_x] = i\frac{h}{2\pi}v_y, \tag{15.95}$$

will be proportional to the corresponding matrices for $\vec{j}$ (in the basis of stationary states where $\vec{j}^2$ and j_z are diagonal matrices—the other quantum numbers n, l, s are fixed and henceforth suppressed):

$$v_i(jm', jm) = c_v j_i(jm', jm), \quad i = x, y, z. \tag{15.96}$$

The proportionality constant c_v depends on the particular kinematical vector quantity $\vec{v}$ considered. Multiplying the matrix of j_i by the matrix of v_i in Eq. (15.96) and summing over i, one finds

$$\vec{j} \cdot \vec{v}\,(jm', jm) = c_v j(j+1)\delta_{m',m}. \tag{15.97}$$

In particular, for the spin operator $\vec{v} = \vec{s}$, we have already calculated (cf. Eq. (15.89))

$$\vec{j} \cdot \vec{s}\,(jm', jm) = \frac{1}{2}(j(j+1) + s(s+1) - l(l+1))\delta_{m',m}, \tag{15.98}$$

so the proportionality constant (sometimes called the "reduced matrix element") becomes in this case

$$c_s = \frac{j(j+1) + s(s+1) - l(l+1)}{2j(j+1)}, \tag{15.99}$$

which, when inserted in Eq. (15.96) for $v_i = s_i$ and $i = z$ gives the desired result Eq. (15.91). The Wigner–Eckart theorem implies that the intuition of the Pauli–Landé vector model can be applied correctly, with the caveat that the squares of angular momentum vectors involve the characteristic replacement $\vec{j}^2 \to j(j+1)$, etc.

15.3.2 The Stark effect: wave-mechanical treatment

Shortly after the introduction of wave mechanics, Schrödinger (1926f) and Epstein (1926), independently of one another, applied the new wave mechanics to the Stark effect.[33] We describe Schrödinger's treatment of the Stark effect in his third paper on

[33] Schrödinger's paper was received by *Annalen der Physik* on May 10 and published July 13, 1926. Epstein's paper is signed July 29 and appeared in *Physical Review* in October 1926. Epstein had moved from Munich to Pasadena in 1921. In his paper, Epstein (1926, p. 695, note 1) cited Schrödinger's first and second "communication" (*Mitteilung*) on wave mechanics (entitled "Quantization as a problem of proper values"), as well as the "equivalence" paper connecting wave mechanics to the Heisenberg–Born–Jordan matrix mechanics, but not the third quantization paper, containing Schrödinger's analysis of the Stark effect. Presumably, the July 13 issue of *Annalen der Physik* had not reached Pasadena by July 29.

wave mechanics, as it is chronologically prior to Epstein's (done entirely independently), which in any event arrives at essentially identical conclusions. However, Epstein's paper does carry the perturbation theory of the energy levels to second order (Schrödinger contents himself with verifying agreement with the first-order Stark effect as given by the old theory). Here, we adopt the coordinate notations of Epstein, as they coincide with the ones we used in Volume 1, in describing the solution of the Stark effect in the old quantum theory. In particular, we use the parabolic coordinates (originally due to Kramers, later used by both Schrödinger[34] and Epstein in their 1926 articles; cf. Eq. (6.70) and Figure 6.5):

$$z = \frac{\xi - \eta}{2}, \quad x + iy = \sqrt{\xi\eta}e^{i\varphi}, \quad r = \sqrt{x^2 + y^2 + z^2} = \frac{\xi + \eta}{2}. \tag{15.100}$$

Classically, one finds in these coordinates a Hamiltonian (cf. Eq. (6.71)):

$$H = \frac{1}{2\mu}\left(\frac{4\xi}{\xi + \eta}p_\xi^2 + \frac{4\eta}{\xi + \eta}p_\eta^2 + \frac{1}{\xi\eta}p_\varphi^2\right) - \frac{2e^2}{\xi + \eta} + \frac{1}{2}e\mathcal{E}(\xi - \eta). \tag{15.101}$$

Here, p_ξ, p_η, and p_φ are the conjugate momenta to the parabolic coordinates ξ, η, φ (in particular, p_φ is the just the usual azimuthal angular momentum), and e, μ are the charge and mass of the electron, respectively. The applied electric field is $\mathcal{E}\hat{z}$. As discussed in Volume 1, parabolic coordinates play a special role in the theory of the Stark effect, as the Hamilton–Jacobi equation (cf. Eq. (6.72)) is separable in these coordinates, with a solution $S(\xi, \eta, \varphi)$ that can be written as a sum of functions $S_\xi(\xi) + S_\eta(\eta) + S_\varphi(\varphi)$. This suggests, via the relation $\psi \simeq e^{iS}$ between the wave function and the Hamilton–Jacobi function (as elucidated by Schrödinger in his second paper), that the wave equation can also be solved, using parabolic coordinates, in terms of a *product* Ansatz for the wave function:

$$\psi(\xi, \eta, \varphi) = \psi_\xi(\xi)\psi_\eta(\eta)\psi_\varphi(\varphi). \tag{15.102}$$

The insertion in Eq. (15.101) of the usual wave mechanical replacement $p_\xi \rightarrow (h/2\pi i)\,\partial/\partial\xi$ (hereafter $(\hbar/i)\,\partial/\partial\xi$), etc., for the canonical momenta, leading to the second-order differential operator version of the Hamiltonian operator, would, if applied directly in Eq. (15.101), lead to a non-self-adjoint operator (with an associated matrix that is not Hermitian), as ξ and p_ξ do not commute with each other and appear in an asymmetric ordering in the classical formula. Schrödinger had already solved the problem of ordering of differential operators in section 4 of his equivalence paper, arriving at a uniquely ordered kinetic energy operator in any coordinate system obtained by a point transformation from Cartesian coordinates (cf. Eq. (14.211)). The application of

[34] In fact, Schrödinger uses (λ_1, λ_2) for (ξ, η). We have retained the earlier notation for consistency with the discussion in Volume 1.

this procedure to the case of a single electron in a Coulomb potential and subject to an applied electric field $\mathcal{E}\hat{z}$ leads to a Hamiltonian operator[35]

$$H = -\frac{2\hbar^2}{\mu}\frac{1}{\xi+\eta}\left(\frac{\partial}{\partial\xi}\xi\frac{\partial}{\partial\xi} + \frac{\partial}{\partial\eta}\eta\frac{\partial}{\partial\eta} + \frac{\xi+\eta}{4\xi\eta}\frac{\partial^2}{\partial\varphi^2}\right) + V, \tag{15.103}$$

with

$$V = -\frac{e^2}{r} + e\mathcal{E}z = -\frac{2e^2}{\xi+\eta} + e\mathcal{E}\frac{\xi-\eta}{2}. \tag{15.104}$$

Note that the ambiguous ordering of $p_\xi = \hbar\,\partial_\xi/i$ and ξ (and p_η and η) in the classical expression Eq. (15.101) has now been resolved uniquely, giving a properly self-adjoint operator, involving symmetrized terms like $p_\xi \xi p_\xi$, etc.[36]

For our purposes, it suffices to solve the Schrödinger equation for the hydrogen atom *without* an external electric field (i.e., setting $\mathcal{E} = 0$) in parabolic coordinates. We then sketch the derivation of the actual formula for the first-order Stark effect, that is, the formula for the energy levels in the presence of an external electric field, which involves the calculation of a matrix element using only the lowest-order wave functions for the stationary states. To emphasize the close mathematical similarity to the calculation of quantized levels in the old theory, we retain the same notation for the separation constants $\alpha_1 = E$, α_2, and α_3 used in the treatment of the Hamilton–Jacobi equation (6.72) in our analysis of the Schrödinger equation, which now reads simply $H\psi = E\psi = \alpha_1\psi$, with H the differential operator given in Eq. (15.103). Inserting this Hamilton operator into $H\psi = \alpha_1\psi$, dividing both sides by ψ and multiplying by $2\mu(\xi+\eta)$, we arrive at the Schrödinger equation:

$$\begin{aligned} 2\mu(\xi+\eta)\alpha_1 &= -\frac{\hbar^2}{\psi}\left(4\frac{\partial}{\partial\xi}\xi\frac{\partial}{\partial\xi} + 4\frac{\partial}{\partial\eta}\eta\frac{\partial}{\partial\eta} + \left(\frac{1}{\xi}+\frac{1}{\eta}\right)\frac{\partial^2}{\partial\varphi^2}\right)\psi \\ &\quad -4\mu e^2 + \mu e\mathcal{E}(\xi^2-\eta^2). \end{aligned} \tag{15.105}$$

Note the similarity between this equation and the Hamilton–Jacobi equation (6.74). The Schrödinger equation, like the Hamilton–Jacobi equation for this system, is separable in parabolic coordinates. In the case of the Schrödinger equation, this means that its solution factorizes as indicated in Eq. (15.102).[37]

[35] The derivation of the form of the Laplacian in parabolic coordinates needed here is easily accomplished using Eq. (6.71): in this case, one finds $\Delta = 4/r^2$ for the determinant of the metric matrix $g = \mathrm{diag}(2\xi/r, 2\eta/r, 1/\xi\eta)$. Inserting this in Eq. (6.71), with $(q_1, q_2, q_3) = (\xi, \eta, \varphi)$, one easily obtains Eqs. (15.103) and (15.104). See Chapter 14, note 89, for the equivalent calculation in spherical coordinates. The volume element in parabolic coordinates is $d^3r (= dxdydz) = \Delta^{-1/2}d\xi d\eta d\varphi = ((\xi+\eta)/4)d\xi d\eta d\varphi$.

[36] Note that in checking the self-adjoint property of H by integration by parts, the $\xi + \eta$ factor in the integration measure (see note 35) must be taken into account. It cancels the factor $1/(\xi+\eta)$ in the kinetic energy term, with the remaining differential operator assuming a symmetric form.

[37] Schrödinger uses Λ_1 and Λ_2 for what we are denoting ψ_ξ and ψ_η, respectively; Epstein uses $M(\xi)$ and $N(\eta)$.

A glance at Eq. (15.105) shows that

$$\frac{1}{\psi}\frac{\partial^2 \psi}{\partial \varphi^2} = \frac{1}{\psi_\varphi}\frac{d^2\psi_\varphi(\varphi)}{d\varphi^2} \tag{15.106}$$

can be written in terms of a function only of ξ and η, hence, is independent of φ, the only independent variable in Eq. (15.106). It must therefore be a constant, independent of ξ, η, and φ.[38] It follows that we can write

$$\psi_\varphi = Ce^{im\varphi}, \tag{15.107}$$

where m is an integer. The constant value of Eq. (15.106), $-m^2$, is chosen as the (negative) square of an integer to ensure the single-valuedness of the dependence $e^{im\varphi}$ of the wave function on the azimuthal angle φ.[39]

After we substitute $-m^2\psi$ for $\partial^2\psi/\partial\varphi^2$ in Eq. (15.105), we are left with an equation that splits into a part that only depends on ξ and a part that only depends on η. The two parts must therefore each be constant, repeating the reasoning used in Eq. (15.106) to detach and solve for the φ-dependence. Denoting these constants by $\mp 2\alpha_2$, as we did in the corresponding Eqs. (6.78)–(6.79) in the old quantum theory, we arrive at

$$-\frac{4\hbar^2}{\psi_\xi}\frac{d}{d\xi}\left(\xi\frac{d\psi_\xi}{d\xi}\right) + \frac{m^2\hbar^2}{\xi} - 2\mu e^2 + \mu e\mathcal{E}\xi^2 - 2\mu\alpha_1\xi = -2\alpha_2\,, \tag{15.108}$$

$$-\frac{4\hbar^2}{\psi_\eta}\frac{d}{d\eta}\left(\eta\frac{d\psi_\eta}{d\eta}\right) + \frac{m^2\hbar^2}{\eta} - 2\mu e^2 - \mu e\mathcal{E}\eta^2 - 2\mu\alpha_1\eta = +2\alpha_2. \tag{15.109}$$

These last two equations can be rewritten as

$$\xi\frac{d^2\psi_\xi}{d\xi^2} + \frac{d\psi_\xi}{d\xi} + \frac{1}{4}\left[\frac{2\mu e^2}{\hbar^2} - \frac{2\alpha_2}{\hbar^2} - \frac{m^2}{\xi} + \frac{2\mu\alpha_1}{\hbar^2}\xi - \frac{\mu e\mathcal{E}}{\hbar^2}\xi^2\right]\psi_\xi = 0, \tag{15.110}$$

$$\eta\frac{d^2\psi_\eta}{d\eta^2} + \frac{d\psi_\eta}{d\eta} + \frac{1}{4}\left[\frac{2\mu e^2}{\hbar^2} + \frac{2\alpha_2}{\hbar^2} - \frac{m^2}{\eta} + \frac{2\mu\alpha_1}{\hbar^2}\eta + \frac{\mu e\mathcal{E}}{\hbar^2}\eta^2\right]\psi_\eta = 0. \tag{15.111}$$

We now solve Eq. (15.110) for $\psi_\xi(\xi)$ for the case that $\mathcal{E} = 0$, following both Schrödinger and Epstein, who consider the field-free case first, as a prelude to the insertion of the

[38] The reasoning is identical to that employed in Section 6.3 in our solution of the classical Hamilton–Jacobi equation.

[39] The separation constant α_3 for φ in the Hamilton–Jacobi equation has here been replaced directly by its quantized value $m\hbar$.

electric field perturbatively. Note that for $\mathcal{E} = 0$ the equations for $\psi_\xi(\xi)$ and $\psi_\eta(\eta)$ differ only by the change of sign of α_2, so only the first need be examined in detail.

It will be convenient to introduce the quantity n defined as

$$na \equiv \frac{\hbar}{\sqrt{-2\mu\alpha_1}}, \tag{15.112}$$

where $a \equiv \hbar^2/\mu e^2$ is the Bohr radius. The notation n is chosen here with malice aforethought: n will turn out to be the principal quantum number. For the time being, though, n is simply a peculiarly defined positive real number. Eq. (15.112) allows us to write

$$\alpha_1 = -\frac{\hbar^2}{2\mu a^2 n^2} \left(= -\frac{2\pi^2 \mu e^4}{n^2 h^2} \right). \tag{15.113}$$

With the help of Eqs. (15.112) and (15.113), Eq. (15.110) for $\mathcal{E} = 0$ can be rewritten as:

$$\xi \frac{d^2\psi_\xi}{d\xi^2} + \frac{d\psi_\xi}{d\xi} + \frac{1}{4}\left[\frac{2}{a} - \frac{2\alpha_2}{\hbar^2} - \frac{m^2}{\xi} - \frac{\xi}{n^2a^2}\right]\psi_\xi = 0. \tag{15.114}$$

To solve this equation[40] we first need to examine its asymptotic behavior. For small ξ, the third term in Eq. (15.114) is dominated by the m^2/ξ piece. Keeping only this contribution in the third term, we find the possible behaviors $\psi_\xi \simeq \xi^{\pm|m|/2}$. For large ξ, keeping only the terms proportional to ξ, we find the possible exponential behaviors $\psi_\xi \simeq \exp(\pm\xi/2na)$.

If we extract the dominant asymptotic behavior at both large and small ξ, the remaining factor (call it $f(\xi)$) should be:

1. non-singular, approaching a non-zero constant at $\xi \to 0$ (thereby selecting the behavior $\psi_\xi \propto \xi^{+|m|/2}$); and
2. power bounded (selecting $\psi_\xi \propto \exp(-\xi/2na)$, up to powers of ξ) at large ξ, in accord with Schrödinger's regularity requirements stated already in his first paper.

We therefore define a new function $f(\xi)$ by setting

$$\psi_\xi(\xi) = \xi^{|m|/2} e^{-\xi/2na} f(\xi), \tag{15.115}$$

[40] We do not treat the equation, as Schrödinger does, using the sophisticated complex analysis methods, due to Schlesinger, which we described in the treatment of the radial Schrödinger equation for the Coulomb problem in Chapter 14 (see also the web resource, *Solving the Radial Schrödinger Equation*). Instead, we employ the power series analysis of asymptotic behavior more familiar to modern students, which is, in fact, the technique employed by Epstein.

where $f(\xi)$ should be finite as $\xi \to 0$ and power bounded for $\xi \to \infty$. Given this factorization, we can write the first derivative of ψ_ξ and ξ times the second derivative of ψ_ξ as:[41]

$$\frac{d\psi_\xi}{d\xi} = \left(\frac{|m|}{2}\xi^{|m|/2-1}f - \frac{\xi^{|m|/2}}{2na}f + \xi^{|m|/2}f'\right)e^{-\xi/2na}, \tag{15.116}$$

$$\begin{aligned}\xi\frac{d^2\psi_\xi}{d\xi^2} &= \left(\frac{|m|}{2}\left(\frac{|m|}{2}-1\right)\xi^{|m|/2-1}f - \frac{|m|\xi^{|m|/2}}{2na}f + |m|\xi^{|m|/2}f'\right.\\ &\quad \left. + \frac{\xi^{|m|/2+1}}{4n^2a^2}f - \frac{\xi^{|m|/2+1}}{na}f' + \xi^{|m|/2+1}f''\right)e^{-\xi/2na}.\end{aligned} \tag{15.117}$$

When these results are inserted into Eq. (15.114), the last two terms in $\frac{1}{4}[\ldots]\psi_\xi$ cancel against the terms $(|m|^2/4\xi)\psi_\xi$ and $(\xi/4n^2a^2)\psi_\xi$ in $\xi\, d^2\psi_\xi/d\xi^2$. The first term in $d\psi_\xi/d\xi$ cancels against the term $-(|m|/2)\psi_\xi/\xi$ in $\xi\, d^2\psi_\xi/d\xi^2$. Dividing the remainder of Eq. (15.114) by $\xi^{|m|/2}e^{-\xi/2na}$ and grouping terms with f, f', and f'', we find:

$$\xi f'' + \left(|m| + 1 - \frac{\xi}{na}\right)f' + \left(\frac{1}{2a}\left(1 - \frac{1}{n}\right) - \frac{\alpha_2}{2\hbar^2} - \frac{|m|}{2na}\right)f = 0. \tag{15.118}$$

The solution of this equation will now be written as a power series expansion ξ. We will see that an acceptable asymptotic behavior for the wave function requires the expansion to terminate, at which point f will become a polynomial. We thus set

$$f(\xi) = \sum_k c_k \xi^k. \tag{15.119}$$

Inserting this expansion into Eq. (15.118), we find:

$$\begin{aligned}\sum_k c_k \Big(k(k-1)\xi^{k-1} &+ (|m|+1)k\xi^{k-1}\\ &- \frac{k}{na}\xi^k + \left(\frac{1}{2a}\left(1 - \frac{|m|+1}{n}\right) - \frac{\alpha_2}{2\hbar^2}\right)\xi^k\Big) = 0.\end{aligned} \tag{15.120}$$

Replacing the summation variable k by $k+1$, we can rewrite the terms of order $k-1$ in ξ in Eq. (15.120) as terms of order k:

$$\sum_k kc_k(k+|m|)\xi^{k-1} = \sum_k (k+1)c_{k+1}(k+1+|m|)\xi^k. \tag{15.121}$$

For Eq. (15.120) to hold, the coefficients of ξ^k must vanish for all k. Inserting Eq. (15.121) into Eq. (15.120), we thus find the following recursion relation for the coefficient c_k:

[41] In Eq. (15.115), ψ_ξ is written as the product of three factors. Using 0s, 1s, and 2s to indicate which factors are differentiated zero, one, and two times, we can schematically write $d\psi_\xi/d\xi$ as $[100 + 010 + 001]\psi_\xi$, and $d^2\psi_\xi/d\xi^2$ as $[(200 + 110 + 101) + (110 + 020 + 011) + (101 + 011 + 002)]\psi_\xi = [200 + 2(110) + 2(101) + 020 + 2(011) + 002]\psi_\xi$.

$$\frac{c_{k+1}}{c_k} = \frac{k - \frac{n}{2}\left(1 - \frac{|m|+1}{n}\right) + \frac{\alpha_2 na}{2\hbar^2}}{na(k+1)(k+1+|m|)}. \tag{15.122}$$

As the polynomial in Eq. (15.119) needs to break off for finite k to get a legitimate wave function,[42] there must be some value n_ξ for k such that $c_{k+1} = 0$. Eq. (15.122) tells us that this quantum number is given by:

$$n_\xi = \frac{n}{2}\left(1 - \frac{|m|+1}{n}\right) - \frac{\alpha_2 na}{2\hbar^2}. \tag{15.123}$$

To find the solution of equation (15.111) for $\psi_\eta(\eta)$ for $\mathcal{E} = 0$, we proceed in the exact same way as we did in Eqs. (15.114)–(15.123) for $\psi_\xi(\xi)$. Given its behavior at small and large η, we write $\psi_\eta(\eta)$ as (cf. Eq. (15.115))

$$\psi_\eta(\eta) = \eta^{|m|/2} e^{-\eta/2na} g(\eta). \tag{15.124}$$

We then derive an equation for $g(\eta)$, analogous to Eq. (15.118) for $f(\xi)$. The solution of this equation will be a polynomial $\sum_k \hat{c}_k \eta^k$ (where the "hat" is used to distinguish the coefficients from those in Eq. (15.119) for $f(\xi)$) that will break off if and only if there is a value n_η for k such that (cf. Eq. (15.123)):

$$n_\eta = \frac{n}{2}\left(1 - \frac{|m|+1}{n}\right) + \frac{\alpha_2 na}{2\hbar^2}. \tag{15.125}$$

Combining Eq. (15.123) and Eq. (15.125), we find

$$\frac{n}{2}\left(1 - \frac{|m|+1}{n}\right) - n_\xi = n_\eta - \frac{n}{2}\left(1 - \frac{|m|+1}{n}\right), \tag{15.126}$$

or

$$n = n_\xi + n_\eta + |m| + 1. \tag{15.127}$$

Recall the result in the old quantum theory for the energy in the field-free limit in parabolic coordinates (see Eqs. (6.92) and (6.95)):

$$\alpha_1 = -\frac{2\pi\mu e^2}{(I_\xi + I_\eta + I_\varphi)^2}, \tag{15.128}$$

$$I_\xi = n_\xi h, \quad I_\eta = n_\eta h, \quad I_\varphi = n_\varphi h \; (= |m|h). \tag{15.129}$$

[42] Otherwise, for large k, $c_{k+1}/c_k \simeq 1/na(k+1)$ and $f(\xi)$ approximates for large ξ the exponential expansion $\sum_k (\xi/na)^k/k! = \exp(+\xi/na)$, reversing the falling exponential in Eq. (15.115).

Comparing this with Eq. (15.113), we see that the difference between the old and the new theory lies in the final term +1 in Eq. (15.127).[43] This extra term obviates the need for a special condition to rule out $|m| = 0$ (cf. Volume 1, p. 294), which in the old theory corresponded to rectilinear orbits passing through the nucleus. As Epstein noted toward the end of his paper on the Stark effect in wave mechanics:

> It will be remembered that the restriction for the azimuthal quantum number [$|m| > 0$] was an additional one, not following from the dynamical conditions. It was introduced by Bohr for the purpose of eliminating plane orbits, moving in which the electrons would sooner or later undergo a collusion [sic] with the nucleus. In our new theory an additional restriction is not necessary (Epstein 1926, p. 708).

We now sketch how the formula for the energy levels in the first-order Stark effect—the third of the great successes of the old quantum theory recounted in Volume 1—is recovered in wave mechanics. The power-series solutions discussed above for the wave functions $\psi_\xi(\xi)$ and $\psi_\eta(\eta)$ turn out to be associated Laguerre polynomials.[44] The full normalized energy eigenfunction solutions $\psi_{n_\xi n_\eta m}(\xi, \eta, \varphi)$ for the state characterized by quantum numbers n_ξ, n_η, m (with principal quantum number $n = n_\xi + n_\eta + |m| + 1$) in the unperturbed case (i.e., a hydrogen atom in zero electric field) take the form

$$\psi_{n_\xi n_\eta m} = C_{n_\xi n_\eta m} \left(\frac{\xi\eta}{n^2 a^2}\right)^{|m|/2} e^{-(\xi+\eta)/2na} L_{n_\xi}^{|m|}\left(\frac{\xi}{na}\right) L_{n_\eta}^{|m|}\left(\frac{\eta}{na}\right) e^{im\varphi}, \tag{15.130}$$

where $C_{n_\xi n_\eta m}$ is an explicitly known normalization constant that we do not give here (see, e.g., Condon and Shortley 1963, p. 399).

Introducing the Stark perturbation operator $H^{\text{Stark}} \equiv \frac{1}{2}e\mathcal{E}(\xi - \eta)$, the procedure of first-order degenerate perturbation theory (see Schrödinger 1926f, sec. 2; cf. also Eq. (65)) instructs us to calculate the matrix of the perturbing operator in the basis of the n^2 degenerate states of identical unperturbed energy corresponding to a given principal

[43] The roots of the extra factor of unity in the transition from the old to the new theory can be traced quite precisely by employing the WKB (or WKB-J) approximation of the Schrödinger equation, due to Wentzel (1926), Kramers (1926), Brillouin (1926), and Jeffreys (1924). It then becomes clear that the WKB method applied to the one-dimensional Schrödinger equations for ψ_ξ and ψ_η lead to quantization conditions equivalent to those of the old quantum theory, with the simple replacement $n_\xi \to n_\xi + 1/2$ (and likewise $n_\eta \to n_\eta + 1/2$). The displacement of previously integer quantum numbers by a fraction (called the *Maslov index*), here 1/2, is typical in most problems involving librational motion in the old theory. In this case, it explains the replacement of the old quantum theory result, $n = n_\xi + n_\eta + n_\varphi$ by $n = n_\xi + 1/2 + n_\eta + 1/2 + |m| = n_\xi + n_\eta + |m| + 1$, Eq. (15.127). See Duncan and Janssen (2014, sec. 6) for a detailed explanation of this critical correction to the Bohr–Sommerfeld rules of the old theory provided by the new quantum mechanics.

[44] The associated Laguerre polynomial $L_p^q(z)$ (p, q integer) satisfies the second-order differential equation $(zd^2/dz^2 + (q + 1 - z)d/dz + p)L_p^q = 0$. Defining $\xi = naz$, we easily see from Eqs. (15.118)–(15.123) that $f(\xi) = L_{n_\xi}^{|m|}(\xi/na)$, up to normalization. We are here using the notation of standard quantum textbooks such as Messiah (1966) or Schiff (1968) for the associated Laguerre polynomials. Other authors, including Schrödinger, employ a slightly different definition. In Condon and Shortley (1963), for example, we find $L_{n_\xi+|m|}^{|m|}$ instead of $L_{n_\xi}^{|m|}$, etc.

quantum number n. The calculation, in which one uses standard properties of Laguerre functions, gives[45]

$$\int \psi^{\star}_{n'_\xi n'_\eta m'}(\xi,\eta,\varphi)\left(e\mathcal{E}\frac{\xi-\eta}{2}\right)\psi_{n_\xi n_\eta m}(\xi,\eta,\varphi)\left(\frac{\xi+\eta}{4}\right)d\xi d\eta d\varphi$$
$$= \frac{3}{2}e\mathcal{E}n(n_\xi - n_\eta)a\delta_{n'_\xi n_\xi}\delta_{n'_\eta n_\eta}\delta_{m'm}. \tag{15.131}$$

The perturbation matrix is, as we see, diagonal in this basis, so the Stark energy shifts can be read off directly and are seen to be identical to the results obtained in the old quantum theory (see Eq. (6.96), second term, with $a = \hbar^2/\mu e^2 = h^2/4\pi^2\mu e^2$).

The treatment of the Stark effect in the new theory, however, has some distinct advantages of the treatment in the old quantum theory. We already encountered one of these. As Epstein (1926, p. 708) noted, the new theory did away with the extra restriction $|m| \neq 0$ required in the old theory to avoid orbits going through the nucleus. The new theory had two additional advantages. First, the problematic non-uniqueness of orbits that the old quantum theory ran into in the treatment of the Stark problem (see Volume 1, p. 293) is beautifully resolved in the new theory. Second, the new theory replaced Kramers' inherently somewhat ambiguous calculations of the intensities of the various Stark components on the basis of the correspondence principle (see Volume 1, p. 298) by a perfectly well-defined procedure leading to results that ended up agreeing better with improved experimental data. To conclude this section, we take a closer look at these two advantages of the new theory over the old one.

As Schrödinger (1926d, pp. 33–34; quoted in Section 14.4.2) had already pointed out explicitly in his second paper, the troublesome ambiguity in choice of orbits in degenerate systems in the old quantum theory turns into the totally unproblematic non-uniqueness of bases in the linear space of proper functions of the wave equation in the new quantum theory. The allowed wave functions arise as *linear aggregates* (in modern terms—linear combinations) of a complete set of basis eigenfunctions of a given energy, and although the specific basis functions specifying the states of a given energy will depend on the coordinate system in which the wave equation is separated, the set of all possible linear combinations of these functions—which, as Schrödinger states, represents the *possible state[s] of vibration* of the electron—is the same regardless of the chosen coordinate system.

The solution of the Stark effect via parabolic coordinates affords a beautiful opportunity to illustrate Schrödinger's point in complete analytic detail. In wave mechanics, the stationary states are associated with eigenfunctions of the Hamilton operator of the

[45] Here we give the most direct version of the first-order calculation, as found in modern texts, in terms of the full three-dimensional wave functions. Both Schrödinger and Epstein calculate the energy shift by analyzing the perturbation, to first order in $\mathcal{E}$, in the eigenvalues of the self-adjoint operators associated with the one-dimensional differential equations Eqs. (15.108) and (15.109) of the separated problem. The perturbation in the desired energy $E = \alpha_1$ is then given as the sum of the two shifts for the ξ and η equation.

system. Thus, to solve the problem of a bound electron in the hydrogen atom in spherical coordinates (r, ϑ, φ), we need to find normalizable (square-integrable) solutions of the time-independent Schrödinger equation,

$$-\frac{\hbar^2}{2\mu}\left(\frac{1}{r^2}\frac{\partial}{\partial r}\left(r^2\frac{\partial\psi}{\partial r}\right)+\frac{1}{r^2\sin\vartheta}\frac{\partial}{\partial\vartheta}\left(\sin\vartheta\frac{\partial\psi}{\partial\vartheta}\right)\right.$$

$$\left.+\frac{1}{r^2\sin^2\vartheta}\frac{\partial^2\psi}{\partial\varphi^2}\right)-\frac{e^2}{r}\psi=E\psi, \tag{15.132}$$

for energy eigenvalues $E < 0$, where the wave function ψ is a function of (r, ϑ, φ). We highlight this choice of coordinates by using the notation $\psi^{\text{spherical}}(r, \vartheta, \varphi)$ for solutions of Eq. (15.132). The negative-energy normalizable solutions of Eq. (15.132) correspond to the discrete energies E_n labeled by the value of the principal quantum number n. The Schrödinger equation can be *separated* in spherical coordinates, which means that Eq. (15.132) has solutions of the form $\psi_r(r)\psi_\vartheta(\vartheta)\psi_\varphi(\varphi)(= R_{n,l}(r)P_{lm}(\vartheta)\Phi_m(\varphi)$, in more usual notation). For each value of the principal quantum number $n = n_r + l$, the angular momentum quantum number $l = n_\vartheta$ can take on the values $0, 1, 2, \ldots, n-1$ and, for each value of l, the azimuthal quantum number m (where $|m| = n_\varphi$) can take on the values $-l, -l+1, \ldots, l-1, l$. For each value of n, there are $\sum_{l=0}^{n-1}(2l+1) = n^2$ degenerate orthogonal solutions of Eq. (15.132). We can conveniently label these solutions with the values of n, l, and m and introduce the notation $\psi_{nlm}^{\text{spherical}}(r, \vartheta, \varphi)$ for them. Any solution of Eq. (15.132) for $E = E_n$ must be a linear combination of the solutions $\psi_{nlm}^{\text{spherical}}(r, \vartheta, \varphi)$ with different values of l and m but a fixed value of n.

In parabolic coordinates (ξ, η, φ) (see Eq. (15.100)), the time-independent Schrödinger equation for the same system is $H\psi = E\psi$ (cf. Eq. (15.103) for H), that is,

$$-\frac{\hbar^2}{2\mu}\left(\frac{4}{\xi+\eta}\frac{\partial}{\partial\xi}\left(\xi\frac{\partial\psi}{\partial\xi}\right)+\frac{4}{\xi+\eta}\frac{\partial}{\partial\eta}\left(\eta\frac{\partial\psi}{\partial\eta}\right)+\frac{1}{\xi\eta}\frac{\partial^2\psi}{\partial\varphi^2}\right)-\frac{2e^2}{\xi+\eta}\psi=E\psi, \tag{15.133}$$

where the wave function ψ is now a function of (ξ, η, φ). We highlight this choice of coordinates by adopting the notation $\psi^{\text{parabolic}}(\xi, \eta, \varphi)$ for solutions of Eq. (15.133).[46] The Schrödinger equation can once again be *separated* in parabolic coordinates, which means that Eq. (15.133) has solutions of the form $\psi_\xi(\xi)\psi_\eta(\eta)\psi_\varphi(\varphi)$ (see Eq. (15.102)). For $E < 0$, there are normalizable solutions only for discrete energies E_n labeled by the principal quantum number $n = n_\xi + n_\eta + n_\varphi + 1$ (see Eq. (15.127)). For any fixed value of n, there are n^2 degenerate orthogonal solutions, labeled by the values of the integer quantum numbers n_ξ, n_η, and $n_\varphi = |m|$ (where m is the same as in spherical coordinates). We introduce the notation $\psi_{n_\xi n_\eta m}^{\text{parabolic}}(\xi, \eta, \varphi)$ for these solutions. For a given value of n,

[46] Note that only two of the three coordinates have actually been changed. The azimuthal angle coordinate φ is the same in both coordinate systems. The φ-derivative terms in Eqs. (15.132) and (15.133) are, in fact, identical, as can readily be established with the help of Eq. (15.100) for the transformation from Cartesian to parabolic coordinates.

all combinations of positive or zero values of n_ξ, n_η, and $|m|$ consistent with $n = n_\xi + n_\eta + |m| + 1$ are possible. The number of such combinations is n^2. Any solution of Eq. (15.133) for $E = E_n$ must be a linear combination of the solutions $\psi^{\text{parabolic}}_{n_\xi n_\eta m}(\xi, \eta, \varphi)$ with different values of n_ξ, n_η, and $|m|$ but a fixed value of n. The only advantage of parabolic coordinates in this context lies in the fact that the Schrödinger equation *continues to be separable* even when the external electric field is switched on, which is *not* the case for spherical coordinates. However, there is *absolutely no distinction*, either of analytic convenience or of conceptual validity, in the use of either system for the pure Coulomb problem.

Any solution $\psi^{\text{spherical}}(r, \vartheta, \varphi)$ of Eq. (15.132) can be immediately converted into a solution $\psi^{\text{parabolic}}(\xi, \eta, \varphi)$ of Eq. (15.133) simply by expressing (r, ϑ, φ) in terms of (ξ, η, φ). The allowed physical states are therefore identical, whichever coordinate system we use, unlike the orbits selected by the quantum conditions in the old quantum theory. However, the individual *separated* solutions (labeled by definite triples of quantum numbers in either coordinate system) are not in one-to-one correspondence. Since we are dealing with different representations of one and the same self-adjoint operator, any solution $\psi^{\text{spherical}}_{nlm}(r, \vartheta, \varphi)$ of Eq. (15.132) must be a linear combination of the solutions $\psi^{\text{parabolic}}_{n_\xi n_\eta m}(\xi, \eta, \varphi)$ of Eq. (15.133). After all, any solution $\psi^{\text{spherical}}_{nlm}(r, \vartheta, \varphi)$—with (r, ϑ, φ) expressed in terms of (ξ, η, φ)—of Eq. (15.132) must also be a solution of Eq. (15.133) for the same energy $E = E_n$. The lack of a one-to-one correspondence between the separated solutions in two different coordinate systems can be regarded as the "residue" in wave mechanics of the problem of the non-uniqueness of the orbits—that is, of quantized elliptical orbits of different shape when the system is solved in spherical or parabolic coordinates—in the old quantum theory. This residue, of course, is no problem at all in wave mechanics.

The explicit separated solutions are as follows. In spherical coordinates, they are

$$\psi^{\text{spherical}}_{nlm}(r, \vartheta, \varphi) = C_{nlm} \left(\frac{2r}{na}\right)^l e^{-r/na} L^{2l+1}_{n-l-1}\left(\frac{2r}{na}\right) P^m_l(\cos\vartheta) e^{im\varphi}, \tag{15.134}$$

while in parabolic coordinates, they are

$$\psi^{\text{parabolic}}_{n_\xi n_\eta m}(\xi, \eta, \varphi) = C_{n_\xi n_\eta m} \left(\frac{\eta\xi}{n^2 a^2}\right)^{|m|/2} e^{-(\xi+\eta)/2na} L^{|m|}_{n_\xi}\left(\frac{\xi}{na}\right) L^{|m|}_{n_\eta}\left(\frac{\eta}{na}\right) e^{im\varphi}, \tag{15.135}$$

where P^m_l and $L^{\cdots}_{\cdots}$ are the associated Legendre and Laguerre polynomials. C_{nlm} and $C_{n_\xi n_\eta m}$ are normalization constants. Dropping these, and the identical exponential radial dependence $e^{-r/na} = e^{-(\xi+\eta)/2na}$ and azimuthal dependence $e^{im\varphi}$ in both cases, we find that

$$\psi^{\text{spherical}}_{nlm}(r, \vartheta, \varphi) \propto r^l L^{2l+1}_{n-l-1}\left(\frac{2r}{na}\right) P^m_l(\cos\vartheta), \tag{15.136}$$

and that

$$\psi^{\text{parabolic}}_{n_\xi n_\eta m}(\xi,\eta,\varphi) \propto r^{|m|}\sin^{|m|}\vartheta L^{|m|}_{n_\xi}\left(\frac{\xi}{na}\right)L^{|m|}_{n_\eta}\left(\frac{\eta}{na}\right). \tag{15.137}$$

As discussed above, the functions in Eq. (15.136) must be linear combinations of those in Eq. (15.137) (and conversely). This is tedious to demonstrate algebraically in complete generality, but easy to see for the special case of maximal azimuthal quantum number, $|m| = l$. In this case, using the addition formula

$$\sum_{n_\xi=0}^{n-|m|-1} L^{|m|}_{n_\xi}\left(\frac{\xi}{na}\right)L^{|m|}_{n-|m|-1-n_\xi}\left(\frac{\eta}{na}\right) = L^{2|m|+1}_{n-|m|-1}\left(\frac{\xi+\eta}{na}\right) = L^{2l+1}_{n-l-1}\left(\frac{2r}{na}\right) \tag{15.138}$$

and

$$P^l_l(\cos\vartheta) = (2l-1)!!\sin^l\vartheta, \tag{15.139}$$

we find very simply that

$$\varphi^{\text{spherical}}_{nll} = (2l-1)!!\sum_{n_\xi=0}^{n-l-1}\varphi^{\text{parabolic}}_{n_\xi,n-l-1-n_\xi,m=l}. \tag{15.140}$$

For example, for the $2p$ states with maximal $|m| = 1$, the sum in Eq. (15.140) degenerates to a single term and we have a one–one correspondence between normalized states in spherical coordinates, which we denote as $|n\,l\,m\rangle$, and normalized states in parabolic coordinates, which we denote as $|n_\xi\; n_\eta\; m)$:

$$|2\,1+1\rangle = |0\,0+1), \quad |2\,1-1\rangle = |0\,0-1). \tag{15.141}$$

For the $2s$ state, $m = l = 0$ and the sum in Eq. (15.140) contains two terms and we have

$$|2\;0\;0\rangle = \frac{1}{\sqrt{2}}\{|1\;0\;0) + |0\;1\;0)\}. \tag{15.142}$$

The remaining state ($2p$ with $m = 0$) is evidently

$$|2\;1\;0\rangle = \frac{1}{\sqrt{2}}\{|1\;0\;0) - |0\;1\;0)\}. \tag{15.143}$$

Once the term $eEr\cos\vartheta = \frac{1}{2}eE(\xi-\eta)$, describing an external field in the z-direction, is added to the Hamilton operator of the system, the problem is no longer separable in spherical coordinates, neither in the old quantum theory $[S(r,\vartheta,\varphi) \neq S_r(r) + S_\vartheta(\vartheta) + S_\varphi(\varphi)]$ nor in wave mechanics $[\psi(r,\vartheta,\varphi) \neq \psi_r(r)\psi_\vartheta(\vartheta)\psi_\varphi(\varphi)]$. However, the problem continues to be separable in parabolic coordinates, both in the old quantum theory [$S(\xi, \eta, \varphi) = S_\xi(\xi) + S_\eta(\eta) + S_\varphi(\varphi)$; cf. Eq. (6.73)] and in wave mechanics [$\psi(\xi, \eta, \varphi) = \psi_\xi(\xi)\psi_\eta(\eta)\psi_\varphi(\varphi)$; cf. Eq. (15.102)]. From the point of view of the old quantum theory, this means that the dynamics *must* be analyzed in parabolic coordinates.

In wave mechanics, the separated eigenfunctions in spherical coordinates (i.e., the $|\cdots\rangle$ states) are no longer eigenfunctions of the new Hamilton operator, but rather linear combinations of them give, at least to first order, the separated eigenfunctions in parabolic coordinates (the $|..)$ states). To account for the first-order Stark effect, the quantum conditions of the old quantum theory had to be imposed in parabolic coordinates. From the point of view of the new quantum theory, this is directly related to the fact that, in standard first-order degenerate perturbation theory, the matrix of the perturbing part of the Hamilton operator (here the term with the external electric field) is diagonal in the basis of the (unperturbed) states $|n_\xi\, n_\eta\, m)$ in parabolic coordinates (cf. Eq. (15.131)), *but not in spherical coordinates.*

The final advantage of the new theory over the old one in the treatment of the Stark effect concerns the intensities of the various Stark components (cf. Duncan and Janssen 2015, pp. 242–250). As shown in Volume 1 (p. 298), Kramers (1919) computed these on the basis of the correspondence principle from averages of the Fourier components of the initial and final orbits of the transition involved. His calculations were generally in good agreement with Stark's experimental findings. In his Nobel lecture, Bohr (1923d, p. 39) included a diagram showing the agreement between Kramers' calculations and Stark's data and commented: "the theory reproduces completely the main feature of the experimental results." In a National Research Council *Bulletin* by Edwin P. Adams on quantum theory, Kramers' results are hailed as "most convincing evidence for the value of [Bohr's] principles" (1923, p. 88).[47]

The following year, the Canadian physicist John Stuart Foster published new data on the intensities in the Stark effect, which once again agreed with Kramers' calculations. Foster had measured the relative intensities of some Stark components of H_ε, the Balmer line corresponding to the transition $n = 7 \rightarrow n = 2$.[48] He wrote:

> At the time Kramers' dissertation was published there were no observations with which to compare his theoretical estimates of the relative intensities of the Stark effect components of H_ε. Two of the plates exposed ... during this investigation show many components

[47] This is also the assessment of several historians. In his book on the Bohr model, Kragh (2012, p. 205) writes: "Kramers arrived at theoretical values for the relative intensities that he modestly described as 'convincing'. In fact the agreement between theory and experiment was nearly perfect" (Kragh 2012, p. 205). Kramers' biographer concurs: "The agreement between theory and experiment is surprisingly good ... Kramers describes this impressive agreement in his customary guarded manner" (Dresden 1987, p. 109). Leone, Paoletti, and Robotti (2004, p. 288) likewise state that "Kramers's theoretical predictions were in excellent agreement with Stark's measurements."

[48] See the biographical memoir by Bell (1966, pp. 150–153) for a discussion of Foster's work on the Stark effect that he started as a graduate student at Yale in the early 1920s and continued until the late 1930s as a professor at McGill. As Bell points out, "[d]uring all of his work on the Stark effect he brooded over the design of Lo Surdo discharge tubes, and he was able to make them behave better than anyone else" (Bell 1966, p. 151). See Leone, Paoletti, and Robotti (2004, p. 288), especially figures 4, 6, 9, and 10, for a comparison between the experimental setups used by Lo Surdo and Stark.

> of this line . . . The relative intensities of the stronger inner components are very close to the predicted values (Foster 1924, p. 675).

In subsequent experiments, however, undertaken with his graduate student Laura Chalk,[49] Foster found relative intensities for some Stark components of H_α and H_β that differed markedly from those reported by Stark and from the theoretical values found by Kramers. The new experimental values, however, generally agreed with the intensities predicted by wave mechanics.[50] Referring to the diagram reproduced in Figure 15.2, comparing Stark's data with the theoretical values given by Kramers (1919), Schrödinger (1926f), and Epstein (1926) for the intensities of some Stark components of H_α (corresponding to the transition $n = 3 \rightarrow n = 2$), Foster and Chalk reported:

> By means of wave mechanics, Schrödinger has made quantitative calculations of the intensities of Stark components in hydrogen which are commonly considered to be an improvement on the earlier estimates based on the correspondence principle. That this is so in the case of H_β was shown recently by the writers in a quantitative experimental investigation [i.e., Foster and Chalk 1926].
>
> The greatest variation of the new theory from Prof. Stark's results, however, occurs in the parallel components of H_α. There are three pairs of such components which have been photographed; and in the original experiments, as well as in the older quantum theory, the outside components were found to be the strongest. This is further supported by the recent calculations of Epstein on wave mechanics. In contrast to these results, Schrödinger finds the greatest intensity for the pair with intermediate displacements. The difference between Schrödinger's calculations and the observations of Stark is obviously rather large to be considered as an experimental error. Yet this is what it appears to be according to numerous plates obtained by the junior author in an extension to the earlier experiments, the new results being in general agreement with the calculations of Schrödinger (Foster and Chalk 1928, p. 830).

The three pairs of parallel Stark components of H_α mentioned by Foster and Chalk and shown in Figure 15.2 correspond to the six solid lines in Figure 6.7, with $\Delta\nu = \pm 2, \pm 3, \pm 4$ (in the appropriate units).

Foster and Chalk suspected that Epstein's calculations only agreed with Stark's old data because Epstein had made some errors. This suspicion was confirmed the following year when Gordon and Minkowski (1929) corrected Epstein's calculations and showed that they led to the same results as Schrödinger's (Condon and Shortley 1963, p. 400).

[49] In 1928, Chalk became the first woman to earn her PhD in physics at McGill.

[50] See the diagrams in Condon and Shortley (1963, p. 401) based on the experimental work of Foster and Chalk (1926, 1928, 1929), Mark and Wierl (1929), and others in the late 1920s, which show that in most cases the quantum-mechanical predictions for the intensities of the Stark components of H_α, H_β, H_γ, and H_δ agree within a few percent with experiment. Condon and Shortley (1963, p. 402) caution that these measurements "do not agree in detail with the theory, but provisionally we shall regard this as due to the large number of variations in physical conditions in the experimental work." The dependence of the results of measurements of the intensities of the Stark components of the Balmer lines on the experimental setup is emphasized by Mark and Wierl (1929, p. 538).

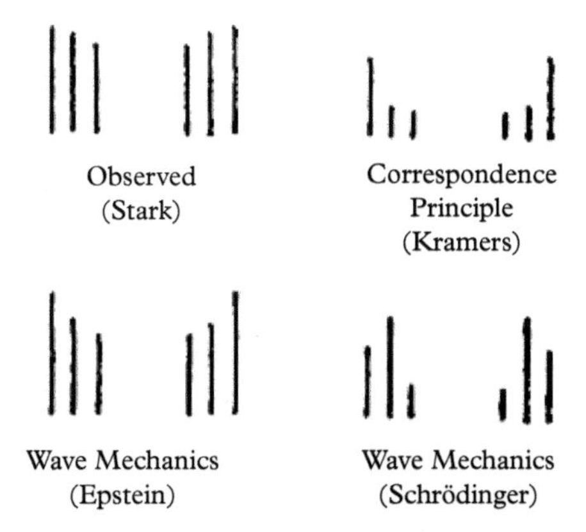

Figure 15.2 *Four sets of values (one experimental, three theoretical) for the intensities of the six parallel Stark components of H_α (Foster and Chalk 1928, p. 830, Fig. 1).*

Foster and Chalk concluded that the intensities found by Schrödinger agreed well with their experimental findings:

> Within the limits of experimental error the new results agree, we believe, with the calculations by Schrödinger. The well-marked agreement in the lines H_α and H_β is of especial importance, since it is in the case of low quantum numbers that the new quantum-theoretical calculations show the greatest departure from the estimates of the older theory (Foster and Chalk 1928, pp. 116–118).

Two years earlier, in 1926, Foster had spent almost a year in Copenhagen on an International Education Board fellowship, which gave him "the chance to publish his most important single paper" (Bell 1966, p. 148). In this paper, communicated by Bohr, Foster (1927) presented both theoretical and experimental results on the Stark effect in helium. We end this section with a quote from Heisenberg's later recollections of Foster's visit and his results on helium, which will serve as a nice segue to the next section:

> I was mainly busy with the practical application of quantum mechanics to the spectrum of the helium atom. Foster's beautiful measurements of the Stark effect in the helium spectrum played an important role in this work. Foster had come to Copenhagen for a short stay from Canada to compare his results with those of the new theory. Most of the discussions took place in Mrs. Maar's week-end cottage ... On the garden benches between the rosebeds ... we spread out the enlargements of Foster's spectral photographs and the measured line positions were compared with the results of the theory. The agreement was perfect, and we were happy to see how many of the most complicated and apparently unconnected details resulted more or less automatically from the formulae of quantum mechanics. Bohr too was glad to note how the Stark-effect once again, just as ten years earlier with the hydrogen atom, proved one of the most beautiful confirmations that one was on the right road to an understanding of the atom (Heisenberg 1967, p. 102).

15.4 The problem of helium

Section 7.4 discussed in some detail the failure of the old quantum theory to come to terms with even the most basic features of the spectrum of helium, the atoms of which possess in their un-ionized state two electrons and therefore represent the simplest atomic system beyond hydrogen, which had been treated with remarkable success by the Bohr–Sommerfeld technology. In particular, two problems of transcendent importance could not be treated with even qualitative success by the old theory: an explanation of the existence of two separate, and (apparently) totally non-communicating—via transitions connected with emission or absorption of light quanta—series of quantized energy levels (or "terms", in the spectroscopic language of the time), later dubbed the "par(a)helium" and "orthohelium" levels; and the calculation of the binding energy of the two electrons in the lowest energy state (*Normalzustand*, or "ground state" in current terminology).

Shortly after the discovery of the terrestrial existence of helium as a gas emitted from cleveite,[51] a mineral containing uranium (the element had first been associated with spectral lines of the sun), spectroscopic measurements carried out by Runge and Paschen (1896) established the existence of two sets of terms, with transitions only occurring between the terms of each set separately (see Figure 15.3).

After many attempts to separate these components chemically (as if they were to be associated with two distinct elements, soon dubbed "Par(a)helium" and "Orthohelium") had failed (Nath 2013, Ch. 12), it was reluctantly accepted that the same atom possessed two sets of states that were completely disconnected from one another in regard to the possibility of radiative transitions. Moreover, the terms of parahelium were all singlets, while those of orthohelium initially appeared to consist of doublets. The arrangement of the levels in each set bore a striking similarity to the spectra of alkali atoms, such as lithium, sodium, etc.(Figure 15.3, left panel)—thus, there were two distinct series of sharp, principal and diffuse lines corresponding to the terms of parahelium and orthohelium, respectively (Figure 15.3, right panel). The separation of corresponding terms for parahelium and orthohelium could be as large as an electron volt, but became smaller as one went to higher values of the principal quantum number. Increase in the resolution of the spectroscopic studies revealed that the apparent doublets of orthohelium were actually triplets, with two members of the triplet very close together, thus having evaded separation in the initial spectroscopic measurements.

The binding energy problem encompassed the issue of the ionization potential of helium (the energy required to remove a single electron from the un-ionized atom in its ground state), which by now had finally (after some initial difficulties—cf. Section 7.4) been pinned down by ultraviolet observations of Lyman (1922) to be 24.6 eV. The first calculation of this quantity by Bohr in 1913, using a model in which the two electrons moved on the same circular orbit with a 180-degree phase difference, had given a value of 28.9 eV, about 4 eV too large. A much more sophisticated non-planar model (now referred to as the "Bohr–Kemble" model) that had become the favorite candidate for

[51] The alpha particles, or helium nuclei, emitted during decays of elements in the uranium chain eventually pick up two electrons and become neutral helium.

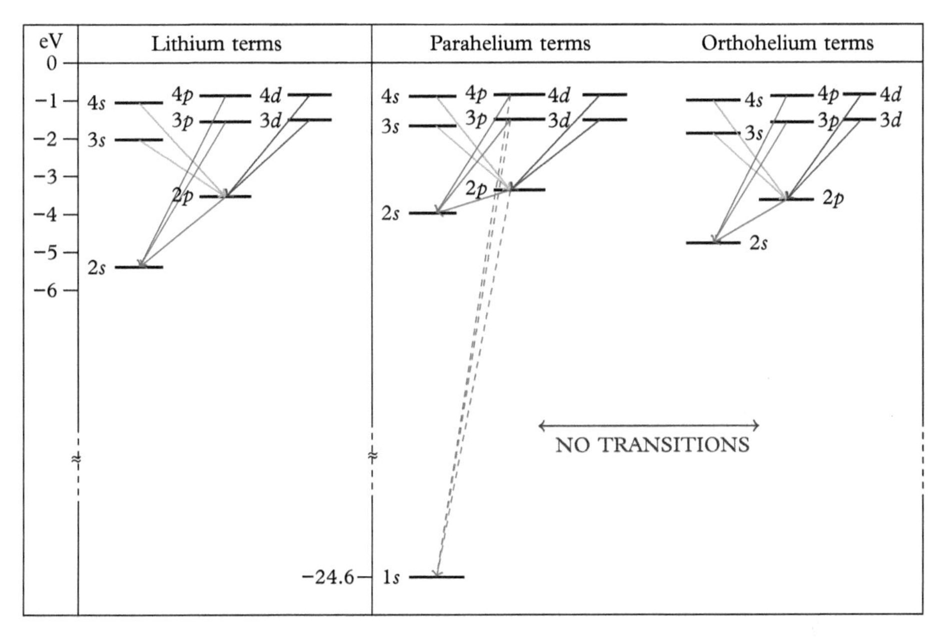

Figure 15.3 *Energy level diagrams for lithium(left panel) and para- and orthohelium (right panel). All three sets of terms display principal (red), sharp (green), and diffuse (blue) series, which are qualitatively similar, with the exception of an additional principal series (dashed red) for parahelium, which is entirely in the ultraviolet. Transitions between para- and orthohelium terms were not observed.*

the ground state configuration was found in virtuosic calculations of Van Vleck (1922) and Kramers (1923a) to be about 20.7 eV, now again about 4 eV off from the measured value, but on the other side!

The advent of the new quantum mechanics, supplemented by the spinning electron hypothesis of Goudsmit and Uhlenbeck (1925), immediately brought the problem of helium once again to the forefront, and Bohr in particular (an early adopter of the electron spin idea) was especially eager to see if the inclusion of electron spin would prove to be useful in untangling the mysteries of helium. To that end, he invited Goudsmit to visit Copenhagen, starting in early February 1926, to work on just this problem. Bohr was convinced that the distinction between para- and orthohelium was to be found in the relative orientation of the spin of the two electrons (aligned or anti-aligned), but it was unclear how an energy separation of the order of an electron volt could arise from a direct magnetic interaction between the electrons, as we shall see shortly. After a month of fruitless (and to Bohr completely unconvincing) calculations, Goudsmit was given a return train ticket to Holland with instructions to stop in Hamburg to inform Pauli of the newly discovered Thomas factor of 2 (Mehra and Rechenberg 1982c, pp. 284–285). It would fall to Heisenberg, who arrived in Copenhagen at the end of April to replace Kramers, as well as to give lectures at the University of Copenhagen, to uncover the physical principles essential to an understanding of the helium spectrum.

15.4.1 Heisenberg and the helium spectrum: degeneracy, resonance, and the exchange force

In early June 1926, Heisenberg submitted a paper to *Zeitschrift für Physik* that began the systematic effort to decipher the spectra of multi-electron atoms, which would be the central focus of theoretical atomic physics for many years to come. The title of this article was "The many-body problem and resonance in quantum mechanics" (*Mehrkörperproblem und Resonanz in der Quantenmechanik*). The abstract of this paper is short and to the point, and worth quoting in full:

> This work attempts to give a foundation for the quantum mechanical treatment of the many body problem. A resonance phenomenon characteristic of the quantum mechanics of the many body problem is thoroughly investigated, and a connection established of the results so obtained with the Einstein–Bose counting [statistics] and the Pauli exclusion of equivalent orbits (Heisenberg 1926b, p. 411).

Heisenberg's paper begins with a fairly effusive acknowledgement of the importance of the new wave mechanics of Schrödinger, which provides a "mathematically significantly more convenient approach to the realm of quantum mechanics" (p. 411). There is a brief discussion of the extent to which the Schrödinger theory can be regarded as a fulfillment of the de Broglie–Einstein program, in which a scalar matter wave is firmly based in conventional three-dimensional space (in analogy to the fields of electromagnetism)—correctly, Heisenberg points out that in the Schrödinger theory, one is forced to work in a d-dimensional space when treating a system with d degrees of freedom.[52] Such a matter wave would only correspond to the de Broglie–Einstein field for a single-point particle with no spin degrees of freedom.[53]

These speculations have relatively little to do with the main thrust of the paper, however, and are not considered further here. More relevant is what Heisenberg has to say about the role of Einstein–Bose statistics and the Pauli exclusion principle in *quantum mechanics* (which at this stage is still used fairly generally to refer specifically to the matrix variant of the theory developed in Göttingen), as opposed to the accommodation of these ideas in Schrödinger's approach, which after all descends in some sense from the work on gas theory described in Sections 13.5 and 14.2.

> The aim of our investigation is the quantum-mechanical treatment of systems consisting of several point masses. Such a treatment seems at first to encounter serious difficulties: the aspects of the de Broglie wave theory that lead to the Einstein–Bose statistics seem to have no analogy in quantum [i.e., matrix] mechanics; supplementary conditions, such as the Pauli exclusion of equivalent orbits, have as such no place in the mathematical scheme of quantum mechanics ... Finally, we may recall a well-known difficulty of the quantitative treatment of spectra: the separation between the singlet- and triplet-system

[52] We use d, rather than Heisenberg's f, to denote the number of degrees of freedom, as he also uses f, confusingly, to represent the term inducing radiative transitions.

[53] Quantum field theory does in fact restore the description of quantum systems in terms of a single operator valued field defined on the physical space–time, not the Schrödinger coordinate space.

in the spectra of the alkaline earths and the helium spectrum is orders of magnitude too large to be interpreted as the magnetic interaction energy of two rotating electrons (Heisenberg 1926b, p. 413).

The aim of the paper can therefore be simply summarized as follows: (i) to find a natural way to incorporate the restrictions imposed by Einstein–Bose statistics, or the Pauli principle (in the cases appropriate for each), in the mathematical framework of matrix mechanics; and (ii) to show that the consistent application of these ideas lead to a better understanding of the spectra of multi-electron atoms, in particular of helium. Heisenberg closes the introductory section of the paper by asserting that a solution of (ii) will follow naturally once (i) is accomplished: at that point, "the above mentioned difficulties resolve themselves automatically" (*lösen sich ... ganz von selbst*; p. 413).

Before we describe Heisenberg's procedure in the rest of the paper, it is useful to explain by a simple calculation the problem of magnetic interaction energy alluded to here, as it explains the failure of Goudsmit, among others, to make any progress on the helium problem in the early months of 1926. The potential energy of two magnetic dipoles, each of magnetic moment μ, and separated by a distance r is roughly (apart from directional dependences of no interest here) given by $H_{\text{mag}} \sim \mu^2/r^3$. For two electrons in an atom, we have $\mu \sim \mu_B = e\hbar/2mc$ and $r \sim a = \hbar^2/me^2$ (the Bohr radius). Some simple algebra thus gives

$$H_{\text{mag}} \sim \frac{e^8 m}{4\hbar^4 c^2} = \frac{1}{4}\left(\frac{e^2}{\hbar c}\right)^4 mc^2 = \frac{1}{4}\alpha^4 mc^2. \tag{15.144}$$

Inserting for the fine structure constant $\alpha = 1/137$ and the electron rest energy $mc^2 \sim 5 \times 10^5$ eV, we find magnetic energies of the order of 10^{-4} eV. This is of the order of the fine structure splittings, more than a thousand times smaller than the observed splitting between singlet and triplet states in helium.

In the first section of the paper, Heisenberg introduces a toy model that is formally very similar in structure to the problem of real physical interest—a helium atom described by the two electron Hamiltonian

$$H = \frac{1}{2m}\vec{p}_1^{\,2} - 2e^2/r_1 + \frac{1}{2m}\vec{p}_2^{\,2} - 2e^2/r_2 + \frac{e^2}{|\vec{r}_1 - \vec{r}_2|}. \tag{15.145}$$

Instead, one considers a two-particle system of similar form, but one that is completely solvable in both classical and quantum mechanics, namely, the system of two one-dimensional simple harmonic oscillators, which are also coupled by a weak harmonic potential:[54]

$$H = \frac{1}{2m}p_1^2 + \frac{1}{2}m\omega^2 q_1^2 + \frac{1}{2m}p_2^2 + \frac{1}{2}m\omega^2 q_2^2 + m\lambda q_1 q_2. \tag{15.146}$$

[54] Heisenberg's Hamiltonian can be rewritten in a form that displays even more clearly the analogy to the two-electron helium system: defining $\tilde{\omega} = \sqrt{\omega^2 + \lambda}$, one has $H = p_1^2/2m + m\tilde{\omega}^2 q_1^2/2 + p_2^2/2m + m\tilde{\omega}^2 q_2^2/2 - m\lambda(q_1 - q_2)^2/2$.

In fact, this Hamiltonian actually represents a system of two *decoupled* harmonic oscillators, with shifted frequencies, as the usual canonical transformation to normal modes reveals. Setting

$$p_1' = \frac{1}{\sqrt{2}}(p_1 + p_2), \quad q_1' = \frac{1}{\sqrt{2}}(q_1 + q_2), \tag{15.147}$$

$$p_2' = \frac{1}{\sqrt{2}}(p_1 - p_2), \quad q_2' = \frac{1}{\sqrt{2}}(q_1 - q_2), \tag{15.148}$$

one finds that the Hamiltonian takes, in terms of the new primed canonical coordinates, the form

$$H = \frac{1}{2m}p_1'^2 + \frac{1}{2}m\omega_1'^2 q_1'^2 + \frac{1}{2m}p_2'^2 + \frac{1}{2}m\omega_2'^2 q_2'^2, \tag{15.149}$$

with

$$\omega_1'^2 = \omega^2 + \lambda, \quad \omega_2'^2 = \omega^2 - \lambda. \tag{15.150}$$

Classically, setting $q_2' = 0$, one obtains a motion in which $q_1(t) = q_2(t)$, so the particles move in tandem in the same direction, while setting $q_1' = 0$ corresponds to motion in opposite directions. Quantum mechanically, the discrete energy levels of the system are labeled with two quantum numbers $n_1', n_2' = 0, 1, 2, \ldots$, with energies

$$E(n_1', n_2') = \frac{h\omega_1'}{2\pi}(n_1' + \frac{1}{2}) + \frac{h\omega_2'}{2\pi}(n_2' + \frac{1}{2}). \tag{15.151}$$

If λ is a positive number, ω_1' is larger than ω_2', and the excitations of the q_1' mode are more energetic than those of the q_2' mode, and one obtains the energy level diagram displayed qualitatively in Figure 15.4(a).

It is apparent that the energy levels displayed in Figure 15.4(a) have been segregated into two sets of alternating columns, denoted by dots and crosses, respectively (following the notation introduced by Heisenberg in Fig. 2 of his paper). The reason for this becomes clear once we posit that our two particles, in addition to being identical in mass, also have equal electric charge. Thus, the system possesses an electric dipole moment $f_{\rm dip}(q_1, q_2) = e(q_1 + q_2)$. The symmetry property $f_{\rm dip}(q_1, q_2) = f_{\rm dip}(q_2, q_1)$ is simply a reflection of the identity of the particles, and will be present in the expressions, of higher order in q_1 and q_2, describing all the higher multipole moments describing the radiative interaction of the system with the electromagnetic field.[55] As all such symmetric functions of the coordinates only involve even powers of q_2', the radiative transitions induced by such moments can only change the quantum number n_2' by an even number.

[55] As a more realistic example, for two identical charged particles in three spatial dimensions, with vector coordinates $\vec{r}_1, \vec{r}_2$ replacing the coordinates q_1, q_2 in Eq. (15.146), the relevant multipole moments would involve the electric dipole vector $e(\vec{r}_1 + \vec{r}_2)$, the magnetic dipole moment vector $e(\vec{r}_1 \times \dot{\vec{r}}_1 + \vec{r}_2 \times \dot{\vec{r}}_2)$, the electric quadrupole moment tensor $Q_{ab} = e(3(r_{1a}r_{1b} + r_{2a}r_{2b}) - \delta_{ab}(r_1^2 + r_2^2)), a, b = 1, 2, 3$, and so on. All the moments are symmetric in interchange of the particle labels 1,2, for obvious reasons.

In fact, the argument is much more general than this electromagnetic example shows—as Heisenberg is at pains to explain, *any* interaction with the external world must involve an interaction energy that is symmetric in the coordinates (or, if velocity dependent, momenta) of particles 1 and 2 if they are truly identical, as interchanging the labeling of identical particles cannot alter the physical situation. Consequently, physical interactions (including other than electromagnetic interactions, collisions with other particles, etc.) can only evoke transitions from dotted states to dotted states (where n_2' remains an even number), or crossed states to crossed states (where n_2' is odd). It would appear therefore that one can obtain a consistent, and closed, dynamics by *restricting oneself to states of either type.* Heisenberg states at this point:

> This indeterminacy of the quantum-mechanical solution seems to me the most essential result of this investigation. It gives just as much freedom to introduce into the system of quantum mechanics the requirements of Bose–Einstein statistics or the Pauli exclusion of identical orbits (Heisenberg 1926b, p. 416).

The connection between the segregation of the states into two sets which do not connect dynamically with each other, and the Bose–Einstein counting or Pauli exclusion principle can hardly be clear to the reader at this point, and is really only clarified later in the paper. The origins of this idea in fact may reasonably be assumed to go back to conversations between Heisenberg and Fermi a few months earlier in Rome (his resonance paper was submitted in June 1926, the trip to Italy having occurred in April).[56] Fermi had already published a discussion of an alternative (to Bose–Einstein) statistics based on the Pauli exclusion principle, in two papers, a short article summarizing his approach in early February (Fermi 1926a), and a longer one in March (Fermi 1926b), in which the famous Fermi–Dirac distribution appears for the first time—albeit for the somewhat peculiar situation of a system of many non-interacting electrons, subject to the Pauli exclusion principle, and bound to a central point by an isotropic harmonic potential. Fermi had already, in his first article, emphasized that, in order to obtain a gas statistics compatible with Nernst's theorem (which required the specific heat to vanish as the temperature was reduced to absolute zero) "it appears necessary to admit some complement to the [quantization] rules of Sommerfeld,[57] for the treatment of systems such as ours which contain indistinguishable elements [particles]" (Fermi 1926a, p. 146). In his two articles in early 1926, Fermi explores the consequences of imposing, as the desired complementary condition, the Pauli exclusion forbidding two electrons to occupy the same quantum state. What is of particular interest in connection with Heisenberg's resonance/helium paper three months later are the comments made by Heisenberg in his oral interviews with Kuhn in February 1963:

[56] See Pérez and Ibáñez (2022), who give a detailed discussion of Fermi's contributions to the statistics bearing his name.

[57] Fermi's articles are both written in the context of the old quantum theory, with particle orbits selected by Bohr–Sommerfeld quantization.

> [W]e did a tour in Italy. On the way back I saw Fermi in Rome, and I think that Fermi told me about his new statistics in agreement with the Pauli exclusion principle. He told me that when one applied Pauli's exclusion principle one got something which is somewhat related to Bose statistics, but which is definitely different from Bose's statistics. It was a kind of complement. He told me that the relation between his statistics and Bose's was something like plus and minus.[58]

Although Heisenberg did not reference either of Fermi's articles in his resonance paper, these intriguing remarks (albeit almost forty years later) should be kept in mind in the discussion that follows of his introduction of alternative symmetry requirements on quantum states, which allowed him to resolve the orthohelium/parahelium puzzle in his June 1926 resonance paper.

The importance of the appearance of two distinct sets of states in his oscillator toy model is that it motivates Heisenberg's next step, which is to explain that the existence of two equally acceptable symmetry requirements on quantum states holds not merely for the extremely special case of two coupled one dimensional harmonic oscillators, but much more generally—indeed, for any quantum mechanical system in which two states, initially degenerate (in virtue of the identity of the particles comprising the system) when only a part of the Hamiltonian is considered, have their energy separated (i.e., the degeneracy is lifted) when an interaction connecting these states to each other is added. It is precisely this generality that allows the results of this paper to be applied immediately to the problem of helium.

The essential feature that makes this conclusion possible is the *linearity* of the quantum-mechanical equations of motion.[59] This is in contradistinction to classical mechanics, where non-linearity is almost always present. As Heisenberg points out, the linearity of the harmonic oscillator problem in classical mechanics, and of more general systems, is also the essential precondition for "proper resonance" (*eigentliche Resonanz*). In a linear system such as the harmonic oscillator, the frequency of the motion is independent of amplitude, but once anharmonicities are present the natural frequency changes as the amplitude increases, which means that an applied field oscillating at the initial resonant frequency will necessarily go off-resonance as energy is transferred to the system and the amplitude increases. That this does not happen in quantum-mechanical systems means that, even for systems (such as atoms) that have a dynamics completely different from that induced by a harmonic potential, the behavior of the system under applied oscillating fields retains, as in dispersion theory, many of the features of the *simple* harmonic oscillator case.

In section 2 of the paper, Heisenberg (1926b) introduces the archetypal model system to be investigated. One posits two "completely identical" (*völlig gleich*, p. 417) systems *a* and *b*, each with *d* degrees of freedom. These could, for example, be the two hydrogen atoms in a hydrogen molecule, or, in the case of immediate interest, the two electrons in a helium atom, in which case $d = 4$, corresponding to the three spatial and one

[58] *American Institute of Physics, Oral Histories*, Interview of Thomas Kuhn with Werner Heisenberg, Session IX, February 27, 1963.

[59] This linearity is really only obvious in the Schrödinger formulation, where the wave equation involves only the first power of the wave function, a point which is not emphasized by Heisenberg.

spin degrees of freedom. Thus, at the lowest order the Hamiltonian of the system is $H_0 = H^a + H^b$, where H^a and H^b are identical functions of the kinematic variables of system a and system b. The stationary-state energies of either system may now be enumerated as $E_n^a = E_n^b = E_n$ where now we number (as in the Bohr hydrogen atom) the states starting at unity, $n = 1, 2, 3, \ldots$ For simplicity, we also assume for the time being that there is no degeneracy: each subsystem has exactly one state with a given energy.[60] One also assumes the presence of an interaction energy term λH^1, where H^1 depends completely symmetrically on the variables of system a and b. For example, H^1 might be, in the helium case, the electrostatic repulsion energy of the two electrons, $e^2/|\vec{r}_a - \vec{r}_b|$.

With the interaction term switched off, one can identify states of the combined system that are associated with any given pair m, n, with $m \leq n$, of the quantum numbers labeling the states of the isolated systems. Evidently, the unperturbed energy of such a state is $E(m,n) = E_m + E_n$. If $m \neq n$, there are two distinct states with the same energy, which we may write as "*mn*" (subsystem a in state m, subsystem b in state n), or "*nm*", where the energies of the subsystems are exchanged. On the other hand the states where $m = n$ are clearly non-degenerate—they are, in spectroscopic language, "singlets".

We now consider the effect of switching on the interaction λH^1, to first order in λ. The shift E^1 for the singlet states is directly given by the diagonal matrix element of the perturbation,

$$E^1(mm) = \lambda H^1(mm, mm). \tag{15.152}$$

For the paired states with $m \neq n$, however, we have a problem in degenerate perturbation theory. The prescription, given first in Born, Heisenberg, and Jordan (1926, Ch. 2, sec. 2) is simple: one forms the 2×2 matrix of H^1 with respect to the unperturbed states *mn* and *nm* (with $m < n$),

$$\mathbf{H}^1 = \begin{pmatrix} H^1(mn, mn) & H^1(mn, nm) \\ H^1(nm, mn) & H^1(nm, nm) \end{pmatrix}.$$

Note that by virtue of the fact that systems a and b are treated completely symmetrically by the interaction term H^1, we must have $H^1(mn, mn) = H^1(nm, nm) = \alpha$, where α is a real number. The Hermiticity property of energy matrices in matrix mechanics also implies that $H^1(mn, nm) = (H^1(nm, mn))^\star$, but it will turn out for the states of interest in helium that these matrix elements may be chosen real, and turn out to be positive (here, H^1 is the electrostatic repulsion term) so we can set $H^1(mn, nm) = H^1(nm, mn) = \beta > 0$. Of course, both the numbers α and β depend on the particular pair of states m, n of the two subsystems we choose to examine, a dependence omitted here for simplicity.

[60] In this section, Heisenberg's notation is particularly inept and confusing, so here and later, we alter it to make the reasoning more transparent.

The next step is to diagonalize $\mathbf{H}^1$ using a 2×2 orthogonal similarity transformation matrix $\mathbf{S}$. In other words, we seek an $\mathbf{S}$ such that $\mathbf{S}^{-1}\mathbf{H}^1\mathbf{S}$ is a diagonal matrix. This is easily found to be

$$\mathbf{S} = \begin{pmatrix} \frac{1}{\sqrt{2}} & \frac{1}{\sqrt{2}} \\ \frac{1}{\sqrt{2}} & -\frac{1}{\sqrt{2}} \end{pmatrix}, \tag{15.153}$$

whence[61]

$$\mathbf{S}^{-1}\mathbf{H}^1\mathbf{S} = \mathbf{S}^{-1}\begin{pmatrix} \alpha & \beta \\ \beta & \alpha \end{pmatrix}\mathbf{S} = \begin{pmatrix} \alpha+\beta & 0 \\ 0 & \alpha-\beta \end{pmatrix} = \begin{pmatrix} E_\bullet & 0 \\ 0 & E_+ \end{pmatrix}. \tag{15.154}$$

This pattern is repeated for every pair of non-identical states m, n (with $m < n$) of the two identical subsystems, so that the initial degeneracy is lifted, and we now have two states of differing energy: the lower state, with energy shift $E_+ = \alpha - \beta$ denoted (somewhat unfortunately, but here following Heisenberg's choice) by a plus symbol; and the higher energy state, with energy shift $E_\bullet = \alpha + \beta$, by a dot. Recalling that the overall spacing of the non-interacting energies E_n may no longer be constant as in the harmonic oscillator case (for the Coulomb problem the spacings get smaller for the higher states), we obtain the qualitative spectrum displayed in Figure 15.4(b), where the dotted states associated with the pair $m \leq n$ are labelled with round parentheses, (mn), while the "plus" labelled states are labelled $[mn]$ (and exist only for $m \neq n$).

The next step is to verify that the two subsets of states of the entire system labelled by "dots" or "pluses" do not communicate dynamically, as in the oscillator problem. We again consider a physical interaction energy term to which the system may be subjected, resulting in transitions between states. As for the charged oscillator, such transitions are associated with the matrix elements $f_{m_1n_1,m_2n_2}$ of the kinematical quantity (e.g., an electric or magnetic dipole moment, or higher multipole) f responsible for the transition. The identity of the subsystems translates into the symmetry property $f_{m_1n_1,m_2n_2} = f_{n_1m_1,n_2m_2}$, as no physical property can be affected by simply switching the labeling of the subsystems. If we once again consider a particular degenerate pair of states of the uncoupled subsystems, labelled by mn, and examine the 2×2 matrix of this perturbing term, we find that its symmetry requires that the 2×2 matrix of f takes exactly the form of the Hamiltonian matrix in Eq. (15.154): namely the two diagonal elements are equal (as $f_{mn,mn} = f_{nm,nm} \equiv \alpha'$), and the two off-diagonal elements are also equal ($f_{mn,nm} = f_{nm,mn} \equiv \beta'$). In making a similarity transform with the orthogonal matrix $\mathbf{S}$ we therefore find that the correspondingly transformed transition energy term is transformed into a 2×2 matrix $\mathbf{f}' = \mathbf{S}^{-1}\mathbf{f}\mathbf{S}$, which is automatically diagonal ($= \text{diag}(\alpha' + \beta', \alpha' - \beta')$): in other words, it connects dotted states to dotted states, and plus states to plus states only. Transitions between a dotted state and a plus state are therefore forbidden, just as in the coupled oscillator case.

[61] Note that here we have the simple property that $\mathbf{S}^{-1} = \mathbf{S}$.

We saw that Heisenberg, in the introduction to his paper, mentioned in quite enthusiastic (if in some respects qualified) terms the Schrödinger formulation of quantum-mechanical systems in terms of a eigenfunction problem for wave functions in coordinate space. After describing the separation of the dotted and plus systems in the language of matrix mechanics, he points out that the same result can be understood in terms of the symmetry or anti-symmetry of the corresponding wave functions for the full system. The eigenfunction associated with subsystem "a" (resp. "b") with energy E_n is thus a complex function $\varphi_n^a(q_1^a, q_2^a, \ldots, q_d^a)$ (resp. $\varphi_n^b(q_1^b, q_2^b, \ldots, q_d^b)$), where the number of degrees of freedom of each subsystem is given by d. For the pair of states mn of the non-interacting system "ab" we therefore have the wave function $\varphi_m^a\varphi_n^b$. The matrix elements of the perturbation H^1 are given in the way identified by Schrödinger, as the integrals

$$\begin{aligned} H^1(mn, mn) &= H^1(nm, nm) \\ &= \int (\varphi_m^a\varphi_n^b)^\star H^1(q_1^a, \ldots, q_d^a, q_1^b, \ldots, q_d^b)\varphi_m^a\varphi_n^b dq_1^a \ldots dq_d^b \equiv \alpha, \end{aligned} \tag{15.155}$$

$$\begin{aligned} H^1(mn, nm) &= H^1(nm, mn) \\ &= \int (\varphi_m^a\varphi_n^b)^\star H^1(q_1^a, \ldots, q_d^a, q_1^b, \ldots, q_d^b)\varphi_n^a\varphi_m^b dq_1^a \ldots dq_d^b \equiv \beta. \end{aligned}$$

The linear transformation corresponding to **S**, which gives the states of definite energy once the interaction H^1 is included, therefore produces energies

$$\begin{aligned} E_{(mn)} &= E_n + E_m + E_\bullet, \text{ with eigenfunction } \frac{1}{\sqrt{2}}(\varphi_m^a\varphi_n^b + \varphi_n^a\varphi_m^b), \\ E_{[mn]} &= E_n + E_m + E_+, \text{ with eigenfunction } \frac{1}{\sqrt{2}}(\varphi_m^a\varphi_n^b - \varphi_n^a\varphi_m^b), \end{aligned} \tag{15.156}$$

where $E_\bullet$ (resp. E_+) is $\alpha + \beta$ (resp. $\alpha - \beta$). The desired "non-existence of inter-combinations between the two subsystems" (which here now means the states denoted by dots or pluses) is now a trivial consequence of the vanishing of an integral over an anti-symmetric integrand. Taking for example a system of two electrons subject to dipole radiation, the function f is just the dipole moment of the pair, $f = e(q_i^a + q_i^b)$ (where $i = x, y, z$ is a spatial index), and dipole transitions from a dotted state (mn) to a plus state $[mn]$ involve the transition matrix element

$$\int (\varphi_m^a\varphi_n^b - \varphi_n^a\varphi_m^b)^\star e(q_i^a + q_i^b)(\varphi_m^a\varphi_n^b + \varphi_n^a\varphi_m^b)dq_1^a dq_2^a \ldots dq_1^b dq_2^b \ldots dq_2^d. \tag{15.157}$$

This integral is equal to the one obtained by interchanging the labels a and b—as this reverses the sign of the integral (and this depends on f being invariant under the interchange of particle labels), we conclude that it must be zero. It must be admitted that the desired decoupling of states is considerably more transparent in the wave function language, at least for physicists, then, as now, more accustomed to the methods of calculus than of matrix algebra.

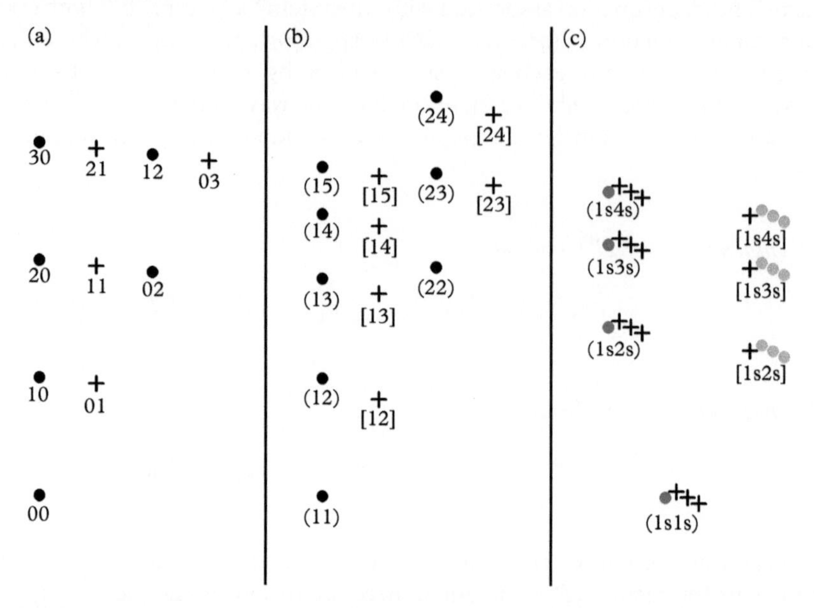

Figure 15.4 *(a) Energy level diagram for system of two coupled simple harmonic oscillators (after Fig. 2, in Heisenberg 1926b). Integer pairs indicate quantum levels of the isolated subsystems. (b) Energy level diagram for system of two coupled identical subsystems. Integer pairs enclosed in round parentheses indicate symmetric states and square brackets anti-symmetric states (after Fig. 3, ibid).(c) Energy level diagram for the s-states of Helium (after Fig. 5, ibid). Plus signs denote states obeying Bose–Einstein statistics, and the colored circles helium s-states (red circles = parahelium, green circles = orthohelium).*

In section 3 of this first paper on the many-body problem in quantum mechanics, Heisenberg (1926b) turns at last to the system of preeminent importance at this stage, the neutral helium atom. The results presented here are of a broad, qualitative nature, but they do resolve the first great puzzle of the helium spectrum, which had remained unexplained for the preceding three decades: the apparently complete absence of radiative transitions between the two sets of terms labelled as para- and ortho- helium energy levels. Heisenberg first points out that the difference in energy between the states with symmetric and anti-symmetric spatial wave functions (given from Eq. (15.154) as $E_\bullet - E_+ = 2\beta$, with β the lower integral in Eq. (15.156), later to be called the "exchange energy") owes its origin to the electrostatic repulsion between the two electrons (as $H^1 = e^2/|\vec{r}_a - \vec{r}_b|$), and therefore should be of the same order as the shifts of alkali valence electron levels observed due to screening effects (also due to this repulsion term). They should therefore be of the order of *fractions* of an electron volt, as observed, and not *thousandths* of an electron volt, as in the case of the fine structure, or from a direct magnetic dipole–dipole interaction of the two electrons.

However, the addition to the argument of the Goudsmit–Uhlenbeck electron spin feature requires a reconsideration of the symmetry arguments presented above, as for any given pair of distinct spatial states of the two electrons (prior to the introduction of the electrostatic repulsion term) there now exist four possible spin terms. "How the calculation leads to results described here," Heisenberg (1926b, p. 422) promises, "will be explained in a paper which will appear shortly." This second paper (Heisenberg 1926c) was submitted about six weeks after the first. In this later paper, the possible spin orientations of the two electrons are described in terms of the projections $m_1\hbar$, $m_2\hbar$ of the spin angular momentum of each electron onto the direction of an external magnetic field (we may as well assume it to be in the z direction). With $m_1, m_2 = \pm\frac{1}{2}$ there are obviously four possible choices for the spin pair (m_1, m_2), which are degenerate in energy as the Hamiltonian has no spin-dependent terms. These four terms can be sorted into a singlet corresponding to the total spin angular momentum $\vec{S}$ being zero (hence, $S_z = (m_1 + m_2) = 0$) and a triplet of terms with $S = 1, S_z = +1, 0, -1$. The terms with $S_z = +1$ must be identified with the pair $(+\frac{1}{2}, +\frac{1}{2})$ (Heisenberg's state "*a*"), while the pair $(-\frac{1}{2}, -\frac{1}{2})$ must be the $S_z = -1$ state (Heisenberg's state "*d*"). These two states are evidently symmetric under interchange of the particle labels (Heisenberg 1926c, p. 512). The spin configuration corresponding to zero total spin $S = 0$ (and, of course, $S_z = 0$) is the anti-symmetric state (labelled "*c*" by Heisenberg), while the spin configuration with total spin $S = 1$ and $S_z = 0$ is the symmetric (state "*b*") combination of the pairs $(+\frac{1}{2}, -\frac{1}{2})$ and $(-\frac{1}{2}, +\frac{1}{2})$:[62]

> Let us consider now again that we have written the Schrödinger eigenfunctions of this problem, following Eq. (15.157), then the [spin] coordinates of both electromagnets [i.e., electrons] in the eigenfunctions belonging to the "Triplet"-states a, b, d will be *symmetric*, while the eigenfunction belonging to the "Singlet" state c will change sign under exchange of the electrons, and is therefore anti-symmetric (Heisenberg 1926c, p. 512).

Returning to the results presented in Heisenberg (1926b, sec. 3), one finds that the states of a two-electron system consisting of a single series of terms, for example the $1s, 2s, 3s, \ldots$ terms corresponding to spatial s-states for both electrons, can be represented as shown in Figure 15.4(c).[63] As expected each pair of electron spatial states is accompanied by four possible spin configurations, which are displayed as a singlet and a triplet. The lowest state $(1s, 1s)$, non-degenerate in the absence of spin, is clearly

[62] It is apparent that the property of being invariant under the permutation of the electron labels cannot be altered by a spatial rotation of the system, so that the state with $S = 1, S_z = 0$ must possess the same symmetry property as the $S = 1, S_z = \pm\frac{1}{2}$ states. The other orthogonal, and therefore anti-symmetric, combination of $(+\frac{1}{2}, -\frac{1}{2}), (-\frac{1}{2}, +\frac{1}{2})$ must therefore correspond to the $S = 0$ singlet.

[63] Heisenberg represents the triplet levels as having slightly different energies, as would be the case via the Zeeman effect if a magnetic field were actually present. With no external field these states would all lie at the same vertical level.

symmetric in regards to the spatial part of its wave function, as are the higher $(1s, ns)$ states in the left column (with $n = 2, 3, \ldots$), while the states on the right are the spatially anti-symmetric combinations $[1s, ns]$. The states represented by black plus signs have either symmetric (resp. anti-symmetric) spin and symmetric (resp. anti-symmetric) spatial behavior (so the overall behavior of the wave function is symmetric under particle interchange), while the colored dots represent states which are spin symmetric, spatially anti-symmetric, or vice versa, so that the overall symmetry of the wave function is anti-symmetric. Of the two systems (dots or pluses) in the figure, as Heisenberg points out

> For the helium spectrum it is an empirical fact, that only one system in the Figure is present, and, as far as we can see here, at least qualitatively agrees with the He spectrum; the other system [i.e., the terms labelled by plus signs] is not realized in Nature. This fact appears to me . . . to indicate the true connection between the quantum-mechanical indeterminacy [i.e., the seemingly arbitrary, but equally consistent, choice of symmetric or antisymmetric solutions] on the one hand and the Pauli rule [exclusion principle] and Bose–Einstein counting on the other (Heisenberg 1926b, p. 422).

In other words, in the case of a two-particle system, the imposition of either the Pauli principle or the Bose–Einstein counting procedure corresponds to the choice of either overall anti-symmetry or overall symmetry in the eigenfunctions of the whole system, reducing the number of allowed states (roughly, by one half). The anti-symmetric choice clearly corresponds to the Pauli rule, as it is impossible to form an anti-symmetric combination of two completely (i.e., spin and spatial) identical one-particle states. We must therefore discard the states in Figure 15.4(c) denoted by black plus signs (whose spin and spatial eigenfunctions are symmetric under particle interchange) as physically irrelevant for the helium problem.

Now it is the *spatial symmetry* that determines whether the arguments presented previously concerning the vanishing of dipole or higher multipole moment matrix elements are valid, so we conclude, with Heisenberg, that *the two systems, labelled by red dots or green dots, do not combine via radiative transitions*. The red dots, which are all spin singlets and spatially symmetric, correspond to parahelium and include the lowest (or ground state) of the atom, as the spatial combination $(1s, 1s)$ must be symmetric. The green dots are all spin triplets and therefore correspond to the orthohelium levels, with spatially anti-symmetric eigenfunctions. With these observations, the long-standing puzzle of the duplicated spectrum of helium can be regarded as definitively resolved. The removal of the other half of the possible states by appeal to the Pauli exclusion principle must still of course be regarded as a starting postulate, the deeper origins of which would only be uncovered once the concepts and machinery of relativistic quantum field theory became available.

This first paper of Heisenberg's on the "many" (here, at most two!)-body problem in quantum mechanics concludes with a brief discussion (to be greatly amplified in Heisenberg 1927a, submitted in December 1926) of the generalization of the results to a system of n identical particles (or sub-systems). In particular, he shows that of the $n!$ automatically degenerate states that can be formed from the products of single particle states by simply relabeling the particle indices, there is always a single distinguished linear combination that is anti-symmetric under interchange of any pair of particles. Such a wave function necessarily obeys the exclusion principle, as if any pair of particles are assigned to identical states, their exchange must not only produce a change of sign, but also leave the wave function of the entire system unchanged, whence we conclude that the latter must vanish. The formula Heisenberg writes for the distinguished linear combination is easily seen to be a determinant with elements composed of products of the single particle wave functions—a point made explicit in Heisenberg (1927a), where the full power of the representation theory of the permutation group is brought to bear on the n-body problem.[64]

Moreover, the first-order degenerate perturbation theory for a perturbing interaction H^1 that is symmetric in the particle coordinates always produces an $n! \times n!$ matrix with this state as an eigenstate. This is simply because the application of the matrix of H^1 to the totally anti-symmetric combination must leave it totally anti-symmetric (as H^1 is symmetric under interchange of any two-particle coordinates). The coupling of the system to radiation also cannot induce transitions from this totally anti-symmetric state to any state violating the Pauli exclusion principle (i.e., with two-particles in identical states), as the transition matrix element would involve the integral of the product of a function anti-symmetric under interchange of the two particles in question (the initial wave function, assumed to satisfy the Pauli principle) and a symmetric function (namely, the multipole moment function f, automatically symmetric, times the final wave function, now assumed symmetric under exchange of the two particles).

Heisenberg's 1926 papers on the many-body problem accomplish the introduction of one of the strangest (and least comprehensible from a classical viewpoint) features of the new mechanics: the need for the introduction of extrinsic symmetry conditions on a quantum mechanical problem which reduce the number of physically admissible states, and hence, alter the "counting" procedure that must be applied in a quantum treatment of any system composed of several *identical* particles. In particular, in these papers Heisenberg succeeds in finding the key to the strange bifurcation of the helium spectrum that had puzzled atomic physicists for almost thirty years. However, there is one aspect of his discussion modern readers may well find puzzling, and for good reason, namely, the emphasis throughout on the central importance of "resonance" behavior in understanding the results. Indeed, the word "Resonanz" appears in the title of Heisenberg (1926b) and Heisenberg (1927a), and in several critical parts of the development.

Classically, we are used to viewing a resonance phenomenon as one in which a periodic system gains energy preferentially when an external influence is applied with a time dependence of the same (or close) frequency as that with which the isolated system oscillates. More generally, we can imagine the transfer of energy as occurring between

[64] For some reason, the anti-symmetric combination of single-particle wave functions relevant to describing the electronic structure of atoms, written in determinant form, has become known as the "Slater determinant", although it was clearly first introduced both by Heisenberg in his articles of 1926, and independently by Dirac (1926b, p. 669).

two independent systems of the same intrinsic frequency. In quantum mechanics, the frequency ν associated with a state is related to the energy E of the state by the Planck relation $E = h\nu$, so we are naturally led to consider *degenerate* systems, of equal (quantized) energy. As Heisenberg (1926b, p. 416) emphasizes at the beginning of section 2 of his paper, quantum mechanics has the additional simplifying feature, in comparison to classical mechanics, of *linearity*. The transfer of energy to a classical periodic system will typically increase the amplitude of the motion, and if non-linearities are present (as they always are to some degree, except in the idealized case of exactly harmonic motion) this will shift the frequency, so that the system is no longer resonant with the applied force. In quantum mechanics, on the other hand, the transfer of electromagnetic energy $E_a - E_b = h\nu_{ab}$ (as given by the Bohr condition) back and forth between an electromagnetic wave and an atom with two quantized states a, b can continue indefinitely at precisely the resonant frequency ν_{ab}, a fact essential for the phenomenological equivalence of quantum dispersion theory with the classical Drude dispersion theory based on simple harmonic oscillators.

For Heisenberg (1926b), however, the use of the term “resonance” is extended to a somewhat different implication, really as a placeholder for concepts not yet available (or fully appreciated) in the Spring of 1926. In effect, whenever an initial degeneracy of states present in the Hamiltonian of a quantum-mechanical problem (typically due to a symmetry[65] effective at the level of approximation corresponding to that Hamiltonian) is lifted by the addition of a new perturbing term to the energy of the system, the new stationary states of the system must really be represented, in the Schrödinger scheme of energy eigenfunctions, as linear combinations of the original set of eigenfunctions, which may have been the convenient choices for the original unperturbed states.

In the toy model of two coupled oscillators, the energy splitting induced by an interaction term (the $q_1 q_2$ term in Eq. (15.146)) results classically in a “beat” phenomenon in which energy can be transferred back and forth between the two originally uncoupled oscillators, and this can be regarded as a “resonance oscillation” (*Resonanzschwebung*) analogous to the transfer of energy between two systems in resonance. Quantum mechanically, this beat phenomenon is associated with the time dependence associated with states formed as linear combinations of energy eigenstates of slightly different energy, as Schrödinger (1926e, p. 755) had already shown. In his papers of 1926, Heisenberg frequently uses the term “resonance” simply in this sense, to imply that the states of interest quantum-mechanically are fused versions of states under a process of continuous interconversion by some perturbation. In the helium atom, this perturbation (a large one!) is the electrostatic repulsion of the two electrons. The peculiar aspect of the problem is that only certain superpositions, or combinations, of the states of the unperturbed problem are allowed once one takes into account either the Pauli principle or the Bose–Einstein counting restrictions. In the case of helium (or larger atoms more generally) the situation is further complicated by the presence of spin, leading to variable symmetry properties of the spin degrees of freedom of the problem, and a further reduplication of the states of the system, whence the appearance of separate para- and ortho-spectra.

[65] In the present case, the symmetry is the exchange symmetry that is always presnt if one considers multiparticle systems composed of truly *identical* particles.

1. *George Uhlenbeck, Hans Kramers, Samuel Goudsmit.*

2. *Copenhagen, 1926. Back row left to right: David Dennison, Ralph Kronig, Bidhu Bhusan Ray. Front row left to right: Yoshio Nishina and Werner Kuhn.*

3. *Llewellyn Hilleth Thomas.*

4. *Outdoors at a cabin in Pontresina, Switzerland. Front row left to right: Otto Hahn, Max Born, Rudolf Ladenburg, Robert Pohl, Fritz Reiche. Back row left to right: Bertha Reiche, Auguste Pohl, Mrs. Pummerer, Else Ladenburg, Edith Hahn, and Hedwig Born.*

5. *Paul Ehrenfest and Pascual Jordan.*

6. *Werner Heisenberg.*

7. *Norbert Wiener and Max Born.*

8. *Paul Dirac.*

9. *Satyendra Nath Bose.*

10. *Louis de Broglie.*

11. *Erwin Schrödinger.*

12. *Arthur Holly Compton, Werner Heisenberg, George Monk, Paul Dirac, Carl Eckart, Henry Gale, Robert Mulliken, Friedrich Hund, Frank C. Hoyt. Chicago 1929.*

13. *John H. Van Vleck.*

14. *John C. Slater.*

15. *Wolfgang Pauli and Max Born.*

16. *Born's tombstone.*

17. *Schrödinger's tombstone.*

18. *Enrico Fermi, Werner Heisenberg, Wolfgang Pauli. Como, 1927.*

19. *Fifth Solvay Conference, 1927.*

20. *Ernst Hellinger.*

21. *David Hilbert.*

22. *John von Neumann.*

We have concluded our discussion of Heisenberg's 1926 papers on multi-particle quantum mechanics with this digression on Heisenberg's insistence on the importance of "resonance" in atomic problems partly because this usage would persist, especially in the quantum chemistry literature, well past the point when the importance of the linear structure of the theory, and its complete formulation in terms of linear vector spaces (and especially, Hilbert space), was fully understood. One frequently encounters references in modern quantum chemistry to *resonance structures, resonance hybrids, resonance contributors, etc.*, in reference to molecular states viewed as linear combinations of more simple configurations in which definite states are assigned to equivalent subsystems. This language can clearly be traced back to Heisenberg's (1926b) seminal paper of June 1926.

15.4.2 Perturbative attacks on the multi-electron problem

In his second paper on helium, entitled "On the spectra of atomic systems with two electrons," Heisenberg (1926c) addresses the problem of calculating from quantum mechanics, necessarily approximately, the energy, or term, levels of parahelium and orthohelium that had been the subject of intense spectroscopic examination since the turn of the century. The complicating feature was of course the presence in the Hamiltonian of the electrostatic repulsion term between the two electrons, $e^2/|\vec{r}_1-\vec{r}_2| \equiv e^2/r_{12}$, without which the system is immediately solvable, yielding Schrödinger eigenfunctions that are just products of hydrogenic single-particle wave functions (with, of course, $Z = 2$ for the nuclear charge). In this paper, matrix-mechanical language has been more or less abandoned and the entire procedure is framed in terms of the Schrödinger formalism, which is manifestly better adapted to the task at hand.

As discussed previously, the resemblance of both the para- and ortho- helium spectra to the spectra of alkali metals had been known for almost three decades, and suggested that both sets of terms corresponded to a situation in which a single valence electron was excited to various higher energy levels while the other electron remained in the lowest (1*s*) state, its effect on the dynamics of the higher valence electron being (to a first approximation) to form, with the nucleus, an atomic "core" of net charge $\approx Z-1(=1$ for helium, 2 for Li^+, etc). Thus, the higher levels should asymptotically approach just the hydrogenic levels, as was the case for alkali metal terms. The deviation from purely hydrogenic behavior was typically expressed as a *spectral defect* δ, defined by parametrizing the total energy of the two electrons, for nuclear charge Z, as

$$E=(-\frac{Z^2}{1^2}-\frac{(Z-1)^2}{(n+\delta)^2})\times 13.6\text{ eV}. \tag{15.158}$$

This phenomenology made it clear that an effective perturbation theory could be formulated for the highly excited states by splitting the exact two electron Hamiltonian,

$$H=\frac{\vec{p}_1^{\,2}}{2m}-\frac{Ze^2}{r_1}+\frac{\vec{p}_2^{\,2}}{2m}-\frac{Ze^2}{r_2}+\frac{e^2}{r_{12}}, \tag{15.159}$$

into an unperturbed part

$$H_0 = \frac{\vec{p}_1^{\,2}}{2m} - \frac{Ze^2}{r_1} + f(r_1) + \frac{\vec{p}_2^{\,2}}{2m} - \frac{Ze^2}{r_2} + f(r_2), \tag{15.160}$$

and a perturbation

$$H^1 = \frac{e^2}{r_{12}} - f(r_1) - f(r_2), \tag{15.161}$$

where the screening function $f(r)$ should be chosen so that $-Ze^2/r + f(r)$ approaches $-(Z-1)e^2/r$ at large r but is dominated by the full nuclear potential $-Ze^2/r$ at small r. Heisenberg chose to set $f(r)$ equal to the continuous function defined by $f(r) = e^2/r$ for $r > r_0$, $f(r) = e^2/r_0$ for $r < r_0$, with r_0 chosen (at least in principle) by the requirement that the perturbation theory in H^1 around H_0 converge as rapidly as possible. This amounts to modeling the screening of the nucleus by the inner electron with a uniform shell of charge $-e$ spread over the sphere of radius r_0.[66] The central potential problem defined by H_0 is analytically tractable (if somewhat messy), and one then hopes that first-order perturbation theory would give an encouraging result for the energy shifts induced by H^1 (expressed, as had become the fashion, in terms of a spectral defect, as in Eq. (15.158)).

The splitting between para- and ortho- energy levels had been found in the first paper to be twice the "exchange" term (as it would come to be called—this terminology does not appear in these first two Heisenberg papers) $\beta = H^1(mn, nm)$ (cf. Eqs. (15.156) and (15.157)), the notation for which in Heisenberg (1926c) has been changed from $H^1(nm, mn)$ in Eq. (15.156)) to $H^1(vw, wv)$. Now v denotes the $1s$ state of the lower electron, while w corresponds to the higher ("valence") electron, occupying a state characterized by principal quantum number n and angular momentum k (modern notation: l). In principle, the wave functions used to evaluate this matrix element of the electrostatic repulsion[67] term $e^2/|\vec{r}_1 - \vec{r}_2|$ should be eigenfunctions corresponding to the shielded non-interacting Hamiltonian H_0 in Eq. (15.160), but in fact Heisenberg simply uses the hydrogenic $1s$ wave function (with the full nuclear charge Z) for the v state, and the hydrogenic n, k state (with the fully screened charge $Z-1$) for the w state in the integral, and thus, his perturbation theory is not fully consistent.

The results of this (as Heisenberg himself admits) first rather crude computation of the helium spectrum are somewhat encouraging, but can hardly be regarded as a convincing *quantitative* confirmation of the new theory. The problem is that the method used from the outset works best precisely in the limit (large orbital angular momentum k)

[66] Below, we briefly discuss Unsöld (1927), who shows that Heisenberg's $f(r)$ is very close to the screened potential one finds by using the continuous charge density (charge times the wave function squared) of the inner electron.

[67] The contributions of the single particle terms $f(r_1)$ or $f(r_2)$ to this matrix element vanish, as one encounters in each case an overlap integral of two orthogonal wave functions, namely φ_v and φ_w, which vanishes. The integral involving the electrostatic repulsion term, which depends on both r_1 and r_2, does not vanish.

where the deviations from pure hydrogenic behavior, precisely the interesting new input coming from the exchange term, become extremely—in fact unmeasurably—small. For the lower states where the spectroscopy yields more accurate results (a spectral defect given to two or three significant digits, for example), first order perturbation theory can only be assumed to give crude agreement. For example, for the splitting of the spectral defect between the ortho-$(1s, 2p)$ and para-$(1s, 2p)$ state (or $2p - 2P$ splitting, in the notation of the time), Heisenberg's calculation gives 0.061 versus the measured value of 0.075. Or, in direct energy terms, the perturbation theory gives an ortho–para energy splitting of 0.21 eV, as against the empirical value of 0.254 eV.

We are clearly not yet in the regime of precise quantitative agreement at the level of, for example, Sommerfeld's fine structure formula of 1916—even though the energy splittings being considered are far greater than those of the fine structure! In particular, Heisenberg's perturbative method was unsuited to the treatment of the *s*-states with angular momentum zero (i.e., all the states displayed in Figure 15.4(c)), for which the electrostatic repulsion term was of similar magnitude to the nuclear potential energy. States of low angular momentum in multi-electron atoms had already caused problems, of course, in the old quantum theory, as they corresponded to the "penetrating (or dipping) orbits" where the electron followed a comet-like highly elliptical orbit, traveling between the outer regions of the atom and the inner parts of the "Atomrumpf", or atomic core. This complicated the calculation of screening effects, as Schrödinger (1921) pointed out.

We can see the residue of these concerns in an interesting paper, submitted to *Annalen der Physik* in mid-December 1926 by Albrecht Unsöld (1927) a student of Sommerfeld's in Munich. The first chapter of this paper is entitled "Penetrating orbits" (*Eindringende Bahnen*). Chief among these low angular momentum states was of course the $(1s, 1s)$ ground state of helium, or in the current spectroscopic notation, the 1S parahelium term, the calculation of the energy of which had proven the ultimate obstacle on which the old quantum theory had foundered (see Section 7.4).

In his paper, Unsöld (1927) made a systematic survey of the adequacy of straightforward perturbation theory[68] for the study of the low angular momentum states of two-electron (He and Li^+) and three-electron (Li, Be^+, B^{++}, and C^{+++}) atoms. Unsöld's perturbation theory, in which the lowest order Hamiltonian H_0 is just the sum of kinetic and nuclear potential energies of the electrons, and the perturbation H^1 is the sum of pair-wise electrostatic repulsion terms, is really an expansion in the inverse atomic number $1/Z$. This is simply because the nuclear Coulomb energy terms in H_0 are proportional to the nuclear charge Z, while the number of electrostatic repulsion terms is fixed (one for the two-electron atoms, three for the three-electron atoms).[69] This perturbation theory—which we emphasize is, unlike Heisenberg's, formally consistent—will

[68] Which, although it had been developed in detail in the Three-Man-Paper, now goes under the name of "Rayleigh–Schrödinger" perturbation theory, inasmuch as the requisite matrix elements are always computed using wave-mechanical methods.

[69] More precisely, given that the mean radial coordinate of an electron is of order a/Z, the kinetic plus nuclear potential energy of an electron is of order $Ze^2 \cdot Z/a \propto Z^2$, while the electrostatic repulsion term is of order $e^2 \cdot Z/a \propto Z$.

therefore give accurate results in the limit of large Z (for a fixed number of electrons) even for the lowest energy states of the atoms, unlike Heisenberg's approach, which required large n and l for accuracy. We once again see the frequent tension where theoretically tractable phenomena are empirically unavailable, and vice versa—but in this case, especially for the three-electron atoms, where higher values of Z (up to $Z = 6$) were available, Unsöld was able to achieve an impressive agreement between theory and experiment.

In chapter 1, section 3 of his paper, Unsöld (1927) calculates the first-order perturbative shifts in two-electron atoms due to the electrostatic repulsion term, for the $(1sns)$ (parahelium) and $[1sns]$ (orthohelium) terms indicated in Figure 15.4(c). This involves the computation of the "direct" term α (cf. Eq. (15.156)) and the "exchange" (*Austausch*) term β, given in this case by the integrals

$$\alpha(1s, ns) = \int e\psi_{1s}^2(r_1)\frac{1}{|\vec{r}_1 - \vec{r}_2|}\, e\psi_{ns}^2(r_2)\, d^3r_1 d^3r_2, \tag{15.162}$$

$$\beta(1s, ns) = \int e\psi_{1s}(r_1)\psi_{ns}(r_1)\frac{1}{|\vec{r}_1 - \vec{r}_2|}\, e\psi_{1s}(r_2)\psi_{ns}(r_2)\, d^3r_1 d^3r_2. \tag{15.163}$$

Here, $\psi_{ns}(r)$ is the real (hence, no complex conjugates) and radially symmetric Schrödinger wave function for a single electron in a hydrogenic atom of nuclear charge Z, with principal quantum number n and angular momentum $l = 0$. Note that the direct term $\alpha(1s, ns)$ has a clear classical interpretation, as the electrostatic potential energy of two charge density distributions $e\psi_{1s}(r)^2$ and $e\psi_{ns}(r)^2$. The "exchange" term $\beta(1s, ns)$ has no such classical interpretation, instead having its origins in the purely quantum mechanical anti-symmetry constraints imposed on the two-particle wave function by the Fermi–Dirac statistics that had emerged earlier in the year.

The shift in the unperturbed energy $-(Z^2/1^2 + Z^2/n^2) \times 13.6$ eV is then just $\alpha + \beta$ for the parahelium (singlet) $(1sns)$ state, and $\alpha - \beta$ for the orthohelium (triplet) state $[1sns]$. Both integrals can be computed in closed form, as Heisenberg had stated in his first paper on helium. For the direct term, Unsöld found

$$\alpha(1s, ns) = Z(\frac{2}{n^2} - \frac{13}{162})(\times 13.6 \text{ eV}), \quad n \geq 2. \tag{15.164}$$

The effect of the first term in brackets is (for large Z) just to produce the expected screening, as

$$-\frac{Z^2}{n^2} + \frac{2Z}{n^2} \sim -\frac{(Z-1)^2}{n^2}. \tag{15.165}$$

The analytic computation of the exchange term for arbitrary n is somewhat messier, although for any given value of n it can be carried through to an analytic result: in particular, for $n = 2$, one obtains $\beta(1s, 2s) = \frac{32}{3^6} Z \times 13.6 \text{ eV} = 0.597$ eV. The ortho–para splitting, which is theoretically given by $2\beta(1s, 2s) = 1.19$ eV, is experimentally 0.83 eV, which gives a good idea of the inaccuracy of first-order perturbation theory for these low s-states.

However, as Unsöld points out, the situation should improve for larger values of the nuclear charge Z. If one now calculates the long-desired *ground state* energy—namely, the energy of the parahelium $(1s1s)$ state—one finds (in this case of a non-degenerate state, there is no exchange term)

$$\alpha(1s, 1s) = \frac{5}{4} Z \times 13.6 \text{ eV}, \tag{15.166}$$

so that the total energy (= binding energy of both electrons) is, to first order,

$$E(1s, 1s) = -(2Z^2 - \frac{5}{4} Z) \times 13.6 \text{ eV}, \tag{15.167}$$

and the ionization energy (needed to remove one electron), given by subtracting from this the energy ($-Z^2 \times 13.6$ eV) of a single electron bound to the nucleus is

$$\mathcal{J} = (Z^2 - \frac{5}{4} Z) \times 13.6 \text{ eV}. \tag{15.168}$$

The comparison with the experimental values is illuminating: for helium ($Z = 2$) one finds from this formula 20.3 eV (experiment gave 24.6 eV), for Li^+ 71 eV (expt. 75.6 eV), and for Be^{++} 149 ev (expt. 154 eV, although this was unavailable to Unsöld). Thus, the error in the result decreases from 17% for neutral helium, to 6% and 3% for Li^+ and Be^{++}, respectively. Evidently the perturbation theory is becoming increasingly accurate as we increase the nuclear charge, as expected, making the relative contribution of the single electrostatic repulsion energy term progressively smaller.

Even more impressive results could be obtained for "lithium-like" atoms with three electrons, with Z values ranging from 3 for neutral lithium to 6 for triply ionized carbon. Term values for such "stripped" atoms (the word appears in English in Unsöld 1927, p. 369) had recently been published by Millikan and Bowen (1926; see also Bowen and Millikan 1924), using "hot spark spectra" in which multiple ionization states of the atoms could be obtained. Here, the states examined corresponded to the helium ground state $(1s1s)$ configuration with a third s-state electron at principal quantum number $n \geq 2$. The results were as usual presented in terms of a spectral defect for the third (valence electron)—the total energy is written as the helium energy (with variable Z) for the two $1s$ electrons (cf. Eq. (15.167)), plus a third energy (which is compared with the term values obtained by Bowen and Millikan), written in terms of a spectral defect δ of the screened valence term,

$$E_{\text{val}}(ns) = -\frac{Z_a^2}{(n-\delta)^2} \times 13.6 \text{ eV}, \quad Z_a \equiv Z - 2. \tag{15.169}$$

The product $(Z - 2)\delta$ is given by Unsöld in terms of the direct and exchange integrals $Z\alpha, Z\beta$ as $n^3(\alpha + \beta/2)$ (neglecting terms that would be relevant at the second order of perturbation theory), and comparison with the extrapolation of the experimental numbers to large Z (here facilitated by data for four increasing values of Z) gave agreement to 2.5% for the 2S and 1.5% for the 3S states, respectively. This was surely regarded as very strong evidence that the new theory was finally giving encouraging signs of quantitatively addressing the complicated dynamics of multi-electron atoms. Nevertheless, there was still the need for a calculational procedure that could handle accurately the empirically more frequently encountered situation of atoms in a low ionization state. The key to doing this was actually in the note added in proof in Schrödinger's first paper on wave mechanics.

15.4.3 The helium ground state: perturbation theory gives way to variational methods

In a short note at the end of his first paper on wave mechanics (submitted on January 27), Schrödinger (1926c) pointed out that the differential eigenvalue equation he had formulated (and successfully applied to the hydrogen atom) could be viewed as equivalent to a variational principle—closely related, but not identical, to the one (lacking any real motivation, and subsequently discarded) with which he had begun his paper, based on the classical Hamilton–Jacobi equation (cf. Eq. (14.63), et seq.). Now Schrödinger introduced a variational principle based on a "Hamiltonian integral" constructed from the classical kinetic energy quadratic form $T(q, p = \partial/\partial q)$ and the wave function $\psi(q)$ (where q and p are shorthand for the generalized coordinates and momenta $q_1, q_2, \ldots, q_n$, $p_1, p_2, \ldots, p_n$ of a system of n degrees of freedom).

For the simple case of a single particle of mass m moving in three dimensions under the influence of a potential energy field $V(\vec{r})$, the new postulate was that the integral

$$I = \int \left\{ \frac{h^2}{8\pi^2 m} (|\vec{\nabla}\psi(\vec{r})|^2 + V(\vec{r})\psi(\vec{r})^2 \right\} d^3r, \tag{15.170}$$

$$= \int \psi(\vec{r}) \left\{ -\frac{h^2}{8\pi^2 m} \vec{\nabla}^2 + V(\vec{r}) \right\} \psi(\vec{r}) d^3r, \tag{15.171}$$

be *stationary* under variations $\delta\psi(\vec{r})$ respecting the normalization condition $\int \psi(\vec{r})^2 d^3r = 1$.[70] The first-order variation of this integral under $\psi \rightarrow \psi + \delta\psi$ (introducing a Lagrange multiplier to impose the normalization condition) is easily found to be

$$\delta(I - \lambda \int \psi(\vec{r})^2 d^3r) = 2 \int \delta\psi(\vec{r}) \left\{ -\frac{h^2}{8\pi^2 m} \vec{\nabla}^2\psi(\vec{r}) + V(\vec{r})\psi(\vec{r}) - \lambda\psi(\vec{r}) \right\} d^3r, \tag{15.172}$$

[70] Note that Schrödinger is still under the impression that one can deal consistently with real wave functions throughout, so the complex conjugate wave function expected by a modern reader is absent.

which we require to vanish for arbitrary choices of the first-order variation $\delta\psi(\vec{r})$. This clearly implies the Schrödinger equation obtained previously in the paper:

$$-\frac{h^2}{8\pi^2 m}\vec{\nabla}^2\psi(\vec{r}) + V(\vec{r})\psi(\vec{r}) = \lambda\psi(\vec{r}), \tag{15.173}$$

with the Lagrange multiplier λ playing the role of the energy eigenvalue E. This (at first sight rather innocuous) reformulation of the variational postulate would become the kernel of an enormously efficacious and calculationally productive technique for extracting quantitative quantum-mechanical predictions for complex atoms, where a direct analytic solution of the Schrödinger equation was out of the question. The Hartree–Fock procedure, which became the workhorse of atomic theory in the following decades, is in fact a direct elaboration and application of this observation of Schrödinger.

The utility of this variational principle for the helium problem, in combination with a general procedure for dealing with a general class of similar variational problems introduced by Ritz (1909) almost twenty years earlier, was demonstrated in the Spring of 1927 by Georg Kellner (1927), a student of von Laue in Berlin, in an article published in *Zeitschrift für Physik* summarizing the results of his PhD dissertation.

First, let us observe that the ground state of an atom plays a special role in Schrödinger's extremal principle, as it represents an *absolute*, rather than merely *relative*, minimum of the integral I in Eq. (15.170). Let us write the wave function of the electrons in a general atomic system as $\psi(q)$, where q represents the collection of coordinate variables—thus, for helium, it would contain the six coordinate variables for the Cartesian (or spherical) coordinates $\vec{r}_1, \vec{r}_2$ of the two electrons. We consider only spin singlet states (as our immediate aim is the calculation of the ground state energy of helium), so the spin degrees of freedom are ignorable. Suppose that the complete set of energy eigenfunctions ψ_n, satisfying

$$H\psi_n(q) = E_n\psi_n(q), \tag{15.174}$$

are known, where for convenience we order the energy eigenvalues $E_1 < E_2 \leq E_3 \leq E_4 \ldots$[71] A general wave function (satisfying the usual boundary conditions, square-integrability, uniqueness, etc.) can be written as a linear combination of these eigenfunctions

$$\psi(q) = \sum_{n=1}^{\infty} c_n\psi_n(q). \tag{15.175}$$

[71] We assume, as is the case for helium, that the ground state is a non-degenerate singlet. Higher states may or may not be degenerate, hence, the less-than-or-equal signs. Note also that a complete set of eigenfunctions can always be chosen to be *real*, by taking linear combinations, as if ψ satisfies $H\psi = E\psi$, so does ψ^*, as the Hamiltonian being considered here, kinetic energy plus nuclear potential energy and electrostatic repulsion energy of the electrons, is real. The argument also neglects the possible existence of a continuous spectrum—for reasons that will be fairly obvious shortly.

At this point we also assume the eigenfunctions are orthogonal:

$$\mathcal{J}_{mn} \equiv \int \psi_n(q)\psi_m(q)dq = \delta_{nm}. \tag{15.176}$$

The usual unit normalization of the wave function, $\int \psi(q)^2 dq = 1$, then corresponds to the constraint $\sum_n c_n^2 = 1$. Schrödinger's integral in Eq. (15.171) can now be rewritten as an infinite sum, as

$$\begin{aligned} I &= \int \psi H \psi dq = \sum_{nm} c_n c_m \int \psi_n H \psi_m dq \\ &= \sum_{nm} c_n c_m \int \psi_n E_m \psi_m dq \\ &= \sum_n c_n^2 E_n. \end{aligned} \tag{15.177}$$

As $\sum_n c_n^2 = 1$, and $E_1 < E_n, n \geq 2$, we have

$$I = \sum_n c_n^2 E_n \geq \sum_n c_n^2 E_1 = E_1, \tag{15.178}$$

with equality achieved only if $c_1 = 1, c_2 = c_3 = \ldots 0$.[72] In other words, if we view Schrödinger's integral I as a function of the coefficients c_n, then the function $I(c_n)$ achieves an absolute minimum exactly when $\psi(q) = \psi_1(q)$, the ground state wave function of the system, at which point the value of the integral gives us precisely the desired ground state energy. It is easy to see that the function $I(c_n)$ is also stationary at all the other (higher than ground state) energy eigenfunctions of the system (i.e., when $c_N = 1, c_n = 0$ for $n \neq N$), but those points represent saddle points in the multi-dimensional space of the c_n, with negative curvature in some directions and positive curvature in others, and not absolute minima.

Now in general (apart from hydrogenic systems), the exact energy eigenfunctions $\psi_n(q)$ of the atomic system are not known. However, if one has to hand a *complete* set of basis functions φ_n, each obeying the appropriate boundary conditions for the problem (in this case, square-integrability), then an arbitrary physically acceptable wave function $\psi(q)$ has a unique expansion

$$\psi(q) = \sum_n c_n \varphi_n(q) \tag{15.179}$$

in terms of this complete function set. The functions φ_n may be chosen to be mutually orthogonal (as are the exact energy eigenfunctions ψ_n discussed above), but all that is

[72] Including a continuous spectrum in Eq. (15.175), in the form of an integral $\int d\alpha c(\alpha)\psi(\alpha; q)$, would simply lead to the supplementary condition $c(\alpha) = 0$ here, as $E_1 < E_\alpha$, that is, the ground state is energetically lower than all continuum states.

really necessary is that they be linearly independent (and, of course, complete). It may in fact be convenient to sacrifice orthogonality for computational efficiency. Given this fact, the integral $I = \int \psi H \psi dq$, where in the integrand the wave function ψ, and hence, the expansion coefficients c_n, appear only quadratically, must therefore be an infinite quadratic form,

$$I(c_n) = \sum_{mn} I_{mn} c_m c_n, \tag{15.180}$$

where I_{mn} is now a real symmetric matrix with numerical elements. This quantity needs to be minimized subject to the constraint $\int \psi(q)^2 dq = 1 = \sum_{mn} \mathcal{J}_{mn} c_m c_n$ (where, in the special case where we are dealing with an orthonormal set of functions, $\mathcal{J}_{mn} = \delta_{mn}$, cf. Eq. (15.176)). The variational problem identified by Schrödinger thus reduces to the extremal condition

$$\frac{\partial}{\partial c_m} \sum_{mn} (I_{mn} c_m c_n - \lambda \mathcal{J}_{mn} c_m c_n) = 0 \Rightarrow \sum_n (I_{mn} - \lambda \mathcal{J}_{mn}) c_n = 0. \tag{15.181}$$

The equation on the right is a generalized eigenvalue problem for λ, which is just the energy eigenvalue—but as it stands, the equation is infinite dimensional and not directly solvable, One may however hope that if the expansion function set is chosen well (and this really means is chosen to imitate as closely as possible the important features of the exact energy eigenfunctions of the system, especially the low energy ones), the convergence in the infinite sums over n is rapid, and one may obtain an adequate approximation by truncating the sums at some value $n = N$. The resulting eigenvalue equation, $\sum_{n=1}^{N} (I_{mn} - \lambda \mathcal{J}_{mn}) c_n = 0$, is then solved by finding the roots of the $N \times N$ determinant of the matrix $I_{mn} - \lambda \mathcal{J}_{mn}$. The lowest root corresponds to an approximate value for the ground state energy, of which it is a rigorous upper bound.

To summarize, the Ritz procedure, as applied by Kellner, involves (a) choice of a set of expansion functions φ_n which reflect at least approximately the physically expected behavior of the actual energy eigenfunctions of the system, and which are (b) computationally "agreeable" (in other words, are functions for which the necessary integrals leading to the I_{mn} and $\mathcal{J}_{mn}$ matrix elements are analytically tractable).

An important property of the ground state wave function that should be enforced from the start was its overall spatial symmetry. It had already been known from the early 1920s that the singlet ground state of helium could not have a permanent magnetic moment, as helium (and indeed, all inert gases) is dia- rather than para-magnetic. The total spin of the two electrons was zero in the ground state, as Heisenberg had shown, so the presence of a non-zero total orbital angular momentum would necessarily lead to a (unobserved) magnetic moment. By now (late 1927), the group theoretic implications of physical quantities like the angular momentum operators would have been clear to theorists like Eugene Wigner, with whom Kellner was in contact (see Keller 1927, p. 96, note 2). In particular, the application of a component of the total orbital angular momentum to the wave function, say εL_i (where $i = x, y$, or z, and ε is infinitesimal), gives the change in the wave function when the coordinate axes are rotated by ε around the ith axis. If the wave function corresponds to a state of vanishing total angular momentum, the change in the wave function $\psi(\vec{r}_1, \vec{r}_2)$ under rotation vanishes, which implies that it is rotationally invariant. It must therefore be constructed from the only rotational invariants available in the two-electron problem, $r_1 = |\vec{r}_1|$, $r_2 = |\vec{r}_2|$, and $\vec{r}_1 \cdot \vec{r}_2 = r_1 r_2 \cos(\gamma)$, where γ is the angle between the coordinate vectors $\vec{r}_1, \vec{r}_2$ of the two electrons. Equivalently, we may write the wave function as a function of only three variables, $\psi(r_1, r_2, x)$, where $x = \cos(\gamma)$, instead of the initial six (the Cartesian components of $\vec{r}_1$ and $\vec{r}_2$).

Kellner does not give this argument directly, but rather appeals to a separation of variables in the Schrödinger equation into r_1, r_2, and x and three other Euler angles α_1, α_2, and α_3. Wave functions can then be constructed as products of functions $\psi(r_1, r_2, x)$ and functions $A(\alpha_1, \alpha_2, \alpha_3)$. The eigenvalue equation for the latter function produces a lowest eigenvalue (zero) when $A(\alpha_1, \alpha_2, \alpha_3)$ is independent of the Euler angles $\alpha_1, \alpha_2, \alpha_3$. Once again, we are led to examine functions that depend only on the radial coordinates of the two electrons and the angle between their coordinate vectors.[73]

The desired function $\psi(r_1, r_2, x)$ ($x = \cos(\gamma)$) can be expanded in a complete set of functions formed by taking the orthogonal radial Laguerre functions, which solve the radial part of a hydrogenic atom problem (one electron, nucleus of charge Z)[74]

$$\psi_n(r) = L'_n(\eta_n)e^{-\eta_n/2}, \quad \eta_n \equiv \frac{2Z}{na}r, \; a = \frac{h^2}{4\pi^2 me^2} (= \text{Bohr radius}), \tag{15.182}$$

together with the Legendre polynomials $P_k(x), k = 0, 1, 2, \ldots$ for the angular variable $x = \cos(\gamma)$, which form a complete set of basis functions for the space of piecewise continuous functions $f(\gamma), 0 \leq \gamma \leq \pi$, orthogonal with respect to the measure $\sin(\gamma)d\gamma$. Thus, one could expand the desired function $\psi(r_1, r_2, x)$ as

$$\psi(r_1, r_2, x) = \sum_{mnk} c_{mnk}\psi_m(r_1)\psi_n(r_2)P_k(x). \tag{15.183}$$

One is free to alter the parameter Z in Eq. (15.182) to some non-integer value $Z^\star$ to account for the screening effect of one electron on the nuclear potential seen by the other, without spoiling the orthogonality of the function set. However, the radial exponential falloff of the higher Laguerre functions, $\psi_n \sim \exp(-Zr/na)$, $n \geq 2$ is physically

[73] Kellner gives a misleading "quick" argument for this conclusion, based on the uniqueness of the ground state, as follows: a non-spherically symmetric ground state wave function could be subjected to an arbitrary spatial rotation and still retain the same energy, which would violate the uniqueness of the wave function. This is misleading as there are many atoms with a non-vanishing total electronic orbital angular momentum in the ground state—for example, boron, with an unpaired $2p$ electron—which therefore have a non-spherically symmetric ground state wave function. It should be noted that the existence of a ground state wave function for the neutral helium atom (i.e., a discrete unique lowest eigenvalue) was only rigorously established by Kato (1951b), whose results also strongly supported the validity of the Ritz–Kellner approach to calculating the ground state energy—not that this had ever been in any doubt for the many physicists employed in atomic theory in the decades following the discovery of quantum mechanics.

[74] Here, $L'_n(\eta) = \frac{d}{d\eta}L_n(\eta)$, with $L_n(\eta)$ the usual Laguerre polynomials.

incorrect for the ground state wave function, where both electrons presumably hug the nucleus closely, approximately as expected for $1s$ states, so that the spreading of the wave function, which would reflect correctly the behavior of states of higher principal quantum number, is here inappropriate. To deal with this problem, Wentzel abandons the orthogonality requirement, by keeping the *same* exponential behavior $\exp(-Z^\star r/a)$ for $n = 1$ for all the higher functions $\psi_n, n \geq 2$. The effective charge $Z^\star$ is chosen by setting the empirically measured binding energy of -78.9 eV to the energy of two identical electrons bound independently (and ignoring their electrostatic repulsion) to a nucleus of screened charge $Z^\star$, namely $-2 \times (Z^\star)^2 \times 13.6$ eV, whence $Z^\star = 1.7$, and the radial function parameter becomes $\eta = 3.4r/a$, independent of principal quantum number. In summary, Kellner's approach is to minimize the Schrödinger integral over functions of the form given in Eq. (15.183), where the sum is truncated of course at some point, using radial functions of the form

$$\psi_m(r) = e^{-\eta/2}L'_m(\eta), \quad \eta = \frac{3.4r}{a}. \tag{15.184}$$

Of course, as these expansion functions are no longer orthogonal (they are, however, linearly independent), the generalized eigenvalue problem Eq. (15.181) must be solved after all the relevant matrix elements $I_{mn}, \mathcal{J}_{mn}$ have been calculated, with $\mathcal{J}_{mn}$ no longer proportional to the identity matrix. The fact that the same exponential factor appears in all the functions turns out to simplify considerably the evaluation of the necessary integrals.

Kellner was able to carry through the Ritz procedure using up to four basis functions (thereby yielding 4×4 matrices in the associated generalized eigenvalue problem), namely, the functions $\psi_{110}, \psi_{111}, \psi_{112}$, and ψ_{220}. Using only the first function, one obtained an ionization potential of 22.9 volts, already appreciably more accurate than the Unsöld result of 20.3 eV, which was 20% off from the experimental value of 24.6 eV. Using all four basis functions increased the ionization energy to 23.75 eV, only 3.5% off from the experimental value, or, considered as an error in the total binding energy (which is greater than the ionization energy by the binding energy of a single electron, $4 \times 13.6 = 54.4$ eV), off by only 0.9%.

The Ritz method was pushed further, and probably to the limits of computational feasibility in the pre-digital-computer years of the late 1920s, by Egil Hylleraas (1928), who was able to reduce the discrepancy between the observed and calculated ionization energy to 0.5% by considering a Ritz expansion truncated to 11 terms. The basis functions employed in a second paper by Hylleraas (1929), which are powers of $s = r_1 + r_2$, $t^2 = (r_1 - r_2)^2$, and $u = r_{12} = |\vec{r}_1 - \vec{r}_2|$ multiplied by an exponential $\exp(-ks/2)$, would become the foundation for modern calculations (see Nakashima and Nakatsuji 2007) of the ground state energy of helium following from the Schrödinger equation, which are now accurate to 40 significant digits! Technical *tours de force* of this kind are, of course, of purely academic interest, as measurements are only available to eight-digit accuracy, and many other physical effects contributing to the fine structure are neglected in the Hamiltonian Eq. (15.159), and make their presence known already in the fourth significant digit.

Nevertheless, the success of the helium calculations of the late 1920s was a stunning confirmation of the power and correctness of the new theory, redeeming completely the embarrassing failure of the old quantum theory to handle atoms with more than one electron. In fact, it could be argued that the new theory was even more powerful than the classical mechanics it was replacing, for which effective general treatments of the three- (or more) body problem have never been found, due to the intrinsically chaotic nature of generic solutions of such problems. The source of this new power can be directly traced to the linearity of the underlying equations of quantum mechanics, which renders them susceptible to a much more extensive array of effective approximate treatments than have ever been available for the analysis of the non-linear equations of classical mechanics.

Part IV

The Formalism of Quantum Mechanics and Its Statistical Interpretation

16

Statistical Interpretation of Matrix and Wave Mechanics

16.1 Evolution of probability concepts from the old to the new quantum theory

The interpretation of the absolute square of the Schrödinger wave function as a probability is practically the first thing a contemporary student learns about the wave-mechanical approach to quantum mechanics in many modern texts. It may thus come as a surprise that the term "probability" (*Wahrscheinlichkeit* in German) appears only twice in the five foundational papers on wave mechanics (the four papers of the "Quantization as an eigenvalue problem" series, and the "equivalence" paper on the connection between wave and matrix mechanics). In both cases, the term occurs in the combination "transition probability", referring to the relation between matrix elements of coordinate operators and intensities of radiation processes (Schrödinger 1982, pp. 59, 82). Admittedly, in the final section of the fourth paper of the tetralogy, "On the physical significance of the field scalar," there is a tantalizing remark, which follows a reminder that the electric charge density associated with an atomic electron is given by $e\psi\overline{\psi}$, the electron charge times the absolute square of the electron's wave function:

> $\psi\overline{\psi}$ is a kind of *weight-function* in the system's configuration space. The *wave-mechanical* configuration of the system is a superposition of many, strictly speaking of *all*, point-mechanical configurations kinematically possible. Thus, each point-mechanical configuration contributes to the true wave-mechanical configuration with a certain *weight*, which is given precisely by $\psi\overline{\psi}$. If we like paradoxes, we may say that the system exists, as it were, simultaneously in all the positions kinematically imaginable, but not "equally strongly" in all (Schrödinger 1926h, p. 120).

Born would later point out in a paper on Ehrenfest's adiabatic theorem that directly addresses the probabilistic import of de Broglie–Schrödinger waves:

> In his latest communication Schrödinger is also concerned with the square of wave amplitudes, introducing the term "weight-function", which already approaches quite closely the terminology of statistics (Born 1926c, p. 168).

Constructing Quantum Mechanics. Anthony Duncan and Michel Janssen, Oxford University Press.
 DOI: 10.1093/oso/9780198883906.003.0016

But, as Born rightly points out, the thrust of Schrödinger's arguments leading to the wave-mechanical formalism, and indeed the source of its intuitive appeal, much in contrast to the abstraction of the Göttingen approach, was the assertion that wave mechanics could be regarded as the "physical optics" underlying the geometrical "ray optics" described by the classical mechanics of particle motion. As such, the wave function would be a direct physical representation of the electron—its "field scalar"—much as the electric and magnetic fields were of electromagnetic radiation.

In fact, the direct and unambiguous importation of statistical/probabilistic concepts into the new quantum mechanics would be accomplished almost exclusively at the hands of Schrödinger's rivals in Göttingen: Born, Heisenberg, Jordan, and behind the scenes, constantly communicating with them and providing crucial stimulation, Pauli in Hamburg. Paul Dirac, visiting Bohr's Institute in Copenhagen in the Fall of 1926 (and overlapping there with Heisenberg) would also independently arrive at very similar results.

The existence of *intrinsically* stochastic processes at the atomic level was, of course, not a novel feature of the quantum-mechanical treatment of atoms in the post-1925 era of matrix/wave mechanics. At the beginning of the century, Rutherford and Soddy (1902) had already recognized that atoms undergoing radioactive decay have a specific probability of random decay per unit time interval, which led them to the famous exponential law of radioactive decay. Planck (1913), in his "second theory" of heat radiation, had explicitly introduced an intrinsic stochastic element in the (quantized) gain or loss of energy of Planckian resonators (see Section 3.6, note 78). Soon thereafter, the notion of randomly occurring quantum jumps was reprised by Bohr (1913a) in the first installment of his famous trilogy (see Section 4.5.1). Again, a few years later, we have the appearance of explicitly probabilistic arguments in Einstein's (1916a, 1917a) introduction of the A and B coefficients (see Section 3.6). In Chapter 10, we discussed how Einstein's stochastic radiation theory, together with Bohr's correspondence principle, played a crucial role in the developments in dispersion theory leading up to matrix mechanics. The BKS theory (cf. Section 10.4) also prominently featured a stochastic element, as emphasized in the (purely qualitative) paper introducing the theory:

> [T]he occurrence of transition processes for the given atom itself, as well as for the other atoms with which it is in mutual communication, is connected with this mechanism by probability laws which are analogous to those which in Einstein's theory hold for the induced transitions between stationary states when illuminated by radiation (Bohr, Kramers, and Slater 1924a, pp. 164–165).

This theory, however, was decisively refuted by the X-ray experiments of Bothe and Geiger (1925a) and Compton and Simon (1925) (see Section 10.4).

In any event, by the summer of 1926, it was increasingly apparent to many of the theorists engaged with the new quantum theories that the nascent formalism(s) would necessarily have to produce a sensible interpretation of the stochastic content of the theory—specifically, an explanation within the theory of the random quantum "jumps" (*Sprünge*) or "transitions" (*Übergänge*) of atomic electrons interacting with radiation.

In the last week of June 1926, Born submitted a short "preliminary" investigation of this question, examined not in the context of radiatively induced transitions of atomic electrons, but for the problem of collisions of a material particle (electron, α-particle, etc.) with an atom in a specified stationary state.[1] The asymptotic behavior of the solution ψ of the Schrödinger equation can be determined perturbatively, starting from the assumption that at zeroth order the incoming particle is a free plane wave in the x-direction, that is, that the wave function $\psi_0 = e^{ikx}$, with wave vector $\vec{k} = (k, 0, 0)$ and momentum $\hbar k$ in the x-direction. The first-order perturbation of the wave function corresponds to a linear superposition of outgoing spherical waves $e^{i\vec{k}'\cdot\vec{r}}$, with coefficients given by a complex amplitude $\varphi(\vec{k}')$.[2] Born (1926a, pp. 865–866) asserts that *the probability of the incoming electron to be scattered from the incoming (x) direction to the outgoing direction specified by the wave vector* $\vec{k}'$ *is determined by the amplitude* $\varphi(\vec{k}')$. In a note added in proof, he famously corrected himself and noted: "More careful consideration shows that the probability is proportional to the square of the quantity $[\varphi(\vec{k}')]$" (Born 1926a, p. 865). This is the first explicit statement of what would eventually come to be called the "Born rule". In a follow-up paper a month later, Born (1926b) considerably amplified these observations in his preliminary note, offering a full examination of scattering amplitudes in both one and three dimensions as well as an exposition of the now famous Born perturbation expansion.[3]

Before we move on to this follow-up paper, we want to draw attention to the following starkly explicit passage in the preliminary note, in which Born expresses his acceptance of both the completeness of the quantum theory as it stood in mid-1926 and the fundamental indeterminism apparently required to interpret solutions of the Schrödinger equation:

> Here the entire problem of determinism raises itself. From the standpoint of our quantum mechanics there doesn't exist a quantity that will causally determine the effect of a collision; but likewise empirically we have up to now no indication that there are inner properties of the atom that determine a definite result of the collision. Should we hope to discover such properties (perhaps, phases of inner atomic motions) later and determine them in individual cases? Or should we believe that there is a pre-established harmony in the inability of both theory and observation to provide conditions for a causal process, based on the non-existence of such conditions? I myself am inclined to abandon determinism in the atomic world. But this is a philosophical question for which physical arguments alone are not decisive (Born 1926a, p. 866).

Born's identification of the squared complex amplitude coefficients appearing in the asymptotic regime (i.e., far from the center of scattering) with scattering probabilities

[1] From a modern perspective, the problem of scattering of photons from atoms is clearly analogous to the problem of scattering of material (massive) particles: indeed, as Born (1926a, p. 863) mentions, Bohr had frequently emphasized the close relation of radiative and collision processes.

[2] We have somewhat simplified Born's notation, which allowed for inelastic transitions in which the atomic state was changed, with a concomitant shift in the energy of the scattered particle. The scattered wave vector $\vec{k}'$ is expressed in Cartesian coordinates as $\vec{k}' = (\alpha, \beta, \gamma)$.

[3] For historical analyses of Born's collision papers and his probabilistic interpretation of the wave function, see, for example, Konno (1978), Wessels (1980), Pais (1982b), Beller (1990), and Im (1996).

received additional support in his study of one-dimensional scattering in his second paper. In that case, the problem simplifies considerably, as there are only two complex numbers to be considered, for an incoming particle from the left to be either reflected from (i.e., backward, to $x \to -\infty$), or "transmitted" through the central potential, emerging as an outgoing particle moving to the right for $x \to +\infty$. The absolute squares of these two complex numbers sum gratifyingly to unity, reinforcing the interpretation of the two squared amplitudes (reflection and transmission coefficient, respectively) as probabilities. In Born's view, the predictive content of the theory was exhausted by these probabilistic statements: nothing whatsoever could be said about the result of the scattering on an event-by-event basis.

The physical interpretation of the Schrödinger wave function in the case where it consisted of a linear superposition of *states of different energy*[4] turned out to cause serious distress for theorists in the Göttingen camp. Many of them at least initially resisted the notion that an electron could be simultaneously in more than a single stationary state—an idea that Schrödinger, on the other hand, was quite willing to accept from the beginning. In his first paper, he speaks of a "potpourri of proper vibrations" (Schrödinger 1926c, p. 11). And in the final section of the "equivalence" paper, he considers a hydrogen atom

> in such a state that the field scalar ψ is given by a series of discrete proper functions
>
> $$\psi = \sum_k c_k\, u_k(x)\, e^{2\pi i E_k t/h} \tag{16.1}$$
>
> (Schrödinger 1926e, p. 60).

In fact, superpositions of states of differing energy would play an essential role in Schrödinger's (1926h) reproduction of the Kramers–Heisenberg dispersion formulas in the fourth installment of his tetralogy. Such superpositions, however, were evidently viewed with profound suspicion by many theorists. In a letter to Pauli on June 26, 1926, Sommerfeld wrote

> We have had Schrödinger here [in Munich], together with Heisenberg. My overall impression is that the "wave mechanics" is indeed a marvelous micromechanics, but that it does not provide in the least measure a solution to the fundamental riddles of quantum theory. At least for now, I no longer believe Schrödinger the instant he begins to calculate with the c_k (the amplitudes of the different simultaneous eigenvibrations) (Pauli 1979, Doc. 141).

Three months later, in a paper on the interpretation of Ehrenfest's adiabatic theorem in wave mechanics, Born addresses this question head-on, after restating Eq. (16.1) in a slightly different notation

[4] In the case of an elastic collision process, the outgoing particle states are all degenerate in energy, differing only in the direction of the scattered particle.

> The question now is what this function [ψ] means physically. Schrödinger is of the opinion, as is clear from his latest communication [i.e., Schrödinger (1926h)], that in a *single* atom "many eigenvibrations can be excited" simultaneously. It is not quite clear to me whether "excitation of an eigenvibration" is equivalent to what one meant, using the old Bohrian language, when saying that the atom "is in a stationary state". In any case the statement that an atom can be at one and the same time in several stationary states not only runs counter to the whole Bohrian theory, but directly contradicts the natural, and so far never disputed, interpretation of the stationary states corresponding to points of the continuous spectrum (Born 1926d, pp. 169–170).[5]

Two paragraphs later, Born definitively nails his flag to the Bohrian mast: "we shall therefore hold onto the Bohrian picture, that an atomic system is always only in *one* stationary state" (Born 1926d, p. 170) The interpretation of the linear combination of Eq. (16.1)—which Born regards a perfectly valid wave function to be subjected to evolution via the time-dependent Schrödinger equation—now acquires an epistemological core:

> In general we shall at any given moment only know that, on the basis of its prehistory and the prevailing physical conditions, a certain probability exists for the atom to be in the nth state. We now assert that, as a measure of this "state-probability" one should choose the quantity
>
> $$|c_n|^2 = \left| \int \psi(x,t)\, \psi_n^\star(x)\, dx \right|^2 \tag{16.2}$$
>
> (Born 1926c, pp. 170–171).

From a modern perspective, it looks as if Born is simply mistaking a pure state for a mixed state at this point. The distinction between pure and mixed states, however, would not be introduced until the following year by von Neumann (1927b; see Section 17.2). Yet, even though this distinction was not available to Born at this point, he did, in this same paper, arrive at some correct conclusions about the behavior of an atom (or more precisely, an ensemble of identical atoms) initially placed in the superposed state Eq. (16.1). Born considered the case where the initial Hamiltonian (with eigenfunctions u_k or ψ_n in Eq. (16.2)) is *temporarily* perturbed and then returns, after a finite time interval T, to its original form. With this calculation, an important step forward is made in the recognition of the peculiar "interference of probabilities" characteristic of quantum theory, a step that would prove of critical importance to Jordan and Dirac in their development of statistical transformation theory.[6]

Born (1926d, p. 172) first considers the special case where the wave function at time zero is given by $\psi(x,0) = \psi_n(x)$, that is, when the particle is with certainty occupying the

[5] There follows a discussion of the clearly visible rectilinear paths of ionized electrons in a Wilson cloud chamber, which correspond to a clear indication of localized particle motion, and not to the superpositions of motion in many different directions, as represented, for example, by an outgoing spherical wave.

[6] Jordan (1927b, p. 812) credits Pauli for the phrase "interference of probabilities" but in the follow-up paper to his initial note on collision phenomena, Born (1926b, p. 804) already talked about the "interference ... of 'probability waves'."

nth discrete stationary state (with energy W_n) of the initial unperturbed Hamiltonian. One then imagines switching on and off a temporary perturbation (say, an imposed external electric field) over a finite time period, with the Hamiltonian returning to its original form at time $t = T$. Born (1926d, p. 173, Eq. (16)) then finds that the wave function, for later times $t > T$, takes the form

$$\psi_n(x,t) = \sum_m b_{nm}\,\psi_m(x)\,e^{2\pi i W_m t/h}, \tag{16.3}$$

where the array of complex amplitudes b_{nm} depends on the details of the perturbation but always satisfies $\sum_m |b_{nm}|^2 = 1$.[7] Born then asserts that the squared amplitude $|b_{nm}|^2$ is *the probability that the system after the termination of the perturbation ($t > T$) finds itself in the state m*. Once again, in any individual case, the end-point of the perturbation is one in which the particle is in a single definite stationary state, despite the manifest appearance in the wave function of many different states. The probabilities simply refer to the frequencies with which various end states are found if the process is repeated many times, starting from state n.

We shall shortly see how Heisenberg attempted to rationalize this peculiar claim in terms of stochastic quantum jumps, but for the time being let us follow Born just a little further. Suppose we begin the process of perturbing the system at time zero with the particle *already in a state of superposition of several stationary states* (perhaps as a result of a prior perturbing influence):

$$\psi(x,0) = \sum_n c_n\,\psi_n(x). \tag{16.4}$$

By the linearity of the time-dependent Schrödinger equation, the result for the wave function for times $t > T$ is simply given as the corresponding linear combination of the solutions for different values of n given in Eq. (16.3), that is,

$$\begin{aligned}\psi(x,t) &= \sum_n c_n \sum_m b_{nm}\,\psi_m(x)\,e^{2\pi i W_m t/h}\\ &= \sum_{mn} c_m\,b_{mn}\,\psi_n(x)\,e^{2\pi i W_n t/h}.\end{aligned} \tag{16.5}$$

Introducing

$$C_n \equiv \sum_m c_m\,b_{mn}, \tag{16.6}$$

[7] The absence of a minus sign in the complex exponential, in accordance with modern usage, derives from Schrödinger's choice of sign in his time-dependent equation, which is basically the equation satisfied by the complex conjugate of the wave function as used nowadays.

we can rewrite the final expression in Eq. (16.5) as

$$\psi(x,t) = \sum_n C_n \psi_n(x)\, e^{2\pi i W_n t/h}. \tag{16.7}$$

By the same logic as used previously, the positive quantity $|C_n|^2$ indicates the probability of now finding the system, having started in the superposition Eq. (16.4) at time zero, post-disturbance ($t > T$) in the definite stationary state n. If the processes of "quantum jumping" from each of the possible m states present in the initial superposed state to state n at time $t > T$ were statistically independent, we should expect, instead of

$$|C_n|^2 = \Big|\sum_m c_m\, b_{mn}\Big|^2, \tag{16.8}$$

the sum of independent probabilities

$$\sum_m |c_m|^2\, |b_{mn}|^2, \tag{16.9}$$

which is in general a different number. Born's conclusion is inescapable:

> The quantum jumps between two states characterized by m and n *cannot* proceed as independent events—for then the above expression [Eq. (16.8)] would read $\sum_m |c_m|^2\, |b_{mn}|^2$ [Eq. (16.9)] (Born 1926d, p. 174).

About three weeks after Born submitted his paper on quantum transitions induced by external disturbances (either temporally limited, or extended adiabatically over an infinite time), Heisenberg completed a short note on fluctuation phenomena in quantum mechanics, which attempted, as the abstract proclaims, "to show that quantum mechanics is always in agreement with the fluctuation formulas required by the theory of discontinuities" (Heisenberg 1926d). By the "theory of discontinuities," Heisenberg meant the Bohrian conception of dynamical evolution of quantum systems as periods of occupancy of stationary states, interrupted stochastically by quantum jumps between these states, with the actual times of the said quantum jumps not predictable in any way, other than as a statistical distribution. The observable quantities, according to Heisenberg, are time averages which are given by the theory and only constrain the frequency of these quantum jumps probabilistically.

As a concrete example of these ideas, Heisenberg considers two harmonically bound particles, which he refers to as "atom a" and "atom b", very weakly coupled by an interaction term. This is the same model that he had exploited so successfully in deciphering the mysterious features of the helium spectrum (Heisenberg 1926b; see Section 15.4.1).

Thus, the full Hamiltonian of the system is (cf. Eq. (15.146))

$$H = H^a + H^b + H_{\text{coupling}}. \tag{16.10}$$

In this model, which can be solved completely in either classical or quantum mechanics, the total energy of the system is conserved, but the energy of each atom varies harmonically, as energy is exchanged slowly between the two atoms by the agency of the weak coupling. One imagines a situation in which atom a is in stationary state n at time zero, while atom b is in state $m \neq n$. In the course of time, the weak coupling between the atoms induces quantum transitions in which atom a jumps from state n to state m, while atom b jumps (simultaneously) from state m to state n, and vice versa. In a basis of states labeled by the stationary states of the uncoupled atoms, the Hamiltonian of the first atom is therefore effectively (as far as needed for the description of the time evolution of the situation described here) the 2×2 matrix

$$\mathbf{H}^a = \begin{pmatrix} E_n & 0 \\ 0 & E_m \end{pmatrix}. \tag{16.11}$$

In order to examine the energy fluctuations of atom a once the system is placed in an eigenstate of the full Hamiltonian H, we must perform an orthogonal transformation S to diagonalize the latter, as we did in Eqs. (15.153) and (15.154). Applying this transformation to the matrix $\mathbf{H}^a$, we find

$$\begin{aligned} \mathbf{S}^{-1}\mathbf{H}^a\mathbf{S} &= \mathbf{S}^{-1}\begin{pmatrix} E_n & 0 \\ 0 & E_m \end{pmatrix}\mathbf{S} \\ &= \frac{1}{2}\begin{pmatrix} E_n + E_m & E_n - E_m \\ E_n - E_m & E_n + E_m \end{pmatrix}. \end{aligned} \tag{16.12}$$

In fact, the same simple algebra can be performed if we replace the energy H^a of atom a by any function $f(H^a)$ (square, cube, etc.): one simply obtains a 2×2 matrix as in Eq. (16.12), with E_n (or E_m) replaced by $f(E_n)$ (or $f(E_m)$).

The diagonal matrix elements of this transformed matrix are interpreted, as usual in matrix mechanics, as time averages (over a sufficiently long interval) of the associated quantity. In the situation described previously, in which atom a is initially in the state n, we see that the subsequent evolution of the full system (given by the Hamiltonian H including the coupling term) is such that the average energy of atom a becomes $(E_n + E_m)/2$. The "discontinuity theory" asserts that this average result obtains as a consequence of random stochastic quantum jumps of atom a between state n and state m, as an energy quantum of magnitude $|E_n - E_m|$ is exchanged between the two atoms. All that we can deduce from the theory is that atom a spends on average half its time in state n and half its time in state m, which then automatically gives the average $(f(E_n) + f(E_m))/2$ for the average over time of any function of the energy of atom a. Thus, mean square fluctuations of the energy of atom a are exactly as expected if the atom a were in either state n or

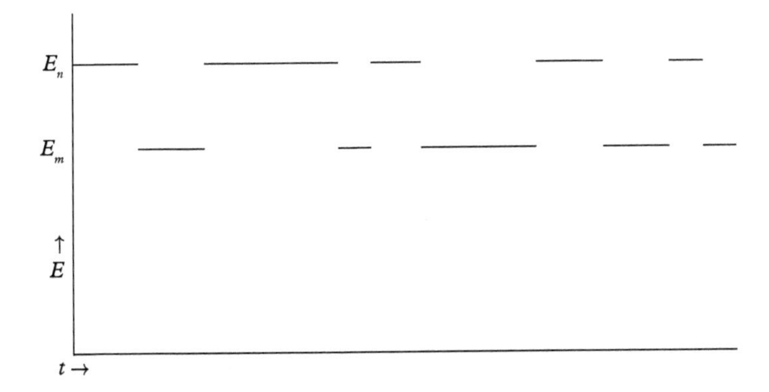

Figure 16.1 *Quantum jumps of an atom coupled weakly to another in Heisenberg resonance model (see Heisenberg 1926d, Fig. 1).*

state m at any given time, and randomly jumping from one to the other. Operationally, Heisenberg concluded, the results obtained are exactly as though atom a behaved as indicated in Figure 16.1:

> The result is therefore that, with respect to all fluctuation quantities in the case discussed here, quantum mechanics arrives at the same results as the discontinuous conception—in other words, it is seen that the fact of discontinuity [i.e., discrete quantum jumps] is naturally incorporated in the system of quantum mechanics (Heisenberg 1926d, p. 504).

What Heisenberg failed to consider is that the agreement of fluctuations as calculated in matrix mechanics (which even in this simple model is only valid in the limit of a very weak coupling of the two atoms) with the fluctuations found in the model of time evolution of the state via discrete quantum jumps depends on restricting oneself to a set of compatible observables (represented by commuting operators): in this case, the energy H^a of atom a, and powers thereof. One can easily see that the values obtained for the mean value of various powers of the kinetic energy K^a of atom a (e.g., the first term in Eq. (15.146)) obtained from the appropriate diagonal matrix elements *do not agree* with the picture of the time evolution displayed in Figure 16.1. This is simply because the kinetic energy of atom a is not diagonal in the n, m basis for the full (kinetic plus potential) energy of atom a, and the discontinuity picture of the figure simply cannot reproduce the required quantum dispersions.

From the examples cited to this point, it seems clear that the theorists in close contact with the Göttingen school, even those (like Born) inclined to adopt (for practical reasons) the methodology of the wave-mechanical approach, were nevertheless anxious to maintain a strong element of discreteness, and *discontinuity*, in their conceptualizations of quantum dynamics. This led to the erroneous picture of particles restricted to definite stationary states and hopping stochastically between them, a picture inimical to the continuous time development of linearly superposed waves intrinsic to the Schrödinger approach. The result of these misconceptions was inevitably to obscure the correct connection between the formal components of the theory (matrix elements, wave functions etc.) and the probabilistic outcomes connected with these mathematical quantities. All of this is perfectly understandable if we keep in mind that a proper understanding of quantum statistics, with a clear distinction between pure and mixed states, the role of measurement in altering the observed quantum state, and the incompatibility of measurements of non-commuting observables would all only come into clear focus with von Neumann's (1927b) synoptic development of the probabilistic foundations of quantum theory a year later, which is discussed in detail in Section 17.2.

The discussions of the precise role of probability concepts in quantum theory by Born and Heisenberg in Summer and Fall 1926 left much unclear and unresolved. However, there was one particular point, raised in Heisenberg's fluctuation paper immediately after his discussion of the coupled two-atom system, that, like Born's recognition of the "interference of probabilities", would prove to be key to the development of Dirac's and Jordan's statistical transformation theory. The orthogonal matrix S—in the two-atom problem a simple 2×2 matrix giving the diagonalizing matrix for the coupled system Hamiltonian (in a basis consisting of a pair of uncoupled stationary states)—is connected directly with probability, in a much more generalized context than the simple toy model with which the paper is primarily concerned. Consider *any* quantum system that can be subject to perturbation, perhaps of finite duration in time (examples of which could be found in Born's papers on collisions and adiabatic processes). Then, Heisenberg states (emphasis in the original):

> *If the perturbed system finds itself in state α, then $|S_{\alpha\beta}|^2$ gives the probability that (in the case for example of collisions, or the sudden removal of the perturbation) the system is found to be in state β* ... By general principles one of course has
>
> $$\sum_{\beta} |S_{\alpha\beta}|^2 = 1. \qquad (16.13)$$
>
> (Heisenberg 1926d, p. 505)

The requirement Eq. (16.13) is indeed essential for the probability interpretation proposed by Heisenberg to be tenable: the system, when examined as to its status *vis-a-vis* the unperturbed Hamiltonian (following removal of the perturbation, or once the colliding particles have receded from one another) must be found to be in *some* state β, and basic properties of probability theory requires the condition Eq. (16.13) to hold.

Heisenberg's fluctuation paper marks the first occurence in print of the remarkable connection between dynamical probabilistic features of the new formalisms (be it in Born's wave-mechanical analysis of collision processes or in Heisenberg's matrix-mechanical analysis of a two-state system) and the kinematical problem of describing a quantum-mechanical system in a variety of different coordinates. But this same connection had already been explored, in a related context, in a long letter from Pauli to Heisenberg of October 19, 1926 (Pauli 1979, Doc. 143). After discussing a variant of Born's one-dimensional potential scattering theory (yielding reflection and transmission coefficients) for a particle constrained to move on a circular loop and a brief discussion of the impossibility of specifying simultaneously the initial conditions of a collision process in both coordinate (q) and momentum (p) space, Pauli goes on to describe a particular kinematical transformation (classically, the canonical transformation interchanging q and p), which allows the Schrödinger formalism to be rewritten in a completely equivalent form, but with wave functions prescribed in momentum, rather than, as usual, in coordinate space. He then continues:

> The second question is: how do the given matrix elements come to determine the collision probabilities. The direction in which one must proceed is, I believe, the following ... All diagonal elements of the matrices (at least, for functions of p only or of q only) can already be kinematically specified. For one can first ask for the probability that, in a specific stationary state of the system, the coordinates q_k of the particles ($k = 1, \ldots, f$) lie between q_k and $q_k + dq_k$. The answer to this is
>
> $$|\psi(q_1 \cdots q_f)|^2 dq_1 \cdots dq_f, \tag{16.14}$$
>
> where ψ is the Schrödinger eigenfunction. (From the *corpuscular* standpoint it is also therefore sensible, that it is specified in a multi-dimensional space [i.e., in the higher-dimensional configuration space of a multi-particle system, and not, as would be the case if the wave function were a physical field—e.g., Schrödinger's scalar field [*Skalarfeld*]—in three spatial dimensions, analogous to the electromagnetic field]). We must regard this probability [function] as in principle observable, just like the intensity of light is a function of space in a standing light wave ... Here one can make a clever mathematical move [*einen mathematischen Witz machen*]: there is a corresponding probability density in *p-space*: for this one sets (in one dimension, for the sake of simplicity) [the momentum and coordinate matrices]
>
> $$p_{ik} = \int p\,\varphi_i(p)\,\overline{\varphi}_k(p)\,dp;$$
>
> $$\frac{2\pi i}{h} q_{ik} = -\int \varphi_i \frac{\partial \overline{\varphi}_k}{\partial p} dp = +\int \frac{\partial \varphi_i}{\partial p} \overline{\varphi}_k dp.$$
>
> where bars denote complex conjugation (Pauli 1979, Doc. 143, p. 347).

Although Pauli does not explicitly say so, it is clear that the new p-space wave functions introduced here, $\varphi_i(p)$, are just the Fourier transforms of the original q-space Schrödinger stationary state wave functions $\psi_i(q)$:

$$\begin{aligned} \varphi_i(p) &= \int \psi_i(q)\, e^{-2\pi i p q/h} dq/\sqrt{h}, \\ \psi_i(q) &= \int \varphi_i(p)\, e^{+2\pi i p q/h} dp/\sqrt{h}. \end{aligned} \tag{16.15}$$

With a little algebra one sees that the matrices defined here are exactly those that one would define using Schrödinger q-space wave functions, as explained in his "equivalence" paper.[8] The probability postulate transfers naturally to p-space, giving now the probability of finding the system (again, one with several particles) *known to be in a stationary state n* to have the momentum variables in a given range (e.g., p_k between p_k and $p_k + dp_k$):

$$|\varphi_n(p_1 \cdots p_f)|^2 dp_1 \cdots dp_f. \tag{16.16}$$

Within a month, the connection between coordinate transformations (of a very general kind) and conditional probabilities would be generalized in spectacular form by Dirac, visiting in Copenhagen at the time and talking frequently to Heisenberg. Via Heisenberg, the news of Dirac's work spread quickly to Pauli and Jordan. The latter, it turned out, had independently arrived at essentially the same results as Dirac. It is to this formalism developed by Dirac and Jordan that we turn next.

16.2 The statistical transformation theory of Jordan and Dirac

Section 16.1 traced the first steps toward the recognition of the peculiar probabilistic aspects of the new quantum theory, especially in some remarkable papers of Born (1926a, 1926b, 1926d), which would eventually lead to a Nobel Prize, and in one of Pauli's many letters during this period (Pauli 1979, Doc. 143). Two new formalisms for quantum theory with a full-fledged probabilistic interpretation were published the following year: (1) the *statistical transformation theory* of Jordan (1927b, 1927g) and Dirac (1927a), amplified three years later in the latter's famous textbook on quantum mechanics (Dirac 1930); (2) the *Hilbert-space formalism* of von Neumann (1927a, 1927b, 1927c), amplified five years later in another classic of quantum mechanics (von Neumann 1932). We discuss the first formalism in this section and the second in Chapter 17.[9]

The generalization of the early contributions to the probabilistic interpretation of quantum theory in these two formalisms went hand in hand with efforts to resolve two other questions that had already drawn considerable attention in the course of 1926.

The first question was how to change variables in the new quantum theory. In the context of matrix mechanics, this question arose in the guise of how to implement canonical transformations. Jordan (1926a, 1926b) and Fritz London (1926a, 1926b) had each published a pair of papers on this topic, expanding on ideas already contained in Born, Heisenberg, and Jordan (1926).[10] In the context of wave mechanics, the question took

[8] Using standard Fourier technology, one easily verifies, for instance, that the coordinate matrix q_{ik} in Pauli's letter is given by $\int q\,\psi_i(q)\,\overline{\psi}_k(q)\,dq$.

[9] Both this section and Chapter 17 closely follow Duncan and Janssen (2013).

[10] For detailed analysis of these papers, see Lacki (2004) and Duncan and Janssen (2009).

the form of how to transform wave functions of one (set of) variable(s) to wave functions of another. For the special case of going from wave functions in coordinate-space to wave functions in momentum-space, (cf. Section 16.1), this issue had already been settled by Pauli in a long letter to Heisenberg of October 19, 1926 (Pauli 1979, Doc. 143; see also Section 16.1).

The second question addressed the relation between the different guises in which the new quantum theory had arrived on the scene, most importantly Schrödinger's wave mechanics and the Göttingen matrix mechanics, but also two other versions closely related to the latter, Dirac's q-number theory (see Section 12.2) and the operator calculus of Born and Wiener (1926) (see Section 14.5). In his "equivalence" paper, Schrödinger (1926e) had already taken some important steps in clarifying the relation between wave mechanics and matrix mechanics. Similar results can be found in a paper by Eckart (1926) and in a letter from Pauli to Jordan of April 12, 1926 (Pauli 1979, Doc. 131).[11] However, it was only with the Dirac–Jordan statistical transformation theory and especially with von Neumann's Hilbert space formalism that the relation between these different forms of quantum mechanics was fully understood.

16.2.1 Jordan's and Dirac's versions of the statistical transformation theory

On Christmas Eve 1926, Dirac, on an extended visit to Bohr's institute in Copenhagen, wrote to Jordan, working with Born in Göttingen:

> Dr. Heisenberg has shown me the work you sent him, and as far as I can see it is equivalent to my own work in all essential points. The way of obtaining the results may be rather different though . . . I hope you do not mind the fact that I have obtained the same results as you, at (I believe) the same time as you (Dirac to Jordan, December 24, 1916, AHQP microfilm).[12]

Dirac's (1927a) paper, "The physical interpretation of the quantum dynamics," had been received by the Royal Society on December 2, 1926, and was published on January 1, 1927. Jordan's (1927b) paper, "On a new foundation [*Neue Begründung*] of quantum mechanics," had been received by *Zeitschrift für Physik* on December 18, 1926, and was published January 18, 1927. We refer to this paper as *Neue Begründung* I to distinguish it from its sequel, "On a new foundation [*Neue Begründung*] of quantum mechanics II," which we refer to as *Neue Begründung* II (Jordan 1927g), submitted June 3, 1927 to the same journal. In this sequel Jordan tried both to simplify and to generalize his theory.[13] What Dirac and Jordan, independently of one another, had worked out and presented in these papers has come to be known as the Dirac–Jordan (statistical) transformation

[11] For discussion, see Section 14.5.
[12] This letter is also quoted and discussed by Mehra and Rechenberg (2000–2001, pp. 72, 83–87).
[13] A short version of *Neue Begründung* I (Jordan 1927c) was presented to the Göttingen Academy by Born on behalf of Jordan on January 14, 1927.

theory.[14] Here, we give an overview of the development of this theory and highlight the differences between the versions of Jordan and Dirac. The rest of Section 16.2 provides a closer look at Jordan's version, examining two papers by Jordan (1927b, 1927g) himself and one written in between (but only published early the following year) by Hilbert, von Neumann, and Nordheim (1928).

Roughly speaking, we can say that the difference between the Dirac and the Jordan version of the theory is fourfold. First, in Jordan's case the emphasis is on the statistical interpretation, while in Dirac's case the emphasis is on transformation properties. Second, Jordan presented the theory in axiomatic form, whereas Dirac did not. Third, Jordan and Dirac used canonical transformations, central to the formalism, in different ways. Finally, Dirac's notation is vastly superior to Jordan's. Our focus is on Jordan's version of the theory but we occasionally help ourselves to elements of Dirac's notation for clarity.[15]

Statistical transformation theory owed much to insights of Pauli. While Pauli is not mentioned in Dirac's (1927a) paper, he is mentioned seven times in the first three pages of *Neue Begründung* I. In a footnote, Jordan (1927b, p. 811) mentions a forthcoming paper by Pauli (1927) on gas degeneracy, in which $|\psi(q)|^2 dq$ is interpreted as the probability of finding the system at a position somewhere in the region $(q, q + dq)$ (where $q \equiv (q_1, \ldots, q_f)$ with f the number of degrees of freedom of the system). This same interpretation of the wave function, in both q- and p-space, can be found in Pauli's letter to Heisenberg of October 19, 1926 (Pauli 1979, Doc. 143; see Section 16.1). As Heisenberg wrote to Pauli on October 28, 1926 (Pauli 1979, Doc. 144), this letter made the rounds in Copenhagen, and Dirac was among the ones who read it. Jordan did not, as he was in Göttingen at the time, but he and Pauli regularly met during this period, even going on vacation together, so there was ample opportunity for Pauli to fill him in on the probabilistic interpretation of Schrödinger wave functions.

Whatever he took from these conversations, Jordan presented a broad generalization of the idea found in Pauli's paper on gas degeneracy and in his letter to Heisenberg. Jordan was after nothing less than a new unified foundation for quantum mechanics—the *Neue Begründung* of the title of the relevant papers—by trying to show that its laws could be derived as "consequences of a few simple statistical assumptions" (Jordan 1927b, pp. 810–811). The central quantities in Jordan's formalism are what he called "probability amplitudes". This echoes Born's (1926b, p. 804) term "probability waves" for Schrödinger wave functions but is a far-reaching generalization of Born's concept. Moreover, Jordan (1927b, p. 811) credited Pauli rather than Born with suggesting the term. He defined a complex probability amplitude, $\varphi(a, b)$, for two arbitrary quantum-mechanical quantities $\hat{a}$ and $\hat{b}$ with fully continuous spectra.[16] When he wrote *Neue*

[14] The term "transformation theory"—and even the term "statistical transformation theory"—is sometimes used more broadly; cf. Jammer (1966, p. 293) and Duncan and Janssen (2013, p. 193, note 51).

[15] For discussions of Dirac's version of the theory, see, for example, Jammer (1966, pp. 302–305), Kragh (1990, pp. 39–43), Darrigol (1992, pp. 337–345), Mehra and Rechenberg (2000–2001, pp. 72–89), Rechenberg (2010, pp. 543–548), and Bacciagaluppi and Valentini (2009, pp. 100–103).

[16] This is the first of several instances where we enhance Jordan's own notation (cf. note 34). In *Neue Begründung* I, Jordan used different letters for quantities and their values. We almost always use the same letter

Begründung I, he clearly labored under the illusion that it would be relatively straightforward to generalize his formalism to cover quantities with wholly or partly discrete spectra as well. In *Neue Begründung* II (Jordan 1927g), he would discover that such a generalization is highly problematic (see Section 16.2.3). Blissfully unaware of these complications for the time being and following Pauli's lead, Jordan (1927b, p. 811) interpreted $|\varphi(a,b)|^2 da$ as the conditional probability $\Pr(a|b)$ for finding a value between a and $a + da$ for $\hat{a}$ given that the system under consideration had been found to have the value b for the quantity $\hat{b}$.[17]

Eigenfunctions $\psi_E(q)$ with eigenvalues E (solutions of the time-independent Schrödinger equation for some Hamiltonian $\hat{H}$ of a one-dimensional system in configuration space) are examples of Jordan's probability amplitudes $\varphi(a,b)$. The quantities $\hat{a}$ and $\hat{b}$ in this case are the position $\hat{q}$ and the Hamiltonian $\hat{H}$, respectively. Hence,

$$|\psi_E(q)|^2 dq = |\varphi(q,E)|^2 dq \tag{16.17}$$

gives the conditional probability $\Pr(q|E)$ that $\hat{q}$ has a value between q and $q + dq$ given that $\hat{H}$ has the value E. This is the special case of Jordan's interpretation given in Pauli's (1927) paper on gas degeneracy (in f dimensions).

As mentioned, Jordan took an axiomatic approach, reflecting the influence of the mathematical tradition in Göttingen (Lacki 2000). In fact, Jordan had been Richard Courant's assistant before becoming Born's. In his *Neue Begründung* papers, Jordan began with a series of postulates about his probability amplitudes and the rules they ought to obey—the formulation and even the number of these postulates varied (see Sections 16.2.2–16.2.4)—and then developed a formalism realizing these postulates.

A clear description of the task at hand can be found in the introduction of Hilbert, von Neumann, and Nordheim's paper on Jordan's theory (to which we return in Section 16.2.3):

> One imposes certain physical requirements on these probabilities, which are suggested by earlier experience and developments, and the satisfaction of which calls for certain relations between the probabilities. Then, secondly, one searches for a simple analytical apparatus in which quantities occur that satisfy these relations exactly (Hilbert, von Neumann, and Nordheim 1928, pp. 2–3).

The quantities satisfying the relations that Jordan postulated for his probability amplitudes are the transformation "matrices" taking pride of place in Dirac's (1927a) version of the theory, where we put "matrices" in scare quotes because their "rows" and "columns" are labeled by continuous variables. We already saw an example of this in Section 16.1, where we examined the interpretation of the transformation matrix $S_{\alpha\beta}$ in

for a quantity and its values and use a "hat" to distinguish the former from the latter. The main exception is the Hamiltonian $\hat{H}$ and the energy eigenvalues E. Dirac (1927a) used primes to distinguish the value of a quantity from the quantity itself.

[17] The now standard notation $\Pr(a|b)$ is not used in the sources we discuss. At this point, we suppress another wrinkle, the "supplementary amplitude" (*Ergänzungsamplitude*; see Section 16.2.2).

Heisenberg's (1926d) note on fluctuations. As Heisenberg wrote to Jordan on November 24, 1926, after Jordan had told him about the paper he was working on:[18]

> I hope that what's in your paper isn't exactly the same as what's in a paper Dirac did here. Dirac's basic idea is that the physical meaning of $S_{\alpha\beta}$, given in my note on fluctuations, can greatly be generalized, so much so that it covers all physical applications of quantum mechanics there have been so far, and, according to Dirac, all there ever will be (AHQP microfilm, quoted in Duncan and Janssen 2013, p. 179).

It is easy to understand why Heisenberg would be pleased with this development. He felt strongly that the interpretation of the quantum formalism should naturally emerge from matrix mechanics without any appeal to wave mechanics. For this reason, he initially disliked Born's probabilistic interpretation of the wave function (as well as, one may add, Bohr's concept of complementarity, with its emphasis on wave–particle duality[19]). What Heisenberg saw as Dirac's generalization of his own work showed that the probability interpretation could either be given in terms of Schrödinger wave functions or in terms of transformation matrices familiar from matrix mechanics. As he went on to explain to Jordan:

> S is the solution of a transformation to principal axes and also of a differential equation à la Schrödinger, though by no means always in q-space. One can introduce matrices of a very general kind, e.g., S with indices $S(q, E)$. The $S(q, E)$ that solves Born's transformation to principal axes in Qu.M. II (Ch. 3, Eq. (13) [a reference to Born, Heisenberg, and Jordan (1926, p. 351); for discussion, see Section 12.3.3] is Schrödinger's $S(q, E) = \psi_E(q)$ (AHQP microfilm, quoted in Duncan and Janssen 2013, p. 179).

This connection between transformation matrices and wave functions is also clearly stated in Dirac's paper:

> *The eigenfunctions of Schrödinger's wave equation are just the transformation functions ... that enable one to transform from the (q) scheme of matrix representation to a scheme in which the Hamiltonian is a diagonal matrix* (Dirac 1927a, p. 635; emphasis in the original).

Wave functions become probability amplitudes in the Jordan–Dirac theory, so the connection between transformation matrices and wave functions also provides a connection between transformation matrices and probability amplitudes. That connection is illustrated with a simple example taken from lecture notes by Dirac.[20] This example shows

[18] Heisenberg made sure, both at the time and in later recollections, to mention his own contribution to the development of the Dirac–Jordan statistical transformation theory. He slipped in a reference to his note on fluctuations in his discussion of the theory in the introduction of a paper on resonance phenomena submitted December 22, 1926, before his note on fluctuations or the papers by Dirac (1927a) and Jordan (1927b) were even published (Heisenberg 1927a, p. 240, note 1). He likewise referred to his "own considerations about fluctuations" when he mentioned the statistical transformation theory in his contribution to the memorial volume for Pauli (Heisenberg 1960, p. 44; cf. Duncan and Janssen 2013, pp. 179–180).

[19] See Section 16.3.

[20] AHQP microfilm, "Notes for Dirac's first lecture course on quantum mechanics, 115 pp. [October 1927 ?]" (Kuhn et al. 1967, p. 32; cf. Darrigol 1992, p. 344).

how one of Jordan's postulates for his probability amplitudes is satisfied when those amplitudes are set equal to the appropriate transformation matrices.

One of the features Jordan saw as characteristic of quantum mechanics, and which he therefore included among his postulates, was that the usual addition and multiplication rules of basic probability theory apply to his probability *amplitudes* rather than to the probabilities themselves. Jordan (1927b, p. 812) used the phrase "interference of probabilities" for this feature. He once again credited Pauli with the name for this phenomenon, even though Born (1926b, p. 804) had already talked about the "interference of ... 'probability waves'" in his paper on quantum collisions. Section 16.2.2 discusses how Jordan's postulate about the addition and multiplication of probability amplitudes eventually boils down to the requirement that the amplitudes $\varphi(a, c)$, $\psi(a, b)$ and $\chi(b, c)$ for the quantities $\hat{a}$, $\hat{b}$, and $\hat{c}$ with purely continuous spectra satisfy the relation

$$\varphi(a, c) = \int db\, \psi(a, b)\, \chi(b, c). \tag{16.18}$$

It is easy to see intuitively, though harder to prove, that this relation is indeed satisfied if these three amplitudes are equated with transformation "matrices".

Dirac (1927a, p. 630) introduced the compact notation (ξ'/α') for the transformation matrix from the α-basis to the ξ-basis (from now on, we drop the scare quotes).[21] The absolute square of this matrix, $|(\xi'/\alpha')|^2$, times $d\xi'$ then gives the probability that the quantity ξ (or $\hat{\xi}$ in our notation) has a value between ξ' and $\xi' + d\xi'$ given that the quantity α (or $\hat{\alpha}$ in our notation) has the value α'. Although the notation follows Dirac's, this formulation of the probability interpretation follows Jordan (1927b, p. 811).[22] The probability amplitudes $\varphi(a, c)$, $\psi(a, b)$ and $\chi(b, c)$ are thus set equal to the transformation matrices (a/c), (a/b), and (b/c), respectively. These matrices relate wave functions in a-, b-, and c-space representation to one another. From

$$\psi(a) = \int db\, (a/b)\, \psi(b), \quad \psi(b) = \int dc\, (b/c)\, \psi(c), \tag{16.19}$$

the usual formula for the product of a matrix and a vector (with an integral rather than a sum over the relevant index), it follows that, for all ψ,

$$\psi(a) = \iint db\, dc\, (a/b)\, (b/c)\, \psi(c). \tag{16.20}$$

[21] Dirac actually used the notation $(.|.)$ in his handwritten letters and manuscripts during this period, but apparently did not mind that this was rendered as $(./.)$ in print. It is tempting to read $(./.)$ as $\langle .|. \rangle$, that is, as the standard bra-ket notation for the inner product of two vectors in Hilbert space. Keep in mind, however, that the Hilbert-space formalism had not been introduced yet. In fact, Dirac would not break up his brackets into bras and kets until well over a decade later (Dirac 1939, cf. Section C.5.5). For discussion, see Borrelli (2010).

[22] For careful discussion of Dirac's (1927a, secs. 6–7, pp. 637–641) own formulation, see Darrigol (1992, pp. 342–343).

Comparing this expression to $\psi(a) = \int dc\,(a/c)\,\psi(c)$, we see that

$$(a/c) = \int db\,(a/b)\,(b/c), \tag{16.21}$$

in accordance with Eq. (16.18), which Jordan presented as a consequence of one of his postulates.[23]

This brings us to an important difference between Jordan's version and Dirac's version of statistical transformation theory. For Dirac, the *transformation* element was primary, for Jordan the *statistical* element was. Most of Dirac's paper is devoted to the development of the formalism that allowed him to represent the laws of quantum mechanics in different yet equivalent ways (Dirac 1927a, secs. 2–5, pp. 624–637). The probability interpretation of the transformation matrices is then grafted onto this formalism in the last two sections (Dirac 1927a, secs. 6–7, pp. 637–641). Jordan's paper begins with some axioms about probability (Jordan 1927b, Pt. I, secs. 1–2, pp. 809–816). It is then shown that those can be implemented by equating probability amplitudes with transformation matrices, or, to be more precise, with integral kernels of canonical transformations (Jordan 1927b, Pt. 2, secs. 1–6, pp. 816–835; see Section 16.2.2).

Heisenberg strongly preferred Dirac's version of the theory over Jordan's. For one thing, he disliked Jordan's axiomatic approach. As he told Kuhn in his interview for the AHQP project in the early 1960s:

> Jordan used this transformation theory for deriving what he called the axiomatics of quantum theory ... This I disliked intensely ... Dirac kept within the spirit of quantum theory while Jordan, together with Born, went into the spirit of the mathematicians (AHQP interview with Heisenberg, session 11, pp. 7–8).

Others were less put off by Jordan's axiomatic approach. As Ehrenfest told Jordan, who still remembered it with amusement when interviewed decades later: "Since you wrote the paper axiomatically, that only means that one has to read it back to front."[24] In that case, one would encounter probability amplitudes in the guise of transformation matrices first, as in Dirac's version of the theory.

Heisenberg had more serious reservations about Jordan's version of the theory, which initially made it difficult for him even to understand *Neue Begründung* I. He expressed his frustration in a letter to Pauli a few weeks after the paper was published. After praising Jordan's (1927d) habilitation lecture, which had just appeared in *Die Naturwissenschaften*,[25] he wrote:

[23] Dirac lecture notes, p. 7 (cf. note 20). At this point in these notes, Dirac actually considered a purely discrete rather than a purely continuous case and thus used sums instead of integrals.

[24] AHQP interview with Jordan, session 3, p. 17; quoted by Mehra and Rechenberg (2000–2001, p. 69) and Duncan and Janssen (2009, p. 360).

[25] It was subsequently translated by Oppenheimer and reprinted in *Nature*.

> I could not understand Jordan's big paper [i.e., *Neue Begründung* I]. The "postulates" are so intangible and undefined, I cannot make heads or tails of them (Heisenberg to Pauli, February 5, 1927; Pauli 1979, Doc. 153).[26]

About a month later, Heisenberg wrote to Jordan himself, telling him that he was working on "on a fat paper" [*an einer dicken Arbeit*][27]

> that one might characterize as physical commentary on your paper and Dirac's. You should not hold it against me that I consider this necessary. The essence from a mathematical point of view is roughly that it is possible with your mathematics to give an exact formulation of the case in which p and q are *both* given with a certain *accuracy* (Heisenberg to Jordan, March 7, 1927, AHQP microfilm; emphasis in the original).[28]

Heisenberg now had a clearer picture of Jordan's approach and of how it differed from the approach of Dirac, who had meanwhile left Copenhagen and had joined Born and Jordan in Göttingen. After registering some disagreements with Dirac, Heisenberg turned to his disagreements with Jordan:

> With you I don't quite agree in that, in my opinion, the relation $\int \varphi(x,y)\,\psi(y,z)\,dy$ [cf. Eq. (16.18) above] has nothing to do with the laws of probability. In all cases in which one can talk about probabilities the usual addition and multiplication of probabilities are valid, *without* "interference." *With* Dirac I believe that it is more accurate to say: all statistics is brought in only through our experiments (Heisenberg to Jordan, March 7, 1927, AHQP microfilm; emphasis in the original).

The reservation expressed in the first two sentences is largely a matter of semantics. Heisenberg did not dispute that the relation in question holds in quantum mechanics (it corresponds to the familiar completeness relation: see note 46). He also did not dispute that this relation describes interference phenomena.[29] What he objected to were the

[26] Even Dirac's (1927) paper, which for modern readers is much more accessible than Jordan's *Neue Begründung* I, was difficult to understand for many of his contemporaries. In a letter of June 16, 1927, Ehrenfest told Dirac that he and his colleagues in Leyden had had a hard time with several of his papers, including the one on transformation theory (Kragh 1990, p. 46).

[27] This is the paper introducing the uncertainty relations (Heisenberg 1927b), which is discussed in Section 16.3.

[28] Heisenberg's (1927b) paper was received by *Zeitschrift für Physik* on March 23, 1927. Shortly before he submitted the paper, he wrote to Jordan for help with a sign error in a derivation in *Neue Begründung* I that he needed for his paper (see note 51). For discussion of Heisenberg's reliance on transformation theory in his uncertainty paper, see Jammer (1966, pp. 326–328), Mehra and Rechenberg (2000–2001, pp. 148–151, 159–161), and Beller (1985, 1999, pp. 91–95).

[29] A simple example, the familiar two-slit experiment, immediately makes that clear. Let the quantity $\hat{c}$ (value c) be the position where photons are emitted; let $\hat{b}$ (values b_1 and b_2) be the positions of the two slits; and let $\hat{a}$ (value a) be the position on the screen where photons are detected. On the assumption (vindicated by modern theory) that the relation also applies to quantities with discrete values (such as $\hat{b}$ in this example), the integral on the right-hand side of Eq. (16.21) reduces to a sum of two terms in this case: $\varphi(a,c) = \psi(a,b_1)\,\chi(b_1,c) + \psi(a,b_2)\,\chi(b_2,c)$. Multiplying the left- and the right-hand side by their complex conjugates, we find that the probability $\Pr(a|c)$ is equal to the sum of the probabilities $\Pr(a\,\&\,b_1|c)$ and $\Pr(a\,\&\,b_2|c)$ *plus* the interference terms $\psi(a,b_1)\,\psi^*(a,b_2)\,\chi(b_1,c)\,\chi^*(b_2,c)$ and $\psi(a,b_2)\,\psi^*(a,b_1)\,\chi(b_2,c)\,\chi^*(b_1,c)$. Of course, as we now know, it is critical that the quantity $\hat{b}$ (which slit the particle went through) is not actually measured in this case.

addition and multiplication rules for probability *amplitudes* from which Jordan derived the relation. This did not materially affect the theory as Jordan (1927b, p. 813) only ever used those rules to derive this particular relation and a closely related one in which we recognize the familiar orthogonality relation (see Eqs. (16.25) and (16.27) and note 46).[30]

The reference to Dirac for Heisenberg's second reservation about Jordan's position is probably to some conversations he had with Dirac in Copenhagen, which also left a clear trace in Dirac's paper. The concluding sentence of Dirac's paper on transformation theory is:

> The notion of probabilities does not enter into the ultimate description of mechanical processes: only when one is given some information that involves a probability ... can one deduce results that involve probabilities (Dirac 1927a, p. 641).[31]

Instead of elaborating on this second reservation, Heisenberg told Jordan that he had just written a 14-page letter about these matters to Pauli and suggested that Jordan have Pauli send him this letter for further details. In this letter, the blueprint for his uncertainty paper, Heisenberg told Pauli:

> One can, like Jordan, say that the laws of nature are statistical. But one can also, and that to me seems considerably more profound, say with Dirac that all statistics are brought in only through our experiments. That we do not know at which position the electron will be at the moment of our experiment, is, in a manner of speaking, only because we do not know the phases [in the classical multiply periodic description via action/angle variables, these are the $\vec{w}$ angle variables, determining the coordinates of the particle at any given time, cf. Eq. (A.167)], if we do know the energy ... and in *this* respect the classical theory would be no different. That we *cannot* come to know the phases without ... destroying the atom is characteristic of quantum mechanics (Heisenberg to Pauli, February 23, 1927; Pauli 1979, Doc. 154, p. 377; emphasis in the original; see also Heisenberg 1927b, p. 177).

So Heisenberg—and with him Dirac—held on to the idea that nature itself is deterministic and that all the indeterminism that quantum mechanics tells us we shall encounter is the result of unavoidable disturbances of nature in our experiments. As he made clear in his habilitation lecture, Jordan (1927d) wanted to keep open the possibility that nature is intrinsically indeterministic, as did Born (1926a, p. 866, 1926b, pp. 826–827).

In his *Neue Begründung* papers, Jordan did not discuss the nature of the probabilities he introduced. He did not even properly define these probabilities. This was done only

[30] Heisenberg elaborated on his criticism of Jordan on this score in his uncertainty paper (Heisenberg 1927b, pp. 183–184, p. 196). Von Neumann (1927a, p. 46) initially followed Jordan (see also Hilbert, von Neumann, and Nordheim 1928, p. 5; cf. note 63), but changed his mind after reading Heisenberg's uncertainty paper (von Neumann 1927b, p. 246) (see Section 17.2).

[31] Quoted and discussed by Kragh (1990, p. 42). Dirac's position, in turn, was undoubtedly influenced by the exchange between Pauli and Heisenberg that led to the latter's uncertainty paper.

by von Neumann (1927b) with the help of the notion of ensembles of systems from which one randomly selects members, a notion developed by Richard von Mises (1928) and published in book form the following year (see Section 17.2).

As far as the issue of determinism versus indeterminism is concerned, Dirac thus stayed closer to classical theory than Jordan. In other respects, however, Jordan stayed closer. Most importantly, Jordan's use of canonical transformations is closer to their use in classical mechanics than Dirac's. In the letter to Jordan from which we quoted at the beginning of this subsection, Dirac clearly identified part of the difference in their use of canonical transformations:

> In your work I believe you considered transformations from one set of dynamical variables to another, instead of a transformation from one scheme of matrices representing the dynamical variables to another scheme representing the same dynamical variables, which is the point of view adopted throughout my paper. The mathematics appears to be the same in the two cases, however (Dirac to Jordan, December 24, 1916, AHQP microfilm).

Traditionally, canonical transformations had been used the way Jordan used them, as transformations to new variables, and not the way Dirac used them, as transformations to new representations of the same variables. Canonical transformations had been central to the development of matrix mechanics. Prior to *Neue Begründung* I, Jordan (1926a, 1926b) had published two important papers on the implementation of canonical transformations in matrix mechanics (Lacki 2004; Duncan and Janssen 2009). Talking about these two papers in his interview with Kuhn for the AHQP project, Jordan said:

> Canonical transformations in the sense of Hamilton–Jacobi were ... our daily bread in the preceding years, so to tie in the new results with those as closely as possible—that was something very natural for us to try (AHQP interview with Jordan, session 4, p. 11).

Jordan's use of canonical transformations in his *Neue Begründung* papers was twofold. First, Jordan tried to show that integral kernels in canonical transformations have all the properties that probability amplitudes must satisfy according to his postulates. Second, he tried to use canonical transformations to derive differential equations for probability amplitudes for arbitrary quantities (such as the time-independent Schrödinger equation for $\varphi(q, E) \equiv \psi_E(q)$) from the trivial differential equations satisfied by the probability amplitude for $\varphi(p, q)$ for some generalized coordinate $\hat{q}$ and its conjugate momentum $\hat{p}$ that he started from.

Both ways in which Jordan relied on canonical transformations in his *Neue Begründung* papers turned out to be problematic and resulted in serious mathematical problems for his version of statistical transformation theory (see Sections 16.2.2–16.2.4). These problems do not affect Dirac's version since he only relied on canonical transformations in a very loose sense. The two versions of the theory, however, do share a number of other mathematical problems, which would only be solved by von Neumann (1927a; see also Chapter 17).

In the rest of this section, we take a closer look at three papers on Jordan's version of statistical transformation theory: *Neue Begründung* I (Jordan 1927b; see Section 16.2.2) Hilbert, von Neumann, and Nordheim (1928; see Section 16.2.3) and *Neue Begründung* II (Jordan 1927g; see Section 16.2.4). We do not cover Dirac's (1927a) paper but highlight some of the differences between his version of the theory and Jordan's. As mentioned, we also use elements of Dirac's notation to enhance Jordan's.

16.2.2 Jordan's "New foundation (*Neue Begründung*) . . . " I

Neue Begründung I (Jordan 1927b) consists of two parts. In Part One (*I. Teil*), consisting of sections 1–2 (pp. 809–816), Jordan laid down the postulates of his theory. In Part Two (*II. Teil*), consisting of sections 3–7 (pp. 816–838), he presented the formalism realizing these postulates. In the abstract of the paper, Jordan announced that his new theory would unify all earlier formulations of quantum theory:

> The four forms of quantum mechanics that have been developed so far—matrix theory, the theory of Born and Wiener, wave mechanics, and q-number theory—are contained in a more general formal theory. Following one of Pauli's ideas, one can base this new theory on a few simple fundamental postulates (*Grundpostulate*) of a statistical nature (p. 809).[32]

Jordan claimed that he could recover both the time-dependent and the time-independent Schrödinger equation as special cases of the differential equations he derived for the probability amplitudes central to his formalism. This is the basis for his claim that wave mechanics can be subsumed under his new formalism. Nowhere in the paper did he show explicitly how matrix mechanics is to be subsumed under the new formalism. Perhaps Jordan felt that this did not require a special argument as the new formalism had grown naturally out of matrix mechanics and his own contributions to it (Jordan 1926a, 1926b). Like Dirac's (1927a) own version of statistical transformation theory, Jordan's version can be seen as a natural extension of Dirac's (1925) q-number theory. It is only toward the end of his paper (sec. 6) that Jordan turned to the operator theory of Born and Wiener (1926). In our discussion of *Neue Begründung* I, we omit this section along with some mathematically intricate parts of sections 3 and 5 that are not necessary for understanding the paper's overall argument. We also do not cover the final section of Jordan's paper, which deals with quantum jumps.[33]

Although we do not describe Jordan's unification of the various forms of quantum theory in any detail, we will cover von Neumann's (1927a) criticism of the Dirac–Jordan way of proving the equivalence of matrix mechanics and wave mechanics as a prelude to his own proof based on the isomorphism of two distinct instantiations of Hilbert space, the space l^2 of square-summable sequences, and the space L^2 of square-integrable

[32] Unless noted otherwise, all page references in this subsection refer to *Neue Begründung* I (Jordan 1927b).

[33] Jordan (1927a) had written a paper on this topic before, which is similar to the considerations in Heisenberg's (1926d) note on fluctuations and in Pauli's letter to Heisenberg of October 19, 1926 (Pauli 1979, Doc. 143) discussed in Section 16.1 (see also Duncan and Janssen 2013, pp. 181–182).

functions (see Chapter 17 and Appendix C). Here, our discussion of *Neue Begründung* I focuses on the portion of Jordan's paper that corresponds to the last sentence of the abstract, which promises a statistical foundation of quantum mechanics. Laying this foundation actually takes up most of the paper (secs. 1–2, 4–5).

Referring to a footnote in a forthcoming paper by Pauli (1927, p. 83, note), Jordan proposed the following interpretation of the energy eigenfunctions $\varphi_n(q)$ (where n labels the different energy eigenvalues) of a system (in one dimension): "If $\varphi_n(q)$ is normalized, then $|\varphi_n(q)|^2 dq$ gives the probability that, if the system is in the state n, the coordinate $\hat{q}$ has a value between q and $q + dq$" (p. 811; "hat" on $\hat{q}$ added). A probability amplitude such as this one, connecting position and energy, can be introduced for any two quantities.

In *Neue Begründung* I, Jordan focused on quantities with completely continuous spectra. Suppressing a slight complication to be addressed shortly (what Jordan calls the "supplementary amplitude" [*Ergänzungsamplitude*]), we can state the basic idea as follows: for two quantities $\hat{x}$ and $\hat{y}$ that can take on a continuous range of values x and y, respectively, there is a complex probability amplitude $\mathcal{A}(x, y)$ such that $|\mathcal{A}(x, y)|^2 \, dx$ gives the probability that $\hat{x}$ has a value between x and $x + dx$ given that $\hat{y}$ has the value y.[34]

Many expressions and equations in Jordan's paper turn into easily recognizable ones in modern quantum mechanics if for $\mathcal{A}(x, y)$ we read $\langle x|y \rangle$, the modern Dirac notation (cf. note 21) for the inner product of the eigenvectors $|x\rangle$ and $|y\rangle$ of the operators $\hat{x}$ and $\hat{y}$ in Hilbert space. In hindsight, one recognizes that a realization of Jordan's postulates can, in fact, be obtained by setting $\mathcal{A}(x, y) = \langle x|y \rangle$. Of course, Jordan and Dirac formulated their statistical transformation theory *before* von Neumann (1927a) introduced Hilbert space. Moreover, even after the introduction of Hilbert space, a number of serious mathematical hurdles had to be cleared before probability amplitudes could be identified with inner products. In particular, for quantities $\hat{x}$ and $\hat{y}$ with continuous spectra, such as the position or momentum of a particle in an infinitely extended region, the "vectors" $|x\rangle$ and $|y\rangle$ are *not* elements of Hilbert space, although an inner product $\langle x|y \rangle$ can be defined in a generalized sense (as a distribution) as an integral of products of continuum normalized wave functions, as is routinely done in modern quantum mechanics.

With the advent (beginning in the late 1930s) of the theory of distributions, the Dirac delta function (which up to that point had been regarded as a mathematical monstrosity) also made sense. Because of these mathematical problems, as we shall see in Chapter 17, von Neumann did not simply set probability amplitudes equal to inner products, but rather took an entirely different approach to formulating the basic rules of quantum

[34] Recall that this is our notation (cf. note 16). There are two key differences between Jordan's own notation and ours: 1. Jordan uses different Greek letters for the amplitudes of different quantities. We use $\mathcal{A}$ for all of them; 2. Jordan frequently uses different letters for a quantity and its value (in other cases he uses the same letter but adds a subscript "0" or a prime to distinguish values from quantities). We use the same letter for a quantity and its value and distinguish between the two by putting a "hat" on the letter when it represents a quantity.

mechanics in his Hilbert space formalism. Here, we occasionally help ourselves to elements of the Hilbert space formalism in Dirac notation, either as a shortcut for Jordan's more cumbersome derivations or to help modern readers recognize elements familiar from modern quantum mechanics in Jordan's formalism. For the most part, however, we deviate from Jordan's own presentation only in that we use an enhanced version of his confusing notation.

In the introduction of his paper (pp. 809–810), Jordan motivated his project by raising the question whether the connection between the time-independent Schrödinger equation and the classical Hamilton–Jacobi equation[35] for a Hamiltonian H that is a function of some coordinate q and its conjugate momentum p could be generalized to equations for a new Hamiltonian that is a function of new coordinates and momenta (Q, P) related to old ones (q, p) through a canonical transformation. He then presented Pauli's probabilistic interpretation of solutions of the time-independent Schrödinger equation and proposed to generalize that interpretation in the way indicated above.

Reflecting the influence of Göttingen's mathematicians, Jordan presented his theory axiomatically. At the end of the first section, he listed two postulates, labeled I and II (p. 811). Only two pages later, in section 2, "Statistical foundation of quantum mechanics," these two postulates are superseded by a new set of four postulates, labeled A through D (pp. 813–814).[36] In *Neue Begründung* II, Jordan (1927g, p. 6) presented yet another set of postulates, three this time, labeled I through III (see Section 16.2.4).[37] The exposition of Jordan's theory by Hilbert, von Neumann, and Nordheim (1928), written in between *Neue Begründung* I and II, starts from six "physical axioms" (see Section 16.2.3). We start from Jordan's four postulates of *Neue Begründung* I, which we paraphrase and comment on below.

Postulate A. For two mechanical quantities $\hat{q}$ and $\hat{\beta}$ that stand in a definite kinematical relation to one another there are two complex-valued functions, an amplitude $\mathcal{A}(q, \beta)$ and a "supplementary amplitude" (*Ergänzungsamplitude*) $\tilde{\mathcal{A}}(q, \beta)$,[38] such that

$$\mathcal{A}(q, \beta)\, \tilde{\mathcal{A}}^{*}(q, \beta)\, dq \tag{16.22}$$

(where the asterisk indicates complex conjugation) gives the probability of finding a value between q and $q + dq$ for $\hat{q}$ given that $\hat{\beta}$ has the value β.

[35] Jordan gives three equations: Eq. (1) is the time-independent Schrödinger equation; Eq. (2) gives the relation $S = \varepsilon \ln \psi$ (with $\varepsilon = \hbar/i$) between the Hamilton–Jacobi function S and the Schrödinger wavefunction ψ (see our Eq. (14.61)); Eq. (3) expresses Eq. (1) in terms of S (see our Eq. (14.60)). Jordan notes that in the limit that $h \to 0$, this last equation turns into the classical Hamilton–Jacobi equation.

[36] In the short version of *Neue Begründung* I presented to the Göttingen Academy on January 14, 1927, Jordan (1927c, p. 162) only introduced postulates I and II.

[37] In his overview of recent developments in quantum mechanics in *Die Naturwissenschaften*, Jordan (1927h, Pt. 2, p. 648), after explaining the basic notion of a probability amplitude (cf. Postulate A), listed only two postulates, or axioms as he now called them, namely "the assumption of probability interference" (cf. Postulate C) and the requirement that there is a canonically conjugate quantity $\hat{p}$ for every quantum-mechanical quantity $\hat{q}$ (cf. Postulate D).

[38] Jordan's own notation for these two quantities is $\varphi(q, \beta)$ and $\psi(q, \beta)$, respectively. We use $\mathcal{A}$ for all amplitudes and $\tilde{\mathcal{A}}$ for all supplementary amplitudes (cf. note 34).

Comments: As becomes clear later on in the paper, "mechanical quantities that stand in a definite kinematical relation to one another" are quantities that can be written as functions of some set of generalized coordinates and their conjugate momenta. In his original postulate I, Jordan wrote that $\mathcal{A}(q, \beta)$ "is independent of the mechanical nature (the Hamiltonian) of the system and is determined only by the kinematical relation between $\hat{q}$ and $\hat{\beta}$" (p. 162; "hats" added).[39] Hilbert, von Neumann, and Nordheim (1928, p. 5) made this into a separate postulate, their axiom V: "A further physical requirement is that the probabilities only depend on the functional nature of the quantities $F_1(pq)$ and $F_2(pq)$, i.e., on their kinematical connection [*Verknüpfung*], and not for instance on additional special properties of the mechanical system under consideration, such as, for example, its Hamiltonian."[40]

It turns out that for all quantities represented, in modern terms, by Hermitian operators, the amplitude $\mathcal{A}(q, \beta)$ and the supplementary amplitude (*Ergänzungsamplitude*) $\tilde{\mathcal{A}}(q, \beta)$ are equal to one another. At this point, however, Jordan wanted to leave room for quantities represented by non-Hermitian operators. This is directly related to the central role of canonical transformations in his formalism. As Jordan (1926a, 1926b) had already found the year before, canonical transformations need not be unitary and therefore do not always preserve the Hermiticity of the conjugate variables one starts from (Duncan and Janssen 2009). The *Ergänzungsamplitude* does not appear in the presentation of Jordan's formalism by Hilbert, von Neumann, and Nordheim (1928; see Section 16.2.3). In *Neue Begründung* II, Jordan (1927g, p. 3) restricted himself to Hermitian quantities and silently dropped the *Ergänzungsamplitude* (see Section 16.2.4).[41] In anticipation of this, we shall simply set $\tilde{\mathcal{A}}(q, \beta) = \mathcal{A}(q, \beta)$ everywhere. In that case, the probability in Eq. (16.22) is given by:

$$\Pr(\hat{q} \in (q, q + dq) \mid \hat{\beta} = \beta) = \left|\mathcal{A}(q, \beta)\right|^2 dq. \tag{16.23}$$

The absolute square of the probability amplitude for quantities with a continuous spectrum therefore corresponds, strictly speaking, to a *probability density.*

Postulate B. The probability amplitude $\mathcal{A}(\beta, q)$ for finding that $\hat{\beta}$ has the value β given that $\hat{q}$ has the value q is the complex conjugate of the probability amplitude $\mathcal{A}(q, \beta)$ for finding that $\hat{q}$ has the value q given that $\hat{\beta}$ has the value β:[42]

$$\mathcal{A}(\beta, q) = \mathcal{A}^{\star}(q, \beta) \tag{16.24}$$

[39] In his AHQP interview with Jordan, Kuhn emphasized the importance of this aspect of Jordan's formalism: "The terribly important step here is throwing the particular Hamiltonian function away and saying that the relationship is only in the kinematics" (session 3, p. 15).

[40] With $\varphi(q, \beta) = \langle q|\beta\rangle$, the statement about the kinematical nature of probability amplitudes translates into the observation that they depend only on the inner-product structure of Hilbert space and not on the Hamiltonian governing the time evolution of the system under consideration.

[41] For detailed discussion of Jordan's notion of an *Ergänzungsamplitude*, see Duncan and Janssen (2013, section 2.4, pp. 217–221).

[42] Jordan writes $\varphi(q, \beta)$ and $\overline{\varphi}(\beta, q)$ for $\mathcal{A}(q, \beta)$ and $\tilde{\mathcal{A}}(q, \beta)$, respectively (cf. notes 34 and 38).

(p. 13, Eq. (11)).[43] This implies a symmetry property of the probabilities themselves: the probability density of finding a value of β for $\hat{\beta}$ given the value q for $\hat{q}$ is equal to the probability density of finding a value q for $\hat{q}$ given the value β for $\hat{\beta}$: $|\mathcal{A}(\beta, q)|^2 = |\mathcal{A}(q, \beta)|^2$.

Postulate C. The probabilities combine through interference. In section 1, Jordan already introduced the phrase "interference of probabilities" (p. 812) to capture the striking feature in his quantum formalism that the probability *amplitudes* rather than the probabilities themselves follow the usual composition rules for probabilities.[44] Let F_1 and F_2 be two outcomes [*Tatsachen*] for which the amplitudes are φ_1 and φ_2. If F_1 and F_2 are mutually exclusive, $\varphi_1 + \varphi_2$ is the amplitude for the outcome "F_1 or F_2". If F_1 and F_2 are independent, $\varphi_1 \varphi_2$ is the amplitude for the outcome "F_1 and F_2".

Consequence. Let $\mathcal{A}(q, \beta)$ be the probability amplitude for the outcome F_1 that $\hat{q} = q$ given that $\hat{\beta} = \beta$. Let $\mathcal{A}(Q, q)$ be the probability amplitude for the outcome F_2 that $\hat{Q} = Q$ given that $\hat{q} = q$. Since F_1 and F_2 are independent, Jordan's multiplication rule tells us that the probability amplitude for "F_1 and F_2" is given by the product $\mathcal{A}(Q, q)\,\mathcal{A}(q, \beta)$. Now let $\mathcal{A}(Q, \beta)$ be the probability amplitude for the outcome F_3 that $\hat{Q} = Q$ given that $\hat{\beta} = \beta$. According to Jordan's addition rule, this amplitude is equal to the "sum" of the amplitudes for "F_1 and F_2" for all different values of q. Since $\hat{q}$ has a continuous spectrum, this "sum" is actually an integral. The probability amplitude for F_3 is thus given by (p. 813, Eq. (13)):

$$\mathcal{A}(Q, \beta) = \int \mathcal{A}(Q, q)\,\mathcal{A}(q, \beta)\,dq. \tag{16.25}$$

Special case. If $\hat{Q} = \hat{\beta}$, the amplitude $\mathcal{A}(\beta', \beta'')$ becomes the Dirac delta function. Jordan introduced the notation

$$\delta_{\beta'\beta''} = \begin{cases} 1 \text{ for } \beta' = \beta'' \\ 0 \text{ for } \beta' \neq \beta'' \end{cases} \tag{16.26}$$

(p. 814. Eq. (16)). In a footnote he conceded that this is mathematically dubious: "This in the case of a continuously variable β mathematically not very correct notation [*nicht sehr korrekte Schreibweise*] may be seen as an expression of a well-known mathematical procedure" (p. 814). In *Neue Begründung* II, Jordan (1927g, p. 5) used the delta function that Dirac (1927a, pp. 625–627) had meanwhile introduced. Here and in what follows we give Jordan (and Dirac) the benefit of the doubt and assume the normal properties of the delta function.[45]

[43] This property follows immediately if we set $\mathcal{A}(q, \beta) = \langle q|\beta\rangle$.
[44] Recall, however, Heisenberg's criticism of this aspect of Jordan's work (see note 30).
[45] For a brief history of the delta function focusing on its role in quantum mechanics, see, for example, Jammer (1966, pp. 301–302, pp. 313–314).

Since $\mathcal{A}(\beta', q)$ is just the complex conjugate of $\mathcal{A}(q, \beta')$, the amplitude $\mathcal{A}(\beta', \beta'')$ is given by (p. 814, Eq. (17)):[46]

$$\mathcal{A}(\beta', \beta'') = \int \mathcal{A}^\star(q, \beta')\, \mathcal{A}(q, \beta'')\, dq = \delta_{\beta'\beta''} \tag{16.27}$$

Postulate D. For every $\hat{q}$ there is a conjugate momentum $\hat{p}$. Before stating this postulate, Jordan offered a new definition of what it means for $\hat{p}$ to be conjugate to $\hat{q}$. If

$$\mathcal{A}(p, q) = e^{-ipq/\hbar} \tag{16.28}$$

(p. 814, Eq. (18), where we replaced Jordan's $\varepsilon \equiv h/2\pi i$ by $\hbar/\imath$) is the amplitude for finding $\hat{p} = p$ given that $\hat{q} = q$, then $\hat{p}$ is conjugate to $\hat{q}$.

Anticipating a special case of the uncertainty principle (cf. note 28), Jordan noted that Eq. (16.28) implies that "[f]or a given value of $\hat{q}$ all possible values of $\hat{p}$ are *equally probable*" (p. 814). His habilitation lecture published about two months before Heisenberg's (1927b) paper contains a statement that is even more suggestive of the uncertainty principle:

> With different experimental setups one can observe different coordinates. However, with a particular setup one can, at best, observe particular coordinates of an atom exactly, while it will then be impossible in that setup to observe the corresponding momenta exactly (Jordan 1927d, p. 108, note 1, in the German version; this note was omitted in the English translation).

The same idea can already be found in the letter from Pauli to Heisenberg of October 19, 1926 (Pauli 1979, Doc. 143). While Jordan may not have seen this letter, he and Pauli discussed these matters at length in person (see Section 16.2.1).

For $\hat{p}$s and $\hat{q}$s with completely continuous spectra, Jordan's definition of when $\hat{p}$ is conjugate to $\hat{q}$ is equivalent to the standard one that the operators $\hat{p}$ and $\hat{q}$ satisfy the commutation relation $[\hat{p}, \hat{q}] \equiv \hat{p}\,\hat{q} - \hat{q}\,\hat{p} = \hbar/i$ (p. 815, Eq. (25)). This equivalence, however, presupposes that we associate $\hat{q}$ with "multiplying by q" and $\hat{p}$ with "$(\hbar/\imath)\,\partial/\partial q$", as we routinely do in Schrödinger's wave mechanics in configuration space. Rather than simply assuming these associations, however, Jordan tried to derive them from his postulates. This is the goal of the manipulations in Eqs. (19ab)–(24) on pp. 814–815, immediately following postulate D, under the subheading "Consequences" (*Folgerungen*).

[46] Setting probability amplitudes equal to inner products of vectors in Hilbert space and using Dirac notation, we see that Eqs. (16.25) and (16.27) turn into the familiar *completeness* and *orthogonality* relations:

$$\langle Q|\beta\rangle = \int \langle Q|q\rangle\langle q|\beta\rangle\, dq, \quad \langle \beta'|\beta''\rangle = \int \langle \beta'|q\rangle\langle q|\beta''\rangle\, dq = \delta(\beta' - \beta'').$$

Since the eigenvectors $|q\rangle$ of the operator $\hat{q}$ are not in Hilbert space, the spectral theorem, first proven by von Neumann (1927a), is required for the use of the resolution of the unit operator $\hat{1} = \int dq|q\rangle\langle q|$ (see Chapter 17 and Appendix C).

Our reconstruction of Jordan's rather convoluted argument will show that the argument, as it stands, fails, but a slightly amended version of it works (cf. Duncan and Janssen 2013, pp. 202–207). The probability amplitude in Eq. 16.28 trivially satisfies the following pair of equations (p. 814, Eqs. (19ab)):

$$\left(p + \frac{\hbar}{i}\frac{\partial}{\partial q}\right) \mathcal{A}(p, q) = 0, \tag{16.29}$$

$$\left(\frac{\hbar}{i}\frac{\partial}{\partial p} + q\right) \mathcal{A}(p, q) = 0. \tag{16.30}$$

Following Jordan (pp. 814–815, Eqs. 20–22), we now define the map T, which takes functions f of p and turns them into functions Tf of Q (the value of a new quantity $\hat{Q}$ with a fully continuous spectrum):

$$T: \; f(p) \; \rightarrow \; (Tf(p))(Q) \equiv \int \mathcal{A}(Q, p) f(p) \, dp. \tag{16.31}$$

In other words (p. 814, Eq. (21)):

$$T \ldots = \int dp \, \mathcal{A}(Q, p) \ldots \tag{16.32}$$

For the special case that $f(p) = \mathcal{A}(p, q)$, we get:

$$(T\mathcal{A}(p, q))\,(Q) = \int \mathcal{A}(Q, p)\, \mathcal{A}(p, q) \, dp = \mathcal{A}(Q, q), \tag{16.33}$$

where in the last step we used Eq. (16.25), Jordan's version of the familiar completeness relation (cf. note 46). So T maps $\mathcal{A}(p, q)$ onto $\mathcal{A}(Q, q)$ (p. 815, Eq. 22):[47]

$$\mathcal{A}(Q, q) = T\mathcal{A}(p, q). \tag{16.34}$$

[47] At this point, Jordan's notation, $\varphi(x, y) = T.\rho(x, y)$, gets particularly confusing as the x on the left-hand side and the x on the right-hand side refer to values of different quantities. The same is true for the equations that follow (Eqs. 23ab, 24ab).

Likewise, we define the inverse map T^{-1}, which takes functions F of Q and turns them into functions $T^{-1}F$ of p:[48]

$$T^{-1}: \ F(Q) \ \rightarrow \ (T^{-1}F(Q))(p) \equiv \int \mathcal{A}^{\star}(Q,p)\, F(Q)\, dQ. \tag{16.35}$$

In other words,[49]

$$T^{-1} \ldots = \int dQ\, \mathcal{A}^{\star}(Q,p) \ldots \tag{16.36}$$

For the special case that $F(Q) = \mathcal{A}(Q,q)$ we get (again by Eq. (16.25), Jordan's version of completeness):

$$\left(T^{-1}\mathcal{A}(Q,q)\right)(p) = \int \mathcal{A}^{\star}(Q,p)\, \mathcal{A}(Q,q)\, dQ = \mathcal{A}(p,q), \tag{16.37}$$

or, more succinctly,

$$\mathcal{A}(p,q) = T^{-1}\mathcal{A}(Q,q). \tag{16.38}$$

Applying T to the left-hand side of Eq. (16.29), we find:

$$T\left(\left(p + \frac{\hbar}{i}\frac{\partial}{\partial q}\right)\mathcal{A}(p,q)\right) = Tp\,\mathcal{A}(p,q) + \frac{\hbar}{i}\frac{\partial}{\partial q}\, T\mathcal{A}(p,q) = 0, \tag{16.39}$$

where we used that differentiation with respect to q commutes with applying T (which only affects the functional dependence on p). Using Eq. (16.38) for the first occurrence

[48] To verify that T^{-1} is indeed the inverse of T, we take $F(Q)$ in Eq. (16.35) to be $(Tf)(Q)$ in Eq. (16.31). In that case we get:

$$\begin{aligned}(T^{-1}Tf)(p) &= \int \mathcal{A}^{\star}(Q,p)\left(\int \mathcal{A}(Q,p')f(p')\,dp'\right) dQ \\ &= \iint \mathcal{A}^{\star}(Q,p)\,\mathcal{A}(Q,p')\,f(p')\,dQ\,dp' \\ &= \int \delta_{pp'}\, f(p')\,dp' = f(p),\end{aligned}$$

where in the last line we used Eq. (16.27), Jordan's version of the familiar orthogonality relation, to set $\int \mathcal{A}^{\star}(Q,p)\,\mathcal{A}(Q,p')\,dQ = \delta_{pp'}$. This proof becomes much easier to read for a modern reader if, for convenience, one sets $\mathcal{A}(Q,p) = \langle Q|p\rangle$.

[49] Jordan used the *Ergänzungsamplitude* to represent T^{-1} in this form (p. 815, note).

of $\mathcal{A}(p,q)$ in this equation and Eq. (16.34) for the second, we can rewrite Eq. (16.39) as (p. 815, Eq. (23a)[50]):

$$\left(T p\, T^{-1} + \frac{\hbar}{i}\frac{\partial}{\partial q}\right)\mathcal{A}(Q,q) = 0. \tag{16.40}$$

Similarly, applying T to the left-hand side of Eq. (16.30), we find:

$$T\left(\left(\frac{\hbar}{i}\frac{\partial}{\partial p} + q\right)\mathcal{A}(p,q)\right) = T\frac{\hbar}{i}\frac{\partial}{\partial p}\mathcal{A}(p,q) + q\,T\mathcal{A}(p,q) = 0, \tag{16.41}$$

where we used that multiplying by q commutes with applying T. Once again using Eqs. (16.34) and (16.38), we can rewrite this as (p. 815, Eq. (23b)[51]):

$$\left(T\frac{\hbar}{i}\frac{\partial}{\partial p}\, T^{-1} + q\right)\mathcal{A}(Q,q) = 0. \tag{16.42}$$

Eqs. (16.40) and (16.42) gave Jordan a representation of the quantities $\hat{p}$ and $\hat{q}$ in the Q-basis. The identification of $\hat{p}$ in the Q-basis is straightforward. The quantity p in Eq. (16.29) turns into the quantity $T p\, T^{-1}$ in Eq. (16.40). This is just what Jordan had come to expect on the basis of his earlier use of canonical transformations. The identification of $\hat{q}$ in the Q-basis is a little trickier. Equation (16.30) told Jordan that the position operator in the original p-basis is $-(\hbar/i)\,\partial/\partial p$ (note the minus sign). This quantity turns into $-T(\hbar/i)\,\partial/\partial p\, T^{-1}$ in Eq. (16.42). This then should be the representation of $\hat{q}$ in the new Q-basis, as Jordan stated right below this last equation: "With respect to [*in Bezug*

[50] There is a sign error in this equation: $-TxT^{-1}$ should be TxT^{-1}.

[51] There is a sign error in this equation: $-y$ should be y. The sign errors in Eqs. (16.40) and (16.42) (Jordan's Eqs. (23ab)) confused Heisenberg, who was using *Neue Begründung* I in connection with his uncertainty paper (see Section 16.3 and note 28). He wrote to Jordan to ask for clarification:

> Today just a quick question, since I have been trying in vain, enduring persistent fits of rage, to derive your Eq. (23ab) from (19ab) in your transformation paper. According to my certainly not authoritative opinion it should be $\rho = e^{+xy/\varepsilon}$ and not $\rho = e^{-xy/\varepsilon}$, for out of $(x + \varepsilon\,\partial/\partial y)\,\rho(x,y)$ [Eq. (16.29)] I always get—God be darned—$(+TxT^{-1} + \varepsilon\,\partial/\partial y)\,\varphi(x,y) = 0$ [Eq. (16.40)]. Now it's possible that I am doing something nonsensical with these constantly conjugated quantities ($\tilde{F}, F^{\times}, F^{\dagger}$: read: F-blurry, F-cross, and F-deceased [*lies: F-verschwommen, F-Kreuz, und F-gestorben*]), but I don't understand anything anymore. Since, however, the quantity $\rho(x,y)$ forms the basis of my mathematics, I am kindly asking you for clarification of the sign (Heisenberg to Jordan, March 17, 1927, AHQP microfilm).

Jordan's response does not seem to have survived but from another letter from Heisenberg to Jordan a week later, we can infer that Jordan wrote back that the expression for $\rho(x,y)$ in *Neue Begründung* I is correct but that there are sign errors in Eqs. (23ab) (our Eqs. (16.40) and (16.42)). In the meantime Heisenberg had submitted his uncertainty paper and replied: "I now fully agree with your calculations and will change my calculations accordingly in the proofs" (Heisenberg to Jordan, March 24, 1927, AHQP microfilm).

auf] the fixed chosen quantity [$\hat{Q}$] every other quantity [$\hat{q}$] corresponds to an operator [$-T(\hbar/i)\,\partial/\partial p\, T^{-1}$]" (p. 815).[52]

With these representations of his quantum-mechanical quantities $\hat{p}$ and $\hat{q}$, Jordan could now define their addition and multiplication through the corresponding addition and multiplication of the differential operators representing these quantities.

Jordan's next step was to work out what the differential operators $T p\, T^{-1}$ and $-T(\hbar/i)\,\partial/\partial p\, T^{-1}$, representing $\hat{p}$ and $\hat{q}$ in the Q-basis, are in the special case that $\hat{Q} = \hat{q}$. In that case, Eqs. (16.40) and (16.42) turn into:

$$\left(T p\, T^{-1} + \frac{\hbar}{i}\frac{\partial}{\partial q}\right) \mathcal{A}(q', q) = 0, \tag{16.43}$$

$$\left(T \frac{\hbar}{i}\frac{\partial}{\partial p}\, T^{-1} + q\right) \mathcal{A}(q', q) = 0. \tag{16.44}$$

On the other hand, $\mathcal{A}(q', q) = \delta(q' - q)$. So $\mathcal{A}(q', q)$ trivially satisfies (p. 815, Eq. (24ab)):

$$\left(\frac{\hbar}{i}\frac{\partial}{\partial q'} + \frac{\hbar}{i}\frac{\partial}{\partial q}\right) \mathcal{A}(q', q) = 0, \tag{16.45}$$

$$(-q' + q)\, \mathcal{A}(q', q) = 0. \tag{16.46}$$

Comparing Eqs. (16.45)–(16.46) with Eqs. (16.43)–(16.44), we arrive at

$$T p\, T^{-1}\, \mathcal{A}(q', q) = \frac{\hbar}{i}\frac{\partial}{\partial q'}\, \mathcal{A}(q', q), \tag{16.47}$$

$$-\, T \frac{\hbar}{i}\frac{\partial}{\partial p}\, T^{-1}\, \mathcal{A}(q', q) = q'\, \mathcal{A}(q', q). \tag{16.48}$$

Eq. (16.47) suggests that $T p\, T^{-1}$, the momentum $\hat{p}$ in the q-basis acting on the q' variable, is just $(\hbar/i)\,\partial/\partial q'$. Likewise, Eq. (16.48) suggests that $-T(\hbar/i)\,\partial/\partial p\, T^{-1}$, the position $\hat{q}$ in the q-basis acting on the q' variable, is just multiplication by q'. As Jordan put it in a passage that is hard to follow because of his confusing notation:

> Therefore, as a consequence of [Eqs. (16.45)–(16.46)], the operator x [multiplying by q' in our notation] is assigned (*zugeordnet*) to the quantity [*Grösse*] Q itself [$\hat{q}$ in our

[52] Because of the sign error in Eq. (16.42) (see note 50), Jordan set $\hat{q}$ in the Q-basis equal to $T((\hbar/i)\,\partial/\partial p)T^{-1}$.

> notation]. One sees furthermore that the operator $\varepsilon\, \partial/\partial x$ [$(\hbar/i)\, \partial/\partial q'$ in our notation] corresponds to the momentum P [$\hat{p}$] belonging to Q [$\hat{q}$] (Jordan 1927b, p. 815).[53]

It is by this circuitous route that Jordan arrived at the usual functional interpretation of coordinate and momentum operators in the Schrödinger formalism. He emphasized that the association of $(\hbar/i)\, \partial/\partial q$ and q with $\hat{p}$ and $\hat{q}$ can easily be generalized (pp. 815–816). Any quantity (*Grösse*) obtained through multiplication and addition of $\hat{q}$ and $\hat{p}$ is associated with the corresponding combination of differential operators q and $(\hbar/i)\, \partial/\partial q$.

As it stands, Jordan's argument fails. We cannot conclude that two operations are identical because they give the same result when applied to one special case, here, the amplitude $\mathcal{A}(q', q) = \delta(q' - q)$ (cf. Eqs. (16.47)–(16.48)). We need to show that they give identical results when applied to an *arbitrary* function. One can easily fix this flaw in Jordan's argument, however, using only the kind of manipulations he himself used at this point (see Duncan and Janssen 2013, pp. 205–207).

With these identifications of $\hat{p}$ and $\hat{q}$ in the q-basis it is straightforward to verify that Jordan's new definition of conjugate variables in Eq. (16.28) reduces to the standard definition, $[\hat{p}, \hat{q}] = \hbar/i$, at least for quantities with completely continuous spectra. Letting the commutator of $(\hbar/i)\, \partial/\partial q$ and q act on an arbitrary function $f(q)$, we find

$$\left[\frac{\hbar}{i}\frac{\partial}{\partial q}, q\right] f(q) = \frac{\hbar}{i}\left(\frac{\partial(q f(q))}{\partial q} - q\frac{\partial f(q)}{\partial q}\right) = \frac{\hbar}{i} f(q). \tag{16.49}$$

Given the association of $(\hbar/i)\, \partial/\partial q$ and q with the quantities $\hat{p}$ and $\hat{q}$ that Jordan meanwhile established, it follows that these quantities indeed satisfy the usual commutation relation (p. 815, Eq. (25)):

$$[\hat{p}, \hat{q}] = \hat{p}\,\hat{q} - \hat{q}\,\hat{p} = \frac{\hbar}{i}. \tag{16.50}$$

This concludes Part I of *Neue Begründung* I. Jordan wrote:

> This is the content of the new theory. The rest of the paper will be devoted, through a mathematical discussion of these differential equations [Eqs. (16.40)–(16.42) and similar equations for probability amplitudes involving other quantities], on the one hand, to proving that our postulates are mathematically consistent [*widerspruchsfrei*] and, on the other hand, to showing that the earlier forms [*Darstellungen*] of quantum mechanics are contained in our theory (p. 816).

We focus on the first of these tasks, which amounts to providing a realization of Jordan's postulates.

[53] Note that Jordan used the term "operator" [*Operator*] *not* for an operator acting in an abstract Hilbert space but for the differential operators $(\hbar/i)\, \partial/\partial x$ and (multiplying by) x and for combinations of them.

At the beginning of section 4 of *Neue Begründung* I, "General comments on the differential equations for the amplitudes," Jordan announced:

> To prove that our postulates are mathematically consistent, we want to give a new foundation of the theory—independently from the considerations in section 2—based on the differential equations which appeared as end results there (p. 821).

He began by introducing the canonically conjugate variables $\hat{\alpha}$ and $\hat{\beta}$, satisfying the commutation relation $[\hat{\alpha}, \hat{\beta}] = \hbar/i$. They are related to the basic variables $\hat{p}$ and $\hat{q}$, for which the probability amplitude, according to Jordan's postulates, is $\mathcal{A}(p, q) = e^{-ipq/\hbar}$ (see Eq. (16.28)), via:

$$\hat{\alpha} = f(\hat{p}, \hat{q}) = T\hat{p}\,T^{-1}, \tag{16.51}$$

$$\hat{\beta} = g(\hat{p}, \hat{q}) = T\hat{q}\,T^{-1}, \tag{16.52}$$

with $T = T(\hat{p}, \hat{q})$ (p. 821, Eq. (1)[54]). Note that the operator $T(\hat{p}, \hat{q})$ defined here is different from the operator $T\ldots = \int dp\,\mathcal{A}(Q, p)\ldots$ defined in Eq. (16.31). The $T(\hat{p}, \hat{q})$ operator defined in section 4 is a similarity transformation operator implementing the canonical transformation from the pair $(\hat{p}, \hat{q})$ to the pair $(\hat{\alpha}, \hat{\beta})$. We will see later that there is an important relation between the T operators defined in sections 2 and 4.

Jordan now posited the fundamental differential equations for the probability amplitude $\mathcal{A}(q, \beta)$ in his theory (p. 821, Eqs. (2ab)):[55]

$$\left\{ f\left(\frac{\hbar}{i}\frac{\partial}{\partial q}, q\right) + \frac{\hbar}{i}\frac{\partial}{\partial \beta} \right\} \mathcal{A}(q, \beta) = 0, \tag{16.53}$$

$$\left\{ g\left(\frac{\hbar}{i}\frac{\partial}{\partial q}, q\right) - \beta \right\} \mathcal{A}(q, \beta) = 0. \tag{16.54}$$

These equations have the exact same form as Eqs. (16.40) and (16.42) (p. 815, Eqs. (23ab)), with the understanding that the operator T is defined differently. As Jordan put it in the passage quoted above, he took the equations that were the end result in section 2 as his starting point in section 4.

We turn to Jordan's discussion of Eqs. (16.53)–(16.54).[56] As Jordan pointed out:

> As is well known, of course, one cannot in general simultaneously impose two partial differential equations on one function of two variables. We shall prove, however, in section

[54] The numbering of equations in *Neue Begründung* I starts over in section 3, the first section of Part Two, and then again in section 4. The final two sections (sections 6 and 7, pp. 831–838) each have their own set of equation numbers.

[55] He introduced separate equations for the *Ergänzungsamplitude* (p. 821, Eqs. (3ab)).

[56] Before we do so, we show that these equations are easily recovered in the modern Hilbert space formalism (Duncan and Janssen 2013, p. 211). The result of the momentum operator $\hat{\alpha}$ in Eq. (16.51) acting on

5: the presupposition—which we already made—*that $\hat{a}$ and $\hat{\beta}$ are connected to $\hat{p}$ and $\hat{q}$ via a canonical transformation (1)* [our Eqs. (16.51)–(16.52)] *is the necessary and sufficient condition for (2)* [our Eqs. (16.53)–(16.54)] *to be solvable* (p. 822, our emphasis).

In section 5, "Mathematical theory of the amplitude equations" (pp. 824–828), Jordan made good on this promise. To prove that the "presupposition" is *sufficient,* he used canonical transformations to construct a simultaneous solution of the pair of differential equations in Eqs. (16.53)–(16.54) (pp. 824–825, Eqs. (9)–(17)). He did this in two steps.

1. He showed that the sufficient condition for $\mathcal{A}(Q, \beta)$ to be a solution of the amplitude equations in the Q-basis, given that $\mathcal{A}(q, \beta)$ is a solution of these equations in the q-basis, is that the pair $(\hat{p}, \hat{q})$ is related to the pair $(\hat{P}, \hat{Q})$ by a canonical transformation.
2. He established a starting point for generating such solutions by showing that a very simple canonical transformation (basically switching $\hat{p}$ and $\hat{q}$) turns the amplitude equations (16.53)–(16.54) into a pair of equations immediately seen to be satisfied by the amplitude $\mathcal{A}(q, \beta) = e^{iq\beta/\hbar}$.

With these two steps Jordan showed that the assumption that $\hat{P}$ and $\hat{Q}$ are related to $\hat{p}$ and $\hat{q}$ through a canonical transformation is indeed *sufficient* for Eqs. (16.53)–(16.54) to be simultaneously solvable. We take a closer look at this part of Jordan's argument.

The proof that this assumption is *necessary* as well as sufficient is much more complicated (pp. 825–828, Eqs. (18)–(34)). The mathematical preliminaries presented in

eigenvectors $|\beta\rangle$ of its conjugate operator $\hat{\beta}$ in Eq. (16.52) is:

$$\hat{\alpha}|\beta\rangle = -\frac{\hbar}{i}\frac{\partial}{\partial\beta}|\beta\rangle.$$

Taking the inner product of these expressions with $|q\rangle$ and using that $\hat{\alpha} = f(\hat{p}, \hat{q})$, we find that

$$-\frac{\hbar}{i}\frac{\partial}{\partial\beta}\langle q|\beta\rangle = \langle q|\hat{\alpha}|\beta\rangle = \langle q|f(\hat{p}, \hat{q})|\beta\rangle.$$

Since $\hat{p}$ and $\hat{q}$ are represented by the differential operators $(\hbar/i)\, \partial/\partial q$ and q, respectively, in the q-basis, we can rewrite this as

$$\langle q|f(\hat{p}, \hat{q})|\beta\rangle = f\left(\frac{\hbar}{i}\frac{\partial}{\partial q}, q\right)\langle q|\beta\rangle.$$

Combining these last two equations, we arrive at Eq. (16.53) (upon replacing $\langle q|\beta\rangle$ by $\mathcal{A}(q, \beta)$ as usual). Likewise, using that $\hat{\beta}|\beta\rangle = \beta|\beta\rangle$ and that $\hat{\beta} = g(\hat{p}, \hat{q})$, we can write the inner product $\langle q|\hat{\beta}|\beta\rangle$ as

$$\langle q|\hat{\beta}|\beta\rangle = \beta\langle q|\beta\rangle = g\left(\frac{\hbar}{i}\frac{\partial}{\partial q}, q\right)\langle q|\beta\rangle,$$

where in the last step we used the representation of $\hat{p}$ and $\hat{q}$ in the q-basis. From this equation we can read off Eq. (16.54).

section 3 (pp. 816–821) are needed only for this part of the proof in section 5, neither parts of which are covered here.

However, we do need to explain an important result that Jordan derived in section 5 as a consequence of this part of his proof (p. 828, Eqs. (35)–(40)): Canonical transformations $T(\hat{p}, \hat{q})$ as defined in section 4 (see Eqs. (16.51)–(16.52)), which are differential operators once $\hat{p}$ and $\hat{q}$ have been replaced by their representations $(\hbar/i)\,\partial/\partial q$ and q in the q-basis, can be written as integral operators T as defined in section 2 (see Eq. (16.31)).

This result is central to the basic structure of Jordan's theory and to the logic of his *Neue Begründung* papers. It shows that Jordan's probability amplitudes do double duty as integral kernels of the operators implementing canonical transformations. As such, Jordan showed, they satisfy the versions of the familiar completeness and orthogonality relations required by his postulate C (see Eqs. (16.25) and (16.27)). To paraphrase the characterization of Jordan's project that we already quoted in Section 16.2.1, Jordan postulated certain relations between his probability amplitudes in Part One of his paper and then, in Part Two, presented "a simple analytical apparatus in which quantities occur that satisfy these relations exactly" (Hilbert, von Neumann, and Nordheim 1928, p. 2). These quantities, it turns out, are the integral kernels of canonical transformations. Rather than following Jordan's own proof of this key result, which turns on properties of canonical transformations, we present (in Appendix C) a modern proof, which turns on properties of Hilbert space and the spectral theorem.

But first, closely following Jordan's own argument in section 5 of *Neue Begründung* I, we show how to construct a simultaneous solution of Eqs. (16.53)–(16.54), the two differential equations for the amplitude $\mathcal{A}(q, \beta)$. Suppose we can exhibit just one case of a canonical transformation $(\hat{p}, \hat{q}) \to (\hat{\alpha}, \hat{\beta})$ as in Eqs. (16.51)–(16.52)), where they manifestly have a unique simultaneous solution. According to Jordan,[57] any other canonical pair can be arrived at from the pair $(\hat{p}, \hat{q})$ via a new transformation function $S(\hat{P}, \hat{Q})$, in the usual way

$$\hat{p} = S\hat{P}S^{-1}, \quad \hat{q} = S\hat{Q}S^{-1}, \tag{16.55}$$

with $S = S(\hat{P}, \hat{Q})$ (p. 824, Eq. (10)). The connection between the original pair $(\hat{\alpha}, \hat{\beta})$ and the new pair $(\hat{P}, \hat{Q})$ involves the composite of two canonical transformations (p. 824, Eq. (11)):

$$\hat{\alpha} = f(\hat{p}, \hat{q}) = f(S\hat{P}S^{-1}, S\hat{Q}S^{-1}) \equiv F(\hat{P}, \hat{Q}), \tag{16.56}$$

$$\hat{\beta} = g(\hat{p}, \hat{q}) = g(S\hat{P}S^{-1}, S\hat{Q}S^{-1}) \equiv G(\hat{P}, \hat{Q}). \tag{16.57}$$

[57] As we discuss below, this assumption is false.

In the new Q-basis, Eqs. (16.53)–(16.54) take the form (p. 824, Eqs. (12ab)):

$$\left\{F\left(\frac{\hbar}{i}\frac{\partial}{\partial Q}, Q\right) + \frac{\hbar}{i}\frac{\partial}{\partial \beta}\right\} \mathcal{A}(Q, \beta) = 0, \tag{16.58}$$

$$\left\{G\left(\frac{\hbar}{i}\frac{\partial}{\partial Q}, Q\right) - \beta\right\} \mathcal{A}(Q, \beta) = 0. \tag{16.59}$$

Jordan now showed that

$$\mathcal{A}(Q, \beta) = \left(S\left(\frac{\hbar}{i}\frac{\partial}{\partial q}, q\right) \mathcal{A}(q, \beta)\right)\Big|_{q=Q} \tag{16.60}$$

(p. 824, Eq. (13)) is a simultaneous solution of the differential equations in the Q-basis (Eqs. (16.58)–(16.59)) as long as $\mathcal{A}(q, \beta)$ is a simultaneous solution of the differential equations in the q-basis (Eqs. (16.53)–(16.54)). Using the operator S and its inverse S^{-1}, we can rewrite the latter pair of equations as (p. 825, Eqs. (14ab)):[58]

$$S\left\{f\left(\frac{\hbar}{i}\frac{\partial}{\partial q}, q\right) + \frac{\hbar}{i}\frac{\partial}{\partial \beta}\right\} S^{-1} S \mathcal{A}(q, \beta) = 0, \tag{16.61}$$

$$S\left\{g\left(\frac{\hbar}{i}\frac{\partial}{\partial q}, q\right) - \beta\right\} S^{-1} S \mathcal{A}(q, \beta) = 0, \tag{16.62}$$

both taken, as in Eq. (16.60), at $q = Q$. Written more explicitly, the first term in curly brackets in Eq. (16.61), sandwiched between S and S^{-1}, is

$$S\left(\frac{\hbar}{i}\frac{\partial}{\partial q}, q\right) f\left(\frac{\hbar}{i}\frac{\partial}{\partial q}, q\right) S^{-1}\left(\frac{\hbar}{i}\frac{\partial}{\partial q}, q\right), \tag{16.63}$$

again taken at $q = Q$. With the help of Eq. (16.56), this can further be rewritten as

$$S(P, Q) f(P, Q) S(P, Q)^{-1}\Big|_{P=\frac{\hbar}{i}\frac{\partial}{\partial Q}} = F\left(\frac{\hbar}{i}\frac{\partial}{\partial Q}, Q\right). \tag{16.64}$$

The second term in curly brackets in Eq. (16.61), sandwiched between S and S^{-1}, is simply equal to

$$S\frac{\hbar}{i}\frac{\partial}{\partial \beta} S^{-1} = \frac{\hbar}{i}\frac{\partial}{\partial \beta}, \tag{16.65}$$

58 This step is formally the same as the one that got us from Eqs. (16.29)–(16.30) to Eqs. (16.40)–(16.42).

as S does not involve β. Using Eqs. (16.60) and (16.63)–(16.65), we can rewrite Eq. (16.61) as

$$\left\{F\left(\frac{\hbar}{i}\frac{\partial}{\partial Q}, Q\right) + \frac{\hbar}{i}\frac{\partial}{\partial \beta}\right\} \mathcal{A}(Q, \beta) = 0, \tag{16.66}$$

which is just Eq. (16.58). A completely analogous argument establishes that Eq. (16.62) reduces to Eq. (16.59). This concludes the proof that $\mathcal{A}(Q, \beta)$ in Eq. (16.60) is a solution of the amplitude equations in the new Q-basis as long as $\mathcal{A}(q, \beta)$, out of which $\mathcal{A}(Q, \beta)$ was constructed with the help of the operator S implementing a canonical transformation, is a solution of the amplitude equations in the old q-basis.

As S is completely general, we need only exhibit a single valid starting point, that is, a pair (f, g) and an amplitude $\mathcal{A}(q, \beta)$ satisfying the amplitude equations in the q-basis (Eqs. (16.53)–(16.54)), to construct general solutions of the amplitude equations in some new Q-basis (Eqs. (16.58)–(16.59)). The trivial example of a canonical transformation switching the roles of coordinate and momentum does the trick (cf. Eqs. (16.56)–(16.57)):

$$\hat{\alpha} = f(\hat{p}, \hat{q}) = -\hat{q}, \quad \beta = g(\hat{p}, \hat{q}) = \hat{p}. \tag{16.67}$$

In that case, Eqs. (16.53)–(16.54) become (p. 825, Eq. 16):

$$\left\{q - \frac{\hbar}{i}\frac{\partial}{\partial \beta}\right\} \mathcal{A}(q, \beta) = 0, \tag{16.68}$$

$$\left\{\frac{\hbar}{i}\frac{\partial}{\partial q} - \beta\right\} \mathcal{A}(q, \beta) = 0. \tag{16.69}$$

Except for the minus signs, these equations are of the same form as the trivial equations (16.29)–(16.30) for $\mathcal{A}(p, q)$, satisfied by the basic amplitude $\mathcal{A}(p, q) = e^{-ipq/\hbar}$. In the case of Eqs. (16.68)–(16.69), the solution is (p. 825, Eq. (17)):

$$\mathcal{A}(q, \beta) = e^{i\beta q/\hbar}. \tag{16.70}$$

This establishes that the canonical nature of the transformation to the new variables is a sufficient condition for the consistency (i.e., simultaneous solvability) of the pair of differential equations in Eqs. (16.58)–(16.59).

Jordan went on to prove the converse, that is, that the canonical connection is also a necessary condition for the consistency of Eqs. (16.58)–(16.59) (pp. 825–828). This is done by explicit construction of the operator S (in Eq. (16.60)), given the validity of Eqs. (16.58)–(16.59). We omit this part of the proof.

Jordan then used some of the same techniques to prove a key result in his theory (p. 828). Here, we help ourselves to the modern Hilbert space formalism and von Neumann's spectral theorem (see Chapter 17 and Appendix C) to elucidate his result. We start from Eq. (16.60), replacing Jordan's probability amplitudes by inner products of vectors in Hilbert space:

$$\langle Q|\beta\rangle = \left(S\left(\frac{\hbar}{i}\frac{\partial}{\partial q}, q\right)\langle q|\beta\rangle\right)\Big|_{q=Q}. \tag{16.71}$$

This equation tells us that the differential operator S maps arbitrary states $|\beta\rangle$ (recall that $\hat{\beta}$ can be any Hermitian operator) expressed in its components $\langle q|\beta\rangle$ in the q-basis onto its components $\langle Q|\beta\rangle$ in the Q-basis. The spectral theorem, which gives us the resolution $\int dq|q\rangle\langle q|$ of the unit operator, tells us that this mapping can also be written as

$$\langle Q|\beta\rangle = \int dq\,\langle Q|q\rangle\langle q|\beta\rangle. \tag{16.72}$$

Schematically, we can write

$$S\left(\frac{\hbar}{i}\frac{\partial}{\partial q}, q\right)\ldots = \int dq\,\langle Q|q\rangle\ldots \tag{16.73}$$

In other words, $\langle Q|q\rangle$, which is just Jordan's probability amplitude $\mathcal{A}(Q, q)$, is the integral kernel for the integral representation of the canonical transformation operator S. Using nothing but the properties of canonical transformations and his differential equations for probability amplitudes (Eqs. (16.53)–(16.54)), Jordan derived an equation of exactly the same form as Eq. (16.73), which we give here in its original notation (p. 828, Eq. (40)):

$$T\left(\varepsilon\frac{\partial}{\partial x}, x\right) = \int dx\,\varphi(y, x)\ldots \tag{16.74}$$

Translated into our notation, this equation is:

$$T\left(\frac{\hbar}{i}\frac{\partial}{\partial q}, q\right)\ldots = \int dq\,\mathcal{A}(Q, q)\ldots \tag{16.75}$$

Jordan claimed that Eqs. (16.53)–(16.54) contain both the time-independent and the time-dependent Schrödinger equations as special cases.[59] The time-independent Schrödinger equation is a special case of Eq. (16.54):

> If in (2b) [our Eq. (16.54)] we take β to be the energy W, and g to be the Hamiltonian function $H(p, q)$ of the system, we obtain the Schrödinger wave equation, which corresponds to the classical Hamilton–Jacobi equation. With (2b) comes (2a) [our Eq. (16.53)] as a second equation. In this equation we need to consider f to be the time t (as a function of p and q) (p. 822).

[59] Cf. Hilbert, von Neumann, and Nordheim (1928, sec. 10, pp. 27–29, "The Schrödinger differential equations"), discussed briefly by Jammer (1966, p. 311).

Actually, the variable conjugate to $\hat{H}$ would have to be *minus* $\hat{t}$. For $\hat{\alpha} = f(\hat{p}, \hat{q}) = -\hat{t}$ and $\hat{\beta} = g(\hat{p}, \hat{q}) = \hat{H}$ (with eigenvalues E), Eqs. (16.53)–(16.54) become:

$$\left\{\hat{t} - \frac{\hbar}{i}\frac{\partial}{\partial E}\right\} \mathcal{A}(q, E) = 0, \tag{16.76}$$

$$\left\{\hat{H} - E\right\} \mathcal{A}(q, E) = 0. \tag{16.77}$$

If $\mathcal{A}(q, E)$ is set equal to $\psi_E(q)$, Eq. (16.77) is indeed just the time-independent Schrödinger equation.

Jordan likewise claimed that the time-dependent Schrödinger equation is a special case of Eq. (16.53):

> if for β we choose the time t [this, once again, should be $-t$], for g [minus] the time $t(p, q)$ as function of p, q, and, correspondingly, for f the Hamiltonian function $H(p, q)$ (p. 823).

This claim is *much* more problematic. For $\hat{\alpha} = f(\hat{p}, \hat{q}) = \hat{H}$ (eigenvalues E) and $\hat{\beta} = g(\hat{p}, \hat{q}) = -\hat{t}$, Eqs. (16.53)–(16.54) become:

$$\left\{\hat{H} - \frac{\hbar}{i}\frac{\partial}{\partial t}\right\} \mathcal{A}(q, t) = 0, \tag{16.78}$$

$$\{\hat{t} - t\} \mathcal{A}(q, t) = 0. \tag{16.79}$$

If $\mathcal{A}(q, t)$ is set equal to $\psi(q, t)$, Eq. (16.78) turns into the time-dependent Schrödinger equation. However, time is a parameter in quantum mechanics and *not* an operator $\hat{t}$ with eigenvalues t and eigenstates $|t\rangle$ (cf. note 90).

This also makes Eqs. (16.76) and (16.79) problematic. Consider the former. For a free particle, the Hamiltonian is $\hat{H} = \hat{p}^2/2m$, represented by $((\hbar/i)\,\partial/\partial q)^2/2m$ in the q-basis. The solution of Eq. (16.77),

$$\mathcal{A}(q, E) = e^{i\sqrt{2mE}q/\hbar}, \tag{16.80}$$

is also a solution of Eq. (16.76) as long as we define $\hat{t} \equiv m\hat{q}\hat{p}^{-1}$, as suggested by the relation $q = (p/m)\,t$. Note, however, that we rather arbitrarily decided on this particular ordering of the non-commuting operators $\hat{p}$ and $\hat{q}$. Using that

$$\frac{\hbar}{i}\frac{\partial}{\partial q}\, e^{i\sqrt{2mE}q/\hbar} = \sqrt{2mE}\, e^{i\sqrt{2mE}q/\hbar} \tag{16.81}$$

and that $\hat{t} \equiv m\hat{q}\hat{p}^{-1}$, we find that, for the amplitude $\mathcal{A}(q,E)$ in Eq. (16.80), $\hat{t}\mathcal{A}(q,E)$ gives

$$mq\left(\frac{\hbar}{i}\frac{\partial}{\partial q}\right)^{-1} e^{i\sqrt{2mE}q/\hbar} = \frac{mq}{\sqrt{2mE}}\, e^{i\sqrt{2mE}q/\hbar}, \tag{16.82}$$

which is indeed equal to $(\hbar/i)\,\partial\mathcal{A}(q,E)/\partial E$, as required by Eq. (16.76):

$$\frac{\hbar}{i}\frac{\partial}{\partial E}\, e^{i\sqrt{2mE}q/\hbar} = \frac{mq}{\sqrt{2mE}}\, e^{i\sqrt{2mE}q/\hbar}. \tag{16.83}$$

So with $\hat{t} \equiv m\hat{q}\hat{p}^{-1}$, both Eq. (16.53) and Eq. (16.54) hold in the special case of a free particle. It is not at all clear, however, whether this will be true in general.

It is understandable that we can get Jordan's formalism to work, albeit with difficulty, for a free particle where the energy spectrum is fully continuous. Recall that, in *Neue Begründung* I, Jordan restricted himself to quantities with completely continuous spectra. As he discovered when in *Neue Begründung* II, he tried to generalize his formalism to quantities with partly or wholly discrete spectra, this restriction is not nearly as innocuous as he made it sound in *Neue Begründung* I.

Consider the canonical transformation $\hat{\alpha} = T\hat{p}\,T^{-1}$ (see Eq. (16.51) that plays a key role in Jordan's construction of the model realizing his postulates. Consider (in modern terms) an arbitrary eigenstate $|p\rangle$ of the operator $\hat{p}$ with eigenvalue p, that is, $\hat{p}\,|p\rangle = p\,|p\rangle$. It only takes one line to show that then $T|p\rangle$ is an eigenstate of $\hat{\alpha}$ with the same eigenvalue p:

$$\hat{\alpha}\,T|p\rangle = T\hat{p}\,T^{-1}\,T|p\rangle = T\hat{p}\,|p\rangle = p\,T|p\rangle. \tag{16.84}$$

It likewise only takes one line to show that if $|\alpha\rangle$ is an eigenstate of $\hat{\alpha}$ with eigenvalue α, $T^{-1}|\alpha\rangle$ is an eigenvector of $\hat{p}$ with that same eigenvalue α:

$$\hat{p}\,T^{-1}\,|\alpha\rangle = T^{-1}\,T\hat{p}\,T^{-1}\,|\alpha\rangle = T^{-1}\,\hat{\alpha}\,|\alpha\rangle = \alpha\,T^{-1}\,|\alpha\rangle. \tag{16.85}$$

In other words, the operators $\hat{\alpha}$ and $\hat{p}$ connected by the canonical transformation $\hat{\alpha} = T\hat{p}\,T^{-1}$ have the same spectrum. This simple observation, more than anything else, reveals the limitations of Jordan's formalism. It is true, as Eq. (16.77) demonstrates, that his differential equations for probability amplitudes (Eqs. (16.53)–(16.54)) contain the time-independent Schrödinger equation as a special case. However, since the energy spectrum is bounded from below and, in many interesting cases, at least partially discrete, we will never arrive at the time-independent Schrödinger equation starting from the trivial Eqs. (16.68) and (16.69) for the probability amplitude $e^{iq\beta/\hbar}$ between $\hat{q}$ and $\hat{\beta}$—recall that $\hat{\beta} = \hat{p}$ in this case (see Eq. (16.67))—and performing some canonical transformation. As Eqs. (16.84) and (16.85) show, a canonical transformation cannot get us from $\hat{p}$s and $\hat{q}$s with completely continuous spectra to $\hat{\alpha}$s and $\hat{\beta}$s with partly discrete spectra. This, in turn, means that, in many interesting cases (i.e., for Hamiltonians with at least partly discrete spectra), the time-independent Schrödinger equation does *not* follow from Jordan's postulates. In Jordan's defense one could note that this criticism is unfair as he explicitly restricted himself to quantities with fully continuous spectra in *Neue Begründung* I. However, when we turn to *Neue Begründung* II in Section 16.2.4, we will see that Jordan had to accept in this second paper that the extension of his general formalism to quantities with wholly or partly discrete spectra only served to accentuate the problem and did nothing to alleviate it.

16.2.3 Hilbert, von Neumann, and Nordheim on Jordan's "New foundation (*Neue Begründung*) . . . " I

In the Winter semester of 1926–1927, Hilbert gave a course entitled "Mathematical methods of quantum theory." The course consisted of two parts. The first, "The older quantum theory," was essentially a repeat of a course he had given under the same title in 1922–1923. The second, "The new quantum theory," covered the developments since 1925. As he had in 1922–1923, Nordheim prepared the notes for this course (see Sauer and Majer 2009, pp. 504–707; the second part takes up pp. 609–707). At the very end (pp. 700–706), we find a concise exposition of the main line of reasoning of Jordan's (1927b) *Neue Begründung* I.

This exposition served as the basis for a paper by Hilbert, von Neumann, and Nordheim (1928). As the authors explain in the introduction (pp. 1–2), "important parts of the mathematical elaboration" were due to von Neumann, while Nordheim was responsible for the final text (Duncan and Janssen 2009, p. 361). The paper was submitted to the *Mathematische Annalen* on April 6, 1927, but only published at the beginning of the volume for 1928. It thus appeared after the trilogy by von Neumann (1927a, 1927b, 1927c), to be discussed in Chapter 17, which rendered much of it obsolete. Here, we cover the main points of this earlier paper by Hilbert, von Neumann, and Nordheim.[60]

The lecture notes for Hilbert's course do not mention Dirac at all, and even though in the paper it is acknowledged that Dirac (1927a) had independently arrived at and published similar results as Jordan, the focus continues to be on the latter. There are only a handful of references to Dirac, most importantly in connection with the delta function and in the discussion of the Schrödinger equation for a Hamiltonian with a partly

[60] For other discussions of Hilbert, von Neumann, and Nordheim (1928), see Jammer (1966, pp. 309–312), Mehra and Rechenberg (2000–2001, pp. 404–411), and Lacki (2000, pp. 295–300, focusing mainly on the paper's axiomatic structure). Jammer talks about the "Hilbert–Neumann–Nordheim transformation theory" as if it were a new version of the theory, going beyond "its predecessors, the theories of Dirac and Jordan" (Jammer 1966, p. 312). However, what Jammer sees as the new element, the identification of probability amplitudes with the kernels of certain integral operators, is part and parcel of Jordan's version of the theory.

discrete spectrum (Hilbert, von Neumann, and Nordheim 1928, p. 8 and p. 30, respectively). Both the lecture notes and the paper stay close to the relevant sections of *Neue Begründung* I, although the notation used is an improvement over Jordan's (not a high bar to clear). For one thing, they use the same Greek letter (φ) for all probability amplitudes (cf. note 34). Here, we adopt the notation of Hilbert and his co-authors, except we continue to use $\mathcal{A}$ for all amplitudes, "hats" to distinguish kinematical quantities from their numerical values, and an asterisk to indicate complex conjugation (they use a bar).

As mentioned in Section 16.2.2, when we discussed postulates A–D of *Neue Begründung* I, Hilbert, von Neumann, and Nordheim (1928, pp. 4–5) based their exposition of Jordan's theory on six "physical axioms."[61] Axiom I introduces the basic idea of a probability amplitude. The amplitude for the probability density that a mechanical quantity $\hat{F}_1(\hat{p}\,\hat{q})$ (some function of momentum $\hat{p}$ and coordinate $\hat{q}$) has the value x given that another such quantity $\hat{F}_2(\hat{p}\,\hat{q})$ has the value y is written as

$$\mathcal{A}(x\,y; \hat{F}_1\,\hat{F}_2). \tag{16.86}$$

Jordan's *Ergänzungsamplitude* still made a brief appearance in the notes for Hilbert's course (Sauer and Majer 2009, p. 700) but is silently dropped in the paper. Section 16.2.2 discussed how amplitude and supplementary amplitude are identical as long as we only consider quantities represented, in modern terms, by Hermitian operators. In that case, the probability density $w(x\,y; \hat{F}_1\,\hat{F}_2)$ of finding $\hat{F}_1 = x$ given that $\hat{F}_2 = y$ is equal to the product of $\mathcal{A}(x\,y; \hat{F}_1\,\hat{F}_2)$ and its complex conjugate. Hilbert, von Neumann, and Nordheim (1928, p. 4) immediately set $w(x\,y; \hat{F}_1\,\hat{F}_2)$ equal to this product,

$$w(x\,y; \hat{F}_1\,\hat{F}_2) = \mathcal{A}(x\,y; \hat{F}_1\,\hat{F}_2)\,\mathcal{A}^{\star}(x\,y; \hat{F}_1\,\hat{F}_2), \tag{16.87}$$

which, of course, will always be real. Although they did not explicitly point out that this maneuver eliminates the need for the *Ergänzungsamplitude*, they put great emphasis on the restriction to Hermitian operators. Sections 6–8 of their paper ("The reality conditions," "Properties of Hermitian operators," and "The physical meaning of the reality conditions") are devoted to this issue.[62]

Axiom II corresponds to Jordan's postulate B and says that the amplitude for finding a value for $\hat{F}_2$ given the value of $\hat{F}_1$ is the complex conjugate of the amplitude of finding that same value for $\hat{F}_1$ given that same value of $\hat{F}_2$. This symmetry property entails that the corresponding conditional probability densities are the same. Axiom III is not among Jordan's postulates. It basically states the obvious demand that when $\hat{F}_1 = \hat{F}_2$, the probability density $w(x\,y; \hat{F}_1\,\hat{F}_2)$ be either 0 (if $x \neq y$) or 1 (if $x = y$). Axiom IV corresponds to Jordan's postulate C and states that the amplitudes, rather than the probabilities themselves, follow the usual composition rules for probabilities

[61] In the lecture notes we find four axioms that are essentially the same as Jordan's four postulates (Sauer and Majer 2009, pp. 700–701).

[62] The importance of the Hermiticity property had already been emphasized in the theory of complex Hermitian forms of infinitely many variables in Hilbert's theory of integral equations: see, for example, Hilbert (1912, p. 162).

(cf. Eqs. (16.25) and (16.27) and note 46):

$$\mathcal{A}(x\,z;\hat{F}_1\,\hat{F}_3) = \int \mathcal{A}(x\,y;\hat{F}_1\,\hat{F}_2)\,\mathcal{A}(y\,z;\hat{F}_2\,\hat{F}_3)\,dy \tag{16.88}$$

(p. 4). Though they did not use Jordan's phrase "interference of probabilities," the authors emphasized the central importance of this particular axiom:[63]

> This requirement [Eq. (16.88)] is obviously analogous to the addition and multiplication theorems of ordinary probability calculus, except that in this case they hold for the amplitudes rather than for the probabilities themselves. The characteristic difference to ordinary probability calculus lies herein that initially, instead of the probabilities themselves, amplitudes occur, which in general will be complex quantities and only give ordinary probabilities if their absolute value is taken and then squared (p. 5).

Axiom V, as mentioned in Section 16.2.2, makes part of Jordan's postulate A into a separate axiom. It demands that probability amplitudes for quantities $\hat{F}_1$ and $\hat{F}_2$ depend only on the functional dependence of these quantities on $\hat{q}$ and $\hat{p}$ and not on "special properties of the system under consideration, such as, for example, its Hamiltonian" (Hilbert, von Neumann, and Nordheim 1928, p. 5). Axiom VI, finally, adds another obvious requirement to the ones recognized by Jordan: that probabilities be independent of the choice of specific coordinate systems.

Before they introduced the axioms, Hilbert, von Neumann, and Nordheim (1928, p. 2) had already explained, in a passage quoted in Section 16.2.1, that the task at hand was to find "a simple analytical apparatus in which quantities occur that satisfy" axioms I–VI. As we know from *Neue Begründung* I, the quantities that fit the bill are the integral kernels of certain canonical transformations, implemented as $T\hat{p}\,T^{-1}$ and $T\hat{q}\,T^{-1}$ (cf. Eqs. (16.51)–(16.52)). After introducing this "simple analytical apparatus" in sections 3–4 ("Basic formulae of the operator calculus," "Canonical operators and transformations"), the authors concluded in section 5 ("The physical interpretation of the operator calculus"):

> *The probability amplitude $\mathcal{A}(x\,y;\hat{q}\,\hat{F})$ between the coordinate $\hat{q}$ and an arbitrary mechanical quantity $\hat{F}(\hat{q}\,\hat{p})$—i.e., for the situation that for a given value y of $\hat{F}$, the coordinate lies between x and $x + dx$—is given by the kernel of the integral operator that canonically transforms the operator $\hat{q}$ into the operator corresponding to the mechanical quantity $\hat{F}(\hat{q}\,\hat{p})$* (Hilbert, von Neumann, and Nordheim 1928, p. 14; emphasis in the original, "hats" added, φ replaced by $\mathcal{A}$).

They immediately generalized this definition to cover the probability amplitude between two arbitrary quantities $\hat{F}_1$ and $\hat{F}_2$. In section 3, they had already derived differential equations for integral kernels $\mathcal{A}(x\,y)$ (Hilbert, von Neumann, and Nordheim 1928,

[63] Heisenberg (1927b, pp. 183–184, p. 196) criticized this aspect of Jordan's work in his uncertainty paper (see cf. note 30 and Section 16.3, where we quote the relevant passages). von Neumann (1927a, p. 46) initially followed Jordan but changed his mind after reading Heisenberg's criticism (von Neumann 1927b, p. 246; see Section 17.2).

pp. 10–11, Eqs. (19ab) and (21ab)). Given the identification of these integral kernels with probability amplitudes in section 5, these equations are just Jordan's fundamental differential equations for the latter (Jordan 1927b, sec. 4, Eqs. (2ab); our Eqs. (16.53)–(16.54)).

In section 4, they also stated the key assumption that any quantity of interest can be obtained through a canonical transformation starting from some canonically conjugate pair of quantities $\hat{p}$ and $\hat{q}$:

> we shall assume that every operator $\hat{F}$ can be generated out of the basic operator $\hat{q}$ by a canonical transformation. This statement can also be expressed in the following way, namely that, given $\hat{F}$, the operator equation $T\hat{q}\,T^{-1}$ has to be solvable. The conditions that $\hat{F}$ has to satisfy for this to be possible will not be investigated here (Hilbert, von Neumann, and Nordheim 1928, p. 12; "hats" added).

This passage suggests that the authors recognized the importance of this assumption but did not quite appreciate that it puts severe limits on the applicability of Jordan's formalism (see our discussion at the end of Section 16.2.2). In the simple examples of canonical transformations ($\hat{F} = f(\hat{q})$ and $\hat{F} = \hat{p}$) that they considered in section 9 ("Application of the theory to special cases"), the assumption is obviously satisfied and the formalism works just fine. In section 10 ("The Schrödinger differential equations"), however, they set $\hat{F}$ equal to the Hamiltonian $\hat{H}$ and claimed that one of the differential equations for the probability amplitude $\mathcal{A}(x\,W; \hat{q}\,\hat{H})$ (where W is an energy eigenvalue) is the time-independent Schrödinger equation. As soon as the Hamiltonian has a wholly or partly discrete spectrum, however, there simply is no operator T such that $\hat{H} = T\hat{q}\,T^{-1}$.

In sections 6–8, Hilbert, von Neumann, and Nordheim (1928, pp. 17–25) showed that the necessary and sufficient condition for the probability density $w(x\,y; \hat{F}_1\,\hat{F}_2)$ to be real is that $\hat{F}_1$ and $\hat{F}_2$ are both represented by Hermitian operators. As pointed out above, they thus implicitly rejected Jordan's attempt to accommodate $\hat{F}$s represented by non-Hermitian operators through the introduction of the *Ergänzungsamplitude*. They also showed that the operator representing the canonical conjugate $\hat{G}$ of a quantity $\hat{F}$ represented by a Hermitian operator is itself Hermitian.

The authors ended their paper on a cautionary note, emphasizing its lack of mathematical rigor. They referred to von Neumann's (1927a) forthcoming paper, "Mathematical foundation [*Mathematische Begründung*] of quantum mechanics," for a more satisfactory treatment of the Schrödinger equation for Hamiltonians with partly discrete spectra. In the concluding paragraph, they warned the reader more generally:

> In our presentation the general theory receives such a perspicuous and formally simple form that we have carried it through in a mathematically still imperfect form, especially since a fully rigorous presentation might well be considerably more tedious and circuitous [*mühsamer und umständlicher*] (Hilbert, von Neumann, and Nordheim 1928, p. 30).

In Chapter 17, we return to this paper by Hilbert, von Neumann, and Nordheim to discuss the role it played in the development of von Neumann's thinking about quantum mechanics.[64]

16.2.4 Jordan's "New foundation (*Neue Begründung*) ... " II

In April and May 1927, while at Bohr's institute in Copenhagen on an International Education Board fellowship, Jordan (1927g) wrote *Neue Begründung* II,[65] which was received by *Zeitschrift für Physik* June 3, 1927.[66] In the abstract he announced a "simplified and generalized" version of the theory presented in *Neue Begründung* I.

One simplification was that Jordan, like Hilbert, von Neumann, and Nordheim (1928), dropped the *Ergänzungsamplitude* and restricted himself accordingly to physical quantities represented (in modern terms) by Hermitian operators and to canonical transformations preserving Hermiticity. He also simplified his notation (cf. note 34).[67] Following Hilbert, von Neumann, and Nordheim, he started using the same Greek letter for all amplitudes (Φ). We continue to use $\mathcal{A}$. Following Dirac (1927a), he started using the same letter for a mechanical quantity and its possible values, using primes to distinguish the latter from the former. When, for instance, the letter β is used for some quantity, its possible (real) values are denoted as β', β'', etc. We continue to use the notation $\hat{\beta}$ for the quantity (and the operator representing that quantity) and the notation $\beta, \beta_1, \beta_2, \ldots$ for its values. As in our discussion of *Neue Begründung* I in Section 16.2.2, we occasionally resort to using the Hilbert space formalism in Dirac notation (cf. Appendix C).

While Jordan's new notational conventions are a welcome improvement, they only affect the cosmetics of the paper. Unfortunately, the generalization of the formalism promised in the abstract to handle cases with wholly or partly discrete spectra turns out to be much more problematic than Jordan suggested and is ultimately untenable. By the end of the paper, Jordan is counting quantities nobody would think of as canonically conjugate, such as different spin components, as pairs of conjugate variables and has abandoned the notion, central to the formalism of *Neue Begründung* I, that any quantity

[64] As Lacki (2000, p. 299) notes in his discussion of Hilbert, von Neumann, and Nordheim (1928): "this was to be Hilbert's last physics paper, and the sources are scarce as to what was his further thinking and attitude towards quantum theory."

[65] Not long after leaving for Copenhagen, Jordan wrote a long letter to Dirac, who was still in Göttingen, touching on some of the issues addressed in *Neue Begründung* II (Jordan to Dirac, April 14, 1927 (AHQP microfilm), quoted in Duncan and Janssen 2013, p. 224, note 96). Unfortunately, we do not have Dirac's reply. In between *Neue Begründung* I and II, Jordan (1927e) published a short paper showing that his theory has the desirable feature that the conditional probability of finding a certain value for some quantity is independent of the scale used to measure that quantity.

[66] Jordan (1927g) was written after (and partly in response to) the paper in which von Neumann (1927a) introduced the Hilbert space formalism. As in Duncan and Janssen (2013), we follow the logical rather than the temporal sequence, and deviate from the temporal order by covering Jordan's paper before von Neumann's, which is covered in Section 17.1.

[67] When Kuhn complained about the "dreadful notation" of *Neue Begründung* I in his interview with Jordan for the AHQP project, Jordan said that in *Neue Begründung* II he just wanted to give a "prettier and clearer" exposition of the same material (session 3, p. 17, quoted in Duncan and Janssen 2009, p. 360).

of interest (e.g., the Hamiltonian) is a member of a pair of conjugate variables connected to some initial pair of $\hat{p}$'s and $\hat{q}$'s by a canonical transformation.

As shown at the end of Section 16.2.2, the canonical transformation

$$\hat{\alpha} = T\hat{p}\,T^{-1}, \quad \hat{\beta} = T\hat{q}\,T^{-1} \tag{16.89}$$

(cf. Eqs. (16.51)–(16.52)) can never get us from quantities with completely continuous spectra (such as position or momentum) to quantities with wholly or partly discrete spectra (such as the Hamiltonian). In *Neue Begründung* II, Jordan (pp. 16–17) evidently recognized this problem although it is not clear that he realized the extent to which this undercuts his entire approach.

The central problem is brought out somewhat indirectly in the paper. As Jordan (pp. 1–2)[68] already mentioned in the abstract and then demonstrated in the introduction, the commutation relation, $[\hat{p}, \hat{q}] = \hbar/i$, for two canonically conjugate quantities $\hat{p}$ and $\hat{q}$ cannot hold as soon as the spectrum of one of them is partly discrete. This means, for example, that action/angle variables $\hat{J}$ and $\hat{w}$, where the eigenvalues of the action variable $\hat{J}$ are restricted to integral multiples of Planck's constant, cannot satisfy the canonical commutation relation.

To prove this claim, Jordan considered a pair of conjugate quantities $\hat{\alpha}$ and $\hat{\beta}$ where $\hat{\beta}$ is assumed to have a purely discrete spectrum. (One runs into the same problem as soon as either $\hat{\alpha}$ or $\hat{\beta}$ has a single discrete eigenvalue.) Suppose $\hat{\alpha}$ and $\hat{\beta}$ satisfy the standard commutation relation:

$$[\hat{\alpha}, \hat{\beta}] = \frac{\hbar}{i}. \tag{16.90}$$

As Jordan pointed out, it then follows that an operator that is some function F of $\hat{\beta}$ satisfies[69]

$$[\hat{\alpha}, F(\hat{\beta})] = \frac{\hbar}{i} F'(\hat{\beta}). \tag{16.91}$$

Jordan now chose a function such that $F(\beta) = 0$ for all eigenvalues $\beta_1, \beta_2, \ldots$ of $\hat{\beta}$, while $F'(\beta) \neq 0$ at those same points.[70] In that case, the left-hand side of Eq. (16.91) vanishes

[68] Unless indicated otherwise, all page numbers in this subsection refer to *Neue Begründung* II (Jordan 1927g).

[69] If the function $F(\beta)$ is assumed to be a polynomial, $\sum_n c_n \beta^n$, which is all we need for what we want to prove, although Jordan considered a "fully transcendent function" (p. 2), Jordan's claim is a standard result in elementary quantum mechanics:

$$[\hat{\alpha}, F(\hat{\beta})] = [\hat{\alpha}, \sum_n c_n \hat{\beta}^n] = \sum_n c_n\, n \frac{\hbar}{i} \hat{\beta}^{n-1} = \frac{\hbar}{i} \frac{d}{d\hat{\beta}} \left(\sum_n c_n \hat{\beta}^n \right) = \frac{\hbar}{i} F'(\hat{\beta}),$$

where in the second step we repeatedly used that $[\hat{\alpha}, \hat{\beta}] = \hbar/i$ and that $[\hat{A}, \hat{B}\,\hat{C}] = [\hat{A}, \hat{B}]\,\hat{C} + \hat{B}\,[\hat{A}, \hat{C}]$ for any three operators $\hat{A}$, $\hat{B}$, and $\hat{C}$.

[70] See Duncan and Janssen (2013, pp. 225–226) for a simple concrete example of such a function.

at all these points, whereas the right-hand side does not. Hence, Eq. (16.91) cannot hold. Since Eq. (16.91) is a direct consequence of Eq. (16.90), the latter cannot hold either.

Much later in his paper, in section 4, Jordan (p. 16) acknowledged that it follows directly from this result that no canonical transformation can ever get us from a pair of conjugate variables $\hat{p}$s and $\hat{q}$s with completely continuous spectra to $\hat{\alpha}$s and $\hat{\beta}$s with partly discrete spectra. It is, after all, an essential property of canonical transformations that they preserve canonical commutation relations. From Eq. (16.89) and $[\hat{p}, \hat{q}] = \hbar/i$ it follows that

$$[\hat{\alpha}, \hat{\beta}] = [T\hat{p}\,T^{-1}, T\hat{q}\,T^{-1}] = T[\hat{p}, \hat{q}]\,T^{-1} = \frac{\hbar}{i}. \tag{16.92}$$

Eq. (16.92) cannot hold for $\hat{\alpha}$s and $\hat{\beta}$s with partly discrete spectra. Hence, such $\hat{\alpha}$'s and $\hat{\beta}$'s cannot possibly be obtained from $\hat{p}$ and $\hat{q}$ through a canonical transformation.

We discuss below how this impossibility affects Jordan's general formalism. When Jordan, in the introduction of *Neue Begründung* II, showed that no quantity with a partly discrete spectrum can satisfy a canonical commutation relation, he presented it not as a serious problem for his formalism, but rather as an argument for the superiority of his alternative definition of conjugate variables in *Neue Begründung* I (Jordan 1927b, p. 814). According to that definition $\hat{p}$ is canonically conjugate to $\hat{q}$ if the probability amplitude $\mathcal{A}(p, q)$ has the simple form $e^{-ipq/\hbar}$ (see Eq. (16.28)). As we saw in Section 16.2.2, Jordan then showed that, for $\hat{p}$s and $\hat{q}$s with purely continuous spectra, this implies $[\hat{p}, \hat{q}] = \hbar/i$, the standard definition of what it means for $\hat{p}$ to be conjugate to $\hat{q}$. In *Neue Begründung* II, Jordan (p. 6) extended his alternative definition to quantities with wholly or partly discrete spectra, in which case the new definition no longer reduces to the standard one.

As Jordan wrote in its opening paragraph, *Neue Begründung* II only assumes a rough familiarity with *Neue Begründung* I. He thus had to redevelop much of the formalism of his earlier paper, while trying to simplify and generalize it at the same time. In section 2 ("Basic properties of quantities and probability amplitudes"), Jordan began by restating the postulates to be satisfied by his probability amplitudes.

He introduced a new notation for these amplitudes. Instead of $\varphi(\beta, q)$, which in our notation is $\mathcal{A}(\beta, q)$, he now wrote (p. 6)

$$\Phi_{\alpha p}(\beta', q'), \tag{16.93}$$

for which we use the notation $\mathcal{A}_{\hat{\alpha}\hat{p}}(\beta, q)$. The primes distinguish values of quantities from those quantities themselves. The subscripts α and p (or $\hat{\alpha}$ and $\hat{p}$ in our notation) denote which quantities are canonically conjugate to the quantities $\hat{\beta}$ and $\hat{q}$ for which the probability amplitude is being evaluated.

It turns out that one has a certain freedom in picking the $\hat{\alpha}$ and $\hat{p}$ conjugate to $\hat{\beta}$ and $\hat{q}$, respectively, and settling on a specific pair of $\hat{\alpha}$ and $\hat{p}$ is equivalent to fixing the phase ambiguity of the amplitude $\mathcal{A}(\beta, q)$ up to some constant factor.[71] So for a given

[71] For detailed discussion of this point, see Duncan and Janssen (2013, pp. 234–235).

choice of $\hat{\alpha}$ and $\hat{p}$, the amplitude $\mathcal{A}_{\hat{\alpha}\hat{p}}(\beta, q)$ is essentially unique. In this way, Jordan (p. 20) could answer, at least formally, von Neumann's (1927a, p. 3) objection in "Mathematical foundation ... " that probability amplitudes are not uniquely determined even though the resulting probabilities are (see Section 17.1). It is only made clear toward the end of *Neue Begründung* II that this is the rationale behind these additional subscripts. Their only other role is to remind the reader that $\mathcal{A}_{\hat{\alpha}\hat{p}}(\beta, q)$ is determined not by one Schrödinger-type equation in Jordan's formalism, but rather by a pair of such equations involving both canonically conjugate pairs of variables, $(\hat{p}, \hat{q})$ and $(\hat{\alpha}, \hat{\beta})$ (p. 20).[72]

Jordan also removed the restriction to systems of one degree of freedom that he had adopted for convenience in *Neue Begründung* I (Jordan 1927b, p. 810). So $\hat{q}$, in general, now stands for $(\hat{q}_1, \ldots \hat{q}_f)$, where f is the number of degrees of freedom of the system under consideration. The same is true for other quantities. Jordan (pp. 4–5) spent a few paragraphs examining the different possible structures of the space of eigenvalues for such f-dimensional quantities depending on the nature of the spectrum of its various components—fully continuous, fully discrete, or combinations of both. He also introduced the notation $\delta(\beta' - \beta'')$ for a combination of the Dirac delta function and the Kronecker delta (which Jordan (p. 5) called the "Weierstrassian symbol").

In *Neue Begründung* II, the four postulates of *Neue Begründung* I are replaced by three postulates—or "axioms" as Jordan now also called them (p. 6)—numbered with Roman numerals. Jordan's new postulates or axioms do not include the key portion of postulate A of *Neue Begründung* I, which states the probability interpretation of the amplitudes. That is relegated to section 5, "The physical meaning of the amplitudes" (p. 19). Right before listing the postulates, however, Jordan mentions that he will only consider "real (Hermitian) quantities" (p. 5), thereby obviating the need for the *Ergänzungsamplitude* and simplifying the relation between amplitudes and probabilities. There is no discussion of the *Ergänzungsamplitude* amplitude in the paper. Instead, Jordan silently dropped it. It is possible that this was not even a matter of principle for Jordan but only one of convenience. Right after listing the postulates, he wrote that the restriction to real quantities is made only "to keep things simple" (*der Einfachkeit halber*, p. 6).

Other than the probability-interpretation part of postulate A, all four postulates of *Neue Begründung* I return, generalized from one to f degrees of freedom and from quantities with completely continuous spectra to quantities with wholly or partly discrete spectra. Axiom I corresponds to the old postulate D. It says that for every generalized coordinate there is a conjugate momentum.

Axiom II consists of three parts, labeled (A), (B), and (C) (p. 6). Part (A) corresponds to the old postulate B (cf. Eq. 16.24)), asserting the symmetry property, which, in our version of Jordan's new notation, becomes:

$$\mathcal{A}_{\hat{\alpha}\hat{p}}(\beta, q) = \mathcal{A}^{\star}_{\hat{p}\hat{\alpha}}(q, \beta). \tag{16.94}$$

[72] These subscripts do not affect the translation of probability amplitudes to inner products in Hilbert space. To make (our discussion of) *Neue Begründung* II easier to follow, the modern reader can still substitute $\langle\beta|q\rangle$ for $\mathcal{A}_{\hat{\alpha}\hat{p}}(\beta, q)$ everywhere.

Part (B) corresponds to the old postulate C (cf. Eq. 16.25)), which gives the basic rule for the composition of probability amplitudes,

$$\overline{\sum_q} \mathcal{A}_{\hat{\alpha}\hat{p}}(\beta, q)\, \mathcal{A}_{\hat{p}\hat{P}}(q, Q) = \mathcal{A}_{\hat{\alpha}\hat{P}}(\beta, Q), \tag{16.95}$$

where the notation $\overline{\sum}_q$ indicates that, in general, we need a combination of integrals over the continuous parts of the spectrum of a quantity and sums over its discrete parts. In Eq. (16.95), $\overline{\sum}_q$ refers to an ordinary integral as the coordinate $\hat{q}$ has a purely continuous spectrum. Eq. (16.95) gives the completeness relation, familiar from modern quantum mechanics, in Jordan's formalism (cf. notes 46 and 72). As in *Neue Begründung* I (see Eq. (16.27)), we arrive at Jordan's version of the familiar orthogonality relations by considering the special case $\hat{Q} = \hat{\beta}$ in Eq. (16.95). This gives (p. 7, Eq . (5))

$$\begin{aligned} \overline{\sum_q} \mathcal{A}_{\hat{\alpha}\hat{p}}(\beta, q)\, \mathcal{A}^{\star}_{\hat{\alpha}\hat{p}}(\beta', q) &= \delta(\beta - \beta'), \\ \overline{\sum_\beta} \mathcal{A}_{\hat{\alpha}\hat{p}}(\beta, q)\, \mathcal{A}^{\star}_{\hat{\alpha}\hat{p}}(\beta, q') &= \delta(q - q'), \end{aligned} \tag{16.96}$$

where $\delta(\beta - \beta')$ can be either the Dirac delta function or the Kronecker delta, as $\hat{\beta}$ can have either a fully continuous or a partly or wholly discrete spectrum.

Part (C) of axiom II is the generalization of the definition of conjugate variables familiar from *Neue Begründung* I to f degrees of freedom and to quantities with wholly or partly discrete spectra. Two quantities $\hat{\alpha} = (\hat{\alpha}_1, \ldots, \hat{\alpha}_f)$ and $\hat{\beta} = (\hat{\beta}_1, \ldots, \hat{\beta}_f)$ are canonically conjugate to one another if

$$\mathcal{A}_{\hat{\alpha},-\hat{\beta}}(\beta, \alpha) = C\, e^{i\left(\sum_{k=1}^{f} \beta_k \alpha_k\right)/\hbar}, \tag{16.97}$$

where C is a normalization constant.[73]

Axiom III, finally, generalizes axiom III of Hilbert, von Neumann, and Nordheim (1928, p. 4), which states the obvious requirement for a quantity $\hat{\beta}$ (and its conjugate $\hat{\alpha}$) with just one component that

$$\mathcal{A}_{\hat{\alpha}\hat{\alpha}}(\beta_1, \beta_2) = \delta(\beta_1 - \beta_2) \tag{16.98}$$

(where, once again, $\delta(\beta_1 - \beta_2)$ can be either the Dirac delta function or the Kronecker delta), to quantities $\hat{\beta} = (\hat{\beta}_1, \ldots, \hat{\beta}_f)$ with an arbitrary number f components.

[73] Contrary to what Jordan (p. 7) suggested, the sign of the exponent in Eq. (16.97) agrees with the sign of the exponent in the corresponding formula in *Neue Begründung* I (Jordan 1927b, p. 814, Eq. 18; our Eq. (16.28)).

We need to explain one more aspect of Jordan's notation in *Neue Begründung* II. As discussed in Sections 16.2.1 and 16.2.2, probability amplitudes do double duty as integral kernels of canonical transformations. Jordan (p. 6) introduced the special notation $\Phi_{\alpha p}^{\beta q}$ (which becomes $\mathcal{A}_{\alpha p}^{\beta q}$ in our notation) to indicate that the amplitude $\mathcal{A}_{\hat{\alpha}\hat{p}}(\beta, q)$ serves as such an integral kernel. We can think of $\mathcal{A}_{\alpha p}^{\beta q}$ as "matrices" with β and q as "indices" that, in general, will take on both discrete and continuous values. This clearly brings out the double role of this quantity. In *Neue Begründung* I, we frequently encountered canonical transformations such as $\hat{\alpha} = T\hat{p}\,T^{-1}$ and $\hat{\beta} = T\hat{q}\,T^{-1}$ (see Eqs. (16.51)–(16.52)). In *Neue Begründung* II, such transformations are written with $\mathcal{A}_{\alpha p}^{\beta q}$s instead of Ts. While bringing out more clearly an important feature of Jordan's formalism, the new notation conceals an important shift in his usage of these transformation equations. This shift is only made explicit in section 4, entitled "Canonical transformations." Up to that point, and especially in section 3, "The functional equations of the amplitudes," Jordan appears to be vacillating between two different interpretations of these canonical transformation equations, the one of *Neue Begründung* I, in which $\hat{\alpha}$ and $\hat{\beta}$ are different from the $\hat{p}$ and $\hat{q}$ we started from, and one, inspired by Dirac (1927a), as Jordan acknowledged in section 4 (pp. 16–17), in which $\hat{\alpha}$ and $\hat{\beta}$ are just different representations of the same $\hat{p}$ and $\hat{q}$.

Before he got into any of this, Jordan (sec. 2, pp. 8–10) examined five examples, labeled (a) through (e), of what he considered to be pairs of conjugate quantities and convinced himself that they qualify as such under his new definition (16.97) by checking that they satisfy the completeness and orthogonality relations in Eqs. (16.95) and (16.96). The examples include familiar pairs of canonically conjugate variables, such as action/angle variables (Jordan's example (c)), but also quantities that we normally would not think of as conjugate variables, such as different spin components (a special case of example (e)).[74] This last example raises the question whether Jordan's definition of conjugate variables was getting too permissive. The main problem with Jordan's formalism, however, was not that he was asking too little of his conjugate variables, but that he was asking too much of his canonical transformations!

Canonical transformations enter into the formalism in section 3 (pp. 13–16), where Jordan introduced a simplified yet at the same time generalized version of Eqs. (2ab) of *Neue Begründung* I for probability amplitudes (our Eqs. (16.53)–(16.54)). They are simplified in that there are no longer additional equations for the *Ergänzungsamplitude*. They are generalized in that they are no longer restricted to systems with only one degree of freedom and, much more importantly, no longer restricted to cases where all quantities involved have purely continuous spectra. Quantities with partly or wholly discrete spectra are now also allowed.

Recall how Jordan built up his theory in *Neue Begründung* I. He posited a number of axioms to be satisfied by his probability amplitudes. He then constructed a model for these postulates. To this end he identified probability amplitudes with the integral kernels for certain canonical transformations. Starting with differential equations trivially satisfied by the amplitude $\mathcal{A}(p, q) = e^{-ipq/\hbar}$ for some initial pair of conjugate variables

[74] For detailed discussion of these two examples, action/angle variables and a pair of spin components, see Duncan and Janssen (2013, pp. 229–230).

$\hat{p}$ and $\hat{q}$, Jordan derived differential equations for amplitudes involving other quantities related to the initial ones through canonical transformations. As we showed at the end of Section 16.2.2, this approach breaks down as soon as we ask about the probability amplitudes for quantities with partly discrete spectra, such as, typically, the Hamiltonian.

Although Jordan (p. 14) emphasized that one has to choose initial $\hat{p}$s and $\hat{q}$s with "fitting spectra" (*passende Spektren*) and that the equations for the amplitudes are solvable only "if it is possible to find" such spectra, he did not state explicitly in section 3 that the construction of *Neue Begründung* I fails for quantities with discrete spectra.[75] That admission is postponed until the discussion of canonical transformations in section 4. At the beginning of section 3, the general equations for probability amplitudes are given in the form (p. 14, Eqs. (2ab)):

$$\mathcal{A}^{\beta q}_{\alpha p}\hat{B}_k - \hat{\beta}_k\mathcal{A}^{\beta q}_{\alpha p} = 0, \tag{16.99}$$

$$\mathcal{A}^{\beta q}_{\alpha p}\hat{A}_k - \hat{\alpha}_k\mathcal{A}^{\beta q}_{\alpha p} = 0, \tag{16.100}$$

where $\hat{A}_k$ and $\hat{B}_k$ are defined as (p. 13, Eq. (1)):

$$\hat{B}_k = \left(\mathcal{A}^{\beta q}_{\alpha p}\right)^{-1}\hat{\beta}_k\,\mathcal{A}^{\beta q}_{\alpha p}, \quad \hat{A}_k = \left(\mathcal{A}^{\beta q}_{\alpha p}\right)^{-1}\hat{\alpha}_k\,\mathcal{A}^{\beta q}_{\alpha p}. \tag{16.101}$$

Jordan (pp. 14–15) then showed that the differential equations of *Neue Begründung* I are included in these new equations as special cases. Since there is only one degree of freedom in that case, we do not need the index k. We can also suppress all indices of $\mathcal{A}^{\beta q}_{\alpha p}$ as this is the only probability-amplitude/transformation-matrix involved in the argument. So we have

$$\hat{A} = \mathcal{A}^{-1}\,\hat{\alpha}\,\mathcal{A}, \quad \hat{B} = \mathcal{A}^{-1}\hat{\beta}\,\mathcal{A}. \tag{16.102}$$

These transformations, however, are used very differently in the two installments of *Neue Begründung*. Although Jordan only discussed this change in section 4, he already alerted the reader to it in section 3, noting that "$\hat{B}$, $\hat{A}$ are the operators for $\hat{\beta}$, $\hat{\alpha}$ *with respect to* $\hat{q}$, $\hat{p}$" (p. 15, our emphasis).

[75] In his "Mathematical foundation . . . ," von Neumann (1927a) had already put his finger on this problem: "A special difficulty with [the approach of] Jordan is that one has to calculate not just the transforming operators (the integral kernels of which are the "probability amplitudes"), but also the value-range onto which one is transforming (i.e., the spectrum of eigenvalues)" (p. 3).

Suppressing all subscripts and superscripts, we can rewrite Eqs. (16.99)–(16.100) as:

$$(\mathcal{A}\hat{B}\mathcal{A}^{-1} - \hat{\beta})\,\mathcal{A} = 0, \tag{16.103}$$

$$(\mathcal{A}\hat{A}\mathcal{A}^{-1} - \hat{\alpha})\,\mathcal{A} = 0. \tag{16.104}$$

Using that (p. 15, Eq. 8)

$$\mathcal{A}\hat{A}\mathcal{A}^{-1} = \hat{\alpha} = f(\hat{p}, \hat{q}), \quad \mathcal{A}\hat{B}\mathcal{A}^{-1} = \hat{\beta} = g(\hat{p}, \hat{q}); \tag{16.105}$$

that $\hat{p}$ and $\hat{q}$ in the q-basis are represented by $(\hbar/i)\partial/\partial q$ and multiplication by q, respectively; and that $\hat{\alpha}$ and $\hat{\beta}$ in Eqs. (16.103)–(16.104) are represented by $-(\hbar/i)\partial/\partial\beta$ and multiplication by β, respectively, we see that, in this special case, Eqs. (16.99)–(16.100) (or, equivalently, Eqs. (16.103)–(16.104)) reduce to (p. 15, Eqs. (9ab))

$$\left(g\left(\frac{\hbar}{i}\frac{\partial}{\partial q}, q\right) - \beta\right)\mathcal{A} = 0, \tag{16.106}$$

$$\left(f\left(\frac{\hbar}{i}\frac{\partial}{\partial q}, q\right) + \frac{\hbar}{i}\frac{\partial}{\partial \beta}\right)\mathcal{A} = 0, \tag{16.107}$$

which are just Eqs. (2b) and (2a) of *Neue Begründung* I, respectively (Jordan 1927b, p. 821; cf. Eqs. (16.54) and (16.53)). This is the basis for Jordan's renewed claim that his general equations for probability amplitudes contain both the time-dependent and the time-independent Schrödinger equations as a special case (see Section 16.2.2). It is certainly true that, if the quantity $\hat{B}$ in Eq. (16.101) is chosen to be the Hamiltonian, Eq. (16.106) turns into the time-independent Schrödinger equation. However, there is no canonical transformation that connects this equation for $\psi_n(q) = \mathcal{A}(q, E)$ to the equations trivially satisfied by $\mathcal{A}(p, q)$, which formed the starting point for Jordan's construction of his formalism in *Neue Begründung* I.[76]

Jordan finally conceded this point in section 4 of *Neue Begründung* II (pp. 16–17). Following Dirac (1927a), he switched to a new conception of canonical transformations. Whereas before he saw canonical transformations such as $(\hat{\alpha} = T\hat{p}\,T^{-1}, \hat{\beta} = T\hat{q}\,T^{-1})$, as taking us from one pair of conjugate variables $(\hat{p}, \hat{q})$ to a *different* pair $(\hat{\alpha}, \hat{\beta})$, he now saw them as taking us from one particular *representation* of a pair of conjugate variables to a *different representation* of those *same* variables. Dirac had already put his finger on this difference between Jordan's original conception of canonical transformations and

[76] This problem does not affect Dirac's version of the theory as Dirac did not try to derive the differential equations for arbitrary amplitudes from a handful of postulates.

his own in the letter to Jordan from which we quoted at the beginning of Section 16.2.2:

> In your work I believe you considered transformations from one set of dynamical variables to another, instead of a transformation from one scheme of matrices representing the dynamical variables to another scheme representing the same dynamical variables, which is the point of view adopted throughout my paper (Dirac to Jordan, December 24, 1916 (AHQP microfilm), quoted in Duncan and Janssen 2013, p. 188).

The canonical transformation in Eq. (1) in section 3 of *Neue Begründung* II (our Eqs. (16.101) and (16.102)) is an example of a canonical transformation in the sense of Dirac. By giving up on canonical transformations in the older sense, Jordan effectively abandoned the basic architecture of *Neue Begründung* I.

This is how Jordan explained the problem at the beginning of section 4 of *Neue Begründung* II:

> Canonical transformations, the theory of which, as in classical mechanics, gives the natural generalization and the fundamental solution of the problem of the integration of the equations of motion, were originally [footnote citing the Three-Man-Paper of Born, Heisenberg, and Jordan (1926)] conceived of as follows: the canonical quantities $\hat{q}$, $\hat{p}$ should be represented as functions of certain other canonical quantities $\hat{\beta}$, $\hat{\alpha}$:
>
> $$\hat{q}_k = G_k(\hat{\beta}, \hat{\alpha}), \quad \hat{p}_k = F_k(\hat{\beta}, \hat{\alpha}). \qquad (1)$$
>
> On the assumption that canonical systems can be defined through the usual canonical commutation relations, a formal proof could be given [footnote referring to Jordan (1926a)] that for canonical $\hat{q}$, $\hat{p}$ and $\hat{\beta}$, $\hat{\alpha}$, Eq. (1), as was already suspected originally, can always be cast in the form
>
> $$\hat{q}_k = T\hat{\beta}_k T^{-1}, \quad \hat{p}_k = T\hat{\alpha}_k T^{-1}. \qquad (2)$$
>
> However, since, as we saw, the old canonical commutation relations are not valid [cf. Eqs. (16.90)–(16.92)], this proof too loses its meaning; in general, one can *not* bring equations (1) in the form (2).
>
> Now a modified conception of canonical transformation was developed by Dirac [footnote citing Dirac (1927a) and Lanczos (1926)]. According to Dirac, it is not about representing certain canonical quantities as functions of other canonical quantities, but about switching, *without a transformation of the quantities themselves*, to a different *matrix representation* (Jordan 1927g, pp. 16–17; emphasis in the original, "hats" added).

In terms of the modern Hilbert space formalism, canonical transformations, in the sense of Dirac, transform the matrix elements of an operator in one basis to matrix elements of that same operator in another. This works whether or not the operator under consideration is part of a pair of operators corresponding to canonically conjugate quantities. Consider the matrix elements of the position operator $\hat{q}$ (with a purely continuous spectrum) in the q-basis, now written *a la* Dirac:

$$\langle q|\,\hat{q}\,|q'\rangle = q'\delta(q' - q). \tag{16.108}$$

Let $\hat{\beta}$ be an arbitrary self-adjoint operator. In general, $\hat{\beta}$ will have a spectrum with both continuous and discrete parts. Von Neumann's spectral theorem (see Section C.5.6) tells us that

$$\hat{\beta} = \sum_n \beta_n\,|\beta_n\rangle\langle\beta_n| + \int \beta\,|\beta\rangle\langle\beta|\,d\beta, \tag{16.109}$$

where sums and integrals extend over the discrete and continuous parts of the spectrum of $\hat{\beta}$, respectively.

We now want to find the relation between the matrix elements of $\hat{q}$ in the β-basis and its matrix elements in the q-basis. Using the von Neumann (purely continuous) spectral decomposition $\int q\,|q\rangle\langle q|\,dq$ of $\hat{q}$, we can write:

$$\begin{aligned}
\langle\beta|\,\hat{q}\,|\beta'\rangle &= \int dq\,\langle\beta|q\rangle\,q\,\langle q|\beta'\rangle \\
&= \int dq\,dq'\,\langle\beta|q\rangle\,q'\,\delta(q' - q)\,\langle q'|\beta'\rangle \\
&= \int dq\,dq'\,\langle\beta|q\rangle\,\langle q|\,\hat{q}\,|q'\rangle\,\langle q'|\beta'\rangle.
\end{aligned}$$

In the notation of *Neue Begründung* II, the last line would be the "matrix multiplication" (cf. Eqs. (16.101) and (16.102)):

$$\left(\Phi^{q\beta}_{q\alpha}\right)^{-1}\hat{q}\,\Phi^{q\beta}_{q\alpha}. \tag{16.110}$$

The translation of this expression into modern notation shows that Jordan's formalism, even with a greatly reduced role for canonical transformations, implicitly relies on the spectral theorem, which von Neumann (1927a) published in "Mathematical foundation ...," submitted just one month before *Neue Begründung* II.

In addition to a response to von Neumann's criticism about the non-uniqueness of Jordan's probability amplitudes (see note 71), *Neue Begründung* II also contains some criticism of von Neumann's approach (p. 20). In particular, Jordan complained that von Neumann showed no interest in either canonical transformations or conjugate variables. As we shall see in Section 17.1, this is simply because von Neumann did not need either for his formulation of quantum mechanics. That formulation clearly did not convince Jordan. In fact, von Neumann's 'Mathematical foundation ... " only seems to have increased Jordan's confidence in his own approach. After his brief discussion of von Neumann's paper, he concluded: "It thus appears that the amplitudes themselves are to be considered the fundamental concept of quantum mechanics" (pp. 20–21).

It is unclear whether Jordan ever came to appreciate the advantages of von Neumann's approach over his own. In the preface of *Anschauliche Quantentheorie*, his textbook on quantum mechanics, Jordan (1936) described the statistical transformation theory of Dirac and himself as "the pinnacle of the development of quantum mechanics" (p. VI) and as the "most comprehensive and profound version of the quantum laws" (p. 171, quoted in Duncan and Janssen 2009, p. 361). He did not discuss any of von Neumann's contributions in this book. However, in *Elementare Quantenmechanik*, the textbook he co-authored with Born, we do find elaborate expositions (Born and Jordan 1930, Ch. 6, pp. 288–364) of the two papers by von Neumann (1927a, 1927b) that will occupy us in Chapter 17.

16.3 Heisenberg's uncertainty relations

On April 30, 1926, Heisenberg arrived in Copenhagen to take up a new position as lecturer at the university, as well as Bohr's primary collaborator and assistant, the role previously occupied by Kramers, who by this point had decamped to the Netherlands, to take up a professorship in Utrecht. Heisenberg remained in Copenhagen for a year and a half (with occasional departures for scientific or personal reasons), where he finally was able to have the kind of in-depth discussions with Bohr that he had so looked forward to on the occasion of his first extended stay in 1924–1925 (see Section 10.5).

This was a period of intense activity on a variety of problems of central importance in atomic physics for Heisenberg. The paper with Jordan essentially settling the question of complex spectra and the anomalous Zeeman effect (Heisenberg and Jordan 1926) had recently been completed. Heisenberg then went on in Copenhagen to write a series of impressive papers on a wide range of topics: on the resonance problem for multi-particle systems in quantum mechanics, resolving the puzzle of ortho/parahelium spectra (Heisenberg 1926b, see Section 15.4.1); on the perturbative evaluation of the helium spectrum (Heisenberg 1926c; see Section 15.4.2); on the nature of quantum fluctuations induced by quantum "jumps" (Heisenberg 1926d, see Sections 16.1 and 16.2.1); and on the symmetry requirements for the states of general multi-electron atoms, fully incorporating the requirements of the exclusion principle, such as the antisymmetry of the electronic eigenfunctions (Heisenberg 1927a, see Section 15.4.1).

The interaction between Heisenberg and Bohr on this occasion, if more intense and direct than a year and a half earlier (where Kramers had served as a frequent, if unwanted, intermediary), turned out to be considerably more contentious.[77] The primary reason for this seems to have been Heisenberg's rejection of the wave-mechanical approach (with the sole exception of its utility in the calculation of transition, or

[77] For discussion of the interaction between Heisenberg and Bohr in the period leading up to the former's uncertainty paper and the latter's Como lecture on complementarity (see Section 16.4), see, for example, Kalckar (1985), the introduction to the first of two volumes of Bohr's collected papers devoted to his writings on the foundations of quantum physics (Bohr 1972–2008, Vols. 6 and 7), Pais (1991, Ch. 14, secs. (c)–(f), pp. 300–320) in his biography of Bohr, Mehra and Rechenberg (2000–2001, secs. II.3, (g)–(h), pp. 151–157), and Kojevnikov (2020, pp. 94–96).

exchange, matrix elements), whereas Bohr had readily accepted Schrödinger's formalism, and was busily exploring the fundamental conceptual issues (quantum jumps, correspondence principle, etc.) with its use. Heisenberg insisted on sticking as closely as possible to the ideas and methodology of matrix mechanics, and by the Fall of 1926 his closest interlocutor on the elucidation of conceptual issues in quantum mechanics was Pauli (in Hamburg), with whom he corresponded frequently.

In Pauli's letter to Heisenberg of October 19, 1926 (Pauli 1979, Doc. 143), cited and quoted already in Sections 16.1 and 16.2.1 in connection with the transformation from wave-functions in q-space to those in p-space, we find an extended discussion of the connection between the (squared) matrix elements of a scattering potential and Born's scattering probabilities. In his discussion of the scattering of particles from a three-dimensional force center, Pauli emphasized that a particular scattering process can be specified by applying boundary conditions to a set of momentum variables (for example, energy of the incoming particle and its angular momentum around the center), but that there then arises the "obscure point" [*dunkle Punkt*] that while "the ps must be taken to be *controlled*" (i.e., specified), the corresponding qs (in this case, time and angular direction) "must be taken to be *uncontrolled*." A little later, we find the following prescient comments, which can reasonably be regarded, in the words of Mehra and Rechenberg (2000–2001, p. 145), as "the crucial stimulus which put Heisenberg straight on the route to the uncertainty relations:"

> So much for the mathematics. The corresponding physics is to a large extent still unclear to me. The first question is, why only the ps, and in any case not *both* the ps *and* the qs, are allowed both to be prescribed with arbitrary precision ... It is always the same business: on account of dispersion, arbitrarily narrow rays cannot exist in the wave optics of the ψ-field, and one cannot simultaneously assign "c-numbers" to the "p-numbers" and to the "q-numbers". One can view the world with the p-eye, and one can view it with the q-eye, but if one tries to open both eyes, one goes crazy [*wird man irre*] (Pauli to Heisenberg, October 19, 1926, Pauli 1979, Doc. 143).

By early February 1927, Heisenberg was writing to Pauli (1979, Doc. 153) that he had returned to the interpretational questions posed by the confluence of probabilistic and kinematic concepts in the Jordan–Dirac statistical transformation theory, stimulated by the need to elucidate the precise meaning of phrases such as "the probability that the electron sits at a certain point" in Jordan's habilitation lecture published in *Die Naturwissenschaften*.[78] These issues were discussed intensively, but unproductively, with Bohr (see Mehra and Rechenberg 2000–2001, p. 151). Apparently, the disagreements stemmed mainly from Bohr's insistence on employing wave-mechanical imagery in the interpretation of the numerous ingenious thought-experiments devised by both interlocutors in an attempt to pin down the slippery character of coordinate and momentum measurements in the new quantum theory, while Heisenberg insisted on starting from the formal equations of quantum mechanics (enhanced by the Dirac–Jordan formalism),

[78] This is Heisenberg's paraphrase of the passage "the probability that the system-point is at a given place in the configuration space" in (Oppenheimer's translation of) Jordan's (1927d, p. 568) article.

as he had by now adopted Einstein's admonition that "the theory determines what can be observed."[79]

In any event, by mid-February, Bohr experienced one of his frequent bouts of exhaustion (cf. Section 10.3, note 30) and went skiing for four weeks in Norway (Kalckar 1985, p. 16), allowing Heisenberg to develop his ideas without further interference. Shortly after Bohr had left, on February 23, Heisenberg sent a long letter to Pauli outlining the main conclusions of what would become the famous uncertainty paper (Heisenberg 1927b). On March 9, he also sent the first draft of the paper (the only copy!) to Pauli, by registered mail, with the request to return it in a few days so that he could discuss it with Bohr when Bohr returned from Norway. Pauli returned the paper promptly, with uncharacteristically few criticisms, and Heisenberg thereupon submitted it at about the time of Bohr's return and before the latter was able to digest its contents and raise objections—which he would shortly, and with some vehemence. This is clear from a "note added in proof" to the paper. This note also helps explain the two-month gap between March 23, when the paper was received by *Zeitschrift für Physik*, and May 29, when the paper finally appeared.[80]

Tensions between Heisenberg and Bohr ran high during these two months.[81] Mehra and Rechenberg (2000–2001, pp. 184) quote from two anguished letters from Heisenberg to Pauli during this period, written May 16 and May 31, 1927, respectively (Pauli 1979, Docs. 163 and 164), and then from another, written June 3, 1927 (Pauli 1979,

[79] In a private conversation at his home following Heisenberg's colloquium on his new theory in Berlin on April 28, 1926, Einstein emphasized to Heisenberg, as recalled many years later by the latter in his autobiography, that he had abandoned his earlier Machian prejudice, according to which theories should be constructed entirely on the basis of directly observable quantities. According to Heisenberg, Einstein asserted: "But on principle, it is quite wrong to try founding a theory on observable magnitudes alone. In reality the very opposite happens. It is the theory which decides what we can observe" (Heisenberg 1971, p. 63). As Mehra and Rechenberg (2000–2001, p. 154) caution, Heisenberg's (1971, p. 77) recollection in his autobiography of his flash of insight as he remembered his conversation with Einstein two years later "does dramatize the situation quite a bit." There is no mention of that conversation in Heisenberg's (1967, pp. 105–106) brief account of how he came to write the uncertainty paper in his reminiscences four years earlier.

[80] In a letter of March 10, 1927, Heisenberg told Bohr, still vacationing in Norway, that he had drafted a paper and sent it to Pauli, but did not provide any further details (Kalckar 1985, p. 16). Bohr returned to Copenhagen around March 18, as can be inferred from a letter he wrote to Kronig that day. Heisenberg's paper was received by *Zeitschrift für Physik* five days later. It is not entirely clear whether Bohr saw the manuscript before Heisenberg submitted it. According to Pais (1991, p. 308) he did: when "Bohr came back from Norway, Heisenberg showed him the paper he had written." This is also what Heisenberg remembered: "When Bohr returned from Norway, I was already able to present him with the first version of a paper along with the letter from Pauli" (Heisenberg 1967, p. 106). Nevertheless, according to Kalckar (1985, p. 16), "Bohr arrived back in Copenhagen ... only to find that Heisenberg had already sent the paper off for publication." He surmises that Heisenberg was "eager to get the manuscript off before every paragraph of it would be overhauled by Bohr." Mehra and Rechenberg (2000–2001, p. 181) side with Pais. Whether or not Bohr laid eyes on the manuscript before it was submitted, it seems clear that he only raised serious objections after the paper had been sent off. As Oskar Klein, who succeeded Heisenberg as Bohr's assistant (see Section 16.4), recalled in an interview in 1968: "Bohr read the paper and was at first very taken with it but when he began to look more closely he became very disappointed" (quoted in Pais 1991, p. 308).

[81] As Heisenberg (1967, p. 104) recalled, "Bohr often came up to my room late at night to talk with me of the difficulties in quantum theory, which tortured both of us ... Our evening discussions quite often lasted till after midnight, and we occasionally parted somewhat discontented, for the difference in the directions in which we sought the solution seemed often to make the problem more difficult" (see also Heisenberg 1971, pp. 76–77).

Doc. 165), after the tension has dissipated. In the first two letters, Heisenberg is particularly concerned about Oskar Klein joining the fray on the side of Bohr, which Heisenberg felt "seemed to make matters worse."[82] In an interview decades later, Heisenberg recalled: "Bohr tried to explain that it was not right and I shouldn't publish the paper. I remember that it ended by my breaking out in tears because I just couldn't stand this pressure from Bohr."[83] As can be inferred from Heisenberg's correspondence during this period,[84] an important part of Bohr's criticism of Heisenberg's paper was aimed at his analysis of the famous "gamma-ray-microscope", which plays a prominent role in the paper (see note 100).

Heisenberg's paper, one of the truly seminal works in the entire development of quantum theory, was entitled "On the visualizable content of quantum-theoretical kinematics and mechanics" [*Über den anschaulichen Inhalt der quantentheoretischen Kinematik und Mechanik*].[85] Its contents are well summarized by the abstract:

> In the present work exact definitions are presented of the words position, velocity, energy, etc. (e.g., of the electron), which also maintain their validity in quantum mechanics, and it is shown that canonically conjugate quantities can only be determined simultaneously with a characteristic uncertainty (§ 1). This uncertainty is the actual reason for the appearance of statistical correlations in quantum mechanics. Its mathematical formulation is possible by means of the Dirac–Jordan theory (§ 2). It is shown, starting from the basic principles obtained in this way, how macroscopic processes can be understood on the basis of quantum mechanics (§ 3). Some particular thought experiments are discussed in elucidation of the theory (§ 4) (Heisenberg 1927b, p. 172).[86]

In the introduction to the paper, Heisenberg explains his use of the word "visualizable" [*anschaulich*] in relation to a scientific theory: one must be able to qualitatively imagine the experimental consequences of the theory (at least, in "all simple cases"), and, secondly, the application of the theory should never lead to contradictions in the prediction of these consequences. According to Heisenberg, such contradictions were present, at least apparently, in the present form of quantum theory—for example, in the conflict between followers of "discontinuity theory" (primarily the Göttingen school, using matrix mechanical methods) and "continuum theory" (Schrödinger and followers of the wave-mechanical approach), as well as in the troublesome duality of particles and waves. It was the hope of the author, expressed at the end of the introductory section, to "clarify these contradictions by a more exact analysis of kinematic and mechanical concepts" (p. 173), and thereby arrive at a visualizable (in the above described sense)

[82] In later recollections, Heisenberg credited Klein with resolving his conflict with Bohr (Heisenberg 1967, p. 106; see also 1971, p. 79).

[83] Interview of Heisenberg by Kuhn, 19 February 1963 (AHQP), quoted by Pais (1991, p. 308).

[84] In addition to the letters cited above: Heisenberg to Pauli, April 4, 1927 (Pauli 1979, Doc. 161) and Heisenberg to Dirac, April 27, 1927 (quoted in Kalckar 1985, pp. 17–19).

[85] For a discussion placing this paper in the development of Heisenberg's views on the foundations of quantum mechanics, see Camilleri (2009a). For critical discussion of the uncertainty principle and uncertainty relations, see Uffink (1990), Uffink and Hilgevoord (1985), and Hilgevoord and Uffink (1988, 2016).

[86] Unless indicated otherwise, all page numbers in the remainder of this section are to Heisenberg's (1927b) uncertainty paper.

understanding of quantum-mechanical laws. This last point is accompanied by a footnote in which Heisenberg gives fulsome acknowledgement to prior work by Bohr and Einstein, but most emphatically to Pauli—citing his article in *Handbuch der Physik* (Pauli 1926b)—"who has contributed in an essential way to the present work."

In section 1 of his article, entitled "The concepts: position, path, velocity, energy" (pp. 174–179), Heisenberg gives a preliminary, and mainly qualitative, discussion of the intrinsic imprecisions in kinematic quantities induced by the constraints of measurement. The discussion of position measurements (how to measure the "position of an electron") foreshadows arguments to be found in any modern elementary text on quantum mechanics. One relies on intuition derived from classical optics: namely, it is impossible to resolve the location of a microscopic object using light of wavelength longer than the desired precision of the location measurement, as a simple consequence of diffraction effects. However, this limitation can easily be overcome, even for an electron, by using light of extremely short wavelength, for example, gamma rays. Thus, the "gamma-ray-microscope" makes its appearance. One illuminates the electron with gamma rays, and deduces its position by observing the scattered light. Of course, we expect that the interaction of the light with the electron will disturb the latter, and so an uncontrolled (by the observer) shift in both the position and velocity/momentum of the electron is to be expected. The *minimum* possible disturbance of the electron manifestly takes place when it interacts with a single photon (still called a "light quantum" [*Lichtquant*] in Heisenberg's paper), which proceeds via a Compton scattering, in which a momentum transfer $p_1 \sim h/\lambda$ from the photon to the electron occurs, where λ is the wavelength of the gamma rays. As pointed out above, the resolution of the measurement of position, in Heisenberg's notation q_1,[87] is determined by just this wavelength λ, whence one finds (given that $q_1 \sim \lambda$)

$$p_1 q_1 \sim h. \tag{16.111}$$

This admittedly rather crude argument would be elaborated into a much more precisely controlled thought experiment by Bohr, taking into proper account the diffraction limit for a microscope with a specified aperture, etc. As Heisenberg points out, the mutually constrained uncertainties expressed in Eq. (16.111) are intimately connected with the fundamental commutation relation $\mathbf{p}\,\mathbf{q} - \mathbf{q}\,\mathbf{p} = h/2\pi i$, a fact which would become apparent in the subsequent section on the Dirac–Jordan formalism.

After discussing the issue of position measurement, Heisenberg goes on to discuss the notion of "path", which classically simply corresponds to the sequence of positions occupied by (say) an electron as time evolves. Heisenberg points out that, with the new understanding of the position concept, the association of a "path" in this sense with a bound atomic electron, say the $1s$ ground state of the electron in hydrogen, is meaningless. For any measurement capable of localizing the electron with a precision q_1 much

[87] Throughout his paper, Heisenberg uses the subscript "1" to denote the uncertainty, or mean deviation, of a measured quantity. We follow this notation in our discussion of his paper. In modern texts, the uncertainty relation would be written $\Delta p \cdot \Delta q \sim h$.

smaller than the typical size $a = \hbar^2/me^2$ (the Bohr radius) of a Bohrian orbit would require a momentum transfer $p_1 \sim h/q_1$ with

$$p_1 \gg \frac{h}{a} = \frac{2\pi me^2}{\hbar}, \tag{16.112}$$

thereby transferring a kinetic energy to the atomic electron

$$\frac{p_1^2}{2m} \gg \frac{4\pi^2 me^4}{2\hbar^2} = 4\pi^2 \cdot 13.7 \text{ eV}. \tag{16.113}$$

The electron would indeed be instantaneously localized (with arbitrarily high precision) within the atom, but would now possess an energy far greater than its binding energy, and would immediately escape, destroying the atom. This "snapshot" of an electron in a bound stationary state is fairly meaningless as an isolated event, but as Heisenberg points out, the repetition of the measurement on an ensemble of identically prepared atoms (all, say, in their $1s$ ground state) would lead, at least in principle, to a probability distribution of the measured q positions predicted precisely by the theory as $\psi_{1s}(q)\overline{\psi}_{1s}(q)$, the absolute square of the Schrödinger energy eigenfunction.[88] Heisenberg is at pains to remind the reader at this point, referring to the Dirac–Jordan theory, that this squared wave function is nothing other than the square of the element $S(1s, q)$ of the transformation matrix going from the q-representation to the (in this case discrete) energy representation, with rows labeled by stationary states $1s$, $2s$, etc. (cf. Section 16.2.1).

At this point Heisenberg comments on an important debate that was still very active among the various camps of quantum theory (see Section 16.2.1). For Born and Jordan, quantum mechanics possessed an intrinsically statistical character, married to an inescapably acausal dynamics, which distinguished it from classical mechanics. Dirac, on the other hand, preferred to assert that the statistical character of the theory was "brought in by our experiments" (p. 177)—by the explicit intervention, in other words, of a macroscopic measuring apparatus. Heisenberg's viewpoint is closer to Dirac,[89] inasmuch as the intrinsic uncertainties he discusses are all manifestly occasioned by quite definite external interventions whereby the atomic system is brought into interaction with some external agency, the effect of which on the atomic system produces an irreducible (and unpredictable) disturbance.

Section 1 continues with a brief discussion of the kinematic concept of speed (equivalent to momentum p via the usual $p = mv$ relation, measurable by a Doppler effect from scattered light). This turns out to be completely analogous to the measurement of position of a $1s$ electron bound in an atom: attempts to pin down the electron momentum to a small fraction of its typical value in the bound state inevitably involve a transfer

[88] This application of what we now call the Born rule was first given by Pauli (see Section 16.1). Heisenberg uses $1S$ instead of $1s$. We use a lower-case s to avoid confusion with the transformation matrix S introduced below.

[89] See the passages quoted in Section 16.2.1 from letters from Heisenberg to Pauli and Jordan of February 23, 1927 and March 7, 1927, respectively.

of momentum sufficient to eject the electron entirely—or "destroy the atom," as Heisenberg put it (p. 177). Repeated measurements will eventually reconstruct a probability distribution in momentum space, this time given by $S(1s, p)\,\overline{S}(1s, p)$ (Heisenberg uses a bar to indicate complex conjugation), the absolute square of the transformation matrix from p space to (discrete) energy space.

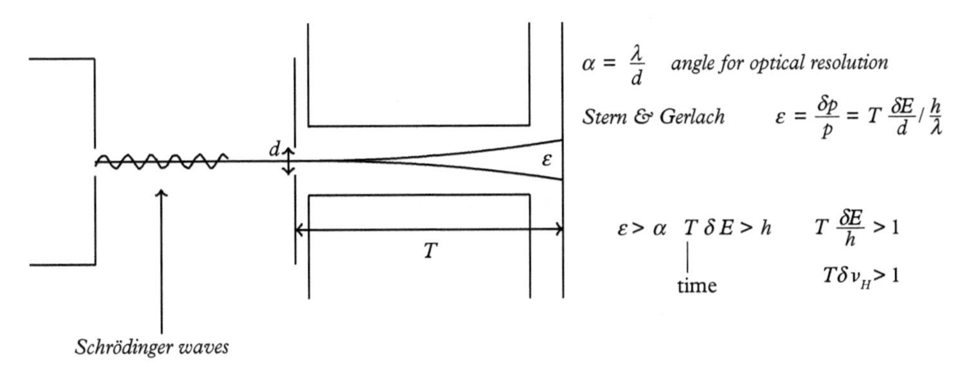

Figure 16.2 *Bohr's Solvay conference drawing of the Heisenberg Stern–Gerlach energy-time thought experiment (Heisenberg 1927b, p. 178, Heisenberg's t_1 and E_1 are Bohr's T and δE, respectively). Some details have been omitted for clarity.*

The final kinematic concept discussed in this first section, energy, requires a distinctly more subtle argument than the one leading to the momentum–position uncertainty relation in Eq. (16.111). Here, the underlying physics is considerably more slippery: energy and time, though canonically conjugate in classical theory, cannot be regarded as conjugate quantities in quite the same way in quantum mechanics, where time the plays the role of a parameter, not an operator.[90] Proceeding as though quantum mechanical matrices can be associated with the classically conjugate action $\mathcal{J}$ and angle w (phase) coordinates, or equivalently (up to proportionality, for conditionally periodic systems) energy E and time t, Heisenberg writes a commutator relation $\mathbf{E}\,\mathbf{t} - \mathbf{t}\,\mathbf{E} = h/2\pi i$ which then, by analogy with the situation for momentum and position, suggests a mutual uncertainty between measurements of energy and time—in this case, for a bound system, interpreting "time" as the temporal moment at which an external agency is brought to bear capable of revealing the energy of the system. Experience with the momentum–position uncertainty relation suggests that the imprecision E_1 of the energy measurement should correlate with the time period t_1 over which the energy measurement is carried out.

The explicit example given by Heisenberg is a thought experiment using an experiment of the Stern–Gerlach type, in which a measurement of the magnetic moment of

[90] Cf. our discussion in Section 16.2.2 of Jordan's attempt to recover the time-dependent Schrödinger equation by considering $\hat{E}$ and $\hat{t}$ as conjugate variables. In 1958, Pauli would prove that no self-adjoint time operator can have the canonical commutator with the Hamiltonian unless the latter has a spectrum coinciding with the entire real line (Pauli 1980). There is extensive literature on this topic: see, for example, Hilgevoord (2002) and Busch (2008).

an atom (contained in a atomic ray of width d, with the atoms moving parallel with momentum $\vec{p}$) is made by applying an inhomogeneous magnetic field perpendicular to the direction of the ray of atoms. The value of the magnetic moment is measured by the perpendicular deflection of the atoms. The energy measurement being made is somewhat indirect insofar as the measured magnetic moment is then used as an indicator for the quantum number of the particular stationary state occupied by the atom—as Heisenberg puts it, we perform a "measurement of quantities which depend only on action variables" (p. 178). The inhomogeneous magnetic field is present only in a short segment of the path of the atoms, which they traverse in a time t_1, the time required for the identification of the atomic state, during which time the field transfers an undeterminable energy of order E_1 to the atoms (see Figure 16.2). The width d of the ray is produced by passing the atoms (just prior to their transit of the field) through an aperture of width d, which then inevitably induces a diffractive-deflection of the order λ/d, where the de Broglie wavelength is $\lambda = h/p$. The deflection induced by the magnetic field needed to distinguish states of different magnetic moment (associated with distinct atomic stationary states) must be at least as large as the diffraction-induced deflection $h/(dp)$. This magnetic deflection (angle ε in Figure 16.2) is equal to the momentum imparted by the field divided by the momentum p of the atom. The momentum imparted by the field, in turn, is equal to the force times t_1, while the force is equal to E_1/d, so the angular deflection caused by the field is equal to $E_1 t_1/(dp)$. It follows that

$$E_1 t_1 \sim h. \tag{16.114}$$

In the final section of the paper (pp. 193–195), Heisenberg returns to yet another thought experiment in which the mutual uncertainty in energy and time is explored, in that case the time "uncertainty" (now written Δt, as in modern discussions) being the observation time required to distinguish two stationary states differing by energy ΔE, with $\Delta E \Delta t \sim h$ (see below). The subtle distinctions in the various operational definitions of time and energy in quantum mechanics (cf. note 90) underly the rather confusing profusion of different physical realizations of the $\Delta E \Delta t \sim h$ uncertainty relation already visible in this first treatment of Heisenberg.

In section 2, entitled "The Dirac–Jordan theory" (pp. 179–184), two main topics come into focus. First, the specific form of the probability amplitude for position given momentum (in Heisenberg's notation $S(q,p) = e^{ipq/\hbar}$) as given by Jordan[91] is used to construct an analytically simple Gaussian wave packet satisfying Eq. (16.111). In fact, this example turns out to saturate the more general form of the uncertainty relation giving the lower bound on the mutual uncertainty of canonically conjugate quantities (not obtained by Heisenberg in this paper). Second, there is an illuminating discussion of a key feature of quantum amplitudes, called "interference of probabilities" by Jordan (see Sections 16.2.1 and 16.2.2), using an ingenious thought experiment involving a sequence of Stern–Gerlach experiments.

[91] In our discussion of Jordan's theory, this amplitude appeared as $\mathcal{A}(q,p)$, the complex conjugate of the amplitude $\mathcal{A}(p,q)$ (see Eqs. (16.28 and (16.24)).

We have already discussed the behavior of wave packets in quantum theory at various points—initially in connection with de Broglie's discussion of the group velocity of "phase waves" (cf. Section 13.2). Here, Heisenberg uses the Dirac–Jordan transformation calculus to exhibit the Fourier transform connection between representations of a state in coordinate and momentum space already described in Pauli's long letter of October 19, 1926 (Pauli 1979, Doc. 143, quoted in Section 16.1). Let η denote the value of some measurable observable (e.g., the energy of an atomic stationary state or of a free electron). Then $S(\eta, q)$ represents, in Heisenberg's language, the transformation matrix from the energy to the coordinate basis. In other words, it is just the Schrödinger energy eigenfunction $S(\eta, q) = \psi_{(E=\eta)}(q)$.[92] Here, the letter η is simply used to indicate that *some particular definite quantum state* (not necessarily an energy eigenstate!) is under consideration—it plays no essential role in what follows.

In the particular case of a free particle localized at some time in a Gaussian wave packet of width q_1 around the central point q', and with mean momentum p', we can write (ignoring normalization factors):[93]

$$S(\eta, q) \propto e^{-\frac{(q-q')^2}{2q_1^2} - \frac{i}{\hbar}p'(q-q')}. \tag{16.115}$$

From the Jordan–Dirac completeness relation (see Eq. (16.25) and note 46 in Section 16.2.2) it follows that

$$S(\eta, p) = \int S(\eta, q)\, S(q, p)\, dq. \tag{16.116}$$

Using $S(q, p) = e^{ipq/\hbar}$, we then find:

$$\begin{aligned} S(\eta, p) &\propto \int e^{-\frac{(q-q')^2}{2q_1^2} - \frac{i}{\hbar}p'(q-q') + ipq/\hbar} dq \\ &\propto e^{ipq'/\hbar} \int e^{-\frac{(q-q')^2}{2q_1^2} - \frac{i}{\hbar}(p'-p)(q-q')} dq \\ &\propto e^{-\frac{q_1^2}{2\hbar^2}(p-p')^2} \cdot e^{ipq'/\hbar}. \end{aligned} \tag{16.117}$$

If we write the Gaussian factor appearing in the last line in the form

$$S(\eta, p) \propto e^{-(p-p')^2/(2p_1^2)}, \tag{16.118}$$

indicating a width, or "uncertainty", in the momentum space wave function of p_1, we can then read off directly the following relation between the widths q_1 and. p_1 in coordinate

[92] Heisenberg already made this identification in a letter to Jordan of November 24, 1926. We already quoted the relevant passage in Section 16.2.1.

[93] To avoid the proliferation of annoying factors of π, we use, as previously, the modified Planck's constant $\hbar = h/2\pi$, which had not at this point reached wide usage.

momentum space, respectively:

$$p_1^2 = \frac{\hbar^2}{q_1^2} \quad \text{or} \quad p_1 q_1 = \hbar. \tag{16.119}$$

The modern expression of this relation would use the variance (root-mean-square deviation) of these Gaussians instead, related to Heisenberg's uncertainties through $\Delta p = p_1/\sqrt{2}$ and $\Delta q = q_1/\sqrt{2}$, so the mutual uncertainty formula for Gaussian wave functions would read

$$\Delta p \cdot \Delta q = \hbar/2. \tag{16.120}$$

In other words, the wider the wave packet in coordinate space, the narrower it is in momentum space, and vice versa.

Heisenberg follows this graphical illustration of the correlation of uncertainties in position and momentum with a prescient discussion of the relation between the choice of representation implied by a given experimental procedure and the "principal axes" that determine the diagonality of the matrix representing any particular physical quantity the experiment attempts to measure. The procedure (which here really means the particular quantum state which the experimenter "prepares" for measurement) in and of itself defines a "direction" which selects only a certain specific set of physical quantities which may be measured without uncertainty. All other observables (if measured repeatedly on identically prepared systems) will give a dispersion of values, and it is the variance of these observations that is being described mathematically by Heisenberg's quantities p_1, q_1, etc. In Chapter 17, we will see how these comments were made completely precise and transparent in von Neumann's (1927a, 1927b) formulation of quantum mechanics in Hilbert space, with the use of projection operators to describe measurements on quantum ensembles.

The thought experiment chosen by Heisenberg (pp. 182–183) to elucidate the probabilistic import of interfering quantum amplitudes—and sharpen the contrast between his (and Dirac's) view of the statistical character of the theory with that advanced by Jordan—has become the model for numerous explanations of quantum amplitudes in modern introductory texts on quantum mechanics.[94] A beam of atoms, all prepared in an identical stationary state n, is passed through two successive Stern–Gerlach magnets, the first one specified by an inhomogeneous field which we denote F_1 (without further specification). The second magnet (with field F_2) is separated by a gap spacious enough to allow the insertion of some apparatus capable of determining the particular atomic state m traveling in the field free region between F_1 and F_2. Both fields are strong enough to induce, by a process that Heisenberg colorfully calls "shaking-action" [*Schüttelwirkung*], atomic transitions, with the amplitude for the transition $n \rightarrow m$ induced by field F_1 written as the (complex) number c_{nm}, while the amplitude for a subsequent

[94] See, for example, McIntyre (2012, pp. 1–10). For a list of other examples, see Janas, Cuffaro, and Janssen (2022, p. 203, note 41).

transition $m \to l$ induced by the second magnet F_2 is written as d_{ml}. As the fields F_1 and F_2 may differ, the c and d matrices may of course be different. The discrete version of Eq. (16.116) implies in this case that the momentum space wave function for an atom initially in the state n undergoes the following transformation in passing through the apparatus (p. 183, Eq (8)):

$$\begin{aligned} S(E_n, p) &\xrightarrow{F_1} \sum_m c_{nm}\, S(E_m, p) \\ &\xrightarrow{F_2} \sum_{m,l} c_{nm}\, d_{ml}\, S(E_l, p) \\ &\equiv \sum_l e_{nl}\, S(E_l, p), \end{aligned} \tag{16.121}$$

where $e_{nl} \equiv \sum_m c_{nm}\, d_{ml}$. It is evident that the relative probability for determining that an atom (initially in state n) proceeding without any disturbance (other than the two magnets) through the whole apparatus to be found (after F_2) in state l is given just by

$$e_{nl}\, \bar{e}_{nl} = \Big|\sum_m c_{nm}\, d_{ml}\Big|^2. \tag{16.122}$$

If we instead imagine an experiment in which the atoms are sent through one by one, and the state of each atom determined in the gap between the two magnets by measurement, then the relative probability of finding the atom to be in state m in the gap is clearly given by $c_{nm}\, \bar{c}_{nm} = |c_{nm}|^2$, while the probability of the atom (known to be in state m) to further proceed through the second magnet and emerge in state l is clearly $d_{ml}\, \bar{d}_{ml} = |d_{ml}|^2$. In this situation, where the state of every atom is "inspected" in the gap between the magnets, the relative probability of finding the emerging atoms in state l is evidently, *by the usual laws of probability theory*,

$$Z_{nl} = \sum_m c_{nm}\, \bar{c}_{nm}\, d_{ml}\, \bar{d}_{ml}, \tag{16.123}$$

which is different from the relative probability in Eq. (16.122). As Heisenberg puts it:[95]

> This expression [i.e., Eq. (16.123)] does not agree with $e_{nl}\, \bar{e}_{nl}$ [our Eq. (16.122)]. Jordan [here Heisenberg cites *Neue Begründung* I] for this reason talks about an "interference of probabilities." I, however, do not wish to accede to this view. For the two experiments, which lead respectively to [$e_{nl}\, \bar{e}_{nl}$] or Z_{nl}, are really physically distinct. In one case the atom experiences no disturbance between F_1 and F_2, in the other it is disturbed by the apparatus, which enables the determination of the stationary state [in the gap] (p. 183).

95 Von Neumann (1927b, p. 246) would take this criticism to heart in the second installment of his 1927 trilogy on quantum mechanics (see note 63 and Section 17.2).

As Heisenberg further explains (somewhat elliptically), this change in the probabilities is physically due to the fact that the gap apparatus inserted to determine the state m necessarily introduces a random phase factor $e^{i\beta_m}$. It is worth taking the time to amplify Heisenberg's argument here. As the phase factors β_m for different states m are random and uncorrelated, the average over an ensemble of measurements of their product must satisfy

$$\langle e^{i\beta_m} e^{-i\beta_{m'}} \rangle = \delta_{mm'}. \tag{16.124}$$

Inserting these phases, and performing an ensemble average (denoted by $\langle \cdots \rangle$) over them in the calculation of the $n \to l$ transition probability, one finds

$$\begin{aligned}\langle |\sum_m c_{nm}\, e^{i\beta_m}\, d_{ml}|^2 \rangle &= \sum_{mm'} \langle c_{nm}\, d_{ml}\, e^{i\beta_m}\, \bar{c}_{nm'}\, \bar{d}_{m'l}\, e^{-i\beta_{m'}} \rangle \\ &= \sum_{mm'} c_{nm}\, d_{ml}\, \bar{c}_{nm'}\, \bar{d}_{m'l}\, \delta_{mm'} \\ &= \sum_m c_{nm}\, \bar{c}_{nm}\, d_{ml}\, \bar{d}_{ml} = Z_{nl}.\end{aligned} \tag{16.125}$$

Thus, the act of ascertaining the state energies in the gap produces a physically distinct alteration in the probabilities after passage through the whole apparatus, by removing the interference terms one would obtain in the squared sum for $e_{nl}\bar{e}_{nl}$. This clearly has nothing to do with any deep, and mysterious, alteration of the rules of probability. Both Dirac and Heisenberg were quite clear on this point.[96]

Section 3, entitled "The transition from micro- to macro-mechanics" (pp. 184–189), begins with a criticism of Schrödinger's (1926g) article of almost the same name (differing only in the qualifier "continuous" before transition) in *Die Naturwissenschaften*. Heisenberg's criticism echoes rather closely Lorentz's criticisms from May 1926 on wave packet spreading (see Section 14.6). Schrödinger had shown that one could construct a superposition of harmonic-oscillator eigenfunctions which, employing high quantum number states, correspond to highly concentrated wave packets bobbing back and forth around the force center *without changing their shape (and in particular their width)* in time. Such objects approach very closely the classical picture of point particles obeying the classical equations of motion—indeed, arbitrarily so, by taking sufficiently high quantum numbers. Heisenberg was at pains to point out (a fact of which Schrödinger was undoubtedly aware) that this appealing behavior was entirely due to the extremely simple character of harmonic oscillator dynamics. One could construct similar wave packets for electrons bound in high orbits of the hydrogen atom, initially concentrated in regions small compared to the Bohr radius of the orbit, and following a Bohrian path, but they would inevitable spread, very soon smearing the electron over the entire volume of the

[96] A perfectly analogous example of the kind of destructive interference Heisenberg considered here is the familiar two-slit experiment, when one introduces, for example, light sources adjacent to each slit to "spy" on the electrons as they pass through the slits.

atom. This was clearly not a fruitful way to imagine a transition from quantum to classical behavior.

Instead, Heisenberg insisted that the only way to understand the emergence of a "classical path" for an electron (say), whether bound in an atom or free, was to imagine a consecutive sequence of approximate position and momentum measurements (each pair of measurements constrained, of course, by the $p_1 q_1 \sim h$ requirement) where one would at least obtain a sequence of approximate $(q(t), p(t))$ values defining a (somewhat blurry) path in classical phase space. Presumably, something analogous to this was at work in a cloud chamber where one could distinguish a definite track for charged particles, the width of which (although macroscopic) did not seem to grow linearly with time as a purely quantum-mechanical treatment (see below) would imply.[97] Presumably, the process Heisenberg describes of sequential approximate p, q measurements, which "relocalize" the evolving wave packet of the particle, are accomplished physically in the cloud chamber by the production of ionized droplets along the path that act as seeds for growth to visible size.

Toward the end of section 3 (pp. 187–188), Heisenberg once again applied the Jordan–Dirac formalism to the calculation of the amplitude $S(\eta, q)$, where the particle state denoted by η now corresponds to a particle described by a wave packet of width q_1 as in Eq. (16.115) at time $t = 0$ and q refers to the position at time t. The calculation, which we omit here, is yet another straightforward exercise in Fourier transformation. The essential result is the famous "spreading of the wave packet": the width, or position uncertainty, grows, for large time, linearly as $q_1(t) \propto \hbar t/mq_1(t=0)^2$. For an electron in the ground state of a hydrogen atom which suddenly "loses" its proton, the resulting wave packet (clearly of width $q_1 \sim a$, with a the Bohr radius), this growth factor can easily be calculated to be about 4×10^{16} per second, leading to a wave packet extending over meters in a mere microsecond. This is evidently not the way charged particles behave in cloud chambers![98]

Section 4, the final section of the paper, entitled "Discussion of some particular thought experiments" (pp. 189–195), is primarily concerned with examining the consequences of the energy-time uncertainty principle. The first example Heisenberg gives is that of determining the time at which "quantum jumps" occur between stationary states:

> The exactness with which such a point of time [of a quantum jump] can be determined is, according to Eq. (2) [i.e., $E_1\, t_1 \sim h$], given by $h/\Delta E$, where ΔE denotes the change in energy in the quantum jump (p. 189).

[97] That Heisenberg was concerned with this issue, which he had discussed intensely with both Einstein and Bohr, is indicated in Heisenberg (1971, p. 77), where he claims to have "concentrated all my efforts on the mathematical representation of the electron path in the cloud chamber," in the period following Bohr's departure for Norway in February 1927.

[98] See also Chapter 14, note 106, for the remarkable case of neutrino wave packet spreading after the transit from supernova SN1987A, which were detected in rapid succession on Earth, despite a huge spreading factor *in the neutrino's rest frame.*

This result is illustrated by consideration of the behavior of an atomic beam passing through a Stern–Gerlach magnet, where the atoms are initially assumed to be in a stationary state with energy E_2, which subsequently decays to a state with a lower energy E_1. The wave function is then a linear combination of two terms with exponential behavior $e^{-iE_2t/\hbar}$ (high-energy state) and $e^{-iE_1t/\hbar}$ (low-energy state), and an examination of the energy is carried out at distinct time intervals Δt. Formally, this amounts to Fourier transforming the wave function from time to energy space, with the time integral extending only over an extent Δt. It is easy to show that the two components of the wave function can only be distinguished if Δt is of the order of (or larger than) $h/\Delta E$. Heisenberg goes on to consider two further examples. The first (pp. 191–193) concerns the phase of electronic motion in resonance fluorescence, where now the electron state oscillates periodically between two stationary states with an energy difference matched to the incident radiation frequency. The second (pp. 193–195) concerns the energy in a rotator model of Ehrenfest and Breit (1922). We do not discuss these examples in further detail here, as they add little to the central conceptual import of the paper.

In the final two pages, Heisenberg summarizes the main points of the article (pp. 195–197):

1. The kinematical concepts of classical theory (momentum, position, energy, etc.) are all exactly definable concepts in the quantum theory. They are however subject to fundamental limitations, such as the mutual uncertainty constraints (between canonically conjugate quantities) embodied in the fundamental relation $p_1\, q_1 \sim h$.
2. There is no need to regard quantum mechanics as unintuitive and abstract, insofar as thought experiments can be devised which allow us to visualize (at least qualitatively) the experimental consequences of the theory in all simple cases.
3. The fundamental completeness relation of the Jordan–Dirac theory,

$$S(q, q'') = \int S(q, q')\, S(q', q'')\, dq', \tag{16.126}$$

   should *not* be regarded as a “probability relation” (cf. our discussion above about “interference of probabilities”).
4. The fact that kinematical quantities like the position or momentum of an electron are represented in the theory by matrices rather than simple numbers leads naturally to the conclusion that their values in certain circumstances cannot be determined exactly, but have a natural dispersion, around a mean value corresponding to a diagonal element of the matrix, with the amount of the dispersion related to the size of the off-diagonal elements of the matrix.
5. The acausal nature of the theory has to be understood in the context of the fundamental experimental limitations imposed by the uncertainty relations. To the extent that quantities can be measured with arbitrary precision, the classical conservation laws (energy, momentum, etc.) continue to hold in quantum mechanics. As regards causality, Heisenberg writes

> In the sharp formulation of the law of causality, "if we know the present exactly, then we can calculate the future", it is not the consequence, but the premise that is false. In principle, we *cannot* get to know all the determining factors [*Bestimmungsstücken*] of the present. Therefore all observation is a selection from a plenitude of [present] possibilities, and a restriction of future ones (p. 197).

On the question of whether the phenomena of quantum mechanics ultimately derive from an underlying causal dynamics, Heisenberg declares himself agnostic (and speculation on the subject "unfruitful and senseless"). The real situation should instead be summarized as follows (cf. note 89):

> As all experiments are subject to the laws of quantum mechanics, and consequently to Eq. (1) [$p_1 q_1 \sim h$], the invalidity of the law of causality on the basis of quantum mechanics can be incontrovertibly established (p. 197).

The paper ends with a "note added in proof" [*Nachtrag bei der Korrektur*][99] of well over half a page (pp. 197–198) in which Heisenberg acknowledges various objections by Bohr that he addressed as he was correcting the page proofs. Probably in an attempt to smooth things over with Bohr after their contentious discussions during this period, Heisenberg credits Bohr's interventions with leading to "an essential deepening and refinement" of his paper and admits that Bohr had shown him that "in some passages in the paper, [he] had overlooked essential points." The main criticism by Bohr mentioned by Heisenberg involves the analysis of the gamma-ray microscope, where Bohr had insisted on treating light both as a wave and a particle.[100] In conclusion, Heisenberg thanks Bohr for sharing the preliminary results of some new investigations, which, Heisenberg announces, "will soon be published in an article on the conceptual structure of the quantum theory" (p. 198). This appears to be a reference to Bohr's famous complementarity paper, which would only be published the following year (Bohr 1928a, 1928b).

16.4 Discussions of the new quantum theory in Como and Brussels 1927

In this section, we turn to two famous conferences in 1927 where many of the founding fathers of quantum theory met and discussed the theory and its interpretation.[101] One is

[99] Quoted and discussed by Mehra and Rechenberg (2000–2001, 185–186). According to Cassidy (2009, p. 170), this note was added as Klein's urging.

[100] For a detailed discussion of Bohr's elaboration of the gamma-ray microscope thought experiment, see Eisberg and Resnick (1985, pp. 66–68), who refer to it as "Bohr's microscope thought experiment". A careful examination of the experimental setup requires, as Bohr showed, application of both the classical wave formula for the resolving power of a microscope (the Abbe formula), as well as the Compton effect momentum transfer formula arising from the light-quantum aspects of the gamma rays used to examine an electron.

[101] For another discussion of the early debates over the interpretation of quantum mechanics, see Jähnert and Lehner (2022).

a conference held in Como, September 11–20, 1927 to mark the centenary of the death of Alessandro Volta; the other is the fifth Solvay conference, held in Brussels, October 24–29, 1927 (with an excursion to Paris for an event to commemorate the centenary of the death of Augustin Fresnel). Whereas quantum mechanics took center stage at the Solvay conference, only one day, September 16, was set aside for it at Como. Accordingly, the talks presented that day take up only one of eight chapters of the proceedings (Como 1928, Vol. 2, Ch. VI, pp. 433–598). Moreover, the apotheosis of the conference was not the lecture by Bohr for which the conference is best remembered today but an address commemorating Volta by Italy's own Guglielmo Marconi (held in Rome rather than Como).

We focus largely on the lecture Bohr gave in Como and essentially reprised in Brussels and on the long gestation period of the famous complementarity paper loosely based on these presentations. This paper, entitled "The quantum postulate and the recent development of atomic theory," only appeared the following year, first in German in the issue of *Die Naturwissenschaften* of April 13, 1928 (Bohr 1928a), a day later in English in a special "Supplement" to *Nature* (Bohr 1928b). In July 1928, a somewhat different English version appeared in the proceedings of the Como conference (Bohr 1928c).[102] A French translation of the German version, finally, was included in the proceedings of the Solvay conference (Bohr 1928d). The final text(s) of Bohr's paper thus only appeared after the period covered in this book. However, through correspondence and through Bohr's presentations in Como and Brussels, many physicists were familiar with its contents well before the paper was published, some (such as Pauli and Heisenberg) more intimately than others. Rather than giving a detailed analysis of the finished paper (see Section 8.9 for some comments on its contents), we focus here on the impressions Bohr's developing ideas made in 1927.[103]

To trace the origins of Bohr's complementarity paper, we go back to July 1926, when Schrödinger, at the invitation of Sommerfeld, presented his new theory in Munich. Heisenberg, based in Copenhagen at the time (see Section 16.3) but visiting his parents in Munich, attended Schrödinger's seminars. He did not like what he heard, tried to raise some objections, but was silenced by Wien, who, as Heisenberg recalled decades later, told him that "one must really put an end to quantum jumps and the whole atomic mysticism and [that] the difficulties [Heisenberg] had mentioned would certainly soon be solved by Schrödinger" (Heisenberg 1967, p. 103; see also Heisenberg 1971, p. 73). As Heisenberg told Pauli at time: "Just as nice as Schrödinger is as a person, just as strange I find his physics ... Schrödinger throws everything 'quantum-theoretical' overboard:

[102] See Bohr (1972–2008, pp. 110–111) for a list of differences between the *Naturwissenschaften–Nature* version (Bohr 1928a, 1928b) and the Como version (Bohr 1928c).

[103] For this section, we have drawn heavily on the following sources (cf. note 77): Kalckar (1985) in the *Collected Works* of Bohr (1972–2008, Vol. 6, pp. 7–53); Pais (1991, Ch. 14, secs. (c)–(f), pp. 300–320) in his biography of Bohr; Mehra and Rechenberg (2000–2001, sec. II.4, pp. 163–199, and sec. II.6 (a)–(f), pp. 232–260), who themselves follow Kalckar (as they acknowledge on p. 165); and Bacciagaluppi and Valentini (2009) on the 1927 Solvay conference. Kalckar and Pais were both great admirers of Bohr (see, e.g., Kalckar 1967, Pais 1967). For a critical review of the history of this period by someone who, unlike these two authors, was not sympathetic to Bohr's views on quantum mechanics, see Beller (1999, especially Chs. 6 and 11).

Photo-electric effect, Franck collisions, Stern–Gerlach effect etc."[104] Heisenberg had already expressed his strong aversion to wave mechanics in another letter to Pauli the previous month. The more he thought about it, he had confided in that letter, "the more repulsive [*desto abscheulicher*] I find it."[105] Of course, Heisenberg was only returning the favor. Section 14.5 discussed how Schrödinger (1926e, p. 46) had declared in print that he had been "repelled" [*abgestoßen*] by Heisenberg's theory. He used even stronger language in private correspondence. In a letter to Wien of February 22, 1926 (from which we already quoted in Section 14.5), he wrote: "the matrix calculus was unbearable [*unerträglich*] to me long before I even had an inkling of my own theory ... I shudder [*mir schaudert*] at the thought that I would one day have to present the matrix calculus to a young student as the true essence of the atom" (von Meyenn 2011, pp. 186–187).

After attending Schrödinger's lectures in Munich, Heisenberg recalled in his scientific autobiography, "I went home rather sadly. It must have been that same evening that I wrote to Niels Bohr about the unhappy outcome of the discussion. Perhaps it was as a result of this letter that he invited Schrödinger to spend part of September in Copenhagen. Schrödinger agreed, and I, too, sped back to Denmark" (Heisenberg 1971, p. 73, quoted in Pais 1991, p. 298)

Whether or not Heisenberg actually wrote this letter,[106] Bohr did invite Schrödinger to Copenhagen. Bohr's reaction to wave mechanics had not been as negative as Heisenberg's. Whereas Heisenberg only conceded that it provided some techniques for simplifying calculations, Bohr was more sympathetic as Schrödinger's wave-imagery fit with his own developing ideas about wave-particle duality. Section 16.3 showed how this contributed to the tensions between Bohr and Heisenberg in early 1927. Like Heisenberg, however, Bohr did not care for Schrödinger's attempts to interpret his wave function as representing some charge density and to do away with quantum jumps (see Chapter 14).

On October 4, 1926, Schrödinger gave a lecture at the Danish Physical Society on the "Foundations of the undulatory mechanics" (Pais 1991, p. 298). But it was in informal discussions during the rest of his visit that Bohr pressed Schrödinger relentlessly on the perceived inadequacies of his interpretation of the theory. In his autobiography, Heisenberg gives a colorful and no doubt somewhat embellished (cf. note 79) account of their discussions:[107]

[104] Heisenberg to Pauli, July 26, 1926 (Pauli 1979, Doc. 142).

[105] Heisenberg to Pauli, June 8, 1926 (Pauli 1979, Doc. 136).

[106] No such letter is to be found in the archives and Heisenberg himself was not sure he wrote it. Following the passage we just quoted about Wien preventing him from criticizing Schrödinger, Heisenberg (1967, p. 103) wrote: "I no longer remember whether or not I wrote to Bohr of this encounter in Munich."

[107] Heisenberg (1971, p. 73) wrote that although Bohr was "normally most considerate and friendly in his dealings with people, he now struck me as an almost remorseless fanatic," who, as Heisenberg (1967, p. 103) put it on another occasion, "was able in such a discussion, which concerned epistemological problems which he considered of vital importance, to insist fanatically and with almost terrifying relentlessness on complete clarity in all arguments."

Schrödinger: "If all this damned quantum jumping were really here to stay then I should be sorry I ever got involved with quantum theory."

Bohr: "But the rest of us are extremely grateful that you did; your wave mechanics has contributed so much to mathematical clarity and simplicity that it represents a gigantic advance over all previous forms of quantum mechanics."

And so the discussion continued day and night. After a few days Schrödinger fell ill, perhaps as a result of the enormous stress ... While Mrs. Bohr nursed him and brought in tea and cake, Niels Bohr kept sitting on the edge of the bed talking at Schrödinger: "But you must surely admit that ... " (Heisenberg 1971, pp. 75–76).

Schrödinger does not mention these histrionics at his sickbed when he wrote to Bohr shortly after his return to Zurich.[108] Yet his long letter confirms the gist of Heisenberg's account. After thanking his host profusely for his hospitality, Schrödinger apologizes for his stubbornness only to resume his vigorous defense of his own views.

We also have Bohr's account of Schrödinger's visit in a letter to Fowler of October 26, 1926 (Bohr 1972–2008, Vol. 6, pp. 423–424; quoted by Kalckar 1985, p. 14–15). This letter starts with "We had great pleasure of the visit of Schrödinger" and ends with "After the discussions with Schrödinger it is very much on my mind to complete a paper dealing with the general principles of the quantum theory." This visit by Schrödinger, then, appears to have been the initial stimulus for Bohr to start writing what would become his complementarity paper.

Initially, Bohr developed his views using Heisenberg as his sparring partner. In Section 16.3, we already covered the period of intense discussions between the two men in late 1926/early 1927. We saw that Heisenberg's breakthrough occurred shortly after Bohr had left Copenhagen for a vacation in mid-February 1927. Bohr apparently had his own epiphany while skiing in the mountains around Gudbrandsdalen in Norway in February–March 1927 (Pais 1991, p. 310). On April 13, 1927, about a month after his return to Copenhagen (see note 80), he presented an early version of his ideas about complementarity in a long letter to Einstein (reproduced in Bohr 1972–2008, Vol. 6, pp. 418–421; English translation: pp. 21–23). A couple of sentences from this letter will give the flavor of Bohr's thinking at the time:

It has of course long been recognized how intimately the difficulties of quantum theory are connected with the concepts, or rather with the words that are used in the customary description of nature, and which all have their origin in the classical theories ... This very circumstance that the limitations of our concepts coincide so closely with the limitations in our possibilities of observation, permits us—as Heisenberg emphasizes—to avoid contradictions (quoted in Kalckar 1985, p. 21).[109]

[108] See Schrödinger to Bohr, October 23, 1926 (Bohr 1972–2008, Vol. 6, pp. 459–461). An English translation of part of this letter is included Kalckar's (1985, pp. 12–13) introduction to this volume.

[109] In this letter, Bohr also touches on a paper by Einstein (1926b) related to what turned out to be fraudulent experiments by Emil Rupp to distinguish between a wave and a particle theory of light (van Dongen 2007a, 2007b).

In this letter, Bohr repeatedly expresses his admiration for Heisenberg's (1927b) uncertainty paper, despite the grief he was giving its author over it during this exact same period (see Section 16.3). After calling it "a very important contribution" [*einen äusserst bedeutungsvollen Beitrag*] at the beginning of his letter, Bohr gets even more effusive in his praise toward the end: "Heisenberg shows in a truly inspired way [*in überaus geistreicher Weise*] how his uncertainty relations may be utilized not only in the actual development of quantum theory, but also for the judgment of its visualizable content."

The period of intense discussion between Bohr and Heisenberg came to an end in the summer of 1927, when Heisenberg left Copenhagen to take up a professorship in Leipzig.[110] New inspiration for Bohr came from a note by Norman Campbell (1927) in *Nature* in the issue of May 26, 1927, in response to Jordan's (1927d) habilitation lecture, which had been published in the issue of April 16 in Oppenheimer's translation (see note 25). Campbell's note was followed by a short reply by Jordan (1927f). All three pieces bore the title "Philosophical foundations of quantum theory." Bohr decided that he wanted to weigh in on this discussion. He started working on a short paper with the same title. Several drafts survive but no article was submitted. At the same time, he produced drafts for the longer paper he had envisioned when writing to Fowler shortly after Schrödinger's visit in October 1926.[111]

Most of these drafts are in the hand of Oskar Klein, who succeeded Heisenberg as Bohr's assistant (Cassidy 2009, p. 172). Unlike his predecessor, Klein was not biased against Schrödinger's wave mechanics, which he had used in some of his own papers (Mehra and Rechenberg 2000–2001, pp. 174–180). Moreover, Klein had sided with Bohr in his dispute with Heisenberg over the latter's uncertainty paper concerning the use of wave imagery (see Section 16.3). Talking about his work with Bohr in 1927 in an interview in 1968, Klein recalls that "Bohr dictated and the next day all he had dictated was discarded and we began anew. And so it went all summer" (quoted in Pais 1991, p. 311).[112]

The term "complementarity" makes it first appearance in a draft of July 10, 1927 (Kalckar 1985, p. 27).[113] Pais (1991, p. 311) suggests that Bohr may have decided to use

[110] On Heisenberg's appointment in Leipzig, see Cassidy (2009, p. 164 and pp. 171–172)

[111] On June 10, 1927, Bohr wrote to Fowler again, telling him: "At present I am busy myself preparing a paper on the more philosophical aspect of quantum theory, which as you know I have planned so long, but for which I think I now have collected the material" (quoted in Mehra and Rechenberg 2000–2001, p. 190). That Bohr was working on these two papers at the same time, Mehra and Rechenberg (p. 190) note, has complicated matters for historians. Kalckar and Pais, in fact, do not distinguish between drafts for one or the other.

[112] See also Klein (1967, p. 89). Pais is quoting from an interview with Oskar Klein by Léon Rosenfeld and Jørgen Kalckar, November 7, 1968 (Niels Bohr Archives). Klein continues: "after a time Mrs. Bohr became unhappy … one time when I sat alone in the little room where we worked she came in crying." This appears to be yet another instance of the impact Bohr's single-minded obsessions could have on those around him. We already recorded its effect on Kramers (see Chapter 10, note 72), Heisenberg (see Section 16.3), and Schrödinger (see above, especially note 107).

[113] This manuscript (in Danish) is presented in facsimile (pp. 63–65), transcription (pp. 59–60), and translation (pp. 61–62) in Bohr (1972–2008, Vol. 6).

this term while sailing with some friends.[114] Both Kalckar (1985, p. 27) and Pais (1991, pp. 310–311) emphasize that the notion of complementarity was the result of Bohr's grappling with the physics (and his discussions about it with Heisenberg and others), and *not* of his reading of various philosophers. More specifically, Kalckar writes:

> Thus Meyer-Abich [1965, pp. 133–140] believes that he can detect influences from *The Principles of Psychology* by William James, whereas Max Jammer [1966, pp. 172–179 and pp. 348–350] is convinced that Bohr was influenced not only by James, but even by Kierkegaard and Harald Høffding. To anyone familiar with Bohr's style of thinking and working these conjectures appear highly unlikely (Kalckar 1985, p. 27).[115]

Bohr kept drafting and redrafting both papers he was working on right up to the time he left for Como on September 7, 1927.[116] He had not found the time to prepare a text for the lecture he was supposed to give there before his departure. The closest thing we have to a manuscript for his lecture is an eight-page document in Bohr's own handwriting and almost illegible in places, still with the same title as the pieces by Jordan and Campbell in *Nature*, and dated September 13, 1927 (Kalckar 1985, p. 29).[117] As Mehra and Rechenberg (2000–2001) point out, this date suggests that Bohr only prepared the text for his lecture while he was already in Como.

Many of the protagonists of our story gathered in Como, among them, in addition to Bohr: Born, Compton, Debye, Heisenberg, Kramers, von Laue, Lorentz, Millikan, Paschen, Pauli, Planck, Rutherford, Smekal, Sommerfeld, Zeeman, and, as Mehra and Rechenberg (2000–2001) add, "the very interested and able mathematician John von Neumann" (p. 192).[118] The most important figures missing were Einstein, Dirac, and Schrödinger, who were invited but could not attend, and Jordan and De Broglie, who do not seem to have been invited (though the latter's brother Maurice was). Jordan was missing at Solvay as well but the other four were present on that occasion. As we mentioned at the beginning of this section, the session dealing with quantum mechanics took place on September 16, 1927 under the title "Theories of structure of matter and of radiation" (Como 1928, Vol. 2, Ch. VI, pp. 433–598). This left room for a paper by Eddington (1928), "The electrical state of a star," that had nothing to do with quantum mechanics.

[114] A picture of Bohr sailing with these friends, Niels Bjerrum, Ole Chievitz, and Holger Hendriksen, can be found in Bohr (1972–2008, Vol. 6, p. 442).

[115] Mehra and Rechenberg (2000–2001, sec. II.4 (b), pp. 166–169) also discuss this issue, taking into account Kalckar's views, in a subsection entitled "The philosophical influences on Niels Bohr."

[116] In a letter of September 6, 1927, Bohr told Yoshio Nishina that he was leaving the next day (Mehra and Rechenberg 2000–2001, p. 192).

[117] This manuscript (in English) is presented in facsimile (pp. 81–88) and in transcription (pp. 75–80) in Bohr (1972–2008, Vol. 6).

[118] In a footnote, Mehra and Rechenberg relate the amusing story that, due to a mixup with the invitations, D. M. Bose of the University of Calcutta instead of S. N. Bose of Dacca University turned up at the meeting. The wikipedia entry on D. M. Bose calls this story into question, noting that D. M. was the better known of the two Boses at the time. In any event, D. M. Bose gave a talk in a session on the structure of matter on September 12, presided over by Lorentz and Quirino Majorana (not to be confused with his nephew Ettore) with a large number of speakers including William Bragg (not to be confused with his son Lawrence), Maurice De Broglie (not to be confused with his brother Louis), Compton, Franck, Gerlach, Rutherford, Smekal, and Stern (Como 1928, Vol. 1, Ch. II, pp. 53–212)

Quirino Majorana chaired the session. In the proceedings there is an eight-page introduction by the Italian mathematician and physicist Paolo Straneo, followed by papers by Born (1928, "On the importance of collision processes for the understanding of quantum mechanics"), Sommerfeld (on the electron theory of metals using Fermi statistics), Levi-Civita (on adiabatic invariants), Debye (1928, "on electric moments"),[119] and von Laue ("On the temperature dependence of X-ray interference"). After the aforementioned paper by Eddington, a paper by Kramers (1928, "The dispersion of light by atoms") and a paper by the Italian priest and physicist Giuseppe Gianfranceschi ("The physical significance of quantum theory"), comes the final paper, the no doubt much-anticipated lecture by Bohr, followed by a discussion with contributions by Born, Fermi, Heisenberg, Kramers, and Pauli.

Unlike the enthusiastic reception of Bohr's Wolfskehl lectures about the correspondence principle five years earlier in Göttingen—the celebrated *Bohr Festspiele* (see Section 9.1.1)—the reception of his Como lecture on the complementarity principle was, in Kalckar's (1985, p. 29) words, "remarkably cool." Pais (1991, p. 315) put it more bluntly: "Bohr's lecture at Como did not bring down the house." Both Pais and Kalckar relate a comment by Rosenfeld, who was present at Como, in an interview in 1968:[120]

> There was a characteristic remark by Wigner after the Como lecture, "This lecture will not induce any one of us to change his own [opinion] about quantum mechanics" (quoted in Pais 1991, p. 315).

Bohr apparently was fond of repeating Wigner's unflattering observation in later years (Kalckar 1985, p. 29). Given that he is remembered today as one of Bohr's staunchest allies,[121] it may come as a surprise that Rosenfeld himself was also unimpressed by the Como lecture, even after reading the presumably much-improved version published a year later. In the same 1963 interview, Rosenfeld went on to say: "In fact, my own view of the Como lecture *when I read it* was that Bohr was just putting in a rather heavy form things which had been expressed much more simply by Born and which were current in Göttingen at the time" (our emphasis). In defense of Bohr, Kalckar (1985, p. 29) points out that "even Rosenfeld at that time was far from appreciating Bohr's message."

As mentioned above, Bohr essentially reprised his Como lecture at the Solvay conference and had a French translation included in its proceedings (Bohr 1928d). Debriefing some of his (former) students on what had transpired in Brussels,[122] Ehrenfest, despite being on Bohr's side in the latter's exchanges with Einstein, complained about the formulation of Bohr's ideas: "Once again the awful Bohr incantation terminology. Impossible for anybody else to summarize." After quoting this passage, Pais (1991, p. 312), like

[119] In this paper, Debye discusses the problems of the Langevin–Debye formula in the old quantum theory and their resolution in the new one (see Section 15.2).

[120] Interview by Kuhn and Heilbron, July 1, 1963 (Niels Bohr Archive).

[121] See, for example, Mehra and Rechenberg (2000–2001, pp. 165–166) and Freire (2015, Ch. 4).

[122] Ehrenfest to Goudsmit, Uhlenbeck, and Dieke of November 3, 1927. For the German original of the long excerpt from this letter quoted by Kalckar (1985, pp. 37–41), see Bohr (1972–2008, pp. 415–418). We quote more passages from this letter when we discuss the Solvay conference.

Kalckar in the case of Rosenfeld, rallies to Bohr's defense. He expresses "deep sympathy" for Ehrenfest's assessment but adds: "I should stress that in later years Bohr himself vastly improved his presentation of the complementarity concept."

Going by the discussion comments following Bohr's lecture in the proceedings of the Como conference, one would not think the lecture was poorly received. Note, however, that the discussion was dominated by those in Bohr's camp (Mehra and Rechenberg 2000–2001, p. 194). Three of the five discussants whose comments are recorded in the proceedings belonged to Bohr's inner circle: Heisenberg, Pauli, and Kramers. The two exceptions were Fermi and Born. The latter was also very sympathetic to Bohr's views. In fact, the comments section in the proceedings begins with Born's strong endorsement of what Bohr had just said: "Prof. Bohr has expressed the views we have formed about the fundamental concepts of quantum theory in such an astute manner [*in so treffender Weise*] that all that remains for me is to add a few comments" (Bohr 1928c, p. 589).[123] Kramers likewise began his comments with: "I shall not be able to add anything fundamental to Professor Bohr's exposition of the physical principles underlying the new quantum mechanics" (Bohr 1928c, p. 591). Here one has to bear in mind that the most prominent members of the opposition to the Copenhagen–Göttingen camp were absent in Como. Bohr would meet with considerably more resistance in Brussels a month later with Einstein, Schrödinger, and De Broglie present.

After the Como meeting, Bohr spent a week at Lake Como, enlisting the help of Pauli to prepare an article based on his Como lecture. He now changed the title to the one under which the article was eventually published: "The quantum postulate and the recent development of atomic theory" [*Das Quantenpostulat und die neuere Entwicklung der Atomistik*]. Back in Copenhagen, he finished the manuscript and on October 11, 1927, submitted it to *Die Naturwissenschaften.* The article made it to the proof stage but not into print. One explanation for this is that Bohr failed to correct the proofs to his own satisfaction in response to extensive feedback from Pauli. Bohr had asked the editors of *Die Naturwissenschaften* to send a copy of the proofs to Pauli, who marked them up and offered additional suggestions for improvements in the letter with which he mailed the annotated copy back to Bohr.[124] Unfortunately, these proofs are no longer extant, but a 12-page typewritten manuscript with an English translation, which Bohr enclosed with a letter to Darwin, has survived.[125] This manuscript was analyzed by Kalckar (1985, pp. 30–32), who concluded that it is already close to the published paper, though closer to the version published in the proceedings of the Como meeting (Bohr 1928c) than to the one published in *Die Naturwissenschaften* and *Nature* (Bohr 1928a, 1928b). The famous

[123] Born concludes his remarks with the statement "that the various mathematical formalisms [of quantum mechanics] can be seen in a unified way as special cases of a general operator theory. Following the approach of Wiener and myself, Jordan and Dirac have developed this calculus and recently Mr. v. Neumann has given it a mathematically rigorous formulation" (Bohr 1928c, p. 589). Born is referring to the statistical transformation theory of Jordan (1927b) and Dirac (1927a) and to von Neumann's (1927a) Hilbert space formalism. To claim that these papers are somehow offshoots of Born and Wiener (1926) is somewhat self-serving.

[124] Pauli to Bohr, October 17, 1927 (Pauli 1979, Doc. 173). See also Bohr (1972–2008, pp. 432–435) and Kalckar (1985, pp. 32–35) for an English translation.

[125] Bohr to Darwin, October 16, 1927. A transcription of the enclosed document can be found in Bohr (1972–2008, pp. 91–98).

exchanges with Einstein at the Solvay conference, he therefore concludes, do not seem to have left important traces in the published paper.[126]

A month after the Como meeting, its most prominent attendees gathered again in Brussels for the fifth Solvay conference: Bohr, Born, Heisenberg, Pauli, Kramers, Lorentz, Compton, Debye, and Planck. This time, however, De Broglie, Dirac, Ehrenfest, Einstein, and Schrödinger were also in attendance. The topic of the meeting was "The quantum theory and the classical theories of radiation,"[127] though its proceedings were eventually published under the title "Electrons and photons" (Solvay 1928), using the term for light quanta introduced by the American chemist Gilbert N. Lewis (1926).

It is interesting to compare the famous group portrait for the conference (see Plate 19) with the one for the first Solvay congress sixteen years earlier (see Volume 1, Plate 9). Several of the leading participants in the 1927 conference were too young to have been at the inaugural one in 1911. This is true not only for Kramers, Pauli, Heisenberg, and Dirac, who were born in 1894, 1900, 1901, and 1902, respectively, but also for Bohr, who had just obtained his doctorate in 1911. Born, sitting next to Bohr in the 1927 picture, while old enough to have attended the 1911 meeting, only developed an interest in quantum theory later. Ehrenfest, diagonally behind Lorentz and Kramers, had already been one of the leaders in quantum theory in 1911 but missed the meeting as he was living in faraway St. Petersburg at the time. He only succeeded Lorentz in Leyden in 1912. Rutherford, who filled in Bohr on what had been discussed at the 1911 meeting (see Volume 1, p. 9 and p. 148), was not invited in 1927. Neither was another important participant in 1911: Sommerfeld. No German physicists were invited to the third and fourth Solvay conferences in 1921 and 1924 and there were lingering concerns about inviting them to the fifth (Bacciagaluppi and Valentini 2009, pp. 6–8). As the group photo shows, a substantial contingent was invited, but not Sommerfeld. Because of this, Planck initially hesitated but eventually accepted the invitation.[128] Jordan was also passed over, as he had been in Como. Comparing the group photos of 1911 and 1927, one cannot help but notice how much some of the participants had aged in the sixteen years that had passed since the first meeting, especially Planck, Curie, Lorentz, and Einstein, sitting next to each other in the front row. Even Bohr, on the far right in the second row, looks old compared to Pauli and Heisenberg diagonally behind him in the third row.

The best-known account of the 1927 Solvay conference is the one by Bohr, written two decades later as part of his contribution to the volume on Einstein in *The Library of Living Philosophers* (Bohr 1949). However, the liveliest account, written right after the event, is undoubtedly the one in the letter from Ehrenfest to three of his (former) students from which we already quoted above (see note 122). Ehrenfest's exuberant report begins:

[126] Mehra and Rechenberg (2000–2001, pp. 195–198) also discuss this manuscript and arrive at essentially the same conclusions as Kalckar.

[127] See Lorentz to Einstein, April 6, 1926, quoted in Bacciagaluppi and Valentini (2009, p. 9), who discuss both the preparation of the conference and the editing of its proceedings in detail (pp. 8–21).

[128] Planck to Born, June 14, 1927 (Mehra and Rechenberg 2000–2001, p. 233).

> Brussels-Solvay was fine! ... BOHR towering completely over everybody. At first not understood at all ... then step by step defeating everybody ... (Poor Lorentz[129] as interpreter between the British and the French who were absolutely unable to understand each other. Summarizing Bohr. And Bohr responding with polite despair.) ... It was delightful for me to be present during the conversations between Bohr and Einstein. Like a game of chess. Einstein all the time with new examples. In a certain sense a sort of perpetuum mobile of the second kind to break the UNCERTAINTY RELATION. Bohr from out of philosophical smoke clouds constantly searching for the tools to crush one example after the other. Einstein like a jack-in-the-box: jumping out fresh every morning. Oh, that was priceless. But I am almost without reservation pro Bohr and contra Einstein. His attitude to Bohr is now exactly like the attitude of the defenders of absolute simultaneity towards him (quoted by Kalckar 1985, pp. 37–38, preserving some of Ehrenfest's idiosyncratic orthography).

Ehrenfest's letter focuses on the informal discussions outside of the official program.[130] This is understandable as most of the talks contained nothing new. This is especially true for the presentations by Born and Heisenberg (1928) on matrix mechanics and the presentation by Schrödinger (1928) on wave mechanics, delivered in the morning and afternoon sessions of October 26, respectively (see Section 8.9 for some comments on these talks). The speakers mainly defended the positions they had staked out before. In terms of understanding the probabilistic aspects of the theory, it is interesting to note that Born and Heisenberg still maintained that if the state of a system is a linear combination of eigenstates of its unperturbed Hamiltonian, it will nevertheless "from the point of view of Bohr's theory ... always be in only *one* quantum state" (Bacciagaluppi and Valentini 2009, p. 386). Thus the energy will have a definite value, even before a measurement is made: we just don't know what it is until we make a measurement (cf. our discussion of Heisenberg's fluctuation paper in Section 16.1; see also Figure 16.1). Schrödinger was perfectly comfortable with the idea that observables do not have definite values in the case of such superpositions, but he still resisted the statistical interpretation of his wave functions. De Broglie's (1928) presentation in the afternoon session of October 25, the day before those by Born, Heisenberg, and Schrödinger, did break new ground. De Broglie introduced what would become pilot-wave theory but the reception of his ideas was decidedly cold.[131]

In the morning session of October 27, it was Bohr's turn. This is where many of those present heard about complementarity for the first time. As we have seen, Heisenberg, Pauli, Schrödinger, and Einstein had all heard about it before. It is not clear exactly what Bohr said on the occasion. Unlike the other speakers, he did not submit a written version of his talk for the conference proceedings. Instead, as we already mentioned, a French translation of his complementarity paper in *Die Naturwissenschaften* and *Nature* was included in the proceedings of the Solvay conference (Bohr 1928d). It seems safe

[129] This was the last Solvay conference Lorentz presided over. He died suddenly in February 1928 before he could finish editing the proceedings of the conference (Bacciagaluppi and Valentini 2009, p. 20).

[130] See Bacciagaluppi and Valentini (2009, p. 18) for the schedule of the conference.

[131] For discussion of these presentations, especially the one by De Broglie, see Bacciagaluppi and Valentini (2009).

to assume, however, that Bohr essentially reiterated what he had said in Como. After this session, most of the conference participants boarded a train to Paris to attend the opening event that evening of a celebration to mark the centenary of the death of Fresnel. They were back in Brussels the following morning and in the afternoon began the general discussion, which continued the next and final day of the conference, October 29.[132]

The uncertainty relations formed the central topic of this general discussion and, if Ehrenfest's letter is any guide, also of the many informal discussions. Most of Ehrenfest's letter deals with the uncertainty relations. Only toward the end of the letter does he broach, with some trepidation, the subject of complementarity:

> In the article in *Naturwissenschaften* you will see how Bohr constantly returns to the "complementary description" of all experience [Ehrenfest once again returns to the uncertainty relations for a few sentences] Bohr says: For the time being we have at our disposal only those words and concepts that yield such a complementary mode of description. But at least we see already that the famous INTERNAL CONTRADICTIONS of quantum theory only arise because we operate with this not yet sufficiently revised language (I know for sure that this last formulation of mine would drive Bohr to COMPLETE DESPAIR). Now read it for yourselves! (Kalckar 1985, pp. 40–41).

Even the complementarity paper, finally published the following year (Bohr 1928a, 1928b), is mostly about the uncertainty relations, especially about Heisenberg's gamma-ray microscope, which Bohr used to show that both the wave and the particle theory of light are needed to fully analyze the situation (cf. note 100).

Whatever Bohr may have meant by complementarity at the time or made out of it later, those listening to him probably had very much the same reaction as the one by Rosenfeld quoted above in connection with the Como conference. Complementarity seemed to be little more than Bohr's attempt to reconcile the particles suggested by matrix mechanics and the waves suggested by wave mechanics by proposing that they were not contradictory but complementary descriptions of situations such as those encountered in the gamma-ray-microscope thought experiment.

In Chapter 17, we will explain how a much deeper understanding of the relation between matrix and wave mechanics was provided by von Neumann's realization months earlier (cf. note 123) that underlying both forms of the theory is the mathematical structure he called a Hilbert space. Even before von Neumann's contribution, those who had absorbed the Dirac–Jordan statistical transformation theory (which, in addition to Jordan and Dirac themselves, would certainly include Heisenberg, Pauli, and Born) may have felt that the wave–particle duality that gets pride of place in Bohr's exposition of complementarity is just one example of a much more general feature of quantum mechanics, that is, that the state of a system can be described in many different ways depending

[132] Transcriptions of comments made during this general discussion are included in the proceedings of the volume. Apparently, stenographed notes were taken during the meeting, which were then edited by Verschaffelt, secretary of the Solvay committee, with help from Kramers, and sent to the participants for their approval (Bacciagaluppi and Valentini 2009, p. 20). In the appendix to their book, Bacciagaluppi and Valentini (2009, pp. 474–497) present Verschaffelt's notes, now in the Niels Bohr Archive. They also provide a transcription of notes by Kramers (pp. 498–501).

on the variable being measured and that the measurement of that variable precludes the simultaneous measurement of variables that do not commute with it (cf. Duncan and Janssen 2013, p. 178). We remind the reader in this context of a passage in Pauli's letter to Heisenberg of October 19, 1926, which we already quoted in Section 16.3: "One can view the world with the p-eye, and one can view it with the q-eye, but if one tries to open both eyes, one goes crazy." In short, one can understand Rosenfeld's assessment (quoted above) that "Bohr was just putting in a rather heavy form things ... which were current in Göttingen at the time."

That said, some participants at the Como and Solvay meetings seem to have gotten a good deal more out of Bohr's groping statements. Unsurprisingly, it was those who had discussed these ideas with him before, especially Pauli and Heisenberg. We illustrate this with two passages from Heisenberg's contribution to the discussion following Bohr's Como lecture (Bohr 1928c, pp. 593–594). In using this example, we are assuming that Bohr's pronouncements in Como and in Brussels can essentially be used interchangeably. Heisenberg's comments, presented rather charitably as elaborations of the views laid out in Bohr's lecture, touch on elements that would become fixtures in the later debate about the interpretation of quantum mechanics (Freire 2022). In the first passage, we recognize what would become the "Heisenberg cut":

> In quantum mechanics, as Prof. Bohr has shown, observation plays a very peculiar role. One could treat the whole world as *one* mechanical system but then we only have a mathematical problem left—access to observations would be blocked. To get to observations, one must therefore cut some subsystem out of the world somewhere and make "statements" or "observations" about this subsystem. One thereby destroys the subtle connections there between the phenomena and at the point where we make the cut [*Schnitt*] between, on the one hand, the system to be observed and, on the other hand, the observer and his instruments, we should expect difficulties for our view (Bohr 1928c, p. 593).

A similar thought can be found on the very first page of the paper Bohr eventually published, but one wonders how many of his listeners would have appreciated this upon first hearing it in Como or Brussels: "an independent reality in the ordinary physical sense can neither be ascribed to the phenomena nor to the agencies of observation. After all, the concept of observation is in so far arbitrary as it depends upon which objects are included in the system to be observed" (Bohr 1928b, p. 580).

In the second passage from Heisenberg's comments at Como to which we want to draw attention, we recognize what would become the collapse of the wave function:

> This discontinuous change of the wave picture [*unstetige Aenderung des Wellenbildes*] in an observation appears to me to be a fundamental feature of quantum mechanics. One has to get serious about the concept of "probability waves". The waves do not have the direct reality we ascribed to the waves of Maxwell's theory in the past. One has to

> interpret them as probability waves and hence expect sudden changes with each new observation (p. 594).[133]

These comments help explain why Heisenberg (1930) in the preface of the published version of his lectures on quantum mechanics in Chicago in 1929 started talking about the "*Kopenhagener Geist der Quantentheorie*" [Copenhagen spirit of quantum theory] (Pais 1991, p. 320).[134]

The most prominent critic of the interpretation that began to coalesce around the ideas of Bohr, Born, Heisenberg, Pauli and others associated with Copenhagen or Göttingen was Einstein. As he told Schrödinger, who was in his corner, half a year after the Solvay conference: "The tranquilizing philosophy [*Beruhigungsphilosophie*]—or religion?—of Heisenberg–Bohr is so cleverly concocted that for the present it offers the believers a soft resting pillow from which they are not easily chased away. Let us therefore let them rest."[135] In the general discussion at Solvay, Einstein already put his finger on a problem with this Heisenberg–Bohr doctrine that is at the heart of his critique of quantum mechanics in the famous EPR paper seven years later (Einstein, Podolsky, and Rosen 1935). Einstein imagines an electron described by a Schrödinger wave function going through a small aperture on a screen. The aperture is at the center of a semi-spherical photographic plate. Einstein's description of the situation is actually accompanied by one of the few figures in the proceedings volume (Solvay 1928, p. 254; p. 440 in Bacciagaluppi and Valentini 2009). The diffracted wave function tells us, on the interpretation preferred by the Copenhagen–Göttingen camp, that there is an equal probability of detecting the electron anywhere on the screen. But if this probabilistic statement is a complete description of the situation, it is mysterious that we find out simultaneously that the electron is present at one point of the semi-spherical screen and not present at some other point. Short of there being some mysterious action at a distance between these points, there is nothing in the quantum formalism, as interpreted by Einstein's opponents, that could explain why the electron is detected at the first and not at the second point. This suggests, Einstein concluded, that quantum mechanics, as understood by the Copenhagen–Göttingen crowd, is incomplete.[136]

[133] In his AHQP interview, Heisenberg claimed that Born's probabilistic interpretation of the wave function owed a debt to the BKS theory in which "virtual radiation" determines the probability of atomic transitions (see Section 10.4, note 79). Heisenberg cautioned, however, that Born strongly disagreed with this claim (see Duncan and Janssen 2013, p. 180, note 17).

[134] See, for example, Howard (2004, 2022) and Camilleri (2009a, 2009b, 2022) for discussion of the twisted history of what came to be called "the Copenhagen interpretation". Although Heisenberg tended to present his views as being the same as Bohr's, he disagreed with Bohr on some key points. Camilleri (2009a, pp. 77–84) identifies two such disagreements. First, whereas Bohr thought the use of classical concepts such as "wave" and "particle" were inevitable to describe even the quantum realm, Heisenberg emphasized the inadequacy of both concepts in the quantum regime. Second, as a result of the early development of quantum field theory (cf. Duncan and Janssen 2008, p. 235), wave–particle duality for Heisenberg came to mean, *not*, as Bohr would have it, that quantum theory calls for both wave and particle concepts, but that quantum theory can be expressed in two completely equivalent ways, either as a theory of particles, satisfying some non-classical statistics, or as a theory of quantized waves.

[135] Einstein to Schrödinger, May 31, 1928 (Klein 1967), quoted for instance in Pais (1991, p. 320).

[136] In his review of Einstein's critique of quantum mechanics, Lehner (2014, pp. 327–331) discusses this 1927 thought experiment and concludes that the EPR paper can be seen as "a direct continuation" of it. The

After the Solvay conference, Bohr kept revising his complementarity paper. He also gave another lecture on the topic, on November 18, 1927, at the Danish Academy.[137] In January 1928, he traveled to Hamburg to discuss the manuscript once again with one of his favorite interlocutors, Pauli (Kalckar 1985, p. 41). By March, Pauli was getting impatient. He informed Bohr that he would not visit Copenhagen again until Bohr had submitted the corrected page proofs. In his reply three days later, Bohr assured Pauli he had meanwhile done so.[138] The complementarity paper, almost two years in the making, would finally see the light of day.

The version of Bohr's (1928b) paper in *Nature* is prefaced by a peculiar one-page editorial (reproduced in facsimile in Kalckar 1985, p. 52). In the penultimate paragraph of this piece, the problem of wave–particle duality is laid out and complementarity is presented as Bohr's solution to it:

> The strange conflict which has been waged between the wave theory of light and the light quantum hypothesis has resulted in a remarkable dilemma. But now we have a parallel dilemma, for a material particle manifests some of the attributes of wave motion. Can these apparently contradictory views be reconciled? According to Bohr, the pictures ought to be regarded not as contradictory but as complementary.

The author(s?) of the editorial were not satisfied with Bohr's solution, as the final paragraph makes clear:

> The new wave mechanics gave rise to the hope that an account of atomic phenomena might be obtained which would not differ essentially from that afforded by the classical theories of electricity and magnetism. Unfortunately, Bohr's statement in the following communication of the principles underlying the description of atomic phenomena gives little, if any, encouragement in this direction ... It is earnestly to be hoped that this is not [the physicists'] last word on the subject, and that they yet may be successful in expressing the quantum postulate in picturesque form (quoted, e.g., in Pais 1991, p. 315).

This disclaimer greatly annoyed Pauli, who, as we have seen, had been heavily involved in the writing of the paper. In a letter to Bohr,[139] he offered the following paraphrase of the offending paragraph:

1927 argument is actually very similar to an argument Einstein gave in a letter to Schrödinger of June 19, 1935, in which he complained that Podolsky had made the EPR paper more complicated than it needed to be (Lehner 2014, pp. 331–332). Imagine two boxes, Einstein wrote to Schrödinger, and imagine we are told that with equal probability a ball is under one of them. Is this a complete description of the situation? According to standard quantum mechanics it is. Quantum mechanics provides no information about which box has the ball under it until we lift up one of the two boxes. Once we do, we find out simultaneously, and this is true no matter how far the boxes are away from each, that it is under one rather than the other. In the thought experiment Einstein presented at the 1927 Solvay conference, we likewise find out simultaneously that an electron hits the photographic plate at one spot and not at any other.

[137] Only a short abstract of this lecture was published (Bohr 1972–2008, Vol. 6, p. 108).

[138] Pauli to Bohr, March 10, 1928 and Bohr to Pauli, March 13, 1928 (Pauli 1979, Docs. 189 and 190), quoted and discussed in Kalckar (1985, p. 43–44).

[139] Pauli to Bohr, June 16, 1928 (Pauli 1979, Doc. 201). See also Bohr (1972–2008, Vol. 6, p. 438–439; English translation: pp. 440–441).

> We British physicists would be awfully pleased if in the future the points of view advocated in the following paper should turn out not to be true. Since, however, Mr. Bohr is a nice man, such a pleasure would not be kind. Since moreover he is a famous physicist and more often right than wrong, there remains only a slight chance that our hopes will be fulfilled (quoted by Kalckar 1985, p. 53).

The debate over these matters continues to this day and is beyond the scope of the present work. In Chapter 17, we return to 1927 and discuss (the first two installments of) a remarkable trilogy by von Neumann (1927a, 1927b, 1927c) in the proceedings of the Göttingen Academy. In the first installment, which appeared in May 1927, before both Como and Solvay, von Neumann introduced the Hilbert space formalism and showed that it provides the structure underlying the various formalisms for quantum mechanics proposed earlier. In the second installment, which did not appear until November 11, 1927, von Neumann greatly improved on the attempts discussed in this chapter to come to terms with the probabilistic aspects of the theory. Two weeks earlier, on October 27, Born had already advertised this paper in the general discussion at the Solvay meeting:[140]

> I should like to point out, with regard to the considerations of Mr. Dirac [who has just made a lengthy comment about probabilities and transformation theory], that they seem closely related to the ideas expressed by my collaborator J. von Neumann, which will appear shortly. The author shows that quantum mechanics can be built up using the ordinary probability calculus, starting from a small number of formal hypotheses: the probability amplitudes and the law of their composition [see Eq. (16.25)] do not really play a role there (Bacciagaluppi and Valentini 2009, p. 448).

Though von Neumann did attend in Como, he was not invited to Brussels. And Born's advertisement of his work seems to have been lost in the general excitement about uncertainty and complementarity.

[140] Born had already promoted the first installment of von Neumann's trilogy in Como (see note 123). At Solvay, Born curiously referred to von Neumann as his "collaborator." As we shall see in Chapter 17, von Neumann came to Göttingen to work with Hilbert, not Born. The title pages of the three papers of von Neumann's (1927a, 1927b, 1927c) trilogy, however, all say that they were "presented" [*vorgelegt*] to the Göttingen Academy by Born.

17

Von Neumann's Hilbert Space Formalism

John von Neumann, the central character in the final scenes of our account of the construction of quantum mechanics, was born (as Neumann Janos Lajos[1]) on December 28, 1903 into a wealthy upper middle-class Budapest family. His father, Miksa (Max), a lawyer and financier, believed deeply in the value of a wide-ranging humanistic education, and the Neumann home boasted a large library with books from floor to ceiling.[2] It was soon apparent that the young Janos possessed remarkable facilities of memory as well as a capacity for complex mental arithmetic. Unlike the case for many child savants, however, von Neumann's early talent in this area would mature into a penetrating and creative mathematical mind. As a student at the Lutheran gymnasium, he received private tutoring at the hands of some of the best Hungarian mathematicians at the University of Budapest, such as Lipót Fejér, Gábor Szegö, and Mihály Fekete.

Despite his insistence on the value of a broad humanistic education, von Neumann's father also emphasized the importance of acquiring skills adapted to an increasingly technological world, and convinced Janos to take a degree in chemical engineering. The necessary background in chemistry—prerequisite to admission to the ETH in Zurich for the desired degree—would be obtained by taking courses (chemistry from Fritz Haber, statistical mechanics with Einstein) at the University of Berlin, where Janos (now Johann) spent most of the two years from September 1921 to September 1923, when he passed (effortlessly) the entrance exam for the ETH. During all of this, von Neumann (having also enrolled before leaving for Berlin in the doctoral mathematics program in the University of Budapest) was working on a remarkable paper on an axiomatic reformulation of set theory which would avoid the famous antinomies with which the standard Cantorian version was beset. This work would be submitted in Budapest as his doctoral

[1] The honorific "von" was added to the family surname (actually, "margittai Neumann" in Hungarian, or "Neumann von Margitta" in German) as a consequence of his father's ennoblement by the Austro–Hungarian emperor Franz Josef in 1913. In Germany, the Hungarian Janos became, for a few years, Johann, which in turn gave way to the anglicized "John" ("Johnny" to his friends) once von Neumann arrived in the United States in 1930.

[2] For these biographical remarks we draw largely on Bhattacharya (2022). For biographical information with more emphasis on his mathematical work, see Israel and Gasca (2009).

Constructing Quantum Mechanics. Anthony Duncan and Michel Janssen, Oxford University Press.
 DOI: 10.1093/oso/9780198883906.003.0017

thesis, which he successfully defended (1926), more or less at the same time receiving his chemical engineering degree from the ETH. From this point on, von Neumann put aside his "practical" education in chemistry, and devoted himself (for the next decade or so) to problems of rich mathematical content, but also enormous relevance to the new quantum theory emerging at just this time.

As in the case of Heisenberg visiting the Bohr Institute in 1924, the Rockefeller Foundation would play a critical role at this stage of von Neumann's career: he was given a postdoctoral fellowship to go to Göttingen, where he could talk to and work with the acknowledged leader in the field of axiomatization and formalization of mathematics, David Hilbert. Indeed, Hilbert had lobbied vigorously in support of von Neumann's application for the fellowship grant, in hopes that he would acquire a brilliant and energetic collaborator in his program of axiomatic formalization of mathematics.[3]

Von Neumann's conversations with Erhard Schmidt in Berlin, who had received his PhD from Hilbert in 1905, had presumably convinced him that Göttingen would be the ideal place to pursue his true love: the resolution of fundamental problems in (at this time) pure mathematics. In fact, when he arrived in Göttingen, the topic that was increasingly occupying Hilbert's attention was the new quantum mechanics. Hilbert had already invited Heisenberg in the Fall of 1925 (after the Three-Man-Paper had been completed) to educate the mathematicians on matrix mechanics. Moreover, as we saw in Chapter 16, Hilbert gave a course entitled "Mathematical methods of quantum theory," which covered both the old and the new quantum theory. Hilbert discussed matrix mechanics, its relation to the theory of infinitely many variables and integral equations (the Hilbert–Hellinger theory, cf. Section 12.3), as well as Schrödinger's wave mechanics. The final part of the course was on Jordan's version of statistical transformation theory, which formed the basis for the paper by Hilbert, von Neumann, and Nordheim (1928) on this theory, discussed in Section 16.2.3.

At this stage, quantum mechanics still existed in several distinct but somehow equivalent manifestations, most importantly matrix mechanics and wave mechanics, the rigorous mathematical substructure of which was as yet unclear, although a formal connection between the two had already been established by Schrödinger (in his "equivalence" paper), Pauli, and Eckart (see Section 14.5), and had been improved upon, albeit in a mathematically suspect way, by Dirac and Jordan with their statistical transformation theory (see Chapter 16). It was a very fortunate coincidence that the relevant mathematical formalism for an "axiomatization" and rigorous formulation of this new physical theory, as von Neumann would show in the next few years, was provided by the rapidly developing "functional analysis", a blend of algebraic and analytic techniques

[3] Although we are concerned primarily with von Neumann's activities in quantum theory in the next two years, one should keep in mind that von Neumann's Rockefeller fellowship, for the academic year 1926–1927, had originally been intended to support his work on the foundations of mathematics. In addition to working on quantum mechanics, von Neumann was clearly still thinking about the consistent axiomatization of set theory, and more generally of mathematics. The final version of his doctoral thesis, "The axiomatization of set theory" (von Neumann 1928), was submitted to *Mathematische Zeitschrift* in July 1927, in the midst of his explorations of the mathematical substructure of quantum mechanics.

that had been pioneered by, among others, Hilbert and his student Erhard Schmidt (see Appendix C).

When von Neumann arrived in Göttingen, the job of bringing Hilbert up to date on the latest developments in quantum mechanics had already been assumed by Lothar Nordheim, a doctoral graduate (1923) of Max Born. Nordheim was familiar with the progress that had been made by both Jordan and Dirac in their statistical transformation theory (see Section 16.2). In particular, building on earlier work by Born and Pauli, they had given a general characterization of the probabilistic content of the theory. At the same time, they had shown a way in which matrix mechanics and wave mechanics could be subsumed under a single formal framework. Nordheim prepared a detailed set of notes in which this theory was presented. The mathematical shortcomings of the Jordan–Dirac formalism, however, must have been glaringly obvious to Hilbert, who enlisted von Neumann in the project of providing a mathematically more palatable account of the results of Jordan and Dirac.

The paper resulting from these efforts is disappointing. It contains little that was not already present, implicitly or explicitly, in the work of Jordan and Dirac (see Section 16.2.3). The general theme—the employment throughout of integral kernels to "rigorize" the idea of matrices with continuous rows and columns—is clearly a product of Hilbert's long years of work on integral equations, but the attempt to unify the operator calculus in this way inevitably leads to the introduction of the mathematically unsavory Dirac delta function, as well as other "improper" (*uneigentliche*) functions, to represent even the simplest observables in operator form.

For example, the integral kernel associated with the identity operator is just the delta function itself, $\delta(x-y)$, while the momentum operator is given by the improper kernel $(\hbar/i)\,\delta'(x-y)$. Given von Neumann's later statements on the subject of such improper functions, it is difficult to believe that he was at all satisfied with the paper, which seems to have been written largely by Nordheim with occasional input from his co-authors. Von Neumann's participation in this project did, however, afford him the opportunity to fully survey the "lay of the land" in the present state of quantum theory, an essential prerequisite to his rigorous formalization of the mathematical substructure of quantum mechanics that would soon appear.

In the balance of this chapter we examine in detail the essential parts of the first two installments of the trilogy on quantum mechanics that von Neumann (1927a, 1927b, 1927c) produced, partly in response to the papers discussed in Chapter 16: Jordan (1927b), Dirac (1927a), and Hilbert, von Neumann, and Nordheim (1928) on statistical transformation theory and Heisenberg (1927b) on the uncertainty relations.[4] Von Neumann's trilogy would provide the backbone of his famous book published five years later (von Neumann 1932).

In the first installment, "The mathematical foundation [*Mathematische Begründung*] of quantum mechanics," presented in a session of the Göttingen Academy of Sciences on May 20, 1927, von Neumann introduced the modern Hilbert space formalism. This

[4] This chapter is based on Duncan and Janssen (2013, secs. 5 and 6). For a shorter version aimed at a broader audience, see Janssen (2019, pp. 142–161).

allowed him to rewrite the probabilities, which Jordan and Dirac had set equal to the absolute squares of probability amplitudes, in a mathematically unobjectionable way in terms of projection operators. As in his paper with Hilbert and Nordheim (see Section 16.2.3), von Neumann endorsed Jordan's assertion that quantum mechanics calls for changes in the basic laws of probability theory. Following Jordan, von Neumann noted that the ordinary multiplication law of probabilities does not hold in quantum mechanics, adding parenthetically that "what does hold is a weaker law corresponding to the 'combination [*Zusammensetzung*] of probability amplitudes' ... which we will not go into here" (von Neumann 1927a, p. 46). As we saw in Section 16.3, the notion von Neumann is referring to here, Jordan's "interference of probabilities", came in for sharp criticism from Heisenberg in his uncertainty paper (Heisenberg 1927b, pp. 183–184, p. 196).

The second installment of the trilogy, "Probability-theoretic construction [*Wahrscheinlichkeitstheoretischer Aufbau*] of quantum mechanics," was presented (along with the third) about half a year later, on November 11, 1927. Von Neumann was far from being idle in the meantime.[5] In July, he submitted a revised version of his dissertation on set theory to *Mathematische Zeitschrift* (von Neumann 1928, see note 3). He also attended the Como conference in September 1927 (see Section 16.4). He did not attend the Solvay conference in late October. In the general discussion at this conference, however (as we mentioned in Section 16.4), Born announced a forthcoming paper by von Neumann,[6] which is clearly "Probability-theoretic construction ... " The author, Born said,

> shows that quantum mechanics can be built up using the ordinary probability calculus, starting from a small number of formal hypotheses: the probability amplitudes and the law of their composition do not really play a role there (Bacciagaluppi and Valentini 2009, p. 448).

The "law of the composition of [probability amplitudes]" refers to Jordan's notion of the "interference of probabilities". As we saw in Section 16.4, Heisenberg (1927b) had criticized this notion in his uncertainty paper. This paper, however, was not published until May 29, nine days after the presentation of the first installment of von Neumann's trilogy. By the time he wrote the second, von Neumann had read and thoroughly absorbed Heisenberg's paper. It is cited right at the beginning of "Probability-theoretic construction ... " (von Neumann 1927b, p. 246). Following Heisenberg's criticism of the notion of "interference of probabilities", von Neumann now distanced himself from Jordan on this point. One of the shortcomings of "Mathematical foundation ... ," he recognized, was that

[5] For discussion of von Neumann's work during this period, see Hashagen (2010).

[6] Born curiously refers to von Neumann as his "collaborator." As we saw above, von Neumann was working with Hilbert in Göttingen, not with Born. The title pages of von Neumann (1927a, 1927b, 1927c), however, all say that these papers were "presented" [*vorgelegt*] to the Göttingen Academy by Born.

> the relation to the ordinary probability calculus was not sufficiently clarified: the validity of its basic rules (addition and multiplication law of the probability calculus) was not sufficiently stressed (von Neumann 1927b, p. 246).

Instead of rederiving Jordan's formula for probabilities in terms of projection operators, von Neumann now started from a careful consideration of statistical ensembles, an approach based on the work of Richard von Mises.[7] In this paper, von Neumann introduced the now familiar density operators to characterize quantum-statistical ensembles as well as the key distinction between pure states and mixed states. He also carefully considered what happens when we select a member from such an ensemble and measure one of its observable quantities. Von Neumann's "Probability-theoretic construction ... " thus settled many of the questions about the role of probabilities in quantum mechanics that we encountered in Chapter 16.

In the third and final installment, "Thermodynamics of quantum-mechanical ensembles," presented in the same session of the Göttingen Academy as the second, von Neumann turned his attention to quantum-statistical mechanics. This is also where he first explicitly introduced the notation involving the trace of products of projection operators for his formula for the probability of finding a particular result upon subjecting a member drawn from some quantum-statistical ensemble to a particular measurement (see note 42).

In Sections 17.1 and 17.2, we cover "Mathematical foundation... " and "Probability-theoretic construction...," respectively. In our discussion of the former, we focus on von Neumann taking aim at the mathematical shortcomings of the theory of Dirac and Jordan, his proof of the equivalence of wave mechanics and matrix mechanics based on the isomorphism between two instantiations of Hilbert space, the space of square-summable sequences (l^2) and the space of square-integrable functions (L^2), and his derivation of the trace formula for probabilities in quantum mechanics. We will not cover the introduction of the Hilbert space formalism, which takes up a large portion of this paper. This material is covered in any number of modern books on functional analysis.[8] In our discussion of "Probability-theoretic construction ...," we likewise focus on the overall argument of the paper, covering the derivation of the trace formula from some basic assumptions about the expectation value of observables in an ensemble of identical systems, the introduction of density operators, and the specification of pure states through the values of a maximal set of commuting operators.

In Section 17.3, we summarize the transition from the statistical transformation theory of Dirac and Jordan, who framed their theory in terms of canonical transformations

[7] Von Mises only published these ideas in book form the following year (von Mises 1928) but he was one of the examiners of von Neumann's *Habilitation* thesis in mathematics in Berlin in December 1927 (Mehra and Rechenberg 2000–2001, p. 402), which explains why von Neumann was familiar with these ideas when he wrote "Probability-theoretic construction ... " earlier in 1927 (cf. note 37).

[8] See Appendix C, "The Mathematics of Quantum Mechanics," where we provide a condensed history of the development of the linear algebraic and functional analytic concepts which form the basis for a rigorous formulation of quantum mechanics. For a complete mathematical treatment see, for example, Fano (1971) and Prugovecki (1981); or, for a more elementary treatment, more than sufficient for our purposes, Dennery and Krzywicki (1967, Ch. 3).

between conjugate variables familiar from classical mechanics, to the Hilbert space formalism of von Neumann, who no longer needed these elements from classical mechanics.[9]

17.1 "Mathematical foundation . . . "

Von Neumann presented his “mathematical foundation (*Mathematische Begründung*) of quantum mechanics” in the session of the Göttingen Academy of May 20, 1927. This is where he first introduced the Hilbert space formalism and the spectral theorem, at least for bounded operators, two contributions that have since become staples of graduate texts in quantum physics and functional analysis (see note 8). The paper is divided into nine parts, comprising 15 sections and two appendices:[10]

1. “Introduction,” sec. I, pp. 1–4;
2. “The Hilbert space,” secs. II–VI, pp. 4–22;
3. “Operator calculus,” secs. VII–VIII, pp. 22–29;
4. “The eigenvalue problem,” secs. IX–X, pp. 29–37;
5. “The absolute value of an operator,” sec. IX (a typo: this should be XI), pp. 37–41;
6. “The statistical assumption [*Ansatz*] of quantum mechanics,” secs. XII–XIII, pp. 42–47;
7. “Applications,” sec. XIV, pp. 47–50;
8. “Summary,” sec. XV, pp. 50–51;
9. “Appendices,” pp. 51–57.

In the introduction, von Neumann lists seven points, labeled α through ϑ (there is no point η), in which he takes stock of the current state of affairs in the new quantum theory and identifies areas where it ran into mathematical difficulties. We paraphrase these points. (α) Quantum theory describes the behavior of atomic systems in terms of certain eigenvalue problems. (β) This allows for a unified treatment of continuous and discontinuous elements in the atomic world. (γ) The theory suggests that the elemental laws of nature are intrinsically acausal and stochastic.[11] (δ) Returning to the formulation of the theory in terms of eigenvalue problems, von Neumann briefly characterizes the different but equivalent ways in which such problems are posed in matrix mechanics

[9] Hence the title of Duncan and Janssen (2013): “(Never) mind your *p*’s and *q*’s.”

[10] Unless noted otherwise, all page numbers in this section refer to von Neumann (1927a).

[11] Parenthetically, he added an important qualification, “(at least the quantum laws known to us)” (von Neumann 1927a, p. 1). He thus left open the possibility that, at a deeper level, the laws would be deterministic again (cf. Chapter 16). Von Neumann’s position at this point was thus basically the same as Jordan’s (see Section 16.2.1). By the time of the second paper of his trilogy, von Neumann had read Heisenberg’s (1927b) uncertainty paper and endorsed Heisenberg’s position that the indeterminism of quantum mechanics is the result of the inevitable disturbance of quantum systems in our measurements (von Neumann 1927b, p. 273; cf. note 41).

and in wave mechanics. (ε) Both approaches have their difficulties. The application of matrix mechanics appears to be restricted to situations with purely discrete spectra. To deal with wholly or partly continuous spectra, one ends up using, side by side, matrices with indices taking on discrete values and "continuous matrices", that is, the integral kernels of the Dirac–Jordan statistical transformation theory, with "indices" taking on continuous values. It is "very hard," von Neumann (p. 2) warned, to do this in a mathematically rigorous way. (ζ) These same problems start to plague wave mechanics as soon as wave functions are interpreted as probability amplitudes. Von Neumann credited Born, Pauli, and Jordan with transferring the probability concepts of matrix mechanics to wave mechanics and Jordan with developing these ideas into a "closed system" (p. 2).[12] This system, however, faces serious mathematical objections because of the unavoidable use of improper eigenfunctions, such as the Dirac delta function, the properties of which von Neumann thought were simply "absurd" (p. 3). His final objection seems mild by comparison but weighed heavily for von Neumann: (ϑ) eigenfunctions in wave mechanics and probability amplitudes in statistical transformation theory are determined only up to an arbitrary phase factor. The probabilities one ultimately is after in quantum theory do not depend on these physically superfluous phase factors and von Neumann therefore wanted to avoid them altogether.[13]

In section II, von Neumann put the different guises in which the eigenvalue problems appear in matrix and in wave mechanics side-by-side. In matrix mechanics, the problem is to find square-summable infinite sequences of complex numbers $\mathbf{v} = (v_1, v_2, \ldots)$ such that

$$\mathbf{H}\mathbf{v} = E\mathbf{v}, \tag{17.1}$$

where $\mathbf{H}$ is the matrix representing the Hamiltonian of the system in matrix mechanics and E is an energy eigenvalue. In wave mechanics, the problem is to find square-integrable complex-valued functions $f(x)$ such that

$$\hat{H}f(x) = Ef(x), \tag{17.2}$$

where $\hat{H}$ is the differential operator, involving multiplication by functions of x and differentiation with respect to x, that represents the Hamiltonian of the system in wave mechanics.

At the beginning of section IV, von Neumann (1927a, pp. 10–11) points out that one way to unify these two approaches is to look upon the discrete set of values $1, 2, 3\ldots$ of the index i of the sequences $\{v_i\}_{i=1}^{\infty}$ in matrix mechanics and the continuous (generally

[12] In this context, von Neumann (1927a, p. 2) referred to his forthcoming paper with Hilbert and Nordheim (1928). Oddly, von Neumann did not mention Dirac at this point, although Dirac is mentioned (alongside Pauli and Jordan) in section XII (von Neumann 1927a, p. 43) as well as in the second paper of the trilogy (von Neumann 1927b, p. 245).

[13] In *Neue Begründung* II, as we saw in Section 16.2.4, Jordan (1927g, p. 8) responded to this criticism by adding subscripts to the probability amplitudes for two quantities $\hat{\beta}$ and $\hat{q}$ indicating a specific choice of the canonically conjugate quantities $\hat{\alpha}$ and $\hat{p}$ (see Eq. (16.93)).

multi-dimensional) domain Ω of the functions $f(x)$ in wave mechanics as two particular realizations of some more general space, which von Neumann called R. Following the notation of his book (von Neumann 1932, sec. 4, pp. 15–16), we call the "space" of index values Z. Eq. (17.1) can then be written as:[14]

$$\sum_{j \in Z} H_{ij} v_j = E v_i. \tag{17.3}$$

"Summation over Z" can be seen as one instantiation of "integration over R", "integration over Ω" as another. In this way, Eq. (17.2) can, at least formally, be subsumed under matrix mechanics. One could represent the operator $\hat{H}$ in Eq. (17.2) by the integral kernel $H(x,y)$ and write

$$\int_{\Omega} dy\, H(x,y) f(y) = E f(x). \tag{17.4}$$

Both the matrix H_{ij} and the integral kernel $H(x,y)$ can be seen as "matrices" H_{xy} with indices $x, y \in R$. For H_{ij}, $R = Z$; for $H(x,y)$, $R = \Omega$. Von Neumann identified this way of trying to unify matrix and wave mechanics as Dirac's way—and, one may add, although he is not mentioned by name at this point, Jordan's way. It is also the formulation adopted in his collaboration with Nordheim and Hilbert.

Von Neumann now rejected this approach. He dismissed the analogy between Z and Ω sketched above as "very superficial, as long as one sticks to the usual measure of mathematical rigor" (p. 11).[15] He pointed out that even the simplest linear operator, the identity operator, does not have a proper integral-kernel representation. Its integral kernel is the improper Dirac delta function: $\int dy\, \delta(x-y) f(y) = f(x)$.

The appropriate analogy, von Neumann (pp. 11–14) argued, is not between Z and Ω, but between the space of square-summable sequences *over* Z and the space of square-integrable functions *over* Ω. In his book, he used the notation F_Z and F_Ω for these two spaces (von Neumann 1932, p. 16).[16] In 1927, he used $\mathfrak{H}_0$ and $\mathfrak{H}$, instead. Today they are called l^2 and L^2, respectively.[17] Von Neumann (pp. 12–13) reminded his readers of the "Parseval formula," which maps sequences in l^2 onto functions in L^2, and a

[14] We replaced von Neumann's (1927a, p. 10) x_is by v_is to avoid confusion with the argument(s) of the functions $f(x)$.

[15] In the introduction of his 1932 book, von Neumann likewise complained about the lack of mathematical rigor in Dirac's approach. After characterizing the approach in terms of the analogy between Z and Ω, he wrote: "It is no wonder that this cannot succeed without some violence to formalism and mathematics: the spaces Z and Ω are really very different and every attempt to establish a relation between them must run into great difficulties" (von Neumann 1932, p. 15).

[16] Jammer (1966, pp. 314–315) also used this 1932 notation in his discussion of von Neumann (1927a).

[17] Earlier in his paper, von Neumann (p. 7) remarked that what we now call l^2 was usually called "(complex) Hilbert space." In a paper on canonical transformations in quantum mechanics, Fritz London (1926b, p. 197) used the term "Hilbert space" for L^2 (Duncan and Janssen 2009, p. 356). See also Sections C.5.2 and C.5.3.

"theorem of Fischer and F. Riesz," which maps functions in L^2 onto sequences in l^2.[18] The combination of these two results establishes that l^2 and L^2 are isomorphic. As von Neumann (p. 12) emphasized, these "mathematical facts that had long been known" could be used to unify matrix mechanics and wave mechanics in a mathematically impeccable manner. With a stroke of the pen, von Neumann thus definitively settled the issue of the equivalence of wave mechanics and matrix mechanics, superseding earlier efforts by Schrödinger (1926e) and others (see Section 14.5)! Von Neumann made it clear that *anything that can be done in wave mechanics, that is, in* L^2*, has a precise equivalent in matrix mechanics, that is, in* l^2. This is true regardless of whether we are dealing with discrete spectra, continuous spectra, or a combination of the two.

In section V, von Neumann (pp. 14–18) introduces abstract Hilbert space, for which he uses the notation $\overline{\mathfrak{H}}$, carefully defining it in terms of five axioms labeled A through E.[19] In section VI, he adds a few more definitions and then states and proves six theorems about Hilbert space, labeled 1 through 6 (pp. 18–22). In section VII, he turns to the discussion of operators acting in Hilbert space (p. 25). This will be familiar terrain for the modern reader and need not be surveyed in any more detail (see Appendix C and note 8).

The same goes for section VIII, in which von Neumann introduces a special class of Hermitian operators. Their defining property is that they are idempotent: $\hat{E}^2 = \hat{E}$. Von Neumann called an operator like this an *Einzeloperator* or *E.Op.* for short (p. 25).[20] They are now known as projection operators. In a series of theorems, numbered 1 through 9, von Neumann (pp. 26–29) proved some properties of such operators. For our purposes, it suffices to know that they are Hermitian and idempotent.

We do need to take a closer look at section IX, where von Neumann (pp. 29–33) uses projection operators to formulate the spectral theorem. Following von Neumann (p. 31), we start by considering a finite Hermitian operator $\hat{A}$ with a non-degenerate discrete spectrum. Order its real eigenvalues: $a_1 < a_2 < a_3 \ldots$. Helping ourselves to modern Dirac notation, as in Section 16.2, we write the associated normalized eigenvectors as $|a_i\rangle$ (with the normalization condition $\langle a_i|a_j\rangle = \delta_{ij}$). Using this notation, we can write the operator $\hat{E}(l)$ introduced by von Neumann at this point as:[21]

$$\hat{E}(l) \equiv \sum_{(i|a_i \leq l)} |a_i\rangle\langle a_i|. \tag{17.5}$$

Although von Neumann, of course, did not think of an *E.Op.* as constructed out of bras and kets, Eq. (17.5) provides a convenient way to show that, contrary to the probability

[18] The paper cites by von Neumann (p. 13, note 15) is Riesz (1907a). In his discussion of von Neumann's paper, Jammer (1966, pp. 314–315) cites Riesz (1907a, 1907b) and Fischer (1907). See Section C.5.3 for a more detailed discussion of these issues.

[19] In his book, von Neumann (1932) adopted the notation 'H. R.' (shorthand for *Hilbertscher Raum*) for $\overline{\mathfrak{H}}$.

[20] As he explains in a footnote, the term *Einzeloperator* is based on Hilbert's term *Einzelform*. For historical discussion, see Jammer (1966, pp. 317–318). See Section C.3 for a brief introduction to projection operators.

[21] Von Neumann initially defined this operator in terms of its matrix elements $\langle v|\hat{E}(l)|w\rangle$ for two arbitrary (square-summable) sequences $\{v_i\}_{i=1}^{\kappa}$ and $\{w_i\}_{i=1}^{\kappa}$ (where we replaced von Neumann's x and y by v and w; cf. note 14). He defined (in our notation): $E(l; v|w) = \sum_{(i|a_i \leq l)} \langle v|a_i\rangle\langle a_i|w\rangle$ (p. 31).

amplitudes of the Jordan–Dirac statistical transformation theory, there is no phase ambiguity in $\hat{E}(l)$. The operator stays the same if we apply an independent complex phase to each eigenvector, by replacing $|a_i\rangle$ by $|a_i\rangle' = e^{i\varphi_i}|a_i\rangle$:

$$|a_i\rangle'\langle a_i|' = e^{i\varphi_i}|a_i\rangle\langle a_i|e^{-i\varphi_i} = |a_i\rangle\langle a_i|. \tag{17.6}$$

The operator $\hat{E}(l)$ has the property:

$$\hat{E}(a_i) - \hat{E}(a_{i-1}) = |a_i\rangle\langle a_i|. \tag{17.7}$$

It follows that:

$$\hat{A} = \sum_i a_i(\hat{E}(a_i) - \hat{E}(a_{i-1})) = \sum_i a_i|a_i\rangle\langle a_i|. \tag{17.8}$$

$\hat{E}(l)$ is piece-wise constant with jumps where l equals an eigenvalue. Hence, we can write $\hat{A}$ as a so-called Stieltjes integral, which von Neumann (pp. 55–57) discussed and illustrated with some figures in appendix 3 of his paper:

$$\hat{A} = \int l d\hat{E}(l). \tag{17.9}$$

As von Neumann (p. 32) notes, these results (Eqs. (17.5)–(17.9)) can easily be generalized from finite Hermitian matrices and finite sequences to bounded Hermitian operators and the space $\mathfrak{H}_0$ or l^2 of infinite square-summable sequences. Since $\mathfrak{H}_0$ is just a particular instantiation of the abstract Hilbert space $\overline{\mathfrak{H}}$, it is clear that the same results hold for bounded Hermitian operators $\hat{T}$ in $\overline{\mathfrak{H}}$. After listing the key properties of $\hat{E}(l)$ for $\hat{T}$,[22] he concludes section IX, saying "We call $\hat{E}(l)$ the resolution of unity [*Zerlegung der Einheit*] belonging to $\hat{T}$" (p. 33).

In section X, von Neumann (pp. 33–37) further discusses the spectral theorem. Most importantly, he concedes that he had not yet been able to prove that it also holds for

[22] As before (see note 21), he first defined the matrix elements $\langle f|\hat{E}(l)|g\rangle$ for two arbitrary elements f and g of Hilbert space. So he started from the relation

$$\langle f|\hat{T}|g\rangle = \int_{-\infty}^{\infty} l d\langle f|\hat{E}(l)|g\rangle,$$

and inferred from that, first, that $\hat{T}|g\rangle = \int_{-\infty}^{\infty} l d\{\hat{E}(l)|g\rangle\}$, and, finally, that $\hat{T} = \int_{-\infty}^{\infty} l d\hat{E}(l)$ (cf. Eq. (17.9)). Instead of the notation $\langle f|g\rangle$, von Neumann (1927a, p. 12) used the notation $Q(f, g)$ for the inner product of f and g (on p. 32, he also used $Q(f|g)$). So, in von Neumann's own notation, the relation he started from is written as $Q(f, Tg) = \int_{-\infty}^{\infty} l dQ(f, E(l)g)$ (von Neumann 1927a, p. 33).

unbounded operators.[23] He only published the proof of this generalization in a paper submitted to *Mathematische Annalen* in December 1928 (von Neumann 1929).[24] The key to the extension of the spectral theorem from bounded to unbounded operators is a so-called Cayley transformation (1929, p. 80). Given an unbounded Hermitian operator $\hat{R}$, introduce the operator $\hat{U}$ and its adjoint

$$\hat{U} = \frac{\hat{R} + i\hat{1}}{\hat{R} - i\hat{1}}, \qquad \hat{U}^\dagger = \frac{\hat{R} - i\hat{1}}{\hat{R} + i\hat{1}}, \tag{17.10}$$

where $\hat{1}$ is the unit operator. Since $\hat{R}$ is Hermitian, it only has real eigenvalues, so $(\hat{R} \pm i\hat{1})|\varphi\rangle \neq 0$ for any $|\varphi\rangle \in \overline{\mathfrak{H}}$, and the inverses displayed in Eq. (17.10) exist. Since $\hat{U}$ is unitary ($\hat{U}\hat{U}^\dagger = \hat{1}$), the absolute value of all its eigenvalues equals 1. $\hat{U}$ is thus a bounded operator for which the spectral theorem holds. If it holds for $\hat{U}$, however, it must also hold for the original unbounded operator $\hat{R}$. The spectral decomposition of $\hat{R}$ is essentially the same as that of $\hat{U}$—one has merely to replace the eigenvalue u of $\hat{U}$ with the corresponding one r of $\hat{R}$, given by $r = i\frac{u+1}{u-1}$. In his book, von Neumann (1932, p. 80) gave Eq. (17.10), but he referred to his 1929 paper for a mathematically rigorous treatment of the spectral theorem for unbounded operators (1932, p. 75, p. 246, note 95, and p. 244, note 78).

Section XI concludes the purely mathematical part of the paper. In this section, von Neumann (pp. 37–41) introduces the "absolute value" of an operator—essentially, the sum of the absolute value squared of all the matrix elements of the operator in any complete orthonormal basis—an important ingredient in his derivation of his formula for conditional probabilities in quantum mechanics (see Eqs. (17.19)–(17.23)).

In section XII, von Neumann (pp. 42–45) finally turnes to the statistical interpretation of quantum mechanics. At the end of section I, he warns the reader that sections II–XI would have a "preparatory character" and that he would only get to the real subject matter of the paper in sections XII–XIV. At the beginning of section XII on the statistical interpretation of quantum mechanics, he writes: "We are now in a position to take up our real task, the mathematically unobjectionable unification of statistical quantum mechanics" (p. 42). He then proceeds to use the spectral theorem and the projection operators $\hat{E}(l)$ of section IX to construct an alternative to Jordan's formula for conditional probabilities in quantum mechanics, which *does not involve probability amplitudes*.

[23] A linear operator $\hat{A}$ in Hilbert space is bounded if there exists a positive real constant C such that $|\hat{A}f| < C|f|$ for arbitrary vectors f in the space (where $|...|$ indicates the norm of a vector, as induced from the defining inner-product in the space). If this is *not* the case, then there exist vectors in the Hilbert space on which the operator $\hat{A}$ is not well defined, basically because the resultant vector has infinite norm. Instead, such unbounded operators are only defined (i.e., yield finite-norm vectors) on a proper subset of the Hilbert space, called the *domain* $\mathcal{D}(\hat{A})$ of the operator $\hat{A}$. Indeed, the Hellinger–Toeplitz theorem (1910) asserts that if $\hat{A}$ is Hermitian, and $\mathcal{D}(\hat{A})$ is the full Hilbert space, then $\hat{A}$ must be bounded. The set of vectors obtained by applying $\hat{A}$ to all elements of its domain is called the *range* $\mathcal{R}(\hat{A})$ of $\hat{A}$. Multiplication of two unbounded operators evidently becomes a delicate matter insofar as the domains and ranges of the respective operators may not coincide. For further details, see Section C.5.

[24] For brief discussions, see Jammer (1966, p. 320) and Mehra and Rechenberg (2000–2001, p. 415).

Recall von Neumann's objections to probability amplitudes. First, Jordan's basic amplitudes, $\varphi(p, q) = e^{-ipq/\hbar}$, which from the perspective of Schrödinger wave mechanics are eigenfunctions of momentum, are not square-integrable and hence, not in Hilbert space. Second, they are only determined up to a phase factor. Von Neumann avoided these two problems by deriving an alternative formula that expresses the conditional probability $\Pr(a|b)$ in terms of projection operators associated with the spectral decomposition of the operators for the corresponding observables $\hat{a}$ and $\hat{b}$. These projection operators are unambiguous and well defined for arbitrary observables, even unbounded ones, whether with discrete, continuous, or mixed eigenvalue spectra.

Von Neumann takes over Jordan's basic statistical Ansatz. Consider a one-particle system in one dimension with coordinate q (von Neumann considered the more general case with coordinates $q \equiv (q_1, \ldots, q_k)$). The probability of finding a particle in some region K if we know that its energy is E_n (i.e., if we know the particle is in the state $\psi_n(x)$ associated with that eigenvalue), is given by (p. 43):[25]

$$\Pr(q \text{ in } K|E_n) = \int_K |\psi_n(q)|^2 dq. \tag{17.11}$$

Next, he considers the probability of finding the particle in some region K if we know that its energy is in some interval I that includes various eigenvalues of its energy, that is, if we only know that, with equal probability, the particle is one of the states $\psi_n(x)$ associated with the eigenvalues within the interval I:[26]

$$\Pr(q \text{ in } K|E_n \text{ in } I) = \sum_{(n|E_n \text{ in } I)} \int_K |\psi_n(q)|^2 dq. \tag{17.12}$$

These conditional probabilities can be written in terms of the projection operators,

$$\hat{E}(I) \equiv \sum_{(n|E_n \text{ in } I)} |\psi_n\rangle\langle\psi_n|, \quad \hat{F}(K) \equiv \int_K |q\rangle\langle q|\, dq, \tag{17.13}$$

that project arbitrary state vectors onto the subspaces of $\mathfrak{H}$ spanned by "eigenvectors" of the Hamiltonian $\hat{H}$ and of the position operator $\hat{q}$ with eigenvalues in the ranges I and K, respectively. The right-hand side of Eq. (17.12) can be rewritten as:

$$\sum_{(n|E_n \text{ in } I)} \int_K \langle\psi_n|q\rangle\langle q|\psi_n\rangle\, dq. \tag{17.14}$$

[25] The left-hand side is shorthand for: $\Pr(\hat{q}$ has value q in $K|\hat{H}$ has value $E_n)$. We remind the reader that the notation $\Pr(.|.)$ is ours and is not used in any of our sources.

[26] The left-hand side is shorthand for: $\Pr(\hat{q}$ has value q in $K|\hat{H}$ has value E_n in $I)$. On the right-hand side, $\sum_{(n|E_n \text{ in } I)}$ is the sum over all n such that E_n lies in the interval I.

We now choose an arbitrary orthonormal discrete basis $\{|\alpha\rangle\}_{\alpha=1}^{\infty}$ of the Hilbert space $\overline{\mathfrak{H}}$. Inserting the corresponding resolution of unity,

$$\hat{1} = \sum_{\alpha} |\alpha\rangle\langle\alpha|, \tag{17.15}$$

into Eq. (17.14), we find

$$\sum_{\alpha} \sum_{(n|E_n \text{ in } I)} \int_K \langle\psi_n|\alpha\rangle\langle\alpha|q\rangle\langle q|\psi_n\rangle \, dq. \tag{17.16}$$

This can be rewritten as

$$\sum_{\alpha} \langle\alpha| \left(\int_K |q\rangle\langle q| \, dq \cdot \sum_{(n|E_n \text{ in } I)} |\psi_n\rangle\langle\psi_n| \right) |\alpha\rangle, \tag{17.17}$$

which is nothing but the trace of the product of the projection operators $\hat{F}(K)$ and $\hat{E}(I)$ defined in Eq. (17.13). The conditional probability in Eq. (17.12) can thus be written as:

$$\Pr(x \text{ in } K|E_n \text{ in } I) = \sum_{\alpha} \langle\alpha|\hat{F}(K)\hat{E}(I)|\alpha\rangle = \text{Tr}(\hat{F}(K)\hat{E}(I)). \tag{17.18}$$

This is our notation for what von Neumann (p. 45) wrote as[27]

$$[\hat{F}(K), \hat{E}(I)]. \tag{17.19}$$

He defined the quantity $[\hat{A}, \hat{B}]$—*not to be confused with a commutator*—as (p. 40):

$$[\hat{A}, \hat{B}] \equiv [\hat{A}^{\dagger}\hat{B}]. \tag{17.20}$$

For any operator $\hat{O}$, he defined the quantity $[\hat{O}]$, which he called the "absolute value" of $\hat{O}$, as (pp. 37–38):[28]

$$[\hat{O}] \equiv \sum_{\mu,\nu} |\langle\varphi_\mu|\hat{O}|\psi_\nu\rangle|^2, \tag{17.21}$$

[27] Since von Neumann (p. 43) chose K to be k-dimensional, he actually wrote: $[\hat{F}_1(\mathfrak{J}_1) \cdot \ldots \cdot \hat{F}_k(\mathfrak{J}_k), \hat{E}(I)]$ (p. 45; "hats" added). For the one-dimensional case we are considering, von Neumann's expression reduces to Eq. (17.19). For other discussions of von Neumann's derivation of this key formula, see Jammer (1966, pp. 320–321) and Mehra and Rechenberg (2000–2001, p. 414).

[28] Using the notation $Q(.,.)$ for the inner product (see note 22) and using A instead of O, von Neumann ([p. 37) wrote the right-hand side of Eq. (17.21) as $\sum_{\mu,\nu=1}^{\infty} |Q(\varphi_\mu, A\psi_\nu)|^2$.

where $\{|\varphi_\mu\rangle\}_{\mu=1}^{\infty}$ and $\{|\psi_\nu\rangle\}_{\nu=1}^{\infty}$ are two arbitrary orthonormal bases of $\overline{\mathfrak{H}}$. Eq. (17.21) can also be written as:

$$[\hat{O}] \equiv \sum_{\mu,\nu} \langle\varphi_\mu|\hat{O}|\psi_\nu\rangle\langle\psi_\nu|\hat{O}^\dagger|\varphi_\mu\rangle = \sum_{\mu} \langle\varphi_\mu|\hat{O}\hat{O}^\dagger|\varphi_\mu\rangle = \mathrm{Tr}(\hat{O}\hat{O}^\dagger), \tag{17.22}$$

where we used the resolution of unity, $\hat{1} = \sum_\nu |\psi_\nu\rangle\langle\psi_\nu|$, and the fact that $\mathrm{Tr}(\hat{O}) = \sum_\alpha \langle\alpha|\hat{O}|\alpha\rangle$ for any orthonormal basis $\{|\alpha\rangle\}_{\alpha=1}^{\infty}$ of $\overline{\mathfrak{H}}$.[29] Using the definitions of $[\hat{A}, \hat{B}]$ and $[\hat{O}]$ in Eqs. (17.20) and (17.22), with $\hat{A} = \hat{F}(K)$, $\hat{B} = \hat{E}(I)$, and $\hat{O} = \hat{F}(K)^\dagger \hat{E}(I)$, we can rewrite Eq. (17.19) as

$$[\hat{F}, \hat{E}] = [\hat{F}^\dagger \hat{E}] = \mathrm{Tr}((\hat{F}^\dagger \hat{E})(\hat{F}^\dagger \hat{E})^\dagger) = \mathrm{Tr}(\hat{F}^\dagger \hat{E}\hat{E}^\dagger \hat{F}), \tag{17.23}$$

where, to make the equation easier to read, we temporarily suppressed the value ranges K and I of $\hat{F}(K)$ and $\hat{E}(I)$. Using the cyclic property of the trace, we can rewrite the final expression in Eq. (17.23) as $\mathrm{Tr}(\hat{F}\hat{F}^\dagger \hat{E}\hat{E}^\dagger)$. Since projection operators $\hat{P}$ are both Hermitian and idempotent, we have $\hat{F}\hat{F}^\dagger = \hat{F}^2 = \hat{F}$ and $\hat{E}\hat{E}^\dagger = \hat{E}^2 = \hat{E}$. Combining these observations and restoring the arguments of $\hat{F}$ and $\hat{E}$, we can rewrite Eq. (17.23) as:

$$[\hat{F}(K), \hat{E}(I)] = \mathrm{Tr}(\hat{F}(K)\hat{E}(I)), \tag{17.24}$$

which is just Eq. (17.18) for $\Pr(x \text{ in } K|E_n \text{ in } I)$ found above.

From $\mathrm{Tr}(\hat{F}\hat{E}) = \mathrm{Tr}(\hat{E}\hat{F})$, it follows that

$$\Pr(x \text{ in } K|E_n \text{ in } I) = \Pr(E_n \text{ in } I|x \text{ in } K), \tag{17.25}$$

which is just the symmetry property imposed on Jordan's probability amplitudes in postulate B of *Neue Begründung* I and postulate II in *Neue Begründung* II (Jordan 1927b, 1927g; see Section 16.2.2 and 16.2.4)

Von Neumann generalized Eq. (17.18) for a pair of quantities to a similar formula for a pair *of sets* of quantities such that the operators for all quantities in each set commute with those for all other quantities in that same set but not necessarily with those for quantities in the other set (p. 45).[30] Let $\{\hat{R}_i\}_{i=1}^{n}$ and $\{\hat{S}_j\}_{j=1}^{m}$ be two such sets of commuting operators: $[\hat{R}_{i_1}, \hat{R}_{i_2}] = 0$ for all $1 \le i_1, i_2 \le n$; $[\hat{S}_{j_1}, \hat{S}_{j_2}] = 0$ for all $1 \le j_1, j_2 \le m$.[31]

[29] Eq. (17.22) shows that $[\hat{O}]$ is independent of the choice of the bases $\{|\varphi_\mu\rangle\}_{\mu=1}^{\infty}$ and $\{|\psi_\nu\rangle\}_{\nu=1}^{\infty}$. Von Neumann (p. 37) initially introduced the quantity $[\hat{O}; \varphi_\mu; \psi_\nu] \equiv \sum_{\mu,\nu} |\langle\varphi_\mu|\hat{O}|\psi_\nu\rangle|^2$. He then showed that this quantity does not actually depend on φ_μ and ψ_ν, renamed it $[\hat{O}]$ (see Eq. (17.21) and note 28), and called it the "absolute value of the operator" $\hat{O}$ (p. 38). Operators with finite absolute value form an important subclass of the bounded operators, called Hilbert–Schmidt operators.

[30] Von Neumann distinguishes between the commuting of $\hat{R}_i$ and $\hat{R}_j$ and the commuting of the corresponding projection operators $\hat{E}_i(I_i)$ and $\hat{E}_j(I_j)$. For *bounded* operators, these two properties are equivalent. If both $\hat{R}_i$ and $\hat{R}_j$ are *unbounded*, von Neumann (p. 45) cautions, "certain difficulties of a formal nature occur, which we do not want to go into here" (cf. note 23).

[31] Unlike von Neumann (see Eqs. (17.19)–(17.20)), we continue to use the notation [.,.] for commutators.

Let $\hat{E}_i(I_i)$ $(i = 1, \ldots, n)$ be the projection operators onto the space spanned by eigenstates of $\hat{R}_i$ with eigenvalues in the interval I_i and let $\hat{F}_j(\mathcal{J}_j)$ $(j = 1, \ldots, m)$ likewise be the projection operators onto the space spanned by eigenstates of $\hat{S}_j$ with eigenvalues in the interval $\mathcal{J}_j$ (cf. Eq. (17.13)). A straightforward generalization of von Neumann's trace formula (17.18) gives the probability that the $\hat{S}_j$s have values in the intervals $\mathcal{J}_j$ given that the $\hat{R}_i$s have values in the intervals I_i:

$$\Pr(\hat{S}_j\text{'s in } \mathcal{J}_j\text{'s} | \hat{R}_i\text{'s in } I_i\text{'s}) = \mathrm{Tr}(\hat{E}_1(I_1) \ldots \hat{E}_n(I_n) \hat{F}_1(\mathcal{J}_1) \ldots \hat{F}_m(\mathcal{J}_m)). \tag{17.26}$$

The outcomes "$\hat{R}_i$ in I_i" are called the "assertions" (*Behauptungen*) and the outcomes "$\hat{S}_j$ in $\mathcal{J}_j$" are called the "conditions" (*Voraussetzungen*) (von Neumann 1927a, p. 45). Because of the cyclic property of the trace, which we already invoked in Eq. (17.25), Eq. (17.26) is invariant under swapping all assertions with all conditions. Since all $\hat{E}_i(I_i)$s commute with each other and all $\hat{F}_j(\mathcal{J}_j)$s commute with each other, Eq. (17.26) is also invariant under changing the order of the assertions and changing the order of the conditions. These two properties are given in the first two entries of a list of five properties, labeled α through ϑ (there are no points ζ and η), of the basic rule (17.26) for probabilities in quantum mechanics (pp. 45–47).

Under point γ, von Neumann noted that projection operators $\hat{F}(\mathcal{J})$ and $\hat{E}(I)$ for "empty" (*nichtssagende*) assertions and conditions, that is, those for which the intervals $\mathcal{J}$ and I are $(-\infty, +\infty)$, can be added to or removed from Eq. (17.26) without affecting the result.

Under point δ, von Neumann follows Jordan, asserting that the ordinary multiplication law of probabilities does not hold in quantum mechanics, excusing this deviation with the rather vague comment that "the dependency relations of our probabilities can be arbitrarily complicated." We remind the reader that von Neumann was unaware at the time of writing this paper of Heisenberg's uncertainty paper, which was published a little more than a week after von Neumann submitted *Mathematische Begündung*. In Section 16.3, we saw that, Heisenberg, in his analysis of quantum-kinematical concepts in the uncertainty paper, clearly explained that the normal multiplication law of probabilities holds perfectly well as long as one is careful to associate actual physical measurements with the "assertions" and "conditions" entering into the conditional probabilities predicted by the theory.

However, as von Neumann notes under point ε: "The addition rule of probabilities is valid" (p. 46). In general, as Jordan (1927b, p. 18) argued in *Neue Begründung* I, the addition rule also fails to hold in quantum mechanics. Instead, Jordan pointed out, the addition rule, like the multiplication rule, holds for the corresponding probability *amplitudes*. Von Neumann departs here from Jordan, asserting the unconditional validity of the addition rule. In specifying the conditional probability, von Neumann takes fully into account the physical constraints of (a) specifying a prepared quantum state (say, selecting particles on a line, in a specified energy range) as the assertion, and (b) determining the presence of the electrons in some preassigned interval of the line, as the condition. The specification of a process in terms of intermediate assertions (or conditions) that are *not measured*, and therefore lead to the non-classical interference effects present in

Jordan's $\int \varphi(x,y)\psi(y,z)dy$ quantum addition law, is thereby avoided.[32] Consider Eq. (17.18) for the conditional probability that we find a particle in some region K given that its energy E has a value in some interval I. Let the region K consist of two disjoint subregions, K' and K'', such that $K = K' \cup K''$ and $K' \cap K'' = \emptyset$. Given that the energy E lies in the interval I, the probability that the particle is *either* in K' *or* in K'', is obviously equal to the probability that it is in K. Von Neumann now notes that

$$\Pr(x \text{ in } K|E \text{ in } I) = \Pr(x \text{ in } K'|E \text{ in } I) + \Pr(x \text{ in } K''|E \text{ in } I). \tag{17.27}$$

In terms of the trace formula (17.18), Eq. (17.27) becomes:

$$\mathrm{Tr}(\hat{F}(K)\hat{E}(I)) = \mathrm{Tr}(\hat{F}(K')\hat{E}(I)) + \mathrm{Tr}(\hat{F}(K'')\hat{E}(I)). \tag{17.28}$$

Similar instances of the addition rule obtain for the more general version of the trace formula in Eq. (17.26).

Under point ϑ, finally, we find the one and only reference in "Mathematical foundation ... " to "canonical transformations", central to the statistical transformation theory of Jordan and Dirac (see Section 16.2). Von Neumann (pp. 46–47) defines a canonical transformation as the process of subjecting *all* operators $\hat{A}$ to the transformation $\hat{U}\hat{A}\,\hat{U}^{\dagger}$, where $\hat{U}$ is some unitary operator. The absolute value squared $[\hat{A}]$ is invariant under such transformations. Recall $[\hat{A}] = \mathrm{Tr}(\hat{A}\hat{A}^{\dagger})$ (see Eq. (17.22)). Now consider $[\hat{U}\hat{A}\,\hat{U}^{\dagger}]$:

$$\begin{aligned} [\hat{U}\hat{A}\hat{U}^{\dagger}] &= \mathrm{Tr}(\hat{U}\hat{A}\,\hat{U}^{\dagger}(\hat{U}\hat{A}\,\hat{U}^{\dagger})^{\dagger}) \\ &= \mathrm{Tr}(\hat{U}\hat{A}\,\hat{U}^{\dagger}\hat{U}\hat{A}^{\dagger}\,\hat{U}^{\dagger}) \\ &= \mathrm{Tr}(\hat{A}\hat{A}^{\dagger}) = [\hat{A}]. \end{aligned} \tag{17.29}$$

Traces of products of operators are similarly invariant. Note that this definition of canonical transformations, unlike the one used by Jordan and Dirac, makes no reference to sorting quantities into sets of conjugate variables.

[32] See Chapter 16, note 29. In the two-slit experiment, for example, the addition law for probabilities of exclusive events, $\Pr(A \text{ or } B) = \Pr(A) + \Pr(B)$, appears to fail unless we realize that the specification of A (resp. B) as "particle arrives at screen after passing through slit 1 (resp. 2)" requires that measurements be performed to ascertain which slit the particle went through, in order to ensure that we are indeed dealing with mutually exclusive processes. As Heisenberg showed in his uncertainty paper, using a similar thought experiment involving atomic beams passing through Stern–Gerlach magnets, the measurements necessary to fix the assertions and conditions appearing in conditional probabilities change the dynamical evolution of the system in just such a way as to restore the validity of the classical probability calculus. In modern language, confusions arise if we are not sufficiently precise in the specification of the *sample space* and/or *event space* for the measured probabilities.

17.2 "Probability-theoretic construction ... "

On November 11, 1927, about half a year after the first installment, von Neumann presented the second and third installments of his trilogy to the Göttingen Academy (von Neumann 1927b, 1927c).[33] The second, "Probability-theoretic Construction [*Wahrscheinlichkeitstheoretischer Aufbau*] of Quantum Mechanics," is important for our purposes; the third, on quantum statistical mechanics, is not.[34] In "Mathematical foundation ... ," as we saw in Section 17.1, von Neumann had simply taken over the basic rule for probabilities in quantum mechanics as stated by Jordan, namely, that probabilities are given by the absolute square of the corresponding probability amplitudes, the prescription now known as the Born rule. In "Probability-theoretic construction ... ," he seeks to derive this rule from more basic considerations.

In the introduction of the paper, von Neumann (p. 245)[35] replaces the old opposition between "wave mechanics" and "matrix mechanics" by a new distinction between "wave mechanics" on the one hand, and what he calls "transformation theory" or "statistical theory", on the other. By this time, matrix mechanics and Dirac's (1925) q-number theory had morphed into the Dirac–Jordan statistical transformation theory. The two names von Neumann used for this theory reflect the difference in emphasis between Dirac (transformation theory) and Jordan (statistical theory). Commenting on the Schrödinger wave function, von Neumann observes: "Dirac interprets it as a row of a certain transformation matrix, Jordan calls it a probability amplitude" (p. 246, note 3). Von Neumann mentions Born, Pauli, and London as those who had paved the way for the statistical theory and Dirac and Jordan as those responsible for bringing this development to a conclusion (p. 245; cf. note 12).[36]

Von Neumann was dissatisfied with the way in which probabilities were introduced in the Dirac–Jordan theory. He lists two objections. First, he notes that the Born rule was not well-motivated:

> The method hitherto used in statistical quantum mechanics was essentially *deductive*: the square of the norm of certain expansion coefficients of the wave function or of the wave function itself was fairly *dogmatically* set equal to a probability, and agreement

[33] The three papers of this trilogy take up 57, 28, and 19 pages, respectively. The first one is thus longer than the other two combined. Note that in between the first and the last two, *Neue Begründung* II appeared (see Section 16.2.4), in which Jordan (1927g) responded to von Neumann's criticism in "Mathematical foundation ... ". Von Neumann did not comment on this response in these two later papers.

[34] For discussion of this third paper, see Mehra and Rechenberg (2000–2001, pp. 439–445). See pp. 431–436 for their discussion of the second paper.

[35] Unless noted otherwise, all page references in this section are to von Neumann (1927b).

[36] He cited the relevant work by Dirac (1927a) and Jordan (1927b, 1927g). He did not give references for the other three authors but presumably was thinking of Born (1926a, 1926b, 1926c), Pauli (1927), and London (1926b). The reference to London is somewhat puzzling. While it is true that London anticipated important aspects of the Dirac–Jordan statistical transformation theory (see Lacki 2004; Duncan and Janssen 2009), the statistical interpretation of the formalism is not among those. Our best guess is that von Neumann took note of Jordan's repeated acknowledgment of London's paper (most prominently perhaps in footnote 1 of *Neue Begründung* I). In his book, von Neumann (1932, p. 2, note 2; the note itself is on p. 238) cited papers by Dirac (1927a), Jordan (1927b), and London (1926b) in addition to the book by Dirac (1930) for the development of transformation theory. In that context, the reference to London is entirely appropriate.

> with experience was verified afterwards. A systematic derivation of quantum mechanics from empirical facts or fundamental probability-theoretic assumptions, i.e., an *inductive* justification, was not given (p. 246; our emphasis).

Secondly, he addresses the relation between quantum probability concepts and ordinary probability theory "Moreover, the relation to the ordinary probability calculus was not sufficiently clarified: the validity of its basic rules (addition and multiplication law of the probability calculus) was not sufficiently stressed" (p. 246). In a footnote, he adds: "For instance, according to Jordan [1927b], the addition and multiplication laws hold for the 'probability amplitudes' and not for their absolute squares, i.e., the probabilities themselves. For an opposing view, see Heisenberg [1927b]." In his uncertainty paper (see Section 16.4), Heisenberg (1927b, pp. 183–184) had criticized Jordan's concept of "interference of probabilities" defined in *Neue Begründung* I as "the circumstance that not the probabilities themselves but their amplitudes obey the usual composition law of the probability calculus" (Jordan 1927b, p. 812). Earlier, von Neumann had endorsed Jordan's formulation (Hilbert, von Neumann, and Nordheim 1928, p. 5; von Neumann 1927a, p. 46). Now, he fully agreed with Heisenberg's attitude: there was nothing at all amiss with the usual rules of probability, as long as they were applied with awareness of the observational constraints imposed in quantum theory by the specification of the "assertions" and "conditions" (in von Neumann's terminology) of the quantum processes in question. The scope and methodology of the paper is succinctly summarized in the introduction:

> In the present work an inductive construction [of quantum mechanics] will be attempted. *In doing this we make the assumption of absolute validity of the usual probability calculus.* It will be shown, not only that this is compatible with quantum mechanics, but (once combined with some not very far-reaching factual and formal assumptions) suffices for a unique derivation [of the theory] (von Neumann 1927b, p. 246; our emphasis).

Von Neumann starts by introducing probabilities in terms of selecting members from a large ensemble of systems.[37] He then presents his "inductive" derivation of his trace formula for probabilities (see Eq. (17.18)), which contains the Born rule as a special case, from two very general, deceptively innocuous, but certainly non-trivial assumptions about expectation values of properties of the systems in such ensembles. From

[37] In his book, von Neumann (1932, p. 158, p. 255, note 156) referred to a book by von Mises (1928) for this notion of an ensemble (*Gesamtheit* or *Kollektiv*) (cf. Lacki 2000, p. 308). Von Neumann may have picked up this notion from von Mises in the period leading up to his *Habilitation* in mathematics in Berlin in December 1927. Von Mises was one of his examiners (Mehra and Rechenberg 2000–2001, p. 402). As von Mises (1928) explained in its preface, his book elaborates on ideas he had presented for "about fifteen years" in various talks, courses, and articles. Von Mises defined a collective (*Kollektiv*) as an ensemble (*Gesamtheit*) whose members are distinguished by some observable marker (*beobachtbares Merkmal*). (Von Neumann used the term *Gesamtheit* for what von Mises called a *Kollektiv*.) One of his examples is a group of plants of the same species grown from a given collection of seeds, where the individual plants differ from one another in the color of their flowers (1928, pp. 12–13). In their discussion of von Neumann's "Probability-theoretic construction ...," Born and Jordan (1930, p. 306) also cited von Mises (1928).

those two assumptions, together with some key elements of the Hilbert space formalism introduced in "Mathematical foundation . . . ," and two assumptions about the repeatability of measurements not explicitly identified until the summary at the end of the paper (p. 271), von Neumann indeed manages to recover Eq. (17.18) for probabilities. As discussed, he downplayed his reliance on the formalism of "Mathematical foundation . . . " by characterizing the assumptions taken from it as "not very far-reaching . . . " He refers to section IX, the summary of the paper, for these assumptions at this point, but most of them are already stated, more explicitly in fact, in section II, "basic assumptions" (pp. 249–252).

Consider an ensemble $\{\mathfrak{S}_1, \mathfrak{S}_2, \mathfrak{S}_3, \ldots\}$ of copies of a system $\mathfrak{S}$.[38] Von Neumann wants to find an expression for the expectation value $\mathcal{E}(\mathfrak{a})$ in that ensemble of some property $\mathfrak{a}$ of the system (we use $\mathcal{E}$ to distinguish the expectation value from the projection operator $\hat{E}$). He makes the following basic assumptions about $\mathcal{E}$ (pp. 249–250):

A. *Linearity*: $\mathcal{E}(\alpha\,\mathfrak{a} + \beta\,\mathfrak{b} + \gamma\,\mathfrak{c} + \ldots) = \alpha\,\mathcal{E}(\mathfrak{a}) + \beta\,\mathcal{E}(\mathfrak{b}) + \gamma\,\mathcal{E}(\mathfrak{c}) + \ldots$ (where α, β, and γ are real numbers).[39]

B. *Positive-definiteness*. If the quantity $\mathfrak{a}$ never takes on negative values, then $\mathcal{E}(\mathfrak{a}) \geq 0$.

To this he added two formal assumptions (p. 252):

C. *Linearity of the assignment of operators to quantities*. If the operators $\hat{S}$, $\hat{T}$, . . . represent the quantities $\mathfrak{a}$, $\mathfrak{b}$, . . . , then $\alpha\hat{S} + \beta\hat{T} + \ldots$ represents the quantity $\alpha\,\mathfrak{a} + \beta\,\mathfrak{b} + \ldots$[40]

D. If the operator $\hat{S}$ represents the quantity $\mathfrak{a}$, then $f(\hat{S})$ represents the quantity $f(\mathfrak{a})$.

In section IX, the summary of his paper, von Neumann once again lists the assumptions that go into his derivation of the expression for $\mathcal{E}(\mathfrak{a})$:

> The goal of the present paper was to show that quantum mechanics is not only compatible with the usual probability calculus, but that, if it [i.e., ordinary probability theory]—along with a few plausible factual [*sachlich*] assumptions—is taken as given, it [i.e., quantum mechanics] is actually the only possible solution. The assumptions made were the following:

[38] "Copies" should be understood here as "specimens of the same kind", *not* as "reproductions of some original". The ensemble may, for example, consist of hydrogen atoms, not necessarily in the same quantum state.

[39] Here, von Neumann appended a footnote in which he looked at the example of a harmonic oscillator in three dimensions. The same point can be made with a one-dimensional harmonic oscillator with position and momentum operators $\hat{q}$ and $\hat{p}$, Hamiltonian $\hat{H}$, mass m, and characteristic angular frequency ω: "The three quantities $[\hat{p}^2/2m, m\omega^2\hat{q}^2/2, \hat{H} = \hat{p}^2/2m + m\omega^2\hat{q}^2/2]$ have very different spectra: the first two both have a continuous spectrum, the third has a discrete spectrum. Moreover, no two of them can be measured simultaneously. Nevertheless, the sum of the expectation values of the first two equals the expectation value of the third" (p. 249). While it may be reasonable to impose condition (A) on directly measurable quantities, it is questionable whether this is also reasonable for hidden variables (see note 45).

[40] In von Neumann's own notation, the operator $\hat{S}$ and the matrix S representing that operator are both written simply as S.

1. Every measurement changes the measured object, and two measurements therefore always disturb each other—except when they can be replaced by a single measurement.
2. However, the change caused by a measurement is such that the measurement itself retains its validity, that is, if one repeats it immediately afterwards, one finds the same result.

In addition, a formal assumption:

3. Physical quantities are to be described by functional operators in a manner subject to a few simple formal rules.

These principles already inevitably entail quantum mechanics and its statistics (p. 271).

Assumptions A and B of section II are not on this new list in section IX. Presumably, this is because they are part of ordinary probability theory. Conversely, assumptions 1 and 2 of section IX are not among the assumptions A–D of section II. These two properties of measurements are guaranteed in von Neumann's formalism by the idempotency of the projection operators associated with those measurements.[41] Finally, the "simple formal rules" referred to in assumption 3 are spelled out in assumptions C–D.

We go over the main steps of von Neumann's derivation of his trace formula from these assumptions. Instead of the general Hilbert space $\bar{\mathfrak{H}}$, von Neumann considers $\mathfrak{H}_0$, that is, l^2 (p. 253; cf. von Neumann 1927a, pp. 14–15; see Section 17.1). Consider some infinite-dimensional Hermitian matrix S, with matrix elements $s_{\mu\nu} = s^{\star}_{\nu\mu}$, representing an Hermitian operator $\hat{S}$. This operator, in turn, represents some measurable quantity $\mathfrak{a}$.

The matrix S can be written as a linear combination of three types of infinite-dimensional matrices labeled A, B, and C. To show what these matrices look like, we write down their finite-dimensional counterparts:

$$A_\mu \equiv \begin{pmatrix} 0 & \cdots & \cdots & \cdots & 0 \\ \vdots & 1 & & & \vdots \\ \vdots & & \ddots & & \vdots \\ \vdots & & & 0 & \vdots \\ 0 & \cdots & \cdots & \cdots & 0 \end{pmatrix}, \tag{17.30}$$

[41] Assumption 2 is first introduced in footnote 30 on p. 262: "Although a measurement is fundamentally an intervention (*Eingriff*), that is, it changes the system under investigation (this is what the "acausal" character of quantum mechanics is based on, cf. [Heisenberg 1927b, on the uncertainty principle]), it can be assumed that the change occurs for the sake of the experiment, that is, that as soon as the experiment has been carried out the system is in a state in which the *same* measurement can be carried out without further change to the system. Or: that if the same measurement is performed twice (and nothing happens in between), the result is the same." Von Neumann (1927c, p. 273, note 2) reiterated assumptions 1 and 2 in the introduction of the final installment of his trilogy and commented (once again citing Heisenberg's uncertainty paper): "1. corresponds to the explanation given by Heisenberg for the a-causal behavior of quantum physics; 2. expresses that the theory nonetheless gives the appearance of a kind of causality."

$$B_{\mu\nu} \equiv \begin{pmatrix} 0 & \cdots & \cdots & \cdots & 0 \\ \vdots & 0 & & 1 & \vdots \\ \vdots & & \ddots & & \vdots \\ \vdots & 1 & & 0 & \vdots \\ 0 & \cdots & \cdots & \cdots & 0 \end{pmatrix}, \tag{17.31}$$

$$C_{\mu\nu} \equiv \begin{pmatrix} 0 & \cdots & \cdots & \cdots & 0 \\ \vdots & 0 & & i & \vdots \\ \vdots & & \ddots & & \vdots \\ \vdots & -i & & 0 & \vdots \\ 0 & \cdots & \cdots & \cdots & 0 \end{pmatrix}. \tag{17.32}$$

The A_μs have 1 in the μ row and the μ column and 0s everywhere else. The $B_{\mu\nu}$s ($\mu < \nu$) have 1 in the μ row and the ν column and in the ν row and the μ column and 0s everywhere else. The $C_{\mu\nu}$s ($\mu < \nu$) have i in the μ row and the ν column and $-i$ in the ν row and the μ column and 0s everywhere else.

The A_μs have eigenvectors (eigenvalue 1) with 1 in the μ row and 0s everywhere else; and infinitely many eigenvectors with eigenvalue 0. The $B_{\mu\nu}$s have eigenvectors (eigenvalue 1) with 1s in the μ and the ν row and 0s everywhere else; eigenvectors (eigenvalue -1) with 1 in the μ row, -1 in the ν row, and 0s everywhere else; and infinitely many eigenvectors with eigenvalue 0. The $C_{\mu\nu}$s have eigenvectors (eigenvalue 1) with $i - 1$ in the μ row, $i + 1$ in the ν row, and 0s everywhere else; eigenvectors (eigenvalue -1) with $-1 - i$ in the μ row, $1 - i$ in the ν row, and 0s everywhere else; and infinitely many eigenvectors with eigenvalue 0.

For the counterpart of $B_{\mu\nu}$ in a simple finite case (with $\mu, \nu = 1, 2, 3$), we have:

$$\begin{pmatrix} 0 & 0 & 1 \\ 0 & 0 & 0 \\ 1 & 0 & 0 \end{pmatrix} \begin{pmatrix} 1 \\ 0 \\ 1 \end{pmatrix} = \begin{pmatrix} 1 \\ 0 \\ 1 \end{pmatrix},$$
$$\begin{pmatrix} 0 & 0 & 1 \\ 0 & 0 & 0 \\ 1 & 0 & 0 \end{pmatrix} \begin{pmatrix} 1 \\ 0 \\ -1 \end{pmatrix} = -\begin{pmatrix} 1 \\ 0 \\ -1 \end{pmatrix}. \tag{17.33}$$

For the counterpart of $C_{\mu\nu}$ we similarly have:

$$\begin{pmatrix} 0 & 0 & i \\ 0 & 0 & 0 \\ -i & 0 & 0 \end{pmatrix} \begin{pmatrix} i-1 \\ 0 \\ i+1 \end{pmatrix} = \begin{pmatrix} i-1 \\ 0 \\ i+1 \end{pmatrix},$$
$$\begin{pmatrix} 0 & 0 & i \\ 0 & 0 & 0 \\ -i & 0 & 0 \end{pmatrix} \begin{pmatrix} -1-i \\ 0 \\ 1-i \end{pmatrix} = -\begin{pmatrix} -1-i \\ 0 \\ 1-i \end{pmatrix}. \tag{17.34}$$

The matrix S can be written as a linear combination of A, B, and C:

$$S = \sum_{\mu} s_{\mu\mu} \cdot A_{\mu} + \sum_{\mu<\nu} \mathrm{Re}\, s_{\mu\nu} \cdot B_{\mu\nu} + \sum_{\mu<\nu} \mathrm{Im}\, s_{\mu\nu} \cdot C_{\mu\nu}, \tag{17.35}$$

where $\mathrm{Re}\, s_{\mu\nu}$ and $\mathrm{Im}\, s_{\mu\nu}$ and the real and imaginary parts of $s_{\mu\nu}$, respectively.

Using von Neumann's linearity assumption (A), we write the expectation value of S in the ensemble $\{\mathfrak{S}_1, \mathfrak{S}_2, \mathfrak{S}_3, \ldots\}$ as:

$$\begin{aligned} \mathcal{E}(S) &= \sum_{\mu} s_{\mu\mu} \cdot \mathcal{E}(A_{\mu}) \\ &\quad + \sum_{\mu<\nu} \mathrm{Re}\, s_{\mu\nu} \cdot \mathcal{E}(B_{\mu\nu}) + \sum_{\mu<\nu} \mathrm{Im}\, s_{\mu\nu} \cdot \mathcal{E}(C_{\mu\nu}). \end{aligned} \tag{17.36}$$

Since the eigenvalues of A_{μ}, $B_{\mu\nu}$, and $C_{\mu\nu}$ are all real, the expectation values $\mathcal{E}(A_{\mu})$, $\mathcal{E}(B_{\mu\nu})$, and $\mathcal{E}(C_{\mu\nu})$ are also real. Now define the matrix U (associated with some operator $\hat{U}$) with diagonal components $u_{\mu\mu} \equiv \mathcal{E}(A_{\mu})$ and off-diagonal components $(\mu < \nu)$:

$$\begin{aligned} u_{\mu\nu} &\equiv \frac{1}{2}\left(\mathcal{E}(B_{\mu\nu}) + i\mathcal{E}(C_{\mu\nu})\right), \\ u_{\nu\mu} &\equiv \frac{1}{2}\left(\mathcal{E}(B_{\mu\nu}) - i\mathcal{E}(C_{\mu\nu})\right). \end{aligned} \tag{17.37}$$

Note that this matrix is Hermitian: $u^{\star}_{\mu\nu} = u_{\nu\mu}$. With the help of this matrix U, the expectation value of S can be written as (p. 253):

$$\mathcal{E}(S) = \sum_{\mu\nu} s_{\mu\nu}\, u_{\nu\mu}. \tag{17.38}$$

To verify this, we consider the sums over $\mu = \nu$ and $\mu \neq \nu$ separately. For the former we find

$$\sum_{\mu} s_{\mu\mu}\, u_{\mu\mu} = \sum_{\mu} s_{\mu\mu} \cdot \mathcal{E}(A_{\mu}). \tag{17.39}$$

For the latter, we have

$$\sum_{\mu\neq\nu} s_{\mu\nu}\, u_{\nu\mu} = \sum_{\mu<\nu} s_{\mu\nu}\, u_{\nu\mu} + \sum_{\mu>\nu} s_{\mu\nu}\, u_{\nu\mu}. \tag{17.40}$$

The second term can be written as $\sum_{\nu>\mu} s_{\nu\mu}\, u_{\mu\nu} = \sum_{\mu<\nu} s^{\star}_{\mu\nu}\, u^{\star}_{\nu\mu}$, which means that

$$\sum_{\mu\neq\nu} s_{\mu\nu}\, u_{\nu\mu} = \sum_{\mu<\nu} 2\,\mathrm{Re}\,(s_{\mu\nu}\, u_{\nu\mu}). \tag{17.41}$$

Now write $s_{\mu\nu}$ as the sum of its real and imaginary parts and use Eq. (17.37) for $u_{\nu\mu}$:

$$\begin{aligned}\sum_{\mu\neq\nu} s_{\mu\nu}\,u_{\nu\mu} &= \sum_{\mu<\nu} \mathrm{Re}\,\{(\mathrm{Re}\,s_{\mu\nu} + i\,\mathrm{Im}\,s_{\mu\nu})\cdot(\mathcal{E}(B_{\mu\nu}) - i\mathcal{E}(C_{\mu\nu}))\} \\ &= \sum_{\mu<\nu} \mathrm{Re}\,s_{\mu\nu}\cdot\mathcal{E}(B_{\mu\nu}) + \sum_{\mu<\nu} \mathrm{Im}\,s_{\mu\nu}\cdot\mathcal{E}(C_{\mu\nu}).\end{aligned} \tag{17.42}$$

Adding Eq. (17.39) and Eq. (17.42), we arrive at

$$\begin{aligned}\sum_{\mu\nu} s_{\mu\nu}\,u_{\nu\mu} &= \sum_{\mu} s_{\mu\mu}\cdot\mathcal{E}(A_\mu) \\ &\quad + \sum_{\mu<\nu} \mathrm{Re}\,s_{\mu\nu}\cdot\mathcal{E}(B_{\mu\nu}) + \sum_{\mu<\nu} \mathrm{Im}\,s_{\mu\nu}\cdot\mathcal{E}(C_{\mu\nu}).\end{aligned} \tag{17.43}$$

Eq. (17.36) tells us that the right-hand side of this equation is just $\mathcal{E}(S)$. This concludes the proof of Eq. (17.38), in which one readily recognizes the trace of the product of S and U:[42]

$$\mathcal{E}(S) = \sum_{\mu\nu} s_{\mu\nu}\,u_{\nu\mu} = \sum_{\mu} (SU)_{\mu\mu} = \mathrm{Tr}(SU) = \mathrm{Tr}(US). \tag{17.44}$$

In other words, U is what is now called a *density matrix*, usually denoted by the Greek letter ρ. It corresponds to a density operator $\hat{U}$ or $\hat{\rho}$.

The matrix U characterizes the ensemble $\{\mathfrak{S}_1, \mathfrak{S}_2, \mathfrak{S}_3, \ldots\}$. Von Neumann (sec. IV, p. 255) now focuses on "pure" (*rein*) or "uniform" (*einheitlich*) ensembles, in which every copy $\mathfrak{S}_i$ of the system is in the exact same quantum state. Von Neumann characterizes such ensembles as follows: one cannot obtain a uniform ensemble "by mixing (*vermischen*) two ensembles unless it is the case that both of these correspond to that same ensemble" (p. 256).[43] Let the density operators $\hat{U}$, $\hat{U}^\star$, and $\hat{U}^{\star\star}$ correspond to the ensembles $\{\mathfrak{S}_i\}$, $\{\mathfrak{S}^\star_j\}$, $\{\mathfrak{S}^{\star\star}_k\}$, respectively. Suppose $\{\mathfrak{S}_i\}$ consists of $\eta \times 100\%$ $\{\mathfrak{S}^\star_j\}$ and $\vartheta \times 100\%$ $\{\mathfrak{S}^{\star\star}_k\}$. The expectation value of an arbitrary property represented by the operator $\hat{S}$ in $\{\mathfrak{S}_i\}$ is then given by (ibid.):

$$\mathcal{E}(\hat{S}) = \eta\,\mathcal{E}^\star(\hat{S}) + \vartheta\,\mathcal{E}^{\star\star}(\hat{S}), \tag{17.45}$$

where $\mathcal{E}^\star$ and $\mathcal{E}^{\star\star}$ refer to ensemble averages over $\{\mathfrak{S}^\star_j\}$ and $\{\mathfrak{S}^{\star\star}_k\}$, respectively. Using Eq. (17.44), we can write this as:

$$\mathrm{Tr}(\hat{U}\hat{S}) = \eta\,\mathrm{Tr}(\hat{U}^\star\hat{S}) + \vartheta\,\mathrm{Tr}(\hat{U}^{\star\star}\hat{S}). \tag{17.46}$$

[42] Von Neumann (1927b, p. 255) only wrote down the first step of Eq. (17.44). It was only in the third installment of his trilogy, that von Neumann (1927c, p. 274) finally introduced the notation trace (*Spur*), which we use here and in Eq. (17.18) in Section 17.1 (using the modern notation "Tr").

[43] The difference between a uniform and a non-uniform ensemble corresponds to the difference between pure states and mixed states. Von Neumann (1927a, p. 43) already implicitly used this distinction in the first installment of his trilogy: Eq. (17.11) describes a pure state and Eq. (17.12) describes a mixed state.

Since $\hat{S}$ is arbitrary, it follows that $\hat{U}$, $\hat{U}^\star$, and $\hat{U}^{\star\star}$ satisfy (ibid.):

$$\hat{U} = \eta\, \hat{U}^\star + \vartheta\, \hat{U}^{\star\star}. \tag{17.47}$$

Von Neumann now proves a theorem pertaining to uniform ensembles (pp. 257–258). That $\hat{U}$ is the density operator for a uniform ensemble can be expressed by the following conditional statement: If ($\hat{U} = \hat{U}^\star + \hat{U}^{\star\star}$) then ($\hat{U}^\star \propto \hat{U}^{\star\star} \propto \hat{U}$). Von Neumann shows that this is equivalent to the statement that there is a unit vector $|\varphi\rangle$ such that $\hat{U}$ is the projection operator onto that vector, that is, $\hat{U} = \hat{P}_\varphi = |\varphi\rangle\langle\varphi|$.[44] Written more compactly, the theorem says:

$$\begin{aligned} &\Big((\hat{U} = \hat{U}^\star + \hat{U}^{\star\star}) \Rightarrow (\hat{U}^\star \propto \hat{U}^{\star\star} \propto \hat{U}) \Big) \\ &\quad \Leftrightarrow \Big(\exists\, |\varphi\rangle,\ \hat{U} = \hat{P}_\varphi = |\varphi\rangle\langle\varphi| \Big). \end{aligned} \tag{17.48}$$

A crucial ingredient of the proof of the theorem is the inner-product structure of Hilbert space. The theorem implies two important results, which, given the generality of the assumptions going into its proof, have the unmistakable flavor of a free lunch. First, pure dispersion-free states (or ensembles) correspond to unit vectors in Hilbert space.[45] Second, the expectation value of a quantity $\mathfrak{a}$ represented by the operator $\hat{S}$ in a uniform ensemble $\{\mathfrak{S}_i\}$ characterized by the density operator $\hat{U} = |\varphi\rangle\langle\varphi|$ is given by the trace of the product of the corresponding matrices:

$$\mathcal{E}(\hat{S}) = \mathrm{Tr}(\hat{U}\hat{S}) = \mathrm{Tr}(|\varphi\rangle\langle\varphi|\hat{S}) = \langle\varphi|\hat{S}|\varphi\rangle, \tag{17.49}$$

which is equivalent to the Born rule.

Von Neumann was still not satisfied. In section V, "Measurements and states," he notes that

> our knowledge of a system $\mathfrak{S}'$, that is, of the structure of a statistical ensemble $\{\mathfrak{S}'_1, \mathfrak{S}'_2, \ldots\}$, is never described by the specification of a state—or even by the corresponding φ [i.e., the vector $|\varphi\rangle$]; but usually by the result of measurements performed on the system (p. 260).

He considers the simultaneous measurement of a complete set of commuting operators and constructs a density operator for (the ensemble representing) the system on the basis

[44] The notation $\hat{P}_\varphi$ (except for the "hat") is von Neumann's own (p. 257).

[45] This is the essence of von Neumann's later no-hidden-variables proof (von Neumann 1932, Ch. 4, p. 171), which was criticized by John Bell (1966, pp. 1–5), who questioned the linearity assumption (A), $\mathcal{E}(\alpha\,\mathfrak{a}+\beta\,\mathfrak{b}) = \alpha\,\mathcal{E}(\mathfrak{a})+\beta\,\mathcal{E}(\mathfrak{b})$ (see note 39). Bell argued, with the aid of explicit examples, that the linearity of expectation values was too strong a requirement to impose on hypothetical dispersion-free states (dispersion-free via specification of additional "hidden" variables). In particular, the dependence of spin expectation values on the (single) hidden variable in the explicit example provided by Bell is manifestly *non-linear*, although the model reproduces exactly the standard quantum-mechanical results when one averages (uniformly) over the hidden variable. For further discussion, see Bacciagaluppi and Crull (2009), Bub (2010), and Dieks (2017).

of outcomes of these measurements showing the measured quantities to have values in certain intervals. He shows that these measurements can fully determine the state and that the density operator in that case is the projection operator onto that state.

Let $\{\hat{S}_\mu\}$ ($\mu = 1, \ldots, m$) be a complete (maximal) set of commuting operators with common eigenvectors, $\{|\sigma_n\rangle\}$, with eigenvalues $\lambda_\mu(n)$:

$$\hat{S}_\mu \, |\sigma_n\rangle = \lambda_\mu(n) \, |\sigma_n\rangle. \tag{17.50}$$

Now construct an operator $\hat{S}$ with those same eigenvectors and completely non-degenerate eigenvalues λ_n:

$$\hat{S}|\sigma_n\rangle = \lambda_n|\sigma_n\rangle, \tag{17.51}$$

with $\lambda_n \neq \lambda_{n'}$ if $n \neq n'$. Define the functions $f_\mu(\lambda_n) = \lambda_\mu(n)$. Consider the action of $f_\mu(\hat{S})$ on $|\sigma_n\rangle$:

$$f_\mu(\hat{S}) \, |\sigma_n\rangle = f_\mu(\lambda_n) \, |\sigma_n\rangle = \lambda_\mu(n) \, |\sigma_n\rangle = \hat{S}_\mu \, |\sigma_n\rangle. \tag{17.52}$$

Hence, $\hat{S}_\mu = f_\mu(\hat{S})$. It follows from Eq. (17.52) that a measurement of $\hat{S}$ uniquely determines the state of the system. As von Neumann states: "In this way measurements have been identified that uniquely determine the state of [the system represented by the ensemble] $\mathfrak{S}'$" (p. 264).

As a concrete example, consider the bound states of a hydrogen atom. These states are uniquely determined by the values of four quantum numbers: the principal quantum number n, the orbital quantum number l, the magnetic quantum number m_l, and the spin quantum number m_s. These four quantum numbers specify the eigenvalues of four operators, which we may make dimensionless by suitable choices of units: the Hamiltonian in Rydberg units ($\hat{H}$/Ry), the angular momentum squared ($\hat{L}^2/\hbar^2$), the z-component of the angular momentum ($\hat{L}_z/\hbar$), and the z-component of the spin ($\hat{\sigma}_z/\hbar$). In this case, in other words,

$$\{\hat{S}_\mu\}_{\mu=1}^4 = (\hat{H}/\text{Ry}, \hat{L}^2/\hbar^2, \hat{L}_z/\hbar, \hat{\sigma}_z/\hbar). \tag{17.53}$$

The task now is to construct an operator $\hat{S}$ that is a function of the $\hat{S}_\mu$s (which have rational numbers as eigenvalues) and that has a completely non-degenerate spectrum. One measurement of $\hat{S}$ then uniquely determines the (bound) state of the hydrogen atom. For example, choose α, β, γ, and δ to be four real numbers, incommensurable over the rationals (i.e., no linear combination of $\alpha, \beta, \gamma, \delta$ with rational coefficients vanishes), and define

$$\hat{S} = \alpha\,\hat{S}_1 + \beta\,\hat{S}_2 + \gamma\,\hat{S}_3 + \delta\,\hat{S}_4. \tag{17.54}$$

One sees immediately that the specification of the eigenvalue of $\hat{S}$ suffices to uniquely identify the eigenvalues of $\hat{H}$, $\hat{L}^2$, $\hat{L}_z$, and $\hat{\sigma}_z$.

Von Neumann thus arrives at the typical statement of a problem in modern quantum mechanics. There is no need anymore for $\hat{q}$s and $\hat{p}$s, where the $\hat{p}$s do not commute with the $\hat{q}$s. Instead one identifies a complete set of commuting operators. Since all members of the set commute with one another, they can all be viewed as $\hat{q}$s. The canonically conjugate $\hat{p}$s do not make an appearance—indeed, their specification is irrelevant for the unique determination of a quantum state of the system.

To conclude this section, we draw attention to one more passage in "Probability-theoretic construction ... ," already quoted in the introduction to this volume (see Section 8.8). Both Jordan and von Neumann considered conditional probabilities of the form

$$\Pr(\hat{a} \text{ has the value } a \,|\, \hat{b} \text{ has the value } b),$$

or, more generally,

$$\Pr(\hat{a} \text{ has a value in interval } I \,|\, \hat{b} \text{ has a value in interval } \mathcal{J}).$$

To test the quantum-mechanical predictions for these probabilities, one needs to prepare a system in a pure state in which $\hat{b}$ has the value b or in a mixed state in which $\hat{b}$ has a value in the interval $\mathcal{J}$, and then measure $\hat{a}$. What happens *after* that measurement? Elaborating on Heisenberg's (1927b) ideas in the uncertainty paper (cf. note 41), von Neumann addresses this question in the concluding section of "Probability-theoretic construction ... ":

> A system left to itself (not disturbed by any measurements) has a completely causal time evolution [governed by the Schrödinger equation]. In the confrontation with experiments, however, the statistical character is unavoidable: for every experiment there is a state adapted [*angepaßt*] to it in which the result is uniquely determined (the experiment in fact produces such states if they were not there before); however, for every state there are "non-adapted" measurements, the execution of which *demolishes* [*zertrümmert*] that state and produces adapted states according to stochastic laws (pp. 271–272, our emphasis).

As far as we know, this is the first time the imagery of a collapse (here a demolition or *Zertrümmerung*) of the state vector occurs in print. The basic idea, however, von Neumann probably picked up from Born or Heisenberg, either in conversation or from lectures and discussions at Como in September 1927 (see Section 16.4, especially the second passage quoted from Heisenberg's comments following Bohr's Como lecture).

17.3 From canonical transformations to transformations in Hilbert space

This section briefly reviews and summarizes the transition from Jordan's version of the Dirac–Jordan statistical transformation theory discussed in Chapter 16 to von Neumann's Hilbert space formalism discussed in this chapter.

The postulates of Jordan's (1927b, 1927g) *Neue Begründung* papers amount to a concise formulation of the fundamental tenets of the probabilistic interpretation of quantum mechanics. Building on insights of Born and Pauli, Jordan (1927b) was the first to state *in full generality* that probabilities in quantum mechanics are given by the absolute square of what he called probability amplitudes. He was also the first to recognize in full generality the peculiar rules for combining probability amplitudes. However, as we have seen, both Heisenberg (1927b, pp. 183–184, p. 196) and von Neumann (1927b, p. 246) criticized his further claim that the basic rules of probability theory no longer hold in quantum mechanics. Moreover, as we saw in Section 16.2, after laying down the peculiar rules for combining probability amplitudes in a set of postulates, Jordan failed to produce a satisfactory formalism realizing those postulates.

In hindsight, it is clear that Jordan was lacking the requisite mathematical tools to do so, namely, abstract Hilbert space and the spectral theorem for operators acting in Hilbert space. Instead, he drew on the canonical formalism of classical mechanics. Jordan was steeped in this formalism, which played a central role in the transition from the old quantum theory to matrix mechanics. He published two papers (Jordan 1926a, 1926b), in which he investigated the implementation of canonical transformations in matrix mechanics (Lacki 2004; Duncan and Janssen 2009). As he put it in his AHQP interview (see Section 16.2.1 for the full quotation), canonically conjugate variables and canonical transformations had thus been his "daily bread" in the years leading up to *Neue Begründung*. Unfortunately, this formalism proved ill suited to the task at hand. As a result, Jordan ran into a number of serious difficulties, which became particularly severe when, in *Neue Begründung* II, he tried to extend his formalism, originally formulated only for quantities with purely continuous spectra, to quantities with wholly or partly discrete spectra. The newly introduced spin variables further exposed the limitations of Jordan's canonical formalism. To subsume these variables under his general approach, Jordan had to weaken his definition of canonically conjugate quantities to such an extent that the concept lost much of its meaning. Under Jordan's definition in *Neue Begründung* II, any two of the three components $\hat{\sigma}_x$, $\hat{\sigma}_y$, and $\hat{\sigma}_z$ of spin angular momentum are canonically conjugate to each other.

All these problems can be avoided if the canonical formalism of classical mechanics is replaced by the Hilbert space formalism, even though other mathematical challenges remain.[46] When Jordan's probability amplitudes $\varphi(a, b)$ for the quantities $\hat{a}$ and $\hat{b}$ are equated with "inner products" $\langle a|b\rangle$ of normalized "eigenvectors" of the corresponding operators $\hat{a}$ and $\hat{b}$, the rules for such amplitudes, as laid down in the postulates of

[46] Even before the introduction of the Hilbert space formalism by von Neumann, Schrödinger recognized that the mathematics of Hilbert and his school, which underlies von Neumann's approach, was better adapted to the new quantum mechanics than the celestial mechanics that Jordan continued to draw on for his approach. In the letter to Wien of February 22, 1926, from which we already quoted in Sections 14.5 and 16.4, Schrödinger talks about "the splendid [*herrliche*] classical mathematics and Hilbert mathematics, the wonderful edifice of eigenvalue theory," adding that he is "happy to have escaped from the dreadful [*schrecklichen*] mechanics with its action/angle variables and perturbation theory, which I never really understood. Now everything is *linear*, everything superposable [*superponierbar*], one calculates as easily and pleasantly as in the old acoustics" (von Meyenn 2011, pp. 184–185, emphasis in the original).

Jordan's *Neue Begründung*, are automatically satisfied. Probabilities are given by the absolute square of these inner products, and Jordan's addition and multiplication rules for probability amplitudes essentially reduce to the familiar completeness and orthogonality relations in Hilbert space. Once the Hilbert space formalism is adopted, the need to sort quantities into $\hat{p}$s and $\hat{q}$s disappears. Canonical transformations, at least in the classical sense as understood by Jordan, similarly cease to be important. Instead of canonical transformations $(\hat{p}, \hat{q}) \to (T\hat{p}\,T^{-1}, T\hat{q}\,T^{-1})$ of pairs of canonically conjugate quantities, one now considers unitary transformations $\hat{A} \to U\hat{A}\,U^{-1}$ of individual Hermitian operators. Such transformations get us from one orthonormal basis of Hilbert space to another, preserving inner products as required by the probability interpretation of quantum theory.

Von Neumann (1927a) introduced the Hilbert space formalism in May 1927 in "Mathematical foundation ... " However, he did not use this formalism to provide a realization of Jordan's postulates along the lines sketched in the preceding paragraph. As we saw in this chapter, von Neumann had some fundamental objections to the approach of Jordan (and Dirac). The basic probability amplitude for $\hat{p}$ and $\hat{q}$ in Jordan's formalism, $\varphi(p, q) = e^{-ipq/\hbar}$ (see Eq. (16.28)), is not a square-integrable function and is thus not an element of the function space L^2 instantiating abstract Hilbert space. The delta function, unavoidable in the Dirac–Jordan formalism, is no function at all. In addition, as we saw earlier, von Neumann objected to the phase-ambiguity of the probability amplitudes.

Jordan's response to this last objection illustrates the extent to which he was still trapped in thinking solely in terms of ps and qs. In *Neue Begründung* II, he eliminated the phase-ambiguity of the probability amplitude for any two quantities by adding two indices indicating a specific choice of the quantities canonically conjugate to those two quantities (see Eq. (16.93)). Von Neumann's response to this same problem was very different and underscores that he was not wedded at all to the canonical formalism of classical mechanics. Von Neumann decided to avoid probability amplitudes altogether. Instead he turned to projection operators in Hilbert space, which he used both to formulate the spectral theorem for a large class of operators[47] in Hilbert space, and to construct a new formula for conditional probabilities in quantum mechanics (see Eq. (17.18) and Eq. (17.26)).

Although von Neumann took Jordan's formula for conditional probabilities as his starting point and rewrote it in terms of projection operators, his final formula is more general than Jordan's in that it pertains both to pure and to mixed states. However, as we have seen, it was not until "Probability-theoretic construction ... ," presented in November 1927, that von Neumann (1927b) carefully defines the difference between pure and mixed states. Here, von Neumann frees his approach from reliance on Jordan's even further and now derives his formula for conditional probabilities in terms of the trace of products of projection operators from the Hilbert space formalism, using a few seemingly innocuous assumptions about expectation values of observables of systems

[47] The spectral theorem holds for *normal* operators, bounded or unbounded, defined as those operators that commute with their adjoint. Self-adjoint (resp. unitary) are evidently normal, as any operator commutes with itself (resp. its inverse).

in an ensemble of copies of those systems characterized by a density operator. He then shows that the density operator for a uniform ensemble is just the projection operator onto the corresponding pure dispersion-free state. Such pure states can be characterized completely by the eigenvalues of a complete set of commuting operators, which led von Neumann to a new way of formulating a typical problem in quantum mechanics. Rather than identifying $\hat{p}$s and $\hat{q}$s for the system under consideration, he realizes that it suffices to specify the values of a *maximal set of commuting operators* for the system. All operators in such sets can be thought of as $\hat{q}$'s. There is no need to find the $\hat{p}$'s canonically conjugate to these $\hat{q}$s. The deep kinematical structure of the new theory—namely, a set of specified, mutually measurable, observables, the collected eigenvalues of which *uniquely* specify a quantum state of the system, and therefore in a sense represent the maximum information which can be associated with a definite quantum state—is determined by the choice of maximal commuting sets of operators in Hilbert space, not by classical phase spaces.

18

Conclusion: Arch and Scaffold

Steven Shapin's *The Scientific Revolution* famously begins with the sentence: "There was no such thing as the Scientific Revolution, and this is a book about it" (Shapin 1996, p. 1). We can use the superposition principle to state its quantum counterpart: there was and there wasn't a quantum revolution, and this is and this isn't a book about it. There clearly *was* a quantum revolution in the sense that the basic formalism for handling problems in spectroscopy and other areas of physics underwent a complete overhaul during the years 1925–1927. Hence, our book definitely *is* about a revolution. However, it is not about a revolution in the sense of the new theory being erected on the burning embers of the old one, the image conjured up by the political connotations of the term "revolution". Our analysis suggests a different image: the new theory was carefully built as an arch on a scaffold provided in large part by the old theory. This is why we used the term "scaffold" in the subtitle of Volume 1 of our book and the term "arch" in the subtitle of Volume 2. In this short concluding chapter, we elaborate and reflect on our use of this metaphor.[1]

18.1 Continuity and discontinuity in the quantum revolution

The imagery of an arch built on a scaffold, discarded once the arch can support itself, captures—even if not perfectly—both continuous and discontinuous aspects of theory change such as the major upheaval in quantum theory in the mid-1920s. Of course, whether one sees continuity or discontinuity in this transition is to some extent a matter of perspective. One would expect historians tracing the events as they unfolded on the ground to see mainly continuity and physicists taking a bird's eye view, comparing the

[1] For more general discussion of the arch-and-scaffold metaphor, see Janssen (2019), on which we have drawn heavily for this chapter. Other authors have referred to an old theory "scaffolding" a new one (Israel 1987, p. 200) or a new theory being "grafted" onto an old one (Lakatos 1970, p. 142) but have not pursued the idea any further. Similar ideas, though not phrased in terms of arches and scaffolds, have been proposed and developed by Jürgen Renn (2006, 2020). Janssen and Renn (2015, 2022) use the arch-and-scaffold metaphor to describe how Einstein found the field equations of general relativity.

Constructing Quantum Mechanics. Anthony Duncan and Michel Janssen, Oxford University Press.
 DOI: 10.1093/oso/9780198883906.003.0018

landscape before and after the transition, to see mainly discontinuity (Midwinter and Janssen 2013, pp. 147–148).

This generalization obviously does not hold for all historians and all physicists. In his book on the Bohr model and its further development, for instance, historian Helge Kragh (2012, p. 368) emphasized discontinuity, writing that matrix mechanics "grew out of what little was left [of the old quantum theory,] its ruins." And in the preface to the first volume of their monumental *The Historical Development of Quantum Theory*, Jagdish Mehra and Helmut Rechenberg (1982a, p. ix) wrote that quantum mechanics "in its conceptions ... made a complete break with the past." The Dutch physicist Hendrik Casimir, who studied with some of the quantum revolutionaries of the mid-1920s, was more careful, using his own building metaphor to do justice to both continuities and discontinuities: "Between 1924 and 1928 [the new quantum mechanics] swept physics like an enormous wave, tearing down provisional structures, stripping classical edifices of illegitimate extensions, and clearing a most fertile soil" (Casimir 1983, p. 51). While causing serious disruption, Casimir's tidal wave did not level the old building—it only washed away parts of it.

Pace Kuhn (1962, p. 137)—who argued that textbooks, to be pedagogically effective, must paper over scientific revolutions and make the new theory (or paradigm) look like a continuation of the old—several quantum textbooks emphasize the discontinuity in the transition from classical to quantum mechanics. In the preface of a popular undergraduate text, for instance, we read that "quantum mechanics is not, in my view, something that flows smoothly and naturally from earlier theories. On the contrary, it represents an abrupt and revolutionary departure from classical ideas" (Griffiths 2005, p. viii). And commenting on the Schrödinger equation in *The Feynman Lectures on Physics*, Richard Feynman wrote: "Where did we get that from? Nowhere. It is not possible to derive it from anything you know. It came out of the mind of Schrödinger, invented in his struggle to find an understanding of the experimental observation of the real world" (Feynman, Leighton, and Sands 1964, Vol. 3, sec. 16-5, p. 16-12).

Unlike the other statements quoted above, Feynman's seems to have found its way into the popular literature. It is echoed, for instance, in the chapter on Schrödinger in Benjamín Labatut's highly acclaimed *When We Cease to Understand the World*. Schrödinger, he writes, "had not derived his equation from pre-existing principles. His thinking had not departed from any known basis. The equation itself was a principle, and his mind had pulled it from nothing" (Labatut 2020, p. 147). Feynman knew better, and was well aware that Schrödinger had not pulled his equation out of thin air: "When Schrödinger first wrote it down, he gave a kind of derivation based on some heuristic arguments and some brilliant intuitive guesses. Some of the arguments he used were even false, but that does not matter; the only important thing is that the ultimate equation gives a correct description of nature" (Feynman, Leighton, and Sands 1964, Vol. 3, sec. 16-1, p. 16-4).

Ours is certainly not the first book on the genesis of quantum mechanics that downplays the discontinuity emphasized in the passages just quoted and instead highlights the continuity in the transition from classical mechanics, via the old quantum theory, to quantum mechanics. In the preface of his classic *The Conceptual Development of Quantum Mechanics*, for instance, Max Jammer (1966, p. vii) wrote that one of his goals was to show "how in the process of constructing the conceptual edifice of quantum mechanics each stage depended on those preceding it without necessarily following from them as a logical consequence." The subtitle of Olivier Darrigol's *From c-Numbers to q-Numbers*, "The classical analogy in the history of quantum theory," points to a similar objective. In the introduction, Darrigol (1992, p. xxii) states his conviction that "to obtain new theories, [theorists (then and now)] extend, combine, or transpose available pieces of theory." One of the mottos he chose for his book comes from unpublished lecture notes by Dirac from 1927: "[T]he new quantum theory requires very few changes from the classical theory, *these changes being of a fundamental nature*, so that many of the features of the classical theory to which it owes its attractiveness can be taken over unchanged into the quantum theory" (pp. xv–xvi, our emphasis). The italicized clause, which gives Dirac's statement a paradoxical flavor, nicely captures the basic idea of *Umdeutung* at the heart of matrix mechanics: Heisenberg (1925c) did not *repeal* classical mechanics, he *reinterpreted* it (see Chapter 11 and Section 12.2 on Dirac's related work).

18.2 Continuity and discontinuity in two early quantum textbooks

Dirac was not the only quantum architect to emphasize continuity in the aftermath of the watershed of the mid-1920s. Sommerfeld (1929) as well as Born and Jordan (1930), for instance, did the same in prefaces to textbooks written only a few years after the quantum revolution. And contrary to Kuhn's claim that textbooks tend to distort history, their assessments, while admittedly self-serving in some respects, reflect the actual course of events reasonably well.

In 1929, Sommerfeld, whom Ehrenfest had sarcastically christened the "pope" of the old quantum theory (Eckert 2013, p. 256), published a book on wave mechanics, which he tellingly called a wave-mechanical *supplement* [*Wellenmechanischer Ergänzungsband*] (Sommerfeld 1929) to the fourth edition of the "Bible" of the old quantum theory, *Atombau und Spektrallinien* (Sommerfeld 1924). The term "supplement" clearly conveys the continuity Sommerfeld saw between the old quantum theory and wave mechanics. The title of the English translation, published the following year, is simply *Wave-Mechanics* (Sommerfeld 1930), but it is prominently mentioned on the book's title page that it is a "supplementary volume to *Atomic Structure and Spectral Lines*," the English translation (Sommerfeld 1923) of the third German edition of *Atombau und Spektrallinien* (Sommerfeld 1922b). In the first paragraph of the preface, Sommerfeld explains:

> The author believes that he is fulfilling a general wish in attempting the present supplementary volume to depict the recent developments which since 1924 to 1926 (thanks to the work of L. de Broglie, Heisenberg, Schrödinger) have transformed the external aspect of atomic physics. The fact that the inner content of the theory, that is, the quantitative assertions that can be tested by experiment, has for the most part survived this

process of regeneration is manifest to those acquainted with the subject. *The new development does not signify a radical change but a welcome evolution of the existing theory, while many fundamental points are* [*clarified*] *and made more precise* (Sommerfeld 1930, p. v, our emphasis).[2]

While Sommerfeld focused on the continuity in the transition from the old quantum theory to wave mechanics, Born and Jordan (who had meanwhile been appointed professor in Rostock) focus on the continuity in the transition from the old quantum theory to matrix mechanics in their 1930 textbook, which they present as the sequel to Born's (1925) treatise on the old quantum theory. In the preface they write:

> We see in the new mechanics the rigorous implementation [*strenge Durchführung*] of the Bohrian program, as developed in the first volume of this book. In our view, the new mechanics does not mean a return to classical pictures and no rejection of Bohrian concepts [*keine Annäherung an die klassischen Vorstellungen und keine Abkehr von den Bohrschen Begriffsbildungen*] (stationary states, quantum jumps, transition probabilities, etc.) but rather the systematic development of the latter into a logically closed system (Born and Jordan 1930, p. VI).

In the introduction, they elaborate:

> So it is not a matter of an "explanation" of the peculiar quantum laws by reducing them to classical pictures. On the contrary, the fundamental primary character of the basic quantum-theoretical assumptions [i.e., those of the old quantum theory of Bohr] has only become clear through the more recent developments. The progress actually consists in *shedding the remnants of the classical point of view* [*Abstreifen der Reste klassischer Betrachtungsweise*]. In this way, a closed theory has emerged which allows a description of all atomic phenomena which is free of contradictions and contains the classical theory as a special limiting case (ibid., p. 1, our emphasis).

This passage, especially the italicized clause, fits nicely with our image of a scaffold discarded once the arch built on that scaffold can support itself.

Sommerfeld and Born and Jordan are not only referring to continuities between different theories in these passages, but also they are referring to different *kinds* of continuity. Sommerfeld is referring to a continuity of empirical content; Born and Jordan to a conceptual continuity, that is, the conceptual continuity between Bohr's old quantum theory and matrix mechanics. At the same time, Born and Jordan seem to be at pains to emphasize the conceptual gap between these two theories and classical theory. This impression, however, is deceptive.

[2] The italicized sentence has "classified" instead of "clarified." The original German text has "Klärungen" (clarifications). In full, the sentence in the original German reads: "Die neue Entwicklung bedeutet nicht einen Umsturz, sondern eine erfreuliche Weiterbildung des Bestehenden mit vielen grundsätzlichen Klärungen und Verscharfungen" (Sommerfeld 1929, p. III). This sentence is quoted (with the final clause rendered as "with many fundamental clarifications and sharpenings") and discussed by Seth (2010, p. 266). He sees a continuity of problems as opposed to a continuity of principles (ibid., pp. 265–267). Midwinter and Janssen (2013, pp. 198–199), using the example of the theory of electric susceptibilities (see Section 15.2), argue that undergirding this continuity of problems is a continuity in mathematical formalism.

One has to keep in mind that Born and Jordan—and this, as we shall see shortly, is true for Sommerfeld as well—made these comments against the backdrop of the rejection by the Copenhagen–Göttingen camp of Schrödinger's attempts to interpret his wave mechanics along the lines of a classical wave theory (see Chapters 14 and 16). In terms of our metaphor, we can say that Schrödinger built his wave mechanics on a scaffold provided by classical wave theory, more specifically, Hamilton's optics (see Section 13.3), and that Born and Jordan are urging their readers to dismantle that scaffold. This does not mean, however, that they see a sharp break between classical and quantum mechanics:

> We are convinced that the real connection between the classical and the new mechanics does not rest so much on the formal resemblance between the wave-mechanical differential equations and boundary-value problems and the methods of classical continuum physics but on the correspondence principle of Niels Bohr (Born and Jordan 1930, pp. V–VI).

This is why they asked Bohr and got his permission to dedicate their book to him.

Sommerfeld was with Born and Jordan (and with his own former student Heisenberg), as he made clear in the preface of his own book:

> I have called this volume a "wave-mechanical" supplement, because for practical manipulation Schrödinger's methods are obviously superior to the specifically "quantum-mechanical" methods [read: the algebraic methods of matrix mechanics]. On the other hand, however, I have left no doubt that the general ideas that have led Heisenberg to enunciate quantum-mechanics are also indispensable for the elaboration of wave-mechanics. The original standpoint of Schrödinger, that transitions are to occur only between co-existing states [i.e., that radiation is produced by beats between different modes of the wave function (see Section 14.4)], is clearly too narrow and does not accurately fit the facts. I have therefore taken over into wave-mechanics the equal treatment of states and transitions, as is done by Heisenberg from the very beginning . . . This, of course, denotes that I am renouncing the more definite wave-kinematic objective, set up by Schrödinger and de Broglie, and am sacrificing pictorial representation to formalism (Sommerfeld 1930, p. v).

Born and Jordan (1930, p. V) also acknowledge the fruitfulness of Schrödinger's approach, which, they explain, proved especially attractive to physicists because of its reliance on the familiar mathematical techniques of differential equations. This is reflected, they note, in the quantum textbooks published in German so far, which all take a wave-mechanical approach. Born and Jordan list Haas (1928), Sommerfeld (1929), De Broglie (1929), and Frenkel (1929). It is therefore of interest, they argue, to see "how far one can get with elementary, that is, mainly algebraic, means" (ibid.). This then is what the term "elementary" stands for in the title of their book, *Elementare Quantenmechanik*. They hasten to add that they do not mean to denigrate wave mechanics. In fact, they inform the reader, they had originally intended to cover wave mechanics as well but the treatment of quantum mechanics in matrix-mechanical terms took up so much space that there was none left for wave mechanics. They promise to make up for this in another volume "as soon as time and energy permit" (ibid., p. VI).

This promissory note gave Pauli the perfect opening line for his scathing review of Born and Jordan (1930): "This book is the second volume in a series in which aim and purpose of the *n*th volume are always made clear through the virtual existence of the $(n+1)$th volume" (Pauli 1930). The term "virtual existence" comes from the preface of *Lectures on Atomic Mechanics*, Vol. I, Born's treatise on the old quantum theory in its waning days. Painfully aware of the inadequacies of this theory, Born had made it clear that his book was

> deliberately conceived as an attempt . . . to ascertain the limits within which the present principles of atomic and quantum theory are valid and . . . to explore the ways by which we may hope to proceed . . . to make this program clear in the title, I have called the present book "Vol. I;" the second volume is to contain a closer approximation to the "final" atomic mechanics . . . The second volume may, in consequence, remain for many years unwritten. In the meantime let its virtual existence serve to make the aim and spirit of this book clear (Born 1925, p. v).[3]

When the virtual $n=2$ volume materialized five years later, Born and Jordan reminded the readers of this passage in its preface:

> This book is the continuation [Born 1925]; it is the "second volume" announced in the preface [of the first volume], the "virtual existence of which was meant to make the aim and spirit of [the first volume] clear." The hope that the veil, which back then was still hanging over the true structure of the laws governing the atom, would soon be lifted was realized in surprisingly fast and thorough fashion (Born and Jordan 1930, p. V).

In his review, Pauli bemoans the one-sidedness of Born and Jordan's presentation of quantum mechanics and the cumbersome derivations of results that can be found much more easily in wave mechanics than in matrix mechanics. A prime example of this is the derivation of the Balmer formula for the hydrogen spectrum. Pauli reminds the reader that, before the advent of wave mechanics, he himself had first derived the Balmer formula in matrix mechanics (Pauli 1926a). Driving the point home, he adds that the reviewer thus "cannot be accused of finding the grapes sour because they are hanging too high for him." The final flourish and best-known part of Pauli's acerbic review is its closing sentence: "The production of the book in terms of print and paper is excellent" (*Die Ausstattung des Buches hinsichtlich Druck und Papier ist vortrefflich*).[4]

[3] We already quoted and discussed this passage at the end of Chapter 1.

[4] Born wrote to Sommerfeld, Pauli's teacher, to complain about this review. "I know full well," he told his Munich colleague, "that the book has major weaknesses . . . but from Pauli's side the nastiness of the attack has other grounds, which are not pretty" (Born to Sommerfeld, October 1, 1930, quoted in von Meyenn 2007, pp. 45–47). Born then explains that he had originally asked Pauli to collaborate with him on the development of matrix mechanics and had approached Jordan only after Pauli had turned him down (cf. Born 1978, pp. 218–219; see Section 12.1). Ever since, Born continues, Pauli "has had a towering rage against Göttingen

So much for "Vol. II" (Born and Jordan 1930). The passage just quoted from the preface of "Vol. I" (Born 1925) is interesting from the perspective of the arch-and-scaffold metaphor. It constitutes a rare instance of a scientist recognizing that he is building a scaffold, rather than an arch. Typically, it is only after a new theory has been built on top of an old one that the latter is treated as a scaffold (Janssen 2019, pp. 107–110). To his credit, Bohr also seems to have been keenly aware of the provisional character of the old quantum theory. In a letter to Sommerfeld of April 30, 1922, for instance, from which we already quoted in Section 9.1.1), he described his work on the theory as a "sincere effort to obtain an inner connection such that one can hope to create a valid fundament for further construction" (Sommerfeld 2004, Doc. 55).

The new building went up faster than either Born or Bohr could have anticipated. When the English translation of Born (1925) was ready for publication in 1927, wave and matrix mechanics had arrived on the scene and the clarification of the relation between the two was well underway (see Chapters 16 and 17). These rapid developments made it questionable to publish a translation of Born's book without any substantive changes. In a special preface to the English edition, Born briefly described the new developments and wrote somewhat defensively:

> Some may be found to ask if, in these circumstances, the appearance of an English translation is justified. I believe that it is, for it seems to me that *the time is not yet arrived when the new mechanics can be built up on its own foundations, without any connection with classical theory*. It would be giving a wrong view of the historical development, and doing injustice to the genius of Niels Bohr, to represent matters as if the latest ideas were inherent in the nature of the problem, and to ignore the struggle for clear conceptions which has been going on for twenty-five years. Further, I can state with a certain satisfaction that there is practically nothing in the book which I wish to withdraw. The difficulties are always openly acknowledged,[5] and the applications of the theory to empirical details are so carefully formulated that no objections can be made from the point of view of the newest theory. Lastly, I believe that this book itself has contributed in some small measure to the promotion of the new theories, particularly those parts which have been worked out here in Göttingen. The discussions with my collaborators Heisenberg, Jordan and Hund which attended the writing of this book have prepared the way for the critical step which we owe to Heisenberg.
>
> It is, therefore, with a clear conscience that I authorise the English translation (Born 1927, p. xi, our emphasis).

While in 1927 Born still wrote that "the time is not yet arrived when the new mechanics can be built up on its own foundations, without any connection with classical theory," only three years later, as we saw above, he and Jordan suggested that the time was now ripe to "shed the remnants of the classical point of view." Of course, we must remember

and has wasted no opportunity to vent it through mean-spirited comments." Born, however, would eventually come to agree with Pauli's criticism of his book with Jordan. In his memoirs, Born (1978, p. 225) calls the authors' self-imposed restriction to matrix methods a "blunder" for which Pauli had rightfully excoriated them (for further discussion, see Duncan and Janssen 2008, p. 641).

[5] At the end of Chapter 7, we quoted the conclusion of Born's book, in which he candidly acknowledges that the old quantum theory only works for hydrogen and already fails for helium.

that they were referring to Schrödinger's classical point of view and that they put great emphasis on the connection of the new quantum mechanics to the old quantum theory via Bohr's correspondence principle.

18.3 The inadequacy of Kuhn's model of a scientific revolution

The passages from the textbooks of Sommerfeld (1929) and Born and Jordan (1930) discussed in Section 18.2 make it abundantly clear that prominent participants in the quantum revolution of 1925–1927 saw a great deal of continuity in the stormy developments of this period. They may have been revolutionaries but they were not iconoclasts in the sense of just smashing the relics of the old guard. Be that as it may, viewed from a distance of almost a century, the transition from classical to quantum mechanics looks decidedly discontinuous. This impression is reinforced by what, despite the barrage of criticism it has been subjected to for over half a century, continues to be the most popular concept of a scientific revolution among historians, philosophers, scientists, and the general public alike, namely, that of a "paradigm shift" introduced in Thomas S. Kuhn's 1962 classic, *The Structure of Scientific Revolutions*. The emphasis on discontinuity in such paradigm shifts, with successive paradigms not only being incompatible but even incommensurable with each other, is what forces Kuhn to explain away passages in textbooks that make the development of science look more or less continuous as concealing revolutions for pedagogical purposes, thereby "truncating the scientist's sense of his discipline's history" (Kuhn 1962, p. 137).[6]

In Kuhn's defense, we note that he was reacting against an equally problematic, purely cumulative picture of the development of science. In 1953, almost a decade before he published *Structure*, Kuhn wrote in an application for a Guggenheim fellowship:

> Science, then, does not progress by adding stones to an initially incomplete structure, but by tearing down one habitable structure and rebuilding to a new plan with the old materials and, perhaps, new ones besides (quoted in Hufbauer 2012, p. 459).[7]

The "adding stones" metaphor with which Kuhn contrasts the "tearing down" metaphor can be found, for instance, in the preface of Rudolf Carnap's *Aufbau*, one of the central texts of logical positivism, the philosophical program against which Kuhn was reacting. In philosophy, Carnap wrote, one ought to proceed as in the natural sciences, where "one stone gets added to another, and thus is gradually constructed a stable edifice,

[6] See Midwinter and Janssen (2013, pp. 197–198) for discussion of a concrete example, a book by Van Vleck (1932), Kuhn's physics PhD adviser at Harvard, on electric and magnetic susceptibilities (cf. Section 15.2).

[7] We follow the discussion of this passage in Janssen (2019, p. 96).

which can be further extended by each following generation" (Carnap 1928, quoted in Sigmund 2017, p. 137).[8]

Neither of these building metaphors for how old theories get replaced by new ones does justice to how theory change typically happens. When scientists are building a new theory, they neither simply *add to* nor simply *tear down* the old theory. The old cumulative picture may be wrong, but so is the alternative picture of a new theory or paradigm built on the ruins of the old one, perhaps with pieces found in the rubble. The advantage of our arch-and-scaffold metaphor is that it leaves room for both addition and subtraction.

In an earlier book, *The Copernican Revolution*, Kuhn (1957, p. 182) actually used the completely different image of an inflection point, a "bend in an otherwise straight road" as he put it, to characterize the "shift in direction in astronomical thought" marked by Copernicus' *De Revolutionibus*. Neither this metaphor of a "bend in the road" nor that of a *gestalt switch*, which Kuhn (1962, p. 85) was fond of using, allow for the sharp break implied by the "tearing down" metaphor of his Guggenheim fellowship application.

As with the Copernican revolution, one will search Kuhn's writings in vain for an account of the quantum revolution in which the new paradigm was erected on the ruins of the old one. And this is not just because, confounding some of his commentators, Kuhn, for the most part, avoided the terminology of *Structure* in his historical writings.[9] In "Reflections on my Critics," his contribution to the proceedings of a 1965 conference in London which pitted him against several well-established philosophers of science, including Karl Popper, Imre Lakatos, and Paul Feyerabend, Kuhn (1970, pp. 256–259) sketched how he saw the transition from the old quantum theory to matrix mechanics. He put great emphasis on the crisis of the old quantum theory (see Chapters 7 and 9), which he justifiably saw as a "case book example" of this key concept of *Structure*,[10] but the way out of the crisis he characterized as "a series of connected steps too complex to be outlined here" (p. 258). He accordingly criticized Lakatos' (1970, pp. 140–154) account of the same episode for introducing "the crisis-resolving innovation like a magician pulling a rabbit from a hat" (Kuhn 1970, p. 258). So for Kuhn, whatever he said in *Structure*, a paradigm shift following a crisis need not always be a wholesale replacement of the old paradigm or an abrupt break with the past.

[8] John Earman has persuasively argued that Carnap and Kuhn have much more in common than these two quotations suggests: he sees evolution, rather than revolution, in going from Carnap to Kuhn (Earman 1993, p. 9).

[9] See Kuhn's (1984, p. 363) response to a set of reviews by Klein, Shimony, and Pinch (1979, see, in particular, p. 437) of his book on black-body theory (Kuhn 1978). This response was reprinted as an afterword to the second edition of this book (Kuhn 1987). For references to the main contributions to the debate over Kuhn's book, see Chapter 2, note 8.

[10] See also an unpublished essay on the "crisis of the old quantum theory" (Kuhn 1966) and the videotape of a 1980 lecture at Harvard based on this essay. In the proceedings of the 1965 London conference, Kuhn (1970, p. 258) wrote: "History of science, to my knowledge, offers no equally clear, detailed, and cogent example of the creative functions of normal science and crisis." In the Q&A following his 1980 lecture at Harvard, he reiterated that the crisis of the old quantum theory is "a textbook example ... as described in *Structure*," adding: "I don't think there are many if any that are that good" (transcribed from the videotape of the lecture). In his AHQP interviews, Kuhn routinely asked his subjects (leading) questions about their awareness of this crisis at the time (Seth 2010, p. 265).

18.4 Evolution of species and evolution of theories

Toward the end of *Structure*, Kuhn (1962, p. 171) used an analogy "that relates the evolution of organisms to the evolution of scientific ideas," with the caveat that the analogy "can easily be pushed too far." The evolutionary biology Kuhn had in mind was almost certainly the population genetics of the so-called Modern Synthesis of the ideas of Darwin and Mendel, which reigned supreme in the early 1960s (Bowler 2003, Ch. 9). This analogy with population genetics may have seriously tripped up Kuhn (Janssen 2019, pp. 164–169).

The application of population genetics to cultural evolution is best known through the last chapter of Richard Dawkins' *The Selfish Gene*, in which selection of *memes*, units of culture, takes the place of selection of genes (Dawkins 1976, Ch. 11). Dawkins did not apply his model of cultural evolution to science but gave no indication that it would not be applicable there as well. Dawkins' model of cultural evolution, like Kuhn's model of scientific revolutions, has come in for harsh criticism. In fact, the paper introducing the arch-and-scaffold metaphor for the evolution of scientific theories (Janssen 2019) appears in a volume entitled *Beyond the Meme*, a collection of essays exploring models of cultural evolution, including the evolution of science, that go beyond Dawkins' model (Love and Wimsatt 2019).[11]

An important disanalogy between the evolution of theories and the evolution of species is that modifications of scientific theories, unlike variations in species that form the input for natural selection, are not just generated at random. Kuhn has little to say about where new scientific ideas come from and some of what he does say might give comfort to those tempted to push his analogy between the evolution of species and the evolution of science beyond its breaking point. Consider, for instance, the following passage in *Structure*: "The new paradigm, or a sufficient hint to permit later articulation, emerges all at once, sometimes in the middle of the night, in the mind of a man deeply immersed in crisis" (Kuhn 1962, p. 90). Combining statements such as these with Kuhn's (1962, Ch. 7) emphasis on the proliferation of different articulations of a paradigm in a period of crisis—in the specific case of the quantum revolution he can legitimately point to "more and wilder versions of the old quantum theory than before" (Kuhn 1970, p. 257)—one may come away with the impression that modifications of theories, not unlike the random variations in species, are typically generated in great profusion and in no particular direction and that the way in which such modifications compete for acceptance by the relevant scientific community is similar to the way variations in species compete for a given ecological niche.

While population genetics may indeed provide a good model for how theories are *accepted*, it is of little help as a model for how theories are *generated*. In fact, population genetics itself has been subjected to similar criticism. The central point of such criticism is that population geneticists tend to ignore that the variations that form the input for natural selection are tightly constrained by the morphology of the evolving species. An

[11] See also Wimsatt and Griesemer (2007) and Caporael, Griesemer, and Wimsatt (2014).

early example of this line of criticism can be found in a famous paper by Stephen Jay Gould and Richard Lewinton.[12] In the abstract, they write:

> An adaptationist programme has dominated evolutionary thought in England and the United States during the past forty years. It is based on faith in the power of natural selection as an optimizing agent. It proceeds by breaking an organism into unitary "traits" [as Dawkins did with the units of culture he called memes] and proposing an adaptive story for each considered separately ... We criticize this approach and attempt to reassert a competing notion (long popular in continental Europe) that organisms must be analyzed as integrated wholes, with *baupläne* so constrained by phyletic heritage, pathways of development, and general architecture that *the constraints themselves become more interesting and more important in delimiting pathways of change than the selective force that may mediate change when it occurs* (Gould and Lewontin 1979, p. 581; our emphasis).

Gould and Lewontin thus championed an approach to biological evolution that deemphasizes the agent of evolutionary change (natural selection) and emphasizes the role of constraints instead (see also Gould 1980). Neil Shubin's (2008) bestseller *Your Inner Fish* gives a good impression of what an account of evolutionary change along these lines looks like. It traces the evolution of various parts of the human body back along our branch of the evolutionary tree. Like Gould and Lewontin, Shubin clearly does not wish to deny that natural selection is the mechanism that "mediates change when it occurs" but he does not discuss that mechanism at all. His focus is squarely on constraints.

A full-blown version of the approach advocated by Gould and Lewontin, which goes by the acronym 'evo-devo' (for evolution and development), has become popular in biology (for discussion of its various aspects, see, e.g., Samson and Brandon 2007). Evo-devo fits much better with the arch-and-scaffold metaphor for theory change in science than population genetics. The arch-and-scaffold metaphor directs our attention to the question of how new theory was *generated.* In our book, we obviously also paid attention to the question of how new theory was *accepted*, but the arch-and-scaffold metaphor highlights how structures in the new theory arose from structures in the old one. This is similar to the kind of evolutionary biology promoted by Gould and Lewontin and practiced by Shubin. They are interested in tracing structures in later species to structures in earlier species and less interested in spelling out the details of the selection process through which those later species displaced their ancestors.[13]

[12] The title of their paper, "The spandrels of San Marco," refers to another architectural metaphor for the evolution of species that can be adapted to the evolution of theories as well, although we will not try to do so here (see Janssen 2019, pp. 170–171).

[13] Shubin even uses the term "scaffolding" at one point: "the scaffolding of our entire body originated in a surprisingly ancient place: single-celled animals" (Shubin 2008, p. 123). Another popular book on evo-devo is Sean B. Carroll's (2005) *Endless Forms Most Beautiful. The New Science of Evo Devo.* Central to Carroll's book is the *development* part of evo-devo, the issue of how regulatory genes control embryonic development. For our purposes, however, it is the *evolution* part of evo-devo central to Shubin's book that matters. This is key to the analogy we want to draw between the evolution of species and the evolution of scientific theories, especially given the backward-looking perspective Shubin's book has in common with ours (see the preface of Volume 1).

18.5 The role of constraints in the quantum revolution

The notion of constraints taking pride of place in evo-devo-type accounts of biological evolution can also be used fruitfully in arch-and-scaffold-type accounts of theory change. The construction of new scientific theories is "constrained by factors such as the empirical and explanatory successes of prior theory, the existing mathematical toolkit, and the culturally specific reservoir of metaphors and analogies available for heuristic purposes (wave and particle imagery for instance)" (Janas, Cuffaro, and Janssen 2022, p. 9). In our account of the genesis of quantum mechanics, we encountered all three types of constraints listed here. We briefly discuss some examples of each one.

The first constraint—especially, that a new theory should account for the data accounted for by the old theory—may be the most obvious one. As Sommerfeld (1930, p. v) wrote in the preface of his wave-mechanical supplement, "the inner content of the [old quantum] theory, that is, the quantitative assertions that can be tested by experiment, has for the most part survived" the transition to the new quantum theory. An occasional 'Kuhn loss' (see Section 15.2) can be tolerated but the bulk of the empirical support for the old theory had better carry over to the new one.

Both Volume 1 and Volume 2 offer clear examples of the second type of constraints, due to what mathematical techniques happen to be available. In Volume 1 (see pp. 24–25 and pp. 223–224), we saw that the further elaboration of Bohr's model of the hydrogen atom by Sommerfeld and others would not have been possible without the techniques from celestial mechanics to which the astronomer Schwarzschild alerted Sommerfeld in correspondence. Another quote from the "Bible" of the old quantum theory underscores this point:

> Up to a few years ago it was possible to consider that the method of mechanics of Hamilton and Jacobi could be dispensed with for physics and to regard it as serving only the requirements of the calculus of astronomic perturbations and mathematics ... [but] since the appearance of the papers [by Epstein (1916a, 1916b), Schwarzschild (1916), and Sommerfeld (1916a, 1915b)] it seems almost as if Hamilton's method were expressly created for treating the most important problems of physical mechanics (Sommerfeld 1923, pp. 555–556, quoted in Hankins 1980, p. 63).

Chapter 17 and Appendix C show how von Neumann used the work of mathematicians such as Hilbert and Hellinger on the still-developing field of functional analysis to recast the basic ideas of the Dirac–Jordan transformation theory in a mathematically and conceptually more satisfactory form. At a more pedestrian level, we saw how Born and Jordan recognized that Heisenberg's arrays of numbers satisfying a non-commutative multiplication rule are nothing but matrices (see Section 12.1).

Illustrating the third type of constraint, we saw how our protagonists struggled to combine wave and particle imagery throughout the period covered in these volumes, from Einstein's (1905, 1909a, 1909b) fluctuation arguments showing that black-body radiation exhibits both wave-like and particle-like behavior (Chapter 3), to De Broglie's

(1924, 1925) and Einstein's (1924, 1925a, 1925b) extension of wave-particle duality to matter (Chapter 13), to discussions of Heisenberg's (1927b) uncertainty principle and Bohr's (1928a, 1928b) complementarity principle (Chapter 16).

These three types of constraints obviously do not exhaust the list of possible constraints. For instance, we saw how physics principles expected to survive the demise of the old theory were put to good use as constraints in the search for its successor. The second law of thermodynamics played such a role in Planck's work on black-body radiation (Chapter 2), energy conservation provided Heisenberg with an important check on what he had wrought at Helgoland (Chapter 11), and Boltzmann's relation between entropy and probability played a central role in Einstein's fluctuation arguments for light quanta and wave–particle duality (Chapter 3). Bohr's correspondence principle presupposed that there were important structural similarities between the classical and the new theory (see especially Sections 5.3 and 10.3).

18.6 Limitations of the arch-and-scaffold metaphor

As Lewinton (1963, p. 230) once cautioned, with a variation on a warning attributed to Thomas Jefferson about liberty, "the price of metaphor is eternal vigilance" (the true origins of both warnings remain unclear). So while useful to paint a picture of the quantum revolution that does justice to the role of constraints and to both continuous and discontinuous elements in the quantum revolution, the arch-and-scaffold metaphor, like all metaphors, has its limitations (otherwise it would be an equivalence and not merely a metaphor). We list three such limitations and illustrate each one with examples from the genesis of quantum mechanics.

First, there is the issue of intent and foresight. If we want to build a physical arch (e.g., a stone-arch bridge), we typically *design* a scaffold to the exact specifications of the arch, though sometimes we may get lucky and find a pre-existing one that will do the job. By contrast, when we say that a new theory is built as an arch on a scaffold provided by the theory it eventually replaced, the scaffold could obviously not have been constructed with the arch in mind. The arch-and-scaffold metaphor thus suggests an element of teleology, giving any historical account that uses the metaphor the bad odor of 'whig history' or 'presentism'. Typically, an earlier theory, which in hindsight is seen as a scaffold for a later theory, started out as an arch in its own right and was seen as such by the scientists who built it. Kramers' dispersion theory, for instance, looked like one of the keystones in the arch of the old quantum theory before it became a scaffold for matrix mechanics. In the case of the old quantum theory (see the passages quoted above from a letter of Bohr to Sommerfeld of 1922 and from the preface of Born's 1925 book), several physicists realized that they were working on a scaffold rather than an arch (without knowing, of course, what the arch would look like), but this appears to be the exception rather than the rule. In any event, the elements of intent and foresight normally involved in building a scaffold are clearly not part of our metaphorical use of the term. As we noted in the preface of Volume 1 (p. v, note 1), our use of the arch-and-scaffold metaphor was inspired by an example of scaffolding from biology: clay crystals that *inadvertently* may have served as a scaffold for RNA molecules (Cairns-Smith 1985, pp. 58–60). As long as we keep in mind that some old theory was not *designed* as a scaffold for some new theory, but *inadvertently* turned into one, we can use the arch-and-scaffold metaphor without running afoul of whiggishness.

We can actually use the arch-and-scaffold metaphor as a pre-emptive defense against the charge of whiggishness (see also Janssen 2019, pp. 112–113). As we also noted in the preface of Volume 1, the position we started from in our analysis of the genesis of quantum mechanics can be compared to that of someone who stands in awe in front of an arch and wonders how it was built. After some investigation, this person discovers that it was done with the help of a scaffold. Similarly, we marveled at the theoretical edifice of quantum mechanics and asked how it was constructed. An important part of the answer, we discovered, is that earlier theory was used as a scaffold. Heisenberg's (1925c) *Umdeutung* paper, once again, is a good example. It can be thought of as an arch built on the scaffold provided by Kramers' dispersion theory (see Chapters 10 and 11). This scaffold has long since been taken down, leaving those encountering the arch (left standing without any visible support) puzzled as to how it got there. Among those who found themselves in that predicament was the late, great Steven Weinberg. Discussing the *Umdeutung* paper in *Dreams of a Final Theory*, he wrote:

> If the reader is mystified at what Heisenberg was doing, he or she is not alone. I have tried several times to read the paper that Heisenberg wrote on returning from Helgoland, and, although I think I understand quantum mechanics, I have never understood Heisenberg's motivations for the mathematical steps in his paper (Weinberg 1992, p. 67).[14]

We hope that our reconstruction in Chapter 11 of how Heisenberg built his arch on Kramers' scaffold has dispelled much of the mystery.

The arch-and-scaffold metaphor can also be used to describe the development of wave mechanics. Whereas the physicists who developed matrix mechanics ended up turning the very theory they were working on into a scaffold for the new one, those who developed wave mechanics fortuitously found their scaffold in earlier work of Fermat and Maupertuis (in the case of de Broglie) and Hamilton (in the case of Schrödinger).[15] As Schrödinger wrote two decades later:

> [Hamilton's] famous analogy between mechanics and optics virtually anticipated wave mechanics, which did not have to add much to his ideas, [but] only take them ... a little more seriously than he was able to take them, with the experimental knowledge of a century ago (Schrödinger 1945, p. 82, quoted in Hankins 1980, p. 64).

Members of the Copenhagen–Göttingen camp, such as Born and Jordan, but also physicists more sympathetic to wave mechanics, such as Sommerfeld, would argue that

[14] This same quotation has been used to motivate several other studies of (the background to) the *Umdeutung* paper: Aitchison et al. (2004), Duncan and Janssen (2007), Blum et al. (2017), and Janssen (2019).

[15] See Sections 13.3 and 14.4.2 and Joas and Lehner (2009) on Schrödinger's use of the optical–mechanical analogy.

Schrödinger was taking the wave-imagery he imported along with Hamilton's mathematics a little too seriously (see the passages quoted above from the prefaces of Born and Jordan 1930 and Sommerfeld 1929, 1930). But Hamilton's optical–mechanical analogy certainly served as a useful scaffold for the construction of wave mechanics.

That a scaffold is typically put up with the arch in mind is responsible for one limitation of the arch-and-scaffold metaphor; that a scaffold is typically taken down once the arch can support itself is responsible for another. Referring to his mathematical publications, Gauss apparently used to say that 'A good building should not show its scaffolding when completed' (quoted in Janssen 2019, p. 103). This aspect of the arch-and-scaffold metaphor applies to the dispersion theory that served as a scaffold for matrix mechanics and it applies more generally to the old quantum theory. Although its planetary picture of atoms remains a fixture in the public imagination and can still be found in early chapters of older quantum textbooks, quantum physicists working at the research frontier quickly left the old quantum theory behind once the arches of matrix and wave mechanics for which it had served as a scaffold could support themselves. The same, however, cannot be said for the further development of quantum theory. The Dirac–Jordan statistical transformation theory can be seen as an arch built on a combination of two scaffolds: one provided by matrix mechanics and one provided by wave mechanics. This unified theory of Dirac and Jordan can, in turn, be seen as the scaffold for the arch of von Neumann's Hilbert space formalism. However, we cannot use the arch-and-scaffold metaphor to describe the relation between these different quantum formalisms if we insist that calling something a scaffold implies that it is discarded once the arch is finished.[16] In this case, all four formalisms—matrix mechanics, wave mechanics, statistical transformation theory, and the Hilbert space formalism—are still part of quantum mechanics as it is used and taught today. In many applications, in fact, it is much more convenient to use wave mechanics than von Neumann's Hilbert space formalism.

The better building metaphor for these further developments may therefore be that of a cathedral built in different styles by successive generations, sometimes with the help of temporary scaffolds, sometimes directly on top of earlier parts of the building under construction. Gould (2002, pp. 1–6) used the metaphor of a cathedral in the first chapter of his magnum opus, *The Structure of Evolutionary Theory*, to describe the development of evolutionary theory from Darwin to the present. Ofer Gal (2021) used the metaphor of building a cathedral to describe the development of science from antiquity to the seventeenth century.

We can likewise think of modern quantum mechanics as a cathedral in which parts of the building that acted as scaffolds for the construction of others are still plainly visible in the current state of the building. Figure 18.1 shows an artist's rendition of what this

[16] What *is* discarded in going from the Dirac–Jordan transformation theory to the Hilbert space formalism is the reliance on the old mathematics of canonical transformations. Instead, the theory is now formulated in terms of the new mathematics of functional analysis. In fact, that is what the title of Duncan and Janssen (2013), "(Never) mind your p's and q's," was meant to convey (see Section 17.3 and note 20 below).

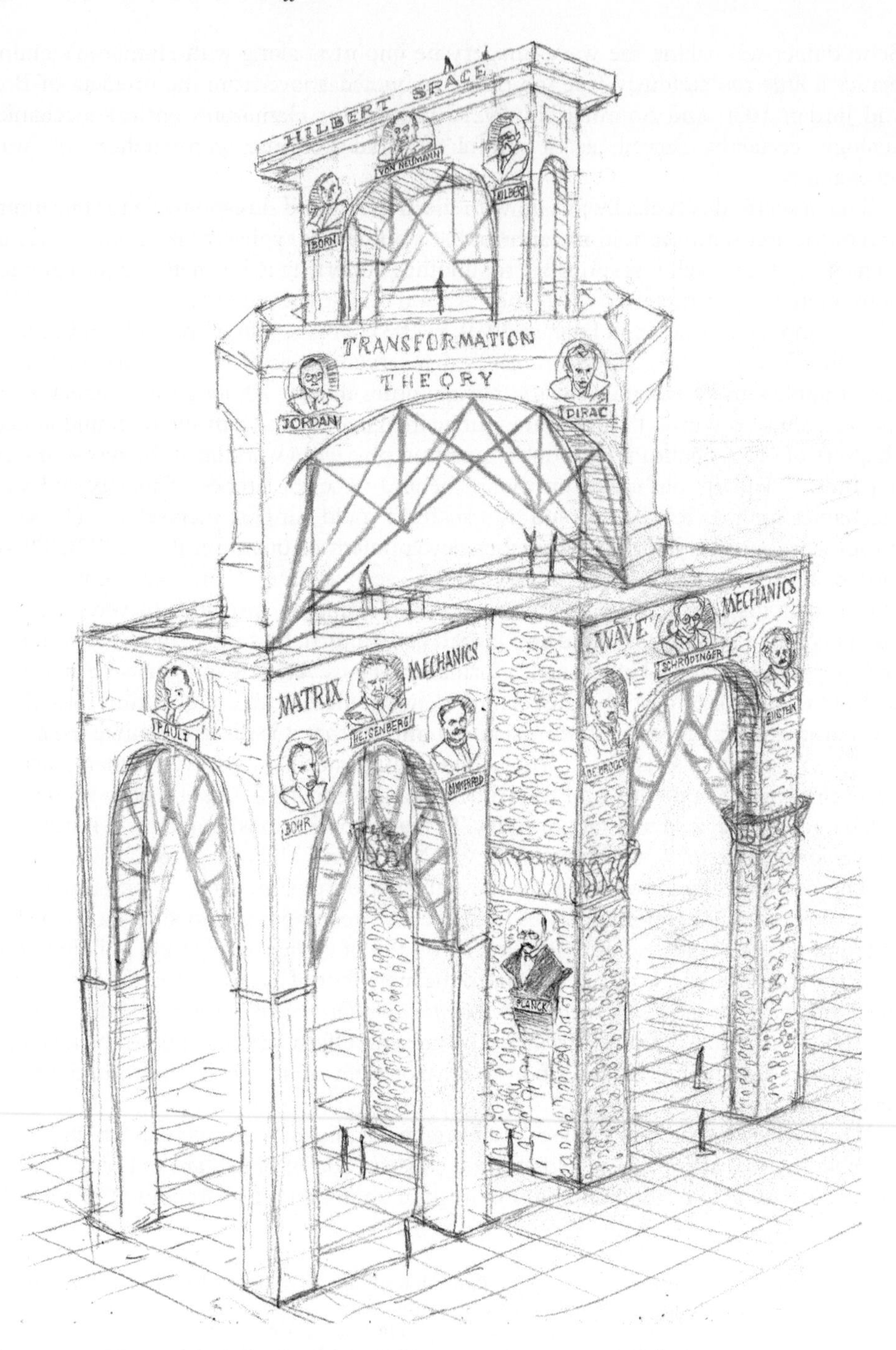

Figure 18.1 *The cathedral of quantum mechanics in the late 1920s, still showing some of the scaffolding used to build it. Drawing by Laurent Taudin (with a nod to M. C. Escher).*

cathedral looked like in the late 1920s, at the end of the period covered in our book. The figure shows the four arches identified above: matrix mechanics, wave mechanics, statistical transformation theory, and the Hilbert space formalism. Some of the scaffolding used to erect these arches is also still visible. The cathedral is adorned with the statues of several of its architects, working in different styles.

One of these architects, Pascual Jordan, used a similar building metaphor (albeit one about a generic building rather than a cathedral) to describe the development of science (including, presumably, the development of quantum theory). In the preface of one of his popular books, he wrote:

> [S]cience might be compared to a building begun with bold courage amidst swamps of unfathomable depth. The absence of a sturdy layer of rocks below the surface makes it necessary to start in places where the connections are relatively strong. Only as the work progresses will it be possible to assess the load-bearing capacity of the chosen foundation more precisely. The construction has to be continued *upwards and downwards at the same time*: the increasing stress on the foundations due to the growing size and weight of the building necessitates constant expansion and reinforcement of the supporting substructure.
>
> An expert's critical examination of the whole building need not show that its sturdiness is doubtful or that it is in imminent danger of collapsing. But it happens quite often that this or that previously untested but seemingly trustworthy part of the foundation actually has only limited and relative strength and has to be secured to lower and stronger parts as construction continues. Sometimes such determinations unexpectedly call for radical changes in the overall planning of the building; but what has already been completed is always the starting point, not only for the construction of new floors or parts of the building, but also to get down to the foundations and strengthen the supporting substructure (Jordan 1947, pp. 8–9, emphasis in the original).

In this passage, especially in the italicized clause, Jordan touches on the third and final limitation of the arch-and-scaffold metaphor to which we want to draw attention. Scaffolds can be used in two different ways: either to erect a new (part of a) building or to prevent an existing (part of a) building from collapsing. This distinction carries over to the metaphorical use of the term "scaffold" as well. We have been using the term in the former sense. Mathematicians, however, if they use the metaphor at all, tend to use it in the latter sense (Janssen 2019, pp. 103–105).

Hilbert, for instance, is recorded as saying in lectures in Göttingen in 1905 (though he used the term "foundation" rather than "scaffold"):

> The buildings of science are not erected the way a residential property is, where the retaining walls are put in place before one moves on to the construction and expansion of living quarters. Science prefers to get inhabitable spaces ready as quickly as possible to conduct its business. Only afterwards, when it turns out that the unevenly laid foundations cannot carry the weight of some additions to the living quarters, does science get

> around to support and secure those foundations. This is not a deficiency but rather the correct and healthy development (Peckhaus 1990, p. 51).[17]

In the introduction of a physics textbook first published in 1900, Paul Volkmann, an associate of Hilbert's, makes a similar point in a passage printed in spaced type for emphasis:

> The conceptual system of physics should not be conceived as one which is produced bottom-up like a building. Rather it is a thorough system of cross-references, which is built like a vault or the arch of a bridge, and which demands that the most diverse references must be made in advance from the outset, and reciprocally, that as later constructions are performed the most diverse retrospections to earlier dispositions and determinations must hold. Physics, in short, is a conceptual system which is strengthened retroactively (Volkmann 1990, pp. 3–4, quoted in Corry 2004, p. 61).

In all likelihood, these remarks by Volkmann are what inspired Jordan's building metaphor. Jordan (1947, p. 9) actually used the phrase "retroactive strengthening" (*rückwirkende Verfestigung*), attributing it to Volkmann.

Jordan's statistical transformation theory provides a good example of the two different uses of the term "scaffold" implied by his own and Volkmann's building metaphors. There are two ways to tell the story of the transition from the Dirac–Jordan statistical transformation theory to von Neumann's Hilbert space formalism in terms of the arch-and-scaffold metaphor. In one version, von Neumann set out to build an arch on the mathematically rickety scaffold provided by the Dirac–Jordan theory. In the other version, von Neumann set out to build a scaffold to prevent the mathematically unsound arch of the Dirac–Jordan theory from collapsing. Either way, we have another illustration of the second limitation of the arch-and-scaffold metaphor: no matter which formalism played the role of the scaffold, it was not taken down after it served its purpose; instead we are left with a composite of arch and scaffold (Janssen 2019, p. 161).

In Section 3.3.3, we saw Millikan combine the two possible uses of the arch-and-scaffold metaphor (though he used the terms "supports" and "underpinnings" instead of "scaffolds"). Commenting on his verification of Einstein's formula for the photoelectric effect but rejecting the light-quantum hypothesis from which Einstein had derived this formula, Millikan wrote that

> we are in the position of having built a very perfect structure and then knocked out entirely the underpinning without causing the building to fall. It stands complete and apparently well tested, but without any visible means of support. These supports must obviously exist, and the most fascinating problem of modern physics is to find them (Millikan 1917, p. 230; see Volume 1, p. 106 for the full quotation).

In other words, the light-quantum hypothesis served as the scaffold for the arch of the formula for the photoelectric effect but this arch was badly in need of a new scaffold to prevent it from collapsing.

[17] The quote is from lecture notes taken by Born. See Rowe (1997, p. 548) for further discussion.

18.7 Substitution and generalization

Given the limitations of the arch-and-scaffold metaphor, it is useful to characterize some of the key moves we encountered in our reconstruction of the genesis of quantum mechanics in less metaphorical terms. Without claiming that this exhausts the full repertoire, we have seen our protagonists make two types of moves when using existing theories as scaffolds to construct new ones, which can be labeled *substitution* and *generalization* (Janssen 2019, pp. 172–173). *Substitution* is when the basic building blocks of (the theory playing the role of) the scaffold are replaced with new ones while the structure built out of them is left intact (think of replacing vacuum tubes by transistors in a logic board). *Generalization* is when a structure exhibited by (the theory playing the role of) the scaffold for a special case is recognized to have broader significance.[18] We close this chapter by sketching some examples of both types of moves discussed in much greater detail in earlier chapters.

The first example of a "substitution" takes us back to Planck's discovery of the law for black-body radiation (see Chapter 2).[19] Recall that, in his pentalogy of the late 1890s (Badino 2015), Planck (1897–1899) derived a relation between, on the one hand, the average energy of radiation of a certain frequency and, on the other hand, the average energy of a charged oscillator with the same (resonance) frequency interacting with that radiation (see Eq. (2.7)). This relation was the centerpiece of a framework that immediately gave Planck a formula for the spectral distribution of the energy in black-body radiation once he had settled on an expression for the entropy of one of these oscillators. Initially, Planck (1900a) believed that the only expression compatible with the second law of thermodynamics was the one that leads to the Wien law. When clear deviations from the Wien law were found for long wavelengths, Planck (1900b) realized this uniqueness argument was wrong. He thereupon replaced the old expression for the entropy of a single oscillator with a new one. When substituted into his general framework, this new expression gives the Planck law. Planck now found himself in the same situation as Millikan a decade and a half later. His new law, like Einstein's formula for the photoelectric effect, stood "complete and apparently well tested, but without any visible means of support." He immediately set out to find such support. Supplying a derivation of his new expression for the entropy of a single oscillator, Planck (1900c, 1901) took the first steps toward quantizing the energy of these oscillators.

Our second example of a "substitution" may well provide the strongest justification for framing our account of the quantum revolution in terms of the arch-and-scaffold metaphor. It is the basic idea behind Heisenberg's (1925c) *Umdeutung* paper (see Chapter 11). In terms of the slogan we used at the beginning of this chapter, Heisenberg did not repeal, but reinterpreted, classical mechanics. He retained the relations between such quantities as position, velocity, and energy in classical mechanics, but

[18] We saw von Neumann make a related move when he recognized that the same structure underlies matrix mechanics and wave mechanics (Chapter 17).

[19] See Lemons, Shanahan, and Buchholtz (2022) for a detailed but concise treatment of Planck and the physics of black-body radiation.

replaced these quantities by arrays of numbers satisfying a non-commutative multiplication rule. Born and Jordan quickly recognized that these arrays of numbers were nothing but matrices and that Heisenberg's multiplication rule was nothing but the standard rule for matrix multiplication. What inspired this move by Heisenberg was the dispersion formula found by Kramers (1924a, 1924b), building on earlier work on dispersion by Ladenburg (1921) and Ladenburg and Reiche (1923) and using a combination of the advanced techniques the old quantum theory had imported from celestial mechanics, elements of Einstein's (1916a, 1917b) quantum theory of radiation, and, most importantly, a sophisticated version of Bohr's correspondence principle. Heisenberg, of course, had been Kramers' co-author of a detailed exposition and extension of the latter's dispersion theory (Kramers and Heisenberg 1925; see Section 10.5).[20]

Heisenberg's breakthrough also illustrates the other maneuver identified above: "generalization". Heisenberg recognized that the kind of argument based on the correspondence principle that Kramers had used to arrive at his dispersion formula could be used much more generally to arrive at a new framework for all of physics. Kramers had first derived a classical formula for the dispersion of light by some multiply periodic system. Appealing to the correspondence principle, he had then turned this classical formula into a quantum formula by replacing amplitudes and frequencies in the Fourier expansion of the motion in one orbit by two-index quantities representing the probability of transitions from one orbit to another and the frequency of the radiation emitted in such transitions. This procedure guaranteed that the quantum formula agreed with the classical formula in the limit of high quantum numbers. Kramers then took the leap of faith that the quantum formula would continue to hold all the way down to the regime of small quantum numbers. Heisenberg realized that, instead of translating the final result of a classical calculation into a quantum formula, one could do the calculation right from the start in terms of the two-index quantities introduced in the translation, subsequently recognized by Born and Jordan as matrices. Heisenberg realized that this could be done not just for dispersion theory but also, at least in principle, for all of physics. Unsurprisingly, the *Umdeutung* paper, in which Heisenberg presented this new framework for all of physics, still bears clear traces of the dispersion theory that had served as the scaffold in its construction. The scaffold was removed only with the further elaboration of Heisenberg's ideas in Born and Jordan's (1925b) Two-Man-Paper and Born, Heisenberg, and Jordan's (1926) Three-Man-Paper.

Dirac's work provides an even-better example of "generalization". By a correspondence-principle argument similar to the one used in dispersion theory and in the *Umdeutung* paper, Dirac arrived at the conclusion that the basic commutation

[20] Another "substitution", which may look obvious in hindsight but was not actually made at the time, is to replace the integral kernels which Jordan used to represent his probability amplitudes by inner products of vectors in Hilbert space (see Sections 16.2 and 17.3). This provides a simple route from the Dirac–Jordan statistical transformation theory to von Neumann's Hilbert space formalism. This route, however, was not available to von Neumann in 1927, as its rigorous formulation depends on advances in mathematics made only decades later, that is, the theory of distributions and the concept of a rigged Hilbert space (cf. Janssen 2019, pp. 154–155). This nicely illustrates the second constraint on theory development discussed above, coming from the available mathematical toolkit.

relation for position and momentum is the direct quantum analogue of Poisson brackets in classical mechanics. His correspondence-principle argument, however, only applied to the extremely restricted class of conditionally periodic systems. Dirac now made the bold generalization that the commutation relations he found on the basis of this argument apply to all systems for which one can write down a Hamiltonian (see Section 12.2, especially note 35).

The most obvious generalization in the development of quantum mechanics is largely implicit in our reconstruction of its genesis. This is the realization that the new framework for dealing with the optical and X-ray spectra of atoms and molecules provides a new framework for all of physics. This is seen most clearly perhaps in the application of the new quantum mechanics, starting shortly after the quantum revolution of 1925–1926, to chemistry (the definitive clarification of the structure of the periodic table based on the new quantum mechanics and Pauli's exclusion principle), nuclear physics, and solid state physics. These developments are beyond the scope of our book. We did, however, briefly touch on the beginnings of another important application of the new quantum mechanics—the quantization of the electromagnetic field (see our discussion of the final section of the Three-Man-Paper in Section 12.3).

Modern textbooks on quantum mechanics continue to devote a fair amount of space to applications in spectroscopy. Solving the Schrödinger equation for the hydrogen atom to recover the Balmer formula and the explanation of the anomalous Zeeman effect through spin–orbit coupling remain textbook staples. But to master the fundamental tenets of modern quantum mechanics one no longer needs the kind of intimate knowledge of the details of atomic spectroscopy that allowed physicists in the 1910s and 1920s to formulate the new theory in the first place. One can think of that specialized knowledge as a scaffold to be discarded now that we are in the safe possession of the arch constructed with its help. It is hard to imagine, however, that this arch would ever have been erected, if the incredible precision of the available spectroscopic data had not forced physicists to develop and accept a theory as foreign to the classical physics they were accustomed to as quantum mechanics inevitably is. In 1913, for instance, the data could not rule out Bohr's easily visualizable model of a helium atom in which two electrons circle the nucleus like planets orbiting the sun. By the mid-1920s much better data made it clear that a far more abstract approach was needed if one wanted to account for the spectrum of ortho- and parahelium or even the binding energy of the ground state of helium.

Appendix

C

The Mathematics of Quantum Mechanics

In the formative years 1925–1927 of the new quantum mechanics, its theoretical practitioners were dealing with a technical deficit in their knowledge base which is in some ways analogous to that faced by the theorists who had, a decade before, undertaken the task of generalizing the tenets of Bohr's 1913 model of the hydrogen atom to deal with a more general class of mechanical models. In that earlier period, the problem was that most physicists were unaware of the elegant and sophisticated techniques of analytical mechanics that had been developed in the nineteenth century by Stäckel and Delaunay, and put into a comprehensive framework in the early twentieth century by Poincaré and Charlier, in their treatises on celestial mechanics. This body of knowledge—specifically, the introduction, where appropriate, and exploitation of action/angle variables—was injected into the arterial system of atomic theorists by Schwarzschild, who was trained as an astronomer and therefore had absorbed these methods years before they became essential tools in the quantum theory.[1]

A similar situation arose in 1925, with Heisenberg's reformulation of the kinematical variables of mechanics as two-index arrays, soon recognized as matrices by his Göttingen colleagues Max Born and Pascual Jordan. Anything more than a passing acquaintance even with finite-dimensional matrices was not to be assumed for the typical theoretical physicist of the time (although Born, as we saw in Chapter 12, was a clear exception to this rule). The situation was even worse when it came to the infinite-dimensional matrices that lay at the heart of Heisenberg's new quantum mechanics. Here again, much of the necessary mathematical framework we now regard as essential for a proper understanding of quantum mechanics had already been developed, in some cases by mathematicians who were colleagues of Heisenberg, Born, and Jordan in Göttingen—sitting right down the hall, as it were. But this material had simply not been necessary for the practitioners of the old quantum theory and had therefore been widely ignored in atomic physics circles.

The algebraic uses of finite matrices, for example, goes back to Sylvester and Cayley in the 1850s (Jammer 1966, p. 204). The study of linear integral equations at the turn of the century led to the introduction of infinite quadratic forms, from which infinite matrices naturally emerged. The relevant mathematical structures for dealing with these infinite entities, which now form the field referred to by mathematicians as "functional analysis", were developed over the first two decades of the twentieth century by David Hilbert, Erhard Schmidt, Ernst Hellinger, Otto Toeplitz, Frigyes Riesz, and others. The work of John von Neumann in the late 1920s (see

[1] For a discussion of the introduction of this technology, and its enthusiastic adoption by the Munich school led by Arnold Sommerfeld, see Volume 1, especially Appendix A, and Section 5.1.4.

Chapter 17), constructing a spectral theory adequate for handling the unbounded operators ubiquitous in quantum mechanics, provided a beautiful completion of the theory of Hilbert space needed for modern quantum mechanics.[2]

In this appendix, we review in a historical context[3] the mathematical ideas that play an essential role in the development of quantum mechanics in the period 1923–1927, that is, up to the formalization given by von Neumann in his 1927 papers, but with a brief glance forward to the spectral theorems proven shortly afterward. We begin with a review of the finite-dimensional linear algebra that provides the basic structures that later are generalized in the theory of functions and infinite-dimensional spaces. Although we try to make the discussion as self-contained as possible, it will be rapid, and some prior knowledge of these matters on the part of the reader will probably be necessary for our treatment to be fully intelligible.

C.1 Matrix algebra

Matrices were introduced in the 1850s by Arthur Cayley as a convenient shorthand for expressing and manipulating linear equations, such as the following, the first equations appearing in Cayley's 1858 *Memoir on the Theory of Matrices*[4]

$$\begin{aligned} y_1 &= A_{11}\,x_1 + A_{12}\,x_2 + A_{13}\,x_3, \\ y_2 &= A_{21}\,x_1 + A_{22}\,x_2 + A_{23}\,x_3, \\ y_3 &= A_{31}\,x_1 + A_{32}\,x_2 + A_{33}\,x_3, \end{aligned} \tag{C.1}$$

which Cayley then rewrites as

$$\begin{pmatrix} y_1 \\ y_2 \\ y_3 \end{pmatrix} = \begin{pmatrix} A_{11} & A_{12} & A_{13} \\ A_{21} & A_{22} & A_{23} \\ A_{31} & A_{32} & A_{33} \end{pmatrix} \begin{pmatrix} x_1 \\ x_2 \\ x_3 \end{pmatrix}. \tag{C.2}$$

In modern notation, we simply write Eq. (C.1) as (using the summation convention for the repeated index, with sums going in this case from 1 to 3)

$$y_i = A_{ij}\,x_j, \tag{C.3}$$

and Eq. (C.2) as

$$\mathbf{y} = \mathbf{A}\,\mathbf{x}, \tag{C.4}$$

[2] References to all the relevant sources will be given below in the process of describing the relevant mathematical concepts.

[3] The reader who is interested in a comprehensive, but very accessible, introduction to the mathematical ideas underlying quantum mechanics, *sans* historical commentary, is referred to the excellent text of Fano (1971). Some general reviews of the early history of functional analysis can be found in Bernkopf (1965, 1968) and Birkhoff and Kreyszig (1984). For another historical discussion of these mathematical ideas and how they entered into the formulation of quantum theory, see Landsman (2022). For a concise history of spectral theory, see Steen (1973).

[4] We have made some inconsequential notational changes, such as replacing (x, y, z) with (x_1, x_2, x_3), relabelling the matrix elements in the modern two-index form, and writing column vectors instead of row vectors.

where the column vectors and the 3×3 (3 rows and 3 columns) matrix $\mathbf{A}$ have been indicated with bold notation.

We find the rule for the composition of matrices straightforwardly by considering the effect of introducing a second set of linear relations for the new variables z_k in terms of the y_i,

$$z_k = B_{ki}\, y_i = B_{ki} A_{ij}\, x_j \equiv C_{kj}\, x_j, \tag{C.5}$$

whence the well-known rule for matrix composition, or multiplication:

$$\mathbf{C} = \mathbf{BA}, \quad C_{kj} = \sum_i B_{ki} A_{ij}. \tag{C.6}$$

We henceforth assume that our matrices are of order n (n rows and n columns) with all indices and sums over indices running from 1 to n. We may also assume, with no additional effort, that our matrices are allowed to contain complex numbers.

The basics of matrix algebra can be summarized in a few lines, as follows:

1. Matrix addition is defined in the obvious way:

$$\mathbf{A} = \mathbf{B} + \mathbf{C} \iff A_{ij} = B_{ij} + C_{ij}, \quad i,j = 1,2,3,\ldots,n. \tag{C.7}$$

2. As the reader may readily verify, associativity holds:

$$\mathbf{A}(\mathbf{BC}) = (\mathbf{AB})\mathbf{C} \quad (\equiv \mathbf{ABC}). \tag{C.8}$$

3. In general, multiplication is *not commutative*:

$$\mathbf{AB} \neq \mathbf{BA}, \tag{C.9}$$

which is the same as saying that at least one element of the matrix difference of the left and right sides, called the *commutator*,

$$[\mathbf{A}, \mathbf{B}] \equiv \mathbf{AB} - \mathbf{BA}, \tag{C.10}$$

is different from zero.

4. The *unit (or identity) matrix* 1 is the matrix defined by the matrix elements $1_{ij} = \delta_{ij}$: it has the property that $1\mathbf{A} = \mathbf{A}1 = \mathbf{A}$ for any matrix $\mathbf{A}$.
5. The *inverse* $\mathbf{A}^{-1}$ of a matrix $\mathbf{A}$ satisfies (if it exists)

$$\mathbf{A}^{-1}\mathbf{A} = \mathbf{A}\mathbf{A}^{-1} = 1. \tag{C.11}$$

If it exists, it is unique. As $\mathbf{B}^{-1}\mathbf{A}^{-1}\mathbf{AB} = 1$, the inverse of $\mathbf{AB}$ is $\mathbf{B}^{-1}\mathbf{A}^{-1}$. The inverse matrix (if it exists), once known, allows us to solve for the variables x_i in Eq. (C.1) in terms of the y_i, as $\mathbf{y} = \mathbf{Ax}$ implies $\mathbf{x} = \mathbf{A}^{-1}\mathbf{y}$, as Cayley was at pains (in a somewhat awkward notation) to emphasize. The inverse of a matrix $\mathbf{A}$ exists if and only if its determinant $\det(\mathbf{A})$ is non-zero. The determinant is defined in terms of the totally antisymmetric symbol $\varepsilon_{i_1 i_2 \ldots i_n}$, which is equal to zero if any two indices coincide, equal to $+1$ if the index set $(i_1, i_2, \ldots, i_n)$ is an even permutation of $(1, 2, 3, \ldots, n)$, and equal to -1 if it is an odd permutation:

$$\det(\mathbf{A}) = \sum_{i_1 i_2 \ldots i_n} \varepsilon_{i_1 i_2 \ldots i_n} A_{1 i_1} A_{2 i_2} \cdots A_{n i_n}. \tag{C.12}$$

The determinant of the product of two matrices conveniently factorizes:

$$\det(\mathbf{AB}) = \det(\mathbf{A})\det(\mathbf{B}). \tag{C.13}$$

For the proof of these assertions, we refer the reader to any of the many standard introductory texts in linear algebra.

C.2 Vector spaces (finite dimensional)

About a decade before Cayley introduced and developed the idea of matrices, Hermann Grassmann, in his treatise "The Linear theory of extension [*lineale Ausdehnungslehre*], a new branch of mathematics" (Grassmann 1844), created the structure we now refer to as a vector space—that is, a set of mathematical objects closed under addition and multiplication by scalar (in Grassmann's case, real) numbers. Grassmann's theory also contained the indispensable concepts of linear independence and dimension, central in any modern presentation of vector space theory. Four decades later, Giuseppe Peano, building on Grassmann's ideas in his book *Calculo geometrico secondo l'Ausdehnungslehre di H. Grassmann* (Peano 1888), formulated the defining characteristics of a linear vector space—closure under addition and scalar multiplication, and existence of a zero vector—in a form recognizable to the modern reader. This work was sadly neglected, perhaps because it was written in Italian, and Peano, working in comparative obscurity in Turin, was not recognized at the time in the top class of European mathematicians. The subject was revived, and made permanently relevant to mathematics and the physical sciences, by Hermann Weyl and others in the 1920s.[5]

Here, we remind the reader of just the essentials needed for an understanding of the development of quantum mechanics. The presentation is based on the concrete spaces $\mathbb{R}^n$ (the space of ordered n-tuples $(x_1, x_2, \ldots, x_n)$ of real numbers) and $\mathbb{C}^n$ (similarly defined, but with complex numbers), rather than a completely abstract presentation, for the simple reason that the Hilbert space of quantum mechanics is effectively an analog of the latter, with n infinite, but with many features passing over without difficulty from the finite- to the infinite-dimensional case.

For the time being, staying with $\mathbb{R}^n$, we have a plethora of widely used notations to indicate the members (vectors) of our vector space. For the time being we stick to the lower case Latin

[5] For example, the basic properties of vector spaces are reprised in the first pages of Courant and Hilbert (1924), which would become the mathematical "Bible" for several generations of twentieth-century physicists.

notation introduced in Section C.1, so that an element $\mathbf{v}$ of $\mathbb{R}^n$ is simply an n-tuple of real numbers $v_1, v_2, \ldots, v_n$, which we can also conveniently display as a column vector, as in Eq. (C.2):

$$\mathbf{v} = \begin{pmatrix} v_1 \\ v_2 \\ \vdots \\ v_n \end{pmatrix}, \quad \mathbf{w} = \begin{pmatrix} w_1 \\ w_2 \\ \vdots \\ w_n \end{pmatrix}, \quad \text{etc.} \tag{C.14}$$

The real numbers $v_1, v_2, \ldots$ etc. are referred to as the "components" of the vector $\mathbf{v}$.

The set of such quantities is evidently closed under scalar multiplication and addition, that is, we may define new vectors by forming

$$\alpha\mathbf{v} + \beta\mathbf{w} = \begin{pmatrix} \alpha v_1 + \beta w_1 \\ \alpha v_2 + \beta w_2 \\ \vdots \\ \alpha v_n + \beta w_n \end{pmatrix}, \tag{C.15}$$

with α and β arbitrary real numbers. Remarkably, from these very simple beginnings, a rich set of structures can be evolved, and then extended, to the case where our n-tuples contain infinitely many components.

A set of (by assumption, non-zero) vectors $\mathbf{v}_1, \mathbf{v}_2, \ldots, \mathbf{v}_m$ are *linearly independent* if no linear combination of them $\alpha_1\mathbf{v}_1 + \alpha_2\mathbf{v}_2 + \ldots + \alpha_m\mathbf{v}$ with at least one non-zero coefficient α_i vanishes (i.e., produces the zero vector, with all zero components). Otherwise, the set is *linearly dependent*. In that case, one of the vectors can be selected to be written as a linear combination of the remaining $m-1$. It is straightforward to show that given n linearly independent vectors $\mathbf{e}_i$ in $\mathbb{R}^n$—called a "basis" set—an arbitrary vector $\mathbf{v}$ can be written in a unique way as a linear combination, $\mathbf{v} = \sum_i \alpha_i \mathbf{e}_i$. A particularly important "canonical" basis, given our definition of the space $\mathbb{R}^n$ in terms of the column vectors in Eq. (C.14), is the set of vectors

$$\mathbf{e}_1 = \begin{pmatrix} 1 \\ 0 \\ \vdots \\ 0 \end{pmatrix}, \quad \mathbf{e}_2 = \begin{pmatrix} 0 \\ 1 \\ \vdots \\ 0 \end{pmatrix}, \quad \text{etc.} \tag{C.16}$$

A linear operator A is defined as a mapping of vectors in $\mathbb{R}^n$ to vectors in $\mathbb{R}^n$ with the property

$$A(\alpha\mathbf{v} + \beta\mathbf{w}) = \alpha A\mathbf{v} + \beta A\mathbf{w}. \tag{C.17}$$

With any basis $\mathbf{e}_i$ (i.e., not necessarily the canonical one indicated in Eq. (C.16)), the knowledge of (the real numbers) A_{ji}, $j, i = 1, 2, \ldots, n$, defined by

$$A\mathbf{e}_i = \sum_j A_{ji}\mathbf{e}_j, \tag{C.18}$$

suffices to determine the action of A on an arbitrary vector $\mathbf{v} = \sum_i v_i \mathbf{e}_i$:

$$A\mathbf{v} = A\sum_i v_i\mathbf{e}_i = \sum_i v_i A\mathbf{e}_i = \sum_{ij} A_{ji} v_i \mathbf{e}_j. \tag{C.19}$$

The jth component of the vector $A\mathbf{v}$ (in the basis $\mathbf{e}_i$) is therefore given by $\sum_i A_{ji}v_i$, just the algebraic construction that Cayley had schematized in matrix notation in Eq. (C.1). We call the set of n^2 numbers A_{ji}, arranged in a square array, the *matrix* $\mathbf{A}$ *of the operator* A *in the basis* $\mathbf{e}_i$. It is important to realize that a given linear operator can be expressed in terms of different matrices (with different matrix elements) simply by changing the set of basis vectors used in Eq. (C.18) to define the matrix elements, just as the same vector $\mathbf{v} = \sum_i v_i\mathbf{e}_i$ can be expressed in terms of a different set of components v_i simply by altering the basis $\mathbf{e}_i$, which can after all be any set of n linearly independent vectors.

The representation of a linear operator in terms of a numerical matrix (given a set of basis vectors) would not be very useful if it were not compatible with the composition rule for the latter. It is straightforward to establish that the successive application of two linear operations leads to a composite linear operator whose matrix is just the product (as defined by Cayley) of the individual matrices associated with each operator:

$$\begin{aligned} BA\mathbf{v} &= B(A\mathbf{v}) \\ &= B\Big(\sum_{ij} A_{ji}v_i\mathbf{e}_j\Big) \\ &= \sum_{ij} A_{ji}v_iB\mathbf{e}_j \\ &= \sum_{ijk} A_{ji}v_iB_{kj}\mathbf{e}_k \\ &= \sum_{ik}\Big(\sum_j B_{kj}A_{ji}\Big)v_i\mathbf{e}_k. \end{aligned} \tag{C.20}$$

We recognize in the parenthesis on the final line the rule for matrix multiplication given in Eq. (C.6).

A non-zero vector $\mathbf{v}$ is said to be an *eigenvector* of the linear operator A if, for some scalar (real or complex) number λ, which we call the associated *eigenvalue*,[6]

$$A\mathbf{v} = \lambda\mathbf{v}. \tag{C.21}$$

If $\lambda = 0$, $\mathbf{v}$ is a *null vector* of A, and from $A_{ij}v_j = 0$ (with not all v_j zero) it is an immediate consequence that the determinant of A must vanish. In fact, for a general eigenvector/eigenvalue pair in Eq. (C.21), we may write

$$(A - \lambda 1)\mathbf{v} = 0, \tag{C.22}$$

which implies, if $\mathbf{A}$ is the $n \times n$ matrix of the linear operator A relative to some (arbitrary) basis,

$$\det(\mathbf{A} - \lambda 1) = 0. \tag{C.23}$$

[6] The corresponding terminology in German was *Eigenvektor/Eigenwert*. The early English terminology was *proper vector/proper value*, but this was soon abandoned in favor of the German terms, which are used ubiquitously in modern texts.

The left-hand side here is a polynomial of degree n in λ, which must possess at least one (possibly complex) root, by the fundamental theorem of algebra.[7] We have therefore the following simple lemma, innocent in appearance, but of great use in the further study of the eigenvalue problem.

Lemma C.2.1. *For an arbitrary linear operator A on $\mathbb{R}^n$ (or $\mathbb{C}^n$), there is at least one non-trivial eigenvalue/eigenvector pair, which may be complex.*

C.3 Inner-product spaces (finite dimensional)

Some additional structure needs to be added to the bare bones of linear vector space theory as described above in order to make it a useful tool for the theory of operators in function spaces which would develop in the first two decades of the twentieth century, and eventually give rise to the area of mathematics now called functional analysis. Specifically, the concept of an *inner product* needs to be introduced. It is the direct descendant of the scalar dot product, which every introductory student of physics learns in encountering the kinematics needed for classical mechanics.

From this point on we assume that we are dealing with the complex vector space $\mathbb{C}^n$ of n-tuples of complex numbers, closed under addition and multiplication by complex scalars. The *inner product* of the vectors $\mathbf{v}$ and $\mathbf{w}$, given by the column vectors indicated in Eq. (C.14), is defined as the (in general complex) number

$$(\mathbf{v}, \mathbf{w}) = \sum_i v_i^* w_i, \tag{C.24}$$

where the asterisk denotes, as usual, complex conjugation. This definition extends in an obvious way the three-dimensional dot product of two (real) three-vectors. For a complex space, the complex conjugation of the components of the left vector ensures that the inner product of a vector with itself is real, positive, and zero if and only if all components of the vector vanish. The reader may easily verify the following properties, which may also be taken as the defining characteristics of the inner product in a more abstract setting:

$$\begin{aligned} &1.\ (\mathbf{v}, \mathbf{v}) \geq 0, \text{ with } (\mathbf{v}, \mathbf{v}) = 0 \text{ if and only if } \mathbf{v} = 0. \\ &2.\ (\mathbf{v}, \mathbf{w}) = (\mathbf{w}, \mathbf{v})^*. \\ &3.\ (\mathbf{v}, \lambda\mathbf{w}) = \lambda(\mathbf{v}, \mathbf{w}). \\ &4.\ (\mathbf{v}, \mathbf{w} + \mathbf{z}) = (\mathbf{v}, \mathbf{w}) + (\mathbf{v}, \mathbf{z}). \end{aligned} \tag{C.25}$$

An inner product of this type automatically implies the existence of a metric (a distance function). First, the norm (or magnitude) $|\mathbf{v}|$ of a vector can be defined as a positive real number by

$$|\mathbf{v}| = \sqrt{(\mathbf{v}, \mathbf{v})}. \tag{C.26}$$

[7] There are in general n complex roots, of course, but these may all coincide. The point is that there is *at least one (in general, complex) value* of λ for which Eq. (C.23) holds, thereby guaranteeing the existence of a pair $(\lambda, \mathbf{v} \neq 0)$ satisfying Eq. (C.21).

In fact, we can also go backwards, and recover the inner product once the norm of every vector is defined, via the following "polarization identity":

$$(\mathbf{v},\mathbf{w}) = \frac{1}{4}\left\{|\mathbf{v}+\mathbf{w}|^2 - |\mathbf{v}-\mathbf{w}|^2 + i|\mathbf{v}-i\mathbf{w}|^2 - i|\mathbf{v}+i\mathbf{w}|^2\right\}, \tag{C.27}$$

which the reader can easily verify by replacing all squared norms with inner products, using Eq. (C.26).

Once a norm has been defined, the "distance" between two elements of the space can be defined simply as $d(\mathbf{v},\mathbf{w}) = |\mathbf{v}-\mathbf{w}|$.[8] A *normalized* vector (or *unit vector*) is one whose magnitude is equal to unity. All of the above can be trivially restricted to the space $\mathbb{R}^n$ by simply using real numbers throughout and dropping the complex conjugate in the definition of the inner product.

The inner product allows us to introduce the critically important concept of *orthogonality*. Just as the vanishing of the dot product of two non-zero three-vectors in real space implies that they are perpendicular to each other, we say that two vectors, $\mathbf{v}$ and $\mathbf{w}$, are orthogonal if their inner product is zero, $(\mathbf{v},\mathbf{w}) = 0$. A set of m mutually orthogonal unit vectors $\mathbf{e}_i$ (with $i = 1, 2, \ldots, m$) are said to form an *orthonormal* set if

$$(\mathbf{e}_i, \mathbf{e}_j) = \delta_{ij}. \tag{C.28}$$

The canonical basis introduced in Eq. (C.16) forms just such an orthonormal set. In $\mathbb{R}^n$ or $\mathbb{C}^n$, any orthonormal set of n vectors is automatically a basis: the linear independence of such a set follows directly by taking the inner product of a linear combination with each element of the basis. If $\sum_j c_j\mathbf{e}_j$ is the zero vector, then $c_i = 0$ for all i:

$$0 = (\mathbf{e}_i, \sum_j c_j\mathbf{e}_j) = \sum_j c_j(\mathbf{e}_i, \mathbf{e}_j) = \sum_j c_j\delta_{ij} = c_i. \tag{C.29}$$

The components c_i of any vector $\mathbf{v} = \sum_i c_i\mathbf{e}_i$ in this basis are just $c_i = (\mathbf{e}_i, \mathbf{v})$. The inner product of any two vectors has a convenient representation in terms of the components of the vectors in an orthonormal basis:

$$\begin{aligned}
(\mathbf{v},\mathbf{w}) &= \left(\sum_k (\mathbf{e}_k,\mathbf{v})\ \mathbf{e}_k, \sum_l (\mathbf{e}_l,\mathbf{w})\ \mathbf{e}_l\right) \\
&= \sum_{kl} (\mathbf{v},\mathbf{e}_k)(\mathbf{e}_l,\mathbf{w})(\mathbf{e}_k,\mathbf{e}_l) \\
&= \sum_k (\mathbf{v},\mathbf{e}_k)(\mathbf{e}_k,\mathbf{w}),
\end{aligned} \tag{C.30}$$

where in the last step we used that $\sum_l(\mathbf{e}_k,\mathbf{e}_l) = \delta_{kl}$. The matrix of any linear operator A in such a basis is easily given in terms of inner products. By definition (cf. Eq. (C.18)),

$$A\mathbf{e}_j = \sum_i A_{ij}\mathbf{e}_i \Rightarrow (\mathbf{e}_i, A\mathbf{e}_j) = A_{ij}. \tag{C.31}$$

[8] The notion of an abstract metric space, characterized by the introduction of a distance function $d(x,y)$, satisfying (i) $d(x,y) = 0$ if and only if $x = y$,(ii) $d(x,y) = d(y,x)$, and (iii) $d(x,z) \leq d(x,y) + d(y,z)$ (triangle inequality), was introduced by Maurice Fréchet, and was an outgrowth of his work on the calculus of variations.

The set of all vectors $\mathbf{w}$ orthogonal to a given vector $\mathbf{v}$ is called the *orthogonal complement of* $\mathbf{v}$. It is a linear subspace, that is, a set of vectors closed under addition and scalar multiplication, as for any two such vectors,

$$(\lambda_1 \mathbf{w}_1 + \lambda_2 \mathbf{w}_2, \mathbf{v}) = \lambda_1(\mathbf{w}_1, \mathbf{v}) + \lambda_2(\mathbf{w}_2, \mathbf{v}) = 0. \tag{C.32}$$

More generally, given a linear subspace V of the full vector space, one can define the orthogonal complement of V, written $\mathbb{C}^n - V$, as the set of all vectors orthogonal to every vector in V.

The introduction of an inner product leads to the very important concept of the *Hermitian adjoint* of a linear operator A. The Hermitian adjoint $A^\dagger$ (often just called the "adjoint") is defined as the operator which, for any two vectors $\mathbf{v}, \mathbf{w}$, satisfies

$$(\mathbf{v}, A\mathbf{w}) = (A^\dagger \mathbf{v}, \mathbf{w}). \tag{C.33}$$

That such an operator exists is established simply by finding its (unique) $N \times N$ matrix, relative to an orthonormal basis $(\mathbf{e}_i), i = 1, 2, \ldots, N$,

$$A_{ij} = (\mathbf{e}_i, A\mathbf{e}_j) = (A^\dagger \mathbf{e}_i, \mathbf{e}_j) = (\mathbf{e}_j, A^\dagger \mathbf{e}_i)^\star = (A^\dagger)^\star_{ji}. \tag{C.34}$$

Thus, the matrix of the adjoint of an operator is found simply by taking the complex conjugate, and then interchanging row and columns, an operation often defined separately as the *transpose*, $A^{\mathrm{T}}_{ij} = A_{ji}$. Hence, for matrices,

$$\mathbf{A}^\dagger = \mathbf{A}^{\mathrm{T}\star}. \tag{C.35}$$

From the property $(\mathbf{AB})^{\mathrm{T}} = \mathbf{B}^{\mathrm{T}}\mathbf{A}^{\mathrm{T}}$, follows the similar rule for adjoints of matrix products,

$$(\mathbf{AB})^\dagger = \mathbf{B}^\dagger \mathbf{A}^\dagger. \tag{C.36}$$

The adjoint operation represents an extension of the process of complex conjugation from complex numbers to operators in a complex vector space. Just as real numbers play a privileged role as a subset of the complex numbers, so do *self-adjoint* operators, defined as operators that are equal to their adjoint:

$$H \text{ is self-adjoint} \iff H = H^\dagger \tag{C.37}$$

A matrix for which $A^\star_{ij} = A_{ji}$ is called *Hermitian*. Evidently, the matrix of a self-adjoint operator relative to an orthonormal basis is Hermitian.[9]

The extent to which this property corresponds to reality for numbers is made clear by the following theorem.

Theorem C.3.1. *Let H be a self-adjoint operator in a complex vector space with inner product. Then (i) all the eigenvalues of H are real, and (ii) the eigenvectors corresponding to two distinct eigenvalues are orthogonal.*

[9] We emphasize that we are discussing finite-dimensional spaces and operators here: the use of the terms "Hermitian" and "self-adjoint" will acquire new subtleties, and will need to be carefully distinguished, in the infinite-dimensional case. In particular, Hermitian operators, characterized by Eq. (C.33), will turn out to be only a small subclass of the more interesting self-adjoint operators that admit a full spectral theory.

Proof. (i) Let λ be an eigenvalue, corresponding to an eigenvector $\mathbf{v}$, that is, $H\mathbf{v} = \lambda\mathbf{v}$. Then,

$$(\mathbf{v}, H\mathbf{v}) = \lambda(\mathbf{v}, \mathbf{v}). \tag{C.38}$$

and, by the self-adjoint property

$$(\mathbf{v}, H\mathbf{v}) = (H\mathbf{v}, \mathbf{v}) = (\mathbf{v}, H\mathbf{v})^\star. \tag{C.39}$$

Since $\lambda = (\mathbf{v}, H\mathbf{v})/(\mathbf{v}, \mathbf{v})$, and both numerator and denominator are real, λ must be real.

(ii) Let λ_1 and λ_2 be any two eigenvalues of H, corresponding to eigenvectors $\mathbf{v}_1$ and $\mathbf{v}_2$, respectively. Then

$$\begin{aligned}(\mathbf{v}_2, H\mathbf{v}_1) &= \lambda_1(\mathbf{v}_2, \mathbf{v}_1) \\ &= (H\mathbf{v}_2, \mathbf{v}_1) = \lambda_2^\star(\mathbf{v}_2, \mathbf{v}_1) \\ &= \lambda_2(\mathbf{v}_2, \mathbf{v}_1),\end{aligned} \tag{C.40}$$

where in going from the second to the third line we have used the previously established reality of the eigenvalues. Subtracting the right-hand sides of the first and third lines, we get

$$(\lambda_1 - \lambda_2)(\mathbf{v}_2, \mathbf{v}_1) = 0. \tag{C.41}$$

Therefore, if the eigenvalues differ, $\lambda_1 \neq \lambda_2$, the associated eigenvectors must be orthogonal, $(\mathbf{v}_2, \mathbf{v}_1) = 0$. □

In fact, a much more powerful assertion can be made concerning self-adjoint operators in $\mathbb{C}^n$: they possess a set of n orthogonal (hence, linearly independent) eigenvectors, which therefore form a basis for the full vector space. This result is based on the intimate connection between the self-adjointness property and the inner product, which asserts that the linear subspace orthogonal to any eigenvector $\mathbf{v}$ (with eigenvalue λ) of H is left invariant under the action of H. Let $\mathbf{w}$ be any vector orthogonal to $\mathbf{v}$. Then

$$(\mathbf{v}, H\mathbf{w}) = (H\mathbf{v}, \mathbf{w}) = (\lambda\mathbf{v}, \mathbf{w}) = \lambda^\star(\mathbf{v}, \mathbf{w}) = 0, \tag{C.42}$$

so $H\mathbf{w}$ is also in the orthogonal complement of $\mathbf{v}$. Our linear operator H can therefore be thought of as restricted to the orthogonal complement subspace (which is $n-1$ dimensional), where it can be represented as an $(n-1) \times (n-1)$ Hermitian matrix, relative to an orthonormal basis in the subspace.

Theorem C.3.2. *Let H be a self-adjoint operator in the n-dimensional complex vector space $\mathbb{C}^n$ with inner product. Then H has n orthogonal eigenvectors.*

Proof. As we saw in lemma C.2.1, an arbitrary linear operator in $\mathbb{C}^n$ has at least one eigenvector—let us call it $\mathbf{v}_1$, with an eigenvalue λ_1, which, by the preceding theorem, must be real. But H can now be restricted to the orthogonal complement of $\mathbf{v}_1$, where it must (again, from lemma C.2.1) have at least one eigenvector $\mathbf{v}_2$ (again, associated with a real eigenvalue λ_2). Proceeding inductively in this way, we can consider the subspace of vectors orthogonal to both $\mathbf{v}_1$ and $\mathbf{v}_2$, which is also invariant under the action of H, and must contain an eigenvector $\mathbf{v}_3$. We end up with a one-dimensional subspace of vectors orthogonal to $\mathbf{v}_1, \mathbf{v}_2, \ldots, \mathbf{v}_{n-1}$, and invariant under the action of H, which must therefore be just the last eigenvector (up to a scalar factor) $\mathbf{v}_n$. □

The set of orthogonal eigenvectors constructed in this proof can immediately be converted into an orthonormal basis (i.e., orthogonal vectors of unit length) by calculating the magnitude of each of the $\mathbf{v}_i$ and dividing by it. Thus, we henceforth assume that

$$(\mathbf{v}_i, \mathbf{v}_j) = \delta_{ij}. \tag{C.43}$$

The matrix of H relative to the *basis of its own eigenvectors* is thus

$$H_{ij} = (\mathbf{v}_i, H\mathbf{v}_j) = \lambda_j(\mathbf{v}_i, \mathbf{v}_j) = \lambda_i\delta_{ij}. \tag{C.44}$$

In other words, in this basis the H_{ij} form a diagonal matrix, which we may call $\mathbf{H}_{\text{diag}}$, with the eigenvalues λ_i sitting on the diagonal line. An alternative notation we sometimes employ in this case is $H = \text{diag}(\lambda_1, \lambda_2, \ldots, \lambda_n)$.

In a general orthonormal basis $\{\mathbf{e}_i\}$ $(i = 1, \ldots, n)$, as we saw earlier, H will have a Hermitian matrix, with $H_{ij} = H^\star_{ji}$, which implies real elements on the diagonal, while elements related by reflection across the diagonal are complex conjugates of each other. The eigenvectors $\mathbf{v}_i$ can, of course, be written as linear combinations of the vectors of the general orthonormal basis $\{\mathbf{e}_i\}$:

$$\mathbf{v}_i = \sum_k U_{ki}\mathbf{e}_k. \tag{C.45}$$

Our diagonal matrix can therefore be written as

$$\lambda_j\delta_{ij} = (\mathbf{v}_i, H\mathbf{v}_j) = (U_{ki}\mathbf{e}_k, HU_{lj}\mathbf{e}_l) = U^\star_{ki}H_{kl}U_{lj}, \tag{C.46}$$

or, in matrix notation,

$$\mathbf{H}_{\text{diag}} = \mathbf{U}^\dagger\mathbf{H}\mathbf{U}. \tag{C.47}$$

Using Eqs. (C.43)–(C.46), we now observe that

$$\begin{aligned} \delta_{ij} &= (\mathbf{v}_i, \mathbf{v}_j) \\ &= \sum_{kl} U^\star_{ki}U_{lj}(\mathbf{e}_k, \mathbf{e}_l) \\ &= \sum_k U^\star_{ki}U_{kj} \\ &= (U^\dagger U)_{ij}. \end{aligned} \tag{C.48}$$

So the matrix $\mathbf{U}^\dagger$ is the inverse of the matrix $\mathbf{U}$. Matrices with this property are called *unitary*. More generally, without reference to a basis, a linear operator U is unitary if its adjoint is equal to its inverse. The eigenvalues of such a U (there must be at least one—lemma C.2.1 again!) are complex numbers of unit modulus, as the following argument shows. Given a unit-length eigenvector $\mathbf{v}$ of U, with $U^\dagger = U^{-1}$, it follows from $U\mathbf{v} = \lambda\mathbf{v}$ and

$$\mathbf{v} = U^{-1}U\mathbf{v} = U^\dagger U\mathbf{v} = \lambda U^\dagger \mathbf{v}, \tag{C.49}$$

that

$$U^\dagger \mathbf{v} = \lambda^{-1}\mathbf{v}. \tag{C.50}$$

From

$$\lambda = (\mathbf{v}, U\mathbf{v}) = (U^\dagger \mathbf{v}, \mathbf{v}) = (\lambda^{-1})^\star, \tag{C.51}$$

it then follows that

$$\lambda\lambda^\star = 1. \tag{C.52}$$

Just as in the case of a self-adjoint operator, a unitary operator will leave the space orthogonal to an eigenvector invariant. Indeed, if $\mathbf{w}$ is any vector orthogonal to an eigenvector $\mathbf{v}$ (with eigenvalue λ) of U, then

$$(U\mathbf{w}, \mathbf{v}) = (\mathbf{w}, U^\dagger \mathbf{v}) = (\mathbf{w}, U^{-1}\mathbf{v}) = \lambda^{-1}(\mathbf{w}, \mathbf{v}) = 0. \tag{C.53}$$

The proof of Theorem C.3.2 can now be repeated essentially word for word to establish the following theorem.

Theorem C.3.3. *Let U be a unitary operator in the n-dimensional complex vector space $\mathbb{C}^n$ with inner product. Then U has n orthogonal eigenvectors. The associated eigenvalues are complex numbers of modulus unity.*

There is a natural one-to-one correspondence between self-adjoint operators and unitary operators in $\mathbb{C}^n$, the generalization of which to infinite-dimensional spaces plays a central role in von Neumann's spectral theory, which we shall later discuss. First, note that if H is a self-adjoint operator, then $H \pm i$ ($= H \pm i1$ where 1 is the identity operator) is invertible, as there are no non-zero vectors $\mathbf{v}$ for which $(H \pm i)\mathbf{v} = 0$ (as the eigenvalues of H are all real). Thus, the operator V, called the *Cayley transform* of H,

$$V \equiv (H - i)(H + i)^{-1} = (H + i)^{-1}(H - i) \tag{C.54}$$

exists, and moreover (cf. Eq. (C.36)),

$$V^\dagger V = (H - i)^{-1}(H + i)(H + i)^{-1}(H - i) = 1, \tag{C.55}$$

so the Cayley transform of a self-adjoint operator is automatically unitary (in finite-dimensional spaces). The generalization of this property to infinite-dimensional operators lies at the heart of the spectral theory that von Neumann developed for unbounded operators.

Theorems C.3.1–C.3.3 all remain valid if we are considering real inner-product spaces based on $\mathbb{R}^n$. The operators analogous to H and U above are now *real, symmetric* matrices satisfying $H_{ij} = H_{ji}$ (or $\mathbf{H} = \mathbf{H}^{\mathrm{T}}$) and *real, orthogonal* matrices O_{ij}, satisfying $O_{ij}O_{ik} = \delta_{jk}$ (or $\mathbf{O}^{\mathrm{T}}\mathbf{O} = 1$), respectively. All of the completeness results obtained above in the complex space $\mathbb{C}^n$ hold here as well in the simpler case of a real vector space. For example, for a real symmetric matrix K_{ij}, there always exists a real orthogonal matrix O_{ij} such that, in analogy to Eq. (C.46),

$$O_{ki}K_{kl}O_{lj} = \kappa_i\delta_{ij}, \quad \text{or} \quad \mathbf{O}^{\mathrm{T}}\mathbf{K}\mathbf{O} = \mathbf{K}_{\mathrm{diag}}. \tag{C.56}$$

However, it should be noted that the eigenvalues of the real orthogonal matrices $\mathbf{O}$ that serve as similarity transformations to effect the diagonalization of a real symmetric matrix are still complex unimodular numbers in general.

Theorem C.3.2 has a generalization of great importance in quantum mechanics to the case where we have two self-adjoint operators, H_1 and H_2, that commute with each other, that is, $H_1H_2 = H_2H_1$. They must have a common eigenvector $\mathbf{v}$, as at least one eigenvector of H_1 with some eigenvalue λ_1 exists, and the space of eigenvalues of H_1 with this eigenvalue is left invariant by H_2, as

$$H_1H_2\mathbf{v} = H_2H_1\mathbf{v} = \lambda_1H_2\mathbf{v}. \tag{C.57}$$

We may therefore restrict H_2 to this space, wherein it must have (lemma C.2.1) at least some eigenvector $\tilde{\mathbf{v}}$, which is therefore a common eigenvector of both H_1 and H_2. We may now restrict both H_1 and H_2 to the orthogonal complement of $\tilde{\mathbf{v}}$, and repeat the argument in this $(n-1)$-dimensional space, to find a second common eigenvector of H_1 and H_2. The result is

Theorem C.3.4. *Let H_1 and H_2 be two self-adjoint operators in the complex vector space $\mathbb{C}^n$ with inner product, which commute with each other, $H_1H_2 = H_2H_1$. Then H_1 and H_2 have n common eigenvectors, which form an orthogonal set.*

Before going on to the infinite-dimensional spaces that are our primary interest in this Appendix, it will be useful to rephrase the results, and to some extent the proof, of Theorem C.3.2 in a form that anticipates the spectral theorems of von Neumann, which are our ultimate objective. Recall that the theorem asserts that a self-adjoint operator H in $\mathbb{C}^n$ possesses exactly n eigenvectors, which can be chosen to form an orthonormal set. In the proof of this assertion, we began by pointing out that there must exist at least one eigenvector (this is true for *any* linear operator) $\mathbf{v}_1$, which we may normalize to unit magnitude, with eigenvalue λ_1. Define a new operator E_1 by its action on an arbitrary vector $\mathbf{v}$, as follows

$$E_1\mathbf{v} = (\mathbf{v}_1, \mathbf{v})\mathbf{v}_1. \tag{C.58}$$

Evidently, E_1 acts on a general vector to produce just the projection of that vector on the one-dimensional subspace spanned by $\mathbf{v}_1$. It is clear that applying E_1 twice gives the same result as applying it once, and in fact, for any integer n,

$$E_1^n = E_1. \tag{C.59}$$

We will call E_1 the *projection operator* onto the one-dimensional space spanned by $\mathbf{v}_1$.[10] Now consider the linear operator

$$H_1 \equiv H - \lambda_1 E_1. \tag{C.60}$$

Since $H\mathbf{v}_1 = \lambda_1 \mathbf{v}_1$ and $E_1\mathbf{v}_1 = \mathbf{v}_1$, H_1 annihilates $\mathbf{v}_1$ and leaves the $n-1$ dimensional space of vectors orthogonal to $\mathbf{v}_1$ invariant. It may therefore be regarded as a linear operator acting just in this latter space, and must have at least one normalized eigenvector $\mathbf{v}_2$, with eigenvalue λ_2 (which might, of course, be the same as λ_1, if there are degeneracies). Repeating the argument above, we define a projection operator E_2 (where $E_2\mathbf{v} = (\mathbf{v}_2, \mathbf{v})\mathbf{v}_2$) on this new eigenvector, and arrive at an operator H_2,

$$H_2 \equiv H_1 - \lambda_2 E_2, \tag{C.61}$$

which acts non-trivially only in the space orthogonal to $\mathbf{v}_1$ and $\mathbf{v}_2$. After n steps we have exhausted the entire space $\mathbb{C}^n$, so.

$$H_n = H - \sum_{i=1}^{n} \lambda_i E_i = 0. \tag{C.62}$$

We can thus write H as:

$$H = \sum_{i=1}^{n} \lambda_i E_i. \tag{C.63}$$

It is apparent from the orthogonality of the eigenvectors that $E_i E_j = 0$ if $i \neq j$, so there is an obvious analogous formula for powers of the operator H:

$$H^N = \sum_{i=1}^{n} \lambda_i^N E_i, \tag{C.64}$$

and for any function $f(H)$ of H constructible as a power series,

$$f(H) = \sum_{i=1}^{n} f(\lambda_i)\, E_i. \tag{C.65}$$

Equation (C.63) is called the *spectral representation* or *spectral decomposition* of the operator H. Its generalization to unbounded operators in infinite-dimensional spaces by von Neumann in 1929 is rightly considered a high point of the new mathematical field of functional analysis, which emerged as a natural generalization of the work of Hilbert and others on integral equations, work which we shall shortly discuss.

[10] The original terminology, introduced by Hilbert (1906a), was *Einzelform*, for the associated quadratic form—this was taken over by von Neumann (1927a) as *Einzeloperator*, or, more succinctly "E-Op". Two years later, von Neumann (1929) adopted the now universally employed term "projection operator" (*Projektionsoperator*).

For now, it will suffice to write this expression in an integral form which is general enough to handle all cases encountered in quantum mechanics. To do this, it is convenient to employ a generalization of the usual Riemann integral, called a *Stieltjes* integral (introduced by T. J. Stieltjes in 1894), which for our applications will take the form

$$\boldsymbol{\int}_a^b f(x)\,dg(x) \equiv \lim_{n\to\infty} \sum_{i=1}^{n} f(\bar{x}_i)\left(g(x_{i+1}) - g(x_i)\right), \tag{C.66}$$

where $f(x)$ is an arbitrary continuous real function and $g(x)$ is an arbitrary bounded, monotonically increasing, but not necessarily continuous real function. The limit is taken with respect to increasingly fine divisions of the integration interval (a, b), with $a < x_1 < \ldots < x_n < b$, and $x_i < \bar{x}_i < x_{i+1}$, just as in the case of the definition of the Riemann integral, for which $g(x) = x$.[11] The important point is that the Stieltjes integral makes perfect sense even when $g(x)$ has jump discontinuities (as long as $f(x)$ does not also have them at the same points, which we here ensure by the simple expedient of assuming $f(x)$ to be continuous).[12]

The eigenvectors $\mathbf{v}_1, \mathbf{v}_2, \ldots, \mathbf{v}_n$ form a complete orthonormal set,

$$\sum_{i=1}^{n} E_i\,\mathbf{v} = \sum_{i=1}^{n} (\mathbf{v}_i, \mathbf{v})\mathbf{v}_i = \mathbf{v}, \tag{C.67}$$

so the projection operators E_i sum to the identity operator (in mathematical lingo, they provide a "partition of unity"). We now construct an operator-valued function $E(\lambda)$ on a single continuous real parameter λ by defining

$$E(\lambda) \equiv \sum_{i, \lambda_i \leq \lambda} E_i, \tag{C.68}$$

where $E(-\infty) = 0$, $E(+\infty)$ is the identity operator, and, for finite real values of λ, $E(\lambda)$ equals the projection operator onto the subspace spanned by eigenvectors for all eigenvalues up to and including the value λ. The assembly of projection operators $E_1, E_2, \ldots, E_n$, or, equivalently, the function $E(\lambda)$ that incorporates these projection operators (and from which they may be recovered), are called the "spectral family associated with the operator H".

With the help of the definition Eq. (C.66),[13] Eq. (C.63) for H can be written as a Stieltjes integral

$$H = \sum_{i=1}^{n} \lambda_i E_i = \boldsymbol{\int}_{-\infty}^{\infty} \lambda\, dE(\lambda), \tag{C.69}$$

[11] Note that we indicate a Stieltjes integral by making the integral sign bold.

[12] It should be emphasized that the way in which the Stieltjes integral generalizes the usual Riemann one is quite different from the Lebesgue generalization, which applies to a far more general class of functions. Nevertheless, the Stieltjes integral is the right tool needed for expressing the most general spectral theorems needed in quantum mechanics.

[13] Strictly speaking, the limit operation in Eq. (C.66) implies a choice of a topology in the space of matrices/operators. As we are working in finite dimensions, one may choose either a "weak" convergence whereby the convergence of the limit is assured for each matrix element separately on the left and right-hand sides of Eq. (C.66), or strong convergence in terms of the norm of the operators, as defined below. In finite-dimensional spaces, these choices are equivalent.

with the obvious generalization to a general function of H (cf. Eq. (C.65)),

$$f(H) = \int_{-\infty}^{\infty} f(\lambda)\, dE(\lambda). \tag{C.70}$$

The expression of a general finite-dimensional self-adjoint operator as a spectral integral rather than a sum seems at this juncture unnecessarily elaborate,[14] but it will turn out that an integral expression of exactly this form (involving a Stieltjes integral) will handle the most general case of importance in quantum mechanics, namely, the unbounded self-adjoint operators ubiquitously present as the representative of physical observables such as position, momentum, and energy, which may possess a continuous as well as a discrete spectrum, and for which the integral form of the spectral representation is unavoidable.

C.4 A historical digression: integral equations and quadratic forms

To the extent to which the Göttingen physicists were familiar with the material presented above on vector spaces and matrices, it would probably[15] have been in the form in which the basic theorems are presented in the first chapter of Courant and Hilbert (1924). Here, the establishment of Theorem C.3.2 (existence of a complete set of orthonormal eigenvectors for a self-adjoint operator H) involves the use of a maximum/minimum principle for the quadratic form associated with the operator. For the time being we follow Courant and Hilbert in restricting the discussion to the real vector space $\mathbb{R}^n$, so our self-adjoint operator is represented by a real symmetric rather than a Hermitian matrix K. We consider an arbitrary *unit* vector $\mathbf{x} = (x_1, x_2, \ldots, x_n)$, where the x_is are the real components of $\mathbf{x}$ in an arbitrary orthonormal basis, with respect to which the matrix elements of K are K_{ij} (with $i, j = 1, 2, \ldots, n$) and satisfy $\sum_i x_i^2 = 1$. The quadratic form associated with K is defined as

$$K(x, x) \equiv (\mathbf{x}, K\mathbf{x}) = \sum_{ij} K_{ij} x_i x_j, \quad K_{ij} = K_{ji}. \tag{C.71}$$

It is easy to see that $K(x, x)$ is a continuous function on a bounded domain, and so must assume its maximum value at some point in or on the surface of the domain.[16] One can then show that the associated vector $\mathbf{x} = \mathbf{v}_1$ is an eigenvector of K, corresponding to some eigenvalue $\lambda_1 = K(\mathbf{v}_1, \mathbf{v}_1)$. One then proceeds to maximize $K(x, x)$ over vectors $\mathbf{x}$ orthogonal to $\mathbf{v}_1$: the result is a new

[14] The spectral integral for a bounded self-adjoint operator in an infinite-dimensional space was first written in this form by F. Riesz (1913).

[15] An exception to this admittedly speculative assertion is Max Born, who was already familiar with the maneuvers of finite-dimensional matrix algebra from his work in 1909 on electron theory and special relativity in the Minkowskian four-dimensional space–time framework. He had also been friends with Otto Toeplitz from his early university days. He explicitly refers to Toeplitz's study of a special class of infinite quadratic forms in his work with von Kármán on the dynamics of crystal lattices (Born and von Kármán 1913, p. 67, note 1). Another of Born's friends from Breslau, Ernst Hellinger, collaborated with Toeplitz in laying out a theory of infinite matrices, connected to the Hilbert theory of quadratic forms (Hellinger and Toeplitz 1910). It is safe to assume that Born was at least aware of these developments in general terms, allowing him to very quickly uncover the connection of matrix mechanics to the extant theory of infinite forms/matrices which he lays out in chapter 3 of the Three-Man-Paper.

[16] It may attain this maximum value at several distinct points of course. One simply picks any maximal point and continues with the inductive procedure described in the rest of this paragraph.

eigenvector $\mathbf{v}_2$, orthogonal to the first, and with an eigenvalue $\lambda_2 \leq \lambda_1$. And so on, until a complete set of orthogonal unit eigenvectors of K are obtained. The analogy of this procedure to the proof given above for Theorem C.3.2 will be clear.

The process of finding a complete set of eigenvectors of the operator K is equivalent to a problem already well known in nineteenth-century analytic geometry: the determination of the principal axis transformation for an ellipsoid in n dimensions, defined as the surface satisfying $K(x,x) = \text{const}$. Here, the problem is to find a set of variables y_p, linearly related to the original $x_i = O_{ip}y_p$, such that the quadratic form resolves into a sum of diagonal terms (squares):

$$\begin{aligned} K(x,x) &= \sum_{ij} K_{ij}x_i x_j \\ &= \sum_{ijpq} O_{ip}K_{ij}O_{jq}y_p y_q \\ &= \sum_{pq} y_p(\mathbf{O}^{\mathrm{T}}\mathbf{K}\mathbf{O})_{pq}y_q \\ &= \sum_{p} \kappa_p y_p^2. \end{aligned} \tag{C.72}$$

This is of course achieved precisely by finding the orthogonal matrix $\mathbf{O}$, which diagonalizes the real symmetric matrix K (cf. Eq. (C.47)):

$$(\mathbf{K}_{\text{diag}})_{pq} = \kappa_p \delta_{pq} = (\mathbf{O}^{\mathrm{T}}\mathbf{K}\mathbf{O})_{pq}, \tag{C.73}$$

where the κ_ps (with $p = 1, 2, \ldots, n$) are the eigenvalues of K.

The quadratic form in Eq. (C.71) can be generalized to a bilinear function of two vectors $\mathbf{x}, \mathbf{y}$ in an obvious way:

$$K(x,y) \equiv (\mathbf{x}, K\mathbf{y}). \tag{C.74}$$

We can rewrite this bilinear function in a form that was of great importance in Hilbert's work by utilizing the existence of a complete set of n normalized eigenvectors $\mathbf{v}_n$ of K, with eigenvalues κ_n, and resolving the vectors $\mathbf{x}$ and $\mathbf{y}$ in the basis of these eigenvectors:

$$\begin{aligned} \mathbf{x} &= \sum_n (\mathbf{x}, \mathbf{v}_n)\,\mathbf{v}_n \equiv \sum_n L_n(x)\,\mathbf{v}_n, \\ K\mathbf{y} &= \sum_n \kappa_n\, L_n(y)\,\mathbf{v}_n, \\ K(x,y) &= \sum_n \kappa_n\, L_n(x)\, L_n(y). \end{aligned} \tag{C.75}$$

This representation of a general bilinear form in terms of eigenvalues and eigenvectors of a symmetric matrix (now called the "spectral resolution" of the form) would play a critical role in Hilbert's algebraic reformulation of the theory of integral equations.

A seemingly unconnected strand of mathematical enquiry occurred in the developing interest in the late-nineteenth and early-twentieth centuries[17] in linear integral equations—equations in which the function to be determined appears linearly, and under an integral sign. The classic example was the Fredholm equation of the second kind, which for a one-dimensional problem takes the form[18]

$$\varphi(s) - \lambda \int_0^1 K(s,t)\,\varphi(t)\,dt = f(s). \tag{C.76}$$

Such integral equations had arisen naturally in the study of certain problems in electrostatic potential theory where the potential is specified on the boundary of a region (in two dimensions) and one wishes to obtain a closed-form expression (as an integral of known functions) for the potential in the interior. Here, the function $K(s,t)$, assumed to be a continuous function of both its variables on the interval $[0,1]$, is called the ***kernel*** of the integral equation. The parameter λ could of course be absorbed in the kernel, but is displayed explicitly to provide a counting mechanism for an expansion of the desired solution for $\varphi(s)$ in powers of λ, presumably convergent if the kernel parameter is sufficiently small. The parameter λ also appears naturally in problems in which periodic motion of the system occurs, as in the vibration of membranes (e.g., the surface of a drum), where the double time derivative becomes the squared frequency.

The algebraic structure connected to this quintessentially analytic problem becomes apparent if we imagine a different sort of approximation, obtained by discretizing the integral as a sum of n terms, so that s and t in Eq. (C.76) are allowed to take the values $s_i = i/n$ and $t_i = j/n$ (with $i,j = 1,2,\ldots,n$). The integral equation then reduces to a finite set of linear algebraic equations:

$$\sum_j (\delta_{ij} - \lambda K_{ij}^{(n)})\,\varphi_j = f_i, \tag{C.77}$$

where $K_{ij}^{(n)} = K(s_i,t_j)/n$ is a $n \times n$ matrix, $\varphi_i = \varphi(s_i)$, and $f_i = f(s_i)$. The solution of the approximated problem clearly amounts to treating φ and f as n-vectors, and inverting the matrix $1 - \lambda\mathbf{K}^{(n)}$:

$$\varphi = \left(1 - \lambda\mathbf{K}^{(n)}\right)^{-1}\mathbf{f}. \tag{C.78}$$

The information contained in the inverse matrix here can also be stored in the associated bilinear form (cf. Eq. (C.74))

$$\overline{K}_n(\lambda;x,y) \equiv \left(\mathbf{x}, \left(1 - \lambda\mathbf{K}^{(n)}\right)^{-1}\mathbf{y}\right). \tag{C.79}$$

Of course, an exact solution of the original integral equation must involve taking the limit $n \to \infty$, and a resolution of the concomitant convergence issues.

[17] For an excellent review of the origins of function spaces in integral-equation theory, see Bernkopf (1965).
[18] The Fredholm integral equation of the first kind is identical to Eq. (C.76), with the $\varphi(s)$ first term missing on the left-hand side. The first problem formulated explicitly as an integral equation of the first kind was Abel's solution of the tautochrone problem in mechanics, in which one seeks to determine the shape of the curve down which a particle will slide in a prescribed time (as a function of initial height). This problem leads to a Fredholm equation of the first kind, with $K(s,t) = \vartheta(s-t)/\sqrt{s-t}$. It was solved by Abel in 1823.

Fredholm's 1903 theory of the integral equation Eq. (C.76) works by analogy to the usual formula for the inverse of a finite matrix involving its determinant and its minor determinants. He was able to develop a generalization of these concepts for the infinite-dimensional case in terms of a determinant defined as an infinite sum, guaranteed to converge if the original kernel $K(s,t)$ is bounded for $0 \leq s, t \leq 1$. The problem was attacked from a different angle in six remarkable papers by David Hilbert, published in the *Göttingen Nachrichten* between 1904 and 1910, and which form the keystone of the area of mathematical work we now refer to as "functional analysis". Hilbert (1912) collected these papers in a single volume with an extensive introduction.

In his first paper, Hilbert explains the overall strategy he would follow in developing a general theory of Fredholm integral equations:

> The method, which I will apply in this first communication, consists in proceeding from an algebraic problem, namely the problem of the orthogonal transformation of a quadratic form of n variables into a sum of squares, to arrive, via the rigorous execution of the $n \to \infty$ limit, at the solution of the transcendental problem at issue (Hilbert 1904, pp. 51–52).

Returning to the discrete approximation in Eqs. (C.77)–(C.78), we note that the algebraic situation is considerably simplified if the kernel $K(s,t)$ is taken to be symmetric, as the associated discretized matrix $K_{ij}^{(n)}$ is then a real symmetric matrix, and we can help ourselves to the powerful algebraic theorems available in the case of Hermitian/symmetric matrices. In Hilbert's terminology, the values of λ for which Eq. (C.76) is satisfied with $f(s) = 0$, that is, for which

$$\varphi(s) = \lambda \int_0^1 K(s,t)\,\varphi(t)\,dt, \quad \int_0^1 K(s,t)\,\varphi(t)\,dt = \frac{1}{\lambda}\varphi(s) \tag{C.80}$$

are called the eigenvalues (*Eigenwerte*) of the kernel $K(s,t)$. These numbers, in the discretized approximation of Eq. (C.77), actually correspond to what in modern terminology would be called the inverse eigenvalues, as is apparent from the preceding equation. Hilbert's theory (with extensions by Schmidt and Hellinger) made it possible to determine the necessary conditions to be satisfied by the kernel $K(s,t)$ so that the finite-dimensional approximations, in which the desired n-dimensional inverse operator of Eq. (C.78) (called the *resolvent* by Hilbert) would converge appropriately to a well-defined integral kernel when n is taken to infinity. In the language of bilinear forms, the problem was to find the resolvent bilinear form $\overline{K}(\lambda; x, y)$ in *infinitely many variables* as a limit $n \to \infty$ of the finite-dimensional $\overline{K}_n(\lambda; x, y)$ approximants in Eq. (C.79), and to determine the form of the corresponding spectral resolution, analogous to Eq. (C.75), in this limit. The use of the word "spectral" (or "spectrum") in this context is historically interesting, as it is introduced at least as early as 1906 by Hilbert, to designate the assembly of eigenvalues of an integral kernel:

> The point spectrum [discrete eigenvalues] together with the accumulation points of the eigenvalues and the line [*Strecken*-i.e., continuous] spectrum taken together are called the *spectrum* of the form K (Hilbert 1906a, p. 172).

The prescient use of this term is remarkable: well before the discoveries of Bohr in 1913, or of modern quantum mechanics in either the matrix or wave version of 1925/26, Hilbert has taken the terminology of an optical spectrum and applied it to exactly the mathematical framework needed later for the understanding of the optical spectra of atomic systems. According to his biographer, Hilbert himself commented on this fortuitous coincidence:

> I developed my theory of infinitely many variables from purely mathematical interests and even called it "spectral analysis" without any presentiment that it would later find an application to the actual spectrum of physics (Reid 1986, p. 183).[19]

In the next section, our discussion of infinite-dimensional spaces shows how a discrete infinite matrix can in fact yield the perhaps unexpected combination of a discrete (enumerable) point spectrum and a continuous spectrum referred to here by Hilbert.

C.5 Infinite-dimensional spaces

Section C.4 showed that Hilbert's study of linear Fredholm integral equations inevitably led him to the consideration of infinite-dimensional real symmetric arrays K_{ij}. The question naturally arose whether the ability to diagonalize such an array by an orthogonal transformation, giving a spectrum of discrete eigenvalues κ_i, guaranteed for finite matrices by Eq. (C.56), would extend to *infinite-dimensional* real symmetric arrays. Hilbert was able to isolate a class of matrices (or quadratic forms) for which this was indeed the case, and was able to generalize the diagonalization process to a wider class of matrices, though the most general case relevant to quantum mechanics (for unbounded self-adjoint operators) would have to wait till the end of the 1920s and John von Neumann's work.

Before going on to the infinite-dimensional case, we state some results from the theory of finite-dimensional self-adjoint operators/matrices in a new way, which smooths the transition to function and Hilbert-space theory. First, note that the result $\mathbf{H}_{\text{diag}} = \mathbf{U}^\dagger\mathbf{H}\mathbf{U}$, which asserts the existence of a unitary matrix $\mathbf{U}$ that effects the diagonalization of an arbitrary Hermitian matrix $\mathbf{H}$, implies $\mathbf{H}\mathbf{U} = \mathbf{U}\mathbf{H}_{\text{diag}}$, or, in terms of matrix elements

$$H_{ik}U_{kj} = \lambda_j U_{ij}. \tag{C.81}$$

The columns of the unitary matrix U_{ij} thus directly provide the components of the eigenvectors of $\mathbf{H}$. It will be convenient to introduce a slightly different (and admittedly superfluous, but nonetheless useful) notation for these eigenvectors, namely

$$\mathbf{H}\mathbf{v}^{(j)} = \lambda_j \mathbf{v}^{(j)}, \quad v_k^{(j)} \equiv U_{kj}. \tag{C.82}$$

Using this notation, we can rewrite Eq. (C.81) as

$$H_{ik}v_k^{(j)} = \lambda_j v_i^{(j)}. \tag{C.83}$$

[19] Reid's book has no notes and no bibliography and no source is given for this remark.

The unitarity conditions $\mathbf{U}^\dagger\mathbf{U} = \mathbf{U}\mathbf{U}^\dagger = 1$, together with Eq. (C.81), are equivalent to the following four results:

1. (Orthogonality) From $\mathbf{U}^\dagger\mathbf{U} = 1$, we have $\sum_i U^\star_{in}U_{im} = \delta_{nm}$ or

$$\sum_i v_i^{(n)\star} v_i^{(m)} = (v^{(n)}, v^{(m)}) = \delta_{nm}. \tag{C.84}$$

2. (Completeness) From $\mathbf{U}\mathbf{U}^\dagger = 1$, we have $\sum_n U_{in}U^\star_{jn} = \delta_{ij}$ or

$$\sum_n v_i^{(n)} v_j^{(n)\star} = \delta_{ij}. \tag{C.85}$$

3. (Resolution of the identity) As an immediate consequence of the completeness relation Eq. (C.85), we have, for any two vectors $\mathbf{w}, \mathbf{z}$,

$$\begin{aligned}(\mathbf{w}, \mathbf{z}) &= \sum_{ij} w_i^\star \delta_{ij} z_j \\ &= \sum_{ij} w_i^\star \Big(\sum_n v_i^{(n)} v_j^{(n)\star}\Big) z_j \\ &= \sum_n \Big(\sum_i w_i^\star v_i^{(n)}\Big)\Big(\sum_j v_j^{(n)\star} z_j\Big) \\ &= \sum_n \big(\mathbf{w}, \mathbf{v}^{(n)}\big)\big(\mathbf{v}^{(n)}, \mathbf{z}\big).\end{aligned} \tag{C.86}$$

4. (Spectral resolution of $\mathbf{H}$) Using $H_{ik}v_k^{(n)} = \lambda_n v_i^{(n)}$, we find, using Eqs. (C.85) and (C.83),

$$\begin{aligned}H_{ij} &= H_{ik}\delta_{kj} \\ &= \sum_n H_{ik} v_k^{(n)} v_j^{(n)\star} \\ &= \sum_n \lambda_n v_i^{(n)} v_j^{(n)\star}.\end{aligned} \tag{C.87}$$

The basic thrust of the initial phase of development of functional analysis, contained in the work of Hilbert, Schmidt, Hellinger, and others, was to study the extent to which results of this type, valid for an arbitrary finite-dimensional complex inner-product space, could be extended to infinite-dimensional spaces. The result was the construction of a remarkable bridge between the algebraic theory of infinite-dimensional spaces and the analytic theory of spaces of functions (of appropriately restricted behavior).

Our very cursory review of the elements of functional analysis of central importance in quantum mechanics is divided into subsections from this point.[20] The headings of these subsections

[20] For an excellent overview of the early history of functional analysis (up to about 1932), see Birkhoff and Kreyszig (1984).

should allow the more mathematically advanced reader to find easily, and skip ahead to, material that may be unfamiliar. Throughout this appendix, we presume that the reader is familiar with the basic concepts of calculus, linear algebra, and some rudiments of real and complex analysis.

C.5.1 Topology: open and closed sets, limits, continuous functions, compact sets

The transition from finite- to infinite-dimensional vector spaces obviously introduces a plethora of new problems connected with the convergence of the infinite summations now required to perform even the simplest algebraic processes (e.g., the multiplication of two matrices). Thus, questions of convergence, or more generally, of the existence of limits, take center stage. These notions all implicitly depend on the existence of an assumed *topology*, which roughly speaking gives us a precise way to talk about the "neighborhood" of a point in a space—namely, the set of points "close" to the given point.[21]

A set of points X becomes a topological space by the simple expedient of selecting a distinguished class of subsets of X, called the "open sets", subject to the restriction that the intersection of any two (or finitely many) such open sets is an open set, and the union of arbitrarily many open sets is an open set. A closed set is simply the complement of an open set in the space. From basic set theory, it follows that finite unions and arbitrary intersections of closed sets are closed. From this innocent starting point a huge number of useful concepts can be derived, essential for the understanding of modern analysis, both real and complex. We restrict ourself here to the most basic and essential ideas.

The idea of a topological space is obviously one of enormous generality. For our purposes, we consider only topologies which are "norm-" or "metric-induced". That is to say, in all the vector spaces of interest, there is an inner product, from which a norm can be defined (cf. Eq. (C.26)), and then in turn, a distance or "metric" function specifying the separation of two distinct vectors as a positive real number:

$$d(\mathbf{v},\mathbf{w}) = |\mathbf{v}-\mathbf{w}| = \sqrt{(\mathbf{v}-\mathbf{w},\mathbf{v}-\mathbf{w})}. \tag{C.88}$$

Define the set $B_\varepsilon(\mathbf{v})$, the open ball of radius ε around the point (vector) $\mathbf{v}$, as the set of all vectors $\mathbf{w}$ satisfying $d(\mathbf{v},\mathbf{w}) < \varepsilon$. The topology we assume for all the normed vector spaces under consideration—both $\mathbb{R}^n$ and $\mathbb{C}^n$, as well as the Hilbert space to be introduced shortly—unless explicitly stated otherwise, defines an open set as any set consisting of an arbitrary union of open balls.

A neighborhood U of a point x in a topological space X is simply a set containing an open set, which itself contains x. A sequence of points $x_1, x_2, \ldots$ is said to converge to a given point x (relative to the given topology) if for every neighborhood U of x, all the points in the sequence past a certain point x_M (where M will in general depend on U) lie in U. It is easy to see that with the norm-induced topology discussed above, convergence of a sequence of vectors $\mathbf{v}_i$ to a limit vector $\mathbf{v}$ is tantamount to saying that for every ε, there exists an integer M such that

$$d(\mathbf{v}_i,\mathbf{v}) < \varepsilon, \quad i \geq M, \tag{C.89}$$

[21] Many of the concepts that lie at the heart of point-set topology were introduced into mathematics by Maurice Fréchet in his seminal doctoral thesis (Frechet 1906). In particular, he defined the notions of compact sets, and of complete and separable spaces, and worked with infinite-dimensional spaces. The concept of a distance function (écart), (or *metric*, in later terminology) was also introduced here.

for which we shall use the more concise and familiar notation

$$\lim_{i\to\infty} \mathbf{v}_i = \mathbf{v}. \tag{C.90}$$

Much progress in mathematics has been made by encountering sequences of mathematical objects which in some sense "ought to converge", but do not in fact do so, under the terms of the given topology. Frequently, this lack of convergence has been "cured" by either changing (typically, enlarging) the underlying set of objects being studied, or by changing the topology used to define convergence. A sequence of points x_i in a metric space (one equipped with a distance function $d(x, y)$, cf. note 8) is called a "Cauchy sequence" if for any positive ε, however small, there is an integer M (depending on ε) such that

$$d(x_i, x_j) < \varepsilon, \quad i, j \geq M. \tag{C.91}$$

The points in a Cauchy sequence get closer and closer to each other the further out we go, so it seems natural to expect that there should be a single limit point x to which the sequence should converge, in the sense of Eq. (C.89). A classic example is the set of rational numbers, in which a Cauchy sequence such as that generated by the decimal expansion of π, for example,

$$3,\ 3.1,\ 3.14,\ 3.141,\ 3.1415,\ 3.14159,\ \ldots \tag{C.92}$$

has no rational number as its limit. The construction of the real numbers can be viewed as the "completion" of the rational numbers by adding to them all Cauchy sequences not convergent within the rationals, each of which defines a new "irrational" number.[22] The resulting space of real numbers is, by construction, *Cauchy-complete*, that is, every Cauchy sequence converges to a point in the space. It is straightforward to show that the same completion process can be carried out on the field of complex numbers $z = r_1 + ir_2$, where r_1 and r_2 are rational numbers, to obtain the normal complex plane C, consisting of complex numbers with both real and imaginary parts taking values in the real numbers. Also, the procedure can be extended to the linear vector space $\mathbf{Q}^n$ consisting of n-tuples of rational numbers, which can be Cauchy-completed into the space $\mathbb{R}^n$ which we have studied before. The completion process for an infinite-dimensional space, which leads to function spaces and Hilbert space, is discussed below.

A few additional concepts from general topology will be indispensable in the following. We introduce them as a collection of definitions, with some obvious consequences of each.

1. Given a subset A of a topological space X, the *closure*, written $\overline{A}$, of A is the intersection of all closed sets containing A. As arbitrary intersections of closed sets are closed, it is apparent that the closure of any set A renders it closed. It is also clear that the closure of $\overline{A}$ is the set $\overline{A}$ itself—no further points are added, by performing the closure twice. Roughly speaking, the closure of a set A is effected by adding to the set all the boundary points (not themselves in A) that are limits of convergent sequences in A. For example, the closure of the open ball $B_\varepsilon(\mathbf{v})$, consisting of vectors $\mathbf{w}$, $|\mathbf{w} - \mathbf{v}| < \varepsilon$ is the closed ball $\overline{B}_\varepsilon(\mathbf{v})$, consisting of vectors $\mathbf{w}$, $|\mathbf{w} - \mathbf{v}| \leq \varepsilon$.
2. A subset A of X is *dense in* X if the closure of A is X: $\overline{A} = X$.

[22] Strictly speaking, a real number must be defined in terms of equivalence classes of Cauchy sequences, as many distinct Cauchy series can converge to the same real number.

3. A topological space X is *separable* if there exists a countable set $A = (x_1, x_2, x_3, \ldots)$ dense in X. Using an example discussed previously, the closure of the rational numbers (now viewed as a subset of the real numbers) leads to the real numbers, so the reals are topologically separable, as the rational numbers form a denumerable set.
4. A function f mapping from one topological space X to another Y (which may be the same as X) is *continuous at the point* $x \in X$ if, for every neighborhood V of $y = f(x)$, there exists a neighborhood U of x such that $f(U) \subset V$. A continuous function (with no point specified) is simply one that is continuous at every point in X.[23] In terms of the familiar open-ball definitions in $\mathbb{R}^n$ (say), this amounts to the usual "epsilon–delta" type of definition: given a point $x \in \mathbb{R}^n$, with $y = f(x) \in \mathbb{R}^n$, then for any ε, however small, we can find δ such that if $d(z, x) < \delta$, then $|f(z) - f(x)| = |f(z) - y| < \varepsilon$—in other words, so that $f(B_\delta(x)) \subset B_\varepsilon(y)$. Intuitively, a continuous function maps small changes into small changes, and convergent sequences into convergent sequences:

$$\text{f continuous} \Rightarrow \lim_{i\to\infty} f(x_i) = f(x) \;\text{ if }\; \lim_{i\to\infty} x_i = x. \tag{C.93}$$

5. A subset B of a topological space is *compact* if every non-constant infinite sequence in B has a convergent subsequence, with a limit point in B.[24] If the subsequence converges in X to a point not in B, we say that B is pre-compact, or relatively compact. In this case, B has compact closure. This concept formalizes in an abstract way a notion of "smallness" for subsets of a topological space. As an example, consider a finite closed interval on the real line $[a, b] : (x, a \leq x \leq b)$. It can be shown (special case of the Bolzano–Weierstrass theorem) that any infinite sequence contained in this interval will contain a subsequence that "bunches up" at some point $x \in [a, b]$. The important feature needed here is that $b - a$ is finite: evidently there are non-constant sequences with no convergent subsequence if we take $b = \infty$, for example. Once the interval is finite, the bunching is easily established: the sequence must have an infinite number of elements in either the left half $a \leq x \leq (a + b)/2$ or the right half $(a + b)/2 \leq x \leq b$ of the original interval. Continuing in this fashion, we establish the existence of infinitely many members of the sequence in successively smaller quarters, one-eighths, etc., of the original interval, eventually zooming in on a point contained in intervals of geometrically vanishing size, all of which have infinitely many members of the sequence. If we consider an open interval (a, b) or half-open interval $(a, b]$, then there are convergent subsequences with limit points corresponding to the missing boundary point, and the interval is pre-compact (with compact closure, the closed interval $[a, b]$). This all generalizes to

Theorem C.5.1. *Any closed, bounded subset of $\mathbb{R}^n$ or $\mathbb{C}^n$ (with the usual open ball metric topology induced by the norm) is compact.*

This simple and intuitive characterization of a compact set will be one of the numerous casualties of a transition from finite-dimensional to infinite-dimensional spaces.

[23] The reader may be familiar with an alternative, and equivalent, global definition that a function is continuous if the preimage $f^{-1}(V)$ of any open set in Y is an open set in X.

[24] To be precise, the definition of compactness given here is usually called "sequential compactness"—it coincides with the more general definition in terms of open coverings for all metric spaces with the topology induced by the metric—in particular, for the Hilbert spaces of concern in this appendix.

C.5.2 The first Hilbert space: l^2

Hilbert spaces were part of the formal equipment of mathematicians for almost two decades before the name "Hilbert space" came into wide usage. The underlying geometrical structure of linear vector spaces was somewhat concealed by the fact that results were framed in terms of quadratic or bilinear forms, a natural consequence of the origins of functional analysis in the study of integral equations, as we saw above.

An early formulation where we can see the germs of the modern viewpoint (eventually formalized abstractly by von Neumann in the late 1920s) is in Erhard Schmidt's 1908 paper on the solution of linear equations with infinitely many variables, where a space consisting of real functions $A(x)$, with x integer (and thus, in modern notation, infinite vectors with components (A_n)) is subjected to the constraint $\sum_n A_n^2 < \infty$, and a (real) inner product defined by $(A; B) = \sum_{x=1}^{x=\infty} A(x)B(x)$ $(= \sum_n A_n B_n)$. In a footnote, Schmidt acknowledges:

> I owe the geometrical meaning of the concepts and theorems developed in this chapter to Kowaljewski. It becomes even clearer if $A(x)$ is defined not as a function but as a vector in a space of infinitely many dimensions (Schmidt 1908, p. 56).

Schmidt's space of square-summable infinite vectors (A_n), extended to complex numbers and with an inner product $(A, B) = \sum_n A_n^\star B_n$, is now referred to as l^2. It is the basic underlying space in terms of which all quantum-mechanical systems can be represented. The space l^2 is also defined and studied in a roughly contemporaneous paper of Maurice Fréchet (1908) entitled "Essay on the analytic geometry of an infinity of coordinates" (*Essai de géométrie analytique a une infinité de coordonnées*).

In 1912, as noted in Section A.4, Hilbert published a volume containing a reprint of his six seminal papers on linear integral equations from the period 1904–1910, with an extensive introduction summarizing the essential elements of his theory. The first section of the introduction covers the algebraic theory of infinite forms, leading up to the spectral theorems for compact, and then, more generally, bounded quadratic forms (definitions to come below). In the second section, this theory is applied to linear integral equations of Fredholm type. Schmidt's "functions" have become "value-systems" (*Wertsysteme*) $x_1, x_2, \ldots$ for which $\sum_n x_n^2 \leq 1$, in other words the "unit ball" in Schmidt's infinite-dimensional space. Hilbert's treatment emphasizes topological aspects of this space, as one might expect from the restriction to vectors (or "value systems") in the unit ball.

The first concept Hilbert introduces in his 1912 compendium is that of boundedness (*Beschränktheit*). We now look at how Hilbert introduces this extremely important concept in the very first section of his 1912 summary, where the language of infinite forms is used. We then translate the concepts into vector-space language, where the relevant elements are vectors, and linear operators acting on these vectors. Hilbert begins:

> A function of infinitely many variables $F(x_1, x_2, x_3, \ldots)$ is said to be bounded, if the absolute value of the n-th restriction of this function $F(x_1, x_2, \ldots, x_n, 0, 0, \ldots)$, for all $x_1, x_2, \ldots$ for which $\sum_p x_p^2 \leq 1$, lies underneath a fixed bound M independent of n. In particular, a *linear form* $a_1 x_1 + a_2 x_2 + \ldots$ is bounded if and only if $a_1^2 + a_2^2 + \ldots$ converges. A *bilinear form* $\sum_{p,q} a_{pq} x_p y_q$ [is said to be bounded if and only if] $\sum_{p,q=1}^{n} a_{pq} x_p y_q$ lies under a bound M independent of n.

A *linear transformation* $x_p = \sum_q a_{pq}y_q$ is said to be bounded, if the associated bilinear form $\sum_{p,q} a_{pq}x_p y_q$ is bounded. Such a transformation takes every value-system [i.e., infinite vector (y_q)] with convergent square sum [i.e., $\sum_q y_q^2 < \infty$] into another such [i.e., $\sum_p x_p^2 < \infty$] (Hilbert 1912, p. IV).

We now translate all of this into modern language, emphasizing the underlying vector-space structure that would finally be brought to the fore by von Neumann in his construction of the mathematical foundations of quantum mechanics. We simultaneously make the fairly trivial extension to complex vector spaces needed in this theory.

Let $\mathbf{v} = (v_1, v_2, \ldots)$ be a vector in the unit ball of the infinite-dimensional complex vector space l^2 (consisting of all vectors (v_p) with complex components and $(\mathbf{v}, \mathbf{v}) = \sum_p |v_p|^2 \leq 1$). A (complex-valued) function F on l^2 is said to be bounded if the restriction of $|F|$ to the n-dimensional subset of the unit ball with $v_p = 0$ $(p > n)$ is bounded by some positive number M, independent of n. Of particular interest are linear functions (in modern language, linear functionals) of the form $F(\mathbf{v}) = (\mathbf{a}, \mathbf{v}) \equiv \sum_p a_p^\star v_p$. This is a bounded function in the sense of the preceding definition in virtue of the *Cauchy–Schwarz inequality*:

Theorem C.5.2. *For any* $\mathbf{v}, \mathbf{w}$ *in* l^2,

$$|(\mathbf{v}, \mathbf{w})|^2 \leq (\mathbf{v}, \mathbf{v})(\mathbf{w}, \mathbf{w}). \tag{C.94}$$

Proof. Let λ be a real variable. Using the positive definite property of the inner product, we must have, for all λ

$$\begin{aligned} 0 &\leq (\mathbf{v} - \lambda(\mathbf{w}, \mathbf{v})\mathbf{w},\ \mathbf{v} - \lambda(\mathbf{w}, \mathbf{v})\mathbf{w}) \\ &= |(\mathbf{w}, \mathbf{v})|^2(\mathbf{w}, \mathbf{w})\lambda^2 - 2|(\mathbf{w}, \mathbf{v})|^2\lambda + (\mathbf{v}, \mathbf{v}). \end{aligned} \tag{C.95}$$

This quadratic function of λ can never go negative, which means that the discriminant cannot be positive:

$$4|(\mathbf{w}, \mathbf{v})|^4 - 4(\mathbf{v}, \mathbf{v})(\mathbf{w}, \mathbf{w})|(\mathbf{w}, \mathbf{v})|^2 \leq 0. \tag{C.96}$$

Dividing by $4|(\mathbf{w}, \mathbf{v})|^2$, we see that this implies Eq. (C.94). □

Applying this theorem to the linear functional $F(\mathbf{v}) = (\mathbf{a}, \mathbf{v})$ we see that F is indeed bounded as long as $\mathbf{a}$ has finite l^2 norm, since if $(\mathbf{a}, \mathbf{a}) = \sum_p |a_p|^2 \equiv M^2 < \infty$ and $\mathbf{v}$ is in the unit ball, then $|F(\mathbf{v})| \leq M$. We leave it as an exercise to the reader that the Cauchy–Schwarz inequality can also be used to establish that l^2 is indeed a linear complex vector space, as we have been assuming, namely, that if $\mathbf{v}, \mathbf{w} \in l^2$ (i.e., $\sum_p |v_p|^2 < \infty, \sum_p |w_p|^2 < \infty$) then $\alpha\mathbf{v} + \beta\mathbf{w} \in l^2$, where α and β are complex numbers. Finally, Schmidt (1908) showed that every Cauchy sequence $\mathbf{v}_n$ in l^2 converges to a vector $\mathbf{v}$ in l^2:

$$\lim_{m,n\to\infty} |\mathbf{v}_n - \mathbf{v}_m| \to 0 \ \Rightarrow\ \exists \mathbf{v} \in l^2,\ \lim_{n\to\infty} \mathbf{v}_n = \mathbf{v}. \tag{C.97}$$

In other words, l^2 is *complete* (cf. discussion in Section A.5.1 on Cauchy completeness of the real numbers). It is the archetypal example of a *Hilbert space* (a linear vector space equipped with an inner product, and complete with respect to the norm induced by that inner product), although that term did not come into wide usage until the 1920s.[25]

Returning to Hilbert's summary of his theory in 1912, we recall that a bilinear form $\sum_{p,q} a_{pq}x_p y_q$ is said to be *bounded* if it takes on absolute values less than some fixed finite bound M when the vectors $\mathbf{x}$ and $\mathbf{y}$ (with components given by the sequences x_p and y_q) lie in the unit ball (i.e., $|\mathbf{x}| \leq 1$ and $|\mathbf{y}| \leq 1$). In modern vector notation, we require

$$|(\mathbf{x}, A\mathbf{y})| \leq M, \quad |\mathbf{x}|, |\mathbf{y}| \leq 1, \tag{C.98}$$

where A is the linear operator with matrix elements a_{pq}. It is clear that the inner product on the left is made largest if the vectors $\mathbf{x}, \mathbf{y}$ are chosen to be unit vectors on the boundary of the unit ball, so Eq. (C.98) is tantamount to

$$\frac{|(\mathbf{x}, A\mathbf{y})|}{|\mathbf{x}| \cdot |\mathbf{y}|} \leq M \tag{C.99}$$

for general non-zero (and not necessarily normalized) vectors $\mathbf{x}$ and $\mathbf{y}$.

Linear operators on l^2 can be provided with a norm (or *metric*) by considering the amount to which an operator "stretches" (i.e., changes the magnitude of) each vector in the space. We shall use the double bar notation, $||A||$, for the norm of a linear operator A to distinguish it from the norm $|\mathbf{v}|$ defined for vectors. The norm $||A||$ of a linear operator A is defined as the maximum stretching induced by the action of A, namely, the positive real number

$$||A|| \equiv \sup_{\mathbf{v} \neq 0} \frac{\sqrt{(A\mathbf{v}, A\mathbf{v})}}{\sqrt{(\mathbf{v}, \mathbf{v})}} = \sup_{\mathbf{v} \neq 0} \frac{|A\mathbf{v}|}{|\mathbf{v}|}. \tag{C.100}$$

It is easy to verify that the usual axioms for a norm are satisfied with this definition ($||A|| \geq 0, = 0 \iff A = 0, ||\alpha A|| = |\alpha| ||A||$ for scalar α, subadditivity of the norm $||A + B|| \leq ||A|| + ||B||$). From the norm a metric can be induced in the usual way and a topology defined on the space of linear operators, allowing us to define concepts such as the (uniform) convergence of a sequence of operators. For the class of operators of finite norm, we define

Definition C.5.1. *If the supremum indicated in the definition Eq. (C.100) of $||A||$ exists (i.e., is finite), we say that the linear operator A is bounded.*

[25] The space of square-summable sequences is referred to explicitly as "Hilbert space" by Hellinger and Toeplitz (1927, p. 1434), where it is denoted R_∞, as well as in von Neumann (1927a), discussed in Chapter 17.

The close connection of this definition to Hilbert's bounded bilinear forms can be seen from the following lemma

Lemma C.5.3.

$$||A|| = \sup_{\mathbf{v},\mathbf{w}\neq 0} \frac{|(\mathbf{v}, A\mathbf{w})|}{|\mathbf{v}|\,|\mathbf{w}|}. \tag{C.101}$$

Proof. First, by using the Cauchy–Schwarz inequality, we find

$$\begin{aligned} |(\mathbf{v}, A\mathbf{w})| &\leq \sqrt{(\mathbf{v},\mathbf{v})}\sqrt{(A\mathbf{w}, A\mathbf{w})} \\ &\leq \sqrt{(\mathbf{v},\mathbf{v})}\sqrt{(\mathbf{w},\mathbf{w})}\,||A|| = |\mathbf{v}|\,|\mathbf{w}|\,||A||. \end{aligned} \tag{C.102}$$

Next, we show that $|(\mathbf{v}, A\mathbf{w})|$ can be made as close to $\sqrt{(\mathbf{v},\mathbf{v})}\sqrt{(\mathbf{w},\mathbf{w})}||A||$ as we wish. From the definition Eq. (C.100) it follows that we can choose the vector $\mathbf{w}$ such that

$$\frac{\sqrt{(A\mathbf{w}, A\mathbf{w})}}{\sqrt{(\mathbf{w},\mathbf{w})}} \geq ||A|| - \varepsilon, \tag{C.103}$$

for arbitrarily small ε. Now choose $\mathbf{v} = A\mathbf{w}$, so that

$$\begin{aligned} (\mathbf{v}, A\mathbf{w}) &= (A\mathbf{w}, A\mathbf{w}) = (\mathbf{v},\mathbf{v}) \\ &= |A\mathbf{w}|^2 = |\mathbf{v}|^2 = |A\mathbf{w}|\,|\mathbf{v}| \\ &= \sqrt{(A\mathbf{w}, A\mathbf{w})}\sqrt{(\mathbf{v},\mathbf{v})} \\ &\geq (||A|| - \varepsilon)\sqrt{(\mathbf{w},\mathbf{w})}\sqrt{(\mathbf{v},\mathbf{v})}. \end{aligned} \tag{C.104}$$

As ε can be chosen arbitrarily small, the supremum definition in Eq. (C.101)) is shown to be equivalent to the original definition Eq. (C.100). □

The concept of a bounded operator plays an essential role for two main reasons. As indicated in the final sentence of the quote from Hilbert above, a bounded linear transformation maps any vector in l^2 into a vector in l^2—in other words, it is defined on the whole Hilbert space. Indeed, if $\mathbf{v} = A\mathbf{w}$ with $|\mathbf{w}| < \infty$, then $|\mathbf{v}| = |A\mathbf{w}| \leq ||A||\,|\mathbf{w}| < \infty$ (as $||A||$ is finite).[26] Secondly, bounded operators correspond precisely, when viewed as functions from the vector space to itself, to the *continuous* functions, which map convergent sequences to convergent sequences. Let $\mathbf{v}_n \to \mathbf{v}$ be such a convergent sequence of vectors. Thus, $|\mathbf{v} - \mathbf{v}_n| \to 0$ for $n \to \infty$. Then

$$\begin{aligned} |A(\mathbf{v} - \mathbf{v}_n)| &\leq ||A||\,|\mathbf{v} - \mathbf{v}_n| \to 0 \quad \text{for } n \to \infty, \\ A\mathbf{v}_n &\to A\mathbf{v} \quad \text{for } n \to \infty, \end{aligned} \tag{C.105}$$

and we have shown that A maps convergent sequences to convergent sequences. Thus, viewed as a function on the vector space (with the norm topology), it is continuous.

[26] We will see later that unbounded operators, which abound in quantum mechanics, are only well defined on a subspace of the full Hilbert space (called the *domain* of the operator).

In his fourth paper in the series on the foundations of integral-equation theory, Hilbert introduced an additional concept, that of a "completely continuous" quadratic form $F(\mathbf{x}) = (\mathbf{x}, K\mathbf{x})$, where K is a linear operator acting on the l^2 vector $\mathbf{x}$. In operator language, Hilbert's definition[27] is equivalent to the more concise and easily digested requirement that the operator K is completely continuous if it maps any sequence of vectors contained in a bounded subset of l^2 (i.e., the vectors in the sequence have bounded norm) into a set with a convergent subsequence. Even more concisely, K is completely continuous (in modern language, *compact*) if it maps any bounded subspace of l^2 into a set with compact closure.[28] It is easy to see that completely continuous/compact operators are automatically bounded, so we have the natural hierarchy *completely continuous operators* $\subset$ *bounded (continuous) operators* $\subset$ *general (including unbounded) linear operators on* l^2. Completely continuous operators/forms are in a sense the most precise infinite-dimensional analogs of finite-dimensional operators/forms, as the following theorem proved by Hilbert in his fourth article demonstrates.

Theorem C.5.4. *If a bounded form K is also completely continuous, then it may always be brought into the following form by an orthogonal transformation*

$$K(\mathbf{x}) = k_1 x_1^2 + k_2 x_2^2 + \ldots, \tag{C.106}$$

where the quantities $k_1, k_2, \ldots$ are the eigenvalues of K, and, in the event there are infinitely many of them,[29] *possess as their only accumulation point the value zero* (Hilbert 1906a, p. 201,Theorem V).

In other words, the matrix of K—real and symmetric, as Hilbert is discussing the real version of l^2 at this point[30]—can be brought into diagonal form by an infinite orthogonal transformation (i.e., the infinite-dimensional version of Eq. (C.56)), with the resulting discrete eigenvalues having the stated properties. In this sense, completely continuous operators provide the natural extension of Theorem C.3.2 to the infinite-dimensional case: they possess a purely discrete (point) spectrum, and their eigenvectors span the entire space. In the same paper, Hilbert provided some *sufficient* conditions for a form K to be completely continuous. For example, it suffices for the sum of the squared matrix elements $\sum_{m,n} K_{mn}^2$ to be finite.

Some explicit examples may be useful here.

1. An operator K whose matrix (in some complete orthonormal basis of l^2) can be brought into the diagonal form $K = \mathrm{diag}(1, \frac{1}{2}, \frac{1}{3}, \frac{1}{4}, \ldots)$ by orthogonal transformation is both bounded (with $||K|| = 1$), and completely continuous. There are an infinite number of non-zero eigenvalues, and they accumulate at zero.
2. An operator K whose matrix can be brought into the diagonal form $K = \mathrm{diag}(1, \frac{1}{\sqrt{2}}, \frac{1}{\sqrt{3}}, \frac{1}{\sqrt{4}}, \ldots)$ by orthogonal transformation is bounded (with $||K|| = 1$), and still

[27] Hilbert defines the form $K(\mathbf{x})$ to be completely continuous if for every weakly convergent sequence of vectors $\mathbf{v}^{(n)}$, with $\mathrm{w} - \lim_{n\to\infty} \mathbf{v}^{(n)} = \mathbf{v}$, $K(\mathbf{v}^{(n)}) \to F(\mathbf{v})$. A sequence of vectors is weakly convergent if the vectors converge to the limit vector component by component.

[28] The closure of a set is simply the set together with all the accumulation (limit) points of sequences in the set.

[29] This proviso takes into account the possibility that the matrix of K is finite-dimensional, that is, can be orthogonally transformed to a matrix with a finite non-zero sub-block, with finitely many non-zero eigenvalues, in which case the eigenvalue 0 appears infinitely many times.

[30] Later in this article (Ch. 12), Hilbert defines complex Hermitian forms and extends his results to this case.

completely continuous, even though it does not satisfy the Hilbert criterion of square-summability ($\sum_n (1/\sqrt{n})^2 = \infty$). This shows that Hilbert's criterion is sufficient for complete continuity, but not necessary.

3. An operator K whose matrix can be brought into the diagonal form $K = \text{diag}(\frac{1}{2}, \frac{2}{3}, \frac{3}{4}, \frac{4}{5}, \ldots)$ by orthogonal transformation is bounded (with $||K|| = 1$), but not completely continuous. Consider the infinite sequence of unit vectors $\mathbf{e}^{(n)}$ corresponding to the basis in which K is diagonal (thus, $\mathbf{e}^{(1)} = (1, 0, 0, \ldots)$, $\mathbf{e}^{(2)} = (0, 1, 0, \ldots)$, etc.). The sequence $K\mathbf{e}^{(n)}$ eventually approaches $\mathbf{e}^{(n)}$, that is, a sequence of mutually orthogonal unit vectors, which clearly do not possess a convergent subsequence.
4. Finally, the matrix H of the Hamiltonian of the simple harmonic oscillator can be brought into the diagonal form $K = \hbar\omega\, \text{diag}(\frac{1}{2}, \frac{3}{2}, \frac{5}{2}, \ldots)$ by orthogonal transformation: it is clearly unbounded ($||H|| = \infty$). The domain of such an operator—the set of vectors to which it can sensibly be applied—is a proper subset of l^2: for example, acting on the perfectly sensible—that is, normalizable—element $\mathbf{v} = (1, \frac{1}{3}, \frac{1}{5}, \frac{1}{7}, \ldots)$ of l^2, it produces the infinite norm vector $\hbar\omega(\frac{1}{2}, \frac{1}{2}, \frac{1}{2}, \ldots)$.

The beginnings of the spectral theory of operators in infinite-dimensional spaces were forged by Hilbert in his fourth communication, in which he was able to expand the theory developed for the very restricted class of completely continuous operators to the much larger set (containing the former) of bounded operators. Of course, the discussion is given throughout in the language of real quadratic forms $K(\mathbf{x}) = (\mathbf{x}, K\mathbf{x}) = \sum_{pq} K_{pq} x_p x_q$, with K_{pq} a real symmetric matrix, and the components x_p of the vector $\mathbf{x}$ chosen so the latter has norm less than or equal to one, $\sum_p x_p^2 \leq 1$. The form is now assumed only to be bounded: thus, for some fixed finite M and all $\mathbf{x}$, $K(\mathbf{x}) \leq M$.

We rephrase Hilbert's fundamental spectral theorem in modern language as a statement about matrix elements of the linear operator K relative to some preassigned orthonormal basis of l^2. To avoid confusion, we also define the eigenvalues λ_p of K in the now usual way, as a (real) number for which there exists a non-zero vector $\mathbf{v}^{(p)} \in l^2$ such that $K\mathbf{v}^{(p)} = \lambda_p \mathbf{v}^{(p)}$ (whereas in the early work of Hilbert and collaborators, the inverse eigenvalue appears on the right-hand side, as we saw earlier). Hilbert discovered that for a general bounded form/matrix the discrete eigenvalues (the "point spectrum", which is the whole story for completely continuous forms) do not in general exhaust the spectrum, which can also contain continuous sections, which Hilbert dubbed the "line spectrum" (*Streckenspektrum*), but which is nowadays more typically called the "continuous spectrum". In other words, the eigenvectors $\mathbf{v}^{(p)}$ associated with the discrete eigenvalues, although they may be chosen to be orthonormal, do not span the full space l^2. We also change the notation slightly to make Hilbert's result appear as the natural extension of our previous finite-dimensional discussion.

Theorem C.5.5. *Let $K(\mathbf{x}) = (\mathbf{x}, K\mathbf{x}) = \sum_{pq} K_{pq} x_p x_q$ be a bounded quadratic form ... Let $\mathbf{v}^{(n)}$ be the eigenvector of K associated with the eigenvalue λ_n [in the point spectrum of K]. The spectral matrix $\sigma_{pq}(\mu)$ [which can be shown to exist as a quadratic form in the x_p: the explicit construction of this form in terms of the continuous eigenfunctions of K would be provided a few years later by Hellinger] is bounded and positive definite*[31] *and its matrix elements are continuous functions of μ.*

[31] The operator A is positive definite if for all $\mathbf{x} \neq 0$, $(\mathbf{x}, A\mathbf{x}) > 0$.

For $\mu > \mu'$ in the continuous spectrum $\sigma(\mu) - \sigma(\mu')$ is positive definite, while for values of μ outside the continuous spectrum, $\sigma(\mu)$ is constant. Then the following equations hold:

$$\delta_{pq} = \sum_n v_p^{(n)} v_q^{(n)} + \int d\sigma_{pq}(\mu), \tag{C.107}$$

$$K_{pq} = \sum_n \lambda_n v_p^{(n)} v_q^{(n)} + \int \mu \, d\sigma_{pq}(\mu). \tag{C.108}$$

(Hilbert 1906a, pp. 189–190, Theorem II).

These two formulas represent the extension to infinite-dimensional bounded symmetric (hence, self-adjoint) operators of the finite-dimensional results Eq. (C.86) (resolution of the identity) and Eq. (C.87) (spectral resolution of K). The finite-dimensional formulas involving purely discrete sums must in general be supplemented by an integral part taking into account the continuous spectrum, if present. Of course, it is also perfectly possible for a bounded operator in l^2 to have a purely discrete spectrum (cf. examples 1, 2, and 3 above), or, indeed, to have no point spectrum and only a continuous spectrum (e.g., the matrix given in Eq. (12.253), with the first two rows and columns excluded). The spectral resolution of K_{pq} in Eq. (C.108) can be trivially extended to functions of this matrix, in particular the resolvent matrix $(1 - \lambda K)^{-1}$ needed to solve the original Fredholm equation (cf. Eq. (C.78)):

$$\left(\frac{1}{1-\lambda K}\right)_{pq} = \sum_n \frac{1}{1-\lambda\lambda_n} v_p^{(n)} v_q^{(n)} + \int \frac{1}{1-\lambda\mu} \, d\sigma_{pq}(\mu). \tag{C.109}$$

The spectral form/matrix is not explicitly defined by Hilbert in terms of the "eigenvectors" (which are not normalizable, hence not really in the Hilbert space l^2). A few years later, Hellinger (1910) would provide a complete specification of $\sigma(\mu)$. However, Hilbert's results in 1906 contain all of the essential properties needed to characterize completely the spectral measure, in particular,[32]

Theorem C.5.6. *With K as in Theorem C.5.5, let $(\mathbf{v}^{(n)}, \mathbf{w}^{(n)})$ together represent a complete orthonormal basis of l^2, with the first set $(\mathbf{v}^{(n)})$ the eigenvectors associated with the eigenvalues of the point spectrum of K. The spectral matrix $\sigma_{pq}(\mu)$ satisfies*

$$\int u(\mu) \, d\sigma_{pr}(\mu) \int u(\mu') \, d\sigma_{rq}(\mu') = \int u(\mu)^2 d\sigma_{pq}(\mu), \tag{C.110}$$

$$\sum_n w_p^{(n)} w_q^{(n)} = \int d\sigma_{pq}(\mu), \tag{C.111}$$

where $u(\mu)$ is an arbitrary real function of the real variable μ (Hilbert 1906a, p. 198, Theorem III).

[32] Again, we are translating Hilbert's terminology based on quadratic forms to more transparent linear algebraic terms.

The proof of these results (involving Hilbert's careful use of limiting procedures) is non-trivial but, in hindsight, unnecessary. They all follow almost trivially once we have the spectral theorem of von Neumann, especially when expressed in the extremely intuitive notation invented by Dirac. The extension of Hilbert's results by Hellinger, leading to an explicit recipe for construction of the spectral measure for the continuous spectrum (if present), was used in the Three-Man-Paper of Born, Heisenberg, and Jordan (1926) to motivate the appearance of continuous spectra in matrix mechanics (cf. Section 12.3.4). We describe these results again below. However, they follow almost trivially once we have the much more transparent notation of Dirac, and the full machinery of the von Neumann spectral theory for self-adjoint operators at our disposal.

C.5.3 Function spaces: L^2

By the first decade of the nineteenth century, as a result of the work of Fourier (based on prior work by Euler and Bernoulli), it had been established that a continuous function $f(t)$ of a single variable that repeats itself with a fixed period T (i.e., $f(t+T) = f(t)$ for all t) can be written as a convergent infinite sum of trigonometric functions:

$$\begin{aligned} f(t) &= a_0 + \sum_{m=1}^{\infty}\left(a_m \cos\left(\frac{2\pi mt}{T}\right) + b_m \sin\left(\frac{2\pi mt}{T}\right)\right) \\ &= \sum_{m=-\infty}^{+\infty} c_m e^{2i\pi mt/T}. \end{aligned} \tag{C.112}$$

That the trigonometric functions appearing in such Fourier series are orthogonal in the integral sense—and here we simplify the algebra by rescaling the period T to 2π and using angular notation—had already been noticed by Euler

$$\begin{aligned} &\int_{-\pi}^{+\pi} \cos(m\vartheta)\cos(n\vartheta)\, d\vartheta = 0, \qquad m \neq n, \\ &\int_{-\pi}^{+\pi} \sin(m\vartheta)\sin(n\vartheta)\, d\vartheta = 0, \qquad m \neq n, \\ &\int_{-\pi}^{+\pi} \sin(m\vartheta)\cos(n\vartheta)\, d\vartheta = 0. \end{aligned} \tag{C.113}$$

By the beginning of the twentieth century, the analogy of these relations to the inner-product structure that naturally appeared in the vector-space theory was clear, and had been extended by the discovery of a plethora of other sets of orthogonal function (Hermite, Legendre, Laguerre, Gegenbauer, etc., polynomials) that could be used to expand continuous functions defined on a finite interval.

The exact nature of the convergence of the trigonometric series appearing in Eq. (C.112) (or other orthogonal polynomial expansions), however, would require an important advance in the theory of integration. For our purposes, in this short sketch of the early history of functional analysis, the essential point is that the type of convergence needed in the construction of the infinite-dimensional function spaces of importance in quantum mechanics was only properly understood at the beginning of the twentieth century, with the introduction by Lebesgue (in his doctoral thesis of 1902) of a new kind of integration, of far greater generality than the Riemann integral, to which students are typically introduced in their first calculus course. Roughly speaking, the Riemann integral calculates the area under the curve representing a function by a vertical slicing operation (integral = area = sum of products of the height of thin vertical slices under the curve times the width of these slices), while the Lebesgue integral calculates the area by a horizontal slicing procedure (Lebesgue integral = sum of products of the total length, or "measure", of horizontal slices where the function has a value in a given interval, times the corresponding value of the function). The precise definition of the Lebesgue measure would take us too far afield, but fortunately a short but excellent exposition of the basic elements of the Lebesgue theory is given by Dennery and Krzywicki (1967, pp. 184–189).

The necessity of introducing a new type of integral is connected with the important property of *completeness* (which we introduced above in connection with the transition from the rational to the real numbers, and also in the definition of the Hilbert space l^2). The reader will recall that we defined a normed vector space to be Cauchy-complete if any Cauchy convergent sequence of vectors $\mathbf{v}^{(n)}$, that is, a sequence satisfying[33]

$$\lim_{m,n\to\infty} |\mathbf{v}^{(m)} - \mathbf{v}^{(n)}| = 0, \tag{C.114}$$

converges to a vector in the space, that is, if there exists a vector $\mathbf{v}$ such that

$$\lim_{n\to\infty} \mathbf{v}^{(n)} = \mathbf{v}. \tag{C.115}$$

Now the space of complex continuous functions $f(x)$ on some closed interval $[a, b]$ (usually denoted $C[a, b]$) is certainly a linear vector space, as linear combinations of such functions are continuous, and it is obviously infinite-dimensional. In the spirit of Euler's orthogonality relations, we can define an inner product of two functions in this space as follows[34]

$$(\mathbf{f}, \mathbf{g}) \equiv \int_a^b f^\star(s)\, g(s)\, ds. \tag{C.116}$$

Notice that we have employed our usual bold notation to indicate that the function is now being treated as an element in a vector space. The reader may easily verify that the inner product defined this way satisfies the basic properties enumerated in Eq. (C.25). With an inner product in hand, we have instantly a natural norm (or magnitude) defined on the function space $C[a, b]$:

$$|\mathbf{f}| \equiv \sqrt{(\mathbf{f}, \mathbf{f})} = \sqrt{\int_a^b |f(s)|^2 ds}. \tag{C.117}$$

Supplied with a complete set of orthogonal functions (now viewed as vectors) in terms of which a general continuous function can be expressed as a linear combination—in other words, with a *complete orthogonal basis*—one might think that we have all that we need for the construction of a full

[33] The precise meaning of Eq. (C.114) is as follows: for any preassigned positive ε, however small, we can find an integer M such that for all $m, n > M$, $|\mathbf{v}^{(m)} - \mathbf{v}^{(n)}| < \varepsilon$. In other words, the members of the sequence are getting arbitrarily close to each other as we follow the sequence.

[34] It is easy to see that all of the desired properties of the inner product, and norm, that follow from this definition are preserved if a real, positive weight function $w(x)$ is included under the integral sign.

theory of linear operators, satisfying the various topological properties discussed in Section C.4 (continous/bounded, completely continous/compact, etc.). The catch is that our space $C[a, b]$, while it is a perfectly fine (infinitely-dimensional) complex inner-product space, is not yet a *Hilbert* space, as it lacks one remaining, but critical, property: Cauchy-completeness. It is easy to construct sequences of continuous functions that are Cauchy convergent but do not converge to a continuous function. Consider, for example, the Fourier expansion for the step function $S(\vartheta)$ defined by the Fourier series

$$S(\vartheta) = \frac{4}{\pi} \sum_{k=0}^{\infty} \frac{1}{2k+1} \sin((2k+1)\vartheta). \tag{C.118}$$

This function is definitely discontinuous, as $S(\vartheta) = -1$ for $-\pi < \vartheta < 0$ and $S(\vartheta) = 1$ for $0 < \vartheta < \pi$. On the other hand, the series of functions corresponding to the partial sums of the Fourier series,

$$S^{(m)}(\vartheta) \equiv \sum_{k=0}^{m} \frac{1}{2k+1} \sin((2k+1)\vartheta), \tag{C.119}$$

is a Cauchy series, as one can show, using the norm defined as above on the interval $(-\pi, +\pi)$, that for $m < n$,

$$|\mathbf{S}^{(n)} - \mathbf{S}^{(m)}| = \frac{16}{\pi} \sum_{k=m+1}^{n} \frac{1}{(2k+1)^2}, \tag{C.120}$$

which is of order $O(1/m)$ as $m \to \infty$. So we have a Cauchy sequence in the space of continuous functions converging to a non-continuous function. Although in this case the resultant non-continuous function is a very simple one (the step function), which is still perfectly integrable by normal (i.e., Riemann–Stieltjes) methods, in general one can show that Cauchy sequences of continuous functions can converge to extremely discontinuous functions for which the Riemann–Stieltjes integral does not exist. The integration theory developed by Lebesgue is nevertheless capable of handling such functions, for example, functions that are continuous except at a countable number of points, or even functions such as the Dirichlet function (defined as taking the value 1 on all rational numbers, 0 otherwise), which is nowhere continuous. We henceforth assume that wherever the integral sign $\int$ appears, it refers to a Lebesgue integral. For further details and a systematic development of the theory, the reader is referred to Dennery and Krzywicki (1967) or any standard modern text on real analysis.

The fundamental result in the theory of function spaces which laid the foundation for much of the subsequent development of functional analysis is the *Riesz–Fischer theorem* of 1907, which reflects several results found independently and published in the same journal (*Comptes Rendus de l'Académie des Sciences*) within a few months. Fischer's (1907) theorem establishes the basic result that allows us to view a function space as a proper (i.e., Cauchy-complete) Hilbert space. Once again, we consider functions defined on an interval $[a, b]$, but now the much larger space of *Lebesgue square-integrable* functions, which we now denote $L^2(a, b)$ (which contains $C[a, b]$, but is much larger!). Defining a norm on this space with Eq. (C.117) (but this time, with Lebesgue integration), a Cauchy sequence of (complex) functions $f_n(x)$ is one satisfying

$$\lim_{m,n\to\infty} \int_a^b |f_n(s) - f_m(s)|^2 dx = \lim_{m,n\to\infty} |\mathbf{f}_n - \mathbf{f}_m|^2 \to 0. \tag{C.121}$$

The convergence relative to the norm defined as a Lebesgue integral of squared functions is frequently referred to (following Fischer) as "convergence in the mean". Fischer (1907) then establishes the following critical result.

Theorem C.5.7 (Fischer). *If a set $f_n(x)$ of Lebesgue square-integrable functions form a Cauchy sequence, in the sense of Eq. (C.121), then there exists a Lebesgue square-integrable function $f(s)$ to which the sequence converges, that is, such that*

$$\lim_{n\to\infty} \int_a^b |f(s) - f_n(s)|^2 \, ds = 0 \tag{C.122}$$

The space $L^2(a, b)$ of Lebesgue square-integrable functions on the interval $[a, b]$, equipped with the norm Eq. (C.117), is complete. Introducing the inner product via Eq. (C.116), we have a complete inner-product space, in other words, a Hilbert space.

The structural identity of the function space $L^2[a, b]$ with Hilbert's sequence space l^2 is guaranteed by two results, one of which, Parseval's identity, goes back to the early theory of Fourier series, while the other, due to Riesz (1907a, 1907b), establishes a direct mapping between the two spaces. Beginning with the latter, we have

Theorem C.5.8 (Riesz). *Let $\{\varphi_m(s)\}$ be an orthonormal (relative to the inner product in Eq. (C.116)) sequence of Lebesgue square-integrable functions on the interval $[a, b]$:*

$$\int_a^b \varphi_m(s)^\star \varphi_n(s) \, ds = \delta_{mn}. \tag{C.123}$$

The complex sequence $\{f_m\}$ is a vector in l^2 (i.e., $\sum_{m=1}^{\infty} |f_m|^2 < \infty$) if and only if there is a (Lebesgue) square-integrable function $f(x)$ for which $\{f_m\}$ are the so-called Fourier coefficients, that is,

$$f_m = \int_a^b \varphi_m^\star(s) f(s) \, ds. \tag{C.124}$$

If we now assume additionally that the orthonormal set $\{\varphi_m(x)\}$ is a complete (sometimes called "maximal") basis, which Hilbert (1912, p. IX) expresses in terms of the following "completeness relation" (*Vollständigkeitsrelation*), also sometimes referred to as Parseval's identity,

$$\sum_m \left| \int_a^b \varphi_m^\star(s) f(s) \, ds \right|^2 = \int_a^b |f(s)|^2 ds, \tag{C.125}$$

then the isomorphism between the l^2 and $L^2(a,b)$ spaces becomes clear. More precisely, in modern language, we say that a one-to-one, onto mapping (bijection) T is a *Hilbert space isomorphism* from l^2 to $L^2(a,b)$ if it preserves the inner product structure,

$$(T\mathbf{f}, T\mathbf{g}) = (\mathbf{f}, \mathbf{g}). \tag{C.126}$$

The mapping

$$T: \ (f_1, f_2, f_3, \ldots) \in l^2 \ \rightarrow \ \sum_m f_m \, \varphi_m(s) \in L^2(a,b) \tag{C.127}$$

provides such an isomorphism, as the structure of the inner product for the sequence space l^2 maps directly to the inner product defined in $L^2(a,b)$ as a Lebesgue integral:

$$\begin{aligned}
(\mathbf{f}, \mathbf{g}) &= \sum_m f_m^\star g_m \\
&= \sum_m \left(\int_a^b f^\star(s)\, \varphi_m(s)\, ds \right) \left(\int_a^b \varphi_m^\star(t)\, g(t)\, dt \right) \\
&= \int_a^b f^\star(s)\, g(s) ds,
\end{aligned} \tag{C.128}$$

where the completeness relation used in going from the second to the third line follows from Parseval's identity Eq. (C.125) once we recall that a polarization identity can be used to write inner products in terms of norms (cf. Eq(C.27)). Riesz's strategy in developing his theory was to first prove the desired properties in the specific case where the orthonormal functions were $\varphi_m(\vartheta) = \frac{1}{\sqrt{2\pi}} e^{im\vartheta}$ on $[-\pi, +\pi]$ (i.e., the functions in the Fourier series in Eq. (C.112)), and then extend them to any other orthonormal basis $\{\psi_p\}$ related to the $\{\varphi_m\}$ by an infinite orthogonal rotation, $\psi_p = \sum_m O_{pm}\varphi_m$.

The Riesz–Fischer theory establishes the intimate connection—indeed, the structural identity—of the infinite-dimensional sequence space l^2 to the function space of Lebesgue square-integrable functions on an interval $L^2(a,b)$. Both satisfy the axiomatic definition of a *Hilbert space*, which would later be given explicitly by von Neumann (1929): an infinite-dimensional linear vector space equipped with an inner product that is Cauchy-complete with respect to the norm induced by the inner product. Early work on Hilbert space often assumed implicitly (as we have above) that these spaces are *separable*, that is, they are spanned by a countable basis of orthogonal vectors.[35] This is the *only* case of relevance to quantum theory: the Hilbert spaces of non-relativistic quantum mechanics, and even the Fock spaces of quantum field theory, are separable Hilbert spaces, and as such, are all structurally the same as the conceptually simple space l^2. This identity lies at the heart of the true equivalence of matrix and wave mechanics: any problem formulated in terms of wave mechanics can be equivalently formulated, and with no loss of generality or physical content, as a problem in matrix mechanics, and vice versa, via the structural isomorphism of the spaces l^2 and L^2.

[35] An equivalent definition of separable space posits the existence of a countable dense subset of vectors in the space.

The intimate connection between function spaces and the algebraic theory of infinite-dimensional matrices and the quadratic forms constructed from them had already been fully appreciated and exploited in 1906 by Hilbert, in the fifth of his series on integral-equation theory:

> As a link and mediating mechanism between the theory of functions and equations with infinitely many variables, as I have developed it in my fourth communication [(Hilbert 1906a)], and the theory of integral equations, which express relations between functions of a single variable s, one requires the existence of some system of infinitely many continuous functions $\varphi_1(s), \varphi_2(s), \ldots$ of the variable s, which in the interval $s = a$ to $s = b$ satisfy the following properties [i.e., orthonormality (Eq. (C.123)) and completeness (Eq. (C.128))] (Hilbert 1906b, p. 442).

With such a complete, orthonormal set of functions in hand, the Fredholm integral equation (cf. Eq. (C.76))

$$\varphi(s) - \lambda \int_0^1 K(s,t)\, \varphi(t)\, dt = f(s), \tag{C.129}$$

can be studied *exactly*—without recourse to the awkward limiting procedure first used by Hilbert, in which the interval $[0, 1]$ was discretized into n subintervals, and then the limit $n \to \infty$ taken—by expanding the functions $\varphi(s)$ and $f(s)$ in an orthonormal basis $\varphi_p(s)$:[36]

$$\varphi(s) = \sum_p \alpha_p\, \varphi_p(s), \quad f(s) = \sum_p a_p\, \varphi_p(s). \tag{C.130}$$

Multiplying the preceding equation by $\varphi_p(s)$ and integrating, we find that

$$\int \varphi_p(s)\, \varphi(s)\, ds - \lambda \int \varphi_p(s)\, K(s,t) \varphi(t)\, ds\, dt = \int \varphi_p(s) f(s)\, ds = a_p. \tag{C.131}$$

It follows that

$$\alpha_p - \lambda \int \varphi_p(s)\, K(s,t) \sum_q \alpha_q \varphi_q(t)\, ds\, dt = a_p, \tag{C.132}$$

or

$$\alpha_p - \lambda \sum_q K_{pq}\, \alpha_q = a_p, \tag{C.133}$$

where $K_{pq} \equiv \int \varphi_p(s)\, K(s,t)\, \varphi_q(t)\, ds\, dt$. The matrix K_{pq} is real symmetric if the real kernel $K(s,t) = K(t,s)$ is symmetric. If the matrix is also bounded (continuous), the spectral theory developed by Hilbert in his fourth paper can be applied to solve Eq. (C.133) for the (infinite) vector $\alpha = (1 - \lambda K)^{-1}\mathbf{a}$ (provided λ^{-1} is not in the spectrum of K), via the spectral resolution Eq. (C.109). Of course, if K is completely continuous, the continuous spectrum is absent (the integral in Eq. (C.109)), and the problem is solved once the discrete eigenvalues λ_n and associated eigenvectors $v^{(n)}$ are determined.

[36] We are returning to the purely real spaces and functions studied by Hilbert, hence, the absence of complex conjugates. We omit the limits 0, 1 on the integrals as the theory extends trivially to any interval $[a, b]$.

C.5.4 The axiomatization of Hilbert space

The foundations of a rigorous, and to this day stable and comprehensive, mathematical framework in which to express the laws of quantum mechanics were laid in 1927 by von Neumann (1927a, 1927b, 1927c). In the first of these, von Neumann introduces the central concept that will provide the stage of action in which quantum-mechanical processes would henceforth be visualized, the infinite-dimensional Hilbert space, explicitly named as such (*Der Hilbertsche Raum*) in the heading of section II of the paper. The serious business of defining and describing the two primary incarnations of such a space—the discrete space l^2 of square-summable sequences, and the function space L^2 of square-integrable functions defined on some interval (or, more generally, some volume element Ω in a higher-dimensional Euclidean space)—begins in section III.[37] After outlining the basic properties of such spaces, much as we have done in the preceding sections, von Neumann gives an explicit list of the axioms defining an abstract Hilbert space—axioms that capture the essential characteristics of both the original Hilbert–Schmidt discrete space l^2 and the function spaces spanned by orthonormal function bases. The axioms conveniently summarize the concepts we have already covered. An abstract complex Hilbert space $\mathfrak{H}$ is defined by von Neumann as follows (von Neumann 1927a, pp. 15–17):

(A) $\mathfrak{H}$ is a linear vector space.

(B) $\mathfrak{H}$ is equipped with an inner product $(\mathbf{f}, \mathbf{g})$, from which a norm, and hence, metric can be defined.

(C) $\mathfrak{H}$ is infinite-dimensional (there are linearly independent sets consisting of arbitrarily many vectors).

(D) $\mathfrak{H}$ contains an everywhere dense sequence (i.e., there is a sequence of vectors $\mathbf{f}_1, \mathbf{f}_2, \ldots$ in $\mathfrak{H}$ such that for any $\mathbf{f} \in \mathfrak{H}$ members of this sequence can be found arbitrarily close to $\mathbf{f}$) (Separability axiom).

(E) $\mathfrak{H}$ is Cauchy-complete (every Cauchy sequence converges to an element of $\mathfrak{H}$).

The statement of axioms (A–E) is followed in section 6 by a number of propositions. The proofs are short and can be found in the paper of von Neumann or in any text on functional analysis (see, for example, Kreyszig 1978). Here, we state the results (somewhat abbreviated), with some minor changes of notation to adapt von Neumann's results to our previous discussion, and with some short comments on the proofs.

1. (a) Every orthonormal set of vectors is either finite or countable. (b) A complete orthonormal set is countable. This depends on the separability axiom, which asserts that there exists a dense countable sequence of vectors $\mathbf{f}_1, \mathbf{f}_2, \ldots$ If $\mathbf{v}_n$ is an orthonormal set, the open balls of radius $1/\sqrt{2}$ around each $\mathbf{v}_n$ are non-intersecting, but must each contain at least one $\mathbf{f}_i$, whence (a). For (b), start with the $\mathbf{f}_i$ and use Schmidt orthogonalization to construct a complete orthonormal sequence of vectors.

[37] Von Neumann uses the notation $\mathfrak{H}_0$ for l^2, $\mathfrak{H}$ for L^2. We continue to use the modern notations for these spaces, but will use von Neumann's $\mathfrak{H}$ for the general abstract space.

2. Let $\mathbf{v}_n$ be an orthonormal set (not necessarily complete!). Then, for every $\mathbf{f}, \mathbf{g} \in \mathfrak{H}$,

$$\sum_{n=1}^{\infty} (\mathbf{f}, \mathbf{v}_n)(\mathbf{v}_n, \mathbf{g}) \text{ is absolutely convergent,}$$
$$\sum_{n=1}^{\infty} |(\mathbf{f}, \mathbf{v}_n)|^2 \leq (\mathbf{f}, \mathbf{f}) \equiv |\mathbf{f}|^2. \tag{C.134}$$

These results follow easily from the positivity of the norm induced by the inner product. The first assertion follows from the absolute convergence of the sums of the right-hand side of the "polarization identity"

$$\begin{aligned}(\mathbf{f}, \mathbf{v}_n)(\mathbf{v}_n, \mathbf{g}) = & \left|\left(\frac{\mathbf{f}+\mathbf{g}}{2}, \mathbf{v}_n\right)\right|^2 \\ & -\left|\left(\frac{\mathbf{f}-\mathbf{g}}{2}, \mathbf{v}_n\right)\right|^2 \\ & +i\left|\left(\frac{\mathbf{f}+i\mathbf{g}}{2}, \mathbf{v}_n\right)\right|^2 \\ & -i\left|\left(\frac{\mathbf{f}-i\mathbf{g}}{2}, \mathbf{v}_n\right)\right|^2.\end{aligned} \tag{C.135}$$

For the second part, we note that

$$\left(\mathbf{f} - \sum_n (\mathbf{f}, \mathbf{v}_n)\mathbf{v}_n,\ \mathbf{f} - \sum_n (\mathbf{f}, \mathbf{v}_n)\mathbf{v}_n\right) \geq 0, \tag{C.136}$$

which reduces to $\sum_n |(\mathbf{f}, \mathbf{v}_n)|^2 \leq |\mathbf{f}|^2 < \infty$.

3. Let $\mathbf{v}_n$ be an orthonormal set (not necessarily complete!). The series $\sum_{n=1}^{\infty} c_n \mathbf{v}_n$ converges to a vector $\mathbf{f} \in \mathfrak{H}$ if and only if $\sum_{n=1}^{\infty} |c_n|^2$ is finite. In this case, $c_n = (\mathbf{v}_n, \mathbf{f})$. The proof amounts to showing that the partial sums appearing in the Cauchy limit test have norms equal to that of a Cauchy sequence of real numbers, which must converge, as the real numbers are complete.

4. Let $\mathbf{v}_n$ be an orthonormal set (not necessarily complete!). For every $\mathbf{f}$, the series

$$\mathbf{f}' = \sum_{n=1}^{\infty} (\mathbf{v}_n, \mathbf{f})\mathbf{v}_n \tag{C.137}$$

converges, and $\mathbf{f}-\mathbf{f}'$ is orthogonal to all the $\mathbf{v}_n$. This is a direct consequence of the preceding two propositions.

5. The following three conditions are both necessary and sufficient for an orthonormal set of vectors $\mathbf{v}_n$ to be complete (namely, there can exist no vector in the space orthogonal to all the $\mathbf{v}_n$):

a. The linear space spanned by $\mathbf{v}_n$ is everywhere dense in $\mathfrak{H}$.

b. For all vectors $\mathbf{f}$,

$$\mathbf{f} = \sum_{n=1}^{\infty} c_n \mathbf{v}_n, \quad c_n = (\mathbf{v}_n, \mathbf{f}). \tag{C.138}$$

c. For all $\mathbf{f}, \mathbf{g} \in \mathfrak{H}$,

$$(\mathbf{f}, \mathbf{g}) = \sum_{n=1}^{\infty} (\mathbf{f}, \mathbf{v}_n)(\mathbf{v}_n, \mathbf{g}). \tag{C.139}$$

That completeness implies (b) is a direct consequence of the preceding proposition (4). We leave it to the reader to check that (b) implies both (a) and (c), which in turn imply completeness.

6. There exist countable complete orthonormal sets. To prove this, we start with a countable everywhere dense set of vectors, which must exist by axiom (D), which are then subjected to Gram–Schmidt orthogonalization.

This section of von Neumann's paper concludes with the comment

> With these results we have in fact shown that every space $\mathfrak{H}$ with the properties (axioms) (A–E) must agree in all its properties with the usual Hilbert space $\mathfrak{H}_0$ [i.e., l^2] (von Neumann 1927a, p. 22).

As discussed, Riesz and Fischer also establish that the function space $L^2(a, b)$ of Lebesgue square-integrable functions on an interval equally well satisfies the five von Neumann axioms, and must therefore "agree in all its properties" with results derivable in the context of the sequence space l^2. Von Neumann is here reviewing known results in a compact and lucid way.

C.5.5 A new notation: Dirac's bras and kets

In his third paper on quantum mechanics, "The physical interpretation of the quantum dynamics," Dirac (1927a) introduced a notation that would eventually be adopted almost universally by physicists using the theory. Although the notation is now regarded as an extremely convenient way to write formulas involving vectors and linear operators in an inner-product space, it is important to realize that Dirac had none of this in mind when developing his version of the Dirac–Jordan statistical transformation theory (see Chapter 16). His primary objective was to generalize the formal properties of physical observables expressed in discrete matrix form in the Heisenberg–Born–Jordan formalism to problems in which the "parameters that label the rows and columns of the matrices may take either discrete values or all values in certain continuous ranges, or perhaps both" (p. 625) The two formal developments that Dirac introduced in this paper to handle the situation where matrices with a continuous range of one or both of the indices occur were: (a) the introduction of the famous Dirac "delta function" $\delta(x)$, which would later be put on a solid mathematical foundation by the development of distribution theory; and (b) an embryonic version of the bra-ket notation, later introduced explicitly as a natural "mathematical syntax" for quantum mechanics (Dirac 1939), which is our chief concern here.

To simplify the discussion, we assume that we are dealing with a system with a single degree of freedom, and that a particular mechanical quantity (say energy or momentum) is represented by its matrix $g(\alpha_1, \alpha_2)$, where the indices α_1 and α_2 can take either discrete or continuous values (or both), typically of some conserved quantity.[38] It might prove advantageous to make a canonical transformation from the original α description to some other kinematical quantity β, so that our original quantity g is now written classically as a function of the βs rather than the αs. This transformation is implemented by a unitary rotation, $G = bgb^{-1}$, where the unitary "matrix" $b(\beta, \alpha)$ implements the transformation from an α description to a β one. Writing out this matrix equation using integrals to denote both the continuous and discrete ranges of the variables, one has

$$G(\beta_1, \beta_2) = \int b(\beta_1, \alpha_1)\, d\alpha_1\, g(\alpha_1, \alpha_2)\, d\alpha_2\, b^{-1}(\alpha_2, \beta_2). \tag{C.140}$$

The transformation matrix $b(\beta_1, \alpha_1)$ would later be simplified to (β_1/α_1), giving birth to the now familiar Dirac bra-ket notation, which is now written with angle rather than round brackets, and a vertical rather than slanted bar: $\langle\beta_1|\alpha_1\rangle$. At this point, the two parts (the "bra" $\langle\beta_1|$, and the "ket" $|\alpha_1\rangle$) of $b(\beta_1, \alpha_1) = \langle\beta_1|\alpha_1\rangle$ are not viewed as separate entities: only later (Dirac 1939) would the notation be used to express state vectors being combined through an inner product to produce the complex number $b(\beta_1, \alpha_1)$. Also, the kinematical quantity expressed by the matrix g (in the α scheme) or G (in the β scheme), is not regarded as an object with an independent representation-free existence, namely, a linear operator on a vector space. We do not find at this point $g(\alpha_1, \alpha_2)$ written as $\langle\alpha_1|g|\alpha_2\rangle$, for example.

The famous delta "function" makes its appearance in this paper as the representative of the identity matrix in situations where the dynamical variable has a continuous spectrum, and must be represented by matrices with continuous rows and columns. Thus, for example, the unitarity condition Eq. (C.84) in the discrete case becomes (with U replaced by b),

$$\int b^\star(\beta, \alpha)\, b(\beta, \alpha')\, d\beta = \delta(\alpha - \alpha'). \tag{C.141}$$

Here, the improper "function" $\delta(\alpha - \alpha')$ is defined by the properties (analogous to those of the Kronecker delta in the discrete case: $\sum_n \delta_{nm} = 1$ for all m and $\delta_{nm} = 0$ for $n \neq m$):

$$\int \delta(\alpha - \alpha')\, d\alpha = 1, \quad \delta(\alpha - \alpha') = 0 \text{ for } \alpha \neq \alpha'. \tag{C.142}$$

For our purposes, we "dissect" the transformation-theory notation of Dirac's 1927 paper, after rewriting (β/α) as $\langle\beta|\alpha\rangle$, into the individual bra $\langle\beta|$ and ket $|\alpha\rangle$ components, now regarded as elements of the dual vector space and the vector space of states, respectively. This is the reinterpretation Dirac (1939) proposed later. It turns out to be a highly compact way to express the spectral measures for the operators of interest in quantum theory.

[38] We have slightly altered Dirac's notation to avoid the unsightly proliferation of primes to which he subjects us.

Thus, our previous notation (whether for finite-dimensional complex inner-product spaces or for Hilbert space) will be replaced as follows:

$$\mathbf{v} \rightarrow |v\rangle, \tag{C.143}$$

$$\alpha\mathbf{v} + \beta\mathbf{w} \rightarrow \alpha|v\rangle + \beta|w\rangle, \tag{C.144}$$

$$(\mathbf{v}, \mathbf{w}) \rightarrow \langle v|w\rangle, \tag{C.145}$$

$$(\mathbf{v}, \mathbf{w}) = (\mathbf{w}, \mathbf{v})^\star \rightarrow \langle v|w\rangle = \langle w|v\rangle^\star, \tag{C.146}$$

$$|\mathbf{v}| \rightarrow \sqrt{\langle v|v\rangle}, \text{ etc.} \tag{C.147}$$

Of course, we emphasize that the concept of a quantum state as identified with a vector in Hilbert space is completely absent from the papers on transformation theory by Dirac (1927a) and Jordan (1927b, 1927g, see also Chapter 16). These ideas are really due to von Neumann. It turns out however, that the spectral theory developed by the latter is most transparently expressed (and certainly in a form instantly recognizable to modern theoretical physicists or quantum chemists) in the notation introduced by Dirac a decade later. Our admittedly anachronistic use of the bra-ket language should thus make the discussion easier to follow.[39]

C.5.6 Operators in Hilbert space: von Neumann's spectral theory

The new quantum mechanics that emerged in late 1925 and early 1926—in the superficially distinct guises of Göttingen matrix mechanics, Schrödinger's wave mechanics, or Dirac's q-number formalism—finally received the rigorous mathematical infrastructure needed to express the underlying unity of the formalism at the hands of John von Neumann. In a series of remarkable papers published in 1927–1929, he established the conceptual framework that has supported all subsequent applications of quantum theory. A summary of the essential mathematical ideas was presented in a paper in the Spring of 1927, "Mathematische Begründung der Quantenmechanik" ("Mathematical Foundation of Quantum Mechanics") (von Neumann 1927a).

The mathematical deficiencies in the formalisms used in quantum theory up to this point were identified by von Neumann in the first two sections of the paper:

1. The appearance of mathematically malodorous constructions such as matrices with a continuous range of rows and columns in problems where a continuous spectrum appears, as in the Three-Man-Paper of Born, Heisenberg, and Jordan (1926) or Dirac's (1927a) paper on transformation theory.
2. The appearance of improper (i.e., non-normalizable) (eigen)functions, and the apparent need for "functions" like Dirac's $\delta(x)$, defined by (according to von Neumann) "absurd" (von Neumann 1927a, p. 3) properties such as Eq. (C.142).
3. The appearance of (apparently unavoidable) unphysical elements in the formalism, for example, the complex phase $e^{i\varphi}$ associated with an eigenfunction, or, in the case of a set

[39] For a very readable account of Hilbert-space theory phrased entirely in Dirac notation, see Dennery and Krzywicki (1967). The much more comprehensive text (Prugovecki 1981) also makes extensive use of Dirac notation.

ψ_k of degenerate eigenfunctions, the freedom of redefinition to a new set by an arbitrary unitary combination $\psi_k \to \hat{\psi}_k = U_{kl}\psi_l$.

Needless to say, von Neumann was hardly the first to recognize these difficulties. He did, however, recognize that the first step towards their resolution was to establish a unified mathematical framework of sufficient generality to re-express all of these problems in a mathematically precise way. In sections II and III of his paper, von Neumann (1927a) reviews the treatment of the energy-eigenvalue problem in both matrix and wave mechanics, employing the two appropriate incarnations of infinite-dimensional Hilbert space, the square-summable sequences comprising l^2 discussed above (for which von Neumann says the name "Hilbert space is common"), and the space of functions (in modern notation $L^2(\Omega)$) whose absolute squares are integrable over some region Ω, that is, $\int_\Omega |\psi|^2 dv$ where dv is a line, surface, or volume element, for problems formulated in 1, 2, 3 or higher dimensions, respectively.[40] This leads to the general axiomatic formulation of Hilbert space in section V of the paper, discussed in Section C.5.4.

The physics of quantum mechanics is, of course, contained in the mechanical quantities, which have undergone a transformation from ordinary functions on phase space in classical theory to matrices/differential operators in the matrix (l^2)/wave (L^2) formulations of the new quantum theory. Von Neumann identifies the critical property, Hermitian symmetry for matrices, self-adjointness for operators, in section III, specifically for the physical quantity *primus inter pares* in quantum theory, the Hamiltonian, representing the energy of the quantum system. For matrices, we must have

$$H_{\mu\nu} = H^{\star}_{\nu\mu}; \tag{C.148}$$

for the Hamiltonian expressed as a differential operator, we must have

$$\int_\Omega (H\psi_2)^{\star}\psi_1\,dv = \int_\Omega \psi_2^{\star} H\psi_1\,dv, \tag{C.149}$$

for all functions ψ_1 and ψ_2 that "vanish (sufficiently strongly) at the boundaries of the integration" (von Neumann 1927a, p. 9). The latter requirements had of course been explored already by Schrödinger, who recognized in them the key to obtaining quantized results from a completely continuous wave problem.

The general eigenvalue problem for a symmetric/Hermitian[41] linear operator T is formulated in section IX. One seeks all real numbers λ (the eigenvalue) and associated (finite norm) eigenvectors $\mathbf{v} \in \mathfrak{H}$ such that

$$T\mathbf{v} = \lambda\mathbf{v}. \tag{C.150}$$

This formulation, as von Neumann points out, is deficient in two ways:

1. The eigenvectors, even if normalized to unit magnitude, are undetermined up to an overall complex phase, if non-degenerate, or worse, if degenerate, to an overall unitary rotation

[40] The discrete l^2 space is denoted $\mathfrak{H}_0$ by von Neumann, the $L^2(\Omega)$ space $\mathfrak{H}$. The general abstract Hilbert space defined axiomatically in section V is called $\overline{\mathfrak{H}}$.

[41] At this point symmetry means Hermiticity, and the crucial distinction between merely symmetric/Hermitian and self-adjoint operators in Hilbert space had not yet been fully appreciated by von Neumann. Unraveling these subtleties would take another two years.

within each subspace spanned by eigenvectors with equal eigenvalues. One wishes to write a representation of T in terms of its eigenvectors and eigenvalues in which such ambiguities are removed.

2. The formulation cannot handle situations in which a continuous spectrum appears, as the requirement that the eigenvectors **v** be in the Hilbert space (i.e., have finite norm) cannot be satisfied in general. The eigenfunction $\psi(x) = e^{2\pi ipx/h}$ of the momentum operator $\frac{h}{2\pi i}\frac{d}{dx}$, for example, squares to unity, which clearly cannot be integrated over $(-\infty, +\infty)$: $\int_{-\infty}^{+\infty} |\psi(x)|^2 dx = \infty$.

The solution of both problems, as von Neumann recognized, is to write the spectral representation of the operator in terms of the projection operators onto the eigenspace associated with eigenvalue λ.[42] We saw how this works for finite-dimensional operators at the conclusion of Section A.3. Consider the projection operator associated with eigenvector $\mathbf{v}_1$ in Eq. (C.58). Under a change of complex phase of the eigenvector, $\mathbf{v}_1 \to e^{i\varphi}\mathbf{v}_1$,

$$E_1\mathbf{v} \to (e^{i\varphi}\mathbf{v}_1, \mathbf{v})e^{i\varphi}\mathbf{v}_1 = e^{-i\varphi}(\mathbf{v}_1, \mathbf{v})e^{i\varphi}\mathbf{v}_1 = E_1\mathbf{v}. \tag{C.151}$$

The reader may easily verify that, in the presence of degeneracy, with n degenerate eigenvectors, $\mathbf{v}_1, \ldots, \mathbf{v}_n$, the projection operator $\sum_{i=1}^{n} E_n$ is invariant under unitary rotations, $\mathbf{v}_n \to U_{nm}\mathbf{v}_m$, for any U satisfying $U^\dagger U = 1$. For finite-dimensional self-adjoint operators, the spectral family $E(\lambda)$ defined in Eq. (C.68) take values for every λ that are independent of the phase normalization and (in the degenerate case) unitary reorientation of the eigenvectors. These comments evidently extend smoothly to the case of infinite-dimensional operators of the Hilbert–Schmidt type (i.e., completely continuous), which have a pure point spectrum. For such operators, the spectral representation Eq. (C.69) (with $n = \infty$) is perfectly valid, with the $E(\lambda)$ defined in terms of an infinite discrete set of projection operators E_i, as in Eq. (C.68).

That the spectral representation of an operator in terms of a Stieltjes integral over a spectral family is the appropriate vehicle for the case of operators with a continuous spectrum is one of von Neumann's most productive insights. It provides the necessary structure for the development of a mathematically rigorous spectral theory of unbounded operators, as well as a very convenient language in which to phrase the statistical content of the theory. Although the proof of the spectral theorem in the unbounded case would only be finished two years later, two important physical examples where the theorem is essentially self-evident are already given in von Neumann (1927a). These are

$$\text{(a)}\ \mathbf{p} = \frac{h}{2\pi i}\frac{d}{dx}\ldots, \quad \text{(b)}\ \mathbf{q} = x\ldots, \tag{C.152}$$

the momentum operator and the position operator for a particle moving on the x-axis. The ellipses indicate that the operators are to be applied to square-integrable functions $\psi(x) \in L^2(-\infty, +\infty)$. We are employing the function-space realization of the Hilbert space, appropriate for a wave mechanical formulation. Also, we are using bold font to distinguish operators from their eigenvalues.

[42] If λ is non-degenerate, the associated eigenspace is the one-dimensional line consisting of all complex multiples of the single eigenvector. If λ is n-fold degenerate, the associated eigenspace is the linear subspace generated by all linear combinations of eigenvectors with eigenvalue λ. Note that these eigenspaces have an invariant meaning independent of how we choose to normalize or orient the individual eigenvectors.

We now show, following von Neumann, how to construct the spectral families $E(\lambda)$, as well-defined (bounded) operators in $L^2(-\infty,+\infty)$, for the momentum and position operators, which have an entirely continuous spectrum. This means that there are *no* normalizable eigenfunctions in either case: for the momentum operator, the eigenfunctions are of the non-square-integrable form $\psi_p(x) = e^{2\pi ipx/h}$ (with eigenvalue p), while for the position operator the eigenfunction $\psi_q(x)$ corresponding to eigenvalue q must satisfy $x\psi_q(x) = q\psi_q(x)$, that is, $\psi_q(x) = 0$ for $x \neq q$, which again cannot be normalized to unity in L^2.

First, for the momentum operator, given a (Lebesgue-)square-integrable function $\psi(x)$, there is a well-defined Fourier transform $\tilde{\psi}(k)$,

$$\tilde{\psi}(p) = \int_{-\infty}^{+\infty} e^{-2\pi ipx/h}\psi(x)\,dx. \tag{C.153}$$

Of course, the original wave function $\psi(x)$ can be reconstructed from its momentum-space version $\tilde{\psi}(p)$ by an inverse Fourier transformation:

$$\psi(x) = \int_{-\infty}^{+\infty} \tilde{\psi}(p)\,e^{2\pi ipx/h}dp/h. \tag{C.154}$$

Define a linear operator $E(p)$ on the function space L^2 as follows:

$$E(p)\psi(x) = \int_{-\infty}^{p} \tilde{\psi}(p')\,e^{2\pi ip'x/h}dp'/h. \tag{C.155}$$

It is apparent from the definition that (i) $E(p) \to 0$ for $p \to -\infty$, and (ii) $E(p) \to 1$ for $p \to \infty$, as expected from any spectral family. From the theory of the Fourier integral,[43] $E(p)\psi(x) \in L^2$ when $\psi(x) \in L^2$. Put otherwise, $E(p)$ is everywhere defined in the Hilbert space. We leave it to the reader to verify that $E(p)$ is bounded (in fact, of norm 1, $||E(p)|| = 1$), self-adjoint, and satisfies the projection operator requirement $E(p)^2 = E(p)$. The spectral representation of the momentum operator, Eq. (C.69), now follows, as, for any $\psi(x)$ in the domain of $\mathbf{p}$,

$$\begin{aligned}
\int_{-\infty}^{+\infty} p\,d(E(p)\psi(x)) &= \int_{-\infty}^{+\infty} p\,d\left\{\int_{-\infty}^{p} \tilde{\psi}(p')\,e^{2\pi ip'x/h}dp'/h\right\} \\
&= \int_{-\infty}^{+\infty} p\,\tilde{\psi}(p)\,e^{2\pi ipx/h}dp/h \\
&= \frac{h}{2\pi i}\frac{d}{dx}\int_{-\infty}^{+\infty} \tilde{\psi}(p)\,e^{2\pi ipx/h}dp/h \\
&= \frac{h}{2\pi i}\frac{d}{dx}\psi(x).
\end{aligned} \tag{C.156}$$

Hence, $E(p)$ is indeed the spectral family for $\mathbf{p} \equiv \frac{h}{2\pi i}\frac{d}{dx}$, as required.

[43] See, for example, Ch. 3, theorems 4.5 and 4.6 in Prugovecki (1981).

The position operator is defined as the multiplication operator

$$\mathbf{q}\psi(x) = x\psi(x). \tag{C.157}$$

We use q to denote eigenvalues of this operator. We define a spectral family for the position operator as

$$\begin{aligned} E(q)\psi(x) &= \psi(x), \qquad x \leq q, \\ &= 0, \qquad x > q, \end{aligned} \tag{C.158}$$

or, more succintly, using the step function,

$$E(q)\,\psi(x) = \vartheta(q - x)\,\psi(x). \tag{C.159}$$

The Stieltjes spectral integral for $\mathbf{q}\,\psi(x)$ now reads

$$\int_{-\infty}^{\infty} q\,d\Big\{E(q)\psi(x)\Big\} = \lim_{n\to\infty}\sum_{i=1}^{n}\overline{q}_i\Big(E(q_{i+1})\psi(x) - E(q_i)\psi(x)\Big), \tag{C.160}$$

where $\overline{q}_i$ is chosen somewhere in the indefinitely decreasing interval (q_i, q_{i+1}). In evaluating the Stieltjes integral as the limit of the discretized sum on the right, we consider x as fixed and having some definite value. In dividing the Stieltjes integral range into a set of intervals, only the single interval satisfying $q_i < x < q_{i+1}$ gives a non-vanishing contribution, namely, $\overline{q}_i\,\psi(x)$. As $n \to \infty$, $\overline{q}_i \to x$ and we have

$$\int_{-\infty}^{\infty} q\,d\{E(q)\psi(x)\} = x\,\psi(x), \tag{C.161}$$

as desired.

It is important to realize that in the spectral representations of the momentum and position operators in terms of their respective spectral families $E(p)$ and $E(q)$, there is no reference anywhere to non-normalizable eigenfunctions: only well-defined operators appear throughout, and they are assumed to act only on vectors in their domain, which, for unbounded operators (such as $\mathbf{p}, \mathbf{q}$), are a proper subset of the full Hilbert space. For example, the position operator for a particle moving on a one-dimensional line can only be applied sensibly to functions such that $x^2|\psi(x)|^2$ is (Lebesgue) integrable from $-\infty$ to $+\infty$. Thus, the perfectly sensible (i.e., square-integrable, hence, element of the Hilbert space) wave function $\psi(x) = C/\sqrt{x^2 + a^2}$ is not in the domain of $\mathbf{q}$, as the wave function $Cx/\sqrt{x^2 + a^2}$ is not square-integrable. The spectral projection operators $E(p)$ and $E(q)$, however, have operator norm 1 and cannot therefore increase the magnitude of any vector upon which they act. They therefore act sensibly on all vectors in L^2: their domain is the entire Hilbert space. Also, the (in 1927 still) highly objectionable Dirac delta function is scrupulously avoided through judicious use of the Stieltjes integral.

Von Neumann was fully aware at this point in early 1927 that, as shown by Hilbert and Hellinger, any *bounded* self-adjoint operator T admits a spectral family of projection operators $E(\lambda)$ such that the operator can be written as the Stieltjes integral representation

$$T = \int \lambda \, dE(\lambda). \tag{C.162}$$

Section C.3 discussed how, despite its continuous appearance, such an integral can handle the case of a discrete (point) spectrum perfectly well. The Hilbert–Schmidt–Hellinger theory (cf. Section C.5.2) extended the intuition of the finite-dimensional case to infinite-dimensional *bounded* self-adjoint operators T, satisfying $|T\mathbf{v}| < C|\mathbf{v}|$ for all $\mathbf{v}$ and some finite real positive C. For the subset of completely continuous operators only a point spectrum was present, but in general one would have both discrete and continuous eigenvalues, with the absolute magnitude of the eigenvalues of either type bounded by C. Hellinger (1910) stated the result, described in some detail in Section 12.3.4 with an explicit example, in a slightly different notation than von Neumann (1927a). In particular, Hellinger did not use projection operators, which would prove so fruitful for von Neumann, both in his mathematical development of the theory of unbounded operators and in his statistical formulation of quantum mechanics. To see the connection between the two, it is best to employ (somewhat anachronistically) the bra-ket notation of Dirac, as the modern student of quantum mechanics will instantly recognize the results of both authors once translated into this notation.

Let us consider the (by 1910) completely solved case of a Hermitian infinite matrix $K_{pq} = K^{\star}_{qp}$ corresponding to a bounded operator on l^2: thus, $|K\mathbf{v}| \leq C|\mathbf{v}|$ for all $\mathbf{v} \in l^2$, with C a finite positive real constant. Hellinger's analysis is complicated by the need to consider degeneracies in the continuous spectrum, which we exclude for simplicity. For definiteness, we assume that the continuous spectrum is contained in the finite interval $(a > -\infty, b < \infty)$ on the real axis. We also assume initially that there are no discrete eigenvalues corresponding to finite norm eigenvectors (i.e., no point spectrum), as in the examples given by von Neumann above for the position and momentum operators (in the L^2 Hilbert space). As a concrete example, the reader can refer to the matrix of Eq. (12.253) with the first two rows and columns (giving the point spectrum) excluded. Here, we use the language of operators in Hilbert space, rather than of bilinear forms, however. As we saw in that case, there exist eigenvectors φ for a continuous range of eigenvalues λ with components $\varphi_p(\lambda)$ $(p = 1, 2, 3, \ldots)$:[44]

$$K_{pq}\,\varphi_q(\lambda) = \lambda\,\varphi_p(\lambda). \tag{C.163}$$

From

$$\begin{aligned} \varphi^{\star}_p(\lambda')\,K_{pq}\,\varphi_q(\lambda) &= \lambda\,\varphi^{\star}_p(\lambda')\,\varphi_p(\lambda) \\ &= \left(K_{qp}\,\varphi_p(\lambda')\right)^{\star}\varphi_q(\lambda) \\ &= \lambda'\,\varphi^{\star}_p(\lambda')\,\varphi_p(\lambda), \end{aligned} \tag{C.164}$$

[44] In Section 12.3.4, we wrote λ and $\varphi_p(\lambda)$ as W and $x_k(W)$, respectively (see Eqs. (12.258)–(12.259)).

we find, restoring explicit summation signs, that

$$(\lambda - \lambda')\sum_{p=1}^{\infty}\varphi_p(\lambda)\,\varphi_p^{\star}(\lambda') = 0. \tag{C.165}$$

It follows that:

$$\begin{aligned} &\text{for } \lambda \neq \lambda': \quad \sum_{p=1}^{\infty}\varphi_p(\lambda)\,\varphi_p^{\star}(\lambda') = 0, \\ &\text{for } \lambda = \lambda': \quad \sum_{p=1}^{\infty}\varphi_p(\lambda)\varphi_p^{\star}(\lambda) = +\infty. \end{aligned} \tag{C.166}$$

If the latter sum were finite, $\varphi_p(\lambda)$ would correspond to a normalizable eigenvector, and λ would be an element of the point spectrum, which we are temporarily assuming absent.

Thus, the eigenvectors $\varphi_p(\lambda)$ are not square-summable (cf. Eq. (12.261)). If we cut off the summation at $p = M$, the function

$$\delta_M(\lambda, \lambda') \equiv \sum_{p=1}^{M}\varphi_p(\lambda)\,\varphi_p^{\star}(\lambda') \tag{C.167}$$

is a finite function that is strongly peaked (at a positive value) for small $\lambda - \lambda'$, with the peak at $\lambda = \lambda'$ going to positive infinity as $M \to \infty$. In the limit we can replace it by a Dirac delta function, with the positive weight $\sigma(\lambda)$ (which is non-zero only for values $a < \lambda < b$ in the continuous spectrum[45]) determined by the integral over λ' (say):

$$\sum_{p=1}^{\infty}\varphi_p(\lambda)\,\varphi_p^{\star}(\lambda') = \sigma(\lambda)\,\delta(\lambda - \lambda'). \tag{C.168}$$

Define the function $\Phi(\lambda)$ as follows

$$\Phi(\lambda) \equiv \sum_p \left| \int_a^{\lambda} \varphi_p(\lambda')\,d\lambda' \right|^2. \tag{C.169}$$

[45] As K is bounded, we must have a, b finite, of course.

The weight $\sigma(\lambda)$ of the delta function is simply related to $\Phi(\lambda)$, as the following calculation shows

$$\begin{aligned}\Phi(\lambda) &= \sum_p \int_a^\lambda \varphi_p(\lambda')\,d\lambda' \int_a^\lambda \varphi_p^\star(\lambda'')\,d\lambda'' \\ &= \int_a^\lambda d\lambda' \int_a^\lambda \sum_p \varphi_p(\lambda')\,\varphi_p^\star(\lambda'')\,d\lambda'' \\ &= \int_a^\lambda d\lambda' \sigma(\lambda') \int_0^\lambda \delta(\lambda' - \lambda'')\,d\lambda'' \\ &= \int_a^\lambda \sigma(\lambda')\,d\lambda'. \end{aligned} \tag{C.170}$$

Thus, we have simply that

$$\sigma(\lambda) = \frac{d\Phi(\lambda)}{d\lambda}. \tag{C.171}$$

Note that the appearance of an essentially arbitrary function $\sigma(\lambda)$ here is simply a reflection of the freedom to independently normalize each continuum eigenvector $\varphi_p(\lambda)$ of the matrix K_{qp} by a λ-dependent factor. In many cases this freedom is employed to set $\sigma(\lambda) = 1$ but, for the time being, we retain the more general normalization to parallel Hellinger's treatment more closely.

We now translate these results into the more efficient and transparent Dirac bra-ket notation. The complete orthonormal basis in l^2 with respect to which the matrix K_{pq} is computed is denoted $|p\rangle$ $(p = 1, 2, 3, \ldots)$.[46] Thus,

$$K_{pq} = \langle p|K|q\rangle, \quad \langle p|q\rangle = \delta_{pq}. \tag{C.172}$$

The continuum eigenvectors will now be re-expressed in Dirac notation as follows

$$\varphi_p(\lambda) \to \langle p|\lambda\rangle, \tag{C.173}$$

so that the complex number $\varphi_p(\lambda)$ is regarded as the (finite!) inner product of the infinite norm vector $|\lambda\rangle$ with the unit norm basis vector $|p\rangle$. Any sequence vector $|v\rangle$ (not necessarily square-summable!) can be expanded in the canonical basis $|q\rangle, q = 1, 2, 3, \ldots$, as

$$|v\rangle = \sum_q |q\rangle\langle q|v\rangle. \tag{C.174}$$

The eigenvalue equation in this notation becomes

$$K|\lambda\rangle = \lambda|\lambda\rangle. \tag{C.175}$$

[46] The vectors $|1\rangle, |2\rangle, \ldots$ can be regarded as the canonical basis $(1, 0, 0, 0, \ldots), (0, 1, 0, 0, \ldots), \ldots$ in the sequence space l^2.

Taking the inner product with the basis vector $|p\rangle$ and using Eq. (C.174) to expand $|\lambda\rangle$ in the basis $|p\rangle$, we find that

$$\lambda\langle p|\lambda\rangle = \langle p|K|\lambda\rangle = \langle p|K|\Big(\sum_q |q\rangle\langle q|\lambda\rangle\Big) = \sum_q K_{pq}\langle q|\lambda\rangle. \tag{C.176}$$

With the help of Eq. (C.173), this translates into the defining equation Eq. (C.163):

$$\sum_q K_{pq}\,\varphi_q(\lambda) = \lambda\,\varphi_p(\lambda). \tag{C.177}$$

Despite the fact that the vector $|\lambda\rangle$ (which makes sense as an infinite sequence vector with finite components $\langle q|\lambda\rangle = \varphi_q(\lambda)$) is not an element of the l^2 Hilbert space ($\sum_q |\langle q|\lambda\rangle|^2 = \infty$), operators constructed from integrals involving $|\lambda\rangle\langle\lambda|$ act perfectly sensibly within l^2. Recalling that the continuous spectrum lies entirely in the interval (a, b), we construct the operator

$$E(\lambda)\begin{cases} = 0 & \text{for } \lambda \le a \\ = \displaystyle\int_a^\lambda |\lambda'\rangle\langle\lambda'|\frac{d\lambda'}{\sigma(\lambda')} & \text{for } a < \lambda < b \\ = \displaystyle\int_a^b |\lambda'\rangle\langle\lambda'|\frac{d\lambda'}{\sigma(\lambda')} & \text{for } \lambda \ge b. \end{cases} \tag{C.178}$$

The meaning of the ket-bra symbols $|\lambda'\rangle\langle\lambda'|$, called the “outer product” of the ket $|\lambda'\rangle$ with the bra $\langle\lambda'|$, is clarified by defining the action of $E(\lambda)$, for example, for $a < \lambda < b$, on an arbitrary vector $|\psi\rangle$ in l^2:

$$E(\lambda)\,|\psi\rangle = \int_a^\lambda |\lambda'\rangle\langle\lambda'|\psi\rangle\frac{d\lambda'}{\sigma(\lambda')}. \tag{C.179}$$

In fact, for any λ, $E(\lambda)$ is a projection operator. First, it is Hermitian, as $\sigma(\lambda)$ is real. To check the projection property, we once again consider $a < \lambda < b$ and calculate

$$E(\lambda)\,E(\lambda) = \int_a^\lambda \frac{d\lambda'}{\sigma(\lambda')}\int_a^\lambda \frac{d\lambda''}{\sigma(\lambda'')}\,|\lambda'\rangle\langle\lambda'|\lambda''\rangle\langle\lambda''|. \tag{C.180}$$

The inner product $\langle\lambda'|\lambda''\rangle$ can be written in terms of the components of these two vectors in the basis $|p\rangle$

$$\langle\lambda'|\lambda''\rangle = \sum_p \langle\lambda'|p\rangle\langle p|\lambda''\rangle = \sum_p \varphi_p(\lambda'')\,\varphi_p^\star(\lambda') = \sigma(\lambda'')\,\delta(\lambda'' - \lambda'), \tag{C.181}$$

where in the second step we used Eq. (C.173) and in the third step we used Eq. (C.168). Substituting this into Eq. (C.180) and carrying out the integration over λ'', we arrive at

$$E(\lambda)\,E(\lambda) = \int_a^\lambda \frac{d\lambda'}{\sigma(\lambda')}|\lambda'\rangle\langle\lambda'| = E(\lambda), \tag{C.182}$$

which proves the projection property. As such a projection operator is necessarily bounded (it can never increase the norm of a vector it acts on), it can be defined on the entire Hilbert space, despite the fact that it is constructed from entities like $|\lambda\rangle$, which are not vectors in the Hilbert space. The matrix of $E(\lambda)$ in our canonical l^2 basis, which of course uniquely specifies the operator, can be constructed from knowledge of the continuum eigenvectors $\varphi_p(\lambda) = \langle p|\lambda\rangle$, as, for $a < \lambda < b$,

$$E_{pq}(\lambda) = \int_a^\lambda \langle p|\lambda'\rangle\langle\lambda'|q\rangle \frac{d\lambda'}{\sigma(\lambda')} = \int_a^\lambda \varphi_p(\lambda')\varphi_q^\star(\lambda') \frac{d\lambda'}{\sigma(\lambda')}. \tag{C.183}$$

The non-trivial part of the Hellinger theory is the proof of completeness, that is, to show, once all continuum eigenvectors $\varphi_p(\lambda) = \langle\lambda|p\rangle$ have been identified and used to construct the spectral family $E(\lambda)$, that

$$E(\lambda) = 1 \text{ for } \lambda > b, \tag{C.184}$$

where b is an upper bound for the continuous spectrum.

Recall that we are explicitly considering here only bounded operators without a point spectrum. If a point spectrum is present, there are (at most denumerably many) orthonormal eigenvectors $|m\rangle$ ($m = 1, 2, \ldots$) with discrete eigenvalues λ_m (i.e., $K|m\rangle = \lambda_m|m\rangle$) and the spectral family is defined (cf. Eq. (C.68)) by including both discrete and continuous parts of the spectrum:

$$E(\lambda) = \sum_{m, \lambda_m \leq \lambda} |m\rangle\langle m| + \int_a^\lambda |\lambda'\rangle\langle\lambda'| \frac{d\lambda'}{\sigma(\lambda')}. \tag{C.185}$$

Here, the integral must be interpreted as equal to zero if $\lambda < a$ and taken over the entire interval (a, b) if $\lambda \geq b$. Again, the final step (Hellinger 1910, Ch. III) of the spectral theorem shows that the projection operators onto the discrete and continuous spectrum exhaust the full Hilbert space. The largest eigenvalue magnitude is just the finite norm $||K||$ of the operator, so if $\lambda \geq ||K||$, $E(\lambda) = 1$ and we get the *spectral resolution (or partition) of the identity*

$$\sum_m |m\rangle\langle m| + \int_a^b |\lambda\rangle\langle\lambda| \frac{d\lambda}{\sigma(\lambda)} = 1. \tag{C.186}$$

If we apply the operator K on the left of Eq. (C.186), and use that $|m\rangle$ is an eigenvector of K with eigenvalue λ_m, we get the spectral theorem for the operator K:

$$K = \sum_m \lambda_m |m\rangle\langle m| + \int_a^b \lambda|\lambda\rangle\langle\lambda| \frac{d\lambda}{\sigma(\lambda)} \equiv K_{\text{disc}} + K_{\text{cont}}. \tag{C.187}$$

The Hellinger proof of this spectral theorem lies in establishing that any bounded Hermitian operator without a point or continuous spectrum (in this case $K-(K_{\text{disc}}+K_{\text{cont}})$) must vanish identically. Once all the discrete and continuous eigenvectors (with components $\langle p|m\rangle$ and $\langle p|\lambda\rangle = \varphi_p(\lambda)$, respectively) have been included in the spectral sum and integral in Eq. (C.187), the remainder must therefore vanish (Hellinger 1910, Ch. III, sec. 8).

Referring to the definition Eq. (C.185) of the spectral family $E(\lambda)$, we see that the operator K can be written very simply as a Stieltjes spectral integral

$$K = \int \lambda \, dE(\lambda). \tag{C.188}$$

The Stieltjes integral takes care effortlessly of both the point and continuous parts of the spectrum of K (cf. discussion at end of Section A.3). The spectral integral representation, at least as established by Hellinger (1910) for bounded self-adjoint operators, can therefore be written in exactly the form surmised by von Neumann (1927a) for unbounded operators.

The form of the Hellinger theory given by Born in the Three-Man-Paper (cf. Section 12.3.3) can be obtained from the formulas above by some minor changes of notation. We illustrate the conversion for the continuum contribution only. The bilinear form associated with the continuum spectrum of K in Eq. (C.187) is

$$\begin{aligned}\sum_{kl} x_k x_l \langle k|K_{\text{cont}}|l\rangle &= \sum_{kl} x_k x_l \int \lambda \langle k|\lambda\rangle\langle\lambda|l\rangle \frac{d\lambda}{\sigma(\lambda)} \\ &= \sum_{kl} x_k x_l \int \lambda \, \varphi_k(\lambda)\, \varphi_l^\star(\lambda) \frac{d\lambda}{\sigma(\lambda)}.\end{aligned} \tag{C.189}$$

Now recall that $\sigma(\lambda) = d\Phi/d\lambda$. This follows from Eq. (C.169), which is just Eq. (12.251) with λ substituted for W, $\Phi(\lambda)$ for $\varphi(W)$, a for W_0, and $\varphi_k(\lambda)$ for $x_k(W)$. We can thus change variables from λ to Φ and introduce a function $y(\Phi)$ by setting

$$\sum_k \varphi_k(\lambda)\, x_k \, d\lambda = y(\Phi)\, d\Phi, \tag{C.190}$$

which entails that

$$y^\star(\Phi) = \sum_l \varphi_l^\star(\lambda)\, x_l \frac{d\lambda}{d\Phi} = \frac{1}{\sigma(\lambda)} \sum_l \varphi_l^\star(\lambda)\, x_l. \tag{C.191}$$

Eq. (C.190) is equivalent to Eq. (12.252), with the changes of notation indicated above. Finally, inserting Eqs. (C.190) and (C.191) into Eq. (C.189), we find

$$\sum_{kl} x_k x_l \langle k|K_{\text{cont}}|l\rangle = \int \lambda(\Phi) y(\Phi) y^\star(\Phi) d\Phi, \tag{C.192}$$

which is the continuum part of Eq. (12.248), again with the indicated changes of notation.

Unfortunately, as Born *et al.* had already realized in their discussion of continuous spectra in the Three-Man-Paper, most of the operators representing physical observables in quantum mechanics (energy, momentum, angular momentum, for example) are unbounded: their spectrum extends to values of arbitrarily large magnitude. They were thus obliged simply to assume the natural extension of Hellinger's results to the unbounded case:

> If we allow ourselves here to carry over the results of Hellinger to the unbounded forms which appear in our case [i.e., in matrix mechanics], this seems to us to be justified by

the fact that Hellinger's methods clearly correspond completely to the physical content of the posed problem (Born, Heisenberg, and Jordan 1926, p. 590).

For the two years following his "Mathematical foundation of quantum mechanics," von Neumann struggled with the problem of extending the clean and beautiful results of the Hilbert–Schmidt–Hellinger theory of bounded forms/operators to the unbounded case needed for quantum mechanics. His efforts would include at least one false start, leading to a paper withdrawn from publication at the last minute,[47] but concluded successfully with the famous paper in *Mathematische Annalen* in 1929, in which the notion of a normal operator (one which commutes with its adjoint—self-adjoint and unitary operators being obvious special cases) is shown to be the critical ingredient in establishing a spectral theorem for unbounded operators analogous to Eq. (C.188) for the bounded case. The developments in functional analysis between 1927 and 1929 leading to von Neumann's spectral theory are strictly speaking beyond the time horizon of this book, but for the curious reader we briefly review the central points of the development.

The difficulties in the case of unbounded operators primarily arise from the need to identify and treat carefully the *domains* of the various operators under investigation. A bounded operator, once defined on a dense subset of the Hilbert space, can be extended uniquely to the entire Hilbert space.[48] However, unbounded operators *cannot* be applied sensibly to all elements of the Hilbert space: an unbounded operator can (indeed must) give vectors of infinite norm when applied to certain finite norm vectors. For example, the Hamiltonian operator of the simple harmonic oscillator,

$$H = \frac{\hbar\omega}{2}\,\mathrm{diag}(1, 3, 5, 7, \ldots), \tag{C.193}$$

applied to the unit norm vector (in l^2),

$$\frac{\sqrt{6}}{\pi}(1, 1/2, 1/3, 1/4, \ldots), \tag{C.194}$$

clearly gives a vector whose squared components sum to infinity. The *domain* of H, written $D(H)$, is just the set of elements $\mathbf{v}$ in l^2 for which $H\mathbf{v}$ has finite norm. The reader can easily check that $D(H)$ is a linear subspace of l^2, but it is clearly a proper subset of the full Hilbert space. Different unbounded operators will have different domains, which evidently complicates the task of defining the sums, products, etc., of such operators.

The first step in dealing with the desired extension of Hellinger spectral theory is to identify the subclass of operators that have the important property of Hermiticity. Hence, the following definition: following Eqs. (C.33)–(C.37) in the finite-dimensional case, we say that an operator A (bounded or unbounded) is *Hermitian* if[49]

$$(\mathbf{v}, A\mathbf{w}) = (A\mathbf{v}, \mathbf{w}), \quad \forall \mathbf{v}, \mathbf{w} \in D(A). \tag{C.195}$$

[47] For an account of Marshall Stone's later description of this premature attempt by von Neumann to formulate a spectral theorem for the unbounded case, see Birkhoff and Kreyszig (1984, p. 309).

[48] Let $\mathbf{v}_n$ be a denumerable dense subset of $\mathfrak{H}$. Any vector $\mathbf{v} \in \mathfrak{H}$ is the limit of a convergent subsequence $\mathbf{v}_i$ of the $\mathbf{v}_n$. As a bounded operator K is continuous, we can define $K\mathbf{v}$ as the limit of the convergent sequence $K\mathbf{v}_i$.

[49] In the mathematical literature, it is more common to encounter the term "symmetric", instead of "Hermitian", even in the complex case. However, we stick to the term "Hermitian", which is also used in the seminal paper by von Neumann (1929).

For such operators, we have the converse of the previously asserted property for bounded operators A generally, namely that $D(A) = \mathfrak{H}$, in the *Hellinger–Toeplitz* theorem:[50]

Theorem C.5.9 (Hellinger–Toeplitz (1910)). *A Hermitian operator defined on the entire Hilbert space $\mathfrak{H}$ is bounded.*

Evidently, the treatment of unbounded Hermitian operators inevitably involves us in questions of definition of domain (and, in the case of products of operators, the *range* of operators, where the range $R(A)$ is the set of vectors obtained by action of A on vectors in its domain $D(A)$).

A few examples here will serve to illustrate the type of domain issues one encounters in dealing with unbounded operators in Hilbert space.

1. In l^2, for the Hamiltonian of the simple harmonic oscillator in Eq. (C.193), $D(H)$ is the set of complex sequences (c_n) for which

$$\sum_{n=1}^{\infty} n^2 |c_n|^2 < \infty. \tag{C.196}$$

Note that such sequences form a dense subset in l^2—in fact, a subset of such sequences, namely all column vectors of complex numbers with at most a finite number of non-zero elements, is already dense in l^2. Indeed, any vector $\mathbf{v} \in l^2$ with components (c_n) such that

$$\sum_{n} |c_n|^2 < \infty \tag{C.197}$$

is the limit (in the l^2 norm) of the vectors $\mathbf{v}_n = (c_1, c_2, \ldots, c_n, 0, 0, 0, \ldots)$.

2. In L^2, the quantum mechanical position operator $\mathbf{q}$ for a particle on an infinite line, defined by

$$\mathbf{q}\psi(x) \equiv x\psi(x), \tag{C.198}$$

where $\psi(x) \in L^2$, has domain $D(q)$ equal to the set of Lebesgue square-integrable functions $\psi(x)$ for which

$$\int_{-\infty}^{+\infty} x^2 |\psi(x)|^2 dx < \infty. \tag{C.199}$$

$D(q)$ contains the set of Lebesgue square-integrable functions of compact support (i.e., which vanish identically for $|x| > C$ for some finite real positive C), which is known to be dense in $L^2(-\infty, +\infty)$.[51]

[50] Original proof in Hellinger and Toeplitz (1910). For a modern proof, see Kreyszig (1978, p. 525).

[51] We can even require the functions of compact support to be infinitely differentiable (see theorem 5.6 in Prugovecki 1981, p. 129).

3. In L^2, the once continuously differentiable functions $\psi(x)$ for which

$$\int_{-\infty}^{+\infty} |d\psi(x)/dx|^2 dx < \infty \tag{C.200}$$

are in the domain $D(p)$ of the momentum operator $\mathbf{p} = (h/2\pi i)\, d/dx$ for a particle on a line. Once again, this set of functions is dense in $L^2(-\infty, +\infty)$. For example, the Hermite functions $H_n(x)$ form a complete orthonormal basis for $L^2(-\infty, +\infty)$, so finite linear combinations of them (clearly in the domain of $\mathbf{p}$) are dense in $L^2(-\infty, +\infty)$.

The Hellinger–Toeplitz theorem tells us that we must necessarily deal in quantum mechanics with operators that are not defined on all the elements of the Hilbert space. However, it is critical that such operators be at least defined on a dense subset of the Hilbert space. This is essential to the spectral theory of von Neumann, which ensures that projection operators onto the eigenspaces of physical observables corresponding to specified ranges of values of these observable are well defined. Such operators form a critical part of the statistical interpretation of the theory, formalized by von Neumann (1927b, 1927c).

Given a densely defined operator A in a Hilbert space, one can define the *adjoint* $A^\dagger$ of the operator as follows.[52] For a given vector $\mathbf{v}$, the action of the adjoint on $\mathbf{v}$ is defined provided there exists a vector $\mathbf{v}'$ such that

$$(\mathbf{v}', \mathbf{w}) = (\mathbf{v}, A\mathbf{w}) \text{ for all } \mathbf{w} \in D(A), \tag{C.201}$$

in which case we define

$$A^\dagger \mathbf{v} = \mathbf{v}'. \tag{C.202}$$

If $D(A)$ were not dense in the Hilbert space, there would exist vectors $\boldsymbol{\Delta}$ of non-zero norm that are orthogonal to all $\mathbf{w} \in D(A)$, and therefore the action of $A^\dagger$ would not be unique, as both $\mathbf{v}'$ and $\mathbf{v}' + \boldsymbol{\Delta}$ would satisfy Eq. (C.201). The set of all vectors $\mathbf{v}$ for which Eq. (C.201) is satisfied is the domain of the adjoint, $D(A^\dagger)$. An important property of the adjoint of any densely defined operator is that it is *closed*—it is defined on the limit $\mathbf{v}$ of any convergent sequence $\mathbf{v}_i$ for which $A^\dagger \mathbf{v}_i$ converges, with $\lim_{i\to\infty} A^\dagger \mathbf{v}_i = A^\dagger \mathbf{v}$.[53]

Now suppose our operator A is Hermitian (i.e., A satisfies Eq. (C.195)). Then $D(A^\dagger)$ obviously contains the vectors in the domain $D(A)$ of A, and on those vectors, A and $A^\dagger$ agree. However, it may happen that the domains of a Hermitian operator and its adjoint are not identical. The special class of operators for which a satisfactory spectral theory exists are those symmetric operators A for which $D(A) = D(A^\dagger)$ and (on the vectors of the common domain) $A = A^\dagger$:

Definition C.5.2. *A densely defined Hermitian operator A is self-adjoint if $D(A) = D(A^\dagger)$ and $A = A^\dagger$.*

A self-adjoint operator is equal to its adjoint, and hence, must be a closed operator.

[52] In the mathematical literature, the notation A^* is more common for the adjoint of the operator A.
[53] This may not be true of the original densely defined operator (see Kreyszig 1978, sec. 10.3).

The importance of self-adjointness of an operator A—as opposed to mere Hermiticity (Eq. (C.195))—lies in the behavior of the Cayley transform $V = (A - i)(A + i)^{-1}$ (cf. Eq. (C.54)). This operator maps us from $R(A + i)$ (the set of vectors $\mathbf{v}$ for which $\mathbf{v} = (A + i)\mathbf{w}$ for some $\mathbf{w} \in D(A)$) to $R(A - i)$. Note the identity, following from the Hermiticity of A, for any $\mathbf{v} \in D(A)$

$$\begin{aligned}((A \pm i)\mathbf{v}, (A \pm i)\mathbf{v}) &= (A\mathbf{v}, A\mathbf{v}) \mp i(\mathbf{v}, A\mathbf{v}) \pm i(A\mathbf{v}, \mathbf{v}) + (\mathbf{v}, \mathbf{v}) \\ &= |A\mathbf{v}|^2 + |\mathbf{v}|^2. \end{aligned} \tag{C.203}$$

Clearly, $\mathbf{v} \neq 0$ implies $(A \pm i)\mathbf{v} \neq 0$, so $A \pm i$ are both one-to-one operators, and $V = (A - i)(A + i)^{-1}$ is a one-to-one mapping of $R(A + i)$ to $R(A - i)$. It is also norm-preserving (i.e., an *isometry*), as for $\mathbf{v} \in R(A + i)$, that is, $\mathbf{v} = (A + i)\mathbf{w}$ with $\mathbf{w} \in D(A)$, $V\mathbf{v} = (A - i)\mathbf{w}$ and

$$|V\mathbf{v}| = |(A - i)\mathbf{w}| = |(A + i)\mathbf{w}| = |\mathbf{v}|. \tag{C.204}$$

However, V is not in general globally defined, and onto, which is necessary if we are to define the inverse and verify that V *is unitary*—that is, $VV^\dagger = V^\dagger V = 1$, with $D(V) = \mathfrak{H}$. In fact, there are many isometric operators that are not unitary. A simple example is the shift operator in l^2 defined by $V(c_1, c_2, c_3, \ldots) = (0, c_1, c_2, c_3, \ldots)$ with $\sum_n |c_n|^2 < \infty$. This operator is one-to-one, but not onto, and has no inverse.

If A is not merely Hermitian, but self-adjoint, then one can show that the domain of definition of V (namely $R(A + i)$) is dense in $\mathfrak{H}$. One can also show that $R(A \pm i)$ is topologically closed and must therefore coincide with the full Hilbert space (von Neumann 1929, p. 80). In this case, the Cayley transform is unitary, bounded, and globally defined. This property lies at the crux of the importance of self-adjoint operators in the spectral theory of unbounded operators.

The whole point of the Cayley maneuver is to transform the problem of determining a spectral resolution for a possibly unbounded self-adjoint operator into that of determining the spectral resolution for a particular type of bounded operator, namely, unitary operators, for whom the domain of definition is the entire Hilbert space, and for which, given the existence of the rigorous spectral theory of Hilbert and Hellinger, there should be a straightforward representation of the form Eq. (C.188). In fact, given that the relation between the Cayley transform V and the self-adjoint operator A is one-to-one and invertible, with $A = i(1 - V)^{-1}(1 + V)$, a complex number $e^{i\varphi}$ in the spectrum of the unitary operator V corresponds to the real number $-\cot(\varphi/2)$ in the spectrum of A:

$$i(1 - e^{i\varphi})^{-1}(1 + e^{i\varphi}) = i\frac{2\cos(\varphi/2)}{-2i\sin(\varphi/2)} = -\cot(\varphi/2). \tag{C.205}$$

The spectral representation for the bounded and globally defined unitary operator V, expressed as usual as a Stieltjes integral that may have jump discontinuities if there is a point spectrum,

$$V = \int_0^{2\pi} e^{i\varphi}\, dE(\varphi), \tag{C.206}$$

can be established by methods going back to Hilbert and collaborators.[54] The spectral family $E(\varphi)$ defined by

$$E(\varphi) = \int_0^{\varphi} dE(\varphi') \tag{C.207}$$

form a set of projection operators with the usual properties: $E(0) = 0$, $E(2\pi) = 1$, $E(\varphi)\,E(\varphi') = E(\varphi)$ for $\varphi < \varphi'$, etc. The associated self-adjoint operator A shares exactly the same spectral family with V, only the eigenvalue needs to be appropriately altered in the spectral integral,

$$A = \int_0^{2\pi} -\cot(\varphi/2)\, dE(\varphi), \tag{C.208}$$

or, in a more familiar form, changing variables from φ to $\lambda = -\cot\varphi/2$,

$$A = \int \lambda\, dE(\lambda), \tag{C.209}$$

where the integral extends over the spectrum (point and continuous) of A.

The careful analysis of the domain and range issues needed to establish the required properties of and connections between the self-adjoint operator A and its Cayley transform extends by two years beyond the time frame of 1923–1927, which we have identified for this second part of our account of the development of quantum theory, and is beyond the scope of our brief review. Moreover, the proof that the operators of physical interest in quantum mechanics are indeed self-adjoint, and hence, compatible with the von Neumann spectral theory, is not always trivial. While straightforward for the coordinate and momentum operators, for which we established the spectral theorem directly above,[55] a rigorous demonstration of the self-adjointness of the Schrödinger Hamiltonian operator for helium (or more general multi-electron atoms) had to wait until 1951, when the necessary techniques were introduced by Tosio Kato (1951a). It should also be admitted that, except for some work on multichannel scattering in the 1960s, where a knowledge of functional analysis became essential, a basic understanding of the spectral theory of operators in Hilbert space, enough to correctly implement the spectral theorem in Eq. (C.187), was sufficient for most quantum physicists post-1930. An amusing anecdote related by the Hungarian-born American mathematician Peter Lax illustrates the point:

> And what do the physicists think of these matters? In the 1960s [Kurt Otto] Friedrichs [cofounder of the Courant Institute, New York] met Heisenberg, and used the occasion to express to him the deep gratitude of the community of mathematicians for having created quantum mechanics, which gave birth to the beautiful theory of operators in Hilbert space. Heisenberg allowed that this was so; Friedrichs then added that the mathematicians have, in some measure, returned the favor. Heisenberg looked noncommittal, so Friedrichs pointed out that it was a mathematician, von Neumann, who clarified the

[54] For a more streamlined modern proof employing trigonometric polynomials, see Prugovecki (1981, sec. III.6).

[55] See, for example, Kreyszig (1978, sec. 10.7) or Prugovecki (1981, p. 224).

difference between a self-adjoint operator and one that is merely symmetric. "What's the difference?," said Heisenberg (Lax 2002, p. 414).

Our brief survey of the development of the mathematical technology of quantum theory must end here. The reader interested in a more comprehensive treatment is encouraged to consult one of the many books treating mathematical issues in quantum mechanics for further details.[56]

[56] From a pedagogical point of view, we recommend the previously cited Kreyszig (1978) and Prugovecki (1981).

Bibliography

Aaserud, Finn, and John L. Heilbron. 2013. *Love, Literature, and the Quantum Atom. Niels Bohr's 1913 Trilogy Revisited.* Oxford: Oxford University Press.

Aaserud, Finn, and Helge Kragh, eds. 2015. *One Hundred Years of the Bohr Atom. Proceedings from a Conference.* Copenhagen: Det Kongelige Danske Videnskabernes Selskab.

Adams, Edwin P. 1923. *The Quantum Theory*. Washington, D. C.: National Research Council.

Aitchison, Ian J. R., David A. McManus, and Thomas M. Snyder. 2004. "Understanding Heisenberg's 'Magical' Paper of July 1925: A New Look at the Calculational Details." *American Journal of Physics* 72: 1370–1379.

Alexandrow, W. 1921. "Eine Bemerkung zur Langevinschen Formel für die Suszeptibilität paramagnetischer Körper." *Physikalische Zeitschrift* 22: 258–259.

Bacciagaluppi, Guido, and Elise Crull. 2009. "Heisenberg (and Schrödinger, and Pauli) on Hidden Variables." *Studies in History and Philosophy of Modern Physics* 40: 374–382.

Bacciagaluppi, Guido, Elise Crull, and Owen J. E. Maroney. 2017. "Jordan's Derivation of Blackbody Fluctuations." *Studies in History and Philosophy of Modern Physics* 60: 23–34.

Bacciagaluppi, Guido, and Antony Valentini. 2009. *Quantum Theory at the Crossroads. Reconsidering the 1927 Solvay Conference.* Cambridge: Cambridge University Press.

Badino, Massimiliano. 2015. *The Bumpy Road. Max Planck from Radiation Theory to the Quantum (1896–1906).* New York: Springer.

Badino, Massimiliano, and Jaume Navarro, eds. 2013. *Research and Pedagogy: A History of Quantum Physics through Its Textbooks.* Berlin: Edition Open Access.

Barut, Asim O., Halis Odabasi, and Alwyn van der Merwe, eds. 1991. *Selected Popular Writings of E.U. Condon.* New York: Springer.

Baym, Gordon. 1969. *Lectures on Quantum Mechanics.* Reading, MA: Addison-Wesley.

Bell, John S. 1966. "On the Problem of Hidden Variables in Quantum Mechanics." *Reviews of Modern Physics* 38: 447–452. Reprinted on pp. 1–13 of John S. Bell, *Speakable and Unspeakable in Quantum Mechanics* (Cambridge: Cambridge University Press, 1987).

Bell, Robert Edward. 1966. "John Stuart Foster, 1890–1964." *Biographical Memoirs of Fellows of the Royal Society* 12: 147–161.

Beller, Mara. 1985. "Pascual Jordan's Influence on the Discovery of Heisenberg's Indeterminacy Principle." *Archive for History of Exact Sciences* 33: 337–349.

——. 1990. "Born's Probabilistic Interpretation: A Case Study of Concepts in Flux." *Studies in History and Philosophy of Science* 21: 563–588.

——. 1999. *Quantum Dialogue. The Making of a Revolution.* Chicago: University of Chicago Press.

Bergmann, Peter Gabriel. 1942. *Introduction to the Theory of Relativity.* New York: Prentice-Hall. Page reference to corrected and enlarged reprint: New York: Dover, 1976.

Bernkopf, Michael. 1965. "The Development of Function Spaces with Particular Reference to their Origins in Integral Equation Theory." *Archive for History of the Exact Sciences* 3: 1–96.

——. 1968. "A History of Infinite Matrices." *Archive for History of Exact Sciences* 4: 308–358.

Bethe, Hans, and Roman Jackiw. 1986. *Intermediate Quantum Mechanics*. Boulder, CO: Westview Press.

Bhattacharya, Ananyo. 2022. *The Man from the Future. The Visionary Life of John von Neumann*. New York: Norton.

Biedenharn, Lawrence C. 1983. "The 'Sommerfeld Puzzle' Revisited and Resolved." *Foundations of Physics* 13: 13–34.

Birkhoff, Garrett, and Erwin Kreyszig. 1984. "The Establishment of Functional Analysis." *Historia Mathematica* 11: 258–321.

Birks, John B., ed. 1962. *Rutherford at Manchester*. London: Heywood. Reprint: New York: Benjamin, 1963.

Blum, Alexander, Martin Jähnert, Christoph Lehner, and Jürgen Renn. 2017. "Translation as Heuristics: Heisenberg's Turn to Matrix Mechanics." *Studies In History and Philosophy of Modern Physics* 60: 3–22.

Bôcher, Maxime. 1907. *Introduction to Higher Algebra*. New York: Macmillan.

——. 1911. *Einführung in die höhere Algebra*. Leipzig: Teubner. German translation of Bôcher (1907).

Bohr, Niels. 1913a. "On the Constitution of Atoms and Molecules (Part I)." *Philosophical Magazine* 26: 1–25. Reprinted in facsimile in Bohr (1972–2008, Vol. 2, pp. 161–185) and in Aaserud and Heilbron (2013, pp. 203–227); also reprinted in Birks (1962, pp. 228–256) and ter Haar (1967, pp. 132–159).

——. 1913b. "On the Constitution of Atoms and Molecules (Part II)." *Philosophical Magazine* 26: 476–502. Reprinted in facsimile in Bohr (1972–2008, Vol. 2, pp. 188–214) and in Aaserud and Heilbron (2013, pp. 228–254).

——. 1914. "Om Brintspektret." *Fysisk Tidsskrift* 12: 97–114. Based on a lecture at the Physical Society of Copenhagen, December 20, 1913. English translation, "On the Spectrum of Hydrogen," in Bohr (1922d, Essay I, pp. 1–19; reprinted in facsimile in Bohr 1972–2008, Vol. 2, pp. 283–301).

——. 1918. "On the Quantum Theory of Line Spectra. Parts I and II." *Det Kongelige Danske Videnskabernes Selskab. Skrifter. Naturvidenskabelig og Matematisk Afdeling. 8. Raekke, IV.1:* 1–100. Reprinted in facsimile in Bohr (1972–2008, Vol. 3, pp. 67–166). Introduction and Part I reprinted in van der Waerden (1968, pp. 95–136). Part III: Bohr (1922f).

——. 1920. "Über die Serienspektra der Elemente." *Physikalische Zeitschrift* 2: 423–469. Based on a lecture at the German Physical Society, Berlin, April 27, 1920. Reprinted in Bohr (1922c, Essay II). English translation, "On the Series Spectra of the Elements," in Bohr (1922d, Essay II, pp. 20–60; reprinted in facsimile in Bohr 1972–2008, Vol. 3, pp. 242–282).

——. 1921a. "Atomic Structure." *Nature* 107: 104–107. Reprinted in Bohr (1972–2008, Vol. 4, pp. 72–82).

——. 1921b. "Constitution of Atoms." Report prepared for the 1921 Solvay Congress. Published in Bohr (1972–2008, Vol. 4, pp. 100–174).

——. 1921c. "Atomic Structure." *Nature* 108: 208–209. Reprinted in Bohr (1972–2008, Vol. 4, pp. 177–180).

——. 1921d. "Atomernes Bygning og Stoffernes fysiske og kemiske Egenskaber." *Fysisk Tidsskrift* 19: 153–220. Based on a lecture on October 18, 1921 before the Physical Society and the Chemical Society of Copenhagen. Reprinted as Bohr (1922a). German translation: Bohr (1922b; reprinted as Essay III in Bohr 1922c). Amended English translation, "The Structure of the Atom and the Physical and Chemical Properties of the Elements," in Bohr

(1922d, Essay III, pp. 61–126). Page references to Bohr (1924a [2nd ed. of Bohr 1922d], Essay III, pp. 61–126; reprinted in facsimile in Bohr 1972–2008, Vol. 4, pp. 263–328).

——. 1922a. *Atomernes Bygning og Stoffernes fysiske og kemiske Egenskaber.* Copenhagen: Jul. Gjellerups Forlag. Reprint of Bohr (1921d). Reprinted in facsimile in Bohr (1972–2008, Vol. 4, pp. 183–256).

——. 1922b. "Der Bau der Atome und die physikalischen und chemischen Eigenschaften der Elemente." *Zeitschrift für Physik* 9: 1–67. German translation of Bohr (1921d). Reprinted as Essay III in Bohr (1922c).

——. 1922c. *Drei Aufsätze über Spektren und Atombau.* Braunschweig: Vieweg. Essay I: German translation of Bohr (1914); Essay II: Bohr (1920); Essay III: Bohr (1922b). English translation: Bohr (1922d).

——. 1922d. *The Theory of Spectra and Atomic Constitution. Three Essays.* Cambridge: Cambridge University Press. Essay I: translation of Bohr (1914); Essay II: translation of Bohr (1920); Essay III: amended translation of Bohr (1921d). 2nd ed.: Bohr (1924a).

——. 1922e. *Seven Lectures on the Theory of Atomic Structure.* English translation of Bohr's Wolfkehl Lectures (*Bohr Festspiele*) in Göttingen, June 12–22, 1922, in Bohr (1972–2008, Vol. 4, pp. 341–419).

——. 1922f. "On the Spectra of Elements of Higher Atomic Number." *Det Kongelige Danske Videnskabernes Selskab. Skrifter. Naturvidenskabelig og Matematisk Afdeling. 8. Raekke, IV.1:* 101–118. Part III of Bohr (1918). Based on a manuscript from 1918 with an appendix added in 1922. Reprinted in facsimile Bohr (1972–2008, Vol. 3, pp. 167–184).

——. 1923a. "L'application de la théorie des quanta aux problèmes atomiques." In *Atomes et Électrons. Rapports et Discussions du Conseil de Physique tenu a Bruxelles du 1er au 6 Avril 1921 sous les auspices de l'Institut International de Physique Solvay,* 228–247. Paris: Gauthier Villars. English translation, "On the application of the quantum theory to atomic problems," in Bohr (1972–2008, Vol. 3, pp. 364–380).

——. 1923b. "The Effect of Electric and Magnetic fields on Spectral Lines." *Proceedings of the Physical Society* 35: 275–302. Based on the Seventh Guthrie Lecture at the Physical Society of London, March 24, 1922. Reprinted in Bohr (1972–2008, Vol. 3, pp. 417–446).

——. 1923c. "Über die Anwendung der Quantentheorie auf den Atombau. I. Die Grundpostulate der Quantentheorie." *Zeitschrift für Phyik* 13: 117–165. Page reference to English translation: Bohr (1924b).

——. 1923d. "On Atomernes Bygning (Nobel lecture presented December 11, 1922)." In *Les Prix Nobel en 1921–1922.* Stockholm: P. A. Norstedt & Fils. Reprinted in Bohr (1972–2008, Vol. 4, pp. 427–465). Page references to English translation: Bohr (1923e).

——. 1923e. "The Structure of the Atom." *Nature* 112: 29–44. Supplement to *Nature,* July 7, 1923. Translation (by Frank C. Hoyt) of Bohr (1923d). Reprinted in Bohr (1972–2008, Vol. 4, pp. 467–482).

——. 1924a. *The Theory of Spectra and Atomic Constitution. Three Essays.* 2nd ed. Cambridge: Cambridge University Press. 2nd ed. of Bohr (1922d). Reprinted in facsimile in Bohr (1972–2008, Vol. 2, pp. 283–301 [Essay I: translation of Bohr 1914], Vol. 3, pp. 242–282 [Essay II: translation of Bohr 1920], and Vol. 4, pp. 257–340 [title page, preface to 1st and 2nd eds., Essay III: amended translation of Bohr 1921d, appendix]).

——. 1924b. "On the Application of the Quantum Theory to Atomic Structure. Part I. The Fundamental Postulates." *Proceedings of the Cambridge Philosophical Society* Supplement: 1–42. Translation of Bohr (1923c). Reprinted in facsimile in Bohr (1972–2008, Vol. 3, pp. 458–499).

Bohr, Niels. 1924c. "Zur Polarisation des Fluorescenzlichtes." *Die Naturwissenschaften* 12: 1115–1117. Reprinted in Bohr (1972–2008, Vol. 5, pp. 145–147 [facsimile], pp. 148–154 [English translation]).

——. 1925. "Atomic Theory and Mechanics." *Nature* 116: 845–852. Based on a lecture of August 30, 1925 at the Sixth Scandinavian Mathematics Congress in Copenhagen. Reprinted in Bohr (1972–2008, Vol. 5, pp. 273–280).

——. 1926. "Atomtheorie und Mechanik." *Die Naturwissenschaften* 14: 1–10. German translation of Bohr (1925).

——. 1928a. "Das Quantenpostulat und die neuere Entwicklung der Atomistik." *Die Naturwissenschaften* 16: 245–257. English translation: Bohr (1928b).

——. 1928b. "The Quantum Postulate and the Recent Development of Atomic Theory." *Nature* 121: 580–590. Supplement to *Nature*, April 14, 1928. Reprinted in facsimile in Bohr (1972–2008, Vol. 6, pp. 148–158). Also reprinted in Bohr (1934, pp. 52–91) [this version is reprinted in Wheeler and Zurek (1983, pp. 87–126)].

——. 1928c. "The Quantum Postulate and the Recent Development of Atomic Theory." In Como (1928, Vol. 2, pp. 565–598), including discussion remarks (pp. 589–598). Reprinted in facsimile in Bohr (1972–2008, Vol. 6, pp. 113–136).

——. 1928d. "Le postulat des quanta et le nouveau développement de l'atomistique." In Solvay (1928, pp. 215–247). Translation of Bohr (1928a).

——. 1934. *Atomic Theory and the Description of Nature.* Cambridge: Cambridge University Press.

——. 1949. "Discussion with Einstein on Epistemological Problems in Atomic Physics." In *Albert Einstein: Philosopher–Scientist.* Edited by Paul Arthur Schilpp, pp. 200–241. Evanston, IL: The Library of Living Philosophers.

——. 1972–2008. *Collected Works.* Edited by Léon Rosenfeld, Finn Aaserud, Erik Rüdiger, et al. Amsterdam: North-Holland.

Bohr, Niels, and Dirk Coster. 1923. "Röntgenspektren und periodisches System der Elemente." *Zeitschrift für Physik* 12: 342–374. English translation in Bohr (1972–2008, Vol. 4, pp. 520–548).

Bohr, Niels, Hendrik A. Kramers, and John C. Slater. 1924a. "The Quantum Theory of Radiation." *Philosophical Magazine* 47: 785–822. Page references to reprint in van der Waerden (1968, pp. 159–176).

——. 1924b. "Über die Quantentheorie der Strahlung." *Zeitschrift für Physik* 24: 69–87.

Born, Max. 1924. "Über Quantenmechanik." *Zeitschrift für Physik* 26: 379–395. Reprinted in Born (1962, Vol. 2, pp. 61–77). English translation in van der Waerden (1968, pp. 181–198).

——. 1925. *Vorlesungen über Atommechanik.* Berlin Heidelberg: Springer. English translation: Born (1927).

——. 1926a. "Zur Quantenmechanik der Stoßvorgänge. (Vorläufige Mitteilung)." *Zeitschrift für Physik* 37: 863–867. Reprinted in Born (1962, Vol. 2, pp. 228–232). English translation, "On the Quantum Mechanics of Collisions," in Wheeler and Zurek (1983, pp. 52–55).

——. 1926b. "Quantenmechanik der Stoßvorgänge." *Zeitschrift für Physik* 38: 803–827. Reprinted in Born (1962, Vol. 2, pp. 233–257). Abridged translation in Ludwig (1968, pp. 206–225).

——. 1926c. "Das Adiabatenprinzip in der Quantenmechanik." *Zeitschrift für Physik* 40: 167–192. Reprinted in Born (1962, Vol. 2, pp. 258–283).

——. 1926d. *Problems of Atomic Dynamics.* Cambridge, MA: The MIT Press.

——. 1926e. *Probleme der Atomdynamik.* Berlin: Julius Springer.

——. 1927. *The Mechanics of the Atom*. London: Bell. Translation of Born (1925).

——. 1928. "Über die Bedeutung der Stossvorgänge für das Verständnis der Quantenmechanik." In Como (1928, Vol. 2, pp. 443–447).

——. 1940. "Prof. Otto Toeplitz." *Nature* 145: 617. Reprinted in Born (1962, Vol. 2, pp. 612–613).

——. 1962. *Ausgewählte Abhandlungen*. 2 Vols. Göttingen: Vandenhoeck & Ruprecht.

——. 1971. *The–Born–Einstein Letters*. New York: Walker. Correspondence between Albert Einstein and Max and Hedwig Born from 1916 to 1955 with commentaries by Max Born.

——. 1978. *My Life. Recollections of a Nobel Laureate*. New York: Charles Scribner's Sons.

Born, Max, and Werner Heisenberg. 1928. "La méchanique des quanta." In Solvay (1928, 143–184). English translation, "Quantum Mechanics," in Bacciagaluppi and Valentini (2009, pp. 372–401).

Born, Max, Werner Heisenberg, and Pascual Jordan. 1926. "Zur Quantenmechanik II." *Zeitschrift für Physik* 35: 557–615. Page references to translation in van der Waerden (1968, 321–385).

Born, Max, and Pascual Jordan. 1925a. "Zur Quantentheorie aperiodischer Vorgänge. I." *Zeitschrift für Physik* 33: 479–505. Reprinted in Born (1962, Vol. 2, pp. 97–123).

——. 1925b. "Zur Quantenmechanik." *Zeitschrift für Physik* 34: 858–888. Page references to Chs. 1–3 are to the English translation in van der Waerden (1968, pp. 277–306). Ch. 4 is omitted in this translation.

——. 1930. *Elementare Quantenmechanik*. Berlin: Springer.

Born, Max, and Theodore von Kármán. 1912. "Über Schwingungen in Raumgittern." *Physikalische Zeitschrift* 13: 297–309. Reprinted in Born (1962, Vol. 1, pp. 231–243).

——. 1913. "Über die Verteilung der Eigenschwingungen von Punktgittern." *Physikalische Zeitschrift* 14: 65–71. Reprinted in Born (1962, Vol. 1, pp. 249–255).

Born, Max, and Wolfgang Pauli. 1922. "Über die Quantelung gestörter mechanischer Systeme." *Zeitschrift für Physik* 10: 137–158. Reprinted in Born (1962, Vol. 2, pp. 1–22).

Born, Max, and Norbert Wiener. 1926. "Eine neue Formulierung der Quantengesetze für periodische und nicht periodische Vorgänge." *Zeitschrift für Physik* 36: 174–187. Reprinted in Born (1962, Vol. 2, pp. 214–227).

Borrelli, Arianna. 2010. "Dirac's Bra-ket Notation and the Notion of a Quantum State." In *Styles of Thinking in Science and Technology*. Edited by Hermann Hunger, Felicitas Seebacher, and Gerhard Holzer, pp. 361–371. Vienna: Verlag der Österreichischen Akademie der Wissenschaften.

Bose, Satyendra Nath. 1924. "Plancks Gesetz und Lichtquantenhypothese." *Zeitschrift für Physik* 26: 178–181.

Bothe, Walther. 1923. "Über eine neue Sekundärstrahlung der Röntgenstrahlen. I. Mitteilung." *Zeitschrift für Physik* 16: 319–320.

——. 1924. "Über eine neue Sekundärstrahlung der Röntgenstrahlen. II. Mitteilung." *Zeitschrift für Physik* 20: 237–255.

Bothe, Walther, and Hans Geiger. 1924. "Ein Weg zur experimentellen Nachprüfung der Theorie von Bohr, Kramers, und Slater." *Zeitschrift für Physik* 25: 44.

——. 1925a. "Experimentelles zur Theorie von Bohr, Kramers, und Slater." *Die Naturwissenschaften* 13: 440–441.

——. 1925b. "Über das Wesen des Comptoneffekts: ein experimenteller Beitrag zur Theorie der Strahlung." *Zeitschrift für Physik* 32: 639–663.

Bowen, Ira S., and Robert A. Millikan. 1924. "The Series Spectra of the Stripped Boron Atom (BIII)." *Proceedings of the National Academy of Sciences* 10 (5): 199–203.

Bowler, Peter J. 2003. *Evolution. The History of an Idea.* 3rd ed. Berkeley, Berkeley: University of California Press.
Breit, Gregory. 1924. "The Quantum Theory of Dispersion." *Nature* 114: 310.
——. 1932. "Quantum Theory of Dispersion." *Reviews of Modern Physics* 4: 504–576.
Brillouin, Léon. 1926. "La mécanique ondulatoire de Schrödinger: une méthode générale de resolution par approximations successives." *Comptes Rendus de l'Academie des Sciences* 183: 24–26.
Bruns, Heinrich. 1895. "Das Eikonal." *Abhandlungen der mathematisch-physikalischen Classe der Königlichen Sächsischen Gesellschaft der Wissenschaften (Leipzig)* 35: 325–435.
Bub, Jeffrey. 2010. "Von Neumann's 'No Hidden Variables' Proof: A Reappraisal." *Foundations of Physics* 40: 1333–1340.
Buchwald, Jed Z. 1985. *From Maxwell to Microphysics. Aspects of Electromagnetic Theory in the Last Quarter of the Nineteenth Century.* Chicago: University of Chicago Press.
Burger, Herman Carel, and Hendrik Berend Dorgelo. 1924. "Beziehungen zwischen inneren Quantenzahlen und Intensitäten von Mehrfachlinien." *Zeitschrift für Physik* 23: 258–266.
Burgers, Johannes M. 1917. "Adiabatic Invariants of Mechanical Systems." *Philosophical Magazine* 33: 514–520.
Busch, Paul. 2008. "The Time-Energy Uncertainty Relation." In *Time in Quantum Mechanics*, 2nd ed. (1st ed. 2002). Edited by J. Gonzalo Muga, Rafael Sala Mayato, and Íñ igno L. Egusquiza, pp. 73–105. New York: Springer.
Cairns-Smith, Alexander Graham. 1985. *Seven Clues to the Origin of Life: A Scientific Detective Story.* Cambridge: Cambridge University Press.
Camilleri, Kristian. 2009a. *Heisenberg and the Interpretation of Quantum Mechanics: The Physicist as Philosopher.* Cambridge: Cambridge University Press.
——. 2009b. "Constructing the Myth of the Copenhagen Interpretation." *Perspectives on Science* 17: 26–57.
——. 2012. "Review of Carson, Kojevnikov, and Trischler (2011)." *Isis* 103: 794–796.
——. 2022. "Orthodoxy and Heterodoxy in the Post-War Era." In Freire (2022, pp. 847–869).
Campbell, Norman. 1927. "Philosophical Foundations of Quantum Theory." *Nature* 119: 779.
Cantor, Geoffrey. 1994. "The Making of a British Theoretical Physicist: E. C. Stoner's Early Career." *The British Journal for the History of Science* 27: 277–290.
Caporael, Linnda R., James R. Griesemer, and William C. Wimsatt, eds. 2014. *Developing Scaffolds in Evolution, Culture, and Cognition.* Cambridge, MA: The MIT Press.
Carnap, Rudolf. 1928. *Der logische Aufbau der Welt.* Frankfurt: Meiner. Page reference to English translation: *The Logical Structure of the World* (Berkeley: University of California Press, 1967).
Carroll, Sean B. 2005. *Endless Forms Most Beautiful. The New Science of Evo Devo.* New York: Norton.
Carson, Cathryn, Alexei Kojevnikov, and Helmut Trischler, eds. 2011. *Weimar Culture and Quantum Mechanics: Selected Papers by Paul Forman and Contemporary Perspectives on the Forman Thesis.* London: Imperial College Press/World Scientific.
Casimir, Hendrik Brugt Gerhard. 1983. *Haphazard Reality. Half a Century of Science.* New York: Harper & Row.
——. 1996. "Levensbericht Ralph Kronig." *Levensberichten en herdenkingen.* Amsterdam: Koninklijke Nederlandse Akademie van Wetenschappen, pp. 55–60.
Cassidy, David C. 1991. *Uncertainty. The Life and Science of Werner Heisenberg.* New York: Freeman.

——. 2009. *Beyond Uncertainty. Heisenberg, Quantum Physics, and the Bomb.* New York: Bellevue Literary Press. 2nd rev. ed. of Cassidy (1991).

Christiansen, Christian. 1870. "Über die Brechungsverhältnisse einer weingeistigen Lösung des Fuchsin; briefliche Mittheilung." *Annalen der Physik* 217: 479–480.

——. 1871. "Ueber das Brechungsverhältniss des Fuchsins." *Annalen der Physik* 219: 250–259.

Clark, George Lindenberg, and William Duane. 1923a. "The Wave-lengths of Secondary X-rays." *Proceedings of the National Academy of Sciences* 9: 413–418.

——. 1923b. "The Wave-lengths of Secondary X-rays (Second Note)." *Proceedings of the National Academy of Sciences* 9: 419–424.

Coben, Stanley. 1971. "The Scientific Establisment and the Transmission of Quantum Mechanics to the United States, 1919–32." *American Historical Review* 76: 442–466.

Como. 1928. *Atti del Congresso Internazionale dei Fisici 11–20 Settembre 1927, Como-Pavia-Roma.* 2 Vols. Bologna: Nicola Zanichelli.

Compton, Arthur Holly. 1921a. "The Magnetic Electron." *Journal of the Franklin Institute* 192: 145–155.

——. 1921b. "The Degradation of Gamma-ray Energy." *Philosophical Magazine* 41: 749–769.

——. 1922. "Secondary Radiations Produced by X-rays, and Some of Their Applications to Physical Problems." *Bulletin of the National Research Council* 4: 1–56.

——. 1923. "A Quantum Theory of the Scattering of X-rays by Light Elements." *Physical Review* 21: 483–502. Reprinted in facsimile in Compton (1973, pp. 382–401) and in Weart (1976, pp. 152–171).

——. 1961. "The Scattering of X rays as Particles." *American Journal of Physics* 29: 817–820.

——. 1973. *Scientific Papers of Arthur Holly Compton: X-ray and Other Studies.* Edited by Robert S. Shankland. Chicago: University of Chicago Press.

Compton, Arthur Holly, and Alfred Walter Simon. 1925. "Directed Quanta of Scattered X-rays." *Physical Review* 26: 289–299. Reprinted in Compton (1973, 508–518).

Condon, Edward U. 1951. "Evolution of the Quantum Theory." *The Scientific Monthly* 72: 217–222. Page reference to reprint in Barut et al. (1991, pp. 199–208).

——. 1962. "60 Years of Quantum Physics." *Physics Today* 15: 37–48. Page reference to reprint in Barut et al. (1991, pp. 262–278).

Condon, Edward U., and George H. Shortley. 1963. *The Theory of Atomic Spectra.* Cambridge: Cambridge University Press.

Corben, Herbert Charles, and Philip Stehle. 1994. *Classical Mechanics.* 2nd ed. New York: Dover.

Corry, Leo. 2004. *David Hilbert and the Axiomatization of Physics (1898–1918). From* Grundlagen der Geometrie *to* Grundlagen der Physik. Dordrecht: Kluwer.

Coster, Dirk. 1922. "Röntgenspectra en de atoomtheorie van Bohr." PhD dissertation, Rijksuniversiteit Leiden.

Courant, Richard, and David Hilbert. 1924. *Methoden der mathematischen Physik (Erster Band).* Berlin: Springer.

Darrigol, Olivier. 1986. "The Origin of Quantized Matter Waves." *Historical Studies in the Physical and Biological Sciences* 16: 197–253.

——. 1988. "Statistics and Combinatorics in Early Quantum Theory, I." *Historical Studies in the Physical Sciences* 19: 17–80.

——. 1991. "Statistics and Combinatorics in Early Quantum Theory, II: Early Symptoms of Indistinguishability and Holism." *Historical Studies in the Physical Sciences* 21: 237–298.

——. 1992. *From c-Numbers to q-Numbers: The Classical Analogy in the History of Quantum Theory.* Berkeley: University of California Press.

Darrigol, Olivier. 2000. *Electrodynamics from Ampère to Einstein.* Oxford: Oxford University Press.
——. 2009. "A Simplified Genesis of Quantum Mechanics." *Studies In History and Philosophy of Modern Physics* 40: 151–166.
——. 2012. *A History of Optics from Greek Antiquity to the Nineteenth Century.* Oxford: Oxford University Press.
Darwin, Charles Galton. 1928. "The Wave Equations of the Electron." *Proceedings of the Royal Society of London A* 118: 654–680.
Darwin, Charles Galton, and Ralph Fowler. 1922. "On the Partition of Energy." *Philosophical Magazine* 44: 450–479.
Davisson, Clinton Joseph. 1916. "The Dispersion of Hydrogen and Helium on Bohr's Theory." *Physical Review* 8: 20–27.
Dawkins, Richard. 1976. *The Selfish Gene.* Oxford: Oxford University Press.
De Broglie, Louis. 1924. "Recherches sur la théorie des quanta." PhD dissertation, Paris. Published as De Broglie (1925). Page references to English translation (2004) by A.F. Kracklauer, available online at fondationlouisdebroglie.org/LDB-oeuvres/De_Broglie_Kracklauer.pdf
——. 1925. "Recherches sur la théorie des quanta." *Annales de Physique* 10: 22–128. Abridged English translation, "Investigations on quantum theory," in Ludwig (1968, pp. 73–93).
——. 1928. "La nouvelle dynamique des quanta." In Solvay (1928, pp. 105–141). English translation, "The New Dynamics of Quanta," in Bacciagaluppi and Valentini (2009, pp. 341–363).
——. 1929. *Einführung in die Wellenmechanik.* Leipzig: Akademische Verlagsgesellschaft.
Debye, Peter. 1909. "Das Verhalten der Lichtwellen in der Nähe eines Brennpunktes oder einer Brennlinie." *Annalen der Physik* 30: 755–776.
——. 1910. "Der Wahrscheinlichkeitsbegriff in der Theorie der Strahlung." *Annalen der Physik* 33: 1427–1434.
——. 1912. "Einige Resultate einer kinetischen Theorie der Isolatoren." *Physikalische Zeitschrift* 13: 97–100.
——. 1915. "Die Konstitution des Wasserstoffmoleküls." *Königlich Bayerische Akademie der Wissenschaften zu München. Mathematisch-physikalische Klasse. Sitzungsberichte:* 1–26.
——. 1923. "Zerstreuung von Röntgenstrahlen und Quantentheorie." *Physikalische Zeitschrift* 24: 161–166. English translation, "X-ray Scattering and Quantum Theory," in Debye (1954, pp. 80–88).
——. 1928. "Über elektrische Momente." In Como (1928, Vol. 2, pp. 515–531).
——. 1954. *The Collected Papers of Peter J. W. Debye.* New York and London: Interscience Publishers.
Dennery, Philippe, and André Krzywicki. 1967. *Mathematics for Physicists.* New York: Harper & Row.
Dennison, David M. 1926. "The Rotation of Molecules." *Physical Review* 28: 318–333.
Dieks, Dennis. 2017. "Von Neumann's Impossibility Proof: Mathematics in the Service of Rhetorics." *Studies in History and Philosophy of Modern Physics* 60: 136–148.
Dirac, Paul Adrien Maurice. 1925. "The Fundamental Equations of Quantum Mechanics." *Proceedings of the Royal Society of London* A 109: 642–653. Reprinted in van der Waerden (1968, pp. 307–320).
——. 1926a. "Quantum Mechanics and a Preliminary Investigation of the Hydrogen Atom." *Proceedings of the Royal Society of London* A 110: 561–569. Reprinted in van der Waerden (1968, pp. 417–427).

———. 1926b. "On the Theory of Quantum Mechanics." *Proceedings of the Royal Society of London* A 112: 661–677.
———. 1927a. "The Physical Interpretation of the Quantum Dynamics." *Proceedings of the Royal Society of London* A 113: 621–641.
———. 1927b. "The Quantum Theory of the Emission and Absorption of Radiation." *Proceedings of the Royal Society of London* A 114: 243–265.
———. 1928. "The Quantum Theory of the Electron." *Proceedings of the Royal Society of London* A A117: 610–624.
———. 1930. *Principles of Quantum Mechanics*. Oxford: Clarendon.
———. 1939. "A New Notation for Quantum Mechanics." *Proceedings of the Cambridge Philosophical Society* 35: 416–418.
Dresden, Max. 1987. *H. A. Kramers: Between Tradition and Revolution*. New York: Springer.
Drude, Paul. 1900. *Lehrbuch der Optik*. Leipzig: S. Hirzel. English translation (by C. R. Mann and R. A. Millikan): *The Theory of Optics*. New York: Longmans, Green, 1902.
———. 1906. *Lehrbuch der Optik*. 2nd edn. Leipzig: S. Hirzel.
Duncan, Anthony. 2012. *The Conceptual Framework of Quantum Field Theory*. Oxford: Oxford University Press.
Duncan, Anthony, and Michel Janssen. 2007. "On the Verge of *Umdeutung* in Minnesota: Van Vleck and the Correspondence Principle." 2 Pts. *Archive for History of Exact Sciences* 61: 553–624, 625–671.
———. 2008. "Pascual Jordan's Resolution of the Conundrum of the Wave-Particle Duality of Light." *Studies in History and Philosophy of Modern Physics* 39: 634–666.
———. 2009. "From Canonical Transformations to Transformation Theory, 1926–1927: The Road to Jordan's *Neue Begründung*." *Studies in History and Philosophy of Modern Physics* 40: 352–362.
———. 2013. "(Never) Mind Your *p*'s and *q*'s: von Neumann versus Jordan on the Foundations of Quantum Theory." *The European Physical Journal H* 38: 175–259.
———. 2014. "The Trouble with Orbits: The Stark Effect in the Old and the New Quantum Theory." *Studies In History and Philosophy of Modern Physics* 48: 68–83.
———. 2015. "The Stark Effect in the Bohr–Sommerfeld Theory and in Schrödinger's Wave Mechanics." In Aaserud and Kragh (2015, pp. 217–271).
———. 2022. "Quantization Conditions, 1900–1927." In Freire (2022, pp. 77–94).
Earman, John. 1993. "Carnap, Kuhn, and the Philosophy of Scientific Methodology." In *World Changes. Thomas Kuhn and the Nature of Science*. Edited by Paul Horwich, pp. 9–36. Cambridge, MA: The MIT Press.
Eckart, Carl. 1926. "Operator Calculus and the Solution of the Equations of Quantum Dynamics." *Physical Review* 28: 711–726.
———. 1930. "The Application of the Group Theory to the Quantum Dynamics of Monatomic Systems." *Reviews of Modern Physics* 2: 305–330.
Eckert, Michael. 2013. *Arnold Sommerfeld. Atomphysiker und Kulturbote 1868–1951. Eine Biographie*. Göttingen: Wallstein. Page references to English translation (by Tom Artin): *Arnold Sommerfeld. Science, Life and Turbulent Times*. New York: Springer, 2013.
———. 2020. *Establishing Quantum Physics in Munich. Emergence of Arnold Sommerfeld's Quantum School*. Cham: Springer.
Eddington, Arthur Stanley. 1928. "The Electrical State of a Star." In Como (1928, Vol. 2, pp. 541–544).

Ehrenfest, Paul. 1913. "Bemerkung betreffs der spezifischen Wärme zweiatomiger Gase." *Verhandlungen der Deutschen Physikalischen Gesellschaft* 15: 451–457. Reprinted in facsimile in Ehrenfest (1959, pp. 333–339).

Ehrenfest, Paul, and Gregory Breit. 1922. "Ein bemerkenswerter Fall von Quantisierung." *Zeitschrift für Physik* 9: 207–210.

Einstein, Albert. 1905. "Über eine die Erzeugung und die Verwandlung des Lichtes betreffenden heuristischen Gesichtspunkts." *Annalen der Physik* 17: 132–148. Reprinted in Einstein (1987–2021, Vol. 2, Doc. 14). English translation in Stachel (2005, pp. 177–197).

——. 1909a. "Zum gegenwärtigen Stand des Strahlungsproblems." *Physikalische Zeitschrift* 10: 185–193. Reprinted in Einstein (1987–2021, Vol. 2, Doc. 56).

——. 1909b. "Über die Entwicklung unserer Anschauungen über das Wesen und die Konstitution der Strahlung." *Physikalische Zeitschrift* 10: 817–825. Reprinted in Einstein (1987–2021, Vol. 2, Doc. 60).

——. 1916a. "Strahlungs-Emission und -Absorption nach der Quantentheorie." *Deutsche Physikalische Gesellschaft. Verhandlungen* 18: 318–323. Reprinted in Einstein (1987–2021, Vol. 6, Doc. 34).

——. 1916b. "Zur Quantentheorie der Strahlung." *Physikalische Gesellschaft Zürich. Mitteilungen* 18: 47–62. Reprinted as Einstein (1917a). Reprinted in Einstein (1987–2021, Vol. 6, Doc. 38).

——. 1917a. "Zur Quantentheorie der Strahlung." *Physikalische Zeitschrift* 18: 121–128. Reprint of Einstein (1916b). Page references to English translation in van der Waerden (1968, 63–77).

——. 1917b. "Zum Quantensatz von Sommerfeld und Epstein." *Deutsche Physikalische Gesellschaft. Verhandlungen* 19: 82–92. Reprinted in Einstein (1987–2021, Vol. 6, Doc. 45).

——. 1918. "Nachtrag." *Preußische Akademie der Wissenschaften* (Berlin). *Physikalisch-mathematische Klasse. Sitzungsberichte:* 478. (Supplement to Weyl 1918). Reprinted in Einstein (1987–2021, Vol. 7, Doc. 8).

——. 1922a. "Über ein den Elementarprozeß der Lichtemission betreffendes Experiment." *Preussische Akademie der Wissenschaften* (Berlin). *Physikalisch-mathematische Klasse. Sitzungsberichte:* 882–883. Reprinted in Einstein (1987–2021, Vol. 7, Doc. 68).

——. 1922b. "Zur Theorie der Lichtfortpffanzung in dispergierenden Medien." *Preussische Akademie der Wissenschaften* (Berlin). *Physikalisch-mathematische Klasse. Sitzungsberichte:* 18–22. Reprinted in Einstein (1987–2021, Vol. 13, Doc. 43).

——. 1924. "Quantentheorie des einatomigen idealen Gases." *Preußische Akademie der Wissenschaften* (Berlin). *Physikalisch-mathematische Klasse. Sitzungsberichte:* 261–267. Reprinted in Einstein (1987–2021, Vol. 14, Doc. 283).

——. 1925a. "Quantentheorie des einatomigen idealen Gases. Zweite Abhandlung." *Preußische Akademie der Wissenschaften* (Berlin). *Physikalisch-mathematische Klasse. Sitzungsberichte:* 3–14. Reprinted in Einstein (1987–2021, Vol. 14, Doc. 385).

——. 1925b. "Zur Quantentheorie des idealen Gases." *Preußische Akademie der Wissenschaften* (Berlin). *Physikalisch-mathematische Klasse. Sitzungsberichte:* 18–25. Reprinted in Einstein (1987–2021, Vol. 14, Doc. 427).

——. 1926a. "Vorschlag zu einem die Natur des elementaren Strahlungs-Emissionsprozesses betreffenden Experiment," *Die Naturwissenschaften* 14: 300–301. Reprinted in Einstein (1987–2021, Vol. 15, Doc. 223).

——. 1926b. "Über die Interferenzeigenschaftendes durch Kanalstrahlen emittierten Lichtes." *Preußische Akademie der Wissenschaften* (Berlin). *Physikalisch-mathematische Klasse. Sitzungsberichte:* 334–340. Reprinted in Einstein (1987–2021, Vol. 15, Doc. 278).

——. 1987–2021. *The Collected Papers of Albert Einstein*. 16 Vols. Edited by John Stachel, Martin J. Klein, Robert Schulmann, Diana Barkan Buchwald, et al. Princeton: Princeton University Press.

Einstein, Albert, Boris Podolsky, and Nathan Rosen. 1935. "Can Quantum-Mechanical Description of Physical Reality be Considered Complete." *Physical Review* 47: 777–780. Reprinted (along with Bohr's response and commentary by Rosenfeld) in Wheeler and Zurek (1983, pp. 138–141).

Eisberg, Robert, and Robert Resnick. 1985. *Quantum Physics of Atoms, Molecules, Solids, Nuclei, and Particles*. 2nd ed. Hoboken, NJ: Wiley.

Epstein, Paul. 1916a. "Zur Theorie des Starkeffektes." *Annalen der Physik* 50: 489–521.

——. 1916b. "Zur Quantentheorie." *Annalen der Physik* 51: 168–188.

——. 1922a. "Die Störungsrechnung im Dienste der Quantentheorie. I. Eine Methode der Störungsrechnung." *Zeitschrift für Physik* 8: 211–228.

——. 1922b. "Die Störungsrechnung im Dienste der Quantentheorie. II. Die numerische Durchführung der Methode." *Zeitschrift für Physik* 8: 305–320.

——. 1922c. "Die Störungsrechnung im Dienste der Quantentheorie. III. Kritische Bemerkungen zur Dispersionstheorie." *Zeitschrift für Physik* 9: 92–110.

——. 1926. "The Stark Effect from the Point of View of Schrödinger's Quantum Theory." *Physical Review* 28: 695–710.

Fano, Guido. 1971. *Mathematical Methods of Quantum Mechanics*. McGraw-Hill.

Farmelo, Graham. 2009. *The Strangest Man: The Hidden Life of Paul Dirac, Mystic of the Atom*. New York: Basic Books.

Fellows, Frederick Hugh. 1985. "J. H. Van Vleck: The Early Life and Work of a Mathematical Physicist." PhD dissertation, University of Minnesota.

Fermi, Enrico. 1926a. "Sulla quantizzazione del gas perfetto monoatomico." *Rendiconti Lincei* 3: 145–149.

——. 1926b. "Zur Quantelung des idealen einatomigen Gases." *Zeitschrift für Physik* 36: 902–912.

Feynman, Richard P., Robert B. Leighton, and Matthew Sands. 1964. *The Feynman Lectures on Physics*. 3 Vols. Reading, MA: Addison-Wesley.

Fierz, Marcus, and Victor F. Weisskopf, eds. 1960. *Theoretical Physics in the Twentieth Century. A Memorial Volume to Wolfgang Pauli*. New York: Interscience Publishers.

Fischer, Ernst. 1907. "Sur la convergence en moyenne." *Comptes Rendus de l'Académie des Science* 144: 1022–1024.

Forman, Paul. 1971. "Weimar Culture, Causality, and Quantum Theory, 1918–1927: Adaptation by German Physicists and Mathematicians to a Hostile Intellectual Environment." *Historical Studies in the Physical Sciences* 3: 1–115. Reprinted in Carson, Kojevnikov, and Trischler (2011, pp. 87–201).

——. 2022. "The Reception of the Forman Thesis in Modernity and Post-modernity." In Freire (2022, pp. 871–886).

Foster, John Stuart. 1924. "Observation of the Stark Effect in Hydrogen and Helium." *Physical Review* 23: 667–684.

——. 1927. "Application of Quantum Mechanics to the Stark Effect in Helium." *Proceedings of the Royal Society of London A* 117: 137–163.

Foster, John Stuart, and Mary Laura Chalk. 1926. "Observed Relative Intensities of Stark Components in Hydrogen." *Nature* 118: 592.

——. 1928. "Observed Relative Intensities of Stark Components of H_α." *Nature* 121: 830–831.

Foster, John Stuart, and Mary Laura Chalk. 1929. "Relative Intensities of Stark Components in Hydrogen." *Proceedings of the Royal Society of London A* 123: 108–118.

Fowler, Ralph. 1925a. "Applications of the Correspondence Principle to the Theory of Line-Intensities in Band-Spectra." *Philosophical Magazine Series 6* 49: 1272–1288.

———. 1925b. "A Note on the Summation Rules for the Intensities of Spectral Lines." *Philosophical Magazine Series 6* 50: 1079–1083.

Franck, James, and Paul Knipping. 1919. "Die Ionisierunsspannungen des Heliums." *Physikalische Zeitschrift* 20: 481–488.

Fréchet, Maurice. 1906. "Sur quelques points du calcul fonctionnel." *Rendiconti del Circolo Mathematico di Palermo* 22: 1–74.

———. 1908. "Essai de géométrie analytique à une infinité de coordonnées." *Nouvelles Annales de Mathématiques* 8: 97–116.

Freire, Olival Jr. 2015. *The Quantum Dissidents: Rebuilding the Foundations of Quantum Mechanics (1950–1990).* Heidelberg: Springer.

———. 2019. *David Bohm: A Life Dedicated to Understanding the Quantum World.* Cham: Springer International Publishing.

———, ed. 2022. *The Oxford Handbook of the History of Quantum Interpretations.* Oxford: Oxford University Press.

Frenkel, Jacov. 1929. *Einführung in die Wellenmechanik.* Berlin: Julius Springer.

Füchtbauer, Christian. 1920. "Die Absorption in Spektrallinien im Lichte der Quantentheorie." *Physikalische Zeitschrift* 21: 322–324.

Füchtbauer, Christian, and W. Hofmann. 1914. "Über Maximalintensität, Dämpfung und wahre Intensitätsverteilung von Serienlinien in Absorption." *Annalen der Physik* 43: 96–134.

Fues, Erwin. 1926a. "Das Eigenschwingungsspektrum zweiatomiger Moleküle in der Undulationsmechanik." *Annalen der Physik* 80: 367–396.

———. 1926b. "Zur Intensität der Bandenlinien und des Affnitätsspektrum zweiatomiger Moleküle." *Annalen der Physik* 81: 281–313.

Gal, Ofer. 2021. *The Origins of Modern Science. From Antiquity to the Scientific Revolution.* Cambridge: Cambridge University Press.

Gearhart, Clayton A. 2010. "'Astonishing Successes' and 'Bitter Disappointment': The Specific Heat of Hydrogen in Quantum Theory." *Archive for History of Exact Sciences* 64: 113–202.

Gimeno, Gonzalo, Mercedes Xipell, and Marià Baig. 2021. "Operator Calculus: The Lost Formulation of Quantum Mechanics. A Mathematical Reconstruction." *Archive for History of Exact Sciences* 75: 283–322.

Glazebrook, Sir Richard Tetley. 1886. "Report on Optical Theories." In *British Association for the Advancement of Science. Report—1885,* pp. 157–261. London: Spottiswoode.

Goldstein, Herbert, Charles P. Poole, and John L. Safko. 2002. *Classical Mechanics.* 3rd ed. San Francisco: Addison-Wesley.

Gordon, Walter. 1928. "Die Energieniveaus des Wasserstoffatoms nach der Diracschen Quantentheorie des Elektrons." *Zeitschrift für Physik* 48: 11–14.

Gordon, Walter, and Rudolph Minkowski. 1929. "Über die Intensitäten der Starkeffektkomponente der Balmerserie." *Die Naturwissenschaften* 17: 368.

Goudsmit, Samuel. 1971. "De ontdekking van de electronenrotatie." *Nederlands Tijdschrift voor Natuurkunde* 37: 386–392. English translation available online at www.lorentz.leidenuniv.nl/goudsmit.html.

———. 1976. "Fifty Years of Spin: It Might as Well Be Spin." *Physics Today* 29: 40–43.

Goudsmit, Samuel, and Ralph de Laer Kronig. 1925. "Die Intensität der Zeemankomponenten." *Die Naturwissenschaften* 13: 90.

Gould, Stephen Jay. 1980. "The Evolutionary Biology of Constraint." *Daedalus* 109 (2): 39–52.

———. 2002. *The Structure of Evolutionary Theory*. Cambridge, MA: Harvard University Press.

Gould, Stephen Jay, and Richard C. Lewontin. 1979. "The Spandrels of San Marco and the Panglossian Paradigm: A Critique of the Adaptationist Programme." *Proceedings of the Royal Society of London* B 205: 581–598.

Gouy, M. G. 1920. "Mesures spectrophotométriques absolues, et applications à la physique solaire." *Annales de Physique* 13: 188–216.

Grassmann, Hermann. 1844. *Die lineale Ausdehnungslehre, ein neuer Zweig der Mathematik*. Leipzig: Otto Wigand. Translation in: Lloyd C. Kannenberg, *A New Branch of Mathematics: The Ausdehnungslehre of 1844 and Other Works* (Chicago: Open Court, 1995).

Greenspan, Nancy Thorndike. 2005. *The End of the Certain World. The Life and Science of Max Born. The Nobel Physicist Who Ignited the Quantum Revolution*. New York: Basic Books.

Griffiths, David J. 1999. *Introduction to Electrodynamics*. 3rd ed. Upper Saddle River, NJ: Prentice Hall.

———. 2005. *Introduction to Quantum Mechanics*. 2nd ed. Upper Saddle River, NJ: Pearson Prentice Hall.

Haas, Arthur Erich. 1928. *Materiewellen und Quantenmechanik*. Leipzig: Akademische Verlagsgesellschaft.

Haber, Fritz, and Walter Zisch. 1922. "Anregung von Gasspektren durch chemische Reaktionen." *Zeitschrift für Physik* 9: 302–326.

Hallo, J. J. 1902. "De magnetische draaiing van het polarisatievlak in de nabijheid van een absorptieband." PhD dissertation, Universiteit van Amsterdam.

Hamilton, William Rowan. 1828. "Theory of Systems of Rays." *Transactions of the Royal Irish Academy* 15: 69–174. Reprinted in Hamilton (1931).

———. 1834. "On a General Method in Dynamics." *Philosophical Transactions of the Royal Society* 124: 247–308. Reprinted in Hamilton (1940).

———. 1835. "Second Essay on a General Method in Dynamics." *Philosophical Transactions of the Royal Society* 125: 95–144. Reprinted in Hamilton (1931).

———. 1837. "Third Supplement to an Essay on the Theory of Systems of Rays." *Transactions of the Royal Irish Academy* 17, part 1: 1–144. Reprinted in Hamilton (1931).

———. 1931. *The Mathematical Papers of Sir William Rowan Hamilton*. Vol. I: *Geometrical Optics*. Edited by Arthur W. Conway and John L. Synge. Cambridge: Cambridge University Press.

———. 1940. *The Mathematical Papers of Sir William Rowan Hamilton*. Vol. II. *Dynamics*. Edited by Arthur W. Conway and John L. Synge. Cambridge: Cambridge University Press.

Hankins, Thomas L. 1980. *Sir William Rowan Hamilton*. Baltimore MD: The Johns Hopkins University Press.

Hartree, Douglas Rayner. 1927a. "The Wave Mechanics of the Atom with a Non-Coulomb Central Field. Part I. Theory and Methods." *Proceedings of the Cambridge Philosophical Society* 24: 89–110.

———. 1927b. "The Wave Mechanics of the Atom with a Non-Coulomb Central Field. Part II. Some Results and Discussion." *Proceedings of the Cambridge Philosophical Society* 24: 111–132.

———. 1928. "The Wave Mechanics of the Atom with a Non-Coulomb Central Field. Part III. Term Values and the Intensities of Series in Optical Spectra." *Proceedings of the Cambridge Philosophical Society* 24: 426–437.

Hashagen, Ulf. 2010. "Die Habilitation von John von Neumann an der Friedrich-Wilhelms-Universität in Berlin: Urteile über einen ungarisch-jüdischen Mathematiker in Deutschland im Jahr 1927." *Historia Mathematica* 37: 242–280.

Heilbron, John L. 1983. "The Origins of the Exclusion Principle." *Historical Studies in the Physical Sciences* 13: 261–310.

Heisenberg, Werner. 1925a. "Über eine Anwendung des Korrespondenzprinzips auf die Frage nach der Polarisation des Fluoreszenzlichtes." *Zeitschrift für Physik* 31: 617–628.

——. 1925b. "Zur Quantentheorie der Multiplettstruktur und der anomalen Zeemaneffekte." *Zeitschrift für Physik* 32: 841–860.

——. 1925c. "Über die quantentheoretische Umdeutung kinematischer und mechanischer Beziehungen." *Zeitschrift für Phyik* 33: 879–893. Page references to translation in van der Waerden (1968, pp. 261–276).

——. 1926a. "Über quantummechanische Kinematik und Mechanik." *Mathematische Annalen* 95: 683–705.

——. 1926b. "Mehrkörperproblem und Resonanz in der Quantenmechanik." *Zeitschrift für Physik* 38: 411–426.

——. 1926c. "Über die Spektra von Atomsystemen mit zwei Elektronen." *Zeitschrift für Physik* 39: 499–518.

——. 1926d. "Schwankungserscheinungen und Quantenmechanik." *Zeitschrift für Physik* 40: 501–506.

——. 1927a. "Mehrkörperprobleme und Resonanz in der Quantenmechanik. II." *Zeitschrift für Physik* 41: 239–267.

——. 1927b. "Über den anschaulichen Inhalte der quantentheoretischen Kinematik und Mechanik." *Zeitschrift für Physik* 43: 172–198. Reprinted in English translation, "The Physical Content of Quantum Kinematics and Mechanics," in Wheeler and Zurek (1983, pp. 62–84).

——. 1930. *The Physical Principles of the Quantum Theory*. Chicago: University of Chicago Press. German edition: *Die physikalischen Prinzipien der Quantentheorie*. Leipzig: S. Hirzel. English translation by Carl Eckart and Frank C. Hoyt.

——. 1960. "Erinnerungen an die Zeit der Entwicklung der Quantenmechanik." In Fierz and Weisskopf (1960, pp. 40–47).

——. 1967. "Quantum Theory and Its Interpretation." In Rozental (1967, pp. 94–108).

——. 1971. *Physics and Beyond: Encounters and Conversations*. New York: Harper & Row.

Heisenberg, Werner, and Pascual Jordan. 1926. "Anwendung der Quantenmechanik auf das Problem der anomalen Zeemaneffekte." *Zeitschrift für Physik* 37: 263–277.

Hellinger, Ernst. 1910. "Neue Begründung der Theorie der quadratischen Formen von unendlich vielen Veränderlichen." *Journal für die reine und angewandte Mathematik* 136: 210–271.

Hellinger, Ernst, and Otto Toeplitz. 1910. "Grundlagen für eine Theorie der unendlichen Matrizen." *Mathematische Annalen* 69: 289–330.

——. 1927. "Integralgleichungen und Gleichungen mit unendlich vielen Unbekannten." *Enzyklopädie der mathematischen Wissenschaften* 2 (3-2): 1335–1597.

Helmholtz, Hermann von. 1892. "Elektromagnetische Theorie der Farbenstreuung." *Sitzungsberichte der Königlich Preußischen Akademie der Wissenschaften* (Berlin), pp. 1093–1109.

——. 1893. "Elektromagnetische Theorie der Farbenstreuung." *Annalen der Physik* 43: 389–405. Reprinted in *Wissenschaftliche Abhandlungen von Hermann von Helmholtz* (Leipzig: Barth, 1895), Vol. 3, pp. 505–525.

Hendry, John. 1981. "Bohr–Kramers–Slater: A Virtual Theory of Virtual Oscillators and Its Role in the History of Quantum Mechanics." *Centaurus* 25: 189–221.

Hertz, Gustav. 1920. "Über die Absorptionsgrenzen in der L-Serie." *Zeitschrift für Phyik* 3: 19–25.

Hilbert, David. 1904. "Grundzüge einer allgemeinen Theorie der linearen Integralgleichungen (Erste Mitteilung)." *Königliche Gesellschaft der Wissenschaften zu Göttingen. Mathematisch-physikalische Klasse. Nachrichten:* 49–91.

——. 1906a. "Grundzüge einer allgemeinen Theorie der linearen Integralgleichungen (Vierte Mitteilung)." *Königliche Gesellschaft der Wissenschaften zu Göttingen. Mathematisch-physikalische Klasse. Nachrichten:* 157–227.

——. 1906b. "Grundzüge einer allgemeinen Theorie der linearen Integralgleichungen (Fünfte Mitteilung)." *Königliche Gesellschaft der Wissenschaften zu Göttingen. Mathematisch-physikalische Klasse. Nachrichten:* 439–480.

——. 1912. *Grundzüge einer allgemeinen Theorie der linearen Integralgleichungen.* Leipzig, Berlin: Teubner.

Hilbert, David, John von Neumann, and Lothar Nordheim. 1928. "Über die Grundlagen der Quantenmechanik." *Mathematische Annalen* 98: 1–30.

Hilgevoord, Jan. 2002. "Time in Quantum Mechanics." *American Journal of Physics* 70: 301–306.

Hilgevoord, Jan, and Jos Uffink. 1988. "The Mathematical Expression of the Uncertainty Principle." In *Microphysical Reality and Quantum Description.* Edited by Alwyn van der Merwe, Gino Tarozzi, and Franco Selleri, pp. 91–114. Dordrecht: Kluwer.

——. 2016. "The Uncertainty Principle." In *The Stanford Encyclopedia of Philosophy* (Winter 2016 Edition). Edited by Edward N. Zalta. Available online at plato.stanford.edu/archives/win2016/entries/qt-uncertainty/.

Hoffmann, Dieter, ed. 2008. *Max Planck: Annalen Papers.* Weinheim: Wiley.

Hönl, Helmut. 1925. "Die Intensitäten der Zeemankomponenten." *Zeitschrift für Physik* 31: 340–354.

Howard, Don. 2004. "Who Invented the Copenhagen Interpretation? A Study in Mythology." *Philosophy of Science* 71: 669–682.

——. 2022. "The Copenhagen Interpretation." In Freire (2022, pp. 521–542).

Hufbauer, Karl. 2012. "From Student of Physics to Historian of Science: T. S. Kuhn's Education and Early Career, 1940–1958." *Physics in Perspective* 14: 421–470.

Hund, Friedrich. 1925. "Zur Deutung verwickelter Spektren." *Zeitschrift für Physik* 34: 296–308.

——. 1961. "Göttingen, Kopenhagen, Leipzig im Rückblick." In *Werner Heisenberg und die Physiker unserer Zeit.* Edited by Fritz Bopp, pp. 1–7. Braunschweig: Vieweg.

——. 1967. *Geschichte der Quantentheorie.* Mannheim: Bibliographisches Institut. Page references to English translation, *The History of Quantum Theory* (New York: Harper & Row, 1974), and to 3rd German ed. (Darmstadt: Wissenschaftliche Buchgesellschaft, 1984).

Hylleraas, Egil A. 1928. "Über den Grundzustand des Heliumatoms." *Zeitschrift für Physik* 48: 469–494.

——. 1929. "Neue Berechnung der Energie des Heliums im Grundzustande, sowie des tiefsten Terms von Ortho-Helium." *Zeitschrift für Physik* 54: 347–366.

Im, Gyeong Soon. 1996. "Experimental Constraints on Formal Quantum Mechanics: The Emergence of Born's Quantum Theory of Collision Processes in Göttingen, 1924–1927." *Archive for History of Exact Sciences* 50: 73–101.

Israel, Giorgio, and Ana Millán Gasca. 2009. *The World as a Mathematical Game.* Basel: Birkhäuser Verlag.

Israel, Werner. 1987. "Dark Stars: The Evolution of an Idea." In *Three Hundred Years of Gravitation.* Edited by Stephen W. Hawking and Werner Israel, pp. 199–276. Cambridge: Cambridge University Press.

Jähnert, Martin. 2016. "Practicing the Correspondence Principle in the Old Quantum Theory: A Transformation Through Application." PhD dissertation, Technische Universität, Berlin.
——. 2019. *Practicing the Correspondence Principle in the Old Quantum Theory*. Cham: Springer.
Jähnert, Martin, and Christoph Lehner. 2022. "The Early Debates About the Interpretation of Quantum Mechanics." In Freire (2022, pp. 135–172).
Jammer, Max. 1966. *The Conceptual Development of Quantum Mechanics*. New York: McGraw-Hill.
Janas, Michael, Michael E. Cuffaro, and Michel Janssen. 2022. *Understanding Quantum Raffles. Quantum Mechanics on an Informational Approach: Structure and Interpretation*. Cham: Springer.
Janssen, Michel. 2019. "Arches and Scaffolds: Bridging Continuity and Discontinuity in Theory Change." In Love and Wimsatt (2019, pp. 95–199).
Janssen, Michel, and Christoph Lehner, eds. 2014. *The Cambridge Companion to Einstein*. Cambridge: Cambridge University Press.
Janssen, Michel, and Matthew Mecklenburg. 2007. "From Classical to Relativistic Mechanics: Electromagnetic Models of the Electron." In *Interactions: Mathematics, Physics and Philosophy, 1860–1930*. Edited by V. F. Hendricks, K. F. Jørgensen, J. Lützen, and S. A. Pedersen, pp. 65–134. Berlin: Springer.
Janssen, Michel, and Jürgen Renn. 2015. "Arch and Scaffold: How Einstein Found His Field Equations." *Physics Today* (November): 30–36.
——. 2022. *How Einstein Found His Field Equations. Sources and Interpretation*. Cham: Birkhäuser.
Jeffeys, Harold. 1924. "On Certain Approximate Solutions of Linear Differential Equations of the Second Order." *Proceedings of the London Mathematical Society* 23: 428–436.
Joas, Christian, and Christoph Lehner. 2009. "The Classical Roots of Wave Mechanics: Schrödinger's Transformations of the Optical-mechanical Analogy." *Studies in History and Philosophy of Modern Physics* 40: 338–351.
Jordan, Pascual. 1924. "Zur Theorie der Quantenstrahlung." *Zeitschrift für Physik* 30: 297–319.
——. 1926a. "Über kanonische Transformationen in der Quantenmechanik." *Zeitschrift für Physik* 37: 383–386.
——. 1926b. "Über kanonische Transformationen in der Quantenmechanik. II." *Zeitschrift für Physik* 38: 513–517.
——. 1927a. "Über quantenmechanische Darstellung von Quantensprüngen." *Zeitschrift für Physik* 40: 661–666.
——. 1927b. "Über eine neue Begründung der Quantenmechanik." *Zeitschrift für Physik* 40: 809–838.
——. 1927c. "Über eine neue Begründung der Quantenmechanik." *Königliche Gesellschaft der Wissenschaften zu Göttingen. Mathematisch-physikalische Klasse. Nachrichten:* 161–169.
——. 1927d. "Kausalität und Statistik in der modernen Physik." *Die Naturwissenschaften* 15: 105–110. Translated (by J. Robert Oppenheimer) as: "Philosophical Foundations of Quantum Theory." *Nature* 119: 566–569 (reaction by Norman R. Campbell [1927] and reply by Jordan [1927f] on p. 779). Page references to English translation.
——. 1927e. "Anmerkung zur statistischen Deutung der Quantenmechanik." *Zeitschrift für Physik* 41: 797–800.
——. 1927f. "Philosophical Foundations of Quantum Theory." *Nature* 119: 779. Reply to Campbell (1927).
——. 1927g. "Über eine neue Begründung der Quantenmechanik II." *Zeitschrift für Physik* 44: 1–25.

———. 1927h. "Die Entwicklung der neuen Quantenmechanik." 2 Pts. *Die Naturwissenschaften* 15: 614–623, 636–649.

———. 1936. *Anschauliche Quantentheorie. Eine Einführung in die moderne Auffassung der Quantenerscheinungen.* Berlin: Julius Springer.

———. 1947. *Das Bild der modernen Physik.* Hamburg-Bergedorf: Stromverlag.

Jordan, Pascual, and Oskar Klein. 1927. "Zum Mehrkörperproblem der Quantentheorie." *Zeitschrift für Physik* 45: 751–765.

Jordi Taltavull, Marta. 2013. "Sorting Things Out: Drude and the Foundations of Classical Optics." In Badino and Navarro (2013, pp. 27–68).

———. 2017. "Transformation of Optical Knowledge from 1870 to 1925: Optical Dispersion between Classical and Quantum Physics." PhD dissertation, Humboldt University, Berlin.

Kalckar, Jørgen. 1967. "Niels Bohr and His Youngest Disciples." In Rozental (1967, pp. 227–239).

———. 1985. "Introduction." In Bohr (1972–2008, Vol. 6, pp. 7–53).

Kangro, Hans, ed. 1972. *Planck's Original Papers in Quantum Physics.* London: Taylor & Francis. Translations: Dirk ter Haar and Stephen G. Brush.

Kato, Tosio. 1951a. "Fundamental Properties of Hamiltonian Operators of the Schrödinger Type." *Transactions of the American Mathematical Society* 70: 196–211.

———. 1951b. "On the Existence of Solutions of the Helium Wave Equation." *Transactions of the American Mathematical Society* 70: 212–218.

Keller, Alex. 1983. *The Infancy of Atomic Physics: Hercules in his Cradle.* Oxford: Clarendon.

Kellner, Georg W. 1927. "Die Ionisierungsspannung des Heliums nach der Schrödingerschen Theorie." *Zeitschrift für Physik* 44: 91–109.

Klein, Martin J., ed. 1967. *Letters on Wave Mechanics.* New York: Philosophical Library. Translation of Przibram (1963).

———. 1970. "The First Phase of the Bohr–Einstein Dialogue." *Historical Studies in the Physical Sciences* 2: 1–39.

Klein, Martin J., Abner Shimony, and Trevor Pinch. 1979. "Paradigm Lost? A Review Symposium." *Isis* 4: 429–440.

Klein, Oskar. 1967. "Glimpses of Bohr as a Scientist and Thinker." In Rozental (1967, pp. 74–93).

Kojevnikov, Alexei. 2020. *The Copenhagen Network. The Birth of Quantum Mechanics from a Postdoctoral Perspective.* Berlin: Springer.

———. 2022. "Quantum Historiography and Cultural History: Revisiting the Forman Thesis." In Freire (2022, pp. 887–908).

Konno, Hiroyuki. 1978. "The Historical Roots of Born's Probabilistic Interpretation." *Japanese Studies in the History of Science* 17: 129–145.

———. 1993. "Kramers' Negative Dispersion, the Virtual Oscillator Model, and the Correspondence Principle." *Centaurus* 36: 117–166.

Kopfermann, Hans, and Rudolf Ladenburg. 1928. "Experimental Proof of 'Negative Dispersion'." *Nature* 122: 438–439.

Kox, Anne J. 2013. "Hendrik Antoon Lorentz's Struggle with Quantum Theory." *Archive for History of Exact Sciences* 67: 149–170.

Kragh, Helge. 1979. "Niels Bohr's Second Atomic Theory." *Historical Studies in the Physical Sciences* 10: 123–186.

———. 1990. *Dirac. A Scientific Biography.* Cambridge: Cambridge University Press.

———. 2012. *Niels Bohr and the Quantum Atom. The Bohr Model of Atomic Structure, 1913–1925.* Oxford: Oxford University Press.

Kramers, Hendrik A. 1919. "Intensities of Spectral Lines. On the Application of the Quantum Theory to the Problem of the Relative Intensities of the Components of the Fine Structure and of the Stark Effect of the Lines of the Hydrogen Spectrum." *Det Kongelige Danske Videnskabernes Selskab. Skrifter. Naturvidenskabelig og Matematisk Afdeling* 8 (3.3): 285–386. Reprinted in Kramers (1956, pp. 3–108).

——. 1923a. "Über das Modell des Heliumatoms." *Zeitschrift für Phyik* 13: 312–341.

——. 1923b. "Das Korrespondenzprinzip und der Schalenbau des Atoms." *Die Naturwissenschaften* 11: 550–559.

——. 1924a. "The Law of Dispersion and Bohr's Theory of Spectra." *Nature* 113: 673–676. Reprinted in van der Waerden (1968, pp. 177–180).

——. 1924b. "The Quantum Theory of Dispersion." *Nature* 114: 310–311. Reprinted in van der Waerden (1968, pp. 199–201).

——. 1926. "Wellenmechanik und halbzahlige Quantisierung." *Zeitschrift für Physik* 39: 828–840.

——. 1928. "La diffusion de la lumière par les atomes." In Como (1928, Vol. 2, pp. 545–557).

——. 1935. "Atom- og kvanteteoriens udvikling i årene 1913–1925." *Fysisk Tidsskrift* 33: 82–96.

Kramers, Hendrik A., and Werner Heisenberg. 1925. "Über die Streuung von Strahlung durch Atome." *Zeitschrift für Physik* 31: 681–707. Page references to translation in van der Waerden (1968, pp. 223–252).

Kramers, Hendrik A., and Helge Holst. 1922. *Bohrs atomteori: almenfatteligt fremstillet*. Copenhagen: Gyldendal Nordisk forlag.

——. 1925. *Das Atom und die Bohrsche Theorie seines Baues. Gemeinverständlich dargestellt*. Berlin: Springer. Translation (with revisions) of Kramers and Holst (1922).

Kramers, Hendrik A., and Wolfgang Pauli. 1923. "Zur Theorie der Bandenspektren." *Zeitschrift für Physik* 13: 351–367.

Kratzer, Adolf. 1923. "Die Feinstruktur einer Klasse von Bandenspektren." *Annalen der Physik* 376: 72–103.

Kreyszig, Erwin. 1978. *Introductory Functional Analysis with Applications*. New York: Wiley.

Kronig, Ralph de Laer. 1925a. "Über die Intensität der Mehrfachlinien und ihrer Zeemankomponenten." *Zeitschrift für Physik* 31: 885–897.

——. 1925b. "Über die Intensität der Mehrfachlinien und ihrer Zeemankomponenten. II." *Zeitschrift für Physik* 33: 261–272.

——. 1926. "The Dielectric Constant of Diatomic Dipole-Gases on the New Quantum Mechanics." *Proceedings of the National Academy of Sciences of the United States of America* 12: 488–493.

——. 1960. "The Turning Point." In Fierz and Weisskopf (1960, pp. 5–39).

Kuhn, Thomas S. 1957. *The Copernican Revolution. Planetary Astronomy in the Development of Western Thought*. Cambridge, MA: Harvard University Press. Page reference to 20th printing, 1999.

——. 1962. *The Structure of Scientific Revolutions*. Chicago: University of Chicago Press. Page reference to the 4th (50th anniversary) ed., 2012.

——. 1966. "The Crisis of the Old Quantum Theory, 1922–1925." Unpublished manuscript. Massachusetts Institute of Technology Archives.

——. 1970. "Reflections on My Critics." In Lakatos and Musgrave (1970, pp. 231–278).

——. 1978. *Black-Body Theory and the Quantum Discontinuity, 1894–1912*. Oxford: Oxford University Press. Reprinted, including Kuhn (1984) as a new afterword, as Kuhn (1987).

——. 1984. "Revisiting Planck." *Historical Studies in the Physical Sciences* 14: 231–252. Page reference to reprint in Kuhn (1987, pp. 349–370).

——. 1987. *Black-Body Theory and the Quantum Discontinuity, 1894–1912.* 2nd ed. Chicago: University of Chicago Press. Reprint of Kuhn (1978) including Kuhn (1984) as a new afterword.

Kuhn, Thomas S., John L. Heilbron, Paul Forman, and Lini Allen. 1967. *Sources for History of Quantum Physics. An Inventory and Report.* Philadelphia: The American Philosophical Society.

Kuhn, Werner. 1925. "Über die Gesamtstärke der von einem Zustande ausgehenden Asborptionslinien." *Zeitschrift für Physik* 33: 408–412. Translation in van der Waerden (1968, pp. 253–257).

Labatut, Benjamín. 2020. *When We Cease to Understand the World.* New York: New York Review of Books. Translated from the Spanish by Adrian Nathan West.

Lacki, Jan. 2000. "The Early Axiomatizations of Quantum Mechanics: Jordan, von Neumann and the Continuation of Hilbert's Program." *Archive for History of Exact Sciences* 54: 279–318.

——. 2004. "The Puzzle of Canonical Transformations in Early Quantum Mechanics." *Studies in History and Philosophy of Modern Physics* 35: 317–344.

Ladenburg, Rudolf. 1921. "Die quantentheoretische Deutung der Zahl der Dispersionselektronen." *Zeitschrift für Physik* 4: 451–468. Page references to translation in van der Waerden (1968, pp. 139–157).

Ladenburg, Rudolf, and Stanislaw Loria. 1908. "Über die Dispersion des leuchtenden Wasserstoffs." *Verhandlungen der Deutschen Physikalischen Gesellschaft* 10: 858–866.

Ladenburg, Rudolf, and Rudolph Minkowski. 1921. "Die Verdampfungswärme des Natriums und die Übergangswahrscheinlichkeit des Na-Atoms aus dem Resonanz- in den Normalzustand auf Grund optischer Messungen." *Zeitschrift für Physik* 6: 153–164.

Ladenburg, Rudolf, and Fritz Reiche. 1923. "Absorption, Zerstreuung und Dispersion in der Bohrschen Atomtheorie." *Die Naturwissenschaften* 11: 584–598.

——. 1924. "Dispersionsgesetz und Bohrsche Atomtheorie." *Die Naturwissenschaften* 12: 672–673.

Lakatos, Imre. 1970. "Falsification and the Methodology of Scientific Research Programmes." In Lakatos and Musgrave (1970, pp. 91–196).

Lakatos, Imre, and Alan Musgrave, eds. 1970. *Criticism and the Growth of Knowledge.* Cambridge: Cambridge University Press.

Lanczos, Cornelius. 1926. "Über eine Feldmäßige Darstellung der neuen Quantenmechanik." *Zeitschrift für Physik* 35: 812–830.

Landé, Alfred. 1922. "Zur Theorie der anomalen Zeeman- und magneto-mechanischen Effekte." *Zeitschrift für Physik* 11: 353–363.

——. 1923a. "Fortschritte beim Zeemaneffekt." *Ergebnisse der exakten Naturwissenschaften* 11: 147–162.

——. 1923b. "Zur Theorie der Röntgenspektren." *Zeitschrift für Phyik* 16: 391–396.

——. 1924a. "Das Wesen der relativistischen Röntgendubletts." *Zeitschrift für Phyik* 24: 88–97.

——. 1924b. "Die absoluten Intervalle der optischen Dubletts und Tripletts." *Zeitschrift für Phyik* 25: 46–57.

——. 1926. "Neue Wege der Quantentheorie." *Die Naturwissenschaften* 14: 455–458.

Landsberg, Grigori S., and Leonid I. Mandelstam. 1928. "Eine neue Erscheinung bei der Lichtzerstreuung in Krystallen." *Die Naturwissenschaften* 16: 557–558.

Landsman, Klaas. 2022. "The Axiomatization of Quantum Theory through Functional Analysis: Hilbert, von Neumann, and Beyond." In Freire (2022, pp. 473–493).

Langevin, Paul. 1905. "Sur la Théorie du Magnétisme." *Journal de Physique* 4: 678–688.

Laue, Max von. 1914. "Die Freiheitsgrade von Strahlenbündeln." *Annalen der Physik* 44: 1197–1212.

Lax, Peter D. 2002. *Functional Analysis*. New York: Wiley.

Lehner, Christoph. 2014. "Einstein's Realism and His Critique of Quantum Mechanics." In Janssen and Lehner (2014, pp. 306–353).

Lemons, Don S., William R. Shanahan, and Louis J. Buchholtz. 2022. *On the Trail of Blackbody Radiation. Max Planck and the Physics of His Era.* Cambridge, MA: The MIT Press.

Leone, Matteo, Alessandro Paoletti, and Nadia Robotti. 2004. "A Simultaneous Discovery: The Case of Johannes Stark and Antonino Lo Surdo." *Physics in Perspective* 6: 271–294.

Leroux, F.-P. 1862. "Recherches sur les indices de réfraction des corps qui ne prennent l'état gazeux qu'a des températures elevées: Dispersion anomale de la vapeur d'iode." *Comptes Rendus de l'Académie des Science* 55: 126–128.

Lewis, Gilbert N. 1926. "The Conservation of Photons." *Nature* 118: 874–875.

Lewontin, Richard C. 1963. "Models, Mathematics and Metaphors." *Synthese* 15: 222–244.

London, Fritz. 1926a. "Über die Jacobischen Transformationen der Quantenmechanik." *Zeitschrift für Physik* 37: 915–925.

———. 1926b. "Winkelvariable und kanonische Transformationen in der Undulationsmechanik." *Zeitschrift für Physik* 40: 193–210.

Lorentz, Hendrik Antoon. 1878. "Over het verband tusschen de voortplantingssnelheid van het licht en de dichtheid en samenstelling der middenstoffen." *Koninklijke Akademie van Wetenschappen te Amsterdam. Wis- en Natuurkundige Afdeeling. Verslagen van de Gewone Vergaderingen:* 1–112. English translation, "Concerning the Relation Between the Velocity of Propagation of Light and the Density and Composition of Media," in Lorentz (1934–1939, Vol. 2, pp. 1–119).

———. 1892. "La théorie électromagnétique de Maxwell et son application aux corps mouvants." *Archives Néerlandaises des Sciences Exactes et Naturelles* 25: 363–552. Reprint: Leyden: Brill, 1892. Also reprinted in Lorentz (1934–1939, Vol. 2, pp. 164–343).

———. 1909. *The Theory of Electrons and Its Applications to the Phenomena of Light and Radiant Heat.* Leipzig: Teubner. 2nd expanded ed. 1916.

———. 1916. *Les théories statistiques en thermodynamique: conférences faites au Collège de France en novembre 1912.* Leipzig: Teubner.

———. 1934–1939. *Collected Papers.* 9 Vols. Edited by Pieter Zeeman and Adriaan D. Fokker. The Hague: Nijhoff.

———. 2008. *The Scientific Correspondence of H. A. Lorentz.* Vol. 1. Edited by Anne J. Kox. New York: Springer.

Love, Alan C., and William C. Wimsatt, eds. 2019. *Beyond the Meme. Development and Structure in Cultural Evolution.* Minneapolis: University of Minnesota Press.

Ludwig, Günther. 1968. *Wave Mechanics.* Oxford: Pergamon.

Lyman, Theodore. 1922. "The Spectrum of Helium in the Extreme Ultraviolet." *Nature* 110: 278–279.

MacKinnon, Edward Michael. 1977. "Heisenberg, Models, and the Rise of Matrix Mechanics." *Historical Studies in the Physical Sciences* 8: 137–188.

———. 1982. *Scientific Explanation and Atomic Physics.* Chicago: University of Chicago Press.

Manneback, Charles. 1926. "Die Dielektrizitätskonstante der zweiatomigen Dipolgase nach der Wellenmechanik." *Physikalische Zeitschrift* 27: 563–569.

Mark, Herman F., and Raimund Wierl. 1929. "Über die relativen Intensitäten der Starkeffektkomponenten von H_β und H_γ." *Zeitschrift für Physik* 53: 526–541.

McIntyre, David H. 2012. *Quantum Mechanics, A Paradigms Approach*. Upper Saddle River, NJ: Pearson.

Mehra, Jagdish, and Helmut Rechenberg. 1982a. *The Historical Development of Quantum Theory*. Vol. 1. *The Quantum Theory of Planck, Einstein, Bohr and Sommerfeld: Its Foundations and the Rise of Its Difficulties*. Berlin: Springer.

——. 1982b. *The Historical Development of Quantum Theory*. Vol. 2. *The Discovery of Quantum Mechanics 1925*. Berlin: Springer.

——. 1982c. *The Historical Development of Quantum Theory*. Vol. 3. *The Formulation of Matrix Mechanics and Its Modifications 1925–1926*. Berlin: Springer.

——. 1982d. *The Historical Development of Quantum Theory*. Vol. 4. Part 1. *The Fundamental Equations of Quantum Mechanics. 1925–1926*. Part 2. *The Reception of the New Quantum Mechanics. 1925–1926*. Berlin: Springer.

——. 1987. *The Historical Development of Quantum Theory*. Vol. 5. *Erwin Schrödinger and the Rise of Wave Mechanics*. Part 1. *Schrödinger in Vienna and Zurich 1887–1925*. Part 2. *The Creation of Wave Mechanics. Early Response and Applications 1925–1926*. Berlin: Springer.

——. 2000–2001. *The Historical Development of Quantum Theory*. Vol. 6. *The Completion of Quantum Mechanics 1926–1941*. Part 1. *The Probability Interpretation and the Statistical Transformation Theory, the Physical Interpretation, and the Empirical and Mathematical Foundations of Quantum Mechanics 1926–1932*. Part 2. *The Conceptual Completion of the Extension of Quantum Mechanics 1932–1941. Epilogue: Aspects of the Further Development of Quantum Theory 1942–1999. Subject Index: Volumes 1 to 6*. Berlin: Springer.

Mensing, Lucy. 1926. "Die Rotations-Schwingungsbanden nach der Quantenmechanik." *Zeitschrift für Physik* 36: 814–823.

Mensing, Lucy, and Wolfgang Pauli. 1926. "Über die Dielektrizitätskonstante von Dipolgasen nach der Quantenmechanik." *Physikalische Zeitschrift* 27: 509–512.

Messiah, Albert. 1966. *Quantum Mechanics*. Vol. 1. Amsterdam: North-Holland.

Meyer-Abich, Klaus Michael. 1965. *Korrespondenz, Individualität und Komplementarität: Eine Studie zur Geistesgeschichte der Quantentheorie in den Beiträgen Niels Bohrs*. Wiesbaden: Franz Steiner Verlag.

Midwinter, Charles, and Michel Janssen. 2013. "Kuhn Losses Regained: Van Vleck from Spectra to Susceptibilities." In Badino and Navarro (2013, pp. 137–205).

Millikan, Robert A. 1917. *The Electron: Its Isolation and Measurement and the Determination of Some of its Properties*. Chicago: University of Chicago Press.

Millikan, Robert A., and Ira S. Bowen. 1926. "Series Spectra of Beryllium, BeI and BeII." *Physical Review* 28: 256–258.

Moore, Walter. 1989. *Schrödinger. Life and Thought*. Cambridge: Cambridge University Press.

Muller, Fred A. 1997–1999. "The Equivalence Myth of Quantum Mechanics." 2 Pts plus addendum. *Studies in History and Philosophy of Modern Physics* 28: 35–61, 219–247 (I–II), 30: 543–545 (addendum).

Münster, Gernot. 2020. "(K)eine klassische Karriere?" *Physik Journal* 19 (June): 30–34.

Nakashima, Hiroyuki, and Hiroshi Nakatsuji. 2007. "Solving the Schrödinger Equation for Helium Atom and Its Isoelectronic Ions with the Free Iterative Complement Interaction (ICI) Method." *Journal of Chemical Physics* 127: 224104.

Nath, Biman B. 2013. *The Story of Helium and the Birth of Astrophysics*. New York: Springer.

Nernst, Walther. 1906. "Über die Berechnung chemischer Gleichgewichte aus thermischen Messungen." *Königliche Gesellschaft der Wissenschaften zu Göttingen. Mathematisch-physikalische Klasse, Nachrichten:* 1–40.

Ornstein, Leonard Salomon, and Herman Carel Burger. 1924a. "Strahlungsgesetz und Intensität von Mehrfachlinien." *Zeitschrift für Physik* 24: 41–47.

——. 1924b. "Intensitäten der Komponenten im Zeemaneffekt." *Zeitschrift für Physik* 28: 135–141.

——. 1924c. "Nachschrift zu der Arbeit Intensität der Komponenten im Zeemaneffekt." *Zeitschrift für Physik* 29: 241–242.

Pais, Abraham. 1967. "Reminiscences from the Post-war Years." In Rozental (1967, pp. 215–226).

——. 1982a. *"Subtle is the Lord ... " The Science and the Life of Albert Einstein.* Oxford: Oxford University Press.

——. 1982b. "Max Born's Statistical Interpretation of Quantum Mechanics." *Science* 218: 1193–1198.

——. 1986. *Inward Bound. Of Matter and Forces in the Physical World.* Oxford: Clarendon.

——. 1991. *Niels Bohr's Times, in Physics, Philosophy, and Polity.* Oxford: Clarendon.

Pauli, Wolfgang. 1921a. "Relativitätstheorie." In *Encyklopädie der mathematischen Wissenschaften, mit Einschluss ihrer Anwendungen.* Vol. 5, *Physik,* Pt 2. Edited by Arnold Sommerfeld, pp. 539–775. Leipzig: Teubner. English translation: London: Pergamon, 1958. Reprinted: New York: Dover, 1981.

——. 1921b. "Zur Theorie der Dieelektrizitätskonstante zweiatomiger Dipolgase." *Zeitschrift für Physik* 6: 319–327. Reprinted in Pauli (1964, Vol. 2, pp. 39–47).

——. 1923. "Über die Gesetzmässigkeiten des anomalen Zeemaneffektes." *Zeitschrift für Phyik* 16: 155–164. Reprinted in Pauli (1964, Vol. 2, pp. 151–160).

——. 1925a. "Über den Einfluss der Geschwindigkeitsabhängigkeit der Elektronenmasse auf den Zeemaneffekt." *Zeitschrift für Phyik* 31: 373–385. Reprinted in Pauli (1964, Vol. 2, pp. 201–213).

——. 1925b. "Über den Zusammenhang des Abschlusses der Elektronengruppen im Atom mit der Komplexstruktur der Spektren." *Zeitschrift für Phyik* 31: 765–783. Reprinted in Pauli (1964, Vol. 2, pp. 214–232).

——. 1925c. "Ueber die Intensitäten der im elektrischen Feld erscheinenden Kombinationslinien." *Det Kongelige Danske Videnskabernes Selskab. Skrifter. Naturvidenskabelig og Matematisk Afdeling* 7 (3): 3–20. Reprinted in Pauli (1964, Vol. 2, pp. 233–250).

——. 1926a. "Über das Wasserstoffspektrum vom Standpunkt der neuen Quantenmechanik." *Zeitschrift für Physik* 36: 336–363. Reprinted in Pauli (1964, Vol. 2, pp. 252–279). Page references to translation in van der Waerden (1968, pp. 387–415).

——. 1926b. "Quantentheorie." In *Handbuch der Physik.* Vol. 23, Pt. 1. Edited by Hans Geiger and Karl Scheel, pp. 1–278. J. Springer-Verlag. Reprinted in Pauli (1964, Vol. 1, pp. 269–548).

——. 1927. "Über Gasentartung und Paramagnetismus." *Zeitschrift für Physik* 41: 81–102. Reprinted in Pauli (1964, Vol. 2, pp. 284–305).

——. 1930. "Review of Born and Jordan (1930)." *Die Naturwissenschaften* 18: 602.

——. 1932. "Les théories quantiques du magnétisme: l'électron magnetique." In *Le Magnetisme. Rapports et Discussions du Sixième Conseil de Physique Tenu à Bruxelles du 20 au 25 Octobre 1930,* pp. 175–338 (Discussion: 239–280). Paris: Gauthier-Villars. Reprinted in Pauli (1964, pp. 502–607).

——. 1946. "Remarks on the History of the Exclusion Principle." *Science* 103: 213–215. Reprinted in Pauli (1964, Vol. 2, pp. 1073–1075).

——. 1947. "Exclusion Principle and Quantum Mechanics." In *Prix Nobel 1946.* Stockholm. Nobel lecture delivered December 13, 1946. Reprinted in Pauli (1964, Vol. 2, pp. 1080–1096).

——. 1964. *Collected Scientific Papers*. 2 Vols. Edited by Ralph de Laer Kronig and Victor F. Weisskopf. New York: Interscience Publishers.

——. 1979. *Wissenschaftlicher Briefwechsel mit Bohr, Einstein, Heisenberg u.a. Band I: 1919–1929/Scientific Correspondence with Bohr, Einstein, Heisenberg, a.o. Volume I 1919–1929*. Edited by Armin Hermann, Karl von Meyenn, and Victor F. Weisskopf. New York: Springer.

——. 1980. *The General Principles of Quantum Mechanics*. Berlin: Springer.

——. 1999. *Wissenschaftlicher Briefwechsel mit Bohr, Einstein, Heisenberg u.a. Band IV, Teil II: 1953–1954/Scientific Correspondence with Bohr, Einstein, Heisenberg, a.o. Volume IV, Part II: 1953–1954*. Edited by Karl von Meyenn. Berlin: Springer.

Pauli, Wolfgang, Léon Rosenfeld, and Victor F. Weisskopf, eds. 1955. *Niels Bohr and the Development of Physics; Essays Dedicated to Niels Bohr on the Occasion of his Seventieth Birthday*. Oxford: Pergamon Press.

Pauling, Linus. 1926. "The Quantum Theory of the Dielectric Constant of Hydrogen Chloride and Similar Gases." *Physical Review* 27: 568–577.

Peano, Giuseppe. 1888. *Calcolo geometrico secondo l'Ausdehnungslehre di H. Grassmann, preceduto dalle operazioni della logica deduttiva*. Torino: Fratelli Bocca Publishers.

Peckhaus, Volker. 1990. *Hilbertprogramm und Kritische Philosophie. Das Göttinger Modell interdisziplinärer Zusammenarbeit zwischen Mathematik und Philosophie*. Göttingen: Vandenhoeck & Ruprecht.

Pérez, Enric, and Joana Ibáñez. 2022. "Indistinguishable Elements in the Origin of Quantum Statistics. The Case of Fermi-Dirac Statistics." *European Physical Journal H* 47: 1–25.

Pérez, Enric, and Blai Pié Valls. 2015. "Ehrenfest's Adiabatic Hypothesis in Bohr's Quantum Theory." In Aaserud and Kragh (2015, pp. 272–289).

Planck, Max. 1897–1899. "Über irreversible Strahlungsvorgänge. 1.–5. Mittheilung." *Königlich Preussische Akademie der Wissenschaften* (Berlin). *Sitzungsberichte:* 57–68, 715–717, 1122–1145 (1897), 449–476 (1898), 440–480 (1899). Reprinted in Planck (1958, Vol. 1, 493–504 (I), 505–507 (II), 508–531 (III), 532–559 (IV), 560–600 (V)).

——. 1900a. "Über irreversible Strahlungsvorgänge." *Annalen der Physik* 1: 69–122. Reprinted in facsimile in Planck (1958, Vol. 1, pp. 614–667) and in Hoffmann (2008, pp. 461–514).

——. 1900b. "Über eine Verbesserung der Wien'schen Spectralgleichung." *Deutsche Physikalische Gesellschaft. Verhandlungen* 2: 202–204. Reprinted in Planck (1958, Vol. 1, 687–689). English translation in Kangro (1972, pp. 35-45).

——. 1900c. "Zur Theorie des Gesetzes der Energieverteilung im Normalspektrum." *Deutsche Physikalische Gesellschaft. Verhandlungen* 2: 237–245. Reprinted in Planck (1958, Vol. 1, pp. 698–706). English translation in Kangro (1972, pp. 35-45).

——. 1901. "Über das Gesetz der Energieverteilung im Normalspektrum." *Annalen der Physik* 4: 553–563. Reprinted in facsimile in Planck (1958, Vol. 1, pp. 717–727) and in Hoffmann (2008, pp. 537–547). Slightly abbreviated English translation in Shamos (1959, pp. 305–313).

——. 1913. *Vorlesungen über die Theorie der Wärmestrahlung*. 2nd ed. (1st ed.: 1906). Leipzig: Barth. English translation: *The Theory of Heat Radiation* (New York: Dover, 1991). Reprinted in Planck (1988, 1–239).

——. 1920. *Die Entstehung und bisherige Entwicklung der Quantentheorie*. Leipzig: Barth. Nobel lecture held in Stockholm, June 2, 1920. Reprinted in Planck (1958, Vol. 3, pp. 121–136). English translation ("The Origin and Development of the Quantum Theory"): Oxford: Clarendon Press, 1922.

Planck, Max. 1925a. "Zur Frage der Quantelung einatomiger Gase." *Preußische Akademie der Wissenschaften* (Berlin). *Physikalisch-mathematische Klasse. Sitzungsberichte:* 49–57. Reprinted in facsimile in Planck (1958, Vol. 2, pp. 584–592).

——. 1925b. "Über die statistische Entropiedefinition." *Preußische Akademie der Wissenschaften* (Berlin). *Physikalisch-mathematische Klasse. Sitzungsberichte:* 442–451. Reprinted in facsimile in Planck (1958, Vol. 2, pp. 593–606).

——. 1958. *Physikalische Abhandlungen und Vorträge.* 3 Vols. Braunschweig: Vieweg.

——. 1988. *The Theory of Heat Radiation.* New York: American Institute of Physics. With an introduction by Allan A. Needell.

Powell, John L., and Bernd Crasemann. 1962. *Quantum Mechanics.* Reading MA: Addison-Wesley.

Prugovecki, Eduard. 1981. *Quantum Mechanics in Hilbert Space.* 2nd ed. New York: Dover.

Przibram, Karl, ed. 1963. *Briefe zur Wellenmechanik.* Vienna: Springer-Verlag. English translation: Klein (1967).

Raman, Chandrasekhara Venkata. 1928. "A New Radiation." *Indian Journal of Physics* 2: 387–398.

Rechenberg, Helmut. 2010. *Werner Heisenberg – Die Sprache der Atome. Leben und Wirken – Eine wissenschaftliche Biographie. Die "Fröhliche Wissenschaft" (Jugend bis Nobelpreis).* Berlin: Springer.

Reich, Karin. 1994. *Die Entwicklung des Tensorkalküls. Vom absoluten Differentialkalkül zur Relativitätstheorie.* Basel: Birkhäuser.

Reiche, Fritz, and Willy Thomas. 1925. "Über die Zahl der Dispersionselektronen, die einem stationären Zustand zugeordnet sind." *Zeitschrift für Physik* 34: 510–525.

Reid, Constance. 1986. *Hilbert—Courant.* New York: Springer. Combines two earlier biographies by the same author: *Hilbert* (New York: Springer, 1970) and *Courant in Göttingen and New York: The Story of an Improbable Mathematician* (New York: Springer, 1976).

Renn, Jürgen. 2006. *Auf den Schultern von Riesen und Zwergen: Einsteins unvollendete Revolution.* New York: Wiley. Revised English translation: Hanoch Gutfreund and Jürgen Renn, *The Einsteinian Revolution. The Historical Roots of His Breakthroughs* (Princeton: Princeton University Press, 2024).

——. 2020. *The Evolution of Knowledge. Rethinking Science for the Anthropocene.* Princeton: Princeton University Press.

Ricci, Gregorio, and Tullio Levi-Civita. 1901. "Méthodes de calcul différentiel absolu et leurs applications." *Mathematische Annalen* 54: 125–201.

Riesz, Frigyes. 1907a. "Über orthogonale Funktionensysteme." *Königliche Gesellschaft der Wissenschaften zu Göttingen. Mathematisch-physikalische Klasse. Nachrichten:* 116–122.

——. 1907b. "Sur les systèmes orthogonaux de fonctions." *Comptes Rendus* 144: 615–619.

——. 1913. *Les systemes d'équations linéaires à une infinité d'inconnues.* Paris: Gauthier-Villars.

Ritz, Walther. 1909. "Über eine neue Methode zur Lösung gewisser Variationsprobleme der mathematischen Physik." *Journal für die reine und angewandte Mathematik* 135: 1–61.

Robertson, Peter. 1979. *The Early Years. The Niels Bohr Institute, 1921–1930.* Copenhagen: Akademisk Forlag.

Ross, Perley Ason. 1923a. "Change in Wave-length by Scattering." *Proceedings of the National Academy of Sciences of the United States of America* 9: 246–248.

——. 1923b. "The Wave-length and Intensity of Scattered X-rays." *Physical Review* 22: 524–525.

Rowe, David E. 1997. "Perspective on Hilbert (Essay Review of books by Herbert Mehrtens, Volker Peckhaus, and Michael-Markus Toepell)." *Perspectives on Science* 5: 533–570.

Rozental, Stefan, ed. 1967. *Niels Bohr. His Life and Work as Seen by his Friends and Colleagues*. New York: Interscience Publishers.
Ruark, Arthur E. 1926. "Review of Van Vleck 1926a." *Journal of the Optical Society of America* 13: 312.
Runge, Carl, and Friedrich Paschen. 1896. "On the Spectrum of Cléveite Gas." *Astrophysical Journal* 3: 4–28.
Russell, Henry Norris, and Frederick Albert Saunders. 1925. "New Regularities in the Spectra of the Alkaline Earths." *Astrophysical Journal* 61: 38–69.
Rutherford, Ernest, and Frederick Soddy. 1902. "The Cause and Nature of Radioactivity, Part I." *Philosophical Magazine* 4: 370–396.
Samson, Roger, and Robert N. Brandon, eds. 2007. *Integrating Evolution and Development. From Theory to Practice*. Cambridge, MA: The MIT Press.
Sauer, Tilman, and Ulrich Majer, eds. 2009. *David Hilbert's Lectures on the Foundations of Physics, 1915–1927*. Berlin: Springer.
Schiff, Leonard. 1968. *Quantum Mechanics*. 3rd ed. New York: McGraw-Hill.
Schirrmacher, Arne. 2019. *Establishing Quantum Physics in Göttingen: David Hilbert, Max Born, and Peter Debye in Context, 1900–1926*. Cham: Springer.
Schlesinger, Ludwig. 1900. *Einführung in die Theorie der Differentialgleichungen mit einer unabhängigen Variabeln*. Leipzig: G. J. Göschensche Verlagshandlung.
Schmidt, Erhard. 1908. "Über die Auflösung linearer Gleichungen mit unendlich vielen Unbekannten." *Rendiconti del Circolo Mathematico di Palermo* 25: 53–77.
Schrödinger, Erwin. 1917. "Die Ergebnisse der neueren Forschung über Atom und Molekularwärmen." *Die Naturwissenschaften* 5: 537–543, 561–567. Reprinted in Schrödinger (1984, Vol. 1, pp. 174–187).
——. 1921. "Versuch zur modellmässigen Deutung des Terms der scharfen Nebenserien." *Zeitschrift für Phyik* 4: 347–354. Reprinted in Schrödinger (1984, Vol. 3, pp. 3–10).
——. 1922. "Über eine bemerkenswerte Eigenschaft der Quantenbahnen eines einzelnen Elektrons." *Zeitschrift für Physik* 12: 13–23. Reprinted in Schrödinger (1984, Vol. 3, pp. 319–323).
——. 1924a. "Gasentartung und freie Weglänge." *Physikalische Zeitschrift* 25: 41–45. Reprinted in Schrödinger (1984, Vol. 1, pp. 319–323).
——. 1924b. "Über die Rotationswärme des Wasserstoffs." *Zeitschrift für Physik* 30: 341–349. Reprinted in Schrödinger (1984, Vol. 1, pp. 332–340).
——. 1924c. "Bohrs neue Strahlungshypothese und der Energiesatz." *Die Naturwissenschaften* 12: 720–724.
——. 1925. "Bemerkungen über die statistische Entropiefunktion beim idealen Gas." *Preußische Akademie der Wissenschaften* (Berlin). *Physikalisch-mathematische Klasse. Sitzungsberichte:* 434–441. Reprinted in Schrödinger (1984, Vol. 1, pp. 341–348).
——. 1926a. "Die Energiestufen des idealen einatomigen Gasmodells." *Preußische Akademie der Wissenschaften* (Berlin). *Physikalisch-mathematische Klasse. Sitzungsberichte:* 23–36. Reprinted in Schrödinger (1984, Vol. 3, pp. 59–72).
——. 1926b. "Zur Einsteinschen Gastheorie." *Physikalische Zeitschrift* 27: 95–101. Reprinted in Schrödinger (1984, Vol. 1, pp. 358–364).
——. 1926c. "Quantisierung als Eigenwertproblem. (Erste Mitteilung)." *Annalen der Physik* 79: 361–376. Reprinted in Schrödinger (1927, pp. 1–16; 1984, Vol. 3, pp. 82–97). Page references to translation, "Quantisation as a Problem of Proper Values (Part I)," in Schrödinger (1982, pp. 1–12). Translation of secs. 1 and 2 in Ludwig (1968, pp. 94–105).

Schrödinger, Erwin. 1926d. "Quantisierung als Eigenwertproblem. (Zweite Mitteilung)." *Annalen der Physik* 79: 489–527. Reprinted in Schrödinger (1927, pp. 17–55; 1984, Vol. 3, pp. 98–136). Page references to translation, "Quantisation as a Problem of Proper Values (Part II)," in Schrödinger (1982, pp. 13–40). Abridged translation in Ludwig (1968, pp. 106–126).

——. 1926e. "Über das Verhältnis der Heisenberg–Born–Jordanschen Quantenmechanik zu der meinen." *Annalen der Physik* 79: 734–756. Reprinted in Schrödinger (1927, pp. 62–84; 1984, Vol. 3, pp. 143–165). Page references to translation, "On the Relation between the Quantum Mechanics of Heisenberg, Born, and Jordan, and that of Schrödinger," in Schrödinger (1982, pp. 45–61). Also translated in Ludwig (1968, pp. 127–150).

——. 1926f. "Quantisierung als Eigenwertproblem. (Dritte Mitteilung: Störungstheorie, mit Anwendung auf den Starkeffekt der Balmerlinien)." *Annalen der Physik* 80: 437–490. Reprinted in Schrödinger (1927, pp. 85–138; 1984, Vol. 3, pp. 166–219). Page references to translation, "Quantisation as a Problem of Proper Values (Part III: Perturbation Theory, with Applications to the Stark Effect of the Balmer Lines)," in Schrödinger (1982, pp. 62–101).

——. 1926g. "Der stetige Übergang von der Mikro- zur Makromechanik." *Die Naturwissenschaften* 14: 664–66. Reprinted in Schrödinger (1927, pp. 56–61; 1984, Vol. 3, pp. 137–142). Translation, "The Continuous Transition from Micro- to Macro-mechanics," in Schrödinger (1982, pp. 41–44).

——. 1926h. "Quantisierung als Eigenwertproblem. (Vierte Mitteilung)." *Annalen der Physik* 81: 109–139. Reprinted in Schrödinger (1927, pp. 139–169; 1984, Vol. 3, pp. 220–250). Page references to translation, "Quantisation as a Problem of Proper Values (Part IV)," in Schrödinger (1982, pp. 102–123). Abridged translation in Ludwig (1968, pp. 151–126).

——. 1927. *Abhandlungen zur Wellenmechanik.* Leipzig: Barth.

——. 1928. "La mécanique des ondes." In Solvay (1928, pp. 185–213). English translation, "Wave Mechanics," in Bacciagaluppi and Valentini (2009, pp. 406–424).

——. 1945. "The Hamilton Postage Stamp: An Announcement by the Irish Minister of Posts and Telegraphs." In *Scripta Mathematica.* No. 2. *A Collection of Papers in Memory of William Rowan Hamilton.* Edited by David Eugene Smith. New York: Scripta Mathematica.

——. 1982. *Collected Papers on Wave Mechanics.* Providence, RI: American Mathematical Society Chelsea Publishing. Third (augmented) English edition.

——. 1984. *Gesammelte Abhandlungen/Collected Papers.* 4 Vols. Vienna, Braunschweig/Wiesbaden: Verlag der Österreichischen Akademie der Wissenschaften, Vieweg.

Schuster, Arthur. 1895. "The Kinetic Theory of Gases." *Nature* 51: 293.

Schwarzschild, Karl. 1916. "Zur Quantenhypothese." *Königlich Preussische Akademie der Wissenschaften* (Berlin). *Sitzungsberichte:* 548–568.

Schweber, Silvan S. 1986. "The Empiricist Temper Regnant: Theoretical Physics in the United States 1920–1950." *Historical Studies in the Physical and Biological Sciences* 17: 55–98.

——. 1990. "The Young John Clarke Slater and the Development of Quantum Chemistry." *Historical Studies in the Physical and Biological Sciences* 20: 339–406.

Senftleben, Hermann. 1915. "Über die Zahl der Emissionszentren der in Flammen leuchtenden Metalldämpfe und die Beziehungen dieser Zahl zur Helligkeit der ausgesandten Spektrallinien." *Annalen der Physik* 47: 949–1000.

Seth, Suman. 2010. *Crafting the Quantum. Arnold Sommerfeld and the Practice of Theory, 1890–1926.* Cambridge, MA: The MIT Press.

——. 2013. "Forman at Forty: New Perspectives on 'Weimar Culture and Quantum Mechanics'." *Metascience* 22: 567–574. Essay review of Carson, Kojevnikov, and Trischler (2011).

Shamos, Morris H., ed. 1959. *Great Experiments in Physics. Firsthand Accounts from Galileo to Einstein*. New York: Dover.
Shapin, Steven. 1996. *The Scientific Revolution*. Chicago: University of Chicago Press.
Shubin, Neil. 2008. *Your Inner Fish. A Journey into the 3.5-Billion-Year History of the Human Body*. New York: Pantheon Books. Reprint (with new afterword): New York: Vintage Books, 2009.
Sigmund, Karl. 2017. *Exact Thinking in Demented Times. The Vienna Circle and the Epic Quest for the Foundations of Science*. New York: Basic Books.
Slater, John C. 1924. "Radiation and Atoms." *Nature* 113: 307–308.
——. 1925. "The Nature of Radiation." *Nature* 116: 278.
Smekal, Adolf. 1923. "Zur Quantentheorie der Dispersion." *Die Naturwissenschaften* 43: 873–875.
Solvay. 1923. *Atomes et électrons. Rapports et discussions du conseil de physique tenu à Bruxelles du 1er au 6 avril 1921*. Paris: Gauthier-Villars.
——. 1928. *Électrons et photons. Rapports et discussions du cinquième conseil de physique tenu à Bruxelles du 24 au 29 Octobre 1927*. Paris: Gauthier-Villars. Institut International de Physique Solvay.
Sommerfeld, Arnold. 1915a. "Zur Theorie der Balmerschen Serie." *Königlich Bayerische Akademie der Wissenschaften zu München. Mathematischphysikalische Klasse. Sitzungsberichte:* 425–458. Reprinted in facsimile in Sommerfeld (2013). English translation: Sommerfeld (2014).
——. 1915b. "Die allgemeine Dispersionsformel nach dem Bohrschen Model." In *Festschrift Julius Elster und Hans Geitel*. Edited by K. Bergwitz. Braunschweig: Vieweg. Reprinted in Sommerfeld (1968, Vol. 3, pp. 136–171).
——. 1916a. "Zur Quantentheorie der Spektrallinien." Pts. I and II. *Annalen der Physik* 51: 1–94. Reprinted in Sommerfeld (1968, Vol. 3, pp. 172–265).
——. 1916b. "Zur Quantentheorie der Spektrallinien." Pt. III. *Annalen der Physik* 51: 125–167. Reprinted in Sommerfeld (1968, Vol. 3, pp. 266–308).
——. 1916c. "Zur Quantentheorie der Spektrallinien. Ergänzungen und Erweiterungen." *Sitzungsberichte der Bayrischen Akademie der Wissenschaften zu München:* 131–182. Reprinted in Sommerfeld (1968, Vol. 3, pp. 326–377).
——. 1918. "Die Drudesche Dispersionstheorie vom Standpunkte des Bohrschen Modelles und die Konstitution von H_2, O_2, and N_2." *Annalen der Physik* 53: 497–550. Reprinted in Sommerfeld (1968, Vol. 3, pp. 378–431).
——. 1919. *Atombau und Spektrallinien*. Braunschweig: Vieweg.
——. 1922a. "Quantentheoretische Umdeutung der Voigt'schen Theorie des anomalen Zeeman-Effektes vom D-Linientypus." *Zeitschrift für Physik* 8: 257–272. Reprinted in Sommerfeld (1968, Vol. 3, pp. 609–624).
——. 1922b. *Atombau und Spektrallinien*. 3rd ed. Braunschweig: Vieweg.
——. 1923. *Atomic Structure and Spectral Lines*. London: Methuen. Translation (by Henry L. Brose) of Sommerfeld (1922b).
——. 1924. *Atombau und Spektrallinien*. 4th ed. Braunschweig: Vieweg.
——. 1929. *Atombau und Spektrallinien. Wellenmechanischer Ergänzungsband*. Braunschweig: Vieweg.
——. 1930. *Wave-Mechanics*. London: Methuen. Translation (by Henry L. Brose) of Sommerfeld (1929).
——. 1968. *Gesammelte Schriften*. 4 Vols. Edited by Fritz Sauter. Braunschweig: Vieweg.

Sommerfeld, Arnold. 2004. *Wissenschaftlicher Briefwechsel. Band 2: 1919–1951*. Edited by Michael Eckert and Karl Märker. Berlin, Diepholz, München: Deutsches Museum. Verlag für die Geschichte der Naturwissenschaften und der Technik.

——. 2013. *Die Bohr–Sommerfeldsche Atomtheorie. Sommerfelds Erweiterung des Bohrschen Atommodells 1915/16 kommentiert von Michael Eckert*. Berlin: Springer. With an introduction by Michael Eckert.

——. 2014. "On the Theory of the Balmer Series." *European Physical Journal H* 9: 157–177. English translation (by Michael Eckert) of Sommerfeld (1915a).

Sommerfeld, Arnold, and Werner Heisenberg. 1922a. "Eine Bemerkung über relativistische Röntgen Dubletts und Linienschärfe." *Zeitschrift für Physik* 10: 393–398.

——. 1922b. "Die Intensität der Mehrfachlinien und ihrer Zeemankomponenten." *Zeitschrift für Physik* 11: 131–154.

Sommerfeld, Arnold, and Helmut Hönl. 1925. "Über die Intensität von Multiplett-Linien." *Preussische Akademie der Wissenschaften* (Berlin). *Physikalisch-mathematische Klasse. Sitzungsberichte*: 141–161.

Sommerfeld, Arnold, and Iris Runge. 1911. "Anwendung der Vektorrechnung auf die Grundlagen der geometrischen Optik." *Annalen der Physik* 35: 277–298.

Sopka, Katherine Russell. 1988. *Quantum Physics in America. The Years Through 1935*. New York: Tomash Publishers/American Institute of Physics.

Stachel, John, ed. 2005. *Einstein's Miraculous Year. Five Papers that Changed the Face of Physics*. Princeton University Press. Centenary Edition. Originally published in 1998.

Steen, Lynn Arthur. 1973. "Highlights in the History of Spectral Theory." *The American Mathematical Monthly* 80: 359–381.

Stehle, Philip. 1966. *Quantum Mechanics*. San Francisco, London, Amsterdam: Holden-Day.

——. 1994. *Order, Chaos, Order. The Transition from Classical to Quantum Physics*. Oxford: Oxford University Press.

Stern, Otto, and Max Volmer. 1919. "Über die Abklingungszeit der Fluoreszenz." *Physikalische Zeitschrift* 20: 183–188.

Stolzenburg, Klaus. 1984. "Introduction. Part 1. The Theory of Bohr, Kramers, and Slater." In Bohr (1972–2008, Vol. 5, pp. 3–96).

Stoner, Edmund C. 1924. "The Distribution of Electrons among Atomic Levels." *Philosophical Magazine* 48: 719–736.

Stuewer, Roger H. 1975. *The Compton Effect: Turning Point in Physics*. Canton, MA: Science History Publications.

——. 2014. "The Experimental Challenge of Light Quanta." In Janssen and Lehner (2014, pp. 143–166).

——. 2018. *The Age of Innocence. Nuclear Physics between the First and Second World Wars*. Oxford: Oxford University Press.

ter Haar, Dirk, ed. 1967. *The Old Quantum Theory*. Oxford: Pergamon.

Thomas, Llewellyn Hilleth. 1926. "The Motion of the Spinning Electron." *Nature* 117: 514.

Thomas, Willy. 1925. "Über die Zahl der Dispersionselektronen, die einem stationären Zustand zugeordnet sind." *Die Naturwissenschaften* 13: 627.

Tzara, Christophe. 1988. "The Neutrino Bursts from the Supernova SN 1987 A Do Not Rule out the Existence of Wave-packet Spreading." *Physics Letters A* 132: 159–160.

Uffink, Jos. 1990. "Measures of Uncertainty and the Uncertainty Principle." PhD dissertation, University of Utrecht.

Uffink, Jos, and Jan Hilgevoord. 1985. "Uncertainty Principle and Uncertainty Relations." *Foundations of Physics* 15: 925–944.

Uhlenbeck, George Eugene. 1976. "Fifty Years of Spin—Personal Reminiscences." *Physics Today* 29: 43–48.

Uhlenbeck, George Eugene, and Samuel Goudsmit. 1925. "Ersetzung der Hypothese vom unmechanischen Zwang durch eine Forderung bezüglich des inneren Verhaltens jedes einzelnen Elektrons." *Die Naturwissenschaften* 13: 953–954.

Unsöld, Albrecht. 1927. "Beiträge zur Quantenmechanik der Atome." *Annalen der Physik* 82: 355–393.

van der Waerden, Bartel Leendert. 1960. "Exclusion Principle and Spin." In Fierz and Weisskopf (1960, pp. 199–244).

——. ed. 1968. *Sources of Quantum Mechanics*. New York: Dover.

van Dongen, Jeroen. 2007a. "Emil Rupp, Albert Einstein, and the Canal Ray Experiments on Wave-Particle Duality: Scientific Fraud and Theoretical Bias." *Historical Studies in the Physical and Biological Sciences* 37 (Supplement): 73–120.

——. 2007b. "The Interpretation of the Einstein-Rupp experiments and Their Influence on the History of Quantum Mechanics." *Historical Studies in the Physical and Biological Sciences* 37 (Supplement): 121–131.

van Oudenaarden, Alexander, Michel H. Devoret, Yu. V. Nazarov, and J. E. Mooij. 1998. "Magneto-electric Aharonov-Bohm Effect in Metal Rings." *Nature* 391: 768–770.

Van Vleck, John H. 1922. "The Normal Helium Atom and its Relation to the Quantum Theory." *Philosophical Magazine* 44: 842–869.

——. 1924a. "A Correspondence Principle for Absorption." *Journal of the Optical Society of America* 9: 27–30.

——. 1924b. "The Absorption of Radiation by Multiply Periodic Orbits, and Its Relation to the Correspondence Principle and the Rayeigh-Jeans Law. Part I. Some Extensions of the Correspondence Principle." *Physical Review* 24: 330–346. Reprinted in van der Waerden (1968, pp. 203–222).

——. 1924c. "The Absorption of Radiation by Multiply Periodic Orbits, and Its Relation to the Correspondence Principle and the Rayeigh-Jeans Law. Part II. Calculation of Absorption by Multiply Periodic Orbits." *Physical Review* 24: 347–365.

——. 1926a. *Quantum Principles and Line Spectra. Bulletin of the National Research Council 10, Pt. 4.* Washington, D. C.: National Research Council.

——. 1926b. "Magnetic Susceptibilities and Dielectric Constants in the New Quantum Mechanics." *Nature* 118: 226–227.

——. 1927a. "On Dielectric Constants and Magnetic Susceptibilities in the new Quantum Mechanics. Part I. A General Proof of the Langevin-Debye Formula." *Physical Review* 29: 727–744.

——. 1927b. "On Dielectric Constants and Magnetic Susceptibilities in the new Quantum Mechanics. Part II. Application to Dielectric Constants." *Physical Review* 30: 31–54.

——. 1928. "On Dielectric Constants and Magnetic Susceptibilities in the new Quantum Mechanics. Part III. Application to Dia- and Paramagnetism." *Physical Review* 29: 587–613.

——. 1932. *The Theory of Electric and Magnetic Susceptibilities*. Oxford: Oxford University Press.

——. 1971. "Reminiscences of the First Decade of Quantum Mechanics." *International Journal of Quantum Chemistry. International Journal of Quantum Chemistry. Symposium No. 5, 1971* (a symposium held in honor of Van Vleck). Edited by Per-Olov Lödwin, pp. 3–20. New York: Wiley.

Voigt, Woldemar. 1908. *Magneto und Elektrooptik*. Leipzig: Teubner.

Volkmann, Paul. 1900. *Einführung in das Studium der theoretischen Physik, insbesondere das der analytischen Mechanik mit einer Einleitung in die Theorie der Physikalischen Erkentniss*. Leipzig: Teubner.

von Meyenn, Karl. 2007. "Jordan, Pauli, und ihre frühe Zusammenarbeit auf dem Gebiet der Quantenstrahung." In *Pascual Jordan (1902–1980). Mainzer Symposium zum 100. Geburtstag*. Preprint 329. Edited by Dieter Hoffmann, Jürgen Ehlers, and Jürgen Renn, pp. 37–46. Berlin: Max Planck Institute for History of Science.

——, ed. 2011. *Eine Entdeckung von ganz außerordentlicher Tragweite. Schrödingers Briefwechsel zur Wellenmechanik und zum Katzenparadoxon*. 2 Vols. Berlin, Heidelberg: Springer.

von Mises, Richard. 1928. *Wahrscheinlichkeit, Statistik und Wahrheit*. Vienna: Springer.

von Neumann, John. 1927a. "Mathematische Begründung der Quantenmechanik." *Königliche Gesellschaft der Wissenschaften zu Göttingen. Mathematisch-physikalische Klasse. Nachrichten:* 1–57.

——. 1927b. "Wahrscheinlichkeitstheoretischer Aufbau der Quantenmechanik." *Königliche Gesellschaft der Wissenschaften zu Göttingen. Mathematisch-physikalische Klasse. Nachrichten:* 245–272.

——. 1927c. "Thermodynamik quantenmechanischer Gesamtheiten." *Königliche Gesellschaft der Wissenschaften zu Göttingen. Mathematisch-physikalische Klasse. Nachrichten:* 273–291.

——. 1928. "Die Axiomatisierung der Mengenlehre." *Mathematische Zeitschrift* 27: 669–752.

——. 1929. "Allgemeine Eigenwerttheorie Hermitescher Funktionaloperatoren." *Mathematische Annalen* 102: 49–131.

——. 1932. *Mathematische Grundlagen der Quantenmechanik*. Berlin: Springer.

Waller, Ivar. 1926. "Das Starkeffekt zweiter Ordnung bei Wasserstoff und die Rydbergkorrektion der Spektra von He and Li^+." *Zeitschrift für Physik* 38: 635–646.

Wasserman, Neil Henry. 1981. "The Bohr-Kramers-Slater Paper and the Development of the Quantum Theory of Radiation in the Work of Niels Bohr." PhD dissertation, Harvard University.

Weart, Spencer R., ed. 1976. *Selected Papers of Great American Phsyicists*. New York: American Institute of Physics.

Weinberg, Steven. 1992. *Dreams of a Final Theory*. New York: Pantheon. Page reference to reprint: New York, Vintage Books, 1994.

Wentzel, Gregor. 1926. "Eine Verallgemeinerung der Quantenbedingungen für die Zwecke der Wellenmechanik." *Zeitschrift für Physik* 38: 518–529.

Wessels, Linda. 1979. "Schrödinger's Route to Wave Mechanics." *Studies in History and Philosophy of Science* 10: 311–340.

——. 1980. "What Was Born's Statistical Interpretation?" In *PSA: Proceedings of the 1980 Biennial Meeting of the Philosophy of Science Association*. Vol. 2. Edited by Peter D. Asquith and Ronald N. Giere, pp. 187–200. East Lansing, MI: Philosophy of Science Association.

Weyl, Hermann. 1918. "Gravitation und Elektrizität." *Königlich Preussische Akademie der Wissenschaften* (Berlin). *Sitzungsberichte:* 465–478, 478–480 ("Erwiderung des Verfassers" to Einstein 1918).

——. 1922. *Space—Time—Matter*. London: Methuen. Translation (by Henry L. Brose) of *Raum–Zeit–Materie. Vorlesungen über allgemeine Relativitätstheorie*. 4th ed. Berlin: Springer, 1921. Page reference to reprint: New York: Dover, 1952.

Wheaton, Bruce R. 1983. *The Tiger and the Shark. Empirical Roots of Wave-Particle Dualism*. Cambridge: Cambridge University Press.

Wheeler, John Archibald, and Wojciech Hubert Zurek, eds. 1983. *Quantum Theory and Measurement*. Princeton: Princeton University Press.

Whittaker, Edmund T. 1951. *A History of the Theories of Aether and Electricity*. 2 Vols. London: Nelson. Reissued in facsimile as Vol. 7 in the series *The History of Modern Physics, 1800–1950*. New York: Tomash Publishers/American Institute of Physics.

Wien, Wilhelm. 1919. "Über Messungen der Leuchtdauer der Atome und der Dämpfung der Spektrallinien. I." *Annalen der Physik* 60: 597–637.

——. 1921. "Über Messungen der Leuchtdauer der Atome und der Dämpfung der Spektrallinien. II." *Annalen der Physik* 66: 229–236.

——. 1924. "Über Messungen der Leuchtdauer der Atome und der Dämpfung der Spektrallinien. III." *Annalen der Physik* 73: 483–504.

Wigner, Eugene. 1927. "Einige Folgerungen aus der Schrödingerschen Theorie für die Termstrukturen." *Zeitschrift für Physik* 43: 624–652.

Wilson, Charles Thomson Rees. 1923a. "Investigations on X-rays and β-rays by the Cloud Method." *Nature* 112: 26–27.

——. 1923b. "Investigations on X-rays and β-rays by the Cloud Method. Part I.—X-rays." *Proceedings of the Royal Society of London* 104: 1–24.

Wimsatt, William C., and James R. Griesemer. 2007. "Reproducing Entrenchments to Scaffold Culture: The Central Role of Development in Cultural Evolution." In Samson and Brandon (2007, pp. 227–323).

Wirtinger, Wilhelm. 1897. "Beitrag zu Riemanns Integrationsmethode für hyperbolische Differentialgleichungen und deren Anwendungen auf Schwingungs-probleme." *Mathematische Annalen* 48: 365–389.

Wood, Robert Williams. 1904. "A Quantitative Determination of the Anomalous Dispersion of Sodium Vapour in the Visible and Ultra-violet Regions." *Philosophical Magazine* 8: 293–324.

Wood, Robert Williams, and Alexander Ellett. 1923. "On the Influence of Magnetic Fields on the Polarization of Resonance Radiation." *Proceedings of the Royal Society of London* A 103: 396–403.

Yourgrau, Wolfgang, and Stanley Mandelstam. 1979. *Variational Principles in Dynamics and Quantum Theory*. 3rd rev. ed. New York: Dover.

Zimmer, Ernst. 1934. *Umsturz im Weltbild der Physik*. Munich: Knorr & Hirth. English translation (by Henry Hatfield): *The Revolution in Physics*. New York: Harcourt, Brace and Company, 1936.

Index